mathematics + art

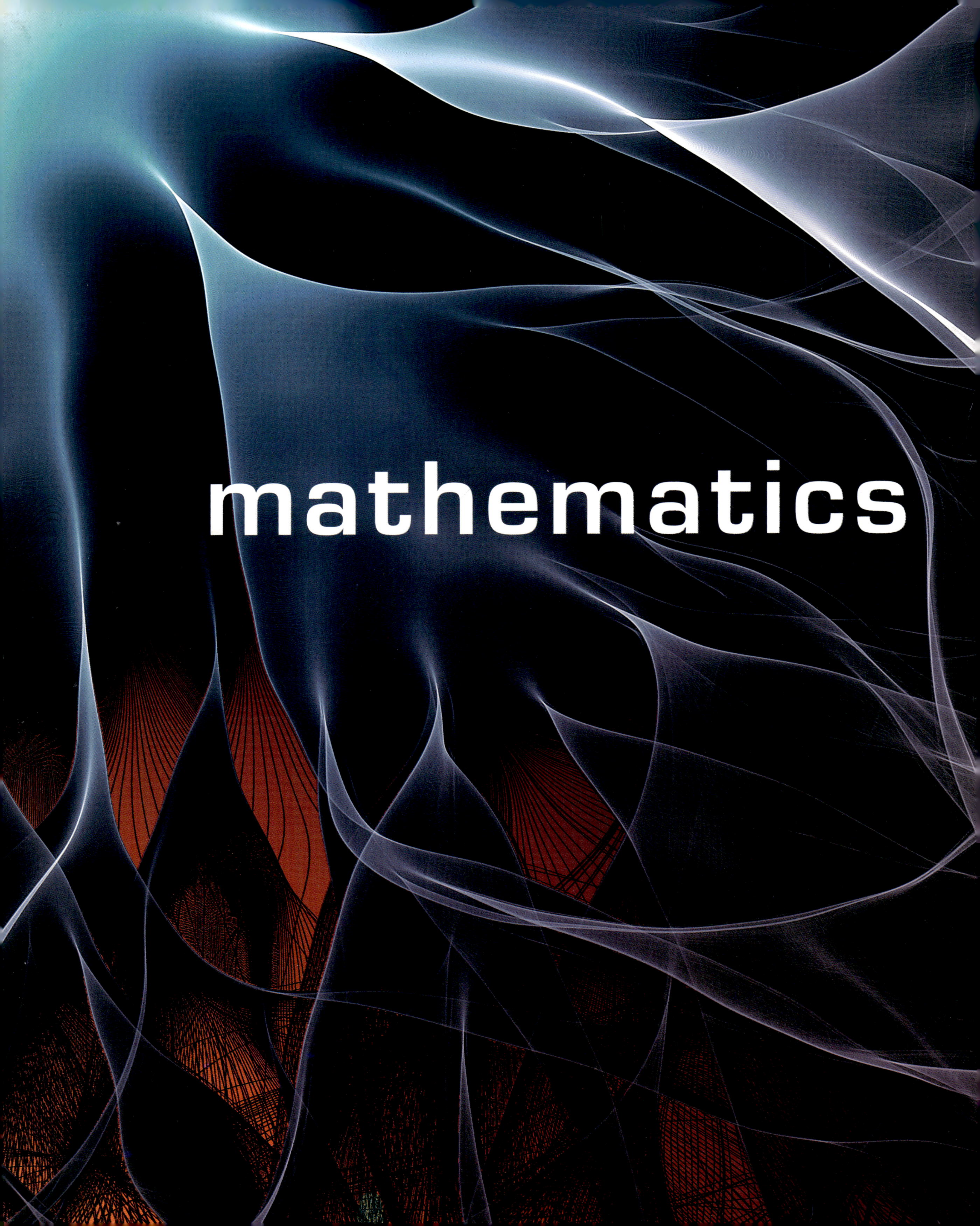
mathematics

+ art A CULTURAL HISTORY

Lynn Gamwell

FOREWORD BY
Neil deGrasse Tyson

PRINCETON UNIVERSITY PRESS
PRINCETON AND OXFORD

FRONT JACKET

Hiroshi Sugimoto (Japanese, born 1948), *Mathematical Form 0009*, 2004. Gelatin silver print, 58 ¾ x 47 in. (149.2 x 119.3 cm). © Hiroshi Sugimoto, courtesy Pace Gallery, New York.

SPINE

Frontispiece to Andreas Speiser, *Die Theorie der Gruppen von endlicher Ordnung* (The theory of groups of finite order; 1927), 4th ed. (Basel, Switzerland: Birkhäuser, 1956). Springer Science and Business Media, Heidelberg. Used with permission.

BACK JACKET

Illustration of the Genesis story of creation in a *Bible moralisée*, French, ca. 1208–15. Egg tempera and gold leaf on vellum. Österreichische Nationalbibliothek, Vienna, Bildarchiv, Codex 2554, fol. 1v.

TITLE PAGE

Eric J. Heller (American, born 1946), *Transport VI*, ca. 2000. Digital print. Courtesy of the artist.

This image, and the one on pages viii–ix, records paths taken by electrons as they flow over the hills and valleys of tiny "landscapes"—surfaces that are measured in microns (one micron equals one millionth of a meter).

PAGES VIII–IX

Eric J. Heller (American, born 1946), *Banyan*, from the *Transport* series, ca. 2000. Digital print. Courtesy of the artist.

Published by Princeton University Press, 41 William Street, Princeton, New Jersey 08540
In the United Kingdom: Princeton University Press, 6 Oxford Street, Woodstock, Oxfordshire OX20 1TW

press.princeton.edu

ISBN 978-0-691-16528-8

Library of Congress Control Number: 2014959232

British Library Cataloging-in-Publication Data is available

Designed by Jason Snyder

This book has been composed in Eurostile and Electra.
Printed on acid-free paper. ∞
Printed in Singapore

10 9 8 7 6 5

For my husband, Charles Brown

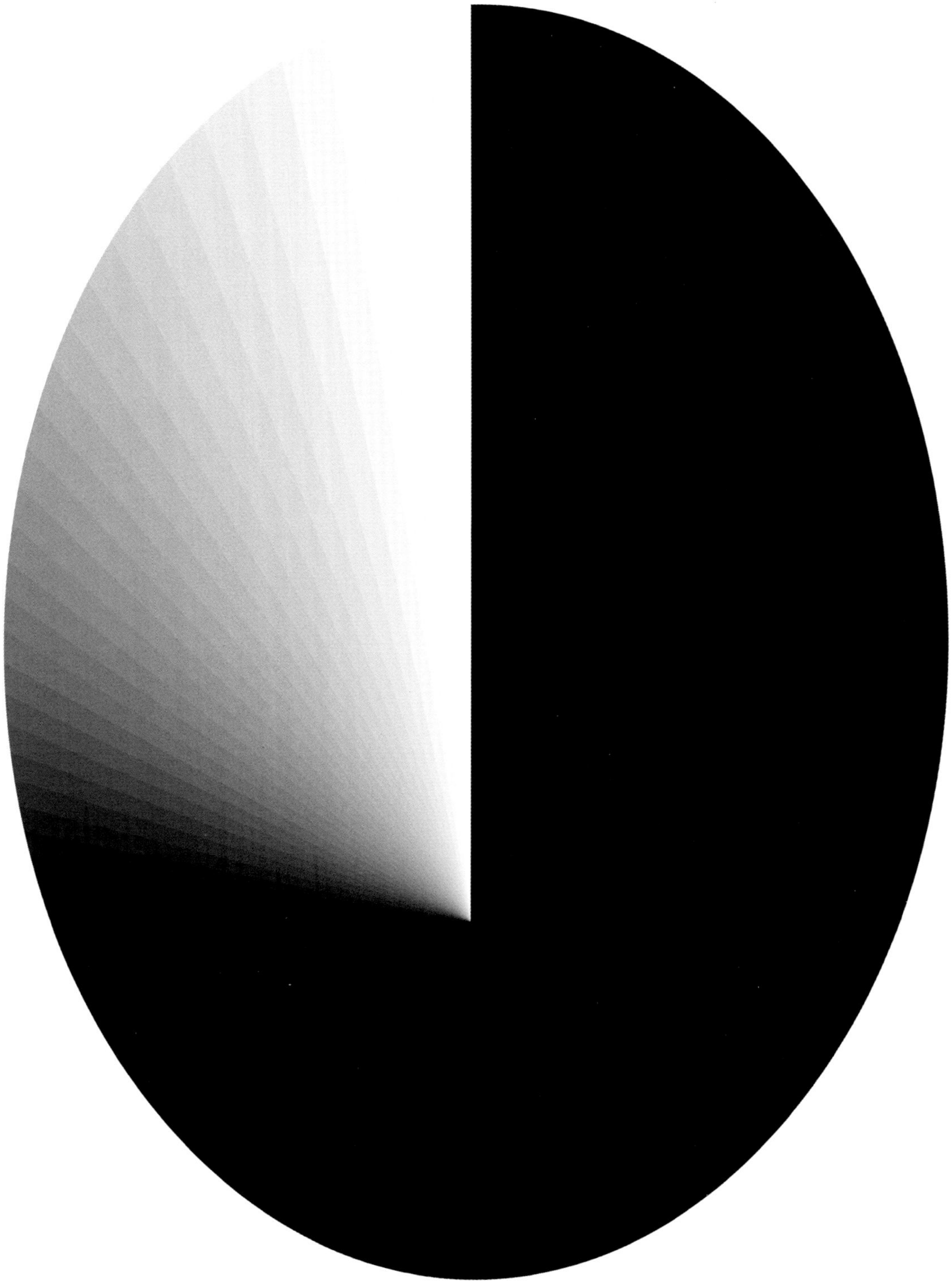

Karl Gerstner (Swiss, b. 1930), *Die Dunkelheit, die sich das Licht gebar* (The darkness, which gave birth to light), 2000. Acrylic on canvas, ca. 40 in. (101.6 cm) high. © Karl Gerstner. Collection of Esther Grether, Basel, Switzerland.

The title of Gerstner's painting is from Goethe's tragedy *Faust* (1790–1832; line 1350). When the Devil appears to Faust in his study, the scholar asks "What is thy name?" The stranger replies that he comes from "the darkness [chaos, irrationality], which gave birth to light [order, reason]." He warns Faust that, in the end, all light will be extinguished, but Faust, overcome by his passion to discover the essence of reality, trades his soul to the Devil in exchange for unlimited knowledge.

Goethe wrote *Faust* during the era of German Romanticism, when intellectuals perceived the Enlightenment's deterministic, clockwork universe as dehumanizing and the cost of knowledge as troublingly high: the soul—human values—can be lost forever. Such Romantic sentiments were never far behind the stoical facade of Germanic culture, which was the birthplace of modern mathematics and abstract art.

Contents

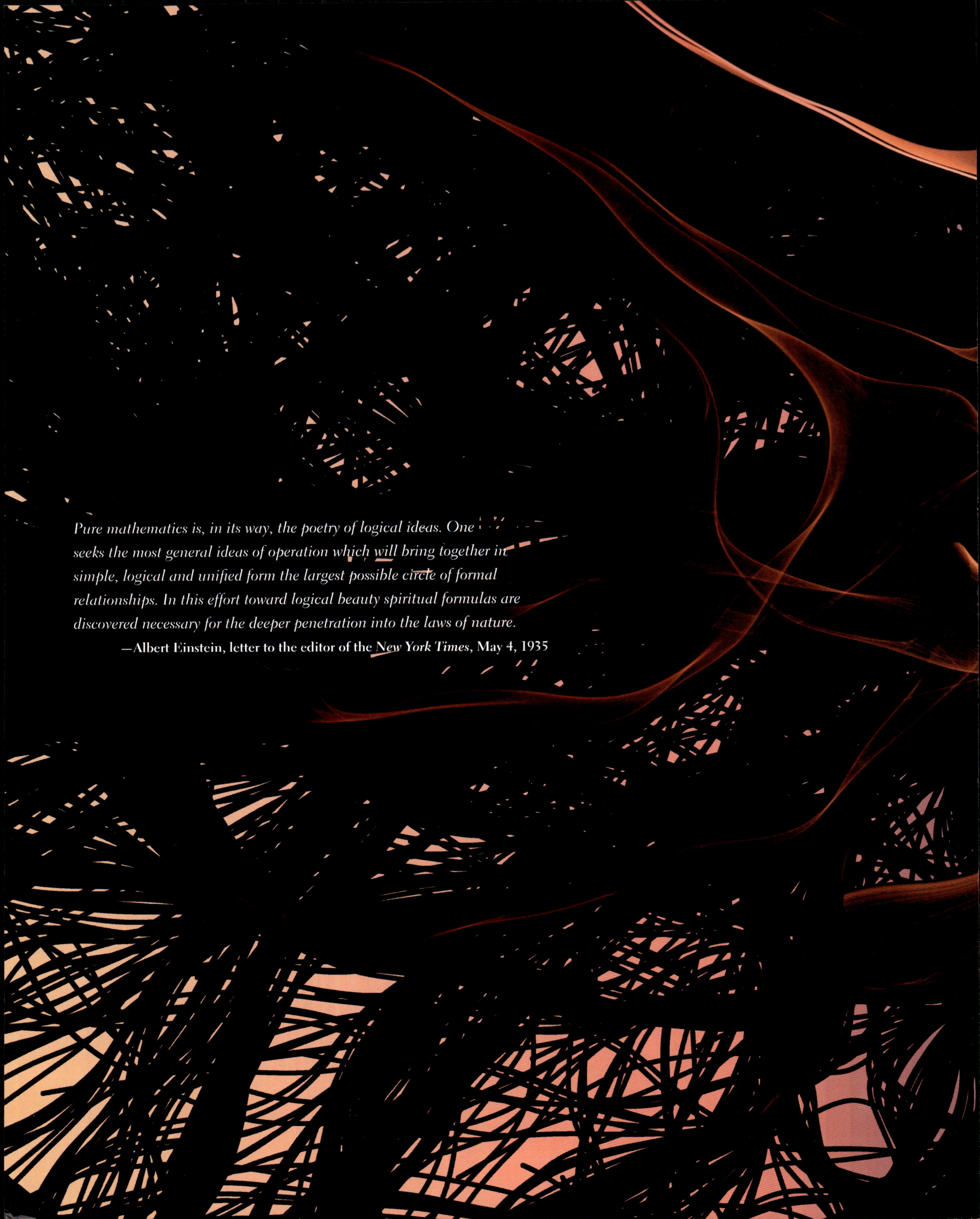

Pure mathematics is, in its way, the poetry of logical ideas. One seeks the most general ideas of operation which will bring together in simple, logical and unified form the largest possible circle of formal relationships. In this effort toward logical beauty spiritual formulas are discovered necessary for the deeper penetration into the laws of nature.

—Albert Einstein, letter to the editor of the *New York Times*, May 4, 1935

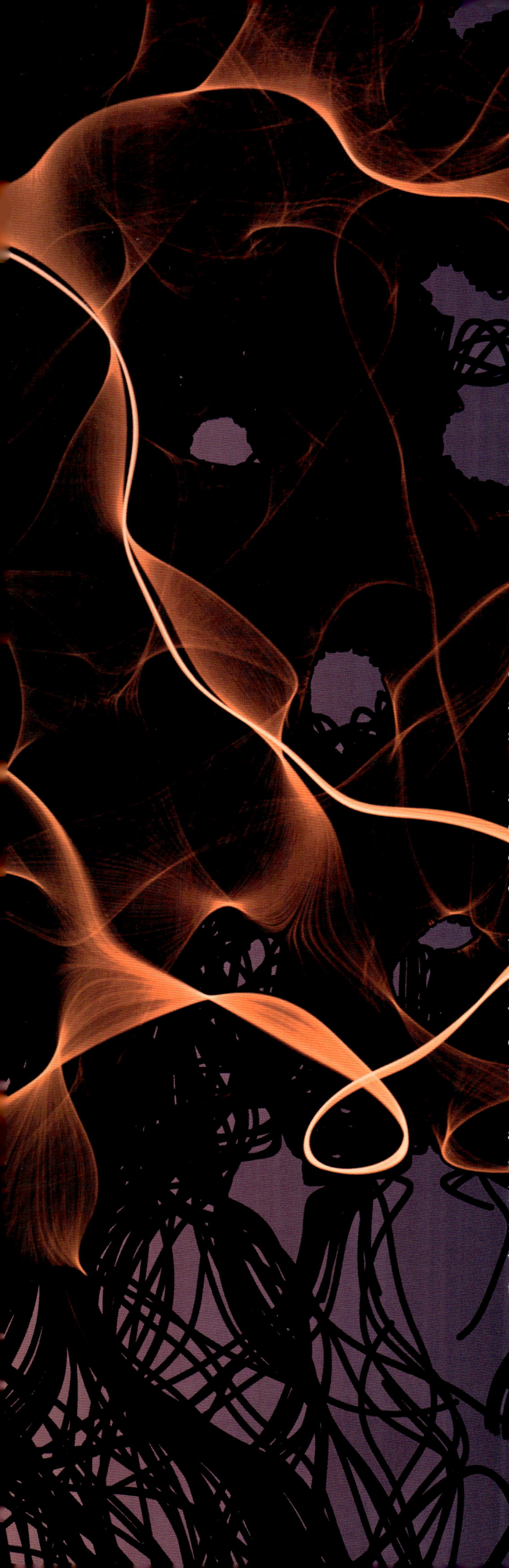

Foreword

Mathematics is the language of the universe.

If we line up the traditional sequence of math instruction, we get arithmetic, algebra, geometry, trigonometry, calculus, and a dozen more branches that excite math enthusiasts. While getting lost in the formulas and concepts, we easily forget—or perhaps were never told—that entire branches of mathematical discovery were driven by the urge to figure out how the world around us works. Isaac Newton invented the basics of calculus just to solve a problem of planetary orbits for which the mathematics of his day was inadequate.

And occasionally the reverse is true. Branches of mathematics have arisen, just out of curiosity, that would be wholly lifted into physics once the need revealed itself. Non-Euclidean geometry, which is basically geometry on curved surfaces, was codified in the early 1800s, a century before Albert Einstein needed it, and was thus available on the shelf for him and others to describe the mass-induced curvature of space-time in his general theory of relativity.

Yes, we conjure symbols in our heads and manipulate them, in an effort to abstract what is real into expressions of pure logic. If we believe that numbers and circles exist only in our heads, then this exercise is stunningly, yet unreasonably, effective at describing the behavior of nature at all scales, from atomic nuclei to the contents and structure of the cosmos itself.

Taking this thought a philosophical step further, we might wonder what numbers and circles are. Where do they come from? Perhaps mathematics does not simply describe the universe, but *is* the universe. Like the invented worlds of immersive video games, generated by thousands of lines of computer code, if we could part the curtains of the cosmos, might we find only equations, furiously calculating all the phenomena we experience?

The value of mathematics to the scientist is clear and present. Given its potency and ubiquity in granting access to the operations of nature, should we be surprised that mathematics has served (and continues to serve) as an irresistible muse for philosophers and artists alike? Isn't it one of the jobs of the artist to help non-artists interpret the world around us, and the world within us? As long as mathematics is what shapes those worlds, the observant artist cannot help but embrace and express this influence on us all.

Neil deGrasse Tyson
New York City, June 2015

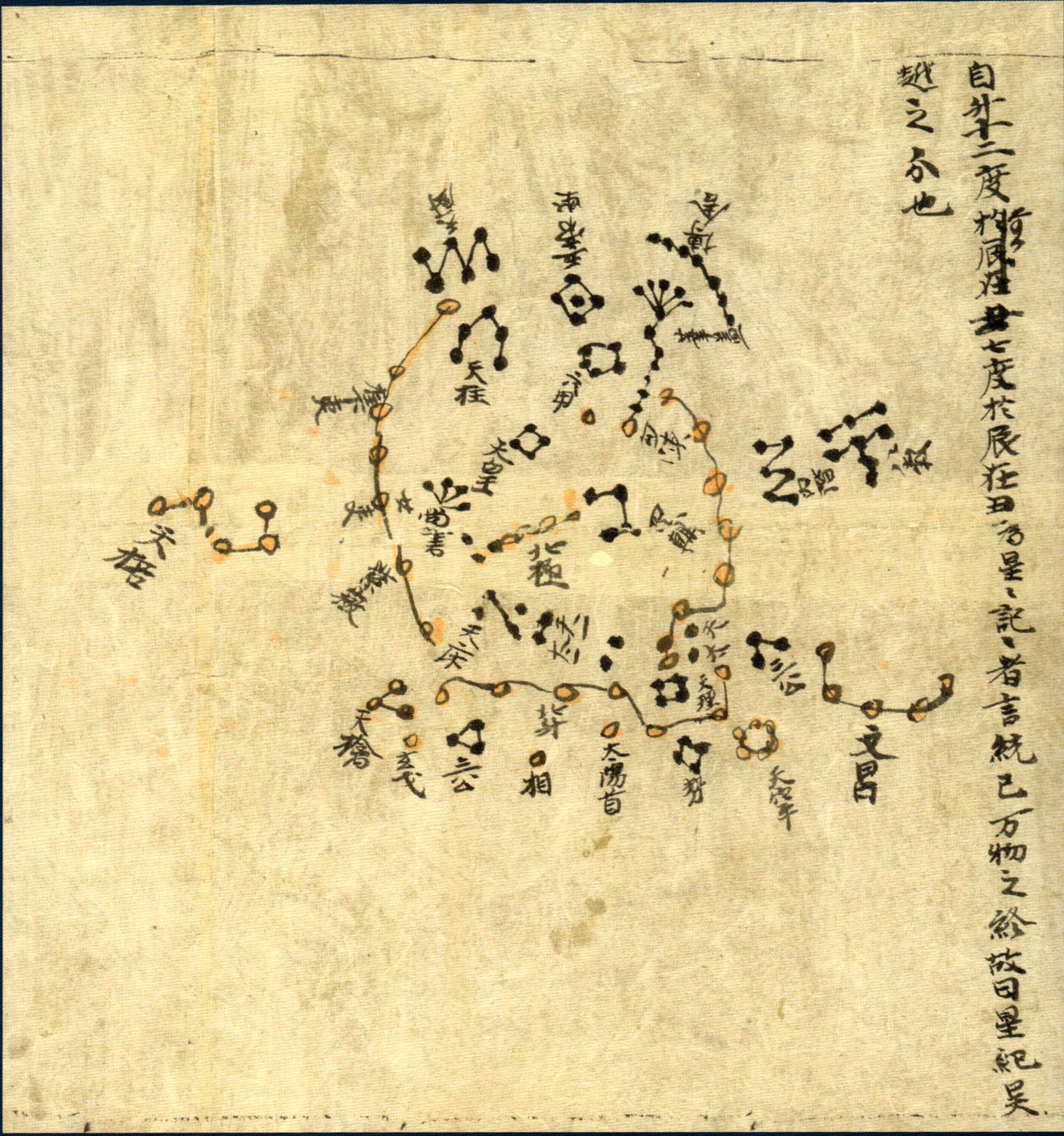
自斗十二度至女七度於辰在丑為星紀星紀者言統紀万物之終故曰星紀吳
越之分也
北極
文昌
相

Preface

Let no one despise symbols! . . . Without symbols we could scarcely lift ourselves to conceptual thinking. Thus, in applying the same symbol to different but similar things, we actually no longer symbolize the individual thing, but rather what they have in common: the concept.

—Gottlob Frege, "On the Scientific Justification of a Conceptual Notation," 1882

NON-SPECIALISTS WHO READ books about mathematics are often disappointed because its secrets are written in a technical language they don't understand. My goal has been to describe in plain English, together with clear symbols and cogent diagrams, the ideas that drive mathematics—numbers, infinity, geometry, and pattern. I have also explored how mathematical symbols, diagrams, and patterns, which are essentially pictorial visualizations of abstract concepts, have inspired artists throughout history, and how architects around the world have used this austere vocabulary to design towering cities, where mathematics—because of its abstractness—is the international language of exact thought.

The practice of mathematics began when early people chipped patterns in rock during the Stone Age and continued in mankind's search for numbers and geometric forms in earth and sky (plate 0-1). From antiquity to the present, a thread running through this history is the distinction between the philosophy and the practice of mathematics. The *philosophy* of mathematics is done by someone like Plato, who asks: What is certain knowledge? (a question in epistemology) or: What are numbers? In what sense do they exist? (questions in metaphysics). The *practice* of mathematics is done by a "working mathematician" like Euclid, who asks: Do the angles of a triangle sum to 180°? (a question in geometry), or: Is there a pattern to the prime numbers? (a question in arithmetic). Occasionally someone does both, such as the Enlightenment scholar Gottfried Leibniz, who wrote a book on metaphysics in the Platonic tradition, *Monadology* (1714), and invented binary notation.

Ancient mathematics relied on simple abstraction and generalization, which was readily absorbed into cultural thought. But over the centuries, as the practice of mathematics became more technical, it was harder for the general public to follow the details, and thus much of the cultural impact of mathematical thought that I describe in this book

0-1. *Dunhuang Star Chart* (detail), AD 649–84. Ink on paper, 9½ × 8 in. (24.4 × 20 cm). © The British Library Board, Or.8210/S.3326 R.2. (8).

This astronomical map records the visible sky around the celestial north pole (the still point in the night sky). The constellation Ursa Major is shown in the lower region, its seven brightest stars forming the Big Dipper. The map is a small section of a star atlas that records 1,339 stars and shows the entire sky visible from the northern hemisphere. Drawn in China during the Tang dynasty (AD 618–907), it is the oldest known star atlas preserved from any civilization in the world.

is of a philosophical nature. But technical mathematics is brought to public attention indirectly when it is applied in the everyday world. For example, in Leibniz's day binary notation was an esoteric concept, known only to a few scholars, but today the use of binary notation in computers is familiar to any educated person. When the practice of mathematics is applied and popularized, mathematical concepts enter the studios and inspire artists.

A second thread running through the history of mathematics is the tension between rationalist accounts of the natural world as operating according to deterministic laws of cause and effect and rebellions against such models (and the mathematics associated with them) as dehumanizing. In antiquity this tension was felt between the Greek rationalist philosopher Democritus, who described a mechanical universe made of inert atoms, and Plato, for whom the cosmos was alive and filled with purpose. In modern times Enlightenment reliance on reason culminated in the invention of calculus and the German Idealism of Immanuel Kant but incited rebellion in Kant's philosophical heirs, the second-generation Idealists Friedrich Schelling and G.W.F. Hegel, who put their faith in feeling and intuition.

I begin (chapter 1) with an overview of mathematics and art from prehistory to the culmination of the classical ideals of rationality, objectivity, and universalizable knowledge during the Enlightenment. I focus on the origins of the Western tradition in Plato, who held that abstract objects (such as numbers and spheres) have an existence independent of human thought. This Platonic outlook has played a dominant role in the philosophy and practice of mathematics and science. I describe the ancient and medieval origins of Platonism in the West, including its religious associations in classical and Christian theology, and variations of Platonism in modern, international secular culture in the context of mankind's ongoing search for knowledge. Charles Darwin's *On the Origin of Species by Means of Natural Selection* (1859) precipitated the decline of organized religion and the emergence of an anti-metaphysical intellectual climate in the West. Mathematicians became wary of Platonism because of its long association with religious dogma, but nevertheless modern mathematics is rooted in Platonism, which is the philosophy held today by most working mathematicians.

Plato declared that there are two worlds. First, there is the natural world of physical objects (such as apples and oranges), which exists in time and space and is known by sense perception—by seeing, hearing, and touching. Second, there is the world of the Forms, including abstract objects (such as numbers and spheres), which exists outside time and space and is known by cognition, intuition, or mystical apprehension. Apples and oranges exist in a "world-out-there," independent of the human mind, as do numbers and spheres, which similarly inhabit a "mathematical-world-out-there." Since numbers and triangles are perfect and eternal, mankind knows these abstract objects (of *mathematics*) with objective certainty and with certitude—a subjective feeling of certainty. Transitory, imperfect plants and animals, earth and sky, embody timeless numbers and perfect geometry, giving nature a fundamental unity. But since natural objects are imperfect embodiments of abstract objects, man's knowledge of nature (of *science*) is inherently partial and changeable.

Plato said, furthermore, that over and above these two worlds there reigns supreme divine Reason—the Good—a mythical Craftsman who created earth and sky by imposing Platonic Forms onto primordial, chaotic matter. The Good was Plato's source of a higher purpose for the cosmos. Opposed to Plato's view were the atomists Democritus and his follower Lucretius, who described a deterministic universe of matter in motion. Lucretius made room for free will (a mind that does not feel "constrained like a conquered thing") by introducing the random swerving of atoms (*De rerum natura* [On the nature of things], first century BC). But Plato found inert atoms in motion (even with an occasional swerve) inadequate to explain the world-out-there, and so he declared the cosmos to be composed of sentient (living, feeling) particles called *monads* (*monad* is Greek for "one"), infused with a World Soul and divine Reason. I highlight the development of the axiomatic method of proof by a follower of Plato, Euclid, in his *Elements* (ca. 300 BC), and Plato's view that art is a *mimêsis* (imitation) of nature. I trace the history of these classical Western views of mathematics and art after the fall of Rome, including their merger with Judeo-Christian theology in the fourth century AD, and the preservation of the Greek texts in Arabic translation by medieval Islamic scholars.

I also review developments in ancient Asian mathematics, focusing on *The Nine Chapters on the Mathematical Art* (246 mathematical problems compiled anonymously before 100 BC), which has served as the textbook for basic education in mathematics throughout the East for over two millennia, much as Euclid's *Elements* has in the West. In opposition to the common view that Greek mathematics was abstract and Chinese case by case, I summarize recent scholarly opinion that Chinese mathematics was abstract without generalization, although the Chinese did not develop a general method of proof.

Following the British scholar Joseph Needham and subsequent Sinologists, I attribute the difference between the Eastern focus on particular examples (considered as abstract paradigms) and the Western abstract, general, axiomatic approach to the ancient cosmologies that formed in these regions. The way people conceive of ultimate reality—whether the ground of being is an unfathomable mystery, a World Soul, monads, or atoms—relates directly to how they think about mathematics. In Chinese thought, the natural world is a harmonious balance of parts that came into being by self-assembly following the Tao (way) of nature that is ultimately inscrutable. Consequently, there never arose within Taoism the assumption that there is a mathematical-world-out-there to explore or laws of nature to discover, and mathematics and science did not advance beyond a basic stage in China during the early dynasties (2100–256 BC) and the imperial era (221 BC–AD 1911). Western creation myths feature a divine person (the Babylonian Marduk, the Hebrew God of Abraham, Plato's Craftsman) who imposed order on chaos and decreed laws of nature. Over the centuries many in the West have looked for the order established by this divine person, leading some to gain knowledge of the laws of nature: ancient Babylonian astronomers discovered the ecliptic, the Enlightenment's Johannes Kepler discovered that the planets move in geometric patterns (plate 0-2), and Isaac Newton discovered the law of universal gravitation.

With this background in place, I turn to the modern and contemporary eras. In chapter 2, I debunk the widely held misconceptions that Euclid's "mean and extreme

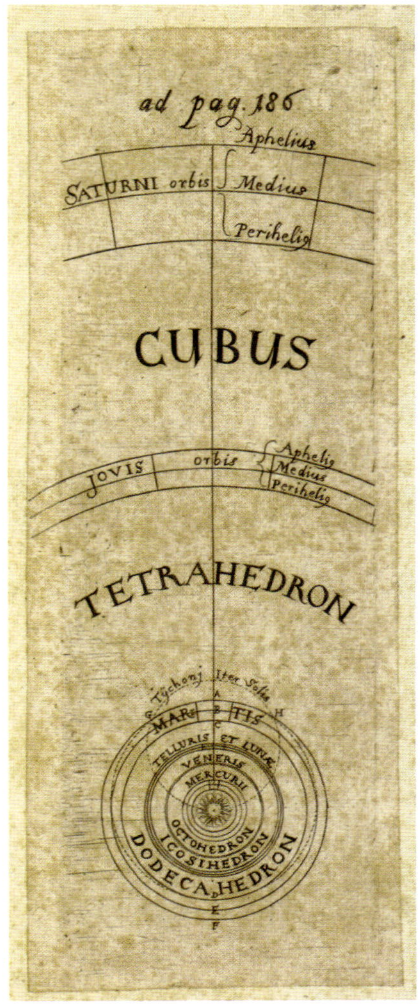

0-2. Diagram of the geometric structure of the solar system by Johannes Kepler in *Harmonices Mundi* (Harmony of the world, 1619); the diagram occurs between pages 186 and 187. Woodcut. Music Division, The New York Public Library for the Performing Arts, Astor, Lenox, and Tilden Foundations.

In AD 1596 the German astronomer Kepler hypothesized that the planets move in circular orbits that circumscribe an invisible cube (*cubus* in Latin), a tetrahedron, and other regular solids. When observation did not conform to this, Kepler sought other geometric forms in the heavens, and by 1609 he had determined that planetary orbits are elliptical.

ratio" (approximately 1.618) is the key to beautiful proportion (the so-called Golden Section) and that it was used in major monuments of art history (such as the Pyramids, the Parthenon, and Leonardo's *Mona Lisa*). I summarize systems of proportion based on the human body that became obsolete after Darwin presented overwhelming evidence that the body does not have a fixed form but evolves over time. Theology had to make way for science as more and more features of the human body and soul yielded to the explanatory power of biology, physiology, and psychology, causing a seismic change from belief in a divine person to denial of the very Creator whose existence had driven Western mathematics for four millennia. This cataclysmic reorientation prompted many attempts to formulate a creed suitable to the secular era.

Chapters 3–10 are, with minor exceptions, centered in Germanic communities, where modern mathematics and abstract art emerged. Groups of scholars and artists in Germany, Austria, Russia, Eastern Europe, Switzerland, Holland, and Scandinavia were "Germanic" in the sense that German was their common language and German Idealism was their shared intellectual heritage. In chapter 3, I describe a distinctly Germanic version of the struggle between reason and intuition that played out as Enlightenment reason (Kant) versus the Romantic imagination (Schelling and Hegel). The battle lines were drawn when René Descartes distinguished between mind and matter and the Enlightenment's pre-eminent philosopher, Kant, declared that one can know only mind (one's own ideas) and not matter (not the moon in the world-out-there). Kant's pronouncement prompted a reaction—German Romanticism—from second-generation German Idealists Schelling and Hegel, the *Naturphilosophen* (nature philosophers). They rejected Kant's solipsism and denied the Enlightenment mind-matter dualism, reviving the ancient Platonic view that everything is made of one sentient substance—monads. The *Naturphilosophen* described a hierarchy of monads at the top of which was an impersonal, supreme intelligence, "Absolute Spirit," which for them was equivalent to the logical structure of all the sentient particles comprising the universe. Another nineteenth-century group of Germanic philosophers known as the *Lebensphilosophen* (life philosophers), Arthur Schopenhauer, Søren Kierkegaard, and Friedrich Nietzsche, argued that the focus of philosophy should be not arid abstractions but the subjective value and purpose of life. Other nineteenth-century rebels turned their backs on deterministic calculus and developed probability theory because, in their opinion, it was a tool for describing the random, uncertain events in human life. I also describe Georg Cantor's invention of set theory as well as his philosophy of the infinite, the culmination of which was a divine "Absolute Infinity," which Cantor proposed in response to the rise of rationalistic science. In the 1910s there was a similar Romantic uprising against soulless calculus in Russia, where the mathematician P. A. Nekrasov set about to use probability theory to "prove" that mankind has free will. Cantor's theory of transfinite numbers was popularized in Moscow by the Russian Orthodox monk and mathematician Pavel Florensky, who inspired Suprematist poets and painters to symbolize infinity.

Cantor's creation of a non-Euclidean arithmetic of infinite sums, together with the discovery by German and Russian mathematicians of non-Euclidean geometries, precipitated a crisis in the foundations of mathematics in the late nineteenth century. In the next three chapters I treat the main responses to this decisive turning point: formalism, logicism, and intuitionism. In chapter 4, I describe David Hilbert's formalist conception of mathematics as an axiomatic system: an internally consistent, self-contained arrangement of abstract, meaning-free, replaceable signs. After detailing how this concept entered Russian literary circles, I describe how Russian Constructivist artists adopted a formalist aesthetic and made paintings and sculptures composed of meaning-free colors and forms arranged within autonomous, imaginary realms. Logicism—the view that mathematics is based on logic—was the premise of modern symbolic logic as developed by the German logician Gottlob Frege and his heir, the British mathematician Bertrand Russell (chapter 5). Logicism developed into British analytic philosophy, which was expressed by the sculptors Henry Moore and Barbara Hepworth, and authors T. S. Eliot and James Joyce. Hilbert and Russell both held modern versions of Platonism (they interpreted Plato's Forms in the context of Cantor's sets), but the leading intuitionist mathematician, the Dutchman L.E.J. Brouwer (chapter 6), declared that abstract objects (circles and triangles) exist only in the human mind. Brouwer, who founded the field of topology, was part of a wave of German Romanticism that swept over Holland at the fin-de-siècle, prompting mathematicians and artists to leave the cities and live in rural communes, connect with nature, and trust their instincts. The amateur mathematician M.H.J. Schoenmaekers, who knew Brouwer and frequented artists' communes, was the conduit of these intuitionist notions to the De Stijl painters Theo van Doesburg and Piet Mondrian.

While early-twentieth-century mathematicians excavated the foundations of their field, scientists, led by Einstein, explored the symmetry of nature, such as the symmetry of mass and energy—mass can be converted into energy, and vice versa ($E = mc^2$). Scientists described such symmetries using the mathematics of group theory (chapter 7). Two mathematicians based in Zurich, Hermann Weyl and Andreas Speiser, wrote popularizations of group theory that inspired Swiss Concrete artists, led by Max Bill, to create art with striking symmetry.

The search for the foundations of mathematics led to the formulation of basic principles (axioms) that underlie arithmetic (1889), geometry (1899), set theory (1908), and logic (1910–13). These successes led Hilbert to suspect that there could be an even lower level to the foundation—a bedrock set of axioms for all branches of mathematics—and he challenged his colleagues to find it (chapter 8). Mathematicians undertook this search in the aftermath of the German defeat in World War I, when there was a Romantic backlash against the exact sciences. In chapter 8 I also describe the advent of quantum physics in the 1920s, and, following the historians of science Max Jammer and Paul Forman, I describe the impact of Romanticism on the so-called Copenhagen interpretation of the subatomic realm. No one then (or now) disagrees about the practice of quantum mechanics (which has produced today's technological world of computers and smartphones), but several of its key founders, including Niels Bohr and his protégé Werner Heisenberg, gave it a philosophical interpretation that remains controversial to this day. Manifesting the prevailing Romantic resistance to reason, these physicists interpreted their data as *inherently* probabilistic and proof that at the most fundamental level of reality, chance rules, as Heisenberg proclaimed: "Quantum mechanics establishes the final failure of causality" (Heisenberg, *Uncertainty Principle*, 1927). I describe the ensuing debate between Germanic physicists who, like Bohr and Heisenberg, declared that nature is fundamentally indeterministic and reality is in the mind of the observer, and the incredulous holdouts for determinism and a world-out-there independent of human observation, including the Frenchman Louis de Broglie and an exasperated Einstein, who exclaimed: "The moon is there even when I'm not looking at it." I end chapter 8 by describing utopian visions—based on both reason and intuition—of artists in Germany in the 1920s and designers associated with the Bauhaus school of design.

Mathematicians who took up Hilbert's challenge never doubted that there *was* one bedrock set of axioms waiting to be discovered. Then in 1931 (chapter 9) a young Viennese logician, Kurt Gödel, proved that such a set of axioms does not—*cannot*—exist because there are inherent limits to artificial, symbolic languages. His Viennese contemporary Ludwig Wittgenstein, who was a fan of both the intuitionist Brouwer and the *Lebensphilosophen*, established a parallel result for natural, spoken language by demonstrating its limits in *Tractatus Logico-Philosophicus* (1921). M. C. Escher and René Magritte were contemporary with these results, and their prints and paintings, like the proofs of Gödel and Wittgenstein, entail paradoxes. I argue, however, that there is no historical evidence to support the common claim that Escher and Magritte were inspired by Gödel and Wittgenstein. Rather, these artists and mathematicians share a common source in nineteenth-century philosophers, such as Nietzsche, who denounced system builders and revelled in enigma. After the proofs of Gödel and Wittgenstein were popularized in the mid-twentieth century, their writings did, however, inspire many artists, such as the American Jasper Johns and the Chinese Gu Wenda.

I begin chapter 10 by observing that Gödel's 1931 theorem was important not only for his surprising result, but also for the new *method* that Gödel invented to achieve his result—a proof by computation—that propelled the development of computers. Rather than do a traditional deductive proof, Gödel encoded mathematical statements in numbers and *computed* his theorem. This new method inspired the young British mathematician Alan Turing to do a thought experiment about a machine that can compute ("On Computable Numbers," 1936). Three years later England declared war on Germany, and Turing joined cryptologists at the British government's secret facility Bletchley Park to build a *real* mechanical computer to decipher Enigma, the German military code. After World War II, Turing joined others to develop these primitive machines into the computer industry.

In chapter 11, I describe how the destruction of Germanic intellectual communities during World War II undermined confidence in Enlightenment ideals of rationality, objectivity, and universalizable knowledge for all who suffered losses. But in countries (including France, Britain, and the Americas) that were not utterly destroyed and which lacked a strong tradition of German Idealism, post-1945 artists continued to express confidence in Enlightenment ideals by creating geometric abstract art that was orderly, executed with detachment, and expressed the power of human reason. The post-war generation of

0-3. Halo of the Cat's Eye Nebula (NGC 6543), 2008. Courtesy of Stefan Binnewies and Josef Pöpsel, Capella Observatory, Mount Skinakas, Crete, Greece.

The Cat's Eye Nebula is a sun-like star that for unknown reasons lost its outer envelope, forming this "halo." The nebula is located 3000 light-years away in the constellation Draco.

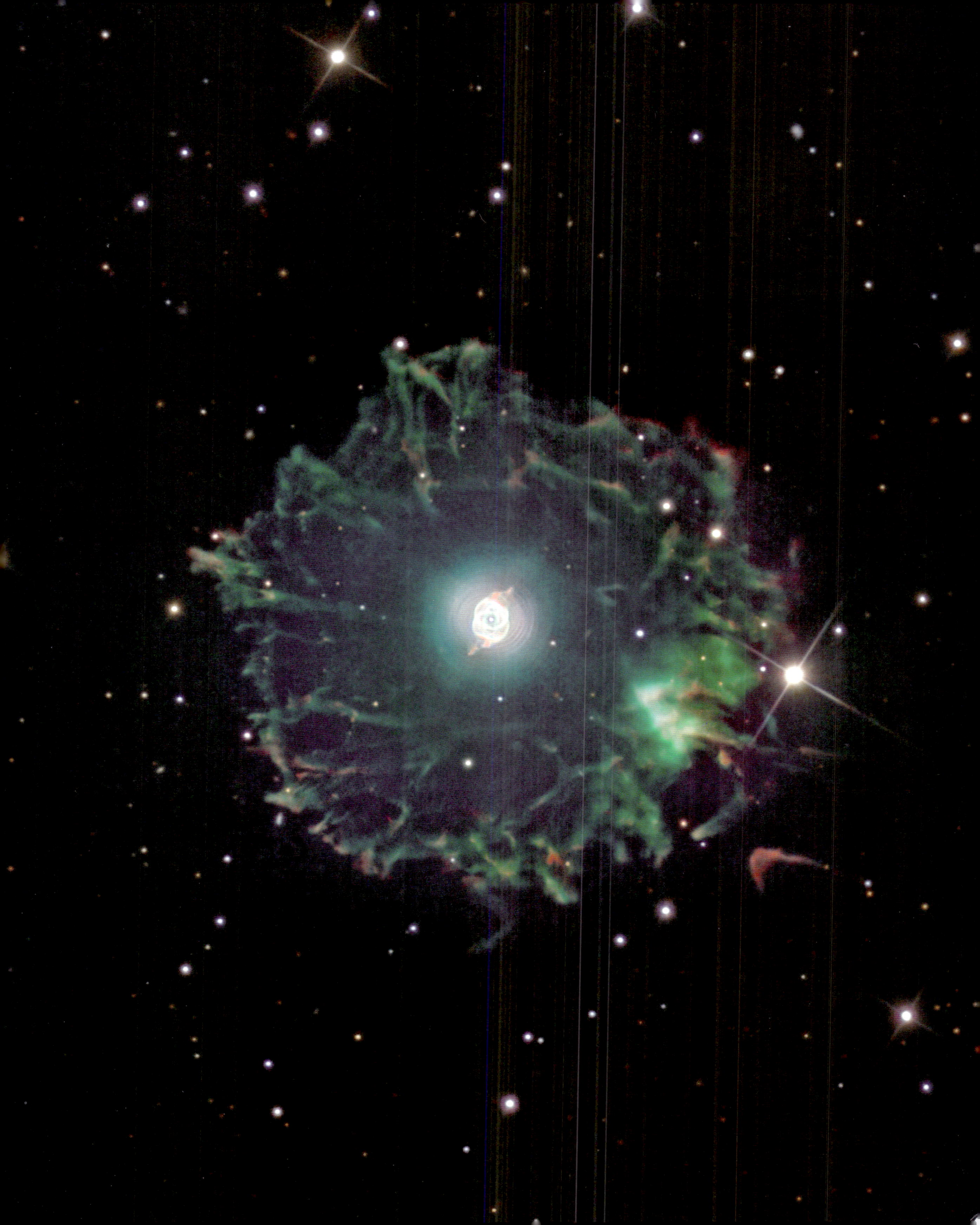

scientists in these countries was filled with assurance that it could find stable, predictable patterns by looking deeper into the subatomic microworld and further out into the cosmic macroworld (plates 0-3 and 0-4). I also describe the rapid development of computers in the post-1945 era in Britain and America (chapter 12), and their adoption as a tool by both mathematicians, as in fractal geometry (1975) and the first computer-assisted proof (1976), and artists, as in digital photography and computer animation for special effects in film.

In contrast, centers of mathematics and science in Berlin and Göttingen were in disarray after 1945, when two German Jews who had spent the war in exile, Theodor Adorno and Max Horkheimer, returned to their homeland and wrote the first in-depth analysis of the loss of confidence in Enlightenment ideals, a condition that came to be called postmodernism (*Dialectic of Enlightenment*, 1947; chapter 13). I conclude the book by observing that mathematics has been largely (although not completely) immune to postmodern critiques of terms such as "truth" and "certainty" because these concepts are so deeply rooted in the history of mathematics. Indeed, the certainty of mathematics gives it a unique status in modern culture, where the interplay between mathematics and the natural world underlies all science and technology.

Almost all the interactions between mathematics and art that I describe are cases of an artist being inspired by a piece of mathematics, and not vice versa (if one excludes mathematicians who strive for aesthetic qualities such as beauty and elegance, as distinct from being inspired by a particular work of art). I have noted the rare exceptions, such as the nineteenth-century French engineer Jean-Victor Poncelet, who generalized Renaissance architect Filippo Brunelleschi's linear perspective into projective geometry. In describing these interactions, I have tried to move beyond vague expressions of intellectual climate and show specific links between the mathematician's study and the artist's studio, such as a historical document or a popularization of mathematics known to the artist. Sometimes I've used a psychological approach in the sense of describing the personality and intellectual environment of a mathematician or artist (when this is known and relevant), and at others I've taken the more sociological approach of considering the mathematician or artist as the product of a particular cultural matrix.

0-4. Center of the Cat's Eye Nebula (NGC 6543), 2004. NASA, ESA, HEIC, and The Hubble Heritage Team (STScI/AURA).

Located at the center of the Cat's Eye Nebula, this cloud-like luminous object was formed by a series of pulses at 1500-year intervals, during which the star gently ejected eleven or more gaseous rings. The outermost visible ring measures 1.2 light-years in diameter.

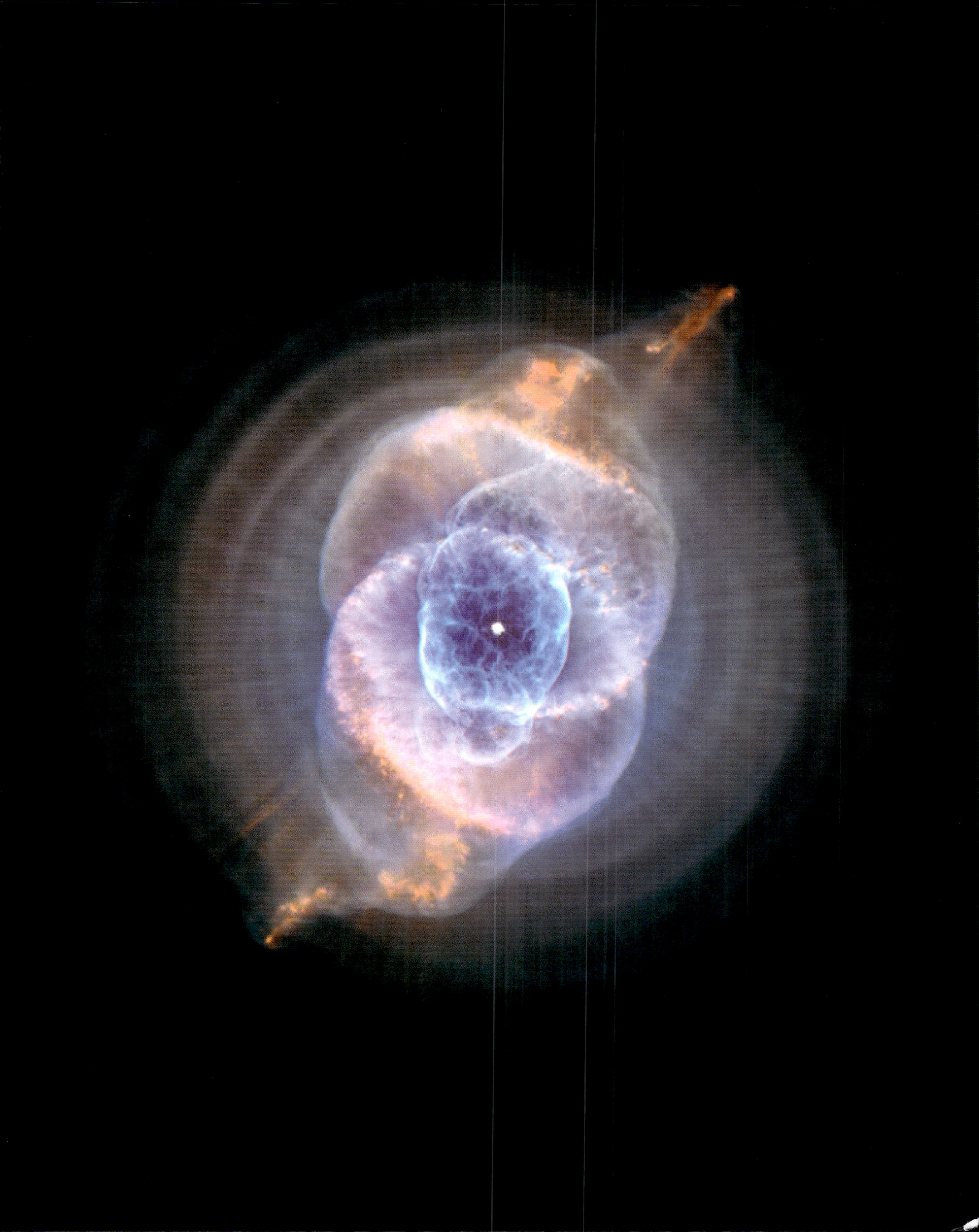

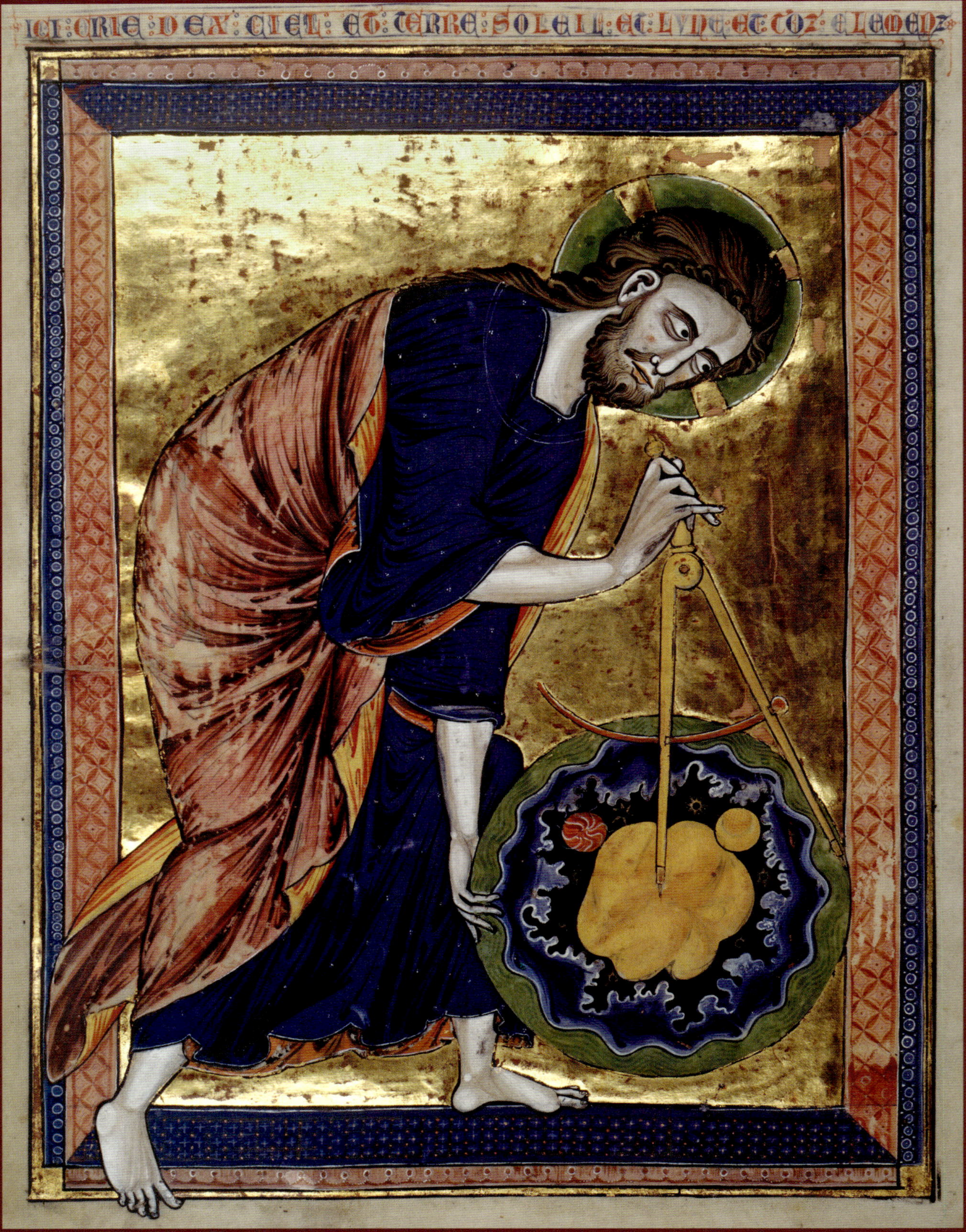
ICI CRIE DEX CIEL ET TERRE SOLEIL ET LUNE ET TOZ ELEMENZ

1
Arithmetic and Geometry

I do not mean by beauty of form such beauty as that of animals or pictures . . . but straight lines and circles, and the plane and solid figures that are formed out of them by turning-lathes and rules and measures of angles; for these I affirm to be not only relatively beautiful like other things, but they are eternally and absolutely beautiful.

—Plato, *Philebus*, 360–347 BC

WHAT ARE NUMBERS? How does mankind know them? What is art? Why do people sing, dance, and draw pictures? Biologists and anthropologists have sought satisfactory answers to these notoriously difficult questions by describing the evolution of *Homo sapiens* into a species that perceives patterns and seeks pleasure.

The ability to recognize quantity—the number of bananas in a bunch or predators in a pack—has an advantage in the struggle for survival. Biologists have demonstrated that many birds and mammals alive today, as well as human infants, have an innate ability to distinguish between small numbers of discrete objects (1, 2, 3, 4, and "many") and to add and subtract these simple sums.[1] *Homo sapiens*' numerical competence is based on neural circuitry for counting and reasoning that they share with animal relatives at least as distant as birds and rodents. This means that when early humans evolved from apes, they inherited the preverbal mechanisms for counting, addition, and subtraction—the basis of arithmetic.

Mankind's ape-like ancestors—and apes still today—could chip the edges of stone to make tools, but the marks they made did not produce symmetrical patterns (plate 1-2). But around 1.4 million years ago, modern man's hominid ancestor *Homo erectus* came down from the trees to live in Africa's tropical grasslands, walked upright, and made the first flat stone tools with balanced halves and symmetrical outlines (plate 1-3).[2] After another million years *Homo sapiens* began chipping three-dimensional symmetrical tools in flaked stone, and, by 300,000 years ago, he carved tools in the round with striking symmetry (plate 1-4). This suggests that the human perceptual/cognitive system evolved in that interval, between 1.4 million and 300,000 years ago. In humans living today, the recognition of flat, two-dimensional shapes (which *Homo erectus* needed to impose a symmetrical outline) does not require the high functioning of the cerebral cortex. However, the perception of left and right and the ability to perceive three-dimensional forms, rotate them in the mind, and judge whether they can be exactly superimposed (which *Homo sapiens*

1-1. Illustration of the Genesis story of creation in a *Bible moralisée*, French, ca. 1208–15. Egg tempera and gold leaf on vellum. Österreichische Nationalbibliothek, Vienna, Bildarchiv, Codex 2554, fol. 1v.

Using a compass, the Creator shaped the cosmos from chaos by drawing its spherical boundary, and then he formed the red sun and golden moon. Next he will use his compass to shape the lump of earth that lies at the center of his cosmos. This is the frontispiece of a *Bible moralisée*, a manuscript containing paraphrases of biblical verses together with moralizing interpretation, which was created for the French royal family by scholarly clergy in Paris in the early thirteenth century. Above the picture the iconographer instructs the royal family that only almighty God (as opposed to mortal astronomers) can encompass the entire cosmos, declaring (in Old French), ICI CRIE DEX CIEL ET TERRE SOLEIL ET LUNE ET TOZ ELEMENZ (Here God creates heaven and earth, sun and moon, and all the elements).

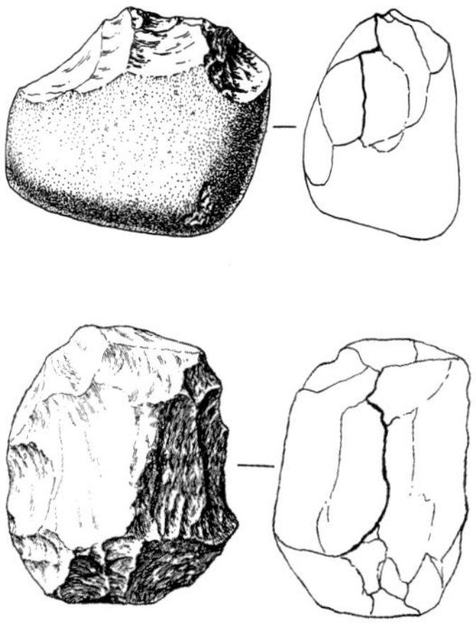

1-2. Asymmetrical tools, 1.8 million years old. Mary D. Leakey, *Olduvai Gorge* (Cambridge: Cambridge University Press, 1971), 3, fig. 9 on 27, and fig. 11 on 29. © 1971 Cambridge University Press. Used with permission.

Front and side views (above) of a side chopper made from a water-worn rounded piece of lava, and a two-edged chopper made from lava (below).

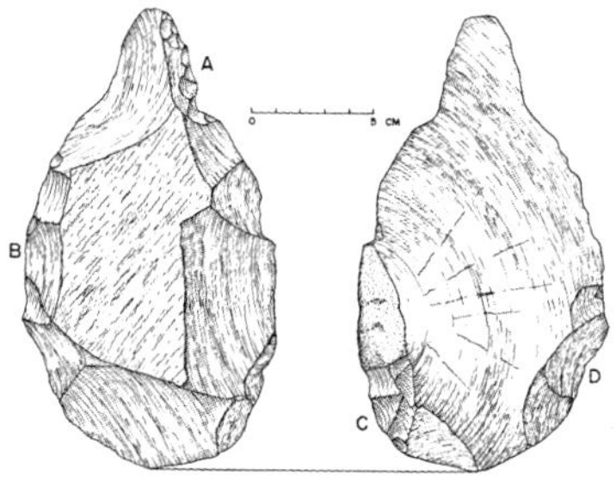

1-3. Hand ax with symmetrical outline, 1.4 million years old. Courtesy of Thomas Wynn.

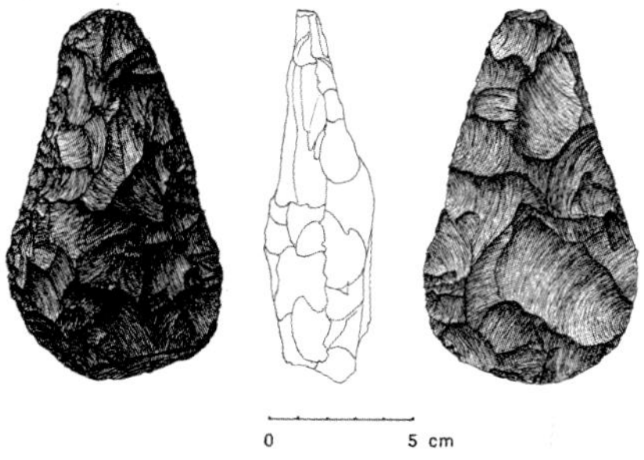

1-4. Symmetrical tool in the round, 150,000–300,000 years old. Courtesy of Thomas Wynn.

needed to carve stone tools in the round) requires a high level of processing in the cerebral cortex, the part of the brain that evolved most recently.[3] Anthropologists hypothesize that the evolution of this spatial perceptual/cognitive system was part of a more general increase in intelligence, and it may have been selected in reproduction, if potential mates saw the ability to make symmetrical tools as a sign of sharp wits and good health.[4] Today newborns have the neurological repertoire to recognize and mentally manipulate simple shapes and forms, and as they grow, children develop the ability to distinguish more complex patterns of shapes and forms—the basis of geometry.

When *Homo erectus* began walking upright, this change in body position prompted not only the development of the perceptual/cognitive system but also the evolution of aesthetic behavior, according to the following anthropological theory.[5] When the womb and birth canal of female bipedal hominids shifted to a vertical orientation (as opposed to horizontal in lower mammals), the result was that as the evolving *Homo sapiens* brain got heavier, the force of gravity pulled the fetus "prematurely" from the womb. Even after birth, the brain of infant *Homo sapiens* continues to grow, its skull bones knitting for another twelve months, during which the mother invests enormous time and energy into caring for the child. When other family members and friends interact with the baby, they elicit a response by speaking in high, lilting sounds or moving rhythmically (clapping, swaying). A responsive infant who joins in this "performance" by smiling and cooing at the pleasant sounds or playing in rhythm with the adult—all of which are *aesthetic* behaviors—receives more attention for longer periods of time from the aunts, uncles, cousins, and neighbors with whom the baby interacts. In other words, the "performing" baby receives more caretaking during its first year of life, and as a result the infant thrives. In this way aesthetic behavior evolved into an innate human trait.

Although the ability to do simple arithmetic was hard-wired into mankind's brain millions of years ago, it was only around 300,000 years ago that the cognitive hardware evolved to perceive and manipulate shapes and forms, which is the basis of geometry, and to possess a preverbal desire for pleasure and human bonding, which is the basis of art. By about 200,000 years ago, the brain of the modern human being—*Homo sapiens*—was fully evolved in terms of its size and shape.

In Africa, the Near East, Europe, and Asia between about 40,000 and 10,000 years ago, human artifacts suddenly show extensive evidence of symbolic thought (plates 1-7 and 1-8). Signs of the use of algorithms (step-by-step procedures to accomplish a task) also appeared, such as carving a pattern of holes in a bone (plate 1-6).[6] These abstract systems must develop in a cultural context, passed on from generation to generation, because although an infant is born with the neurological circuitry to perceive three-dimensional form, to seek pleasure, and to compose sentences, a child does not spontaneously produce mathematics, art, and language. To develop these abstract cognitive skills, a child needs to imitate others in a community. Algorithms were used to apply decorative patterns throughout the Neolithic era between around 10,000 and 2500 BC (plate 1-5), when complex societies formed permanent communities, and the evolutionary drive shifted from biology to culture.

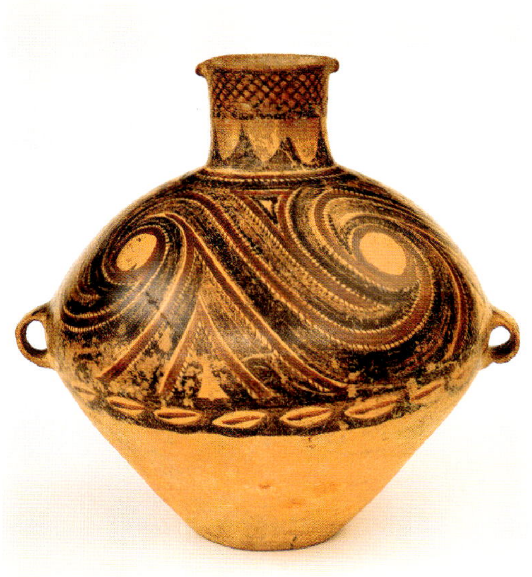

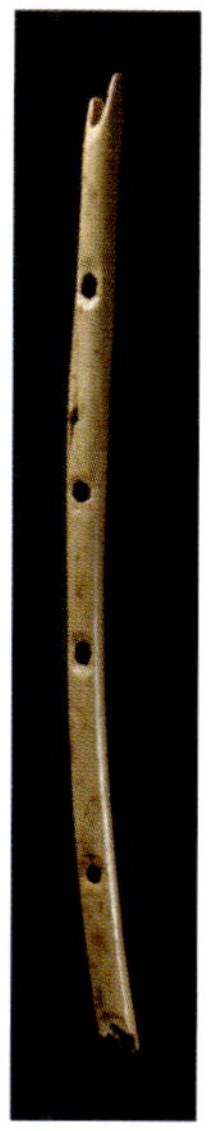

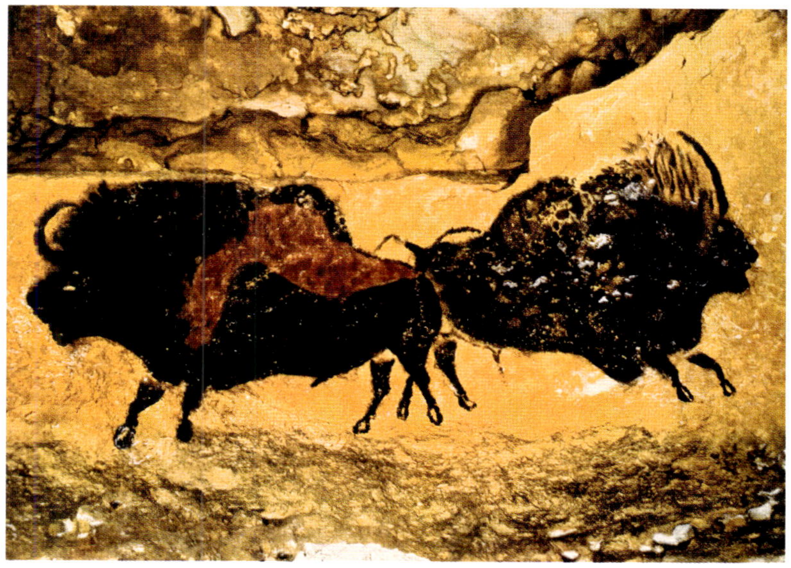

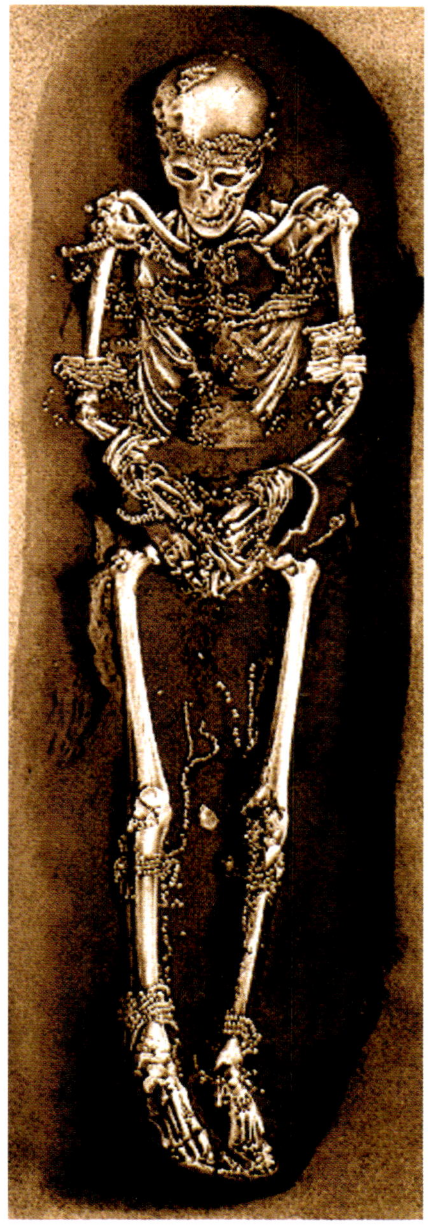

ABOVE LEFT

1-5. Storage jar, Gansu Yangshao culture, Banshan type, Gansu or Qinghai Province, China, ca. 3000–1000 BC. Earthenware painted with red and black slip, 15⅝ in. (40 cm) high, diameter 13¾ in. (35 cm), without handles. Asia Society, New York, Mr. and Mrs. John D. Rockefeller III Collection of Asian Art, 1979.125.

ABOVE CENTER

1-6. Flute found in the Hohle Fels cave in the Danube River valley near Ulm, Germany, 43,000–42,000 years old. Bone, ca. 8½ in. (21.6 cm) long.

One of the first anatomically modern humans in central Europe made this flute by carving a row of finger holes in a bone from the wing of a griffon vulture, a large bird of prey. Thin and hollow but very strong, bird bones are especially suited to making flutes. At one end of the bone the musician cut a double-notched, V-shaped hole, which is assumed to be the mouthpiece, suggesting that this flute was played in a vertical position, like a modern recorder. This and other flutes (bones with a similar pattern of holes) found in nearby caves along the Danube River are the oldest known musical instruments in the world.

ABOVE RIGHT

1-7. Bison painted on the wall of the Upper Paleolithic cave at Lascaux, France, ca. 15,000 BC.

RIGHT

1-8. Burial of a Cro-Magnon man in Sunghir, Russia, 30,000–28,000 years old. Institute of Archaeology, Moscow.

This early modern man was laid to rest in a garment sewn with 3,000 beads, each individually carved from ivory. He was a member of a Cro-Magnon tribe of humans—*Homo sapiens*—who lived in Europe and Russia.

ANCIENT FOUNDATIONS OF MATHEMATICS: ABSTRACTION AND GENERALIZATION

By 3000 BC, in settlements in the Nile, Tigris-Euphrates, and Yellow River valleys, scribes kept numerical records by making marks, such as | for one, || for two, ||| for three, and so on. To avoid a profusion of marks, others were invented for larger numbers, such as the Egyptian ∩ for ten, so that twelve could be written ||∩. Other marks—perhaps a simple picture—signified objects, such as ⌓ for a loaf of bread (in the Egyptian hieroglyph ||⌓

The Tomb of Menna

Menna, whose title was Scribe of the Fields of the Lord of the Two Lands, was overseer of all agricultural activities on the royal estates, which he observes seated at the upper left of the top and middle images. Menna watches farmers plant, harvest, and transport crops. In the center of the top register of the middle image, his surveyors measure the fields preliminary to assessing taxes that the farmers must pay under threat of being beaten for underpayment (top right). Surveyors are at the front and back of the rope (upper register, and in the color detail), each carrying a coiled rope around his shoulder, and they pull taut a rope that is knotted at regular intervals for measuring units of distance. The front of the rope is partly coiled, ending in a knot in the hand of the lead surveyor, and the rope ends in a knot that dangles from the hand of the rear surveyor.

Egyptians measured with a unit of length—a cubit—that was originally based on the length of a forearm, although it was longer than most people's forearms, measuring about 20 1/2 in. (52 cm); the cubit was subdivided into seven palms and four fingers. The knots in this example appear to be 3 cubits apart (a little over 5 ft.). The surveyors are accompanied by a boy with a bag (probably for the rope) and two scribes, who carry pallets on which to record the size of the field of grain shown in the background. An old man with a staff and two boys walk alongside the survey team, which is met by a couple carrying food and drink.

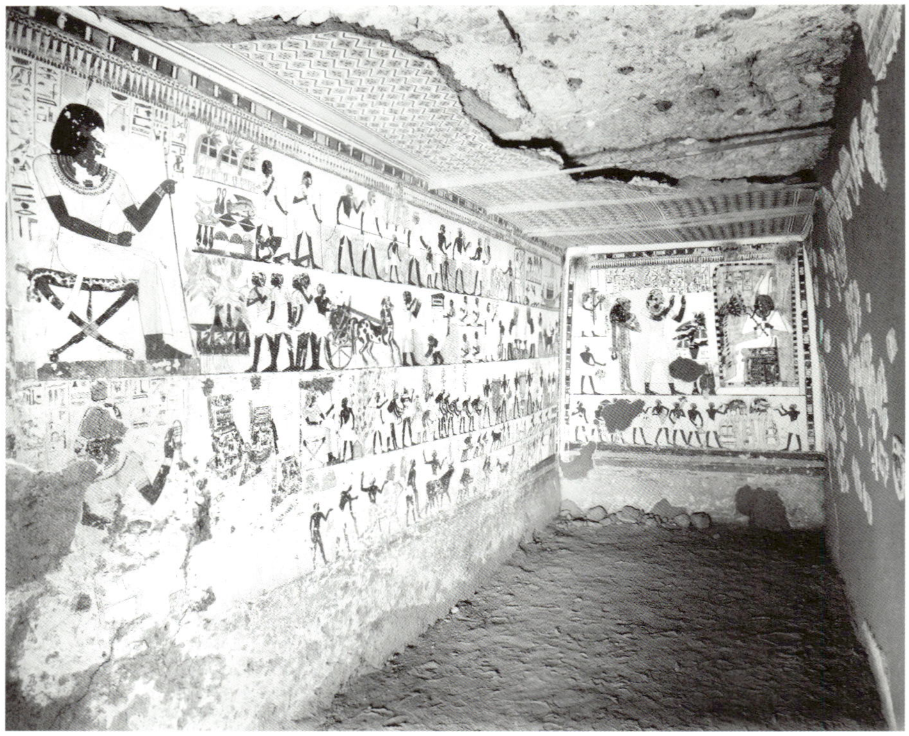

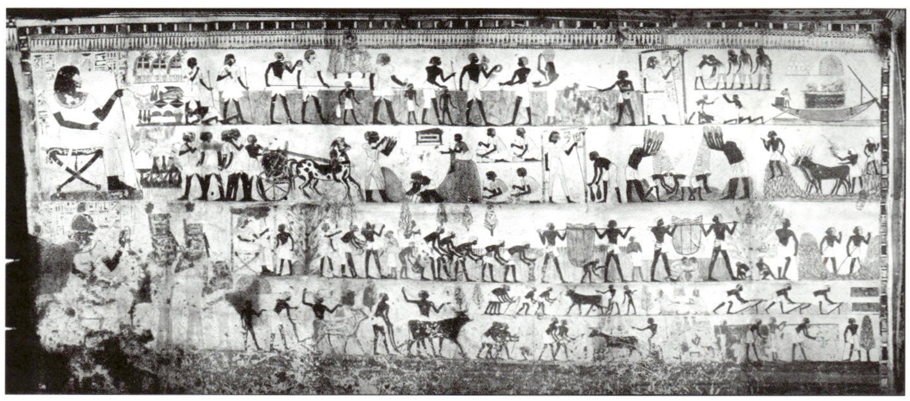

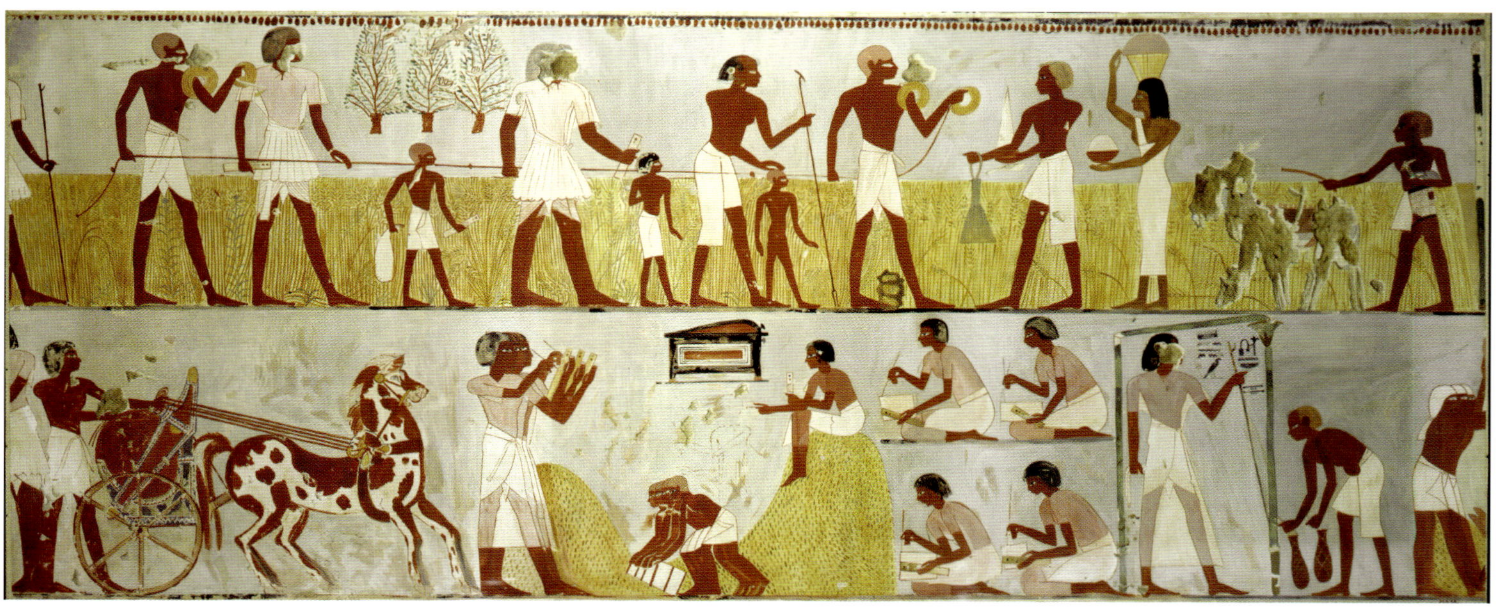

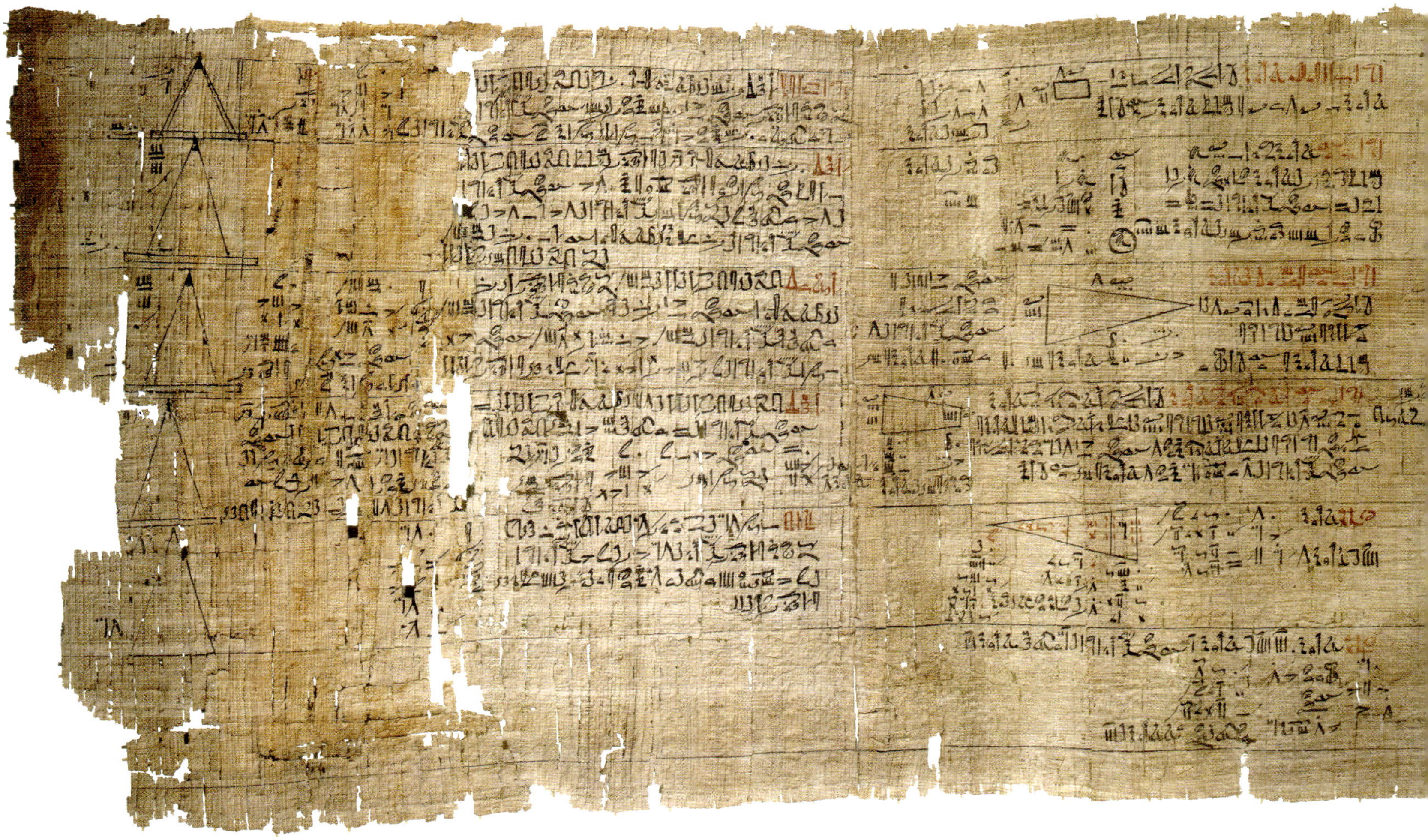

OPPOSITE

1-9. Tomb of Menna, Sheikh Abd el-Qurna, Thebes, Eighteenth Dynasty, late fifteenth–early fourteenth century BC.

Above: Black and white photograph of the Tomb of Menna, taken around 1920 during the Metropolitan Museum of Art Expedition (1907–35). Metropolitan Museum of Art, New York, inv. no. T807.

Middle: Black and white photograph of the left wall, taken around 1920 during the British Expedition (1902–25), led by Robert L. Mond. Griffith Institute, University of Oxford, inv. no. Mond 11.22.

Below: As a member of the Metropolitan Museum of Art Expedition, the job of the American artist Charles K. Wilkinson was to document the ancient paintings in color. Around 1920, Wilkinson did this painting of harvest scenes in the upper register of the left wall of the Tomb of Menna. Tempera on paper, $29\frac{15}{16} \times 73\frac{1}{4}$ in. (76 × 186 cm). Metropolitan Museum of Art, New York, Rogers Fund, 1930.

ABOVE

1-10. Rhind papyrus scroll, ca. 1650 BC. Ink on papyrus, height $12\frac{1}{2}$ in. (31.7 cm). This section of the scroll includes problems 49–55 (right) and problems 56–60 (left). © The Trustees of the British Museum, acc. no. 10057r.

The Rhind papyrus scroll, which was found at Thebes, is written in hieratic (an abridged form of hieroglyphics) and includes geometric figures that are among the oldest known. Egyptologists estimate that the geometric knowledge embodied in these figures is at least one thousand years old, dating to the time of the pyramid-builders of the Old Kingdom.

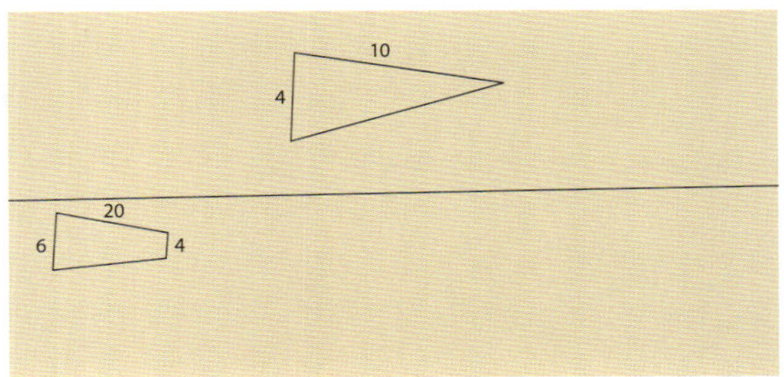

1-11. Diagram of problems 51 and 52 from the Rhind papyrus.

These problems give a lesson in how to calculate the area of a plot of land. The upper figure is a triangle with a base of 4 units and a height of 10. Assuming the height and base are perpendicular (despite the slight deviation in the Rhind papyrus figure), the accompanying text calculates the area of the right triangle as 20 units, which corresponds to the value given today using the modern formula: $A = \frac{1}{2}bh$, which reads "A equals one-half of b times h, where A is the area, b is the base, and h is the height." In the figure below, students of surveying were taught how to calculate the area of a truncated triangle (a trapezoid), with a base of 6, height of 20, and a second base (a cut-off side) of 4. The calculated result of 100 corresponds to the value obtained by the modern formula for finding the area (A) of a trapezoid:

$$A = \left(\frac{b_1 + b_2}{2}\right)h$$

where b_1 and b_2 are the lengths of the two bases and h is the height of the trapezoid.

RIGHT

1-12. Imhotep (Egyptian, ca. 2650–2600 BC), *Step Pyramid of Zoser*, Saqqara, Egypt, 2630–2611 BC.

This first Egyptian pyramid, the Step Pyramid at Saqqara, is the oldest surviving monumental stone structure in the world. The architect Imhotep designed the tomb by placing truncated pyramids—mastabas—of decreasing size on top of each other. The original base mastaba was first extended laterally, and then the "steps" were added vertically. Students of the Rhind papyrus were taught the surveyor's rule for calculating the area of a trapezoid, and scribes have recorded that Egyptian builders knew the related rule for finding the volume of a truncated pyramid. Thus when Old Kingdom workers began constructing tombs in the form of a mastaba, they were able to estimate the amount of material needed.

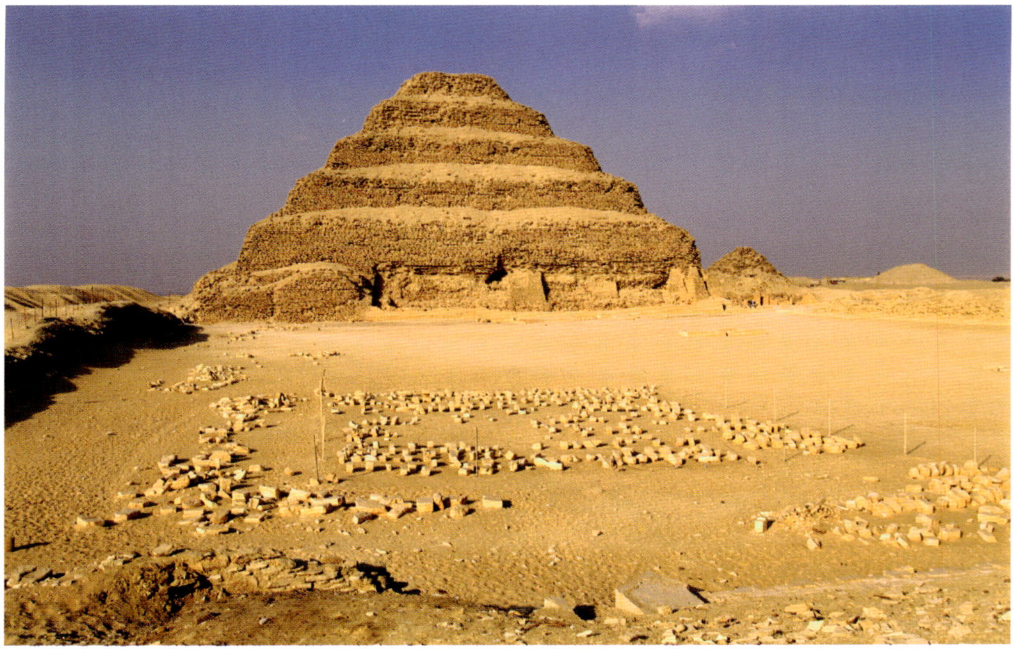

BELOW

1-13. A typical Old Kingdom mastaba tomb has the form of a truncated pyramid.

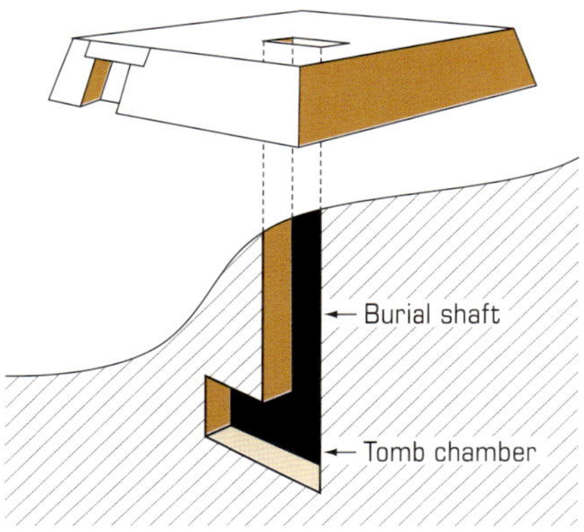

A new light flashed upon the mind of the first man (be it Thales or some other) who demonstrated the properties of the isosceles triangle.

—**Immanuel Kant**, ***The Critique of Pure Reason***, **1781**

for "two loaves"). The predominance of multiples of ten in ancient numerical systems is undoubtedly because human hands, on which counting began, have ten fingers.

The Egyptians and other ancient cultures employed whole numbers (1, 2, 3, . . .), ratios of numbers, written today as either 1 : 2, 1 : 3, 1 : 4, . . . (the ratios of one to two, one to three, one to four, . . .) or as 1/2, 1/3, 1/4, . . . (the fractions one-half, one-third, one-fourth, . . .), and the operations of addition, subtraction, and multiplication, the basis of arithmetic. Ancient peoples also began using triangles and rectangles to measure land and to set out right angles for the corners of buildings. For example, Egyptian surveyors kept careful records of the parcels of farmland along the Nile, so that each year after the floodwaters receded they could redivide the river valley using geometry (from the Greek for "measure the earth"). Surveyors could calculate the area of plots of land and measure the volume of architectural forms (plates 1-9, 1-10, 1-11, 1-12, and 1-13).

The abstraction of counting and measuring from rules of thumb to a systematic arrangement of general solutions or proofs—the step into pure mathematics—took place in Greek culture between the sixth and fourth centuries BC. Egyptian surveyors could find the area of a *particular* plot of land, but they did not move beyond mere calculation into the realm of abstract thought and develop a proof of how to find the area of *all* triangles or *all* rectangles.[7] In the sixth century BC, the Ionian Greek philosopher Thales of Miletus (in present-day Turkey) made that critical move beyond considering only particulars and generalizing to *all*. Thales brought Egyptian surveying methods to Ionia, where, according to ancient sources, he "attacked some problems in a general way"[8] and demonstrated properties of *any* right triangle and *any* triangle with two equal sides, so-called isosceles (from the Greek for "equal-legged") (plate 1-14).[9] In the sixth century BC Anaximander, a follower of Thales who was also from Miletus, declared that the structure of the cosmos is based on proportion and symmetry and that key facts about the natural world, such as the position of the sun, moon, and stars, can be expressed in numbers.[10] This Ionian Greek outlook marks the beginning of the scientific worldview in which nature is understood as embodying a mathematical structure that is discernable by human reason.

What led mankind to take this leap of the imagination from a particular object to its underlying form? The Greek philosophers' sense that the natural world embodies a mathematical structure is grounded in ancient myths that celebrate the victory of a celestial, divine mind over chaos; the world came into being when reason (spirit) imposed structure and laws on the formless void of primordial matter. The Babylonians believed that the divine Marduk ripped the goddess of chaos, Tiamat, in half to form earth and sky, positioned the sun, moon, and stars, and then "ordained the year into sections and he divided it; for the twelve months he fixed three stars" (*Enûma Eliš*, 1500–1100 BC). The god of the Hebrew prophets "laid the foundations of the earth," and in order to contain the rivers and the sea he "set a bound that they may not pass over," putting in place a body of laws to govern the natural world (Psalms 104:5 and 9, ca. 1000 BC; see plate 1-1, chapter frontispiece).[11] In the late eighth century BC, the Greek poet Hesiod told how heaven and earth separated themselves out from a primordial abyss—the chasm of Chaos—that was closed by Eros (sexual desire), creating harmony in the cosmos (*Theogony*, ca. 700 BC).

Babylonian, Hebrew, and Greek people set out to find the structure that had been imposed on formless chaos, and gradually, piece by piece, they found order in the movement of the stars and patterns in numbers. Since these ancient peoples all looked at the same natural world and contemplated the same mathematical-world-out-there, albeit from different cultural perspectives, their observations of nature and their insights about mathematics translated from culture to culture, and over the centuries a cumulative body of knowledge formed that is the basis of Western science and mathematics.

In the late sixth century BC the citizens of Athens on the Greek mainland threw off rule by aristocrats who had inherited their authority and tyrants who had seized power, and, led by Cleisthenes, they invented a new form of government—democracy (from the Greek for "government by the people")—that is historically linked to a scientific approach because of its anti-authoritarian assumption that all humans are equal and its appeal to human reason. In the political reforms that began under Cleisthenes in 508–507 BC, Greek male citizens were selected to hold public office by lottery (on the assumption that all were equally capable). Greek citizens practiced *isonomia* ("equal before law"); there were no judges in the courts but rather a jury of citizens, who were selected by drawing lots from a large pool. Public policy was set at meetings of the Assembly, a public forum open to all citizens, each of whom could speak and cast a vote to decide the issue. As in science and mathematics, truth in Greek democracy was determined by reasoned argument, not on the basis of governmental authority or because it was the custom of the land.

In this political climate, the next generation of Greek citizens began the artistic and literary achievements of the Golden Age of Athens, including the building of the Parthenon, the sculpture of Phidias, and the plays of Aeschylus, Sophocles, and Euripides. Then, in the late fifth and fourth centuries, the scientific approach that had been initiated by Thales and Anaximander in Ionia was developed in Athens to an extraordinary level of abstraction and generalization by Socrates, Plato, and Aristotle, culminating in Euclid.

1-14. Thales's theorems.

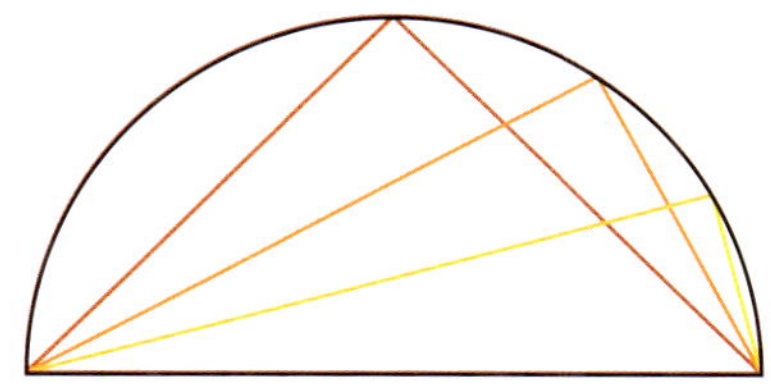

Any angle inscribed in a semicircle makes a right triangle.

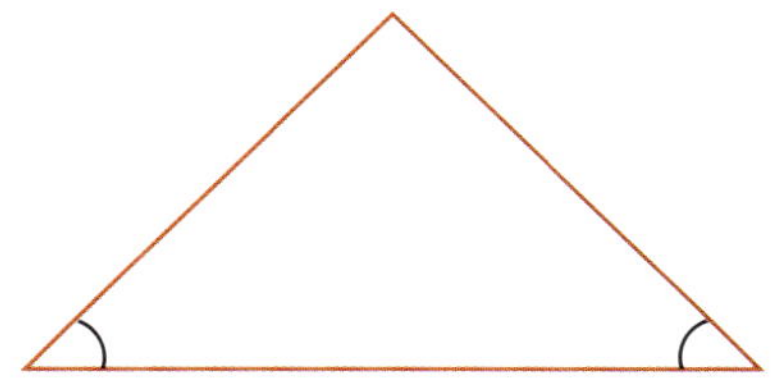

The base angles of an isosceles triangle are equal.

Majority rule is called by the fairest of terms: Equality before the Law. Next, it requires something the tyrant never allows: people hold office by lot, they are accountable for the actions of their administration, and their deliberations are held in public. I propose, therefore, that we abolish the monarchy and increase the power of the people, for in the many is all our strength.

—**Herodotus,**
***The Histories*, 450–420 BC**

Does any say there are gods in heaven? No! there are none if man will not be fool enough to credit the old tale. Let not my words guide your judgment; see for yourselves. I say that tyranny slays its thousands and despoils their goods, and men who break their oath cause cities to be sacked; and, doing so, they are happier than men who walk quietly in the ways of piety from day to day.

—**Euripides, *Bellerophon*, ca. 430 BC**

Greek religion of the sixth and fifth centuries BC centered on the worship of mythological heroes such as Heracles, nature deities such as Zeus, god of sky and thunder, and civic deities, such as Zeus's daughter Athena Parthenos (Athena the virgin), the goddess of Athens. But during the Golden Age, the Athenian understanding of the cosmos in terms of the unpredictable whims of these personal gods gave way to the concept of the natural world as the product of impersonal forces that were discernable by the human mind. After the outbreak of the Peloponnesian War (431–404 BC), confidence in Athena Parthenos was seriously undermined by decades of civil war that devastated Athens and caused poverty throughout the Greek islands.

With faith in Greek mythology shaken, an undercurrent of skepticism surfaced among Greek philosophers. Already in the late sixth century BC, Heraclitus had declared that there is no fixed, unchanging truth—"Everything flows"—from which his follower, Protagoras, inferred that each person's view is an equally valid description of the state of affairs: "Man is the measure of all things." Protagoras openly expressed doubt about the existence of Zeus and Athena: "As to the gods, I have no means of knowing either that they exist or that they do not exist. For many are the obstacles that impede knowledge, both the obscurity of the question and the shortness of human life."[12] The statesman and poet Critias (a relative of Plato) went so far as to suggest that a politician had invented the gods of Greek mythology to increase his power over citizens: "Although the laws kept them from open deeds of violence, men went on doing them in secret; and then it was, I believe, that some clever and sagacious man first invented for mortals the fear of the gods, so there might be something to frighten the wicked, even though their acts or words or thoughts were secret. . . . And for the dwelling of the gods he chose the place that would have the most startling effect on men, . . . the round sky above us, where he saw the lightning and the dreadful crash of thunder."[13]

In this atmosphere of doubt and suspicion about Heracles, Zeus, and Athena, some citizens of Athens perceived the scientific worldview as a threat to the Olympian pantheon. An example is the treatment of the Ionian philosopher Anaxagoras, who travelled from his homeland in Clazomenae (present-day Turkey) to Athens in the fifth century BC, and lectured in public about finding the structure of reality using reason. He described the cosmos as inert matter that had been put in motion by a purposive Mind (*Nous*), after which the natural world operated in predictable ways that could be discerned by man and described in physical (not divine) terms: the sun is an incandescent stone, and the moon is a mass of earth.[14] Fearing that such doctrine posed a serious threat to the authority of the sacred gods of Greek mythology, civic leaders arrested Anaxagoras and exiled him from Athens.[15] Fifty years later the Athenian state put the philosopher Socrates on trial, charging him with not believing "the sun and moon are gods, like other men. . . . since he says the sun is a stone and the moon made of earth."[16] Refusing to flee Athens, Socrates drank the fatal hemlock in 399 BC.

For with what aim did ye insult the gods, and pry around the dwellings of the moon?

—**Aristophanes, *Clouds*, later fifth century BC**

In addition to the scientific approach to understanding the natural world, a second tradition developed within Greek culture in so-called mystery cults, in which knowledge of the world was gained less by reason than in a moment of mystical insight, or intuition. Reason and intuition were used in both the scientific and mystical approaches, the difference being one of emphasis. From their observations of the cycles of nature—day and night followed by dawn, summer and winter followed by spring—members of the mystery cults of Dionysus, Orpheus, and Pythagoras believed that human life was also cyclical—life and death followed by rebirth. The Dionysus cult observed the life cycle of the grapevine, which produces vibrant, luscious grapes, whose dismembered bodies were fermented, producing intoxicating spirits that gave a taste of afterlife in the underworld. Returning to sobriety, they poured wine on the soil, giving libation to begin the cycle anew. The cycle of life-death-rebirth was also part of the cult of Orpheus, the legendary musician who descended into Hades in search of his beloved Eurydice and then returned to the land of the living.

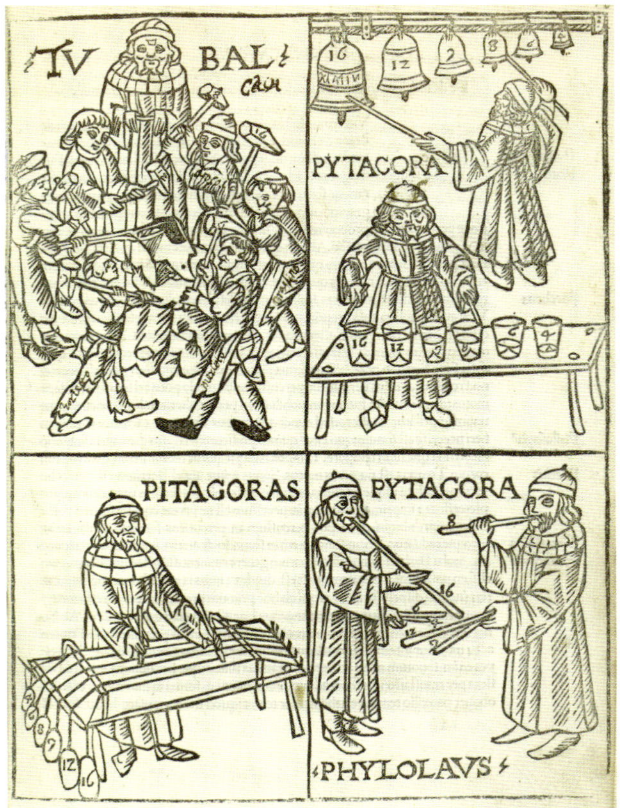

1-15. Harmonic ratios in music, from Franchinus Gaffurius, *Theorica Musicae* (Milan: Phillipas Montegatius,1492), 18. Woodcut. Music Division, The New York Public Library for the Performing Arts, Astor, Lenox, and Tilden Foundations.

In this Renaissance book on music theory, Tubal-cain, the biblical forger (upper left), looks on as smiths make music by striking an anvil with a variety of hammers, while Pythagoras (upper right) strikes graduated bells and glasses filled with diminishing amounts of water. Pythagoras is shown on the lower left playing an instrument that has been tuned by hanging graduated weights from its six strings. As the weights increase from four to sixteen units, the strings are stretched ever tighter, producing increasingly higher pitches when plucked. On the lower right, Pythagoras plays a duet with his student Philolaus, whose flute is twice as long. Together they sound an octave, producing within them a satisfying feeling of resolution.

Pythagoras of Samos lived in the late sixth century BC on the Italian peninsula, where he was the legendary leader of a mystery cult whose members withdrew from society and adopted a strict ascetic lifestyle. Unlike other leaders of mystery cults, Pythagoras made pronouncements about the hidden meaning of numbers, and his cult had followers in the fifth and fourth centuries BC who made important contributions to mathematics.[17] Pythagoras himself wrote nothing, but his ideas survive in the writings of his followers. For example, Philolaus of Croton in southern Italy argued that the cosmos is made of unlimited continua such as fire, water, space, and time, which are limited by numbers, shapes, and forms that join together to form "harmonia"—a harmonious whole (*On Nature*, fifth century BC). To illustrate the process, Philolaus gave a musical example that the Pythagoreans had discovered: if a musician slides a stick up and down a vibrating string, it produces a continuum of sound (plate 1-15 lower left). But this unlimited continuum is limited according to ratios of whole numbers because the main musical consonances—an octave, a fifth, and a fourth—are expressible in ratios of the smallest whole numbers—1 : 2, 2 : 3, and 3 : 4, respectively. If one plucks a string and then divides it in half by pressing it in the middle, plucking the string again produces an octave and the listener experiences an agreeable feeling of finality—a consonance. Thus, according to Philolaus, numerical relations underlie harmony in both music and the soul (plates 1-15 and 1-16). A combination of tones not in a ratio of whole numbers—a dissonance—is experienced by the listener as unresolved. Since the natural world and man's inner being (the soul) are structured by numbers, we can, according to Philolaus, only gain true knowledge of the world and of ourselves by studying numbers.

Whereas philosophers who pursued a scientific approach put their trust in human reason, Greeks who joined mystery cults believed their lives were controlled by the irrational forces of fate—blind and without purpose. Since it was hopeless to even try to exercise rational control of their lives, cult members followed figures such as Dionysus, god of wine, and abandoned themselves to ecstasy and ritual madness. The suffering that they endured in their lives was alleviated by the promise that their lives had a hidden meaning—a mystery—which would be revealed to them in the form of a secret (a riddle, paradox, or absurd

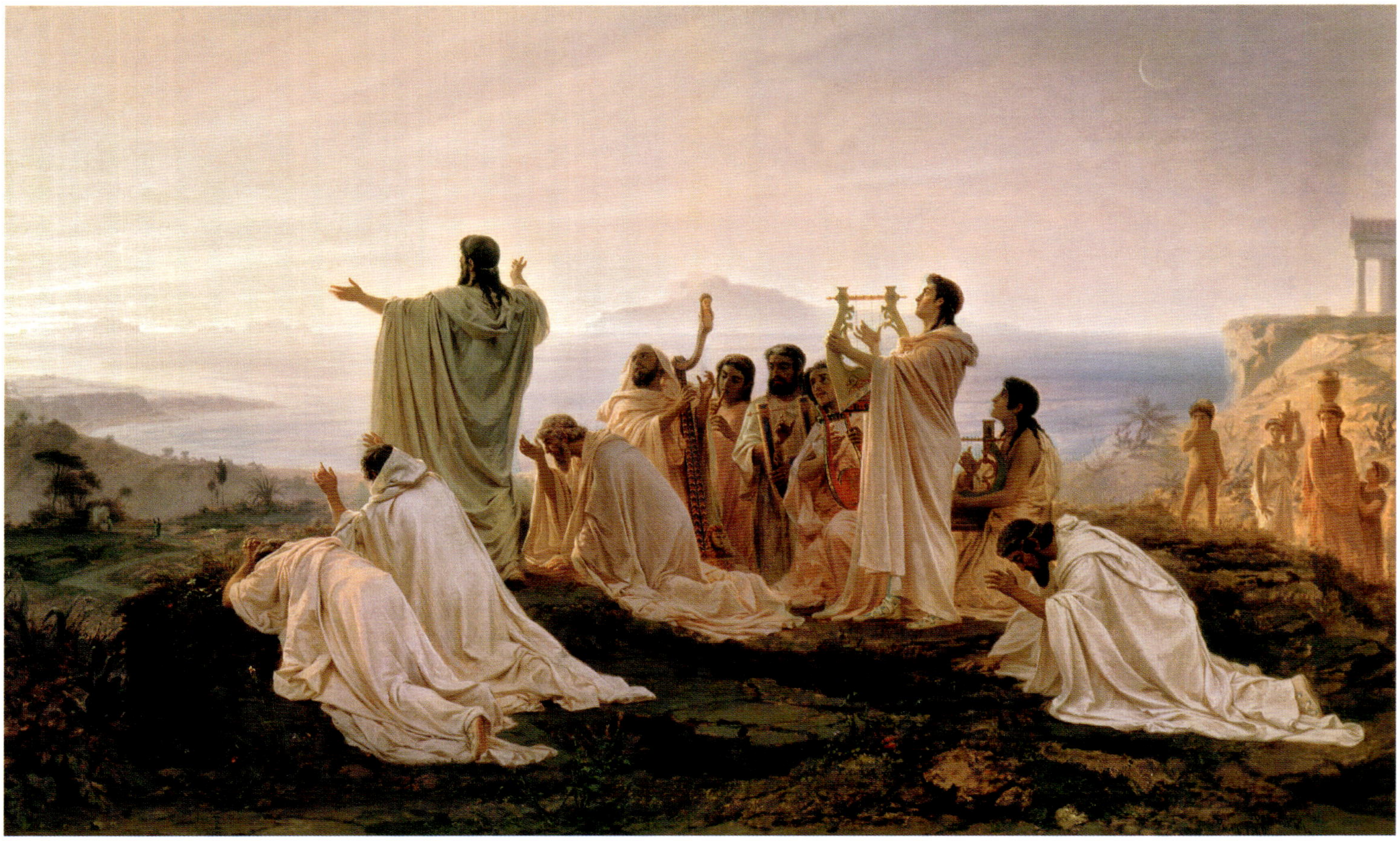

1-16. Fyodor Bronnikov (Russian, 1827–1902), *Pythagorean Hymn to the Rising Sun*, 1869. Oil on canvas, 39¼ × 63½ in. (99.7 × 161 cm). State Tretyakov Gallery Moscow.

The Russian Romantic artist Fyodor Bronnikov painted members of the Pythagorean mystery cult playing stringed instruments in harmony with cyclical rhythms that they observe in nature: the sun rises in the east, follows an arched path overhead, and sets in the west.

All things that are known have number. For it is not possible that anything whatsoever be understood or known without this.

—**Philolaus, *On Nature*, fifth century BC**

phrase) when they were initiated into the cult. They would understand the mystery not by reason but in an epiphany—a flash of insight. Initiates were also promised that their virtues would be rewarded in a life after death.

Mystery cults probably began as spring planting and fall harvest festivals, such as the Eleusinian cult of Demeter, goddess of agriculture, and her daughter Persephone, the goddess of fertility, who was abducted into the underworld by Hades, ruler of the gloomy land of the dead. In response to the pleadings of Demeter to her husband Zeus, the god of thunder allowed the goddess of fertility to return to her mother each spring, but in the fall Persephone had to go back for a cold season in the dark underworld. Because the Athenian members of the Eleusinian cult who paid tribute to Demeter and Persephone in the sixth century BC also followed Zeus and Athena, the cult did not provoke the wrath of public officials, whose only concern was that the power of the Olympian pantheon be preserved. When many Athenians became destitute in the later fifth century BC during the Peloponnesian War, the Eleusinian cult shifted its focus from partaking in a bountiful fall harvest to living in paradise after death, a shift that was repeated during troubled times in later mystery cults.[18]

Many mystery cults promised eternal life and featured figures such as Persephone and Orpheus, who returned alive after visiting the underworld. But whereas in the Eleusinian, Dionysian, and Orphic cults irrational absurdities were transformative, the secret phrases that moved the Pythagoreans were about numbers. Members of the Pythagorean cult had made the stunning and surprising discovery that music has a mathematical basis, and Philolaus drew an analogy, unique to Pythagoreanism, between musical harmony and the harmony of the soul. The most accomplished of all the Pythagorean mathematicians was Philolaus's student Archytas, a contemporary of Plato who was active in the early fourth century BC. Archytas took his teacher's numerical analysis of musical ratios to a new level of complexity by analyzing the numerical ratios of several entire scales, based on his observation of how musicians of his day actually tuned their stringed instruments.[19]

Plato combined the rational, scientific approach of Socratic philosophy with practices of the Pythagorean mystery cult. Plato had learned the rational approach from Socrates, and his early dialogues feature Socrates demonstrating the limits of knowledge gained by sense perception of the natural world. Plato was inspired to go beyond fallible, sensory knowledge of the transitory, imperfect everyday world to attain certain, reasoned knowledge of numbers and spheres in a timeless, perfect world of abstract objects. Plato was also a profoundly political philosopher who as a youth had fought against Sparta in the Peloponnesian War; he knew Socrates from boyhood, and he attended Socrates's trial. Plato was distrustful of the democratic practices that allowed a jury selected by lottery to condemn his wise and virtuous teacher. In addition to certainty, Plato wanted justice, which he found in the Pythagorean promise of a life after death, when scores would be settled. By combining Socratic and Pythagorean methods, Plato achieved a philosophy that met his goals: certainty and justice.

Plato made several trips to the Italian peninsula, where he visited the tyrant Dion in Syracuse and became embroiled in local politics. Plato also became acquainted with the mathematician Archytas, who was a leading politician in the neighboring city of Tarentum and a member of a Pythagorean community there. When the son and successor of Dion, Dionysius II, threatened Plato in 361 BC, it was Archytas who sent a ship to rescue him.

Plato entered into an intellectual exchange with Archytas, from whom he learned the analogy between musical harmony and the harmony of the soul. The related view of the harmony of the spheres, which is traditionally ascribed to Pythagoras, may have originated with Plato himself because it first appeared in the *Republic*[20] and in *Timaeus*,[21] both of which Plato wrote after returning from Italy. In these dialogues Plato articulated his mature doctrine of two worlds: the concrete, transitory world that is perceived by the senses and the abstract, eternal realm known by reason. This is a distinction that the Pythagoreans did not make (they studied numbers only in the sensible world before them), which led Plato to criticize Archytas for focusing only on particular examples in music—"heard harmonics" based on how musicians actually tuned their lyres—and for not pushing beyond particulars to the most general form of the mathematical ratios in music (*Republic*, 530d–531c).

In the *Republic*, Plato also formulated his mature political philosophy. He defined the just person as one with moral excellence and an ordered soul, which is the unity of parts within a well-proportioned whole (*Republic*, 443e–444b, and 462a–b). In other words, the ordered soul has a mathematical ratio, and by analogy a harmonious society, according to Plato, is also the unity of parts within a whole. Thus a person who aspires to be a wise ruler—a philosopher king—must follow a course of study that includes ten

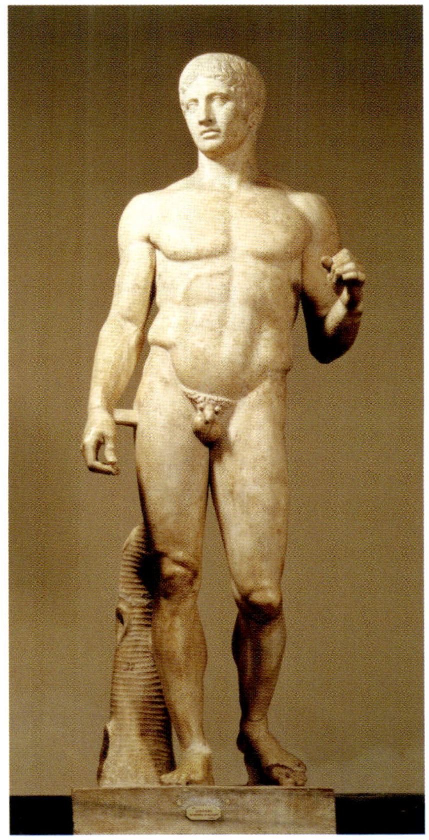

1-17. Polykleitos of Argos, *Doryphoros*, Roman copy after Polykleitos's original bronze, ca. 440 BC. Marble, 6 ft. 11 in. (2.12 m) high. Museo Archeologico Nazionale, Naples.

Polykleitos wrote down the perfect proportions for the human body in a treatise, the *Canon*, which was lost after the fall of Rome, although fragments of the document survive, including "beauty comes about little by little through many numbers" (*Die Fragmente der Vorsokratiker*, ed. and trans. Hermann Diels and Walter Krantz, 6th ed. [Berlin: Weidmann, 1951–52], 40B2). Galen of Pergamum recorded that Polykleitos made a sculpture to illustrate his proportional system (*De placitis Hippocratis et Platonis*, second century AD, bk. 5, 3.16–17), and scholars today believe that this Roman statue is almost certainly a marble copy of Polykleitos's original bronze *Doryphoros*, which was "a statue that artists call the Canon, since they draw their outlines from it as if from a law" (Pliny the Elder, *Naturalis historia*, AD 77–79, bk. 31, 19.55).

years of pure mathematics (*Republic*, 537e). Plato went on to argue that being just is in the best interests of the virtuous person, even if virtue brings no ordinary rewards or, quite the opposite, it elicits punishment as it had Socrates, because man is immortal and his virtue will be rewarded after death. Plato ended the *Republic* by having Socrates tell a story about a slain soldier, Er, who, while lying on his funeral pyre, awakens from death and tells of his journey to an afterlife, where people are judged by their deeds. The just are led to a wondrous place in the sky and the wicked are cast into a terrible underworld. Socrates concluded: "If you will believe with me that the soul is immortal and able to endure all good and ill, we shall keep always to the upward way and in all things pursue justice with the help of wisdom. Then we shall be at peace with Heaven and with ourselves, both during our sojourn here and when, like victors in the Games collecting gifts from their friends, we receive the prize of justice."[22]

In addition to seeking the beauty of music in ratios, the Greeks also looked to mathematics for the aesthetic key to the visual arts, as in the canon of perfect proportions for the male body created by the sculptor Polykleitos (plate 1-17).[23] Indeed, over the centuries many artists in the classical tradition have tried to create a mathematical system that would insure beauty (see chapter 2).

PLATO'S FORMS

They make use of visible figures and discourse about them, though what they really have in mind is the originals of which these figures are images: they are not reasoning, for instance, about this particular square and diagonal which they have drawn, but about the Square and the Diagonal; and so in all cases. The diagrams they draw and the models they make are actual things, which may have their shadows or images in water; but now they serve in their turn as images, while the student is seeking to behold those realities which only thought can apprehend.

—**Plato, *Republic*, 380–367 BC**

At Plato's Academy in Athens, generations of philosophers learned that one can have certain knowledge only of mathematical objects and other Forms (see sidebar on opposite page). This view of mathematical objects became central to Western mathematics and is the common outlook of working mathematicians today. Plato held that physical objects such as apples and oranges exist in time and space, and they are known by sense perception—by seeing or touching them. He held that abstract objects such as squares and cubes exist outside time and space, and they are known by cognition. Apples and oranges exist in a world-out-there, independent of the human mind, as do squares and cubes, which similarly inhabit a mathematical-world-out-there. Since squares and cubes are perfect and eternal, mankind knows them with objective certainty and with certitude—a subjective feeling of certainty.[24]

According to Plato, mathematical objects are real, but since they are immaterial, they do not interact causally with the natural, material world. The mathematical-world-out-there has an existence distinct from our exploration of an area of mathematics, such as geometry, or our invention of the means to study it, such as an algorithm. Human exploration and invention are part of the history of culture, but mathematical reality exists outside history—outside time and space—and thus the world of Platonic mathematical forms is independent of the evolution of animals who can count or rotate a figure in the mind. When an amoeba met two protozoa in the primordial seas, there were three creatures, and their sum was a prime number, even though there were no minds evolved enough to recognize the timeless truth that three is divisible only by itself and one.

Plato's view is the basis of the classical mathematical outlook. Plato's approach also

Plato's Academy

Plato converses with his students in this mosaic, framed by festoons of fruits with comic faces. Done in the first century BC for the home of a wealthy Greek citizen living in Pompeii, the gateway with vases symbolizes the entry into Plato's Academy, over which Plato is said to have written:

ΑΓΕΩΜΕΤΡΗΤΟΣ ΜΗΔΕΙΣ ΕΙΣΙΤΩ
(Let no one unversed in geometry enter here.)

The single olive tree reminds the learned residents of Pompeii that the Academy was located in a grove of olive trees near Athens, and the walls of the Acropolis can be seen in the upper right. Seven bearded philosophers are discussing a geometric object—a gridded globe that sits in a box on the ground in the center of the group. Scholars generally agree that Plato is the third figure from the left, seated holding an open papyrus scroll. He points at the globe with a long brown stick, while another seated figure gestures at it. Above the group is a sundial mounted on a column, and, like the dial's pointer, the figures and the globe cast shadows.

BELOW

1-18. *Plato's Academy*, first century BC. Mosaic, from the Villa of T. Siminius Stephanus in Pompeii, Italy. Museo Archeologico Nazionale, Naples, inv. no 124545.

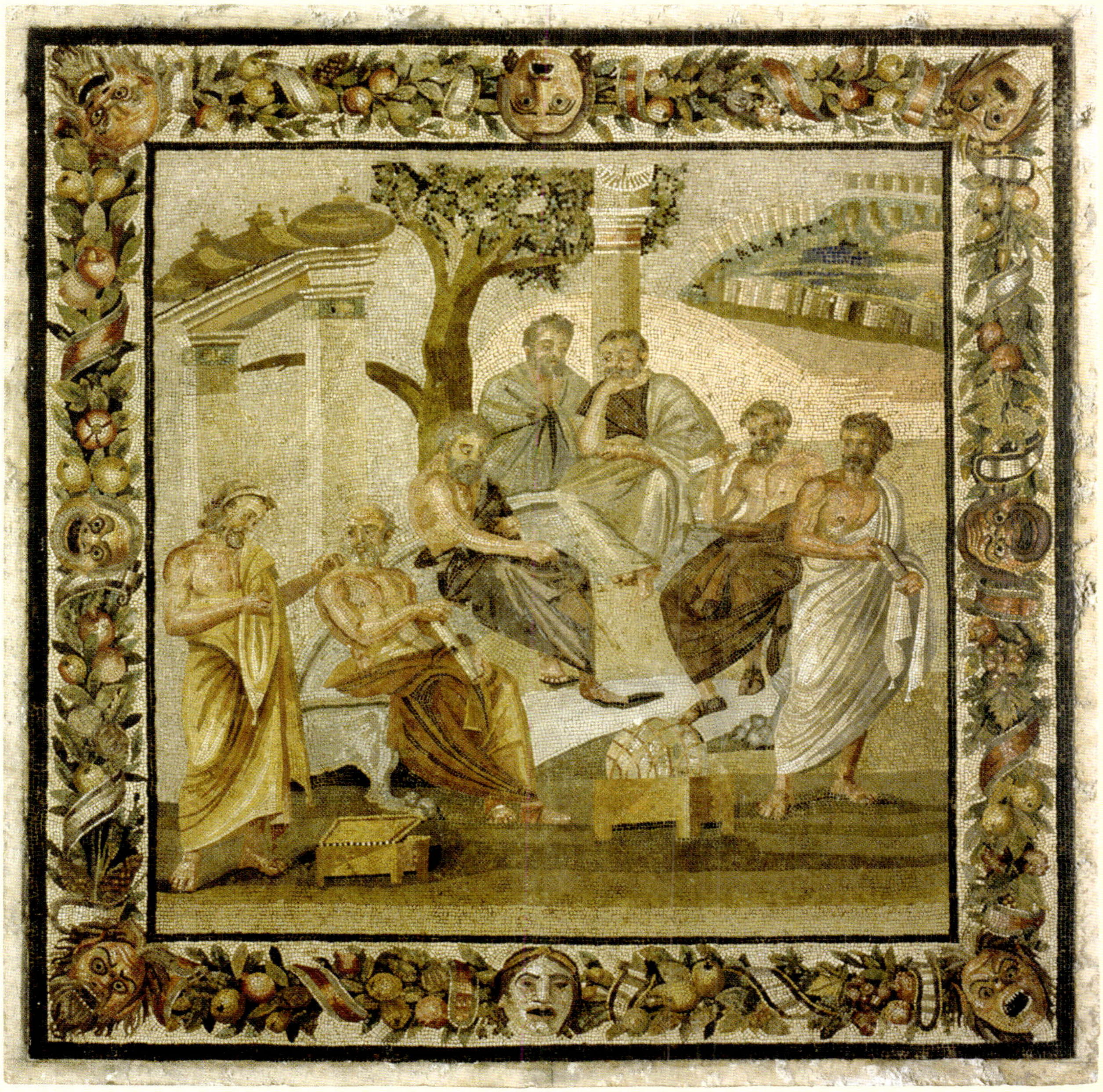

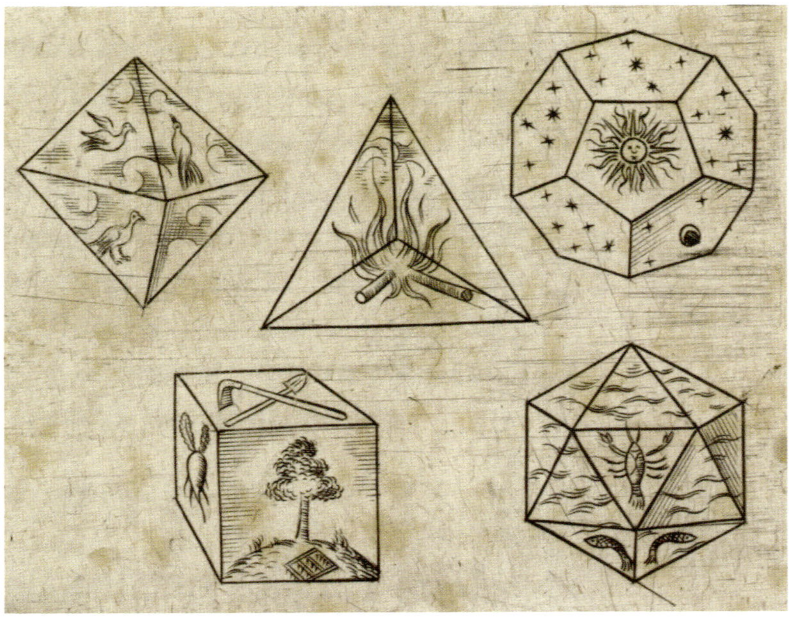

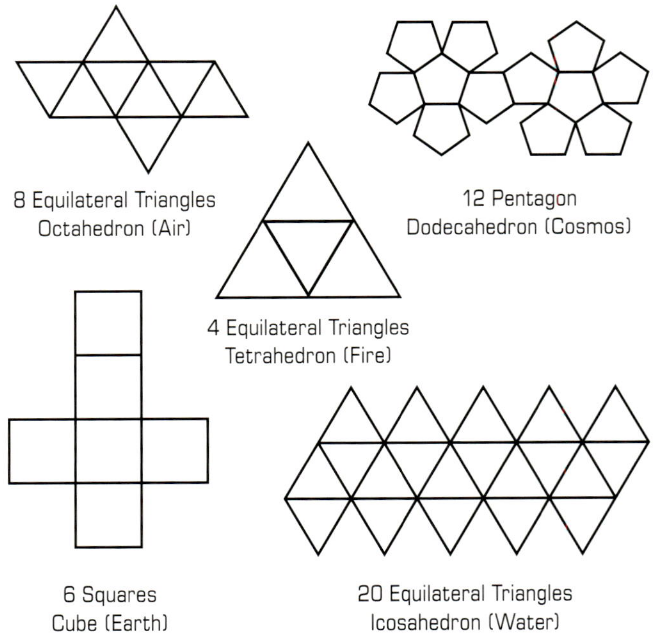

1-19. The Platonic solids (above right) and Johannes Kepler (above left), *Harmonices Mundi* (Harmony of the world, 1619), between 52–53. Woodcut. Music Division, The New York Public Library for the Performing Arts, Astor, Lenox, and Tilden Foundations.

Regular, identical polygons form the faces of the five Platonic solids. Although some figures were known before Plato, he first assembled them as a group—the five regular polyhedrons—in his dialogue on cosmology, *Timaeus* (366–360 BC), in which he described the creation of heaven and earth from elements associated with the solids. Kepler drew this diagram to illustrate Plato's association of four of the solids with earth, air, fire, and water, and the most spherical solid—the dodecahedron—with the entire universe.

underlies the related classical scientific outlook: nature has a fundamental unity and it embodies mathematical patterns. Since the natural world is an imperfect embodiment of numbers and geometric forms, man's knowledge of earth and sky is partial and changeable. Plato also defined visual art as an imitation of nature, and—as a copy of an imperfect embodiment of the Forms—twice removed from timeless perfection.

Plato declared, furthermore, that over and above the natural and mathematical worlds there reigns a supreme, divine Reason—the Good—which was for him the source of a higher purpose for the cosmos. Earlier philosophers in both the scientific and mystical traditions had introduced concepts of a divine Mind: the scientific Ionian Anaxagoras had postulated that the cosmos was set in motion by a purposive Mind, after which the sun, moon, and stars moved in a completely impersonal, mechanical way, and a follower of the mystical Pythagoras, Parmenides of Elea, believed that a divine Intellect—"the One"—had limited the primordial void by generating the numbers, 1, 2, 3, and 4, which sum to 10, from which the One created the natural world. But these pre-Socratic philosophers had not solved the basic problem of any cosmology that makes a divine Mind the source of life on earth: how can an immutable, perfect Mind create (cause) the fluctuating, flawed physical world? Plato solved this in the mystical tradition by making the Good not pure intellect—not an abstract structure of ideas—but a mythical *person*. In other words, Plato deified Mind/Reason, making the Good a divine person with *emotions*—feelings and desires—that connect Reason to the natural world.

Plato introduced his concept of divine Reason in *Timaeus* by telling the story of a mythical Creator of the world, who Plato named the *Demiurge* (Greek for "craftsman"). Plato's Creator/Craftsman "was good" and "desired that all things should come as near as possible to being like himself," and thus he created the earth and heavens by imposing perfect, ideal patterns—the Platonic Forms—onto primordial, chaotic matter: "Desiring, then, that all things should be good and, so far as might be, nothing imperfect, the god took over all that

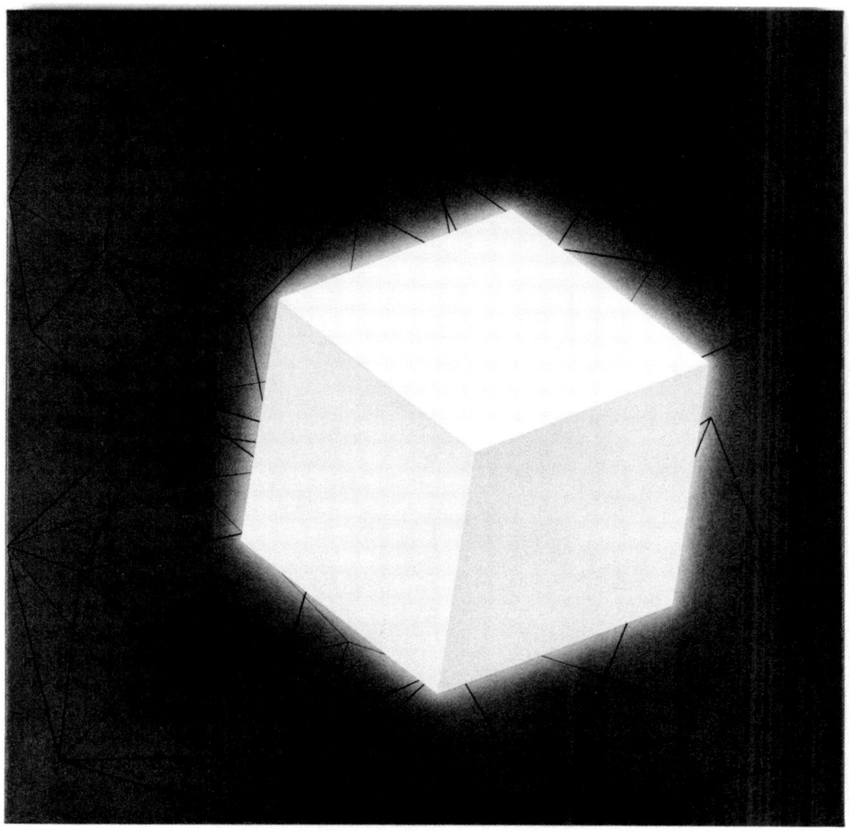

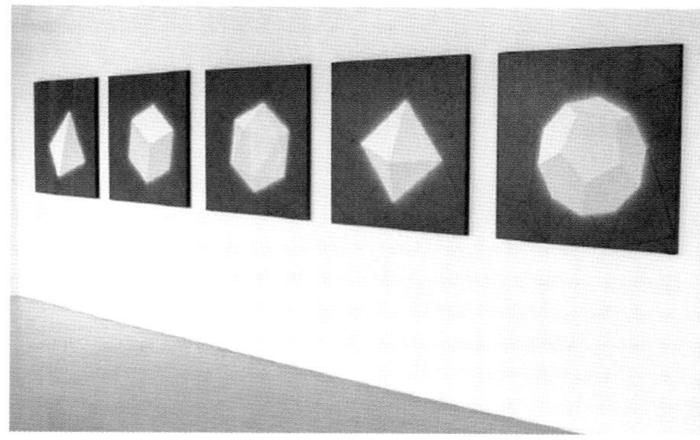

1-20. Rune Mields (German, b. 1935), Installation view (above) of *The Platonic Solids*: (from left to right) *Tetrahedron, Hexahedron [cube], Icosahedron, Octahedron, Dodecahedron*, and *Hexahedron-Earth* (left), 2002. Water-base paint on linen, 39⅜ × 39⅜ in. (100 × 100 cm) ea. Courtesy of the artist and the Galerie Angelika Harthan, Stuttgart. © 2014 VG Bild-Kunst, Bonn/Artists Rights Society, New York.

A hexahedron is a polyhedron with six faces, of which the cube (shown on the left) is an example. In this painting, the contemporary German artist Rune Mields captures Plato's sense that abstract objects are ethereal and unearthly by portraying the cube as luminous and floating in darkness.

is visible—not at rest, but in discordant and unordered motion—and brought it from disorder into order, since he judged that order was in every way the better."[25] Thus Plato's divine Craftsman equated the Good with harmony, measure, and order. To fulfill his desire for goodness, the Craftsman created the natural world as an imperfect copy of the Forms, which included beauty, equality, and bigness, as well as the five regular polyhedrons, the Platonic Solids (plates 1-19 and 1-20). After imposing geometric Forms on matter, the Craftsman also imposed the ultimate Form—the Good—to insure that the cosmos had a purpose.[26]

DEMOCRITUS'S MECHANICAL UNIVERSE

Ionian Greeks initiated the scientific worldview, but in the late fifth century BC it was a member of a mystery cult, the Pythagorean Democritus, who formulated the naturalistic outlook in its most influential form—atomism. The number one had a special place in Pythagorean thought because it limited the boundless, primordial void; understanding the structure of numerical relations was the key to discerning the harmonic order of the cosmos. According to the Pythagoreans, everything in heaven and earth was composed from sentient particles that were made from one substance—monad (from the Greek for "one")—that combines in increasingly complex patterns to form the natural world. Parmenides, following Pythagoras, believed that the cosmos was made of this one substance and that the cosmos was a unified whole—"the One." Parmenides reasoned, furthermore, that if the cosmos is a unity, then change is impossible. How then can we account for the fluctuating world around us? A follower of Parmenides, Democritus answered that nature

Opinion says hot or cold, but the reality is atoms and empty space.

—Democritus, late fifth–early fourth century BC

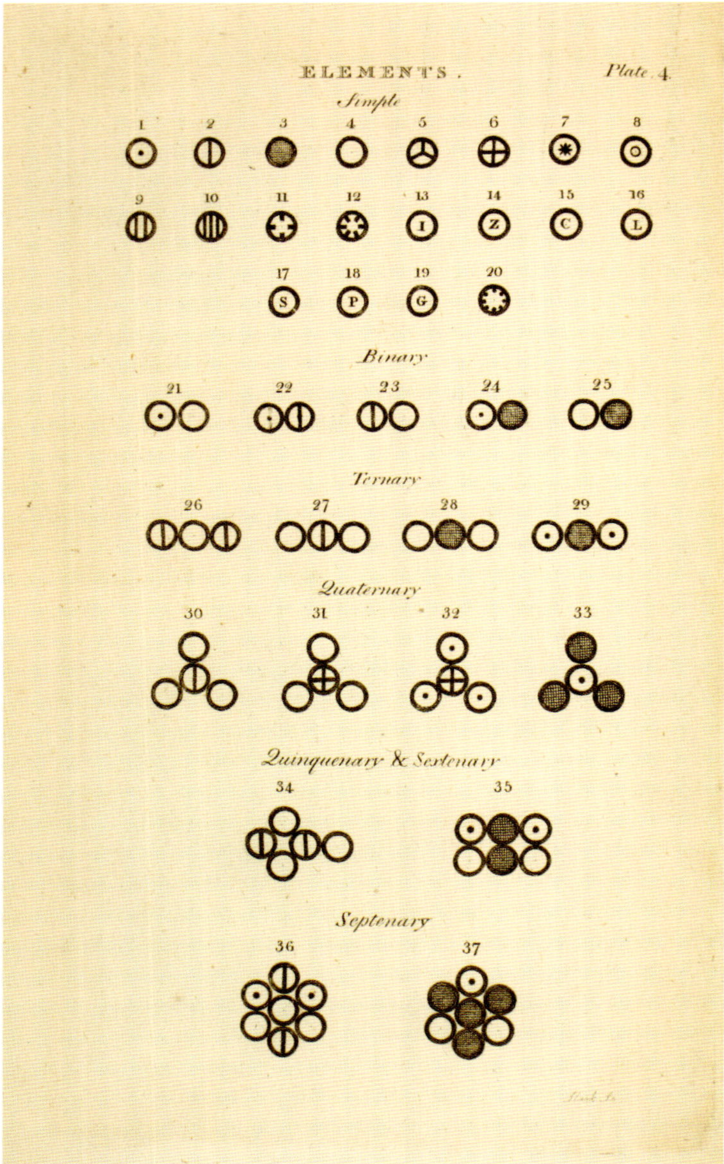

1-21. "Elements" in John Dalton, A *New System of Chemical Philosophy* (Manchester, England: Bickerstaff, Strand, and London, 1808), plate 4. Chemical Heritage Foundation, Othmer Library of Chemical History, Philadelphia.

In the early nineteenth century, the British chemist John Dalton help set the stage for modern atomic theory when he revived Democritus's philosophical view that matter is composed of discrete irreducible units—atoms—which are identical and unchanging, as Dalton wrote: "ultimate particles of all homogeneous bodies are perfectly alike in weight, figure, etc.; every particle of hydrogen is like every other particle of hydrogen" (143).

is composed of minute, indivisible, eternal particles—atoms—which do *not* change but simply rearrange themselves within a void. Democritus also introduced a crucial change from his fellow Pythagoreans: Parmenides's monads were sentient (alive) but Democritus's atoms were inert (lifeless). Democritus and other atomists tried to account for all reality in ways that were consistent with immutable atoms moving in empty space. For example, after excluding purpose from the cosmos, Democritus held that the human mind (psyche, soul) was composed of spherical atoms, as was fire, making both warm and lively.[27] In other words, atomists held that every changing phenomenon is explained by processes within a fundamentally unchanging, unified world with universal, timeless laws. As such, atomism is a scientific outlook that has had a powerful impact on Western thought (plate 1-21). Although ancient atomists expressed interest in mathematics, they did not develop a view of abstract, mathematical objects that related to their atomist explanation of the natural world.

Plato also held that everything is made of one substance arranged in different patterns, but his cosmos was not made of inert bits of matter but, like the Pythagoreans and Parmenides, his building-blocks were living, sentient particles (monads). Because the entire cosmos is made of this shared matter (monads), man is a microcosm of the universe and animated by a common spirit, the World Soul, which, according to the Pythagoreans and Plato, is supernatural and eternal.[28] Many people have been inspired to unite with this living cosmos, and there is a long tradition of pantheism associated with it. This ancient form of pantheism asserts that the transitory natural world is a unity, because everything is made of the same substance, and that nature is also divine, because everything—from the smallest pebble to the farthest star—is animated by divine Reason.

On the other hand, there is no pantheist tradition associated with atomism because people have evidently not been inspired to unite with the mechanical motion of dead particles. Also, since the atomist cosmos entails neither purpose nor a divine Mind, arrangements of atoms are value-free (neither beautiful nor good). Over the centuries, Democritus's mechanical model of a cosmos composed of inert matter in motion has been the main competitor to Plato's organic, living, purposeful cosmos, infused with a World Soul and divine Reason. Mathematical descriptions of the natural world as embodying patterns were given by the ancients in both the scientific approach of atomism, and the combined rational and mystical approach of Platonism, but only Platonism linked mathematics to the divine.

THE AXIOMATIC METHOD: EUCLID'S *ELEMENTS*

The philosopher Aristotle, who studied at Plato's Academy for ten years, agreed with Plato that circles and triangles exist independent of human minds, but he declared that they do not exist independent of their embodiments in particular objects.[29] Aristotle argued that a basic truth (a "first principle") is not discovered by contemplation of a world beyond the senses—the realm of the Forms—but rather by a process of induction that begins in the everyday, visible world. By observing sensible objects in one area of knowledge (such as earth and sky), one can generalize to the first principles of the field, the core propositions from which all the other truths can be deduced. He also declared that the cosmos was not created by Plato's divine Craftsman but was set in motion by another divine person, Aristotle's Prime Mover.[30]

Whether the bedrock of certain knowledge was Plato's immediate intuition of the Forms or Aristotle's slow process of induction of first principles, once the foundation of certainty was in place, both Plato and Aristotle gained further knowledge by the dialectic method of reasoning from premises to a conclusion. In Plato's dialogues in the early to mid-fourth century BC, he took steps to insure the correctness of philosophical arguments by making the premises (the assumptions) of an argument explicit and by standardizing the method of argument (the deduction). Once the assumptions were clearly stated, they could be examined and, if they were wrong, they could be reconsidered. Euclid of Alexandria was a follower of Plato who was active in the late fourth and early third centuries BC. He adopted Plato's goal of clarity and order in philosophical reasoning, and applied it to reasoning about numbers and geometry, developing a proof from premises for mathematical reasoning—the axiomatic method (see sidebar below).[31]

A proof in Euclidean geometry can be characterized as what (since the nineteenth century) has been called a "thought experiment"—an experiment carried out in the imagination—about an abstract object, such as a circle or triangle, which is represented by a

Euclid's Axiomatic Method

Euclid's project in *Elements*, ca. 300 BC, was to write an axiomatic system consisting of a vocabulary, common notions, and axioms (or postulates). The axioms are the basic assumptions of the system. The mathematical terms used in the common notions and axioms are defined in the vocabulary. From the axioms, theorems are proved and once established, the theorems can be used in future proofs.

Vocabulary

Euclid wrote twenty-three definitions, including the following:

1. A point is that which has no part.
2. A line is breadthless length.
3. The extremities of a line are points.

23. Parallel straight lines are straight lines which, being in the same plane and being produced indefinitely in both directions, do not meet one another in either direction.

Common notions

1. Things which are equal to the same thing also equal one another.
2. If equals be added to equals, the wholes are equal.
3. If equals be subtracted from equals, the remainders are equal.
4. Things which coincide with one another equal one another.
5. The whole is greater than the part.

Axioms (postulates)

Let the following be postulated:

1. To draw a straight line from any point to any point.
2. To produce a finite straight line continuously in a straight line.
3. To describe a circle with any center and any distance.
4. That all right angles are equal to one another.
5. That, if a straight line falling on two straight lines makes the interior angles on the same side less than two right angles, the two straight lines, if produced indefinitely, meet on that side on which are the angles less than the two right angles.

1-22. Euclid's proof of the Pythagorean theorem, which states: *In right-angled triangles the square on the side subtending [lying across from] the right angle is equal to the squares on the sides containing the right angle* (*Elements*, book 1, proposition 47).

Euclid's proof is a classic example of abstraction and generalization. In modern times his proof is known as the "windmill" proof because of the shape of the figure. He drew a triangle with a right (90°) angle and then constructed squares on each of its three sides (x, y, and z). Euclid's goal was to show that the areas of the two smaller squares equals the area of the largest square, in algebraic notation, $x^2 + y^2 = z^2$.

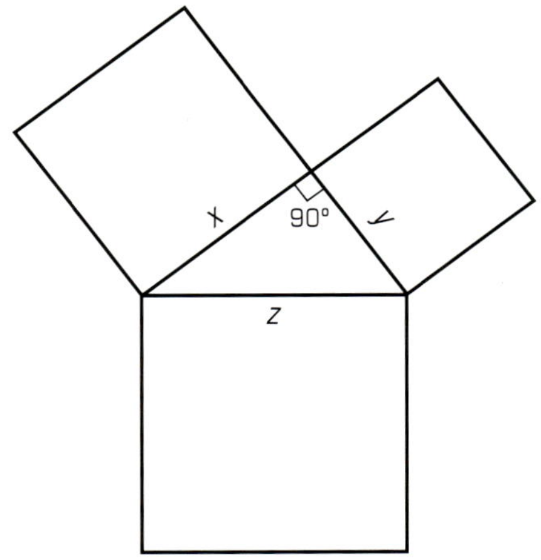

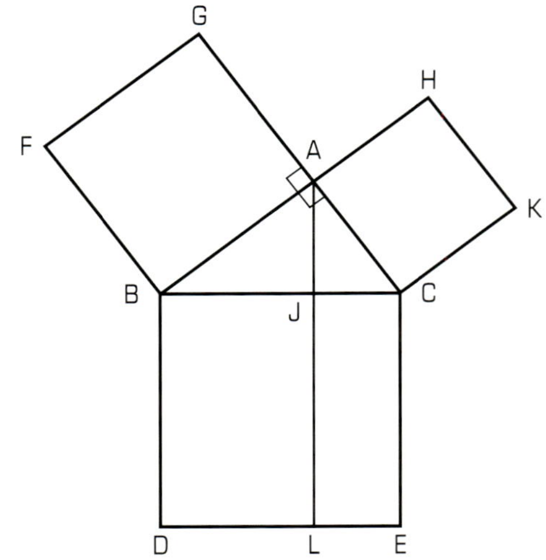

Euclid began by establishing that GAC is a straight line (a truth he will need later in the proof) because GAB and BAC are right angles. Then he drew a straight line down from point A to form the parallelograms BJLD and JCEL. His strategy was to show that the area of BJLD equals the area of the parallelogram on the upper left (the square FGAB), and that JCEL equals the parallelogram on the upper right (the square AHKC).

Beginning on the left, Euclid constructed the triangles FBC and DBA. Angles FBC and DBA are equal because they were formed by adding the same angle (ABC) to a right angle. The lengths of the sides that contain the equal angle are also equal because FB and BA, as well as BC and BD, are sides of squares. Therefore, these triangles are equal using the (previously proven) theorem Proposition 4: *If two triangles have two sides equal, and the angle contained by these sides is equal, then the triangles are equal.*

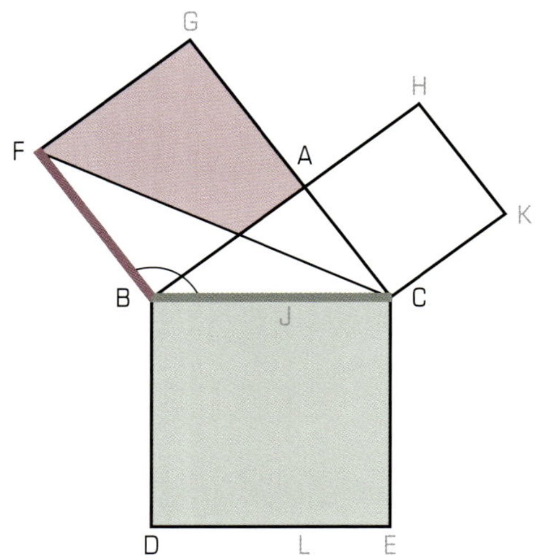

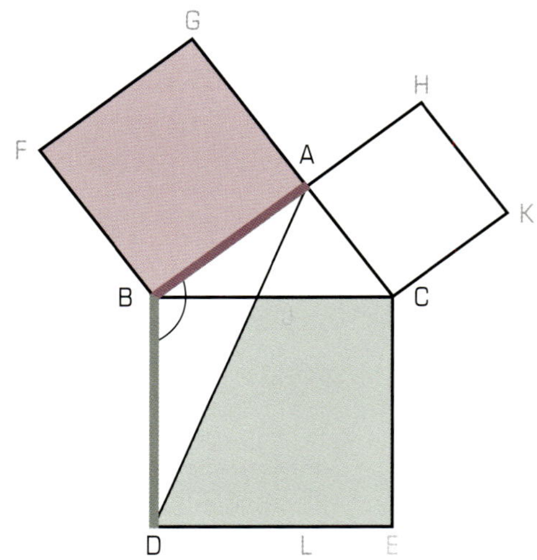

Euclid then deduced that the area of each of these equal triangles is *half* of the area of the parallelogram with which it shares a base. He accomplished this by establishing above that GAC is a straight line and by drawing AL as a straight line. This means that the two pairs of triangles (FBC and DBA) and parallelograms (FGAB and BJLD) lie within the same parallels, and he can use Proposition 41: *If a parallelogram has the same base with a triangle and is in the same parallels, then the parallelogram is double the triangle.*

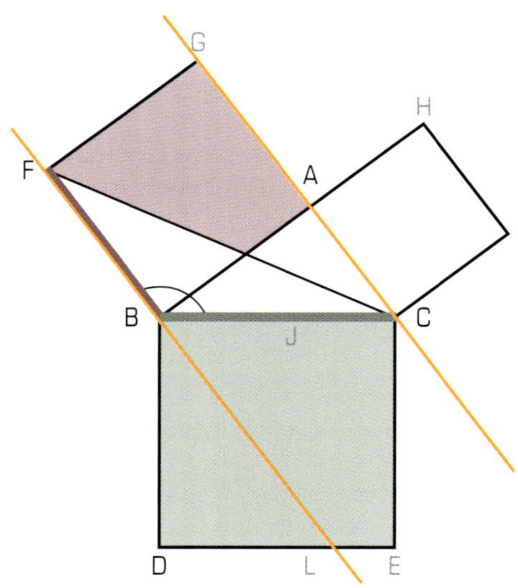

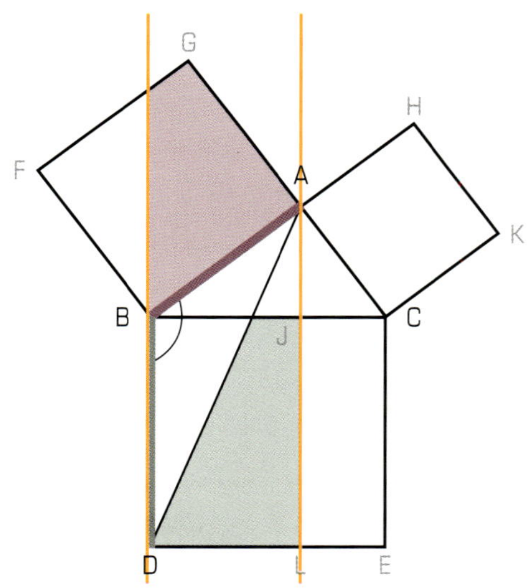

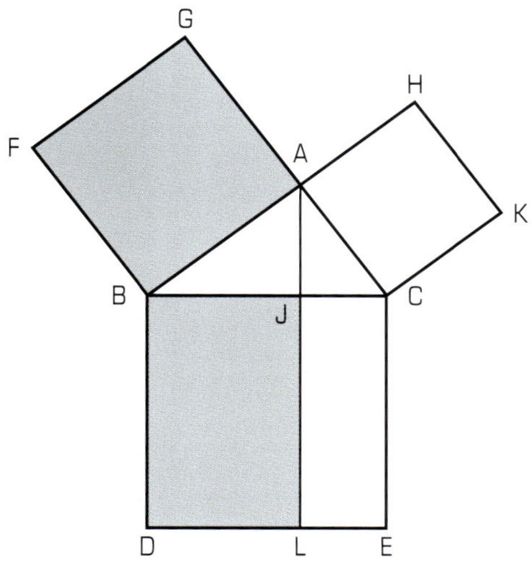

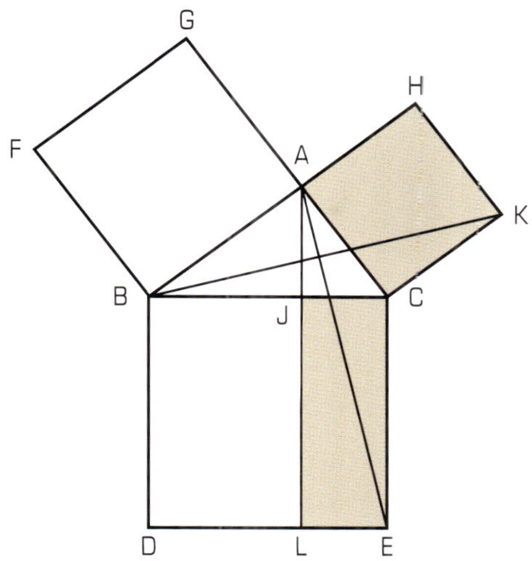

If the halves are equal, then so are the wholes, so parallelograms BJLD and FGAB are equal. Euclid proved the same equality for the parallelograms on the right side of the figure.

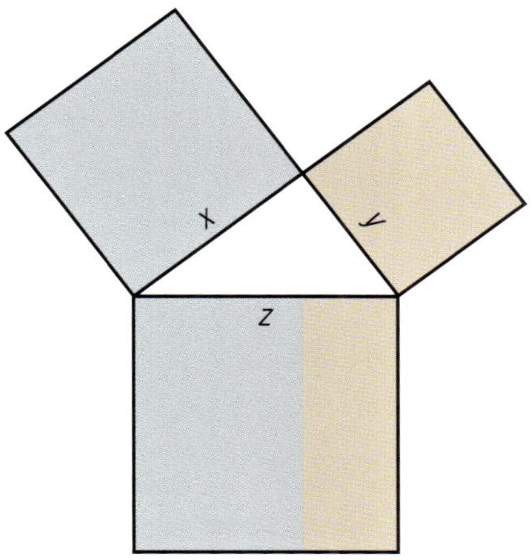

Thus the area of the whole bottom square equals the sum of the areas of the upper squares; in modern notation, $x^2 + y^2 = z^2$ Q.E.D. (*quod erat demonstrandum*, Latin for "which was to be demonstrated").

BELOW

1-23. Ben Shahn (Lithuanian-born American, 1898–1969), *At the Pythagorean School*, 1953. Ink and wash, 8½ x 19 in. (21.6 x 48.3 cm). Collection of Anita and Arthur Kahn, New York.

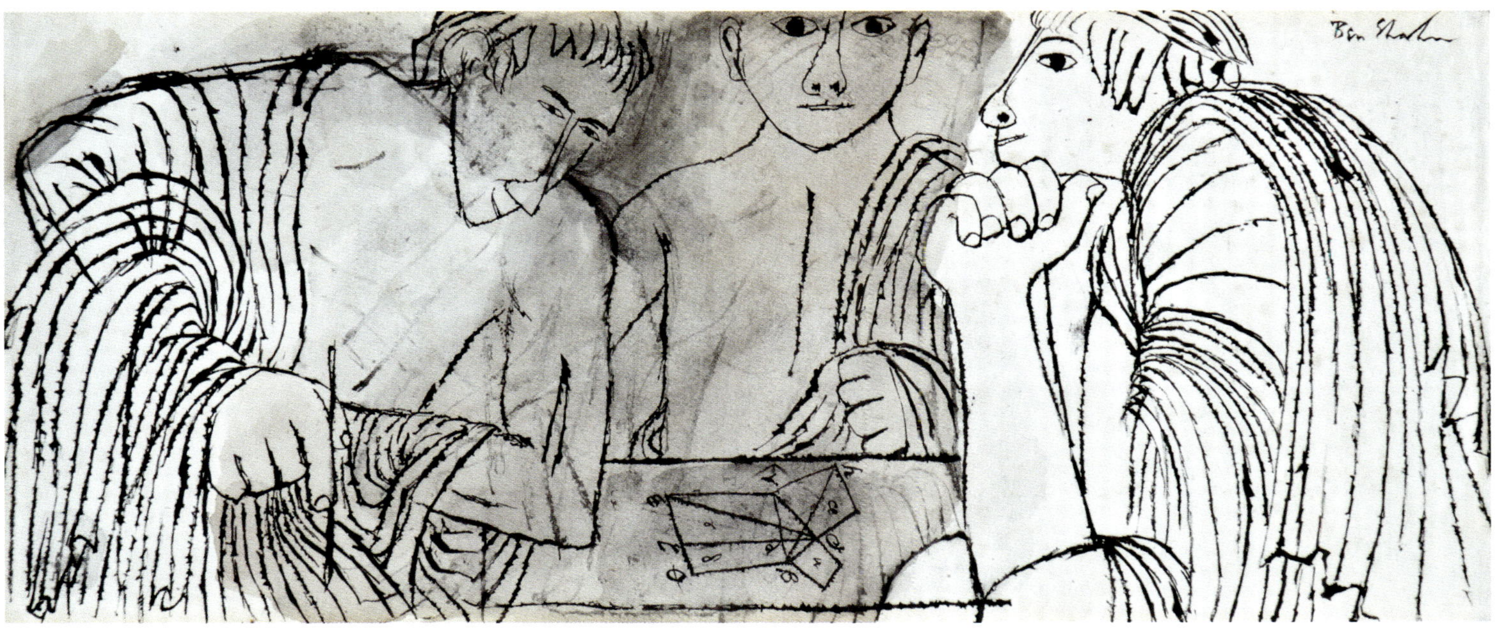

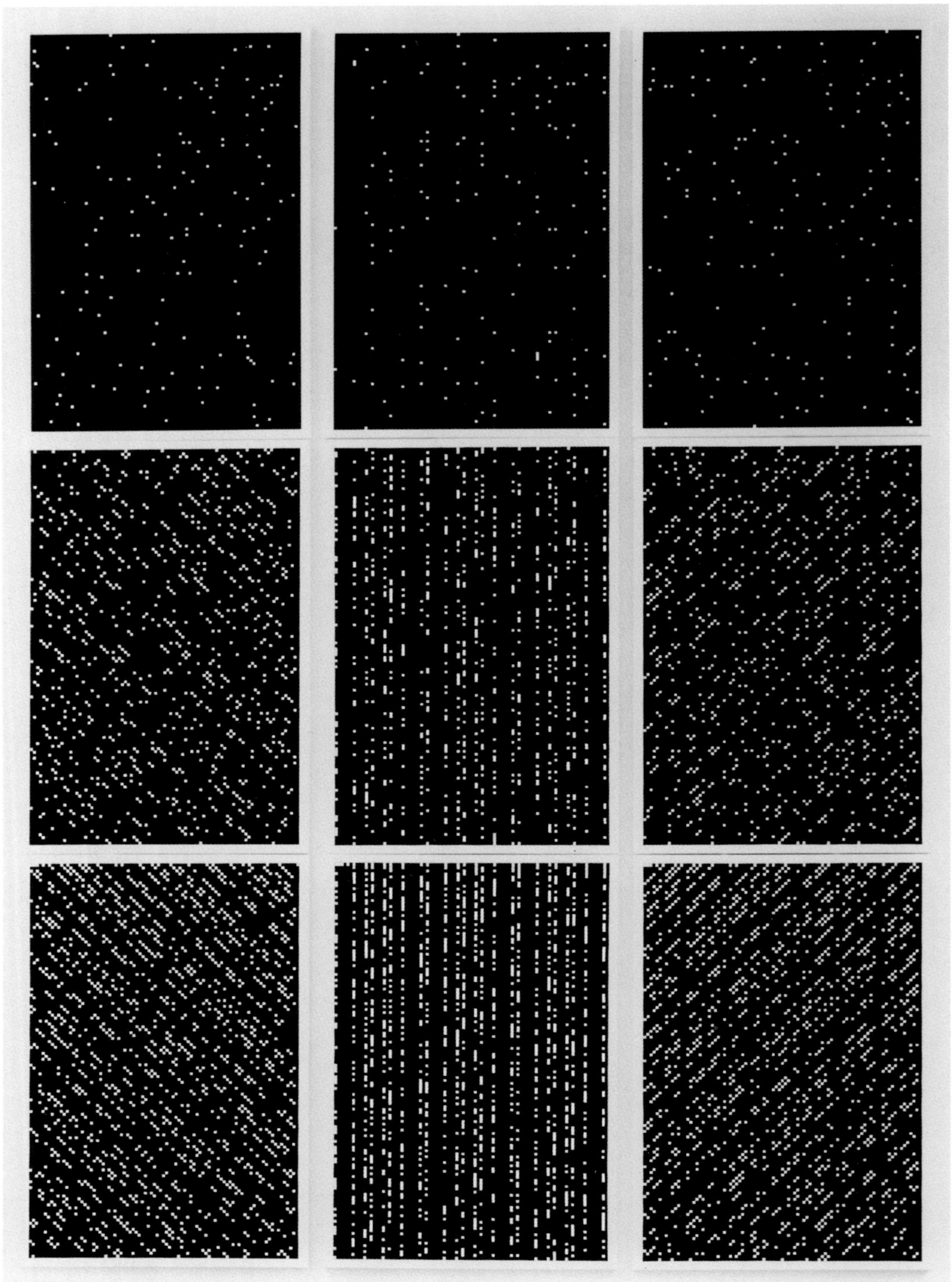

1-24. Rune Mields (German, b. 1935), *The Sieve of Eratosthenes*, 1971. Ink on linen. Courtesy of the artist and the Galerie Angelika Harthan, Stuttgart. © 2014 VG Bild-Kunst, Bonn/Artists Rights Society, New York.

In the Greek city of Alexandria in the third century BC, Eratosthenes of Cyrene, who was director of the city's great classical library, taught a method for finding prime numbers. Beginning with the first whole number greater than 1, circle 2 and cross out every second number after 2. Circle 3 and cross out every third number after 3. Skip 4 because it is already crossed out (which means it is composite), circle 5, and cross out every fifth number after 5. Continue in this way; the procedure is a tool—a sieve—for separating circled prime numbers from crossed-out composites. For another of Eratosthenes's discoveries, see the sidebar on page 29.

Since antiquity, mathematicians have observed patterns in the prime numbers. If we arrange the whole numbers in rows of ten (as in the diagram), then the primes form columns of circles between columns of crossed-out even numbers and multiples of 5. In the lower central panel of Mields's work, she began with 1,2,3, . . . and arranged the whole numbers in rows of 90, using white squares to symbolize prime numbers and black for composites. Since 90 is even, the prime numbers form vertical white bands. The lower-left panel has 89 to a row and the lower-right panel has 91, yielding slanted bands. The central panel in the middle row begins on a higher number, and the central top panel begins on an even higher number; both panels are similarly arranged in rows of multiples of 10. Together the array of nine panels visualizes the pattern of prime numbers (white squares), which occur less frequently as the whole numbers increase. Despite the observation of many patterns in the primes, an understanding of an *overall* pattern—the ability to predict the *next* prime number—had so far eluded mathematicians.

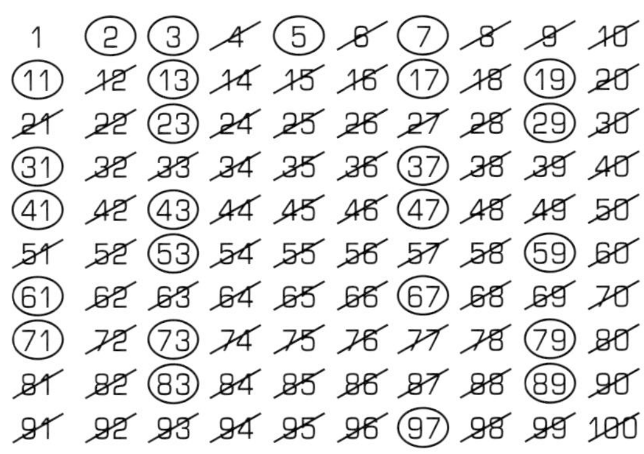

Composite Numbers and Prime Factorization

As we look at the whole numbers greater than 1, we see that some, such as 12, can be written as the products of whole numbers ($12 = 3 \times 4 = 3 \times 2 \times 2$), whereas others, such as 5, can't be expressed as a product. An integer greater than 1 that can be expressed as a product of whole numbers is called *composite*; one that can't is called *prime*. The first ten prime numbers are:

2, 3, 5, 7, 11, 13, 17, 19, 23, 29

Any whole number greater than 1 is either a prime or composite. If it is composite, it can be expressed as a product of two smaller integers, called its *factors*. (A factor divides another number exactly, with a remainder of zero). Each of these factors is either prime or composite, and if composite, it can be decomposed further as a product of integers greater than 1.

Euclid showed that this process can be continued until the composite number is expressed as a product of primes. For example:

Composite number		Prime factors
6	=	2×3
15	=	3×5
46	=	2×23
19,110	=	$2 \times 3 \times 5 \times 7 \times 7 \times 13$

Euclid proved that any composite number can be decomposed (factored) into its primary building-blocks (its prime factors) in just one way (*Elements*, ca. 300 BC, book 7, propositions 30 and 32). The modern statement of the proof is called the fundamental theorem of arithmetic (Carl Friedrich Gauss, *Disquisitiones Arithmeticae*, 1801).

diagram. Euclid's proofs describe how to construct this diagram, which is a visualization of the core concept being asserted by the theorem. Like Plato, Euclid understood these diagrams as symbolizing both ideal Forms that exist outside time and space, and the Forms that are embodied in the transitory natural world. Thus it was crucial to Euclid—and to his many colleagues whose proofs he included in the *Elements*—that the founding premises of his axiomatic system be *true* so that his deductions would yield true conclusions about both the timeless Forms and the natural world.

Euclid began by formulating premises that were self-evidently true, in other words, premises that any rational person intuitively knew were correct based on experience or the reasoning power of the mind. His point was to set these premises down in precise written form so that they could be examined, agreed upon, and used. The cornerstones of his mathematical structure were definitions ("The extremities of a line are points"), postulates ("That all right angles are equal to one another"), and common notions ("The whole is greater than the part").[32] He proceeded to prove each proposition—theorem (from the Greek for "speculation")—each step in every proof based on either a premise (a definition, common notion, or axiom) or a previously proven theorem. Like a mason laying courses of stone over bedrock, Euclid proved each theorem using a deductive structure that preserved validity at every level and constructed mathematics like a grand towering edifice. For example, Euclid proved that in *every* right triangle, the area of the squares on the shorter sides, is equal to the area of the square on the longer side—the Pythagorean theorem (in modern notation, $x^2 + y^2 = z^2$, where x and y are the shorter sides and z is the hypotenuse; plates 1-22 and 1-23).[33]

Euclid's Proof that there is an Infinity of Prime Numbers

Euclid's argument shows how, given a finite collection of prime numbers, there will always be some prime number that is not in the collection. As an illustration he used the collection of three primes A, B, and C, but the method can be applied to any finite collection of primes.

Euclid said: Consider the number $N = A \times B \times C + 1$. Either N is a prime or it is composite. If N is prime, then it is not in the collection because it is larger than A, B, or C. If N is composite, then it may be written as a product of prime factors. But any prime factor of N must be other than A, B, or C because there is a remainder of 1 when N is divided by any of the primes in the original collection. Thus any finite collection of primes will not include them all (*Elements*, ca. 300 BC, book 9, proposition 20).

For example, consider the collection of the first three primes: 2, 3, 5:

$$N = 2 \times 3 \times 5 + 1$$
$$= 30 + 1 = 31$$

31 is prime (and it is larger than any prime in the original collection).

Now consider the collection of the first seven primes: 2, 3, 5, 7, 11, 13, 17:

$$N = 2 \times 3 \times 5 \times 7 \times 11 \times 13 \times 17 + 1$$
$$= 510{,}510 + 1 = 510{,}511.$$

510,511 is composite (its prime factors are 19 × 97 × 277, none of which are in the original collection).

The thirteen books of the *Elements* cover plane geometry (rectangles, triangles, and circles), numbers (whole numbers and ratios), and solids (cubes, pyramids, and spheres). Euclid singled out prime numbers (ones divisible only by themselves and one) as numerical building-blocks (see sidebar on page 21 and plate 1-24). He went on to prove that no finite list can include all the prime numbers; in other words, they are infinitely many (see sidebar above).

GEOMETRIC MODELS OF THE COSMOS

Babylonian astronomers living in the Tigris-Euphrates valley kept detailed, dated records of the visible sky beginning around 700 BC, although these catalogues are based on much earlier records.[34] The Babylonians discerned patterns in the motion of celestial bodies from which they could predict the dates of the summer and winter solstices (plate 1-25). By the fifth century BC, they had realized, furthermore, that, like the sun, the moon rises in the east and sets in the west, and it follows a path at night that is similar to the sun's path in daylight (plate 1-26). This broad path is called the "ecliptic" because when the moon and sun are moving near each other, one can be eclipsed by the other, an event the Babylonians learned to predict. The Babylonians also noticed that constellations of stars accompany the sun, moon, and planets, following the same path each year through the ecliptic (plates 1-27 and 1-28). Having observed this pattern, the Babylonians divided the ecliptic into twelve equal parts (one for each moon), establishing the constellations of the zodiac (plates 1-29, 1-30, and 1-31).[35]

Plato downplayed the role of observation when he set before his students the problem of understanding the design of the universe in mathematical terms, but the theoretical model, he insisted, must in the end explain the observed facts. Eudoxus of Cnidus, who studied philosophy with Plato and mathematics with Archytas, conceived the first sophisticated geometric model of the cosmos, with a stationary earth at its center, that accounted for the apparent retrograde motion of planets (plate 1-32).[36] Eudoxus's contemporary Aristarchus maintained that the sun, not the earth, is the center of the universe, with the earth and other planets revolving around it. But Aristarchus was forgotten and the Eudoxus model survived for a millennium because Aristotle endorsed its basic principles.

After Euclid's axiomatic method was so successful in mathematics, some Greeks in the Hellenistic age came to see deduction as the *only* route to knowledge. They made the mistake of deducing facts about the natural world from first principles and axioms without confirming them with observation. The conviction that genuine knowledge comes only from contemplation also retarded progress in science and engineering because Greek philosophers looked down on the application of pure mathematics to practical tasks that required manual labor, which they considered vulgar. But in the third century BC, Archimedes of Syracuse overcame this prejudice and treated any axiom that described the natural world not as a timeless truth but as a *hypothesis* from which he could predict facts. He then used trial and error to confirm (or refute) the prediction. In about 260 BC he proved a basic theorem about the balance of levers; given the weight of a stone, he could calculate the amount of force needed to move it, the length of the lever needed, and the most efficient spot to position the second stone behind the lever (plate 1-33). Archimedes's contemporary, Eratosthenes of Cyrene, who was a librarian at the great classical library in Alexandria, demonstrated the extraordinary power of mixing mathematics with observation when he correctly calculated the circumference of the earth by using one of Euclid's theorems together with his observation of shadows cast by a stick (see sidebar on page 29).

In the second century BC, Hipparchus of Rhodes codified the celestial records of the Babylonians and added his own extensive observations (plate 1-34). In the *Elements* Euclid had shown how to compute the missing parts of a triangle; for example, if two angles and the side between them are known, the rest of the triangle can be constructed in only one way. Triangles are of particular interest to astronomers because when the "missing part" of the triangle is in the sky, it can be used to measure the distance to a heavenly body. Astronomers also studied angles because they give a way to measure the distance between two objects in the sky by measuring the angle to which you must turn your head to look at one object and then the other. If you make that angle part of a right triangle, then the sides have fixed ratios to each other. Hipparchus founded the field of trigonometry (from the Greek for "to measure a triangle") when he made up precise tables relating the ratios of angles to sides in triangles.

Using trigonometry, Hipparchus calculated the distance from the earth to the moon by first recording the moon's position against the apparently fixed stars from different points of view. When your point of view changes, a near object seems to change position relative to a distant object—a phenomenon called parallax. Near objects appear to move more and far objects less, relative to the viewer. After Hipparchus had determined the parallax of the moon, using trigonometry he could calculate that the moon was at a distance equal to 30 times the earth's diameter. About 50 years earlier, Eratosthenes had determined that

TOP RIGHT

1-25. The sun appears to move through the sky each day in a great half circle, rising in the east and setting in the west. In summer the days grow longer as the sun rises earlier and sets later, reaching its zenith overhead on the longest day—the summer solstice. Near the equator on that day, the vertical blade (gnomon) of a sundial casts no shadow.

In winter the days shorten and the sun's path moves closer to the horizon, causing a gnomon to cast a long shadow. The shortest day is the winter solstice.

BOTTOM RIGHT

1-26. The moon moves from east to west, rising and setting each day in a path that approximates the one taken by the sun. The moon goes through phases, growing from a thin crescent to a full moon, then shrinking back down to a crescent and disappearing one night, becoming the "dark moon," or "new moon." The Babylonians recorded that planets also move in the paths followed by the sun and moon. The orbits of the sun, moon, and planets are all more or less tipped relative to each other, forming a rough plane of the ecliptic overhead but not a narrow path.

OPPOSITE

1-27. The four diagrams show the sky as it appeared over Babylonia (present-day Iraq) in the late fifth century BC. The sky looks a little different today because of the slow change of direction of the earth's axis—so-called axial precession—which traces out an approximate circle in the sky every 26,000 years.

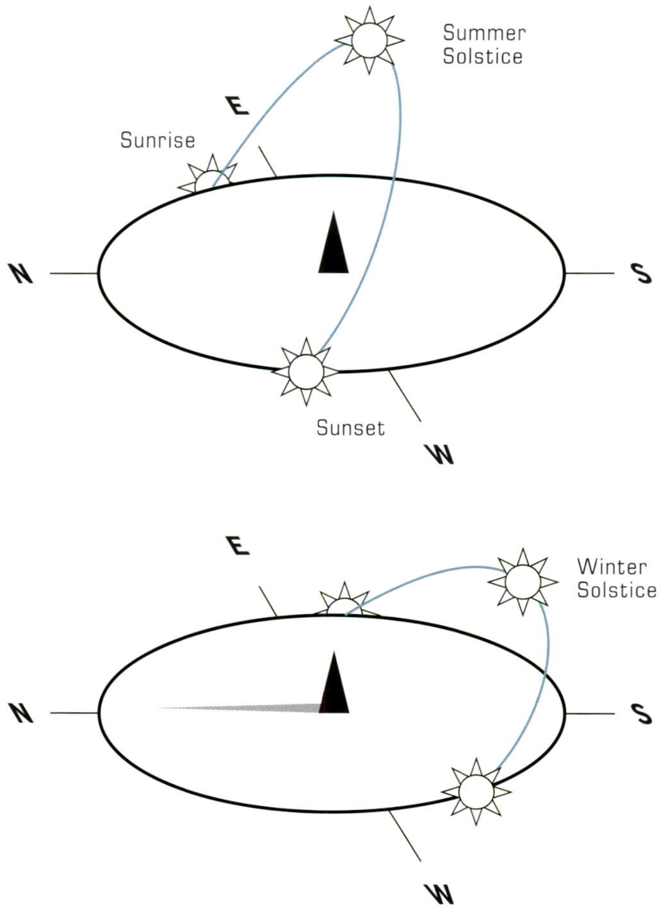

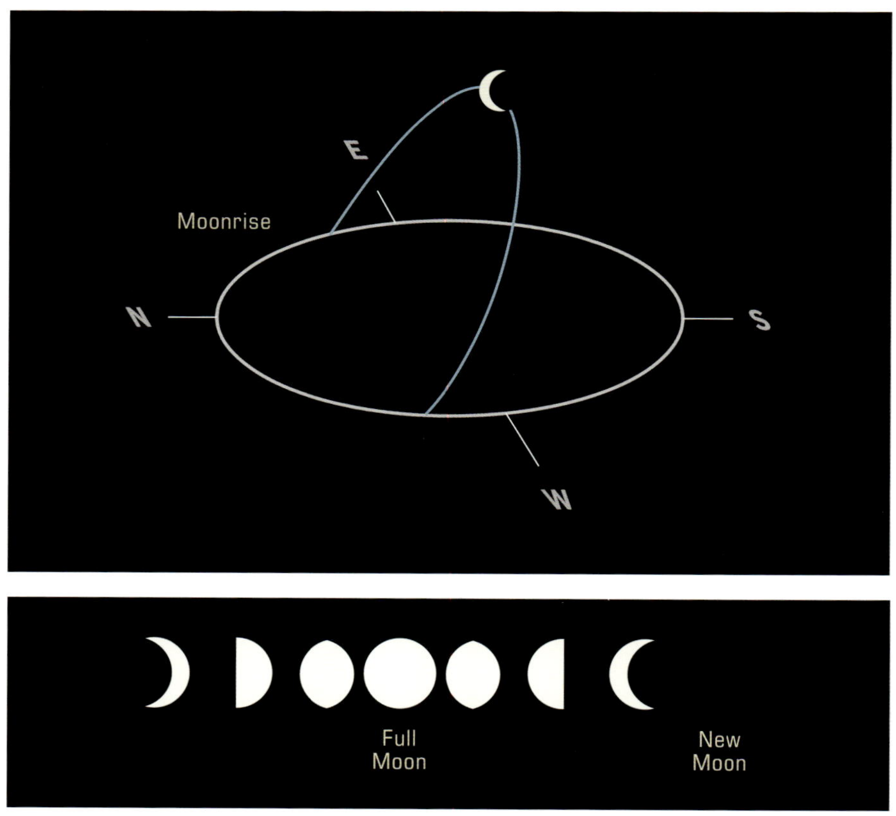

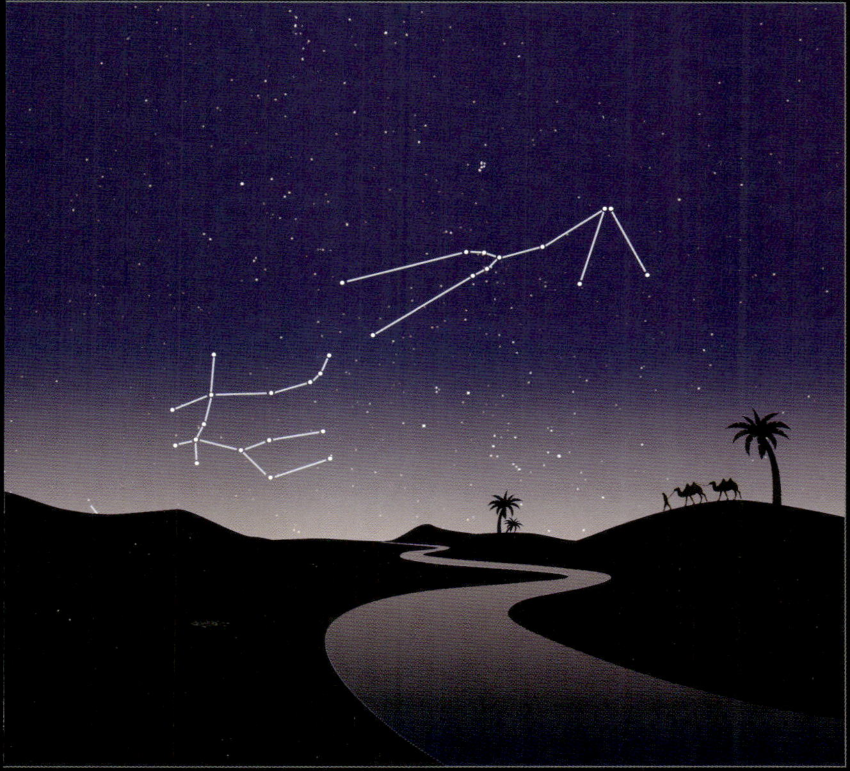

A When Babylonian astronomers gathered to watch the sunrise the morning after the new moon in late June, just before dawn they saw a group of stars—the constellation Gemini—at the horizon.

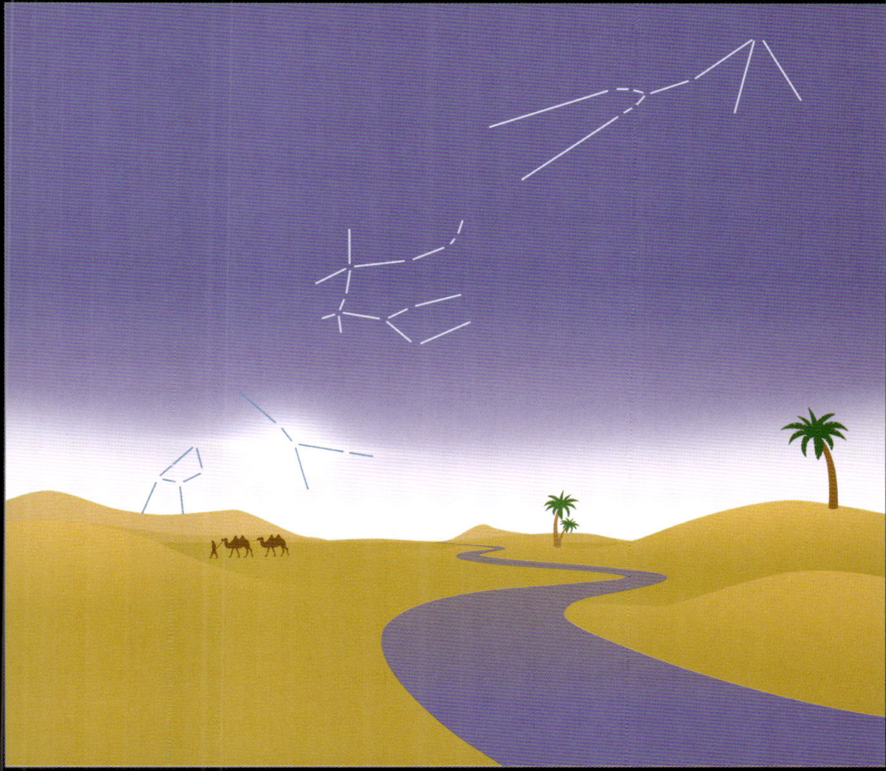

B Then during the next hour, Gemini slowly moved up and to the right and the sun rose “in Cancer” as the stars became invisible in daylight.

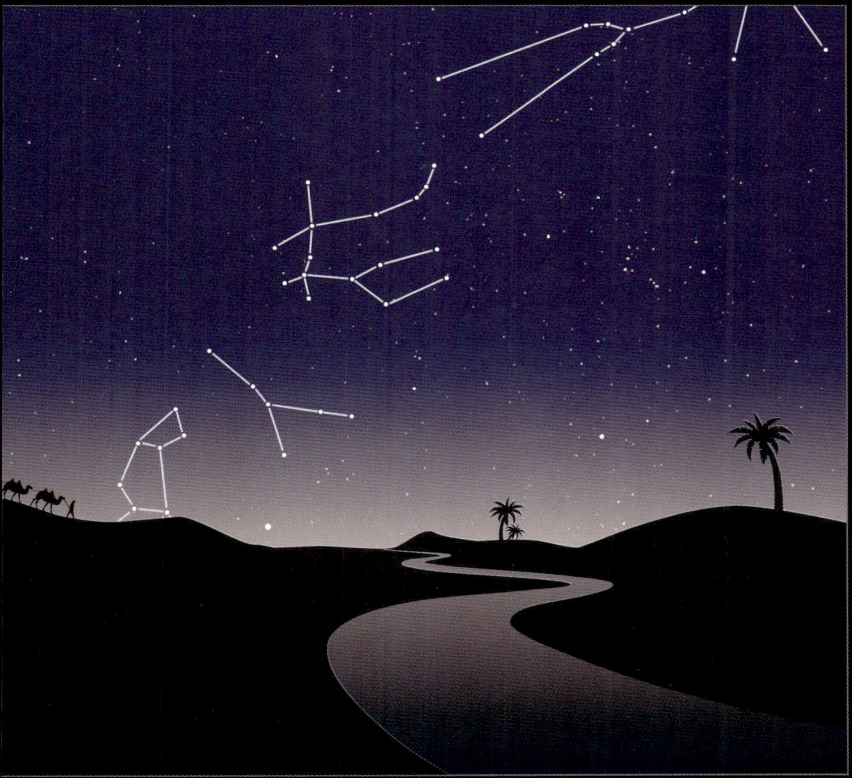

C About 30 days later, on the morning after the next new moon in late July, the Babylonians observed that Cancer was now at the horizon at dawn.

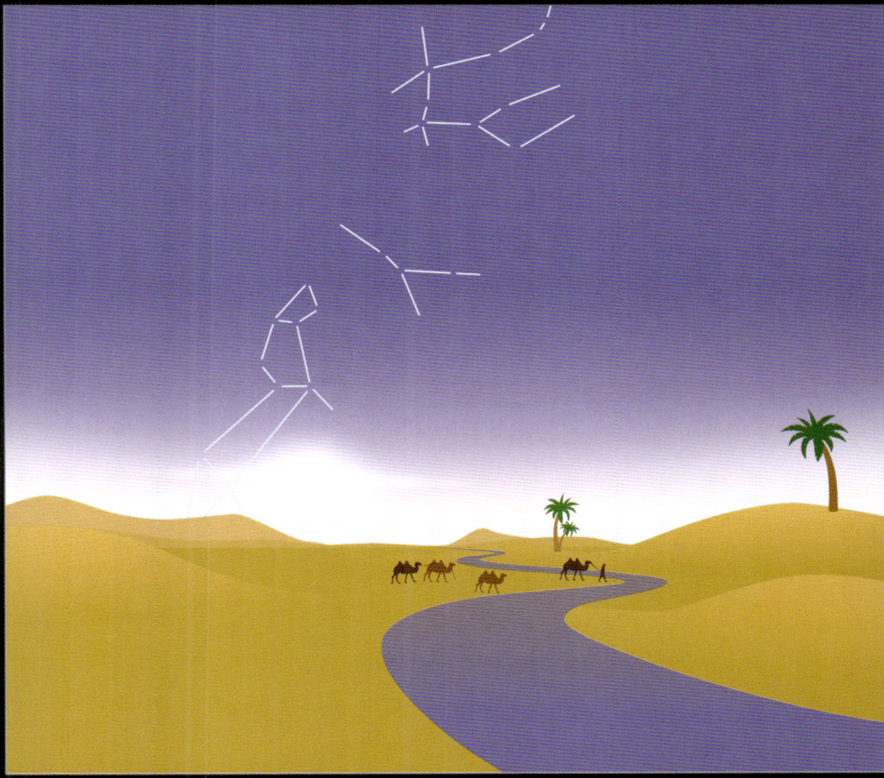

D Then during the next hour, Cancer followed Gemini up and to the right and the sun rose in the constellation Leo.

RIGHT

1-28. The Babylonians noticed that the pattern of constellations accompanying the rising sun repeats after twelve new moons. Thus they divided the circular band of stars in the ecliptic into twelve constellations—the signs of the zodiac—one for each moon. They also divided the circle into 360 units, presumably approximating the days of the year—12 "months" (from Greek for "moons") × 30 days. (This diagram shows the order of the constellations in the night sky; it is not intended to show how they would actually look from a celestial perspective.)

The Babylonian observation of the cyclical nature of time survives today in the circular face of a clock, where the division of an hour into 60 minutes and each minute into 60 seconds is a remnant of their base-60 counting system.

OPPOSITE

1-29. Zodiac.

This diagram shows the constellations arranged counter-clockwise as they appear in the night sky (moving, for example, from Gemini in May, to Cancer in June, to Leo in July). Established in ancient times, these dates are somewhat different today because of the earth's axial precession. (For example, sunrise today enters the constellation Cancer in late July rather than late June.) The pictures of the twelve signs are woodcuts made for the first printed edition of the Italian astronomer Guido Bonatti's manuscript *Liber astronomiae* (Book of astronomy, 1277), which was produced as part of *Decem tractatus astronomiae* (Ten treatises on astronomy) in Augsburg in 1491 by the Bavarian printer Erhard Ratdolt (who also produced the first printed edition of Euclid's *Elements*; see plate 2-19 in chapter 2). In the center of this zodiacal diagram is earth shown as in the Ptolemaic model of the cosmos in plate 1-35.

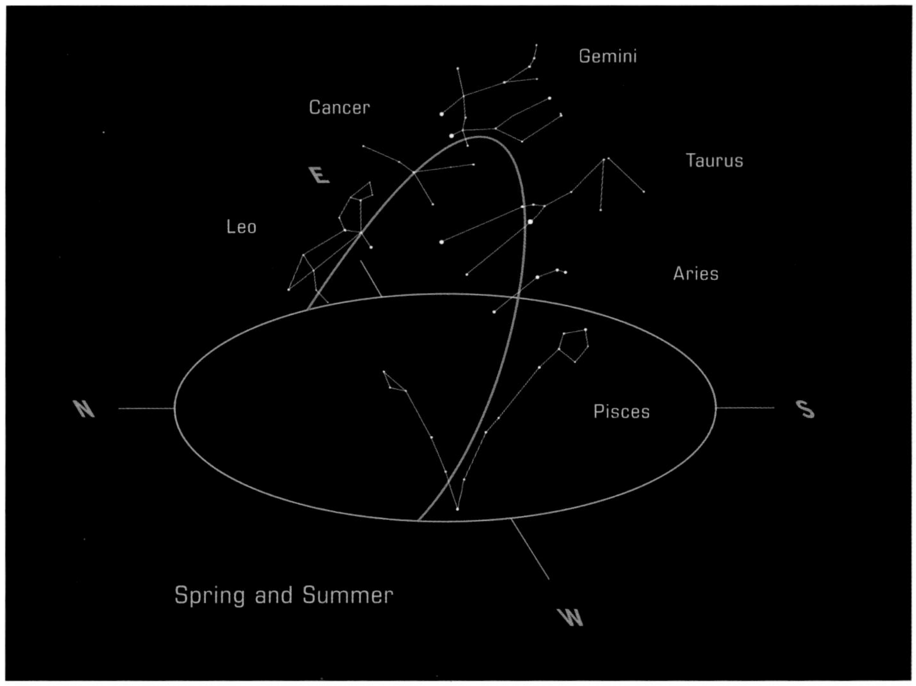

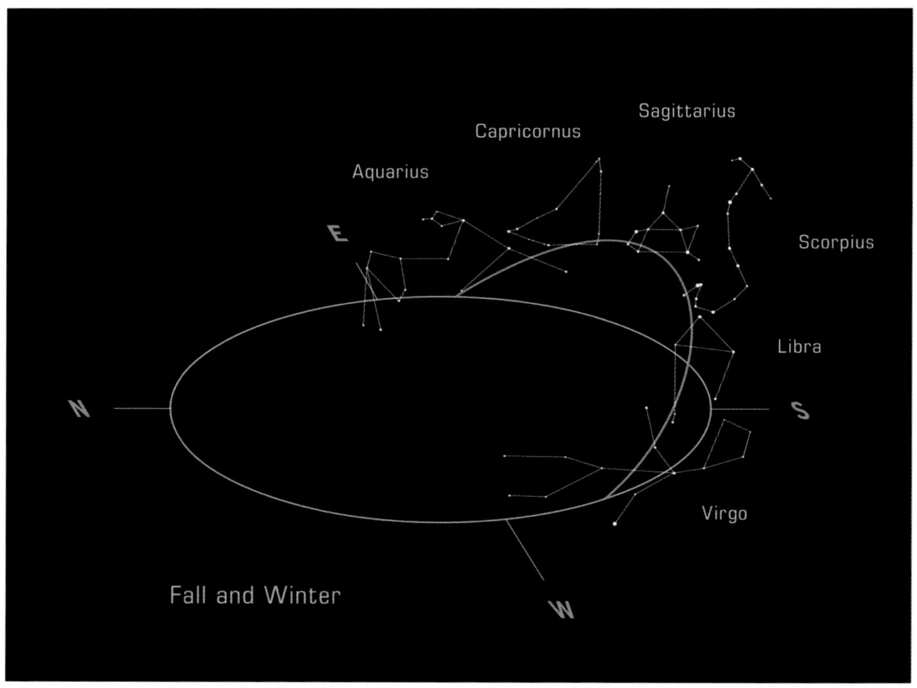

the earth's circumference is about 25,000 miles, which gives it a diameter of about 8000 miles. This meant that the moon is 30 × 8000, or about 240,000, miles away, which is correct and provided astronomers with the first evidence that the cosmos is very large.

In the fourth century BC, Eudoxus had proposed a model of the cosmos as a series of concentric spheres; the stars were embedded on the inside surface of the largest sphere, and transparent spheres carried the sun, moon, and planets, all revolving around a stationary, spherical earth at its center. Hipparchus added a mechanism—an epicycle—that could account for the apparent retrograde motions of the planets (plate 1-36).

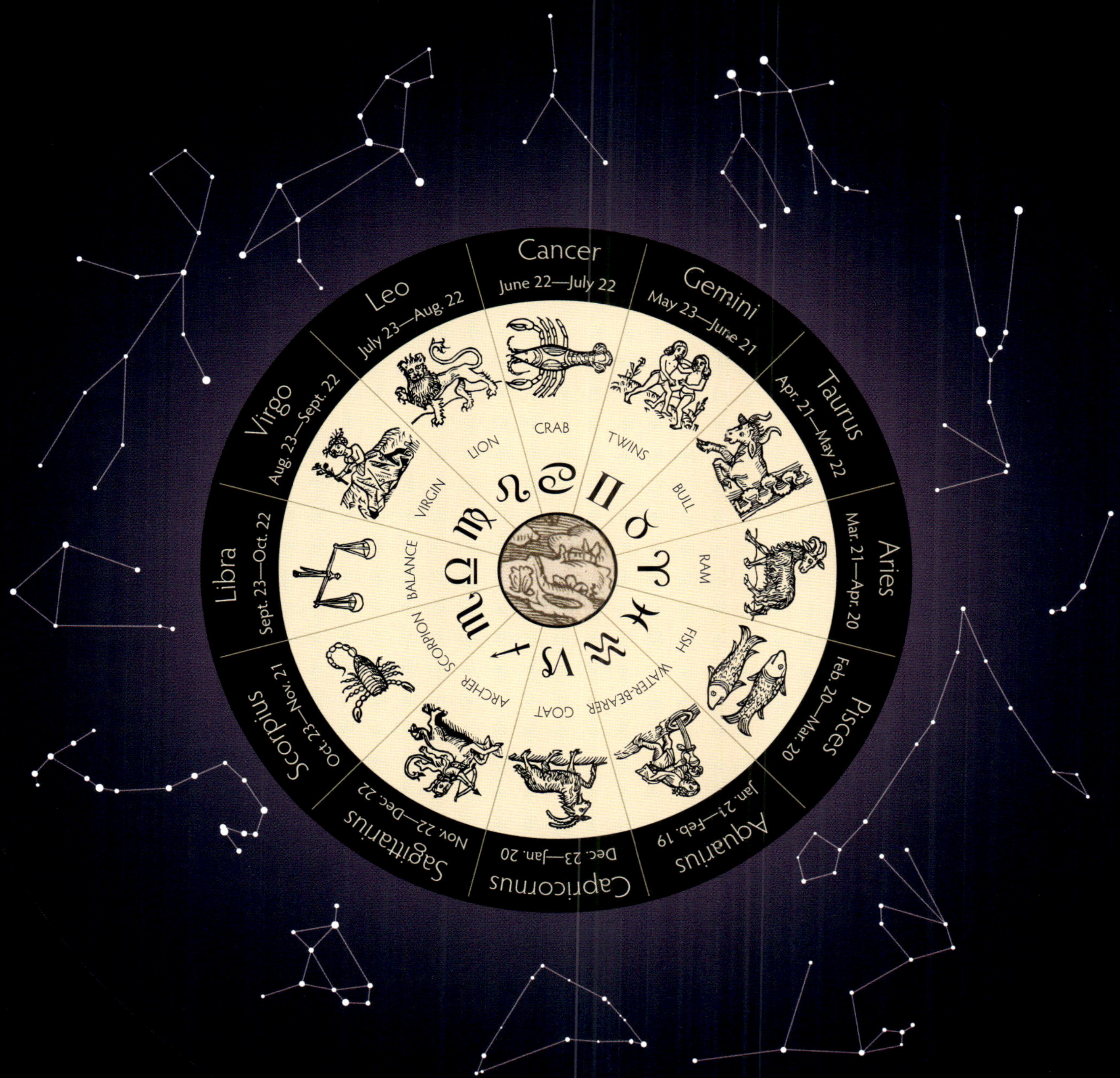
Cancer
June 22—July 22
Gemini
May 23—June 21
Taurus
Apr. 21—May 22
Aries
Mar. 21—Apr. 20
Pisces
Feb. 20—Mar. 20
Aquarius
Jan. 21—Feb. 19
Capricornus
Dec. 23—Jan. 20
Sagittarius
Nov. 22—Dec. 22
Scorpius
Oct. 23—Nov. 21
Libra
Sept. 23—Oct. 22
Virgo
Aug. 23—Sept. 22
Leo
July 23—Aug. 22
CRAB
TWINS
BULL
RAM
FISH
WATER-BEARER
GOAT
ARCHER
SCORPION
BALANCE
VIRGIN
LION

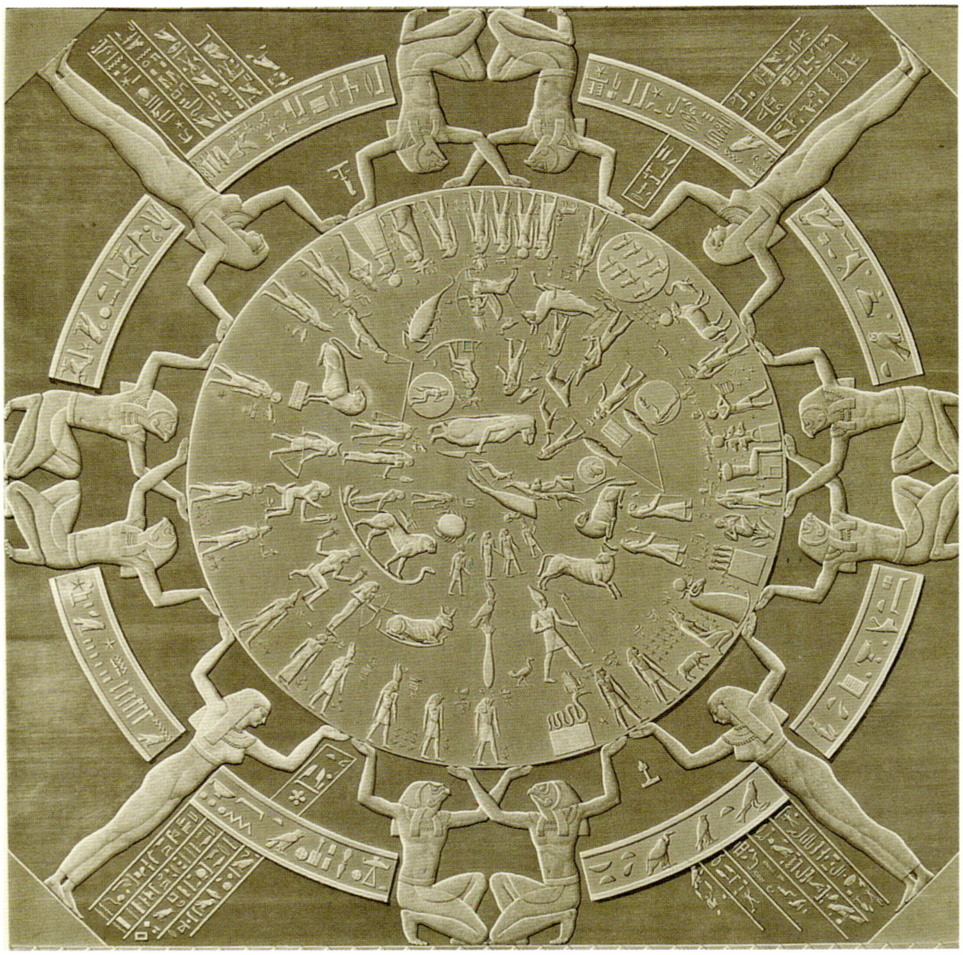

TOP RIGHT

1-30. Zodiac from the Temple at Dendera, Egypt, mid-first century BC. Copper engraving after the original stone carving, in *Description de l'Égypte, ou recueil des observations et des recherches qui ont été faites en Égypte pendant l'expédition de l'armée française* (Paris: Panckoucke, 1821–29). Special Collections, Binghamton University Libraries, Binghamton University, State University of New York.

The Egyptians adopted the zodiac from the Babylonians during the Ptolemaic era (305–30 BC) when they were under Greco-Roman domination. This bas-relief, carved into the ceiling of the Temple complex at Dendera, shows the circular sky supported by four goddesses—standing at north, south, east, and west—and eight kneeling falcon-headed figures. Around the perimeter of the circular sky stand thirty-six decans (each a ruler over a 10-day period) who together comprise the 360-day year. The Egyptian decans surround the twelve constellations of the Babylonian zodiac, which occupy a circular band near the center of the ceiling. Cancer (the crab) is visible below the center slightly to the left, and Leo (the lion) is, in turn, below and to the left of Cancer, riding on a snake-like sled. Representatives of planets and constellations also crowd around the zodiacal circle.

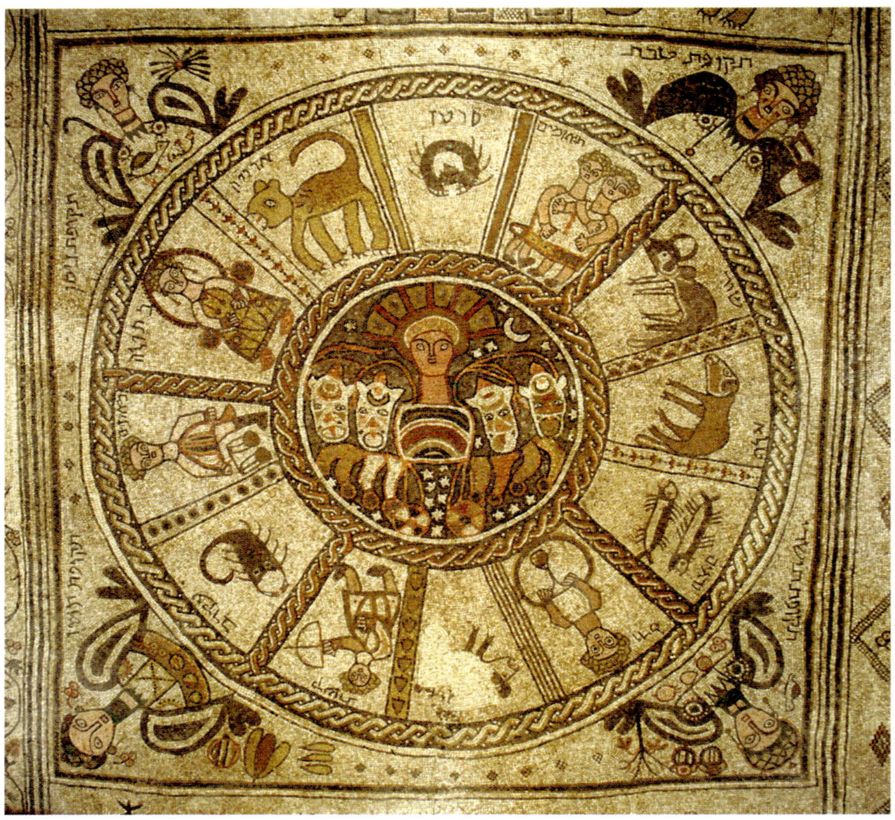

BOTTOM RIGHT

1-31. Zodiac in the synagogue of Beth Alpha, Jezreel Valley, Israel. Mosaic, sixth century AD.

Made of mosaic, this zodiac forms part of the floor of a synagogue built in the land of Israel during the sixth century AD, when the Byzantine Empire dominated the Near East. In a mix of Jewish and Greco-Roman cultures, the twelve signs, labeled in Hebrew, surround the classical sun god Helios, who rides through the starry heavens in a four-horse chariot, wearing a crown of sunbeams. The winged figures in the corners symbolize the longest and shortest days of the year—the summer and winter solstices—and the days when light and dark are approximately equal—the fall and spring equinoxes.

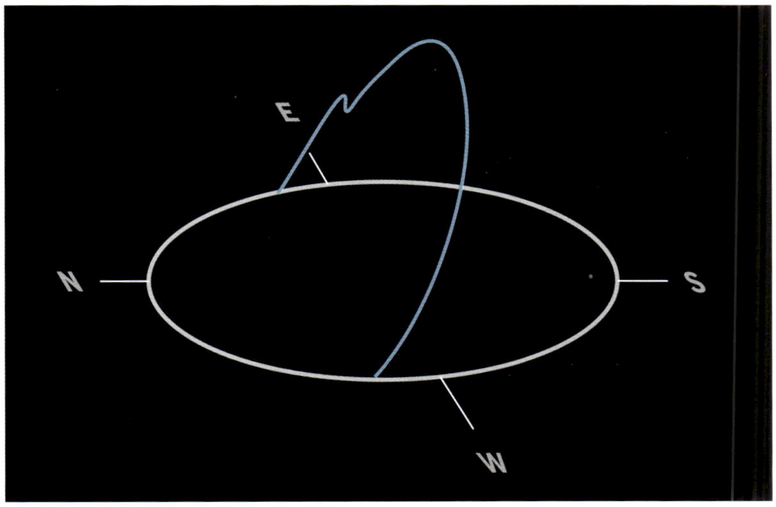

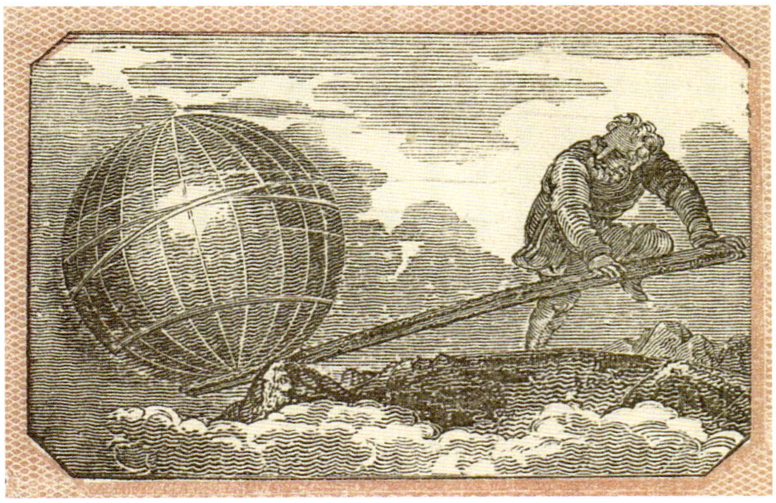

TOP LEFT

1-32. Apparent retrograde motion of a planet.

Like the sun, the planets appear to rise in the east and set in the west. But each planet has a distinct pattern of retrograde motion during which it appears to stop briefly, reverse direction, and then resume its progression in the path of the ecliptic.

BOTTOM LEFT

1-33. "Give me a place to stand, and I can move the earth," a remark of Archimedes quoted by Pappus of Alexandria in *Synagoge*, (ca. AD 340), bk. 8. Engraving, cover of *Mechanics Magazine* 2 (London, 1824). Rare Book and Manuscript Library, University of Pennsylvania.

Eratosthenes's Measurement of the Circumference of the Earth

To estimate the circumference of the earth, Eratosthenes used Euclid's theorem that states that when a line crosses two parallel lines, it creates two sets of angles, and the interior alternate angles are equal.

Eratosthenes knew that the earth is a sphere and he correctly surmised that the sun's rays striking earth are virtually parallel.

Eratosthenes lived in Alexandria in the third century BC. He heard that in the city of Syene (today Aswan), located about 500 miles south of Alexandria, a gnomon of a sundial casts no shadow when the sun is directly overhead at high noon on the summer solstice. He observed that in Alexandria on the same day at noon, a gnomon casts a shadow, and he measured the angle of the sun's rays as 7°. By Euclid's theorem, he knew that the apex of a triangle—which lies in the center of the earth—whose base is the distance between Alexandria and Syene, is also 7°.

From the Babylonians, Eratosthenes knew that a circle has 360 units, thus he figured that the 7° angle at the center of the earth was 7/360 or about 1/50 of a full circle. Figuring that the 500 miles separating Alexandria and Syene is also 1/50 of a circle, Eratosthenes calculated the circumference of the earth to be 50 × 500 miles, or 25,000 miles. (I have used miles in place of the *stadia* used by Eratosthenes, but allowing for the uncertainties, the calculations are as described.) The actual circumference at the equator is about 24,902 miles. Given his knowledge of circles and the circumference of earth, Eratosthenes then calculated the diameter of the earth to be about 8000 miles, which is also quite accurate.

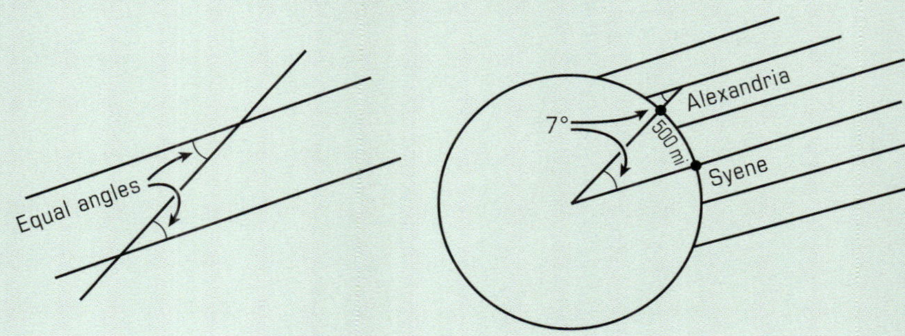

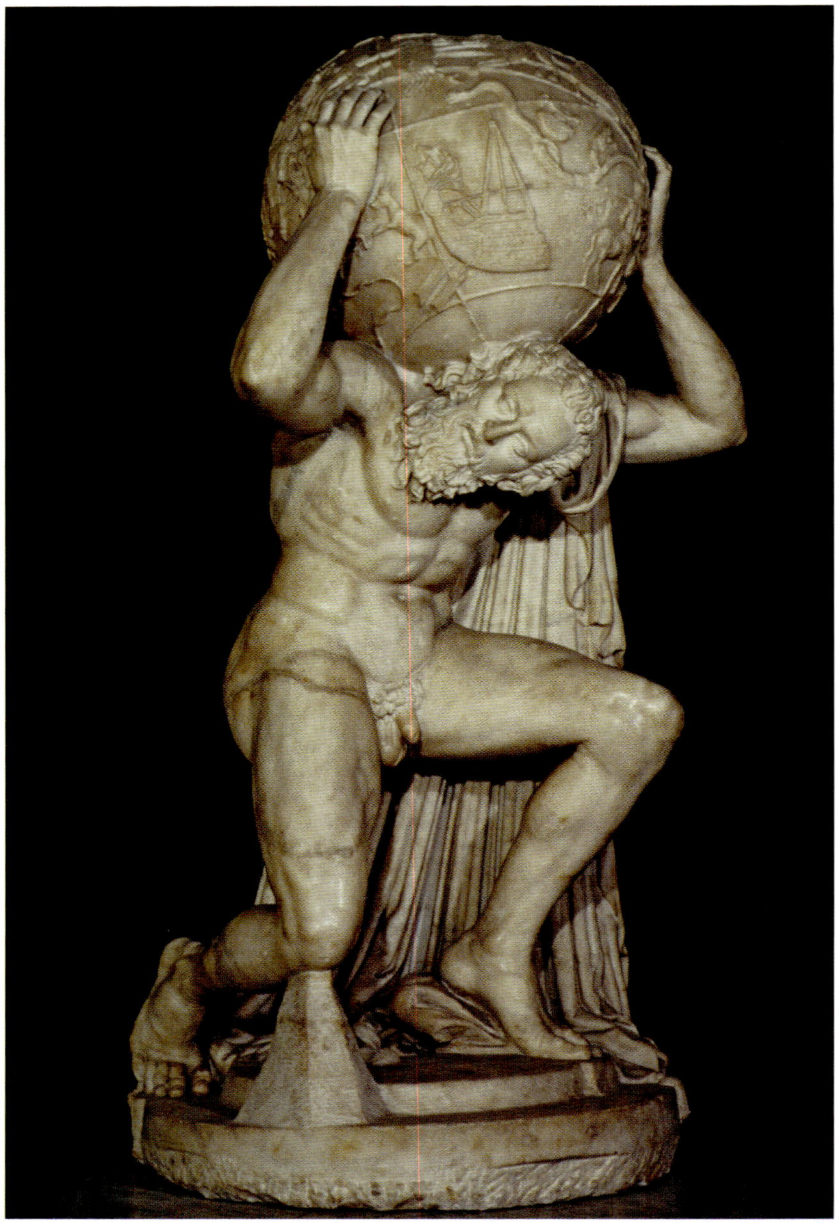

1-34. *The Farnese Atlas*, AD 150, Roman copy of a Greek original of the late second century BC. Marble, height 7 ft. (2.13 m). Museo Archeologico Nazionale, Naples.

Hipparchus completed a star catalogue in 129 BC for which he devised a co-ordinate system that included the celestial equator and the ecliptic. His catalogue and the celestial globes he made were famous in the ancient world but were lost after Rome fell. In 2005 the American astronomer Bradley E. Schaefer demonstrated that the celestial globe hoisted by Atlas in this statue is a copy of one of Hipparchus's lost globes. The sculptor of the original Greek statue had to have referred to a globe by Hipparchus, who was his contemporary, because the position of the constellations matches the sky in 125 BC, and the globe is inscribed with Hipparchus's signature co-ordinate system.

The astronomer Ptolemy of Alexandria, a follower of Hipparchus in the second century AD, reasoned that if the heavens were the realm of perfection, then heavenly bodies would be spherical and move in circular paths at a constant speed (plate 1-35). Ptolemy summarized the geocentric models of Eudoxus and Hipparchus in his *Mathematical Compendium in Thirteen Volumes*, along with his own findings and all the extensive astronomical data collected by the Greeks to which he had access in the classical library in Alexandria. After the decline of Greek civilization, the fall of the Roman Empire, and the destruction of the library in Alexandria,[37] the original Greek writings of Euclid and Ptolemy continued to be copied and studied in the East in the Byzantine Empire until it gradually weakened after the rise of Islam.

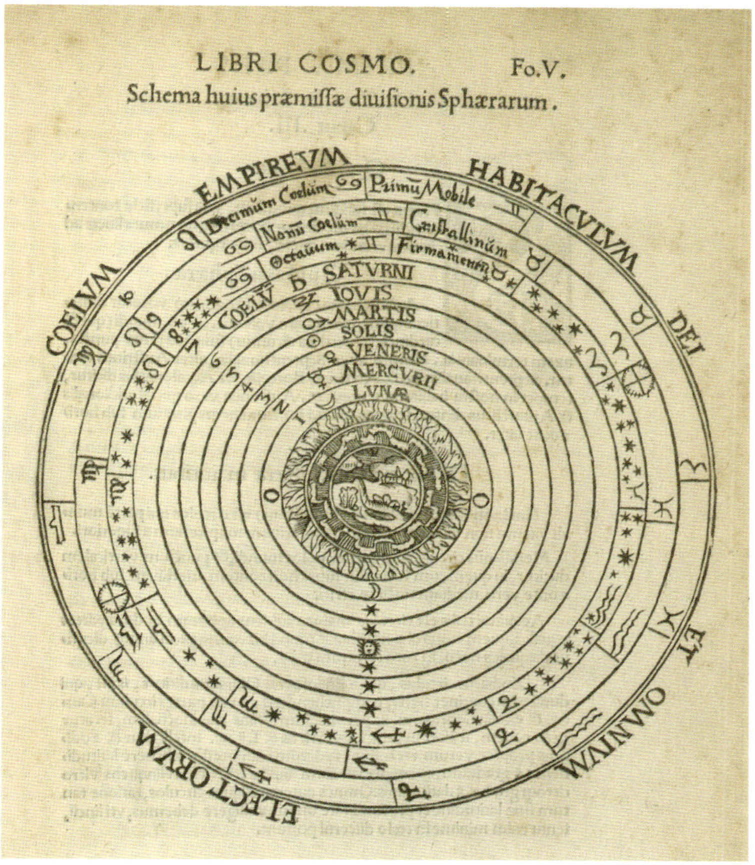

ABOVE

1-35. Ptolemaic model of the cosmos, in Peter Apian, *Cosmographia* (Antwerp: Arnoldi Berckman, 1539), 5. Rare Books Division, The New York Public Library, Astor, Lenox, and Tilden Foundations.

At the center of Ptolemy's cosmos was a stationary earth, surrounded by concentric transparent spheres (made of Aristotle's æther) carrying the moon (orb 1), the inner planets (2–3), the sun (4), and the outer planets (5–7). Then came the orb of fixed stars (8), the firmament (*Firmamentu*), which was divided into twelve parts, each with its constellation of the zodiac, such as ♊ (Gemini, shown above), moving counter-clockwise to ♋ (Cancer), and ♌ (Leo). The firmament was the end of the Aristotelian-Ptolemaic cosmos because of Aristotle's dictum that neither space, nor time, nor void, could exist beyond the outermost boundary of the physical world. When the early Christians adopted the Greek model, they added two layers of heaven (*Coelum*) above the firmament to express their creation story: "God made the firmament, and divided the waters which were under the firmament from the waters which were above the firmament: and it was so" (*Genesis* 1:7). One of these watery orbs was crystalline (*Cristallinum*; 9) and the other was the first to move (*Primu Mobile*; 10). Outside the last sphere the early Christians placed COELVM EMPIREVM HABITACVLVM DEI ET OMNIVM ELECTORVM (the highest heavenly dwelling place of God and the elect).

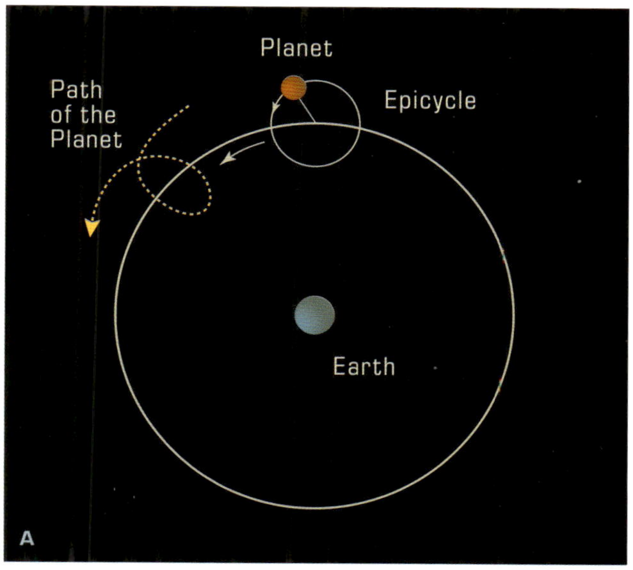

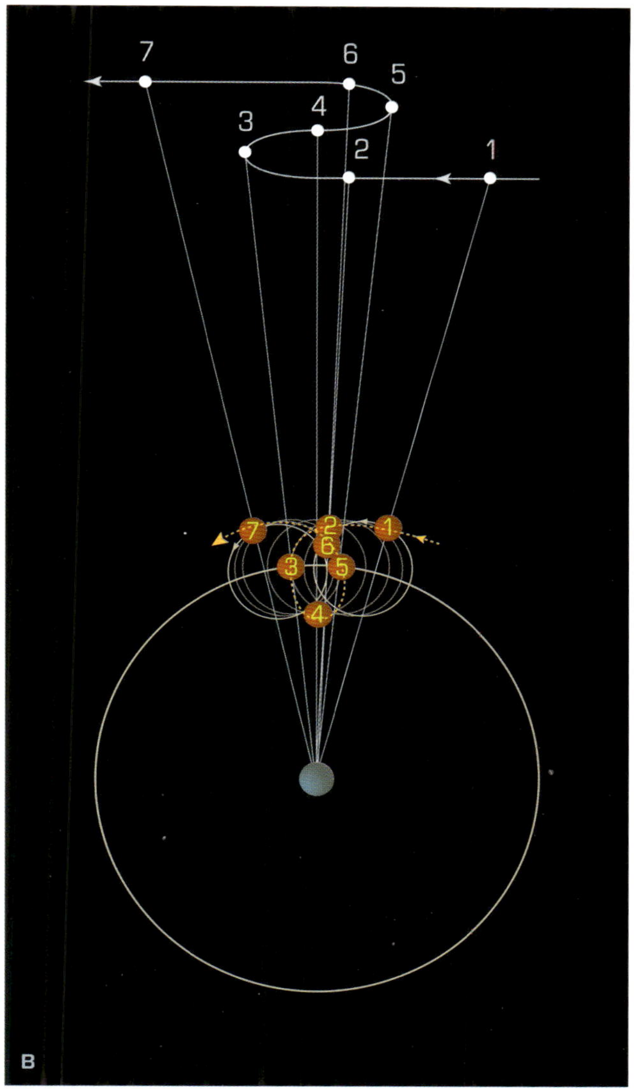

RIGHT, TOP AND BOTTOM

1-36. Hipparchus on the apparent retrograde motion of Mars.

A Hipparchus hypothesized that each planet revolves on a small circle—an epicycle—the center of which moves along a greater circle around an assumed stationary earth.

B The motion of Mars (numbered 1–7 in yellow) along this combination of epicycle and greater circular orbit would, according to Hipparchus, account for its apparent retrograde motion (numbered 1–7 in white).

ISLAMIC MATHEMATICS: ALGEBRA AND ARABIC NUMERALS

Knowledge of Euclid and Ptolemy was eventually lost to the medieval West, but Islamic scholars preserved their writings in Arabic translations from Byzantine Greek texts (plate 1-39). In the ninth century, caliphs established the House of Wisdom in Baghdad as a place for scholars to acquire and translate foreign texts in mathematics and philosophy, especially from classical Greece, and to publish their own original ideas. Ptolemy's thirteen-volume work is known today by the name they gave it, *Almagest* (Arabic for "the greatest").[38] Building on Greek mathematics, Islamic peoples made significant innovations, including the adoption from Hindu mathematics of the zero and positional notation (plate 1-37).

Euclid lacked a notation for dealing with arithmetical operations in general. About AD 825, the Islamic mathematician, al-Khwārizmī, saw how letters could be substituted

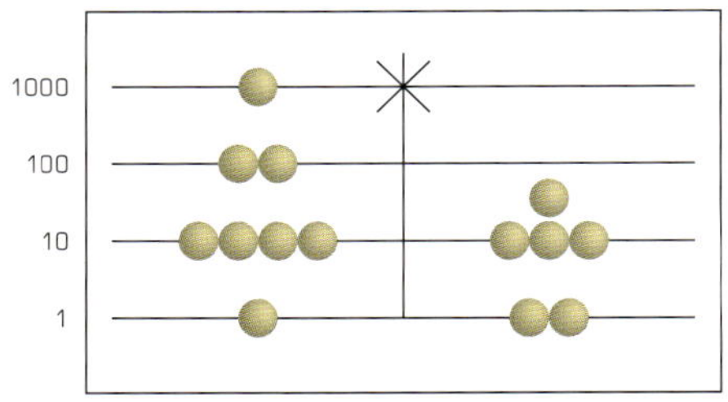

ABOVE

1-37. A counting board—a type of abacus (a tool for calculation)—has movable counters (beads or stones, sometimes strung on wires) which indicate ones, tens, hundreds, and thousands. Although used centuries before Indian mathematicians invented the zero and the decimal point, the concept of base 10, positional notation is implicit in counting boards. In this example and in plate 1-38, placing a bead between lines indicates half the value (5, 50, or 500); thus the beads form the numbers 1241 (above on the left) and 82 (right).

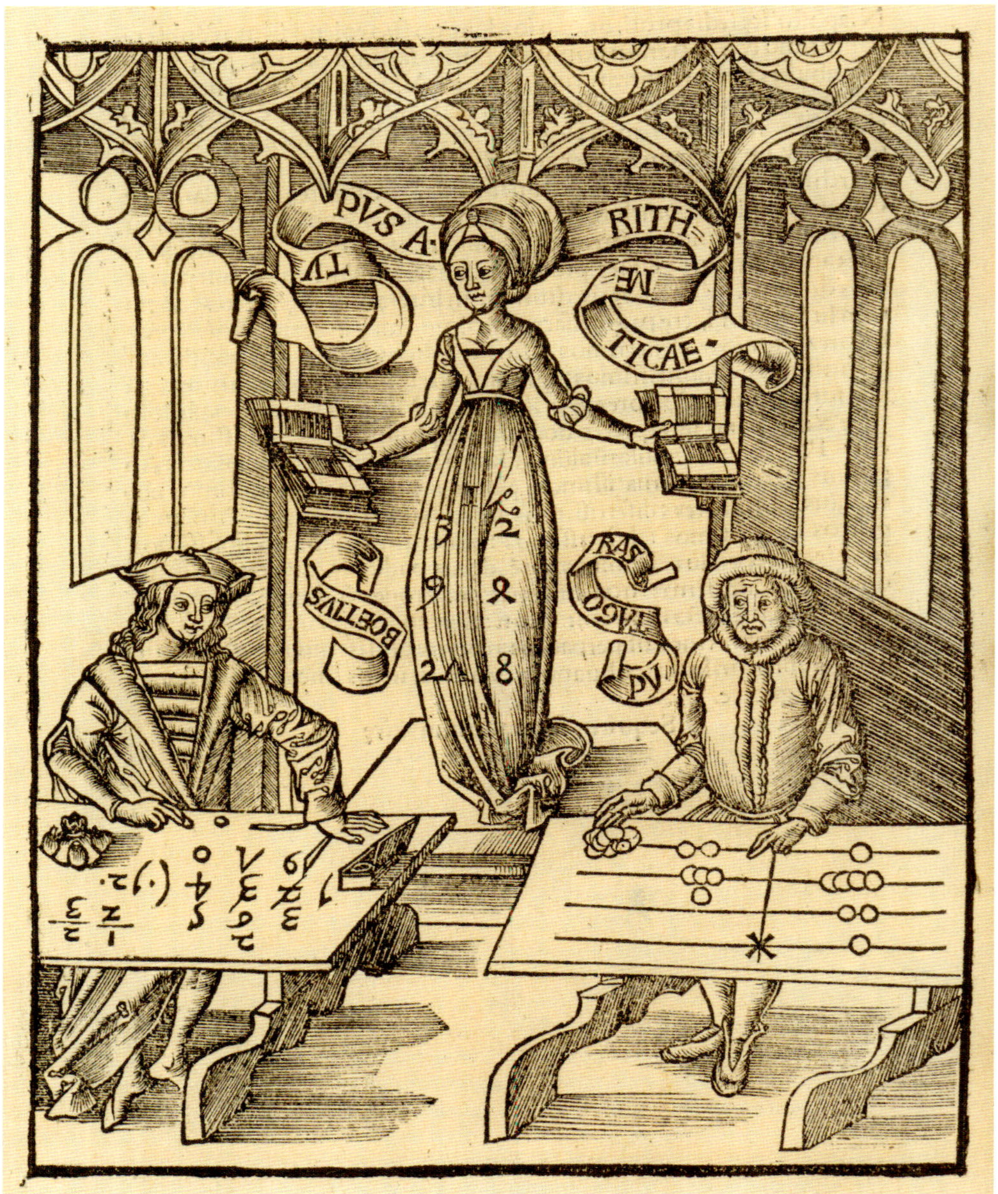

RIGHT

1-38. *Typus Arithmeticae* (Personification of arithmetic), in Gregor Reisch, *Margarita Philosophica* (Freiberg, 1503). Houghton Library, Harvard University, Typ 520.03.736.

The personification of Arithmetic presides over a contest between the early Christian scholar Boethius, author of books on mathematics and music, who smiles as he calculates with Arabic numerals, while a sad Pythagoras uses an old-fashioned counting board, on which he has formed the numbers 1241 and 82. Dressed in the latest fashion, Arithmetic wears a gown embroidered with the Arabic numeral 1 in the centerfold of her skirt, from which descend the progressions 3, 9, 27 and 2, 4, 8.

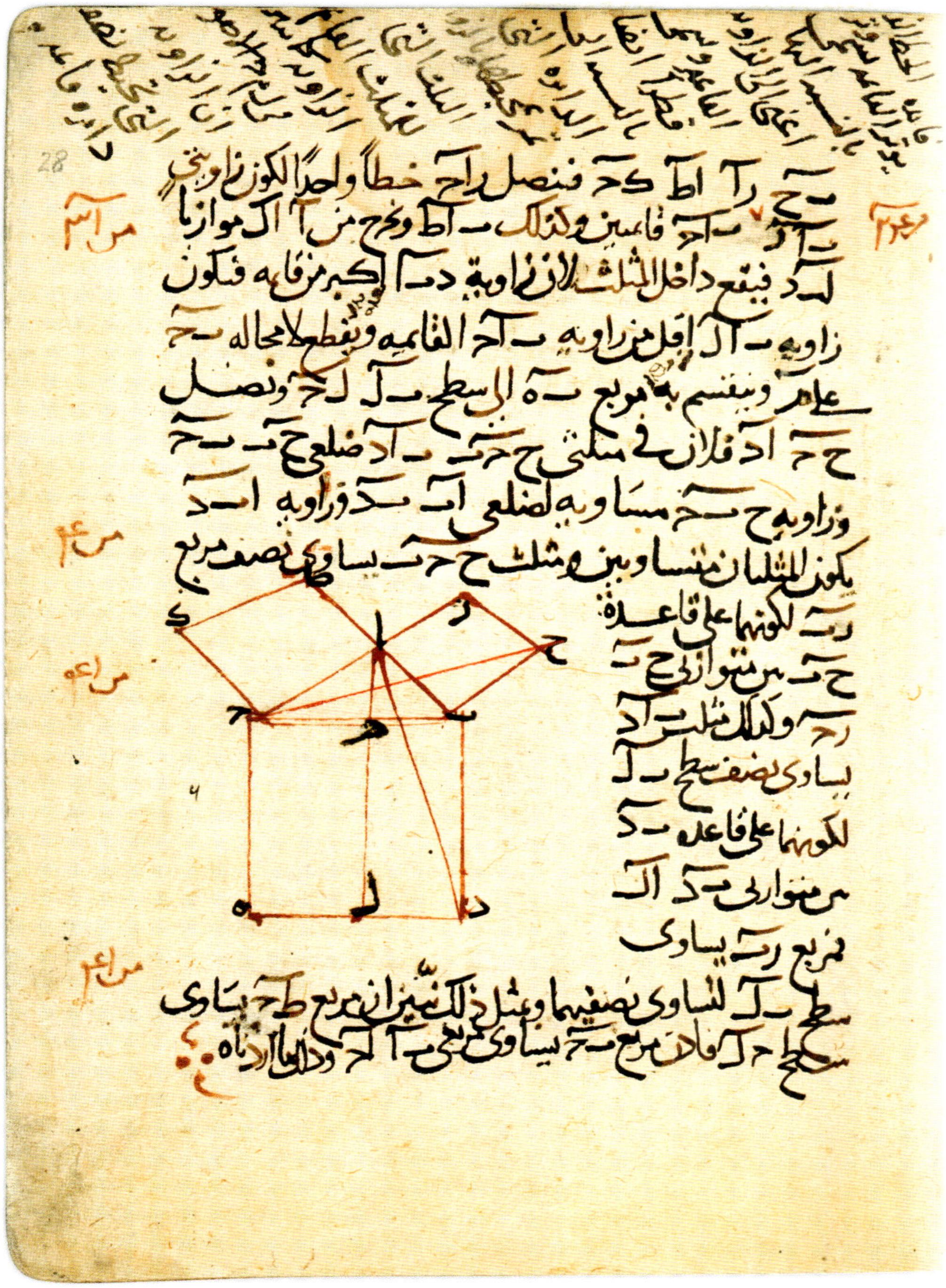

That fondness for science,
. . . that affability and condescension which God shows to the learned, that promptitude with which he protects and supports them in the elucidation of obscurities and in the removal of difficulties,—has encouraged me to compose a short work on calculating by (the rules of) Completion and Reduction, confining it to what is easiest and most useful in arithmetic.

—Al-Khwārizmī, ***Book of Addition and Subtraction According to the Hindu Calculations***, ca. AD 825

for numbers to express general properties. For example, $x + y = y + x$ expresses the fact that the sum of two numbers is independent of their order. The x and y are variables and can be replaced by any pair of numbers. Thus al-Khwārizmī generalized arithmetic and invented algebra (from the Arabic *al-jabr* for "restoration"),[39] which allows us to ask: For what x does $2 + x = 5$?

Al-Khwārizmī wrote the influential *Book of Addition and Subtraction According to the Hindu Calculations,* in which he demonstrated the ease of adding long columns of numbers using the glyphs of the Hindu system (commonly referred to as "Arabic numerals," although they are of Indian origin; see plate 1-38). Al-Khwārizmī also promoted the use of a

TOP LEFT

1-39. Euclid's *Elements*, 300 BC, trans. Nasir al-Din al-Tusi, AD 1258. Ink on vellum, 6⅞ in. (17 cm) high. © The British Library Board, MS Add. 23387, fol. 28.

In the thirteenth century the Islamic scholar Nasir al-Din al-Tusi wrote the definitive Arabic version of Euclid's *Elements*, combing his translation from Byzantine Greek with his expository comments.

1-40. Reza Sarhangi (Iranian-born American, b. 1952) and Robert Fathauer (American, b. 1960), *Būzjānī's Heptagon*, 2007. Digital print, 13 × 13 in. (33 × 33 cm). Courtesy of the artists.

In this collaborative piece, two contemporary mathematicians, Reza Sarhangi and Robert Fathauer, pay homage to the Islamic mathematician Abū al-Wafā' Būzjānī (AD 940–98), who was active in Baghdad in the century after al-Khwārizmī, when the House of Wisdom was still a thriving center of scholarship. Būzjānī contributed to trigonometry, which he applied to astronomy while working at the observatory in Baghdad; he also wrote the practical text, *On Those Parts of Geometry Needed by Craftsmen*, which contained the construction of a regular heptagon (a polygon with seven equal sides and angles), shown in the center portion of this print. Around the perimeter of the heptagon Sarhangi and Fathauer wrote Būzjānī's name seven times in Farsi, the language of Persia (modern Iran).

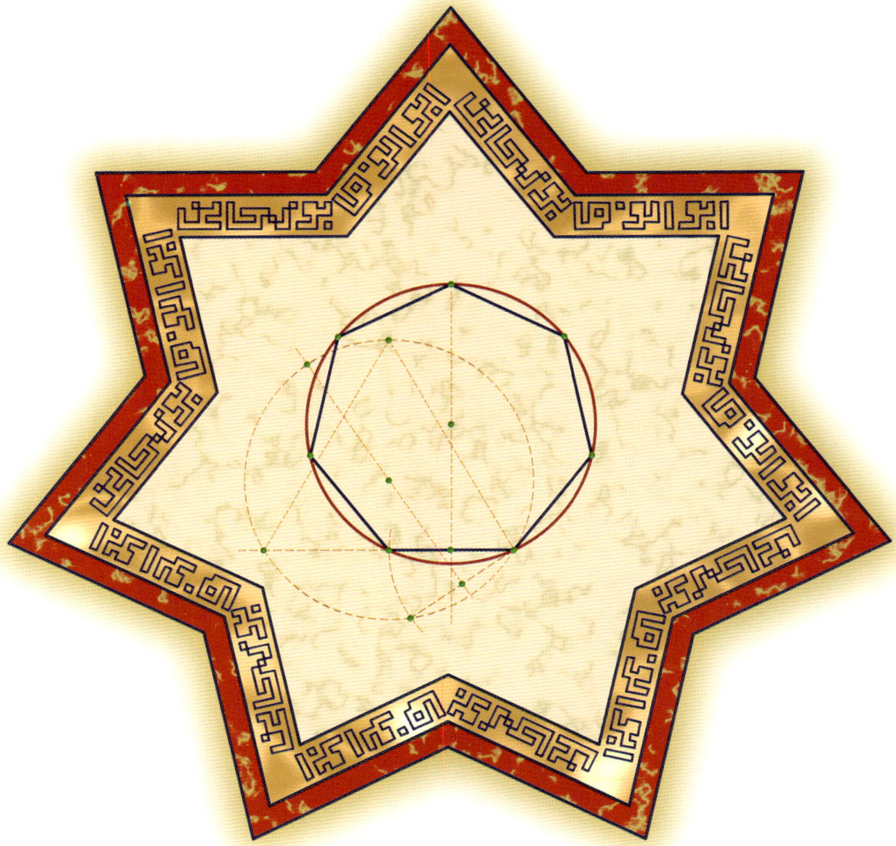

decimal point while working at the House of Wisdom, which remained an intellectual hub until the thirteenth century (plate 1-40). It would take centuries before Westerners abandoned Roman numerals in favor of Arabic numerals and decimal points, which by around 1500 were used throughout the Near East, North Africa, and Europe. By this time the Early Renaissance architect Filippo Brunelleschi had developed linear perspective based on the seven-volume study of optics by Ibn al-Haytham (latinized as "Alhazen"; AD ca. 965– ca. 1040), who gave the first correct description of vision (see plate 2-7 in chapter 2).

NEOPLATONISM AND EARLY CHRISTIANITY

In tracing the cultural history of mathematics, it is helpful to keep in mind the distinction between the philosophy and the practice of mathematics. Plato was engaged in the *philosophy* of mathematics when he asked: Can I achieve certitude by thought alone (a question in epistemology)? He also asked: What is a sphere? In what sense does a perfect sphere exist (topics in metaphysics)? On the other hand, Euclid was a "working mathematician" who *practiced* mathematics when he asked: Are all right angles equal? (a question in geometry), and Are the prime numbers finite or infinite? (a question in arithmetic). Both the philosophy and the practice of mathematics have inspired artists, writers, theologians, and others in the broader culture. But over the centuries, as the practice of mathematics became more technical and its details became harder for the general public to follow, philosophical topics entered the popular imagination in greater proportion.

When Plato deified Reason—making the Good a divine person with feelings and desires—he introduced values (ethics and aesthetics) into mathematics.[40] There are good and bad (harmonious and discordant) ratios, and, according to Plato, the goodness (in an ethical sense) of both the individual and society is equivalent to a unity of parts in harmonious ratios (*Republic*, 443e–444b, and 462a–b). Plato's introduction of ethics and aesthetics into mathematics strikes today's reader as bizarre because in modern biology value judgments are understood as mind-dependent—they are imposed by minds that have evolved to the level of making aesthetic and ethical judgments. When an amoeba met two protozoa in the primordial sea, these creatures did not experience each other as being beautiful or good—they simply were.

The intermixing of mathematics (the Forms) and values (the Good), which began with Plato, was extended within Western thought after classical philosophy merged with Judeo-Christian theology in the fourth century AD. Like Plato, the theologians he inspired discussed topics in the *philosophy* of mathematics and did not *practice* mathematics.

Scholars in the Hebrew and Greek traditions translated Plato's divine Reason into the cultural matrix of late Hellenistic Greek and early Christian societies, intermixing the God of Abraham and the Good, two concepts that were originally very different. The Old Testament prophets were monotheists who believed that there was one omnipotent Creator and lawgiver, and this God of Abraham came to be worshipped within Jewish, Christian, and Islamic cultures. Plato, on the other hand, was a polytheist, who believed that the sun, planets, and the Good were all divine. Unlike the God of Abraham, Plato's Craftsman was not omnipotent but had to coax chaotic matter to take on a Form, and thus the natural world was always an imperfect embodiment of the Forms. The Israelites worshipped the God of Abraham in the Temple of Solomon, and the Greeks built the Parthenon to honor the goddess Athena. Plato, on the other hand, did not worship the Good; he built an Academy where the Good was an object of knowledge.

According to Plato, one attained certain knowledge of the Forms by immediate intuition: "It is not something that can be put into words like other branches of learning; only after long partnership in a common life devoted to this very thing does truth flash upon the soul, like a flame kindled by a leaping spark, and once it is born there it nourishes itself thereafter."[41] One is able to connect from the mundane to the abstract and otherworldly by way of mathematics, especially proportion. When one hears the consonance of notes played on a lyre (strings vibrating in a harmonic ratio in the visible world), one knows the harmony of the cosmos because the strings vibrate in consonance with the faraway realm. When one studies the (mundane) geometric figure of a triangle, one knows the (otherworldly) Form of Triangle. Studying numbers and circles is a stepping-stone to certain knowledge of the mathematical Forms, which, in turn, is the path to knowledge of the Good.

When Plato died in 347 BC, his nephew Speusippus succeeded him as director of the Academy. Speusippus developed Platonic doctrine by postulating that if goodness is mathematical harmony and proportion and, as the Pythagorean Parmenides had argued, if the natural world is created from numbers that are generated from One, then the Good *is* One.[42] Speusippus made several attempts to establish that his uncle had a Pythagorean pedigree of

Perhaps there is a pattern set up in the heavens for one who desires to see it and, seeing it, to establish one in himself.

—Plato, ***Republic***, 380–367 BC

early origin, and his linking Plato's thought with the eminent Parmenides of the fifth century BC may have been such an attempt. No one in Platonic circles adopted this view, and it lay dormant until the third century AD when the Neoplatonist philosopher Plotinus revived it. The strongest of Plato's followers, Aristotle, had also developed his own concept of divine Mind—the Prime Mover—such that, like the One, it was a perfect, immutable person, who, after setting the cosmos in motion, withdrew from the natural world.

After dividing reality into a lower human world and a higher divine realm, Plato urged his followers to focus their attention upward by studying pure mathematics and spending long periods in contemplation: "A man whose thoughts are fixed on true reality has no leisure to look downwards on the affairs of men, to take part in their quarrels, and to catch the infection of their jealousies and hates. He contemplates a world of unchanging and harmonious order, where reason governs and nothing can do or suffer wrong; and, like one who imitates an admired companion, he cannot fail to fashion himself in its likeness."[43] Aristotle similarly suggested that man's highest aspiration should be contemplation of the Prime Mover: "If reason is divine, then, in comparison with man, the life according to it is divine in comparison with human life. But we must not follow those who advise us, being men, to think of human things, and, being mortal, of mortal things, but must, so far as we can, make ourselves immortal, and strain every nerve to live in accordance with the best thing in us; for even if it is small in bulk, much more does it in power and worth surpass everything."[44]

These quotations from Plato and Aristotle put us at the threshold of the medieval monasteries of Europe, where the pious focused their prayers upward after the merger of classical and Christian thought in late antiquity. The Neoplatonic philosopher Plotinus, following the lead of Speusippus, equated Plato's Form of the Good with Parmenides's concept of the One. Plotinus first read Plato's dialogue about Parmenides, in which Plato points out an ambiguity in the concept "the One" that Parmenides had overlooked. Is the One a completely uniform entity or a unified collection of parts? Plato described the two alternatives and their consequences. In the first hypothesis, the One is completely undifferentiated and has no attributes—not even its own existence can be applied to it: "The One in no sense *is* . . . . And if a thing is not, you cannot say that it '*has*' anything, or that there is anything '*of*' it. Consequently, it cannot have a name or be spoken of, nor can there be any knowledge or perception or opinion *of* it."[45] Plato similarly stated the logical consequences that follow from the second hypothesis. With these alternatives in mind, Plotinus then equated the Good with the One as described in Plato's first hypothesis, implying (incorrectly) that Plato himself held the first hypothesis to be true. In Plotinus's day confidence in human reason had plummeted because Greece was a Roman province, suffering under foreign rule and threatened by barbarian invasions. Thus Plotinus struck a chord among fellow Greeks when he emphasized the inability of the human mind to find words to describe the One, falling back on Plato's assertion that the core truths of his philosophy could not be put into words and strengthening the association of divine Mind with ineffability.[46]

In the fifth century AD, Proclus, a follower of Plotinus and one of the last Greek philosophers, taught his Athenian students that the individual soul ("psyche" or "mind") unites with the One by studying geometry. In a commentary on Euclid's *Elements*, Proclus advised the seeker of wisdom to concentrate on geometric diagrams ("figures") of a circle and triangle that are reflections of pure idea—the Forms. Through this process the mind (soul) comes to know (unite with) pure abstract reality.[47]

In the late Roman and early Christian era, scholars continued to study ancient texts based on a curriculum—arithmetic, geometry, music, and astronomy (the *quadrivium*, Latin combination of *quadri* and *via* meaning "the place where four roads meet")—first described by the Pythagoreans and adopted by Plato for his Academy. The seven liberal arts were completed in the first century BC when the Roman scholar Marcus Terentius added to these four subjects the three skills necessary for the scholar to articulate what he had learned: grammar, rhetoric, and logic (the *trivium*).

Early Christian and medieval theologians did not *practice* mathematics, but they studied ancient texts in search of their theological implications. Their *philosophical* debates about the ontological status of abstract objects—their attempts to answer the question: In what sense do numbers and spheres exist?—formed the background of the philosophy of mathematics in modern times. The answers they gave to this question formed three traditions, beginning with realism, which was proclaimed in the early Christian era by Saint Augustine, according to whom abstract objects exist in a Platonic realm. In the later middle ages the French theologian Peter Abelard proclaimed conceptualism, the view that abstract objects exist only in the mind. Then in the early fourteenth century the English friar and logician William of Ockham declared that abstract objects are mere signs or names, an outlook known as nominalism. These scholastic traditions—realism, conceptualism, and nominalism—formed the basis for modern debates on the same topics under other names—logicism (see chapter 5), intuitionism (see chapter 6), and formalism (see chapter 4).[48]

Augustine of Hippo (Saint Augustine) was educated in the seven liberal arts, reading especially Plato and Plotinus. He lived in North Africa in Hippo Regius (today Annaba, Algeria) contemporary with the Desert Fathers (Christian ascetics who, beginning in the third century AD, withdrew to a monastic life in the deserts of Egypt in the northwestern Nile delta). Augustine converted to Christianity and became a priest and then a bishop, establishing a school of theology that combined classical and Christian worldviews: "If those who are called philosophers, especially the Platonists, have said things which are indeed true and well accommodated to our faith, they should not be feared; rather, what they have said should be taken from them as from unjust possessors and converted to our use."[49] Augustine described how studying pagan texts is not an end in itself but a means to know the Christian divinity: "The truly learned are those who, not allowing all the different realities to distract them, attempt their unification into a simple, true and certain whole. Having done so, they can soar on to divine realities not rashly and by faith alone, but contemplating, understanding, and retaining them"[50] (plate 1-41).

Augustine championed Plato's view that one can know timeless, necessary truth—abstract objects—by thought alone. Following Plato, Augustine declared that one perceives the physical world with the eyes, but one contemplates the divine realm of mathematics with the soul (the psyche or mind).[51] Augustine contemplated numbers and forms because he believed that divinity shows itself through geometric forms in nature: "Surveying the earth

Alypius: Haven't you brought before our eyes that venerable and almost divine science of mathematics rightly attributed to and cultivated by Pythagoras? You have expounded to us both his rule of life and not so much the pathways to science as its very fields and crystal clear seas. . . .
Augustine: I must admit you are right.

—**Saint Augustine, *De Ordine* (On order), AD 386**

1-41. Sandro Botticelli (Italian, ca 1445–1510), *Saint Augustine in his Study*, 1480. Fresco, 60 × 44 in. (152 × 112 cm). Church of Ognissanti, Florence.

Wearing clerical robes, Augustine is shown as a saint/scholar in his study, looking heavenward surrounded by the tools of classical learning: an earth-centered Ptolemaic armillary sphere (upper left) and a weight-driven clock (upper right). Books are displayed on a shelf, with an open treatise on geometry on the right. The art historian Richard Stapleford's translation of a phrase in this book's otherwise unreadable text suggests that the overall subject of the painting is the conversion of Augustine ("Intellect and Intuition in Botticelli's *Saint Augustine*," *Art Bulletin* 76, no. 1 [1994], 69–80; the translation is on 73–74). Botticelli depicts the books as he saw them in the Medici family library and probably followed the Medici's practice of binding books by subject matter; a blue binding (colored with the costly pigment lapis lazuli) was used for sacred books, and dark red leather for poetry. On the table, a double-sided lectern holds a sacred book on one side and a book of verse on the other. Thus Botticelli presents Augustine in a setting that balances Christianity and classicism, the sacred and the secular, intuition and reason. The Renaissance chronicler Giorgio Vasari praised Botticelli for capturing Augustine's pious gaze and scholarly concentration: "In the countenance of the Saint he has clearly manifested that power of thought and acuteness of perception which is, for the most part, perceptible in those reflective and studious men who are constantly occupied with the investigation of exalted subjects and the pursuit of obtuse inquiries" ("Sandro Botticelli," *Lives of the most Eminent Painters, Sculptors, and Architects*, 1568).

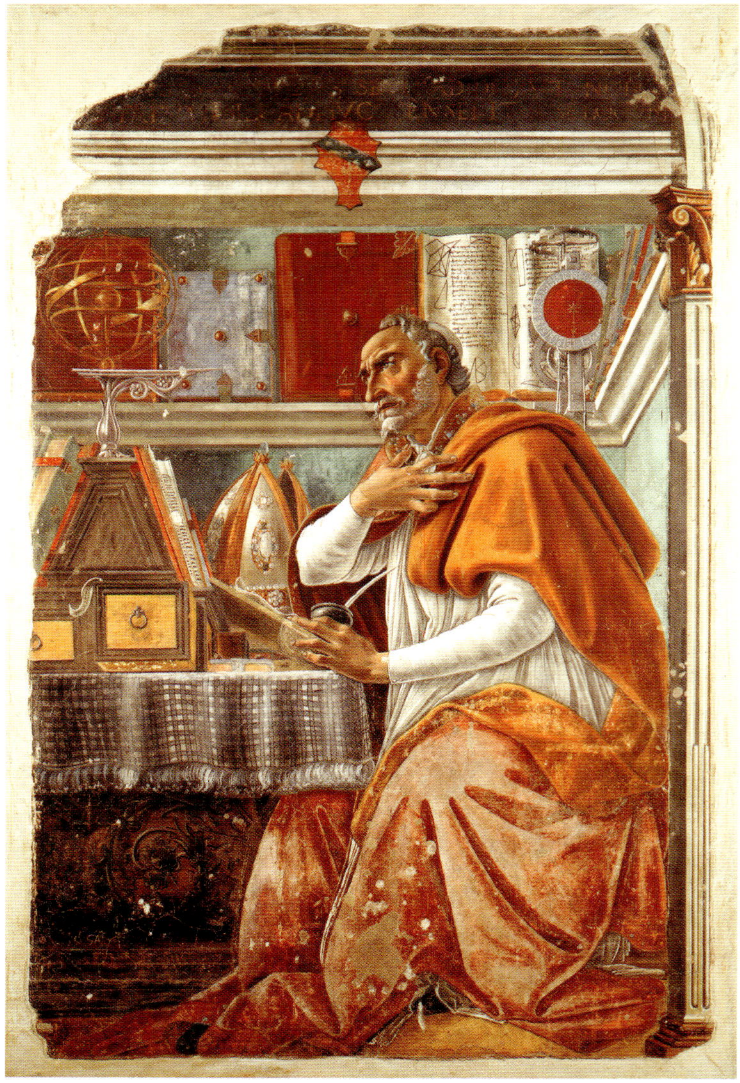

We have not yet succeeded in finding the correct logical analysis of what we mean by our ethical and religious expressions. Now when this is urged against me I at once see clearly, as it were in a flash of light, not only that no description that I can think of would do to describe what I mean by absolute value, but that I would reject every significant description that anybody could possibly suggest, ab initio, *on the ground of its significance. That is to say: I see now that these nonsensical expressions were not nonsensical because I had not yet found the correct expressions, but that their nonsensicality was their very essence.*

—Ludwig Wittgenstein,
"A Lecture on Ethics," 1929

and the heavens, it [reason] found that nothing but beauty pleased it: within beauty forms, within forms proportion, within proportion number. It asked itself where in the real world this or that straight, curved, or other line as conceived by the intelligence might be. It found reality far inferior. Nothing real stood comparison with what the mind could see. It analyzed these forms one by one and arranged them into a discipline which it called *geometry*."[52] Although words are inadequate to describe the Good, a circle—the most symmetrical geometric shape—is, according to Augustine, the best figure to convey the highest level of reality.[53]

Since antiquity thinkers who do not recognize knowledge beyond rational thought have complained that Plato and his followers were inconsistent. On the one hand, Plato said that words cannot describe the Good and Plotinus similarly found the One ineffable, but, on the other hand, they used words to describe these divine entities, albeit in language that is paradoxical and poetic. Regardless, the belief that ultimate personal reality is ineffable took deep root in Western thought.

ASIAN MATHEMATICS: GENERALIZATION WITHOUT ABSTRACTION

One of the oldest mathematical books, the *Zhoubi suanjing* (The arithmetical classic of the gnomon and the circular paths of heaven), was assembled anonymously during the Han dynasty (206 BC–AD 220), although it contains much earlier astronomical calculations, instructions on the use of triangles for measuring distance, and a proof of the Gougu theorem (the Greek Pythagorean theorem; see the sidebar below).

The most important book on mathematics in ancient China was *Jiuzhang suanshu* (The nine chapters on the mathematical art), which served as the textbook for basic education in mathematics throughout Asia for over two millennia, much as Euclid's *Elements* has in the West. *The Nine Chapters* contains 246 mathematical problems in nine chapters, which were written before 100 BC and compiled anonymously between about 200 BC and AD 50, with commentary added in the third and seventh centuries AD. The chapters are on mathematical techniques and applications, such as "Rectangular Fields," on land surveying using plane figures, and "Small Millet and Husked Rice," on exchange rates

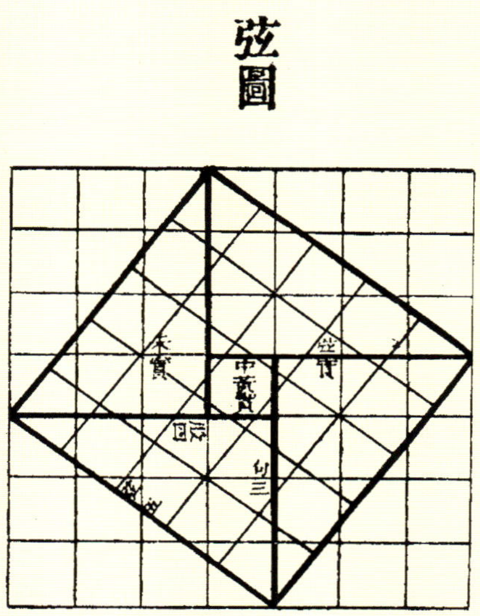

Gougu Theorem

In China, right-angled triangles are *gougu* (Chinese for "hook and thigh"), in reference to the two arms of a carpenter's square. The Chinese approach to the proof is based on a triangle with specific side lengths of 3, 4, and 5. The essence of the proof is this diagram, which is discussed in an accompanying dialogue between the Zhou dynasty ruler Chou Kung and the mathematician Shang Kao, both active in the eleventh century BC.

LEFT

1-42. Gougu theorem (Pythagorean theorem), in the *Zhoubi suanjing* (The arithmetical classic of the gnomon and the circular paths of heaven), Han dynasty, 206 BC to AD 220. Needham Research Institute, Cambridge, England.

The colors in the diagram below show how the proof works. The large square is formed by adjoining four right triangles with sides 3, 4, and 5, leaving the white square in the center to accommodate the difference in length of the sides of the triangles. In modern notation the theorem states that $x^2 + y^2 = z^2$, where z is the hypotenuse. The large square has an area of 25 (z^2).

By moving two of the triangles, as indicated by the arrows, two smaller squares are formed that have the areas 9 and 16, showing that $x^2 + y^2 = z^2$.

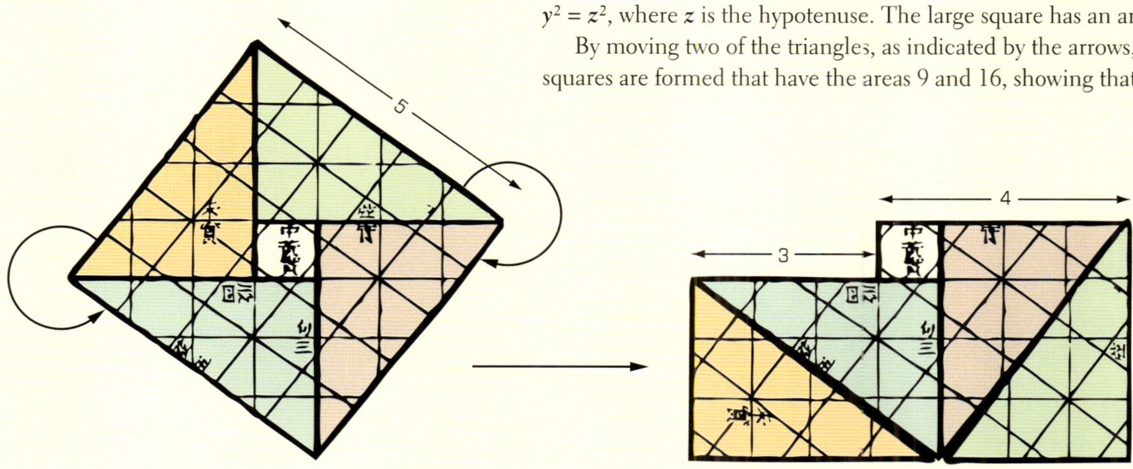

using rules of proportion. Each chapter begins with a simple problem and then progresses to more and more complicated applications.

The Chinese were especially good at numerical combinations, as in the Luoshu ("magic square") diagram, in which the numbers in the rows, columns, and diagonals all add up to the same sum. According to tradition, the diagram originated when a turtle crawled out of the Luo River with the diagram on its shell, and the miraculous animal gave the Luoshu to Yu the Great, the legendary emperor who inaugurated dynastic rule in China by founding the Xia dynasty (ca. 2070–1600 BC). The magic square of order three (with three numbers to a row) symbolized the ancient Chinese view of reality that entails two opposing forces, yin and yang, together with five elements: metal, fire, water, wood, and earth (plates 1-43, 1-44, and 1-45). This system is similar to attempts by other ancient cultures to categorize things in nature, such as the Pythagorean description of the world in terms of pairs of opposites (limited-unlimited, odd-even, one-many, male-female, etc.), and Aristotle's four elements (earth, air, fire, and water), plus the fifth "quintessence" (æther). Although the date of the origin of the diagram is lost in legend, the Luoshu of order three was definitely known by at least the fifth century BC, and Chinese mathematicians developed more complex magic squares of higher orders and with three dimensions ("magic cubes") in the thirteenth century AD.

The vast scope of the Chinese contribution to mathematics began to be brought to the attention of scholars in the East and West in 1954 when the British scholar Joseph Needham released the first volume of his history of Chinese mathematics, science, and technology (*Science and Civilization in China*, 1954+). Needham, a politically active humanist, launched the publication of his massive research project during the Cold War in the hope that awareness of mathematics and science, with cumulative bodies of knowledge that cross international boundaries, would unite peoples of the world.[54] By isolating social, political, and economic factors, Needham tried to explain why people in the East and West had discovered the same truths of mathematics and science in different epochs. Needham suggested that Chinese mathematics was focused on practical applications such as land surveying, calendar making, and record keeping because China had been a bureaucratic society under the rule of warlords and kings in the early Dynasties (2100–256 BC) and emperors, beginning with Qin Shi Huang in the imperial era (221 BC–AD 1911).

Since Needham's volumes began appearing over a half-century ago, it has become common to contrast the Chinese practical approach of applying rules case-by-case, which led to their large collections of algorithms, as opposed to the Greek theoretical approach of generalizing from a particular case to all, which led to Euclid's deductive method of proof. In connection with the first translations of *The Nine Chapters* into English (1999) and French (2004), recent scholars have presented a more nuanced picture.[55] The French translator of *Nine Chapters*, Karine Chemla, has argued that the Chinese text includes discussion of the validity of the practical rules, which can be considered a kind of informal proof. In Greek mathematics generalization is always linked with abstraction, but Chemla argued that these terms can be distinguished. She described Chinese mathematics as having generalization *without* abstraction, in the sense that a particular case is presented as a paradigm, implying a more general conclusion, even though the mathematician never stated the rule in abstract terms.[56] Indeed, there is no written evidence that the Chinese devised rigorous proofs or that they organized their paradigms into a mathematical system that is comparable to Euclid's axiomatic method. They did, however, give clear examples, such as the diagram accompanying the Gougu theorem, which may tell us more than an explicit "proof." An example of a general conclusion that can be drawn from such a diagram occurs at the end of the dialogue about the Gougu theorem. After establishing the theorem, Shang Kao concludes: "He who understands the earth is a wise man, and he who understands the heavens is a sage. Knowledge is derived from the straight line [the shadow]. The straight line is derived from the right angle [or gnomon]. And the combination of the right angle with numbers is what guides and rules the ten thousand things."[57] In Chinese thought, "the ten thousand things" refers to everything in the universe. Post-Needham scholars from East and West, whether they have applauded or denounced Needham, have followed in his footsteps by discussing the history of Asian mathematics in a broad historical context.[58]

In addition to isolating social, political, and economic factors that impacted the development of mathematics in China, Needham also considered Chinese religious and philosophical beliefs because the way people conceive of ultimate reality relates directly to how they think about mathematics. In none of the religions or philosophies that developed in ancient Asia—neither in the creation myths nor in the cosmologies—is there a celestial person who imposed order and enforced laws of nature. Therefore, according to Needham and subsequent scholars, a key reason why Asian mathematics and science did not advance beyond rudimentary levels is that there never arose within Confucianism, Taoism, and Hinduism/Buddhism an assumption that there is an eternal, transcendent mathematical-world-out-there to explore or that there are laws of nature to discover.[59]

In the sixth century BC, Confucius preached an ethical philosophy of correct behavior towards one's fellow man. As a moral doctrine, it had no mathematical component per se, but Confucius's writings are filled with descriptions of clusters of individual people working together in strict hierarchy and harmonious order.[60] The philosophy of Confucius traditionally includes the magical system known as the *I-Ching*, which is described in the divination manual *Zhouyi* (Book of changes). It began as omens and proverbs about animals, the weather, and human emotions, but by the time of Confucius, it had coalesced into a method of discovering the hidden meaning in events and of foretelling the future. The *I-Ching* generates sixty-four patterns of lines (plates 1-46, 1-47, and 1-48). There are several different ways of putting the patterns in an order, and the one that entered the Confucian canon and continues to be used into modern times is traditionally ascribed to Fu Xi, a legendary sovereign who ruled China in the twenty-ninth century BC, about two hundred years before the Egyptian Step Pyramid at Saqqara. Although the *I-Ching* began in antiquity as an attempt to discern numerical patterns in the natural world, as did the Pythagorean/Platonic system, the *I-Ching* degenerated into a way to classify new events by assigning them a number and considering the event explained by the accompanying text. Thus the *I-Ching* became a dead end and discouraged further exploration of nature, but its abstract format lent the system an aura of authority that has attracted users in Asia for over two millennia.

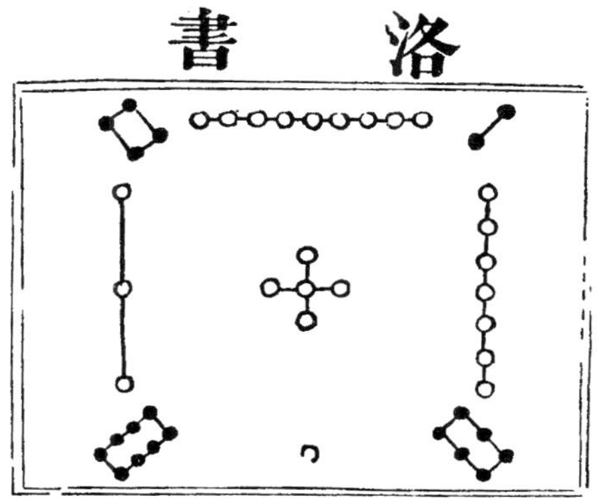

1-43. Luoshu, from Zhu Xi, *Zhouyi* (twelfth century AD), reproduced in *Yitu mingbian* (Clarification of the diagrams in the book of changes), by Hu Wei (1706 AD), chap. 1. Needham Research Institute, Cambridge, England.

The Luoshu (magic square) represents the ancient Chinese picture of reality in which there were opposite forces—yin, the female principle (identified with even numbers, shown as black dots) and yang, the male principle (odd numbers, shown as white dots).

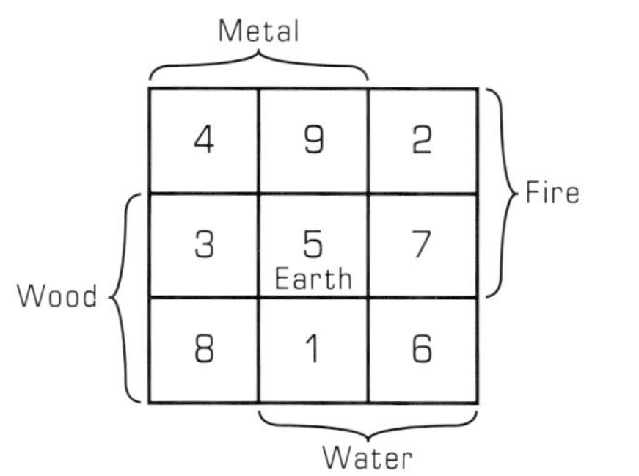

1-44. This diagram of the Luoshu (above) shows the Chinese organization of the elements around the edges of the magic square in yin-yang (even and odd) pairs: metal, fire, water, wood, with 5 in the center representing earth.

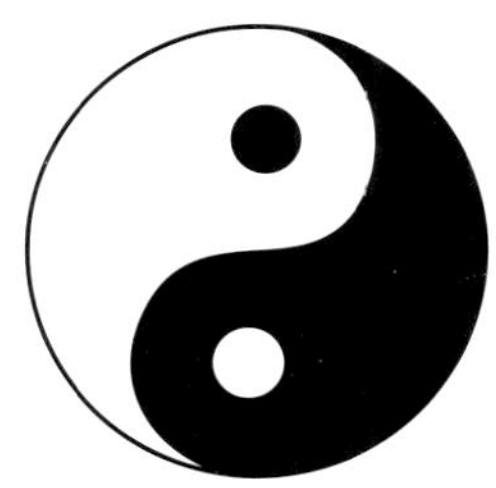

1-45. Yin-yang.

Yin is the feminine or negative principle (characterized by passivity, coldness, and darkness) and yang is the masculine or positive principle (characterized by activity, warmth, and light). The phenomenal world comes into being through the division and fusion of these two cosmic forces.

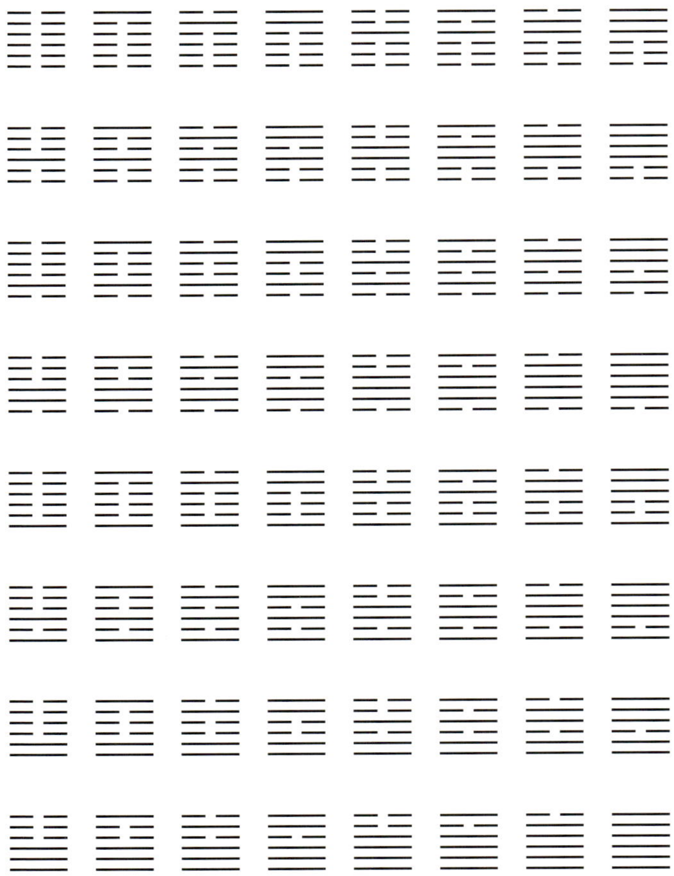

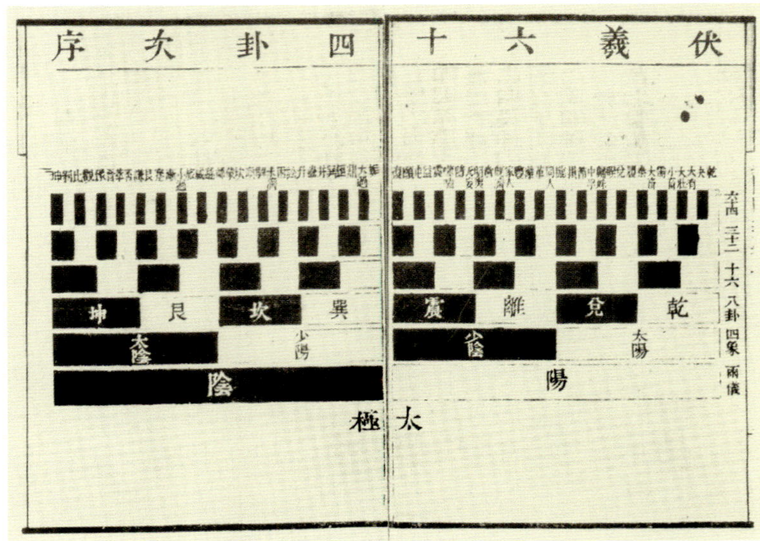

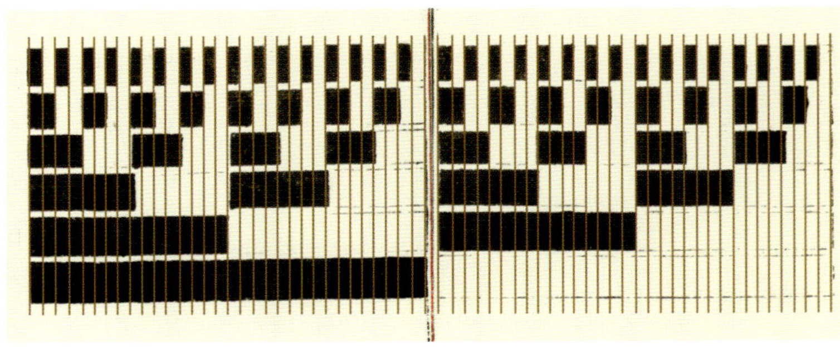

ABOVE LEFT

1-46. *I-Ching* diagram in the Fu Xi sequence.

The *I-Ching* consists of 64 figures (so-called hexagrams), each made up of six parallel (whole or divided) lines. Each figure is accompanied by a brief text, which was written by Confucius. Traditional consultation of the *I-Ching* required an elaborate, lengthy ceremony in which the manipulation of special sticks led to one of the 64 patterns. In modern times it became permissible to forgo the ceremony and quickly get a number by tossing coins, which became common throughout Asia in the brisk business of roadside fortune-telling. Once a number is determined, the text is interpreted by the fortune-teller to determine its meaning for the client's question. For example, the text associated with the first hexagram (the six whole lines in the lower right) foretells "what is great and originating, penetrating, advantageous, correct and firm" (*I-Ching*, trans. James Legge, 1899). Each of the 64 texts are written in a similar nonspecific style so that it needs to be interpreted to fit a particular occasion.

ABOVE RIGHT, TOP

1-47. Segregation table of the symbols of the Book of Changes, in Zhu Xi, *Zhouyi* (twelfth century AD), reproduced in *Yitu mingbian* (Clarification of the diagrams in the book of changes), by Hu Wei (AD 1706), chap. 7. Needham Research Institute, Cambridge, England.

ABOVE RIGHT, BOTTOM

1-48. Diagram of the segregation tables that yields the 64 hexagrams.

The sequence of 64 hexagrams is read beginning in the lower-right corner, from right to left, from bottom to top. This arrangement is traditionally ascribed to the legendary founder of China, Fu Hsi, and the pattern is based on the segregation table shown in plate 1-47. The lowest (white) level symbolizes the cosmic formless void (the *t'ai chi*), which separated at level 1 into the yin (darkness) and yang (light), although each contains its opposite in a secluded state, which emerges when dark and light divide again at the next level. The division into yin and yang continues with no logical end to the process; it traditionally ends at level 6, which has 64 divisions.

This diagram shows how to generate the 64 hexagrams, reading white for a whole line and black for a divided line. We first draw lines to indicate 64 vertical columns, and then, beginning in the lower-right corner, we read the pattern in the first column (from bottom to top). We then move left and read the pattern in the second column. We continue moving left, reading the pattern in all 64 columns, giving us Fu Hsi's sequence of hexagrams.

OPPOSITE

1-49. Fan Kuan (Chinese, fl. 990–1020), *Travellers among Mountains and Streams*, ca. 1000–20. Hanging scroll, ink and colors on silk, 81¼ × 40¾ in. (206.3 × 103.3 cm). National Palace Museum, Taipei, Taiwan, Republic of China.

In the upper-right corner, Fan Kuan wrote that he depicted himself among the travellers moving along the road (in the lower right) below the vast mountains.

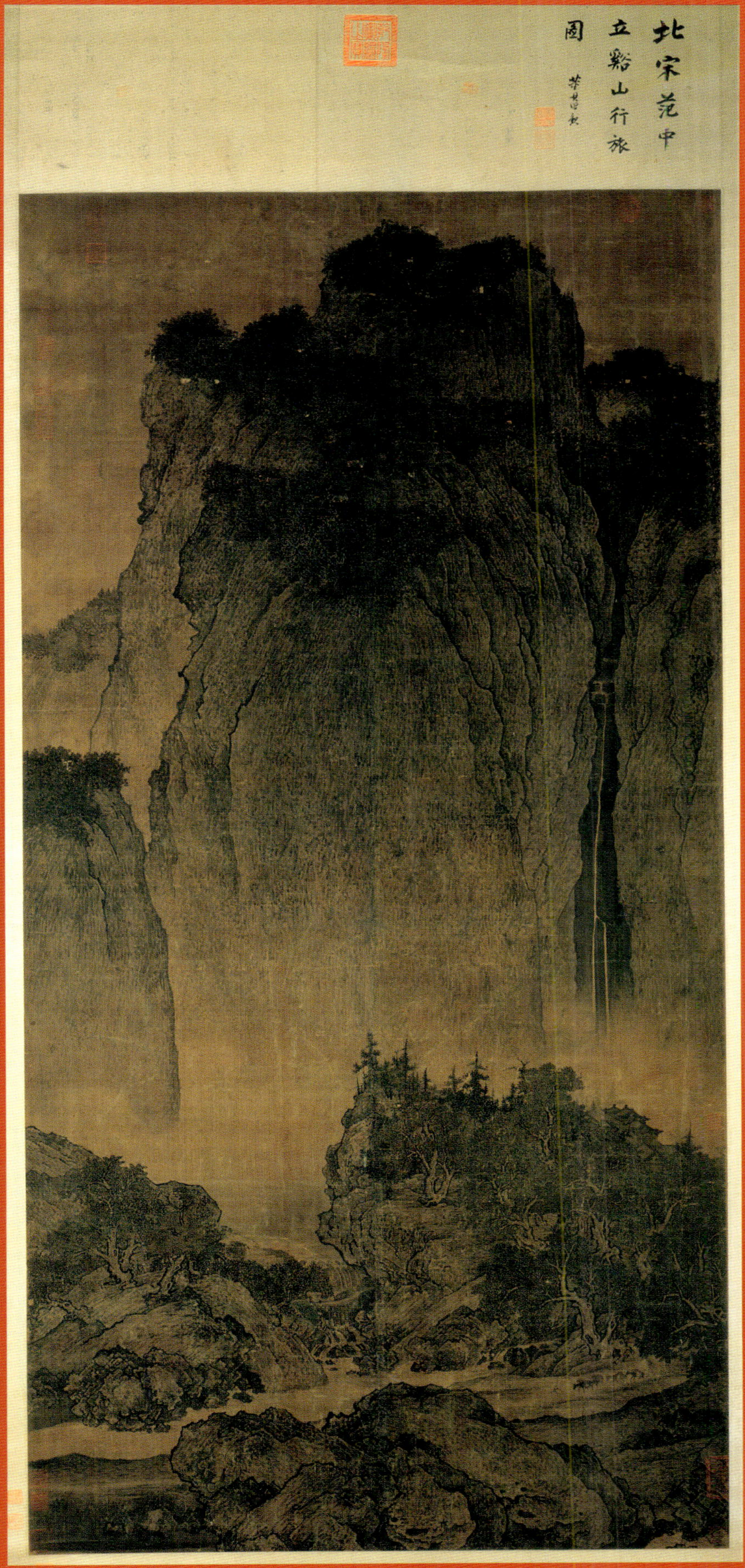

Man takes his law
from the Earth:
Earth takes its law
from heaven;
Heaven takes its law
from the Tao;
The law of the Tao is
its being what it is.

—Lao Tzu, *Tao tê ching*
(The canon of the way and of
virtue), fourth century BC

Taoism arose during a time of political instability in China (the Warring States period (ca. 475–221 BC), when people withdrew from the urban conflict to live in peace and harmony with nature in the countryside. According to the legendary founder of Taoism, Lao Tzu, who probably lived in the fourth century BC, contemporary with Euclid, heaven and earth developed spontaneously from a formless void according to nature's own rules without the aid of a supernatural person. "Tao" literally translates as "way," but it came to have the broader meaning "order of nature." To know the Tao is to be in harmony with the universe as an organic, unified, self-sufficient system in which humans are a small part, an outlook that is beautifully expressed in landscape painting of the Song dynasty (AD 960–1279; see plates 1-49 and 1-50). Taoism has no creation myth in which a superior mind imposed order that mankind's finite mind could hope to discover. Nature is very complex and, according to Taoism, there are no simple patterns to discover; indeed, ultimate reality is unknowable. A follower of Lao Tzu, the philosopher Chuang Tzu, suggested that the natural world operates by some impersonal, deterministic, mechanical device—a hidden metal coil—a "secret spring": "How ceaselessly heaven revolves! How constantly Earth abides at rest! And do the Sun and the Moon contend about their respective places? Who presides over and directs these things? Who binds and connects them together? Who is it that, without trouble or exertion on his part, causes and maintains them? Is it perhaps, that there is some secret spring, in consequence of which they cannot be but as they are?"[61]

In India, Hinduism arose roughly contemporary with the development of Greek mathematics. Indian philosophers distinguished between the unreal, ephemeral physical world of the senses (the world-out-there) and a perfect, infinite realm of pure mind—a genderless cosmic spirit (the Brahma)—which humans know by thought alone. But unlike Plato's realm of the Forms, the Hindu Brahma (pure, unembodied, impersonal mind) is a primordial, formless void. Thus parallel to the Western tradition of knowing the mathematical-world-out-there by thought alone, there developed a Hindu tradition of knowing a formless, cosmic spirit (the Brahma) by meditation.

The Hindu Brahma was described in many texts that began appearing in Sanskrit around 800 BC, which are collectively called the Upanishads. These texts were also the basis for a priesthood and structure of social classes that became hereditary, leading to India's caste system. The Hindu prince Buddha, who lived in Nepal in the late sixth and early fifth centuries BC, renounced his wealth, lived a self-disciplined, austere life, and became a reformer who declared that enlightenment is not determined by class but is open to all. Each person is born, dies, and is reincarnated in the physical world, but anyone can escape cyclical material existence and unite eternally with the spirit of the cosmos—the Brahma—through meditation and ascetic living. Buddha's exhortations encouraged his followers to live apart from the everyday world, and Buddhist monks established monasteries, first in India, and then, following the Silk Route east to China, where the Chinese merged their concept of Tao with the Brahma, resulting in (Taoist) communication with nature as a key to (Buddhist) enlightenment.

Trade routes also took Taoism and Buddhism west to Persia, where their impact on Greek thought cannot be ruled out. Despite the key difference that Western religions

1-50. Tatsuo Miyajima (Japanese, b. 1957), *Keep Changing, Connect with Everything, Continue Forever*, 1998. LED, IC, electric wire, plastic, aluminum panel, iron, 113¼ × 151³⁄₁₆ × 5⅛ in. (288 × 384 × 13 cm). Museum of Contemporary Art, Tokyo, courtesy of the artist and Shiraishi Contemporary Art Inc.

Like the Chinese landscape painters of the Song dynasty, the contemporary Japanese artist Tatsuo Miyajima expresses a Taoist philosophy: *Keep Changing, Connect with Everything, Continue Forever.* The viewer stands before a sparkling, blinking grid of lights. Each light-emitting diode (LED) in the grid is connected to another LED such that when one goes off another goes on. The viewer intuits that there is an underlying pattern, but it is too complex to discern. As such, the artwork symbolizes the natural world as a unity of dynamic, fluctuating parts—trees, rocks, clouds—within a harmonious whole.

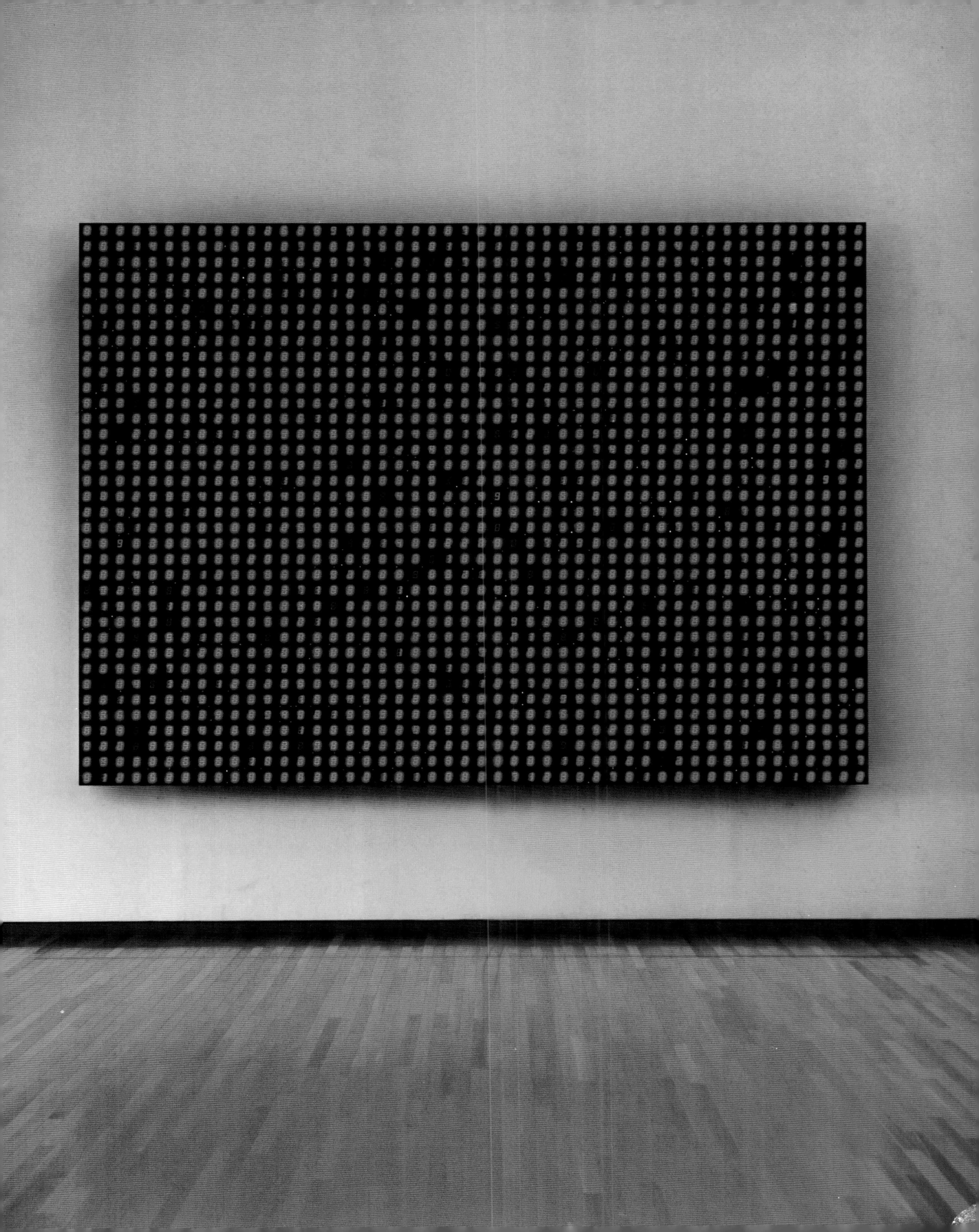

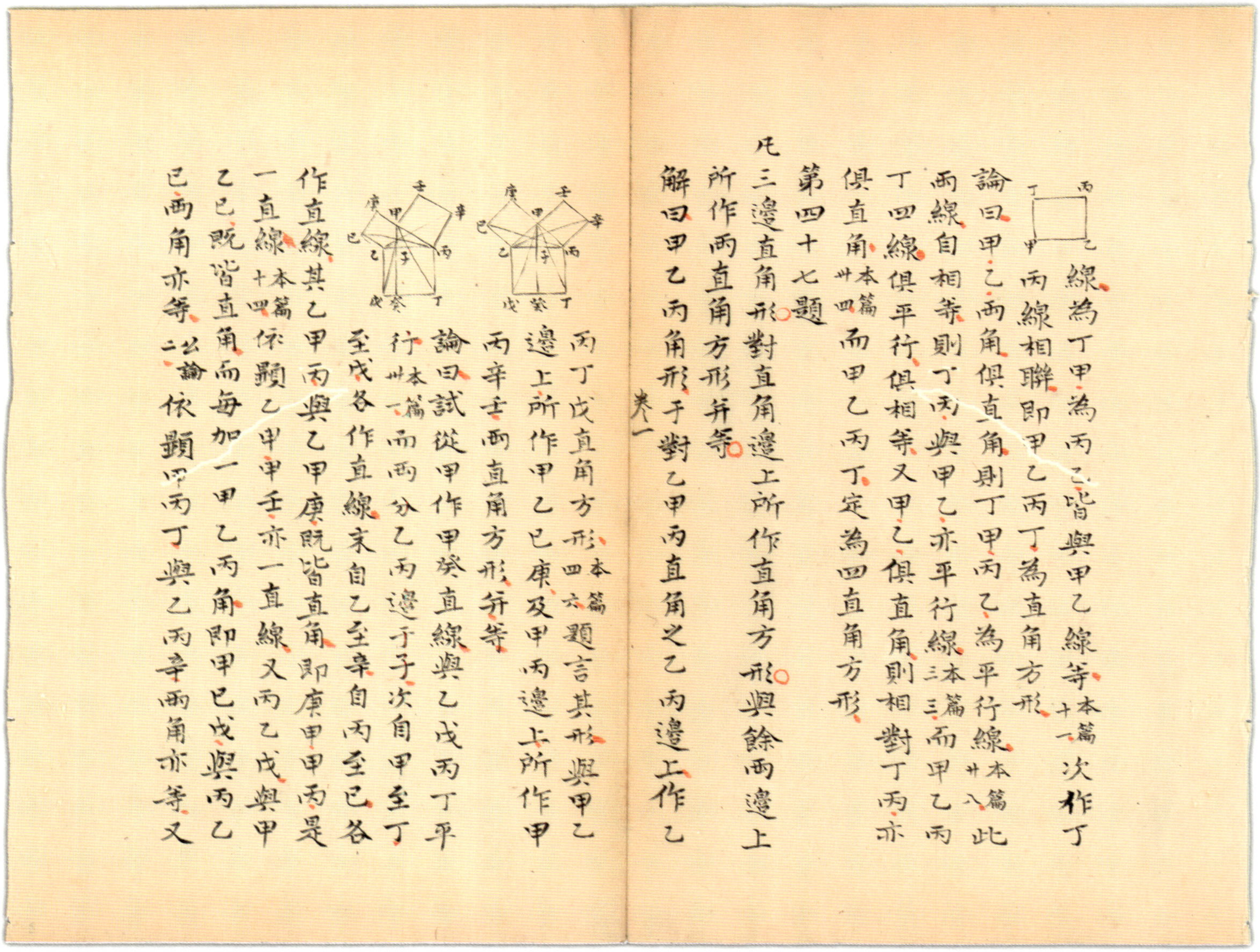

1-51. Euclid's *Elements*, ca. 300 BC, trans. Matteo Ricci and Xu Guangqi, AD 1607, nineteenth-century copy of the 1607 manuscript, book 1, proposition 47. National Library of Australia.

About 1500 years after Chinese mathematicians proved the Gougu theorem (plate 1-42), they were introduced to the Greek version of the proof by the Jesuit missionary Matteo Ricci and the Chinese scholar Xu Guangqi, who translated the first six books of Euclid's *Elements* into Chinese in 1607.

feature a personal deity, scholars have noted similarities between the Western mystical tradition—Greek mystery cults, Neoplatonism, and early Christianity—and the nature mysticism, meditation practices, and belief in reincarnation of Taoism and Buddhism. There is a vast literature on this topic and the weight of scholarly opinion is that the impact of Eastern thought on the Western mystical tradition is unlikely but possible, because there is ample evidence of ancient contact between East and West.[62] The Greek historian Herodotus reported that Greek philosophers travelled to Egypt in the fifth century BC where they learned the (Indian) doctrine of the transmigration of the soul; since Herodotus was a contemporary of Socrates, one can assume that the pre-Socratic philosophers to whom he refers include the Pythagoreans, who taught the doctrine.[63] After Alexander the Great invaded the Indian subcontinent in 327 BC, there were extensive references in classical literature to Indian thought. Greek and Roman philosophers were especially curious about Indian ascetics, the so-called gymnosophists ("naked philosophers"), who shunned the bodily comforts of food and clothing in pursuit of enlightenment. When the third-century AD Greek biographer, Diogenes Laërtius, compiled all the information known to him about Greek and Roman philosophers, he began by naming the gymnosophists among the originators of philosophy.[64] The Neoplatonic philosopher Porphyry reported in the third century AD that his teacher Plotinus, who was educated in Alexandria, tried to gain access (through travel or texts) to the wisdom of the East.[65] Although direct lines of influence and priority are difficult to establish, it is nevertheless common for historians to view the ancient Greek mystery cults, early Christianity, Buddhism, and Taoism in a broad context of Mediterranean and Eastern thought.

With the rise of Islam in the Middle East in the seventh century AD, the East-West trade routes were severed and the doors to commercial and intellectual commerce closed. After trade resumed during the Renaissance, Jesuit missionaries went to China and Japan in the late sixteenth century, where they learned of the technology of Song dynasty printing and Ming porcelain that were unsurpassed in the West. They, in turn, introduced (or reintroduced) Asian peoples to Euclid's *Elements* (plate 1-51).

THE REBIRTH OF LEARNING IN THE WEST

Trust in the power of human reason had been undermined when Greek lands became Roman colonies during the early Christian era, and when Rome fell to the barbarians, confidence in reason was utterly lost. Around AD 500, the Eastern Orthodox monk and mystic Pseudo-Dionysius the Areopagite (so-called because he wrote under the pseudonym Dionysius) formulated an extreme version of ineffability in which the One is so transcendent and so unknowable that the divine can be approached only indirectly by way of negation. This so-called negative theology states what the One is *not*. The seeker embarks on an inner journey—*via negativa* (Latin for "negative way")—to cleanse the mind of all preconceptions, because only then will the sacred truth reveal itself (plate 1-52). Like Augustine, Pseudo-Dionysius thought in mathematical analogies, comparing the infinite God (the

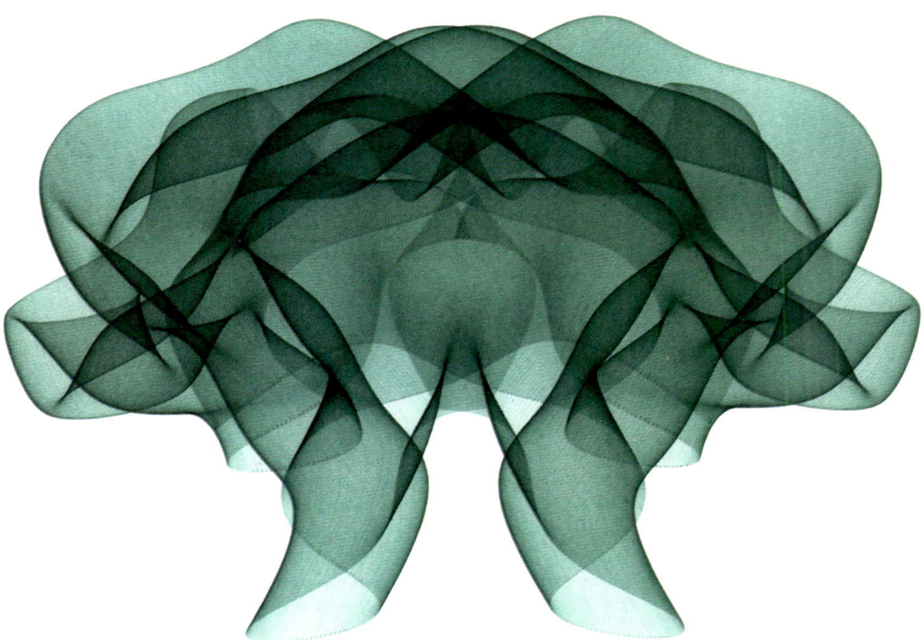

1-52. Roman Verostko (American, b. 1929), *The Cloud of Unknowing*, 1998. Pen-plotted drawing, 11½ × 14½ in. (29.2 × 36.8 cm). Courtesy of the artist.

The contemporary American artist Roman Verostko made this computer-generated drawing in homage to the anonymous, fourteenth-century English monk who wrote *The Cloud of Unknowing*, a spiritual guide to contemplation in the mystical tradition of Pseudo-Dionysius. The monk described pure being as having no definite image or positive form and thus, ultimately, beyond comprehension. Verostko studied at the Benedictine Seminary of Saint Vincent in Pennsylvania, completing a BA in philosophy in 1955, after which he took courses in computer programming and fine arts, while maintaining ties to the seminary and his interest in philosophy. In 1995 Verostko co-founded the Algorists, an international group of artists who make art that is generated—like a medieval chant and a computer graphic—by repetition following simple rules (algorithms).

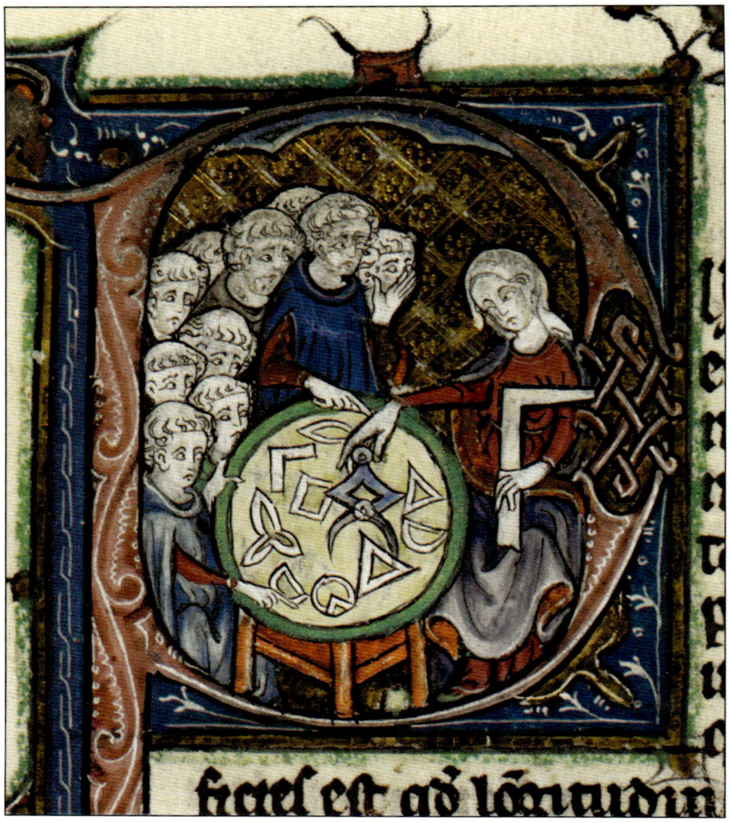

ABOVE AND OPPOSITE

1-53. First page and detail of a hand-copied manuscript of Euclid's *Elements* (300 BC), ca. 1309–16, based on a twelfth-century Latin translation from an Arabic version of the Greek text by Adelard of Bath, Paris, France. Egg tempera and ink on vellum. The British Library, Burney 275, f. 293r. © The British Library Board.

This is a fourteenth-century copy of the oldest surviving Latin translation of Euclid's *Elements*, which was done around 1120 by the English monk Adelard of Bath, who worked from an Arabic translation from Greek. The text in the upper margin announces the subject "PMVS GEOMETE EVCLID" (*Primus geometriae Euclidis* [First geometry of Euclid]), with geometric figures drawn in the right margin. The first line of text reads, "*Punctus est cuius pars non est*" (A point is that of which there is no part), and Geometry, personified as a woman, makes her debut in the illuminated "P" (see detail). She teaches a group of monks using Euclid's tools—a set-square (a straight-edge) and dividers (a compass). Two attentive students point at figures, which she measures with her compass. But the attention of the monk who illuminated Adelard's translation wandered from geometry to the tracery borders, where he drew birds and a rabbit. Below, grotesques use their tails to balance as they fight a dual, and, above, a grotesque with hoofed feet shoots an arrow at a gray fox as his dog gives chase.

One) to the number one, which from his theological perspective he conceived of as a monad from which all numbers can be generated: "For even in a monad, every number pre-exists in the form of a unit, and the monad holds every number in itself singly."[66] Pseudo-Dionysius found that his spiritual ascension to the immaterial, invisible realm was aided by seeing, hearing, or touching beautiful objects in the material world, in what he termed an "anagogical approach" (*anagogicus mos*): "For it is not possible for our mind to be raised to that immaterial representation and contemplation of the Heavenly Hierarchies, without using the material guidance suitable to itself, accounting the visible beauties as reflections of the invisible comeliness."[67] Visible objects, such as pictures and geometric forms, which are seen with the eyes—by light—were especially transportative because they mirror divine light: "Every good gift and every perfect gift is from above, and cometh down from the Father of Lights."[68]

Confidence in reason began to be restored as a rebirth of learning spread throughout the West in the early 1100s. Ancient classical texts had been lost to the West, but interest in them was stimulated when the first Latin translations were made from Arabic translations of Euclid's *Elements* and Ptolemy's *Almagest*, which had been preserved at the House of Wisdom in Baghdad (plate 1-53). In Chartres, near Paris, builders in the town laid the foundations for a new cathedral. Under Thierry of Chartres, who became chancellor of the cathedral's school in 1142, the town became a renowned center for the study of the liberal arts and the revival of classical learning. Thierry recommended a thorough reading of classical texts as preparation for study of sacred scripture: "There are, moreover, four kinds of reasoning that humanity possesses to obtain knowledge of the Creator: arithmetic, music, geometry and astronomy"[69] (plates 1-54 and 1-55). Thierry's writings are filled with theological references to mathematics, and like Pseudo-Dionysius, he identified God with the One (the "unity"): "Since unity creates each number—and numbers are infinite—unity necessarily has no end to its power. Unity is therefore omnipotent."[70]

Architects throughout the medieval West interpreted the ceilings of their Romanesque churches as symbolizing the vaults of heaven, just as Byzantium's Hagia Sophia (AD 532-37) rose from a mundane square base to a divine spherical dome. But in France in the early twelfth century, the dark, cave-like interiors of Romanesque buildings, with their massive stone vaults supported by thick walls, began giving way to the airy naves of

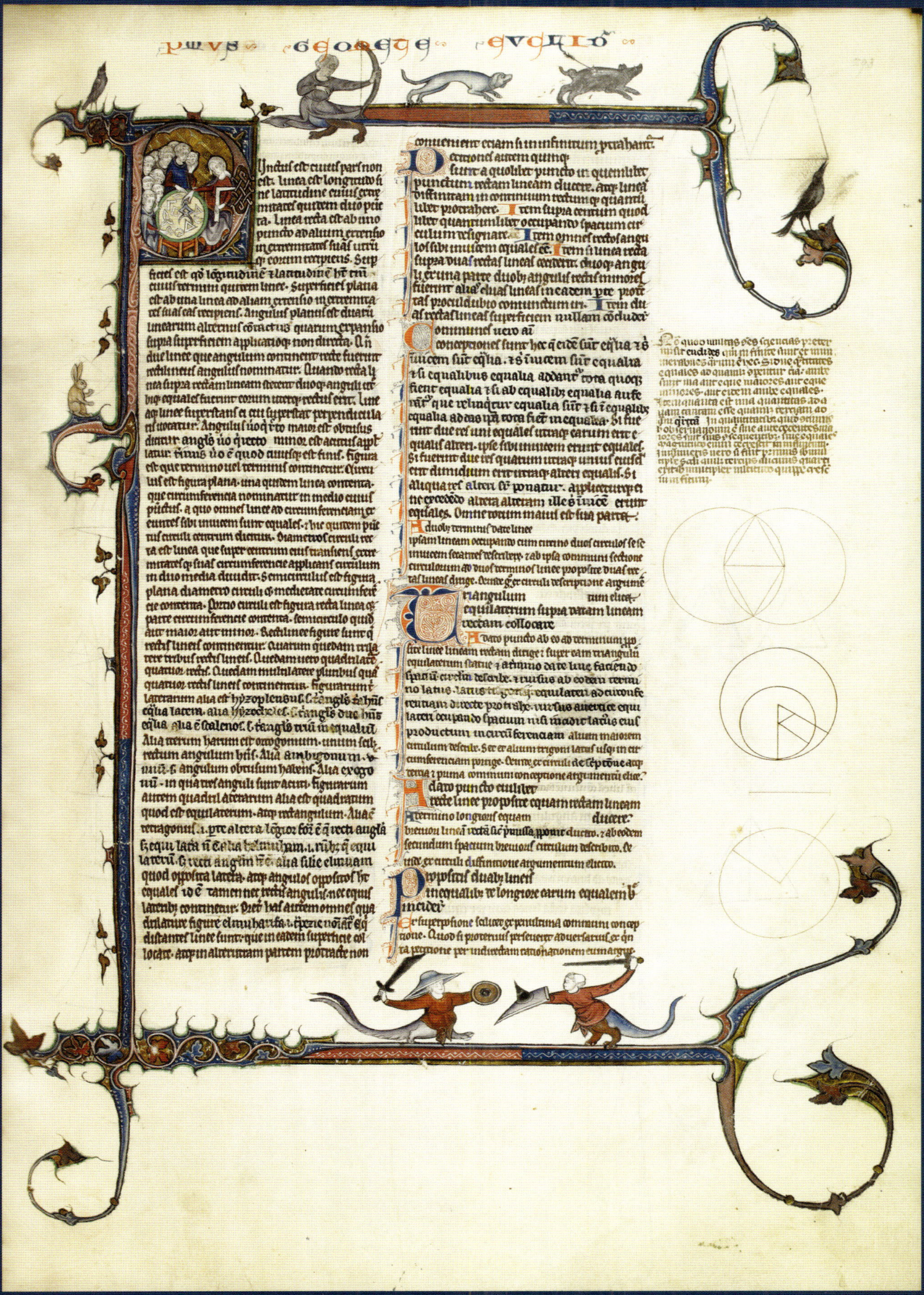

RIGHT

1-54. West façade, Chartres Cathedral, France, twelfth century.

OPPOSITE

1-55. South Portal of the west façade, Chartres Cathedral, AD 1145–55.

Thierry of Chartres, chancellor of the cathedral, supervised the construction of the west façade. In the archivolts of the south (right) portal, the builders placed sculptures personifying the seven liberal arts that were at the heart of Thierry's teaching, including logic (Aristotle, A), geometry (Euclid, C), and music (Pythagoras, G). Chartres is the first and only Romanesque or Gothic cathedral on which the seven liberal arts appear, although they are depicted in less prominent sacred settings, such as in stained glass windows and illuminated manuscripts.

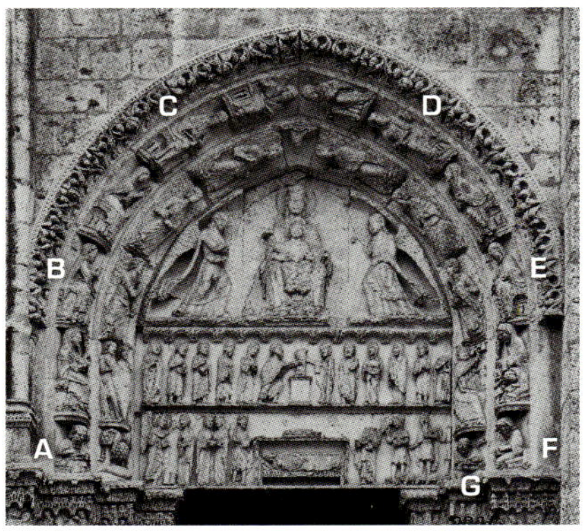

A Aristotle (Dialectic)
B Cicero (Rhetoric)
C Euclid (Geometry)
D Boece (Arithmetic)
E Ptolomy (Astronomy)
F Donatus (Grammar)
G Pythagoras (Music)

When we say that the divinity is the form of being for all things, we do not say that divinity is a form that must exist in matter, just as [we say] of a triangle or quadrangle or something similar.

—Thierry of Chartres, twelfth century AD

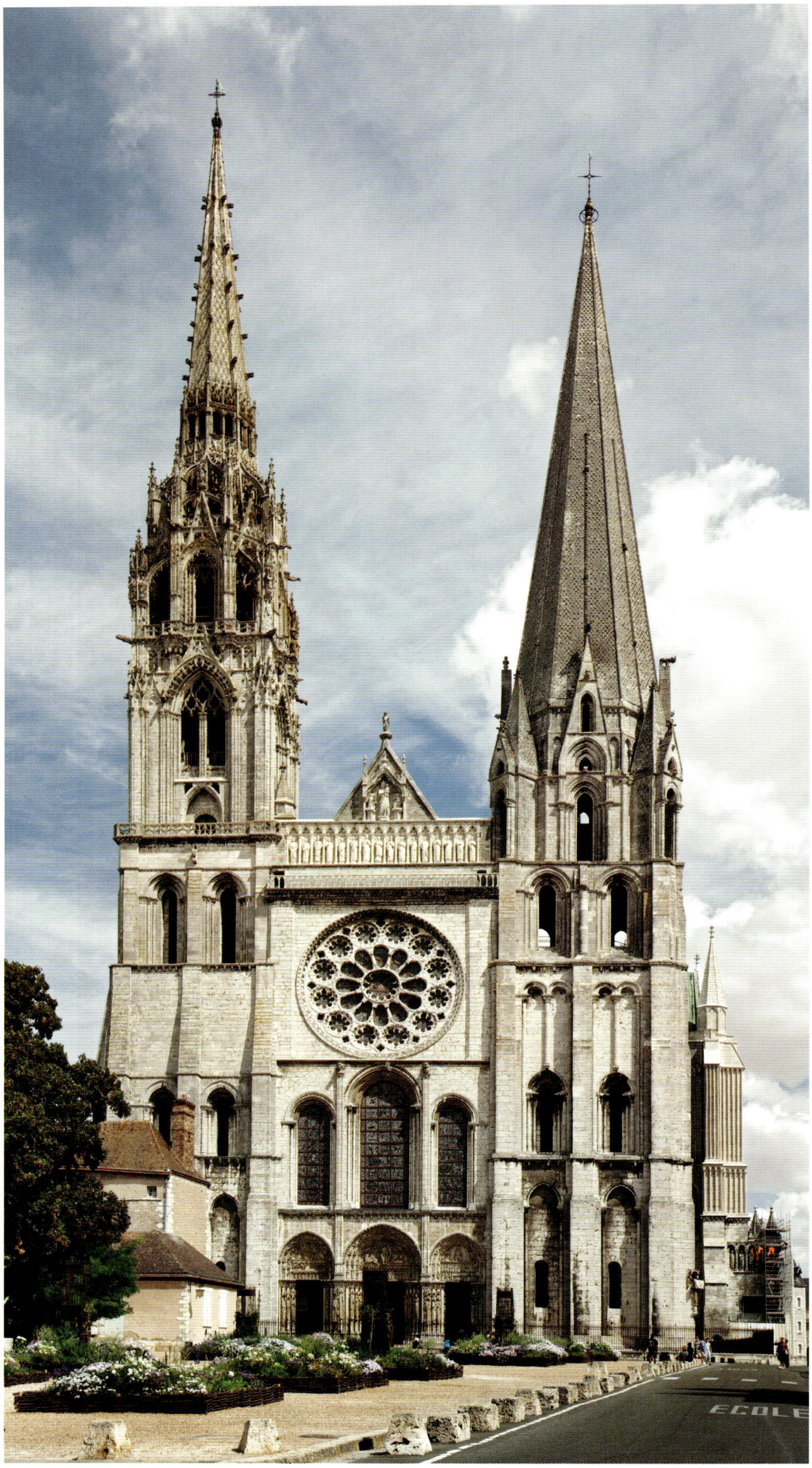

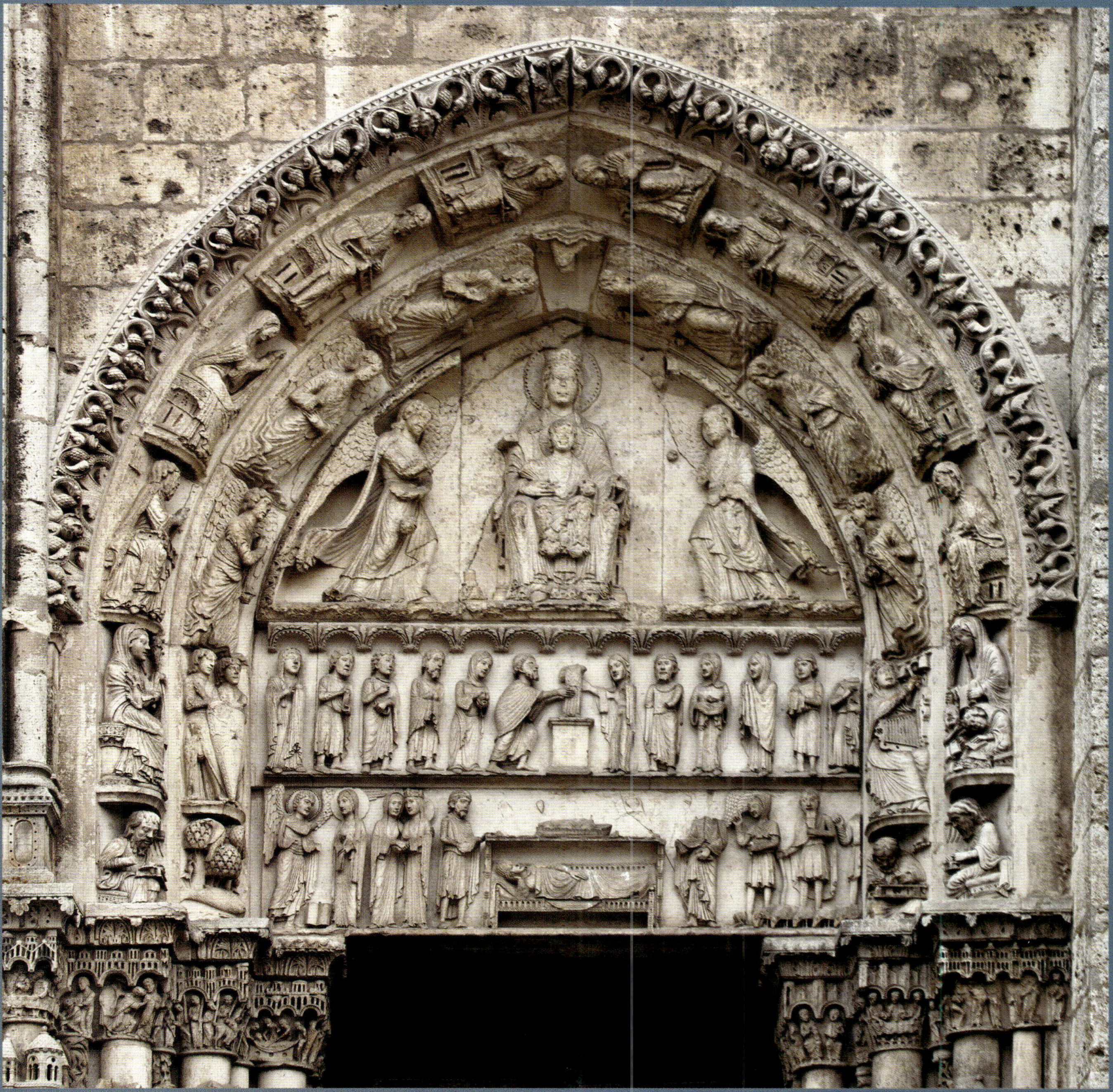

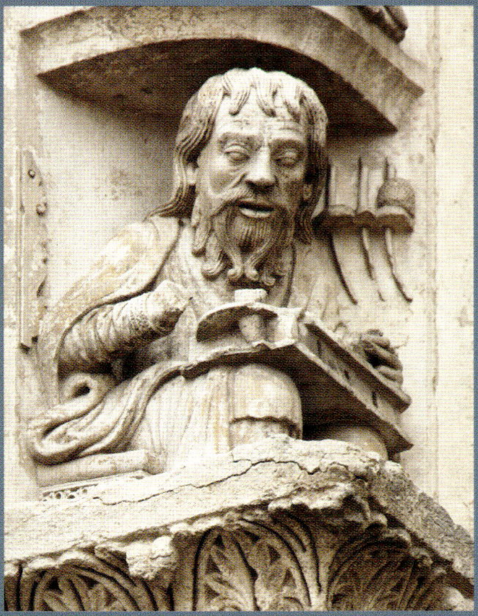

A

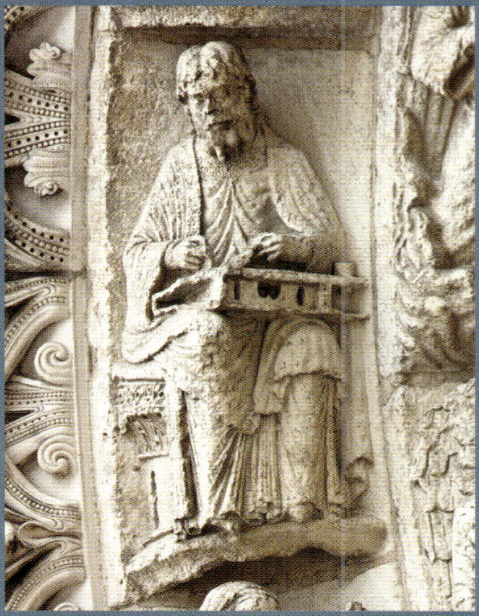

C

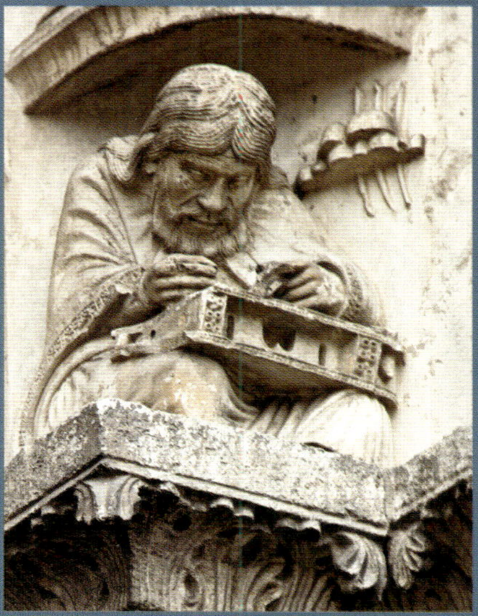

G

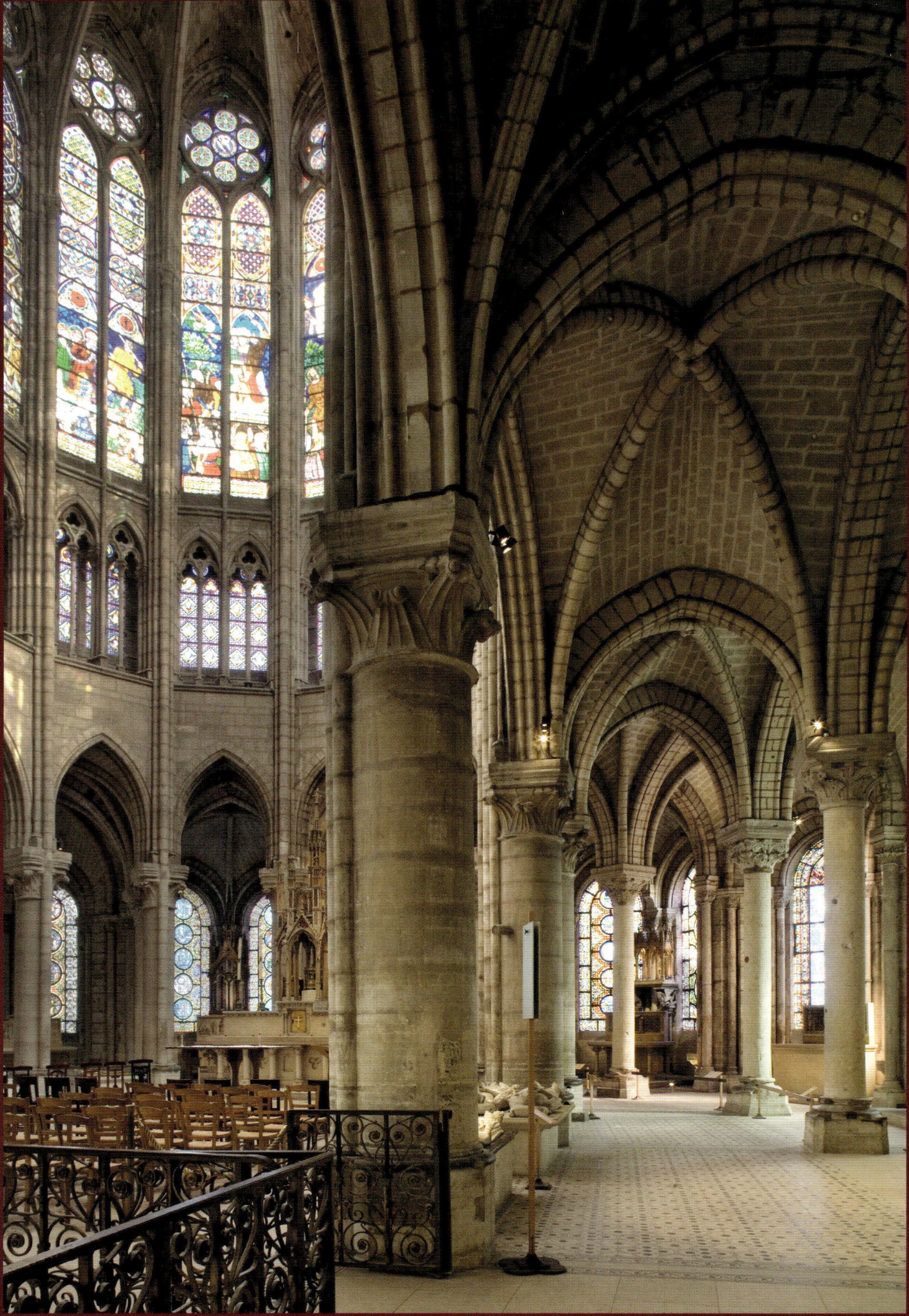

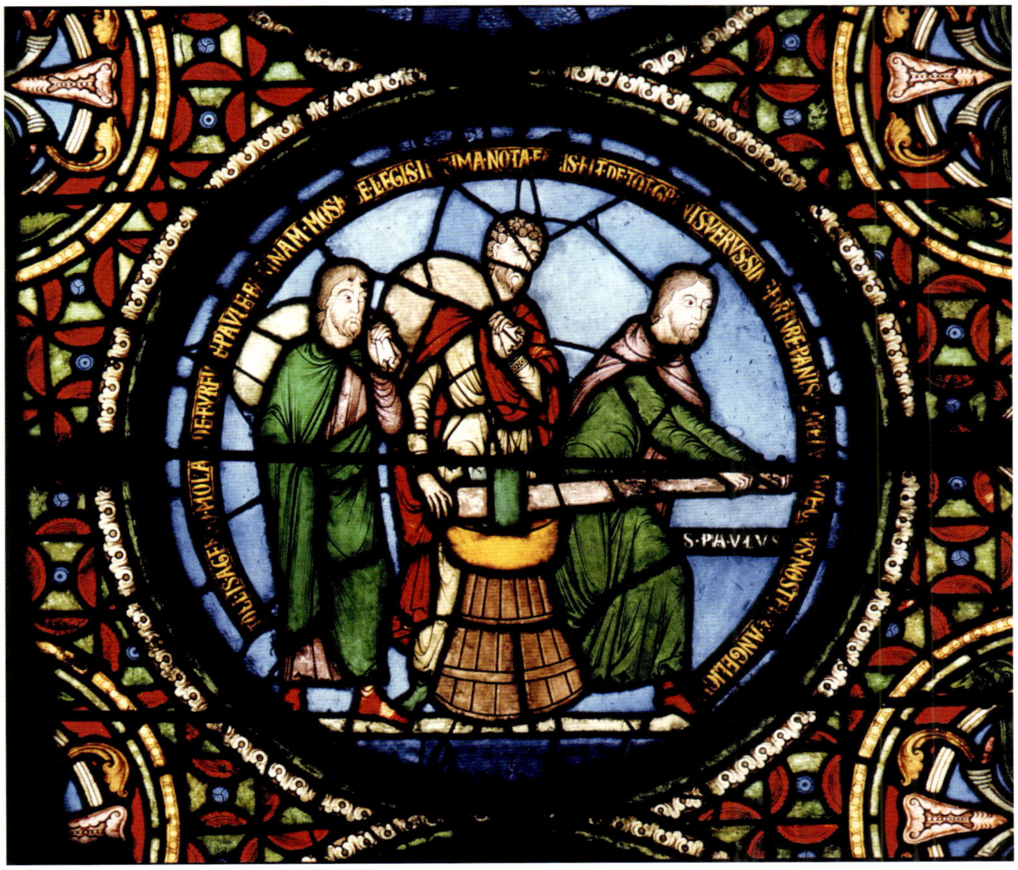

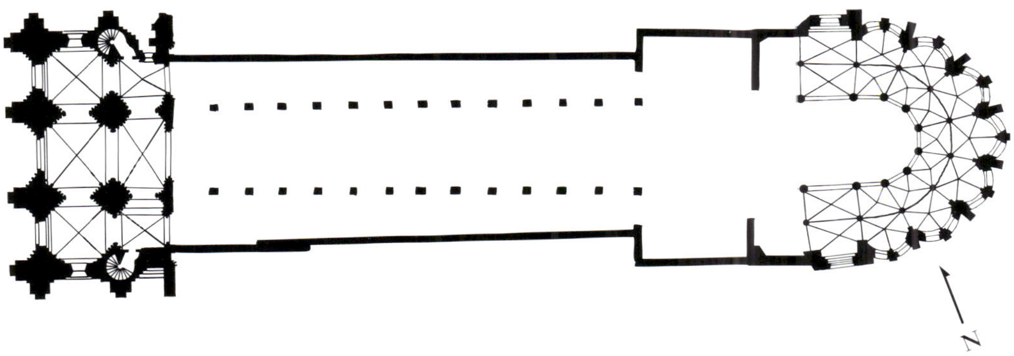

OPPOSITE

1-56. Ambulatory, Saint-Denis, Paris, 1140–44.

After Abbot Suger's ambulatory was completed in the twelfth century, the higher parts of the choir, which are visible in the upper left (the glazed triforium, clerestory, and chevet vault), were built in the thirteenth century and later.

TOP LEFT

1-57. Saint Paul, ambulatory of Saint-Denis, 1130s–1140s. Stained glass.

Two Old Testament prophets carry sacks of grain to a follower of Jesus, the Apostle Paul, who is on the right turning a millstone. Paul grinds their coarse grain to flour, making it more digestible, just as Paul made the Old Testament Judaic Law of Moses more palatable to Christians by reinterpreting the Ten Commandments in his many Epistles (letters), which are preserved in the New Testament.

BOTTOM LEFT

1-58. Plan, Saint-Denis, Paris, as it existed in the twelfth century.

At the opening of the twelfth, century the church of Saint-Denis had heavy masonry walls and the nave was covered with a barrel vault in the Romanesque style. In 1135–40 Abbot Suger began its reconstruction on the west end, building two bays to form a new entry and vestibule. He then turned to the east end and commenced building a new choir, beginning on the ground floor with the ambulatory, the plan of which is shown here. Built in 1140–44, the light-weight rib-vaulting, large expanses of stained glass, and the unified space of the ambulatory (as seen in plate 1-56) began the Gothic style.

Gothic cathedrals with light-weight vaults resting on arched ribs and flying buttresses. Since the walls were not needed for support, they could be pierced through with stained-glass windows, allowing light to stream in. This dramatic shift began in northern France in the 1130s to 1140s in the basilica of Saint-Denis near Paris during a rebuilding program under the guidance of Abbot Suger (plates 1-56 and 1-58).

As it happened, an original Greek manuscript by Pseudo-Dionysius, the titular patron saint of Saint-Denis, had been deposited in the abbey's library, where a Latin translation was done.[71] Thus Suger came to read the author of the *via negativa* and was inspired to use visible, material objects—the stone and glass of the architecture—to transcend to higher levels of invisible divine reality. Reflecting Dionysius's anagogical approach, Suger declared that contemplating the stained-glass windows in the choir transported the faithful from the (imperfect) material to the (perfect) immaterial realm: "We caused to be

RIGHT

1-59. Gerhard Richter (German, b. 1932), *Window for Eternity*, 2007. Stained glass, south transept, Cologne Cathedral, which was begun in 1248.

OPPOSITE

1-60. Thomas Struth (German, b. 1954), *Cologne Cathedral, Cologne 2007*, 2007. Chromogenic print, edition of ten, 78⅞ × 63½ in. (200.34 × 161.29 cm). Courtesy of the artist and Marian Goodman Gallery, New York. © Thomas Struth.

The city of Cologne was heavily bombed during World War II, but the Allied pilots spared her medieval cathedral, although the stained glass window in the south transept shattered. In 2007 the Cologne archdiocese commissioned a new window from German artist Gerhard Richter to replace the clear glass fitted after the war. Using a colorful palette of timeless shapes—squares and rectangles—Richter created what he called a "window for eternity."

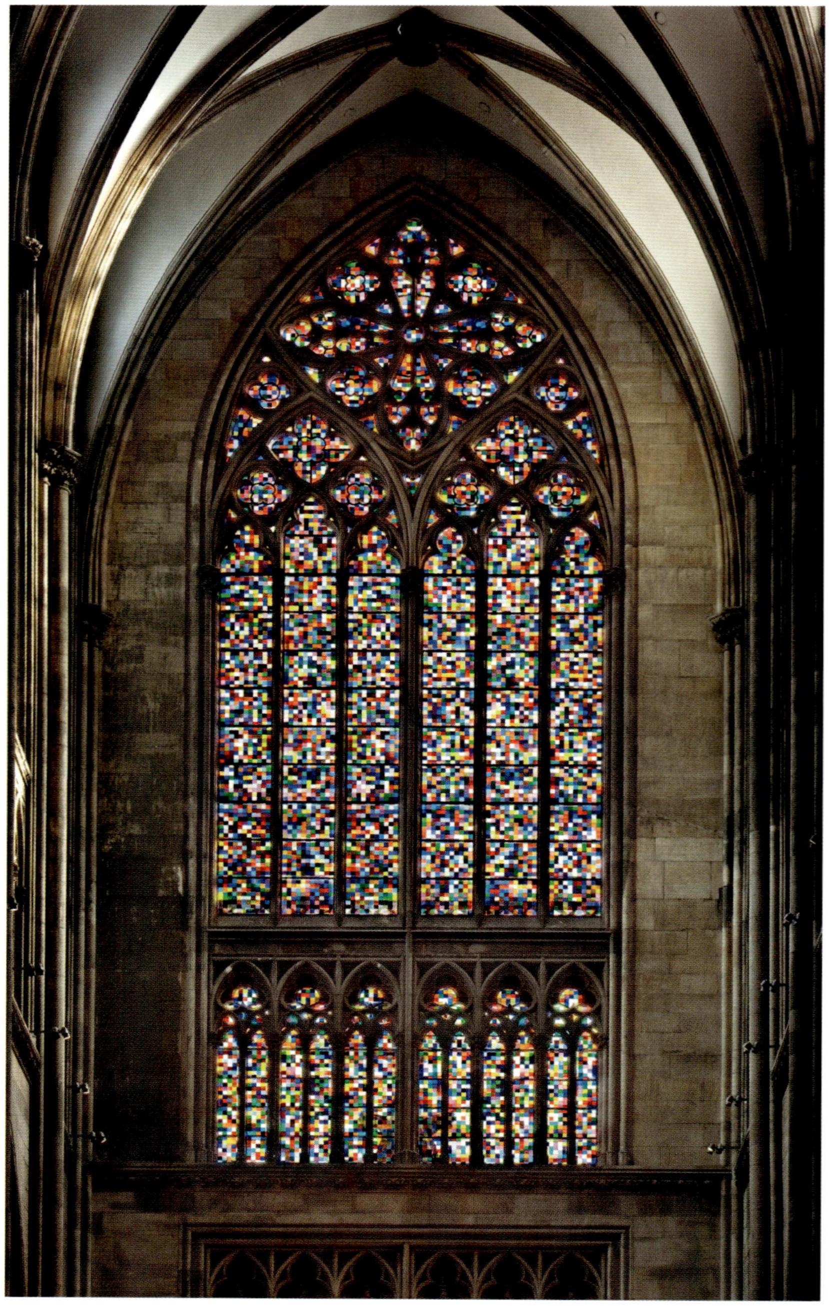

painted, by the exquisite hands of many masters from different regions, a splendid variety of new windows, both below and above. . . . One of these, urging us forward from the material to the immaterial, represents the Apostle Paul turning a mill, and the Prophets carrying sacks to the mill"[72] (plate 1-57). Suger found that looking at the stone pillars and arches was similarly transformative.[73]

At Saint-Denis the plan of the choir with radiating chapels may have been inspired by another ancient linkage of mathematics to the divine: the idea that heavenly bodies move only in perfect circles and epicycles, that is, in the planetary orbits recorded in Ptolemy's *Almagest*.[74] When the cathedral at Chartres burned in 1194, the architects who rebuilt it adopted the Gothic style of Saint-Denis for its north tower. The Gothic style would culminate in the thirteenth century in the cathedrals of Amiens and Cologne (plates 1-59 and 1-60).

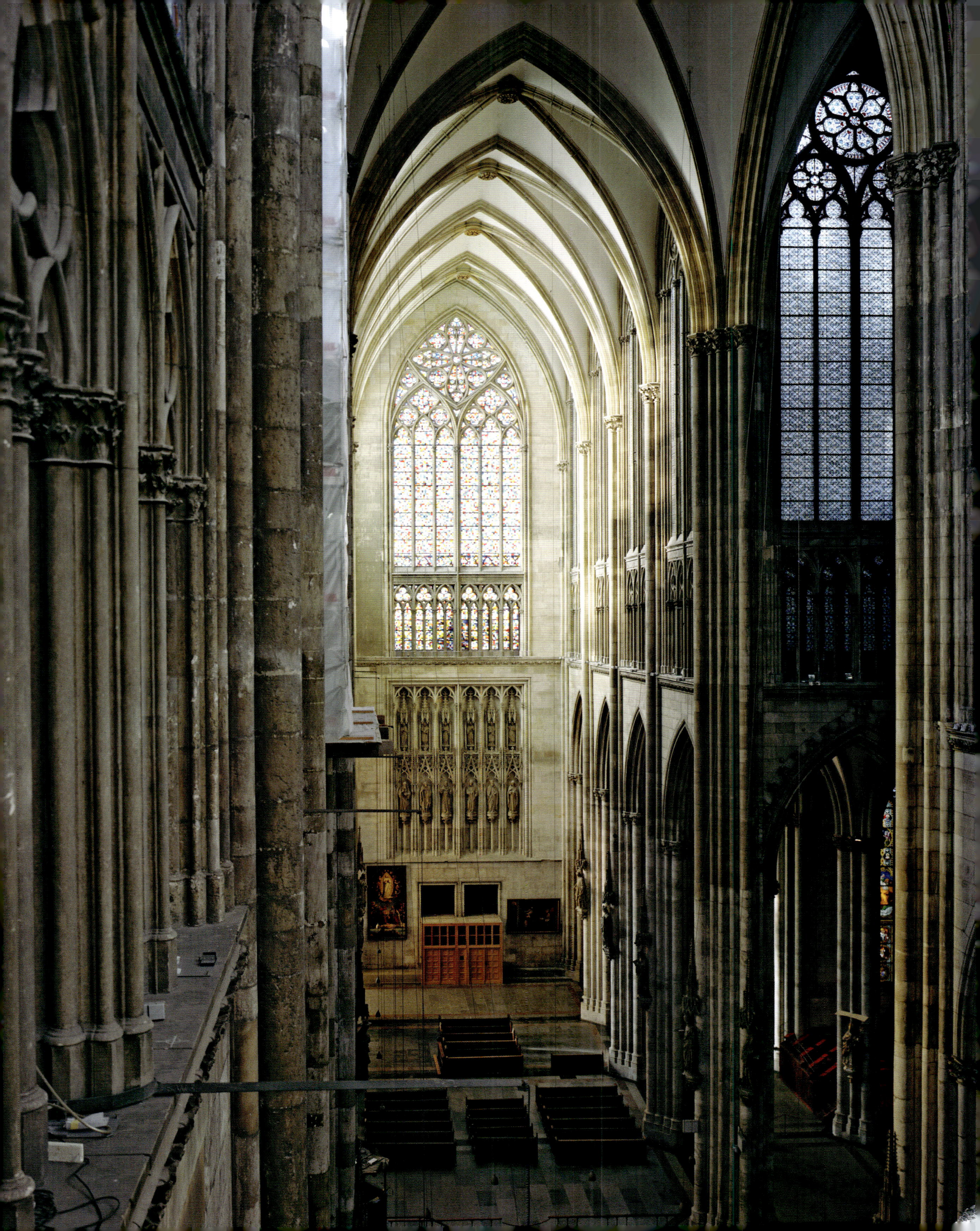

TOP

1-61. Raphael (Italian, 1483–1520), *School of Athens*, Stanza della Segnatura, Vatican, 1508–11. Fresco. © Musei Vaticani, Vatican City.

BOTTOM LEFT

1-62. Pythagoras's writing slate in Raphael's *School of Athens*.

BELOW

1-63. Pythagorean musical ratios.

Pythagoras presents a diagram of a four-string lyre, across whose top is written ΕΠΟΓΛΟΩΝ ("tone"). Above each string its tuning is marked in proportion—VI, VII, VIIII, XII—giving the intervals of an octave, fifth, and fourth, which are written between the strings in Greek letters that spell diapason (1 : 2), diapente (2 : 3), and diatessaron (3 : 4). Below the Pythagorean lyre, the numerical relation of the musical intervals forms a triangle that sums to ten, a figure sacred to the Pythagoreans. Although once common in Western music, Pythagorean tuning fell out of use after the eighteenth century, when musicians divided the twelve notes of the octave so that all the intervals would sound in tune.

I
} 1:2 octave
II
} 2:3 fifth
III
} 3:4 fourth
IIII
X

BOTTOM RIGHT

1-64. Euclid drawing on a slate in Raphael's *School of Athens*.

As his model for Euclid, Raphael used his contemporary, Donato Bramante, the architect who was rebuilding Saint Peter's Basilica. Euclid/Bramante draws two overlaid equilateral triangles. With this symmetrical figure (which does not appear in Euclid's *Elements*) Raphael may have been suggesting that ancient geometry was a source for Bramante's symmetrical floor plan for St. Peter's (see plate 2-6 in chapter 2).

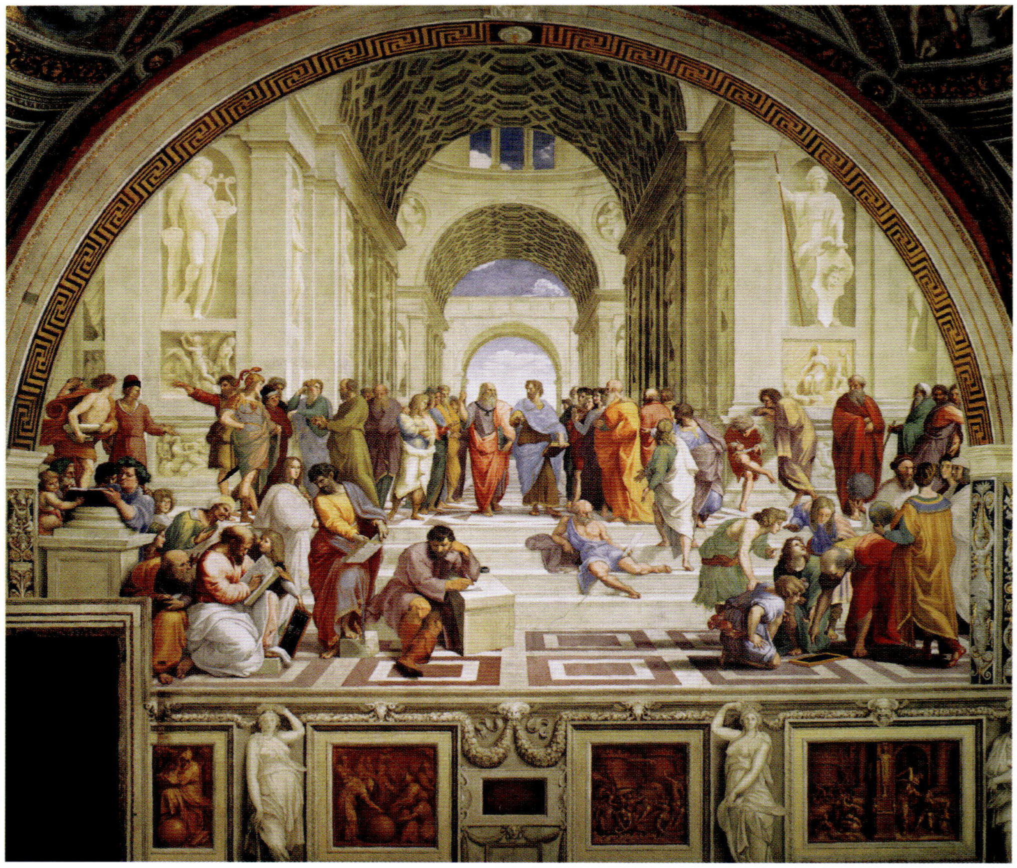

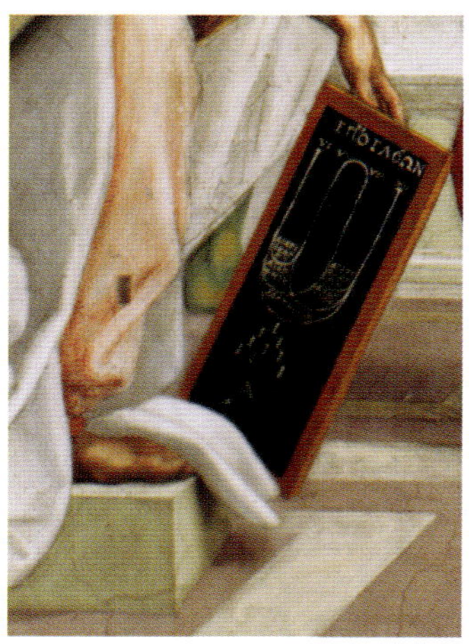

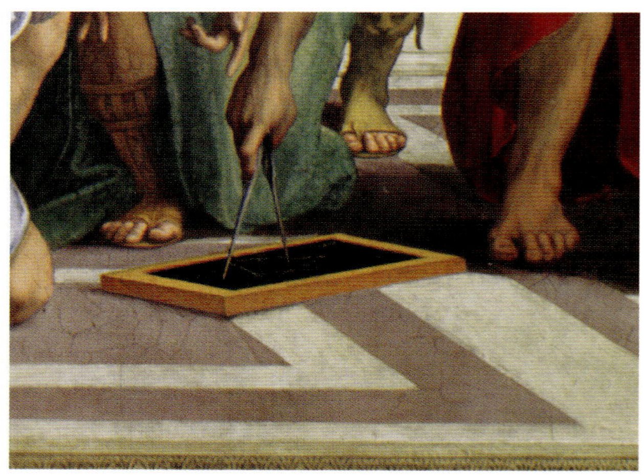

Christian Platonism found expression in the Vatican in Rome at the height of the Italian Renaissance, when the painter Raphael summarized the classical sources of Christian theology in his *School of Athens* (1508–11; see plate 1-61). In the center of the fresco Plato (in the red toga) and Aristotle (in blue) debate how to attain certain knowledge—whether in contemplation of the Forms (Plato points up) or by generalizing from observations of the natural world (Aristotle gestures down). The contest is flanked on the left by Pythagoras representing arithmetic (plates 1-62 and 1-63) and on the right by Euclid, the spokesman for geometry (plate 1-64).

LAWS OF SCIENCE: GALILEO'S LAWS OF MOTION AND KEPLER'S LAWS OF PLANETARY MOTION

Raphael depicted Plato and Aristotle carrying (Renaissance-style) bound copies of their writings: the pages of Plato's book are inscribed TIMEO (*Timaeus*), his dialogue on cosmology, and Aristotle carries ETICA (*Nicomachean Ethics*), in which he wrote: "What is just in the political sense can be subdivided into what is just by nature and what is just by convention. What is by nature just has the same force everywhere and does not depend on what we regard or do not regard as just. . . . Now, some people think that everything just exists only by convention, since whatever is by nature is unchangeable and has the same force everywhere—as, for example, fire burns both here and in Persia—whereas they see that notions of what is just change."[75] This distinction between eternal law and convention formed the basis of Christian ethics, which Thomas Aquinas summarized in his thirteenth-century *Summa Theologiae*. For Aquinas, Aristotle's laws of nature (eternal laws, *lex aeterna*) superseded all man-made, conventional laws: "Human law has the nature of law insofar as it partakes of right reason; and it is clear that, in this respect, it is derived from the eternal law. But insofar as it deviates from reason, it is called an unjust law, and has the nature, not of law, but of violence."[76] In the Renaissance the Italian Galileo Galilei and the German Johannes Kepler set about to comprehend the eternal laws governing the natural world. With confidence, Galileo and Kepler moved the laws of nature from the province of divine will to human reason, spurring the development of Western science and technology.

In antiquity followers of Plato and Aristotle had deduced terrestrial and celestial facts from laws or first principles and stopped there (since, in their opinion, the facts had been proven). But Galileo and Kepler revived Archimedes's method of treating each law of physics or astronomy as a *hypothesis* from which they made *predictions* of facts and took the crucial additional step of confirming (or refuting) each prediction with observation and hands-on experimentation. Galileo dramatically disproved Aristotle's "self-evident" first principle that heavy objects fall faster than light ones, which had remained unchallenged for a millennium, by dropping two balls of equal size but different weights from the Leaning Tower of Pisa. An assembled crowd witnessed their concurrent impacts.

Galileo formulated hypotheses about motion beginning in 1589 in a series of experiments in which he used a water clock (water flowing into a glass from a large vessel) to measure the speed of a brass ball rolling down an inclined plane. He understood the universe as a clockwork composed of mass that is inert unless acted on by a force. He found that a motionless ball would remain at rest or, if it were moving, the ball would continue rolling in a straight line at a constant rate, unless it was acted on by another force. He conjectured that falling bodies steadily increase their speed as they fall, and so he designed an experiment in which he could measure the distance the ball rolled in even intervals of time. The results allowed him to predict that the distance covered is proportional to the square of the elapsed time (in 2 seconds a ball rolls down an inclined plane 4 feet, in 3 seconds it rolls 9 feet, in 4 seconds 16 feet, and so on). With his water clock and falling

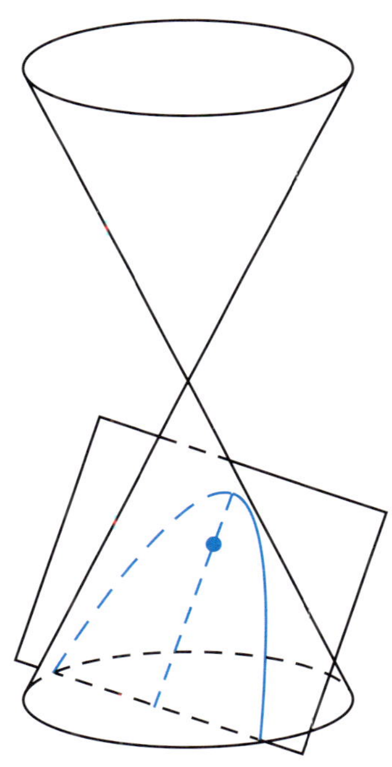

1-65. Parabola.

In *Conica* (Latin for "conical"; ca. 200 BC), Apollonius of Perga described ways of slicing ("sectioning") a cone to produce different curves, so-called conic sections. Slicing through the cone with a plane tipped as shown here produces a parabola.

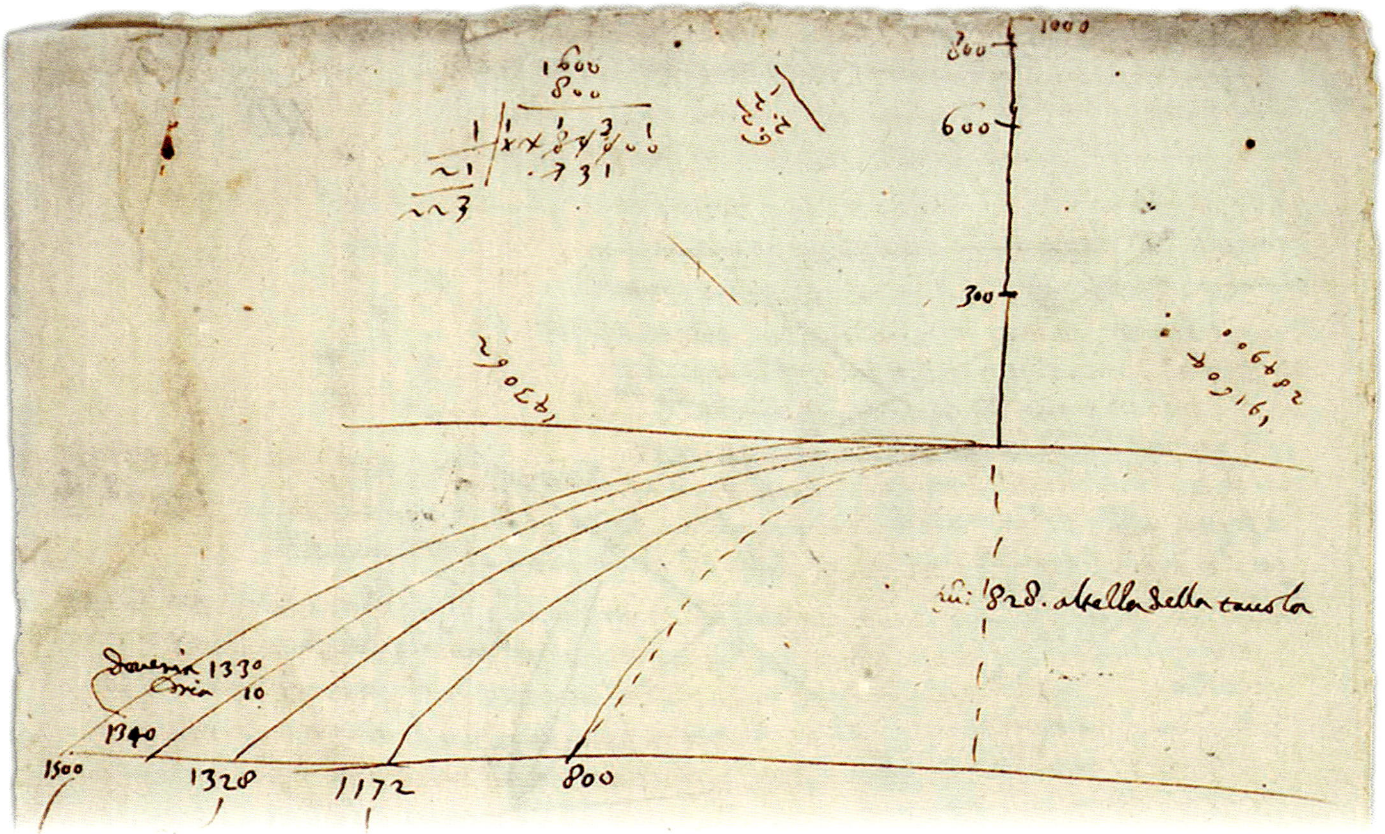

ABOVE

1-66. Galileo Galilei (Italian, 1564–1642), notes on the parabolic paths of projectiles, 1608. Biblioteca Nazionale Centrale di Firenze, Manoscritti Galileiani 72, f. 116v.

In the summer of 1608, Galileo drew this diagram of the parabolic paths of projectiles, which he measured by rolling a brass ball down an inclined plane to launch it horizontally off the end of a table. Measuring in *punti* (one *punto* or "point" equals 0.94 mm), Galileo set up his apparatus on the table, the top of which stood 828 punti (about 30 in., noted on the dotted vertical line, lower right) above the floor. He first tipped the plane up to 300 *punti*; the ball rolled down the plane, continued a short distance on the tabletop, and flew off the table, hitting the floor at 800 *punti*. Then he tipped the inclined plane up to the heights recorded on the solid vertical line (600, 800, 1000 *punti*), and noted that the greater the ball's momentum, the farther it flew in its in parabolic path (hitting at 1172, 1328, 1500 *punti*). For a detailed description of Galileo's calculations on this manuscript page, see Stillman Drake, *History of Free Fall* (Toronto: Wall and Thompson, 1989), 51–65.

OPPOSITE

1-67. Artemisia Gentileschi (Italian, 1593–1652), *Judith Slaying Holofernes*, ca. 1620. Oil on canvas, 78 × 63 in. (199 × 162 cm). Ufizzi, Florence.

Artemisia Gentileschi and Galileo met in 1611, after which they carried on a friendly correspondence for decades. In this painting the artist depicts the beautiful Jewish widow Judith and her maid after they, under cover of darkness, have entered the camp of the Assyrian general Holofernes, whose army is poised to destroy their town of Bethulia at dawn. After charming the drunken Holofernes, Judith grabs his sword and decapitates him (Judith 13: 7–8). In keeping with Galileo's study of the motion of projectiles, Gentileschi painted the conquered general's blood gushing out in parabolic paths.

weights, Galileo made the passage of time a quantity that could be precisely measured.

In Galileo's day people thought that when you fired a cannon ball, it travelled in a straight line and then fell straight down. In considering the path of a projectile, Galileo *combined* the two motions, a uniform motion straight ahead and an accelerating motion straight down, and found that together they formed the arc of a parabola (plate 1-66). The ancient Greek Apollonius of Perga had worked out the properties of a parabola with no application in mind (*Conica*, ca. 200 BC; see plate 1-65). A millennium later, Galileo showed that a parabola describes the path of a projectile. Galileo showed his drawings of parabolas to his friend the court painter Artemisia Gentileschi, who incorporated his findings into her painting *Judith Slaying Holofernes* (ca. 1620; see plate 1-67).[77]

In addition to his mathematical description of motion, Galileo also analyzed building materials, laying the foundation for modern structural engineering. He determined (correctly) that the breaking point of a cantilevered beam is a function of its height times width; using

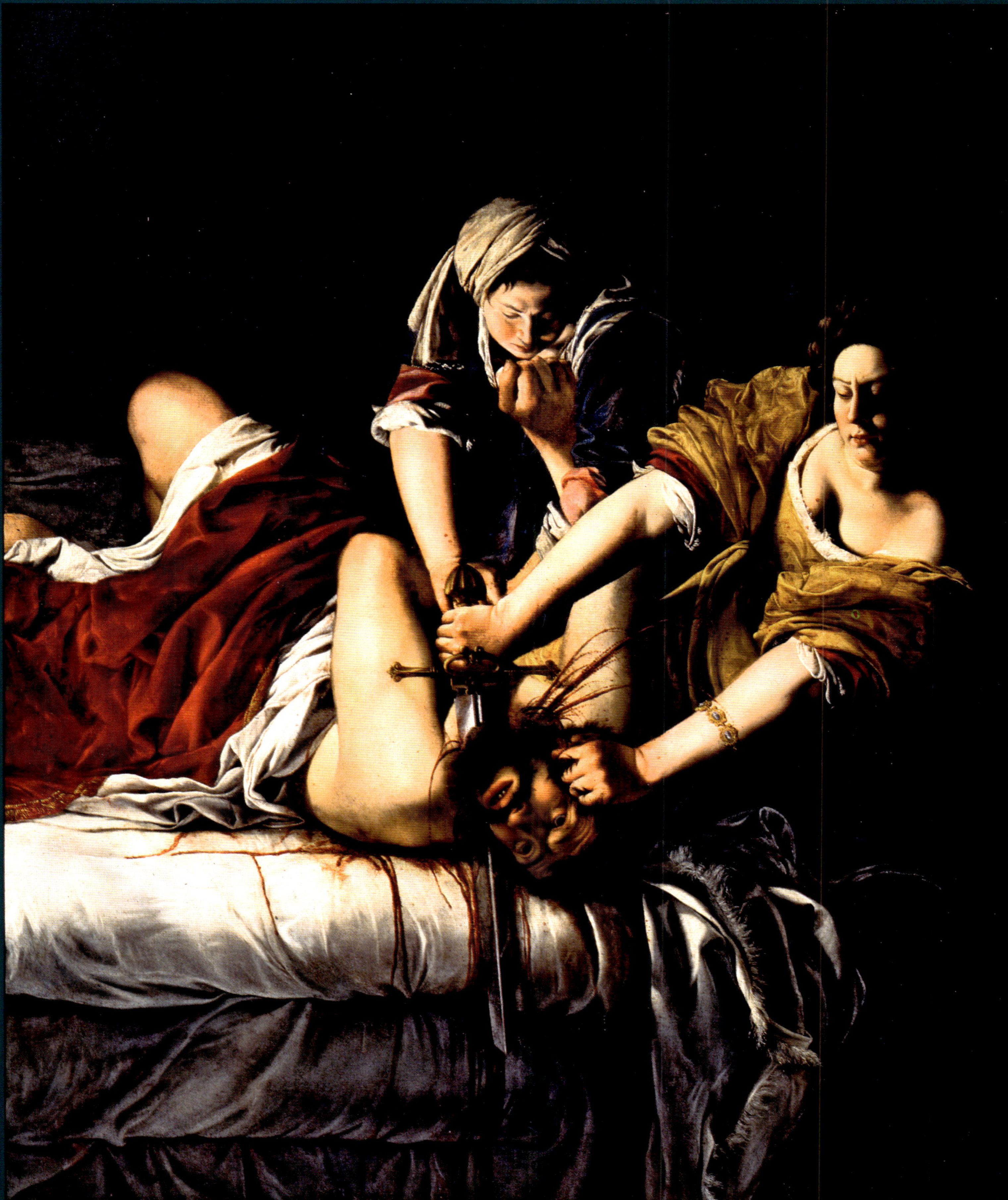

What is the Stress Line of an Arch?

Galileo made the first recorded attempt to determine the curve formed by a hanging chain. He came up with the wrong answer, but more importantly, he had asked the right question. Others soon discovered that a hanging chain forms not a parabola, but a similar-looking catenary curve, as shown in this diagram by the English natural philosopher Robert Hooke, who made an analogy between the shape of a hanging chain (in tension loaded by its own weight) and the shape of a stress line in a masonry arch (carrying its load in compression) (*A Description of Helioscopes*, 1675). The Scottish mathematician David Gregory gave the first mathematical description of the catenary curve, commenting: "In a vertical plane, but in an inverted situation, the chain will preserve its figure without falling, and therefore will constitute a very thin arch. . . . And when an arch of any other figure is supported, it is because within its thickness some *catenaria* is included." David Gregory, "Catenaria," *Philosophical Transactions* 231 (1697), 637. In other words, if a catenary curve lies within (the central third of) the stone blocks that form an arch and its buttresses, then the arch will stand.

Before the seventeenth century, all previous stone architecture—from the massive arches of the Roman Coliseum to the soaring vaults of Cologne Cathedral—had been built by trial and error without the mathematical description of the stress line in an arch that Galileo, Hooke, and Gregory discovered. Consequently, many old arch-based stone vaults do not contain a catenary curve, and therefore they have come tumbling down, as did the dome of Hagia Sophia in Byzantium (collapsed AD 558) and the choir of Beauvais cathedral in France (collapsed 1284). The Great Fire of 1666 destroyed central London, reducing St. Paul's cathedral to smoldering rubble. The architect of the new cathedral, Christopher Wren, followed the advice of his friend Robert Hooke to construct the 102-foot-diameter dome to contain a catenary curve. Therefore the dome of St. Paul's, and all subsequent masonry vaults built to this standard, remains stable to this day.

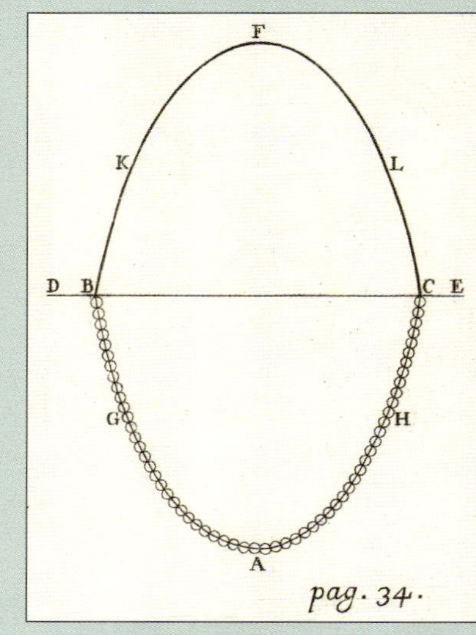

1-68. Robert Hooke's hanging chain, in Giovanni Poleni, *Memorie istoriche della gran cupola del Tempio Vaticano* (Padova: Nella stamperia del Seminario, 1748), 34, fig. 12. Art and Architecture Collection, Miriam and Ira D. Wallach Division of Art, Prints, and Photographs. The New York Public Library, Astor, Lenox, and Tilden Foundations.

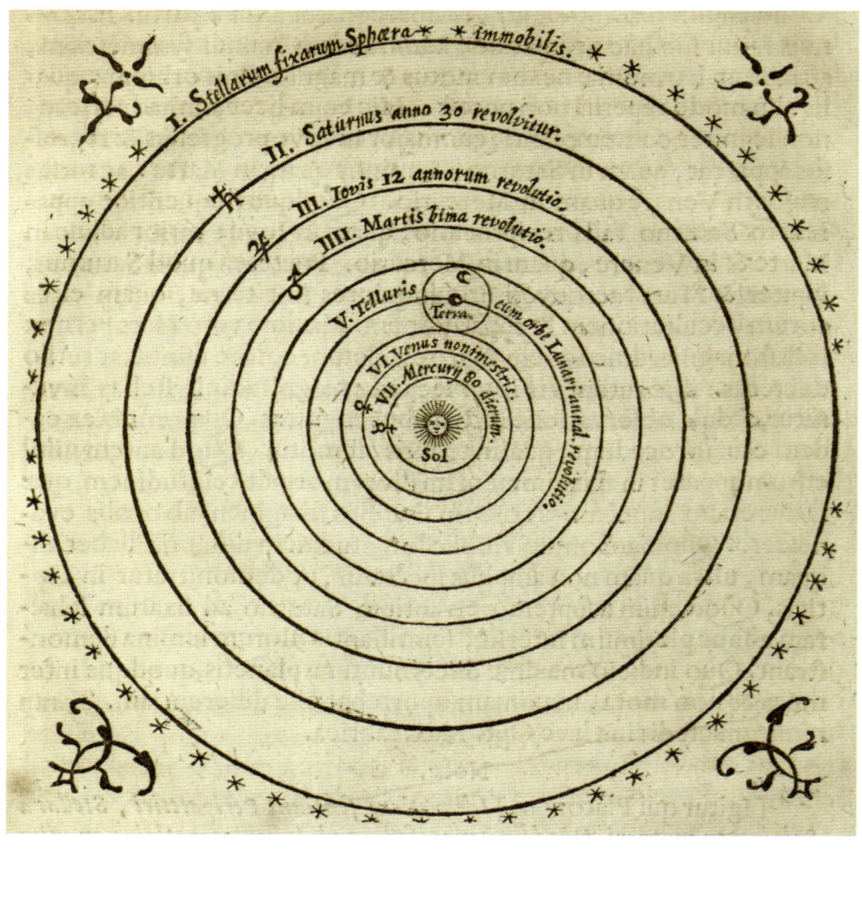

1-69. Nicolaus Copernicus, Model of the cosmos, in *De Revolutionibus Oribium Coelestium* (On the revolutions of the celestial orbs, 1543), 21. Science, Industry & Business Library, The New York Public Library, Astor, Lenox, and Tilden Foundations.

Copernicus placed the sun (*Sol*) at the center of the cosmos, surrounded by concentric orbs that rotate carrying Mercury, Venus, Terra (earth with its moon), Mars, Jupiter, and Saturn. The outermost orb was the immobile sphere of the fixed stars (*Stellarum fixarum sphaera immobilis*). For those accustomed to Christian adaptations of the Aristotelian-Ptolemaic model (plate 1-35), not only had Copernicus demoted earth to being a planet, but gone was "the highest heavenly dwelling place of God."

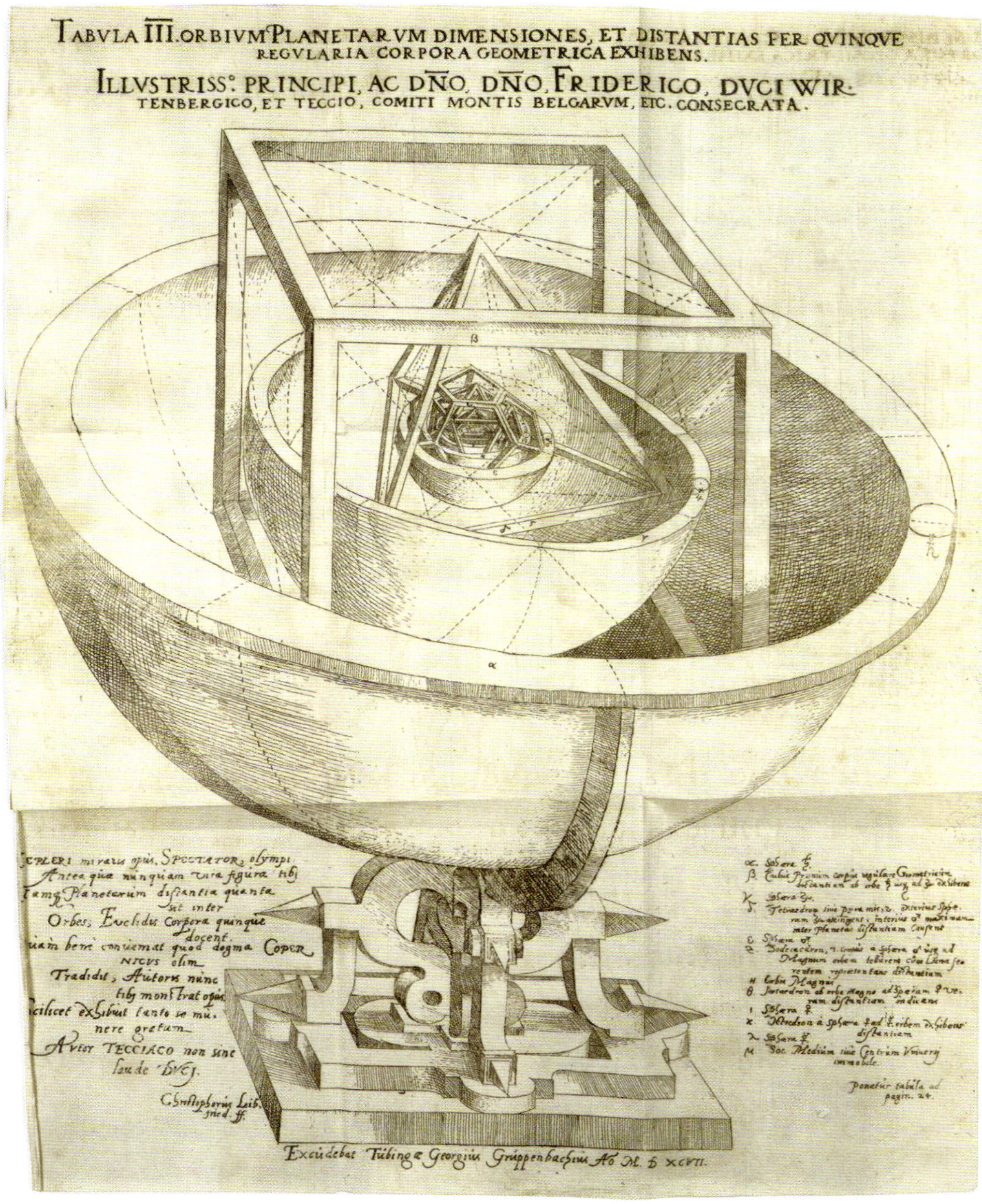

1-70. Johannes Kepler, Model of the cosmos, in *Mysterium Cosmographicum* (Mysterious map of the cosmos) (Tvbingae: Georgius Gruppenbachius, 1596), plate 3. Rare Books Division, The New York Public Library, Astor, Lenox, and Tilden Foundations.

Since there were six (known) planets and five Platonic solids, Kepler reasoned that these numbers are related. Perhaps there is a solid "between" each pair of planets such that each of the six planets orbits around the sun on a sphere, the diameter of which is determined by the forms of the five solids. Saturn orbits on a sphere that circumscribes a cube. Inscribed within the cube is the orbit of Jupiter, which circumscribes a tetrahedron, which in turn forms the orbit of Mars. Between Mars and Earth is a dodecahedron, between Earth and Venus an icosahedron, and between Venus and Mercury, an octahedron. Kepler's calculations assumed circular orbits with epicycles (see plate 0-2 in the Preface). After publishing the treatise in 1596, he could not get his model to match observation, so he abandoned it.

Galileo's formula, architects could calculate the breaking point of any beam of a given material from its rectangular cross-section (*Dialogues Concerning Two New Sciences*, 1638). In the same *Dialogues*, Galileo also speculated about the curve of a hanging chain and hypothesized (incorrectly) that it forms a parabola (see sidebar at top of opposite page).

In the early sixteenth century the Polish astronomer Nicolaus Copernicus replaced another of Aristotle's first principles and made the sun, not the earth, the center of the universe (plate 1-69). Inspired by Copernicus, Kepler determined to prove that the planets move around the sun in circular orbits. To test this hypothesis, he built a Copernican model of the solar system as concentric solids, around which revolved the planets, (*Mysterium Cosmographicum* [Mysterious map of the cosmos] 1596, plate 1-70).

After gaining access to the 20-year, detailed records of planetary movement compiled at the Prague observatory of Tycho Brahe, the indefatigable Kepler discerned a pattern in

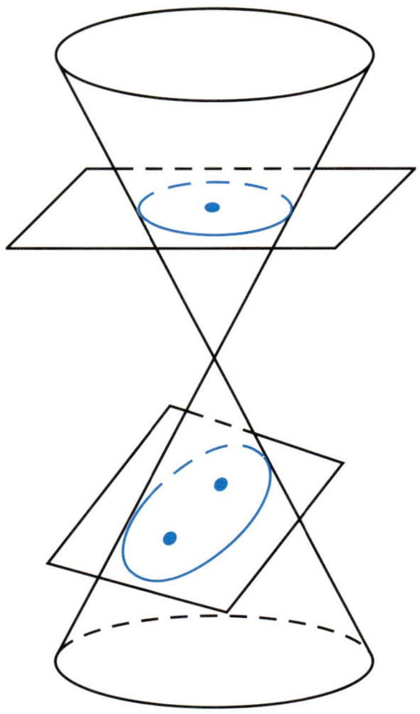

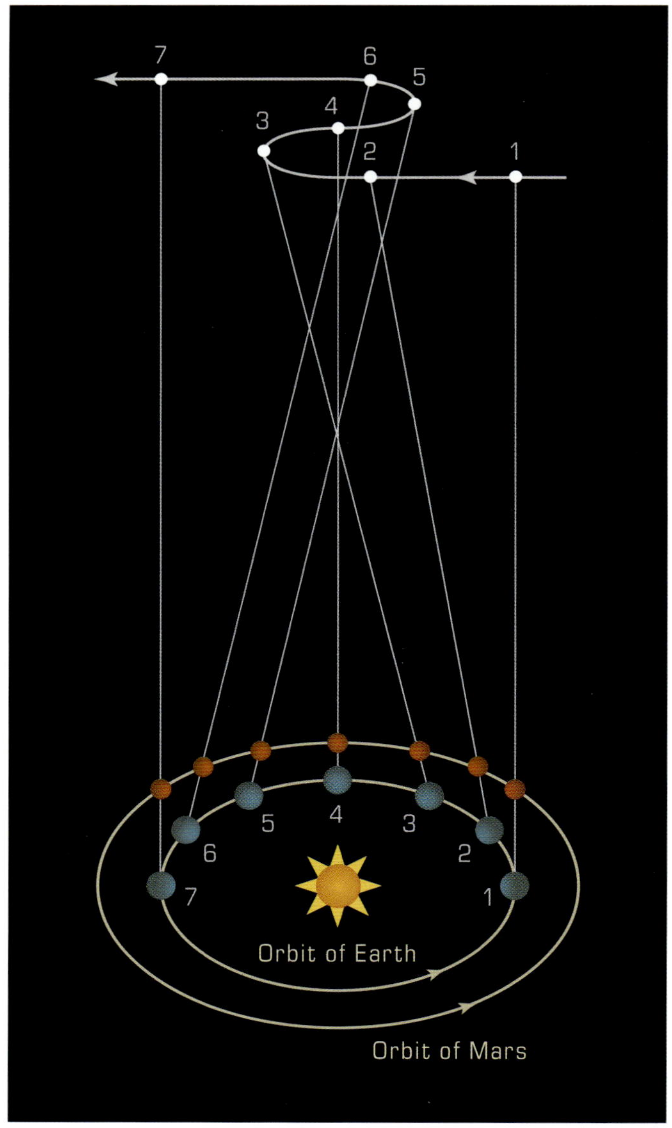

ABOVE

1-71. Ellipse.

Apollonius's method of slicing through a cone can produce an ellipse (all points equidistant from two points ("foci"). The circle is a special case, when the two foci are identical; a circle is all points equidistant from one point (one "focus"; *Conica*, ca. 200 BC).

RIGHT

1-72. Kepler on the apparent retrograde motion of Mars.

Since Earth completes its orbit in a shorter time than Mars, when Earth periodically overtakes the outer planet, Mars will appear to stop and drift backward relative to the constellations. Once Earth has passed Mars, the red planet will appear to resume its forward motion.

Brahe's data that was consistent with each of the planetary orbits being not a circle but an ellipse, another of Apollonius's conic sections (plate 1-71). Arriving at a new theory, Kepler placed the sun at one focus of the ellipse (see sidebar on opposite page). He proposed that each planet sweeps out an equal area in an equal time as it moves along its orbit, speeding up as it approaches the sun and slowing down as it recedes. Kepler also discovered a precise connection between the size of the planet's orbit and the time it takes to go around the sun. He called this ratio his "harmonic law" because "the motions of the heavens are nothing but a kind of perennial harmony (in thought not sound)."[78] In other words, the orbiting planets make a celestial pattern that is analogous to music—a (metaphorical) harmony of the spheres.

The heliocentric model of Kepler, along with later refinements by others, explained geometric features of the visible sky that had eluded previous astronomers. The solar system is like a disc, with the elliptical orbits of the planets lying close to the plane of the ecliptic. The planets rotate on their axes and revolve around the sun, most moving in the same counter-clockwise direction (as seen from an imaginary viewpoint above the North Pole). The sun appears to move westward through the zodiac each year, but it is really earth that is moving in its annual revolution around the sun (plate 1-73). During the night the starry dome above appears to rotate around a fixed point in the northern sky, the North Star, because of the earth's rotation around its axis; the terrestrial North Pole points to the celestial North Star (Polaris). Depending how far north or south of the equator the observer is located, the ecliptic will appear tipped at different angles in the sky (plate 1-74). Kepler's model also accounts for the observation that planets will from time to time appear to reverse their direction of motion as viewed from earth against the background of the constellations (plate 1-72).

Kepler's marvelous achievement is a particularly fine example of the truth that knowledge cannot spring from experience alone but only from the comparison of the inventions of the intellect with observed fact.

—Albert Einstein, "Johannes Kepler," 1930

Kepler's Three Laws of Planetary Motion

In 1609 Kepler discovered the first two laws of planetary motion, which apply to a single planet orbiting the sun. It took him another decade to discover the third law, which he recorded on May 15, 1618. The importance of the third law is that it binds the planets together into one system—the solar system. Kepler, who thought that the solar system was about 6000 years old, described his euphoria ("sacred frenzy") at discovering the third law: "What 25 years ago I already had a vague premonition of; what 16 years ago I determined as the whole aim of research; what I devoted the best part of my life to—*that* I have at last brought into the light. Now, since the dawn eighteen months ago, since the broad daylight three months ago, and since a few days ago, when the full sun illuminated my wonderful speculations, nothing holds me back. I yield freely to the sacred frenzy. . . . The die is cast, and I am writing the book—to be read either now or by posterity, it matters not. It can wait a century for a reader, as God Himself has waited 6000 years for a witness." Johannes Kepler, *Harmonices Mundi* (Harmony of the world, 1619) trans. E. J. Aiton, A. M. Duncan, and J. V. Field (Philadelphia: American Philosophical Society, 1997), bk. 5.

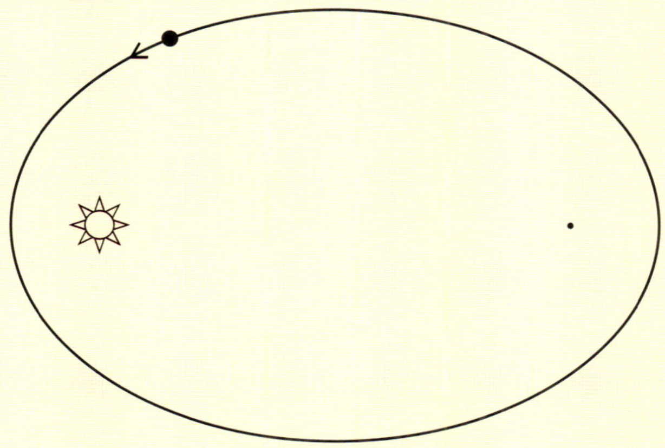

1st Law

A planet moves in an ellipse with the sun at one focus.

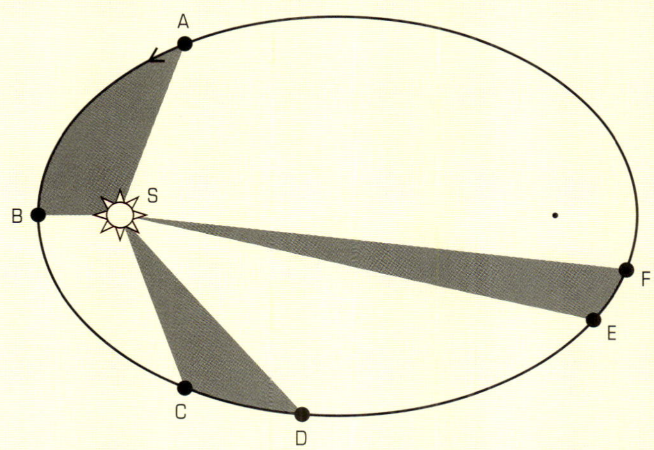

2nd Law

A planet sweeps out equal areas in equal times. A planet speeds up as it approaches the sun and slows down as it moves away such that it takes the planet the same time to move in its orbit from point A to B, C to D, and E to F. Additionally, the shaded areas ABS, CDS, and EFS are all equal.

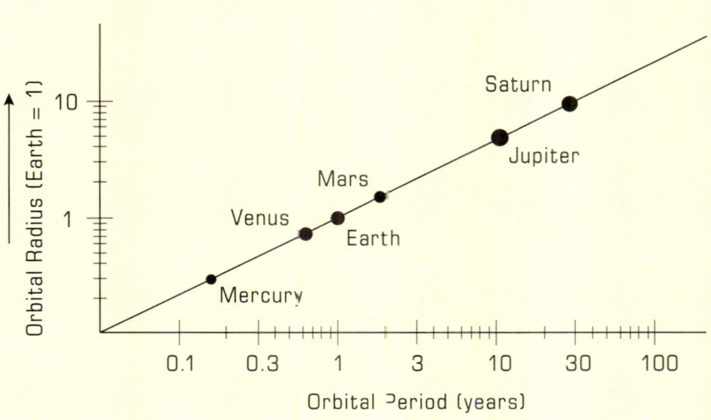

3rd Law

The average radius of a planet's orbit is in a precise ratio to the time it takes that planet to orbit the sun. For each planetary orbit, if you divide the average radius cubed by time squared, you get the same number. In other words, T^2/R^3 is a constant, the same for every planet, despite their vastly different sizes, speeds, and orbits.

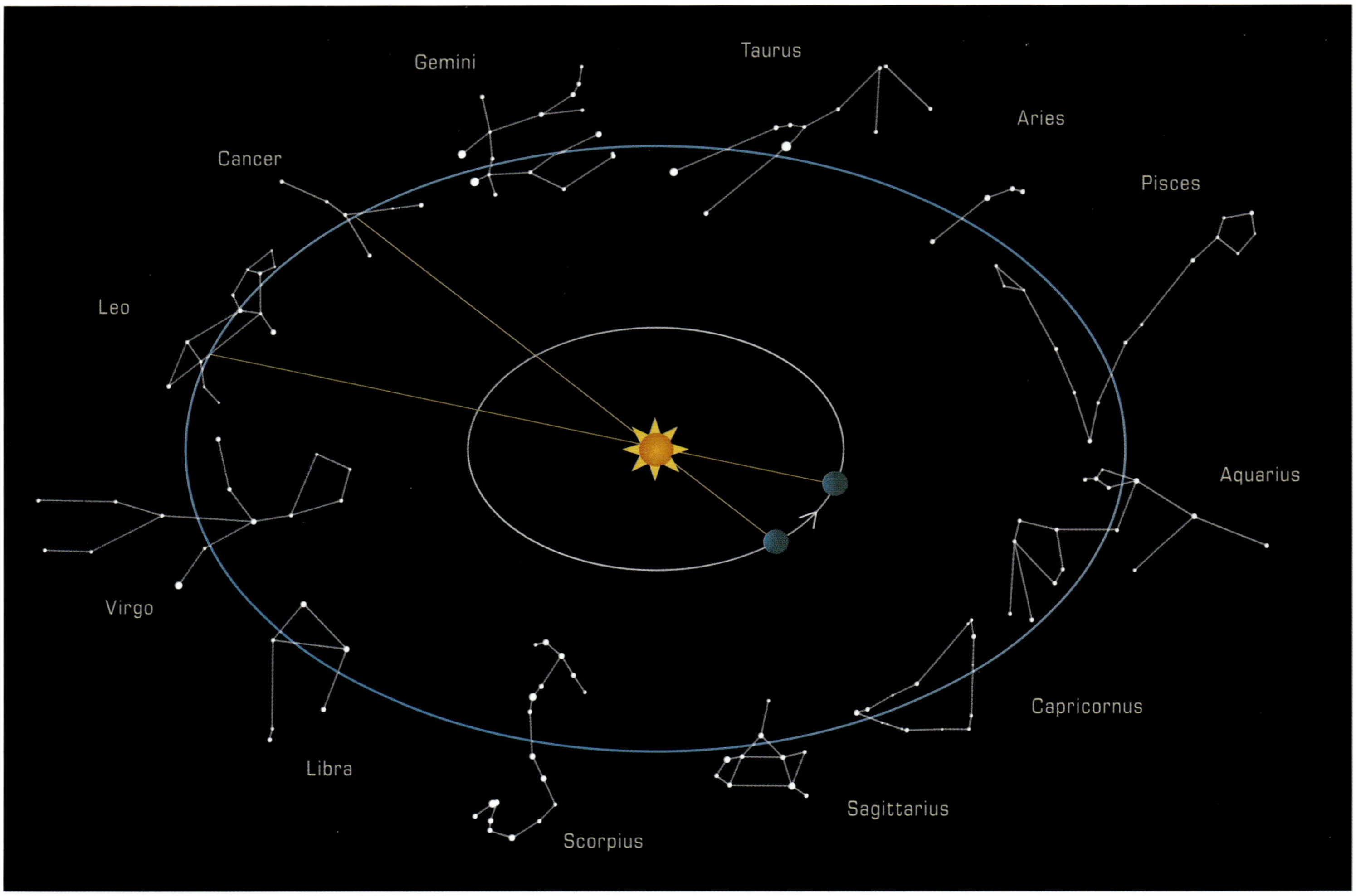

ABOVE

1-73. Kepler's heliocentric model.

The ecliptic is an imaginary plane that cuts through the sky in the paths followed by the sun, moon, planets, and the constellations of the zodiac (shown here in their correct order but not how they would look from a celestial perspective.)

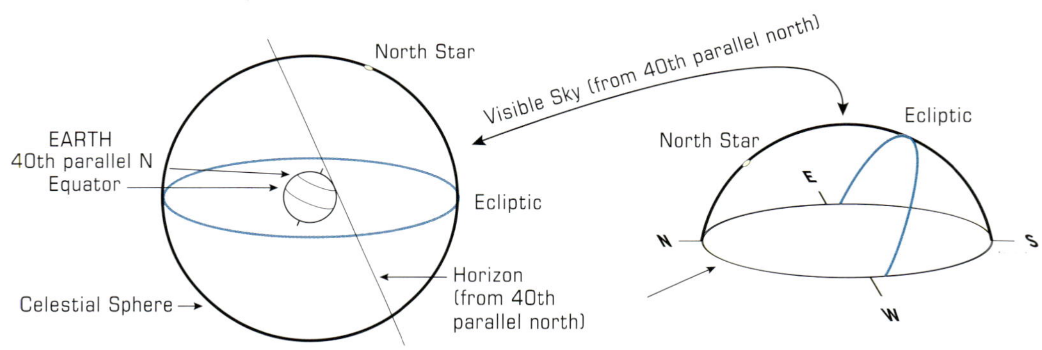

RIGHT

1-74. The visible sky.

This diagram shows the visible sky as seen from the fortieth parallel in the northern hemisphere, a circular latitude that passes through Spain, Greece, Turkey, China, North Korea, Japan, and the northern United States. The whole celestial sphere is the sky visible from anywhere on earth imagined as projected on the inside of a sphere.

Kepler and Galileo thus proclaimed core ideas of the modern scientific worldview: nature has a mathematical structure and man can discover it. Beginning in the Enlightenment and continuing to this day, if scientists are not able to understand the physical nature of a phenomenon, they begin their investigation by trying to discern a pattern in the data and describing it mathematically.

NEWTON'S LAW OF UNIVERSAL GRAVITATION

Galileo did not know what made brass balls fall to the ground, and Kepler could only speculate about what force kept the planets in their orbits. The terrestrial and celestial paths are both conic sections. Are they somehow related? Do they have a common cause? To answer these questions, Newton synthesized Galileo's results into three laws of motion, incorporated Kepler's findings, and then proved that the parabolic paths of earthly projectiles and the elliptical paths of planets and the moon are caused by the same force—gravity.

According to tradition Newton got the idea of universal gravitation when he saw an apple fall to the ground. He suspected that the moon and apple were pulled towards the earth by the same force. If so, why didn't the moon fall to the ground? Newton did a thought experiment and imagined that a cannon ball was shot from the top of a mountain (plate 1-75). If the cannon were charged with a small amount of powder, the ball would follow an almost parabolic path and land in a nearby valley. If more gunpowder were used, the ball would travel farther before hitting the earth, but with enough powder the cannon ball would never strike the earth because as its path curved, so does the surface of the earth; it would go into orbit around the earth. Newton speculated that the moon was like that last cannon ball. Carried forward by its momentum and pulled downward by the earth's gravity, the moon moves in its nearly circular elliptical orbit around the earth.

Newton generalized beyond particular balls, planets, and the moon and declared a general law of nature: every particle of matter in the universe (from a grain of sand to the farthest star) is attracted by the force of gravity to every other particle; in other words, gravity holds the universe together. Newton described gravity mathematically when he asserted that the attraction of bodies is directly proportional to the product of their mass and inversely proportional to the square of the distance between them. Today we write Newton's law of universal gravitation in algebraic form:

$$F = G\left(\frac{m_1 m_2}{d^2}\right)$$

where F is the force of gravity, m_1 and m_2 are the masses of the two bodies, d is the distance between their centers, and G is the gravitational constant.[79] This mathematical description of the force of gravity gave scientists a powerful tool. By replacing the variables with facts (e.g., the mass of the moon), the path of a projectile can be worked out. If the facts are correct, then the path can be known with certainty. Equipped with Newton's law of universal gravitation, in 1968 American engineers needed no new physics to plot the course of a rocket and land a man on the moon.

Galileo's description of falling bodies, Kepler's laws of planetary motion, and Newton's law of universal gravitation underscore a key distinction in science and mathematics that was blurred in antiquity but is a hallmark of modern thought: laws of science (like Newton's law of gravity) describe observed (physical) data, whereas axioms of mathematics (such as $x + y = y + x$) are assumptions made within a specific mathematical system. If the axioms of mathematics are valid in the system and deductions are correctly made, then the

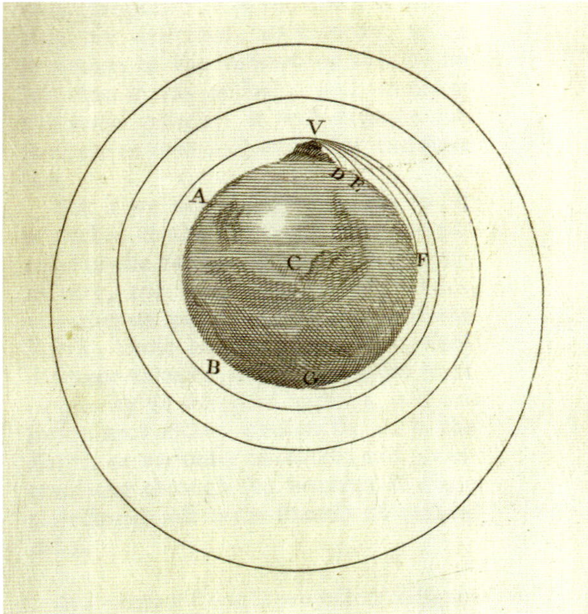

1-75. Paths of a projectile shot from a high mountain at several different velocities; in Isaac Newton, *A Treatise on the System of the World* (1728), an unnumbered illustration between 6 and 7. Science, Industry & Business Library, The New York Public Library, Astor, Lenox, and Tilden Foundations.

conclusions will hold. Galileo, Kepler, and Newton's works are also prime examples of the use of mathematics in the scientific method: first they formulated a mathematical description of carefully observed phenomena, then they used this scientific law to predict observable events, and finally they subjected this law to confirmation (or refutation) by observation. In most cases, later scientists eventually describe the physical mechanism that produces the observed phenomenon, but occasionally mathematics describes a phenomenon whose physical mechanism remains unknown.[80] Galileo, Kepler, and Newton each gave a *mathematical* description of the cause of *physical* effects—falling balls and rotating planets—but they could not give a *physical* description of the cause—gravity.

What *is* gravity? What reaches out from the moon, travels across a void, and tugs on earth's oceans? Newton could precisely calculate the *effect* of gravity, but the physical nature of the force itself remained a mystery to him: "I have not yet been able to discover the cause of these properties of gravity from phenomena and I frame no hypotheses."[81] But something—call it a "force" or "gravitation"—causes ocean tides and planetary motion, and Newton declared that, even if he could not explain gravity physically, he could at least measure it and describe its action mathematically: "It is enough that gravity really does exist and acts according to the laws I have explained, and that it abundantly serves to account for all the motions of the celestial bodies."[82] In other words, the universe certainly acts *as if* there were a force of gravity. To stress that he gave a mathematical description of how gravity affects bodies, not a physical explanation of the mechanism of gravity, Newton named his 1687 treatise *Philosophiæ Naturalis Principia Mathematica* (Mathematical principles of natural philosophy; see plate 1-76) and he stated at the outset: "I have design to give only a mathematical notion of the forces, without considering their physical causes."

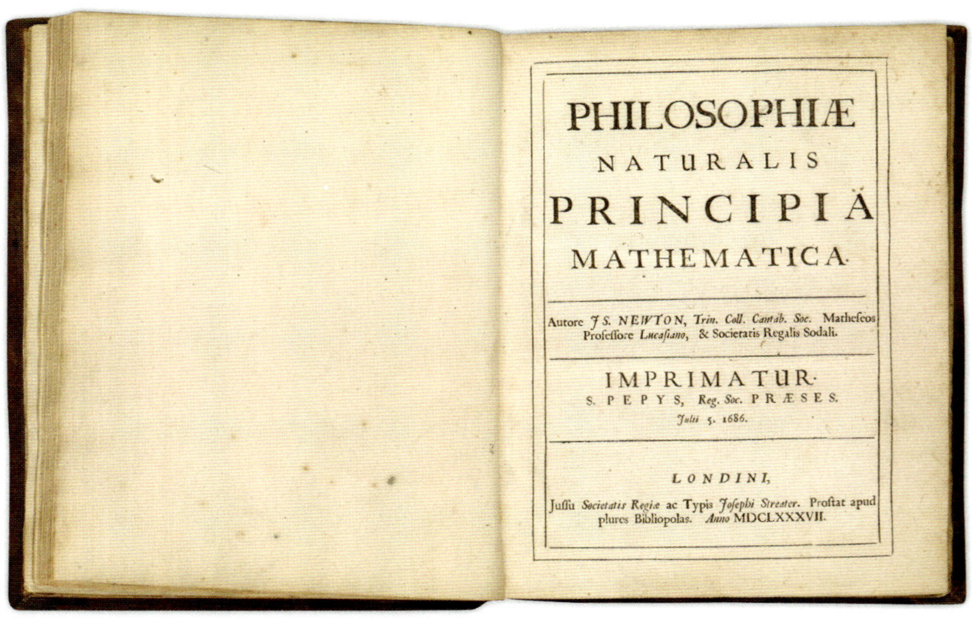
PHILOSOPHIÆ
NATURALIS
PRINCIPIA
MATHEMATICA.

Autore J S. NEWTON, *Trin. Coll. Cantab. Soc.* Matheseos Professore *Lucasiano*, & Societatis Regalis Sodali.

IMPRIMATUR.
S. PEPYS, *Reg. Soc.* PRÆSES.
Julii 5. 1686.

LONDINI,
Jussu *Societatis Regiæ* ac Typis *Josephi Streater*. Prostat apud plures Bibliopolas. *Anno* MDCLXXXVII.

1-76. Isaac Newton, *Philosophiæ Naturalis Principia Mathematica*, ed. Edmond Halley (London: Royal Society, 1687). Rare Books Division, The New York Public Library, Astor, Lenox, and Tilden Foundations.

THEISM, ATHEISM, AND SECULARISM

The rise of science that began during the Enlightenment precipitated major changes in religious beliefs in the West, which affected the traditional association of mathematics and divinity. Throughout history, wherever there was theism (belief in a divine person), there has been atheism (disbelief in the existence of that person). But widespread atheism did not develop in the Christian West until the nineteenth century, after Charles Darwin presented overwhelming evidence that mankind was not created by the God of Abraham but "descended from a hairy, tailed quadruped, probably arboreal in habits."[83] Since Kepler, Galileo, and Newton lived before Darwin and worked primarily in astronomy, their work did not conflict in any immediate or obvious way with the Judeo-Christian-Islamic tradition that a divine Creator imposed order on the cosmos. (Galileo's infamous 1633 conviction for heresy by the Catholic Inquisition resulted not from charges related to atheism—all the participants were devout theists—but from an explosive mix of Pope Urban VIII's rigid enforcement of dogma and Galileo's confrontational presentation of his research.)[84] Indeed, Kepler, Galileo, and Newton each saw himself as discovering geometric patterns in nature that had been put there by a Christian God.

The German Protestant Kepler concluded his discussion of geometry by declaring: "Contemplation of these axioms . . . is lofty, Platonic, and analogous to the Christian faith, looking towards metaphysics and the theory of the soul."[85] The Roman Catholic Galileo declared that God wrote two books: the Bible, written in Hebrew and Greek, and the book of nature, written in mathematics. The latter, Galileo asserted, "stands continually open to our gaze. But the book cannot be understood unless one first learns to comprehend the language and read the letters in which it is composed. It is written in the language of mathematics, and its characters are triangles, circles, and other geometric figures without which it is humanly impossible to understand a single word of it; without these, one wanders about in a dark labyrinth."[86] In other words, according to Galileo, the natural world was in his territory as a mathematician, and Urban VIII's theologians had no jurisdiction over it. Galileo claimed further that since mathematics gives certainty, his knowledge of triangles and circles is of the same high quality as divine knowledge: "Of . . . the mathematical sciences alone; that is, geometry and arithmetic, . . . Divine intellect knows infinitely more propositions, since it knows all. But with regard to those few which the human intellect does understand, I believe that its knowledge equals the Divine in objective certainty, for here it succeeds in understanding necessity, beyond which there can be no greater sureness."[87] The Baroque architect Francesco Borromini—echoing his contemporaries Kepler and Galileo—designed churches as microcosms of the divine geometry of the cosmos, with complex terrestrial patterns below and simple heavenly forms above (plates 1-77, 1-78, and 1-79).[88]

According to Newton, who was a member of the Church of England, cosmic events happen within an ultimate frame of reference—Absolute time and Absolute space—which Newton identified with the eternity and omnipresence of God: "The most perfect system of the sun, planets, and comets could only proceed from the counsel and dominion of an

And in his hand
He took the golden
Compasses, prepar'd
In Gods eternal store,
to circumscribe
This Universe, and all
created things:
One foot he center'd, and
the other turn'd
Round through the vast
profunditie obscure,
And said, thus farr extend,
thus farr thy bounds,
This be thy just Circumference,
O World.

—**John Milton, *Paradise Lost*, 1667**

Geometry . . . is coeternal with God, and by shining forth in the divine mind supplied patterns to God . . . for the furnishing of the world, so that it could become best and most beautiful and above all most like to the Creator.

—**Johannes Kepler, *Harmonices Mundi* (Harmony of the world), 1619**

RIGHT

1-77. Cutaway interior perspective section and plan of Francesco Borromini's Sant'Ivo alla Sapienza, Rome, engraving by Domenico Barrière in *Opera del Cavaliere Francesco Borromino* (Rome: Giannini, 1720), vol. 1, plate 8. The Pierpont Morgan Library, New York.

Built to fit into a small courtyard off a busy street in Rome, the lower part of this symmetrical church is comprised of an array of niches set within pilasters that are linked together by a stringcourse, creating intricate patterns that reflect the complexity of urban life.

BELOW

1-78. Diagram of the plan and dome of Borromini's Sant'Ivo.

The underlying geometry of Borromini's architecture is not obvious in its final design, but—as with the sacred geometry of the cosmos—the mathematical structure has to be discovered. The convex, concave, and straight lines that form the plan and dome of Sant'Ivo are produced by an equilateral triangle with circles at the three apexes and the centers of the three sides.

OPPOSITE

1-79. Dome of Francesco Borromini's Sant'Ivo alla Sapienza, Rome, 1642–60.

Standing in the center of the small sanctuary, the viewer looks up into simple geometric forms covered with stars and filled with light. Atop the six windows are cherubim (two-winged angels who excel in knowledge) and above, seraphim (six-winged angels who excel in love) encircle the base of the lantern. The burst of golden rays in the center of the lantern is a divine symbol of the Holy Spirit and also a reminder that radiant celestial light—the sun—is at the center of Kepler and Galileo's cosmos.

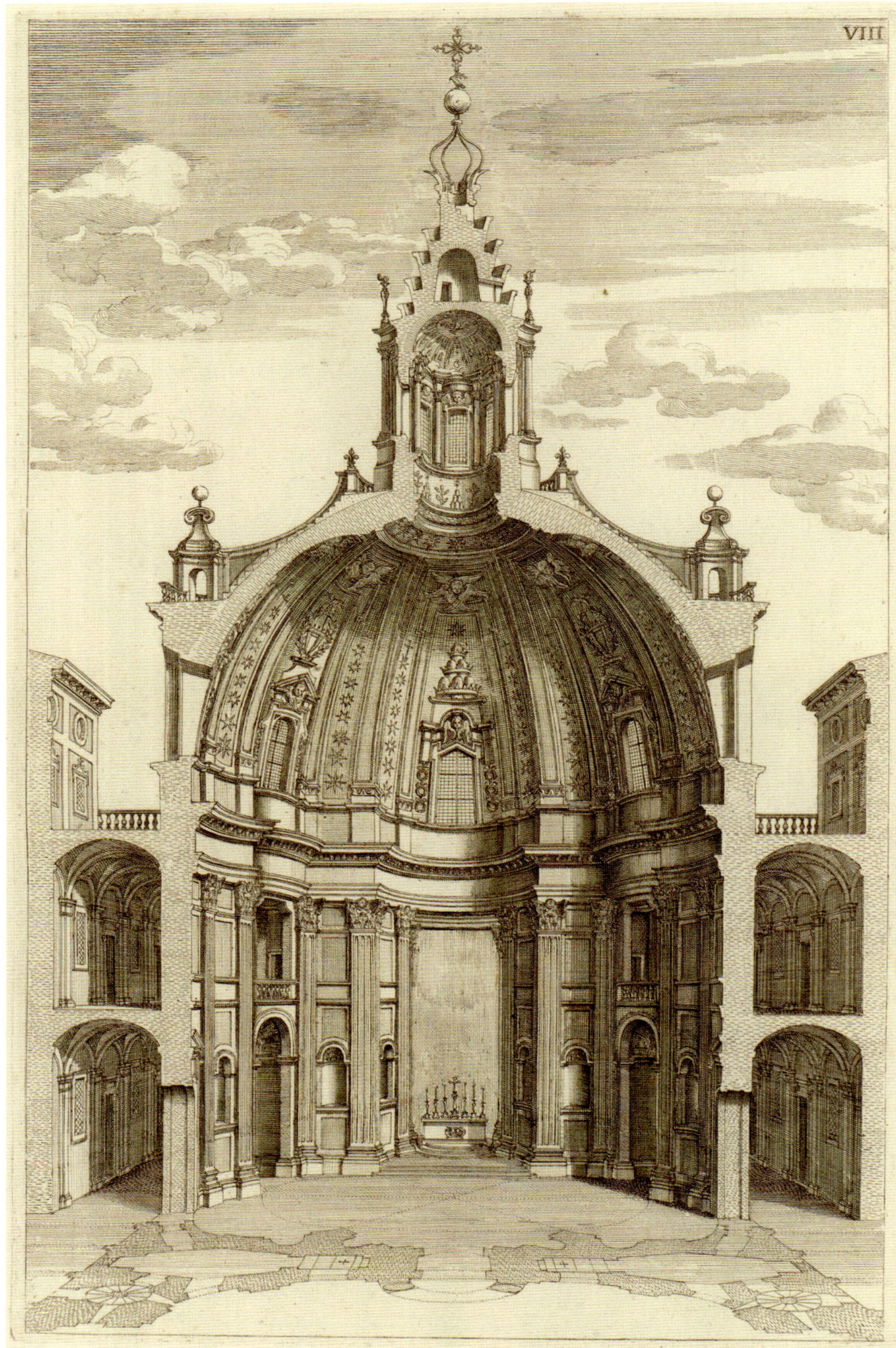

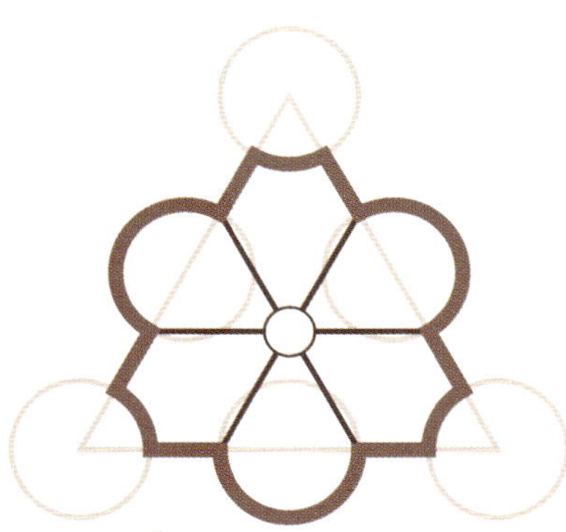

intelligent and powerful Being. . . . The Supreme God is a Being eternal, infinite, absolutely perfect."[89] Newton's goal was to describe that divine mathematical structure, but his discovery of the law of universal gravitation set the stage for the rapid development of the first secular, scientific culture in human history and the separation of mathematics from its traditional association with the divine. After Darwin's publication of *On the Origin of Species* (1859) and *The Descent of Man* (1871), scientists began calling into question

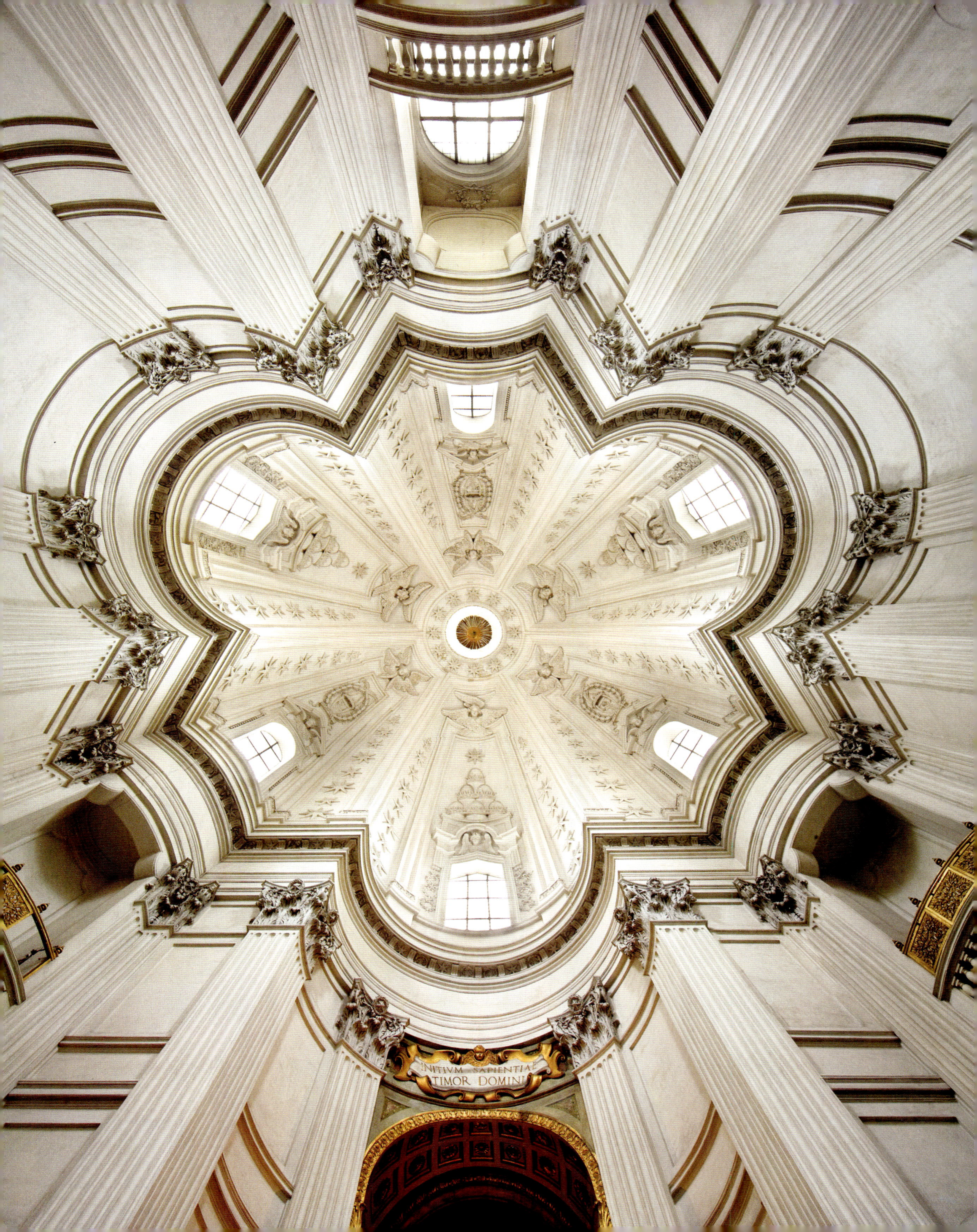
INITIVM SAPIENTIAE
TIMOR DOMINI

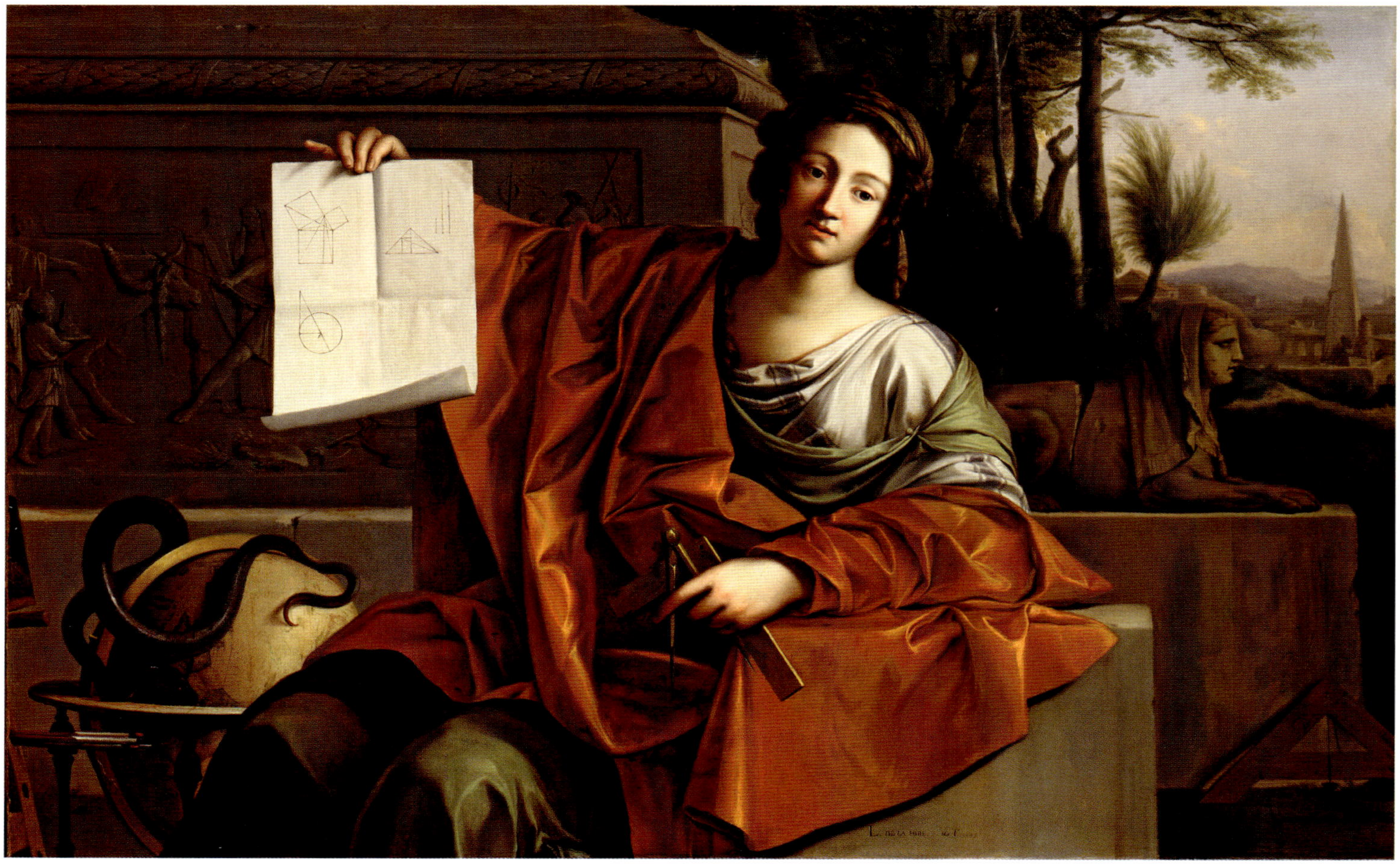

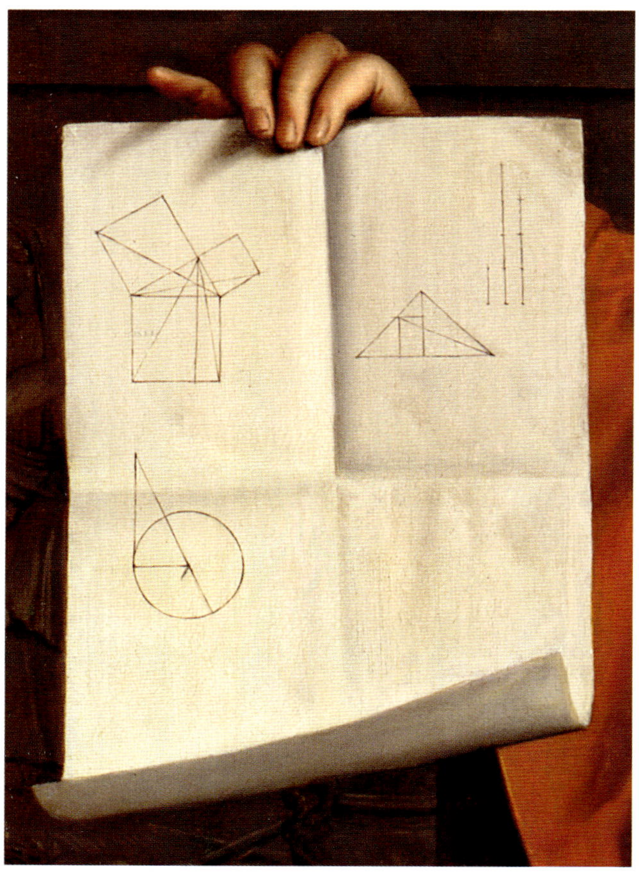

1-80. Follower of Laurent de la Hyre, *Allegory of Geometry*, after 1649. Oil on canvas 40 × 62½ in. (101.6 × 158.6 cm). Toledo Museum of Art, purchased with funds from the Libbey Endowment, gift of Edward Drummond Libbey, 1964.124.

This personification of Geometry, wearing classical attire and holding a (right-angled) straightedge and a compass, signals the dawn of the Enlightenment in the mid-seventeenth century. She presents a paper unfolded to reveal three figures from Euclid's *Elements*: book 1, proposition 47 (the Pythagorean theorem; upper left), book 2, proposition 9 (upper right), and book 3, proposition 36 (lower left). The plumb line leaning against the base of the Egyptian-style sphinx in the background suggests the (practical) Egyptian background of (abstract) Greek geometry. The globe, with its lines of longitude and latitude, suggests the application of geometry to mapping and navigation during the seventeenth- and eighteenth-century age of discovery. The symbolism of the snake on the globe is unknown.

the existence of Newton's Absolute time and Absolute space, as well as a Supreme Being (see chapter 3). As mathematics slowly separated from religion in the nineteenth and twentieth centuries, it nevertheless retained its privileged place in secular society because it gives certain knowledge of abstract objects—numbers, circles, and spheres. Nothing has more profoundly shaped human culture than mankind's cumulative knowledge of the interplay between pure mathematics and the structure of the natural world, which underlies all science and technology.

In addition to changes in religious belief, the rise of science is also linked historically with the rise of democracy and with secularism—the political doctrine that morality and justice should be based only on the well-being of mankind in the present life, without regard for considerations based on the belief in a god or in a life after death. After being invented in Athens in the late sixth century BC, the Hellenistic Greeks continued democratic forms of government, which the Romans adopted for their Republican Constitution that outlined the separation of powers, checks, and balances. But democracy in the ancient world ended in 44 BC when the Roman Senate appointed Julius Caesar as perpetual dictator, and, after his assassination, Rome became an Empire led by its first hereditary emperor, Augustus, the adopted son of Julius Caesar. After the fall of Rome, there was rule by a powerful few—kings, popes, emperors, and dictators—until Enlightenment thinkers focused the spotlight again on human reason, undermining monarchies and reviving democratic ideals (plates 1-80 and 1-81). The English, American, and French revolutions were the political starting points for the rise of modern mathematics, science, and secularism: as democratic reforms swept the West in the aftermath of 1642, 1776, and 1789—and as they sweep North Africa and the East today—the educated public became wary of authority and declared that there are no preordained natural laws. The truths of mathematics are established by reasoned arguments and proofs that can be verified by anyone who cares to follow the details, and the truths of nature are determined by experiments that can be observed by all people equally. Democratic revolutions—in Boston, Paris, Cairo, and Beijing—are also the spiritual starting point for modern culture because they have inspired the passionate quest for freedom and individualism that lies at the core of modern views of the human condition.

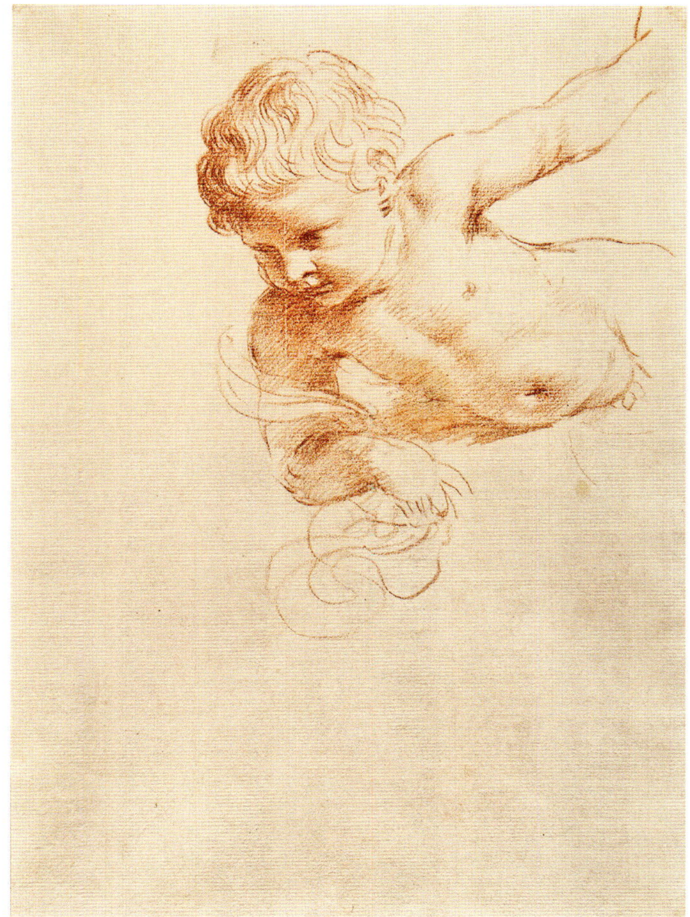

1-81. Giovanni Francesco Barbieri, called Guercino (Italian, 1591–1666), drawing of an angel for *The Assumption of the Virgin*, 1650. Red chalk on paper, 12 × 8⅔ in. (30.5 × 22.2 cm). Private collection.

Drawn by the Italian Baroque artist Guercino, this cherub is poised between the sacred and secular. Swooping down from a supernatural realm, he brings angelic tidings, but this little boy is also of this world—wingless, with shaggy hair and a chubby torso.

Today people with a secular outlook no longer hold a Platonic view of *religion* (they do not believe that nature is guided by divine Reason), but most are Platonists in the sense that they hold a classical view of *science* (they believe that nature embodies mathematical patterns) and of *mathematics* (they regard mathematical objects as eternal and perfect). In the modern era the natural human desire remains strong to connect mentally—via intuition—with the transcendental realm of the mathematical-world-out-there, in the sense of knowing numbers and geometric forms and giving them meaning. Today mathematics continues to inspire great art because abstract objects are, as Plato said, "eternally and absolutely beautiful."

Common notion: Things which are equal to the same thing are also equal to one other.

—**Euclid, *Elements*, ca. 300 BC**

We hold these truths to be self-evident, that all men are created equal.

—**Thomas Jefferson, *The Declaration of Independence*, July 4, 1776**

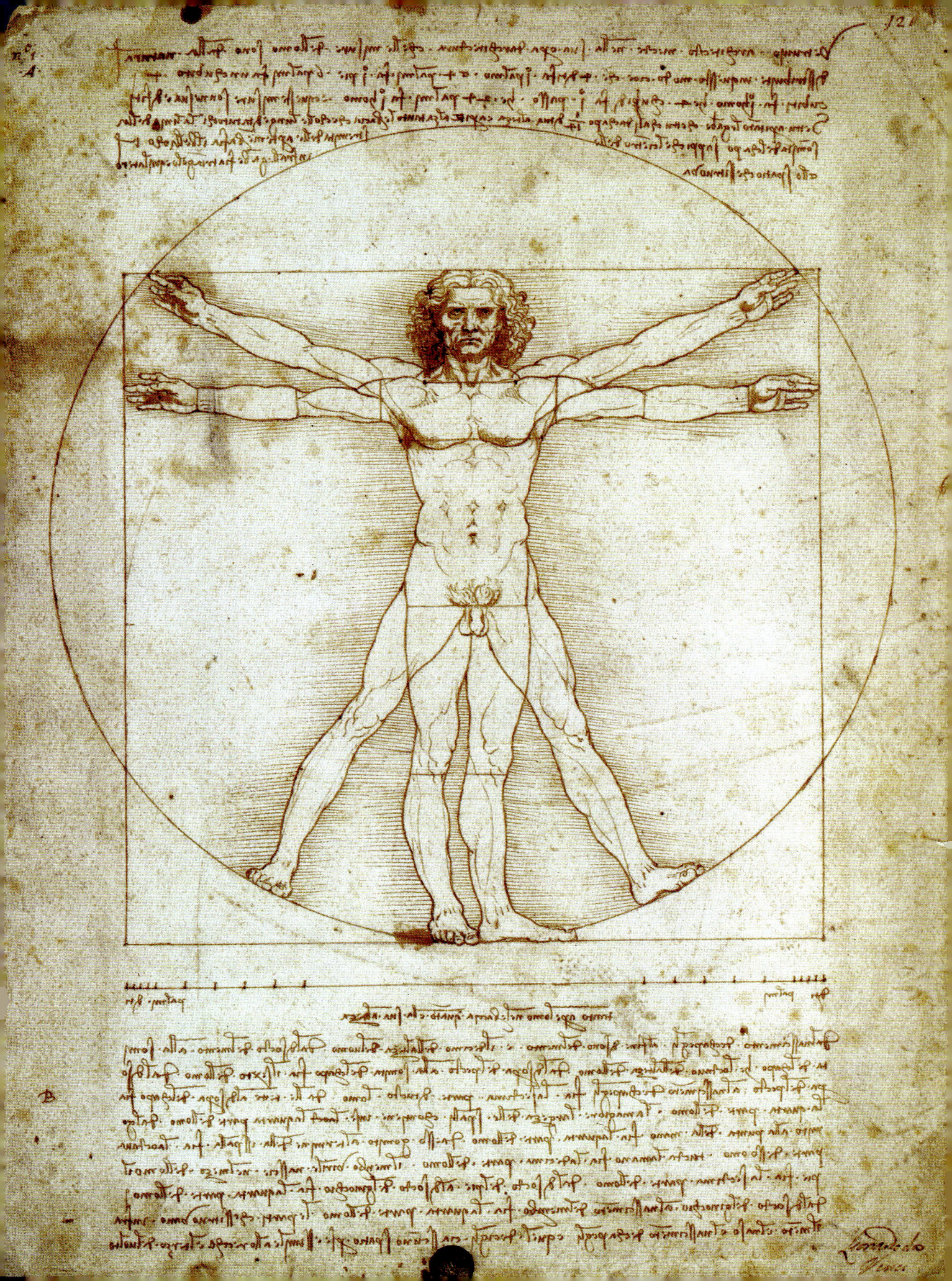

2
Proportion

Our proportion cannot be determined by a number that one can know, nor expressed by any rational quantity, but is always mysterious and secret, and qualified by mathematicians as irrational.

—Luca Pacioli, *Divine Proportion*, 1509

Cells and tissue, shell and bone, leaf and flower, are so many portions of matter, and it is their obedience to the laws of physics that their particles have been moved, molded and conformed. Their problems of form are in the first instance mathematical problems.

—D'Arcy Wentworth Thompson, *On Growth and Form*, 1917

In classical times the Pythagoreans and Plato described the universe as having an elegant geometric order. When they discovered the numerical basis of musical harmony, they concluded it to be a reflection of cosmic harmony, and they established the classical ideal of beauty as the harmonic relation of parts within a whole. Ancient and Renaissance systems of proportion were based, in part, on measurements of the human body, whose harmonious parts were seen as an immutable embodiment of beauty in the image of its divine Creator.

In the *Elements* Euclid showed how to construct many irrational ratios that cannot be expressed as a fraction of whole numbers. The Renaissance friar Fra Luca Pacioli singled out one of Euclid's irrational ratios—the division of a line into extreme and mean ratio—and proclaimed that it was a metaphor for the Almighty because divine nature is irrational in the sense of being beyond the understanding of rational mortals. Thus Pacioli associated Euclid's ratio with theology. But neither Pacioli nor the ancients associated the ratio with art or beauty.

That association was not made until the early 1800s, when German mathematicians first referred to Euclid's division (sectioning) of the line as "golden"; adopted from them and popularized by Adolf Zeising's *New System of Human Proportions* (1854), the term became central to the (false) historical claim that ancient, medieval, and Renaissance artists and architects had used the ratio to determine ideal proportions. Popularized in the era of German Romanticism when the rise of Enlightenment science was seen as endangering mankind's privileged position as

2-1. Leonardo da Vinci (Italian, 1452–1519), *Vitruvian Man*, 1492. Pen and ink with wash over metalpoint, 13½ × 10 in. (34.4 × 25.5 cm). Accademia di Belle Arti, Venice.

Leonardo illustrates Vitruvius's text on ideal proportions: "The center and midpoint of the human body is, naturally, the navel. For if a person is imagined lying back with outstretched arms and feet within a circle whose center is at the navel, the fingers and toes will trace the circumference of this circle as they move around. But to whatever extent a circular scheme may be present in the body, a square design may also be discerned there. For if we measure from the soles of the feet to the crown of the head, and this measurement is compared with that of the outstretched hands, one discovers that this breadth equals the height, just as in areas which have been squared off by use of the set square." Vitruvius, *On Architecture* (first century BC), III.i.3, in *Vitruvius, Ten Book on Architecture*, trans. Ingrid D. Rowland (Cambridge: Cambridge University Press, 1999), 47. To illustrate that the body can be inscribed in both a circle and a square, Leonardo created this composite image, which required him to drop the square below the circle so that their centers do not line up. In the text above the drawing, Leonardo listed Vitruvius's calculations of bodily proportion, and, below, he added a few of his own measures, marked in faint lines on the man's head, arms, and legs. (The text is in Leonardo's mirrored, right-to-left handwriting.) Below the drawing, Leonardo divided a line into units: at both ends he marked off six palms (*palmi*) and the outermost palms he divided further into four fingers (*diti*).

embodying divinity, the "golden mean" offered the public an irresistible idea: one simple ratio—with Euclid's pedigree—placed man at the center of creation with perfect proportions that unify the whole cosmos, nature, and art.

Less than a decade after Zeising's *New System*, Charles Darwin moved *Homo sapiens* from this pedestal and placed him on the tree of life with proportions that change over time. By 1900 Art Nouveau architects and designers had ceased looking for one ideal human proportion that was forever fixed, but instead sought organic patterns in plants and animals that result from a long process of trial and error—evolution by natural selection.

As the scientific worldview took hold and ushered in an era of doubt about traditional Western Judeo-Christian-Islamic theology, the nostalgic concept of a so-called Golden Ratio took a tenacious hold on the imagination of scholars in Germanic academia. In the late nineteenth century influential art historians, including Jacob Burckhardt and Heinrich Wölfflin, followed Zeising's lead and claimed (falsely) that artists and architects used the Golden Section to design great monuments of art history, including the pyramids in Egypt and classical architecture of the Italian Renaissance and Baroque. By 1900 the ratio had been renamed "phi" on the (mistaken) belief that the Greek sculptor Phidias had used it to design the sculpture for the Parthenon in Athens. It was under this guise that German architects in the circle of Peter Behrens, Walter Gropius, and the Bauhaus school of design, wrote the Golden Ratio into textbooks on modern architecture, where it remains a misconception that continues to be taught to this day.

2-2. Albrecht Dürer (German, 1471–1528), man from *Vier Bücher von menschlicher Proportion* (Four books on human proportion; Nuremberg, 1528). Trustees of the National Library of Scotland, Edinburgh.

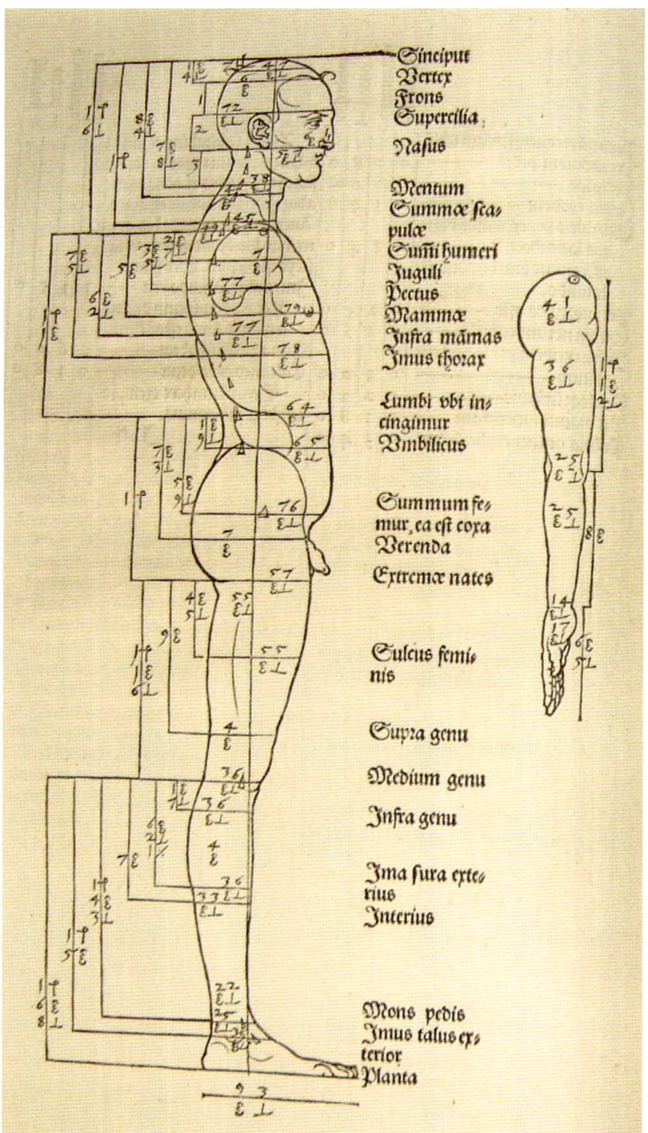

PROPORTION IN CLASSICAL ART

The classical Greek sculptor Polykleitos (fifth century BC) described ideal male proportions in a treatise that is lost, but a marble copy of the bronze statue he made to illustrate his ideal survives (plate 1-17 in chapter 1). Like Polykleitos, the Roman architect Vitruvius (first century BC) saw the human body as a paradigm for the harmonious relation of parts to a whole. He described the relation of the face, hand, and head to height: "For nature composed the human body in such a way that the face, from the chin to the top of the forehead and the lowermost roots of the hairline should be one-tenth [of the total height of the body], the palm of the hand from the wrist to the tip of the middle finger should measure likewise, the head from the chin to the crown one-eighth" (*On Architecture*, first century BC; see plate 2-1, chapter frontispiece).[1] This simple rule—a human is eight heads high—was applied in workshops throughout the Middle Ages.

In the Early Renaissance, the Italian architect Leon Battista Alberti challenged Vitruvius's use of the head by making the foot the relevant unit of measure; an ideal man, he said, was six feet tall (*On*

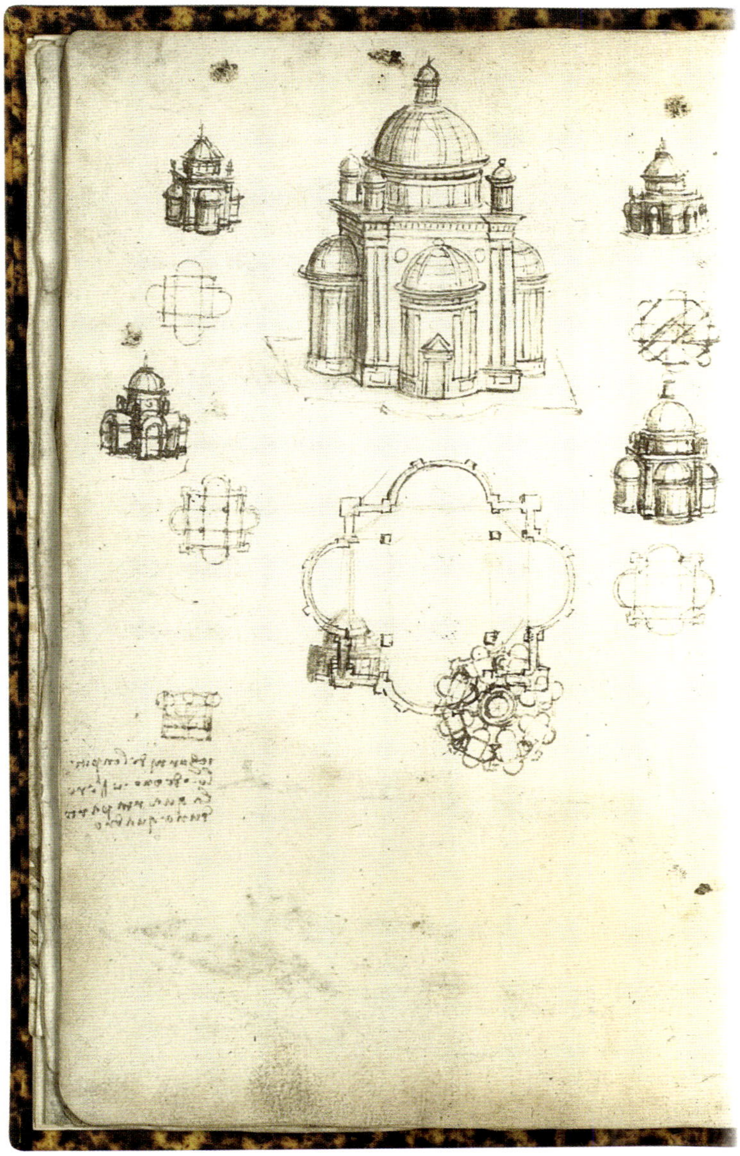

LEFT

2-3. Leonardo da Vinci (Italian, 1452–1519), designs of churches, 1487–90. Codex Ashburnham (complement to MS B), fol. 93v. Bibliothèque de l'Institut de France, Paris.

Leonardo da Vinci designed many symmetrical churches, none of which were built. In 1975 the American mathematician George E. Martin proved that, given the vocabulary available to an Italian Renaissance architect, Leonardo had systematically determined all the possible ways of adding chapels to preserve the symmetry of the plan; George E. Martin, "Leonardo's Theorem," in *The Foundations of Geometry and the Non-Euclidean Plane* (New York: Springer, 1975), 386–92.

BELOW

2-4. Constructions of polygons within a circle, from Leon Battista Alberti, *De Re Aedificatoria* (On the art of building; Florence: Nicolaus Laurentii, 1485), 207, 208. Art and Architecture Collection, Miriam and Ira D. Wallach Division of Art, Prints, and Photographs. The New York Public Library, Astor, Lenox and Tilden Foundations.

Alberti completed his manuscript for *De Re Aedificatoria* in 1452, just when Johannes Gutenberg was printing the first Bible on a press with movable type in Mainz, Germany. Gutenberg's technology spread rapidly to Florence, where this first edition of Alberti's book was printed with these woodcut illustrations in 1485.

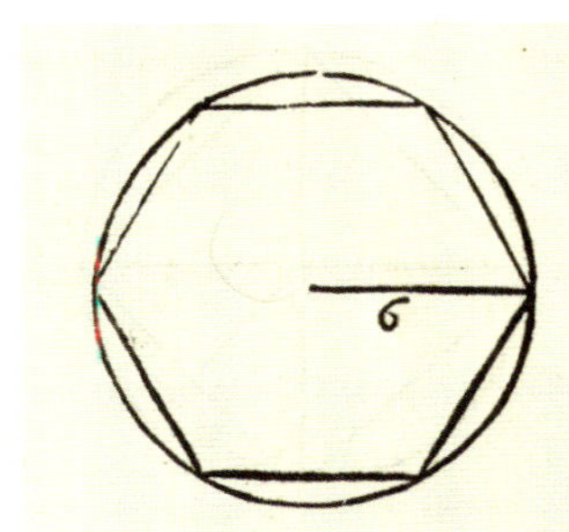

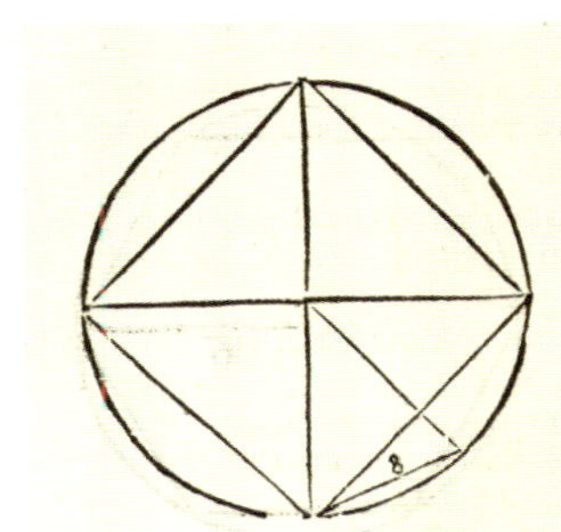

Sculpture, ca. 1464). During the High Renaissance, the German Albrecht Dürer adopted Alberti's foot as his unit after laboriously measuring more than two hundred bodies to confirm the 1 : 6 ratio, which remained standard until the nineteenth century (*Four Books on Human Proportions*, 1528; see plate 2-2).

Alberti especially favored buildings with a symmetrical plan—either circular or with four equal arms—because Vitruvius had declared that the ideal human body could be inscribed in a circle. The circle, Alberti said, was the most perfect shape, although other polygons that could be inscribed in a circle were acceptable as floor plans (plate 2-4). Leonardo da Vinci did drawings for churches as semi-regular polygons (plate 2-3), and Donato Bramante's original design for the most important architectural commission of the High Renaissance, the rebuilding of Saint Peter's basilica in Rome, had a symmetrical plan (plate 2-6).

Beyond stating that architecture should reflect ideal human proportions, Vitruvius did not specify exactly *what* the proportions of a building should be. Renaissance architects measured ancient monuments in the hopes of retrieving the original schemas, but their

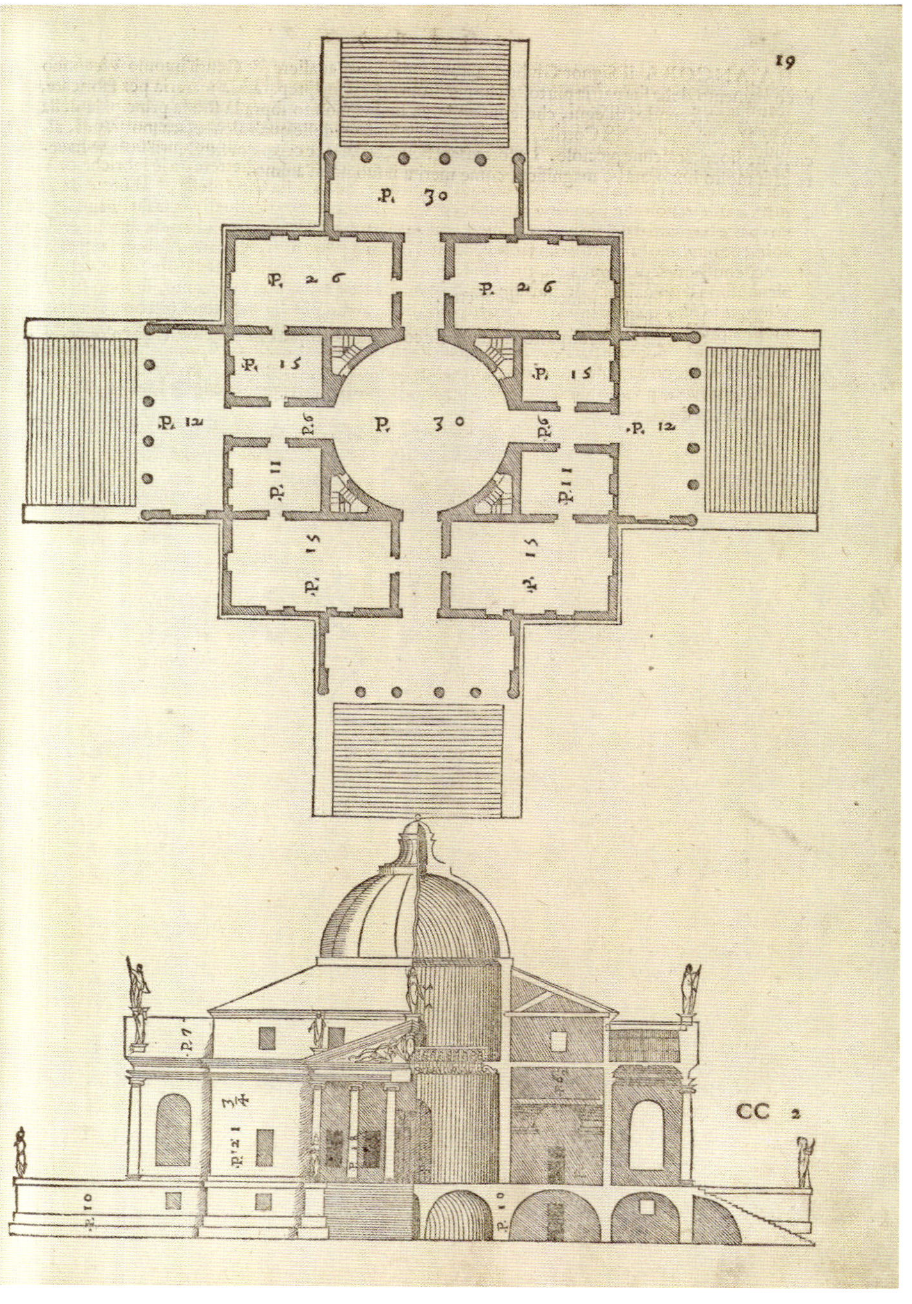

RIGHT

2-5. Andrea Palladio (Italian, 1508–1580), plan and section of Villa Capra "La Rotonda," Vicenza, Italy, begun 1566, completed in 1585 after Palladio's death by Vincenzo Scamozzi, in Palladio's *I Quattro Libri dell'Architettura* (The four books of architecture; Venice: Dominico de' Franceschi, 1570), bk. 2, 19. Spencer Collection, The New York Public Library, Astor, Lenox, and Tilden Foundations.

Alberti's most influential follower, Andrea Palladio, designed stone and brick buildings with symmetrical plans and façades, including villas, palaces, public buildings, and churches. His classical style was widely imitated until the invention of steel in the nineteenth century caused a decline in masonry architecture.

BELOW

2-6. Donato Bramante (Italian, ca. 1444–1514), *Plan for Saint Peter's Cathedral*, Rome, designed 1505, begun 1506.

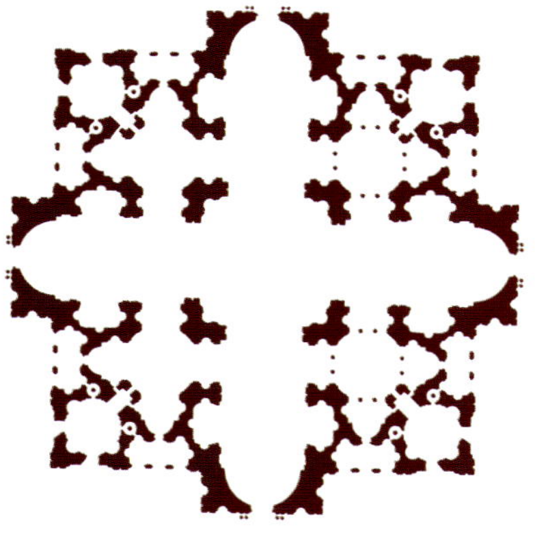

results were inconclusive. Renaissance designers debated how accurate the proportions of buildings should be: should they use only ratios that come out even so they can be measured exactly, the so-called musical ratio (such as 1 : 2, an octave)? Or was it acceptable to introduce incommensurable ratios that can only be estimated, such as the relation of the diameter of a circle to its circumference (the irrational number pi)? Both approaches to ratios were well known to Early Renaissance mathematicians. Alberti proposed a modular approach, in which a building would be designed as an aggregate of whole units. His system, used throughout the Italian Renaissance, became the basis of classical revivals into the nineteenth century (plate 2-5).

LINEAR PERSPECTIVE

An Early Renaissance architect, Filippo Brunelleschi, invented linear perspective as a tool to draw three-dimensional buildings on a two-dimensional surface while accounting for apparent distortion relative to a viewpoint. Alberti codified Brunelleschi's discovery in his treatise *De Pictura* (On painting, 1435), where he showed that lines from the viewer's eye to an object form a visual pyramid, which is intersected by a vertical "picture plane" that can be imagined as being like a piece of glass. The Renaissance theory was based on an eleventh-century treatise on optics by the Islamic mathematician Ibn al-Haytham (Alhazen to the Latin-speaking world), who was the first to correctly model vision as the eye's passive reception of light-rays emanated from glowing sources (fires, stars) or reflected from objects (apples, the moon; plate 2-7). Brunelleschi's contribution was to locate the viewer's eye in three-dimensional space. When he was a young apprentice, horizon lines and vanishing points were already used to account for the position of the viewer's eye up-down and left-right. It was Brunelleschi's invention of a way to measure the rate of diminution in the third dimension—as the viewer moves forward and back—that created the revolution. Early Renaissance artists no longer painted saints floating in golden mist in a faraway place; linear perspective gave them the tool to depict Jesus and the apostles existing right here, right now before their eyes in the natural world (plate 2-8 and the sidebar on pages 78–79).

Piero della Francesca was known in his day as a mathematician, writing the treatises *De prospectiva pingendi* (On perspective for painting), *Trattato d'abaco*

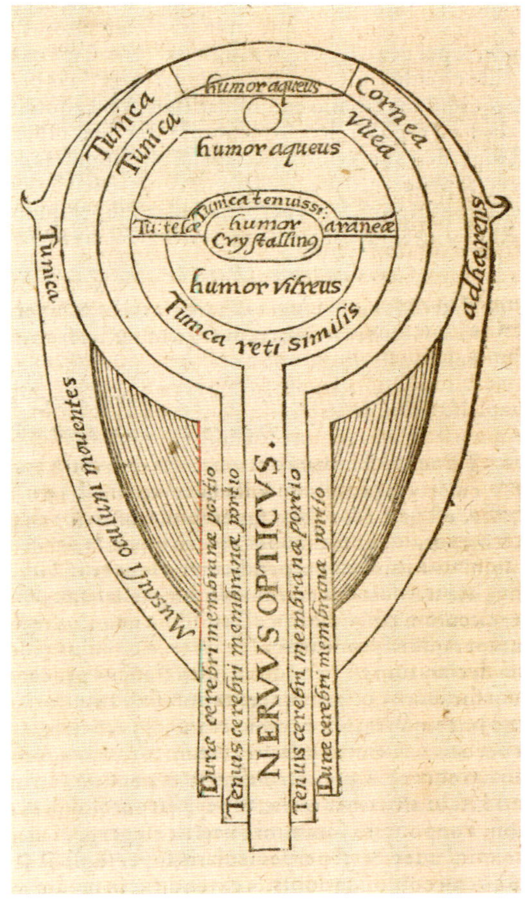

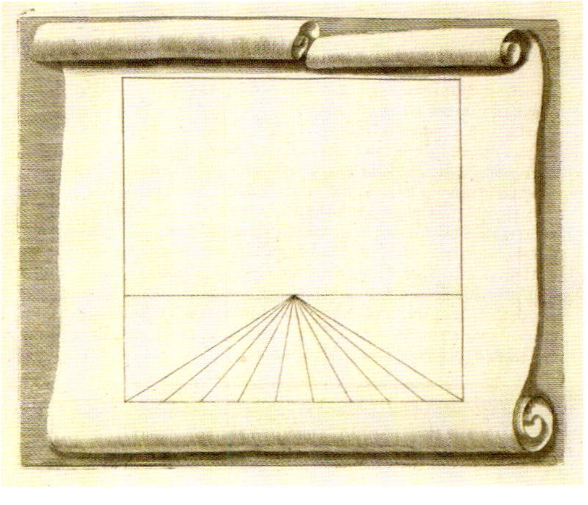

TOP RIGHT

2-7. Anatomy of the eye in Ibn al-Haytham (latinized as Alhazen), *Kitāb al-Manāẓir* (early eleventh century), translated into Latin as *Opticæ Thesaurus: Alhazeni Arabis Libri Septem*, ed. Friedrich Risner (Basel, Switzerland: Episcopios, 1572), 1:6. Rare Books Division, The New York Public Library, Astor, Lenox, and Tilden Foundations.

Ibn al-Hatham declared that vision is caused by light entering the eye through the cornea, pupil, iris, lens, and, after passing through the aqueous humor, striking the retina, which is connected to the optic nerve. In addition to his work on vision, he described *camera obscura* phenomena and rainbows. His seven-volume *Kitāb al-Manāẓir* was translated into Latin as *Opticæ Thesaurus* (Book of optics) by around 1200, and copies of the manuscript were widely circulated in Italy in the early fifteenth century. For the twenty extant copies, see D. C. Lindberg, *A Catalogue of Renaissance and Medieval Optical Manuscripts* (Toronto: Pontifical Institute of Medieval Studies, 1975), 17–18. This first printed edition of Ibn al-Hatham's treatise was compiled by the sixteenth-century German mathematician Friedrich Risner.

BOTTOM RIGHT

2-8. Leon Battista Alberti's *De Pictura* (On painting; 1435), Italian translation by Cosimo Bartoli as *Della Pittura* (Paris: Giacomo Langlois, 1651), 17. Art and Architecture Collection, Miriam and Ira D. Wallach Division of Art, Prints, and Photographs. The New York Public Library, Astor, Lenox, and Tilden Foundations.

Alberti's wrote his 1435 studio manual by hand and did not illustrate it. This diagram, which was added by an anonymous illustrator to an early (1651) printed edition, shows that receding parallel lines meet at a vanishing point on the horizon.

How to Draw a Foreshortened Grid

In what (since the nineteenth century) is called a "thought experiment," Alberti told the artist to imagine looking at an object (such as a cube) with one eye closed. The lines of sight emanating from the open eye create a so-called visual pyramid. A transparent "picture plane" intersects the visual pyramid (lower left), resulting in a projected image of the cube. If the viewer moves, the picture changes because the distortion of the cube is determined by the viewer's position. Thus to draw objects correctly, the artist must consider the position of the viewer in three-dimensional space (up-down, left-right, forward-backward).

A Up-down When standing at the seashore, a person sees where the earth meets the sky at the horizon. Because the viewer is standing on a sphere (the earth), the horizon is always at the viewer's eye level; if the viewer moves up, the viewer sees farther and the horizon appears higher, and if the viewer moves down, the horizon is lower. In a painting the up-down position is indicated by the horizon, which is drawn at the eye level of the viewer (who is outlined in grey).

Left-right Imagine we are drawing a floor with a grid pattern of 2 × 2-ft. square tiles. First we draw the right and left edges of the tiles. As the tiles get farther away, they appear smaller. If the floor is level, the parallel lines of the grid will converge at a point on the horizon (the "vanishing point"). This point is directly in front of the viewer's eye; if the viewer moves left, the point moves left, if the viewer moves right, so does the vanishing point. In this example, the artist has placed the vanishing point for a viewer standing dead center.

B and C Forward-back We know that the tiles get smaller as they get recede into the background, but how much smaller? Walking around on a tile floor, Brunelleschi noticed that the *rate* at which the square tiles get smaller changes as the viewer comes closer or backs away. If the viewer is close, the tiles diminish rapidly from large to small; if the viewer steps back, the rate of diminution is more gradual. To calculate the rate in a painting (shown on the right in B), Brunelleschi instructed the artist to extend the baseline of the painting to one side, and mark it off in 2-ft. units, like the gridded floor. Then imagine turning the picture plane 90° and taking a bird's-eye view of the (painting's) illusionistic space and the (viewer's) real space, as shown in C. Imagine that there is a viewer, who is seen in profile in B and from above in C. In profile the viewer stands 6 ft. tall and is looking at the tile floor. Draw sight-lines from the viewer's eye to the edge of first tile (2 ft. away), the second (4 ft. away), the third (6 ft. away), and so on.

Now we are ready to determine a rate of diminution. We first decide how far back the viewer is standing from the picture plane. If the viewer is standing back 8 ft., then we draw a line (the picture plane) intersecting the visual pyramid at 8 ft. (as shown in B). This is tantamount to moving the picture plane to the 8-ft. mark (shown in C). This intersection of the visual pyramid gives us a rate of diminution. We extend horizontal dotted lines into the painting to complete the tile floor. If we move the picture plane closer to the viewer—to, say, 4 ft.—the rate of diminution will increase dramatically; move it farther away, and the rate will decline.

D Using linear perspective, artists can correctly foreshorten any shape, such as the patterned tile floor, by first gridding it and then projecting the grid. They can also calculate the correct size of any object in the illusionistic space, such as the two men. If artists know the object's real size (its measurement at the picture plane, i.e., at the boundary between real and illusionistic space), then they can calculate its size in the illusionistic space by using properly foreshortened units. This exposition of Brunelleschi and Alberti's method is based on Alberti's verbal description of linear perspective in his (unillustrated) studio manual; see Leon Battista Alberti, *De Pictura* (On painting; 1435), trans. John R. Spencer (New Haven, CT: Yale University Press, 1956), especially 57–58 and the translator's notes on 110–17.

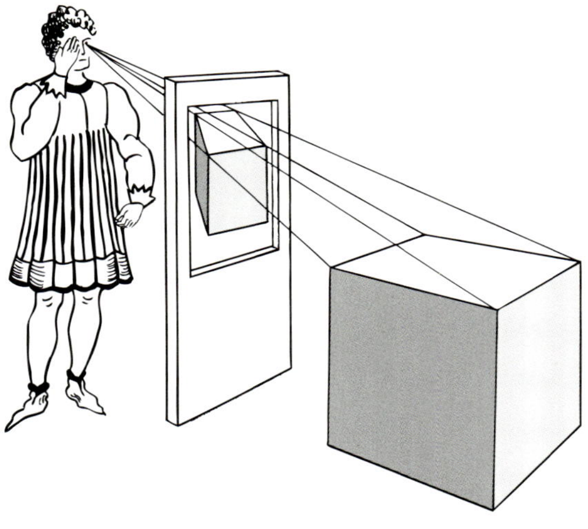

A

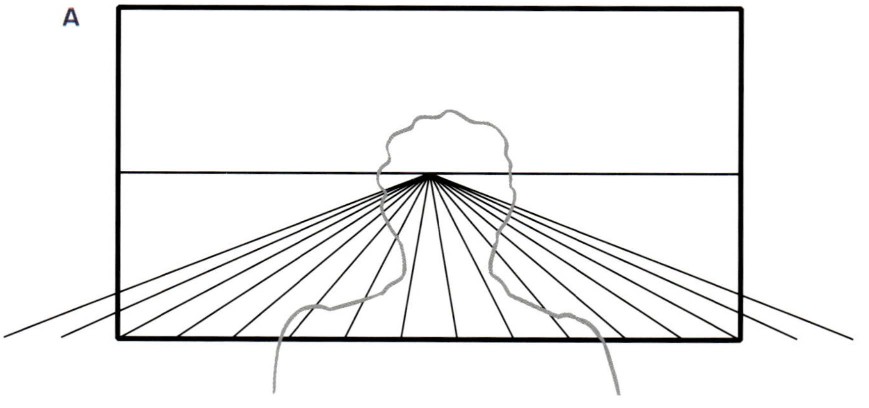

B

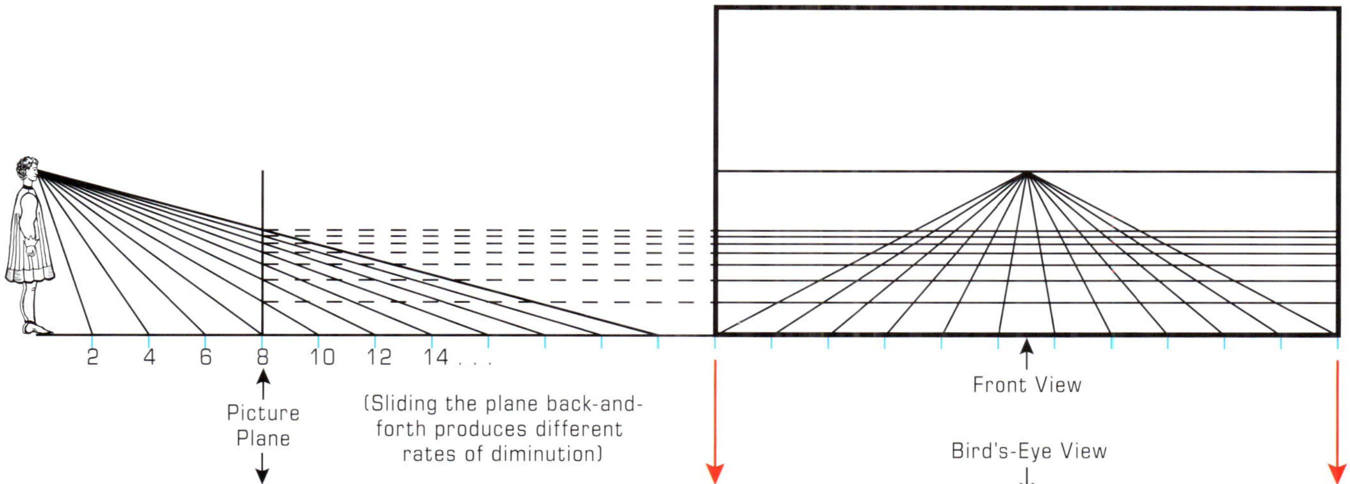

C

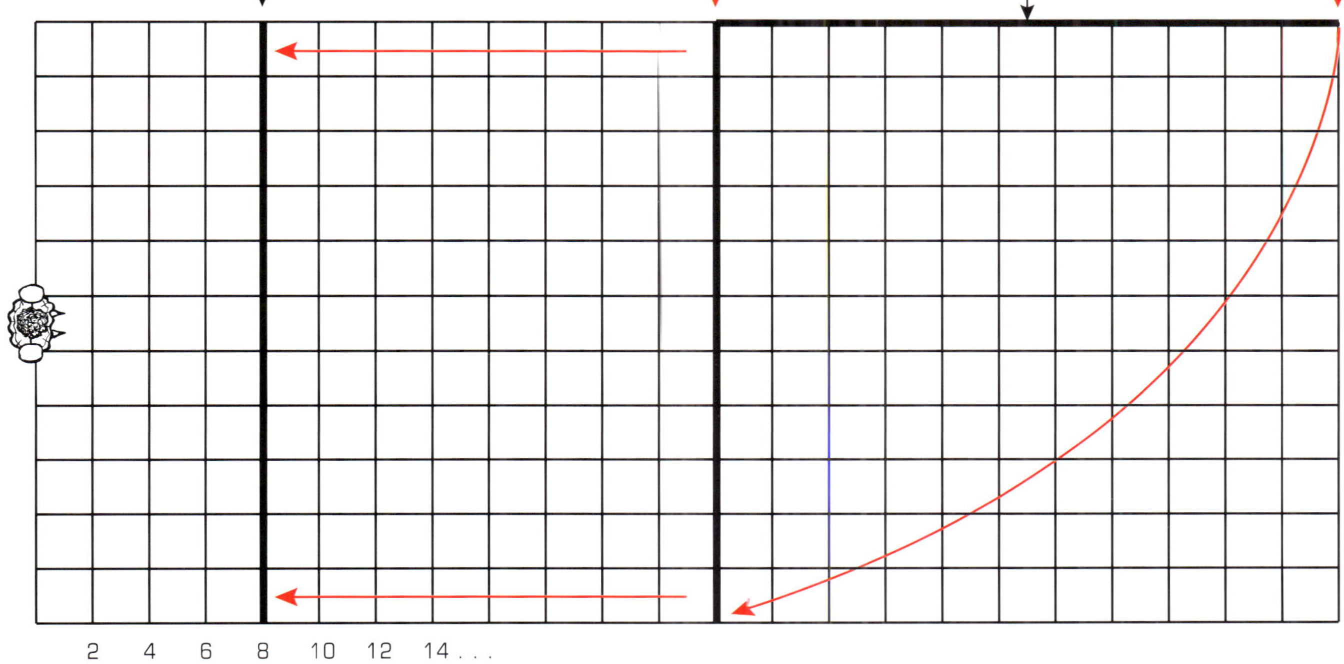

D

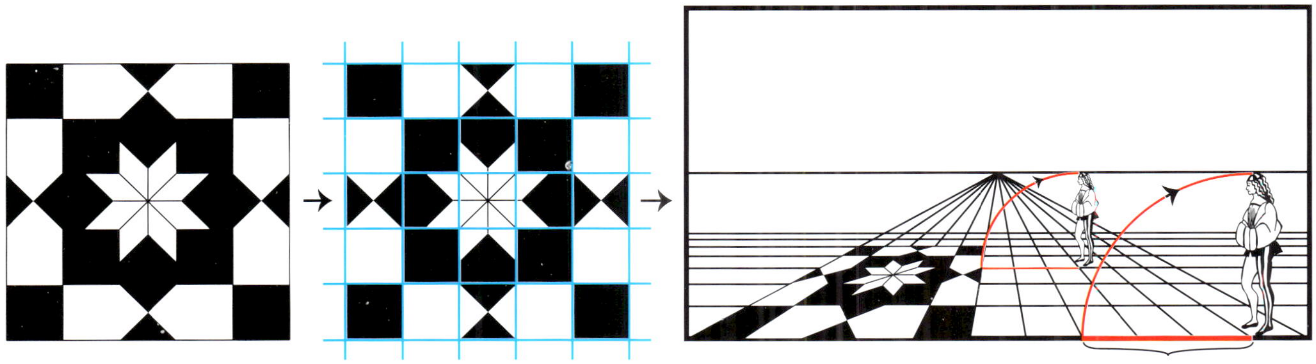

Pattern for a tiled floor

Gridded pattern

Projection of pattern

This grid is drawn in a 2-foot unit so three squares equals the height of a 6-foot man

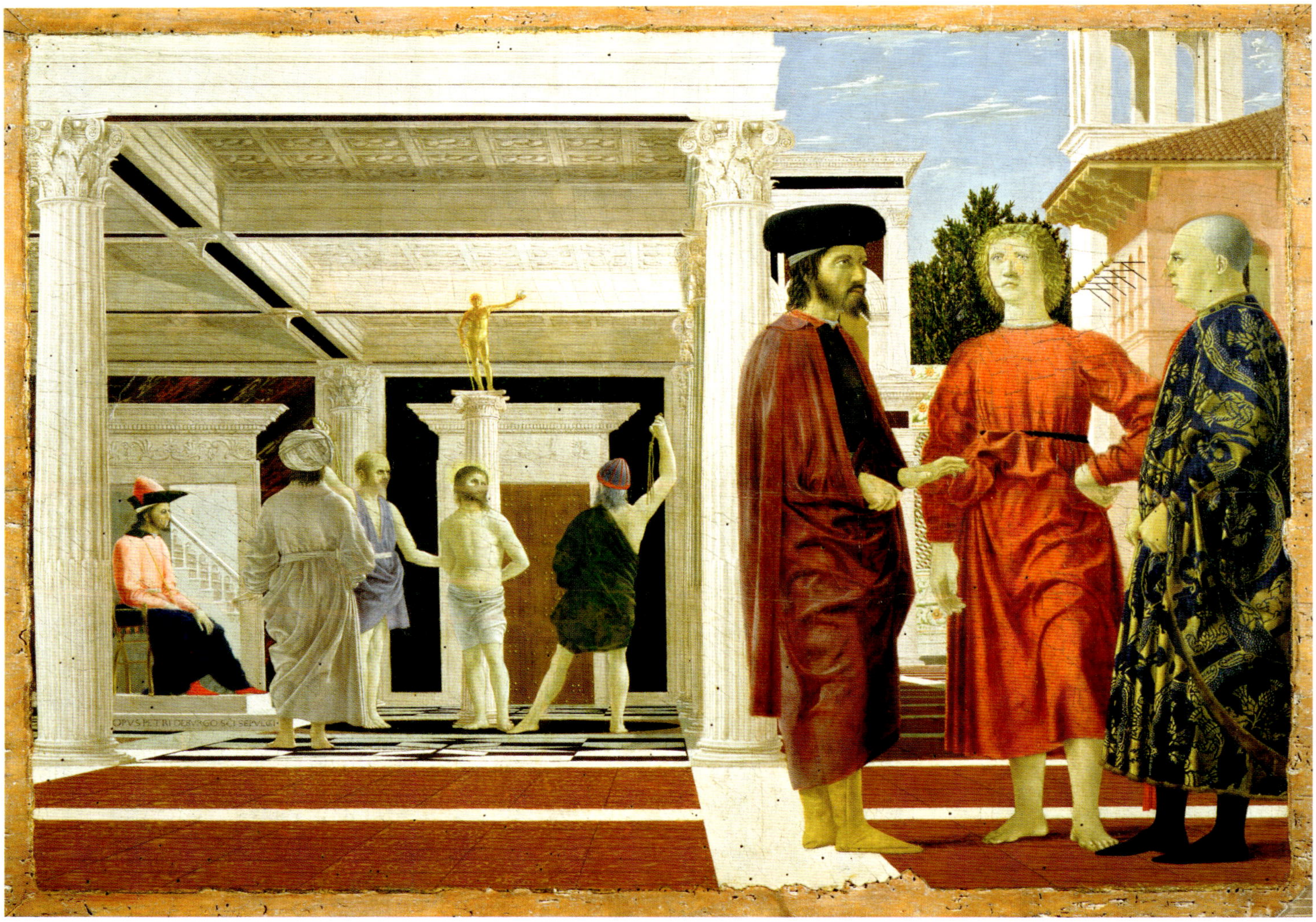

2-9. Piero della Francesca (Italian, ca. 1415–92), *The Flagellation of Christ*, ca. 1455–60. Oil and tempera on panel, 23 × 32⅛ in. (58.4 × 81.5 cm). Galleria Nazionale delle Marche, Urbino.

Piero's perspective skills are manifest in the tile floor, the coffered ceiling, the staircase in the left background, and the buildings in the right rear. As a result of Piero's precise calculation of diminution, the heads of the eight figures are exactly to scale. The reasoned stillness of the scene contrasts with its violent subject matter—Christ being flogged on the eve of his crucifixion. A seated Roman official witnesses the torture, which does not draw the attention of the three standing figures on the right, whose identities are unknown. In 1953 Rudolf Wittkower and B.A.R. Carter did a detailed reconstruction of the architectural space of the painting, from which they concluded that "Piero used perspective in this picture for portraying his three-dimensional design with mathematical accuracy, and furthermore, this design is infused with mathematical symbolism." They are referring to the squaring of the circle (the ancient challenge of constructing a square with the same area as a given circle; Christ stands on a circle within a square). For other examples of mathematical puzzles used symbolically in the mystical tradition of Piero's contemporary Nicolas of Cusa, whose treatises Piero read, see Wittkower and Carter, "The Perspective of Piero della Francesca's *Flagellation*," *Journal of the Warburg and Courtauld Institutes* 16 (1953): 292–302.

(Abacus treatise), and *De quinque corporibus regularibus* (On the five regular solids). *De prospectiva pingendi* was a studio manual on perspective with meticulous instructions for drawing flat patterns (such as tile floors), simple solids (such as a house), and complex solids (such as a human head). Written around 1480, Piero's treatise was the first workshop manual on perspective of its kind, and it circulated widely in studios throughout the sixteenth century. Piero's paintings, such as *The Flagellation of Christ* (ca. 1455–1460; see plate 2-9), display exact perspective calculations for very many elements and a strong sense of order. His *Trattato d'abaco* was an elementary textbook, probably written for a friend or patron, which contained mainly arithmetic along with some algebra and geometry (plate 2-11).[2] Piero dedicated his treatise *De quinque corporibus regularibus* to the Duke of Urbino, Guidobaldo I, who assumed the title in 1482, so the work must date to the last decade of Piero's life. Concerned only with geometry, the treatise contains several original problems that require constructing a sphere with a volume equal to a given solid.

In keeping with the classical preference for symmetrical design, Alberti suggested placing the vanishing point in the center of a painting to achieve left-right, up-down symmetry, as Leonardo da Vinci did in *The Last Supper* (plate 2-10; see also the sidebar on page 82). In less than a century Italian linear perspective spread first to northern Europe and then throughout the West (plate 2-12). The German artist Hans Holbein showed his skill with linear projections by creating a picture, in which part can be seen correctly only when viewed from an extreme slant, an example of so-called anamorphosis (plate 2-14). The diffusion of linear perspective was so complete that by 1600 any European artist of ambition was measured by his application of the technique. Today artists around the world continue to use Renaissance linear perspective to create fantastic illusions of space (plates 2-16 and 2-18), sometimes using mirrors to heighten the illusion (plates 2-15 and 2-17).

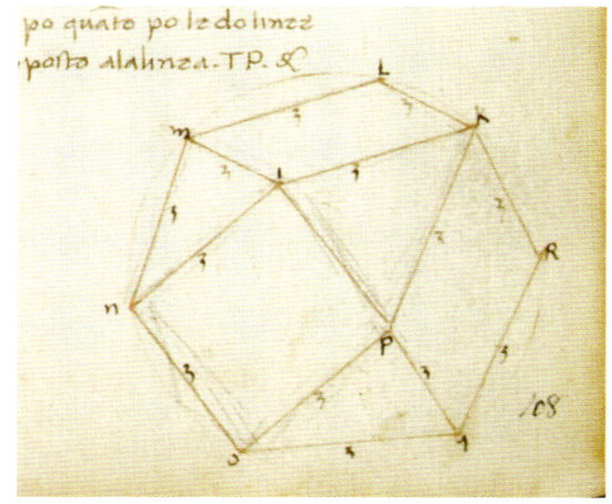

ABOVE

2-11. Piero della Francesca (Italian, ca. 1415–92), *Trattato d'abaco* (Abacus treatise), mid-fifteenth century. Biblioteca Medicea Laurenziana, Florence, MS Ashburnham 359, 108r.

In this manuscript Piero described four polyhedra, which are credited to Archimedes based on a text by Pappus of Alexandria (fifth century AD), who recorded that Archimedes described the three-dimensional solids by listing the number of three-, four-, and five-sided faces of which they are composed. In this drawing Piero took the additional step of *visualizing* one of Archimedes's semi-regular solids (a cuboctahedron, composed of 8 triangles and 6 squares) by illustrating the arrangement of the faces around the corners of the solid.

BELOW

2-10. Leonardo da Vinci (Italian, 1452–1519), *The Last Supper*, 1494–98. Tempera and oil on plaster, 15 ft. 1 in. × 28 ft. 10 in. (4.6 × 8.8 m). Santa Maria delle Grazie, Milan.

When artists began using linear perspective, they were able to draw figures to scale within architectural settings, which created the problem of how to keep a cavernous architectural background from distracting the viewer from the narrative unfolding in the foreground. Leonardo's solution was to create a deep space with plain dark rectangles (suggesting hung tapestries) on the right and left walls, and to put the vanishing point of their receding lines on Jesus, focusing the viewer's attention on the foreground drama. Leonardo also used the light-filled background windows to frame Jesus and Peter, who leans forward, shaking his finger. Thus when looking into the depths of the landscape, the viewer's eye meets the silhouetted actors and returns to the foreground, where Jesus stuns his disciples (provoking Peter's defiant gesture) when he says, "One of you will betray me."

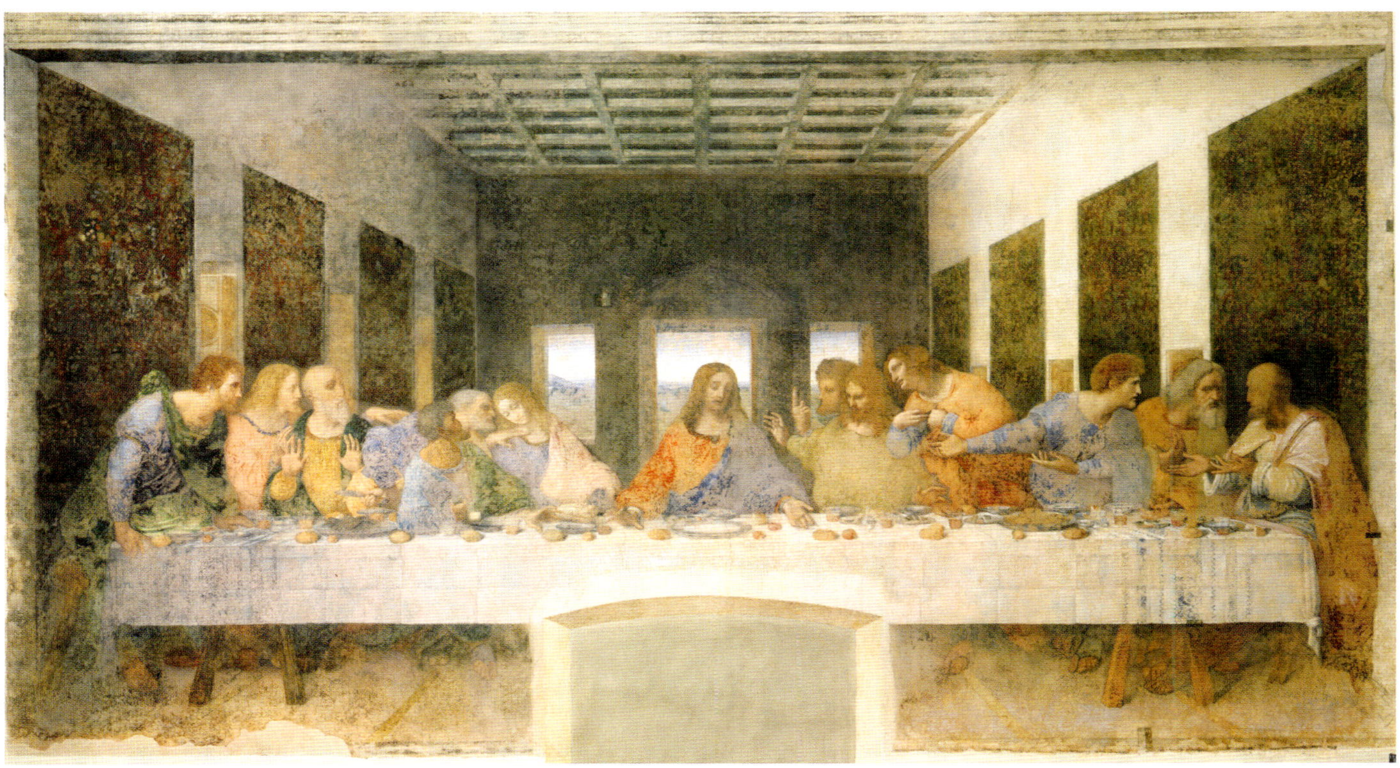

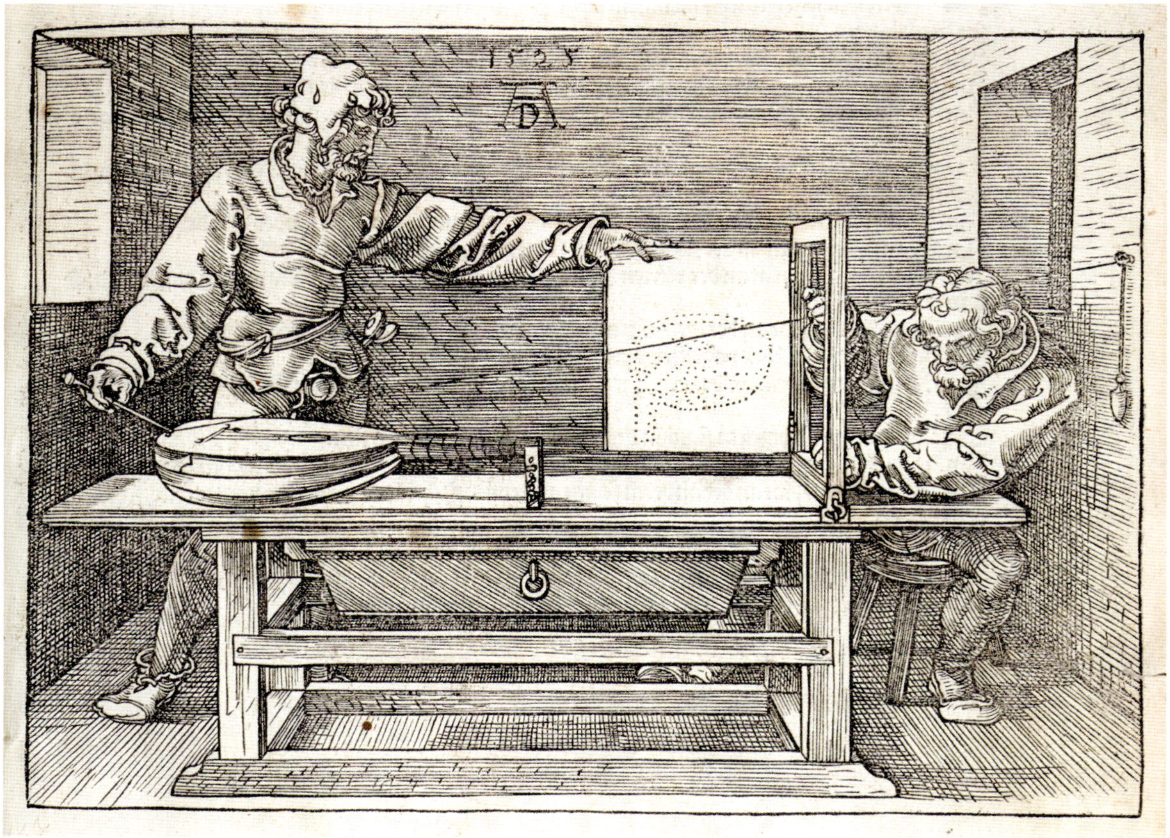

To make clear my exposition in writing this brief commentary on painting, I will take first from the mathematicians those things with which my subject is concerned. When they are understood, I will enlarge on the art of painting.

—**Leon Battista Alberti,** ***De Pictura*****, 1435**

ABOVE

2-12. Albrecht Dürer (German, 1471–1528), demonstration of how to draw a lute in linear perspective, *Underweysung der Messung mit dem Zirckel und Richtscheyt* (Treatise on measurement with compass and straightedge; Nüremberg, 1525).

Dürer learned the principles of Renaissance perspective during his visits to Italy in 1494–95 and 1505–7. Returning home to Nuremberg after his second visit, Dürer published this manual, in which he shows how to draw a lute. First establish the viewpoint of the picture by placing a hook in the right wall, from which a string (sight-line) is pulled through a transparent grid to points on the lute. After each point on the lute is determined, the standing man lets the taut string go slack and swings the picture shut, allowing the seated man to mark that point on a gridded picture plane. Repeat this procedure to form an outline of the lute, which will be correctly foreshortened relative to the given viewpoint.

Critique of Linear Perspective

After completing the *Last Supper*, Leonardo abandoned linear perspective because he had discovered an inconsistency in Brunelleschi and Alberti's theory. The picture plane is flat, but if the viewer's visual field is defined as all points equidistant from a viewpoint (the eye), then the picture plane should be spherical. To contrast a flat and a spherical picture plane, Leonardo drew this bird's-eye-view diagram of three columns (a, b, and c) being looked at by a viewer from one point (h). The viewer's sight-lines are intersected by two picture planes, one flat (d–e) and one curved (f–g). If the picture plane is flat, per Brunelleschi and Alberti, then the columns will (incorrectly) appear larger (wider) as they get farther away. If the field is curved, per Leonardo, then the columns (correctly) appear to get smaller (narrower) as they recede in space. See Carlo Pedretti, "Leonardo on Curvilinear Perspective," *Bibliothèque d'humanisme et Renaissance* 25 (1963): 69–87.

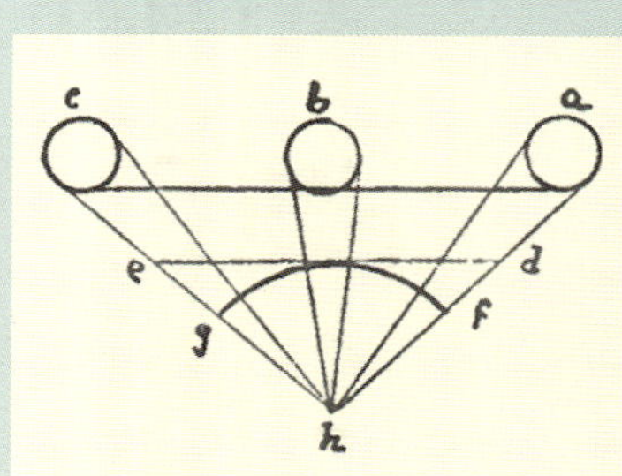

2-13. Leonardo da Vinci (Italian, 1452–1519), diagram of perspective, 1513–14, as copied by Jean Paul Richter in 1881. In his publication of the original Italian text and English translation on facing pages, the historian Richter redrew Leonardo's letters *a–h* to correct (reverse) their mirror image in the original (Institut de France, Paris, MS E, fol. 16a); *Scritti Letterari di Leonardo da Vinci/The Literary Works of Leonardo da Vinci* (London: Sampson Low, Marston, Searle, and Rivington, 1881), 1:63. University of Toronto Library.

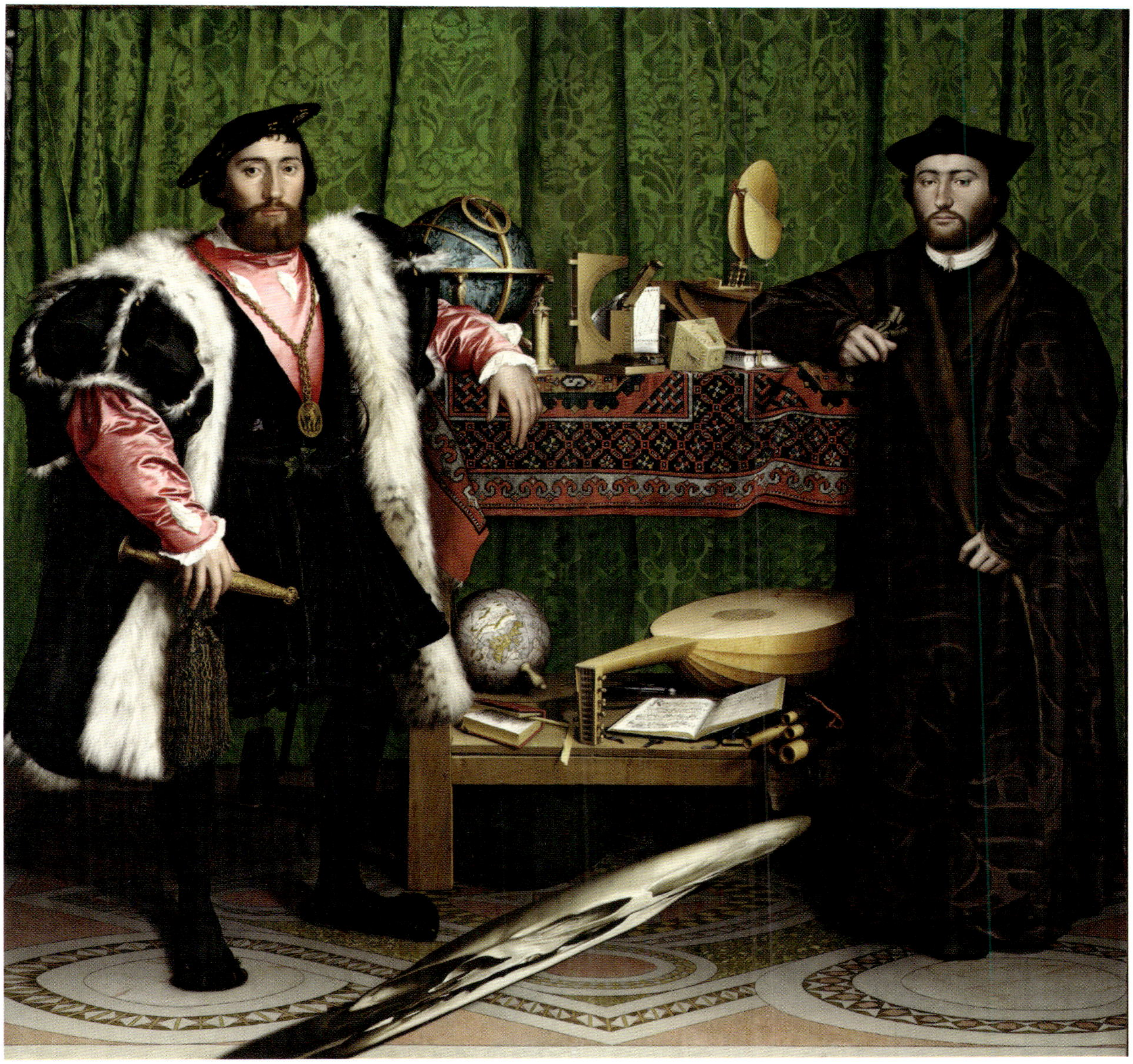

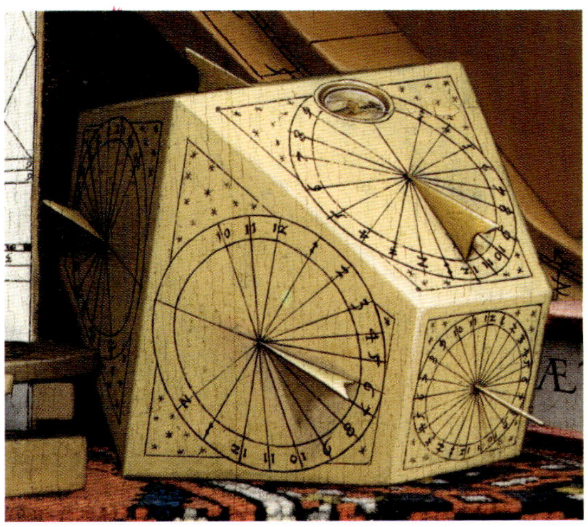

2-14. Hans Holbein (German, ca. 1460–1534), *The Ambassadors*, 1533. Oil on oak panel, 81 × 82½ in. (207 × 209.5 cm). National Gallery, London.

Holbein included an anamorphic painting of a skull in the bottom foreground of this double portrait. To see the skull undistorted, hold the page horizontally at eye level and, keeping one eye closed, view it from the lower-left corner of the painting. Commissioned by the French Dinteville family to hang in their English country estate, Holbein's painting shows Jean de Dinteville, the French ambassador to England, attired in a red silk shirt and an ermine-lined coat, holding a dagger, and his friend Georges de Selve, bishop of Lavaur, in regal damask clerical robes. The painting is in the Northern Renaissance tradition of a *vanitas*, in which the skull is a grim reminder of mortality amidst the transitory pleasures of art and science. On the bottom shelf is a lute and a case of flutes next to an open Protestant Reformation hymnal, a reminder of the contemporary political and religious turmoil caused by the 1533 remarriage of Henry VIII (Holbein's patron), which caused a schism with Rome. To the left, a book on arithmetic is held open with a T-square used as a bookmark, behind which is a terrestrial globe. On the tabletop is (from left to right) a celestial globe and several devices for measuring: a pillar-shaped, pocket-size sundial (a "shepherd's dial"), a wooden universal equinoctial dial with a plumb bob hanging on its left side, a white quadrant used to make angular measurements of celestial bodies, and a multifaceted sundial (see detail). The tall circular object in the back was designed for the Renaissance (Ptolemaic) astronomer to determine the position of a star or planet relative to the ecliptic.

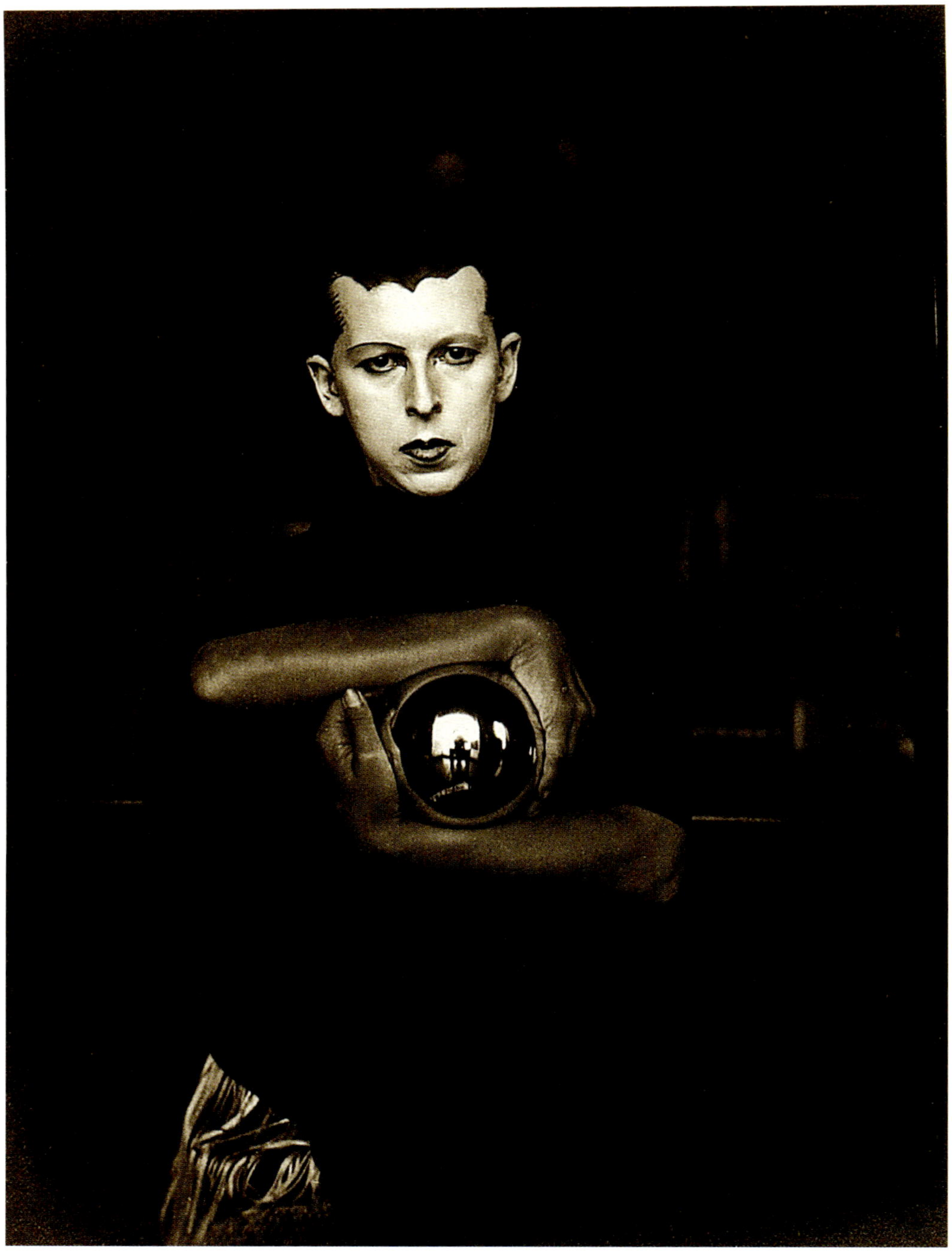

ABOVE

2-15. Claude Cahun (French, 1894–1954), *Self-Portrait*, ca.1927. Gelatin silver print, 4⅝ × 3½ in. (11.7 × 8.8 cm). The Detroit Institute of Arts, Albert and Peggy de Sale Charitable Trust, and the DeRoy Photographic Acquisition Endowment Fund.

Born Lucy Schwob, at age twenty-five the artist adopted the sexually ambiguous name "Claude Cahun" and entered Surrealist circles, photographing herself variously as a dandy, a nymph, or an androgyne, as in this riveting self-portrait with masculine hairstyle and feminine makeup. The viewer looks at Cahun and she stares back, holding up a convex mirror that records a spherical projection of the room. In the mirror the viewer sees a reflection of the camera and—metaphorically—a picture of himself or herself that has been "captured" in Cahun's eye/lens/mirror, reminiscent of Jan Van Eyck's use of a convex mirror in *The Arnolfini Portrait* (1434; National Gallery, London).

OPPOSITE

2-16. Sylvie Donmoyer (French, b. 1959), *Still Life with Magic Square*, 2011. Oil on canvas, 26 × 20 in. (66 × 50.8 cm). Courtesy of the artist.

This contemporary French artist has painted puzzles and geometric objects against the backdrop of two sixteenth-century publications on mathematics, Albrecht Dürer's *Melencolia I* (1514), and Wenzel Jamnitzer's *Perspectiva Corporum Regularium* (Perspective of regular solids; 1568). Dürer's engraving has been interpreted as being about Plato's dialogue on the impossibility of defining Beauty (*Hippias Major,* ca. 380–367 BC). The seated angel holding a compass (personifying Geometry) has failed to construct the Platonic symbol of a perfect universe—a dodecahedron—producing instead the large irregular polygon (on the left), over which Dürer placed a motto expressing Geometry's gloomy attitude towards Plato's (abandoned) search for Beauty: *Melencolia I* (see Patrick Doorly, "Dürer's *Melencolia I:* Plato's Abandoned Search for the Beautiful," *Art Bulletin* 86, no. 2 [2004]: 255–76.) Like Dürer, Jamnitzer lived in Nuremberg, where he was a goldsmith in the court of the German emperor. In 1568 he published this book of geometric patterns that he based on Plato's dialogue on cosmology, *Timaeus* (366–360 BC), and Euclid's *Elements* (ca. 300 BC), illustrating his text with 120 geometric forms based on the five Platonic solids.

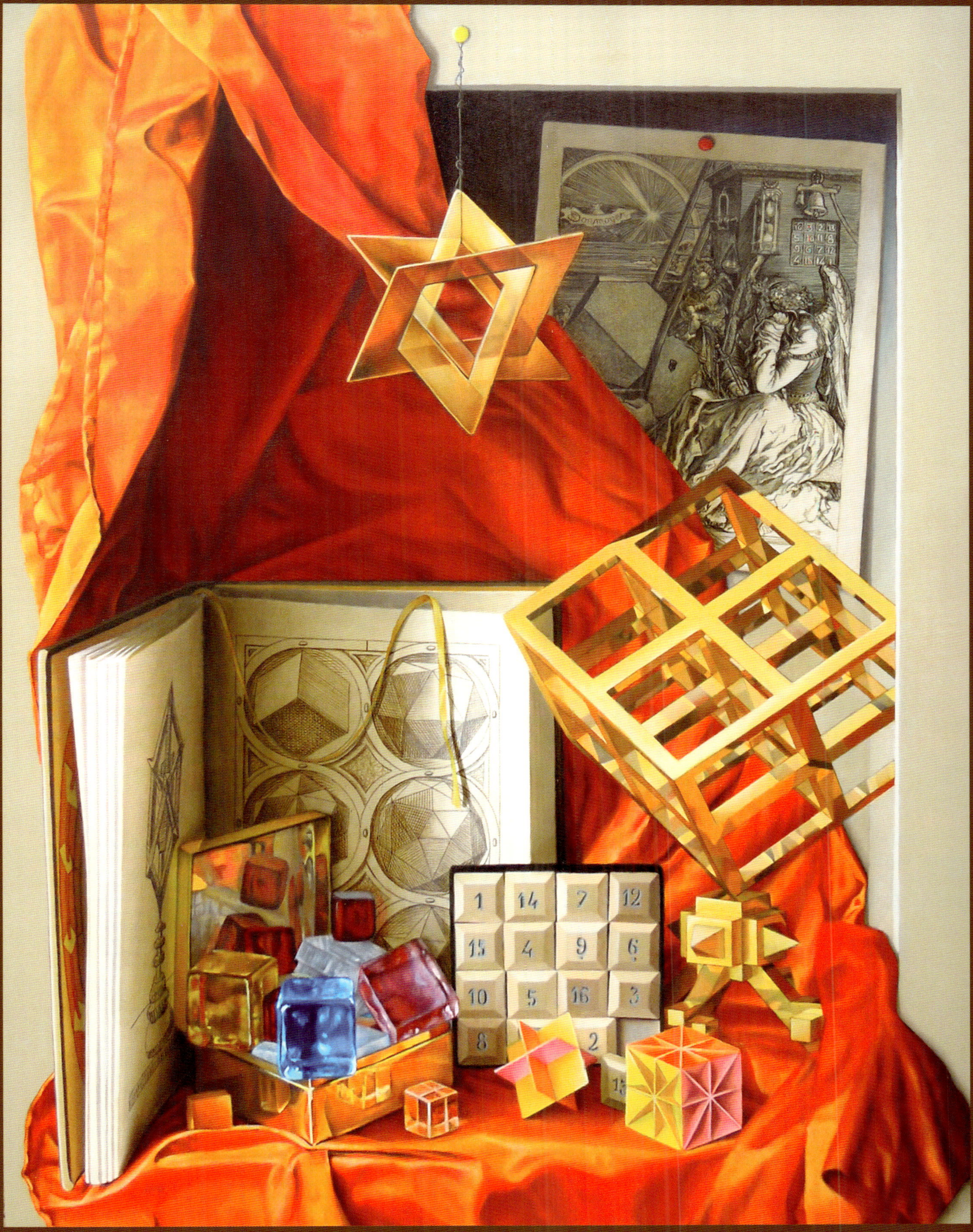
1 14 7 12
15 4 9 6
10 5 16 3
8 2

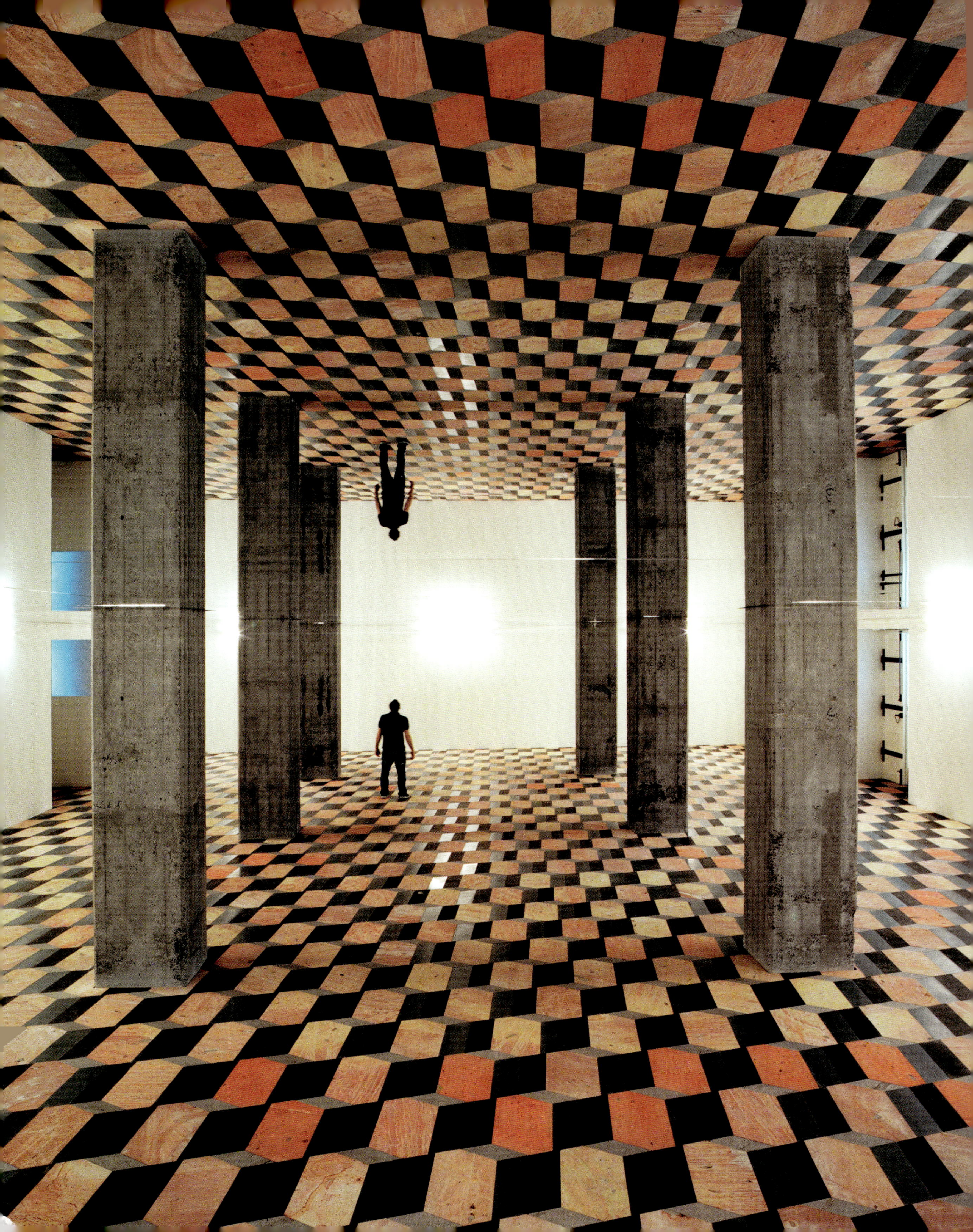

OPPOSITE

2-17. Olafur Eliasson (Danish-born Icelandic, b. 1967), *Frost Activity*, 2004. Mirrored foil, aluminum, limestone, and basalt installation in the Reykjavik Art Museum, Iceland. Courtesy Reykjavik Art Museum.

The contemporary Icelandic artist Olafur Eliasson tiled the floor of the Reykjavik Art Museum with basalt and limestone in the pattern of a reversible geometric figure—a Necker cube—that appears to simultaneously protrude from and intrude into the floor. Drawn in linear perspective such that faces of the cube remain parallel (a so-called parallel projection), the figure is named for the Swiss crystallographer Louis Necker who first published it in 1832, although the identical pattern of reversible cubes has been used to tile floors (such as the floor of the House of the Faun, Pompeii, Italy, 200 BC) since antiquity. Eliasson covered the ceiling with mirrored foil that reflects the floor, engulfing the viewer in the optical illusions.

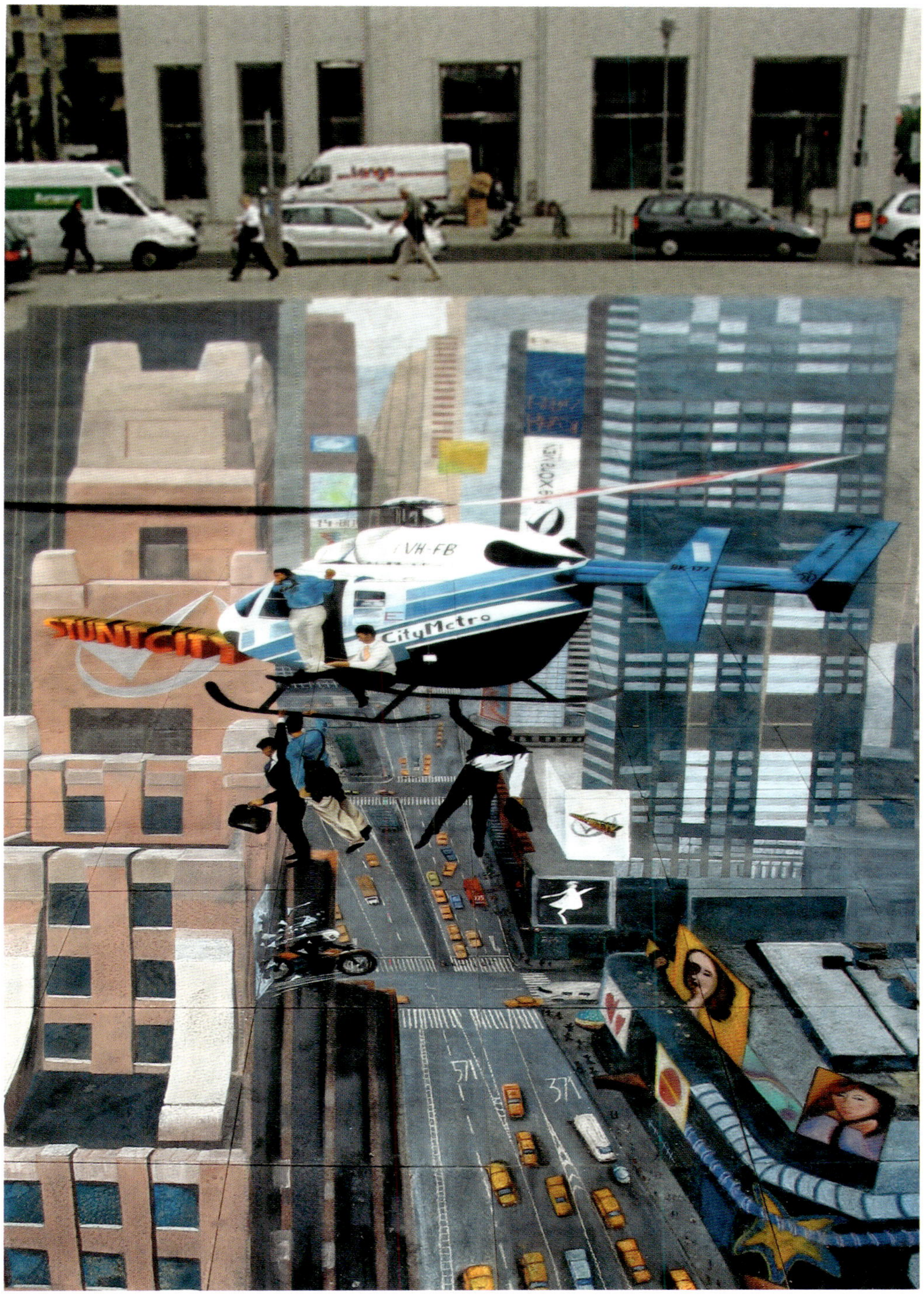

ABOVE

2-18. Edgar Müller (German, b. 1968) and Manfred Stader (German, b. 1958), *Stunt City*, May 2005. Graffiti, chalk on cement, Potsdamer Platz, Berlin, done after an advertisement created by the agency DLKW Lowe, London. Courtesy of the artists.

Drawing with colored chalk on cement, the German graffiti artists Edgar Müller and Manfred Stader created this illusion that the bottom has dropped out of Berlin's Potsdamer Platz. The viewer sees yellow cabs in the street below and brave stunt men, briefcases in hand, hitching a ride to work on an urban helicopter.

THE DIVINE PROPORTION

In the *Elements*, Euclid drew a line and divided it in such a way that the ratio of the larger to the smaller segment is equal to the ratio of the entire line to the larger segment.[3] Mathematically, the ratio can be calculated as a constant, which is approximately 1.618 (see the sidebar below). Euclid assigned no particular significance to it, describing it as an "extreme and mean ratio." During the Renaissance mathematicians developed a new body of *philosophical* literature devoted to the ratio, although their use of the ratio in the *practice* of mathematics was unchanged.

In the thirteenth century the Italian mathematician and astronomer Johannes Campanus did a revision of a twelfth-century Latin translation from Arabic of Euclid's *Elements*. In his commentary on one of the theorems, Campanus made the first recorded philosophical association to the extreme and mean ratio: "Admirable therefore is the power of a line divided according to the ratio with a mean and two extremes; since very many things worthy of admiration of philosophers are in harmony with it, this principle or maxim proceeds from the invariable nature of superior principles, so that it can rationally unite solids that are so diverse, first in magnitude, then in the number of bases, then too in shape, in a certain irrational symphony."[4] Campanus is referring to the ratio's use in the construction of certain geometric figures, including the pentagon and the dodecahedron. Because these diverse geometric objects share this irrational ratio (their constructions have it in common),

OPPOSITE

2-19. The opening page of the first (AD 1482) printed edition of Euclid's *Elements* (300 BC), published as *Elementa Geometriae*, which was based on a twelfth-century Latin translation of an Arabic version of the Greek text (probably the one done by Adelard of Bath) that was revised in 1260 by Johannes Campanus of Novara. Rare Books Division, The New York Public Library, Astor, Lenox, and Tilden Foundations.

In the 1440s Johannes Gutenberg invented a printing press that used movable type, and printers faced the problem of how to reproduce mathematical diagrams. The Bavarian printer Erhard Ratdolt (1442–1528) was active in Venice when he produced this first printed edition of Euclid's *Elements*. In his dedication of this book to his patron, the Doge Mocenigo of Venice, Ratdolt stressed the importance and difficulty of printing the geometric figures, but he did not divulge his method, which is unknown. Other Renaissance editions of Euclid soon followed, with mathematical figures printed as woodcuts. The figures are (from top to bottom): a line, a point, a plane; angles (acute, right angle, perpendicular and non-perpendicular lines); the circle and its parts (diameter, semicircle); triangles (equilateral, right, and obtuse), and the rectangle, square, and parallelogram.

Euclid's Division of a Line into Extreme and Mean Ratio

A line can be divided such that the ratio of the larger segment to the smaller is equal to the ratio of the whole line to the larger segment (Euclid, *Elements*, 300 BC, book 6, proposition 30). This division is called "extreme and mean ratio" because it has three parts: two "extremes" (the smallest segment, y, and the largest, $x + y$), and a "mean" (the in-between length, x).

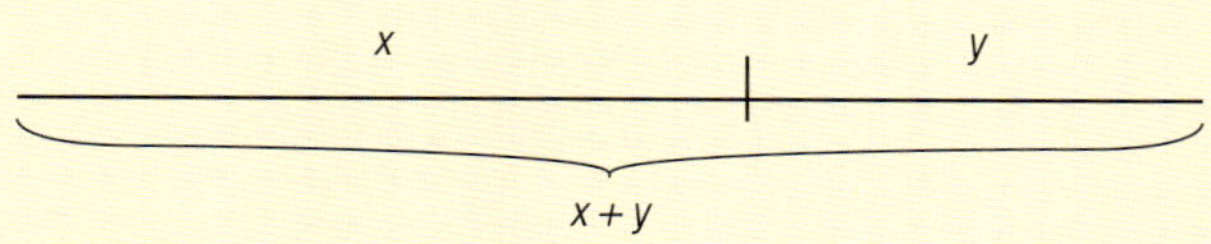

In algebraic notation, the ratio is written

$$x/y = (x + y)/x$$

which reads "the ratio of x to y is the same as the ratio of $(x + y)$ to x."

The lengths x and y are incommensurate; if you divide the larger segment by the smaller, it never comes out even. The larger segment is a little more than half again the smaller, but not exactly, and the ratio is a non-repeating decimal and goes on forever: 1.6180339887498948482 . . . .

The ratio is called "irrational" because it is inexpressible as a ratio of two whole numbers. In other words, a pair of whole numbers, such as 8 and 5, may *approximate* the ratio (8 divided by 5 equals 1.6), but it doesn't equal it exactly. Similarly, the ratio of 8 to 5 (1.6) is *approximately* equal to the ratio of 13 to 8 (13 divided by 8 equals 1.625).

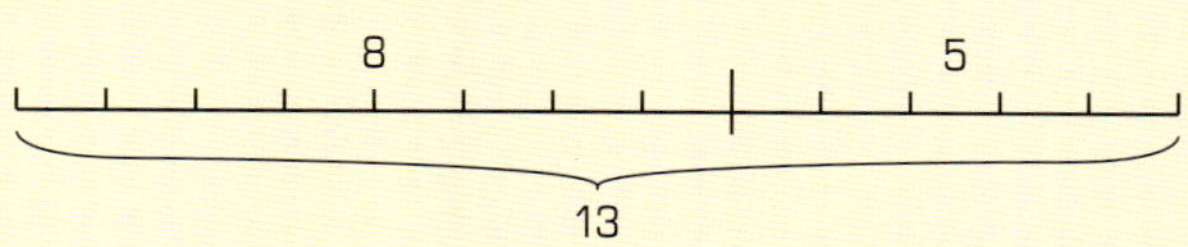

Preclarissimus liber elementorum Euclidis perspi/cacissimi: in artem Geometrie incipit quā foelicissime:

De principijs p se notis: ꝛ pmo de diffini/tionibus earundem.

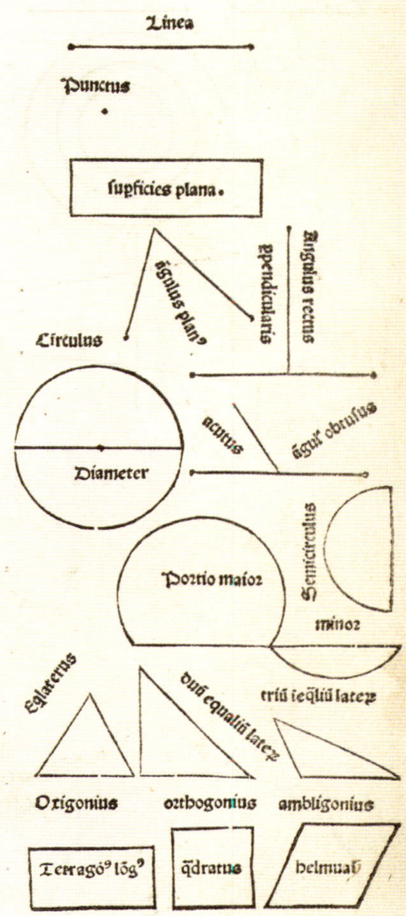

Punctus est cuius ps nō est. ¶ Linea est lōgitudo sine latitudine cuiꝰ quidē extremitates sūt duo pūcta. ¶ Linea recta ē ab vno pūcto ad aliū breuissima extē/sio i extremitates suas vtrūqꝫ eoꝝ reci/piens. ¶ Supficies ē q̄ lōgitudinē ꝛ lati/tudinē tm̄ hꝫ: cuiꝰ termi quidē sūt linee. ¶ Supficies plana ē ab vna linea ad a/liā extēsio i extremitates suas recipiēs ¶ Angulus planus ē duarū linearū al/ternus ꝯtactus: quaꝝ expāsio ē sup sup/ficiē applicatioqꝫ nō directa. ¶ Quādo aūt angulum ꝯtinēt due linee recte rectilineꝰ angulus noiat̃. ¶ Qn̄ recta linea sup rectā steterit duoqꝫ anguli vtrobiqꝫ fuerit eq̄les: eoꝝ vterqꝫ rectꝰ erit ¶ Lineaqꝫ linee supstās ei cui supstat ꝑpendicularis vocat̃. ¶ An/gulus vō qui recto maior ē obtusus dicit̃. ¶ Angulꝰ vō minor re/cto acutꝰ appellat̃. ¶ Terminꝰ ē qd̄ vniuscuiusqꝫ finis ē. ¶ Figura ē q̄ t̄mino vl' termīs ꝯtinet̃. ¶ Circulꝰ ē figura plana vna q̄dem li/nea ꝯtēta: q̄ circūferentia noiat̃: in cuiꝰ medio pūctꝰ ē: a quo oēs linee recte ad circūferētiā exeūtes sibiīuicē sūt equales. Et hic quidē pūctꝰ cētrū circuli dr̃. ¶ Diameter circuli ē linea recta que sup eiꝰ centꝝ trāsiens extremitatesqꝫ suas circūferētie applicans circulū i duo media diuidit. ¶ Semicirculus ē figura plana dia/metro circuli ꝛ medietate circūferentie ꝯtenta. ¶ Portio circu/li ē figura plana recta linea ꝛ parte circūferētie ꝯtēta: semicircu/lo quidē aut maior aut minor. ¶ Rectilinee figure sūt q̄ rectis li/neis cōtinent̃ quarū quedā trilatere q̄ tribꝰ rectis lineis: quedā quadrilatere q̄ q̄tuor rectis lineis. q̄dā mltilatere que pluribus q̄ꝫ quatuor rectis lineis continent̃. ¶ Figurarū trilaterarū: alia est triangulus hn̄s tria latera equalia. Alia triangulus duo hn̄s eq̄lia latera. Alia triangulus triū inequalium laterū. Harꝫ iterū alia est orthogoniū: vnū .s. rectum angulum habens. Alia ē am/bligonium aliquem obtusum angulum habens. Alia est oxigoni/um: in qua tres anguli sunt acuti. ¶ Figurarū autē quadrilateraꝝ Alia est q̄dratum quod est equilaterū atqꝫ rectangulū. Alia est tetragonꝰ longꝰ: q̄ est figura rectangula: sed equilatera non est. Alia est helmuaym: que est equilatera: sed rectangula non est.

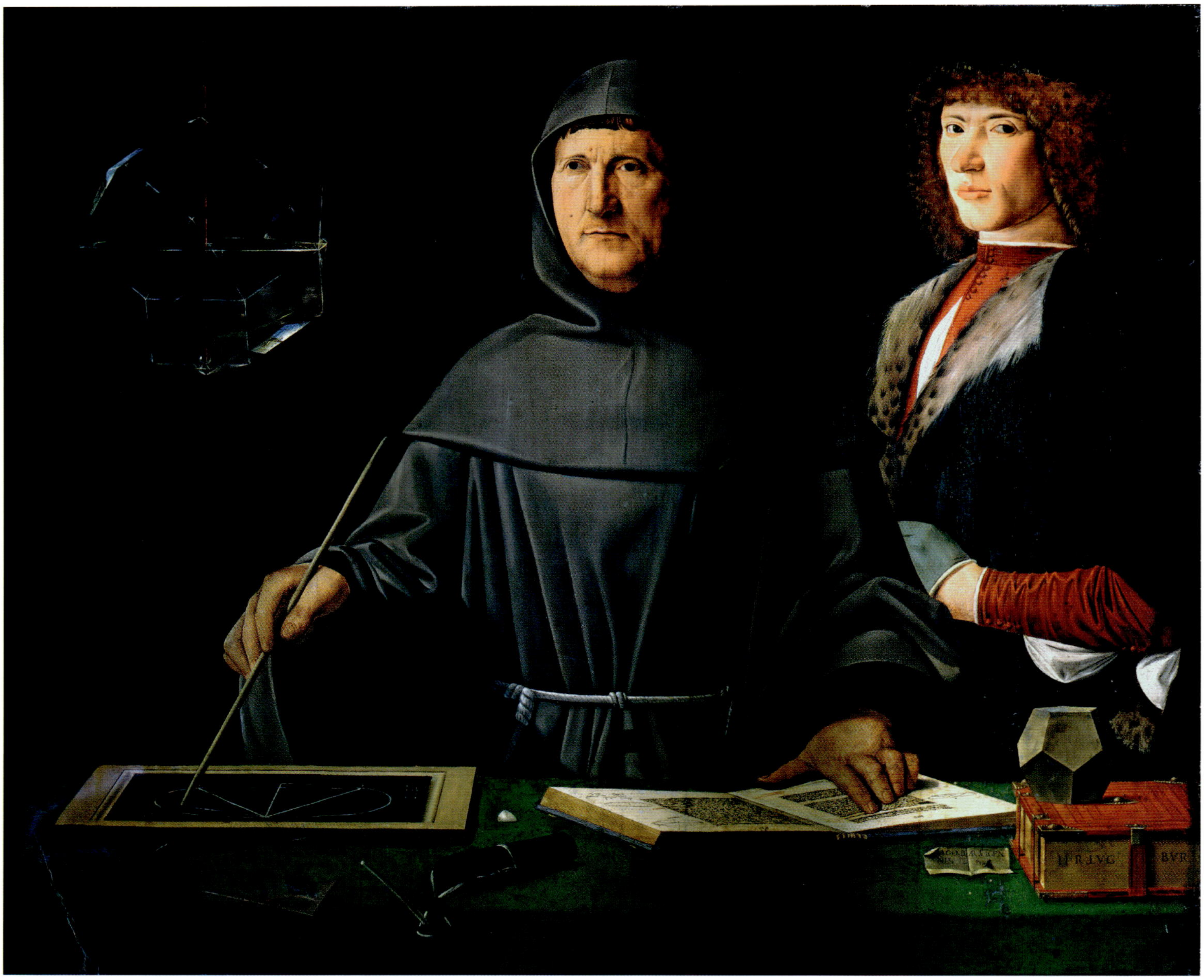

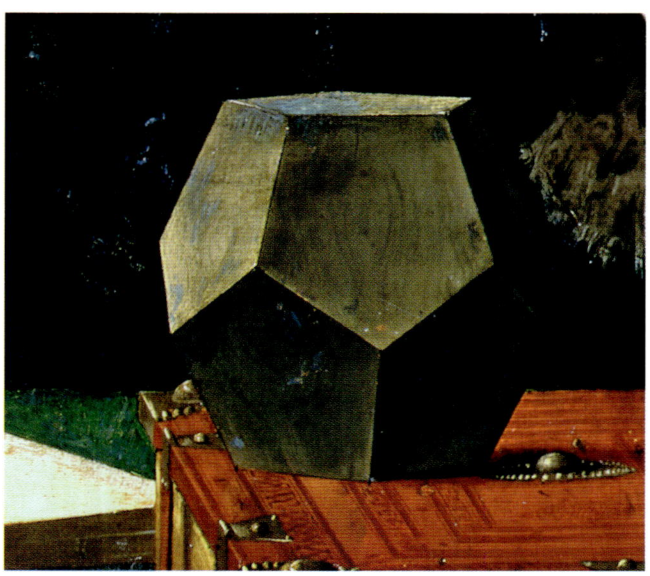

ABOVE

2-20. Jacopo de Barbari (Italian, ca. 1460–1516), *Luca Pacioli*, 1495. Oil on canvas, 39 × 47¼ in. (99 × 120 cm). Museo e Gallerie di Capodimonte, Naples.

The Franciscan monk Fra Luca Pacioli points to a geometric figure on a chalkboard, on the edge of which is written EUCLIDES. On the table are his tools of geometry: a protractor, compasses, a pen, a piece of chalk, a sponge to clean the board, and (unidentified) books. Suspended on the left is one of Archimedes's semi-regular solids, composed of eight triangles and eighteen square faces (a rhombicuboctahedron), made of leaded glass panes and half filled with water. Pacioli's well-clad companion is unidentified.

LEFT

2-21. Solid dodecahedron, detail of Jacopo de Barbari, *Luca Pacioli*, 1495.

The dodecahedron, one of Plato's regular solids, is composed of twelve pentagons that are constructed using Euclid's division of a line into extreme and mean ratio.

Campanus concluded that the ratio created harmony (an "irrational symphony") from disparate parts.

In 1482 Erhard Ratdolt produced the first printed edition of Euclid's *Elements*, using Campanus's revised translation (plate 2-19). Sales of the first edition were brisk, prompting the rapid production of several other editions based on Campanus's translation, including a 1509 edition by the Italian Franciscan friar and mathematician Fra Luca Pacioli. Perhaps inspired by Campanus's musings on the extreme and mean ratio, Pacioli declared that the ratio was a symbol for the Almighty, making a play on words between the mathematical, musical, and theological meanings of "irrational" (*irrationalis* in Latin). The ratio is irrational in the sense that it is not expressible as a ratio of whole numbers; thus, according to Pacioli, the ratio is a symbol for the Divine, who is beyond reason and whose name is not expressible in words. Pacioli further declared that the dodecahedron, whose pentagonal faces are constructed using the ratio, symbolizes heaven (plates 2-20 and 2-21), an association he found in Plato's account of the divinely created universe (*Timaeus*, 366–360 BC).

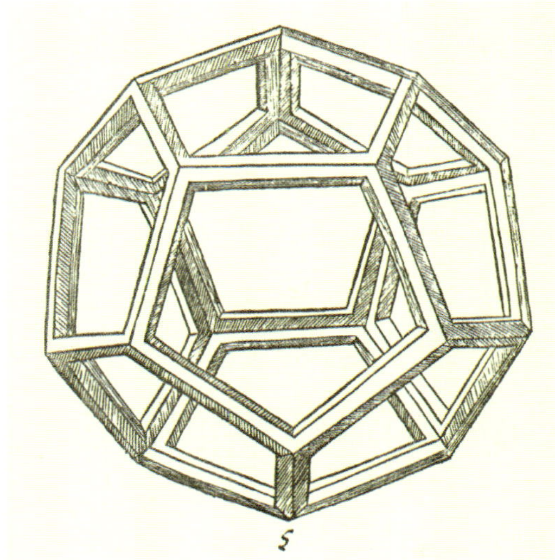

2-22. Open dodecahedron, woodcut illustration by Leonardo da Vinci in Luca Pacioli, *Divina Proportione* (Venice: Paganius, 1509), plate 28.

Pacioli published these views on the divine proportion in *Divina Proportione* (1509), along with his Italian translation from Latin of a mathematical treatise on regular solids by his teacher, Piero della Francesca.[5] Pacioli assembled the book in Milan, where he went in 1496 to serve at the Sforza court. There he met Leonardo da Vinci, who had settled there in 1482. By a fortunate coincidence Pacioli arrived just when Leonardo was seeking a geometry teacher to help him understand linear perspective, which he wanted to use to design a mural that Ludovico Sforza had commissioned from him for the Convent of Santa Maria della Grazie (*The Last Supper*, begun 1495; see plate 2-10). In exchange for Pacioli's tutoring, Leonardo drew geometric diagrams for Pacioli's book, including an open dodecahedron (plate 2-22). Pacioli's lengthy praise for the sublime properties of the divine ratio all centered on its theological symbolism. He neither associated it with beauty nor advocated its use by artists, and there is no evidence that Leonardo ever used it. In his chapter on architecture, Pacioli explicitly concurred with Alberti that designers should use only whole units (only ratios that come out even); in other words, according to Pacioli, artists should *not* use the divine ratio.

THE GOLDEN SECTION

A common misperception holds that artists have used Euclid's ratio since antiquity, but it was not associated with art and beauty until the mid-nineteenth century in Germany. The German mathematician Martin Ohm was the first to call Euclid's ratio the "Golden Section" (*goldener Schnitt*) in the second (1835) edition of his *Die reine Elementar-Mathematik* (Pure elementary mathematics), and the term was repeated in other mathematical publications in the 1840s.[6] The German psychologist and scholar Adolf Zeising adopted the term from these mathematical sources, and in 1854 he announced that the ratio is the key to achieving the harmonious relation of parts within a whole. Zeising began his treatise with a detailed history of the literature on proportion by philosophers and artists, mentioning dozens of authors including Plato, Plotinus, Alberti, Leonardo, Kant, and Hegel. But Zeising omitted Campanus and Pacioli, which suggests that he was evidently unaware of the only writers who had singled out Euclid's ratio as divine.

Zeising declared that the Golden Section "underlies the formation of all beauty and wholeness in nature and in the pictorial arts, and from the beginning it provided the model for all representations and formal relations, whether cosmic or individual, organic or inorganic, acoustic or optical, which found its most perfect realization, however, in the human figure" (*A New System of Human Proportions*, 1854).[7] In other words, according to Zeising the Golden Section is a law of nature and innate in human beings, and therefore (consciously or unconsciously) artists have used it throughout history. To prove his point Zeising claimed to have discovered the ratio in various examples of art and architecture, including a classical statue and a Gothic cathedral (plate 2-23). But whether or not an artist used the Golden Section is a *historical* question that has to be supported by historical evidence, such as an artist's statement or an architect's drawing. One cannot prove that the Golden Section was used by measuring statues and buildings because one can "find" any ratio in any physical object if, like Zeising, one cherry-picks the data and ignores all evidence to the contrary.[8] There is not a scrap of historical evidence that any artist or architect used the Golden Section before the late nineteenth century.[9]

But if the ratio is innate, then artists and architects could have used it unconsciously and there would be no historical evidence (in the form of artist's statements or drawings, which are made when someone is *conscious*). Is the ratio innate? In the 1870s the founder of experimental psychology, Gustav Fechner, tested Zeising's hypothesis that humans

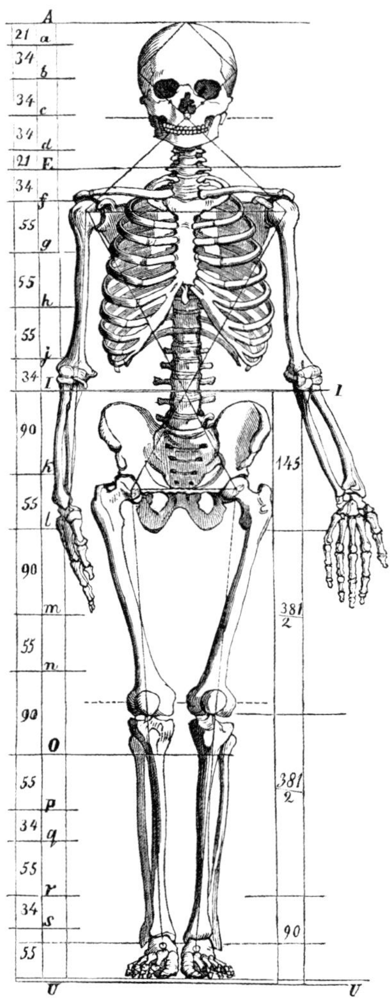

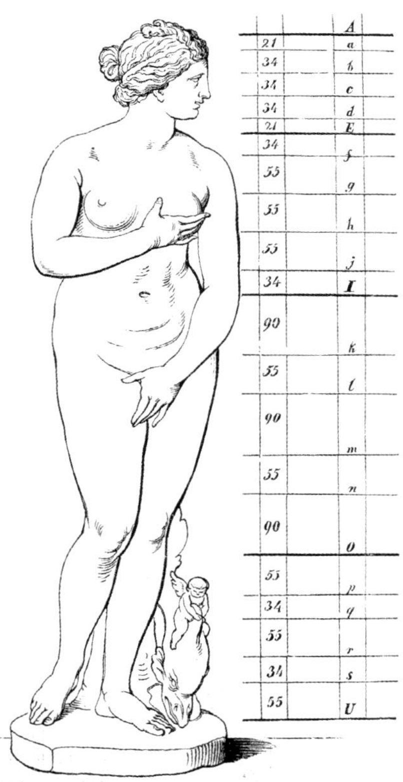

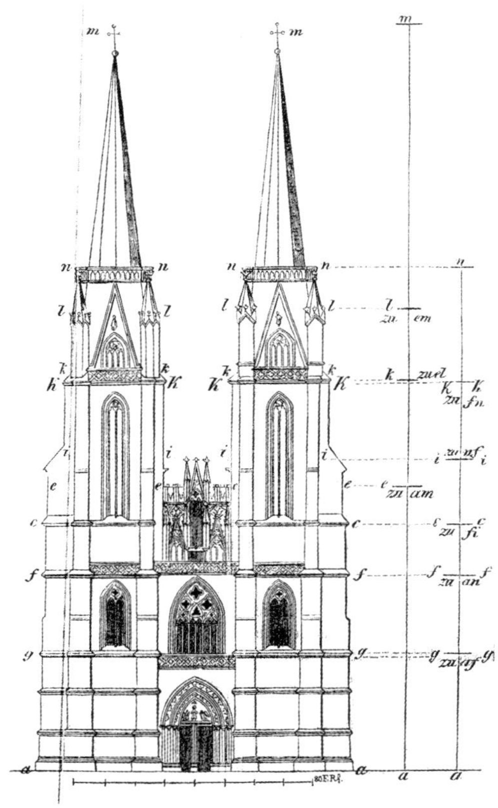

2-23. The proportion of a man (upper left), the Hellenistic or Roman statue of the *Medici Venus* (lower left), and St. Elizabeth's Church, Marburg (right), in Adolf Zeising, *Neue Lehre von den Proportionen des menschlichen Körpers* (A new system of human proportions; Leipzig, Germany: R. Weigel, 1854), 214, fig. 49; 284, fig. 89; 407, fig. 165.

Zeising's book is a classic example of a historian treating a theory not as a hypothesis but as a foregone conclusion. Zeising measured the skeleton, statue, and the church façade looking for the Golden Ratio; when he had a "hit," he marked it, ignoring the parts that didn't fit his theory. The numbers in the columns—and, in general, the distances in Zeising's illustrations—are in the Fibonacci series (1, 1, 2, 3, 5, 8, 13, 21, 34, 55, 89, [assuming that he rounded 89 to 90]). For a definition of the Fibonacci series, see the sidebar on page 104.

get more pleasure from viewing shapes that are in a ratio of 1.618. In one experiment, Fechner showed subjects divided lines; he determined that subjects did *not* show a preference for lines divided in the ratio 1.618 (the "Golden Section"). In another experiment, Fechner showed subjects rectangles; his results established that 35 percent (less than half) of his respondents preferred a *range* of rectangles that *included* those whose sides were in the ratio 1.618 (the "Golden Rectangle"; *Zur experimentellen Aesthetik* [On experimental aesthetics], 1871). In other words, Fechner's experimental data did *not* support (much less prove) the hypothesis that the ratio 1.618 gives greater aesthetic pleasure. Nevertheless, Fechner's (nonexistent) "proof" of the innate aesthetic appeal of the ratio 1.618 entered the psychological literature, where it continues to be debated to this day.[10]

The myth of the Golden Section became a fixture of the burgeoning field of art history in the second half of the nineteenth century when Germanic scholars declared that it was used in major monuments, such as the Great Pyramid of Khufu at Giza.[11] The historian of architecture August Thiersch, who was a professor at the prestigious university of engineering, the Technische Hochschule in Munich, applied Zeising's theory to the design of Classical and Renaissance architecture. In "Die Proportionen in der Architektur" (Proportion in architecture; 1883). Thiersch declared that what makes perfect proportions is the repetition of forms (the Golden Rectangle and others) within the building. To demonstrate his point Thiersch used "regulating lines" (construction lines) to analyze the Parthenon and Saint Peter's Basilica.[12] But, like Zeising, Thiersch gave no *historical* evidence to support this historical claim, and he "found" certain rectangles in the Parthenon and Saint Peter's by singling out ones that confirmed his thesis and ignoring others (plates 2-24 and 2-25).

The Swiss scholar Jacob Burckhardt, who dominated German-language art history from his posts at the University of Basel and the Eidgenössische Technische Hochschule (ETH) in Zurich from 1844 to 1893, wrote to Thiersch that he was "convinced and carried away [*überzeugt und mitgerissen*]" by his essay.[13] Buckhardt's student, the young Swiss art historian Heinrich Wölfflin, hailed Thiersch's thesis as the *Lex Thierschica* (Latin for "Thiersch's law") and contributed *reverse* regulating lines to the analysis of architecture, such as his study of

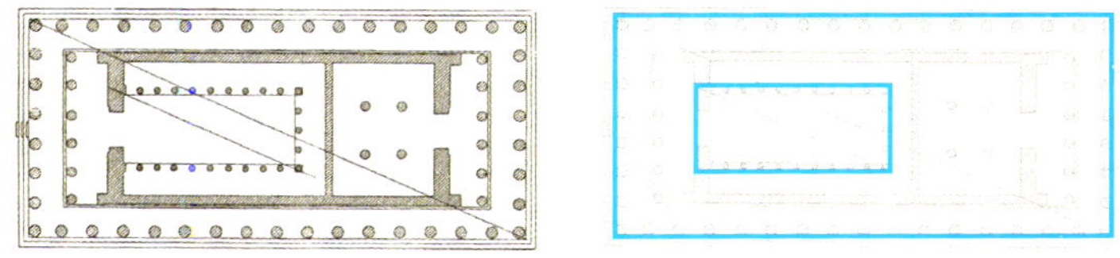

2-24. Plan of the Parthenon, Athens (left), in August Thiersch, "Die Proportionen in der Architektur" (Proportions in architecture), ed. Josef Durm, *Handbuch der Architektur* 4, no. 1 (Darmstadt, Germany: Diehl, 1883): 46, fig. 15.

Thiersch drew "regulating lines" on the plan of the Parthenon (left) to single out two rectangles and allegedly demonstrate that the architects Iktinos and Kallikrates achieved their classic design by using two rectangles—one on the stylobate and one in the cella—of the same proportions (right).

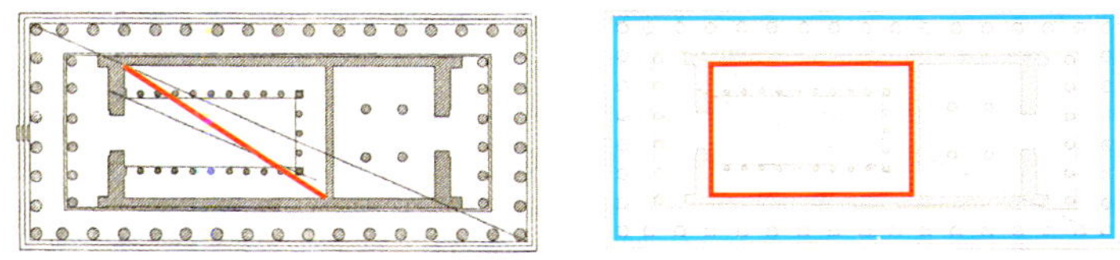

This is one of numerous rectangles that Thiersch ignored.

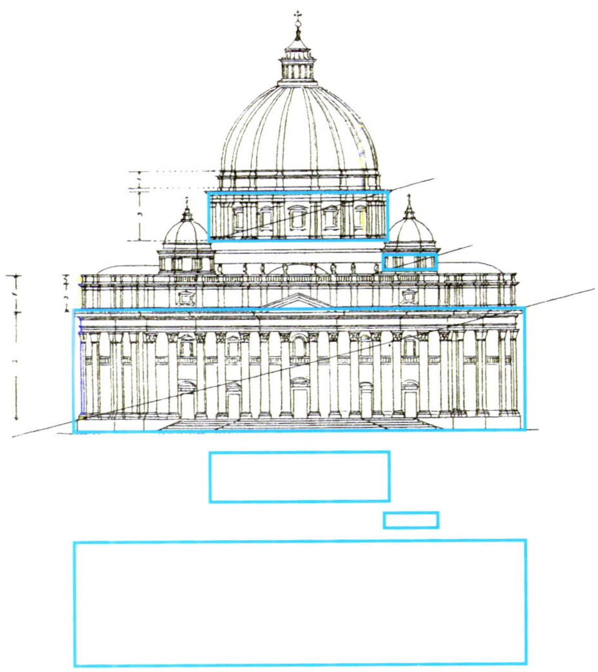

2-25. Saint Peter's Basilica, Rome, in August Thiersch, "Die Proportionen in der Architektur" (1883): 68, fig. 60; reproduced in Jacob Burckhardt, *Geschichte der Renaissance in Italien*, 3rd ed. (Stuttgart: Ebner and Seubert, 1891), 100.

Using regulating lines, Thiersch allegedly demonstrated that the architects of Saint Peter's designed the façade using rectangles with the same proportions. The rectangles he singled out, while ignoring the rest, are shown. On the left-hand side of his diagram, Thiersch also purported to demonstrate that the ratio 1 : 3 is repeated in several horizontal divisions in façade.

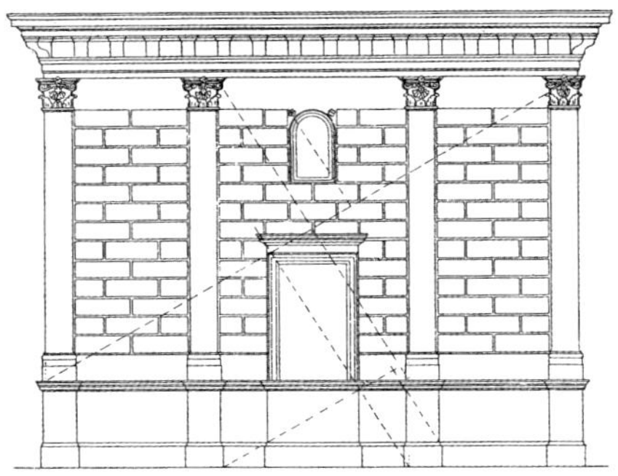

2-26. Top story of the Palazzo Riario (known as the "Cancelleria"), Rome, in Heinrich Wölfflin, *Renaissance und Barock* (Renaissance and Baroque; Munich: T. Ackermann, 1888), 54.

Thiersch used diagonal regulating lines to demonstrate that horizontal rectangles were in proportion to each other. Expanding on this idea, Wölfflin showed that if one drew lines in the opposite direction on the doors and windows (the vertical rectangles), the lines would meet Thiersch's lines at 90° angles. This, Wölfflin claimed, was an objective measure of a well-proportioned building.

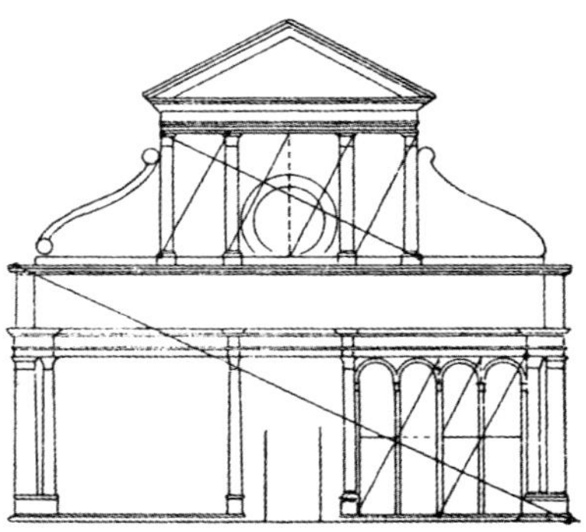

2-27. Santa Maria Novella, Florence, in Heinrich Wölfflin, "Zur Lehre von den Proportionen" (On the theory of proportions), *Deutsche Bauzeitung* 23 (1889), 278.

In an 1888 letter to Wölfflin, Burckhardt praised this diagram by Wölfflin of the façade of Santa Maria Novella in Florence, as well as a similar Wölfflin diagram for the Villa Farnesina in Rome: "What beautiful results! The homogeneity of the wings of the Farnesina with the main building! Then, proving the *Lex Thierschia* on the façade of Santa Maria Novella! And how much you imply, and still promise more"; Burckhardt to Wölfflin, Dec. 12, 1888, in Jacob Burckhardt, *Briefe*, ed. Max Burckhardt (Basel, Switzerland: Schwabe, 1980), 9:168.

the proportions of the Cancelleria in Rome (plates 2-26 and 2-27).[14] In his 1888 art history classic, *Renaissance und Barock* (Renaissance and Baroque), Wölfflin wrote: "Harmony of proportion . . . far from being arbitrary, is determined by the Golden Section, a proportion which we experience as 'pure,' that is, as completely satisfying and therefore completely natural."[15] A few years later Burckhardt expanded the section on proportion in the third (1891) edition of his own classic *Geschichte der Renaissance in Italien* (History of the Renaissance in Italy) to include Thiersch's thesis, and he reproduced Thiersch's Renaissance diagrams (plate 2-25).[16] Burckhardt wrote: "Purely mathematical means will never lead to immutable principles. . . . In recent times, however, a law of proportion had been discovered, which holds good under widely different conditions. . . . [The law is based on] the Golden Section, that constant proportion taught by Euclid . . . [which] is as a rule a constant proportion and similarity among geometric figures, as Euclid demonstrated. . . . Just as that fundamental law of architectural proportion first made its appearance in Greek and Roman buildings, so it was to revive and rise to new importance at the beginning of the Renaissance."[17] Echoing Zeising's contention that the Golden Section is innate, Burckhardt concluded: "Whether architects followed it in practice and then in theory, or the other way round, or indeed consciously at all, may remain undecided. That they observed it is certain, for it shines out from the finest monuments of the Italian Renaissance."[18]

In fact, there is no historical evidence that Italian Renaissance architects used the ratio 1.618.

Wölfflin succeeded Burckhardt as professor at Basel and then taught at universities in Berlin and Zurich, becoming one of the most influential art historians of the early twentieth century. Although in 1914 Wölfflin would still praise Thiersch for formulating the law of the harmony of parts within a whole,[19] his enthusiasm for the Golden Section cooled in the first two decades of the twentieth century after he studied Gestalt psychology, which prompted him to begin analyzing the shift of style from Renaissance to Baroque in terms of five transformations (as described by the mathematics of group theory; see chapter 7). In his discussion of the Cancelleria (plate 2-26) in his *Kunstgeschichtliche Grundbegriffe* (Principles of art history, 1915), Wölfflin mentioned neither the *Lex Thierschica* nor the Golden Section but described the proportions of the building in terms of Alberti's classical definition of beauty as the unity of parts within a whole.[20]

In the early twentieth century the Golden Section was nicknamed phi (φ) for "Phidias," and Thiersch's ideas on proportion entered modernist circles through Walter Gropius, the founding director of the Bauhaus (see chapter 8). As a twenty-year-old student at the Technische Hochschule in Munich, Gropius attended Thiersch's lectures in 1903.[21] Thiersch's son, the architect Paul Thiersch, worked from 1906 to 1907 in the Dusseldorf office of the architectural firm of Gropius's teacher, Peter Behrens. Ernst Neufert, who was a graduate of the Bauhaus and Gropius's assistant on construction of the Bauhaus in Dessau, featured the Golden Section and a diagram based on Zeising in his classic manual for architecture, *Bauentwurfslehre* (1936; see plate 2-28), which has gone through multiple editions; it is still in print today as a standard textbook used to teach the myth of the Golden Section to a new generation of architects. Indeed, the 1936 diagram in plate 2-28 adorns the cover of the latest (2002) paperback edition of Neufert's book, which is translated as *Architects' Data* (Oxford, England: Blackwell, 2002). In another recent example, the Golden Section is the centerpiece of Kimberly Elam's textbook for graphic design, *Geometry of Design: Studies in Proportion and Composition* (New York: Princeton Architectural Press, 2001).[22] These latter-day adopters are both perpetuating the myth of the implementation of the Golden Section throughout all time and now devising systems that are in fact based on Euclid's ratio.

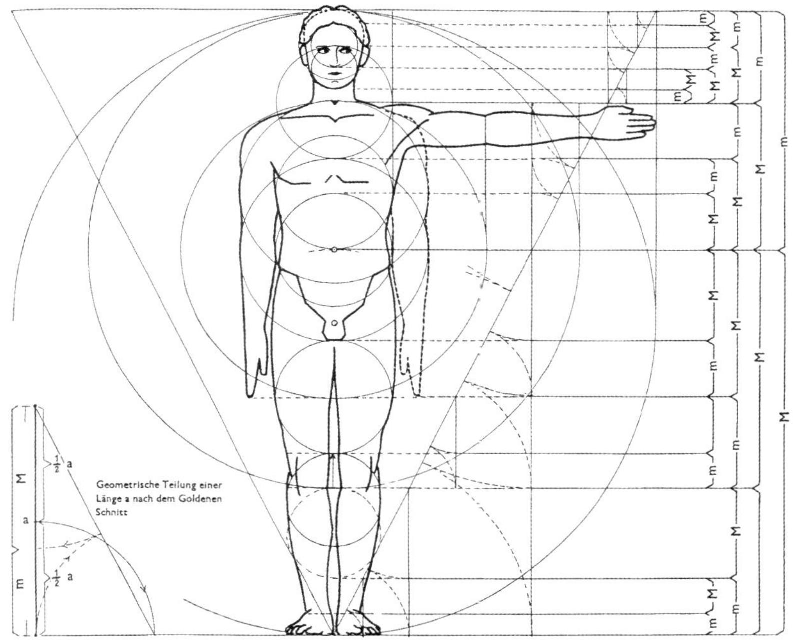

2-28. "Measurement of the Proportion of Man, based on the Research of A. Zeising," in Ernst Neufert, *Bauentwurfslehre*, (Berlin: Bauwelt, 1936), 23. Used with permission.

The text in the lower left reads "Geometrische Teilung einer Länge nach dem Goldenen Schnitt" (Geometric division of a length according to the Golden Section). On the right, a line is divided into the Golden Section, with *m* indicating the smaller section and *M* the larger. The ratio is drawn on the body, which is divided at the navel, and repeated in many smaller proportions. Neufert urged architects to use the Golden Ratio because it puts their buildings in the tradition of others who had (allegedly) used it, including the architects of the Egyptian pyramids and Leonardo da Vinci.

Zeising's doctrine found another advocate in the German Catholic theologian Peter Lenz (Father Desiderius). In 1868 at the Benedictine monastery of Beuron, near Tuttlingen, Germany, Lenz and other monk-artists established a school where monks would be trained to create altarpieces, frescoes, and mosaics. Manifesting the ecumenical spirit of late-nineteenth-century Catholicism, Lenz held that divinity is revealed in a moment of mystical insight in all religions of the world, and that eternal truths are best symbolized by timeless geometric forms. In Berlin in the 1860s Lenz studied collections of Egyptian sculpture, measuring the statues in search of their proportions. It was during this time that he discovered Zeising's *New System of Human Proportions*, which he incorporated into his canon, combining Christian and Egyptian iconography (plate 2-29).[23] From Lenz's correspondence in the early 1870s with fellow monk P. Johannes Blessing, who was researching the mathematical basis of harmony in music, it is clear that these Benedictine clerics were interested in determining an underlying mathematical structure for liturgical art.[24]

Late-nineteenth-century French artists who were interested in mathematics and science, such as Georges Seurat, did not adopt the Golden Section from German sources, which were known to them. Seurat's friend, Charles Henry, who was author of "Introduction à une esthétique scientifique" (Introduction to a scientific aesthetic; 1885) and one of the first to realize that Pacioli, Zeising, and Fechner were all talking about the same ratio,[25] stated emphatically: "The Golden Section and harmonic proportions are completely ignored by contemporary artists."[26] Rather, it was artists interested in *sacred* themes—the French Nabis—who adopted the Golden Section because of its associations with divinity. The aesthetic aims of the Nabis artists were diverse, but several of them, including the group's leader Paul Sérusier, made a strong connection between their religious beliefs and their art. They were drawn to the French theosophist Édouard Schuré, who saw common pantheist themes in the teachings of all great spiritual leaders, including Moses, Pythagoras, Jesus, and Buddha.[27] This merger of world religions can be seen in Nabi works such as Paul Ranson's *Christ and Buddha* (ca. 1890; Triton Foundation, the Netherlands).

The Dutch artist Jan Verkade lived in France in the circle of Paul Gauguin and the Nabis before entering the Abbey of Beuron, where he became a Benedictine monk

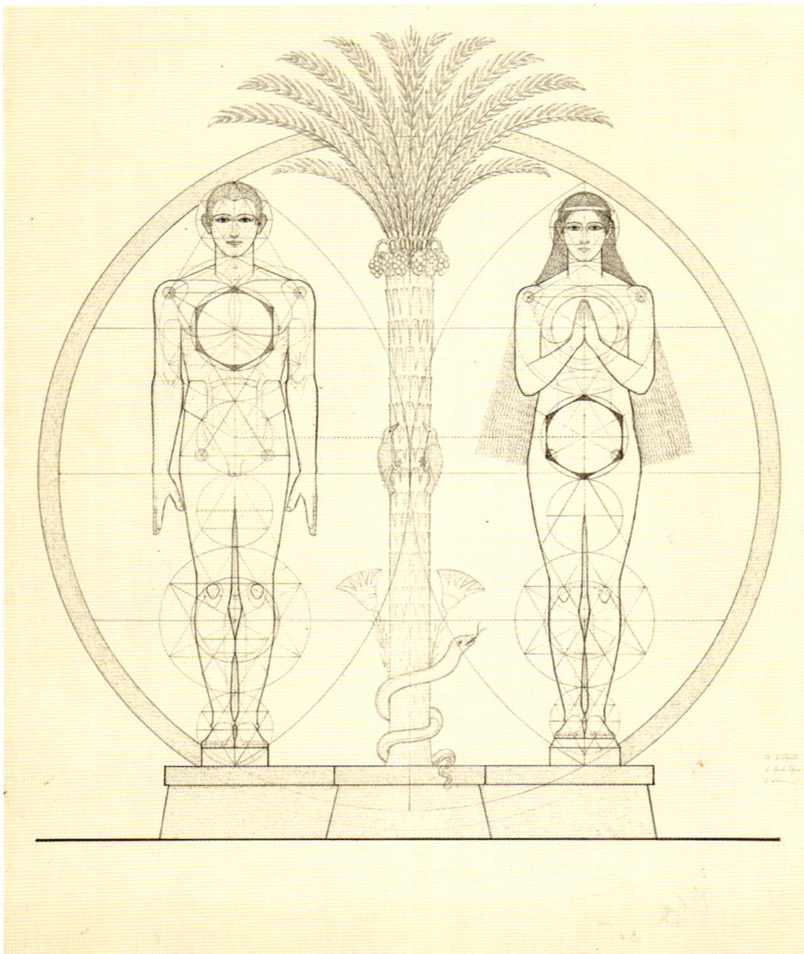

2-29. Peter Lenz (Father Desiderius), *System of Proportion for a Man and Woman*, 1871. Pencil and ink on paper. Archive of the Benedictine Abbey of St. Martin of Beuron, Germany.

and studied with Peter Lenz. The intense expression of piety and brilliant decorative patterns of work produced at the Beuroner Kunstschule (Beuron art school) attracted Sérusier, who in the early 1890s left Brittany in northwest France and travelled to the monastery to study color theory and systems of proportion with Lenz.[28] Sérusier visited Beuron again in 1896 and 1904, accompanied by another Nabi artist, Maurice Denis (plate 2-30). Denis described the impact of the monastery on his friend: "Sérusier doesn't abandon nature or the Breton people [of Brittany], but he introduces a new element from Beuron, a mathematical element in his aesthetic, which will not cease to occupy his thoughts."[29] In 1910 Sérusier painted a golden, glowing geometric form, floating in the heavens above, in a Platonic mix of geometry and divinity that is rare in the modern secular era (plate 2-31; compare plate 1-20 in chapter 1).

Peter Lenz's use of geometry made a deep impression on Denis,[30] who called for an art that would look beyond appearances and reveal an underlying, abstract order, writing in 1890: "Remember that a picture—before being a battle horse, a nude woman, or some anecdote—is essentially a flat surface covered with colors assembled in a certain order."[31] In 1914 Denis purchased and restored a seventeenth-century priory in Saint-Germain-en-Laye near Paris, and in 1919 he established a French version of the Beuron school, the Ateliers d'Art Sacré (Studios for sacred art). In 1921 Sérusier included the Golden Section as part of the artist's "universal language" in his influential studio manual *ABC de la peinture* (ABC of painting). In this context he emphasized geometrical laws of beauty and the Golden Section as an inspirational, philosophical ideal, rather than as a practical way to actually construct a painting.[32]

When I dare to speak of the mathematics in art, people smile as if I were a fool. In our society, people see mathematics and art as opposites, just as they see science and religion.

—Paul Sérusier, letter to Jan Verkade, August 25, 1902

In Paris, enthusiasm for the Golden Section was stoked by the charismatic Romanian writer Matila Ghyka, whose books (1927, 1931) were a rehash of Zeising, Thiersch, Cook, and others. Ghyka's contribution to this literature was his (correct) observation that Pacioli's divine proportion and Zeising's Golden Section are different names for the same ratio. He also claimed (incorrectly) that Leonardo had used the ratio (*Esthétique des proportions dans la nature et dans les arts* [Aesthetics of proportion in nature and the arts], 1927).

The Spanish painter Salvador Dalí fell under the spell of Ghyka when they met in 1948 in Los Angeles, where Ghyka was chairman of the Fine Arts Department at the University of Southern California.[33] Dalí was one of the few avant-garde artists to receive traditional academic training at the renowned École des Beaux-Arts in Paris, where he learned to paint in oil glazes like the Old Masters. Throughout the 1930s, Dalí berated his fellow

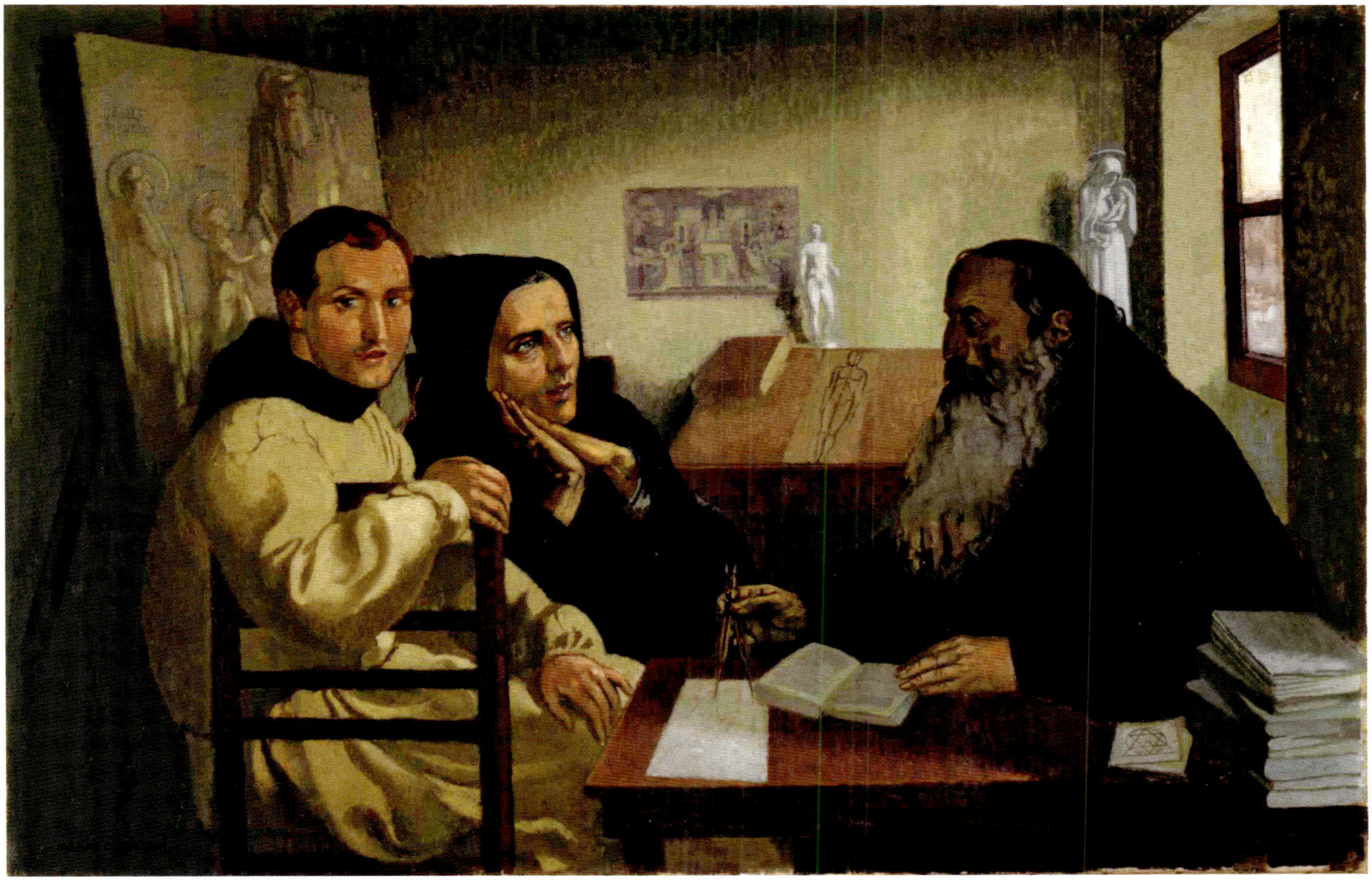

ABOVE

2-30. Maurice Denis (French, 1870–1943), *Moines de Beuron* (Monks of Beuron), 1904. Oil on canvas, 38³⁄₁₆ × 57⅞ in. (97 × 147 cm). Musée départemental Maurice Denis-Le Prieuré, Saint Germain-en-Laye, France.

During his visit with the artist Jan Verkade at the Abbey of Beuron, Denis painted this portrait of three monks seated in a studio: Peter Lenz, the old man with white beard, draws a geometric diagram with a compass to instruct the hooded Franquesa Adalbert, a native of Barcelona, and Verkade in the white cassock. In the background there is a drawing table, a sketch tacked up on the back wall, a plaster cast, and a sculpture of the Madonna (on the right). The light from the window falls on a large easel painting of Saint Benedict and two young disciples, echoing Denis's subject—a Benedictine monk with two apprentices.

RIGHT

2-31. Paul Sérusier (French, 1864–1927), *Golden Cylinder*, ca. 1910. Oil on cardboard, 15 × 11 in. (38.2 × 27.7 cm). Musée des Beaux-Arts, Rennes.

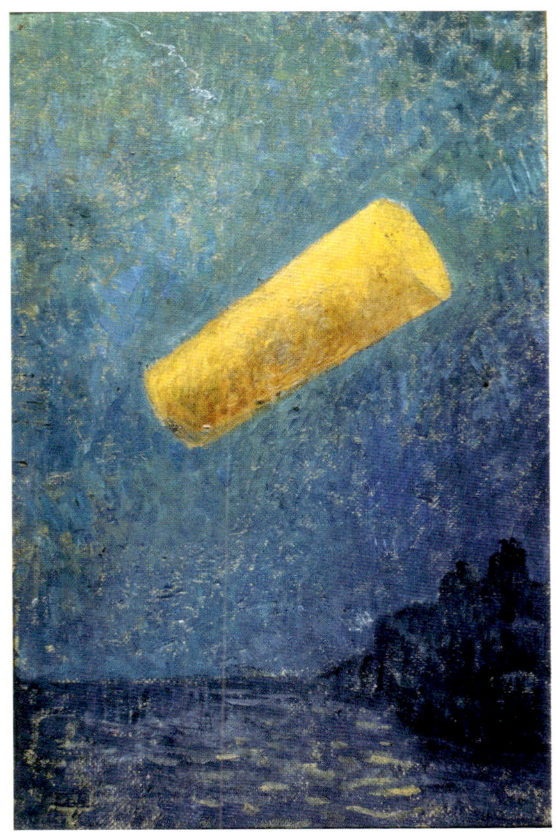

Surrealists because they lacked his technical skills. Dalí met Ghyka when the artist was writing a studio manual: "Now at 45, I want to paint a masterpiece and save Modern Art from chaos and laziness. This book is consecrated to this crusade and I dedicate it to all the young, who have faith in true painting."[34] Dalí was drawn to Ghyka's claim that the Old Masters, especially Leonardo, held a mathematical key for achieving beauty in their art. Ghyka gave Dalí copies of his books, and the men began a correspondence and friendship. Dalí added a summary of Ghyka's thesis to his studio manual.[35]

In the late 1940s Dalí had shifted the focus of his art away from the Freudian unconscious to nuclear physics, doing paintings that mixed the mythical with the atomic, such as *Three Sphinxes of Bikini* (1947; Morohashi Museum of Modern Art, Fukushima, Japan), which refers to the Bikini Atoll in the south Pacific, where nuclear weapons testing began in 1946. In his study of physics Dalí had already learned that solid matter is composed of charged particles held together by electromagnetic forces such that, from a subatomic perspective, matter is mainly empty space. Dalí associated physical dematerialization with spirituality (*Anti-Matter Manifesto*, 1958), and from Ghyka he learned that he could also associate mathematics with spirituality. Dalí also read Pacioli,[36] where he discovered and adopted the link of mathematics with the divine (which had escaped Ghyka). Having read Pacioli's book, Dalí clearly knew that Leonardo had simply drawn geometric diagrams to illustrate the book and that the painter of the *The Last Supper* neither used nor advocated the Golden Section. Nevertheless, Dalí let himself be carried away by the myth that Leonardo had used the Golden Section in his art. He determined to "paint a masterpiece to save Modern Art" to honor Leonardo, whose technical abilities he sought to keep alive. Dalí would symbolize timeless truth by embodying the ratio, which, according to Pacioli, "is always mysterious and secret, and qualified by mathematicians as irrational."

The Golden Section [is] . . . called the divina proporzione *by Luca Pacioli in his memorable book, the most important of all aesthetic treatises.*

—**Salvador Dalí, *50 Secrets of Magic Craftsmanship*, 1948**

Dalí based his monumental *Sacrament of the Last Supper* (1955; see plate 2-32) on Leonardo's *Last Supper* (plate 2-10), the painting on which Leonardo was working when he studied with Pacioli and the only major artwork in which Leonardo used linear perspective. Starting with a rectangular canvas in the proportions of 1.618, Dalí determined the pyramidal forms of the apostles using a diagram from Ghyka that is based on the Golden Section (plate 2-33). The geometrically structured background is a copy of Leonardo's illustration for Pacioli of the (open) dodecahedron (plate 2-22), the shape that symbolized the universe in Plato's *Timeaus*.[37]

During the Roman Catholic sacrament of Holy Communion, the bread and wine (shown on the table in Dalí's painting) are (literally) transformed into the body and blood of Christ. Dalí symbolized this transubstantiation by making the body of Christ fade to transparency in the center of the painting—the "vanishing point" of the perspective construction. Above is the heavenly dodecahedron in which a figure hovers with outstretched arms, a gesture that suggests both the crucifixion and the ascension—in other words, the dual physical/spiritual nature of Jesus as the mortal, finite son of an eternal, infinite God. Dalí called his style "nuclear mysticism," meaning a combination of the current scientific worldview with traditional religious iconography.

Dalí used the Golden Section not to ensure beauty and perfect proportions, as Zeising and Ghyka recommended, but to symbolize the divine, as Pacioli advocated. As such, *The Sacrament of the Last Supper* is a rare example of sacred geometry in modern art—Dalí's nostalgic homage to an Old Master painting created in an age of faith. But Dalí also created jarring mixtures of the sacred and profane, as in his painting of the Madonna as his wife Gala (*The Madonna of Portlligat*, ca. 1950; Fukuoka Art Museum, Fukuoka, Japan). In the

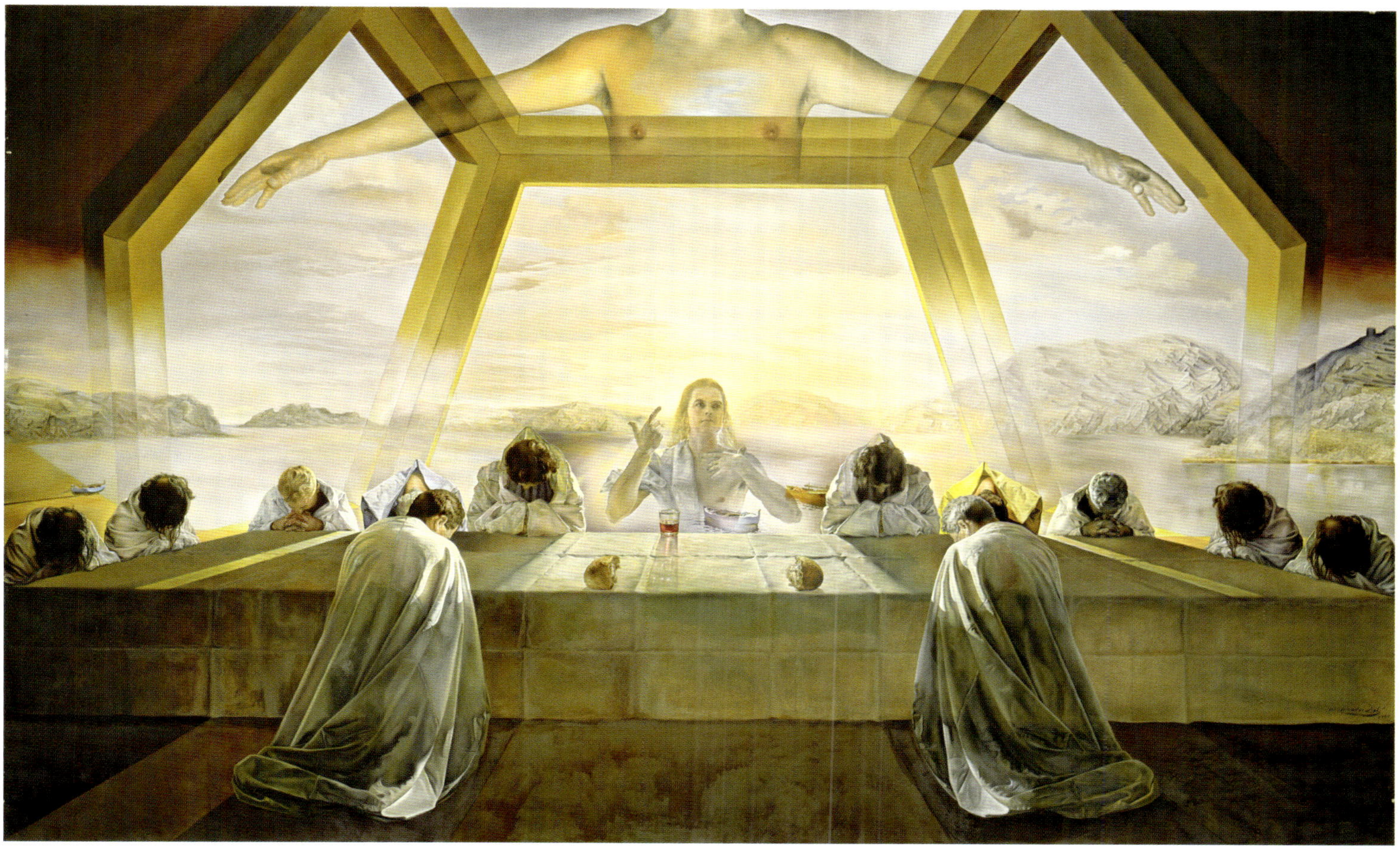

end, Dalí's lack of a spiritual anchor contributed to the "chaos" he sought to dispel as he and other artists grappled to create art appropriate for the nuclear age.

Another disciple of Ghyka, the Swiss architect Le Corbusier, began his career using simple geometric forms, as in *Maison La Roche*, Paris, 1923–25 (plate 2-34). After acquiring Ghyka's 1927 book, he used the ratio 1.618 not to design a building but to confirm his design after the fact (for example, *Villa at Garches*, 1927).[38]

In 1946 Le Corbusier designed a tool, the Modulor—similar to a tape measure that has British units on one side and metric on the other—which could be used by both British and European architects to lay out numerical patterns in a 1.618 ratio because it indicates

ABOVE

2-32. Salvador Dalí (Spanish, 1904–89), *The Sacrament of the Last Supper*, 1955. Oil on canvas, 65⅝ × 105⅛ in. (166.7 × 267 cm). National Gallery of Art, Washington, DC, Chester Dale Collection, 1963.10.115. © 2014 Salvador Dalí, Fundació Gala-Salvador Dalí/Artists Rights Society, New York.

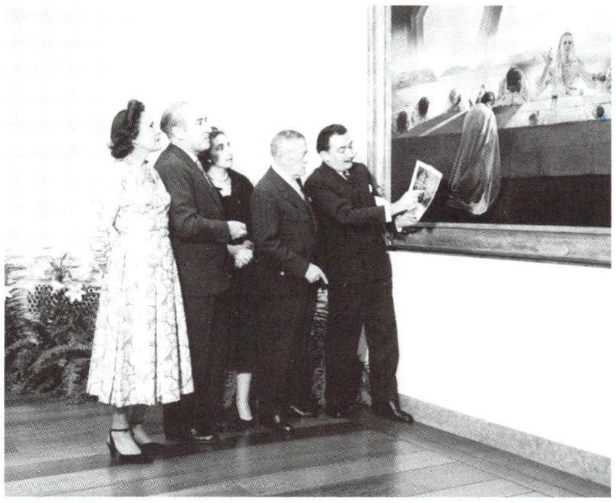

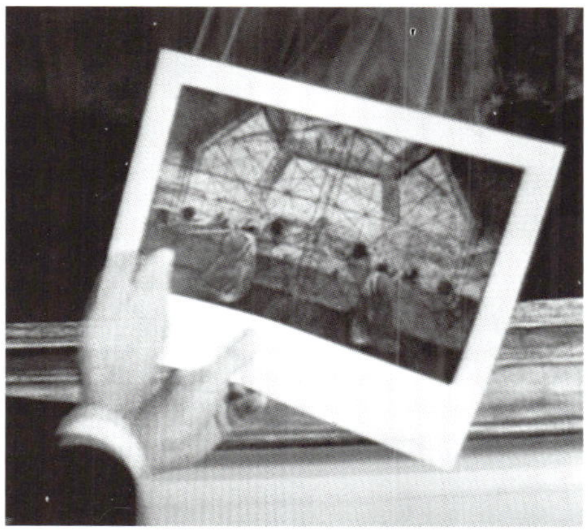

BELOW, LEFT AND RIGHT

2-33. Dalí showing his proportion diagram after Ghyka, standing before *The Sacrament of the Last Supper*, 1955, in the National Gallery of Art, Washington, DC.

TOP RIGHT

2-34. Le Corbusier (Charles-Édouard Jeanneret; Swiss, 1887–1965) and his cousin Pierre Jeanneret (Swiss, 1896–1967), *Maison La Roche*, Paris, 1923–25. Foundation Le Corbusier, Paris. © 2014 F.L.C./ADAGP, Paris/Artists Rights Society, New York.

BOTTOM RIGHT

2-35. Le Corbusier (Charles-Édouard Jeanneret; Swiss, 1887–1965), *The Modulor*, 1945, in Matila Ghyka, "Le Corbusier's Modulor and the Concept of the Golden Mean," *Architectural Review* 103, no. 614 (February 1948): 39–42; the diagrams are on 40. Used with permission. © 2014 F.L.C./ADAGP, Paris/Artists Rights Society, New York.

Le Corbusier designed a proportional system, which he called "The Modulor," by giving one Fibonacci series of numbers based on the height of an average man (1.829 m, AB in the diagram on the left) and a second series based on the height of a man with his arm raised (2.26 m, AC). He then went on to give a geometric "proof" that four parts of the human body—feet, navel, head, and arm—are related in the Golden Ratio.

One can divide a line into Euclid's extreme and mean ratio by a geometric construction. In the figure in the middle, the length AD (distance from the man's feet to his navel) is given. Construct a square with sides AD. Take the midpoint of AD and construct a segment of a circle as shown, which determines the point B (the top of the man's head). Then AD : DB = AB : AD.

In the figure on the right, begin with A'C' (the height of the man with raised arm). D' (the navel) is the midpoint. Construct the two squares to form a rectangle and connect its corners with a diagonal, as shown. Using the upper-left corner of the rectangle as the center, draw an arc of a circle from C' to the diagonal, which determines a point on the diagonal. From that point draw an arc of a circle (with A' at its center) that determines the point E on A'C'. Then A'E : EC' = A'C' : A'E.

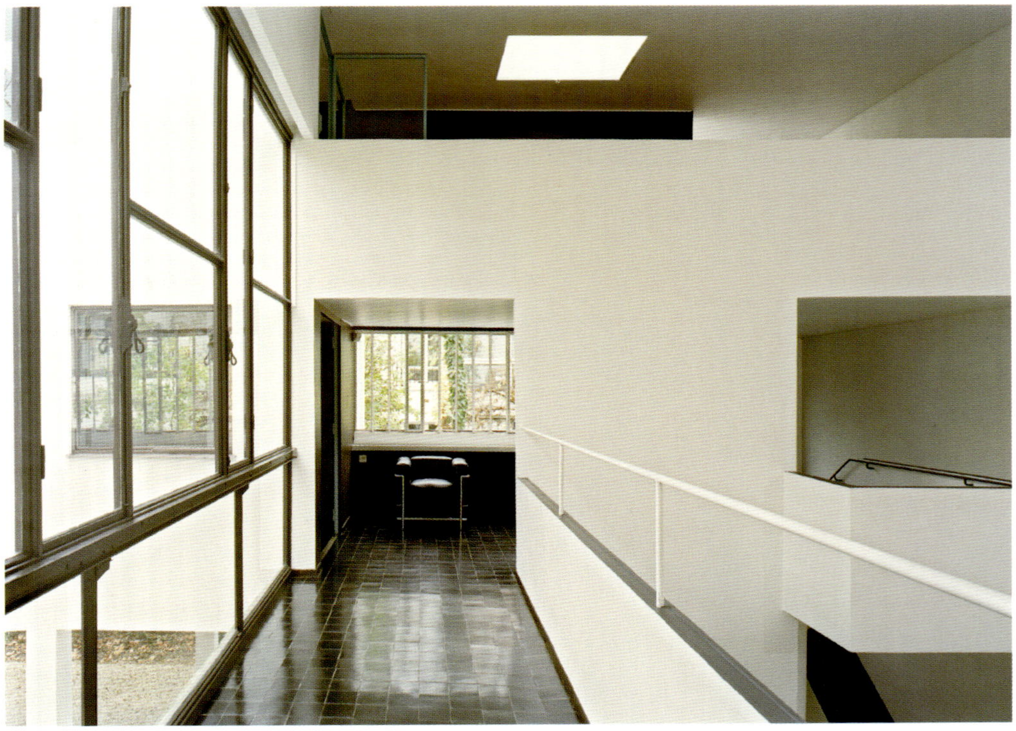

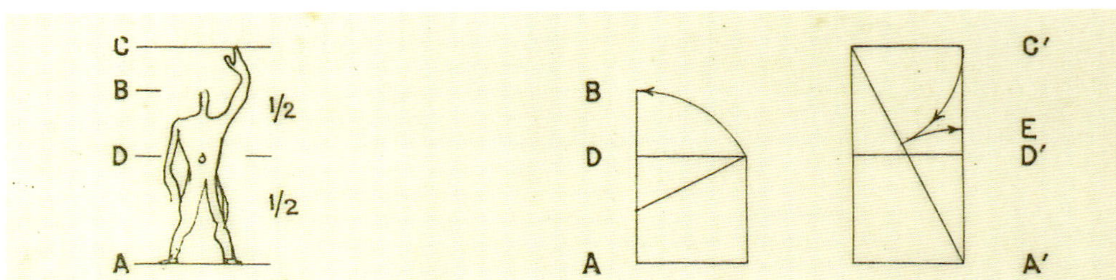

numbers in the Fibonacci series in both measuring systems. According to Le Corbusier, numbers in the Fibonacci series correspond to the proportions of an average human body at key points (the feet, navel, head, and a raised hand; see plate 2-35).[39] Le Corbusier recommended that designers use the tool to determine, for example, the height and width of the doors in a building, because this would automatically incorporate the ratio of 1.618 into the structure. The tool was not adopted by practicing designers and today it is largely forgotten, but in the mid-twentieth century it was featured, along with Alberti's *De re aedificatoria* (On the art of building, 1485; see plate 2-4), in a major 1951 exhibition in Milan on the architectural history of proportion.[40]

Like Dalí, Le Corbusier attributed supernatural power to the ratio 1.618. In his dedication of the apartment building Unité d'Habitation in Marseilles (1947–52), Le Corbusier described the Modulor as if "he" were the personification of architecture: "Dedicated to the Modulor who has ruled [*reglé*] the construction of the entire building. . . . As testimony of our gratitude to the Modulor, who reduced to fifteen the measurements of the entire building, giving it the grace of a harmonious scale."[41] Throughout his long career, Le Corbusier remained convinced that the Golden Section held the key to good design.

PROPORTION AFTER DARWIN

Inspired by Charles Darwin's publication of *On the Origin of Species* in 1859, late-nineteenth-century authors of popular science praised engineering by plants and animals, in which human designers could find models. In England, for example, the clergyman J. G. Wood declared "there is scarcely an invention of man that has not its prototype in Nature" (*Nature's Teachings: Human Invention Anticipated by Nature*, 1877). In Germany the Hungarian-born botanist Raoul Francé, in books such as *Die Pflanze als Erfinder* (The plant as an inventor, 1920), advised architects to study nature to understand form. Francé wrote on protozoa and other marine microorganisms in the 1890s, and he published the first of his popular illustrated booklets on biology for the series *Kosmos*. Francé focused on the geometric structures of biological systems that had evolved to perform certain functions, such as a poppy flower whose form evenly distributes pollen, or marine microorganisms whose spiral shapes propel them through water (plates 2-36, 2-37, and 2-38). Francé called such natural structures "biotechnical" to distinguish them from man-made equivalents. He proposed that, through evolution, nature had selected seven basic geometric forms to solve such technical problems as supporting weight and moving fluids: a crystal, sphere, cone, plane, strip, rod, and spiral (see plate 8-15 in chapter 8).

Darwin's theory of evolution also encouraged nineteenth-century British botanists to seek natural laws governing plant growth, and they determined that incremental growth within a confined space can result in a pattern that lies within the Fibonacci series of

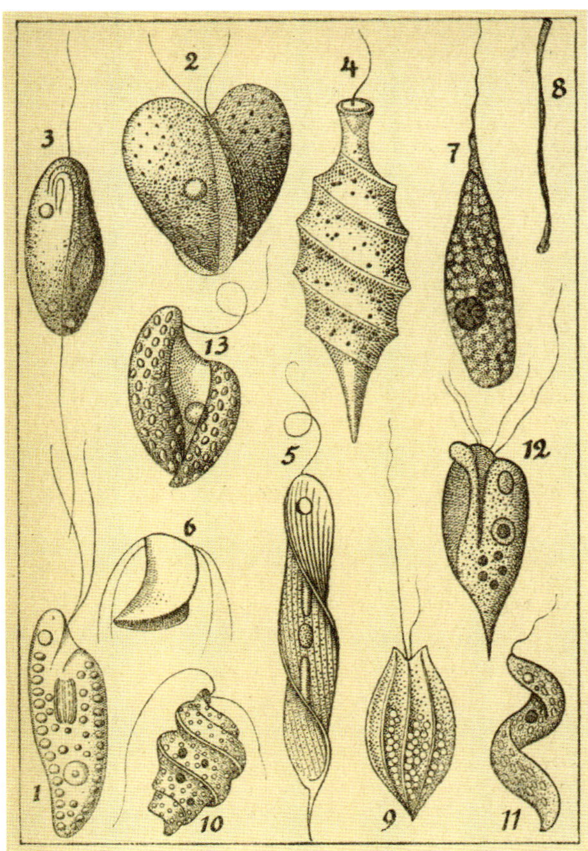

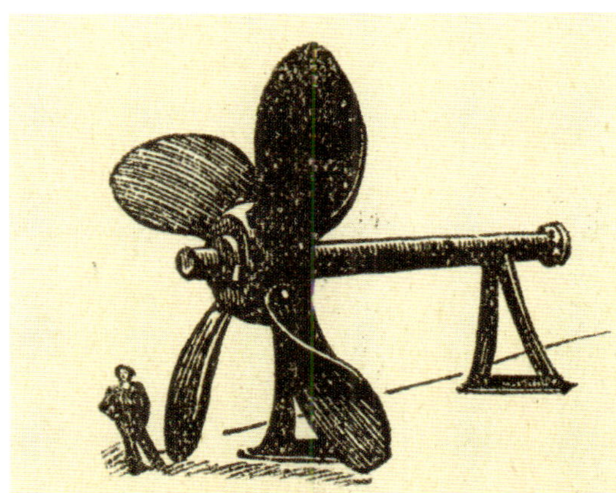

BELOW, LEFT AND TOP RIGHT

2-36. Microscopic marine organisms with flagella and a propeller for a ship, in Raoul Francé, *Die Pflanze als Erfinder* (The plant as an inventor; 1920), 32, plate 10. General Research Division, The New York Public Library for the Performing Arts, Astor, Lenox, and Tilden Foundations.

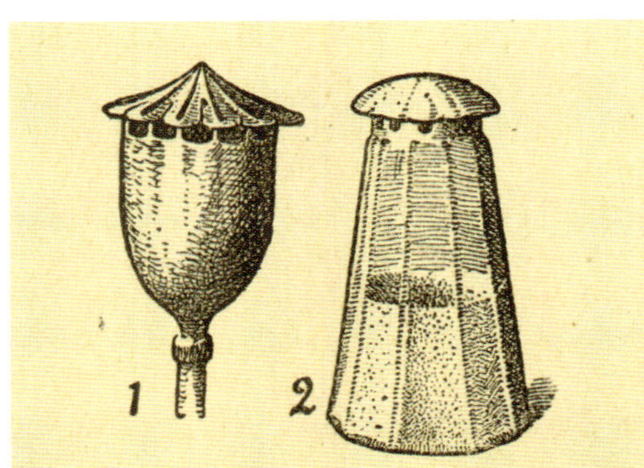

BELOW, BOTTOM RIGHT

2-37. A poppy and a salt shaker, in Raoul Francé, *Die Pflanze als Erfinder* (The plant as an inventor; 1920), 8, plate 1. General Research Division, The New York Public Library for the Performing Arts, Astor, Lenox, and Tilden Foundations.

numbers (see sidebar on page 104), which itself converges towards 1.618 (see sidebar below). Examples of this growth pattern in plants, called phyllotaxis, can be found in sunflowers, daisies, pineapples, pinecones, and broccoli.[42] The British botanists Arthur Harry Church (*On the Relation of Phyllotaxis to Mechanical Laws*, 1901), and D'Arcy Wentworth Thompson (*On Growth and Form*, 1917) explained how phyllotaxis, as well as patterns in the shells and horns of animals, result from mechanical forces and the geometric constraints of growing organisms (plates 2-39, 2-40, 2-41, and 2-42).

The (Arithmetic) Fibonacci Series and the (Geometric) Division into Extreme and Mean Ratio

Johannes Kepler was the first to note similarities between Fibonacci's (arithmetic) series and Euclid's (geometric) division of a line into an extreme and mean ratio. For example, if each number in the Fibonacci series is divided by the number preceding it, the result is around 1.618, becoming a closer approximation as the numbers increase (Kepler, *Harmonices Mundi*, 1619).

Similarly, if a series of rectangles is made with sides in the order of the Fibonacci series, the sides approximate the ratio 1.618. This series can be inter-nested to yield a spiral pattern, which converges toward a self-similar spiral based on the mean-extreme ratio.

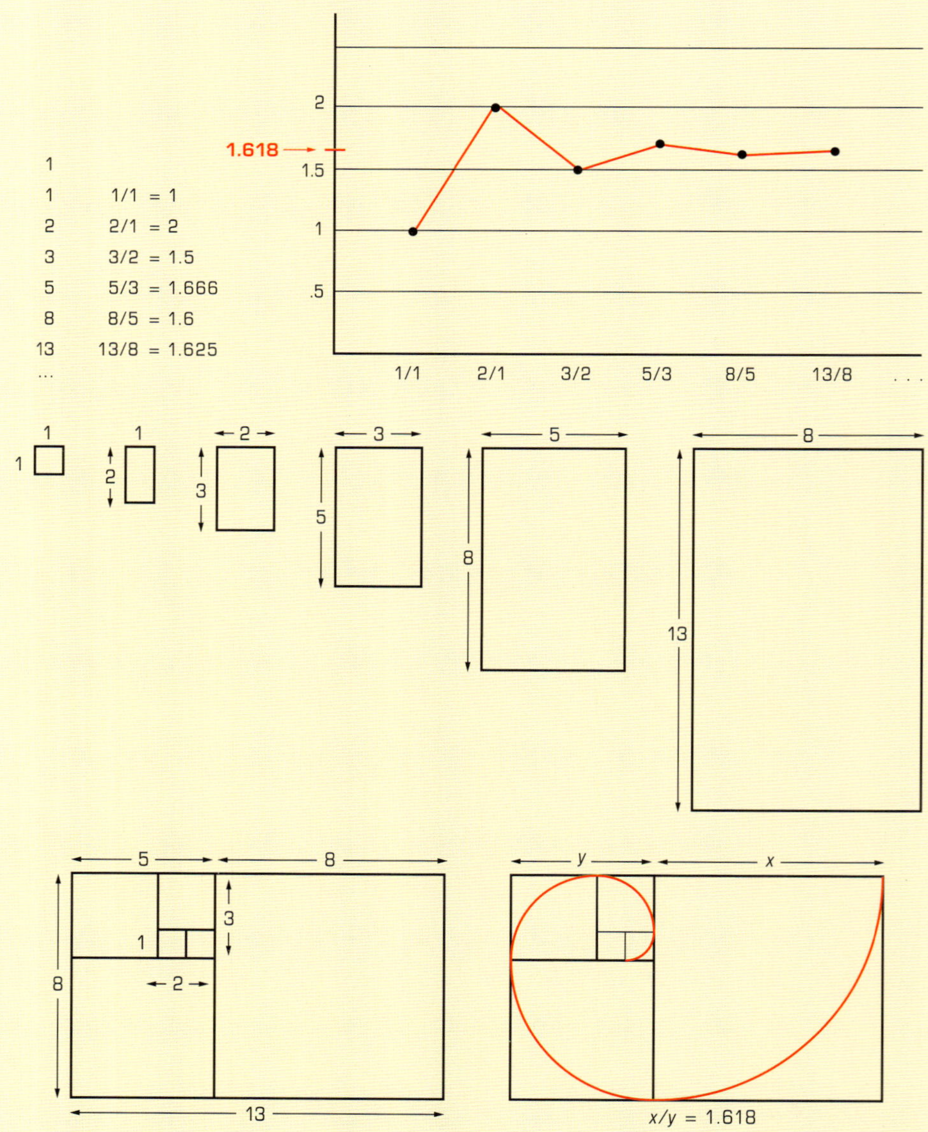

OPPOSITE

2-38. Franz Xaver Lutz (Austrian, b. 1941), *Volumes of a Fish and a Zeppelin*, 1993. Glaze painting. Courtesy of the artist, and © Klaus Tschira Stiftung, Heidelberg.

Trained as both an engineer and an artist, Franz Xaver Lutz works in the tradition of Raoul Francé and Ernst Haeckel. Lutz has described this work: "Long before technology, nature has found the ideal form. . . . The fish and the Zeppelin needed the best form to minimize the resistance of currents."

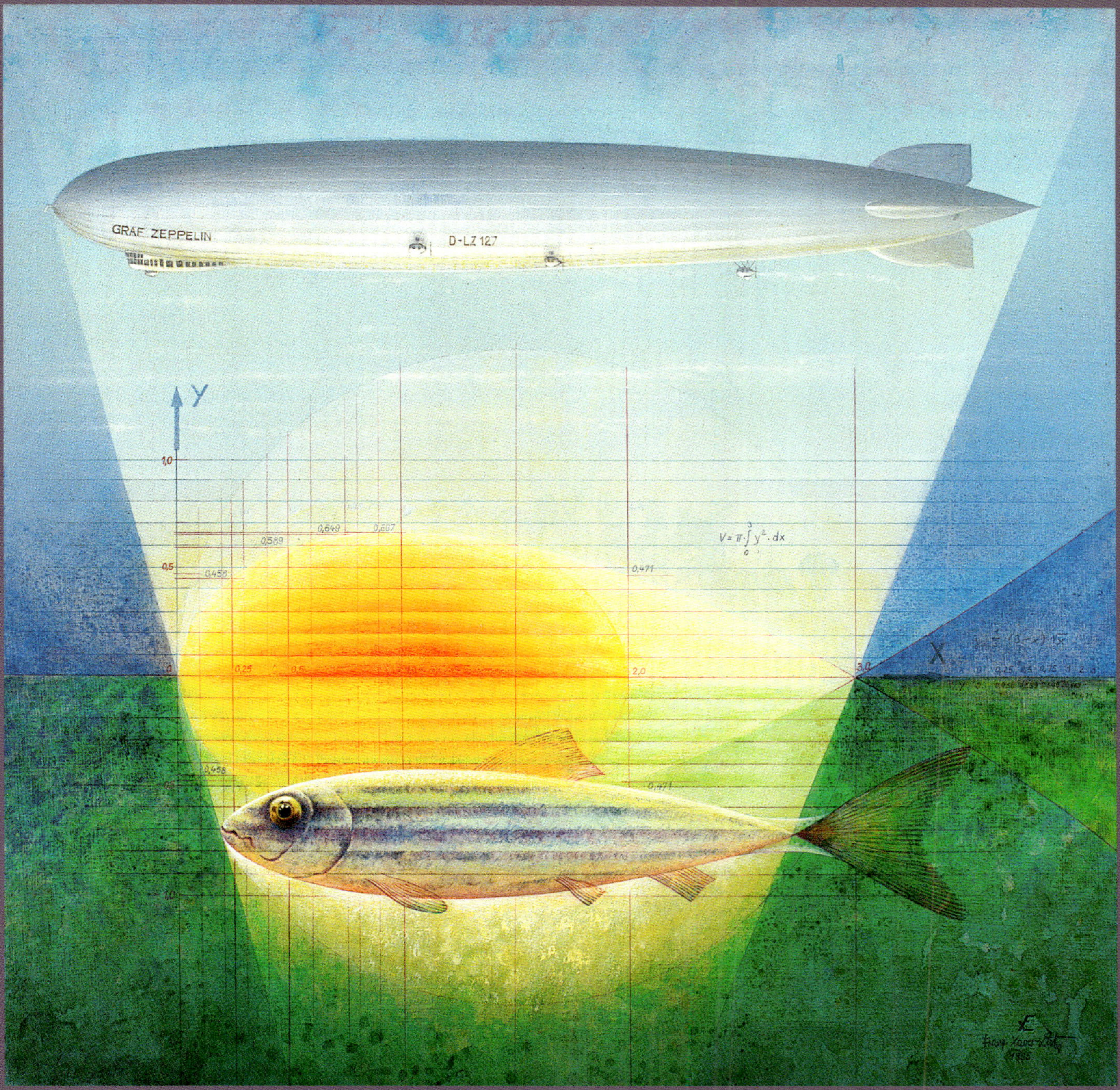
GRAF ZEPPELIN
D-LZ 127
Y
X

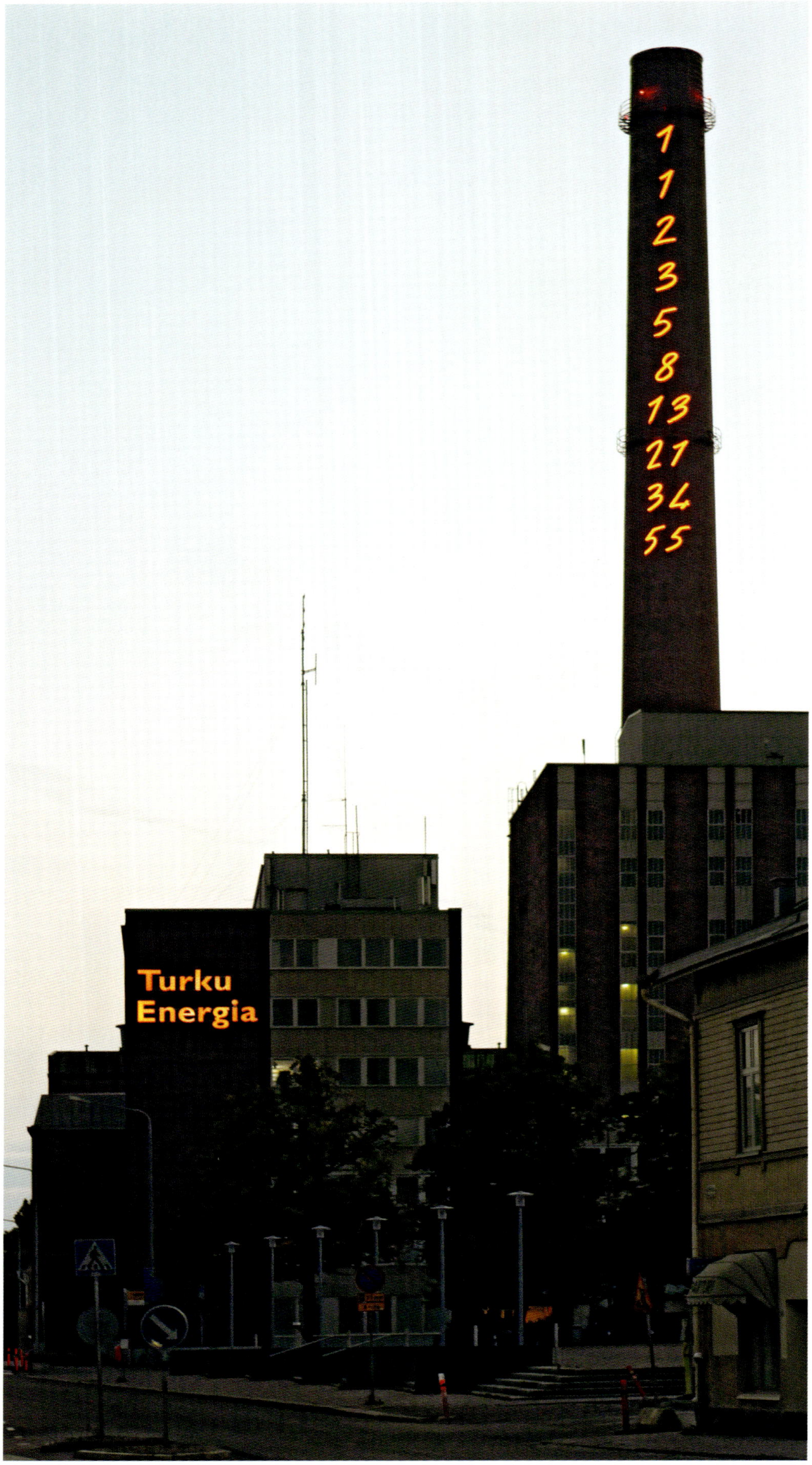

The Fibonacci Series

The medieval mathematician Leonardo Fibonacci (ca. 1170–1250) lived in the port city of Pisa. As the child of a trader with strong ties to Islamic North Africa, Fibonacci learned both Hindu-Arabic and Roman numerals. After he became a mathematician, he advocated the adoption of the Arabic system in the West. In a book that popularized Arabic numerals and the decimal number system, Fibonacci argued that it was easier to calculate in Arabic numerals (*Liber abaci* [Book of the abacus], 1202). As an example of calculation, he gave the series (which became known as the "Fibonacci series") formed by starting with 0 and 1 and then making the next number the sum of the previous two: 0 + 1 = 1, 1 + 1 = 2, 1 + 2 = 3, 2 + 3 = 5, and so on, making the series: 1, 1, 2, 3, 5, 8, 13, 21, . . . Despite Fibonacci's efforts, it took centuries for westerners to abandon Roman numerals in favor of the Arabic system, but by around 1500 Arabic numerals were used throughout the Near East, North Africa, and Europe (see plate 1-38 in chapter 1).

LEFT

2-39. Mario Merz (Italian, 1925–2003), *Flight of Numbers*, 1994. Neon numerals, each 6 ft. 6 in. (2 m) high, on the chimney of the Turku Energia Power Plant in Turku, Finland, 1994. Courtesy of Turku Energia.

The series that bears Fibonacci's name appears in the sculpture of his countryman, the contemporary Italian artist Mario Merz, who suggests the immaterial nature of numbers by making the numerals in light—the orange-red glow from neon tubes—installed here on the chimney of the Turku Energia power plant in Finland.

OPPOSITE

2-40. Mario Merz (Italian, 1925–2003), *Fibonacci Tables*, 1974–76. Charcoal, acrylic paint, metallic paint, and luminous gas-filled tubes on canvas, 105⅛ × 150½ in. (267 × 382 cm). Tate, London.

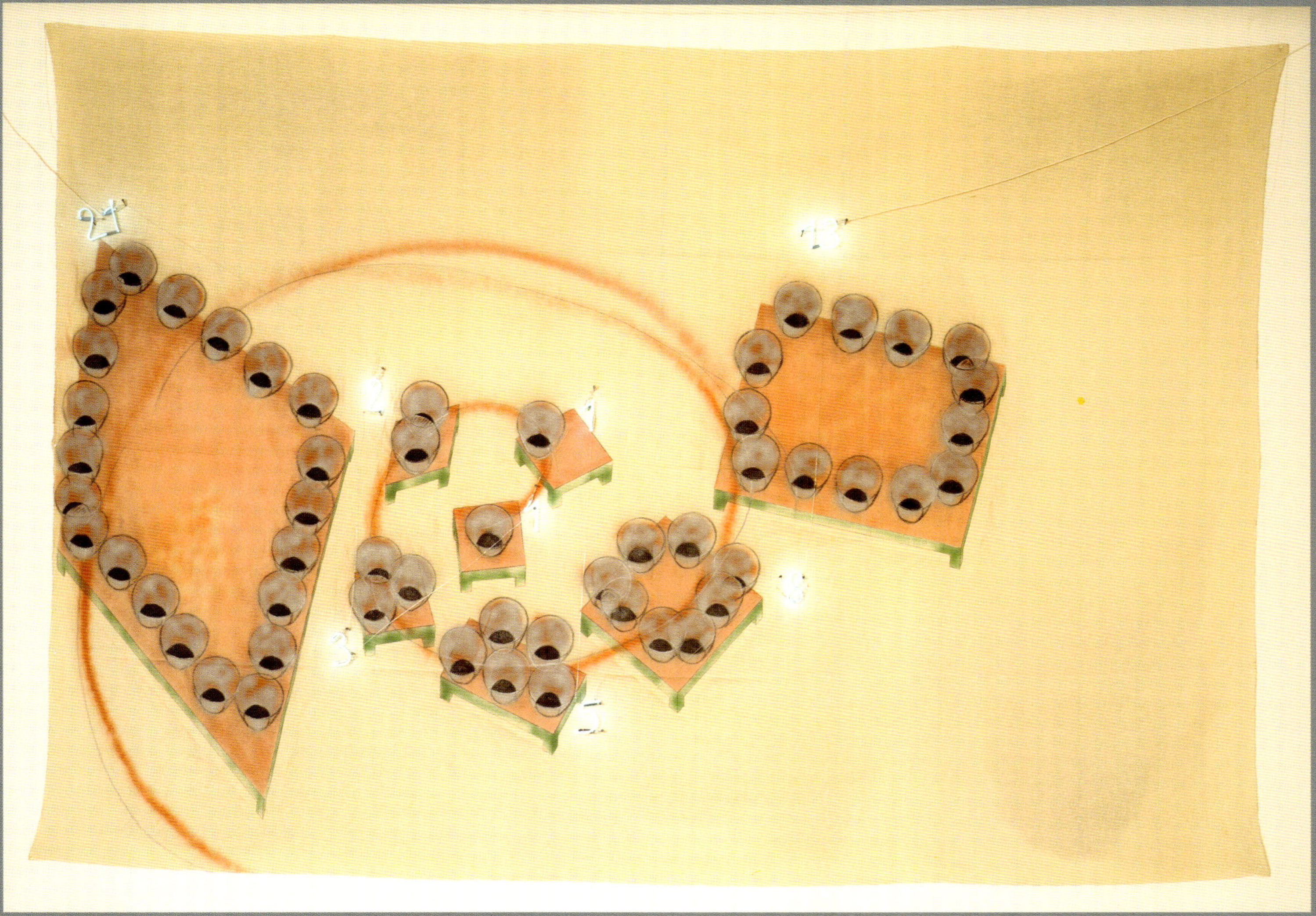

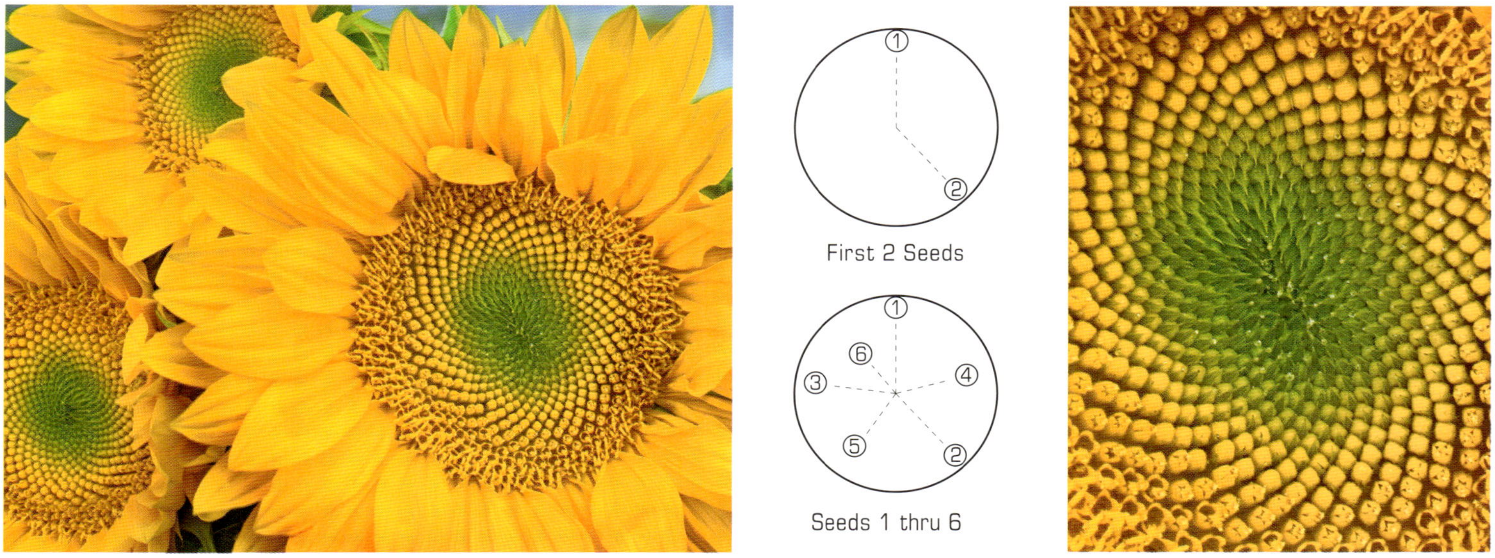

2-41. Phyllotaxis in sunflowers.

Phyllotaxis is a growth pattern in plants that results from the continuous addition of equal units (seeds, petals, or twigs) around an axis within a confined space. In some (not all) plants, this growth pattern results in a pattern in which the units are in a ratio of 1.618 . . . (about 1.62). Whether or not the growth pattern conforms to 1.62 depends on how the first two units divide the confined space. The first two seeds that enter the center of a sunflower divide the circle into two "pieces of pie" which are in the ratio 1.62 (the area of the larger "piece" is 1.62 the area of the smaller "piece"). Such a positioning of the first two units is programmed into the DNA of each botanical species with phyllotaxis.

As additional sunflower seeds enter into the divided space, they move to the place where they are least crowded, thus making a pattern of "phyllotaxis" (from the Latin word for "leaf" and the Greek for "arrangement"). The pattern is usually composed of two systems (of seeds, petals, or twigs) going in opposite directions. Because of the initial division of the space into the ratio 1.62, the two systems remain in this proportion as the plant grows. Thus the total number of spirals in the two systems (those turning clockwise and counter-clockwise) are in the ratio 1.62, and they are consecutive Fibonacci numbers.

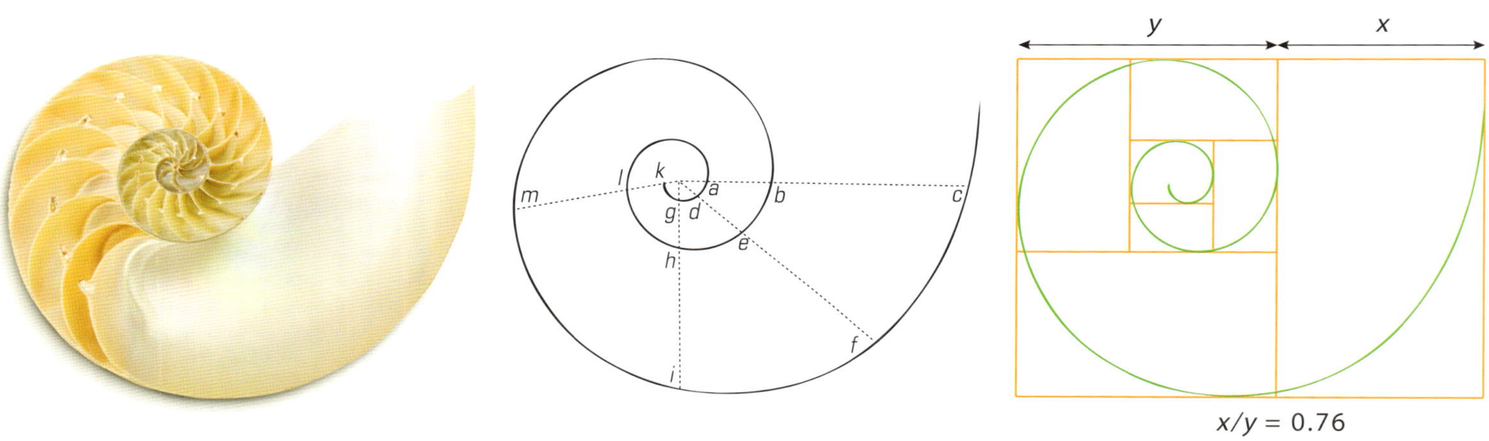

2-42. Nautilus shell (left) and diagram of the spiral of the nautilus (middle), in D'Arcy Wentworth Thompson, *On Growth and Form* (Cambridge: Cambridge University Press, 1917), fig. 367.

In animals, when a growth pattern (as seen in shells and horns) results from the addition of equal units, sometimes the elements that compose the pattern are in a ratio of about 1.62 (as in plants with phyllotaxis) and sometimes not, as in the nautilus shell. The ratio of the shell's width to its opening (as diagrammed by Thompson) is not 1.62 but 0.76 (right).

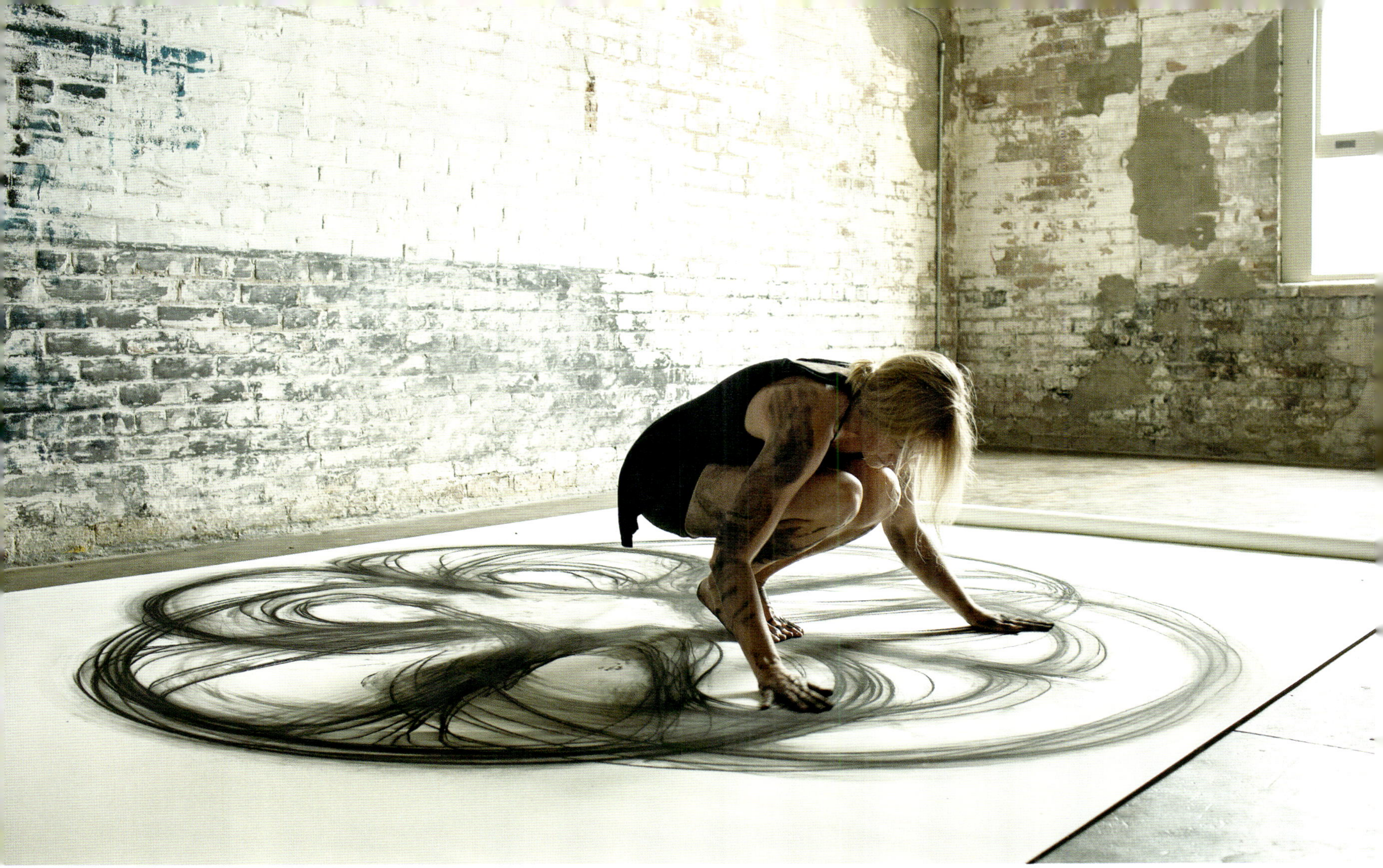

Before Darwin, artists sought ideal proportions in classical images such as Leonardo's picture of Vitruvius's timeless, immobile man, who reflects divinity by fitting within a perfect circle (plate 2-1). But in today's multicultural climate, artists—schooled in evolutionary biology—know that each person occupies a unique place on the tree of life, and the body changes over a lifetime (plate 2-43).

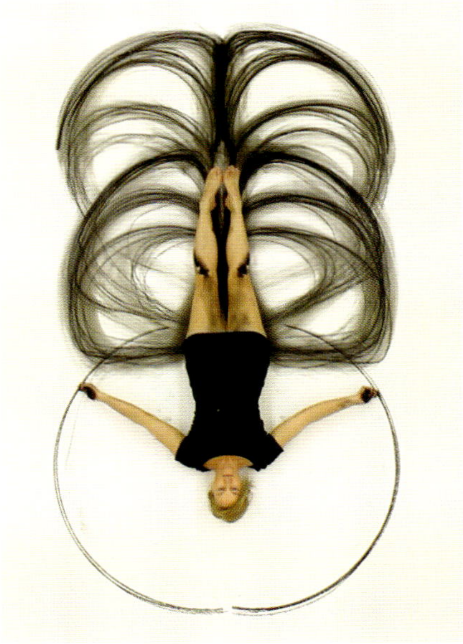

Systems of proportion based on an ideal, immutable human body slowly fell into disuse, and the myth of the Golden Section is fading into history as artists and architects come to understand patterns and proportions in nature as the result of a long process of evolution by natural selection.[43]

ABOVE AND RIGHT

2-43. Heather Hansen (American, b. 1970), kinetic drawings from the series *Emptied Gestures*, 2013. Charcoal on paper, 140 × 140 in. (355.6 × 355.6 cm). Courtesy of the artist and Ochi Gallery, Ketchum, ID.

Heather Hansen records the movement of her hands, feet, and body with charcoal while lying on a large piece of paper. Her dynamic gestures are thus abstracted—*emptied*—into meaning-free patterns, which—since they are unique to her—incorporate slight variations from perfect bilateral symmetry.

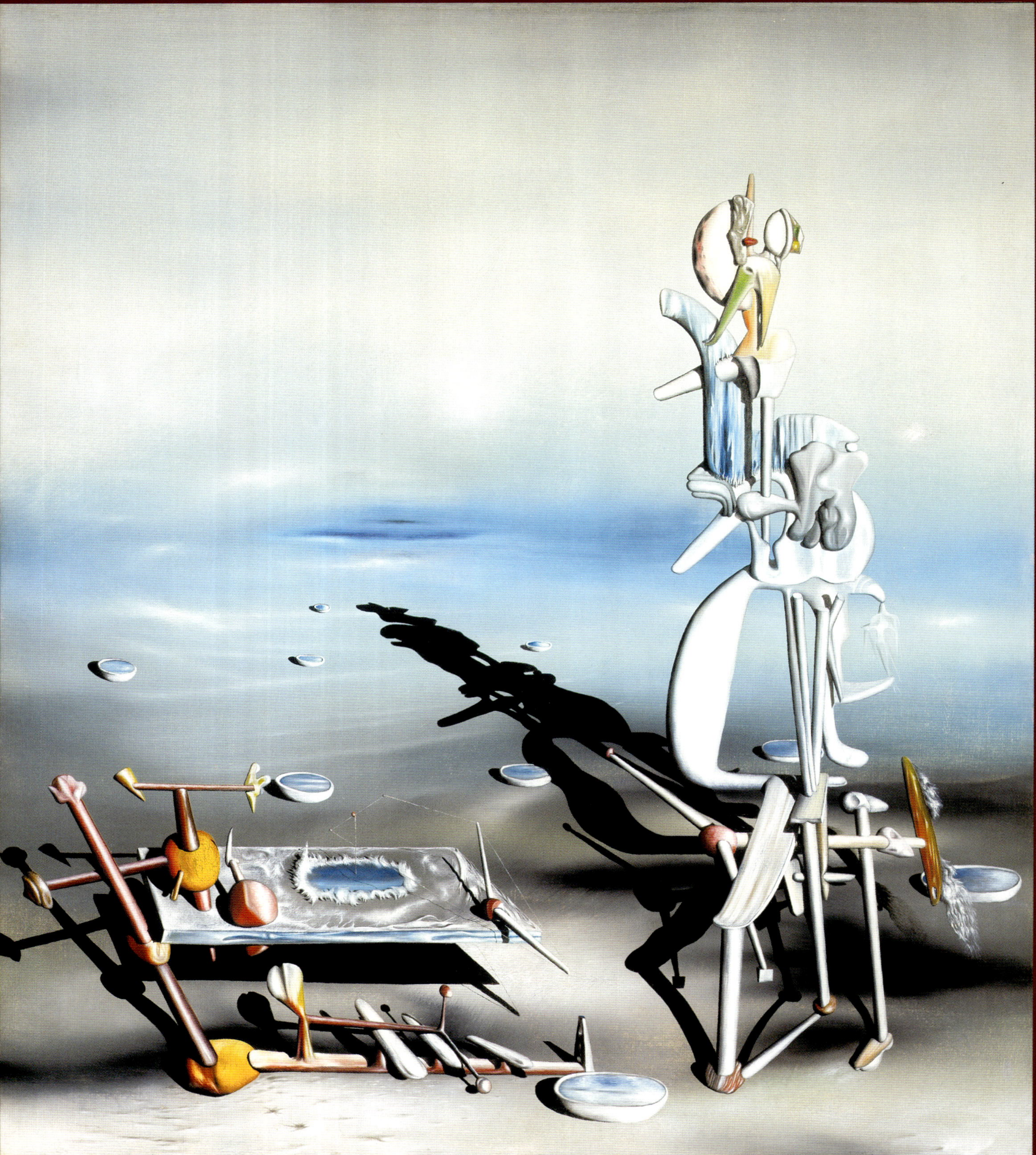

3
Infinity

Apart from the journeys carried out in the imagination or in a dream, for safe strolling or wandering it is absolutely necessary to have a solid base and ground as well as a smooth path, a path which doesn't stop anywhere, but must always remain passable, wherever the journey leads.
—**Georg Cantor, "Contributions to the Theory of the Transfinite," 1887**

Greek philosophers were wary of infinity—the absence of limits. Instead, they made use of the concept of "not finite." Euclid, for example, showed that the totality of prime numbers could not be exhausted in a finite list. When the medieval scholastics discussed concepts such as eternity, most heeded Aristotle's advice to consider a potential infinity (the days following one after another into the future), but not an actual infinity (all days existing together at the same moment in time). However, one Early Renaissance scholastic, the Catholic cardinal Nicholas of Cusa, was drawn to the idea of an actual infinity precisely because it was shrouded in mystery, and he used it to describe the infinite nature of the Christian God.

In the seventeenth century, Isaac Newton and Gottfried Leibniz devised calculus to measure bodies moving through an infinite number of points in space or time. They showed how to carry certain infinite processes to completion, and their specific calculations were precise, but the definitions they used to establish the limit of an infinite series remained vague. It was in the late nineteenth century that the Russian-born German mathematician Georg Cantor gave a precise mathematical definition of the size of an infinite set and described how to perform arithmetic with infinite numbers. This transfinite arithmetic arrived in the mathematics world as a completely new and surprising idea. Cantor observed that when an element is added to an infinite set, the number associated with the number of elements in the set (the cardinal number of the set) is unchanged. Unlike Euclid's arithmetic, in Cantor's transfinite arithmetic the whole is *not* greater than the part.

While some Greek philosophers mused about infinity, others pondered the nature of art. Plato had defined poetry and painting: "Poetic *mimêsis*, like the kind in a painting, is the imitation of appearance" (*Republic*, 596e–602c); in other words, a poem or a painting are depictions ("imitations") of an everyday object such as a tree, which is itself an imperfect copy ("appearance") of the real Tree (its perfect Form). Plato was unenthusiastic

3-1. Yves Tanguy (French, 1900–55), *Divisibilité indéfinie* (Indefinite divisibility), 1942. Oil on canvas, 40 × 35 in. (101.6 × 88.9 cm). Albright-Knox Art Gallery, Buffalo, NY, Room of Contemporary Art Fund, 1945. © 2015 Estate of Yves Tanguy/Artists Rights Society, New York

The French Surrealist Yves Tanguy symbolized a dream in which a bony figure towers over white bowls filled with a blue substance, rendered with realistic shades and shadows, which become ever smaller—dividing indefinitely—as they move towards a misty horizon.

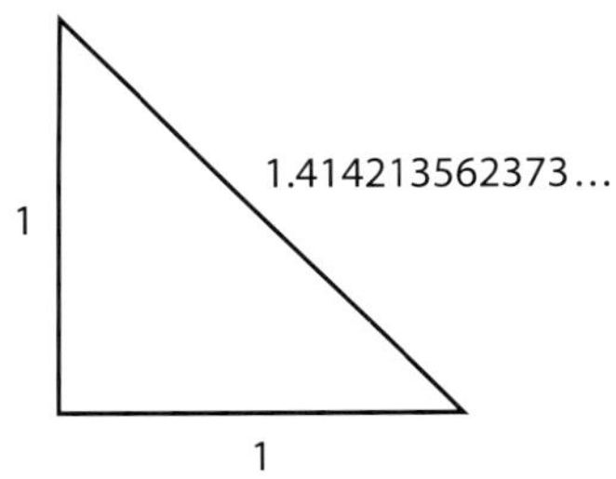

3-2. An irrational number: $\sqrt{2}$.

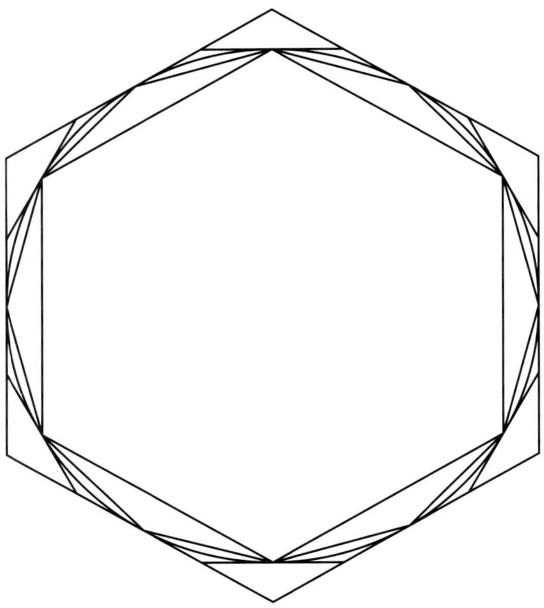

3-3. Archimedes's method of exhaustion for estimating pi, third century BC.

In the fourth century BC, Eudoxus of Cnidus developed a method for approximating the area of a curved surface by filling it with polygons (such as rectangles) whose areas he could measure using Euclidean geometry. As he increased the number of sides of the polygons, they would fill up—"exhaust"—the curve, giving him a closer and closer approximation. In the following century, Archimedes adopted this method to estimate the area of a circle by inscribing and circumscribing it with six-sided figures—hexagons. He could calculate the areas of the inner and outer hexagons, and he reasoned that the area of the "bound" circle was about halfway between. To get a closer approximation, he doubled the number of sides from six to twelve—dodecagons. To get even closer he doubled them again to twenty-four, and so on. To ensure that his lower and upper bounds were correct, Archimedes used either exact values or rounded up or down to an approximate rational number. When the polygons had reached ninety-six sides, he calculated their areas and showed that the value of π is between $3\frac{1}{7}$ (approximately 3.1429) and $3\frac{10}{71}$ (approximately 3.1408), which is consistent with its actual value of about 3.1416.

about mimesis because it was two steps removed from the eternal Forms, hastening to add "its products rank far below truth." Aristotle adopted mimesis from Plato, but since he did not believe that Forms exist independent of their embodiments in particular objects, he did not regard representational art as inferior.

Plato's negativity having been neutralized by Aristotle, the definition of art as mimesis—an imitation of the natural world—became fundamental to classical Western aesthetics. Plants and animals were understood as static, divine creations, but this outlook changed dramatically after Darwin's publication of *Origin of Species* (1859), which described nature as a web of dynamic forces without predetermined purpose or meaning. By the 1890s artists were responding to this new view of nature using a vocabulary of color and forms—in and of themselves—to symbolize invisible organisms, the energy of natural processes, and hidden psychological forces (plate 3-1).[1] Mathematics has been abstract since its inception because it is about patterns of numbers and geometric forms, but for two millennia art had been understood in Western culture as an imitation of the natural world. Abstract art arose as an expression of the scientific worldview, which is only a few hundred years old, and non-objective art—art that does not depict an object—is a uniquely modern phenomenon.

Many modern abstract artists have felt a strong affinity with mathematics precisely because of its abstraction. An early example of such a confluence of art and mathematics occurred in Moscow in the first decades of the twentieth century. After Cantor had published his theory of infinite sets, he went on to speculate about its philosophical implications, and he gave infinite sets a religious interpretation, identifying infinite numbers with a Supreme Being, whom he named "Absolute Infinity." Many of Cantor's colleagues in German academia criticized his work on infinity, but Cantor's practice of mathematics, as well as his philosophical and theological views, were welcomed in Moscow, where the mathematician Pavel Florensky, the poet Aleksei Kruchenykh, and the painter Kazimir Malevich recorded their aspirations to expand their minds to achieve consciousness of the Absolute.

In the nineteenth century, mathematicians and scientists debated whether man's actions are determined by laws of strict cause and effect (like the orbits of the planets) or guided by free will that introduces a degree of unpredictability (like the flow of turbulent water). Is man predestined by fate? Or is he free to choose a path and responsible for his actions? Although both positions are consistent with monotheism, Europeans and Russians with a Judeo-Christian outlook favored free will, and one Russian mathematician, Pavel Nekrasov, tried to settle the debate using probability theory.

The Greeks adopted from Egyptians the whole numbers (1, 2, 3, . . .) as well as fractions, such as 1/2, 3/4, and 5/16, which were acceptable to them because they can be expressed as ratios of whole numbers. Whole numbers and fractions together compose the *rational* numbers—all those expressible as a ratio—and the Pythagoreans, who gave a privileged place to whole numbers, naturally enough thought that the rational numbers were all that existed.

But consider a right triangle with the two short sides equal to 1: what is the length of the hypotenuse? By the Pythagorean theorem ($x^2 + y^2 = z^2$), the square of the hypotenuse equals the sum of the squares of the two sides. The square of each side equals 1, so the square of the hypotenuse is 2, and its length is the square root of 2 (the number which, when multiplied by itself, equals 2).[2] The square of the fraction 7/5 (1.4 in decimal notation) almost equals 2, since 1.4 × 1.4 = 1.96; 1.41 is a little closer (1.41 × 1.41 = 1.9881). One is getting very close with the decimal 1.414213562373, which, times itself, equals 2.000000001131102011499 (plate 3-2). But there is no fraction at all, no finite decimal, no matter how many decimal places it has, which, times itself, equals *exactly* 2. In the third century BC, Euclid named a number that cannot be written as a quotient of two integers "irrational," which has the pejorative overtone of lacking in human reason. The ancients discovered additional examples of irrational quantities: the ratio of the diagonal of a square to one of its sides and the ratio of a circle's circumference to its diameter (pi). The Babylonians used 3 1/8 as an estimate for pi, and the Greek mathematician Archimedes calculated that it was between 3 1/7 (approximately 3.1429) and 3 10/71 (approximately 3.1408). But pi cannot be expressed as a ratio of integers, and its decimal expansion goes on forever without any repeating pattern (3.14159 . . .). The numerical value of pi can be approximated more and more closely only by various infinite processes (plates 3-3 and 3-4).

In the fifth century BC, Zeno formulated several paradoxes about infinity as criticisms of atomism, the philosophy put forth by Democritus that divides space and time into discrete units (atoms). If Achilles gives a tortoise a head start on a racetrack that is composed of an infinite number of points in space, then Achilles cannot win, even though he runs much faster. To overtake the tortoise, Achilles would first have to reach the point where the tortoise started, by which time the animal would have moved forward. Again and again Achilles must reach the new point to which the tortoise advanced, by which time the tortoise has advanced another fraction of the diminishing distance separating them, and so on ad infinitum (plates 3-5 and 3-1, chapter frontispiece). The Greeks knew

3-4. Archimedes's method of exhaustion, from his collected works in Greek and Latin, *Opera Omnia* (Basel, Switzerland: Hervagius, 1544). Houghton Library, Harvard University.

The concept of a limit is implicit in Archimedes's method, which is shown in this diagram by an anonymous illustrator for an early printed edition of Archimedes's work. This diagram shows the general process of doubling the sides to get closer to the bound circle (the "limit"), although Archimedes began with a hexagon, not a square (which will not get to ninety-six by successive doubling of its four sides).

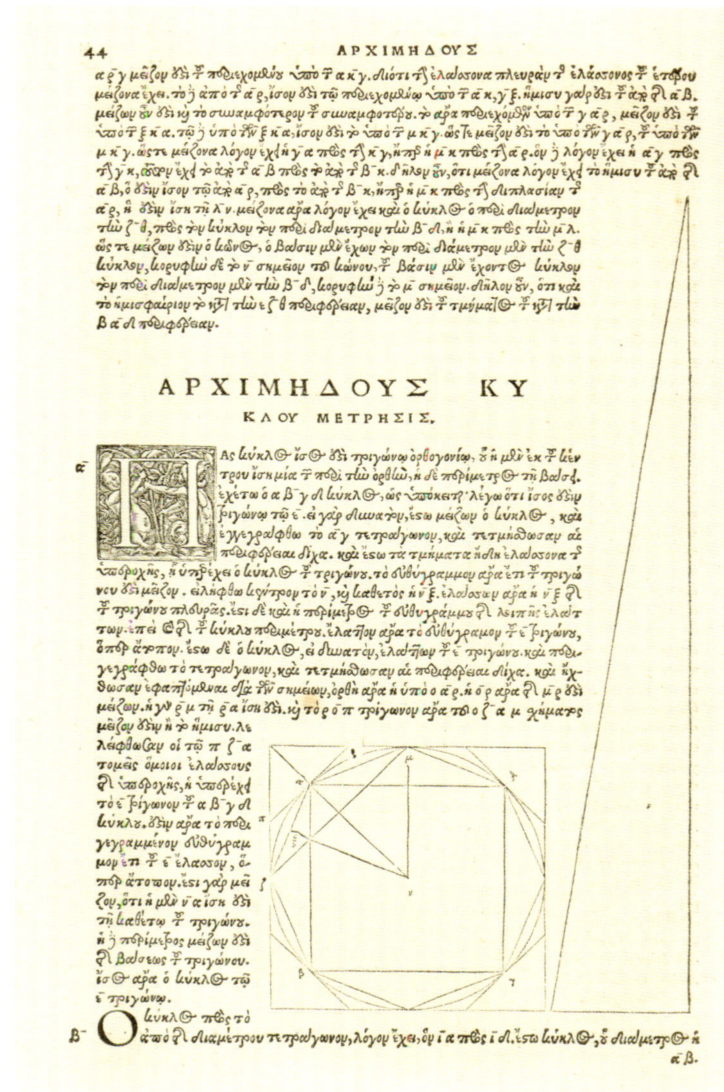
44 ΑΡΧΙΜΗΔΟΥΣ

ΑΡΧΙΜΗΔΟΥΣ ΚΥΚΛΟΥ ΜΕΤΡΗΣΙΣ.

3-5. Zeno's paradox of Achilles and the tortoise.

Achilles cannot catch the tortoise on a racetrack made up of a (potential) infinity of distinct points in space. Inherent in each of Zeno's paradoxes of time and space is the distinction between the discrete and the continuous. The core of this (spatial) paradox is the impossibility of treating space that people move in as composed of an infinity of points that are discrete and separate.

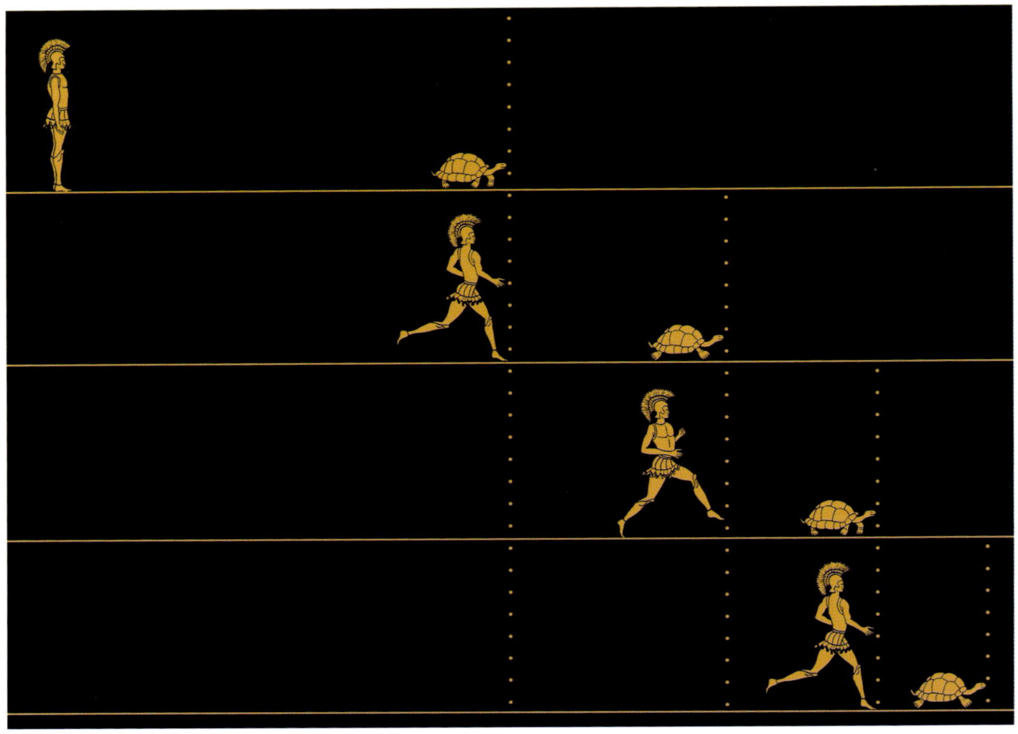

that (in the real world-out-there) Achilles could catch the tortoise and win the race, but they had no mechanism for resolving a conundrum like this that involves an infinite number of steps. Thus Greek mathematicians avoided the topic of infinity whenever possible.

THE MERGER OF GEOMETRY AND ALGEBRA

In the ninth century AD the Islamic mathematician al-Khwārizmī invented algebra, which proved to be an essential tool for measuring objects that are moving through time and space. By the mid-fifteenth century, sailors were venturing onto the open seas where there were no landmarks to guide them, and the fledgling science of positional navigation had produced charts with which sailors could find their location according to two co-ordinates: latitude, the distance north or south of the equator, and longitude, the distance east or west of an agreed point (such as Greenwich, England). But these charts did not accurately represent the earth's roundness; sailors needed navigational maps on which a straight line would correspond to a compass heading. In 1569 the Flemish mapmaker Gerardus Mercator devised a mathematical method for presenting a globe on a flat surface that allowed navigators to chart their courses by drawing the straight lines they needed for their compass headings (plates 3-6 and 3-7).

In the seventeenth century René Descartes took the first step towards measuring moving objects by adopting the cartographer's co-ordinate system and making it into an algebraic ("Cartesian") co-ordinate system (see sidebar at the top of page 114). Descartes noticed that, just as the locations of a sailing ship can be specified by its changing co-ordinates, so the solutions to an equation with two variables, such as $y = 2x$, can be given as a list of ordered pairs of numbers. By representing the pairs as points (by "graphing the

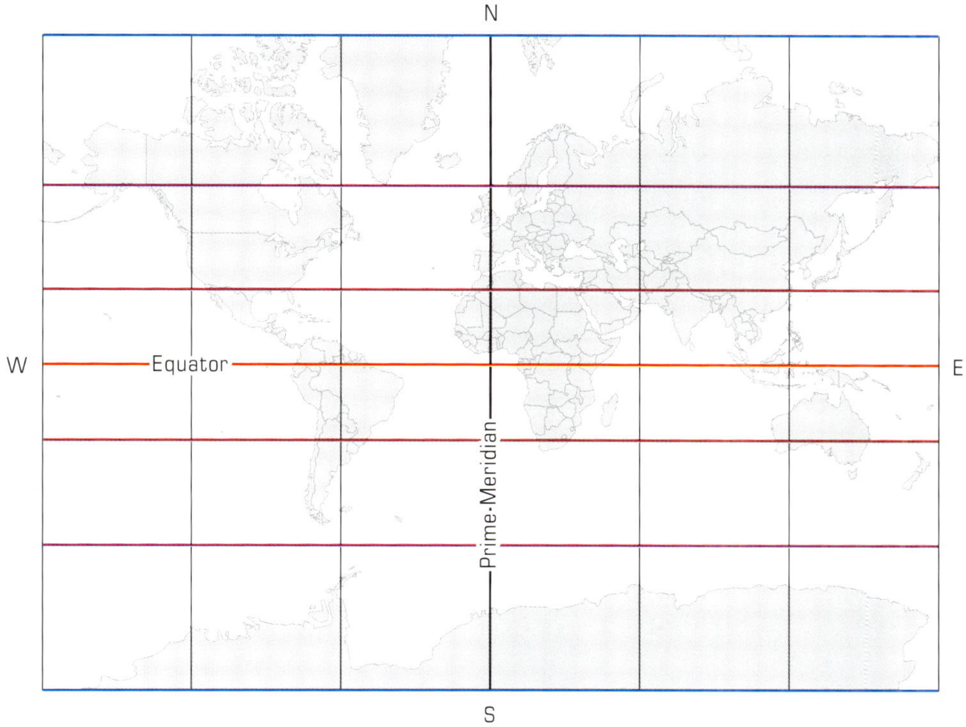

3-6. Mercator map of the world.

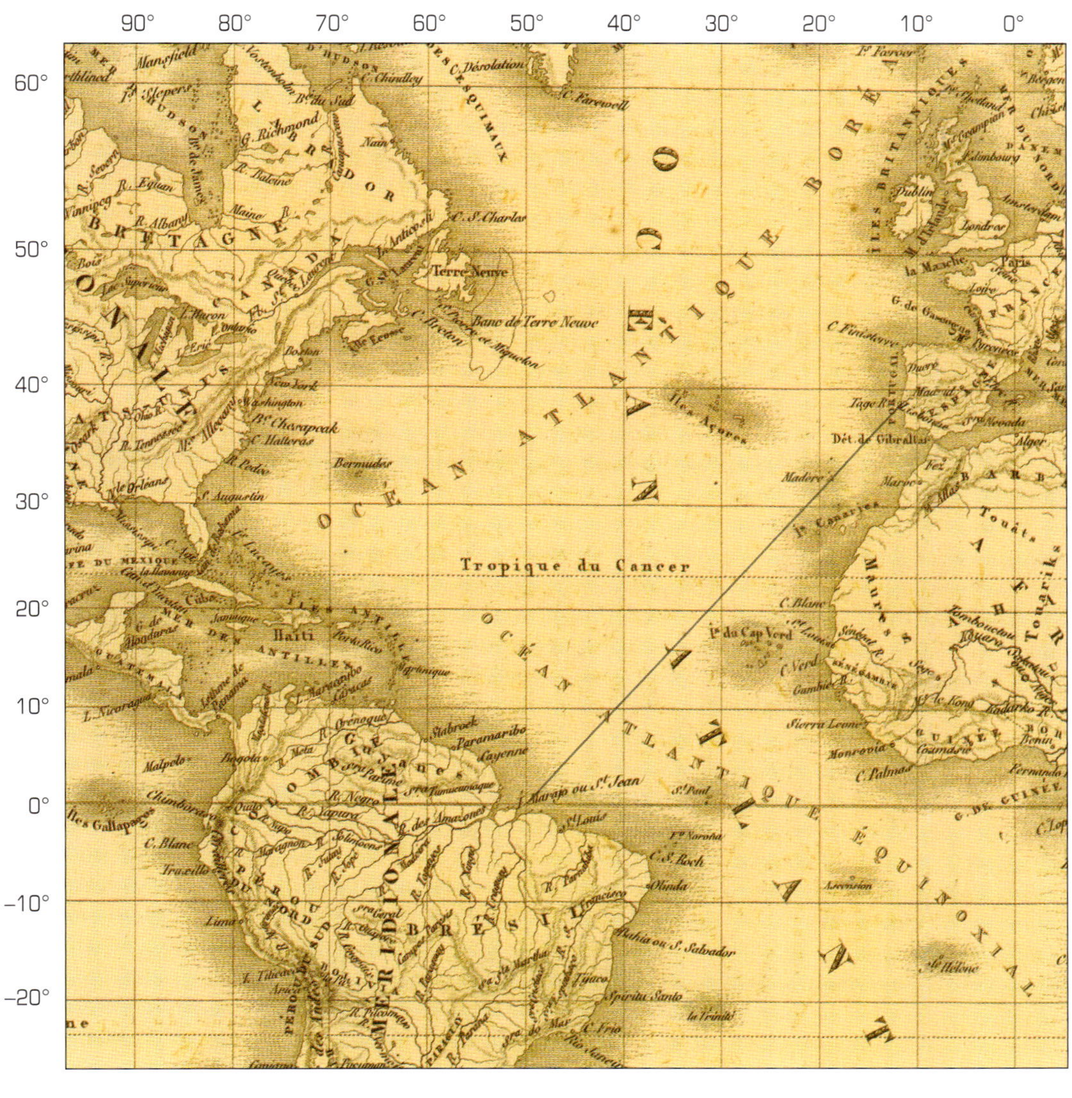

3-7. Chart of a voyage shown on a detail of "Mappemonde aux la projection de Mercator" (World map in a Mercator projection), by Adrien-Hubert Brué, *Atlas universel* (Paris: Ch. Picquet, 1842), 15. Hand-colored engraving, 15¼ × 20½ in. (39 × 52 cm). David Rumsey Map Collection, inv. no. 2741.015. © 2000 by Cartography Associates.

Although Mercator's map distorts areas, especially towards the Poles, he calculated that—crucially—it preserves angles, which means that any straight line on this map is equivalent to a line whose direction is constant as measured by the needle on a compass. Thus a navigator on the open seas could plot a course by drawing a straight line, such as this route from Brazil to Portugal, and stay on course by steering his ship true to a compass heading. The French cartographer who drew this map of the world patriotically made the prime meridian run through Paris.

From Brazil to Portugal

Latitude	Longitude
0°	50° W
10° N	42° W
20° N	33° W
30° N	22° W
38° N	12° W

Cartesian Co-Ordinate System

Descartes adopted the cartographer's co-ordinate system to "map" the solutions to algebraic equations. The solutions to an equation with two variables can be given as a list of ordered pairs of numbers, as shown in the sidebar below. By representing the pairs as points and then connecting the dots, Descartes produced a geometric picture (a map, or graph) of the equation.

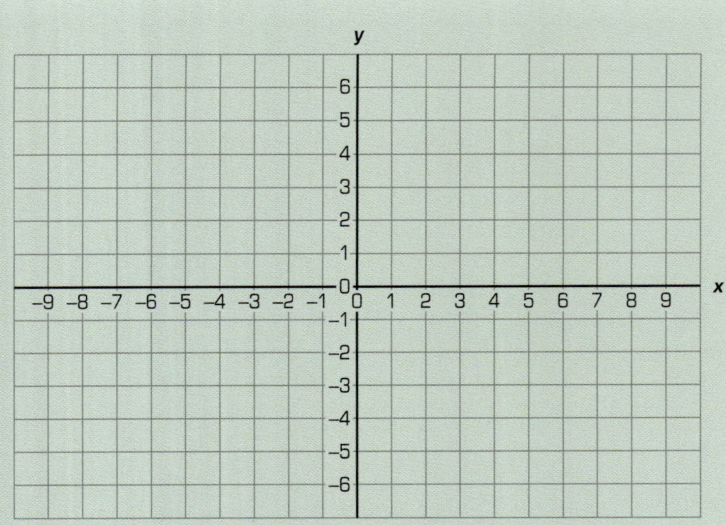

Cartesian coordinate system

Graphs of a Linear and a Quadratic Equation

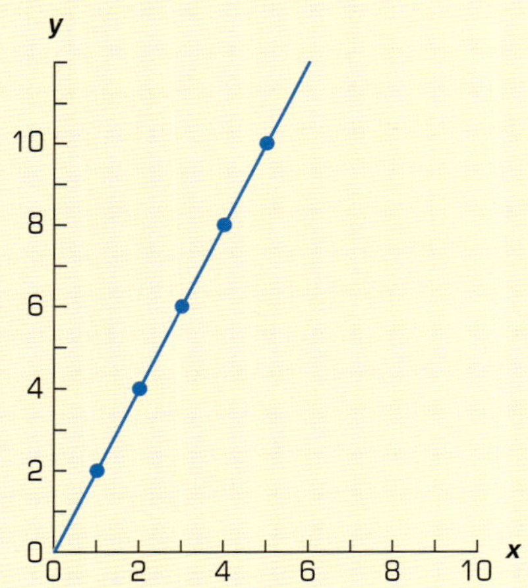

$y = 2x$

x	y
1	2
2	4
3	6
4	8
5	10
⋮	⋮

A Linear equation

In equations such as this ($y = 2x$), in which neither x nor y is raised to a higher power, the graph is a straight line—a "linear" equation.

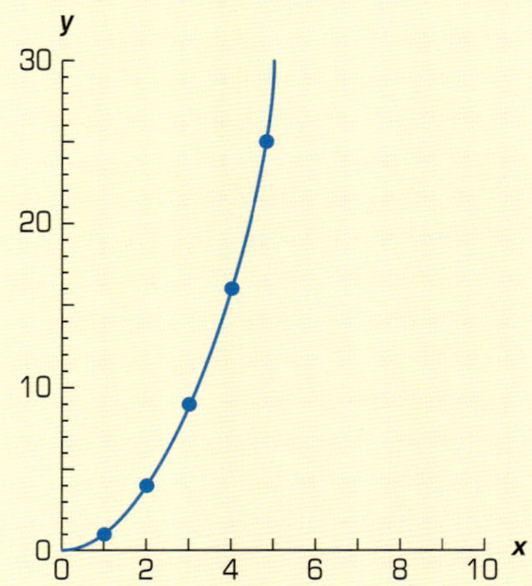

$y = x^2$

x	y
1	1
2	4
3	9
4	16
5	25
⋮	⋮

B Quadratic equation

If one of the variables is squared, as in $y = x^2$, then the equation is "quadratic," and its graph is a curve (in this case, a parabola).

Parabola and Ellipse

A Parabola

A parabola is the set of all points equidistant from a line (the directrix) and a fixed point (the focus of the parabola), which is not on the line. In this diagram, the distance d_1 from the focus to a point on the parabola is the same as the distance d_2 from that point to the line.

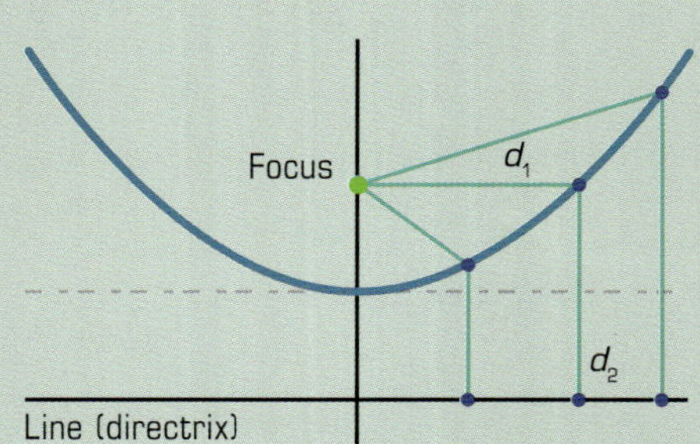

B Ellipse

An ellipse is the set of points such that the sum of the distances from any point on the ellipse to its two fixed foci is constant. In other words, for any point on the ellipse, the sum of the distances from that point to the foci ($d_1 + d_2$) is the same.

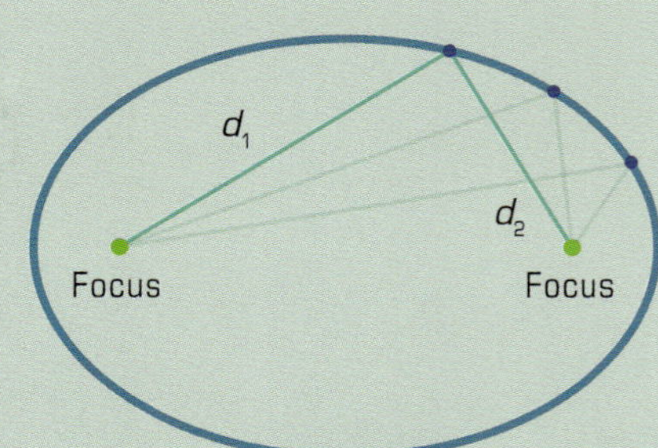

equation") and then connecting the dots, Descartes produced a geometric picture of the changing—"moving"—value of y relative to x. In an appendix to *Discourse on Method* (1637), Descartes went on to generalize the relations between equations and their corresponding graphs, showing, for example, that for all equations (such as $y = 2x$) in which neither x nor y is raised to a higher power, the graph is a straight line (a "linear" equation). On the other hand, if one of the variables is squared (a "quadratic" [square] equation), such as $y = x^2$, then its graph is a curve (see sidebar in the middle of the opposite page).

Conversely, Descartes noted that the geometer's lines, angles, and curves could be plotted on a graph, their points given as a list of paired numbers (co-ordinates); these, in turn, could be expressed as a formula relating the x and y co-ordinates. In late antiquity the Greeks had studied curves formed by a plane intersecting a cone (conic sections), including a parabola and an ellipse (see plates 1-65 and 1-71 in chapter 1 and the sidebar at the bottom of the opposite page). Apollonius of Perga (first century AD) summarized theorems on these curves in his treatise *Conics*, much as Euclid had summarized theorems about curves that can be drawn with a compass (arcs of a circle) in his *Elements*. Descartes showed that curves can be graphed and that their paired values on the x and y co-ordinates translate into an equation. Indeed, using Descartes' new tool, conic sections could be graphed and expressed as numbers (points with paired values on the x and y co-ordinates; the quadradic equation in the sidebar in the middle of the opposite page describes a parabola). In short, Descartes had united arithmetic and geometry and created a new branch of mathematics, algebraic (or "analytic") geometry, in which algebraic equations can be solved geometrically and geometric theorems can be proven using algebra.

CALCULUS: MEASURING CHANGING PHENOMENA

Seventeenth-century mathematicians were especially concerned with measuring the motion of the moon because of its importance to navigators on the open sea. Using a Mercator map, sailors could estimate a ship's latitude by observing the sun's zenith by day and pole star (North Star or South Star) at night. But to measure the ship's longitude—its position east or west of a line of longitude (a meridian)—required both observing the position of the moon and knowing the exact time. Gravity-based water clocks and pendulums are inaccurate on a swaying ship. (The accurate determination of longitude was possible only after the invention in the mid-eighteenth century of the marine chronometer, which is unaffected by rough seas. Before this invention, navigators relied on the position of the moon.) Euclid's geometry of timeless, perfect spheres was inadequate to describe the motion of the moon, which orbits in an elliptical path, constantly speeding up and slowing down under gravitational tugs from earth and its oceans. In 1669 Isaac Newton invented a new mathematical tool—calculus—to describe moving bodies. Leibniz independently formulated an alternative version of calculus, and the two men have gone down in history as its co-discoverers. The following exposition continues to tell Newton's tale of discovery, while describing their approaches in more modern terms by incorporating later refinements and notation.

Measuring a Changing Phenomenon

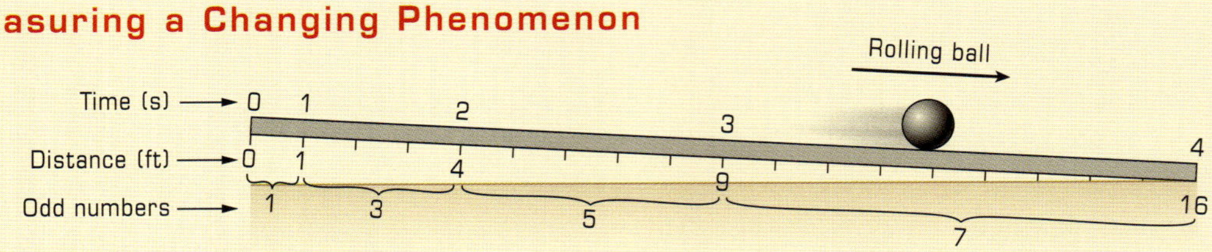

To measure a changing phenomenon, first plot its course on a graph.

Example of a solid brass ball rolling down a plank: in the spirit of Galileo's work, one can adjust the slope of a plank so that a brass ball rolls down the slope slowly enough that its speed can be measured using a water clock. In this example, the slope of the plank is very gentle, only one vertical unit for 16 horizontal units. One can set things up so that a ball released at the top of the plank at time 0 will roll 1 ft. along the plank in the first second. In 2 seconds it rolls 4 ft., in 3 seconds it rolls 9 ft., and, in general, the distance it travels (in feet) is equal to the square of the elapsed time (in seconds): $D = T^2$. The distances between the successive positions are the successive odd integers: 1, 3, 5, 7, 9, 11, 13, ...

We then draw a graph of the distances of the brass ball along the plank over time.

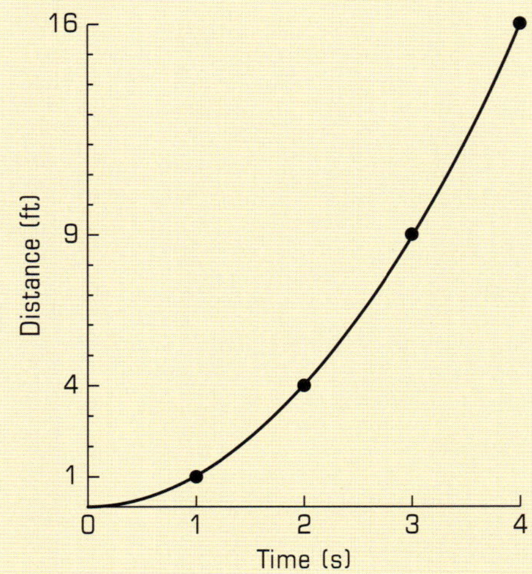

Newton first adopted Descartes's co-ordinate system to represent changing phenomenon over time as a curve on a graph, and he defined an instant as a point on the graph (see sidebar above). Newton then defined the *rate* of change at any instant (at any point) as equal to the slope of a tangent to the curve at that point. An instant of time and a point in space are ideal entities, and, in practice, absolute precision is impossible when measuring the location of a continuously moving body. Newton's key insight was to introduce a "limit" as a way of getting as close to the instant (the point on the graph) as necessary, given the task at hand. For example, to keep a ship on course, a navigator might need to know its position at midnight. Newton gave a way to specify the degree of accuracy needed; at midnight, give or take a second, might be close enough, or give or take a 10th of a second, or a 100th, and so on (see sidebar at top of opposite page). Newton's calculus also gave a method to measure the total *amount* of change by calculating the area under a curve on a graph (see sidebar at bottom of opposite page). This method recalls Archimedes's method of exhaustion, in which the concept of a limit is implicit. Today classical (Newtonian) physics, Einstein's theory of relativity, and twentieth-century descriptions of the subatomic realm (quantum physics) are all rooted in calculus.

In developing calculus, Newton and Leibniz had resurrected age-old problems about infinity. If the exact co-ordinates of the moon at midnight are known, to what point does it move in the next instant? Between 12:00 midnight and 12:01, there are an infinite number of fractions of a second—12:001, 12:0001, . . . ad infinitum. In an attempt to solve the problem, Leibniz's approach was to imagine an expanse in space or duration of

Differential Calculus

Differential calculus measures the slope of a curve at a point, which is equivalent to measuring the rate of change at that instant. Finding the slope of the curve at a point is the same as finding the slope of the tangent to the curve at that point.

To determine the slope at a point (I), draw a second point (J) a little farther along the curve. In this example, the points are 3 seconds apart. Connect the points. This gives a rough approximation of the slope of the curve at I.

The amount of change in distance (shown on the y co-ordinate) during the run (the distance travelled between 1 and 4 seconds, shown on the x co-ordinate) is 15. This is called the "rise" of the curve. The average distance that the brass ball rolled over the elapsed time equals rise/run = 15 ft./3 seconds = 5 ft./second.

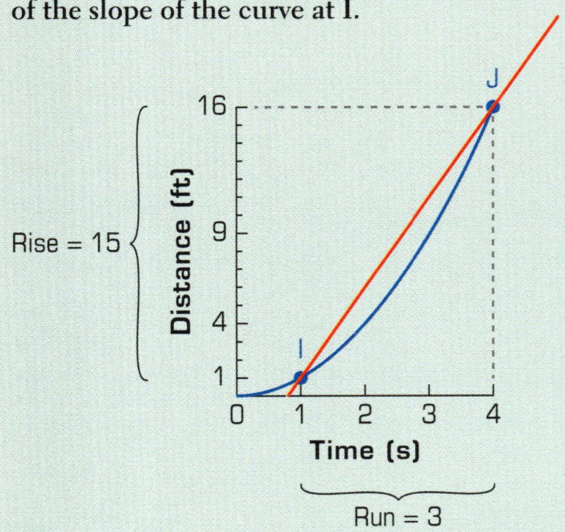

$$\frac{\text{Rise}}{\text{Run}} = \frac{16-1}{4-1} = \frac{15}{3} = \frac{5}{1} = 5 \text{ ft/1 s}$$

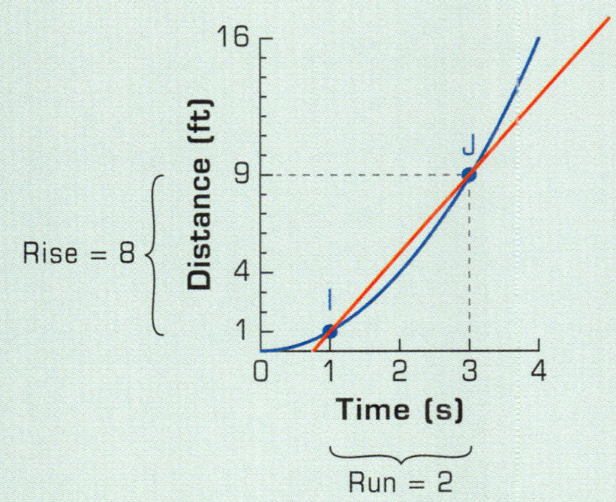

$$\frac{\text{Rise}}{\text{Run}} = \frac{9-1}{3-1} = \frac{8}{2} = \frac{4}{1} = 4 \text{ ft/1 s}$$

To get a more accurate tangent, move J closer to I.

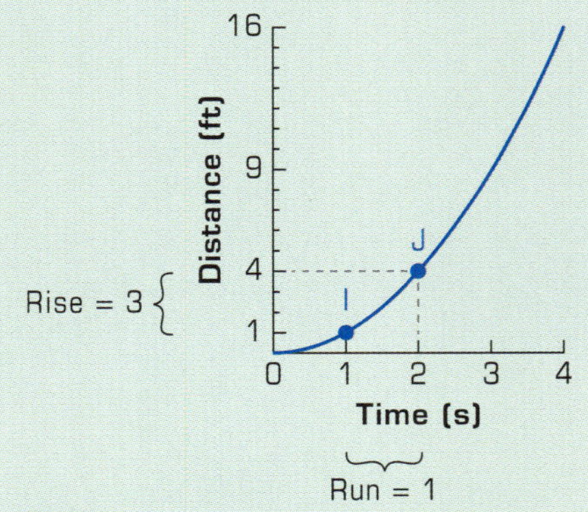

$$\frac{\text{Rise}}{\text{Run}} = \frac{4-1}{2-1} = \frac{3}{1} = 3 \text{ ft/1 s}$$

The limiting case is when J approaches I, that is, when r (the run) approaches zero.

The general (ideal) formula is:

$$\lim_{r \to 0} \frac{\Delta y}{\Delta x} = \text{Slope at I}$$

where Δx stands for "the change in x" and Δy stands for "the change in y." This reads "the slope of the tangent at I (in other words, the rate of change at I) equals the ratio between the change of position on the y co-ordinate and the change on the x co-ordinate, as r approaches zero."

Integral Calculus

Integral calculus measures the area under a curve, which is equivalent to summing up the total amount of change. The basic idea is to fill in the area with regular polygons such as rectangles, because their areas can be measured using Euclidean geometry.

The area of a rectangle equals its width times its height. To approximate the total area, add the areas of all the rectangles. The general (ideal) formula is:

$$\lim_{\text{WIDTH} \to 0} \sum_{k} \left(width_k \times height_k \right) = \text{Area}$$

This reads, "The limit, as the width of the rectangles approach zero, of the sum (denoted by Σ, the Greek letter sigma) of all the widths times the heights of the (k number of) rectangles, equals the area under the curve."

The width can be cut in half to get an even closer approximation, and so on, ad infinitum.

In practice, a width is set that is appropriate to the degree of accuracy required for the task. As the width approaches ever closer to zero, the area of the rectangles approaches with ever-greater accuracy the area under the curve.

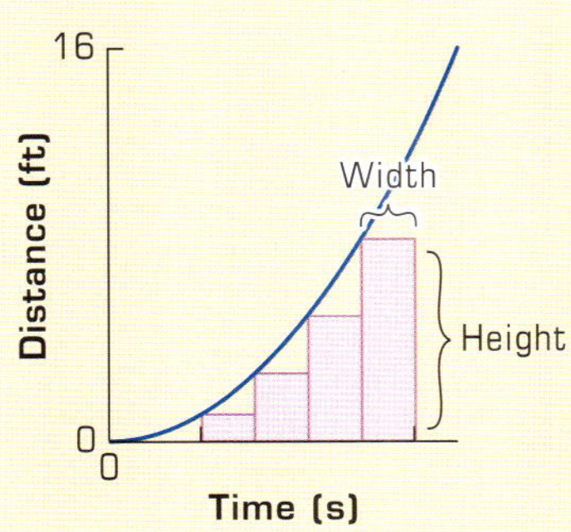

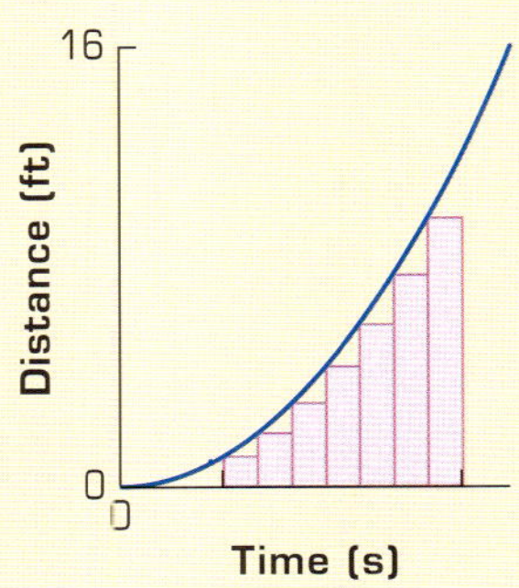

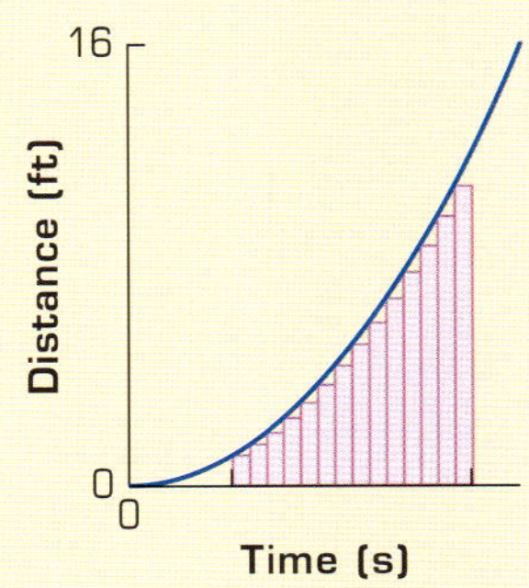

But he who can digest a second or third fluxion, . . . need not, me thinks, be squeamish about any point in divinity. . . . With what appearance of reason shall any man presume to say, that mysteries may not be objects of faith, at the same time that he himself admits such obscure mysteries to be the object of science?

—**Bishop George Berkeley, *The Analyst: a Discourse Addressed to an Infidel Mathematician*, 1734**

time—an "infinitesimal"—that is smaller than any given number but greater than zero, Newton worked with limiting ratios that he called "fluxions" (to suggest something moving in "flux"). In other words, an infinitesimal was a not-quite-dimensionless point in space or time; an engineer could set its value given the task at hand. Although their approaches were somewhat different, Newton's and Leibniz's results were the same. Critics mocked the fluxion as a phantom notion, but Newton, fully aware of its vagueness, used the term because with calculus a navigator could calculate the location of his ship and avoid crashing into rocks in the middle of the night.

MULTIDIMENSIONAL GEOMETRIES

In the century after Newton and Leibniz adopted Descartes's co-ordinate system to represent an event unfolding in time, the German mathematician Carl Friedrich Gauss generalized Descartes's system to make it an even better tool. Gauss noticed that a line could be considered a one-dimensional "space" in the sense that location on a line can be described as a point on that one co-ordinate. A plane with two Cartesian co-ordinates could be thought of as a two-dimensional "space" described by its x and y co-ordinates. Gauss suggested that a z co-ordinate could be added to describe a three-dimensional space, such as a cube, by specifying points as ordered triples on the x, y, and z co-ordinates.

It was an easy step for Gauss to add another variable—a w co-ordinate—for time and to describe points moving in three-dimensional space over time as ordered quadruples in a four-dimensional system.[3] In the mid-nineteenth century the Scottish scientist James Clerk Maxwell used a four-dimensional geometry to describe the positions of molecules moving in a three-dimensional volume of gas over time (*Kinetic Theory of Gas*, 1866). Theoretically there could be "spaces" with any number of co-ordinates—n-dimensional spaces—in which points could be described as ordered n-tuples. By the 1840s several mathematicians had developed such systems and were describing a "four-dimensional space" or, more generally, an "n-dimensional space."[4] These systems, because they use Euclid's five axioms (see the sidebar on page 17 in chapter 1), are considered generalizations of Euclid's *Elements* and are called multidimensional Euclidean geometries. An n-dimensional geometry can be used to record data in any system with multiple variables. For example, a six-dimensional geometry can record the weather with variables for altitude, north-south, east-west, time, humidity, and temperature.

We ought then to regard the present state of the universe as the effect of its anterior state and as the cause of the one to follow. Given for one instant an intelligence which could comprehend all the forces by which nature is animated and the respective situation of the beings who compose it—an intelligence sufficiently vast to submit these data to analysis—it would embrace in the same formula the movements of the greatest bodies of the universe and those of the lightest atoms; for it, nothing would be uncertain and the future, as the past, would be present to its eyes.

—**Pierre-Simon Laplace, *Philosophical Essay on Probabilities*, 1812**

PROBABILITY THEORY

Maxwell not only used a four-dimensional geometry to describe his kinetic theory of gas, but in the same 1866 publication he introduced statistical methods into the physics laboratory to measure atoms and molecules. Although city managers and actuaries had adopted statistics soon after its invention in the seventeenth century, scientists of the day did not see its relevance to their deterministic universe, where, in their view, random events did not occur. When the atomic theory of matter was developed in the nineteenth century, the

French mathematician and astronomer Pierre-Simon Laplace tried to retain macroworld precision of measurement in the microworld, but Maxwell found it impractical to keep track of each molecule of water in a cloud of steam—thereby bringing probability theory, which underlies all statistics, into the halls of science.

The theory of probability had been invented in the seventeenth century when the Frenchman Antoine Gombaud, Chevalier de Méré, became interested in an old gaming puzzle—the problem of the points. Two players contribute equal sums to a prize fund and agree to play a game of chance; the winner will be declared after a specified number of rounds, say the best of five. If the game is interrupted before there is a winner, how do they divide the prize money? Chevalier de Méré sought help from his friend the mathematician Blaise Pascal, who came up with a mathematical method for dividing the stakes. After an exchange of letters commencing in 1654 with his countryman, the mathematician Pierre de Fermat, Pascal determined that it was possible to calculate the average of the outcomes of any large number of random events—probability theory. The toss of an individual coin is unpredictable (it can come up either heads or tails), and each toss is a separate event. But the average of the outcome of many tosses is predictable (the ratio of the number of heads to the total number of tosses is very likely to be close to one-half). Pascal visualized the branching tree of possibilities of such random events as a growing triangle (plates 3-8, 3-9, and 3-10).

Mathematicians in Pascal and Fermat's day worried that natural language (everyday speech) was too ambiguous and mutable to use in science, and so they developed the first artificial languages. In 1685 Leibniz founded modern logic when he invented a set of

The data of the statistical method as applied to molecular science are the sum of large numbers of molecular quantities. In studying the relations between quantities of this kind, we meet with a new kind of regularity, the regularity of averages, which we can depend on quite sufficiently for all practical purposes, but which make no claim to that character of absolute precision which belongs to the laws of abstract dynamics.

—**James Clerk Maxwell, "Molecules,"** ***Nature*****, 1873**

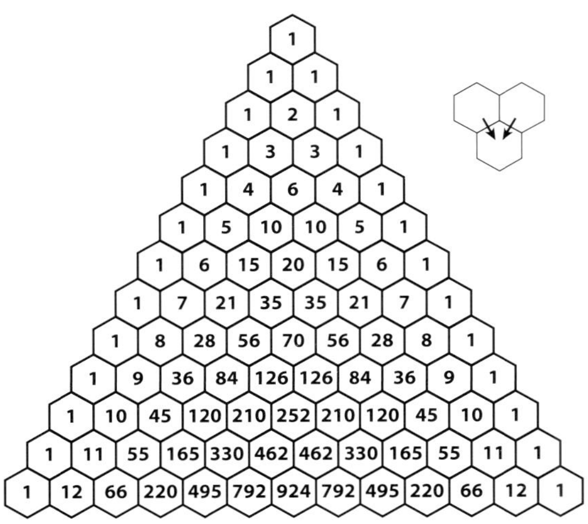

3-8. Pascal's triangle, seventeenth century AD.

Each pair of numbers in the triangle sums to a third located below the pair. Pascal worked out its patterns in the seventeenth century when he was calculating odds for games of chance, such as tossing a coin, in which each toss has two possible outcomes, in his *Traité du triangle arithmétique* (Treatise on the arithmetic triangle) 1653.

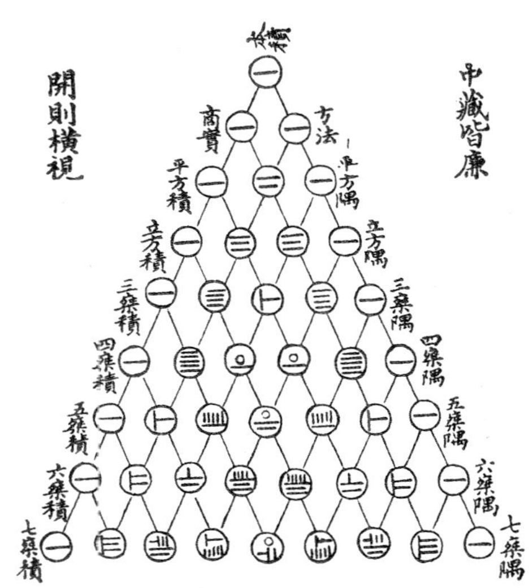

3-9. Yang Hui's triangle in Zhu Shijie's *Siyuan yujian*, AD 1303. Needham Research Institute, Cambridge, England.

The identical triangular pattern, which is named for Pascal in the West, is known in China as Yang Hui's triangle after the Song dynasty mathematician Yang Hui (AD 1238–98). In the columns of characters in the upper right, Yang Hui describes the structure of the diagram, and on the left he tells the reader to view it with the apex of the triangle pointing up (as shown). In Persia (present-day Iran), it is known as Khayyám's triangle after the work of an even earlier mathematician, Omar Khayyám (1048–1131), who described the same pattern and also wrote poetry, a selection of which was translated into English as *The Rubáiyát of Omar Khayyám*.

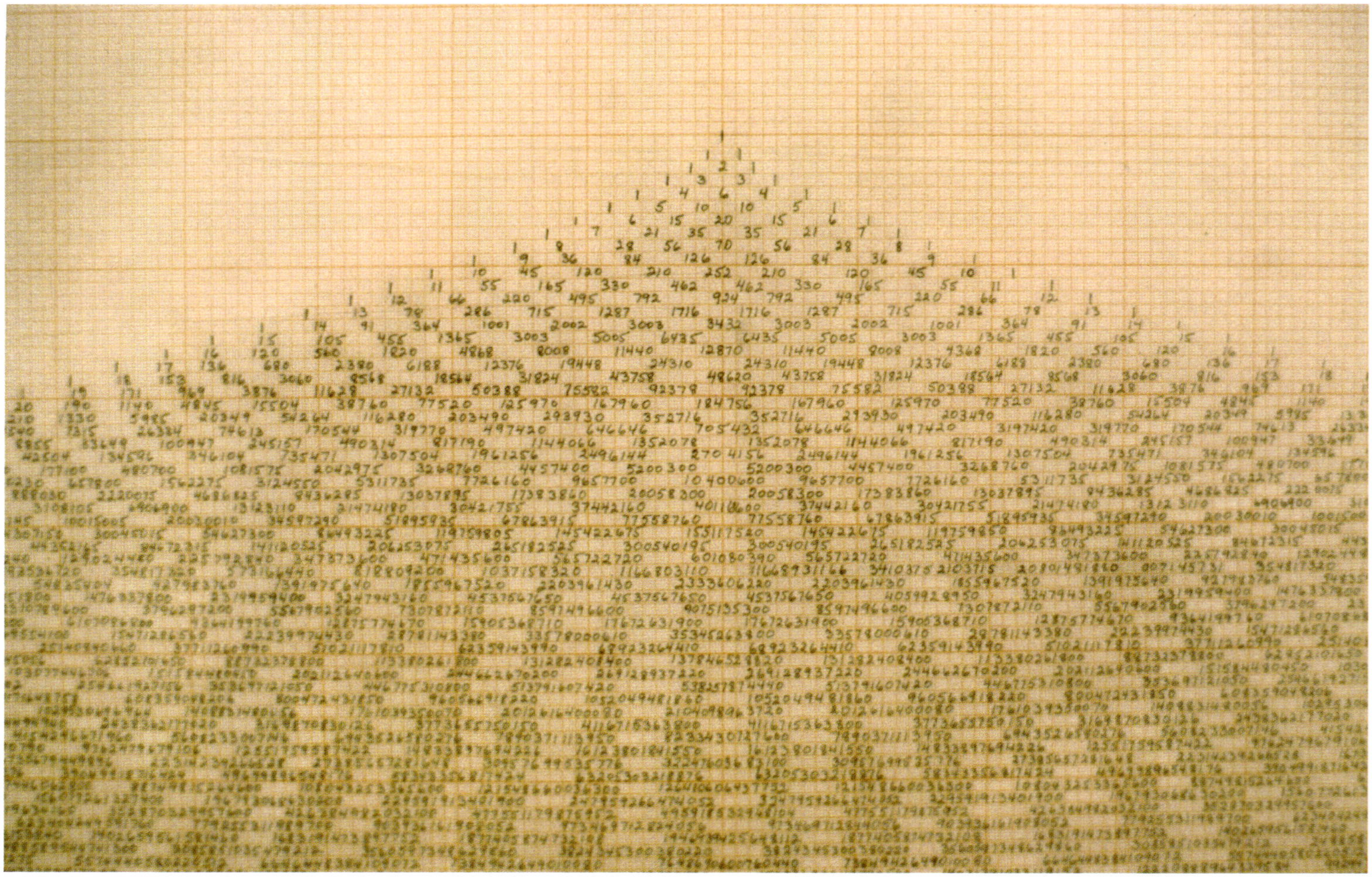

3-10. Agnes Denes (Hungarian-born America, b. 1931), *Pascal's Triangle*, no. 3, from *The Pyramid* series, 1974. India ink on paper; the whole drawing measures 15 in. × 16 ft. (38.1 cm × 4.87 m). Collection of The Ohio State University, 1980.017.000. Courtesy of the artist and the Wexner Center for the Arts.

precise symbols and mechanical rules for manipulating them so that by calculation alone he could determine the truth or falsity of propositions (*The Art of Discovery*, 1685; see chapter 5). In his system, propositions came out either (100 percent) true or (100 percent) false, but Leibniz realized that he could not calculate the truth (or falsity) of statements such as "It will rain tomorrow" because they are not 100 percent true or false. Inspired by Pascal's work, Leibniz recommended that a theory of probabilistic reasoning be developed, and he sent his friend, the Swiss mathematician Jacob Bernoulli, a copy of Pascal's triangle. Between 1684 and 1689, Bernoulli abstracted and generalized Pascal's triangle to yield a theorem—the law of large numbers (*Ars Conjectandi*, published posthumously in 1713), according to which the relative frequency of an event (such as tossing a coin) tends to stabilize as the number of repetitions increases. Bernoulli conceived of the probability of coin tosses, repeated under uniform conditions, as the long-term stability of relative frequencies (50 percent heads and 50 percent tails), and more generally, arithmetic averages arising from repeated trials. Bernoulli wanted to apply probability theory in the real world, such as establishing actuarial tables (life insurance) and assessing taxes; the term "statistics" (German *Statistik*, French *statistique)* derives from the Latin *statisticus*, "of the state." Thus Bernoulli developed a rigorous *probabilistic* logic to use when there is only partial evidence from which to determine that a statement is true (or false) to a degree of probability, as when predicting the weather or making a moral judgment.[5]

CANTOR'S ACTUAL INFINITY AND SET THEORY

As nineteenth-century scientists began exploring the microworld of heat, light, and electricity, they needed mathematical tools to describe another kind of moving phenomena—an oscillating wave. The French mathematician and physicist Joseph Fourier defined a mathematical series that describes the transfer of heat through a solid object, such as a frying pan. Fourier's results were later extended to apply to all periodic, oscillating phenomena, including the motion of sound-waves through the air. Fourier showed that any complex wave pattern (such as the sound produced by playing a chord on a piano) can be analyzed into "pure" waves (sounds produced by striking tuning forks). Fourier defined a mathematical series that decomposes any periodic function (such as the mathematical description of a complex sound-wave) into a set of simple oscillating functions (a set of mathematical descriptions of the pure tones) of which the periodic function is composed (*Memoire on the Propagation of Heat in Solid Bodies*, 1807). A Fourier series can be pictured geometrically on a graph (as in plate 3-11), or described algebraically using trigonometric functions (functions that describe angles).

Cantor's primary motivation for developing set theory came from his work on Fourier series. Cantor began by asking: When must two Fourier series be identical in every way? His work on infinity—transfinite numbers and transfinite arithmetic—was a by-product of his study of the trigonometric functions related to this question. As part of this research, Cantor defined one of the most basic ideas in mathematics—a number—and showed how a number results from counting objects in a set (a manifold of particulars; plate 3-13).

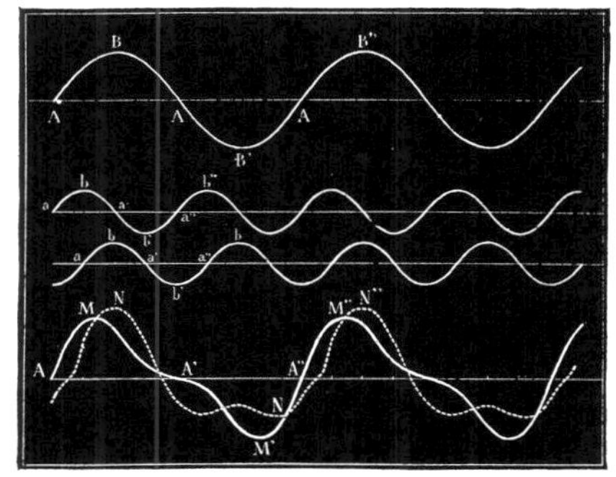

ABOVE

3-11. Pure and complex sound-waves, in Amédée Guillemin, *Le Monde Physique* (Paris: Hachette, 1882), 1:635.

The upper three sound-waves are "pure" sounds that can be produced, for example, by striking three keys on a piano: one base tone (with long waves) and two higher pitches (with shorter waves). The three pure sounds are combined into a complex wave pattern by striking the three keys together, as shown below. Fourier discovered a method for decomposing any periodic function (such as this complex sound-wave) into the simple periodic functions (the three pure sounds) of which it is composed.

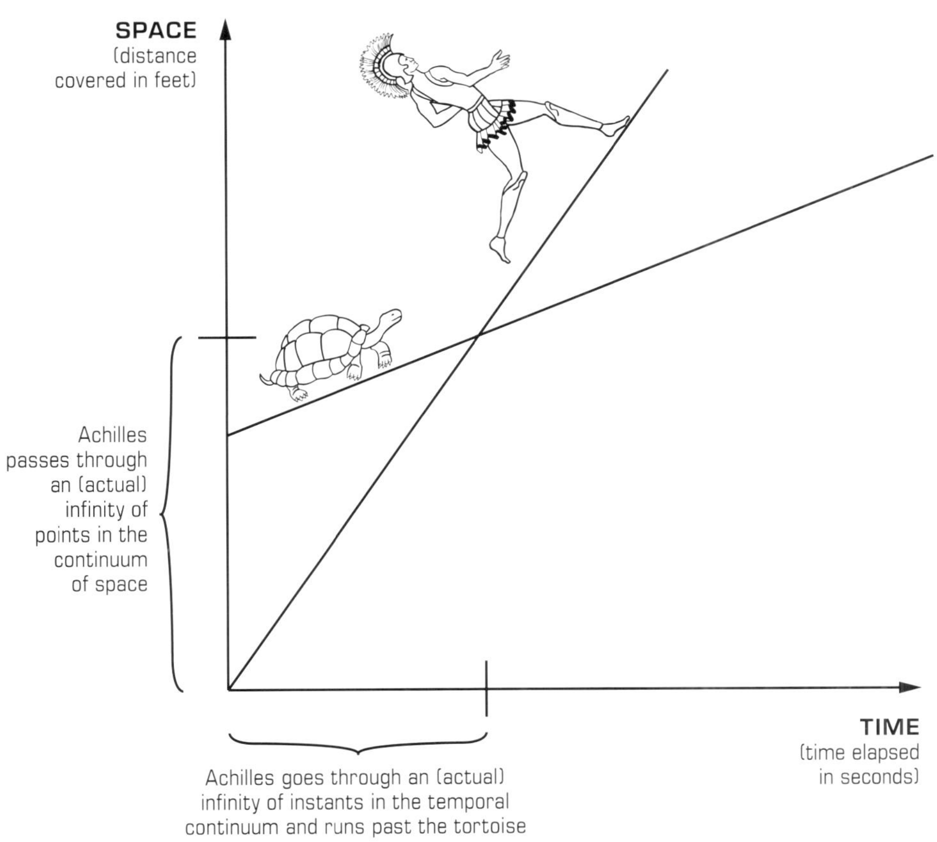

LEFT

3-12. Achilles running past the Tortoise.

Achilles had overtaken the Tortoise, and had seated himself comfortably on his back.

"So you've got to the end of our race-course?" said the Tortoise. "Even though it does consist in an infinite series of distances? I thought some wiseacre or other had proved that the thing couldn't be done?"

"It can be *done," said Achilles. "It* has *been done."*

—**Lewis Carroll, "What the Tortoise Said to the Achilles,"** ***Mind*****, 1895**

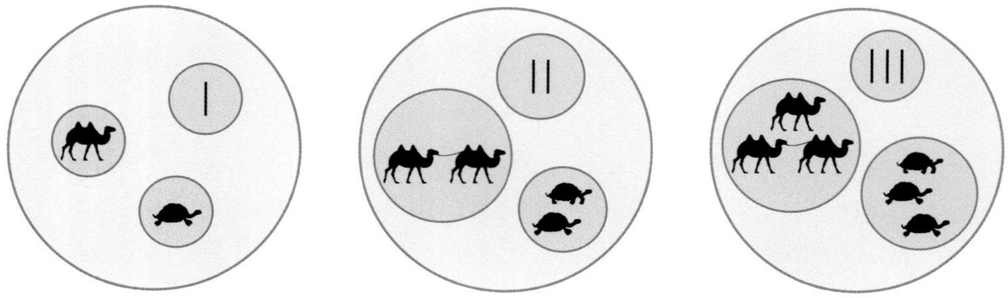

3-13. Cantor's numbers as sets.

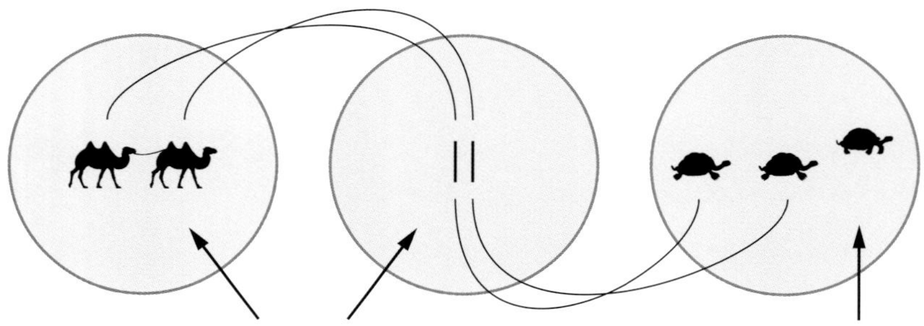

3-14. Cantor's method of comparing the cardinal number of a set, which is the number of elements in the set.

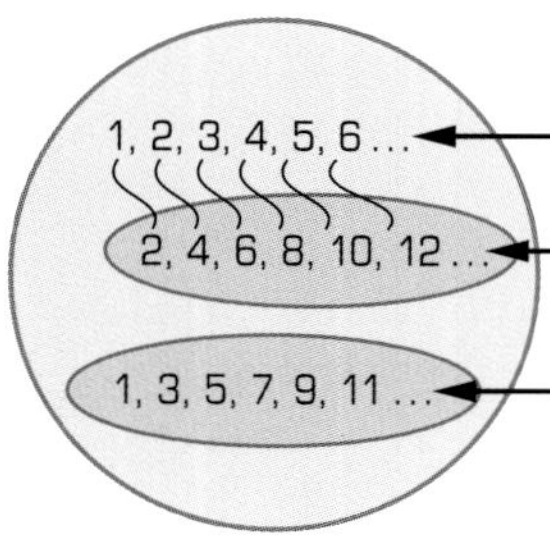

3-15. Infinite (or "transfinite") sets of equal size.

Earlier in the nineteenth century, the English mathematician George Boole and others had discussed classes and sets of objects,[6] but it was Cantor who had the key insight of what came to be called set theory: the size of different sets can be compared by arranging a one-to-one correspondence between the members of one set and another (plates 3-14 and 3-16). He went on to compare infinite sets by matching the numerical series 1, 2, 3, . . . one-to-one with the series 2, 4, 6, . . . demonstrating that the set of whole numbers and the set of even numbers is the same (infinite) size (plates 3-15 and 3-19). Indeed, Cantor defined an infinite set (such as the whole numbers) as a set that can be put into a one-to-one correspondence with one of its proper subsets (such as the even numbers).

Zeno and Euclid understood that the whole is greater than any of its parts. Cantor pointed out that this principle is true for finite sets (there are more human beings than there are Greeks) but false for infinite sets (the set of whole numbers can be matched one-to-one with the set of even numbers). The Greek atomist's space and time were collections of individual points in space and isolated events in time (potential infinites); Zeno showed that this view led to paradoxes if we think of the points as appearing one after another, ad infinitum, in the diminishing distance separating Achilles and the tortoise on a racetrack. Cantor's infinite set of points in space (or moments in time) avoids Zeno's paradox because each infinite set is a completed whole (an actual infinity). The idea of passing to a limit and completing an infinite series is the key to resolving Zeno's paradox, and a graph (a Cartesian co-ordinate system) that separates space and time is the mechanism to resolve the puzzle and show how Achilles overtakes the tortoise (*Foundations of a General Theory of Manifolds*, 1883; see plate 3-12).

I have found . . . the doctrine of sets, the Mengenlehre, *which postulates and explores the immense numbers that an immortal man would not reach even if he spent his eternities counting, and whose imaginary domains have the letters of the Hebrew alphabet as ciphers. It was not given to me to enter into that delicate labyrinth.*

—**Jorge Luis Borges,**
***La cifra* (The cipher), 1981**

Cantor went on to prove that there are different *kinds* of infinity, in the sense that certain infinite series can be shown to have different sizes. He first demonstrated that the rational numbers (the whole numbers together with fractions) are denumerable (they can

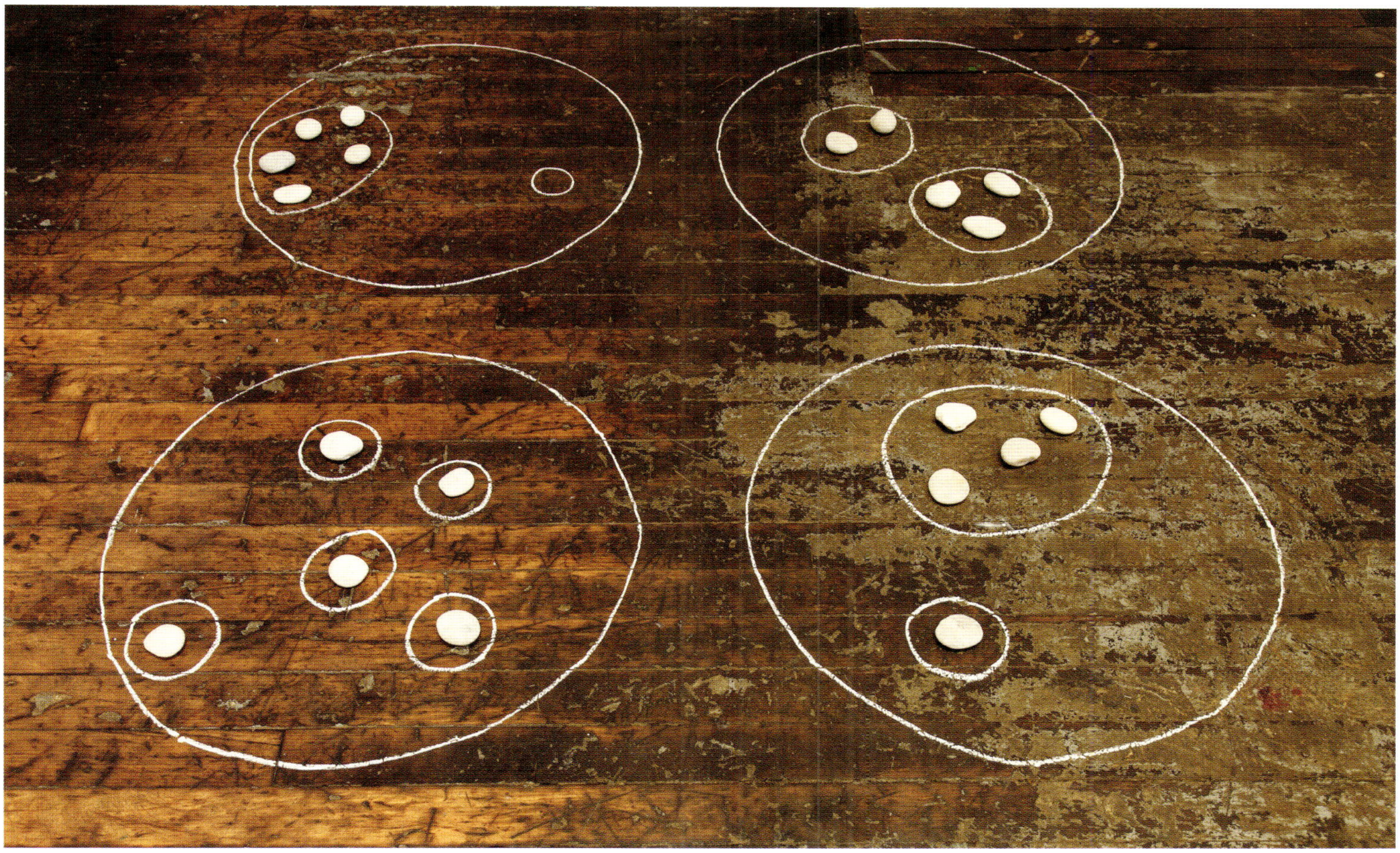

be counted); he drew a diagram in which every rational number would appear, and then he gave a way to "count" them by following the arrows and matching each rational with an integer (plate 3-18). Thus the set of all integers and the set of all rational numbers are the same size, because there is a one-to-one correspondence between the two sets.

The cardinal number of a set is the number of elements in the set. For example, the cardinal number of the set containing the letters in the word "square" is six. Any finite set has a finite cardinal number, and an infinite set has an infinite cardinal. Cantor discovered infinite sets that have different cardinal numbers. He used a Hebrew aleph (ℵ) to denote the cardinal number of an infinite set, and the subscript to indicate the set's cardinality (its "level" of infinity). Cantor called the cardinal number of the set of counting numbers $\aleph_0$ (which reads "aleph-zero" or, more commonly, "aleph-null"). Aleph-null is the smallest infinity. The counting numbers are equal to the even numbers, and so the cardinal number of the set of even numbers is also $\aleph_0$.

The combined set of rational and irrational numbers is called the set of "real numbers." If the real numbers are arranged on a geometric line segment such that numerical differences correspond to distances between points in the one-dimensional space of the line, we have the *real number axis*. There are no gaps on the real number axis—it is *dense*—which means that between any two points (which correspond to numbers), there is always another point (another number). In other words, the real number axis is a line on which points are continuous: Cantor called the real number axis "the continuum."

3-16. Mel Bochner (American, b. 1940), *Five by Four*, 1972. Stones and chalk on floor, size determined by installation. Courtesy of the artist.

The American artist Mel Bochner drew four sets on the floor; each set has subsets that contain a total of five stones (clockwise from the upper left): 5 + 0 = 2 + 3 = 4 + 1 = 1 + 1 + 1 + 1 + 1. The stones in the four sets can be put in one-to-one correspondence; thus the four sets have equal cardinality.

$\aleph_0 + 1 = \aleph_0$

—**James Joyce,**
note for *Ulysses*, 1922

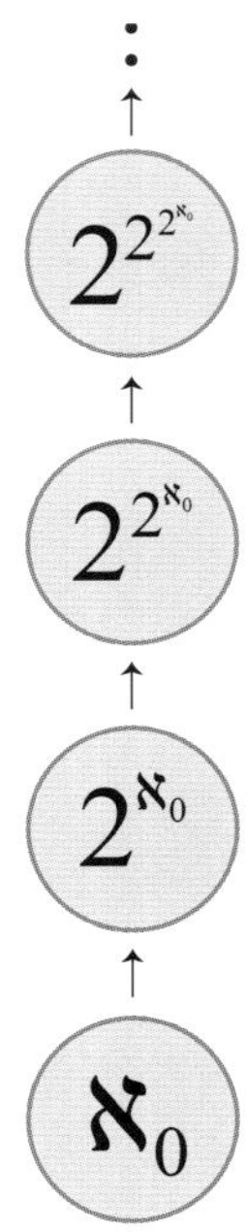

RIGHT

3-17. Cantor built this ascending "ladder" of infinite sets with ever greater cardinality by beginning with a set containing $\aleph_0$ elements and then repeatedly raising 2 to the power of its (infinite) predecessor.

BELOW

3-18. Cantor's proof that the rational numbers are denumerable.

Cantor drew this diagram, on which every rational number (every number expressible as a ratio) would appear. Then he "counted" them by following the arrows.

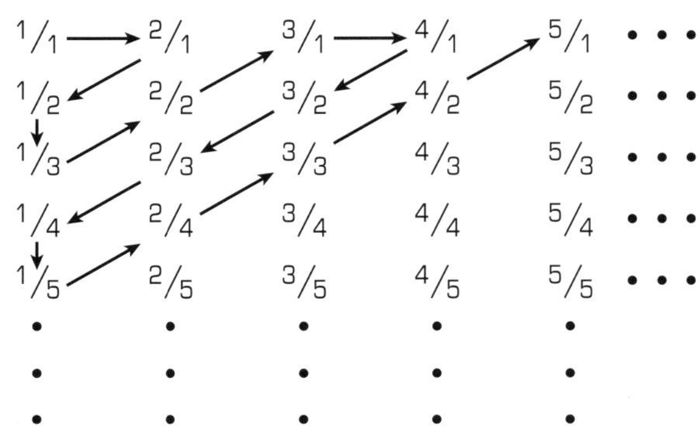

Cantor proved that the set of real numbers exists at a higher level of infinity than the whole "counting" numbers (1, 2, 3, . . .) because it cannot be put in one-to-one correspondence with the whole numbers—it is non-denumerable; the real numbers are uncountably many. Cantor gave the set of real numbers the name "cardinal number $\aleph_1$ (aleph–one)." Cantor believed that $\aleph_1$ was the first transfinite cardinal number greater than $\aleph_0$. His conjecture came to be known as the "continuum hypothesis," and despite spending the rest of his life trying to prove it, he could not. Decades later the conjecture was resolved by the work of the mathematicians Kurt Gödel and Paul Cohen, who showed that the hypothesis is undecidable using the tools of set theory that were available to Cantor.[7]

Cantor also considered what kind of number would result if he raised a number such as 2 to the power of an aleph—in other words, if he made an aleph the exponent, as in $2^{\aleph}$, which reads "two times itself an aleph (infinite) number of times" ($2 \times 2 \times 2 \times \ldots$). Cantor proved that when 2 is raised to the power of $\aleph_0$, it generates an infinite number with a higher cardinality than $\aleph_0$. In other words, $2^{\aleph_0}$ generates a number that cannot be put in one-to-one correspondence with $\aleph_0$, and so $2^{\aleph_0}$ is larger than $\aleph_0$. When 2 is raised to the power of the infinite set with the larger cardinal number ($2^{\aleph_0}$), it generates an even larger infinite number, and so on, generating sets of ever-higher (infinite) cardinality, ad infinitum (plate 3-17).

INFINITY AND "THE ABSOLUTE"

During the Renaissance, Catholic theologians and philosophers began using the term "Absolute" to name both the Judeo-Christian God and some kind of universal reality that underlies the world-out-there. The name Absolute stressed that the Supreme Being and the ground-of-being exist independently of any other being (as opposed to having a relative, dependent existence). The words "Absolute" and "infinity" became associated because the Being and the ground-of-being both have infinite properties.

OPPOSITE

3-19. Paul Klee (Swiss, 1879–1940), *Equals Infinity*, 1932. Oil on canvas mounted on wood, 20¼ × 26⅞ in. (51.4 × 68.3 cm). Museum of Modern Art, New York, acquired through the Lillie P. Bliss Bequest. © 2014 Artists Rights Society, New York.

Klee painted symbols for "equals infinity" using a variation of ∞ which was introduced as the mathematical symbol of infinity in the seventeenth century by John Wallis, who named it "lemniscate" (from the Latin *lemniscus*, "ribbon").

The fifteenth-century German Catholic cardinal Nicholas of Cusa formed the theological concept of an infinite Absolute when he applied philosophical ideas about infinity to divinity. Aristotle had pointed out that the number of days in the future was only a *potential* infinity, since all the days did not exist at the same moment. He warned that an infinite collection of things existing together at the same time—an *actual* infinity (such as the infinite number of points marked off on Zeno's racetrack)—could not exist. Most medieval scholastics adhered to Aristotle's advice and did not refer to an actual infinity in their discussions of eternity, but in the early Renaissance Nicholas of Cusa made an actual infinity the centerpiece of his theology. Nicholas described God as an actual infinity, a spiritual entity that was at every moment omnipotent, omniscient, and omnipresent.

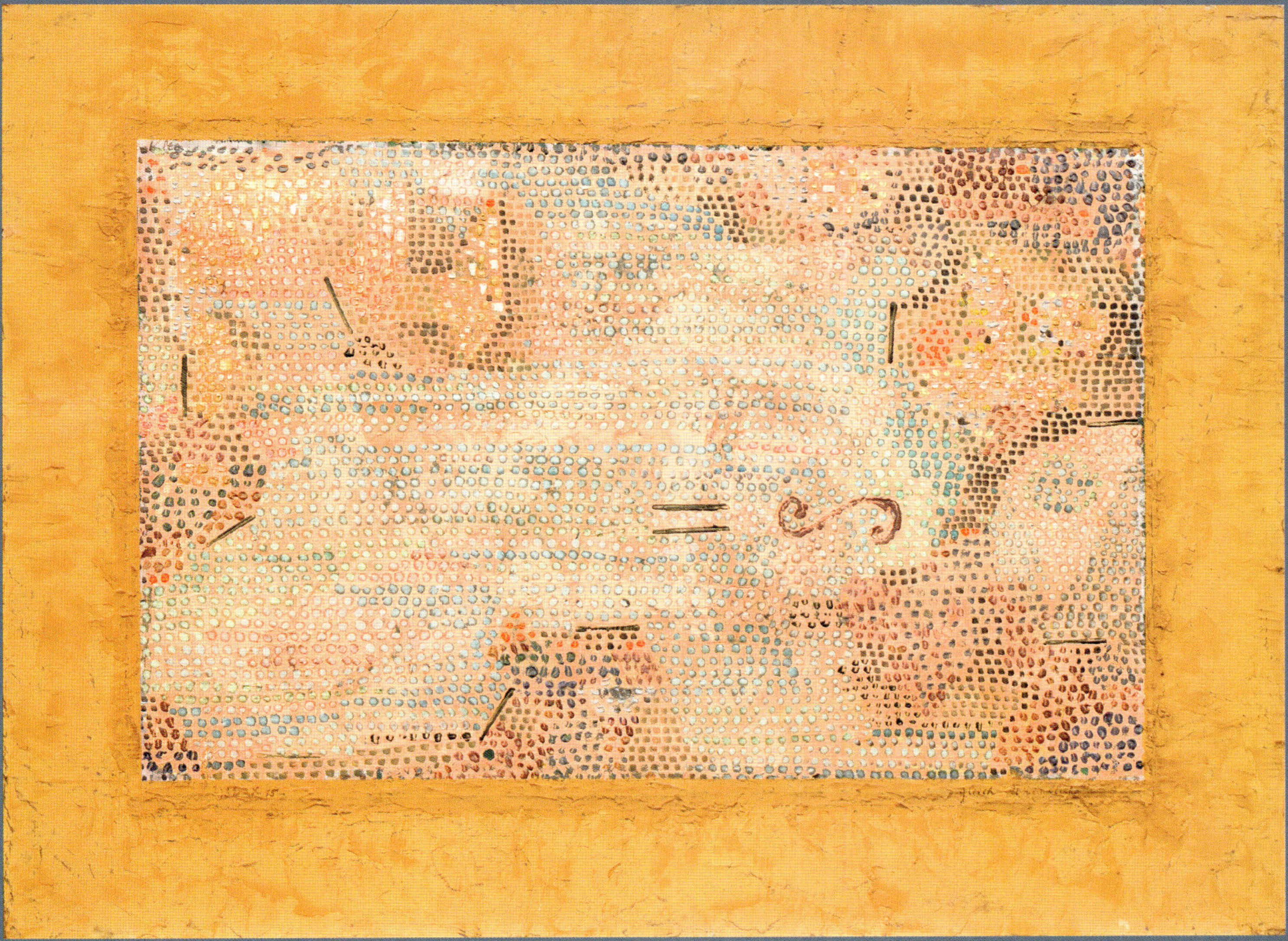

3-20. A man looking beyond the horizon, in Camille Flammarion, *L'Atmosphere: Météorologie Populaire* (Paris: Hachette, 1888), 163.

In this nineteenth-century popular science book on meteorology, an anonymous engraver depicted a medieval cleric who, like Bruno, imagined that beyond the fixed stars in the dome of heaven were other suns, other worlds, and, according to a vision of the Hebrew prophet Ezekiel, a spirit-filled chariot with "a wheel intersecting a wheel" (shown in the upper left-hand corner) that could magically roll in four directions (Ezekiel 1: 16–21).

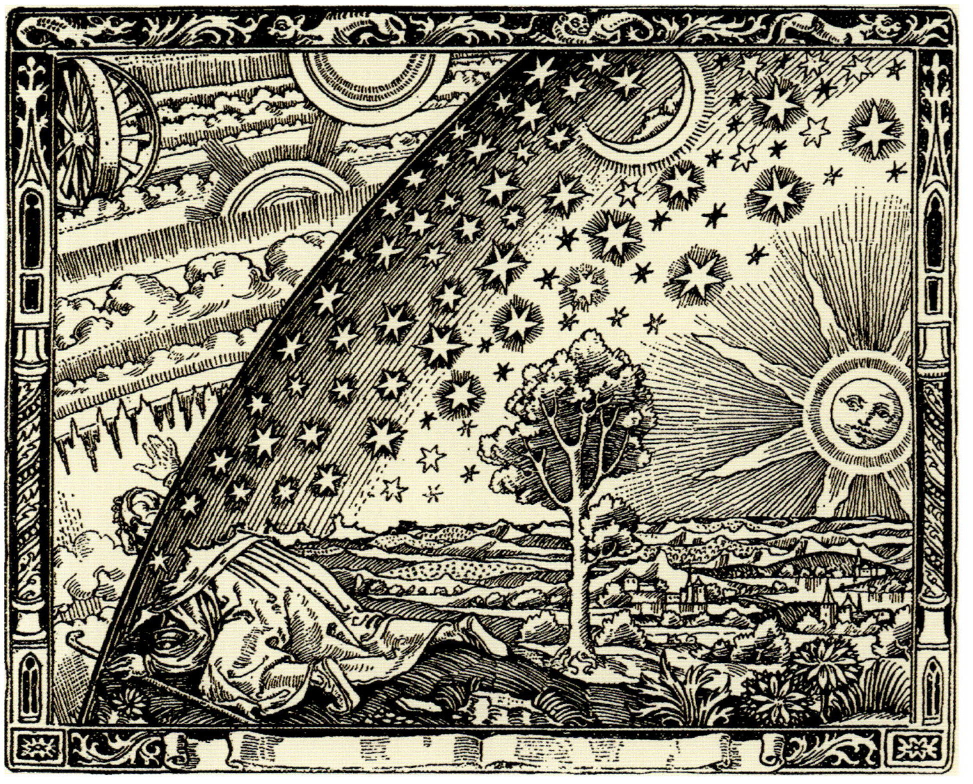

Referring to the greatest number (the maximum), he named this spiritual entity "Absolute Maximality."[8]

Everything we know about mathematics implies that there is no greatest number. Nicholas's claim that there is a greatest counting number is a reminder that he used mathematical analogies to illustrate his theological concepts, but he neither practiced nor contributed to mathematics. Rather, it was as a theologian that Nicholas suggested using mathematical figures as aids to conceiving of Absolute Maximality by first studying finite numbers and forms and then moving on to considering infinite sums and extensions.[9] For conceiving of the Maximum, he suggested the metaphor of an infinite sphere, which would have an infinite diameter and a center everywhere. The paradoxical nature of this object (for example, its diameter and radius, both being infinite, are equal) was, according to Nicholas, in keeping with the ultimate incomprehensibility of the Maximum.[10] Nicholas turned Aristotle's denial of actual infinity into a description of man's spiritual journey towards acceptance of his finiteness and his inherent ignorance of Absolute Maximality. In his finite time on earth with his mortal mind, Nicholas asserted, man can gain only a fragmentary understanding of infinity (*De Docta Ignorantia* [On learned ignorance], 1440). Nicholas also mused that Absolute Maximality created an infinite cosmos filled with myriad worlds, and a century after Nicholas, the Italian Dominican friar Giordano Bruno similarly argued that the Supreme Being created a cosmos that extends without limit beyond the apparent boundary of the night sky (*De l'Infinito Universo et Mondi* [On the infinite universe and worlds], 1584; see plate 3-20).[11] The Catholic inquisitors convicted Bruno of many heresies, including belief in the existence of a plurality of worlds, for which they burned him at the stake in 1600.

For the intellect is to truth as [an inscribed] polygon is to [the inscribing] circle. The more angles the inscribed polygon has the more similar it is to a circle. However, even if the number of angles is increased ad infinitum, *the polygon never becomes equal [to the circle] unless it is resolved into an identity with the circle.*

—**Nicholas of Cusa,**
De Docta Ignorantia
(On learned ignorance), 1440

In the decade after Bruno's death, Galileo invented the telescope and began the transformation of cosmology from theological guesswork to astronomy. But nascent scientists, such as Newton, continued to use theological terms such as "Absolute." Newton described an ultimate frame of reference for cosmic events—Absolute time and Absolute space—which he considered immutable and divine.[12] Absolute time was an *actual* infinity; Newton imagined time as an infinitely long line, stretching back to the past and forward to the future with all the seconds marked off. Absolute space was an inert, stationary framework. For practical purposes, Newton's Absolute space was the framework of the celestial dome of fixed stars within which the planets and other bodies moved.

Although Kepler, Galileo, and Newton each believed that they were discovering a mathematical structure that had been imposed on nature by a Judeo-Christian God, their new outlook was called *natural* philosophy because their laws describe natural (not supernatural) forces. Kepler's laws of planetary motion, Galileo's laws of terrestrial motion, and Newton's law of universal gravitation each describe a completely mechanical, impersonal model of a cosmos composed of inert matter in motion (in the tradition of the ancient Greek atomist Democritus). For many of their contemporaries, this raised the problem of how to preserve free will and values (ethics and aesthetics) in a deterministic world.

The topic had already been addressed in antiquity by the atomist Lucretius (active mid-first century BC), a follower of Democritus. To make room for free will (a mind that does not feel "constrained like a conquered thing"[13]) in the deterministic world of matter in motion, Lucretius introduced a *clinamen* (Latin for "slant"), the random swerving of atoms (*De rerum natura* [On the nature of things], first century BC). Descartes, a contemporary of Kepler and Galileo, proposed another solution in the seventeenth century: matter in motion (orbiting planets, brass balls rolling down an inclined plane) follows deterministic laws of the new natural philosophy, but mankind's mortal mind and Eternal Mind are not material but spiritual. According to Descartes, mind exists outside and independent of the body, operating according to nondeterministic, ethical rules that are imposed from a supernatural realm (*Meditations on First Philosophy*, 1641). If mind and matter are different substances, then how do they interact? How does the (mental) thought that one wants to turn a page of this book cause the (material) hand to move? The Enlightenment's pre-eminent philosopher, Immanuel Kant, showed that the question does not have an answer because Descartes had introduced an unbridgeable chasm between mind and matter. Strictly speaking, one knows mental sensations (seeing rain, feeling wetness) that arise from one's "sensible intuitions" (eyes, touch), and one is unable to perceive the material "transcendental object" (rain falling in the world-out-there).

Not only are the drops of rain mere appearances, but . . . even their round shape, nay even the space in which they fall, are nothing in themselves, but merely modifications or fundamental forms of our sensible intuition, and . . . the transcendental object remains unknown to us.

—**Immanuel Kant**, ***Critique of Pure Reason***, 1781

Like Newton, Leibniz was a Christian who was uneasy with Democritus's mechanical cosmos, but he wanted to avoid the chasm and solipsism introduced by Descartes's mind-matter dualism. Leibniz's solution was to dismiss Democritus and revive the Pythagorean/Platonic organic, living, purposeful cosmos, which was infused with a World Soul and divine Reason. According to the Pythagoreans, everything in heaven and earth is composed from sentient particles—monads (*monad* is Greek for "one"; see chapter 1). Leibniz similarly declared that the universe is not made of two things—mind and matter—but of

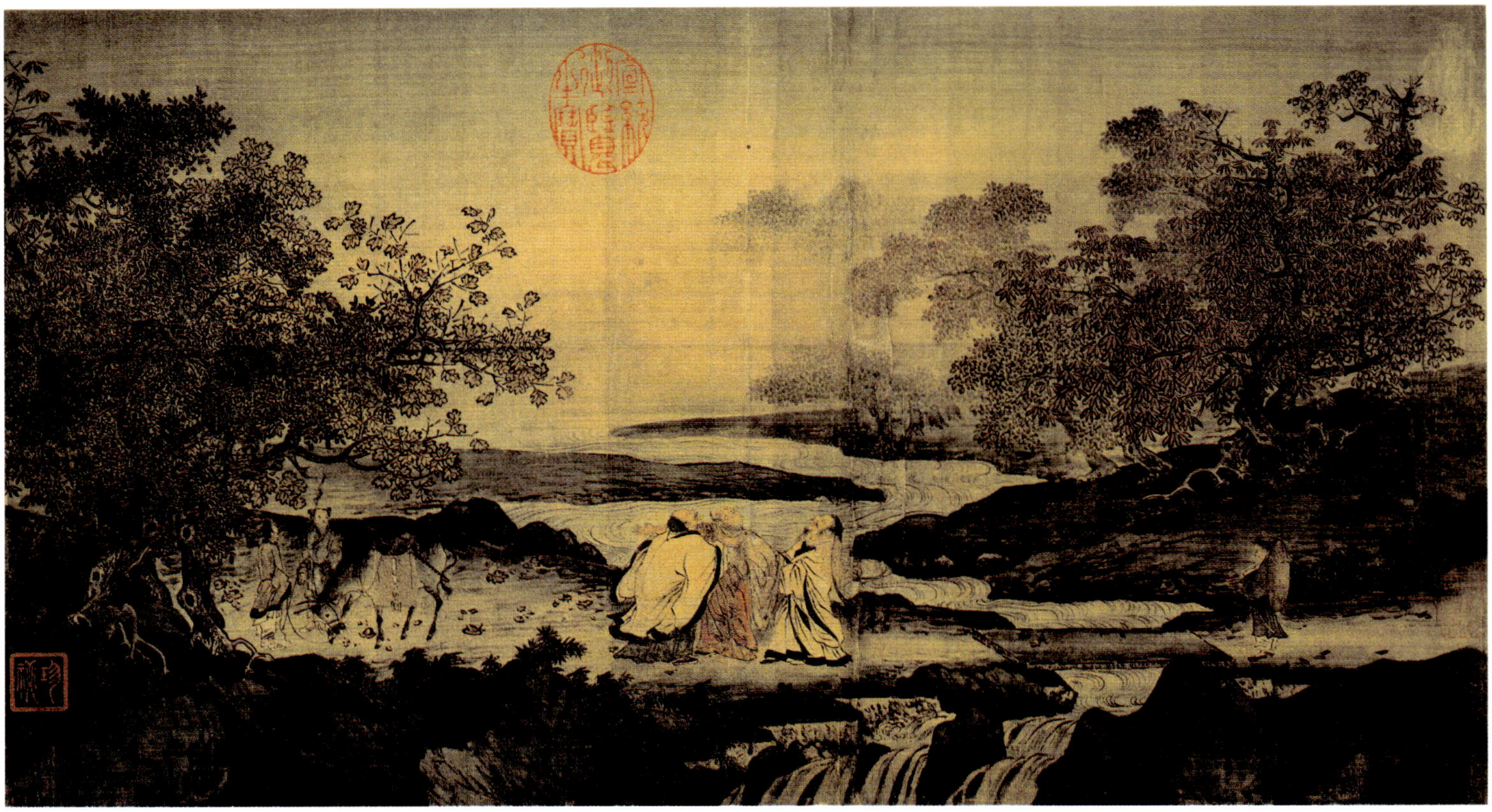

3-21. *Three Men Laughing by the Tiger Stream*, Song dynasty (AD 960–1279). Ink and colors on silk. National Palace Museum, Taipei, Taiwan, Republic of China.

At the opening of the Song dynasty, government scholars made a conscious attempt to merge China's three religions to achieve a more all-embracing outlook that gave practical advice as well as a cosmic vision of man's place in the universe. The motto of the merger was "Confucianism, Taoism, and Buddhism are one," and it encouraged communication with nature as the road to enlightenment.

This Song dynasty painting depicts three scholars representing the three faiths: from left to right, the Taoist scholar Lu Xiujing (AD 407–77), the Buddhist monk Huiyuan (AD 334–416), and Tao Hongjing (AD 456–536), a Taoist scholar and court official who gave advice based on Confucian principles. The friends have just passed through a forest filled with tigers, heedless of the danger because they were lost in congenial conversation. When they reach a clearing, they realize what they have just done and share a laugh at their good fortune. Six centuries later, when the Jesuits arrived in China, they gave this Confucianist-Taoist-Buddhist merger the name Neo-Confucianism, by which it is known in the West.

one kind of substance—monads—sentient particles that are both mental and material. For Leibniz, monads are arranged in an increasingly complex hierarchy from rocks to plants to animals to man to the human mind and, ultimately, to divine Mind. The divine Absolute is, according to Leibniz, the infinite totality of all monads (everything in the universe; *Monadology*, 1714).

Leibniz wrote his *Monadology* at a time when Europeans were first being introduced to Chinese philosophy. The trading ships that sailed to the Far East, guided by magnetic compasses and Mercator maps, carried along Jesuit missionaries who went in search of souls. The political/philosophical outlook that the Jesuits encountered in China was a combination of Confucianism, Taoism, and Buddhism that had been forged around the eleventh century (plate 3-21). Since the Catholic priests were most interested in Confucianism (sidelining Taoism and Buddhism as pagan superstitions) the Jesuits called Chinese philosophy Neo-Confucianism. Because Neo-Confucianism intermixed Confucianist, Taoist, and Buddhist doctrine, Western readers, such as Leibniz, did not always distinguish the three traditions but considered them together as Chinese philosophy.

Leibniz studied Chinese philosophy in depth after meeting a Jesuit missionary to China, Joachim Bouvet, in 1687. Leibniz received a letter from Bouvet in 1701 that included a copy of the *I-Ching* (plate 3-22).[14] Years earlier Leibniz had developed a system of binary notation for expressing all number in terms of two digits (0 and 1), and now he was astonished to read Bouvet's observation of a similar pattern in the Chinese hexagrams of the *I-Ching* (see plates 1-46, 1-47, and 1-48 in chapter 1). Leibniz became

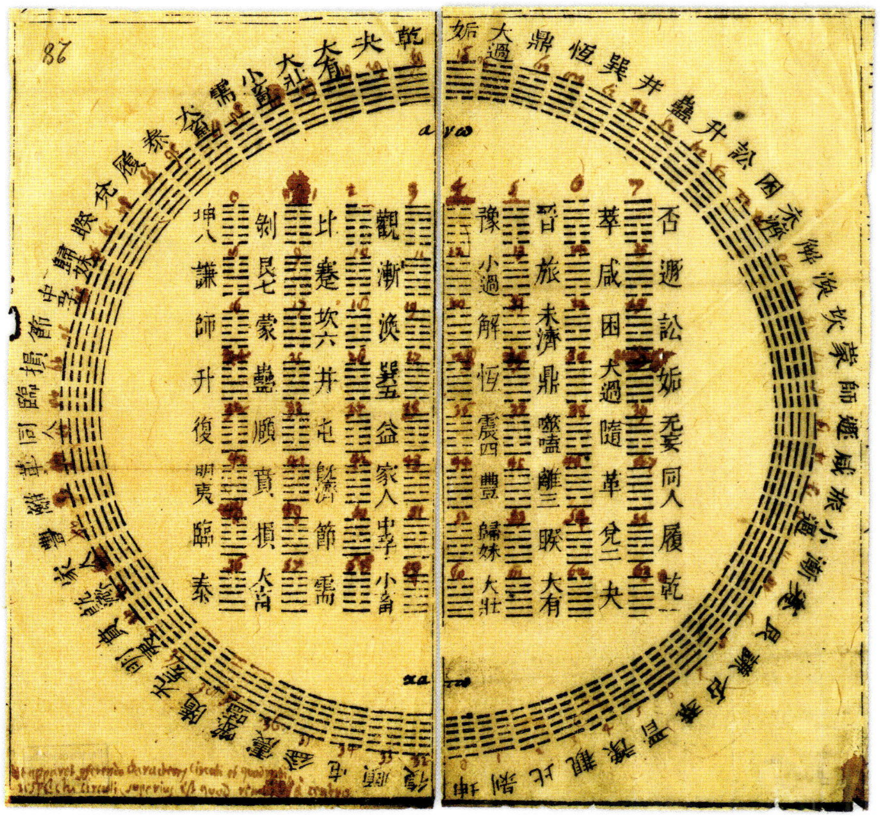

3-22. Gottfried Leibniz's copy of the *I-Ching* (Book of changes). Gottfried Wilhelm Leibniz Bibliothek-Niedersächsische Landesbibliothek, Hannover, LBr 105, fol. 27f.

From Leibniz's numerals written in brown ink above the hexagrams, it is clear that he read this diagram in the Western style from left to right and top to bottom, substituting 0 for the divided lines and 1 for the unbroken lines, which gave him the sequence of the first 64 numerals (including zero) in binary notation (000000 = 0, 000001 = 1, 000010 = 2, 000011 = 3, . . . , 111111 = 63). Living in China, Bouvet presumably read in the Eastern style from right to left and bottom to top, but the pattern reads the same in this reverse direction if you substitute 0 for the unbroken lines and 1 for the divided, as Bouvet evidently did.

convinced that the legendary Chinese emperor Fu Xi had invented binary notation,[15] but there is no evidence to support this. Ancient Chinese scholars had simply found a natural way to organize the hexagrams that have a structural isomorphism to binary notation.

Leibniz also noticed a similarity between the Western Pythagorean/Platonic outlook that he was reviving and Eastern Taoist elements in Neo-Confucianism.[16] Leibniz praised the Chinese for holding a fully naturalist, organic view of nature, rather than describing plants and animals as inert machines (Democritus's mechanical cosmos or Descartes's matter) that needed a supernatural god (*Deus*) or Descartes's "mind" to bring them to life: "Thus one can even find satisfaction with modern Chinese interpreters, and commend them, since they reduce the governance of Heaven and other things to natural causes and distance themselves from the ignorance of the masses, who seek out supernatural miracles—or rather supercorporeal ones—as well as seek out Spirits like those of a *Deus ex machina*."[17] For the Pythagoreans, Plato, and the Taoists, natural interactions occur not because of mechanical cause and effect, as in Newton's clockwork, but from the resonance of like things, as in the sounds of a stringed instrument.

All things reject what is different and follow what is akin to them. Thus it is that if the (respective) vital energies are similar, then they coalesce: if notes harmonize, they make a chord.

—**Dong Zhongshu,** ***Things of the Same Genus Energize Each Other*****, second century BC**

A contemporary of Leibniz, the Dutch Jewish philosopher Baruch Spinoza, formulated a pantheist form of monadology according to which everything is made of one substance, which Spinoza called "God/Nature," and the universe embodies an order that is completely predictable and which the human mind can discover. In his major work *Ethics Demonstrated in a Geometrical Manner* (1677) Spinoza put forth his doctrine in the format of Euclid's *Elements*, as a proof drawn from premises, axioms, and definitions.

Every portion of matter can be conceived as a garden full of plants, and as a pond full of fish. But each branch of a plant, each limb of an animal, each drop of its humors, is still another such garden or pond.

—**Gottlieb Leibniz,** ***Monadology*****, 1714**

FROM ENLIGHTENMENT RATIONALISM TO ROMANTIC *NATURPHILOSOPHIE* AND *LEBENSPHILOSOPHIE*

To Cusa, we can, indeed, articulately trace, word and thing, the recent philosophy of the Absolute. The term Absolute (Absolutum) *he everywhere employs.*

—**William Hamilton, *Discussions of Philosophy and Literature*, 1853**

Like Leibniz and Spinoza, the second-generation German Idealists G.W.F. Hegel and Friedrich Schelling, who are known by their German name *Naturphilosophen* (nature philosophers), denied the Enlightenment dualism between mind and matter and revived the ancient classical view that everything is made of the same substance in the form of sentient particles—monads. For Schelling and Hegel, at the top of the hierarchy of monads was an impersonal mind, a cosmic spirit or "Absolute Spirit," which for them was equivalent to the logical structure of the universe. Newton's Absolute time and Absolute space were akin to temporal and spatial reference co-ordinates; the Absolute of the *Naturphilosophen* also provided a framework, but it was more metaphorical—they considered it their ground-of-being—and, like Nicholas of Cusa's Absolute divinity, Schelling and Hegel considered the Absolute an *actual* infinite whole of everything—all monads arranged in a structure—considered as a spiritual entity, of which the human mind can know only a finite part. Schelling and Hegel also recognized parallels between ultimate reality in Taoism (Lao Tzu's Tao) and a philosophy of nature in which Plato's One and Hegel's Absolute Spirit are ineffable, as Hegel stated in a lecture of 1816: "To the Chinese what is highest in the origin of things is nothing, emptiness, the altogether undetermined, the abstract universal, and this is also called *Tao* or reason. . . . When the Greeks say that the absolute is one, or when men in modern times say that it is the highest existence . . . nothing has been expressed except this same negation."[18]

In the late nineteenth century the Canadian psychiatrist Richard Bucke formulated a secular version of Hegel's Absolute Spirit and the Buddhist Brahma when he described a higher "cosmic" consciousness. Bucke was an educated physician who framed his version in terms of late-nineteenth-century Darwinian evolution. Reminding his reader that the evolution of species is ongoing, Bucke declared that human beings had evolved to the level of self-consciousness (including reason, intuition, and imagination), and he predicted that the next stage in evolution would produce humans with a more complex level of mental functioning—"cosmic consciousness" of ultimate reality (*Cosmic Consciousness: A Study in the Evolution of the Human Mind*, 1901).

Other philosophers, the *Lebensphilosophen* (life philosophers)—Arthur Schopenhauer, Søren Kierkegaard, and Friedrich Nietzsche—argued that there are limits to what can be captured in conceptual systems such as Kant's grand overview (*Weltanschauung*) in his critiques of knowledge. According to the *Lebensphilosophen*, philosophers should not discuss abstract theories but rather focus their attention on immediate and particular experiences that give value and purpose to life. Schopenhauer, Kierkegaard, and Nietzsche also resisted the rise of positivism, the nineteenth-century view that society should be built on scientific principles that give certain "positive" knowledge. Suspicious of reason, they relied on their intuitions, stressing the wholeness and unity of knowledge, and were open to Taoist meditative practices; references to Eastern philosophy occur frequently in the writings of Schopenhauer.[19]

Eastern and Western thought also intermixed in theosophy (a contraction of "theology" and "philosophy"), a pseudo-science with roots in Russia, where the occultist Helena Petrovna Blavatsky interwove the German Idealist view of the human mind as evolving to knowledge of Absolute Spirit with the Buddhist doctrine that each individual attempts to escape the material body and unite with the Brahma (*The Secret Doctrine*, 1888). Blavatsky named her approach "theosophy" and adopted the framework of Buddhism. Nineteenth-century scholars dismissed her as an amateur who distorted Asian religion through an occultist lens, but theosophy popularized Asian thought in Blavatsky's native Russia and, through her disciples, in the West.

After learning about Gauss's multidimensional geometries that describe four-dimensional objects in the mathematical-world-out-there, the British mathematician Charles H. Hinton declared that four-dimensional objects exist in the *physical* world-out-there ("What is the Fourth Dimension?" 1884). One of Blavatsky's disciples, the Russian theosophist Peter Ouspensky, seized on Hinton's hypothetical fourth *spatial* dimension as being the occult realm of the Absolute. But we live in a three-dimensional world; how can one *perceive* a fourth *spatial* dimension? Ouspensky declared that to perceive the four-dimensional Absolute, one had to develop ("evolve") a new kind of occult perception that was a form of higher consciousness, which, borrowing Bucke's term, Ouspensky called "cosmic consciousness" (*Tertium Organum*, 1911). Between the 1880s and 1919, theosophy inspired some early modern artists, including Kruchenykh and Malevich, to create symbols of Ouspensky's four-dimensional Absolute.[20]

CANTOR'S PHILOSOPHY OF THE INFINITE

Cantor matured in the second half of the nineteenth century, when the scientific worldview of a mechanical clockwork universe was on the rise and the *Naturphilosophen* vision of an organic, purposeful cosmos composed of mind-matter monads and culminating in Absolute Spirit was seen as an old-fashioned, Romantic philosophy. Cantor was a deeply religious Lutheran who was part of a resurgence of Romanticism in German intellectual circles in response to a perceived threat from soulless, mechanistic science. Besides engaging in the practice of mathematics, Cantor also sought answers to philosophical questions, as he indicated by giving his founding text of set theory (*Foundations of a General Theory of Manifolds)* the subtitle *A Mathematico-Philosophical [mathematisch-philosophischer] Investigation into the Theory of the Infinite.*[21] In this landmark publication Cantor positioned himself in the philosophical tradition of Plato's organic, living, purposeful cosmos, stating that he shared the fundamental belief of Plato, Spinoza, and Leibniz in "the *unity of all to which we ourselves belong*."[22] Cantor expressed appreciation for Plato's main competitor, Democritus, because his atomic model of a mechanical cosmos composed of inert matter in motion had produced powerful results (Newton's law of gravitation), despite leading to philosophical problems (Descartes's mind-matter dualism). But Cantor vowed to create the mathematical tools to describe the Platonic organic cosmos: "For alongside (or in place of) the mechanical explanation of nature (which inside its proper domain has all the aids and advantages of mathematical analysis at its disposal, but whose one-sidedness and insufficiency have been strikingly exposed by Kant) there has until now not even been the start of an organic explanation of nature that would attempt to go further and that would be armed with the same mathematical rigor; this organic explanation can, I believe, be initiated only by taking up afresh and continuing the works and endeavors of those thinkers" (Plato, Spinoza, and Leibniz).[23] In other words, classical mathematics uses Euclidean straight lines, spheres, and ellipses to give a "mechanical explanation" of smooth natural and man-made objects, such as the moon, the solar system, and the Pyramids of Giza. Cantor vowed to develop the tools of set theory to give "an organic explanation of nature"; in other words, to account for the irregular, complex forms of rough objects such as mountains, plants, and animals, which had not been described in the Euclidean tradition.

The year after the publication of *Foundations* in 1883, Cantor fell into a deep depression that required hospitalization, and he suffered from bouts of depression the rest of his life.[24] That same year the editor of the Swedish journal *Acta Mathematica*, Gösta Mittag-Leffler, a mathematician and close friend of Cantor, asked him to speculate on possible applications of transfinite set theory to research in the sciences.[25] Cantor responded in the fall of 1884 by expressing his hope that he could give a mathematical description of the organic, living cosmos: "I am investigating the application of set theory to the study of organisms, to which we cannot apply traditional mechanical principles. . . . For this we need new mathematical resources, which can essentially be found in the set theory that I have already developed."[26] He went on to confide to his friend Mittag-Leffler that this application of set theory had been the passion driving his work: "The idea of a more precise explanation of the nature of all organic forms has occupied me for 14 years. It is the real reason [*die eigentliche Veranlassung*] why I have pursued this tedious and thankless investigation of set theory and why I never lost sight of it."[27]

Reflecting nineteenth-century cosmology, Cantor described the natural world as consisting of matter and the æther. Like the geologist Charles Lyell and the biologist Charles Darwin, Cantor described matter as composed of basic units, such as atoms, cells, or monads (to use Leibniz's term, as Cantor did).[28] Nineteenth-century scientists imagined the æther as a medium that filled Newton's Absolute space, pervading the whole universe—a volume that was at rest but susceptible to vibratory motion. The idea of the æther was developed because it seemed to physicists studying electricity, magnetism, and light that periodic, oscillating waves (travelling through earth's atmosphere and outer space) needed something to "wave." Cantor hypothesized that matter—the set of all the *discrete* material atoms, cells, or monads that matter comprises—was equal to $\aleph_0$ (aleph-null): "If we think of the example of all living cells at a given time in the universe, which expands infinitely in all directions, this is certainly the epitome of infinitely many individuals; one can raise the question of what is the 'power' (cardinality) of this collection, and I can prove rigorously that the cardinality is the first ($\aleph_0$), and no higher."[29] Furthermore, Cantor hypothesized that the æther—the set of all the *continuous* æthereal atoms or monads that comprise the æther ("a *colossal* playing field for *continuous* motion"[30])—was of a higher infinity and equal to $\aleph_1$ (aleph-one): "the *totality of the material atoms* is of the *first power* ($\aleph_0$), and the *totality of æther atoms*, is of the second power ($\aleph_1$)."[31]

Monads and æther have been tossed into the dustbin of science, but the desire to give a mathematical description of complex organic forms survives today. Indeed, Cantor's words foretell a sentiment expressed a century later by the Polish-born French mathematician Benoît Mandelbrot: "The number of distinct scales of length of natural patterns is for all practical purposes infinite. The existence of these patterns challenges us to study those forms that Euclid leaves aside as being 'formless' to investigate the morphology of the 'amorphous.'"[32] Mandelbrot's approach was to look not at amorphous objects such as mountains, plants, and animals but at the simple, repeatable processes that made them (the growth pattern of crystals, the division of cells), and he invented a new branch of mathematics—fractal geometry—that describes such processes (*The Fractal Geometry of Nature*, 1977; see chapter 12). Mandelbrot credited Cantor with inventing the first example of fractal geometry, a transfinite set (so-called Cantor dust) that is created by the repeated application of a simple algorithm (plate 12-32 in chapter 12).[33] Cantor worked with paper and pencil and lacked a tool with which to compute rapidly—the computer—which Mandelbrot put to good use to investigate algorithms that have been applied billions and billions of times in the course of evolution.

While Cantor was applying set theory to the organic cosmos, his alephs caught the attention of Catholic theologians, who were approaching topics of infinity and related Christian doctrine in the spirit of Pope Leo XIII's sweeping encyclical *Aeterni Patris* (1879), which encouraged the revival of medieval scholastic philosophy in the style of Thomas Aquinas, and supported the advancement of the natural sciences. Although Cantor had been baptized Lutheran, he was open to a dialogue with Catholic scholars. Among the theologians who became interested in reconciling Catholic dogma with Cantor's work was Constantin Gutberlet, a German priest and leader in the dissemination of neoscholastic thought. Gutberlet initiated a correspondence with Cantor over the relevance of transfinite sets to the doctrine of the infinity of God, and Cantor became keenly interested in the theological implications of his work.[34] Critics had questioned the existence of alephs, but in 1896 Gutberlet supplied an answer: infinite sets exist eternally in the mind of God because "in Absolute Mind the entire sequence is always in actual consciousness."[35] Reading such declarations, the Jesuit cardinal Johannes Franzelin expressed concern that the identification of the deity with Cantor's transfinite numbers was a kind of pantheism, which had become so widespread with the rise of the scientific worldview that Pope Pius IX had been prompted to formally condemn it (*Syllabus of Errors*, 1864). Cantor responded to the cardinal's concern by defining a level of infinity reserved for the Judeo-Christian God, which included the entire upward hierarchy of set theory (the ascending ladder of alephs) together with everything in the universe. Borrowing a term from Cusa, Newton, and Hegel, Cantor named the set "Absolute Infinity" (*Absolut Unendlichen*), thereby placing God in a region that transcends all finite and infinite sets.[36]

Georg Cantor, the creator of set theory, was in fact educated by the scholastics.

—**Felix Klein, *The Development of Mathematics in the Nineteenth Century*, 1926**

MATHEMATICS IN MOSCOW: A WELCOME FOR CANTOR

Scholars and theologians since the time of Aristotle had repudiated the concept of actual infinity because of the logical paradoxes it posed. Cantor's radical challenge to this tradition was made even more acute by his description of heretofore unknown, higher levels of infinity. Cantor's teacher, the German mathematician Leopold Kronecker, thought alephs were figments of Cantor's imagination, and he bitterly opposed transfinite numbers: "God made the integers; all else is the work of man."[37] While Western Europeans struggled with Cantor's concepts, Russian mathematicians found it easier to accept the new sets. Russian artists and poets would join in finding expression of the Absolute.

Since its founding in 1864, the prestigious Moscow Mathematical Society was a center for research on topics linking mathematics, philosophy, and theology, giving mathematical objects a mystical interpretation in the tradition of the Russian Orthodox Church. Members opposed classical mathematics, especially the precise, predictable measurements of calculus, which they associated with Newton's mechanistic, determinist cause-and-effect clockwork universe. They pursued so-called free mathematical methods such as probability theory to describe stochastic natural and cultural phenomena. They supported the tsar, associating the monarchy with Christianity and free will, and they took a dim view of the secular, deterministic Marxists.[38] Moscow mathematicians were proud of their countryman Nikolai Lobachevsky for devising a non-Euclidean geometry of an imaginary world (*Imaginary Geometry*, 1826; see chapter 4).

As president of the society from 1891 to 1903, Nikolai Bugayev was the leading mathematician in Moscow in the later nineteenth century. As a philosopher, Bugayev followed Hegel and Cantor in endorsing Plato's living cosmos and its modern expression in Leibniz's monadology.[39] As a mathematician, he studied so-called discontinuous functions, which describe change that is irregular and unpredictable (plate 3-23), to which he gave a philosophical interpretation: "Discontinuity is a manifestation of independent individuality and autonomy. Discontinuity intervenes in questions of final causes and ethical and aesthetic problems."[40] Bugayev was a firm believer in free will, and he carried on a theological dialogue with the fathers of the Russian Orthodox Church.[41]

The truths proposed by the exact sciences do not deny but affirm, based on a firm foundation, our idealistic aspirations towards unity and harmony. . . . In response to ruthless nature, man, with his more profound conception of the workings of the universe, characterizes his destiny in the following terms: "With head held high, the cry of Freedom!"

—Nikolai Bugayev, "Mathematics and the Conception of the World from the Viewpoint of Scientific Philosophy," lecture given at the International Mathematical Congress in Zurich, 1897

In his capacity as a professor of mathematics at Moscow University, Bugayev trained generations of mathematicians to intermix the practice and philosophy of mathematics. Dmitrii Egorov, Bugayev's student from 1887 to 1891, completed his thesis on discontinuous functions in 1903 before joining the mathematics faculty of Moscow University. Egorov and his students went on to make the university a center for the study of functions, as well as set theory. Egorov was also a lifelong elder (lay leader) of the Russian Orthodox Church.[42]

In 1900 the young Pavel Florensky enrolled in Moscow University and studied with Bugayev.[43] Florensky became friends with his teacher's son, the poet Boris N. Bugayev, who adopted the pseudonym Andrei Bely and became a leading poet of the Symbolist generation, whose members turned away from the harsh realities of the everyday world to lose themselves in dreams and fantasies. Bely's first major piece of creative writing, the prose poem *The Northern Symphony* (1902), combined verbal and musical sounds in the Symbolist style. Through his friendship with Bely, Florensky entered Symbolist literary circles, and in 1904 he published the first exposition in Russian of Cantor's *Transfinite Arithmetic*, not in an academic journal but in a periodical of Symbolist poetry published by the Religious and Philosophical Society of Moscow. In the article Florensky, then a twenty-two-year-old graduate student in mathematics, described how Cantor had denoted the total (infinite) number of members in the set of whole numbers (1, 2, 3, . . .) with a Hebrew aleph (ℵ), and then stated these rules for the addition and multiplication of infinite numbers:

$$\aleph + 1 = \aleph$$
$$\aleph \times \aleph = \aleph$$

Florensky not only gave a lucid exposition of transfinite arithmetic for a general audience, but went on to outline his own vision of interpreting symbols in mathematics, art, and poetry in terms of the mathematics of discontinuity ("On Symbols of the Infinite: An Essay on the Ideas of G. Cantor," *Novyi Put'*, 1904).

Florensky turned down a graduate fellowship in mathematics from Moscow University in 1904 to become a Russian Orthodox priest, enrolling at the Moscow Theological Academy, where he wrote a master's thesis on Russian Orthodox theology, *The Pillar and Ground of the Truth* (1914). Florensky's extensive knowledge of Western philosophy and theology is on display in this learned text, as is his command of the mathematics of his time, including erudite discussions of not only Cantor but also the Italian number theorist Giuseppe Peano and the British logician Bertrand Russell. Florensky's theme, in the tradition of negative theology (which traces its origin to the monastic communities of the Desert Fathers of Egypt in the third century AD), was that mystical intuition yields a paradoxical truth, which can be written "A and ¬A" (A and not-A); in other words, as an assertion and its negation.

In modern logic, which developed in Germany and England during this era (see chapter 5), anything can be inferred from contradictory premises. But if A and ¬A implies everything, then the inference relation between the premises and conclusion of a deduction dissolves into triviality. There have, however, been attempts to define systems of logic that can accommodate inconsistency in the premises, treating inconsistent information as paradoxical but, nevertheless, informative. The motivation for developing so-called paraconsistent logic is to study belief systems that are inconsistent but non-trivial, including religious beliefs such as Florensky's, the distorted view of reality of patients with mental disorders, and inconsistent theories that have been held in the history of science.[44]

Florensky's unusual use of mathematics in a work of religious philosophy got the attention of Russian intellectuals. Florensky's thinking reflected the common distrust of reason among fin-de-siècle Symbolist artists and poets for whom truths about life (about persons, the mind, a personal God) were beyond reason: "The *rational* is at the same time *unexplainable*. To explain A is to reduce it to 'something else,' to not-A, to that which is not A and which therefore is not-A. . . . Life, flowing and non-self-identical, might be reasonable; it might be transparent for reason (we have not yet found out if this is the case). But precisely for this reason, life would be non-conformable with rationality, opposed to rationality. It would rip apart the limitedness of rationality."[45] Florensky also discussed what he called "love" between a

3-23. Continuous and discontinuous functions.

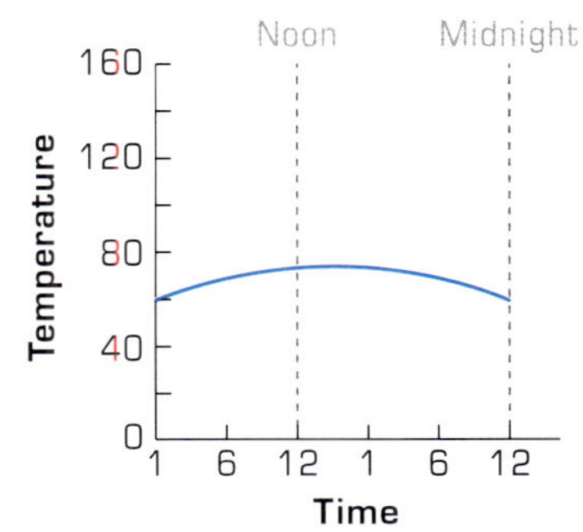

Continuous function: the air temperature is a function of the time of day. On a given day, as the air temperature goes from a low around midnight to a high in midday, it passes through all the degrees in between.

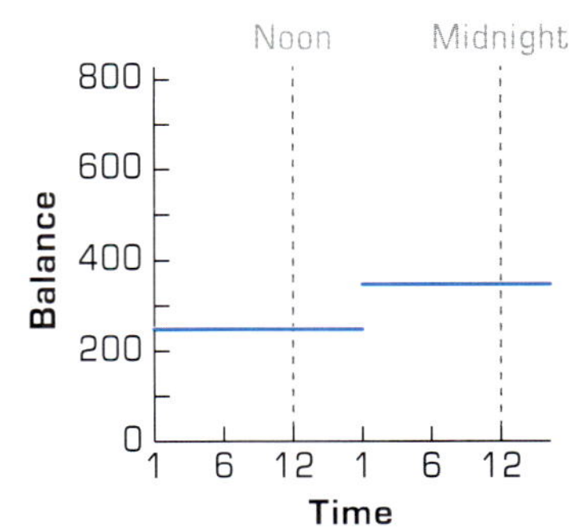

Discontinuous function: a bank balance is a function of when a deposit is made. Let us suppose that a woman makes a deposit of $100 into her bank account at precisely 1:00 p.m. The balance increases $100 at 1:00 p.m. without passing through all the intermediate amounts.

human being (finitude) and a Supreme Being (infinity), which can be expressed only by writing a new logic—a "supralogic" for communicating with a supernatural realm.[46]

Egorov's student Nikolai Luzin underwent a spiritual crisis in the bloody aftermath of the failed Russian revolution of 1905, and he considered giving up mathematics. But reading Florensky's *The Pillar and Ground of the Truth* inspired him to convert to Russian Orthodoxy and to study the Neoplatonist Plotinus (third century AD).[47] For Plotinus, ultimate spiritual (mental) reality—the One—was indescribable by mankind's finite mind *because* it was infinite. In the subsequent medieval mystical tradition of negative theology, which was vibrant in Luzin's day in the Russian Orthodox Church, the infinite Supreme Being was ultimately ineffable but could be described negatively by stating what it is not, although that description would always be partial and incomplete.

Cantor and Luzin were both Platonists who believed that infinite sets exist independent of the human mind.[48] Although the finite human mind might not be able to fully comprehend infinite Mind and infinite sets, humans can describe them indirectly and partially.[49] Luzin's contribution to Cantor's set theory was to describe certain sets in what became known as the Moscow school of descriptive set theory.

SUPREMATISM

The German scientist Gustav Fechner had invented the tools of experimental psychology to prove that the brain has an "evolving consciousness" (*Psychophysics*, 1860) and that animal drives to survive and reproduce cause humans to seek various pleasures, including hearing beautiful sounds (wordless music) and seeing color and form (abstract designs; *Aesthetics*, 1876).[50] Russian scientists, poets, and painters adopted from Fechner the idea that the pleasure of seeing color and form causes the mind to evolve to ever-higher levels of consciousness. In such a state of consciousness, they imagined that they would attain knowledge of the infinite objects—transfinite arithmetic and Absolute Infinity—which they had read about in Florensky's publications on Cantor.[51] To express their knowledge of this higher, Absolute realm, avant-garde writers composed poems with unintelligible words and artists made non-objective paintings that depict nothing at all.

In Saint Petersburg, Nikolai Kulbin, a neurologist and a self-taught painter, founded a group, including painter Wassily Kandinsky and poet Velimir Khlebnikov, which recognized a privileged role for artists in the evolution of the mind.[52] Kulbin named the group Triangle to suggest spiritual ascension to an apex, an image Kandinsky adopted in his book *Concerning the Spiritual in Art* (1911), which espoused the same evolutionary idea. Painter David Burliuk assembled a rival avant-garde group of artists and poets around the goal of creating something completely new by mixing the style of Italian Futurism with Ukrainian folklore. Burliuk gave his group the name Hylaea, an ancient Greek name for Scythia, a region that was part of Ukraine at the opening of the twentieth century.

The Triangle group included the poet Aleksei Kruchenykh, who concurred with Kulbin and Kandinsky that artists were on the vanguard of an expanding consciousness, and the role of art was to describe this higher reality with paradoxical statements that defy common sense. To avoid ordinary speech, Kruchenykh wrote lines of verse by stringing words together—like beads on a string—without syntax (grammatical structure), and he peppered his poems with unintelligible coined words. In 1912 Kruchenykh was the driving force behind a small book of verse, *Mirskontsa* (World backwards), with texts by himself and Khlebnikov and illustrations by artists Mikhail Larionov, Olga Goncharova, and Vladimir Tatlin. To thwart the reader's expectations, he used pages of varying size and paper, layouts with alternating horizontal and vertical formats, text and images in varying colors, and handwritten, typeset, and rubber-stamped words.

According to Kruchenykh, Burliuk upped the ante by suggesting that he write a poem with *all* invented words.[53] Kruchenykh obliged, and in 1913 he wrote "Dyr bul shchyl," an untranslatable arrangement of letters on a page, above which Kruchenykh informed the perplexed reader that he had written three poems in a new language in which "words have no definite meaning" (plate 3-24). Because the new language described a realm beyond common sense and mundane logic, Kruchenykh named it *zaum* (Russian for "transrational"[54]).

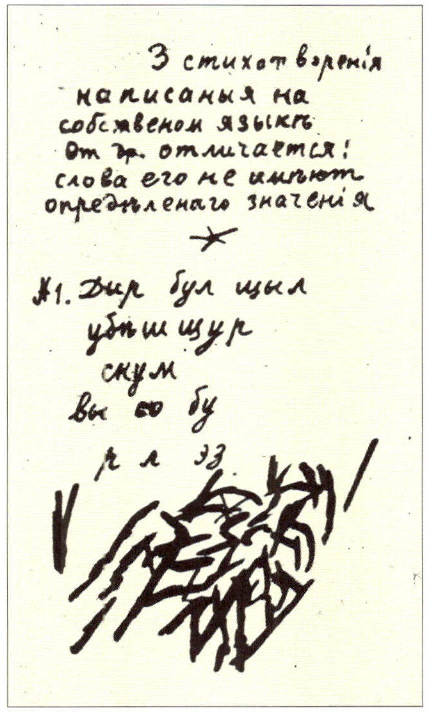

3-24. Aleksei Kruchenykh (Russian, 1886–1968), "Dyr bul shchyl" (poem), and Mikhail Larionov (Russian, 1881–1964) (drawing), handwritten text with lithographs, in *Pomada* (Moscow, 1913), n.p.

top: Kruchenykh wrote in the text above the star:

> *3 poems written in*
> *my own language*
> *different from others:*
> *its words do not have*
> *a definite meaning*

middle: Kruchenykh's poems with his made-up words.

bottom: Mikhail Larionov's drawing.

For whom was Kruchenykh writing these poems? He wrote for other avant-garde poets—he was a poet's poet—and, according to his late-nineteenth-century biological worldview, the poets in his audience possessed expanded (more highly evolved) minds. Kruchenykh did not communicate in the ordinary sense—he deliberately chose obfuscating, made-up words. He composed his poems to reach a small, elite art audience whose brains, to his way of thinking, had evolved enough to perceive an actual infinity—the Absolute. In other words, the subject matter of his verse was neither nonsense nor an occult realm but rather an alleged higher "transrational" level of reasoning.

As a member of avant-garde literary circles in Moscow, Kruchenykh would have been familiar with Florensky's exposition of Cantor's transfinite arithmetic and Kulbin's description of an evolving consciousness. He also read Ouspensky's theosophical description of the Absolute as an occult realm where space has four dimensions. According to Ouspensky, one could perceive the Absolute in an expanded state of consciousness by "intuitive logic—the logic of infinity, the logic of ecstasy" in which everything is both "A and not-A, or Everything is All."[55] In a flourish of bravado, Ouspensky named his book *Tertium Organum* (Third instrument, 1911), putting it in the tradition of Aristotle's treatise on deductive logic *Organon* (Instrument, fourth century BC), and the English philosopher Francis Bacon's study of inductive logic *Novum Organum* (New instrument, 1620).

In 1913 Kruchenykh described his subject matter in the essay "New Ways with Words," ending by quoting the phrase "higher intuition" from Ouspensky:

> Previous poets arrived at the word through rational thought, we arrived at unmediated comprehension through the word.
>
> In our art we already have the first experiments of the language of the future. Art marches in the avant-garde of psychic evolution.
>
> At present there are three units in our psychic life: sensation, representation, concept (and the idea), and a fourth unit is beginning to take shape—"higher intuition" (P. Ouspensky's *Tertium Organum*).[56]

The goal of his verse, Kruchenykh stated, was to describe his own expanded consciousness of a transrational realm, as predicted by evolutionary biology according to neurologists like Kulbin, theologians like Florensky, and theosophists like Ouspensky. Kruchenykh wrote in his 1913 manifesto "Declaration of the Word as Such": "INTRODUCING NEW WORDS, I bring a new content WHERE EVERYTHING begins to slide (the conventions of time, space, etc. Here my view coincides with N. Kulbin's)."[57] The idea of writing with an invented language quickly spread throughout Russian literary circles, and avant-garde poets invented dozens of *zaum* writing styles during the next decade.[58]

Kruchenykh's "Dyr bul shchyl" inspired Malevich to develop a painterly counterpart to *zaum* that would describe his expanded vision of a transcendental realm, which was essentially a mixture of Cantor and Florensky's Absolute Infinity, perceivable with a more evolved brain, and Ouspensky's four-dimensional space of the Absolute, with its Buddhist overtones.[59] In his early Cubist and Futurist work done around 1913–15, Malevich's thinking was based on theosophy, as manifest in the references to a transrational realm in the titles of his works, but by 1915 he had dropped the theosophist vocabulary and adopted more theological/philosophical terminology, referring to spiritual and cosmic themes.[60] In avant-garde circles it was unfashionable to make pious expressions of Russian Orthodox faith, as Florensky did, but avant-garde artists did regard painters of traditional Russian icons as their kin in the brotherhood of artists (plate 3-25).[61] In the context of discussing his oil paintings of landscapes and peasants from 1911 to 1913, Malevich compared realist paintings of similar subjects by a nineteenth-century group known as the Wanderers, with traditional Byzantine-style Orthodox icon paintings of Christian subjects done in egg tempera and gold leaf on wood: "Through the art of the icon painter, I understood the emotional character of peasant art, which I had loved before but I never understood its emotional depth until I studied icons. . . . All the naturalism of the Wanderers was overturned by the icon painters who achieved a high level of technical skill without using aerial and linear perspective. They used color and form to convey a purely emotional perception of their theme."[62] Born near Kiev to Polish immigrant parents, Malevich was baptized Roman Catholic, but from an early age he, like any Russian, was familiar with the Orthodox icon—a physical object that does not simply symbolize a non-physical, spiritual reality but, according to the devout person, *is* spirit. The painters of icons used pictures of saints to express faith in a supernatural realm; Malevich used geometric shapes to symbolize *zaum* and express the spirit of the new age of science and technology. Standing at the crossroads of the sacred and the secular, Malevich retained the mysticism of the Desert Fathers.

To stress the analogy with *zaum* (transrational), Malevich called his style "algoism" (meaning alogical, or opposed to logic).[63] He began by taking images out of their usual contexts and juxtaposing unrelated pictures, as in *Cow and Violin* (1913, State Russian Museum, Saint Petersburg). He then associated pictures with scraps of text, as in *Englishman in Moscow* (1914; see plate 3-26). The large syllables at the top and bottom easily piece together to read: ЗАТМЕНІЕ ЧАСТиЧНОЕ (partial eclipse). Smaller brown letters on the right spell СКАКОВОЕ ОБЩЕСТВО (of the racing society). The pictures of a man and a fish do not have an obvious association to each other or to this text. But

3-25. *Our Lady of Vladimir*, early twelfth century. Tempera and gold leaf on wood panel, 41 × 27⅛ in. (104 × 69 cm). State Tretyakov Gallery, Moscow.

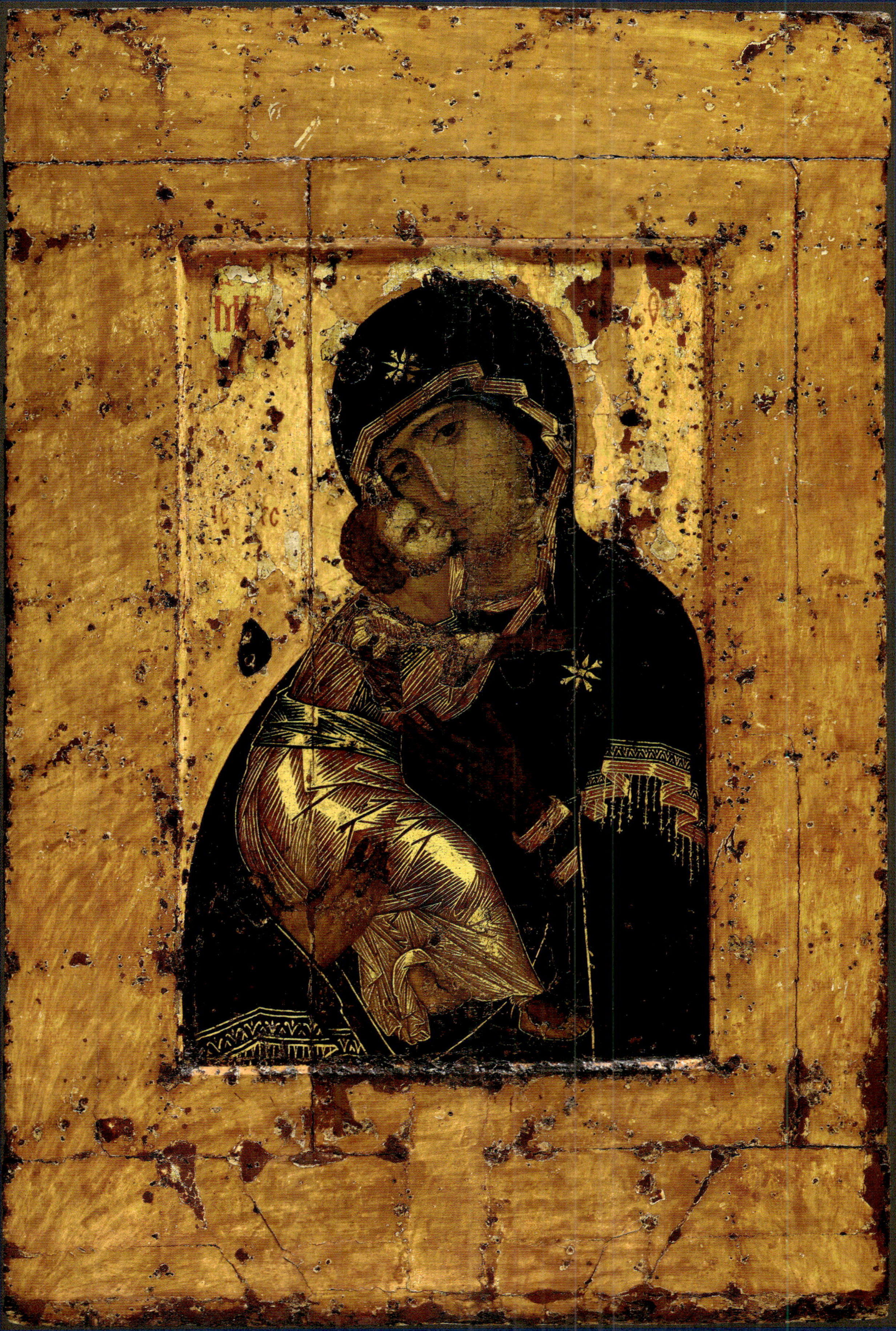

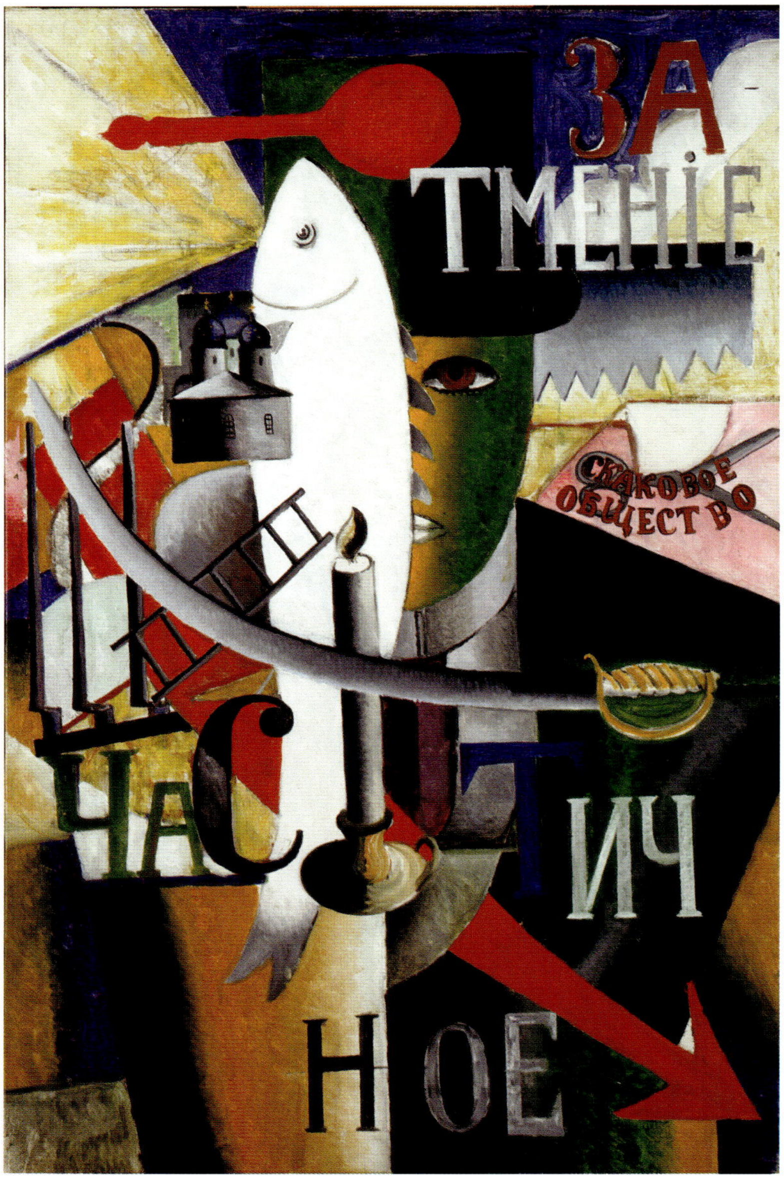

3-26. Kazimir Malevich (Russian, 1879–1935), *An Englishman in Moscow*, 1914. Oil on canvas, 34⅝ × 22½ in. (88 × 57 cm). Stedelijk Museum, Amsterdam.

so far Malevich has not achieved a visual counterpart to Kruchenykh's *zaum* because he is still composing with pictures and words that have conventional associations (to eclipses and Englishmen). But he soon removed all references to the world-out-there, writing: "Naturalism did not stand up to my criticism, and I sought other possibilities—not outside painting, but within the very core of the sentiment of painting, with the expectation that painting itself would sooner or later generate the form from its own painterly qualities, and would escape the electromagnetic attraction of objects and non-pictorial associations. This position led me further and further away from an academic study of nature, from naturalism, from illusionism. My knowledge of icons convinced me that it is neither a question of mastering anatomy and perspective, nor of reproducing nature exactly, but that it is necessary to have an intuition of art and of the reality of art. In other words, I saw that the reality that needed to evolve into ideal form came from the depths of aesthetics."[64]

In 1915 Malevich took the step into completely invented visual marks that are neither pictures nor words. For a performance called *Victory over the Sun*, Malevich designed the costumes and sets, Khlebnikov wrote the preface to the libretto by Kruchenykh, and the musician Mikhail Matyushin (a violinist with the Saint Petersburg Court Orchestra since 1881) composed the score. In the summer of 1913 the collaborators issued a proclamation declaring their intention "to destroy the archaic movement of thought according to the laws of causality, toothless common sense, the 'symmetrical logic,'"[65] and Kruchenykh produced a booklet with the *zaumy* title "Let's Grumble!" which contained his short dramatic sketch for the performance. The play opens with two strong men onstage, one of whom declares:

> *Sun you gave birth to passions*
> *And burned with fiery rays*
> *We will wrap you in a dust veil*
> *And board you up in a concrete house!*[66]

Later, the chorus details the destruction of the sun:

> *We had pulled up the sun by its roots.*
> *They have become permeated with arithmetic and are greasy.*[67]

Malevich illustrated the cover of Kruchenykh's libretto with numerals askew in a web of lines (plate 3-27) because there is no logical arithmetic in the play, which tells of the defeat of the sun (symbol of en*light*enment). Malevich designed the curtain and each of the backdrops in the format of a square-within-a-square, which he filled with images and design motifs in his algoist-style, such as a fanciful building with a window, chimneys, and a spiral staircase (plate 3-28). But one was different; the backdrop for the defeat of the sun had no picture or text but was a completely invented mark—a diagonal line within a square (plate 3-29). The play ends with a celebration of victory over illumination, with the chorus chanting:

> *We are free*
> *The sun is crushed*
> *Hello darkness!*[68]

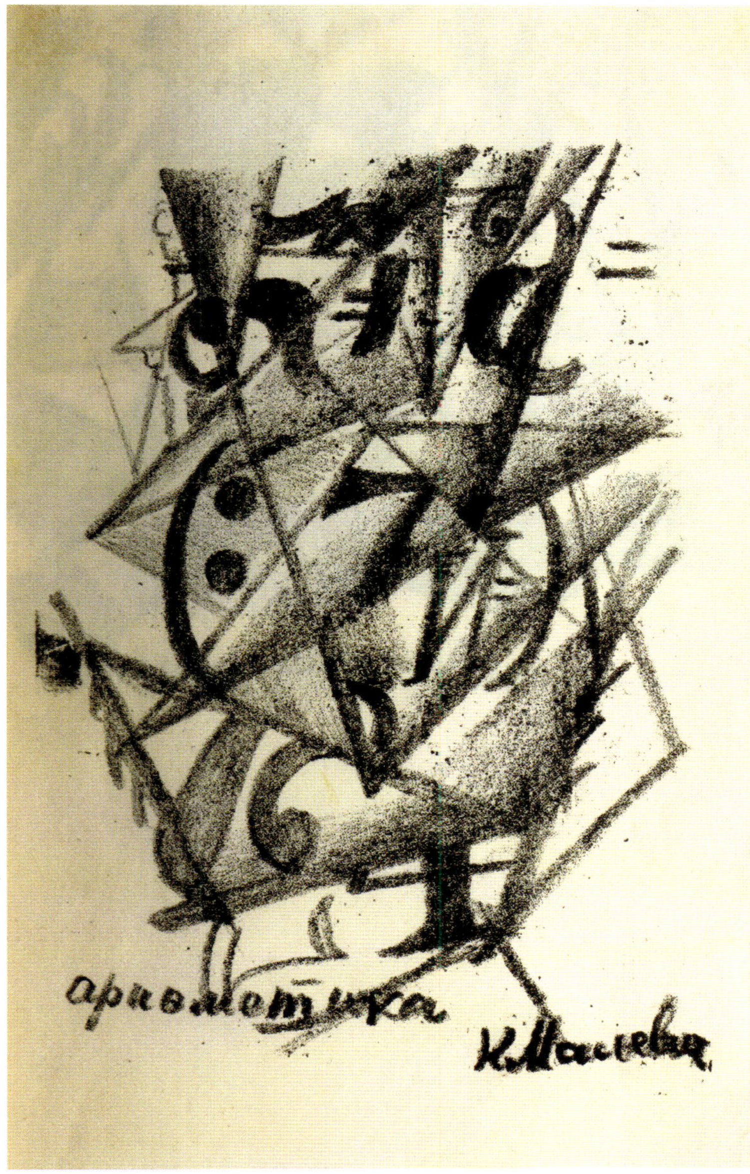

3-27. Kazimir Malevich (Russian, 1879–1935), *Arithmetic–The Science of Numbers*, illustration in Aleksei Kruchenykh, *Vozropsshchem* (Saint Petersburg, Russia, 1913). Lithograph, 5 × 3½ in. (12.7 × 9 cm). Library of Congress, Washington, DC.

In the spring of 1915, Malevich went on to develop an entire non-objective style with invented marks, composing with colored rectangles as in *Black Square* (1915; see plate 3-30), in which he achieved a visual counterpart to Kruchenykh's *zaum*. The art historian Charlotte Douglas has documented that Malevich immediately tried to predate his 1915 works by association to the 1913 performance of *Victory over the Sun*. Indeed, his lifelong, unflattering habit of predating his works to fit a glorified career myth has led to difficulty in establishing an accurate chronology of his work.[69] Proclaiming that his non-objective style has supremacy over other forms of abstract art, Malevich named the style Suprematism.

Malevich declared that by removing references to the world-out-there, he had achieved consciousness of an inner world, which he equated with pure feeling: "The ascent to the heights of non-objective art is arduous and painful. . . . The contours of the objective world—'everything we loved and by which we lived'—becomes lost to sight. . . . The black square on the white field was the first form in which non-objective feeling came to be expressed. The square = feeling; the white field = the void."[70] Severing all ties to the everyday, perceivable world-out-there in the works he created in the spring of 1915, Malevich believed he had also achieved knowledge of a transcendental, supernatural realm, which—borrowing signs from mathematics—he symbolized with polygons (mainly rectangles). Florensky wrote a treatise on icons (*Iconostasis*, 1922) in which he described the experience of meditating on an icon and reaching the point where he no longer sees a physical object but intuits a transcendental reality: "The material iconostasis [a wall of

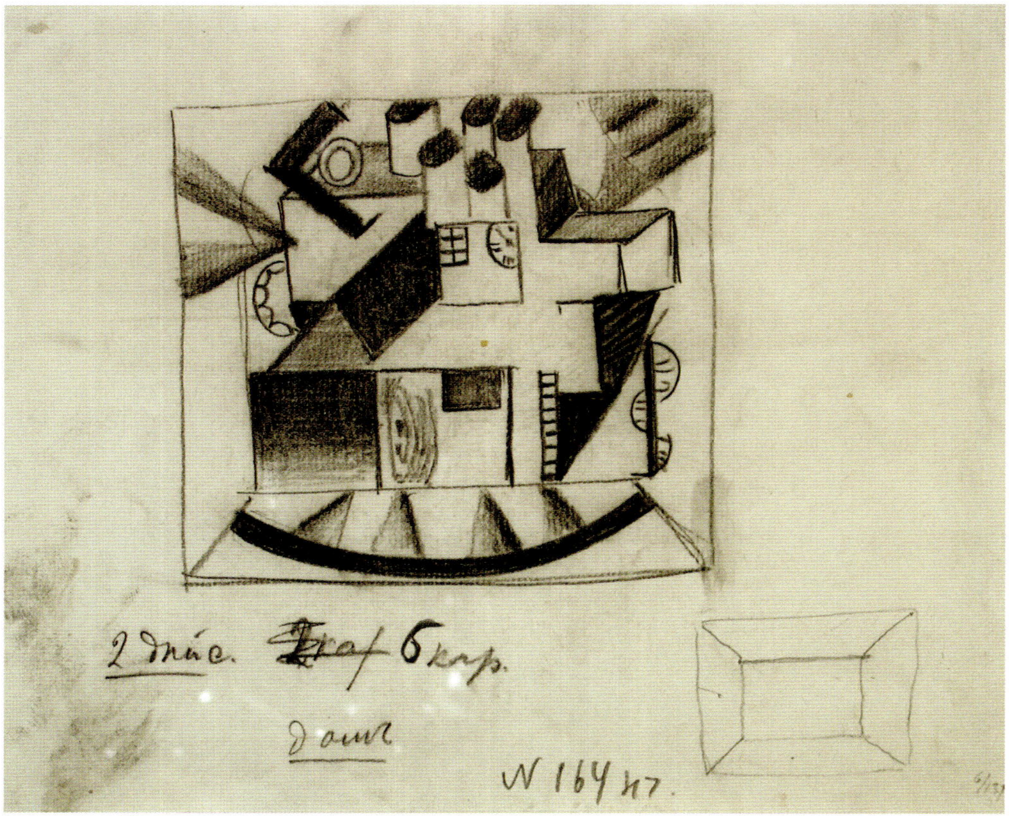

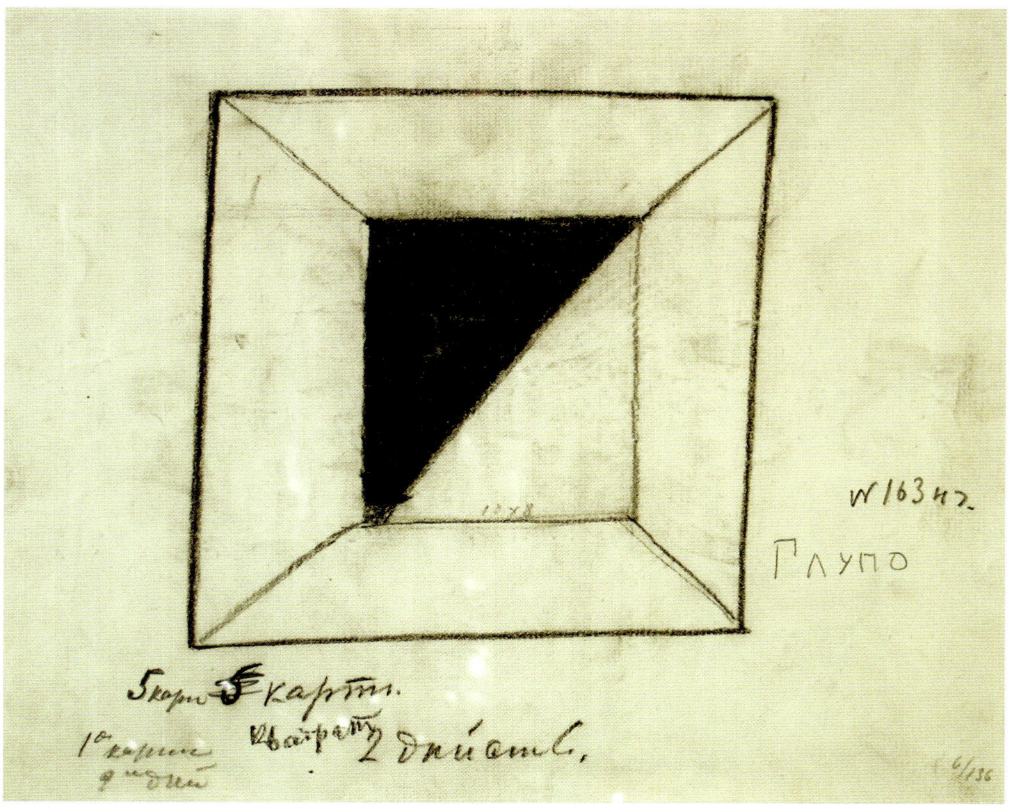

TOP LEFT

3-28. Kazimir Malevich (Russian, 1879–1935), set design for act 2, scene 6, *Victory over the Sun*, 1913, by Mikhail Matyushin and Aleksei Kruchenykh. Pencil on paper, 8½ × 10¾ in. (21.4 × 27.2 cm). Saint Petersburg State Museum of Theater and Music.

BOTTOM LEFT

3-29. Kazimir Malevich (Russian, 1879–1935), set design for act 2, scene 5, *Victory over the Sun*, 1913, by Mikhail Matyushin and Aleksei Kruchenykh. Pencil on paper, 8½ × 10⅞ in. (21.5 × 27.5 cm). Saint Petersburg State Museum of Theater and Music.

OPPOSITE

3-30. Kazimir Malevich (Russian, 1879–1935), *Black Square*, 1915. Oil on canvas, 31¼ × 31¼ in. (79.5 × 79.5 cm). © State Tretyakov Gallery, Moscow.

To paint this square, Malevich applied a uniform layer of black oil paint, but the surface has cracked over the years revealing a web of colors underneath, suggesting that he painted *Black Square* over an earlier work. Indeed, this was confirmed by a 1990 X-ray of the painting; see Andréi Nakov, *Kazimir Malewicz: Catalog raisonné* (Paris: Adam Biro, 2002), 205.

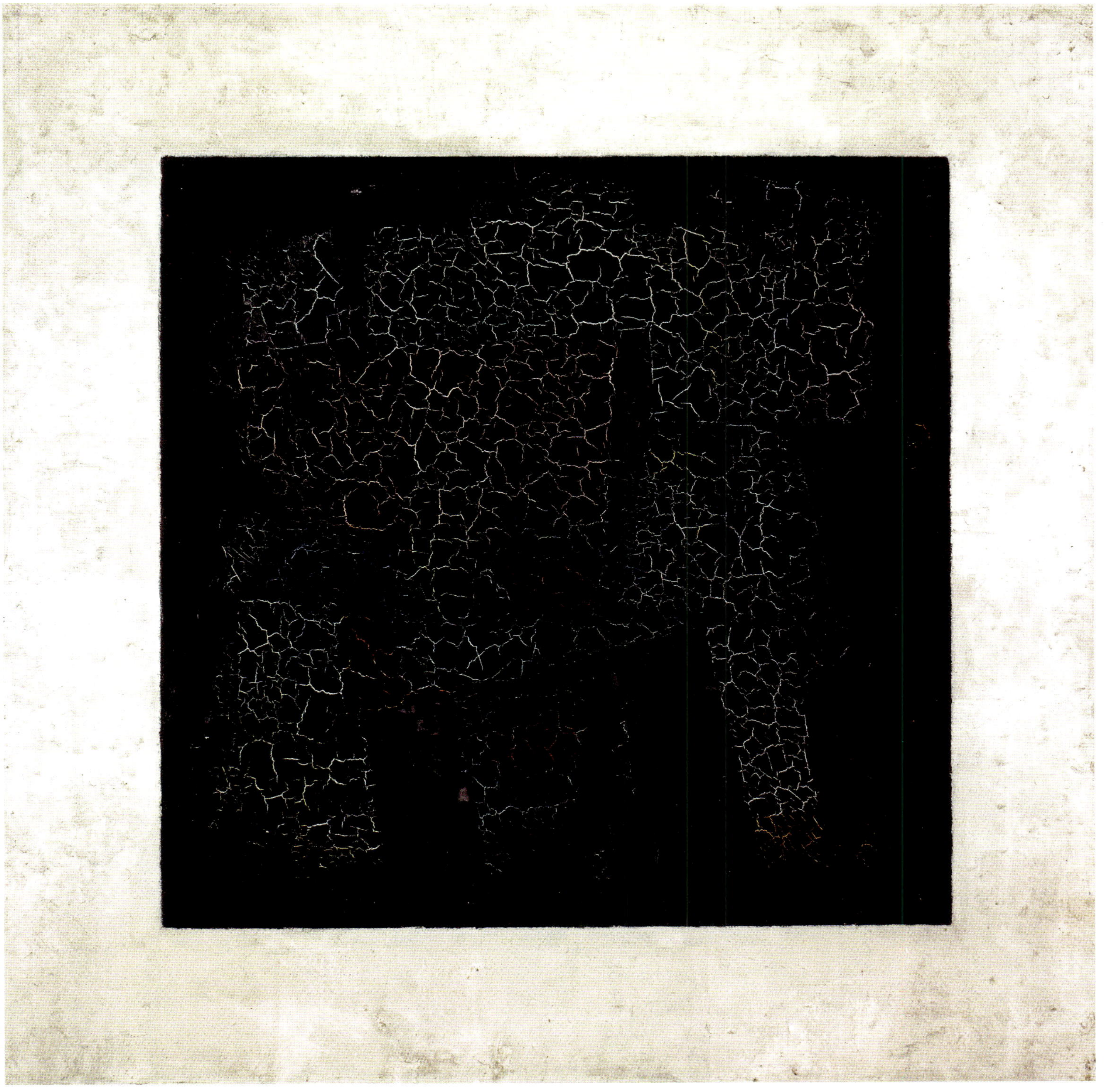

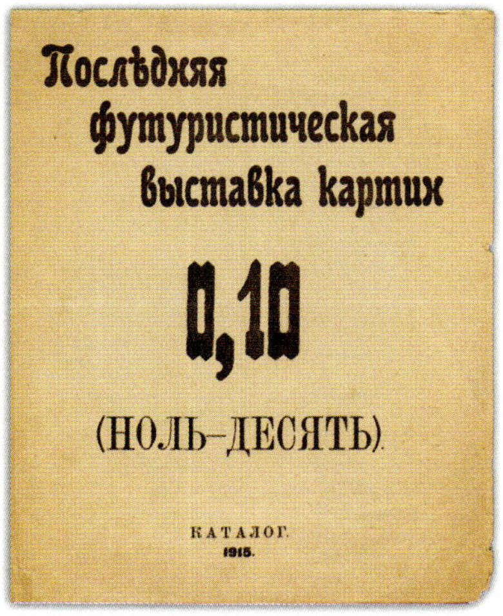

3-31. 0,10 (**НОЛЬ-ДЕСЯТЬ**) **Посльдняя футуристическая выставка картих** (0.10 [zero-ten], the last Futurist exhibition of paintings), 1915. Cover of the exhibition brochure, 11⅞₁₆ × 9 in. (29 × 22.8 cm). Museum of Modern Art, New York, acquired through the generosity of the Orentreich Family Foundation.

The reason for writing "НОЛЬ-ДЕСЯТЬ" (Russian for "zero-ten") below "0,10" is unclear, and it may simply be an error; 0,10 in Russian (written 0.10 in English), is read "one-tenth" (not "zero-ten").

icons] does not, in itself, take the place of the living witnesses, existing *instead* of them; rather, it *points toward* them, concentrating the attention of those who pray on them—a concentration of attention that is essential to the developing of spiritual sight. . . . When this happens, the material iconostasis will self-destruct in that vast obliteration which will destroy the whole range of this world—and which will even destroy faith and hope—and then we will contemplate, in pure love, the immortal glory of God."[71] Recalling Florensky and echoing the title of a 1819 book about the nature of reality by the German Romantic philosopher Arthur Schopenhauer (*The World as Will and Idea*),[72] Malevich wrote that he, like the Desert Fathers, was drawn into an arid transcendental realm: "I was gripped by a kind of timidity bordering on fear when it came to leaving 'the world as will and idea,' in which I had lived and worked and in the reality of which I had believed. But a blissful sense of liberating non-objectivity drew me forth into the 'desert,' where nothing is real except feeling . . . and so feeling become the substance of my life."[73]

In his first public presentation of the new style, Malevich showed thirty-nine Suprematist paintings in the exhibition *0.10 (zero-ten), The Last Futurist Exhibition* (plates 3-31 and 3-32), which opened in Petrograd (today Saint Petersburg) in December 1915. Malevich hung the paintings on the walls from top to bottom (like an iconostasis—a wall of icons). In an Orthodox Russian home, icons are hung high in the corner, often in a group draped with a cloth and lit with candles. Evidently believing that *Black Square* was the most important of his Suprematist paintings, Malevich hung it high in this holy corner of the room, which he described in the early 1920s: "I see the justification and true significance of the Orthodox corner in which the image stands, the holy image, as opposed to all other images and representations of sinners. The holiest occupies the center of the corner, the less holy occupies the walls on the sides. The corner symbolizes that there is no other path to perfection except for the path into the corner."[74]

In a review of the *0.10* exhibition, the critic Alexander Benois accused Malevich of blasphemy for hanging *Black Square* "high above, right under the ceiling, in the 'holy place.' . . . Without a doubt, this is the 'icon' which the Futurists propose as a replacement for the Madonna."[75] There is a sense in which Benois, who was a sophisticated artist and writer, was correct; Malevich did propose the *Black Square* as an icon for the modern secular age, and just as the church used icons to market the faith, Malevich made the *Black Square* his logo, doing four more versions of the painting and using it in many graphic contexts.[76] And, in a final straddling of the sacred-secular divide, Malevich arranged for a version of the *Black Square* to be hung over him in death (plate 3-33), and a copy of the painting placed in his coffin to accompany him to the netherworld.

Despite Malevich's self-serving dogmatism, one must concede that he was right when he declared that he had created a modernist icon, a point that was reiterated in 2007 at a major exhibition of art inspired by Malevich's *Black Square*, including over a hundred artworks done between 1915 and the 2000s by artists from around the world, organized by the Kunsthalle in Hamburg, *Das schwarze Quadrat: Hommage an Malewitsch* (The black square: Homage to Malevich).[77] The icons of the Russian Orthodox Church depict saints of a bygone era, whereas Malevich's *Black Square* symbolizes the spirit of a new age.

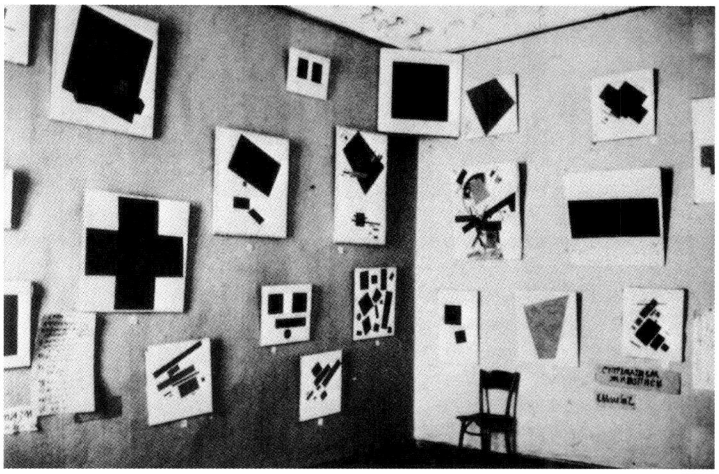 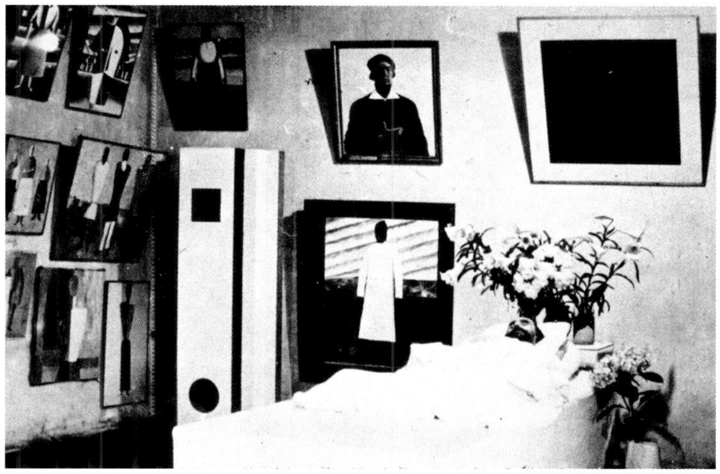

Although the Suprematist's "Absolute" has gradually lost its territory to science, Malevich's (demythologized) *Black Square* continues to resonate with artists today because we live in a secular age in which squares, circles, and numbers comprise the international language spoken in mathematics, science, and technology.

FREE WILL AND PROBABILITY

In an attempt to settle the debate between free will and determinism, the Belgian mathematician Adolphe Quetelet extended Bernoulli's law of large numbers to apply to human behavior; man has free will to make unique choices, but these unpredictable choices cancel each other out when present in large numbers. What remains are statistical regularities that reflect social customs: "Moral phenomena, when observed on a great scale, resemble in some ways physical phenomena. . . . The greater the number of individuals observed, the more do individual peculiarities, whether physical or moral, become effaced [*s'effacent*], and leave prominent general facts in virtue of which society exists and is preserved"[78] (*On Man and the Development of His Faculties, or an Essay on Social Physics*, 1835). In developing so-called social physics, Quetelet saw himself as using statistics to protect free will within the deterministic law of large numbers, and he defended himself again charges of determinism: "How far I am from a blind fatalism, which would regard man as unfit to exercise free will."[79] Quetelet plotted his data showing the statistical regularity of human and social data on a graph (plate 3-34) that revealed the same pattern as Pascal's diagram of inert physical data (plate 3-8).

Quetelet's link of probability and free will was adopted in Russia in the circle of Bugayev, especially by his younger colleague Pavel Nekrasov, who succeeded Bugayev as president of the Moscow Mathematical Society. In 1902 Nekrasov published a massive research project dedicated to the study of "free" mathematics (probability theory), which he linked with free will and Russian Orthodox Christianity, as opposed to "deterministic" mathematics (calculus), which he associated with fatalism and godless Marxism. Nekrasov saw himself as the successor to Quetelet's social physics, but going beyond Quetelet, who had *recorded* actions that he viewed as free choice (such as a groom choosing a

ABOVE LEFT

3-32. *0.10 (zero-ten), The Last Futurist Exhibition of Paintings*, Petrograd, opened in December 1915. Photograph. David King Collection, London.

In this exhibition, *0.10 (zero-ten)*, Malevich showed paintings in his new Suprematist style, composed primarily of colored rectangles; the exhibition catalogue lists thirty-nine works by Malevich, of which twenty-one are shown in this photograph.

ABOVE RIGHT

3-33. Malevich lying in state in his Leningrad apartment, May 17–18, 1935. Photograph. David King Collection, London.

This presentation of Malevich's corpse under a version of *Black Square* hung on the wall is a variation on the display of the deceased in an Orthodox home, where the coffin points into the holy corner where the icon is hung.

Like the dead stroke of
the pendulum clock
Nature devoid of God follows
slavishly the law of gravity.

—Friedrich Schiller, "The Gods of Greece," 1788

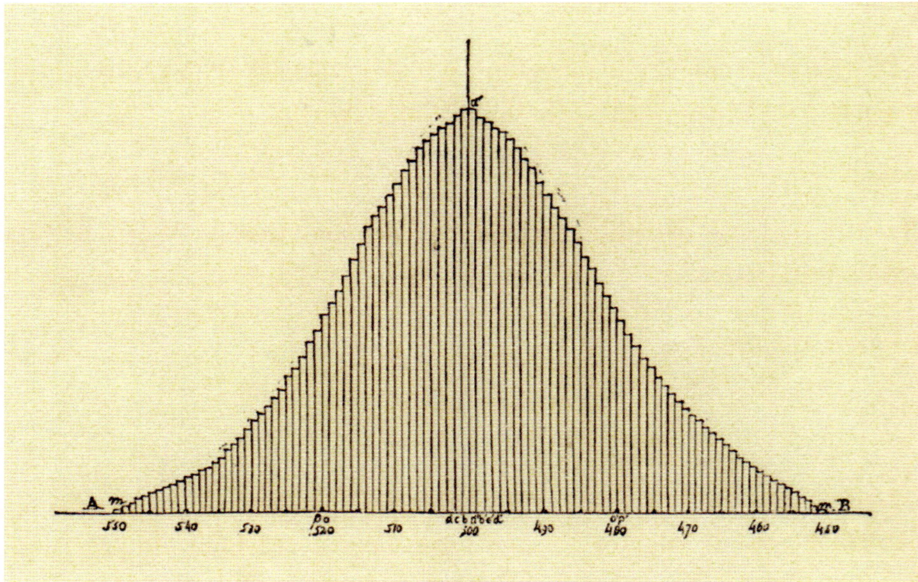

3-34. Normal distribution curve in Adolphe Quetelet, *Lettres sur la théorie des probabilités, appliquée aux sciences morales et politiques* (Brussels: M. Havez, 1846), 396.

Quetelet produced this diagram by plotting data he obtained from the *Edinburgh Medical and Surgical Journal*, which listed the chest circumferences of 5738 Scottish soldiers, whose chests ranged from a puny 33 in. (3 soldiers) to a robust 48 in. (1), with most in the middle range of 39 in. (1075) and 40 in. (1079). After calculating the frequency of each circumference (within a range of 0 to 100%), Quetelet plotted the data for his two terms (circumference and frequency) to produce this binomial (two-term) distribution, which shows that most soldiers are in the middle and the extremes quickly fall off towards zero. The limit of Quetelet's binomial distribution is expressed as a symmetrical "bell-curve" shape, a so-called normal distribution curve.

bride), Nekrasov claimed to have *proven* the existence of free will using statistics.[80] Several Russian and German mathematicians were quick to point out to Nekrasov that a mathematical theorem can't be used to prove any feature of the human condition.[81] Nevertheless, this century-old link of probability theory to free will took deep root in the German Romantic psyche.

We are the quickly dying flame
And again a burning fire.

—**Nikolai A. Vasiliev, "The Longing for Eternity," ca. 1900**

A contemporary of Florensky, Nikolai A. Vasiliev, the son of the Kazan University mathematician Aleksandr V. Vasiliev, began his career as a Symbolist poet (*The Longing for Eternity*, ca. 1900). Beginning around 1910 the younger Vasiliev devoted himself to logic and developed the earliest paraconsistent logic of the modern era, which he titled *Imaginary Logic* (1912) in honor of Lobachevsky's *Imaginary Geometry* (1826).

Vasiliev was drafted into the army after the outbreak of World War I; he fell into a deep depression that led to his dismissal from military service. He then worked as a professor of mathematics at Kazan University until 1922, when the Bolsheviks forced him to resign, after which he declined rapidly and entered a mental hospital, where he died in 1940. Vasiliev's work on paraconsistent logic was rediscovered in the 1960s, and today it is considered a precursor of many-valued (so-called fuzzy) logics, which are designed to handle partial truths by letting the truth-value of a proposition range between 0 (false) and 1 (true), depending on its degree of probability.[82]

In the first two decades of the twentieth century, there had been simmering animosity between members of the Moscow Mathematical Society, who supported the tsarist monarchy, and mathematicians who were politically active Marxists. After the Communist victory in the 1917 Revolution and the Stalinist repression of counter-revolutionaries that began in the 1920s, members of the Moscow Mathematical Society were at high risk because they gave mystical interpretations to certain mathematical objects (discontinuous functions, statistics, non-Euclidean geometry), opening the door to what the Bolsheviks viewed as the evils of metaphysics, theology, and spirituality, as Friedrich Engels, the co-author of *The Communist Manifesto*, wrote in 1883: "The ordinary, metaphysical mathematicians boast with enormous pride of the absolute irrefutability of the results of their science. But these results also include imaginary magnitudes which thereby require a certain reality. When one has become accustomed to ascribe some kind of reality outside our minds to $\sqrt{-1}$ [the square root of minus one, a so-called imaginary number], or to the fourth dimension, then it is not a matter of too much importance if one goes a step further and also accepts the spirit world of the mediums."[83] In a show of power, in 1933 Stalin's Marxist ideologues arrested Florensky (still defiantly wearing his monk's robe) for alleged anti-government sentiments, sent him to a labor camp, and executed him by firing squad on December 8, 1937.[84]

THE END OF THE ABSOLUTE

Nineteenth-century astronomers thought that the planets moved in their orbits within a stationary volume of æther, Newton's Absolute space, while a universal clock ticked off seconds in Absolute time. In 1887 the American physicists Albert Michelson and Edward Morley tried to measure the earth's motion relative to the æther, but after repeated attempts with more and more precise instruments, they concluded it could not be detected. In 1905 Albert Einstein supplied a new overall theory of the universe that accounted for Michelson and Morley's negative results. Their experiment had failed because there is no Absolute space—no æther at rest—against which to define velocity. All objects in the universe move relative to each other within co-ordinate systems that Einstein described, and he defined speed relative to an observer. Nor is there Absolute time in Einstein's universe, because nothing can travel faster than the speed of light. If a body approaches this cosmic speed limit, time (that is, the local time on the moving body) slows down (*Special Theory of Relativity*, 1905).

God is dead. God remains dead. And we have killed him.

—**Frederick Nietzsche, *The Gay Science*, 1882**

In the 1910s, Kruchenykh and Malevich aspired to expand their minds to achieve consciousness of the theosophist Ouspensky's four-dimensional space of the Absolute. But such artistic efforts ceased after a 1919 expedition led by the British astronomer Arthur Eddington confirmed Einstein's space-time universe, in which the fourth dimension is *time* (*General Theory of Relativity*, 1916). This reorientation of the Russian avant-garde was part of a general shift in Western thought away from belief in some form of supernatural four-dimensional realm to reliance on the explanatory power of science (observation and experiment) to account for events in the natural world. By the early 1920s artists were showing less interest in numerology and the occult (alchemy, astrology, extra-sensory perception, and "spiritualism" in the sense of the communication with the "spirits" of deceased persons), and paying more attention to the mathematics associated with science (astronomy, chemistry, and physics).

Can there, will there be major philosophy, literature, music and art of an atheist provenance?

—**George Steiner, *Grammars of Creation*, 2001**

This expansion of the territory of science left intact the classical views of the mathematical-world-out-there (perfect and eternal) and the natural world (imperfect and transitory), but it changed forever mankind's understanding of mind. In place of a supernatural, infinite spirit or psyche that is known by revelation—Plato's Form of the Good, the Judeo-Christian-Islamic God, and Hegel's Absolute Spirit, Cantor's Absolute Infinity—the new

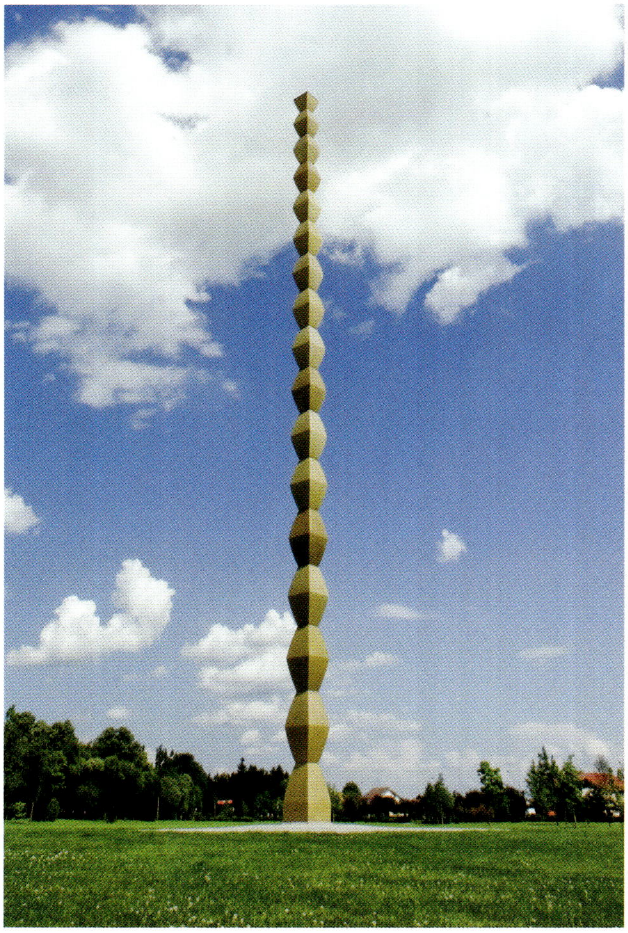

ABOVE

3-35. Constantin Brancusi (Romanian, 1876–1957), *Endless Column*, 1938. Cast iron, 96 ft. 3 in. (29.33 m) high. Târgu Jiu Complex, Romania. © 2014 ADAGP, Paris/Artists Rights Society, New York.

As a memorial to Romanian soldiers who died in World War I, Brancusi created *Endless Column* in Târgu Jiu, the capital of the region of Romania native to Brancusi. Made from seventeen rhombus-shaped modules, the highest is only the bottom half of the diamond-shaped form, suggesting that the column is incomplete but may be continued upward in the viewer's imagination.

OPPOSITE

3-36. *Tribute in Light*, 2013, 88 searchlights in two 50 × 50-ft. (15.2 × 15.2 m) squares, ground zero of the World Trade Center, New York.

Tribute in Light—twin ascending beams of light—began as an impromptu memorial to those who died in the twin towers of the World Trade Center in New York on September 11, 2001. Several teams of architects and designers have organized a reassembly of the tribute to mark the anniversary of the loss.

science of the mind (psychology) declared that there is only natural, finite mind that gives to each person an inner world, which is known by introspection.[85] This shift in territory caused a seismic reorientation in Western culture from belief in a personal lawgiver to denial of the existence of the very Creator whose presence had driven Western mathematics for four millennia.

This cataclysmic change prompted many attempts to formulate a—choose your word—spiritual or psychological creed for the secular era, including revivals of the pantheist tradition that sees the sacred in nature. For example, in search of a form of Protestantism that could coexist with science, the Lutheran theologian Friedrich Schleiermacher described a pantheist outlook in his influential *On Religion: Speeches to Its Cultured Despisers* (1799) that entailed mankind's contemplation of infinity: "In religion everything strives to expand the sharply delineated outlines of our personality and gradually to lose them in the infinite in order that we, by intuiting the universe, will become one with it as much as possible. . . . To be one with the infinite in the midst of the finite and to be eternal in a moment, that is the immortality of religion."[86] Some twentieth-century Western artists looked east and adopted Taoist and Buddhist creeds precisely because they express the mystical in terms of the natural and, since Taoism and Buddhism posit no celestial lawgiver, they merge seamlessly with the scientific worldview (see chapter 10).

One age-old feature of mind (spirit, psyche) survived the shift in territory to secularism: as ancient and medieval man could not describe the One in words or symbols, so modern man finds his inner world ineffable. In the early twentieth century, the founder of phenomenology, Edmund Husserl, declared that the human mind is theoretically capable of an interminable association of ideas (see chapter 8), but an individual's association of ideas is finite because man is mortal. Mankind's ongoing fascination with the topic of infinity may stem from the contrast between infinity and his finitude. Many monuments to the dead are designed around the contrast between the mortality of the flesh and the apparent infinity of time or space (plates 3-35 and 3-36).

When Darwin published *Origin of Species* in 1859, he knew that there must be some physical mechanism to pass traits from generation to generation, but he could not find it. The mechanism was discovered in his lifetime but lost and then rediscovered in 1900, beginning the field of genetics. By the 1940s there was conclusive evidence that DNA is the carrier of genetic information, in 1953 the double-helix structure of the molecule was confirmed, and by the 1970s the order of the four nucleotides (the base pairs that comprise the genetic code) was being sequenced. In 2003 the Human Genome Project was complete, followed by the launch of the study of the epigenome

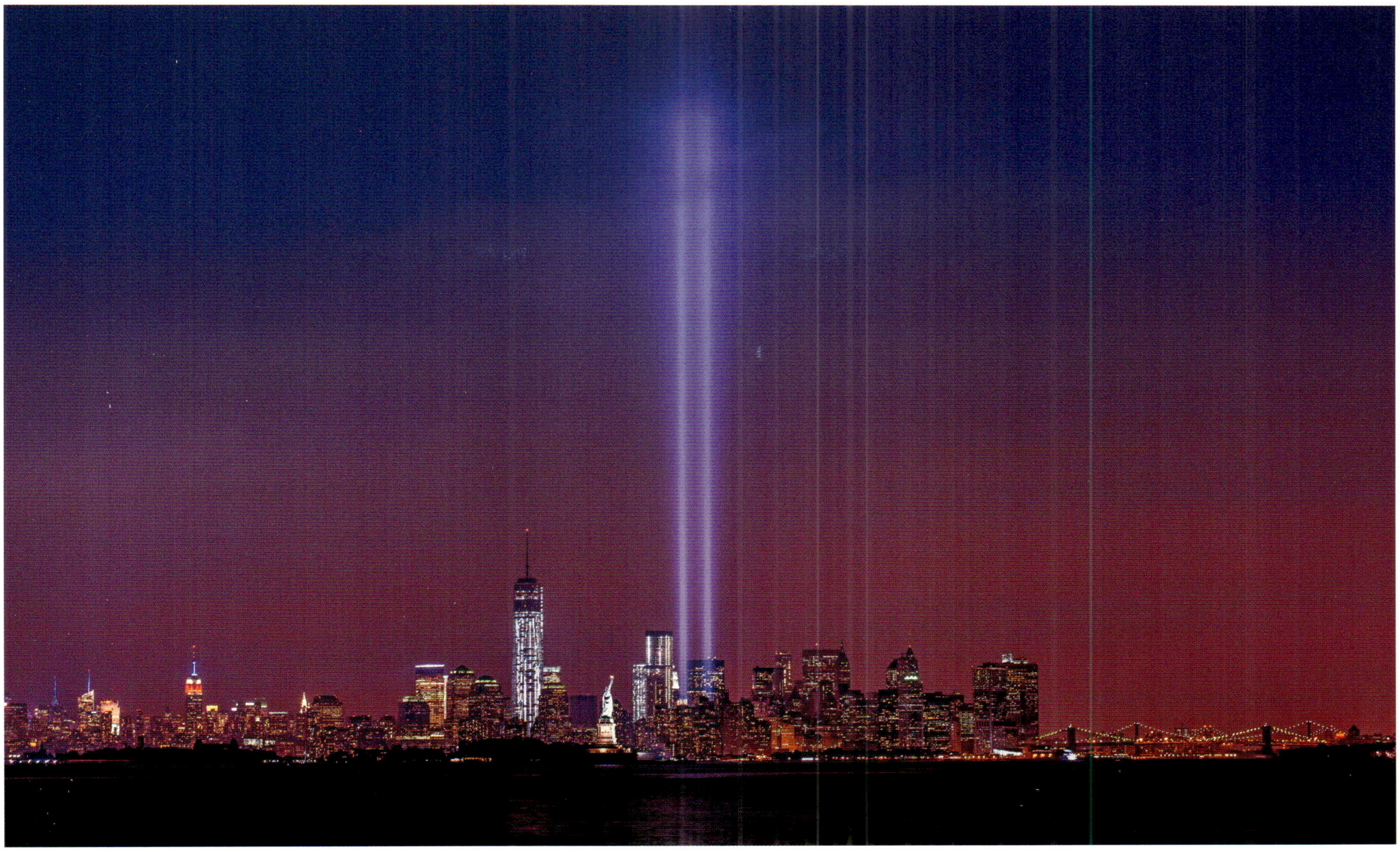

("upon the gene") that seeks to identify the molecules that activate and silence the genes, providing the missing link between a person's (fixed, unchanging) genes and his or her (variable, ever-changing) environment. As the understanding of the mechanism of evolution increased, scientists determined that it was impossible for an individual to evolve a brain during a single lifetime that had significantly more cerebral complexity than the person's parents (and thus be capable of "higher consciousness"), and brain research shifted focus from higher consciousness to so-called altered consciousness (altered by diet, drugs, stress, exercise, or some other cause that has a chemical effect on the central nervous system).

As evolutionary biology developed in the early twentieth century, mathematicians and artists in Germanic culture referred less and less frequently to an evolving consciousness of Absolute Spirit. But expressions of belief in some form of Absolute ceased only after World War II destroyed the community of scholars who studied German Idealism, which was the intellectual underpinning for modern versions of the Absolute. Meanwhile, early-twentieth-century mathematicians and physicists entered the subatomic realm, and they grappled with formulating descriptions of moving particles and forces, with the link between probability theory and free will in the back of many minds.

When I consider the short duration of my life, absorbed in the eternity before and after— memoria hospitis unius diei praetereuntis *[Latin for "the memory of a guest who passed by one day"]—the little space which I fill, and I see the abyss that is engulfed in the infinite immensity of spaces which I ignore and which ignore me, I am frightened and astonished at being here rather than there; for there is no reason why I should be here at this time. . . . The eternal silence of these infinite spaces terrifies me.*

—Blaise Pascal, *Pensées*, 1669

4
Formalism

The procedure of the axiomatic method, as it is expressed here, amounts to a deepening of the foundations of the individual domain of knowledge—a deepening that is necessary for every edifice that one wishes to expand and to build higher while preserving its stability.

—**David Hilbert, "Axiomatic Thought," 1918**

The painting, aiming at complete uniformity, should be the result of its innate qualities (flat surface and square borders).

—**Władysław Strzemiński, *Unism in Painting*, 1928**

WHEN EUCLID WROTE the *Elements* (ca. 300 BC), his fifth postulate was:

> If a straight line falling on two straight lines makes the interior angles on the same side less than two right angles, the two straight lines, if produced indefinitely, meet on that side on which are the angles less than the two right angles.

From the start this axiom raised questions among mathematicians. It was not self-evident like Euclid's other four axioms, but no one was able to prove it using them. Then in the early nineteenth century, two young mathematicians, the Russian Nikolai Lobachevsky and the Hungarian János Bolyai, discovered that, by assuming the negation of the fifth axiom, they could develop non-Euclidean geometries that were completely rational alternatives to Euclid. The nineteenth-century German physiologist Hermann von Helmholtz dealt another blow to Euclidean geometry when he demonstrated that experience and experiment—not intuition—determine whether the world we live in is best described by a Euclidean or non-Euclidean geometry. Helmholtz showed that although the perception of space is innate, as Immanuel Kant had proclaimed (*Critique of Pure Reason*, 1781), neither Euclidean nor non-Euclidean geometry is *a priori* (Latin for "from what is before"). Rather, a man determines the geometry of the space he lives in by experience—by moving around and observing his world. By the late nineteenth century the ties binding geometry to the world were loosening, and mathematicians' confidence in what had once seemed obvious had been shaken.

This moment precipitated a crisis in the foundations of mathematics, and mathematicians reacted by trying to isolate what ensures the certainty of their science. I will follow historians' usual practice and divide their attempts into three groups—formalism, logicism, and intuitionism. These groups differed in their relation to the modern version

4-1. Aleksandr Rodchenko (Russian, 1891–1956), *Red* (*Чистый красный цвет* [Pure red color]), 1921. Oil on canvas, $24\frac{5}{8} \times 20\frac{11}{16}$ in. (62.5 × 52.5 cm). © A. Rodchenko and V. Stepanova Archive, Moscow. © Estate of Aleksandr Rodchenko/RAO, Moscow/VAGA, New York.

Rodchenko's search for the foundations of art prompted him to create the first monochrome paintings of modern art: *Red* (shown here), *Blue*, and *Yellow* (plates 4-16 and 4-17). When Rodchenko put the three paintings on display, he declared: "At this exhibition is the first announcement by me of art made with three primary colors"; *5 × 5 = 25* (Moscow, 1921), n.p.

of Platonism, in which mathematics describes perfect, timeless, abstract objects (Cantor's sets) that exist independently of the human mind. The leader of formalism, the German mathematician David Hilbert, and the leader of logicism, the British logician Bertrand Russell, both held this modern version of Platonism, and they shared the goal of putting traditional mathematics (geometry, arithmetic, etc.) on a secure foundation. To this end they invented new, innovative methods; Hilbert examined the abstract form of geometry and Russell focused on the logical structure of arithmetic (see chapter 5). But the leader of intuitionism, the Dutch mathematician L.E.J. Brouwer, rejected Platonism, maintaining that mathematical objects are creations of the human mind (see chapter 6).

Meanwhile, in the late nineteenth century, painters and designers had precipitated their own crisis by developing the first abstract, completely non-objective art. Since antiquity Western paintings had been pictures of something, but an abstract painting is just color and form. Is it art? Like the modern mathematicians, modern artists became preoccupied with examining the foundations of their field, and their various responses led to different artistic styles. Of the many definitions of art that modern artists proposed, a key dividing line was whether or not art is a mimesis (imitation) of the natural world. Some insisted that the classical understanding of art as mimesis should be preserved, while others argued that representation is superfluous and only formal properties (color and form) are essential to painting and sculpture. Occurring during the rise of science, these debates about mimesis versus non-objective art reflected issues from the laboratory that permeated popular culture, producing some artistic styles based on *vision* and others based on *theory*.

When the scientific worldview emerged in the nineteenth century, French laboratory science developed from a philosophical tradition of Empiricism and was based on observation (for example, Louis Pasteur *first collected data* and observed whether or not it had a pattern, upon which he could *then* base a theory). Science in Germanic lands, on the other hand, developed from Idealist philosophy and was highly theoretical (for example, Gustav Fechner *first formulated a theory*, and *then* he designed experiments that would allow him to make the observations he needed to confirm or refute his theory). Through the popularization of science, these attitudes entered artists' studios. Reflecting French observational science, modern art in Paris was based on vision. Ultimately, Impressionist landscapes, Fauvist portraits, and Cubist still life paintings are all imitations of objects seen in the natural world, albeit very abstracted imitations—they are "pictures." Reflecting Germanic theoretical science, artists from Munich to Moscow symbolized invisible ideas (such as August Endell's pure feeling and Malevich's Absolute Spirit) or, taking an opposite theoretical approach, created meaning-free signs (such as Russian Constructivist Aleksandr Rodchenko's compositions in color and form), thus initiating formalism in the visual arts.[1]

As Rodchenko, Vladimir Tatlin, and others became aware of formalist mathematics, they applied mathematical concepts to their own search for the foundations of art. The painter Władysław Strzemiński wrote a cogent statement of the formalist master narrative of modern art as an inexorable reduction of art to its essence—pure form. Of the three approaches to the foundations of mathematics—formalism, logicism, and intuitionism—formalism has had by far the most profound and lasting impact on the visual arts, first on

Russian Constructivism in the 1910s, then throughout the West in the 1920s, then globally after 1945. The adoption of formalism by visual artists is an example of their interest in the *philosophy* (as opposed to the practice) of mathematics. When artists adopted the philosophical goal of reductionism, they produced very simple paintings and sculptures; in the practice of mathematics, on the other hand, the formalist impulse led mathematicians to extreme complexity (as in formal axiomatic systems), which is absent in the art.

Today we use the word "formalism" to name both the mathematical and the artistic impulse to isolate form (abstract structure), but the term has a different origin in the two fields. In the visual arts, words that translate into English as "form" have been used since antiquity to name the shape of an object (as opposed to its color), as when Plato described the creation of earth by his Craftsman: "Wherefore he made the world in the form of a globe, rounded as from a lathe, having its extremes in every direction equidistant from the center" (Plato, *Timaeus*, 366–360 BC, 33b).[2] The word "form" came into common use in artistic circles only with the rise of abstract art in the nineteenth century, as when John Ruskin wrote: "There is evidently capability of separating color and form, and considering either separately. Form we find abstractedly considered by the sculptor" (*Modern Painters*, 1846).[3] In the second decade of the twentieth century, poets and critics in Russian literary circles adopted the term "formalism" (формализм in Russian, *Formalismus* in German) from the burgeoning science of linguistics, in which researchers named their approach to language formalism to stress that they studied the form (syntax) of language as opposed to its meaning (semantics). By the mid-twentieth century, formalism was used throughout the arts to name any approach that stressed abstract structure (as opposed to imitation, interpretation, or meaning).[4]

Mathematics has always been about form, but the word "formalism" is of modern origin, invented in 1912 by Brouwer to derisively mock Hilbert's approach to the foundations of mathematics: "Formalism [is a] meaningless series of relations to which mathematics are reduced" ("Intuitionism and Formalism," 1912).[5] Although Hilbert himself avoided the term, his colleagues adopted "formalism" to name Hilbert's approach, and the label stuck.

Many modern mathematicians have sensed a resonance between abstract art and their field. Some mathematicians have engaged in dialogues with artists (see chapters 7 and 11), and others have created abstract art themselves, especially after the invention of computer graphics and three-dimensional printing gave them these tools (see chapter 12), much as the invention of the camera spurred the widespread creation of realist art by amateurs a century before. Mathematicians have also commonly applied *philosophical* ideas about aesthetics (beauty, purity) to their mathematical practice. However, since artists generally use very elementary arithmetic and geometry, abstract art usually holds little interest for working mathematicians as far as the *practice* of mathematics is concerned.

Although mathematics has always been abstract and many earlier mathematicians had identified mathematics as a study of formal structures, Hilbert was the first to pursue the full implications of treating geometry as a formal arrangement of abstract, meaning-free, replaceable signs within an internally consistent, self-contained structure. Sensing a common purpose with formalist mathematicians (as well as linguists), Russian Constructivist artists and poets in Moscow, Saint Petersburg, and Kazan reduced their visual and verbal vocabularies to meaning-free signs and composed them as autonomous structures.

The early-twentieth-century search for the foundations of mathematics and art turned the spotlight on the topic of "meaning." A mathematician draws a square on a blackboard and an artist paints a square on a canvas. What determines the meaning of these squares? Do they both refer to the same Platonic Form? Such philosophical questions had been asked since antiquity, but the emergence of modern formalism in mathematics and linguistics, as well as the development of various styles of abstract art, has engendered a massive literature by linguists, psychologists, and philosophers about all types of languages, including natural (spoken), artificial (mathematical), and visual (pictorial) languages. This research into signs and symbols came to be called "formalism" in Germanic intellectual circles, "structuralism" in France, and "semiotics" in Anglo-American universities. These concurrent approaches to understanding symbols and language have long, intertwined histories, and for the purposes of this book it not necessary to trace the threads of the debates, but we will occasionally encounter the ongoing discussions of meaning within formalist/structuralist/semiotic circles.

NON-EUCLIDEAN GEOMETRIES

Over the centuries many mathematicians came to feel that Euclid was wrong to treat the fifth postulate as an axiom and that he should have deduced it from his other four axioms. Many scholars, including the eleventh-century Persian mathematician Omar Khayyam, offered proofs of it, but flaws were always found in their deductions.[6]

Because Euclid's fifth postulate is complicated, several mathematicians wrote equivalent simpler versions of it. The best-known alternative is: through a point can be drawn only one line parallel to a given line. Named after the eighteenth-century Scottish mathematician John Playfair, this alternative formulation is called "Playfair's axiom," but it is best known by its nickname, "the parallel postulate."

In the early nineteenth century, several mathematicians at the University of Göttingen, including Carl Friedrich Gauss, took up the problem. After unsuccessful attempts to prove the parallel postulate, Gauss began to suspect that the fifth axiom does not follow from Euclid's other four axioms after all. In other words, it seemed as if the parallel postulate might be an independent assertion, which meant that alternative geometries could be developed using only the first four axioms. So radical an idea was this that Gauss did not publish it, but he mused about his hunch in lectures to his students and in correspondence with his mathematician friends. Meanwhile, undeterred by the radical nature of the discovery, the young Russian mathematician Nikolai Lobachevsky worked out the mathematics of a non-Euclidean geometry and published it (*Imaginary Geometry*, 1826). Soon after, in 1831 the young Hungarian János Bolyai presented his own formulation of a non-Euclidean geometry as an appendix to a textbook written by his father, the mathematician Farkas Bolyai.[7]

I am becoming ever more convinced that human reason can neither prove nor disprove the necessity of Euclidean geometry.

—**Carl Friedrich Gauss, letter to Wilhelm Olbers, April 28, 1817**

Euclidean and Non-Euclidean Geometries

All geometries share Euclid's first four axioms but differ in the fifth. (For a list of Euclid's axioms see the sidebar on page 17 in chapter 1.)

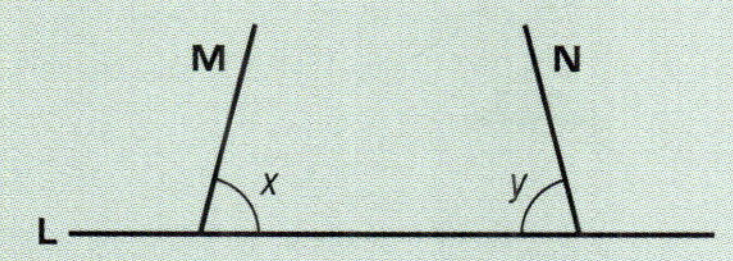

A Euclid's geometry describes a space with zero curvature. Euclid's fifth axiom states: "That, if a straight line falling on two straight lines makes the interior angles on the same side less than two right angles, the two straight lines, if produced indefinitely, meet on that side on which are the angles less than the two right angles." In the diagram, if $x + y$ is less than 180°, the lines will intersect at a point above the bottom line.

This axiom results in plane geometry and describes a flat surface. The angles of a triangle in Euclidean geometry sum to 180°.

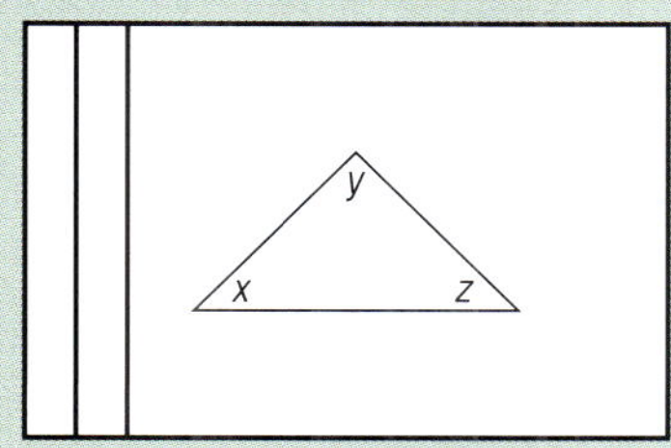

$x + y + z = 180°$

B Gauss's, Lobachevsky's, and Bolyai's non-Euclidean geometry describes a space with negative curvature. Regarding Euclid's fifth postulate in the form that Playfair had given it, Gauss, Lobachevsky, and Bolyai replaced it with: "Through any point lying outside a line, many (an infinite number of) lines can be drawn through the point that do not intersect that given line."

This axiom results in hyperbolic geometry and describes the surface of a pseudosphere, which has a curvature that is everywhere equal and negative. The angles of a triangle in hyperbolic geometry sum to less than 180°.

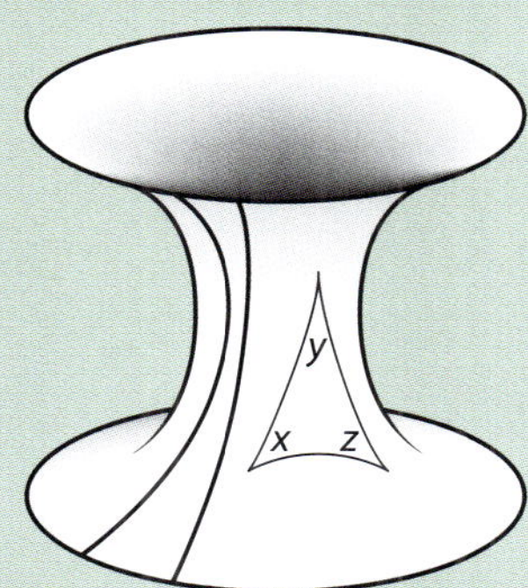

$x + y + z < 180°$

C Riemann's non-Euclidean geometry describes a space with positive curvature. Riemann's fifth axiom is: "Through any point lying outside a line, any line drawn through the point will intersect the given line."

This axiom results in elliptical geometry and describes, for example, the surface of a sphere, which has constant positive curvature. The angles of a triangle sum to more than 180°.

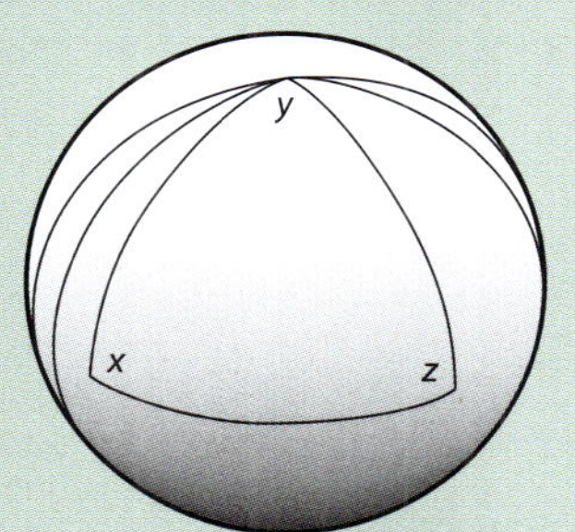

$x + y + z > 180°$

Gauss, Lobachevsky, and Bolyai each attempted to prove the parallel postulate by a so-called proof by contradiction. Suppose that you want to prove an assertion, call it A. In a proof by contradiction you assume the opposite of what you want to prove, $\neg$A (not-A), and then you derive a contradiction. Since only a false premise can lead to a contradiction, it follows that $\neg$A is false and therefore A is true.

Using Euclid's first four axioms as premises together with a denial of the fifth, Gauss, Lobachevsky, and Bolyai each independently looked at the possibility that through a given point many (an infinite number of) lines can be drawn parallel to a given line.[8] But instead of deriving a contradiction, they found themselves proving theorem after theorem in a self-consistent structure. Thus it was that Gauss, Lobachevsky, and Bolyai each developed a non-Euclidean geometry in which Euclid's first four postulates hold but the fifth does not.

I have created a new world from nothing.

—János Bolyai, letter to his father, November 3, 1823

In 1854, the year before Gauss's death, his student and successor at Göttingen, Bernhard Riemann, wrote his doctoral thesis on another non-Euclidean geometry; Riemann assumed that through a given point *no* line can be drawn parallel to the given line; in other words, any line intersects the given line. As the work of Gauss, Lobachevsky, Bolyai, and Riemann showed, the parallel postulate can be neither proven nor disproven—it is *undecidable*—within Euclid's system. Thus Euclid's parallel postulate is independent from the four other axioms and can be replaced by other possibilities, dividing geometry into Euclidean and non-Euclidean geometries. The situation in the mid-nineteenth century was similar to what happened after the discovery, about a century later, that Cantor's continuum hypothesis (his conjecture that the continuum = $\aleph_1$) is undecidable. Cantor's continuum hypothesis could be replaced by other possibilities, and as a result today set theory is divided into Cantorian set theory, which assumes that the next transfinite number greater than $\aleph_0$ is the continuum, and non-Cantorian, which assumes that there are an infinite number of transfinite numbers between $\aleph_0$ and $\aleph_1$.

THE PERCEPTION OF SPACE: HELMHOLTZ'S LEARNED GEOMETRY

Immanuel Kant hypothesized that the laws of Euclidean geometry describe an innate spatial scaffolding in the human brain that forms the intuitive basis of geometry.[9] When a newborn child opens his or her eyes, images enter the brain's spatial framework, where they are arranged over time into the child's perception of a three-dimensional world. Kant held that a similar temporal scaffold in the mind provides the intuitive basis for arithmetic.[10] According to Kant, knowledge of pure mathematics (geometry and arithmetic) was unfailing—a priori—because the intuitions of space and time have their origin in the human mind.

Geometry, however, proceeds with security in knowledge that it is completely a priori, *and has no need to beseech philosophy for any certificate of the pure and legitimate descent of its fundamental concept of space.*

—**Immanuel Kant, *Critique of Pure Reason*,** 1781

Helmholtz, schooled in both German Idealism and physiology, set about to answer in the laboratory the age-old question of how one knows the world. His study of vision and the organs of sense perception (*Handbook of Physiological Optics*, 1856–67) inspired the Impressionists to turn from depicting still lives and landscapes in the world-out-there and to focus instead on their subjective experience of color, capturing with rapid brushstrokes their fleeting impressions of light falling on the retina. Helmholtz also explained hearing ("On the Physiological Causes of Harmony in Music," 1857), which encouraged Jugendstil designers to make an analogy between color and music, both wave-like stimuli (light-waves and sound-waves) that cause nerves to fire. While studying the visual perception of space in the 1860s, Helmholtz concluded that experience must determine whether the natural world is best described by a Euclidean or non-Euclidean geometry.

To understand how human beings know they are in a three-dimensional world, Helmholtz did a thought experiment: he imagined a world composed of a flat plane (a two-dimensional surface with zero curvature), populated with intelligent flat creatures. The inhabitants of this flat world would develop a plane geometry of circles, squares, and triangles whose angles sum to 180°; but, Helmholtz argued, they would not invent spheres

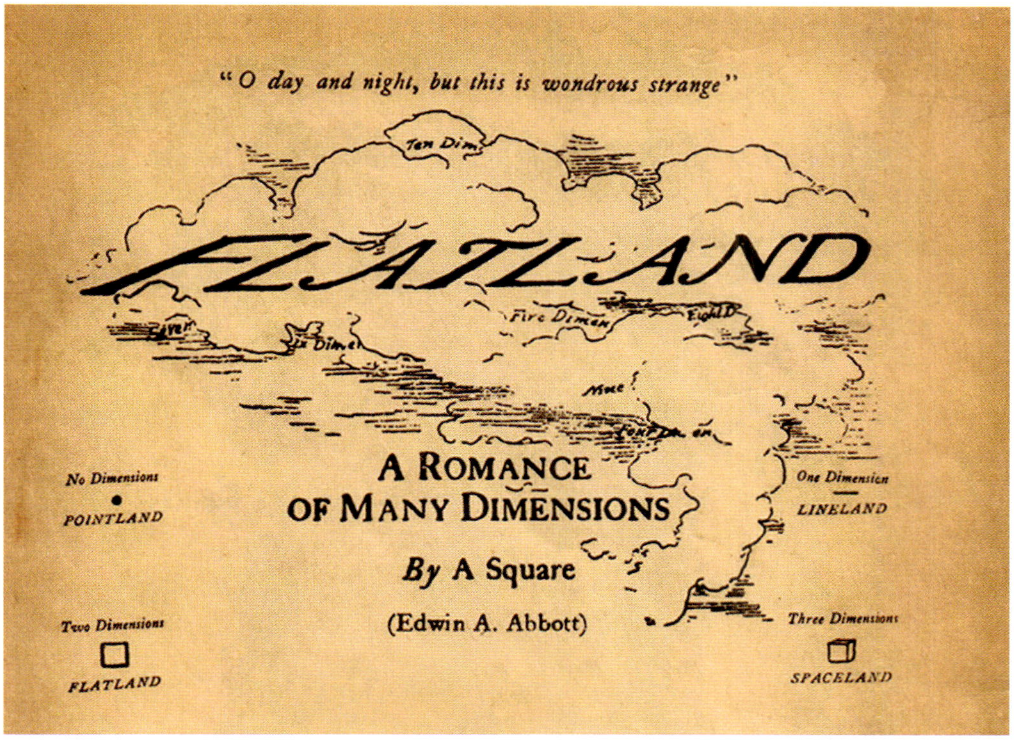

4-2. Edwin A. Abbott (pseud. A. Square), *Flatland: A Romance of Many Dimensions*, illus. by the author (London: Seeley & Co., 1884), cover.

In 1876, Helmholtz gave a popular lecture, "The Origin and Meaning of the Geometrical Axioms," which inspired Edwin A. Abbott to write a satire of Victorian social classes titled *Flatland*. Abbott's story opens in a two-dimensional world populated by polygons whose social rank is determined by their number of sides. The narrator is a lowly commoner named "A. Square." The ruling class includes hexagons and octagons, but the highest rank is reserved for those with an infinite number of sides—Circles—who reign as divine priests of Flatland. Narrow-minded and authoritarian, Circles act as thought police and forbid Flatlanders from imagining a world with three dimensions.

Abbott's satire inspired the artist Dorothea Rockburne to create a metaphorical challenge to Flatland by taking a flat sheet of vellum and folding it, producing "neighborhoods" across the folds that defiantly rise in depth (see plate 4-18). Rockburne based this artwork on topology, the study of surfaces (see pages 229–33 in chapter 6).

and cubes because they had no experience of three-dimensional space. He then imagined a different world that was shaped like the surface of an egg, a two-dimensional surface with positive curvature; the flat creatures living there, he suggested, would develop a kind of spherical geometry in which the angles of triangles lying on the surface of their curved world sum to more than 180°.[11]

From this reasoning, Helmholtz deduced that human beings had developed three-dimensional geometry because they live in a three-dimensional world. He confirmed this hypothesis through laboratory tests of the spatial cues that humans use to judge form and distance. In early 1868 Helmholtz's friend Ernst Schering, professor of mathematics at Göttingen and editor of Gauss's works, told him that Bernhard Riemann, who had died in 1866, had written on this very topic in his 1854 doctoral thesis, which had just been published posthumously as "Über die Hypothesen, welche der Geometrie zugrunde liegen" (On the hypotheses underlying geometry, 1854/1867).[12] The astonished Helmholtz was unaware of this, so Schering sent him a copy of Riemann's 1854 thesis. Helmholtz responded by writing up his own results and, tipping his hat to Riemann, echoed the mathematician's title by naming his essay "Über die Tatsachen, die der Geometrie zugrunde liegen" (On the facts underlying geometry, 1868; see plate 4-2). Helmholtz wrote in the opening paragraphs: "I must confess that the publication of Riemann's work has cancelled the priority of a series of my own works on the subject, but this is of no importance to me because when I see that so distinguished a mathematician should have thought this unusual and discredited subject worthy of his attention, when I find that he and I are companions [*Gefährten*], I am assured of the correctness of my chosen path."[13]

Helmholtz in his physiology laboratory had independently reached the same conclusion that Riemann had by mathematical reasoning: Euclidean geometry is not the only

The axioms of geometry . . . represent no relations of real things. When thus isolated, if we regard them with Kant as forms of intuition transcendentally given, they constitute a form into which any empirical content whatever will fit, and which does not therefore in any way limit or determine beforehand the nature of the content. This is true, however, not only of Euclid's axioms, but also of axioms of spherical and pseudospherical geometry.

—Hermann von Helmholtz, "The Origin and Meaning of the Geometrical Axioms," 1876 possible description of space. Indeed, when Albert Einstein generalized his 1905 *Special Theory of Relativity* to include gravity, he described space as deformed by the gravitational fields of massive bodies, such as stars and galaxies. Thus "straight lines" (light-waves travelling the shortest distance between two points in outer space) follow a curved path, and Einstein adopted Riemann's non-Euclidean geometry to describe the warped space-time of the universe (*General Theory of Relativity*, 1916).

PLATO'S FORMS AS CANTOR'S SETS

Since antiquity Plato's view of mathematical objects has been central to Western mathematics, and it remains today the most common outlook of working mathematicians. In this view, abstract objects such as numbers and spheres exist independently of the human mind. Mathematical objects are real, but since they are immaterial and exist outside time and space, they do not interact causally with the natural, material world. Mankind knows these abstract objects by cognition. Mathematicians describe an objective reality (the mathematical-world-out-there), and the mathematical science is cumulative. Since numbers and triangles are perfect and eternal, mankind's knowledge of them is certain.

Over the centuries there have been many variations of Platonism as it was adopted into Hellenistic, Christian, and Islamic cultures (see chapter 1). In the late nineteenth century, the German mathematician and philosopher Gottlob Frege created the modern version of Platonism when he restated Plato's view that numbers are Forms in terms of Cantor's set theory (*The Foundations of Arithmetic*, 1884). Frege showed how arithmetic could be done with numbers defined in the context of set theory. How are these sets formed? Frege defined a number such as "three" as the set of all things named by the predicate "are three." He defined "equality" between two numbers using Cantor's method for determining the equivalence of sets; two numbers are equal if, and only if, there is a one-to-one correspondence between the members of their sets.

Hilbert was a young professor of mathematics at the University of Göttingen when he came forward in the 1890s to repair the damage done to the foundations of mathematics after Euclid's *Elements* was dislodged. Following Frege, Hilbert was a modern Platonist for whom mathematics describes perfect, timeless, mind-independent, abstract objects—sets.

HILBERT'S FORMAL AXIOMS FOR GEOMETRY: EUCLID'S HIDDEN ASSUMPTIONS MADE EXPLICIT

To put the new set-theoretic version of Platonism on a secure foundation, Hilbert began with Euclidean geometry. In writing *Elements*, Euclid's goal had been to compile proofs by his predecessors, his contemporaries, and himself such that each theorem was deduced, ultimately, from his basic axioms, vocabulary, and common notions. Although it was assumed for two millennia that Euclid had achieved this goal, modern mathematicians, beginning with Gauss and culminating with Hilbert, determined that many of Euclid's assumptions—indeed, his most profound—were left unstated.

An example of a hidden assumption occurs in Axiom 4, which states: All right angles are equal. If Axiom 4 seems self-evident, it is because we are assuming, as Euclid did, that space is everywhere the same and that the geometric properties of a right angle are invariant, no matter where they lie or what their orientation is. Indeed, the homogeneity of space is the most powerful and basic assumption underlying Euclidean geometry.[14]

Euclid evidently placed Axiom 4 directly before Axiom 5 because the parallel postulate is meaningful only if all right angles are equal.[15] Hilbert attributed Euclid's flaw (Axiom 5) to his unwritten assumption (in Axiom 4) that space is homogenous (everywhere the same) and his related assumption that geometry describes the natural world we live in on earth where space *is* homogenous because it is everywhere in the gravitational field of one body—planet earth. However, if we look up at the night sky and gaze beyond our solar system, we are observing space that is not homogenous but warped by the gravity of myriad stars. As Helmholtz argued, the actual geometry here on earth is determined by science (experience and experiment), not mathematics; for an intelligent two-dimensional creature living on the surface of an egg, all right angles are *not* equal. The egg-dweller cannot slide a (rigid) right angle across the surface of his world because the curvature of the surface of an egg (unlike a sphere) varies from one point to another.

Thus Hilbert began his rebuilding of geometry by cutting the ties binding geometry to the world (planet earth and the solar system). He then wrote axioms that are "formal" in the sense that they describe only the abstract structure of a system, using meaning-free symbols from which any reference to the world has been dropped. Hilbert used the axiomatic method of proof as a way to isolate key assertions and arrange them into a logical framework, as Euclid had done, but without reference to the world. Hilbert felt that, once geometry had been formalized, he could prove that the system was consistent (paradox-free) in a way that was clear and obvious to those who cared to follow the details.

Hilbert's approach had been anticipated by the Italian mathematician Giuseppe Peano, who in 1889 wrote a set of formal axioms for arithmetic that were especially simple and elegant; from the second axiom—Every natural number has as a successor—the infinity of counting numbers can be generated ("Peano's Postulates," in *Arithmetices principia* [The principles of arithmetic], 1889). Now Hilbert axiomatized Euclidean geometry by writing twenty axioms that explicitly state Euclid's hidden assumptions and describe only mathematical objects existing in an abstract (formal) structure. He did not provide a vocabulary in which he defined the individual terms of his geometry such as "point," "line," and "plane," as Euclid had done. Rather, he described an abstract system that could apply to anything, beginning his geometry by simply asserting, "Let us consider three distinct systems of things," and he named the things "*points*, *straight lines*, and *planes*."[16] Elsewhere Hilbert commented: "It must be possible to replace in all geometric statements the words *point*, *line*, *plane* by *table*, *chair*, *mug*."[17] Hilbert meant that, since the terms in his vocabulary were undefined terms (meaning-free marks), one can substitute any other terms into the abstract scaffolding—the "theory-form"—without affecting the validity of deductions in the structure. If an assertion about the relation of points is true in classroom geometry, then it can be translated into a parallel assertion about tables that is true in

Peano's Postulates

Italian mathematician Giuseppe Peano's formal system included three undefined primitives—zero, number, successor—a definition of equality, and five axioms. He wrote the axioms in a formal symbolism; informally their meaning is as follows:

1. **Zero is a natural number.**
2. **Every natural number has a successor.**
3. **Zero is not the successor of any number.**
4. **Distinct numbers have distinct successors.**
5. **If a set of numbers contains zero and the successor of every number in the set, then this set contains every number.**

4-3. Geometry as an abstract system.

Tavern in Munich, ca. 1832. Engraving by Schinlels. Bibliothèque des Arts Décoratifs, Paris.

The waiters in this bustling tavern serve their patrons, confident that if tables 1, 2, 3 are in a straight line and table 2 lies between 1 and 3, then 2 also lies between 3 and 1. Their certitude is grounded in the following axiom: "If A B and C are points of a straight line and B lies between A and C, then B lies also between C and A" (David Hilbert, *Grundlagen der Geometrie* [Foundation of geometry; 1899], Axiom 2:1).

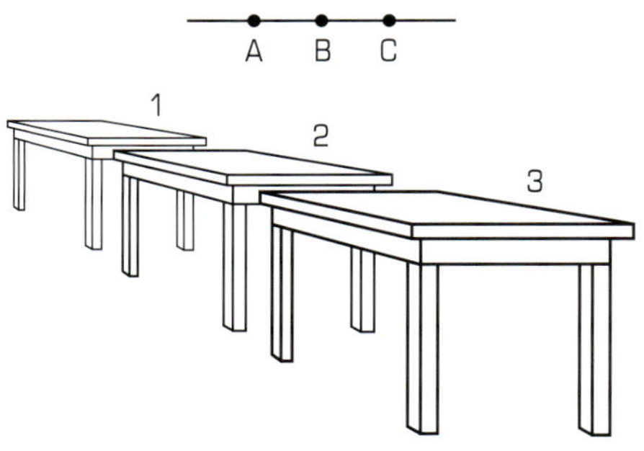

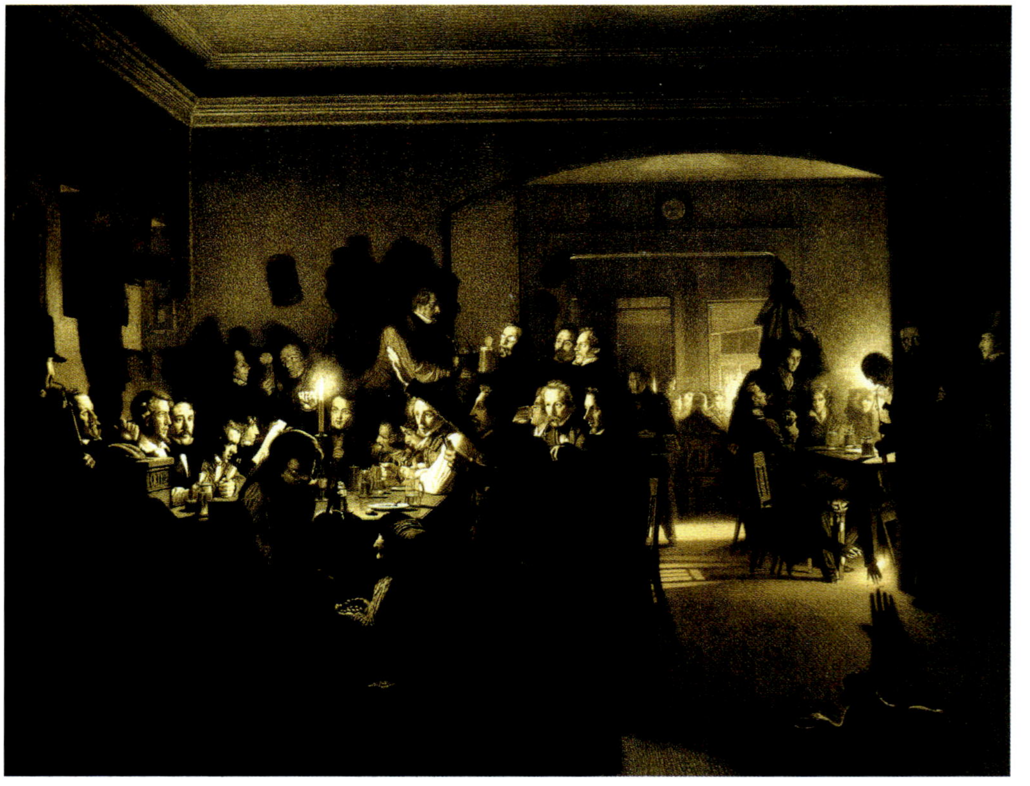

tavern geometry (plate 4-3). Hilbert also defined the mutual relations between points, straight lines, and planes in terms of axioms, which he arranged in five groups: axioms of connection, order, parallels, congruence, and continuity.[18]

The game of chess—the oldest game of pure reasoning (with no element of chance)—is another example. Chess is played with thirty-two pieces on a board of sixty-four squares in the same way that a deductive proof is written in a specific vocabulary; the starting position of the chess pieces is analogous to the axioms stated at the beginning of a proof; the rules of the game are like the rules of inference; and any legal arrangement of the pieces on the board corresponds to a proven theorem. A formalist moves white pieces across a grid following abstract rules with the goal of eliminating one particular black piece, while his opponent might interpret the grid as a battlefield where his black army is trying to capture the white king.

Today the computer is another example of a formal axiomatic system. Written in a vocabulary of zeros and ones, the computer program is analogous to the axioms and rules of inference. Given a starting position and an input (such as data entered on a keyboard), the outcome is completely determined by mechanical rules. Indeed, the computer was invented by the mathematician Alan Turing in the context of debates about the scope and limits of formalism (see chapter 10).

The key difference between Euclid's and Hilbert's axiomatic structures is that Euclid proposed axioms as necessary truths, and he sought a correct description of the natural world-out-there and the idealized realm of the mathematical-world-out-there. Hilbert, on the other hand, proposed axioms as *hypotheses* (which could be true or false in an interpretation) and he wished to show that his set of axioms form an internally consistent structure.

By severing the bonds between mathematics and the world, Hilbert became the first mathematician to consider geometry as a fully formal arrangement of abstract, meaning-free, replaceable signs within an internally consistent, self-contained structure (*Grundlagen der Geometrie*, 1899; English translation in 1902 as *The Foundation of Geometry*). With his formal axioms of geometry, Hilbert established himself as the world's leading figure in formalist mathematics and the University of Göttingen as its home base.

CONSISTENCY ENTAILS EXISTENCE

When Frege created the modern version of Platonism in terms of Cantor's set theory, he translated the age-old philosophical question, "Are Plato's Forms just concepts or do they have an independent existence?" into its modern version, "Are Cantor's sets mind-dependent or do they have an existence independent of human thought?" Frege was a Platonist who answered that sets have an independent existence, arguing that the language of mathematics refers to abstract objects (numbers, circles, sets), and very many mathematical theorems are true. A sentence cannot be true unless it refers to a real state of affairs; therefore, according to Frege, true mathematical theorems commit us to the existence of the mathematical objects to which they refer, such as numbers, circles, and sets.[19]

In his founding text of set theory (*Foundations of a General Theory of Manifolds*, 1883), Cantor had distinguished between an "immanent reality" (an idea in the mind) and a "transient reality" (an object that is not mind-dependent). Any well-defined, logically consistent idea (an immanent reality in the mind) corresponds, according to Cantor, to an abstract object (a transient reality) in the mathematical-world-out there: "I have no doubt that these two sorts of reality always occur together in the sense that a concept designated in the first respect as existent always also possesses in certain, even infinitely many, ways a transient reality. . . . Mathematics, in the development of its ideas, has *only* to take account of the *immanent* reality of its concepts and had *absolutely* no obligation to examine their *transient* reality."[20] In other words, according to Cantor, after Euclid formulated a well-defined, logically consistent mathematical concept, such as his definition of "sphere" as "all points equidistant from a point in three-dimensional space," he did not need to take the additional step of proving the existence of the sphere as a mathematical object. The consistency of Euclid's definition *is* proof that the sphere exists. Where is it? According to Plato, Euclid, Cantor, and Frege, the sphere exists in the mathematical world-out-there, independent of the human mind.

A few years after Cantor formulated set theory, mathematicians began discovering paradoxes in it. For example, after Frege defined numbers in terms of Cantor's set theory, in 1902 the young Bertrand Russell discovered a paradox by considering the set of all sets that are not members of themselves (see chapter 5). But if there are paradoxical concepts (in Cantor's immanent reality), then there are paradoxical abstract objects (in transient reality). How can this be? It *can't* be, and so mathematicians began refining set theory to rid it of paradoxes. In so doing Hilbert, following Cantor, isolated *consistency* as the defining feature of a concept (in immanent reality) that established the existence

The mathematician cannot create things at will, any more than the geographer can; he can only discover what is there and give it a name.

—**Gottlob Frege, *The Foundations of Arithmetic*, 1884**

of a mathematical object (in transient reality), as Hilbert stated in a 1903 letter to Frege: "What is decisive is the recognition that the axioms that define the concept are free from contradiction."[21] For a modern, set-theoretic Platonist, *consistency entails existence*.

In the 1960s the American philosopher Paul Benacerraf raised an objection to the existence of abstract mathematical objects when he argued that, by definition, they are inaccessible to sense perception. In other words, abstract mathematical objects are unobservable and hence unknowable by the scientific method ("What Numbers Could Not Be," 1965). But in a series of publications in the 1970s and 1980s, two American philosophers, W. V. Quine and Hilary Putnam, countered that mathematical objects are indispensible to science in the sense that numbers, geometry, statistics, and calculus are applied throughout the natural sciences; therefore, mathematical objects must exist. This argument is known as "the Quine-Putnam indispensability argument" for modern mathematical Platonism.[22]

To be is to be the value of a bound variable.

—W. V. Quine, "On What There Is," 1961

In the Platonic tradition, knowledge of mathematical objects has been understood as a flash of insight—"only after long partnership in a common life devoted to this very thing does truth flash upon the soul, like a flame kindled by a leaping spark."[23] In a modern secular context, knowledge of abstract objects by cognition came to be called "*intuition*."

Today mathematical Platonism is frequently called "*realism*" because numbers are real (they *really* exist); I will continue to use the traditional term, "*Platonism*." Although Cantor's work on infinity was initially met with resistance and some have quibbled about the details of Frege's arithmetic, today the overwhelming majority of mathematicians around the world have thoroughly adopted this modern version of numbers and arithmetic within the context of sets and set theory.

There are variations of mathematical Platonism in the modern era, depending on what abstract objects are considered real, existing entities. At the conservative end, most would say that the set of natural numbers 1,2,3, . . . (Cantor's aleph-null) exists. A more liberal Platonist might say that an infinite set of higher cardinality (such as Cantor's aleph-one) also exists. But only a radical "absolute Platonist" (*absolute Platonismus*, to borrow a phrase from Hilbert's colleague, Paul Bernays)[24] would believe that there is a set that contains all (imaginable) mathematical objects and, furthermore, that it can itself be treated as a set. The latter is a set, like Cantor's Absolute Infinity, from which most Platonists recoil because it threatens to produce paradoxes. For the purposes of this book, it is not necessary to examine in detail the extensive literature on the ontological status of abstract objects. Suffice it to say that, with a few exceptions that will be noted, modern mathematicians are set-theoretic Platonists, maintaining versions of the classical Platonic outlook, which are based on Cantor's set theory and Frege's arithmetic. The main exception is the group of mathematicians assembled around L.E.J. Brouwer under the banner of intuitionism, who believe that numbers and forms exist only in the mind—mathematicians create them by thought alone (see chapter 6).

THE BANISHMENT OF METAPHYSICS

If abstract mathematical objects exist, what kind of "existence" is this? In the ancient and medieval Platonic traditions, abstract objects resided in the divine realm of the Forms or in the Kingdom of Heaven, but modern, secular, set-theoretic Platonists do not make theological associations with Forms or sets (in any literal sense). As we will encounter repeatedly in this book, there are strong anti-metaphysical tendencies in modern secular thought. So what is metaphysics?

Aristotle wrote about the physical world of earth, air, fire, and water in *Physics*, which he followed with another fourteen books on first causes and first principles of the cosmos; on questions about being, time, and space; and on the subject of theology—"divine things." A century after Aristotle's death, an anonymous editor named these fourteen books "metaphysics" ("after physics"). In the Middle Ages metaphysics continued to refer to questions about first causes and divine things. With the rise of science in the Enlightenment, physics became the new, quantitative field that it is today, leading to a territory shift. Certain topics, such as time and space, moved from metaphysics to physics, while others moved from physics to the realm of metaphysics, which was expanded to include questions about free will and determinism, and also now encompasses value theory (ethics and aesthetics). Today the new "modern" metaphysics, in its most general sense, refers to any attempt to give an account of ultimate reality.[25]

Modern intellectuals who dismiss metaphysics typically use the term in its medieval sense of meaning questions about "divine things" that are above nature ("supernatural")—things that exists beyond the mundane world studied by Newton and Einstein. Taking metaphysics to mean "above and beyond physics" is encouraged by modern coinages such as "metalanguage" and "metamathematics."

If metaphysics is taken to mean "supernatural," then the 2000-year-old association of Platonism with religion is a reason the anti-metaphysical mood is so strong in modern mathematics and science—as it is throughout secular culture. Modern-day secularists desire to break this lingering linkage. For example, in the 1950s Bertrand Russell reflected that in his youth he had made a metaphorical association (which he later disavowed) between Platonism and religion: "I wanted certainty in the kind of way in which people want religious faith. I thought that certainty is more likely to be found in mathematics than elsewhere."[26] More recently the immateriality of abstract mathematical objects had (unwanted) religious associations in the mind of American mathematician Peter Renz: "Belief in the existence of such objects is an act of faith on a par with belief in an unknowable deity,"[27] and the British mathematician E. Brian Davies has complained that intuitive knowledge of "a mathematical realm outside the confines of space and time . . . has more in common with mystical religions than with modern science."[28] Indeed, the founder of set-theoretic Platonism, Cantor himself, gave sets a theological interpretation—culminating in Absolute Infinity—perhaps provoking the antipathy of secular mathematicians for the attribution of metaphysical (supernatural) existence to abstract objects such as sets. The strongly anti-metaphysical stance among secular mathematicians was stoked by the repeated attacks they faced from two camps: adherents of German Romantic *Naturphilosophie*, with its supernatural-sounding beings such as Hegel's Absolute Spirit, and followers of Nietzsche's *Lebensphilosophie*, which focuses on human life (ethics) and human feelings (aesthetics), both topics that nineteenth-century philosophers moved into the domain of metaphysics because they do not yield to the scientific method of measurement and analysis.

Today mathematicians typically use set theory in practice but distance themselves from philosophical issues pertaining to metaphysics (in both its medieval and modern versions). But if people no longer describe numbers and spheres as created by Plato's Craftsman or the Judeo-Christian-Islamic Creator, then where did they come from? According to modern intuitionists, mathematics is a creation of the human mind. Modern Platonists, on the other hand, contemplate numbers and spheres in the mathematical-world-out-there *as if* they are given independently of any human (or mechanical) ability to define or construct them, while hastening to add that they do not intend to suggest that abstract objects have any kind of metaphysical (supernatural) existence. But this ploy does not *avoid* metaphysics (in its modern sense). To assert that one takes abstract objects as a given *is* a metaphysical assertion, which acknowledges that the origin of abstract objects is *unknown*. Indeed, the origin of mathematical objects is one of the great mysteries of modern metaphysics.[29]

For the practical person who is allergic to metaphysics, a philosophical consideration of the origin and existence of abstract objects may seem irrelevant and a bit of an embarrassment. What's the point? Who cares? A recent book with the telling title *Towards a Philosophy of Real Mathematics* argues that philosophical questions of the kind raised by Hilbert about the foundations of mathematics are irrelevant to "real" mathematics, by which the author means the *practice* of mathematics.[30] They are not irrelevant to persons who strive to integrate philosophy and practice in their intellectual lives. Many ancient and medieval people were captivated by philosophical questions about abstract objects that may be unanswerable or even unintelligible, and today certain secular mathematicians and artists continue to ask these questions, expressing a strong desire to connect mentally—via intuition or mysticism—with the mathematical-world-out-there. No one doubts that mathematical research is driven by practical questions, such as Cantor's question about Fourier series and Hilbert's about Euclidean axioms. But philosophical questions also drive research. Doesn't the banishment of metaphysics from modern mathematics threaten to repress the passion to unlock the door to ultimate reality that also motivates and sustains scientific inquiry?

The most beautiful experience we can have is the mysterious. It is the fundamental emotion that stands at the cradle of true art and science.

—Albert Einstein,
"What I Believe," 1930

FORMALIST AESTHETICS: THE AUTONOMY OF MATHEMATICS AND ART

Almost a century before Hilbert, the German philosopher Johann Friedrich Herbart had articulated the formalist outlook as an expression of the scientific worldview. He proposed that the goal of each field of learning, including mathematics and the fine arts, was to isolate the central concepts of its subject matter, after which philosophers like him would unify the concepts (*On Philosophical Studies*, 1807).[31] Herbart was a very influential academic from his post at the University of Königsberg, where from 1809 he occupied Kant's chair in philosophy, after which he taught philosophy at Göttingen from 1834 until his death in 1841. For the young Bernhard Riemann, a student of Gauss at Göttingen in the 1840s, Herbart's analytical attitude was the inspiration for developing a philosophy of mathematics that focused on isolating basic concepts, an approach that Hilbert later adopted.[32]

Herbart, Riemann, and Hilbert all worked in the philosophical tradition of German Idealism, whose founder, Immanuel Kant, had urged scholars to forge a comprehensive, unified worldview (*Weltbild*) or philosophy of life (*Weltanschauung*) that includes mathematics, science, and values (aesthetics and ethics), each a unified field within one grand structure.[33] Kant wrote three treatises about how one gains knowledge of the three fields: for mathematics, *Critique of Pure Reason* (1781); for science, *Critique of Practical Reason* (1788); and he gave aesthetic knowledge its fullest treatment in *Critique of Judgment* (1790). Kant defined the aesthetic realm as autonomous and declared that the truth of an aesthetic judgment, such as "The starry sky is beautiful," is independent of both logic and moral considerations. Aesthetic knowledge is an immediate, subjective intuition.[34]

In 1900 Hilbert described his vision of the unity of mathematics: "Mathematical science is in my opinion an indivisible whole, an organism whose vitality is conditioned upon the connection of its parts. For with all the variety of mathematical knowledge, we are still clearly conscious of the similarity of the logical devices, the relationship of the ideas in mathematics as a whole and the numerous analogies in its different departments."[35] Hilbert also held the related view of science, in which nature has a fundamental unity and it embodies mathematical patterns: "The organic unity of mathematics is inherent in the nature of this science, for mathematics is the foundation of all the exact knowledge of natural phenomena."[36]

Herbart also applied the formalist approach to the arts, viewing each art form as a unified, autonomous field; as a scientist looks at data to discover underlying mathematical patterns, so artists focus on the basic elements of their medium, such as the architect's form and structure and the musician's melody and rhythm. For an aesthetic judgment to be pure, according to Herbart, it must be based only on what is essential to the work of art (such as the pattern of notes and tempo of music) and not on any accidental or variable feature (such as the meaning of the lyrics), as Herbart wrote in 1831: "What have the old artists, who developed the forms of the fugue, or the older ones, who created the architectural orders, intended to express? Absolutely nothing; their thoughts were focused on the essence of each art. The ones who rely on meaning reveal their dread of the essence of each art and their preference for outer glitz." [37]

Herbart inspired the Czech music critic Eduard Hanslick to write the first formalist treatise on music: "Music consists of successions and forms of sound, and these alone constitute the subject" (*The Beautiful in Music*, 1854). A few years later, Helmholtz explained *why* human beings experience certain mathematical ratios of sound as beautiful. In antiquity the Pythagoreans had discovered that when a listener experiences a pair of sounds (such as an octave) as a harmonious consonance, the notes have a precise mathematical ratio (in this case, 1:2; see chapter 1); sounds lacking such ratios are experienced as an unpleasant dissonance. Helmholtz sought an explanation for this subjective experience of harmony in the organs of sense perception—the ear and auditory cortex—and he confirmed that the Pythagorean mathematical ratios are an innate feature of the human neurological circuitry. In his classic essay on music of 1857 ("On the Physiological Causes of Harmony in Music"), Helmholtz explained that humans (unconsciously) perceive "successions and forms of sound" as consonant (pleasant) because the sound patterns resonate with innate structures in the nerves of the human ear and auditory cortex, concluding his essay:

> I have endeavored to lay bare the hidden law, on which depends the agreeableness of consonant combinations. It is in the truest sense of the word unconsciously obeyed, so far as it depends on the upper partial tones, which, though felt by the nerves, are not usually consciously present to the mind. Their compatibility or incompatibility, however, is felt without the hearer knowing the cause of the feeling he experiences. These phenomena of agreeableness of tone, as determined solely by the senses, are of course merely the first step towards the beautiful in music. For the attainment of that higher beauty which appeals to the intellect, harmony and disharmony are only means, although essential and powerful means. In disharmony the auditory nerve feels hurt by the beats of incompatible tones. It longs for the pure efflux of the tones into harmony. It hastens towards that harmony for satisfaction and rest. Thus both harmony and disharmony alternately urge and moderate the flow of tones, while the mind sees in their immaterial motion an image of its own perpetually streaming thoughts and moods. Just as in the rolling ocean, this movement, rhythmically repeated, and yet ever varying, rivets our attention and hurries us along. But whereas in the sea, blind physical forces alone are at work, and hence the final impression on the spectator's mind is nothing but solitude—in a musical work of art the movement follows the outflow of the artist's own emotions. Now gently gliding, now gracefully leaping, now violently stirred, penetrated or laboriously contending with the natural expression of passion, the stream of sound, in primitive vivacity, bears over into the hearer's soul unimagined moods which the artist has overheard from his own, and finally raises him up to that repose of everlasting beauty, of which God has allowed but few of his elect favorites to be the heralds. But I have reached the confines of physical science and must close.[38]

Following the tradition of Herbart and Helmholtz, Hilbert manifested the formalist aesthetic impulse when he described axiomatic structures in aesthetic terms like "simplicity" and "purity," placing great value on finding the fewest axioms, stating them as rigorously as possible, and valuing structures that are well-ordered, consistent, and harmonious.[39] This reductionist impulse towards purity persisted when the formalist approach was imported into other fields in Russia: first linguistics, then literature and the visual arts.

FORMALISM IN RUSSIAN MATHEMATICS AND LINGUISTICS

In the late nineteenth and early twentieth centuries, Moscow mathematicians under the leadership of Nikolai V. Bugayev gave philosophical and religious interpretations to numbers and geometric forms (see chapter 3). But in the 1880s, at Kazan University, the formalist approach gained ground when Aleksandr V. Vasiliev (the father of Nikolai A. Vasiliev) co-ordinated the publication of the first Russian edition of collected works of Nikolai Lobachevsky, who had lived in Kazan, which became a center for the study of Lobachevsky's non-Euclidean geometry. Meanwhile in Saint Petersburg a group of mathematicians around Pafnutii L. Chebyshev established their city as the world's center for research on probability theory. They did not link their sophisticated statistical methods to the research into chance phenomenon conducted in Western laboratories, such as James Clerk Maxwell's 1866 kinetic theory of gas, but rather they studied probability theory as an uninterpreted formal structure. But by the second decade of the twentieth century followers of Chebyshev saw the need for probabilistic methods in the new subatomic physics, and they applied their statistical methods in this domain.[40]

In the early twentieth century German and Russian linguists were also adopting a formalist approach. Earlier linguists had compared basic sounds, grammar, and vocabulary of old languages and their modern descendents, with the goal of organizing languages on a family tree. This biological metaphor was strengthened after Charles Darwin presented species organized on an evolutionary tree (*Origin of Species*, 1859) and focused attention on language as a feature distinguishing humans from other species (*Descent of Man*, 1871; see plates 4-4 and 4-5).[41] But at the fin-de-siècle, linguists studying dead languages adopted a new metaphor from formalist mathematics and began looking at

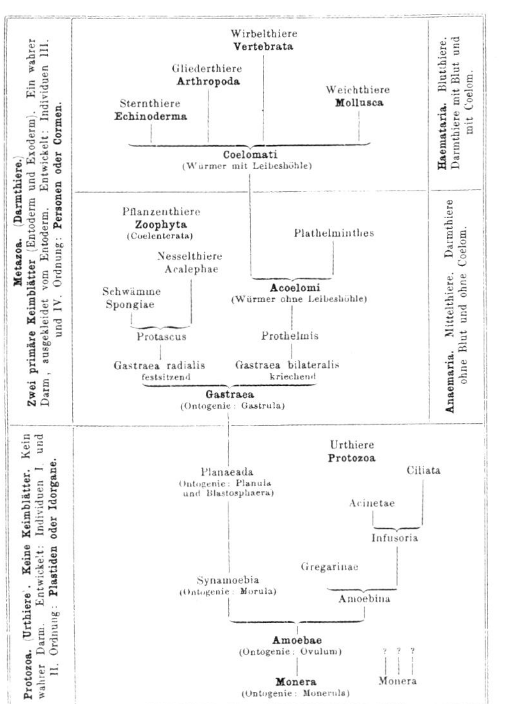

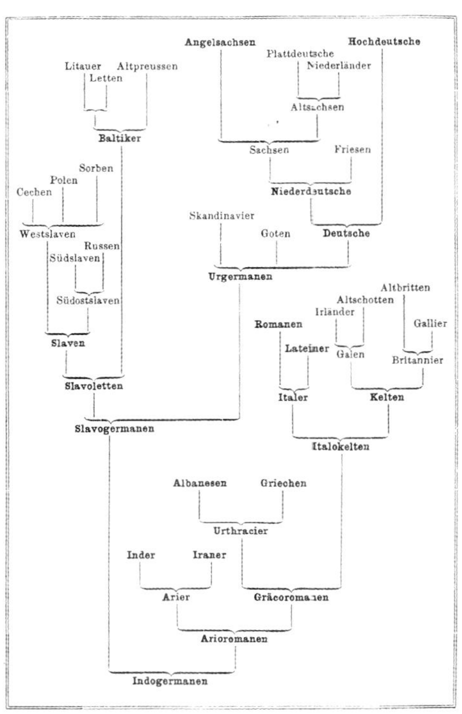

FAR LEFT

4-4. Tree of life, in Ernst Haeckel, *Anthropogenie, oder Entwicklungsgeschichte des Menschen* (Anthropogenic theory, or the history of human development; Leipzig, Germany: Engelmann, 1874), 473, plate 14.

The German biologist Ernst Haeckel was a follower of Charles Darwin.

LEFT

4-5. Tree of languages, in Ernst Haeckel, *Anthropogenie, oder Entwicklungsgeschichte des Menschen* (Anthropogenic theory, or the history of human development; Leipzig, Germany; Engelmann, 1874), 360, plate 10.

Nineteenth-century linguists hypothesized that Western and Near Eastern languages evolved from a common language spoken in lands stretching from the Indian subcontinent to Europe in preliterate times, a lost Indo-European (*Indogermanen*) mother tongue.

languages as formal structures in which the symbols (the pictures or words) are considered as meaning-free variables and patterns within an uninterpreted text. Initially linguists applied the formalist approach only to the remains of dead languages, such as a clay tablet covered with marks. Rather than trying to figure out the meaning (the semantics) of the marks in the lost tongue, formalist linguists set the more limited goal of studying the pattern (the syntax) of untranslated marks on the tablet.

In Germany in the 1870s, linguists at the Leipzig University expanded this formalist approach to describe living languages. The so-called Young Grammarians focused on the units of sound—phonemes—that comprise a spoken tongue, such as the English phonemes "o" (as in "no"), "oi" (as in "loin"), "p" (as in "pit"), and "th" (as in "thus"). In selecting phonemes, they picked out the part of language that lends itself to being treated as a meaning-free variable. The Young Grammarians theorized that the laws governing phonetic changes in spoken languages over time (for example, from Old High German to modern German) were *completely* deterministic, like a rule of inference in geometry.

After studying with the Young Grammarians and completing a doctorate at the Leipzig University in 1870, the Polish linguist Jan Baudouin de Courtenay became a professor of Sanskrit and Indo-European languages from 1875 to 1883 at Kazan University, where mathematicians were already using a formalist approach to study non-Euclidean geometry. At Kazan, Baudouin began his lifelong search for the deterministic laws predicted by the Young Grammarians. For example, in the evolution of Romance languages, Baudouin asserted that *b always* changes to *v* when followed by a vowel. As unlikely as it might seem that one could explain changes in a spoken language by laws that operated, like rules of deductive inference, outside history, Baudouin's approach was widely used in the late nineteenth century to reconstruct lost Indo-European words and to explain changes in spoken languages.[42] Over the years Baudouin softened the strict determinism of his phonetic laws, writing in 1910 that their predictive power was probabilistic, more like weather forecasts.[43]

While Baudouin was establishing the formalist approach to language in Russia in what became known as the Kazan school of linguistics, the rise of formalist linguistics in France and Switzerland, where it was called *structuralisme* (French for "structuralism"), was associated with the work of another student of the Young Grammarians, the Swiss linguist Ferdinand de Saussure, who completed a doctorate at the Leipzig University in 1880. Saussure then taught in Paris and Geneva, and after his death in 1913 his students Albert Sechehaye and Charles Bally published their notes from lectures Saussure had given in Geneva in what became the founding text of structuralist linguistics, *Cours de linguistique générale* (Course in general linguistics, 1916).

In 1900 Baudouin moved to the University of Saint Petersburg, whose mathematics department had already established a formalist approach. From his post as professor of linguistics until 1918, Baudouin exerted great influence on young Russian poets. Entering Russian literary circles, Baudouin viewed avant-garde poetry as an example of rapidly changing phonemes in action.[44]

FORMALIST CRITICISM AND LITERATURE

Formalist poets and novelists typically emphasize the structure (syntax) of language, but rarely maintain that a poem is *only* a structure of sound (phonemes) because sound alone does not make a language. Syntax (grammatical structure) and semantics (meaning) together constitute language.

At the opening of the twentieth century, the leading mathematician in Moscow, Nikolai V. Bugayev, had given his mathematical equations esoteric philosophical meaning, but ironically it was his son Boris N. Bugayev who was the founding figure of Russian formalist literary criticism, which stripped poetry of its meaning. After adopting the pseudonym Andrei Bely, he went on to become a leading Symbolist poet. Bely had learned from his father to give mathematics a broad cultural interpretation, and the young poet applied the formalist approach to literary criticism by stressing the abstract structure of poetry. Like a formalist linguist, Bely focused not on meaning but on meaning-free sound patterns, analyzing the meter of lines of verse and the pattern of stressed and unstressed syllables. He defined rhythm as the pattern of the deviations from standard meter, which he revealed in a chart that produced geometric figures (plate 4-6).[45] For Bely, the geometric figure represented the formal (rhythmic) essence of the poem, and he went on to compare the styles of dozens of Russian poets in terms of the metric patterns embodied in such geometric figures, as in plate 4-7. Bely's accomplishment was essentially to confirm, by statistical means, the aesthetic judgments that other critics made on the basis of their taste, acquired from years of experience: poets who follow the metric rules too closely (such as Tolstoy and early Pushkin) produce predictable sing-song patterns of verse, whereas poets who violate the rules (late Pushkin) create a fresher, richer rhythm (*Symbolism*, 1910).[46]

In 1914 the eighteen-year-old Roman Jakobson enrolled in the linguistics department at Moscow University and entered Bely's literary circle. Jakobson began composing poems by assembling phonemes from various languages. These poems are rare examples of *completely* formal poetry in the sense that they are composed with units of meaning-free phonemes. The poems are comparable to wordless music as a kind of verbal art. Are they literature? If literature is made from language (syntax plus semantics), then they are not. Jakobson published his formalist poems in Futurist periodicals under the pseudonym Roman Aljagrov during his student years, from 1914 to 1920.

In 1915 Jakobson organized the Moscow Linguistic Circle, with which Aleksei Kruchenykh and Velimir

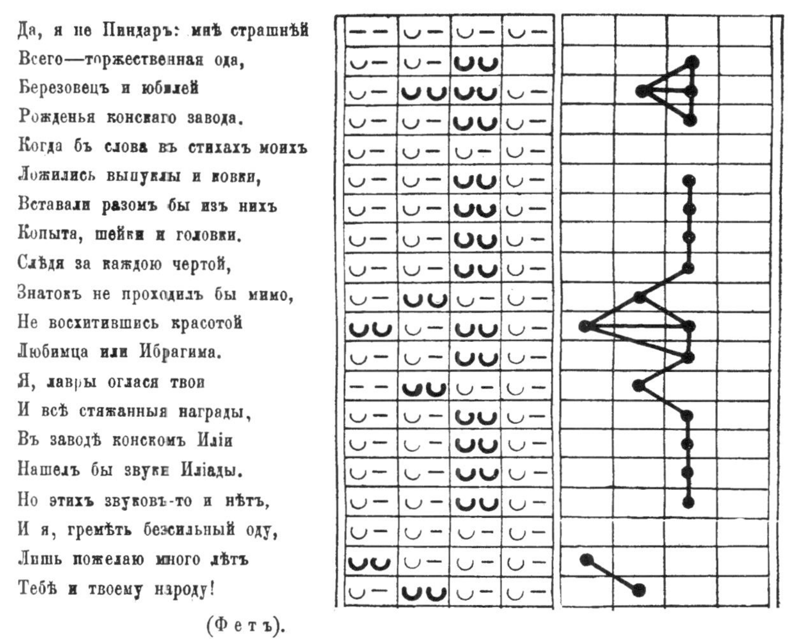

4-6. Andrei Bely's chart of the stress pattern in lines of verse, in *Символизм* (Symbolism; Moscow: Musaget, 1910), 260.

Since antiquity, Western verse has been composed in conventional formats built up from an "iam" (a "metric foot") that consists of an unstressed syllable, υ, followed by a stressed, — (υ—, as in toDAY). Late-nineteenth- and early-twentieth-century Russian verse was written in iambic tetrameter, in which each line has the stress pattern υ—υ—υ—υ—. In this chart Bely recorded where the poet Afanasy Fet, who wrote the lines of verse on the left, broke the rules by putting a stressed syllable in place of an unstressed syllable. In the four left-hand columns, the stress pattern for each line is recorded, and in the four right-hand columns, Bely marked each violation of iambic tetrameter with a dot. He then connected the dots to reveal a geometric figure.

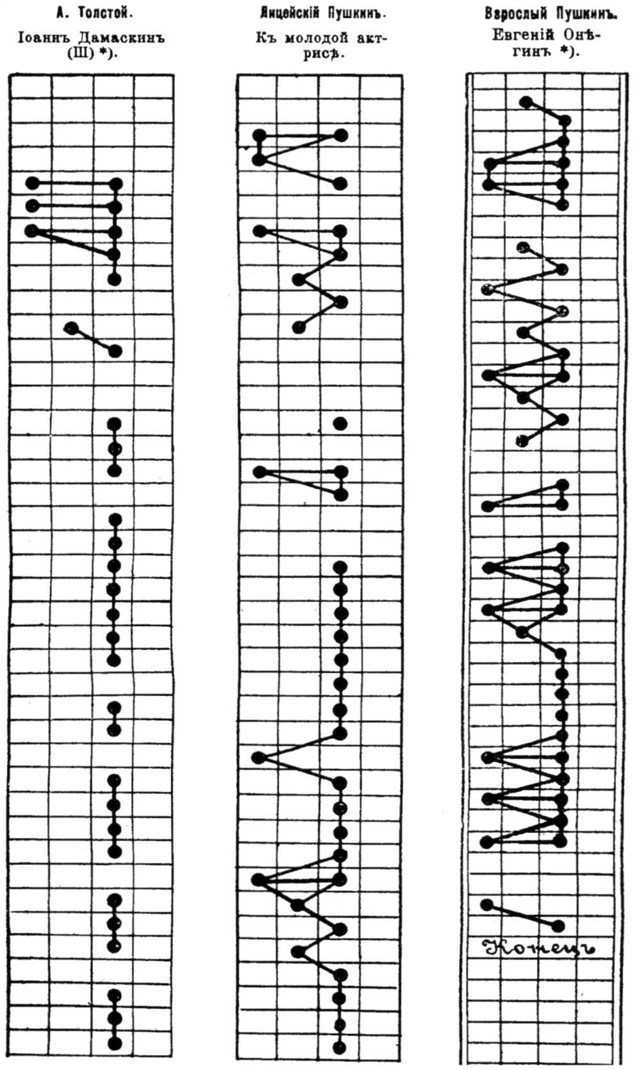

4-7. Andrei Bely's analysis of rhythmic patterns (left to right) in Tolstoy's "John Danascene," Pushkin's early poem "To a Young Actress," and Pushkin's late "Eugene Onegin," in *Символизм* (Symbolism; Moscow: Musaget, 1910), 268 and 270.

As Bely describes these geometric figures (the formal essence of three poems), the figures seem to take on a life of their own: "Comparing the examples of rich rhythms [late Pushkin, on the right] with the examples of poor rhythms [Tolstoy, left, and early Pushkin, middle], we see that the rhythmic figures for the rich rhythms are distinguished by greater complexity. The lines here are broken rather than straight, and simple figures join together here and there to form more complex figures. . . . Pushkin's rhythm describes a broken line, where we find angles, right angles, a 'roof' [horizontal line], and a figure . . . resembling the letter *M* (formed by a pattern of half accents on the second and third feet). The abundance of right angles already looks ahead to the later Pushkin." But wait a minute! What do these poems mean? In Bely's formalist approach, the meaning (semantics) is not considered.

Khlebnikov were associated, and a sister organization in Saint Petersburg, which included critics Osip Brik and Viktor Shklovskii. Jakobson's close friend Shklovskii wrote "The Resurrection of the Word" (1914), a key early text in Russian formalist literary criticism, in which he emphasized physical aspects of words (their sounds and rhythms). He distinguished between the natural world-out-there (the practical, everyday world) and the imaginary world (self-contained and autonomous) created by a piece of literature. Shklovskii declared, furthermore, that the words in a poem refer only to the imaginary world, excluding from consideration the intentions of the author and the tradition of past literature. Russian formalists such as Jakobson and Shklovskii also discounted the impact of social forces, which were the central concern of their bitter rivals, the Marxist critics. But a language only has one semantics, complex as it may be; if the formalist poet and the neighborhood grocer both speak Russian, then the words in the poem refer, if indirectly, to the everyday, natural world, where their meaning evolved. Thus, in the end, Jakobson and Shklovskii's formalist approach is an example of *emphasizing* the uniqueness of literary meaning without completely separating words from the world-out-there.

One of the first poets to adopt a formalist approach was Khlebnikov, who entered Kazan University in 1903, majoring in mathematics. He studied non-Euclidean geometries and natural science while continuing to write, publishing his first poem in 1908. Then from 1908 to 1911, Khlebnikov studied Sanskrit and Slavic languages at the University of Saint Petersburg in the linguistics department led by Baudouin.

While a student, Khlebnikov met the poets Aleksei Kruchenykh and Vladimir Mayakovsky, the musician Mikhail Matyushin, the painter and critic David Burliuk, and the artists Wassily Kandinsky, Elena Guro, and Vladimir Tatlin, as well as the neurologist and amateur painter Nikolai Kulbin. Khlebnikov's first published collaboration with artists, a prose poem with illustrations by Guro and Matyushin, appeared in about 1910. Khlebnikov focused on the sound of words, not their syntax or semantics. After finishing his university studies he began creating a "new universal language" from a structure of sound elements (phonemes) of his own design, similar to those Baudouin and other linguists were using to re-create the lost Indo-European language. Living during the Russian Revolution and World War I, Khlebnikov set the utopian goal that his new common tongue would unite people in peace and harmony. He was not nostalgic for the language of Adam and Eve in the Garden or of a united humanity before the Tower of Babel. Rather Khlebnikov dreamt of a futuristic tongue in which peoples of the globe, united in a scientific worldview, would speak a language like mathematics, in which pictographs would refer unambiguously to concepts common to all.

The goal is to create a common written language shared by all the peoples of this third satellite of the Sun, to invent written symbols that can be understood and accepted by our entire star, populated as it is with human beings and lost here in the universe.

—Velimir Khlebnikov, "Artists of the World!" 1919

What would give the universal language its precision? Khlebnikov suggested an invented alphabet—"a common system of hieroglyphs for the people of our planet"[47]—that consisted of nineteen consonants. (He considered vowels incidental and included them only as an aid to pronunciation.) To each consonant, Khlebnikov assigned a geometric symbol; *B* in the Cyrillic alphabet (which is *v* in the Latin alphabet), for example, was "the turning of one point around another, either in a full circle or only a part of one, along an arc";[48] in other words, it was a pie-shaped segment of a circle. He called on artists to create a typeface for his invented letters: "The task of artists who work with paint is to provide graphic symbols for the basic units of our mental processes. . . . The artist's task would be to provide a special sign for each type of space. Each sign must be simple and clearly distinguishable from all the rest. It might be possible to resort to the use of color, and to designate M with dark blue, B with green."[49]

Like Jakobson and Shklovskii, Khlebnikov believed that the poem exists in its own autonomous, self-contained world, where the words are not used as a practical means of communication but have a literary meaning. The imaginary world of his poems was internally consistent—a paradox-free, autonomous realm, like a non-Euclidean geometry: "If the living language that exists in the mouths of a people may be likened to Euclid's geomeasure [geometry], can the Russian people not therefore permit themselves a luxury other peoples cannot attain, that of creating a language in the likeness of Lobachevsky's geomeasure [geometry], of that shadow of other worlds? Do the Russian people not have a right to this luxury?"[50]

To this end Khlebnikov made up so-called self-spun (coined) words, as in the concluding lines of the poem "начало" ("Beginning," 1908), in which he describes singing in a relaxed mood:

> *Ну же, звонкие поюны,* (*Go on, ringing warbling,*)
> *Славу легких времирей!* (*Long live easy moments!*)

The meaning of Khlebnikov's made-up words are easy to understand from their context. For example, the word "поюны" is coined from a combination of the Russian for "I sing" ("пою") with a made-up ending that turns the verb into a noun. Alternatively, one could translate "поюны" as "singing" or "song." Similarly, the word "времирей" is spun from the real Russian word "врем" (the genitive of "время" for "time") with a made-up suffix.

The word itself is regarded as something like an atom, with its own processes and own structure. Khelbnikov is not a word collector, not a property owner, and not a wise guy seeking to startle. He looks upon words as a scientist.

—Roman Jakobson, "Modern Russian Poetry: Velimir Khlebnikov," 1921

In his poem "Incantation to Laughter" (1908–9), Khlenikov made up words by adding prefixes and suffixes to the Russian word "смех" (laughter), and used words that sound similar to "смех," together with coined words that mimic the sounds of laughter (the Russian equivalent of English sounds like ho, ho, ha, ha, hee, hee). In the eleven lines of the poem, Khlebnikov repeated the self-spun words in various forms, over and over again as in a chant—an incantation—to evoke the sound of laughter.

Like many fin-de-siècle students of cross-cultural research, Khlebnikov envisioned a language of the future that would be universal and unambiguously clear. With this dream unrealized, Khlebnikov died in poverty in 1922 at age thirty-six. Meanwhile, in 1920, during the turbulent aftermath of the October 1917 Revolution, Jakobson left Moscow for Prague, where Herbart's formalist philosophy pervaded the thinking of linguists and art critics. Already in the 1850s Herbart had inspired the Czech critic Eduard Hanslick to write the first formalist analysis of music, and in 1926 Jakobson joined with his Czech colleagues to form the Prague Linguistic Circle, an international group of linguists and literary scholars that remained a vibrant center of formalist debate for about a decade.

RUSSIAN CONSTRUCTIVIST ART: TATLIN AND RODCHENKO

Drawn to Khlebnikov's poetry, the young Vladimir Tatlin was the first visual artist to adopt the poet's formalist approach.[51] Tatlin entered Burliuk's avant-garde circle in Saint Petersburg in 1908 and began painting figures and still lifes in the semi-abstract style of Russian Futurism, but he soon dropped depiction to pursue the formalist essence of art. In two 1912 essays, Burliuk had described the path extending Khlebnikov's formalist approach to the visual arts, observing that painters of his day were focusing on the essence of their medium: "Painting has begun to pursue only painterly objectives. It has begun to live for itself."[52] For painters, that essence was colored marks on a surface: "Painting is a colored space. . . . The component elements into which the essential nature of painting can be broken down are: I. line, II. surface, III. color, IV. objecthood (*faktura*)."[53] By "objecthood (*factura*)," Burliuk meant the painting or sculpture as a physical object. He singled out Khlebnikov ("the finest representative" of modern poetry) for pursuing the same goal for verse, reducing poetry to its essence in self-spun words and sounds.[54]

In the summer of 1914 Tatlin visited Pablo Picasso in his Paris studio. Inspired by Picasso's relief sculpture made from sheet metal and wire, such as *Guitar* (1914; Museum of Modern Art, New York), Tatlin shifted from oil painting to reliefs, which took the shape of "counter-reliefs" (*Corner Counter-Relief*, 1914–15; see plate 4-8),[55] stressing the physical essence of sculpture in a way that is analogous to Khlebnikov's poetry. Tatlin arranged simple planes and volumes cut from sheet metal (analogous to Khlebnikov's universal vocabulary) so that the "grammar" holding them together—the rods piercing the sheet metal and the wire stretched under tension—is exposed. Tatlin gave each corner counter-relief an impersonal, industrial appearance and precise structure that can be comprehended by anyone who looks at it carefully. Russian artists coined the term "Constructivism" in 1921 to name this method; Tatlin "constructed" this sculpture. Like Hilbert's "theory-forms" and Bely's metric graphs, Tatlin's counter-reliefs distilled sculpture to its formal essence of signs (planes, angles, forms) held together by forces (balance, tension, gravity). The essence of a work of visual art, according to formalism, is its set of properties as a physical object—its objecthood, as Burliuk stated in 1912. In the essay "On *Faktura* and Counter-Reliefs," critic Viktor Shklovskii stressed that Tatlin had reduced sculpture to its essence as a physical object: "The

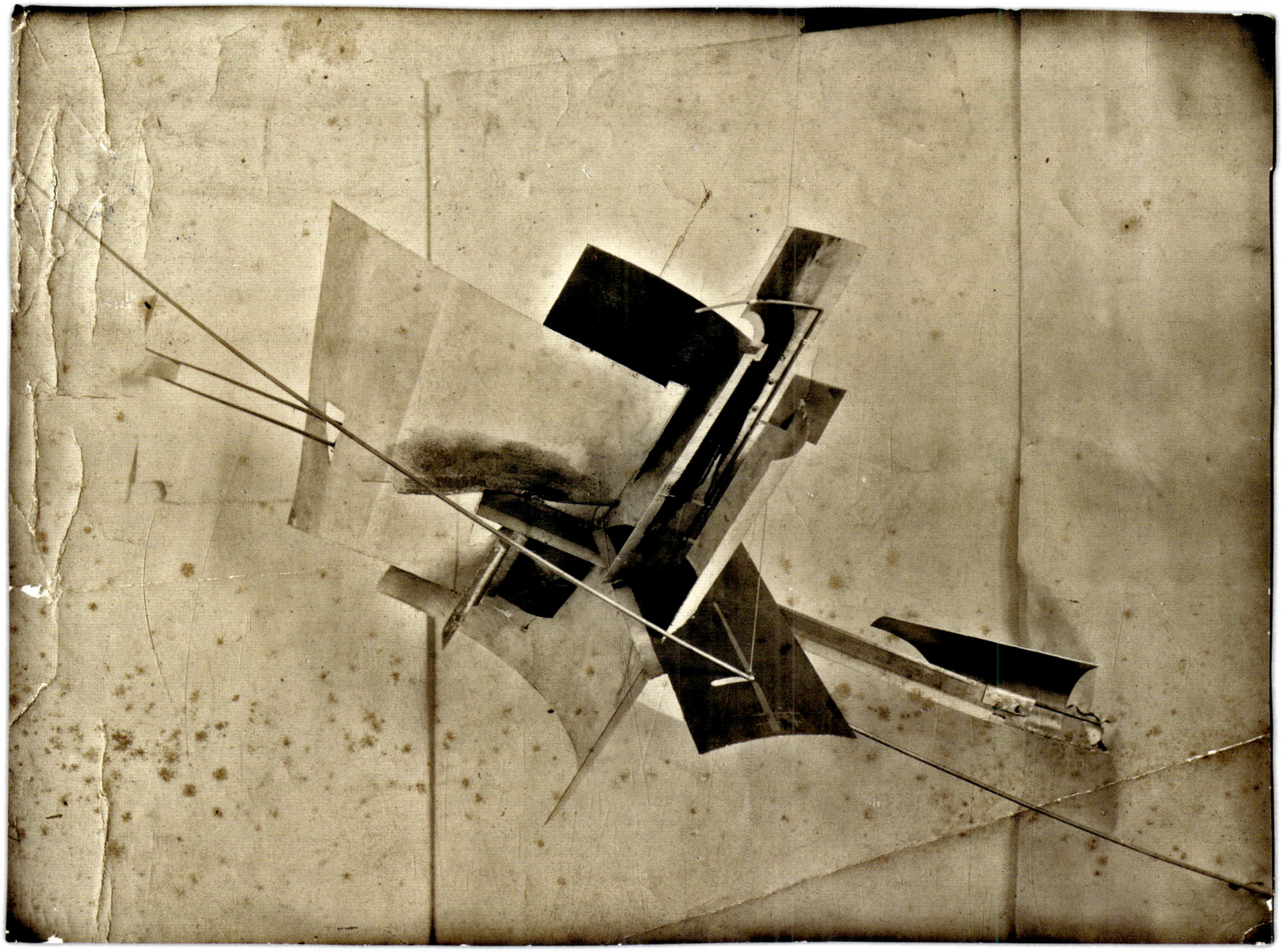

4-8. Vladimir Tatlin (Russian, 1885–1953), *Corner Counter-Relief*, 1914–15. Wood, iron, cable. Photograph of the lost piece taken in 1915 when on display at the exhibition *0.10* in Saint Petersburg. Russian State Archive of Literature and Art, Moscow, inv. no. 998-1-3623-3.

Tatlin's counter-reliefs are lost and survive only in period photographs such as this.

whole effort of a poet and a painter is aimed first and foremost at creating a continuous and thoroughly palpable object, an object with a *faktura*."[56]

Thus in Russia by the second decade of the twentieth century, there emerged two opposing views of the meaning of art: Suprematism and Constructivism. In Moscow, the poet Aleksei Kruchenykh and the Suprematist Kazimir Malevich—in resonance with Nikolai V. Bugayev's philosophical-theological interpretations of mathematics—turned to paradoxical, illogical signs to represent the inexpressible Absolute Spirit. In Saint Petersburg the poet Khlebnikov and the Constructivist Tatlin—inspired by the meaning-free structures of formalist mathematics and linguistics—aimed to created paradox-free, internally consistent art. Visitors to the 1915 Saint Petersburg exhibit *0.10 (zero-ten)*, *The Last Futurist Exhibition of Paintings* could compare the different approaches taken by Malevich and Tatlin. For those hoping to attain the Russian theosophist Peter Ouspensky's cosmic consciousness, Malevich's *Black Square* painting (see plate 3-30 in chapter 3) symbolized the most meaningful thing in the universe—the Absolute. But the art critic Sergei

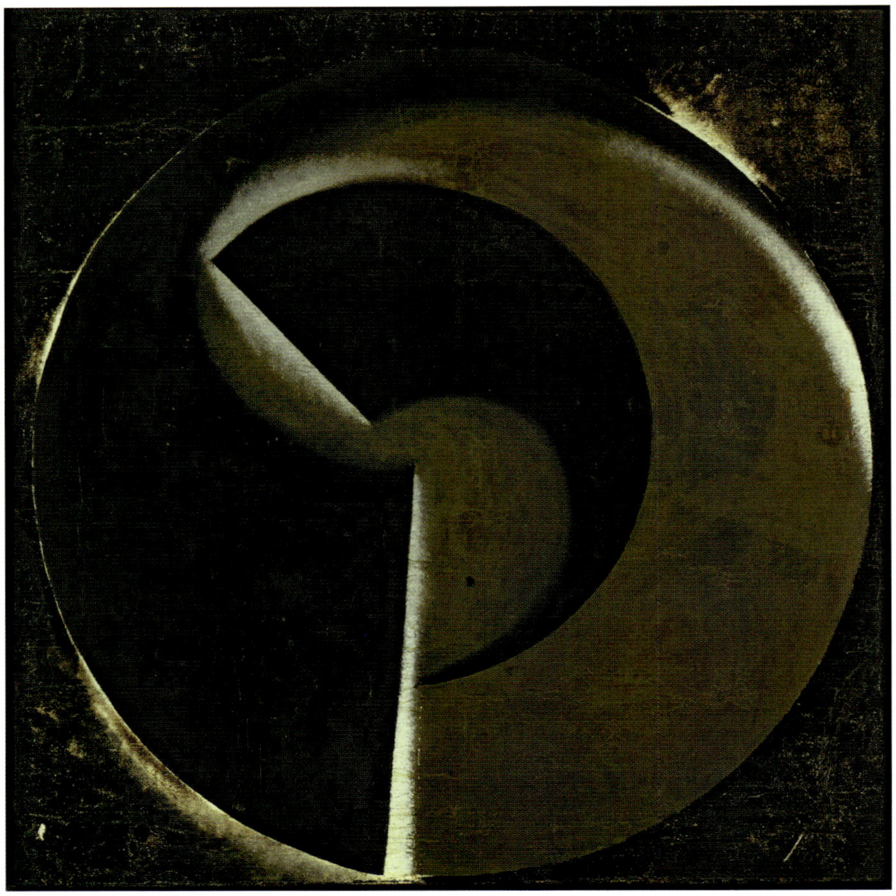

4-9. Aleksandr Rodchenko (Russian, 1891–1956), *Non-Objective Painting no. 80 (Black on Black)*, 1918. Oil on canvas, 32¼ × 31¼ in. (81.9 × 79.4 cm). Museum of Modern Art, New York, gift of the artist, through Jay Leyda. 114.1936. © Estate of Aleksandr Rodchenko/RAO, Moscow/ VAGA, New York.

Isakov (who had a degree in physics and mathematics from Moscow State University) described Tatlin's *Corner Counter-Relief* only in terms of its physical properties: "It is obvious to anyone that before him are the results of some serious, thought-consuming effort to resolve an extraordinarily difficult problem: material and tension."[57]

Aleksandr Rodchenko learned about formalism while a student at the School of Fine Arts in Kazan, where he attended lectures by David Burliuk and performances of poems by Mayakovsky in 1914.[58] The next year Rodchenko moved to Moscow and, inspired by Tatlin's corner counter-reliefs, adopted a Constructivist aesthetic. Like a geometer, Rodchenko drew with a compass and ruler, showing these drawings in the 1916 exhibition *The Store*, organized by Tatlin.

After the October 1917 revolution, avant-garde artists and writers found themselves redefining their role in the new Communist society. Although Rodchenko supported the Bolsheviks and was active in these debates,[59] he concurrently continued to produce his own art as an autonomous formal structure composed of elements arranged according to rules. Between 1917 and 1921 he single-mindedly worked out the implications of this formalist view of visual art, with special attention to reducing the art object to its physical essence. Rodchenko was not trained in mathematics or linguistics, but he had quickly absorbed the formalist outlook of Tatlin, Khlebnikov, and others. It was in Rodchenko's hands that the formalist aesthetic reached its logical conclusion of painting and sculpture reduced to meaning-free signs.

4-10. Kazimir Malevich (Russian, 1879–1935), *Suprematist Composition: White on White*, 1918. Oil on canvas, 31¼ × 31¼ in. (79.4 × 79.4 cm). Museum of Modern Art, New York.

In the 1918 series *Black on Black* (plate 4-9), Rodchenko reduced his palette to black and grays. In 1919 he exhibited paintings from the series at the *Tenth State Exhibition: Non-Objective Creation and Suprematism* in Moscow alongside Malevich's series of white paintings, including *White on White*, also of 1918 (plate 4-10). This dramatic contrast recalls Tatlin's pairing with Malevich in 1915. Malevich saw *White on White* as a step on the road to Absolute Spirit; Rodchenko understood *Black on Black* as an arrangement of meaning-free forms. Rodchenko specifically linked his concept of line to Cantor's concept of the continuum (the real numbers arranged on a line segment such that numerical differences correspond to distances between points). There are no gaps in the continuum; it is dense (between any two points, there is always another point). The infinite number of points in the continuum is equal to any smaller segment of the line, as Rodchenko wrote: "The number of points located on an entire straight line is the same as the number of points on any of its sections; and this means that our entire universe, with all its endless planets, suns, Milky Ways—consists of the same number of points as any, even the smallest, section of a straight line. The entire universe could be created from a piece of straight line, just by placing the points in a different order."[60]

Rodchenko's secular view of infinity underlines his differences from Malevich, for whom infinity was associated with a theosophical cosmic consciousness of "the heights of non-objective art" (*The Non-Objective World*, 1927).[61] Between 1918 and 1920, Rodchenko based his style on line alone as the fundamental building-block from which he made his

TOP RIGHT

4-11. Aleksandr Rodchenko (Russian, 1891–1956), *Spatial Construction no. 12*, ca. 1920. Painted plywood and wire, 24 × 33 × 18½ in. (61 × 83.7 × 47 cm). Photograph taken ca. 1920. © A. Rodchenko and V. Stepanova Archive, Moscow. © Estate of Aleksandr Rodchenko/RAO, Moscow/VAGA, New York.

In his *Spatial Construction* series, Rodchenko cut a flat circle, square, and triangle into concentric bands, which expand into a mobile when hung.

BOTTOM LEFT

4-12. Aleksandr Rodchenko (Russian, 1891–1956), *Spatial Construction no. 11*, ca. 1920. Painted plywood and wire. Photograph of the lost piece taken ca. 1920. © A. Rodchenko and V. Stepanova Archive, Moscow. © Estate of Aleksandr Rodchenko/RAO, Moscow/VAGA, New York.

BOTTOM RIGHT

4-13. Aleksandr Rodchenko (Russian, 1891–1956), *Spatial Construction no. 13*, ca. 1920. Painted plywood and wire. Undated photograph. © A. Rodchenko and V. Stepanova Archive, Moscow. © Estate of Aleksandr Rodchenko/RAO, Moscow/VAGA, New York.

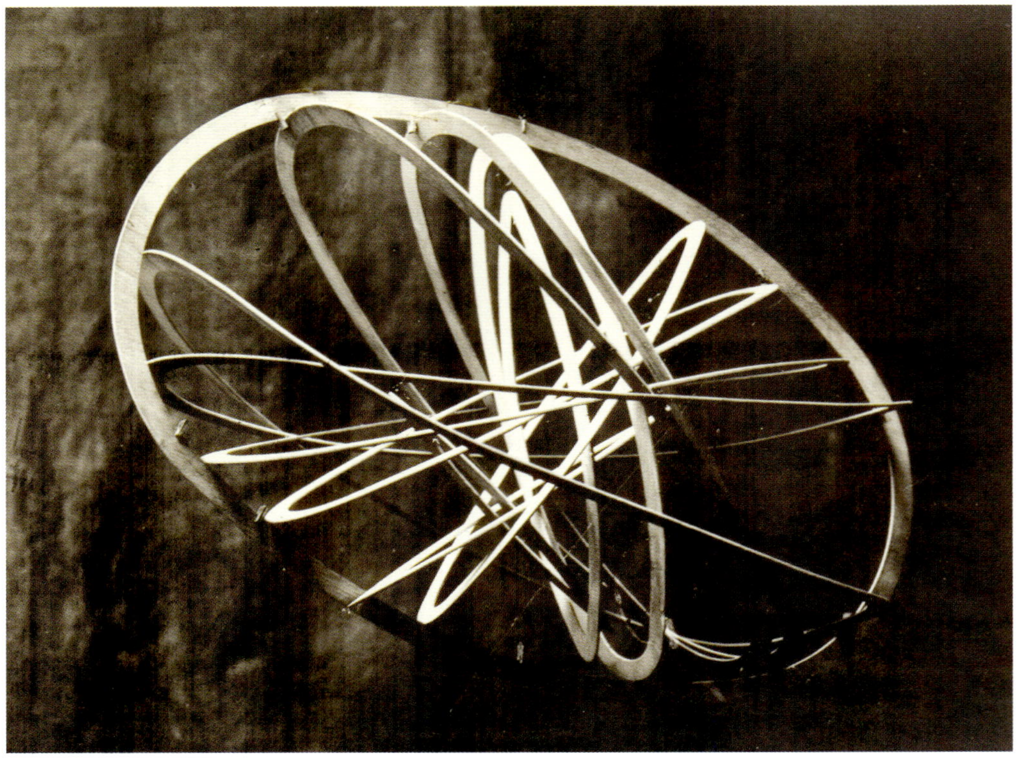

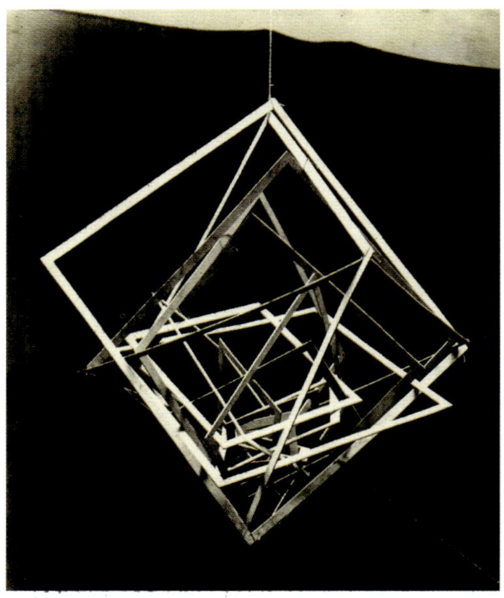

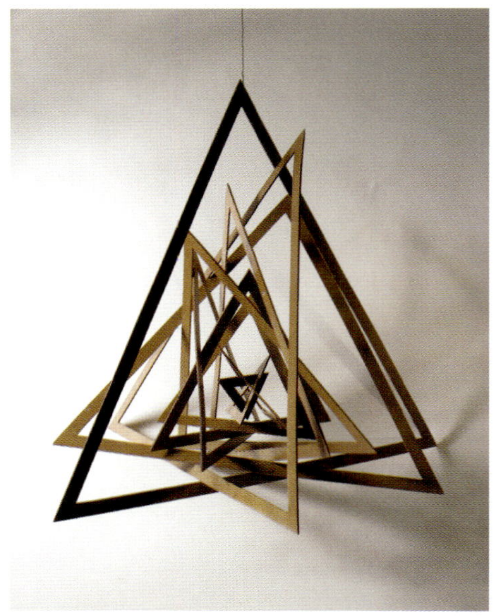

paintings. Naming it "linearism," Rodchenko began investigating basic geometric forms in a series of shapes, including a circle, square, and triangle (plates 4-11, 4-12, and 4-13). Made of wood and covered with silver metallic paint, these works reflect light as they turn. In 1921 he also made a series of geometric sculptures with repeating modules (plates 4-14 and 4-15).[62] Such studies could have applications in the design of architecture, industrial objects, or other practical items, which Rodchenko would soon design—but not before he bid a dramatic farewell to pure form with his monochrome paintings *Red, Blue,* and *Yellow*.

In his earlier 1918 *Black on Black* series, Rodchenko arranged gray shapes on black backgrounds. But in 1921 he painted a canvas in one hue in a uniform, all-over surface,

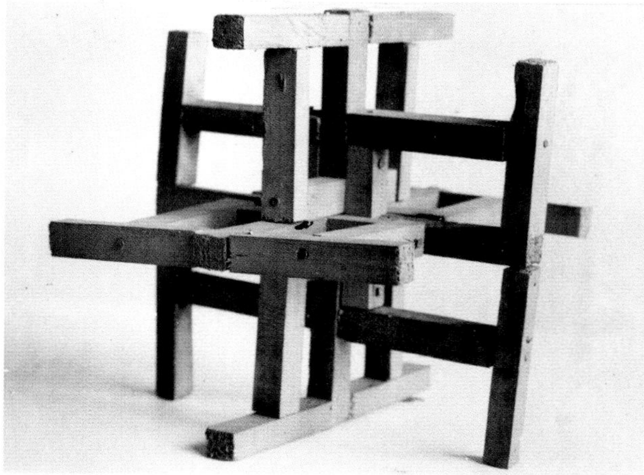

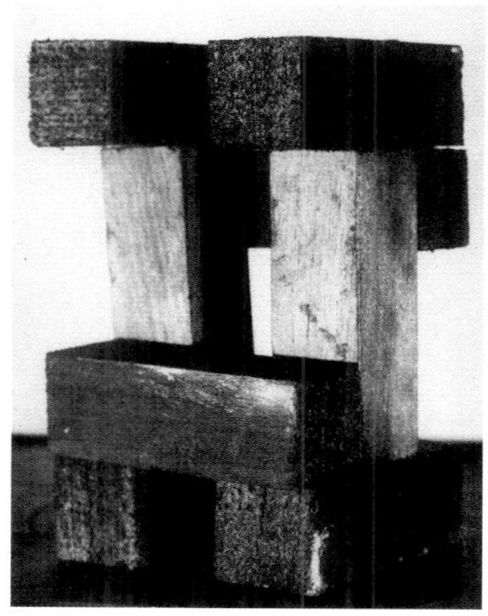

FAR LEFT

4-14. Aleksandr Rodchenko (Russian, 1891–1956), *Spatial Construction no. 20*, 1920–21. Photograph of the lost piece taken in 1924. © A. Rodchenko and V. Stepanova Archive, Moscow. © Estate of Aleksandr Rodchenko/RAO, Moscow/VAGA, New York.

Rodchenko made *Spatial Construction no. 20* from identical T-shaped units.

LEFT

4-15. Aleksandr Rodchenko (Russian, 1891–1956), *Spatial Construction no. 21*, 1920–21. Photograph of the lost piece taken ca. 1920. © A. Rodchenko and V. Stepanova Archive, Moscow. © Estate of Aleksandr Rodchenko/RAO, Moscow/VAGA, New York.

FOLLOWING SPREAD, LEFT

4-16. Aleksandr Rodchenko (Russian, 1891–1956), *Blue* (***Чистый синий цвет*** [Pure blue color]), 1921. Oil on canvas, 24⅝ × 20¹¹⁄₁₆ in. (62.5 × 52.5 cm). © A. Rodchenko and V. Stepanova Archive, Moscow. © Estate of Aleksandr Rodchenko/RAO, Moscow/VAGA, New York.

FOLLOWING SPREAD, RIGHT

4-17. Aleksandr Rodchenko (Russian, 1891–1956), *Yellow* (***Чистый желтый цвет*** [Pure yellow color]), 1921. Oil on canvas, 24⅝ × 20¹¹⁄₁₆ in. (62.5 × 52.5 cm). © A. Rodchenko and V. Stepanova Archive, Moscow. © Estate of Aleksandr Rodchenko/RAO, Moscow/VAGA, New York.

producing a monochromatic painting that he named *Red* (**Чистый красный цвет** [Pure red color]; see plate 4-1, chapter frontispiece). He similarly produced monochrome paintings *Blue* and *Yellow* (plates 4-16 and 4-17) and showed them in the exhibition *5 × 5 = 25* in Moscow as three individual paintings (not as a triptych[63]). In these works Rodchenko reduced painting to what was, in his formalist opinion, its essence—a flat, painted rectangular surface. He declared them to be his last paintings: "I took painting to its logical conclusion and exhibited three canvases: red, blue, and yellow, asserting that: Everything is over."[64] Rodchenko's gesture confirms his self-conception as a researcher excavating the foundations of art. The task had a natural end-point. Rodchenko finished the excavation when he isolated the formal features of painting in 1921, just as Hilbert ended his search for the foundations of geometry when he wrote the formal axioms of geometry in 1899.

The critic Nikolai Tarabukin, who had witnessed the development of Constructivism in Moscow from Tatlin's 1915 counter-reliefs to Rodchenko's 1921 monochrome canvases, agreed with Rodchenko that a conclusion had been reached. Tarabukin described Rodchenko's *Red* as "a smallish, almost square canvas painted entirely in a single red color. This canvas is extremely significant for the evolution of artistic forms which art has undergone in the last ten years. It is not merely a stage that can be followed by new ones but it represents the last and final step of a long journey, the last word, after which painting must become silent, the last 'picture' made by an artist."[65] Some of Rodchenko's remarks have encouraged an interpretation of his work as nihilist, expressing hopelessness, negativity, or death,[66] but the artist's own description of this key moment confirms his emotionally neutral formalist intentions.

By the early 1920s artists had been producing non-objective paintings for decades, but their canvases still contained illusions of shallow spaces and overlapping planes as well as symbolic associations to the artist's feelings or to a philosophical Absolute. Rodchenko reached the end of the formalist reduction by removing the last vestige of representation; his monochrome *Red* is nothing but itself. In making art as a self-contained autonomous system, Rodchenko laid the foundation for formalist art to this day (plates 4-18, 4-19, and 4-20).

How could drawing be of itself and not about something else? —**Dorothea Rockburne, 1973**

ABOVE

4-18. Dorothea Rockburne (Canadian, b. 1932), *Drawing Which Makes Itself: Neighborhood*, 1973. Vellum, pencil, colored pencil, and felt-tip pen on wall. The Museum of Modern Art, New York.

In the 1950s at Black Mountain College in North Carolina, Rockburne learned the formalist aesthetic from her teacher Max Dehn, a German refugee who had earned a doctorate in mathematics with David Hilbert at the University of Göttingen. Dehn had been a graduate student in 1900 when his teacher stood at the podium of the International Congress of Mathematics in Paris and dramatically challenged his audience to solve twenty-three problems, thus setting a research agenda for twentieth-century mathematics. Dehn was the first mathematician to solve one of these so-called Hilbert's problems; today about five still await solution (the number depends on how one defines "solved"). Dehn went on to become a professor of mathematics at Goethe University in Frankfurt until he, being Jewish, was forced out by the Nazis in 1935, after which he and his wife fled Europe via the trans-Siberian railroad escape route to the Sea of Japan. There they boarded a ship and sailed to Japan and on to America, arriving penniless in 1942. Dehn and his wife survived for several years by moving from job to job, until in 1945 he secured a permanent position teaching Greek, Latin, philosophy, and a course he titled "Geometry for Artists" at Black Mountain College.

From Dehn, Rockburne learned to design with a vocabulary of meaning-free points, lines, and planes. In 1973 she made a drawing on the wall of the Bykert Gallery in New York by holding a large sheet of carbon paper against the wall and then folding it and scoring the edge, thus transferring a carbon line to the wall. She repeated this process using black and red carbon paper to produce a pattern. In the drawing shown above, she diagonally folded a piece of vellum, resulting in an X pattern of folds. Rockburne did not draw the carbon lines and the folds but rather—in the formalist tradition of Hilbert and Dehn—she designed an impersonal, algorithmic process of scoring and folding that would (like a formal axiomatic system) *generate* the pattern. In other words, the *drawing made itself*.

OPPOSITE

4-19. Anthony McCall (British, b. 1946), *Breath*, 2004. Video projector, computer, digital file, haze machine; one cycle: 6 minutes. Courtesy of the artist.

In 1973, McCall aimed a 16 mm projector in a darkened room and directed the audience's attention not to an image on a screen but to the projected cone of light, which he declared is the essence of film (*Line Describing a Cone*, 1973). McCall later adopted digital technology to create a so-called vertical film such as this, by projecting linear patterns from above. As light descends in this darkened space, which is filled with a light mist of water vapor, light-waves strike water molecules in the fog and create a volume. In this projection (*Breath*), the pattern on the floor is composed of two elliptical curves and one wavy line. The curves move slowly toward and away from each other, expanding and contracting the volume in a 6-minute cycle, while the wave moves slowly through the curves. Uncomfortable with religious associations to light streaming down from above, McCall aspires to make meaning-free films. But when he titled this conic form *Breath*, he gave it inescapable associations to mathematical objects embedded in a living, breathing cosmos infused with meaning in the Platonic tradition (Plato, *Timaeus*, 366–360 BC).

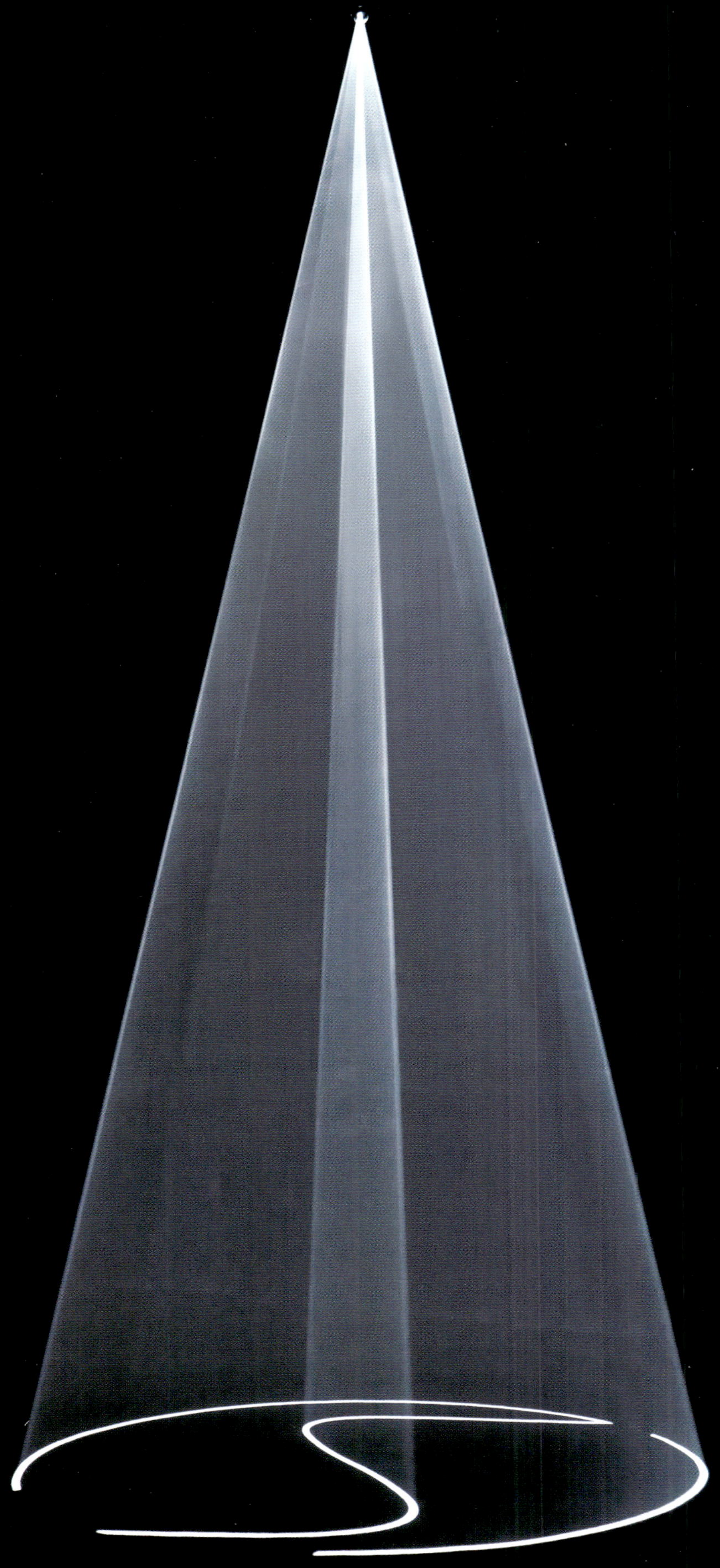

I was always searching for the ultimate film, one that would be nothing but itself.

—Anthony McCall, 2008

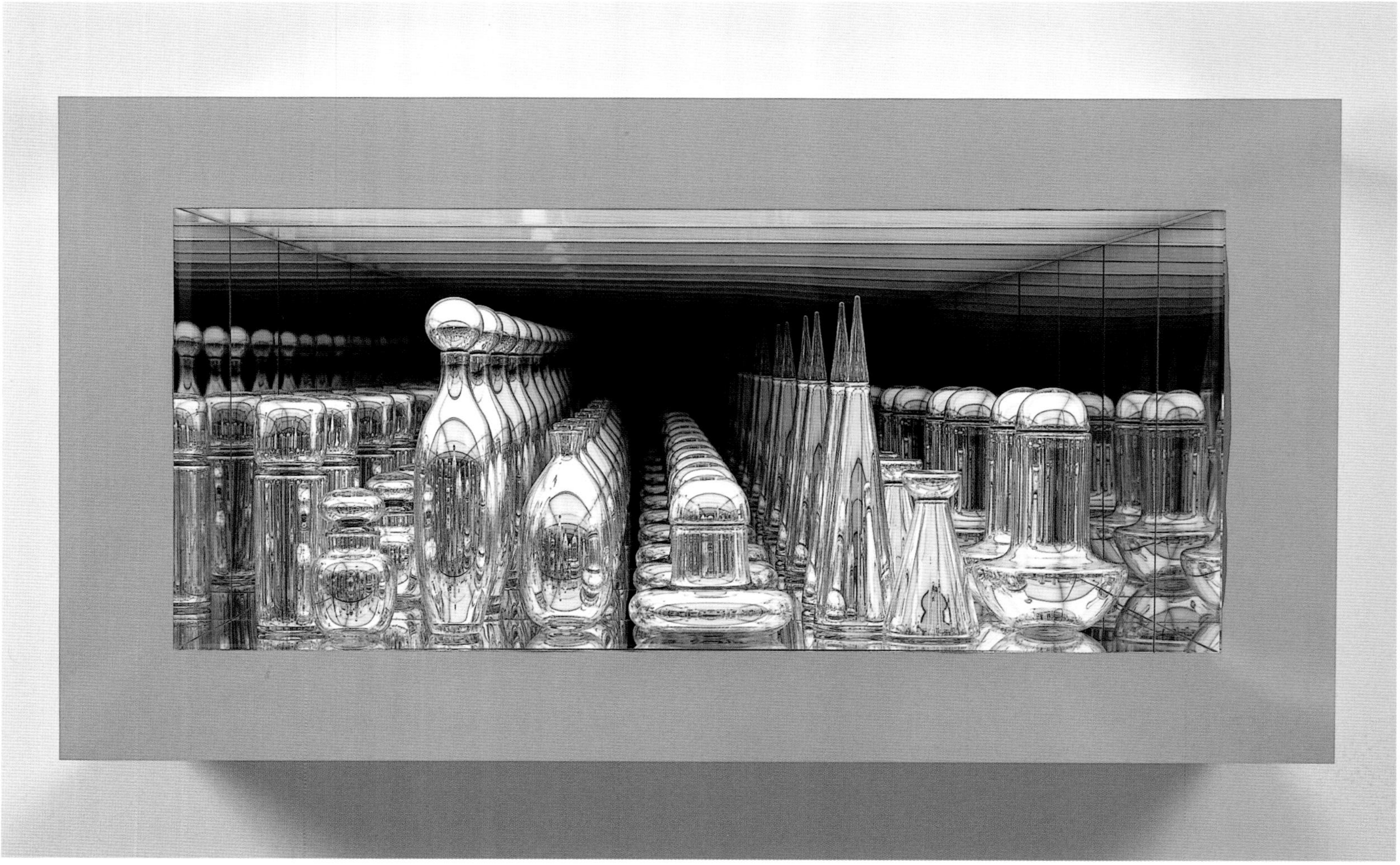

ABOVE AND OPPOSITE

4-20. Josiah McElheny (American, b. 1966), *Czech Modernism Mirrored and Reflected Infinitely*, 2005; and detail. Mirrored glass case with hand-blown mirrored glass objects, 18 ½ × 56 ½ × 30½ in. (47 × 143.5 × 77.5 cm). © Josiah McElheny. Donald Young Gallery, Chicago, and Andrea Rosen Gallery, New York.

Like Hilbert's *Foundation of Geometry* (1899) and Rodchenko's *Red* (1921), McElheny's sculpture is an internally consistent, self-contained arrangement of abstract, meaning-free, replaceable signs. The forms of these eight lidded bottles are based on vessels designed in Czechoslovakia between the 1930s and 1990s, which are similar to ideal, pure design forms used throughout twentieth-century Germanic culture. McElheny blew the bottles and lids in clear glass and then transformed them into mirrors by pouring liquid silver nitrate into them and out again, coating their insides with a thin film. Set within a mirrored box and illuminated from above, the bottles reflect all light to infinity. To ensure that this imaginary world is autonomous, McElheny covered the front of the box with a special "one-way" mirror that is both partially reflective and partially transparent. When one side is brightly lit and the other side dark, the mirror allows viewing from the darkened side only. Lit from above, the bottles stand on the bright side; light bouncing off them is reflected back into the box (as seen in the detail of the second bottle from the left). Standing on the dark side in a dimly lit art gallery, the viewer sees through the one-way mirror as if it were ordinary window glass. But this is not ordinary glass. The viewer does not see his or her reflection in the bottles or in the mirror at the back of the box because light coming from the dark side rebounds from the one-way mirror back into the viewer's space, leaving the pristine bottles—like formalist aesthetics—uncontaminated within a self-enclosed world.

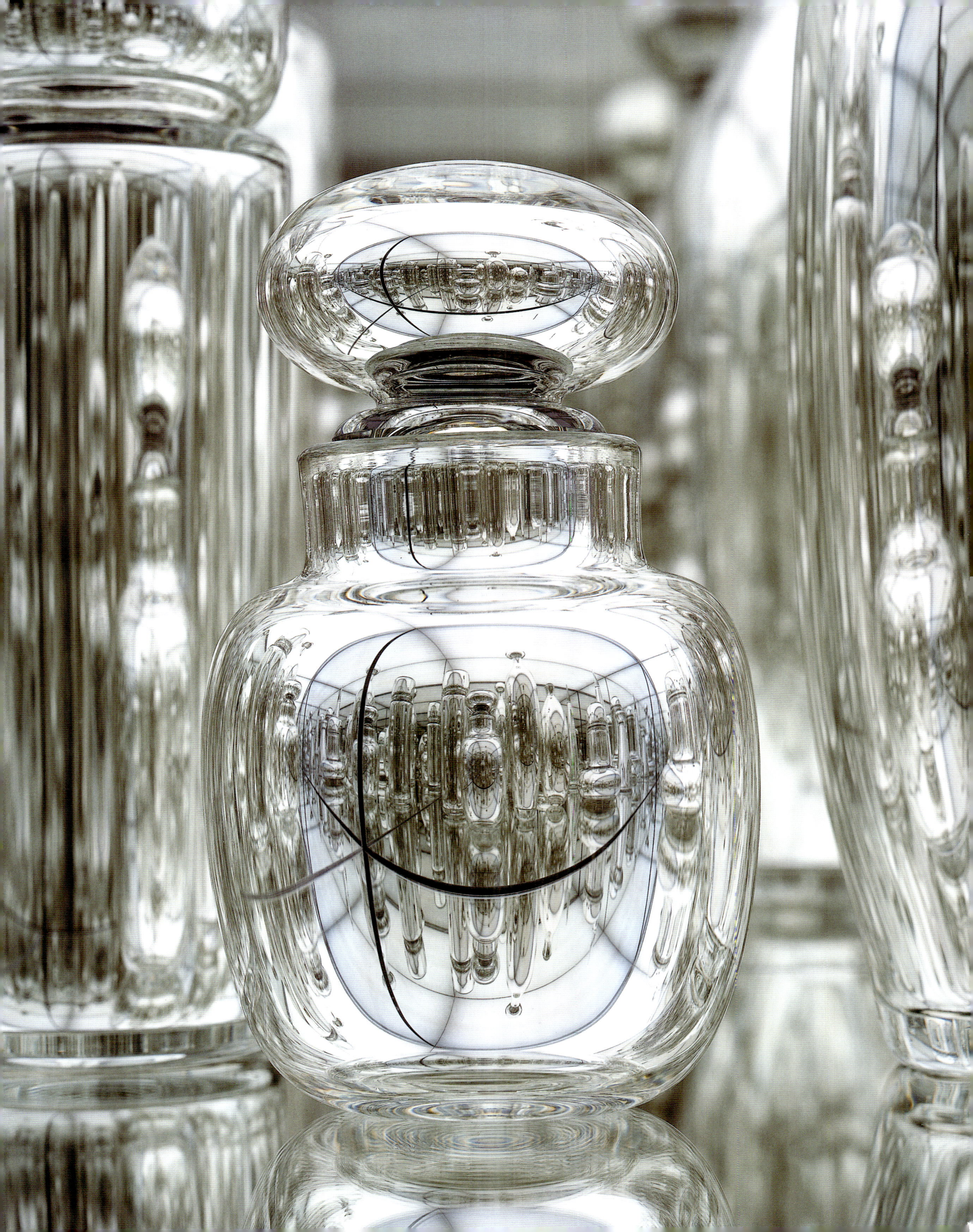

UNISM IN POLAND: STRZEMIŃSKI AND KOBRO

The Russian painter Władysław Strzemiński articulated what became known as the formalist master narrative of visual art. A well-educated colleague of Rodchenko and Tatlin, Strzemiński was born into an aristocratic Russian family from Minsk, studied engineering at a prestigious tsarist military academy, and graduated as an officer in 1911. Then during World War I, he lost an arm and a leg in a grenade explosion. While recovering in a Moscow hospital, Strzemiński met Katarzyna Kobro, an art student and, as expected of the daughters of wealthy families, a volunteer nurse's aide in the war effort. By 1918 Strzemiński was also studying art and, with Kobro, entered avant-garde circles in Moscow and Vitebsk, where they became closely allied with Constructivism. Following Rodchenko, Strzemiński reduced painting to its essence: a flat surface with an all-over "unified" composition—Unism—as in his *Unistic Composition 10* (plate 4-21).[67]

Unable to find employment in Russia, Strzemiński and Kobro immigrated to Poland, where they became citizens in 1924, the year that Polish artists joined together to form Blok, a Constructivist group with its own periodical and collective exhibitions. Strzemiński and Kobro saw the role of the artist as essentially a creator of abstract structures that could then be applied to a practical task. Other members of Blok joined the Polish Communist party and saw themselves as engineers, designing functional objects (architecture, furniture, posters, and typography) for mass production in order to promote the social revolution. This difference in outlook among its members eventually led to the dissolution of Blok in 1926, but Constructivist themes persisted in the artistic communities that thrived for another decade in Warsaw, Kracow, and Łódź.

By 1925 Kobro had developed a sculptural version of Unism. To produce objects that were unified wholes, she built her sculptures from modular units, and to achieve the anonymity of a mathematician, she arranged the units according to precise ratios, including the ratio of approximately 1.618, which was regarded by many artists of her era as aesthetically pleasing (plate 4-22). Kobro and Strzemiński published a booklet, *Composition of Space: Calculations of a Spatio-Temporal Rhythm* (1931), in which they described how to use the ratio 1.618 (the Golden Ratio; see chapter 2) to create design layouts, using illustrations of their own work (plate 4-23).

Painting and sculpting in the spirit of researchers solving problems, Strzemiński and Kobro worked in series. In a 1934 essay on Polish Constructivism, Strzemiński described their working method as the arrangement of units within a whole: "The fundamental method that applies to the contemporary technique is the normalization, the standardization of the models and their production in series. This implies that one works with the aid of uniform elements, that one uses a mathematical method for calculations."[68] Like Rodchenko, Strzemiński and Kobro applied their designs to the production of practical objects: Strzemiński, designed a typeface (plate 4-24) and Kobro a nursery school (4-25).

Early versions of the formalist master narrative of modern art had been articulated in the 1910s by the Russian formalist critics Viktor Shklovskii, David Burliuk, and Nikolai Tarabukin. Working in a Germanic intellectual tradition, their outlook reflected the

4-21. Władysław Strzemiński (Russian-born Polish, 1893–1952), *Unistic Composition 10,* 1931. Oil on canvas. Muzeum Sztuki w Łodzi, Poland.

In *Unistic Composition 10,* Strzemiński maintained the flatness of the surface by applying the paint with a palette knife in thick impasto. Rather than looking through a painted surface as if it were a window into another world, the viewer is stopped at the surface by this opaque pattern, with ridges of paint thick enough to cast shadows. Although this speckled surface is not completely monochrome—the right side is slightly darker, creating a barely perceptible phantom shape—Strzemiński nonetheless created a uniform appearance by choosing pigments of rose, orange, and ochre that are very close in hue (position on a color wheel), value (degree of lightness or darkness), and intensity (amount of pigment).

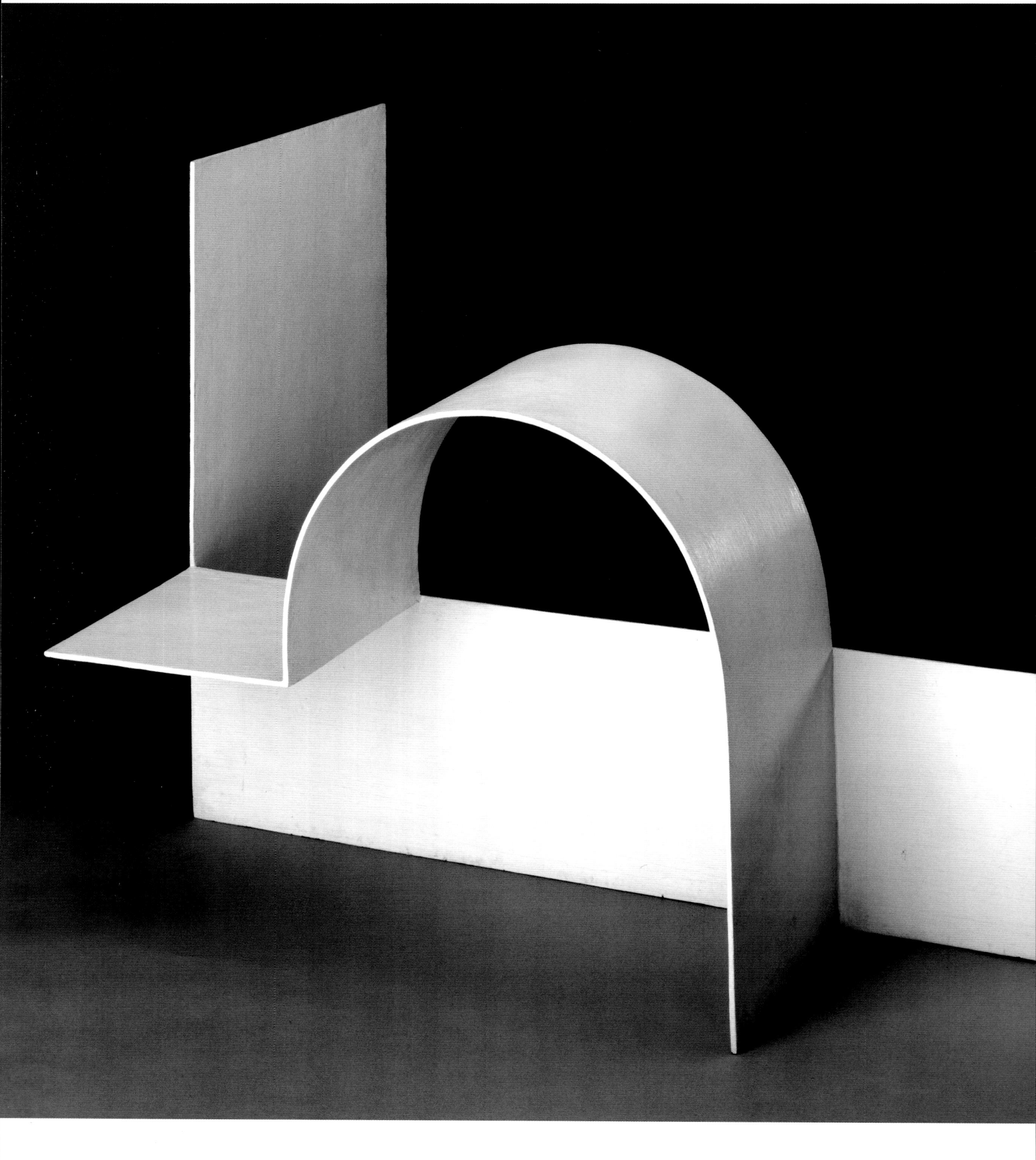

LEFT

4-22. Katarzyna Kobro (Russian-born Polish, 1898–1951), *Space Composition* 3, 1928. Painted steel, 15¾ × 25¼ × 15¾ in. (40 × 64 × 40 cm). Muzeum Sztuki w Łodzi, Poland.

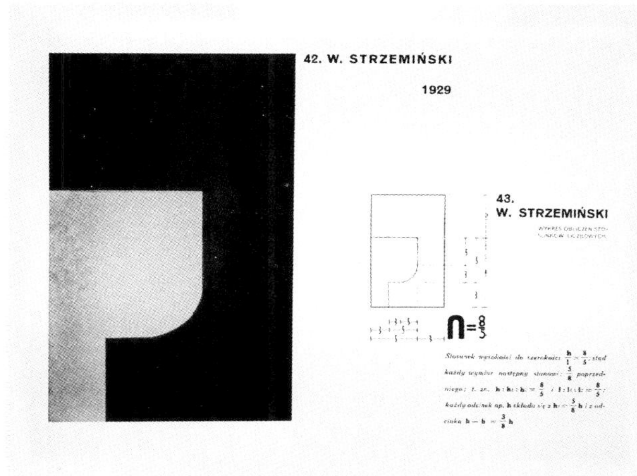

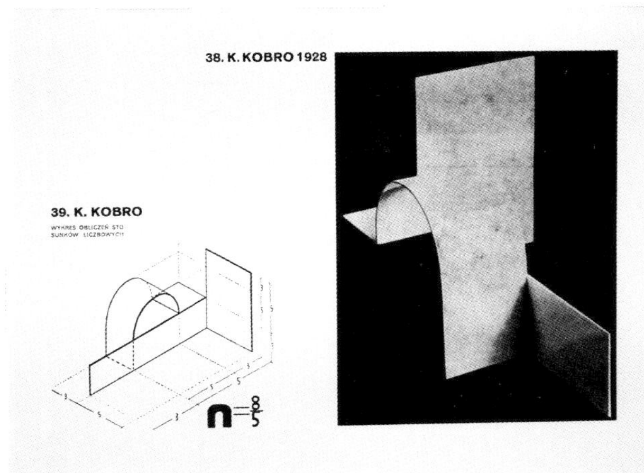

ABOVE RIGHT AND RIGHT

4-23. Katarzyna Kobro (Russian-born Polish, 1898–1951) and Władysław Strzemiński (Russian-born Polish, 1893–1952), *Composition of Space: Calculations of a Spatio-Temporal Rhythm* (Łódź: a.r., 1931), fig. 39, 42.

In their booklet *Composition of Space*, Kobro and Strzemiński described how they used the Golden Ratio, which they symbolized by ∩ and expressed as 8/5 (the ratio 8 to 5; 8 divided by 5 equals 1.6, which approximates the irrational number they had in mind, 1.618 . . .). In Kobro's sculpture, the horizontal bar at the base (8 units wide) and the rectangular plane on the right (5 units high) are in an 8 to 5 ratio. Strzemiński also used the ratio to design his paintings.

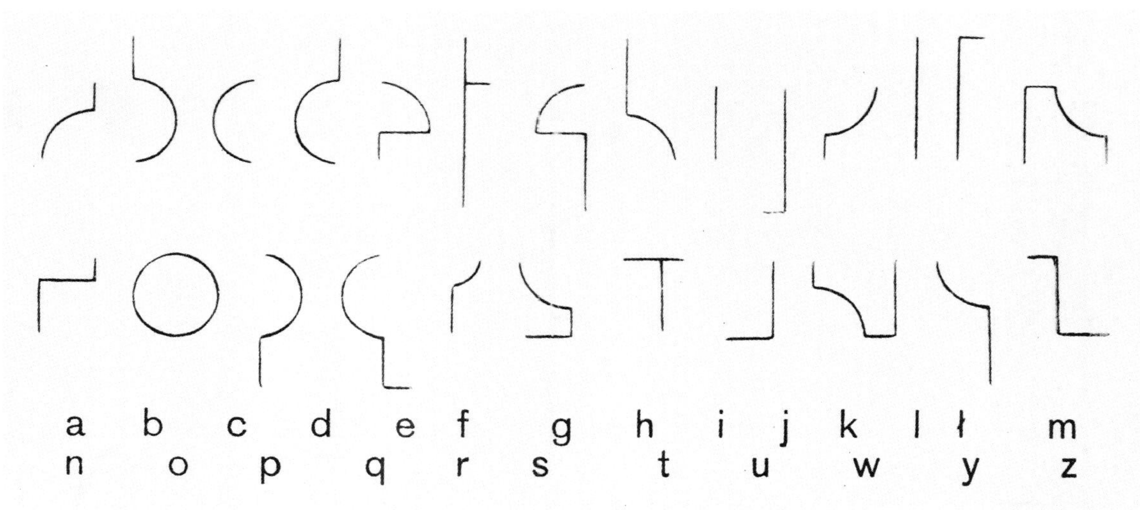

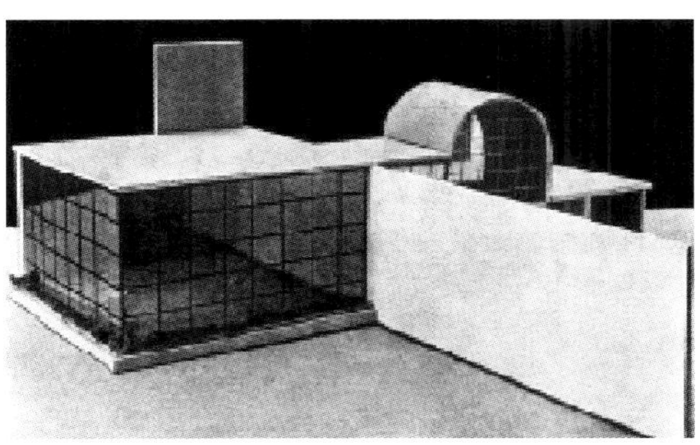

ABOVE

4-24. Władysław Strzemiński (Russian-born Polish, 1893–1952), alphabet, ca. 1930. Ink on paper. Muzeum Sztuki w Łodzi, Poland.

LEFT

4-25. Katarzyna Kobro (Russian-born Polish, 1898–1951), design for a nursery school, ca. 1932. Muzeum Sztuki w Łodzi, Poland.

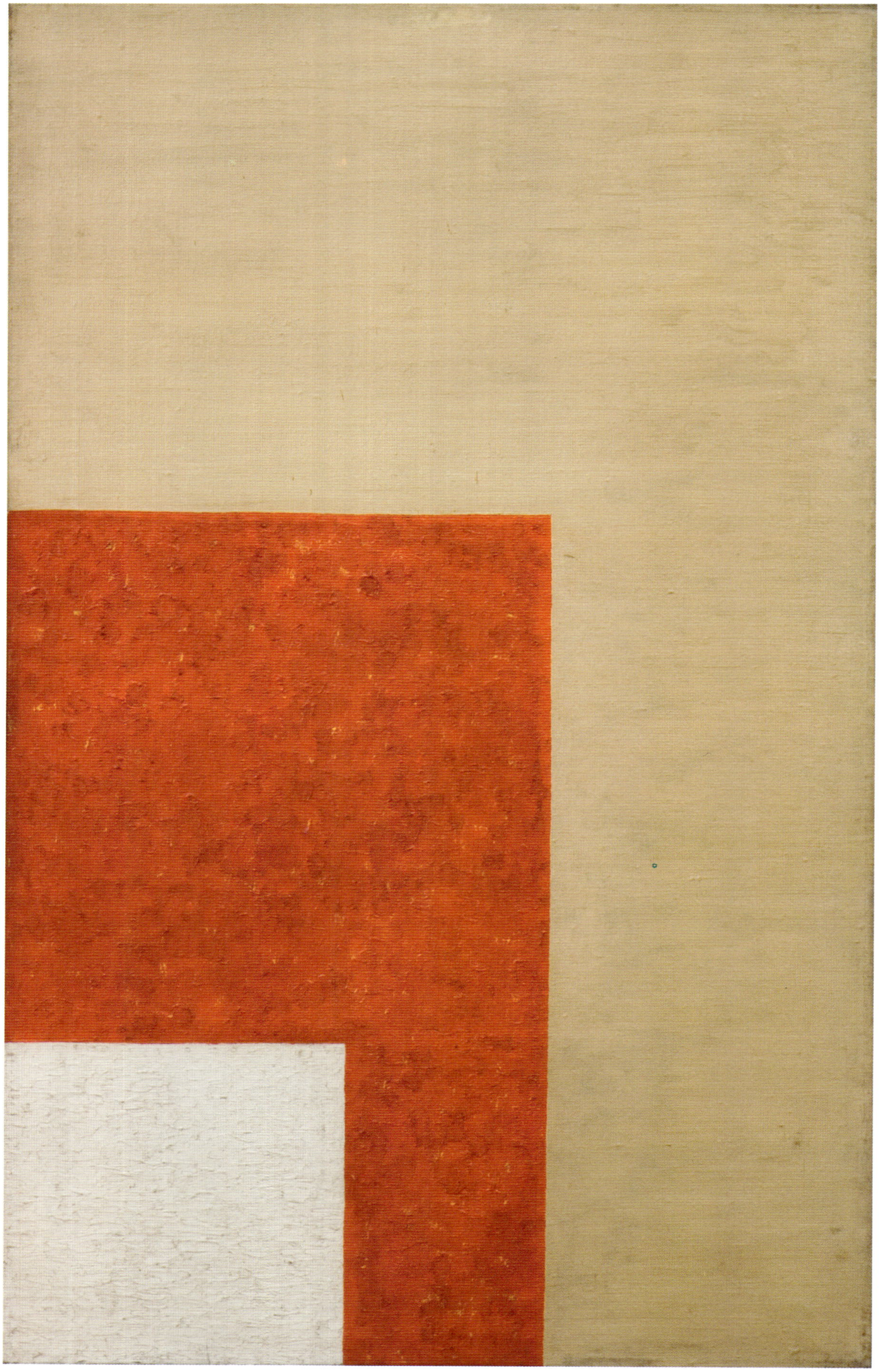

4-26. Władysław Strzemiński (Russian-born Polish, 1893–1952), *Architectural Composition 9c*, 1929. Oil on canvas, 38 × 23⅝ in. (96 × 60 cm). Muzeum Sztuki w Łodzi, Poland.

formalist aesthetics of Herbart. They described what a formalist mathematician like Hilbert called "axiomatizing" a field—determining its basic assumptions (axioms). Carrying out this reductionist process of analysis, Shklovskii, Burliuk, and Tarabukin concluded (as Herbart had) that, from a formalist perspective, the essence of an art form is its physical (formal) properties: for painting, this is a flat surface with an overall composition; for sculpture it is a simple geometric form.

Then in 1924 Strzemiński wrote a cogent summary of the formalist master narrative, describing the progress of painting (putting paint on a surface) from the Renaissance to modern art as the relentless reduction to a flat surface on a rectangular format—a canvas stretched over a rectangular framework.[69] According to the formalist aesthetic, the illusion of space created by Renaissance artists using linear perspective was "incompatible with its essence." Strzemiński saw Impressionist artists as the first to be conscious of the picture plane as a flat surface because they used a pattern of dabs of hues that "connects the disorder of the graphic and spatial forms through an identical tension of color." The Cubists and Futurists moved closer to a flat surface by suggesting a shallow, illusionistic space, and "dynamic tension was spread throughout the whole picture." But "the forms of Cubism and Futurism . . . were gliding and for that reason they were not sufficiently united with the surface." Strzemiński laid the blame for these fluctuating "gliding" planes on the Cubists and Futurist preoccupation with the peripheral task of mimesis—making pictures (landscape, portraits, still lifes). For Strzemiński, the key to further reduction of the painting was to stop depicting the natural world and to begin making completely abstract, non-representational art. "It is time for plastic art to resort to its own means [and declare] a real autonomous existence in the plastic arts. . . . [Art can only reach its essence] when a work of art is plastically self-sufficient; when it constitutes an end in itself and does not look for its justification to values that subsist beyond the picture." Strzemiński identified Malevich as a pioneer of non-representational painting but asserted that Malevich did not achieve total flatness because his compositions of colored forms still contained "too many slanting directions" (balancing one form with another). Rodchenko had achieved flatness by applying uniform color in his monochrome canvases and Strzemiński similarly covered his canvas with a uniform impasto. When Strzemiński did make divisions in the painted surface, as in *Architectural Composition 9c* (plate 4-26), he lined them up with the edges of the canvas. An overall surface and flatness was the goal of the modernist master narrative for painting: "a picture should not be an interaction of forms, but a simultaneous phenomenon."

FORMALISM, MEANING, AND THE MEANING OF FORMALISM

Ever since Hilbert founded formalism in the 1890s, mathematicians who view him as *nothing but* a formalist have attacked him for removing meaning from mathematics and reducing it to a meaningless mechanical process. For example, in a review of Hilbert's *Grundlagen der Geometrie* (Foundation of geometry, 1899), the leading French mathematician of his day, Henri Poincaré, bemoaned: "According to him, it should be possible to reduce reasoning to purely mechanical rules, and, in order to create a geometry, it should be enough to apply these rules slavishly [*servilement*] to the axioms without knowing what the axioms mean."[70] A recent example of this persistent negative attitude towards Hilbert's formalism is a 2006 comment by the British-born American mathematician Freeman Dyson: "The great mathematician David Hilbert . . . espoused a program of formalization, which aimed to reduce the whole of mathematics to a collection of formal statements using a finite alphabet of symbols and a finite set of axioms and rules of inference. This was reductionism in the most literal sense, reducing mathematics to a set of marks written on paper, and deliberately ignoring the context of ideas and applications that give meaning to the marks."[71] These misguided criticisms confuse formalism as a tool—Hilbert's use of formalism to study abstract structure—with formalism as a tedious project of doing deductions by a mechanical process (which is something that machines, not mathematicians, actually do). The criticism overlooks that Hilbert followed a formalist path only *on occasion* (in the 1890s and the 1920s) to reach a *specific goal*: putting classical Platonic mathematics on a firm foundation.[72]

In mathematics, as in any scientific research, we find two tendencies present. On the one hand, the tendency towards abstraction seeks to crystallize the logical relations inherent in the maze of material that is being studied, and to correlate the material in a systematic and orderly manner. On the other hand, the tendency toward intuitive understanding fosters a more immediate grasp of the objects one studies, a live rapport with them, so to speak, which stresses the concrete meaning of the relations.

—**David Hilbert, *Geometry and the Imagination*, 1932**

Critics of Hilbert often confuse the terms "meaning-free" and "meaningless," moving from a dispassionate description of a structure whose variables are uninterpreted (meaning-free) to an ad hominem characterization of such structures as worthless nonsense (meaningless). Once constructed, a formal axiomatic structure—like any piece of pure mathematics—can be *given* meaning and *related* to the world by being assigned an interpretation. Hilbert called his abstract structures "theory-forms," and he thought they were not complete, or very interesting, until they were filled with meaning. Inspired by Einstein's theory of relativity of 1905, Hilbert spent a decade working on mathematical physics, interpreting his meaning-free variables as subatomic particles and forces.[73]

Rodchenko, Tatlin, Strzemiński, and Kobro have also been criticized for emptying art of its content and reducing painting and sculpture to decorative patterns.[74] Such critics often confuse the difference between formalist and representational art with the form-content distinction, concluding incorrectly that non-representational art is *nothing but* form (with no content). Like Hilbert's formalist mathematics, any artistic form can be *given* meaning. With the political upheaval of the Russian revolution and its aftermath, artists adapted their work to the needs of Communism under Vladimir Lenin. Rodchenko *applied* his colored forms to practical design tasks to make everyday life more efficient and pleasant, designing posters, textiles, and tableware. After 1920 he taught the basic course in construction at VKhUTEMAS (Russian acronym for a name that translates as Higher Artistic-Technical Workshops), which was founded by Lenin to train young artists to apply the avant-garde abstract vocabulary of color and form to functional objects and to work with industry to ensure their smooth production. Rodchenko was joined on the VKhUTEMAS faculty by Naum Gabo, a Russian artist educated in Germany who returned to his homeland in 1917 to design public art for the new revolutionary society, where science and mathematics would be held in high esteem (plate 4-27). Avant-garde artists had not been politically active before the revolution, and after October 1917 they struggled to redefine

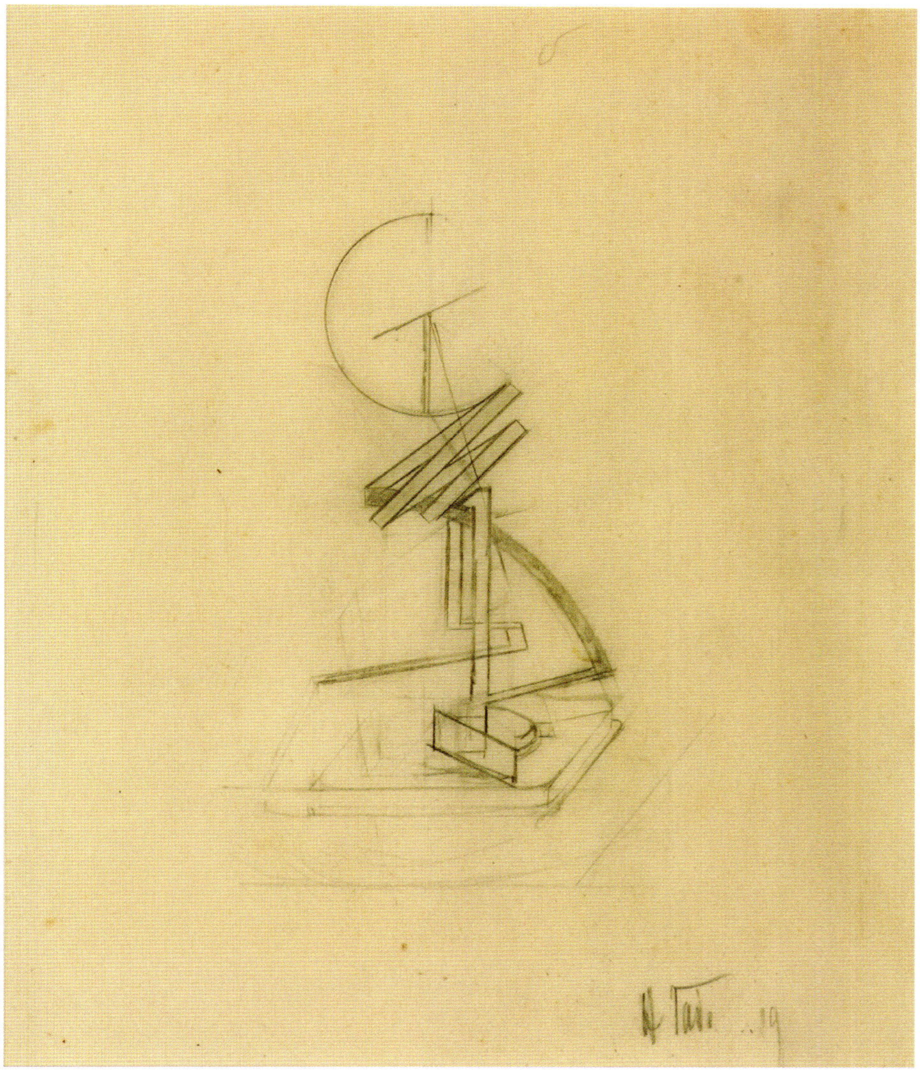

4-27. Naum Gabo (1890–1977), *First Sketch for a Monument for an Institute of Physics and Mathematics*, 1919. Graphite on paper, 13 × 9 in. (33 × 22.9 cm). Tate, London, inv. no. T02156. © Nina and Graham Williams.

Gabo designed this (unrealized) monument for the exact sciences in the new Russia, which the revolution promised would be based on human reason, with tsarist privilege a thing of the past.

their role in the new government, which wanted artists to work in factory-like settings as anonymous constructors of useful objects, not as solo studio practitioners expressing individual personalities ("easelism").[75] But the erasure of authorship did not void meaning; a well-crafted, affordable, functional object could say a lot, as Rodchenko noted on a trip he made to Paris in 1925 to build the interior of a Workers' Club for a display in the Soviet section of the Exposition Internationale des Arts Décoratifs et Industriels Modernes (plate 4-28). The Worker's Club was the first appearance of the output of VKhUTEMAS in the West. Rodchenko, who had never travelled outside Russia, was overwhelmed by the luxurious, expensive consumer goods in the shops of Paris—silk scarves, jewelry, and couture attire. He wrote home to his wife, the artist Varvara Stepanova, that Parisians seemed to be slaves to these objects. Rodchenko's Workers' Club, with its simple forms crafted in plain wood, carried a message for the decadent westerners: "The light is from the East . . . not only in the liberation of the working class. The light is from the East—in the new relation to the person, to woman, to objects. Objects in our hands should also be equal, should also be comrades, and not these black, gloomy slaves like they have here" (plate 4-29).

The plumb line in our hand,
eyes as precise as a ruler, in
a spirit as taut as a compass,
we construct our work as the
universe constructs its own,
as the engineer constructs his
bridges, as the mathematician
his formula of the orbits.

—Naum Gabo and Antoine Pevsner, *The Realist Manifesto*, 1920

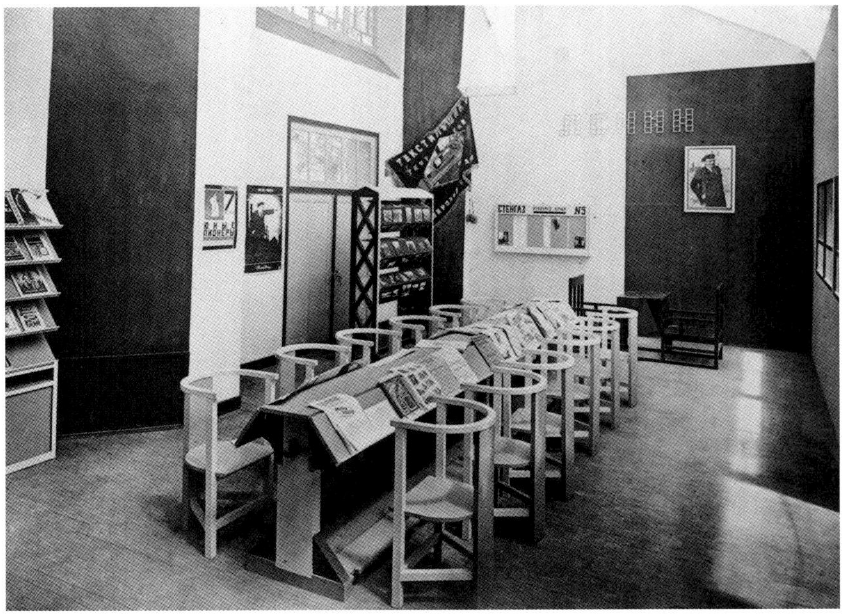

ABOVE

4-28. Aleksandr Rodchenko (Russian, 1891–1956), model for a Worker's Club, 1925, made for the Soviet section of the Exposition Internationale des Arts Décoratifs et Industriels in Paris. Photograph. Museum of Modern Art, New York. © Estate of Aleksandr Rodchenko/RAO, Moscow/VAGA, New York.

OPPOSITE

4-29. Aleksandr Rodchenko (Russian, 1891–1956), *Woman with Leica*, 1934. Photograph. © A. Rodchenko and V. Stepanova Archive, Moscow/© Estate of Aleksandr Rodchenko/RAO, Moscow/VAGA, New York.

After Lenin's death in 1924, government support for the arts diminished, and Rodchenko turned to photography to express "the new relation to the person, to women" in black and white compositions with striking abstract design.

The tables and chairs of Rodchenko's Workers' Club "will take on meaning, will become people's friends and comrades."[76]

The Soviet pavilion at the 1925 Exposition Internationale displayed another key example of applied Constructivism, Tatlin's architectural model for a building complex for the new Communist government (*Monument to the Third International*, known as "Tatlin's Tower," 1919; see plate 4-30). Lenin intended the complex to function as the center of world government after a (supposedly imminent) global Communist revolution that would sweep away old monarchies, wealthy bankers, and the organized church, clearing the way for the equal distribution of wealth in a secular society guided by science. Tatlin symbolized these utopian ideals as timeless truths by making the buildings as perfect forms—a cylinder, a cone, and a cube; yet this government was not static but dynamic, housed in buildings that revolved around an axis within a spiral steel framework that was oriented to the stars.[77] As the art critic Nikolai Punin wrote in 1919: "The spiral is the ideal expression of liberation: its base is set in the earth, it flees from the ground and becomes a symbol of the suspension of all animal, earthly, and groveling interests."[78] But the model was scrapped and hopes for a bright future were dashed after the visionary Lenin's death. The megalomaniac Joseph Stalin succeeded him, ushering in a bleak period of deprivation and suppression of thought in the Soviet Union. In 1927 Malevich was allowed to travel to Berlin for an exhibit of his paintings and for meetings at the Bauhaus with Walter Gropius and László Moholy-Nagy, but his trip was cut short and he was recalled to Moscow because of increasing repression of avant-garde activity by Stalin's arts administration.[79] The Russian people had traded their tsar for a Communist dictator, and in neighboring China there was another Communist revolution—the Chinese Revolution of 1911—during which citizens traded their emperor for a dictator and, like the Russians, found themselves in a society without individual rights. But people did not forget that for a brief moment after October 1917 art had become an instrument of social change; Rodchenko and Tatlin had applied abstract color and form to practical tasks to make the world a better place (plate 4-31).

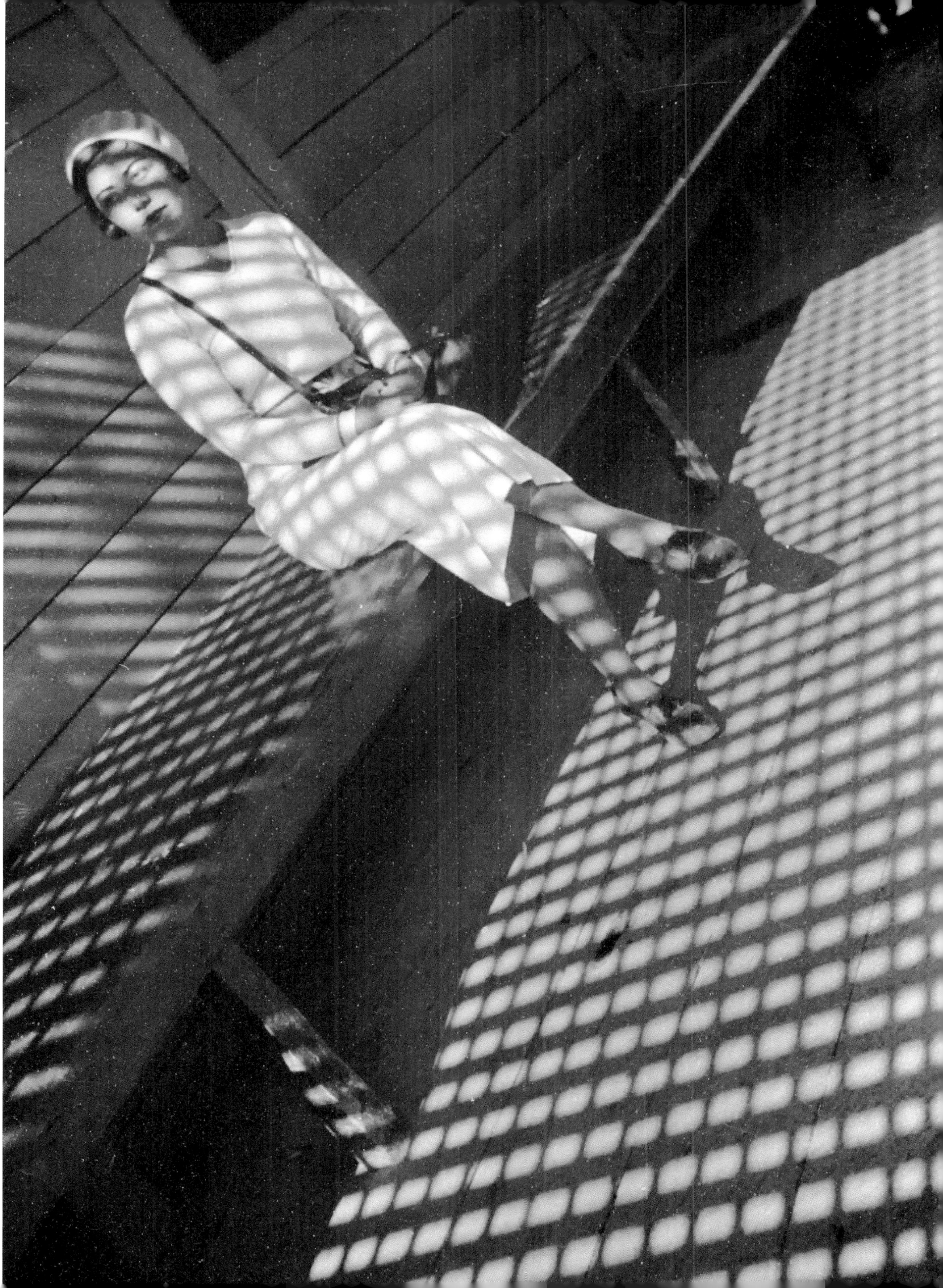

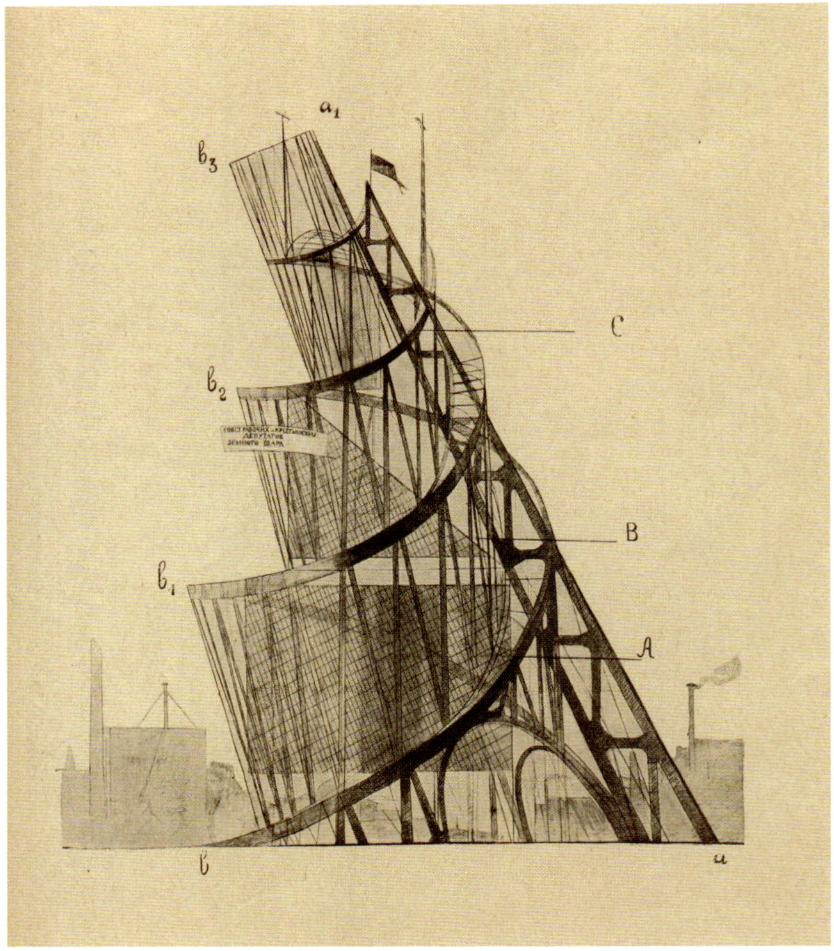

RIGHT

4-30. Vladimir Tatlin (Ukrainian, 1885–1953), illustration of the (unrealized) *Monument to the Third International* (Tatlin's Tower) in Nikolai Punin, *Pamiatnik III Internatsionala* (Saint Petersburg: Izo NKP, 1920). Museum of Modern Art, New York, gift of the Judith Rothchild Foundation.

As the earth rotates on its axis once every twenty-four hours, the constellations in the Northern Hemisphere appear to turn around Polaris (the North Star). The external steel framework of Tatlin's Tower is in the form of a spiral which points to Polaris, so that the tower and Polaris remain motionless relative to each other. Just as all celestial bodies revolve around Polaris, so, in the eyes of Lenin, all terrestrial action would revolve around the tower that housed his Communist headquarters in Moscow, which would soar upward one-third taller than the Eiffel Tower in Paris and twice the height of the Empire State Building in New York. Tatlin designed three steel-and-glass office buildings within the spiral framework, each moving in harmony with the cosmos: at the bottom (A) was a cylindrical auditorium, moving imperceptibly to complete its annual rotation; in the middle (B) a cone-shaped office structure rotated monthly; and on top (C) was a cube housing an information center (with telegraph, telephone, and radio) that rotated daily. On top of that there was an open-air screen on which news bulletins would be projected, and during overcast weather the latest news would be projected onto the clouds.

OPPOSITE

4-31. Ai Weiwei (Chinese, b. 1957), *Fountain of Lights*, 2007. Steel and glass crystals on a wooden base, 23 ft. (7 m) high, installation view at the Louisiana Museum of Modern Art, Humlebæk, Denmark. Collection Faurschou Foundation, Copenhagen, Denmark.

When Ai Weiwei was an infant, his father, a poet, was sent with his family to a labor camp in northwest China, where Ai grew up. When Ai was sixteen, his parents were allowed to return to Beijing, where he studied film and then continued his education in New York between 1981 and 1993, after which he has lived in China. Ai acted as artistic consultant for the Beijing National Stadium (known as the "Bird's Nest") for the 2008 Olympics, but after displaying a list of 5385 children who died when their poorly constructed public schools collapsed in the 2008 Sichuan earthquake, Ai was detained and severely beaten while in police custody. He created *Fountain of Lights* as homage to Tatlin's Tower and as a radiant beacon of hope.

Hilbert was a Platonist who invented the tools of formalism in the context of the foundational crisis. Paul Bernays, Hilbert's closest assistant and collaborator in the 1920s and early 1930s, observed in 1935 that most working mathematicians are Platonists.[80] But most mathematicians are not prepared to defend their Platonism and may, when pressed, retreat to the formalist position of professing belief only in abstract structure. (This is the source of the popular description of the mathematician-on-the-street as "Platonist on weekdays, formalist on Sundays").[81] Far from being only a formalist, Hilbert used formalism as a tool to provide a foundation for the modern version of Plato's view of mathematics as the cognition of perfect, timeless abstract objects (Cantor's sets). As Hilbert declared: "No one should be able to drive us out of the paradise that Cantor has created for us."[82]

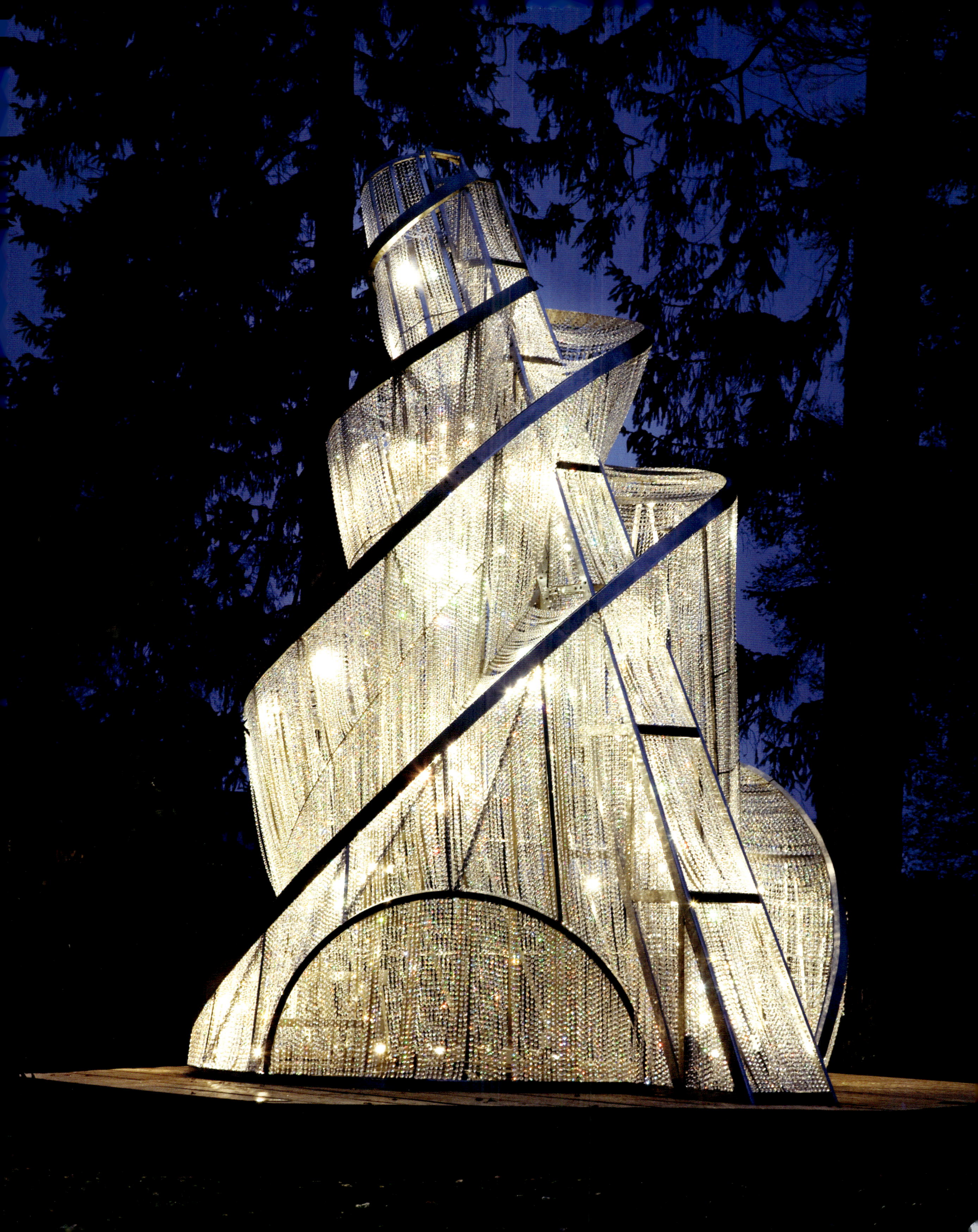

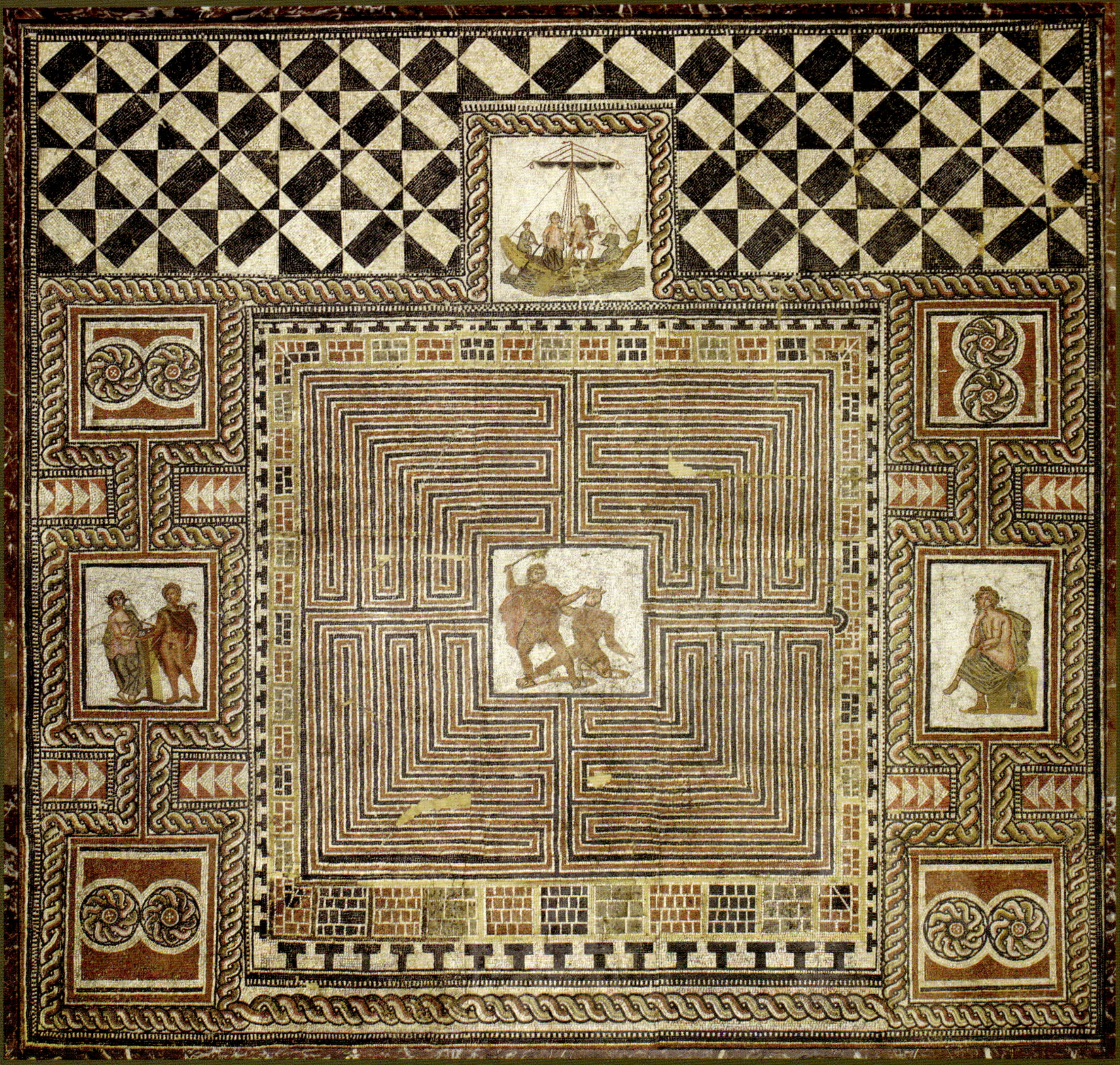

5

Logic

Mathematics, rightly viewed, possesses not only truth, but supreme beauty—a beauty cold and austere, like that of sculpture, without appeal to any part of our weaker nature, without the gorgeous trappings of painting or music, yet sublimely pure, and capable of a stern perfection such as only the greatest art can show.
—Bertrand Russell, "The Study of Mathematics," 1907

With the discovery of non-Euclidean geometries, mathematicians were forced to reexamine the foundations of their science. Formalist mathematicians, led by David Hilbert, tried to secure the foundations of mathematics by severing the links between geometry and the natural world; for them, geometry is an autonomous axiomatic structure. The German mathematician Gottlob Frege and his younger English colleague Bertrand Russell were, like Hilbert, modern Platonists, but unlike Hilbert, they believed the traditional governors of verbal arguments, the laws of logic that describe rational thought, provide the best foundation for mathematics. To secure a basis for modern Platonism, Frege and Russell would first reduce mathematics to logic and then prove that logic is consistent (paradox free). Thus they looked back not to Euclid but to Aristotle, who wrote the first laws of logic—syllogisms—and to the Enlightenment philosopher Gottfried Leibniz, who wrote a logical calculus that he called his *filum Ariadnes*—a thread of Ariadne (plate 5-1). Building on this tradition, Frege invented new tools that set the stage for modern logic.

For Frege and Russell, a mathematical symbol such as + in arithmetic was not meaning-free but represented a mathematical operation (plus) that captured the meaning of the joining (union) of sets as well as the meaning of "and" in everyday speech. Hilbert had written mechanical rules for manipulating the + symbol without reference to the words "plus" or "and." Frege and Russell instead *focused* on the meaning of "plus" in arithmetic and the usage of "and" in informal conversation as well as more technical discourse and then tried to capture these meanings in their *equally mechanical* rules. In other words, Frege and Russell sought to write mechanical rules for + that *merit* an interpretation as "plus" or "and." Their approach, intermixing questions about the artificial language of mathematics and natural spoken language, is called "*logicism*."

The game of chess provides an analogy for the distinction between formalism and logicism. A logician understands the rules of the game as symbolizing the movements of

5-1. *Cretan labyrinth*, first (or fourth) century AD. Mosaic, 18 × 15 ft. (5.48 × 4.57 m). Kunsthistorisches Museum, Vienna.

This mosaic covered the floor of a Roman villa near present-day Salzburg, Austria. Ariadne, the daughter of Minos, king of Crete, is shown on the left, giving a thread to the Athenian hero Theseus before he enters the labyrinth, in exchange for his promise to take her away with him on his ship. In the center of the mosaic Theseus (personifying reason) slays the Minotaur (bestial passion) and then retraces his path and exits the labyrinth guided by the thread of Ariadne (symbolizing logic). Theseus keeps his promise and takes Ariadne with him when he sets sail from the island of Crete (above), only to abandon her on the nearby island of Naxos (right).

soldiers in two armies fighting to capture an opponent's king. Like a foot soldier, the pawn is allowed to advance only step by step (one square per move), but the knight on horseback can jump over other pieces. For the formalist, the game pieces are meaning-free, and their movements are governed by a set of abstract rules. Regardless of approach, however, the rules of the game are equally mechanical. In other words, the distinction between formalism and logicism is *philosophical* and does not have practical consequences for how the game is played.

Frege and Russell (like Hilbert) imagined the abstract objects in the mathematical-world-out-there in terms of Cantor's sets. By the first decade of the twentieth century, the search for the logical structure of assertions about numbers had become the centerpiece of mathematics at Cambridge University in England under Russell's leadership. In collaboration with the philosopher G. E. Moore, Russell expanded the explanatory scope of logic from how one contemplates mathematical objects to also include how one knows the world by direct experience, both the everyday world of plants and animals, and the scientific world of atoms and forces. Russell and Moore translated the age-old philosophical question of how one knows the world into the question of how one *talks* about the world, and they proceeded to analyze the logical structure of *assertions*. Thus began the dramatic transformation of Anglo-American philosophy written in lucid English prose—the tradition of the British Empiricists David Hume, John Locke, and John Stuart Mill—into philosophy that is largely unreadable by the educated public because it is written in a highly technical style and mathematical symbolism peculiar to this school of philosophy—so-called logical empiricism or analytic philosophy. Through Moore and Russell's association with the art critic Roger Fry, the logistic outlook became known to writers, artists, and intellectuals who lived in the Bloomsbury district of London. Artists such as Vanessa Bell and Duncan Grant sought to reveal the underlying form of the visual world, and creative writers such as Virginia Woolf designed their stories on a logical framework. In the 1920s and 1930s, logic permeated the wider circle of British art and literature, manifest in the sculpture of Henry Moore and Barbara Hepworth and the writings of T. S. Eliot and James Joyce.

THE DEDUCTIVE LOGIC OF ARGUMENT: ARISTOTLE'S SYLLOGISMS

In the fourth century BC, Aristotle did for oral arguments what Euclid had done for mathematical proofs. Aristotle made the first recorded attempt to work out a system of valid reasoning—logic (from the Greek *logos* for "word")—by developing a prescribed form of argument, the syllogism (*Organon* [Instrument], fourth century BC). Someone wishing to argue a point first states two facts (premises that share a common term), which are assumed to be true (rather than quarrelling endlessly about basic assumptions). The conclusion of the argument follows from the two premises, as in the following example: All men are mortal. Socrates is a man. Therefore, Socrates is mortal. This syllogism has the form:

All A are B; C is an A; therefore, C is a B.

The truth of the conclusion follows necessarily from the truth of the premises. If the argument has a false premise (for example, All men are Egyptian) although the structure of the argument remains valid, the truth of the conclusion is not guaranteed. Just as Euclid had generalized beyond a particular right triangle to the abstract form of *all* right triangles, so Aristotle generalized beyond particular arguments to the valid structure of *all* arguments of his prescribed format.

For two millennia Aristotle's syllogisms provided users a deductive method of going from general premises to a valid conclusion about a particular case. Despite criticisms for being too rigid, Aristotle's syllogisms survived virtually intact into the early nineteenth century.

LEIBNIZ'S UNIVERSAL LANGUAGE WITH A CALCULUS OF REASON

If logic (words) were to be used in the pursuit of knowledge, how could one be sure those words would be understood? The Western idea of speaking an unambiguous language that is understood by all is rooted in the biblical account of Adam giving each animal a unique name (plate 5-2). After Adam and Eve ate the forbidden fruit and were expelled from Paradise, their descendents still spoke a common language, but it was lost when the Creator decided to "confound their language, that they may not understand one another's speech" in response to the hubristic construction of a tower "whose top may reach unto heavens" which, in a divine play on words, was renamed "Babel" (from the Hebrew *balal* for "confuse"; Genesis 11:1–9; see plate 5-3).

In the classical era Plato's teacher Cratylus attempted to create a true and perfect language in which a name would capture the essence of what it named by somehow imitating or indicating what it denoted. Plato offered as an example the resemblance of certain sounds to things: "the letter ρ [rho] is expressive of rapidity, motion."[1] (In English, an example would be the rushing sound of the letter *r* expressing motion in the words "race," "rush," and "run"). Pointing out that naming (and hence knowledge) of the everyday, natural world was impossible "if everything is changing and nothing persists," Plato suggested that Cratylus's names denoted the stable, timeless Forms, which are the objects of true knowledge.[2] But Cratylus, a follower of Heraclitus who believed that "everything flows," pushed his mentor's idea to an extreme in which total flux means that everything lacks any kind of identity. Unable to finds words that have stable meanings, Cratylus "finally did not think it right to say anything but only moved his finger."[3]

Cratylus . . . criticized Heraclitus for saying that it is impossible to step twice into the same river; for he thought one could not do it even once.

—Aristotle, ***Metaphysics***, fourth century BC

In the seventeenth century a growing awareness of the importance of precise language in scientific discourse prompted Gottfried Leibniz and his contemporaries to invent the first artificial languages intended to overcome the ambiguity, imprecision, and mutability of natural, spoken language. Of the many attempts, Leibniz's efforts have had the most lasting effect.

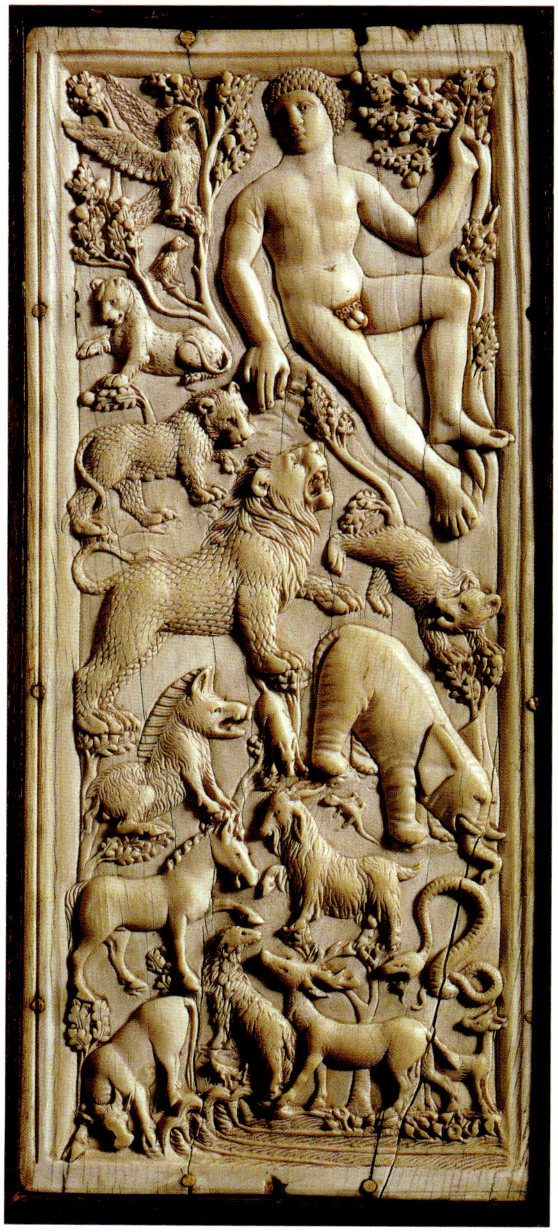

5-2. *Adam naming the animals*, ca. AD 400. Ivory. Museo Nazionale del Bargello, Florence.

This ivory carving illustrates the Biblical text: "And out of the ground the Lord God formed every beast of the field and every fowl of the air; and brought them unto Adam to see what he would call them: and whatsoever Adam called every living creature, that was the name thereof" (Genesis 2:19).

When only a teenager, Leibniz set himself the goal of organizing all scientific knowledge into a vast system, and for the rest of his life he was driven to build this structure, for which he invented two tools. The first was his *linga characteristica universalis*—a language of universal characters. In his youthful enthusiasm Leibniz attempted to assign a prime number to each elementary concept in every field of learning. He planned to multiply together the prime numbers associated with the elementary concepts to form composite numbers corresponding to complex concepts. An advantage of this encoding is that a complex concept can be analyzed into its elementary concepts by factoring its corresponding composite number. There could be no ambiguity in deciphering this code because, as Euclid had proven, every composite number has a unique set of prime factors (see the sidebar on page 21 in chapter 1). The disadvantage of Leibniz's *linga characteristica universalis* as a practical tool, however, is that factoring large composite numbers is notoriously difficult. (It is this very difficulty that today insures the security of systems that use prime numbers to encrypt information in a numerical code and transmit it electronically.)

If Leibniz had completed his *linga characteristica universalis*, he would have created a mythic Adamic language in which the structure of the characters mirrors the structure of reality. But by his mid-twenties he had concluded that the isolation of every elementary concept in all fields of learning was beyond the power of the human mind.[4]

Leibniz's second tool was his *calculus ratiocinator*—a calculus of reasoning. This set of mechanical rules would determine the truth of a proposition in terms of the prime and composite numbers of the elementary and complex concepts contained in the proposition's subject and predicate. For an assertion to be true, the predicate concept had to be contained in the subject; factoring the composite number corresponding to the predicate concept must yield prime factors corresponding to the elementary concepts in the subject. Leibniz's insight was that if he could invent a set of precise symbols and mechanical rules for manipulating them, he could then determine the truth or falsity of a proposition by calculation alone. Using metal gears, Leibniz intended to actually build a calculating machine, for which he invented a machine-friendly language, the binary number system, for expressing all number in terms of two digits (0 and 1).

Leibniz's language of universal characters was utopian in the extreme, but his calculus of reason—*calculus ratiocinator*—foretold the future of modern logic. Leibniz compared logic to the means by which the rational Theseus escaped the legendary labyrinth on the island of Crete, wherein lived the Minotaur, an irrational monster with the head of a bull and the body of a man. Anyone who entered the labyrinth became utterly lost in its baffling passageways, only to be killed by the beast. According to Greek mythology, the warrior Theseus slayed the Minotaur and escaped the maze with the help of Ariadne, daughter of the Cretan king, who gave Theseus a thread—a *filum Ariadnes*[5]—to mark his route. Symbolism had been used in logic since its inception—Aristotle substituted letters for the terms

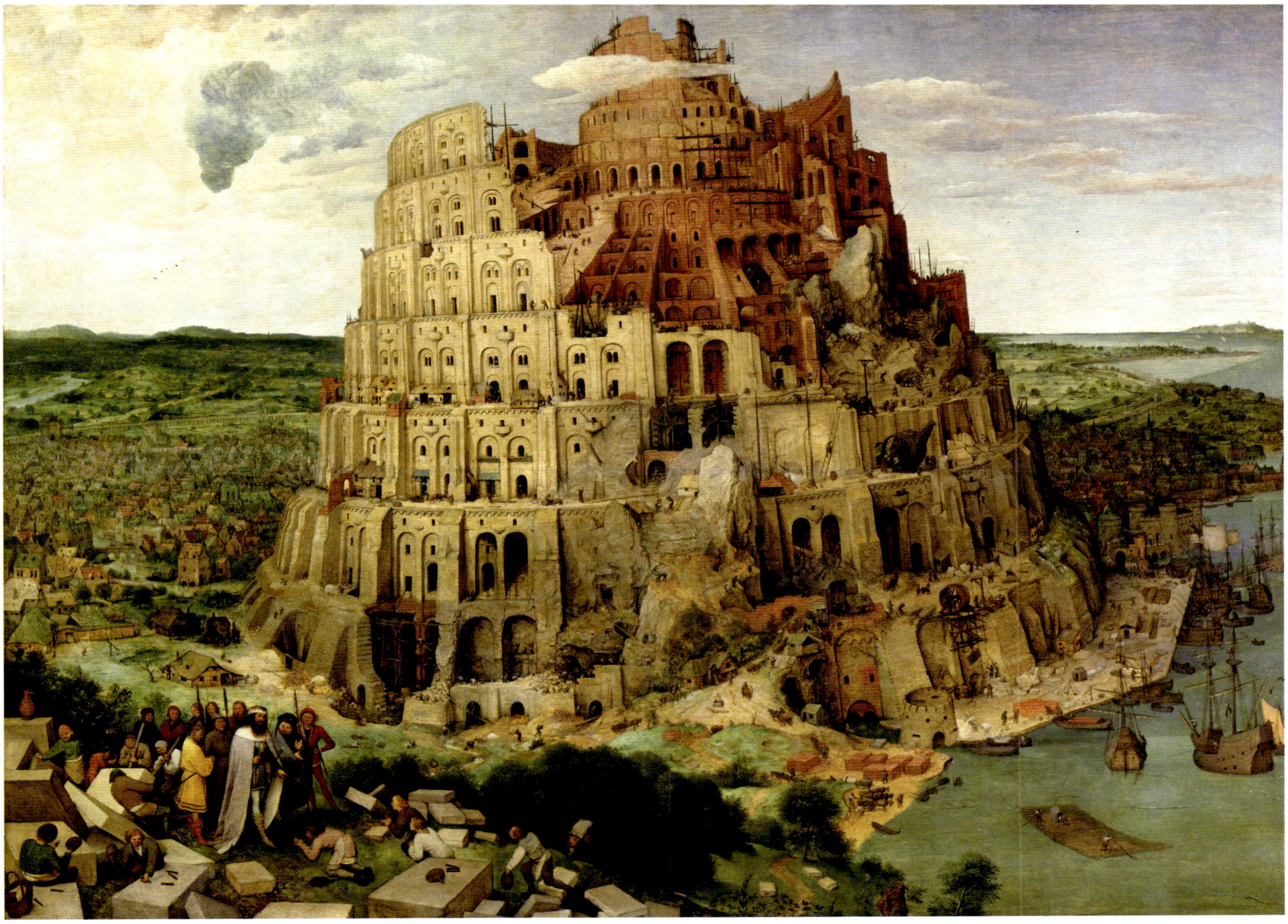

in his syllogisms—but Leibniz made it easier for the mind to grasp patterns of inference by refining the symbolism of logic. Although his calculus was fragmentary, Leibniz inspired others, and, in time, the improved symbolism revealed errors in traditional logic and made generalizations possible. In the end, Leibniz's logical calculus provided the *filum Ariadnes* that led to the development of symbolic logic in the nineteenth century.

5-3. Pieter Bruegel the Elder (Flemish, ca. 1525–69), *The Tower of Babel*, 1563. Oil on oak panel, 44⅞ × 61 in. (114 × 155 cm). Kunsthistorisches Museum, Vienna.

> *God did not create man from the word, and he did not name him. He did not wish to subject him to language, but in man God set language, which had served* him *as a medium of creation, free. God rested when he had left his creative power to itself in man. This creativity, relieved of its divine actuality, became knowledge. . . . The paradisiacal language of man must have been one of perfect knowledge. . . . Even the Tree of Knowledge cannot conceal the fact that the language of Paradise was fully cognizant. . . . The Fall marks the birth of the human word, in which name no longer lives intact. . . . After the Fall, which, in making language mediate, laid the foundations for its multiplicity, linguistic confusion could be only a step away.*
>
> —**Walter Benjamin, "On Language as Such and on the Language of Man," 1916**

SYMBOLIC LOGIC

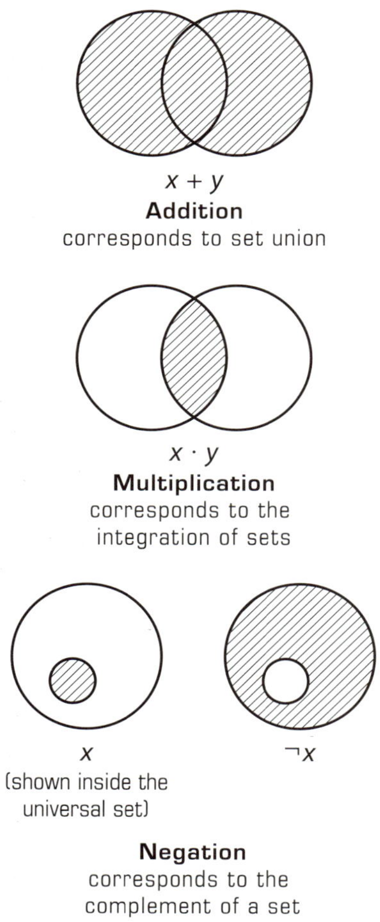

Logic manuals aimed at the general public were popular in nineteenth-century England. Written with the conviction that British citizens could avoid acting on irrational impulses by teaching themselves to be reasonable, these textbooks taught syllogisms and induction, along with informal methods such as argument by comparison or analogy. John Stuart Mill's A *System of Logic* (1843), written in lucid prose, was for a century the most widely-read popular manual on reasoning in the English language. A British contemporary of Mill, George Boole, who was largely self-taught in mathematics, designed the first successful *calculus ratiocinator* by reducing parts of logic to algebra.

Aristotle had focused on terms that occur in propositions (statements that are the premises and conclusion of a syllogism). Boole's contribution was to recognize that there is a structural similarity between a simple, two-element "Boolean" algebra and a logic that uses letters to symbolize propositions, so-called propositional logic (plate 5-5). He adopted Leibniz's binary number system, having the numeral 1 symbolize a true proposition, and 0 a false proposition, and went on to show that one could manipulate the algebraic symbols and yield valid results. Boole published his calculus of reasoning in *The Mathematical Analysis of Logic* (1847) and *The Laws of Thought* (1854), the founding texts of symbolic logic.

Boole's method was essentially a restatement of propositional logic in a two-element *algebra*, and in 1882 the British mathematician John Venn similarly expressed logical relations in *geometry* by picturing sets and propositions as intersecting ellipses (plate 5-4). Since young people learned geometry in school, these diagrams could be enlisted to develop the reasoning power of the mind (plate 5-6). Lewis Carroll, a contemporary of Venn and a lecturer in mathematics at Oxford, developed an alternative to the representation of classes as ellipses, proposing instead intersecting rectangles. Viewing logic as a game, Carroll wrote a booklet of recreational mathematics and invented a board game played with nine pieces on rectangles representing classes (*The Game of Logic*, 1886). According to Carroll, playing his games would develop "clearness of thought—the ability to *see your way* through a puzzle, the habit of arranging your ideas in an orderly and get-at-able form—and, more valuable than all, the power to detect *fallacies*."[6] Carroll published his books on geometry and trigonometry under his given name Charles L. Dodgson, but his books on logic appeared under the pseudonym of the creator of Alice,

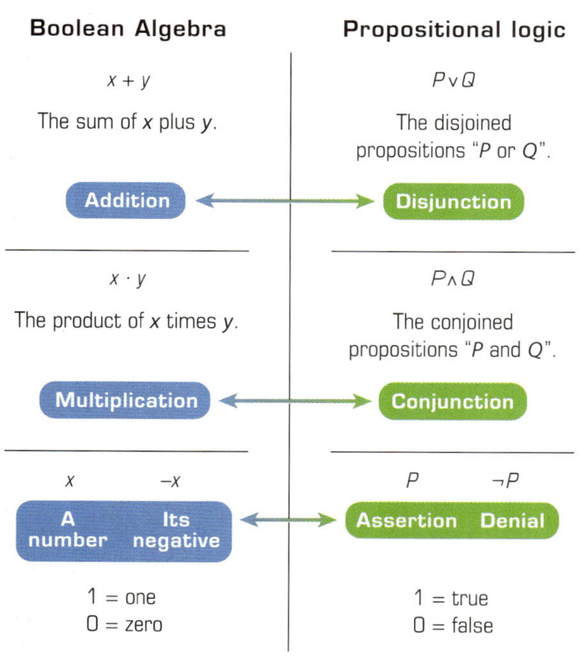

TOP LEFT

5-4. Venn diagrams.

BOTTOM LEFT

5-5. Comparison of Boolean algebra and propositional logic.
A Boolean algebra is a simple, two-element arithmetic of zero and one.

OPPOSITE

5-6. Winslow Homer (American, 1836–1910), *Blackboard*, 1877. Watercolor on paper, 19¾ × 12¾ in. (49 × 32 cm). National Gallery of Art, Washington, DC.

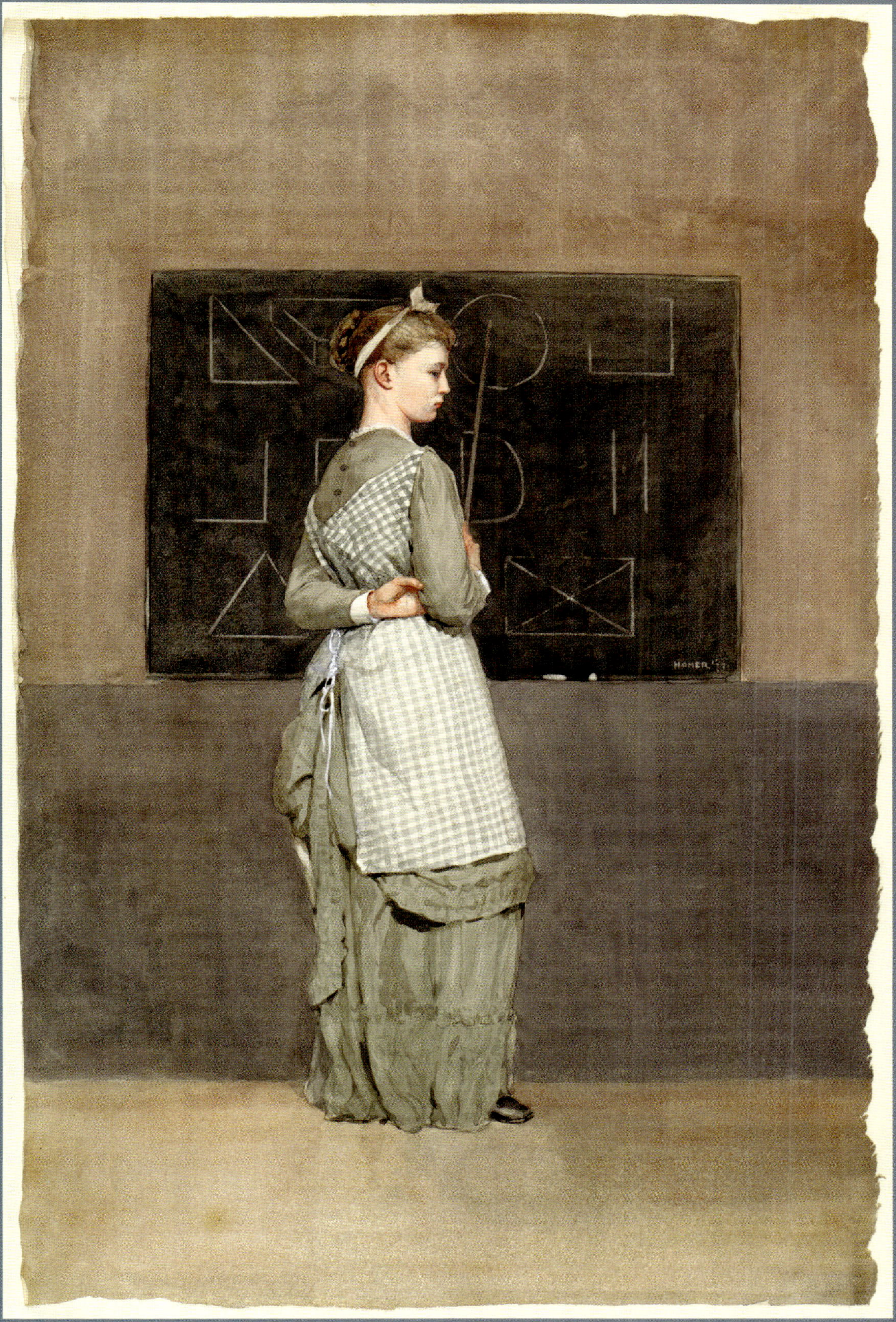

Oh that the words which
make the thoughts obscure
From which they spring, as
clouds of glimmering dew
From a white lake blot
heavens blue portraiture,
Were stripped of their thin
masks and various hue,
And frowns and smiles and
splendors not their own,
Till in the nakedness
of false and true
They stand before their lord
each to receive its due!

—Percy Bysshe Shelley,
"Ode to Liberty," 1820

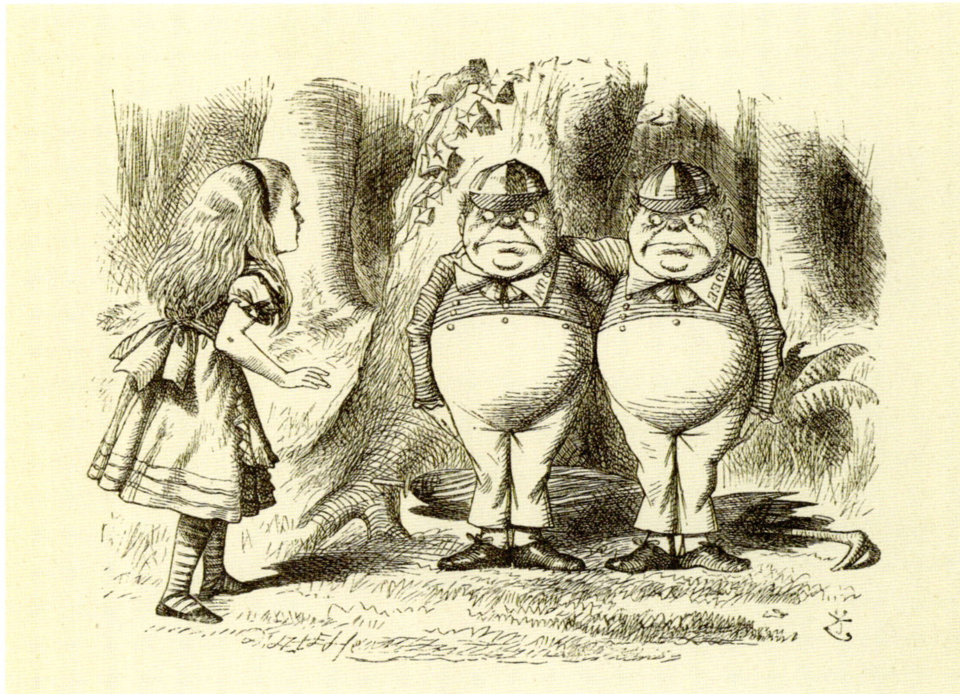

RIGHT

5-7. "Tweedledum and Tweedledee," in Lewis Carroll, *Through the Looking Glass and What Alice Found There*, illus. John Tenniel (London: Macmillan, 1871), 67. Rare Books Library, Columbia University, New York.

"I know what you're thinking about," said Tweedledum: "but it isn't so nohow."

"Contrariwise," continued Tweedledee, "if it was so, it might be; and if it were so, it would be; but as it isn't, it ain't. That's logic."

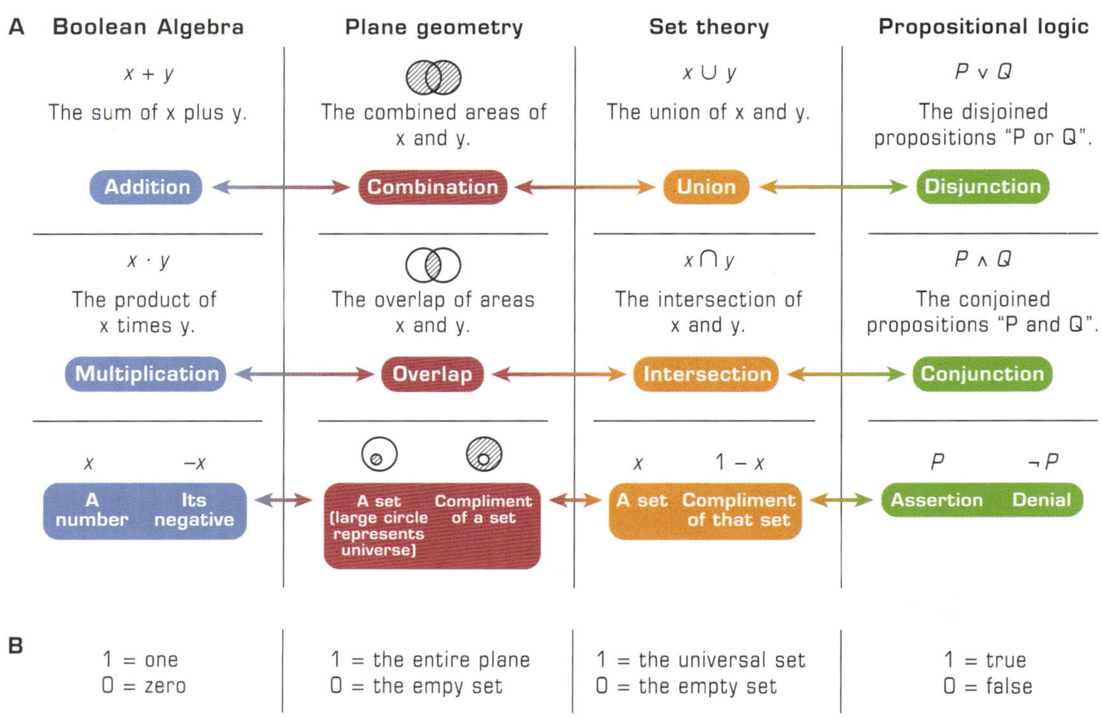

BELOW

5-8. Comparison of Boolean algebra, geometry, sets, and propositional logic.

who, after moving through a looking glass into a game of chess, encountered humorous logical conundrums (plate 5-7). The result of the efforts of Boole, Venn, and Carroll was a demonstration of the structural similarity of Boolean algebra, geometry, set theory, and propositional logic (plate 5-8A), and the possibility of using a binary number system with any of them (plate 5-8B). A half-century later when computing machines began to be powered by an electrical current that has two states, on and off, the machine language was written in Boolean algebra (binary code), with 1 = on and 0 = off.

FREGE'S ARITHMETIC AND PREDICATE LOGIC

Gottlob Frege developed a symbolic system to improve traditional logic. His notation expressed the internal (logical) structure of a sentence by symbolizing its predicate, such as "is mortal" (hence his system is called "predicate logic"). Although he focused on predicates, Frege was not concerned with the subject-predicate *grammar* of assertions but rather with their *logical* structure. The distinction between "all" and "some" of traditional (Aristotelian) logic was carried forward by Frege in his use of "quantifiers." Today logicians use ∀, which reads "for all," and ∃, which reads either "for some" or "there exists."[7] Frege also went beyond traditional logic by developing a way to symbolize relationships, such as *x* loves *y*.

Syllogisms, Propositional Logic, and Predicate Logic

A Syllogisms

Aristotle used letters (A, B, C) to symbolize categories.

	This has the form:
All Greeks are human.	All A are B
All humans are mortal.	All B are C
Therefore,	
All Greeks are mortal.	All A are C

B Propositional Logic

Letters such as *P* and *Q* are used to symbolize propositions, which are assertions that are either true or false.

Sample propositions:	*In symbols:*	*This reads:*
All Greeks are mortal.	*P*	*P*
Socrates is Greek.	*Q*	*Q*
Some roses are red.	*R*	*R*

Example of an inference:

If Greeks are mortal and Socrates is Greek, then some roses are red.	$(P \wedge Q) \to R$	If *P* and *Q*, then *R*

In this example, *P*, *Q*, and *R* are all true statements, so $(P \wedge Q) \to R$ is true. A disadvantage of propositional logic is that is does not allow one to get inside the propositions and show how they interrelate logically. This problem was solved in part by predicate logic.

C Predicate Logic

Frege saw that a notation for "all" and "some" would make logic more useful. Others took up the idea and introduced symbols ∀ for "all" and ∃ for "some" (or "there exists").

Example:	*This reads:*
$\forall x$	All *x*, or For all *x*.
$\exists x$	Some *x*, or There exists an *x*.

To analyze assertions, Frege symbolized their structure: lowercase letters are variables (*x*, *y*, *z*) and constants (*a*, *b*, *c*); capital letters are predicates (*G*, *H*, *P*, *Q*). Thus the proposition "A person is Greek" is symbolized by $\exists x\ (Px \wedge Gx)$, which is read "Something is *P* [a person] and is *G* [Greek]." Although this part of Frege's notation was retained, most was not adopted, and the following examples are written in the notation devised by Giuseppe Peano, which is currently standard.

Example of a deduction:	*In symbols:*	*In words:*
All Greeks are human.	$\forall x\ (Gx \to Hx)$	For all *x*, if *x* is *G*, then *x* is *H*.
All humans are mortal.	$\forall x\ (Hx \to Mx)$	For all *x*, if *x* is *H*, then *x* is *M*.
Therefore,		
All Greeks are mortal.	$\forall x\ (Gx \to Mx)$	For all *x*, if *x* is *G*, then *x* is *M*.

Another example:	*In symbols:*	*In words:*
All Greeks are mortal.	$\forall x\ (Gx \to Mx)$	For all *x*, if *x* is *G*, then *x* is *M*.
Socrates is Greek.	Gs	*s* is *G*
Therefore,		
Socrates is mortal.	Ms	*s* is *M*

Connectives and rules of inference used in both propositional and predicate logic:

Connectives:	*This reads:*	*Examples:*	*This reads:*
$\wedge$	and	$P \wedge Q$	*P* and *Q*
$\vee$	or	$\forall x\ (Fx \vee Gx)$	For all *x*, *x* is *F* or *x* is *G*
$\to$	if, then	$P \to Q$	If *P*, then *Q*
$\leftrightarrow$	if and only if	$P \leftrightarrow Q$	*P* if and only if *Q*
$\neg$	not	$\exists x\ (\neg Fx)$	Some *x* lacks *F*ness

Sample rules of inference:	*This reads:*
$((P \to Q) \wedge P) \to Q$	If (if *P* then *Q*) and *P*, then *Q*.
$\forall x\ ((Fx \to Gx) \wedge Fx) \to \forall x\ (Gx)$	For all *x* (if *x* is *F* then *x* is *G*) and *x* is *F*, then all *x* is *G*.
$((P \vee Q) \wedge \neg P) \to Q$	If (*P* or *Q*) and not-*P*, then *Q*.

Relations in Predicate Logic

Relations

The relation between two (or more) objects can be expressed by using two (or more) variables or constants.

Example:	*In symbols:*	*This reads:*
Everyone fears everyone.	$\forall x, y\ (Fxy)$	For all x and y, x has the relation F to y.
Someone loves someone.	$\exists a, b\ (Lab)$	There are a and b such that a has the relation L to b.

The quantifiers ∀ and ∃ can be used to clarify the structure of assertions. For example, certain propositions contain an ambiguity that can be laid bare using quantifiers:

Ambiguous assertion: Everyone loves someone.

Alternative logical structures:	*This reads:*
$\forall x\ \exists y\ (Lxy)$	For all x there exists some y such that x loves y.

In other words, everyone has a (possibly different) beloved.

$\exists a\ \forall x\ (Lxa)$	There is some a such that for all x, x loves a.

In other words, there is some adorable person who is loved by all.

Arithmetic is Reducible to Logic

The goal of the logistic project was to reduce all of mathematics to logic. Russell and Whitehead proved that at least arithmetic is reducible to logic (*Principia Mathematica*, 1910–13), although they had to resort to three axioms (of infinity, reducibility, and choice), whose logical status has been the subject of debate.

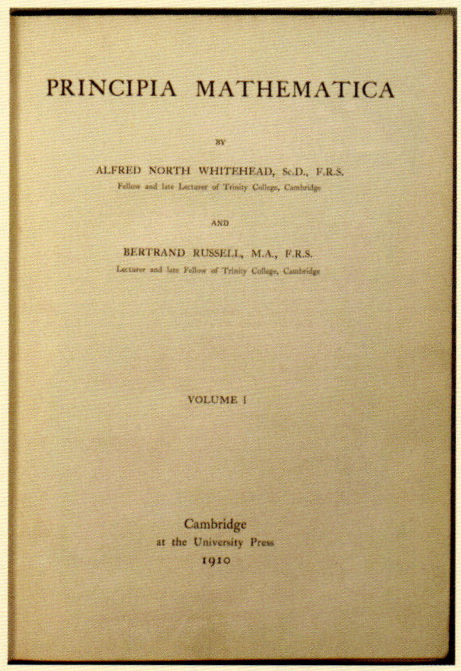
PRINCIPIA MATHEMATICA

BY

ALFRED NORTH WHITEHEAD, Sc.D., F.R.S.
Fellow and late Lecturer of Trinity College, Cambridge

AND

BERTRAND RUSSELL, M.A., F.R.S.
Lecturer and late Fellow of Trinity College, Cambridge

VOLUME I

Cambridge
at the University Press
1910

RIGHT

5-9. Alfred North Whitehead and Bertrand Russell, *Principia Mathematica* (Cambridge: Cambridge University Press, 1910). Institute Archives and Special Collections, Massachusetts Institute of Technology.

In Frege's predicate logic, the variables range over (refer to) individuals, such as numbers. Frege considered predicate logic the lowest level of a hierarchy of formal languages; predicate logic is "first-order logic" (see sidebar on opposite page). At the next, second-order level, the domain of the variables is expanded to range over sets of individuals, such as odd numbers and even numbers. In third-order logic, the variables can range over sets of sets, and so on.

Frege had decided that no geometry—Euclidean or non-Euclidean—should be built into the foundations of mathematics because geometric knowledge was rooted in intuitions about space that are learned from experience. But arithmetic, especially knowledge

Frege's Hierarchy of Formal Languages

N-order logic (and so on . . .)

Domain: sets of sets of sets and so on, rising N levels above that of individual elements

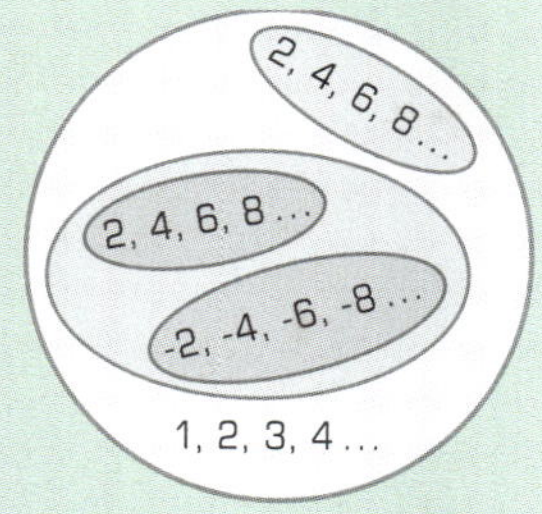

Third-order logic

Domain: individuals, sets of individuals, and sets of sets of individuals

Second-order logic

Domain: individuals and sets of individuals

True assertion in second-order logic:

$$\exists a\ (Oa)$$

which reads "There is a set *a*, such that every individual (number) in the set has *O* (Oddness)." In other words, "There is a set of odd numbers."

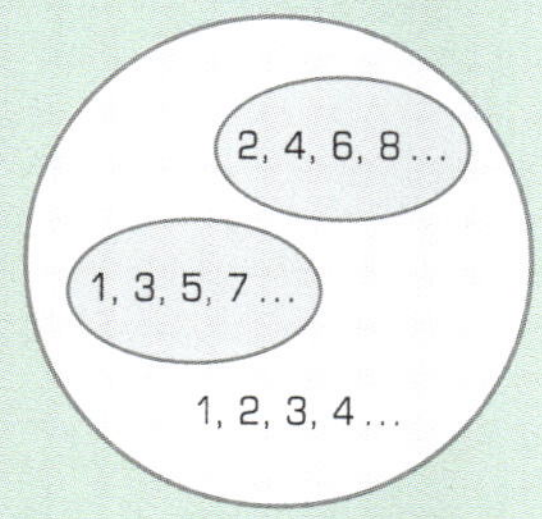

First-order logic

Domain: individuals

True assertion in first-order logic:

$$\exists a, \exists b\ (Ta \wedge Fb)$$

which reads "There exists an *a* and *b* such that *a* has *T* (twoness) and *b* has *F* (Fourness)." In other words, "There are a two and a four."

Description of the language.

Representation of the kind of elements in the domain

of the natural counting numbers (1, 2, 3, . . .), was, in Frege's opinion, grounded in pure reasoning (logic). So he began his work on the foundations of arithmetic by analyzing the language that one uses when talking about numbers in Cantor's set theory (see chapter 3). What is the number one? It is the set of all sets with one member.

Frege's consuming goal was to prove that all branches of mathematics could be derived from the laws of logic alone. Since numbers and arithmetic are at the core of mathematics, if he could show that numbers and numerical patterns can be constructed from logical primitives (such as the symbols for "and" and "not"), the rest would follow. Using Cantor's set theory, Frege first showed that numbers (understood as sets) are equal if and only if there is a one-to-one correspondence between the members of the sets. Frege followed this definition of "equality" with a quote from the eighteenth-century philosopher David Hume expressing a similar view, and hence Frege's principle of numerical equality came to be called "Hume's Principle." Then Frege found that he could derive Peano's arithmetic postulates (see the sidebar on page 159 in chapter 4) from this one single principle—the principle of numerical equality (Hume's Principle)—using only predicate logic (Frege's Theorem, *Basic Laws of Arithmetic*, 1893). Although some have argued that Hume's Principle is not a truth of logic, it is nevertheless a fundamental definition, and Frege's Theorem gave support to the view that arithmetic *is* logic and that logic is the new foundation of mathematics.

THE LINGUISTIC TURN

In developing predicate logic, Frege always returned—with great insight and subtlety of thought—to spoken language as his guide to the logical structure of an assertion. In addition to grounding arithmetic in logic, Frege also attempted to restore confidence in mathematical intuition, which had taken a blow when the parallel postulate was shown to be independent of Euclid's other four axioms. Hilbert had emptied the meaning out of geometric terms such as "point" and "line," but Frege instead appealed to the meaning of arithmetic terms in everyday speech. What does it mean to say that there are three apples in the bowl? In a style that recalls Plato's search for the meaning of terms such as "circle" and "goodness" in his dialogues, Frege proceeded to analyze the subtle nuances in assertions about numbers in everyday language, going from an ordinary person's observations of instances of a universal, to the isolation and grasping of the universal principle itself.

Since it is only in the context of a proposition that words have any meaning, our problem becomes this: to define the sense of a proposition in which a number word occurs.

—**Gottlob Frege,** ***Foundations of Arithmetic*****, 1884**

When Frege turned the spotlight on how people talk about numbers, he initiated what has come to be called the "linguistic turn" in the search for the foundations of mathematics, and he fathered a new discipline, the philosophy of language.[8] Frege's development of predicate logic was the biggest advance in formal deductive reasoning since Aristotle. It and the linguistic turn were the catalysts for so-called logical positivism, a philosophy of language written in a technical style of logical symbolism, which flourished in Germanic intellectual circles in the 1920s and 1930s, and for the analytic philosophy of language that dominates English-speaking academia to this day.

BRITISH ANALYTIC PHILOSOPHY

Meanwhile, at Cambridge University, the mathematician Alfred North Whitehead had begun to study the relation of logic to mathematics. Inspired by Boole's symbolic (algebraic) logic, Whitehead set out in 1891 to design a symbolic system that would encompass all forms of reasoning (*A Treatise on Universal Algebra*, 1898). During the 1890s Whitehead had a bright mathematics student, Bertrand Russell, who also studied philosophy with fellow undergraduate G. E. Moore. After graduation Russell and Moore remained at Cambridge on fellowships and met regularly to plot the overthrow of their common enemy, the German Idealism that had invaded the philosophy departments at Cambridge and Oxford.

The worldview of the second-generation German Idealists—including Hegel's goal of attaining knowledge of Absolute Spirit—had been adopted in the late nineteenth century by the British philosopher F. H. Bradley, a student and then a fellow at Oxford. In the Hegelian spirit, Bradley declared that everyday perceiving and measuring may present to the mind a contradiction (a thesis and antithesis, such as that an object appears both square and oblong when seen from different angles), but at the highest level of consciousness (the synthesis), one knows a harmonious whole—Absolute Spirit—in which all contradictions are resolved (*Principles of Logic*, 1883). Cambridge professor of philosophy J.M.E. McTaggart was also a follower of Hegel (*Commentary on Hegel's Logic*, 1910), but he became troubled because Bradley's ascent to the Absolute culminated in knowledge of an impersonal (atheistic) Spirit. So, in 1910 McTaggart began a massive, multivolume treatise in which he offered to the pious an alternative Absolute, one composed of all immortal souls (the contents of the minds of all humans who have ever lived) bound together eternally by the force of love (*The Nature of Existence*, 1921–27).

As students of McTaggart, Russell and Moore both fell under the spell of German Idealism and Romantism, which had ignited powerful emotions throughout German-speaking regions. But the flame was not sustained in Russell and Moore's more dispassionate British souls, and by the late 1890s they were urging a return to the hallmarks of British philosophy and science: the formulation of concepts that are crystal clear and reliance on what you can see with your own eyes—"seeing is believing." Russell was drawn to Frege's focus on language, which resonated with the analytic methods he and Moore were devising. While studying Frege's theory that numbers are sets, Russell found a contradiction. In 1902, just as Frege was going to press with the second volume of his *Basic Laws of Arithmetic*, he received a letter in which Russell asked whether Frege had considered "the set of all those sets not members of themselves."[9] Does this set include itself? If it does, then it doesn't. If not, then it does. Frege was devastated that this lethal paradox could be derived from his published axioms, which he was forced to modify (plate 5-10).[10]

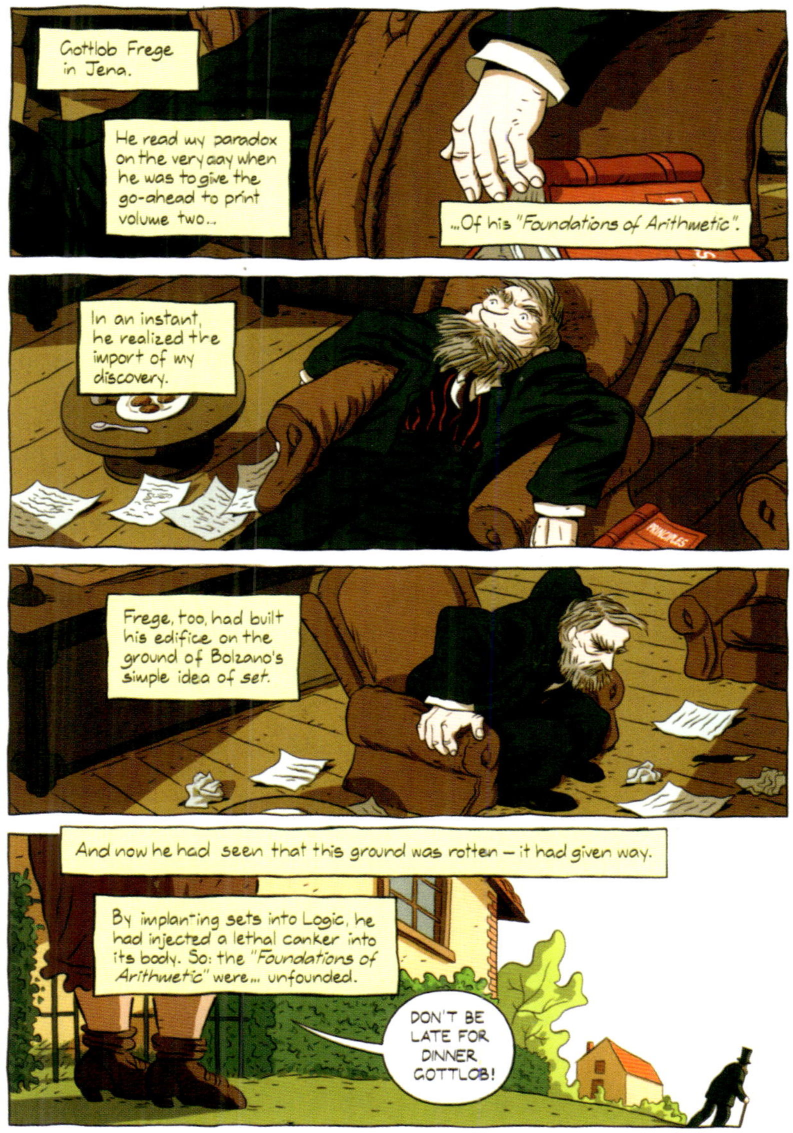

5-10. Frege at home in Jena in 1902 after receiving Russell's letter, in *Logicomix* by Apostolos Doxiadis and Christos H. Papadimitriou (text), Alecos Papadatos (art), Annie di Donna (color) (New York: Bloomsbury, 2009), 170. Used with permission.

Narrated by a fictional Bertrand Russell, this episode of the graphic novel *Logicomix* portrays the fateful moment on the eve of Frege's publication of the second (1903) volume of his *Grundgesetze der Arithmetik* (translated by M. Furth as *Basic Laws of Arithmetic*; vol. 1, 1893). Bernard Bolzano was a Czech mathematician who developed ideas about set theory that made him a forerunner of Georg Cantor. (Evidently using poetic license to link this event more dramatically to their "foundations" theme, Doxiadis and Papadimitriou give the title of Frege's 1893 publication as *Foundations of Arithmetic*, which is the title of an English translation by J. L. Austin of an earlier book by Frege, *Die Grundlagen der Arithmetik*, 1884).

In the definition "the set of all those sets not members of themselves," it was self-reference—the reference to the set in its own definition—that led to the paradox. The logical dangers of self-reference had been noted since antiquity when, in the first century AD, Apostle Paul, referring to Epimenides of Crete, wrote "A prophet of their own said 'All Cretans are liars'" (Titus 1:12). There is a whole range of possible sets, beginning at the conservative end with those that any mathematician would allow, such as the set of natural numbers. Russell's paradox underlines the danger that lurks in more radical sets, such as a set that contains all (imaginable) mathematical objects and can itself be treated as a set. When Russell outlined his own attempt to reduce mathematics to logic in *Principles of Mathematics* (1903), he disallowed any set whose definition entails self-reference.[11]

In the same year that Russell brought out his *Principles*, Moore published *Principia Ethica* (Principles of ethics), in which he proposed an ethics and aesthetics based on an appreciation of "the pleasures of human intercourse and the enjoyment of beautiful objects."[12] To Frege and Russell's list of perfect mathematical objects (such as Circle), Moore added his own version of an ideal Platonic ethical concept, Goodness. Moore's thesis was that, like a perfect circle, pure goodness is never perceived in the world but is known only by intuition. With Russell's *Principles of Mathematics* and Moore's *Principia Ethica*, the analytic method was established in British philosophy.

RUSSELL AND WHITEHEAD'S *PRINCIPIA MATHEMATICA*

The search for new foundations of mathematics had led some to ask if mathematics was deducible from a few laws of thought, but it was not until Frege's invention of predicate logic that it became technically possible to try to actually do the deduction. After Russell discovered the paradox in Frege's system, many mathematicians set about to refine Frege's apparatus (predicate logic) in order to construct arithmetic from logical axioms in a paradox-free structure. Russell and Whitehead together accomplished the massive task of deducing arithmetic from five rules of inference using three primitive logical terms: P (proposition), $\vee$ (or), and $\neg$ (not).[13] To accomplish their task, they assumed three additional axioms that some critics found not to be true as a matter of logic alone: the Axiom of Infinity, the Axiom of Choice, and the Axiom of Reducibility.[14] Despite these reservations, no one doubts that *Principia Mathematica* (1910–13) accomplished the deduction of arithmetic from logic, thus demonstrating the power of predicate logic and initiating its widespread use in diverse fields outside mathematics, including linguistics, economics, and computer science.

Russell and Whitehead confidently titled their book *Principia Mathematica*, echoing the title of a book by an earlier Cambridge mathematician, in which all motion on earth and sky was reduced to a single force (gravity)—Isaac Newton's *Philosophiæ Naturalis Principia Mathematica* (1687). Russell and Whitehead proved that logic—the laws of human reason—are the foundations of mathematics in the sense that arithmetic—numbers and numerical patterns—can be constructed from a very few logical primitives and assumptions.

LOGICAL ATOMISM

After Russell and Whitehead had accomplished the practical task of building the edifice of mathematics from a few logical building-blocks, Russell and Moore applied the new logical tools to other topics in philosophy, beginning with the ancient question: how does one know the world? Taking sense perception to be primary, they began by analyzing the facts of perception (such as perceiving an apple) into component parts (seeing red, touching a round object, smelling sweetness), which they called "sense-data" (today philosophers call such mental experiences "qualia"). Russell and Moore first began writing on this topic at the turn of the last century, when there were lively discussions about the structure of matter at Cambridge, where J. J. Thomson had discovered the electron in 1897. Russell and Moore borrowed a term from the physics laboratory and called each sense-datum an "atomic fact" and its description an "atomic proposition." They proceeded to apply the tools of logic to these components in the hope of solving problems in the theory of knowledge. In what is known as a correspondence theory of truth, for a simple sentence to be true there must be a one-to-one correlation between an atomic proposition (a statement about sense-data) and an atomic fact (an event in the-world-out-there). Additionally, a complex (molecular) sentence is true if it is properly constructed from true atomic propositions.[15]

In the early 1930s the Polish logician Alfred Tarski, using the tools of logic devised by Frege, Russell, and Whitehead, published a very precise definition of the correspondence theory of truth that became widely adopted. Hoping to avoid the paradoxes that result when a sentence is allowed to refer to its own truth or falsity, Tarski employed a hierarchy of languages similar to Frege's, creating an infinite hierarchy of formal languages in which each language is the domain of a metalanguage immediately above it. One can only speak of the truth or falsity of a sentence using its metalanguage one level up ("The Concept of Truth in Formalized Languages," 1933).

"It is snowing" is true if and only if it is snowing.

—Alfred Tarski, "The Concept of Truth in Formalized Languages," 1933

LOGIC AND THE ARTS: ROGER FRY'S FORMALIST CRITICISM

Discomfited by the repercussions of Britain's increasing industrialization, English artists of the late nineteenth century turned their backs on the smokestacks and steam engines that blighted London's air and water and on the poorly crafted goods churned out by factories. James McNeill Whistler and artists of the Aesthetic Movement, which peaked in the 1870s and 1880s, instead painted decorative urban and pastoral genre scenes, retreating to an aesthetic bunker—art for art's sake—that isolated art from the rest of society. Designers in the fin-de-siècle Arts and Crafts movement, led by William Morris, borrowed motifs from Gothic and Renaissance architecture and hand-crafted wood furniture with brass hardware, and printed colorful wallpaper from carved woodblocks.

In opposition to these nostalgic sentiments, the Cambridge-educated Roger Fry vowed to modernize British art, and to this end he imported avant-garde examples from Paris for exhibitions he organized in the 1910s.[16] Fry, who moved in Russell and Whitehead's intellectual circle, saw the French styles through the lens of analytic philosophy. As an undergraduate, Fry had joined Cambridge's elite academic club "The Apostles"; membership was for life and after graduating in 1889 Fry kept his ties to the club, where he met Whitehead, Russell, and Moore.[17] In 1894 Russell married the American Alys Pearsall Smith, whose sister, the art historian Mary Smith, was the wife of the American art historian Bernard Berenson. The Russells, Berensons, and Fry and his wife Helen Coombe had regular social contact.[18] In 1903 Fry helped found the scholarly art journal *The Burlington Magazine*, to which he contributed numerous articles on art and aesthetics,[19] and between 1906 and 1910 Fry was curator of painting at the Metropolitan Museum of Art in New York.

Fry also belonged to a group that met for regular discussion in the Bloomsbury district of London. Members included Clive Bell, who had studied philosophy with Moore at Cambridge, his wife the painter Vanessa Bell, her sister Virginia Woolf, and the artist Duncan Grant. Woolf read Moore's *Principia Ethica* in 1908,[20] and after the release in 1910 of the first volume of Russell and Whitehead's *Principia Mathematica*, the Bloomsbury group discussed the implications of this ground-breaking study. Although *Principia Mathematica* is written in the technical vocabulary of symbolic logic, in which none of the members of the Bloomsbury group were trained, they knew the general themes of the book from Russell's introduction to its first volume (1910) and from prominent reviews of the book, such as the one that appeared on the front page of the *Times Literary Supplement* on September 7, 1911. The (unnamed) reviewer began: "Perhaps twenty or thirty people in England may be expected to read this book," but he went on to give a lucid, non-technical summary of the book's thesis and import.[21] Those in the Bloomsbury group, such as Fry, who took the time to learn a little of the notation could easily appreciate the power of predicate logic to analyze arguments about aesthetics and ethics that were at the heart of Bloomsbury.

In keeping with analytic philosophy, Fry developed an approach to art that was based on logic in the sense that he analyzed the underlying structure of a painting or sculpture. But Fry mixed logic with a more intuitive approach, as did Russell himself, who described his own intellectual style as a combination of intuition (mysticism) and logic in a 1901 essay entitled "Mysticism and Logic."[22] Russell explained how he, as a logician, knew the mathematical-world-out-there by mystical intuition (pure cognition). After having an insight into this ultimate order of reality, Russell described how he then used logical reasoning to craft a proof to be sure that his insight was sound knowledge.

Fry encountered a similar mix of intuition and logic within the British aesthetic tradition of John Ruskin, who wrote in *Modern Painters* (1846) that artists (using their reason) compose with abstract forms and also (using intuition) create spiritual content with a strong moralist overtone (which are not present in Russell).[23] Like both Ruskin and Russell, Fry believed that one can gain knowledge by intuition, speculating that certain artists may intuit an ultimate order of reality: "I think the artist might if he chose take a mystical attitude, and declare that the fullness and completeness of the imaginative life he leads may correspond to an existence more real and more important than any we may know in mortal life."[24]

But how can immaterial, supernatural truths be embodied in material, mundane paintings? In a 1905 essay "Mantegna as a Mystic," Fry described how the Italian Renaissance artist embodied the mystical idea of the immaculate conception, the union of spirit and matter, divinity and mortality, "the word made flesh" (John 1:14): "Mysticism in its higher developments can scarcely become the inspiration for a figurative art. It tends always towards the elimination of differences and distinctions, to regard everything but the Divine unity as illusory and phenomenal, while art, on the other hand, lives entirely upon the distinctions of one object from another; it is bound to emphasize the significance of things, not their essential insignificance in comparison with absolute reality. So it is only

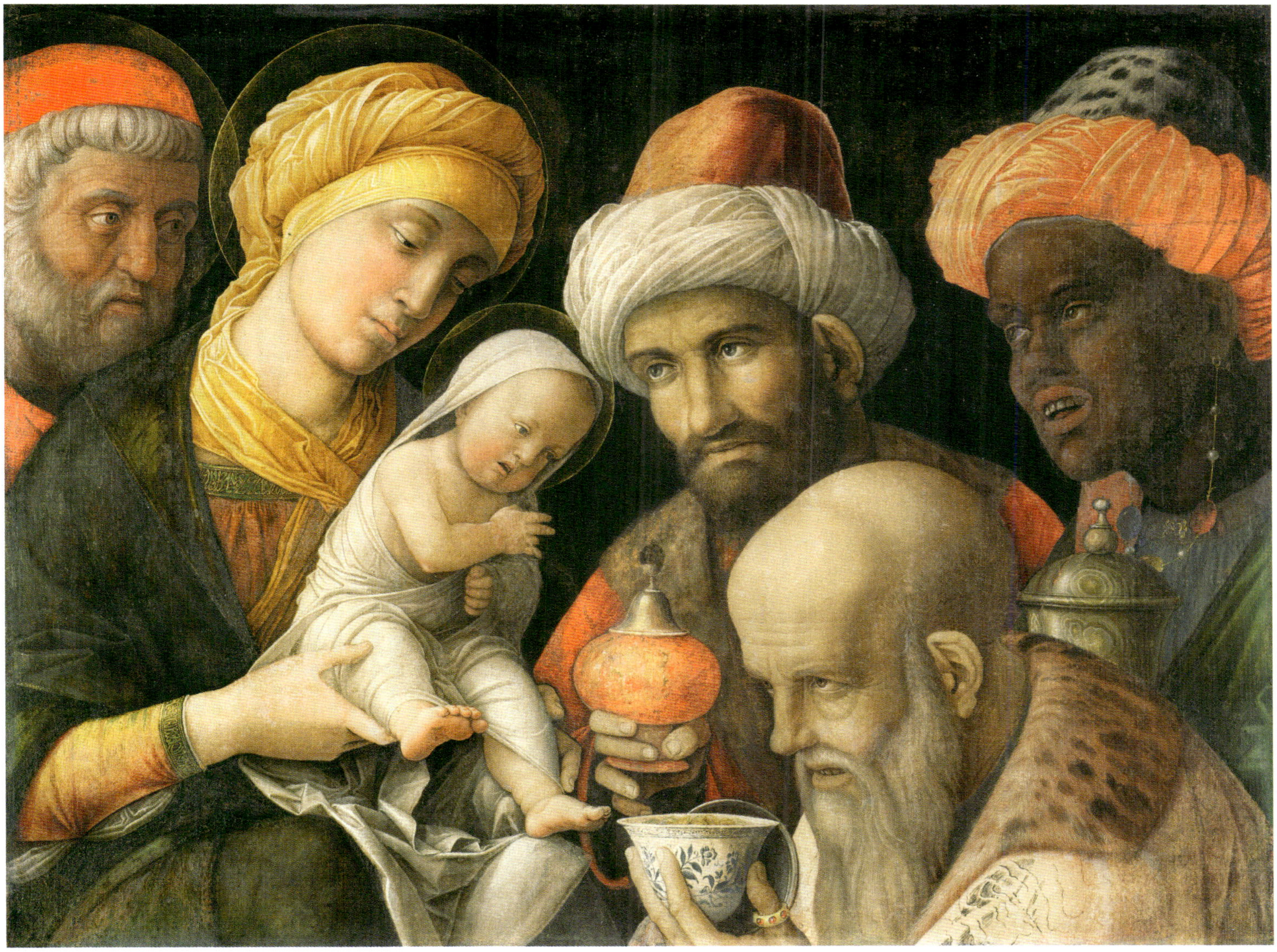

by analogy that one can talk of mysticism in art; it is only by metaphors that artists can convey any sense of mystically apprehended truths. Nevertheless, we feel with regard to a few artists that they do make matter transparent, so that the dimly guessed spiritual realities behind it become clearer for us."[25] For Fry, those few artists included Albrecht Dürer and Mantegna: "Far from being vague and misty in their presentation of form, they are the most precise and definite of all artists. . . . As we see from Dürer's *Melencolia* and Mantegna's Madonnas . . . this mystical idealism [is] consistent with a frank and rigorous realism"[26] (plate 5-11; for Dürer's *Melencolia*, see plate 2-16 in chapter 2).

French and British realism (portraits, landscapes, still lifes) could, according to Fry, also embody the supernatural realm in a way suitable to a secular age, by using logic. In November 1910 Fry opened the first presentation in London of the French modern art of Édouard Manet, Claude Monet, and Paul Cézanne (*Manet and the Post-Impressionists*). He

5-11. Andrea Mantegna (Italian, 1431–1506), *The Adoration of the Magi*, 1495–1500. Distemper on linen, 19⅛ × 25¹³⁄₁₆ in. (54.6 × 70.7 cm). J. Paul Getty Museum, Los Angeles, inv. 85.PA.417.

Fry compared Mantegna's use of muted tones to communicate the "intensity and strangeness of mystical thought" to the technique of Song dynasty Taoist landscape painters such as Fan Kuan (plate 1-49 in chapter 1): "Mantegna, by negation of all the more brilliant and seductive qualities of paint, by reduction of expression to its simplest terms of flat slightly contrasting tones, bounded by contours of the utmost purity and perfection, has found precisely the method to give to his figures their mysterious spiritual life. We may feel the more certain that this instinct was right when we reflect that independently he hit upon almost identically the same technique, and the same general effect of tone, even the same quality of surface, as the great religious painters of China. And in their work too, by similar means, a similar intensity and strangeness of mystical thought becomes apparent"; Roger Fry, "Mantegna as a Mystic," *The Burlington Magazine* 8, no. 32 (1905): 87–98; the quote is on 97–98.

followed in 1912 with an exhibit featuring Pablo Picasso and Henri Matisse (*Second Post-Impressionist Exhibition*). In his description and analysis of this work, Fry adopted terms from British analytic philosophy. In the catalogue to the 1910 exhibition, he described Monet as concerned with "appearances" (sense-data) and Cézanne with "design" (logical form; see plate 5-12): "Cézanne, when rendering the novel aspects of nature to which Impressionism was drawing attention . . . showed how it was possible to pass from the complexity of the appearance of things to the geometrical simplicity which design demands."[27]

Fry essentially translated Russell's logical atomism into an aesthetic theory of how paintings and sculpture represent the world. When Cézanne looked at a still life in his studio, his first task was to analyze the arrangement of the fruit, dish, tablecloth, and knife into its atomic facts, including seven red and three green circles, many purple circles, a thin brown rectangle (the apples, grapes, and knife) on a white plane (the dish and cloth). From these facts Cézanne created an isomorphism in a "painterly" atomic proposition (the painting "asserts" that there are ten apples on a cloth). The artist analyzed his sense-data into basic elements of design (forms, colors), but, according to Fry, it was Cézanne's imposition of logical structure on these elements that produces an emotional impact on the viewer: "We may, then, dispense once and for all with the idea of likeness to Nature, of correctness or incorrectness as a test, and consider only whether the emotional elements inherent in natural form are adequately discovered."[28] What causes the emotions associated with Fry's "emotional elements" of design? Schooled in Moore's *Principia Ethica*, in which logical, rational order is equated with Goodness, Fry gave design a moral overtone that was absent in France; seeing a Cézanne still life in Paris brought sensual pleasure but seeing it in London was a morally uplifting experience.

In his 1912 London exhibition, Fry, now assisted by Bell, included work by the Cubists Picasso and Georges Braque, and the Fauvists Matisse and André Derain. These artists, according to Fry, created abstract, logical form for contemplation as opposed to mere appearances (sense-data) used for practical purposes: "They wish to make images which by the clearness of their logical structure, and by their close-knit unity of texture, shall appeal to our disinterested and contemplative imagination with something of the same vividness as the things of actual life appeal to our practical activities."[29] Fry's examples of "logical structure" included a Picasso Cubist oil sketch of a woman's head and Matisse's *Dance* (1909, Museum of Modern Art, New York).

After seeing the French modernists, Vanessa Bell and Duncan Grant emphasized form in their compositions. For example, in *Virginia Woolf* (ca. 1911–12), Bell reduced the figure, the chair, and the background to flat planes of color in the Fauvist style (plate 5-13). Bell and Grant also experimented with total abstraction (Vanessa Bell, *Abstract Painting*, ca. 1914; Tate, London, and Duncan Grant, *Abstract*, 1915; Anthony d'Offray Gallery, London). Although Fry preferred representational pictures and was not enthusiastic about abstract art, for him the emotional element of design, not imitation of nature, was the aesthetic core of visual art.[30]

In 1913 Fry was too busy to accept a commission to write a book about modern art, so he passed the job on to Clive Bell, who summarized their joint outlook and coined

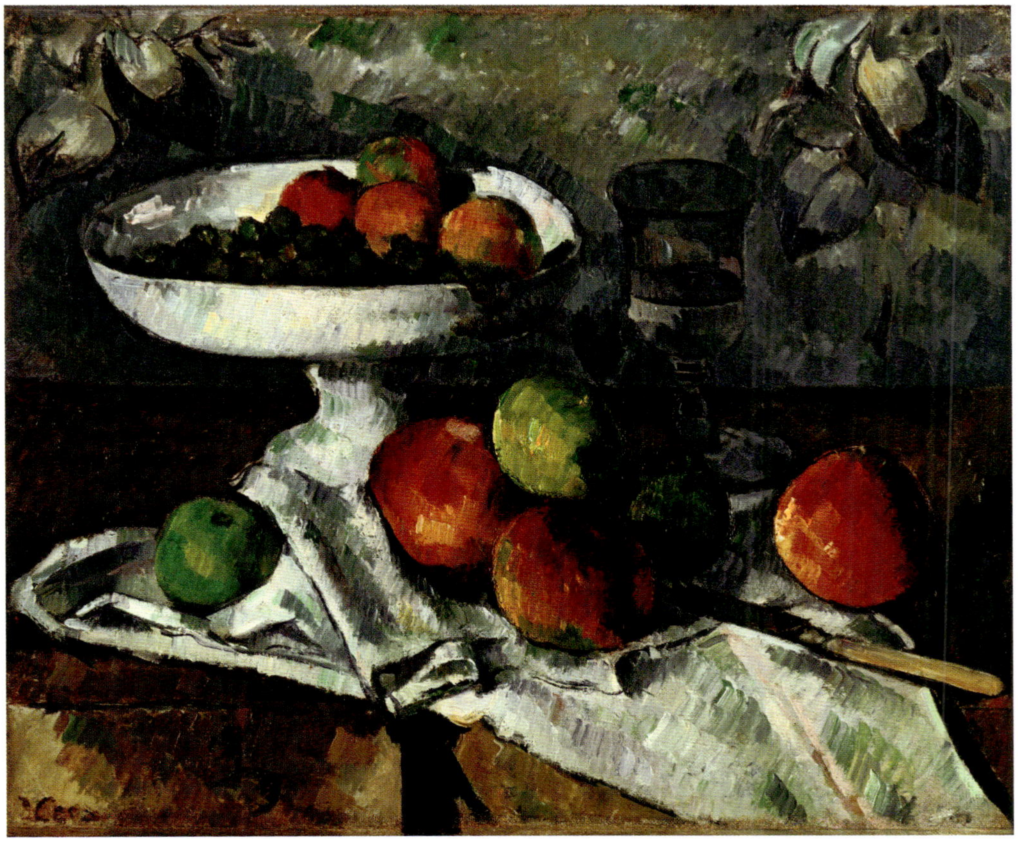

5-12. Paul Cézanne (French, 1839–1906), *Still Life with Fruit Dish*, 1879–80. Oil on canvas, 18¼ × 21½ in. (46.4 × 54.6 cm). Museum of Modern Art, New York, fractional gift of Mr. and Mrs. David Rockefeller.

the term "significant form" to name Fry's "emotional elements of design" (*Art*, 1914).[31] In this widely read treatise, Bell reduced Fry's nuanced rhetoric to simplistic dogma with pronouncements such as: "The representative element in a work of art may or may not be harmful. It is always irrelevant."[32] As Fry put it politely in his review of the book, Bell possessed "an assurance denied to me."[33] Fry would later edit a collection of his essays into his own treatise, in which he declared that sense-data and their representation on a canvas are part of *vision*; the arrangement of these visual and pictorial atoms into logical, molecular scenes constitutes *design* (*Vision and Design*, 1920). Just as for Frege and Russell the essence of mathematics was its logical form, for Fry the essence of art was its design (rational order).

Fry's art criticism came to be known as formalism, the same term adopted by his contemporaries the Russian literary formalists. It is appropriate that they used the same term because, just as Hilbert, Frege, and Russell all arrived (by different routes) at the same basic idea of an axiomatic structure, similarly Bely, Shklovskii, and Fry all (for different reasons) ended up emphasizing form over content. In resonance with Hilbert's focus on mathematics as a formal axiomatic structure, the Russian/German formalist artists and poets composed with shapes and sounds that do not represent the natural world-out-there. Inspired by Frege/Russell's focus on logic and spoken language, British formalist artists used forms to depict ("picture") the natural world, as seen in the figures and landscapes of the Bloomsbury Group, and, beginning in the 1920s, the sculptures of Henry Moore, and Barbara Hepworth.

VIRGINIA WOOLF

She asked him what his father's books were about. "Subject and object and the nature of reality," Andrew had said. And when she said Heavens, she had no notion what that meant. "Think of a kitchen table then," he told her, "when you're not there."

—**Virginia Woolf, *To the Lighthouse*, 1927**

Because of its implications for language, Frege and Russell's logic and analytic philosophy were of great interest to early-twentieth-century British novelists, poets, and critics. But the work of these writers also reflected developments in psychology that were at odds with analytic philosophy. Frege and Russell used words and symbols to represent the observable, natural world-out-there, whereas the neurologists Jean-Martin Charcot in Paris and Sigmund Freud in Vienna asked language to describe an inner world.

It was psychology's emphasis on the interior that inspired a new literary form, the "inner monologue," which purported to record a person's complete flow of conscious thought during a specified period of time. In Paris, Édouard Dujardin (in his novel *Les Lauriers sont coupés*, 1888) and in Vienna, Arthur Schnitzler (in his play *Paracelsus*, 1899) wrote tales in which ordinary people report their thoughts as they go about their uneventful lives, an approach that the America philosopher and psychologist William James termed "stream of consciousness" (*Principles of Psychology*, 1890; see plate 8-6 in chapter 8).[34] Schnitzler's inner monologues by members of dysfunctional Viennese families made him a favorite of Freud.[35]

In London Virginia Woolf mixed the French-Austrian inner monologue with the logical atomism of British analytic philosophy. Her characters report their subjective experiences as objective facts, giving their private feelings a cold, clinical quality.[36] According to logical atomism, the "external" world-out-there is constructed from reports by various people of their "internal" (subjective) sense-data. Analogously, Woolf constructed her fictional world from the viewpoints of characters who report their thoughts and emotions as atomic facts. For example, in a story about the Ramsay family sailing to visit a lighthouse near their seaside home, Woolf constructed an objective event—a dinner party—by reporting each guest's inner monologue. This section of the novel begins, "But what have I done with my life? thought Mrs. Ramsay, taking her place at the head of the table . . . ."[37] Observations and musings of the dinner guests include (referring to Mrs. Ramsay): "How old she looks, how worn she looks, Lily thought, and how remote"[38] and (describing the polite chatter of the women at the table) "What damned rot they talk, thought Charles Tansley, laying down his spoon in the exact middle of his plate" (*To the Lighthouse*, 1927).[39]

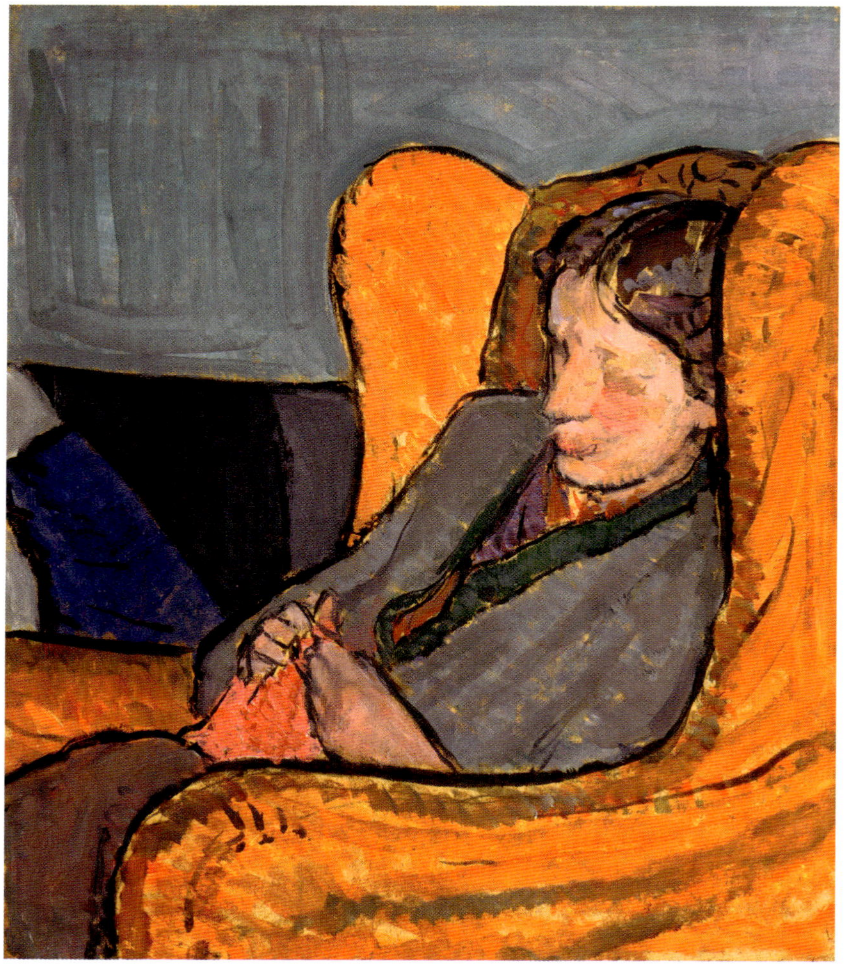

5-13. Vanessa Bell (English, 1879–1961), *Virginia Woolf*, 1912. Oil on board, 15¾ × 13⅜ in. (40 × 34 cm). © National Portrait Gallery, London.

BRITISH SCULPTURE: HENRY MOORE AND BARBARA HEPWORTH

At the opening of the twentieth century, sculptors still drew inspiration from the classical tradition, imitating nature as the Greeks had done. Henry Moore, a young art student at the Leeds School of Art, might have followed that well-worn path but for a chance encounter, in 1921, with Fry's writings emphasizing the emotional elements of design in nature: "Roger Fry's *Vision and Design* was the most lucky discovery for me. I came on it by chance while looking for another book in the Leeds Reference Library. In his essay 'Negro Sculpture,' Fry stressed the 'three-dimensional realization' that characterized African art and its 'truth to material.' Moreover, Fry opened the way to other books and to the realization of the British Museum. That was really the beginning."[40]

Moore developed a vocabulary of organic forms as opposed to pure geometry: "A straight line, a pure curve, solid geometric forms, a perfect sphere, cone, cylinder, cubes,

Abstract qualities of design are essential to the value of a work, but to me of equal importance is the psychological, human element. If both abstract and human elements are welded together in a work, it must have a fuller, deeper meaning.

—Henry Moore, 1934

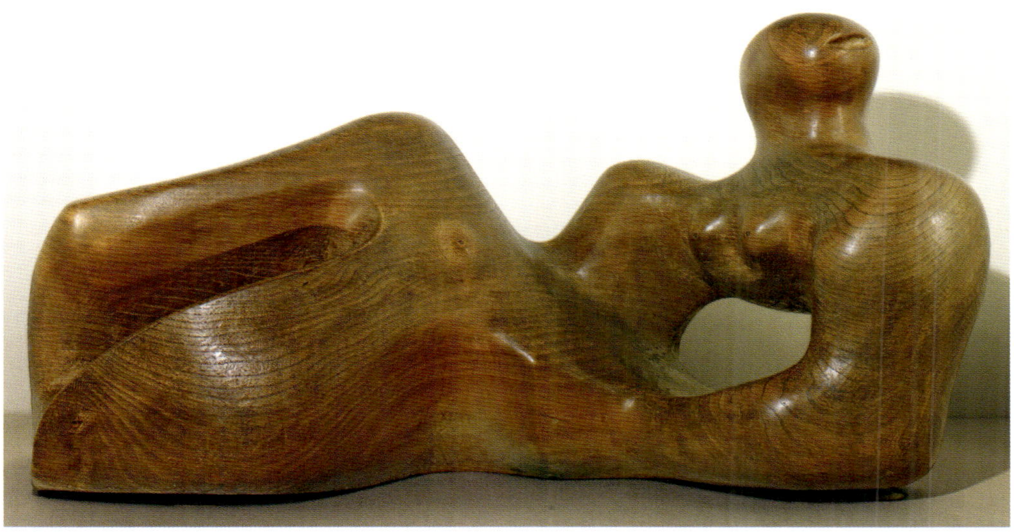

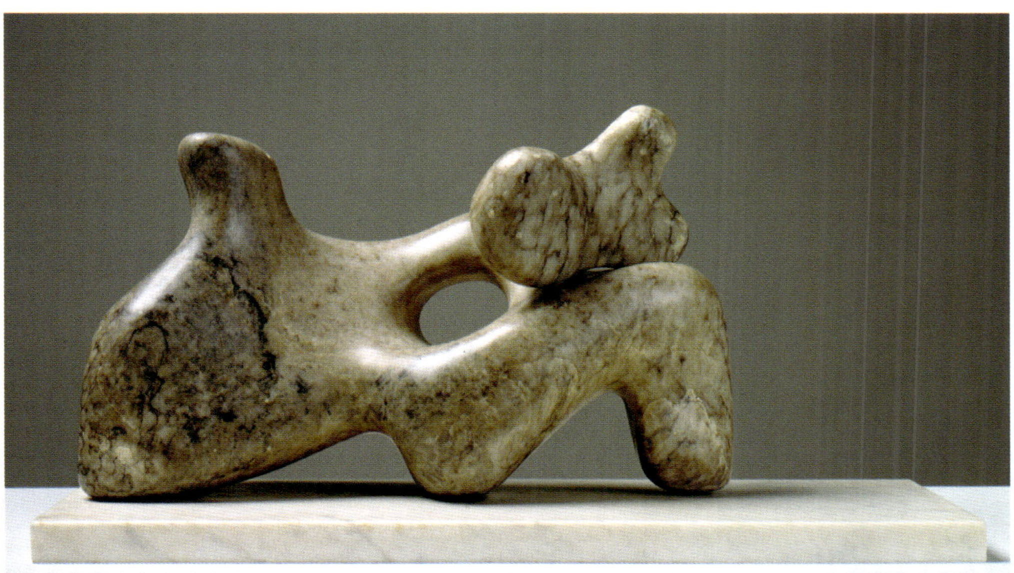

TOP LEFT

5-14. Henry Moore (English, 1898–1986), *Reclining Figure*, 1935-36. Elmwood, 19 × 36¾ × 17½ in. (48.3 × 93.3 × 44.4 cm). Albright-Knox Art Gallery, Buffalo, New York, Room of Contemporary Art Fund, 1939. © The Henry Moore Foundation, 2015.

Moore used local materials such as British elmwood with its distinctive grain, oriented horizontally to emphasize the figure's reclining pose.

BOTTOM LEFT

5-15. Barbara Hepworth (English, 1903–75), *Figure*, 1933. Grey alabaster, slate base, 7⅛ × 5¾ × 21 in. (19 × 14.6 × 6.3 cm). Tate, London.

Hepworth achieved her oft-stated goal of "unity of form and material" by carving these curving forms in harmony with the mottled pattern in the marble.

RIGHT

5-16. Nathaniel Friedman (American, b. 1938), *Trefoil Torso: Canyon Light* (and detail), 1995. Vermont marble, 20 × 14 × 4 in. (50.8 × 35.6 × 10.2 cm). Courtesy of the artist.

Moore and Hepworth inspired Nathanial Friedman to explore the aesthetic qualities of three-dimensional form and void in this sculpture of a marble torso. Friedman, who is both a mathematician and an artist, has described his approach to sculpture: "Like mathematics, sculpture is about form and space. But it is also about light and color, which I learned about from studying the work of Barbara Hepworth and Henry Moore, who inspired me to carve out of stone a space that light can enter. I carved the inside of a trefoil knot at the center of this sculpture, and then I rippled the edges, which reminds me of a canyon. The range of direct and reflected light falling into this hollowed out space depends on the angle of the sun, and the variety is, for all practical purposes, endless."

OPPOSITE TOP

5-17. Barbara Hepworth (English, 1903–75), *Three Forms*, 1935. Marble. © Tate, London.

OPPOSITE BOTTOM

5-18. Ben Nicholson (English, 1894–1982), *1934* (relief), 1934. Oil on wood, 28¼ × 38 × 13 in. (71.8 × 96.5 × 33.2 cm). © Tate, London. © 2014 Angela Verren Taunt/All rights reserved/DACS, London/Artists Rights Society, New York.

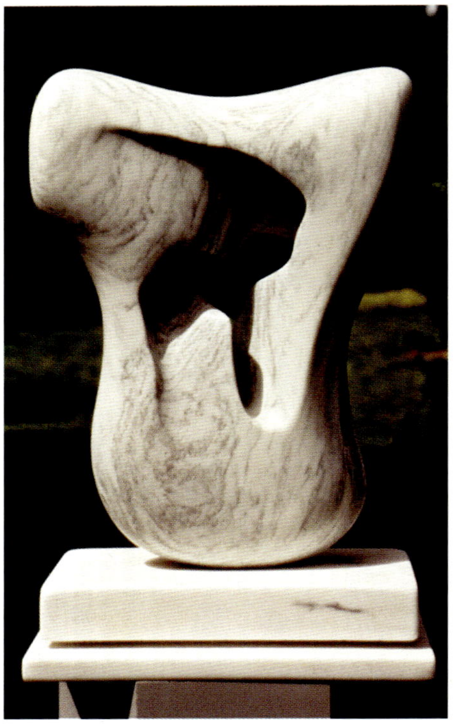

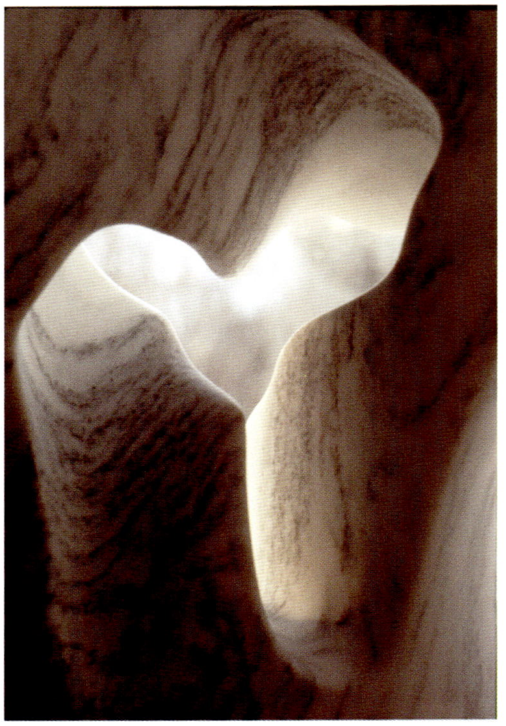

are thought to be beautiful (Plato) but they can be best made by machines, not artists. Art shows the fallible, human, variations from the perfect."[41] By the mid-1930s Moore was conceiving the figure in simple, smooth forms, like driftwood that had been tossed in the sea (*Reclining Figure*, 1935–36; see plate 5-14). Following Fry, the artist declared that abstract organic forms have more "emotional significance" than realistic depictions of their subject: "The removal of the Greek spectacles from the eyes of the modern sculptor . . . has helped him to realize again the intrinsic emotional significance of shapes instead of mainly seeing a representational value."[42] In shunning the classical tradition, Moore looked instead to pre-Columbian and African sculpture, and to modern artists such as Cézanne, because he gave his subject matter a geometric structure.

Moore was the seventh child of a self-educated miner, and he was an open-minded, curious reader his whole life. Barbara Hepworth was the daughter of a well-to-do, highly educated civil engineer who encouraged her to attend the university. She went instead to the Leeds School of Art and harbored an antipathy for the Cambridge-educated Bloomsbury group.[43] Thus Hepworth did not read Fry or Moore but found inspiration in nature and in the work of other sculptors, beginning with Moore. Having met in 1920 as students at Leeds, they continued their studies together at London's Royal College of Art from 1921 to 1924. By 1929 both artists had settled in the Hampstead area of London, regularly visiting one another's studios. In *Figure* (1933), Hepworth carved a reclining mother and child, which, like Moore's figures, were reduced to simple undulating forms—shoulders, heads, legs—which suggest the rolling hills in a landscape (plates 5-15 and 5-16).[44]

In London in the 1930s, Hepworth and her husband, the painter Ben Nicholson, were introduced to the stark geometry of the émigré Constructivists László Moholy-Nagy and Naum Gabo, and plain white squares, circles, and spheres began to appear in their

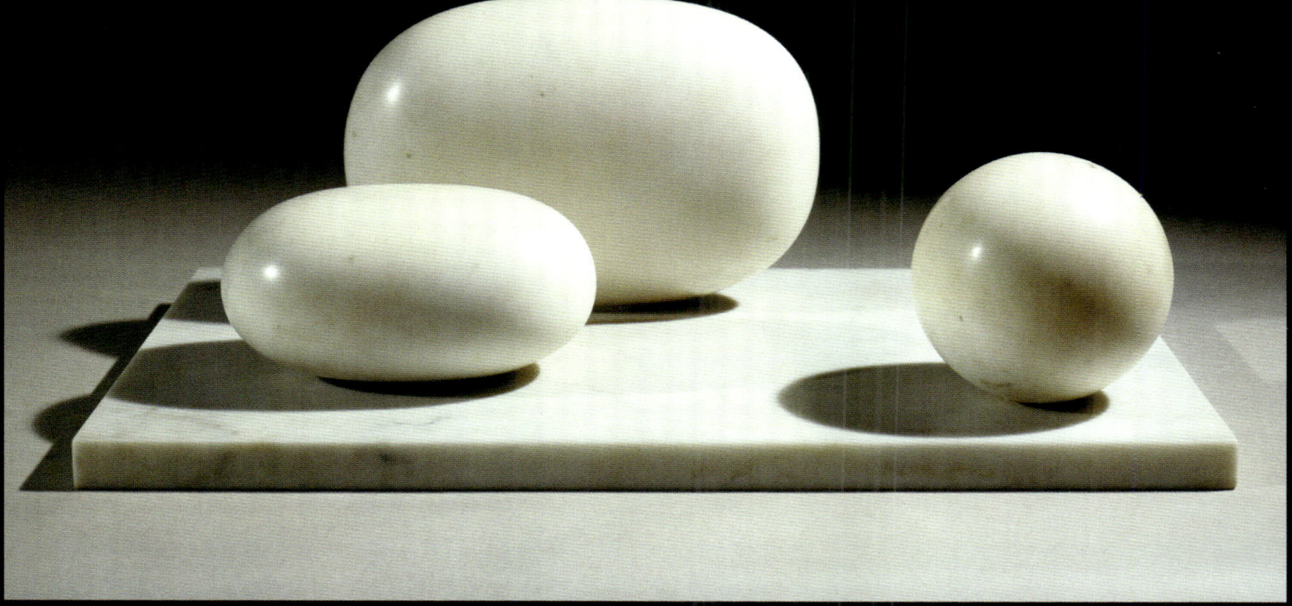

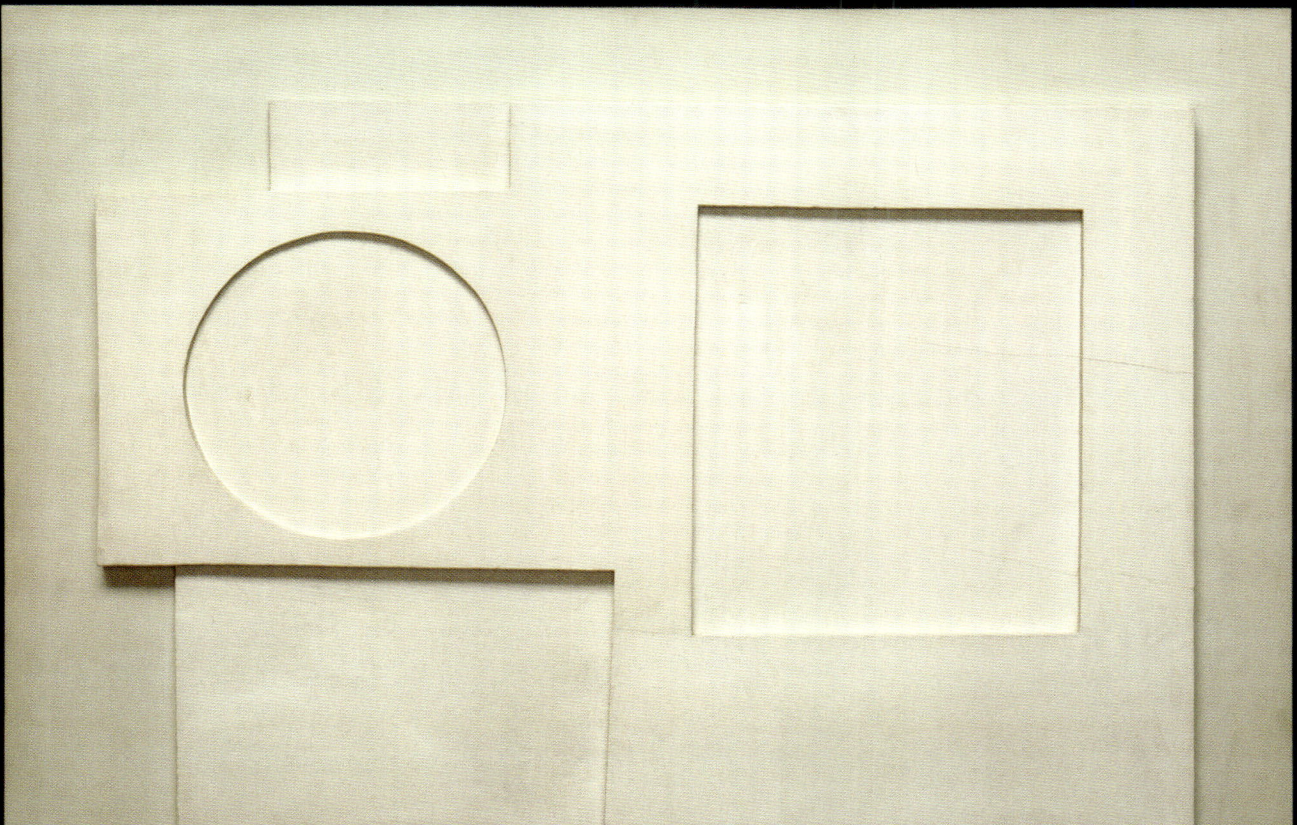

work.[45] In 1933 Hepworth and Nicholson visited the Paris studios of Jean/Hans Arp and Constantin Brancusi, whom Hepworth followed into complete abstraction (*Three Forms*, 1935; see plate 5-17). In the French periodical *Art Concret*, Nicholson read "Towards White Painting" (1930) by the Dutch artist Theo van Doesburg, in which the leader of the De Stijl style of geometric abstraction advocated painting in only white: "WHITE! This is the spiritual color of our times. . . . [which is] the era of perfection, purity, and certitude."[46] Nicholson created his first all-white collage in 1934, and for half a century, he would make paper collages and wood reliefs only in white (plate 5-18).

Among the reviewers of Henry Moore's sculpture in the 1930s was Herbert Read, the most powerful English-language critical voice in the generation after Fry. Read, a curator of ceramics at the Victoria and Albert Museum in London, learned about the expressive use of form when his museum work took him to Germany. There he met the art historian Wilhelm Worringer, whose doctoral dissertation was *Abstraktion und Einfühlung* (Abstraction and empathy, 1907). Worringer's immediate source was psychologist Theodor Lipps's theory of empathy: viewers respond to art out of compassion when they recognize emotions in the art object that they share. Worringer theorized that in times of peace and prosperity artists represent the natural world but in times of fear they withdraw from reality.

The worship of the lowest, most unreasonable levels of the unconscious life which underlies so many of the methods of modern art is absurd. It is only in those higher levels which come into close contact with our conscious intellectual and organized life that the fruitful synthesis of both science and art occur.

—**Roger Fry, review of Herbert Read's *Art Now* (1933), in *Burlington Magazine*, 1934**

Worringer's thesis struck a cord in Read because it helped him understand the surge of abstraction in Britain after World War I.[47] Read was also open to using Freudian psychology, which had gained prestige in Britain during the war as a treatment for shell-shocked soldiers. Read used psychoanalysis to interpret art from the burgeoning German Expressionist and international Surrealist movements. In his influential books *The Meaning of Art* (1931) and *Art Now* (1933), Read praised the form in Moore's sculpture.[48] But Read understood form not as an embodiment of reason (Fry), but as an expression of history (Worringer) and the artist's feelings (Freud). Fry, unmoved by German Expressionism, loathed all art that was not a product of human reason. He dismissed German Expressionism as "a kind of artistic bad manners,"[49] bemoaning the historical and psychoanalytic approaches to form championed by Read, who, ironically, served as editor of Fry's repository of aesthetic etiquette, *The Burlington Magazine*, from 1933 to 1939.

BRITISH LITERARY FORMALISM: T. S. ELIOT AND NEW CRITICISM

The American T. S. Eliot, who became a naturalized British citizen at age thirty-nine, first encountered British analytic philosophy as a graduate student in philosophy at Harvard University, where in 1914 he enrolled in a course on symbolic logic taught by a visiting professor, Bertrand Russell. Lacking a background in mathematics and the exact sciences, Eliot did not excel in the course, but he met Russell again later that year in England, and, inspired by the clarity of Russell's thought, Eliot read further in analytic philosophy.[50] In June 1915, Eliot, now twenty-six years old, married an English woman, Vivienne Haigh-Wood, against the wishes of his parents, who cut him off financially. Russell offered the young couple a room in his London flat and financial support.[51] In these years of Eliot's early poetry ("The Love Song of J. Alfred Prufrock," 1917), he began adopting ideas from analytic philosophy, repeatedly expressing his admiration for Russell's intellect.[52]

As Fry had done for art criticism, Eliot adopted logical atomism to explain how poetic language represents its subject matter. Words name things in the natural world-out-there. Words also name feelings in the inner world, and these emotions were objective facts for Eliot (as they were for Woolf). According to Eliot, a work of art correlates to the emotion it evokes; that is, the poem or play is a physical object (words on a page, players on a stage), the form of which is analogous (isomorphic) to the emotion it conveys. As Eliot declared: "The only way of expressing emotion in the form of art is by finding an 'objective correlative'; in other words, a set of objects, a situation, a chain of events which shall be the formula of that *particular* emotion; such that when the external facts, which must terminate in sensory experience, are given, the emotion is immediately evoked."[53] Like Frege and Russell's meaning of a term like "atom" (which names the same object for all scientists) a piece of literature, in Eliot's view, evokes the *same* emotion in every reader's inner world. This was Eliot's way of introducing the precision of mathematical (artificial) expression into spoken (natural) language. He credited Russell with helping to make English into "a language in which it is possible to think clearly and exactly on any subject. The *Principia Mathematica* is perhaps a greater contribution to our language than they are to mathematics."[54]

Although his friendship with Russell ended in 1918,[55] Eliot continued to find inspiration in analytic philosophy.[56] When Eliot brought out "The Waste Land" (1922), he took the extraordinary step of publishing the poem bristling with footnotes to its extensive allusions and literary references. As these footnotes attest, Eliot intended the poem to have the *same* emotional content for everyone because he viewed his poetry as an objective correlation to the expression of universal emotions in world literature and not as an expression of his own inner world or that of his characters.

In 1927 Eliot converted to the Anglican faith and renounced secularism, including the scientific worldview that was, in his opinion, epitomized by symbolic logic. Analytic themes disappeared from his writing, and he turned his attention to human salvation in poems such as "Ash Wednesday" (1930).

Meanwhile a British contemporary of Eliot, I. A. Richards, had been studying at Cambridge with G. E. Moore, and absorbing logical atomism. Richards would teach English literature at Cambridge from 1922 until 1939; he later taught at Harvard University until his retirement in 1963. He developed a method for analyzing a poem or novel—called "close reading"—which taught students to break down a text into its smallest units (its "atoms") and then to analyze the meaning of the parts. Ever since Aristotle's *Poetics*, philosophers had stressed the formal properties of literature and diction, but only after the emergence of British analytic philosophy did poets and literary critics analyze form to such a minute degree. In his *Meaning of Meaning* (1923), co-authored with the linguist C. K. Ogden, Richards distinguished eight kinds of meaning, including emotive (poetic phrases expressing feelings) and referential (scientific terms denoting objects). Richards showed his training in analytic philosophy by relentlessly drawing distinctions among subtler and subtler shades of meaning.

Anglo-American close reading grew in the 1920s and 1930s, led in the United States by the poet and critic John Crowe Ransom, who from 1937 taught at Kenyon College in Ohio, where he founded the magazine *The Kenyon Review* and published the book that gave close reading its name (*The New Criticism*, 1941). After World War II, New Criticism dominated academic circles throughout the English-speaking world, and close reading was taught in undergraduate literature classrooms in the 1940s and 1950s. The generation that matured in the volatile political climate of the 1960s demanded a criticism that included history and politics, and adherents of New Criticism shifted their focus to a close reading of the "discourse" of race, gender, and power.

JAMES JOYCE

The Frege/Russell version of logic also encouraged the Irish writer James Joyce to treat a novel as an axiomatic structure in the sense that he established a vocabulary and assembled the novel according to rules.[57] In *Ulysses* (1922), Joyce recorded the inner monologues of Dublin residents Leopold Bloom, his wife, Molly Bloom, and his young acquaintance Stephen Dedalus during one ordinary day, June 16, 1904. The reader of *Ulysses* has already met Dedalus in Joyce's autobiographical novel, *A Portrait of the Artist as a Young Man* (1916), in which the character expressed what makes a work of art beautiful:

> "Beauty expressed by the artist . . . awakens, or ought to awaken, or induces, or ought to induce, an aesthetic stasis, an ideal pity or an ideal terror, a stasis called forth, prolonged and at last dissolved by what I call the rhythm of beauty."
>
> "What is that exactly?" asked Lynch.
>
> "Rhythm," said Stephen, "is the first formal aesthetic relation of part to part in any aesthetic whole or of an aesthetic whole to its part or parts or of any part to the aesthetic whole of which it is a part."[58]

In other words, beauty is the harmonic relation of parts within a whole. Joyce described his method for creating *Ulysses* as like making a "mosaic,"[59] and from surviving drafts, notebooks, manuscript pages, and corrected galleys, it is possible to see that he began with a draft of his characters' thoughts, to which he then added chains of associations. When Joyce had written about half of the novel, he devised a schema to impose order on his amorphous manuscript. The schema is a method for organizing the material, but it does not provide a key to the meaning of the novel, which is about thought processes.

Joyce organized the book into eighteen episodes that correspond to events in Homer's *Odyssey* (eighth century BC); to each episode Joyce assigned a location at a specific hour, an organ of the body, an art, and a symbol. Each episode is written in a different literary style, which has a loose metaphorical association to the episode. After devising this schema midstream, Joyce backtracked to earlier chapters to make them conform to this format, resulting in references that often read as tacked-on links to the assigned organ, art, symbol, and technique. But as Joyce continued, he used the scaffold as he wrote, and his "axioms" generated the imaginary world of the book's later episodes. This schema amounts to a set of axioms for the novel in the sense of being a set of consistent rules, unrelated to any known world (neither Bloom's Dublin, nor Odysseus's Aegean, nor the world of the reader). Like Russell and Whitehead's axioms, Joyce's axioms define a self-contained world.[60]

Joyce gave his characters dense subjective experiences, with complex musings, and layers upon layers of associations recorded in documentary detail. But his characters remain oddly opaque—like chatty strangers. For example, writing in the third person in the style of an encyclopedia, Joyce describes myriad particulars of Leopold and Molly Bloom's adulterous marriage, without offering insight into *why* this man and woman betray each other.

Most of Joyce's references to mathematics occur in the seventeenth episode, "Ithaca," which corresponds to Odysseus's homecoming in the *Odyssey*. Bloom and Dedalus arrive at Bloom's home after midnight, in a scene associated with a skeleton (organ), science (art), and comets (symbol), and told in the style of a catechism (literary technique). In writing "Ithaca," Joyce culled data from many books on science and mathematics, taking extensive notes on Russell's *Introduction to Mathematical Philosophy* (1919).[61] In keeping with the "catechism" technique, Joyce described the actions of Bloom and Stephen in the prescriptive question-and-answer format of an elementary Catholic lesson book. For example, after asking, "How did they take leave, one of the other, in separation?" (in other words, how did Stephen and Bloom go their separate ways at the end of their evening together), Joyce described their parting with mathematical precision (standing with their shoulders parallel to each other, they embraced and their arms converged): "Standing perpendicular at the same door and on different sides of its base, the lines of their valedictory arms, meeting at any point and forming any angle less than the sum of two right angles."[62]

I am writing "Ithaca" in the form of a mathematical catechism.

—James Joyce, letter to Frank Budgen, February 1921

Thus Joyce added texture to his "Ithaca" episode with concepts and terms borrowed from mathematics without considering the meaning of these terms in any depth. Authors through the ages have reported interior monologues, enriched their descriptions with detail, and imposed structure on their plots. Joyce is the novelist most representative of the age of free association in psychology, the dawning of the information age, and the century of axiomatic mathematics because he combined in *Ulysses* the denseness of his characters' unedited thoughts, the minuteness of archival detail about their surroundings, and the rigidity of the schema with which he generated the imaginary world of Leopold Bloom, Molly Bloom, and Stephen Dedalus.

Frege and Russell's ideal mathematical objects inspired sculptors Moore and Hepworth to disregard the transitory appearance of individuals and capture the underlying form of the human figure. British poets and novelists similarly echoed a mathematics that was based on formal specifications of how numbers and sets behave, as contrasted with a mathematics based on what numbers and sets are. Inspired by this syntactic approach, the writers Woolf, Eliot, and Joyce describe how their characters act and what they say, but not who they *are* or how they *feel*.

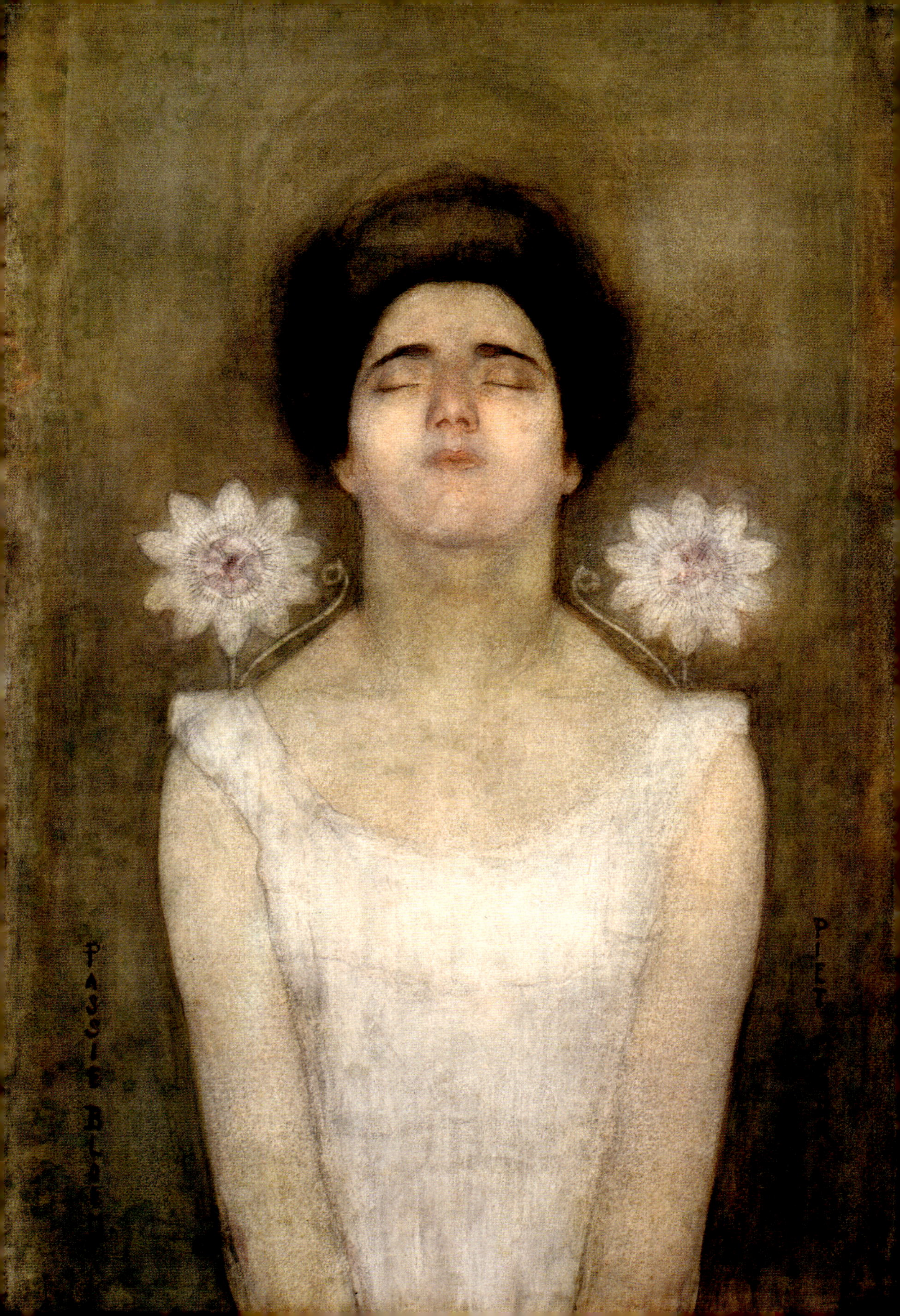
PASSIE
PIET

6
Intuitionism

Only he who recognizes that he has nothing, that he cannot posses anything, that absolute certainty is unattainable, who completely resigns himself and sacrifices all, who gives everything, who does not know anything, who does not want anything and does not want to know anything, who abandons and neglects everything, he will receive all.

—L.E.J. Brouwer, ***Life, Art, and Mysticism***, 1905

At the opening of the twentieth century, Dutch mathematicians and artists became swept up in a resurgence of German Romanticism. Responding to the deterministic materialism that they associated with science and technology, intellectuals, physicians, painters, and poets left Amsterdam to settle in countryside communes where they could reconnect with nature, follow their instincts, and—wary of logic and reason—seek knowledge by intuition. In one thatched hut lived Holland's leading twentieth-century mathematician, L.E.J. Brouwer, who spent long periods in silent meditation thinking about numbers, a practice that became central to *intuitionism*. Brouwer understood numbers, circles, and triangles as creations of the human mind, and he became an outspoken critic of modern Platonism. Today Brouwer's approach is also called "constructivism" because those who adopt it construct mathematical objects with numbers. In addition to engaging in the philosophy of mathematics, Brouwer practiced mathematics by developing topology, which is concerned with the properties of a geometric shape that is continuously deformed.

Dutch artists in Brouwer's extended circle shared his trust in intuition and meditation (plate 6-1); the conduit between Brouwer's thinking and that of the artists was an erstwhile cleric and amateur philosopher of mathematics, M.H.J. Schoenmaekers, who knew Brouwer and frequented artistic circles. Inspired by Schoenmaekers, Piet Mondrian and Theo van Doesburg used geometric forms and primary colors to symbolize mathematical objects that are embodied in the natural world.

6-1. Piet Mondrian (Dutch, 1872–1944), *Passion Flower*, ca. 1901. Watercolor on paper, 28½ × 18⅞ in. (72.5 × 47.5 cm). Gemeentemuseum, The Hague.

Throughout his long career Mondrian was on a spiritual journey. His early paintings were an expression of the religion of his parents, who were followers of John Calvin, a sixteenth-century leader of the Protestant Reformation who preached an ascetic lifestyle. When Mondrian was twenty-nine, he painted this woman meditating between two passion flowers, which are named for the suffering ("passion") of Jesus before his crucifixion; the flower's three stamens represent the nails in Jesus's hands and feet, and its tendrils represent his crown of thorns. In his thirties Mondrian exchanged Calvinism for theosophy, and by age forty he had replaced theosophy with a belief in a spiritual force in nature that he symbolized using a vocabulary of geometric shapes and primary colors.

THE DUTCH WALDEN

The psychiatrist and writer Frederik van Eeden was inspired to found a commune by reading the American political philosopher Henry David Thoreau's account of his solitary life on the shores of Walden Pond near Concord, Massachusetts (*Walden, or Life in the*

Woods, 1854). Thoreau had put into practice an American version of German Idealism, New England Transcendentalism, which was described by the essayist and poet Ralph Waldo Emerson: one should live a simple life close to nature and attain knowledge of the highest spiritual realm (Absolute Spirit) through intuition. Emerson was ordained into the clergy of Unitarianism, a Protestant sect that is pantheist in the sense that it asserts that the deity is a unity (as opposed to a trinity), and that nature is divine. Emerson adopted the term "transcendentalist" from Kant, according to whom space and time are transcendental (a priori) preconditions for experiencing the world, but Emerson gave the term the vaguer meaning of anything that is *intuitive*: "Whatever belongs to the class of intuitive thought, is popularly called at the present day Transcendental."[1] Van Eeden opened his commune in 1898 and named it "Walden" in honor of Thoreau.

Residents of the Dutch Walden rejected their legacy of strict Calvinism and turned to spiritual outlooks with strong mystical traditions such as early Christianity and Buddhism, aspiring to unite all world religions into one global philosophy. They adopted a vegetarian diet from India, practiced meditation, and read Plato's dialogues, the Bible, and Buddhist/Taoist doctrine. Before establishing the commune, van Eeden had practiced psychiatry, and as leader of this vibrant community of artists and intellectuals, he focused debate on topics relevant to the human mind: the causes of mental health and illness, the sources of artistic creativity, how to attain knowledge by meditation, and, after Brouwer joined the group, the power of the mind to create mathematical objects.

A near contemporary of Sigmund Freud, van Eeden had been a medical student in the early 1880s when neurologists were divided over whether hysteria (a nervous disorder) had physical or psychological causes. Van Eeden travelled to Paris to attend lectures on hysteria given at the Salpêtrière Hospital by the leading neurologist of the day, Jean-Martin Charcot, who held that hysteria had physical causes (although they were unknown to him). Charcot declared that he could, however, treat its symptoms (such as convulsions) with psychological methods, including the power of suggestion and hypnosis. After graduating from medical school, van Eeden, together with the young Dutch neurologist Albert Willem van Renterghem, returned to Paris to study with Charcot; they also travelled to Nancy to hear lectures by Charcot's rivals, Ambroise-Auguste Liébeault and Hippolyte Bernheim. The Nancy school held that the causes of hysteria were *purely* psychological, and they also used suggestion and hypnosis in its treatment.

Having studied both sides of the issue, Van Eeden and van Renterghem joined the camp of the Nancy school. They returned to Amsterdam in 1887 and established the first clinic in Europe that offered treatment for psychological disorders,[2] the Instituut Liébeault, and the Nancy school flourished in Holland.[3] In 1889 van Eeden and van Renterghem travelled to Paris to the Exposition Universelle, which had the new Eiffel Tower as its centerpiece, and attended an international congress on hypnotism, with Charcot named honorary chairman. Van Eeden and van Renterghem gave presentations on their research at the Instituut Liébeault, and they were joined at the podium by the Italian criminologist Cesare Lombroso, the philosopher and psychologist William James, and the father of psychoanalysis, Sigmund Freud.[4]

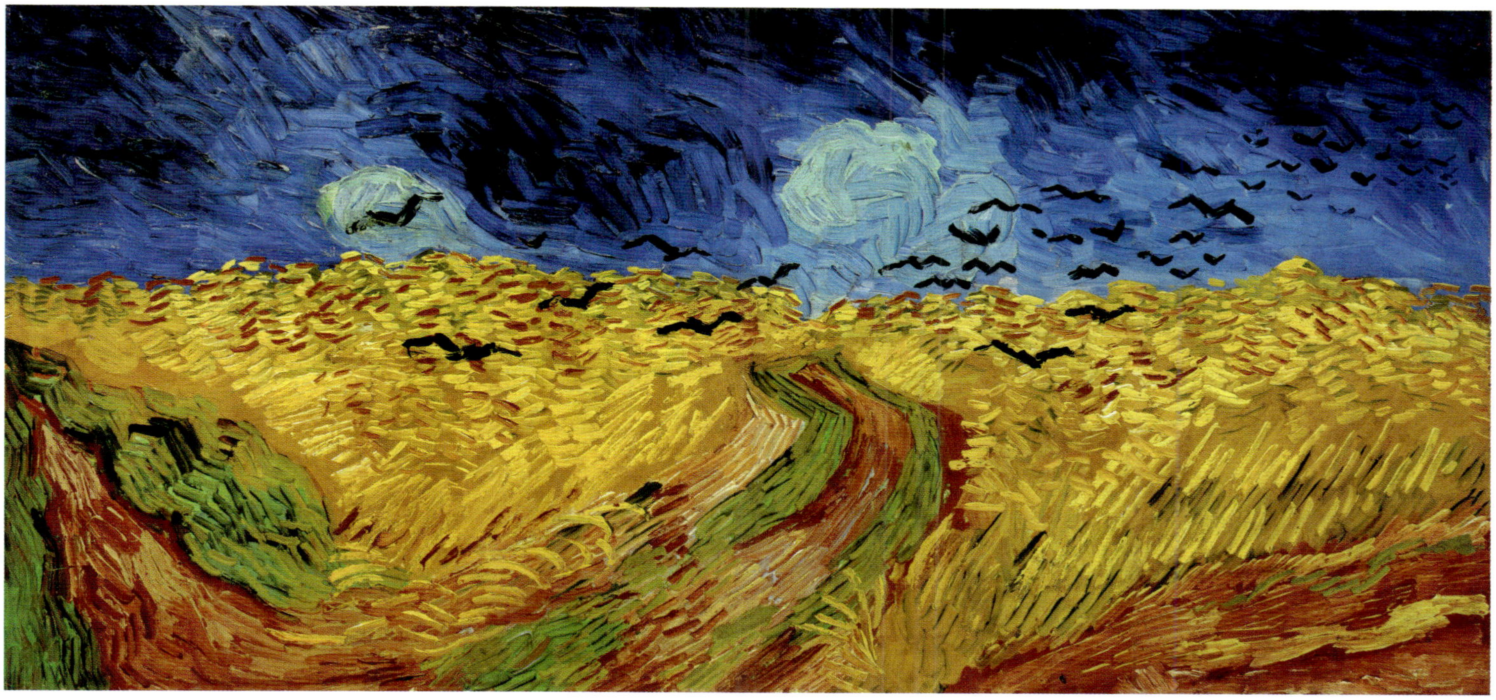

6-2. Vincent van Gogh (Dutch, 1853–1890), *Wheatfield with Crows*, 1890. Oil on canvas, 19.9 × 40.6 in. (50.5 × 103 cm). Van Gogh Museum, Amsterdam (Vincent van Gogh Foundation), in. no. s149V/1962 F779.

Van Gogh painted this landscape in mid-July 1890, during the last weeks of his life.

After the publication of Darwin's *Origin of Species* (1859), neurologists had begun to investigate whether mental illness resulted from a hereditary physical deterioration of the brain. This concept of mental degeneracy was naturally of interest to physicians such as Lombroso who studied criminal behavior. Lombroso restated the ancient association of creativity and madness in evolutionary terms by declaring that all creative persons suffer from a degenerative psychosis (*L'uomo di genio* [Man of genius], 1888). This enormously popular book, which went through dozens of editions between 1888 and 1900 in German, French, and English translations, led certain fin-de-siècle artists to aspire to be fashionably crazy and art critics to gleefully dismiss abstract art as a symptom of psychosis.[5] But van Eeden adopted a more sober tone in an essay he wrote about Vincent van Gogh four months after the Dutch artist's death by suicide on July 29, 1890, at age thirty-seven (plate 6-2). Reflecting the views of Lombroso, van Eeden described how van Gogh's mental illness was manifest in his art as passionate zeal (*gepassioneerden ijver*), for which van Eeden expressed deep admiration.[6]

After treating patients for seven years at his Amsterdam clinic, van Eeden resigned in 1893 to devote himself to his commune and to his literary career, but he would maintain a lifelong interest in psychology, especially in dreams.[7] Two years after opening the Dutch Walden, he attended an international congress on psychology in Paris, where he heard a lecture by an Indian Buddhist, Jagadisha Chattopadhyaya, who strengthened van Eeden's determination to bridge the gap between Western and Eastern thought: "Only by combining the two civilizations can mankind reach that true culture that has been lacking until now."[8]

Intellectuals who took part in discussions at the Dutch Walden included the mathematician Gerrit Mannoury and his student L.E.J. Brouwer. Trained in second-generation German Idealism of the *Naturphilosophen*, Mannoury professed a mystical pantheist vision of an organic cosmos culminating in Hegel's Absolute Spirit. Mannoury was also drawn to Eastern thought, writing a book comparing Absolute Spirit with the cosmic spirit of the Buddhist Brahma (*Het Boeddhisme: Overzicht van leer en geschiedenis* [Buddhism: An overview of its teachings and history], 1907). To understand the impact of Eastern thought on Dutch culture in the early twentieth century, it is helpful to keep in mind that in around the eleventh century AD, the Chinese had merged their native concept of Tao (the way of nature) with the Indian Brahma, resulting in (Taoist) communication with nature as a key to (Buddhist) enlightenment. In other words, the Buddhism that was read by nineteenth- and twentieth-century Westerners (such as Mannoury) had strong Taoist (naturalist) elements.

Mannoury taught his outlook to Brouwer along with generations of students at the University of Amsterdam, where he was professor of mathematics from 1902 to 1937. He and Brouwer used the term "mind" in both Idealist and Buddhist/Taoist senses. Referring to Kant's a priori intuition of the mind's spatial and temporal scaffolding, they held that numbers are known—immediately and with certitude—by reflection on one's own mind. But according to Mannoury and Brouwer, any encounter with divine Mind (the Idealist Absolute Spirit or the Buddhist Brahma) cannot be communicated in words or symbols—such thoughts are ineffable. Mannoury and Brouwer extended this approach to their (ultimately inexpressible) knowledge of mathematical objects such as numbers and triangles. How can a university professor teach students mathematics if he cannot use language? What does he do in the classroom? Mannoury spoke in the style of an exhortation to his disciples about how to achieve the same state of mind in order to have the same insight (as opposed to writing a formal proof on the blackboard).[9] Mannoury and Brouwer held, moreover, that it was impossible to communicate to another person any profound insight about mathematics, as Brouwer wrote in 1905: "Never has anyone been able to communicate his soul by means of language."[10]

Little confusion is possible about the meaning of words such as "equal" and "triangle," but even then two different people will never think of them in exactly the same way.

—**L.E.J. Brouwer, *Life, Art, and Mysticism*, 1905**

Van Eeden gave the word "mind" yet another meaning—the unconscious mind—which, as a psychiatrist, he viewed as the ultimate source of true knowledge. Like Mannoury and Brouwer's distrust of natural and artificial languages, van Eeden took an equally dim view of the ability of words to describe man's inner world or the workings of the unconscious mind. In 1897 van Eeden summed up his views of the limits of language in a treatise written as a series of theorems in the style of Euclid's *Elements*.[11]

BROUWER'S PRIMORDIAL INTUITION OF TIME

Brouwer read van Eeden's utopian writings with great interest,[12] and while he was still a graduate student he bought a strip of land near Amsterdam, where he built a cottage and commenced an ascetic life. He exercised vigorously, ate a vegetarian diet, and dedicated a period each day to silent meditation. In the coming decades, after academic success brought Brouwer wealth and invitations to join the mathematics faculties at prestigious universities in Göttingen and Berlin, he neither left his hut nor changed his rigorous routine.

Convinced that knowledge of mathematics is inherently incomplete and ineffable, Mannoury taught his students to remain ever skeptical (*Methodological and Philosophical Comments on Elementary Mathematics*, 1909). Brouwer absorbed his teacher's Romantic anti-intellectualism, interest in Eastern thought, and view that mathematical objects are mind-dependent. While still a student, Brouwer published the philosophy of mathematics to which he adhered for the rest of his long life (*Life, Art, and Mysticism*, 1905). The central theme of this youthful treatise, which was inspired by the medieval Christian mystics Jakob Böhme and Meister Eckhart, as well as by Buddhism, was that one attains knowledge by intuition.[13]

Brouwer described how counting is based on a person's experience of events over time: "By a move of time a present gives way to another present sensation in such a way that consciousness retains the former one as a past sensation, and moreover, through this distinction between present and past, recedes from both and from stillness, and becomes *mind*."[14] When lost in pure thought in a "state of stillness," one has a "primordial intuition of time" without the aid of words or symbols. Then the mind begins "its transition from its deepest home to the exterior world in which we cooperate and seek mutual understanding."[15] Brouwer's method of doing mathematics was to return to that "state of stillness," where he explored, in silent meditation, the order and pattern of numbers. He used no words or symbols because he held that his core insights were impossible to capture in any kind of notation.

Proper to man is a faculty which accompanies all his interactions with nature, namely the faculty of taking a mathematical view of his life. . . . The basic phenomenon therein is the simple intuition of time, in which repetition is possible in the form: "thing in time and again thing," as a consequence of which moments of life break up into sequences of things which differ qualitatively. These sequences thereupon concentrate in the intellect in mathematical sequences.

—L.E.J. Brouwer, ***Foundations of Mathematics***, 1907

TOPOLOGY

Between 1909 and 1913 Brouwer turned his attention from philosophical speculation to the practice of mathematics and became a founder of topology, which is a generalization of the linear perspective that Brunelleschi and Alberti formulated in the early fifteenth century to project geometric forms onto a plane and to calculate the distortions of the forms relative to a viewpoint (see the sidebar on pages 78–79 in chapter 2). Brouwer began with the work of the French mathematician and engineer Jean-Victor Poncelet, who was a squadron commander during Napoleon's 1812 invasion of Russia. During the French army's disastrous retreat from Moscow (plate 6-3), Poncelet was left for dead and Russian soldiers captured him. While recovering in a Russian prison camp, Poncelet occupied his mind by generalizing linear perspective into projective geometry. Poncelet began by contemplating all kinds of projections—including shadows cast on a wall and architecture drawn in linear perspective (plate 6-4). He asked what each object has in common with its projection (conceived of as its "shadow"). He considered not only projections onto a Renaissance picture plane (which is placed upright, perpendicular to the plane of the horizon), but also onto planes that are tipped or rotated (plate 6-5). Poncelet noticed that a straight line always casts a shadow that is a straight line, although its length is not preserved. The shadow of an elliptical object might be circular or elongated, but the shadow remains an ellipse; a tangent to a curve casts a shadow that is tangent to the shadow of the curve. In general, Poncelet studied geometric features that remain invariant under transformations (*Traité des propriétés projectives des figures* [Treatise on the projective properties of figures], 1822).

Brouwer generalized Poncelet's projective geometry to projections onto surfaces that are stretched or distorted into any shape—so-called rubber-sheet geometry—provided that

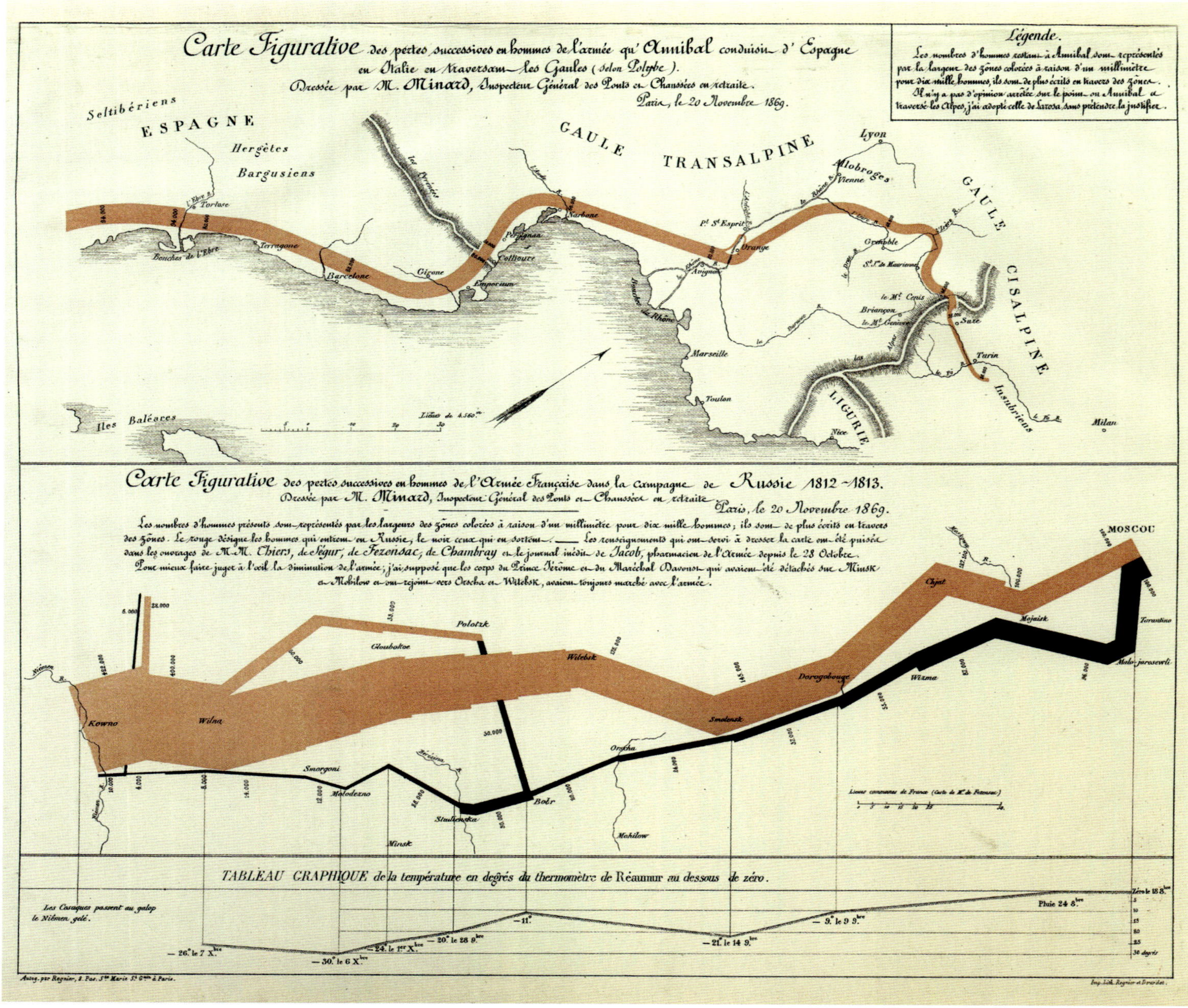

6-3. A pair of flow maps, ca. 19¾ × 23⅜ in. (50× 60 cm), by Charles Joseph Minard in his *Tableaux graphiques et cartes figuratives, 1844–1870* (Paris: Régnier et Dourdet, 1870), 28. École Nationale des Ponts et Chaussées, Paris, fol. 10975.

In an early example of graphic visualization of numerical data, the French engineer Charles Minard made this pair of flow maps showing the diminishing number of men in two of the deadliest military campaigns in human history: the Second Punic War (218–201 BC), in which Hannibal from Carthage, a city of "Punics" (from the Latin name for settlers from Phoenicia), tried to defeat Rome (above), and Napoleon's Russian campaign (1812–13), in which the French general tried to seize Moscow (below). According to Minard's legend (upper right), 1 mm in the width of the brown and black lines represents 1000 soldiers. Minard showed Hannibal's army going from 94,000 to 26,000 men on its march from Spain to Italy, the most precipitous drops occurring during his crossing of the Pyrenees and the Gallic (French) Alps. Most of Hannibal's 37 African elephants (not represented on the graph) also died traversing this rocky terrain at high altitudes and freezing temperatures. Minard showed the French Grande Armée going from 422,000 when it crossed the Neman River into Russia to 10,000 men when it returned; below, he charted the decreasing temperature during Napoleon's winter retreat. Today historians put Hannibal's losses much higher and Napoleon's slightly lower.

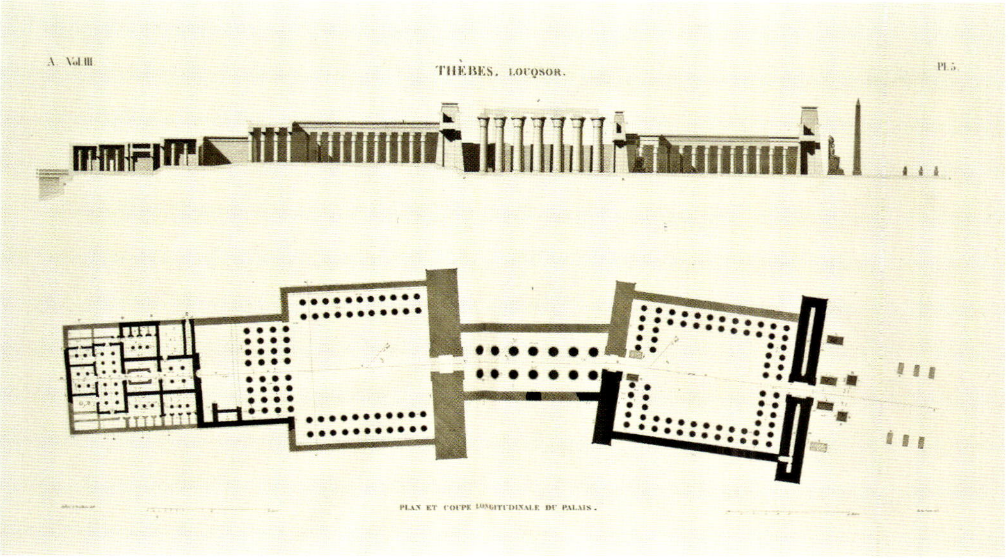

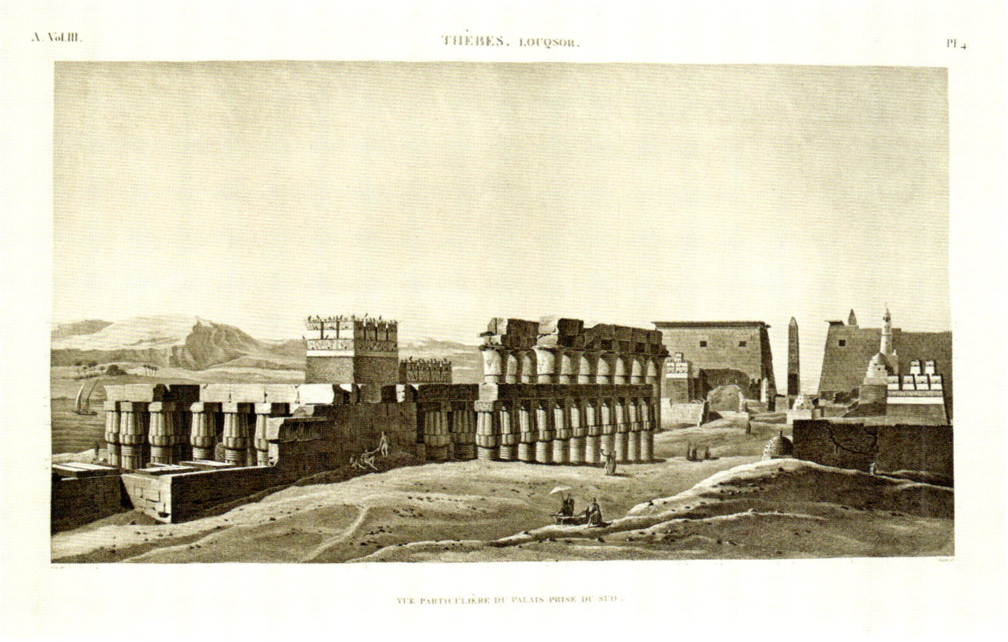

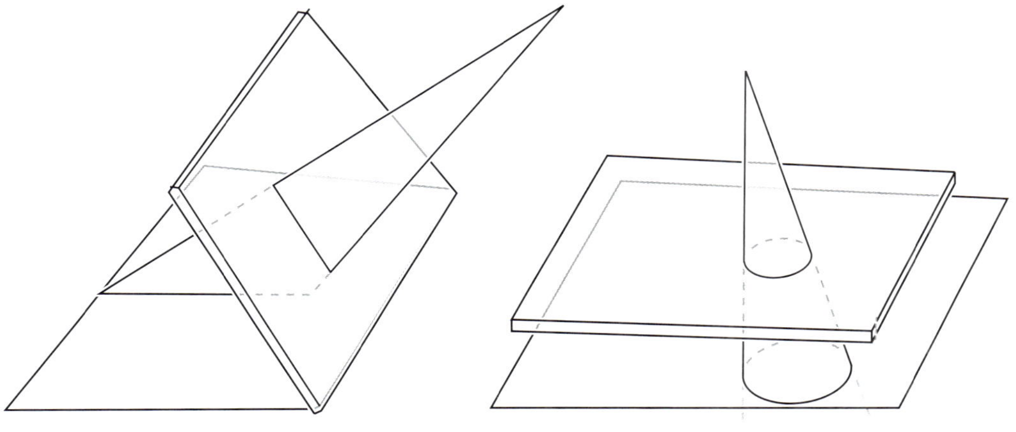

TOP AND CENTER LEFT

6-4. Temple of Luxor at Thebes, in *Description de l'Égypte, ou recueil des observations et des recherches qui ont été faites en Égypte pendant l'expédition de l'armée française* (Paris: Panckoucke, 1821–29), sec. A, vol. 3, plate 4-5. Special Collection, Binghamton University Library, State University of New York.

After Napoleon's army conquered a country, military artists and engineers were brought in to document seized art and architecture. Using linear perspective, Napoleon's artists drew this cross-section, plan, and three-dimensional projection of the Temple of Luxor at Thebes, which was occupied during the French campaign in Egypt (1798–1801). Jean-Victor Poncelet had the bad luck to be assigned to Napoleon's campaign in Russia.

BOTTOM LEFT

6-5. Projective geometry.

A line segment projects to a line segment; an ellipse projects to an ellipse.

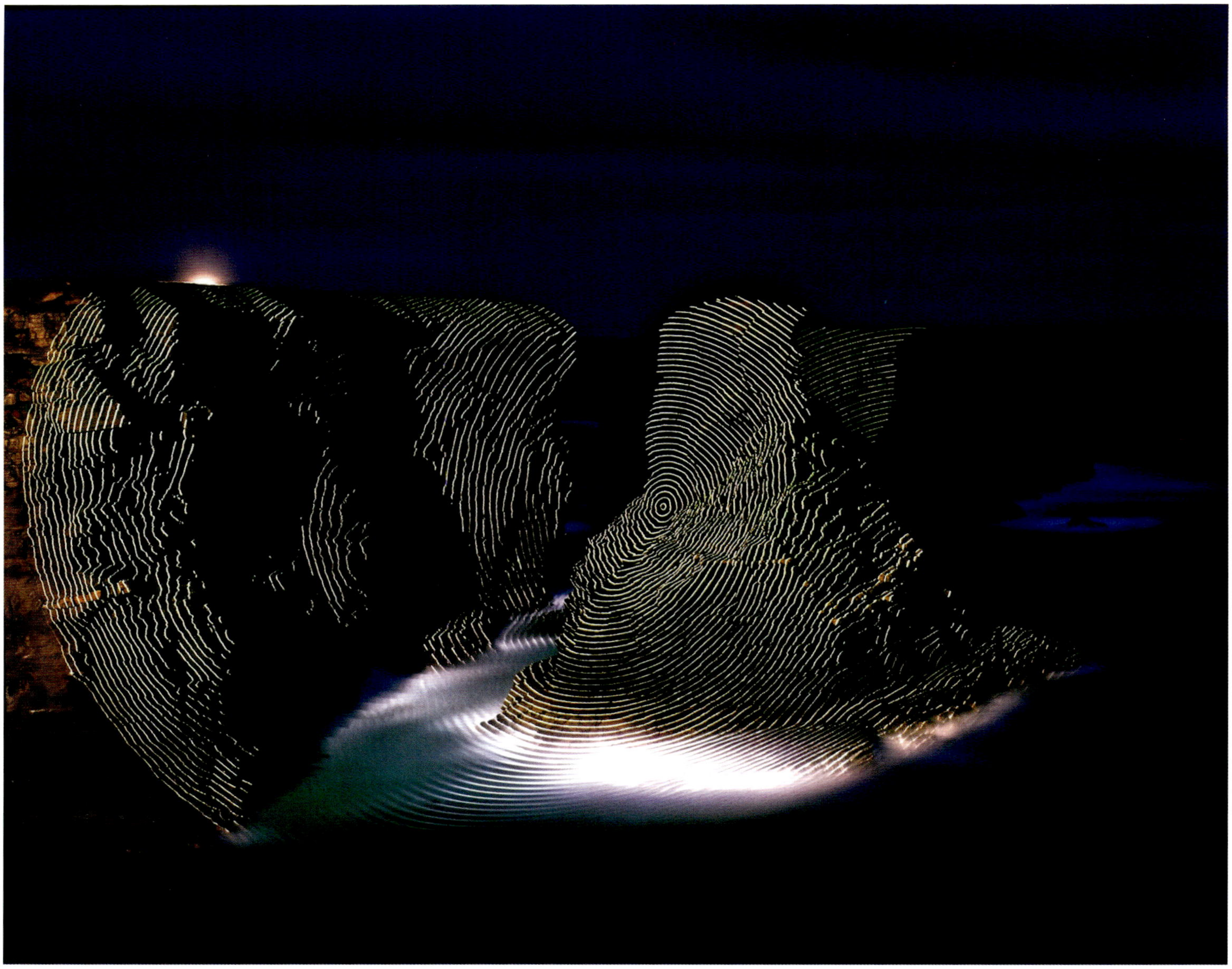

ABOVE

6-6. Jim Sanborn (American, b. 1945), *Kilkee County Clare, Ireland*, 1997. Large-format projection, digital print, 30 × 36 in. (76.2 × 91.4 cm). Courtesy of the artist.

The contemporary artist Jim Sanborn created this artwork by projecting a pattern of concentric circles onto a large rock formation at night from about ½ mile away. He then took this photograph with a long exposure at moonrise.

BELOW

6-7. Topology.

Brouwer's topology describes what properties of a surface remain unchanged when a surface is distorted in the sense of being stretched and pulled (like a piece of rubber) but not cut or torn. In an admissible distortion, as in the model in this diagram, a point in the original plane must transform into one, and only one, point after the plane has been stretched, and neighboring points must transform into neighboring points. Both the transformation and its inverse (being stretched in the opposite direction) must have these properties. William James anticipated Brouwer's "rubber-sheet geometry" in his 1890 model of the stream of consciousness; see plate 8-6 in chapter 8.

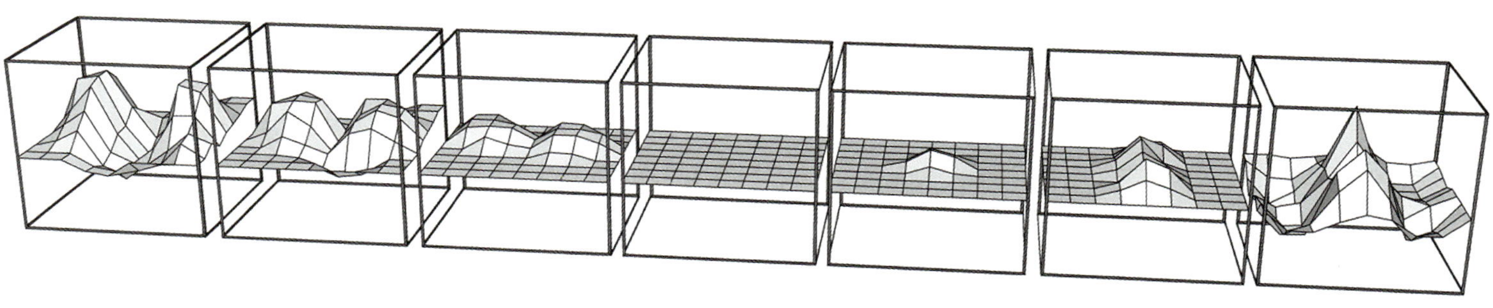

the plane remains continuous (with no holes or tears; see plates 6-6 and 6-7). Brouwer also generalized Poncelet's three-dimensional geometry to any number of dimensions. Thus Brouwer created a geometry that describes figures in all imaginable continuous spaces in worlds with all possible dimensions: the new field of topology. The interchanges between mathematics and the arts are generally one-sided in the sense that they are almost always a case of an artist finding inspiration in mathematics, and not vice versa. This is a rare exception in which the architect Brunelleschi invented linear perspective, after which the mathematicians Poncelet and Brouwer generalized perspective into projective geometry and topology.

After finishing his practical work on topology, Brouwer returned to philosophical topics in the foundations of mathematics and developed his intuitionist approach. According to Brouwer, the counting numbers (1, 2, 3, . . .) are the only mathematical objects that one knows intuitively with immediacy and certainty. Numbers are the building-blocks from which the rest of mathematics can be constructed, in the sense that one can show how to construct any (legitimate) mathematical object using numbers. For example, one can generate the first ten numbers by counting to ten and then construct the whole numbers by giving a finite procedure (add one to the last number) to generate the endless series.

Brouwer threw out many areas of mathematics because they were not built from numbers following some finite procedure. Brouwer disavowed "P or $\neg P$" (P or not-P) as an axiom, asserting that it assumes all mathematical statements can be either proven or disproven, which in his view is not allowed unless you actually have a proof of P or a proof of $\neg P$. Suppose, for example, that P asserts: the decimal expansion of π contains an infinite number of sevens. Most mathematicians would agree that P is true. But according to Brouwer, the only way to prove P or $\neg P$ is to either prove that there are an infinite number of sevens in the decimal expansion of π or to prove that there are only a finite number of sevens in the decimal expansion of π. And any such proof would have to meet strict constructivist standards.

SEMIOTICS AND SIGNIFICS

When World War I broke out in 1914, van Eeden corresponded with Freud about the causes of armed conflict, and Freud went on to write an essay on the topic ("Thoughts for the Times on War and Death," 1915).[16] Van Eeden became convinced that one cause of the hostilities was a breakdown of communication between the warring nations of Europe, so he organized the intellectuals in his circle around the task of creating an international spoken language to promote world peace. Similar motivations had prompted the Polish scholar L. L. Zamenhof to construct "Esperanto" from root words in the Romance languages (*Fundamento de Esperanto*, 1905) and the Italian mathematician Giuseppe Peano to create "Interlingua" with a simple grammar and vocabulary drawn from Latin, French, German, and English (*Vocabulario de Interlingua*, 1915). Van Eeden's group based its approach on the American philosopher and mathematician Charles Sanders Peirce, who in the 1860s had declared that all knowledge of the world-out-there comes from patterns of signs such as words, pictures, and symbols. In what he named "semiotics" (from the Greek for "significant"), Peirce analyzed each sign into a triad: the sign itself (the "signifier"), what it denotes (the "signified"), and the connection between the two (the "interpretation"; "Some Consequences of Four Incapacities," 1868). By analyzing the meaning of terms, Peirce hoped to make language more precise. Van Eeden knew Peirce's semiotics second hand from Peirce's follower, the British writer Victoria Welby, who wrote about what she termed "significs" (the "science of meaning").[17] Welby urged the study of the English language as an ever-changing system in order to avoid ambiguity and misunderstanding (*What is Meaning?* 1903).[18]

Van Eeden's circle included M.H.J. Schoenmaekers, a student of theology who was ordained into the Catholic clergy in 1900. But within two years Schoenmaekers had dissolved his vows and was searching for a new spiritual outlook. He was drawn to the Dutch Walden of van Eeden, who invited Schoenmaekers to help create the universal language that he called "Significs" in honor of Welby.[19]

In 1915 van Eeden met Brouwer, who by then was the world-famous founder of topology, and van Eeden recruited him to help design Significs.[20] Although Brouwer had long maintained that communication between humans at anything but a superficial level is impossible, he resigned himself to language and joined the effort, leaving the extreme isolation of his thatched hut to be at the center of van Eeden's lively entourage of artists and intellectuals. Together the group, which also included Mannoury, read Brouwer's *On the Foundation of Mathematics* (1907), which they hailed as the first attempt to put the Significs program into practice because it sought the most elementary components of thought (numbers). In 1915 Brouwer invented a new verbal code—so-called Intuitive Significs—in which there are basic words, as numbers are basic in mathematics, to which Brouwer gave a political interpretation: "No one can be good or happy in the activity of this unholy society. Those in search of happiness outside this activity do not find any stimulus to active thought in the words of present languages, which in the last analysis are no more than signs of command in society's regulation of labor. . . . Intuitive Significs concerns itself with the creation of new words forming a code of basic means of mutual understanding for the systematic activities of a new and holier society."[21]

During World War I van Eeden also began plans for an (unrealized) Academy of Higher Wisdom, which would teach students to create a harmonious society: "A university, truly independent, not influenced by some sect or party, but universal, seeking unity in all religions and adding to natural science, deepening it with all those functions of the human mind that hitherto were not methodically studied in Western universities. That includes mysticism, occultism, and the philosophy of religion and arts."[22] Van Eeden wanted Schoenmaekers to join the faculty: "He is the right teacher for our Academy of Higher Wisdom."[23] But van Eeden's plans were short-lived due to a breakdown in negotiations with city officials at Amersfoort, where he hoped to locate the school.

Like van Eeden, artists also established utopian communities. The Dutch Symbolist Jan Toorop led an artists' colony near rural Domburg, which van Eeden frequented, becoming good friends with the artist. In 1891 van Eeden bought a drawing from Toorop, whose work he deemed "definitely good" (*zeker goed*).[24]

As a young man Piet Mondrian did pictures that reflected his Calvinist upbringing, such as *Passion Flower* (1901; see plate 6-1, chapter frontispiece), which presents a woman with her eyes closed as if in prayer or meditation. In 1909 Mondrian, together with the painters Cornelis Spoor and Jan Sluyters, had an exhibition at the Stedelijk Museum in Amsterdam. Mondrian showed mainly landscapes, probably also *Passion Flower*,[25] and certainly a similar painting that depicts a girl in thoughtful contemplation (*Devotion*, 1908; Gemeentemuseum, The Hague).[26] In a review of this exhibit in the popular morning newspaper *De Telegraaf* on January 8, 1909, the critic C. L. Dake reflected Lombroso's association of creativity with madness when he described Mondrian's landscapes, including *Red Tree* (1908–10; Gemeentemuseum, The Hague) as "visions of an insane individual."[27] That same day van Eeden wrote in his diary that Mondrian "has totally lost his balance, and now suffers from acute decadence [*acuut verval*]."[28] A month later van Eeden published his own review of the Mondrian-Spoor-Sluyters exhibition in which he repeated his diagnosis.[29] But by this time Freud was formulating a psychoanalytic view of art ("Creative Writers and Daydreaming," 1908), and as Freudian psychoanalysis swept the Netherlands, the view of art as the (therapeutic) expression of repressed desires replaced Lombroso's view of art as a (pathological) symptom of a degenerate psychosis. After visiting Freud in Vienna in 1914, van Eeden adopted Freudian aesthetics and developed an appreciation of abstract art, especially the paintings of Janus de Winter.[30]

Mondrian moved away from Protestantism after reading a German translation of the French theosophist Édouard Schuré's history of religions, *The Great Initiates* (1889) with an introduction by fellow theosophist Rudolph Steiner.[31] In the 1890s Steiner had edited a collection of scientific writings by the German Romantic philosopher and poet Goethe, and, like many heirs to the *Naturphilosophen* spiritual outlook, Steiner understood man as ascending a ladder of knowledge culminating in intuition of Absolute Spirit. In keeping with theosophy's founder, Helena Petrovna Blavatsky, Steiner added the Buddhist idea of cyclical reincarnations, each with a more evolved consciousness. In his 1909 book *Theosophy*, Steiner asserted that thinking about mathematical patterns helps the seeker ascend the ladder, evidently having in mind the calming effect of ritualistic behavior (which is not, of course, unique to mathematics): "Mathematics with its strict laws, which do not yield to the course of ordinary sensible phenomena, form a good preparation for the seeker of the Path. . . . His life-thought must itself be a copy of the unchanging mathematical methods of stating premises and forming conclusions. He must strive wherever he goes and whatever he does to think after this manner. Then the intrinsic lawfulness of the spirit world will flow into him."[32]

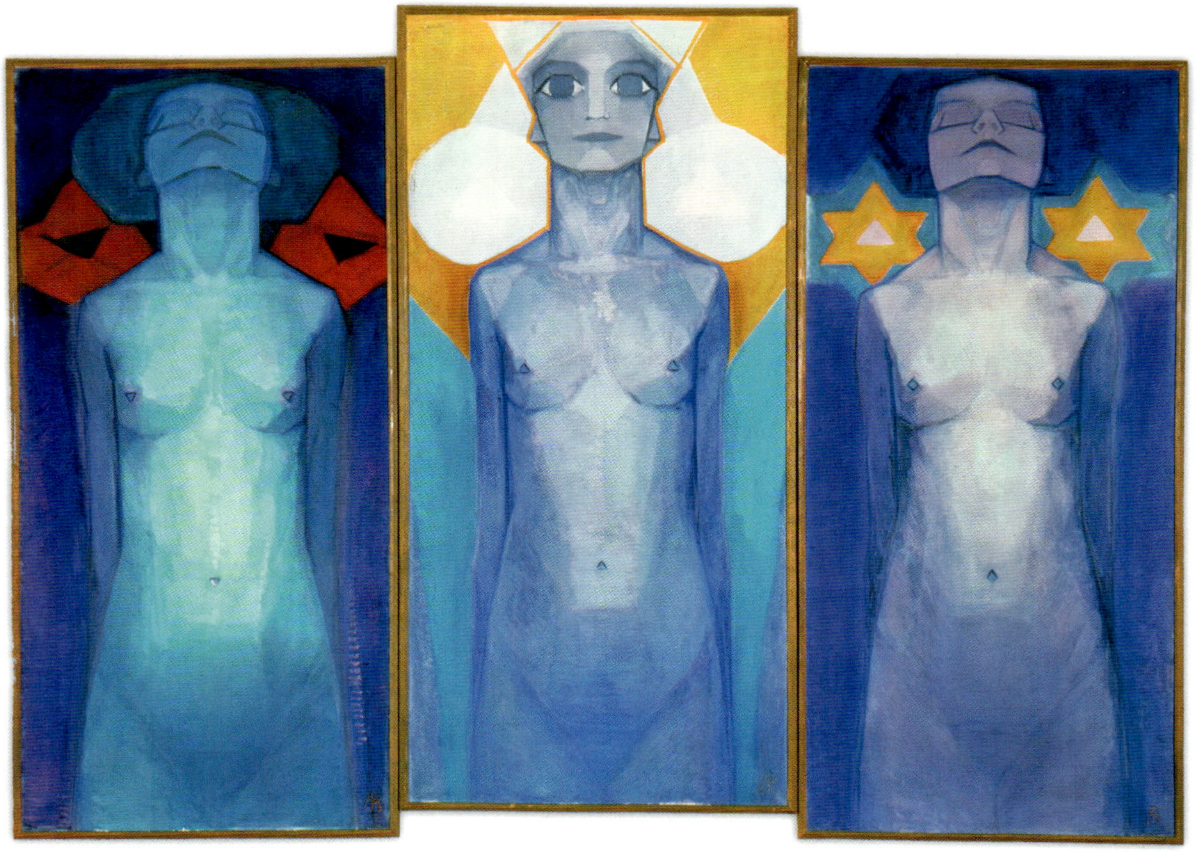

ABOVE

6-8. Piet Mondrian (Dutch, 1872–1944), *Evolution*, ca. 1911. Oil on canvas, central panel 72 × 34 ½ in. (183 × 87.6 cm), side panels 70 × 33½ in. (178 × 85 cm). Gemeentemuseum, The Hague.

BELOW

6-9. Alex Grey (American, b. 1953), *Theologue: The Union of Human and Divine Consciousness Weaving the Fabric of Space and Time in Which the Self and Its Surroundings Are Embedded*, 1986. Acrylic on linen, 60 × 180 in. (152 × 457 cm). Courtesy of the artist. © 1986 Alex Grey.

Today latter-day theosophists and New Age thinkers such as American artist Alex Grey achieve expanded states of consciousness through a combination of meditation and medication. The release in 1953 of chlorpromazine (trade name Thorazine), the first drug effective against severe mental illness, led to the widespread use of various mind-altering drugs in the following decade of the 1960s. In this painting Grey records his pantheist vision of the unity of the self with all of nature that he attains using yoga, meditation, and the drug psilocybin (extracted from mushrooms), which was shown in a 2006 study to enable people to have similar visions in research conducted at Johns Hopkins School of Medicine. R. Griffiths, W. Richards, U. McCann, and R. Jesse, "Psilocybin Can Occasion Mystical Experiences Having Substantial and Sustained Personal Meaning and Spiritual Significance," *Psychopharmacology* 187 (2006): 268–83.

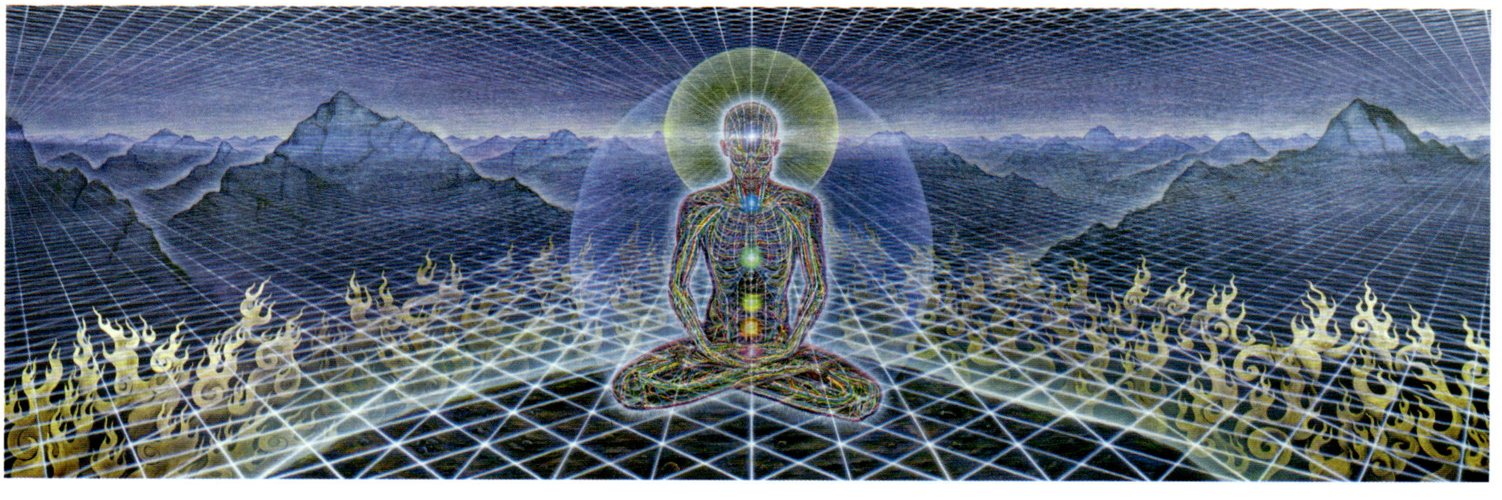

Mondrian moved to Domburg in 1908 and stayed annually until 1916, becoming close friends with Toorop. In March 1908 Steiner gave a series of lectures in Holland, which were published in a book that Mondrian acquired.[33] According to Steiner, the positive mystic, with eyes open and in full mental awareness, attains clairvoyant vision of "higher spheres" in which objects are surrounded with "astral shells," or auras.[34] In May 1909 Mondrian joined the Theosophical Society of Amsterdam.[35] Mondrian summarized this theosophical outlook in *Evolution* of 1910–11 (plate 6-8). A woman evolves spiritually from meditating in an earthly setting (on the left) surrounded by red amaryllis flowers (recalling his 1900 woman surrounded by passion flowers in plate 6-1), to meditating (on the right) while surrounded by six-pointed stars (emblems of the Theosophical Society). In pantheist traditions as old as Pythagoras and Lao Tzu, the goal of knowledge (of the One, the Tao) is a unity of opposites (symbolized by the two triangles in the six-pointed star): male and female, yin and yang, mind and matter. In the words of Blavatsky: "The triangle with its apex pointing upward indicates the male principle, downward the female; the two typifying, at the same time, spirit and matter."[36] In the center panel the woman is now one of Steiner's positive mystics; eyes open, she has reached enlightenment, symbolized by the small triangles and white forms surrounding her head (plates 6-8 and 6-9).

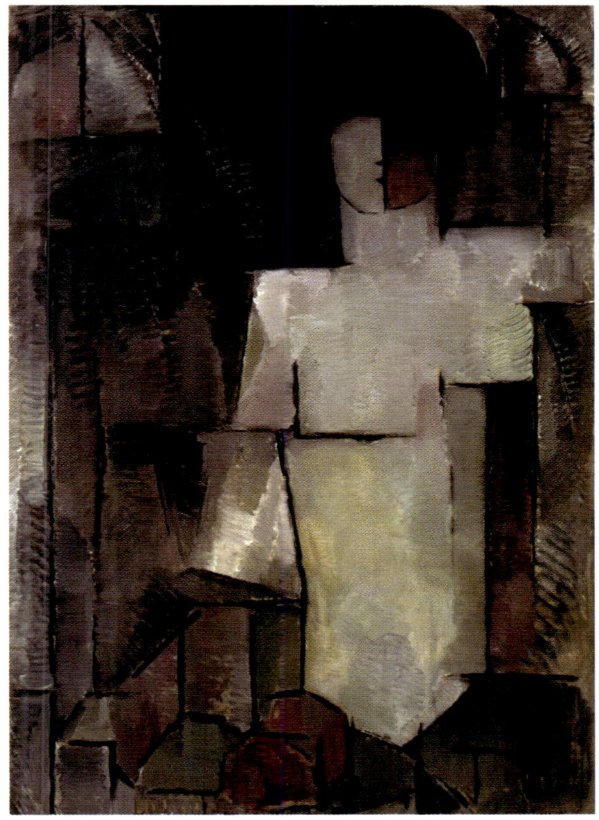

6-10. Piet Mondrian (Dutch, 1872–1944), *The Large Nude*, 1912. Oil on canvas, 55⅛ × 38⅝ in. (140 × 98 cm). Gemeentemuseum, The Hague. © 2015 Mondrian/Holtzman Trust.

Steiner eventually broke with Blavatsky after he came to feel that her theosophy was too occultist. Then he adopted the methods of Western science and in 1913 founded his Anthroposophy Society for the *empirical* study of spiritual matters (combining research methods of anthropology with philosophy).

In 1911 Mondrian saw Cubist work by Picasso and Braque at an exhibition in Amsterdam, and he became more familiar with Cubism during 1912–14, when he shuttled back and forth between Domburg and Paris. Mondrian painted a woman in the Cubist style, devoid of Symbolist trappings and reduced to drab geometric planes (plate 6-10). Picasso and Braque depicted (visible) landscapes, portraits, and still lifes in the realist tradition, and like Cézanne they never crossed the line into total abstraction. When Mondrian arrived in Paris, he had a different goal—the expression of (an immaterial) Absolute Spirit. When he remarked that the Cubists did not take their style to its logical conclusion,[37] he was projecting his goal onto them. Mondrian went on to use Cubist-style planes to reach the logical conclusion of *his* goal, the expression of the Absolute.

In 1914, while writing a treatise about his philosophy of art, Mondrian summarized the outlook that he developed during the Symbolist era, including using geometric forms to symbolize "the spiritual": "Two roads lead to the spiritual: the road of doctrinal teaching, of direct exercise (meditation, etc.) and the slow but certain road of evolution. One sees in art the slow growth of spirituality, of which the artists themselves are unconscious. . . . To approach the spiritual in art, one will use as little of reality as possible, for reality is opposed to the spiritual. Thus the use of elementary forms is quite logical. Since these forms are abstract, we find ourselves confronted by an art that is abstract."[38]

If art transcends the human sphere, it cultivates the transcendent element in mankind, and art, like religion, is a means for the evolution of mankind.

—Piet Mondrian, note in a sketchbook, 1913–14

SCHOENMAEKERS

Schoenmaekers had taken up theosophy, and like Blavatsky and Steiner, he tried to enlist science to give prestigious support to his occult views about alchemy and astrology and his Buddhist understanding of reaching enlightenment through thought alone. In Schoenmaekers's day astronomers and physicists studied the two forces known to them, gravity and electromagnetism. When Newton wrote his *mathematical* description of gravity in the seventeenth century, the *physical* nature of this force, which acts across vast distances to hold the solar system together, was a mystery to him. By the early nineteenth century scientists were investigating two other forces that act on matter at very close range, electricity and magnetism. Then in 1865 James Clerk Maxwell gave a *mathematical* description of the two forces that accounted for all observations and demonstrated that electricity and magnetism are actually two aspects of one force—electromagnetism. Maxwell, like Newton, felt that a mathematical description was incomplete without a *physical* description of the phenomenon. Maxwell expressed the conviction that in the future "a mature theory, in which physical facts will be physically explained, will be formed by those who, by interrogating Nature herself, can obtain the only true solution of the questions which the mathematical theory suggests."[39] But in Schoenmaekers's day gravity and electromagnetism were described *only* by mathematics (as, indeed, they still are today; an explanation of why these physical forces behave as they do has not yet been found).

Schoenmaekers seized on Newton and Maxwell's mathematical theories and declared that they had overlooked Mother Nature's spiritual dimension (*Het nieuwe Wereldbeeld* [The new image of the world], 1915). Van Eeden declared that Schoenmaekers "interprets life's phenomena in a scientific and a poetic-religious way, simultaneously. . . . I heard him talk, and thought: he would have pleased father Goethe. He picks up where Kepler and Newton left off."[40] Kepler and Newton considered the planets as spheres of inert, lifeless matter that move only when acted on by a physical force. According to van Eeden, Schoenmaekers thought about not only physical forces such as gravity but also mental (spiritual) forces that exert power over and above where Kepler and Newton had "left off."

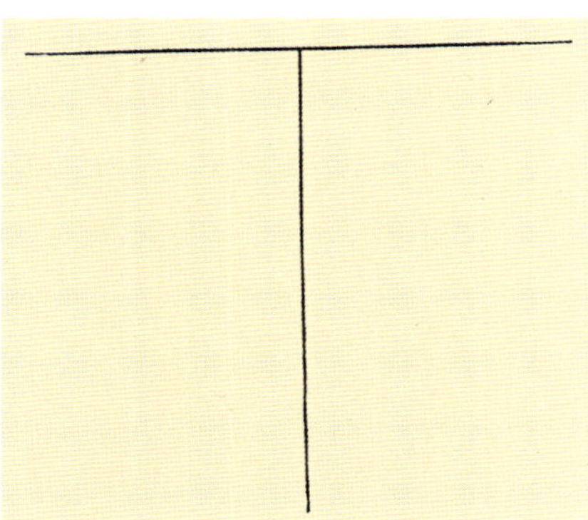

6-11. "The Cross," in M.H.J. Schoenmaekers, *Beginselen der Beeldende Wiskunde* (Principles of plastic mathematics; Bussum, the Netherlands: van Dishoeck, 1916), 72.

At the fin-de-siècle, electromagnetism was the talk of the science world and inventions such as light bulbs and the telegraph were changing everyday life. In the physicist's descriptions of the balance in nature of positive and negative electrical forces, as well as north and south magnetic poles, Schoenmaekers found scientific confirmation of his philosophical view of nature as a unity of opposing forces (horizontal/vertical, male/female, yin/yang, spirit/matter). In his 1916 book *Beginselen der Beeldende Wiskunde* (Principles of plastic mathematics) Schoenmaekers visualized the balance as a pair of horizontal and vertical lines (plate 6-11): "The line is 'horizontal' in its essence. The radius is 'vertical' in its essence. . . . The horizontal is in character line: lean, yielding, lying, ongoing, passive line. The vertical is in character radius: tight, tough, upright, upward, expanding, active radius." The passive (feminine) horizontal and the active (masculine) vertical form a unified whole: "Opposites are separate parts of one reality. They are all real in relation to one

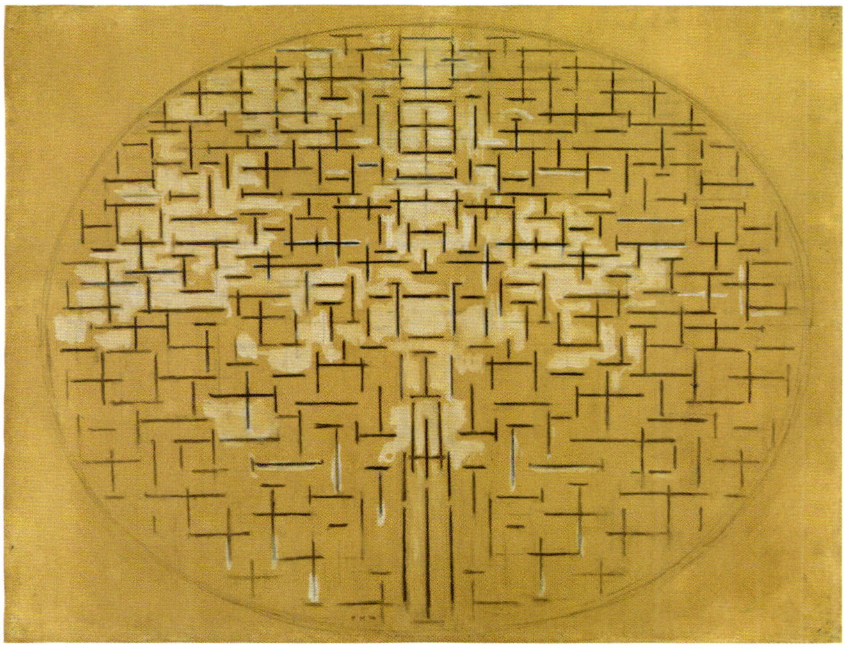

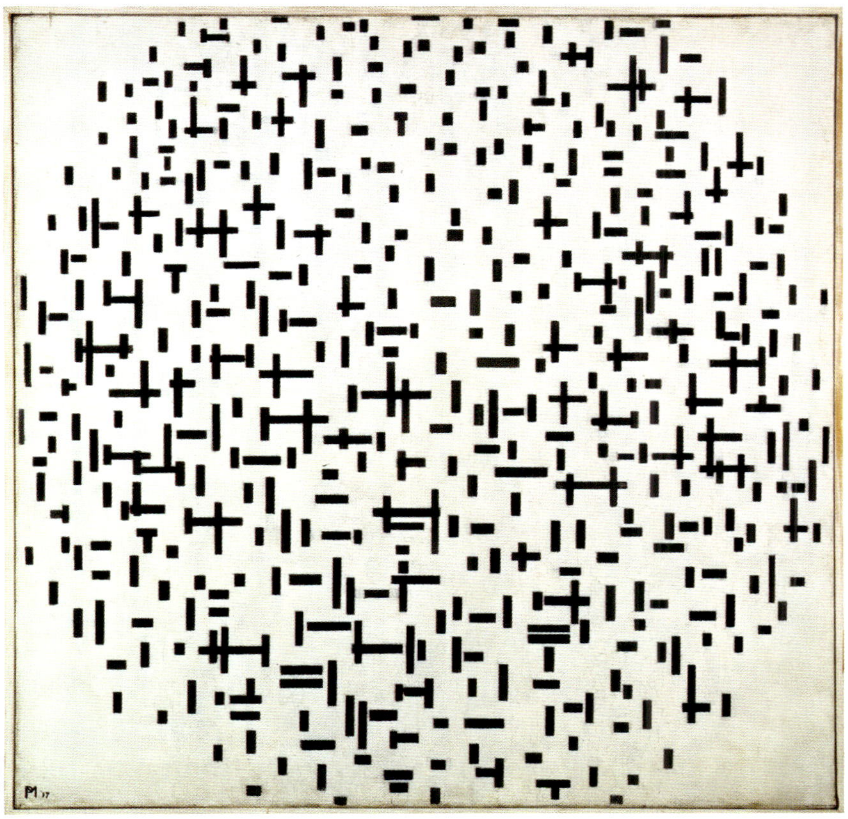

TOP LEFT

6-12. Piet Mondrian (Dutch, 1872–1944), *Pier and Ocean 5 (Sea and Starry Sky)*, 1915. Charcoal and gouache on paper, 34⅝ × 44 in. (87.9 × 111.7 cm). Museum of Modern Art, New York, Mrs. Simon Guggenheim Fund. © 2015 Mondrian/Holtzman Trust.

BOTTOM LEFT

6-13. Piet Mondrian (Dutch, 1872–1944), *Composition in Line* (second state), 1916–17. Oil on canvas, 42½ × 42½ in. (108 × 108 cm). Kröller-Müller Museum, Otterlo, the Netherlands.

another. . . . In this way, the woman is only really woman in relation to the man. And the man is only really man in relation to the woman."[41]

In 1914 Mondrian wrote about his philosophy of art in an essay (to which in 1917 he gave the title "Neo-plasticism in Painting"); while he was revising it[42] into a book-length manuscript during 1915–16, he met Schoenmaekers, who was his neighbor in Laren. Mondrian probably recognized Schoenmaekers's name from his contributions to *Eenheid*, the theosophical weekly to which he subscribed. Mondrian soon adopted Schoenmaekers's spiritual interpretation of the forces of nature as well as his horizontal and vertical lines. After their meeting, Mondrian began describing "the spiritual" in more scientific terms as the forces in nature, or, as he put it, the *force* of nature (in the singular): "that incomprehensible force which is universally active, and that we therefore call the *universal*."[43] To express this force, the artist developed a style that was increasingly more mathematical in the sense of being constructed in plane geometry, with straight lines and 90° angles. In 1915 Mondrian painted *Pier and Ocean 5* with horizontal and vertical lines to suggest the sea and vertical lines in the center front to indicate a pier (plate 6-12). Mondrian abstracted the image further and painted the lines straighter in *Composition in Line* (1916–17; see plate 6-13). With only the slightest reference to the perceived world, Mondrian composed this work, following Schoenmaekers, as a balance of horizontal and vertical lines, symbolizing the cosmos as a unity of horizontal/vertical and female/male.

As a member of van Eeden's circle, Schoenmaekers shared Brouwer's trust in intuition and meditation as the path to finding certain knowledge. Schoenmaekers was the bridge from Brouwer to Mondrian, who also put his faith in intuition and meditation. But Schoenmaekers and Mondrian differed from Brouwer in their philosophical opinions about the origin of abstract objects. For Brouwer, numbers and forms are mind-dependent creations of his intellect, but Schoenmaekers and Mondrian were essentially Platonists who believed that horizontal and vertical lines are eternal forms that are embodied in the natural world.

DE STIJL

Once Mondrian adopted a mathematical vocabulary of horizontal and vertical lines from Schoenmaekers, these elements remained at the core of De Stijl (Dutch for "the style"), which Mondrian developed together with several other artists under the leadership of Theo van Doesburg. Eleven years younger than Mondrian, van Doesburg was a self-educated artist who from about 1909 painted in an Expressionist style. He read Hegelian philosophy, adopted the goal of expressing Absolute Spirit, and in 1912 became a contributor with Schoenmaekers to *Eenheid*. Although Holland was neutral during World War I, young men were called up to defend her borders, and in 1914 van Doesburg was stationed at the Dutch-Belgian front. There he read Wassily Kandinsky's *Concerning the Spiritual in Art* (1911), whose thesis—that art expresses a spiritual realm—he enthusiastically endorsed. Van Doesburg met fellow soldier Antony Kok, a poet, to whom he confided his dream of founding an art magazine. On leave in Amsterdam in the fall of 1915, van Doesburg saw a group exhibition in Amsterdam that included Mondrian's paintings; in his review for *Eenheid*, van Doesburg wrote that Mondrian's work gave him a feeling of tranquility and spirituality.[44] In November 1915 he sent his first letter to Mondrian, beginning their collaboration.

In February 1916 van Doesburg met Mondrian at his studio in Laren, where they were joined by Schoenmaekers and the composer Jakob van Domselaer; the three already knew van Doesburg's writings in *Eenheid*. The day after the visit, van Doesburg wrote to Kok: "I got the idea that van Domselaar and Mondrian are completely under the spell of Dr.

Schoenmaekers's principles. He has recently published a book about *Plastic Mathematics*. Schoenmaekers has a mathematic approach. He considers mathematics to be the only pure yardstick for our feeling. He feels a work of art should therefore always have a mathematical foundation. Mondrian utilizes this principle by expressing his emotions in the two purest forms: the horizontal and the vertical line."[45] At this time van Doesburg was trying his hand at writing art criticism in a psychoanalytic style. In a pamphlet for an exhibition of an artist favored by van Eeden, Janus de Winter, van Doesburg described his canvases with simple planes of color as the therapeutic expression of repressed desires (Theo van Doesburg, *De Winter en zijn werk: Psychoanalytische studie* [De Winter and his work: A psychoanalytic study], 1916).[46] Van Doesburg took along a few paintings by de Winter to the meeting at Mondrian's studio. Schoenmaekers signaled that he had moved beyond theosophy by dismissing de Winter's art as "simply representations of astral visions" ("astral" was a theosophical term for an occult supersensible substance), and Mondrian concurred that the paintings were "beautiful but not very spiritual" (the term "spiritual" now associated with science).[47]

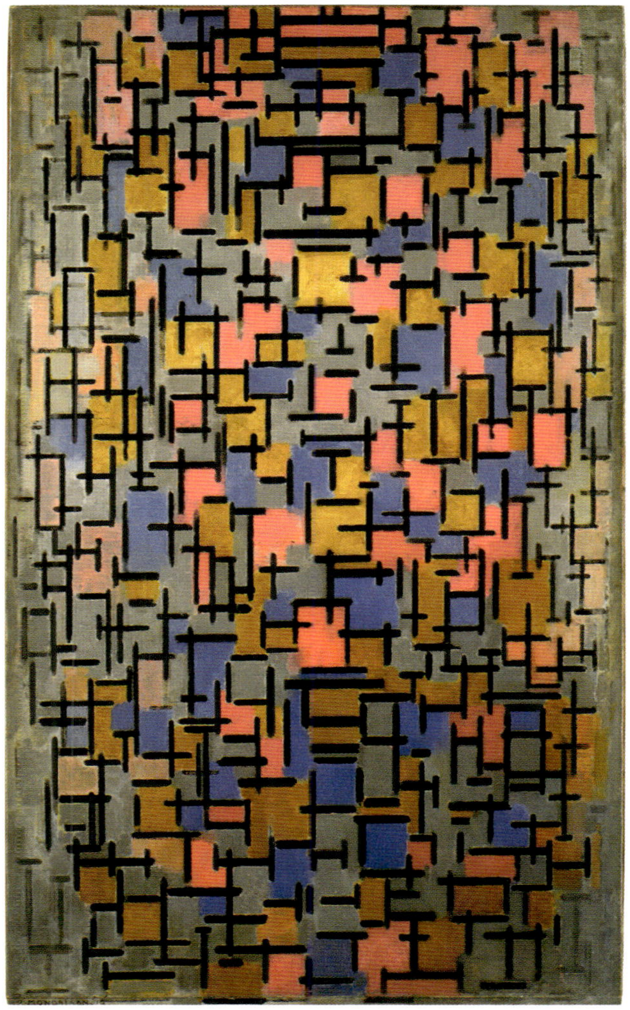

6-14. Piet Mondrian (Dutch, 1872–1944), *Composition*, 1916. Oil on canvas with wood, 46⅞ × 29⅝ in. (119 × 75.1 cm). Solomon R. Guggenheim Museum, New York, Founding Collection, 49.1229.

After adopting Schoenmaekers's mathematical vocabulary, Mondrian began drawing his lines with a ruler in work such as *Composition* of 1916 (plate 6-14), which is non-representational in the sense that it symbolizes an invisible reality (a universal force in nature) as opposed to a painting that is an abstraction from something observed (such as the sea). Mondrian's shift from a Symbolist to a mathematical vocabulary reflected a change within German-speaking theosophy itself, away from an overtly Eastern Buddhist philosophy (championed by Blavatsky) and towards the mysteries of Western science (described by Steiner). In a 1921 letter to Steiner, Mondrian expressed admiration for his writings and stated that his own "Neo-plasticism" was "the art of the future for Anthroposophy and Theosophy."[48] De Stijl expressed a cosmic spirit (the Absolute, the Tao, the Brahma) as well as the scientific worldview in which mathematics describes the natural forces that hold the cosmos together.

By early 1917 van Doesburg himself was composing in a pattern of horizontal and vertical lines and then creating variations by doing a mirror reflection and rotating the pattern, as he remarked, "just like J. S. Bach" (*Composite III*, 1917; see plate 6-15). Artists in the De Stijl circle applied their designs to practical projects: van Doesburg executed *Composite IV* in stained glass for a building designed by the architect J.J.P. Oud, and Bart van der Leck, who was also trained in glass-making, created abstract paintings whose designs he had woven into rugs (plate 6-16).

In 1915–16 Mondrian and van Doesburg had discovered a valuable nugget of mathematical symbolism in the murky sea of Schoenmaekers's thought, but the artists became disenchanted with him while preparing the first issue of *De Stijl*, which appeared in October 1917. Van Doesburg withdrew his invitation to Schoenmaekers to contribute to the

TOP RIGHT

6-15. Theo van Doesburg (Dutch, 1883–1931), *Composition III*, 1917. Stained glass. Kröller-Müller Museum, Otterlo, the Netherlands.

BOTTOM RIGHT

6-16. Bart van der Leck (Dutch, 1876–1958), *Composition no. 4*, 1918. Oil on canvas, 22 × 18 in. (56 × 46 cm). Kröller-Müller Museum, Otterlo, the Netherlands. © 2014 Pictoright Amsterdam/Artists Rights Society, New York.

BELOW

6-17. Georges Vantongerloo (Belgian, 1886–1965), *Construction in a Sphere*, 1917. Painted wood, 6⅝ × 6⅝ × 6⅝ in. (17 × 17 × 17 cm), and construction drawings. In Vantongerloo, *L'art et son avenir* (Antwerp, Belgium: De Sikkel, 1924), 11. © 2014 ProLitteris, Zurich/Artists Rights Society, New York.

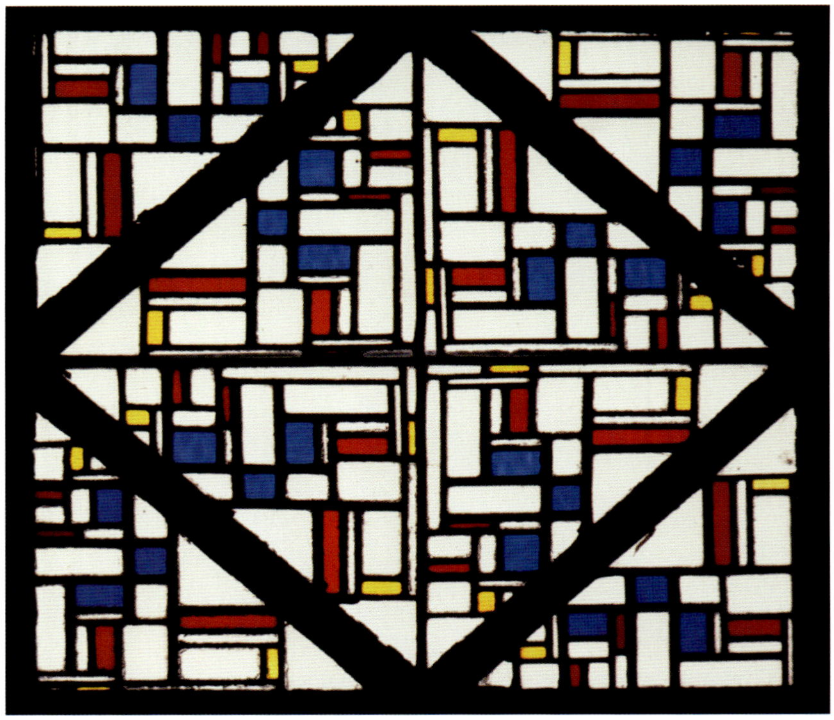

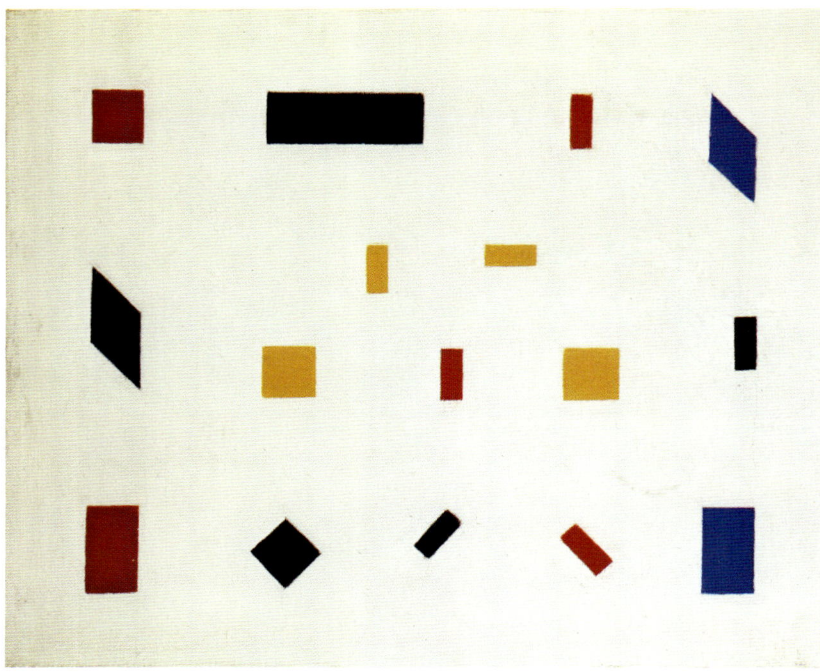

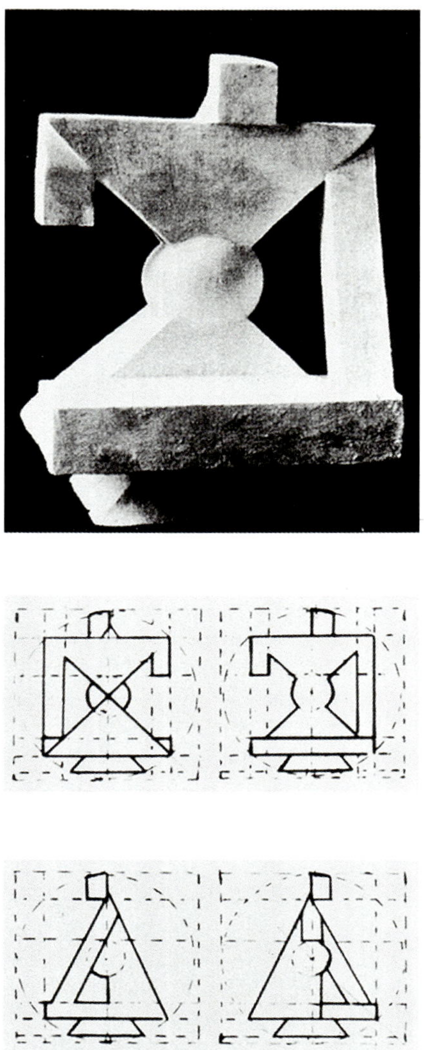

magazine, and Mondrian complained: "Write we must, you and I, . . . Schoenmaekers could be of use, if only he wasn't such a rotten sort of guy—also with him I don't believe it is real."[49]

After visiting Mondrian in Laren in February 1916, van Doesburg had met Oud and added architecture to his plans for the new magazine.[50] Van Doesburg sought a sculptor to complete the group, and in 1918 he met the Belgian artist Georges Vantongerloo and gave him a copy of Schoenmaekers's 1916 book, even though he and the author were no longer on friendly terms. Vantongerloo, who had been sculpting in the style of Rodin, adopted

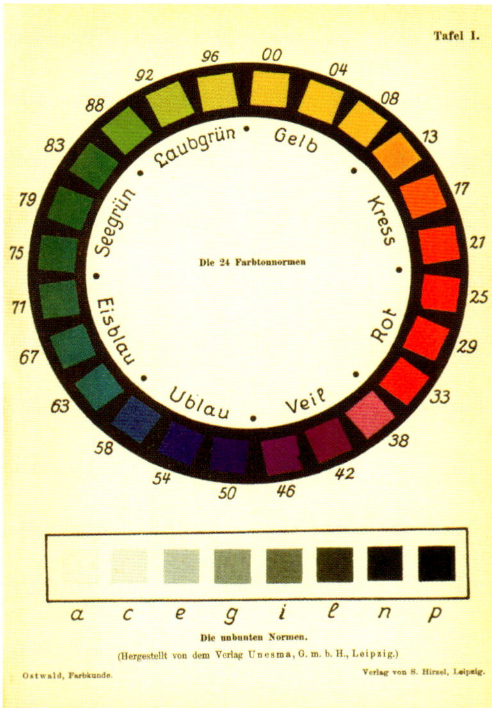

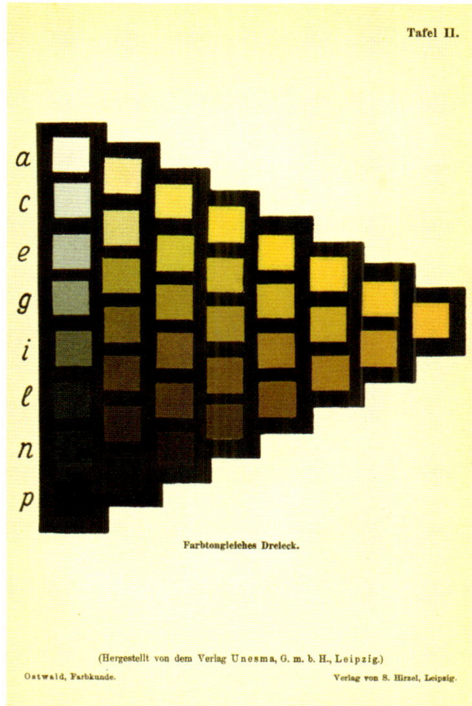

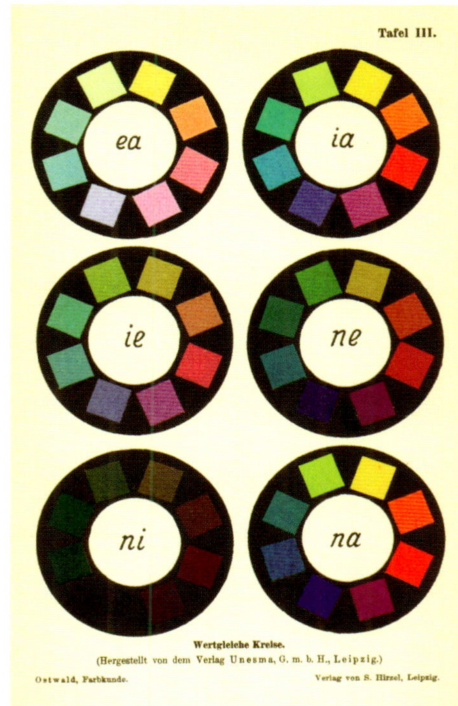

ABOVE LEFT

6-18. *The Grey Scale* and *The 24 Standard Colors*, in Wilhelm Ostwald, *Farbkunde* (Color tidings; Leipzig, Germany: S. Hirzel, 1923), plate 1.

Ostwald established a scale of eight values from white to black (shown at the bottom of this chart, marked a, c, e, g, i, l, n, p), hoping to capture the minimal difference in perceivable intensity at each step by increasing the proportion of white to black in a constant ratio (with no "jumps"). He also divided the color wheel into 100 hues, as indicated by the numbers on the circumference of the wheel on this chart, which has 24 main hues presented as swatches. Ostwald began publishing books on his classification system for mixed colors in 1916 with *Die Farbenfibel* (Color primer; Leipzig, Germany: Unesma).

ABOVE CENTER

6-19. *Triangle with the Same Hue*, in Wilhelm Ostwald, *Farbkunde* (Color tidings; Leipzig, Germany: S. Hirzel, 1923), plate 2.

Ostwald produced a three-dimensional color solid by first arranging his 24 main hues around the equator and then placing his value scale on the vertical axis, with white at the top and black at the bottom. He then showed each hue at full saturation at the tip of the wedge (shown here is no. 8, orange) and then mixed with varying proportions of black, white, and gray, as indicated by the letters. Each triangular slice of Ostwald's color solid has 36 squares of mixed paint, which can be designated by three numbers—the proportion of white, black, and the hue—that total 100 percent.

ABOVE RIGHT

6-20. *Circles with the Same Value*, in Wilhelm Ostwald, *Farbkunde* (Color tidings; Leipzig, Germany: S. Hirzel, 1923), plate 3.

Ostwald also offered his opinion that a pleasing color harmony can be achieved by juxtaposing hues that occur on his color wheel at the intervals of 3, 4, 6, 8, or 12. For example, in the lower-right circle (*na*), the 8 hues occur at intervals of 3 on his color wheel (reading clockwise: 00, 13, 25, 38, 50, 63, 75, 88). Ostwald was of the opinion that juxtaposed hues were harmoniously balanced only if they were close in value. The pure hues shown in *na* differ in value, from light yellow to dark bluish-purple. To equalize the hue and achieve harmony, the artist must, according to Oswald, adjust the value of the pure hues by adding white and black. The three circles on the left-hand side of this chart show the same array of hues (as in *na*) but at close values, from a light pastel above to dark tones below. (There is no experimental confirmation of Ostwald's assertions about color harmonies, which were based on his opinion.)

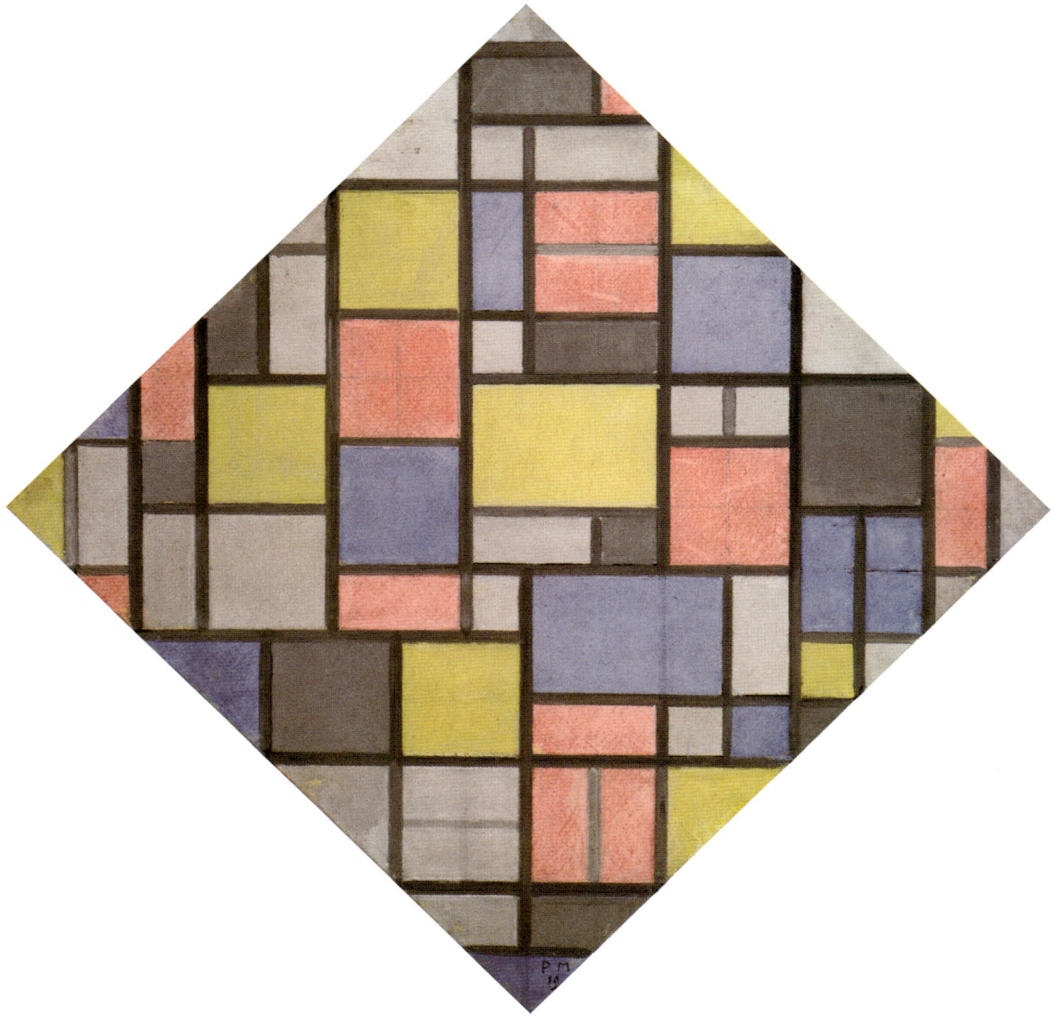

6-21. Piet Mondrian (Dutch, 1872–1944), *Composition with Grid 6: Lozenge Composition with Colors*, 1919. Oil on canvas, 19¼ × 19¼ in. (49 × 49 cm). Kröller-Müller Museum, Otterlo, the Netherlands.

The abstract—like the mathematical—is actually expressed in and through all things . . . the new painting achieved of its own accord a determined plastic expression of the universal, which, although veiled and hidden, is revealed in and through the natural appearances of things.

—Piet Mondrian, "Neo-plasticism in Painting," *De Stijl*, 1917

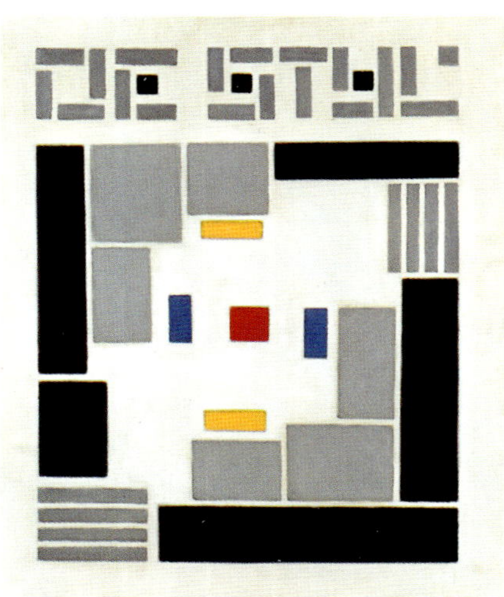

6-22. Vilmos Huszár (Hungarian, 1884–1960), *Composition De Stijl*, logo designed in 1917 for the cover of the first issue of *De Stijl*; painting after 1921. Oil on canvas, 23⅝ × 19¾ in. (60 × 50 cm). Gemeentemuseum, The Hague.

the De Stijl aesthetic and did a series of four symmetrical works that each fit inside a cube or sphere (plate 6-17). Vantongerloo published a series of articles that incorporated Schoenmaekers's outlook, in which he stated that his goal was to balance volume and void in his sculpture as an expression of the unity of positive and negative forces in nature ("Reflections," *De Stijl*, 1918).

De Stijl artists learned that color, too, could be approached mathematically. The German chemist Wilhelm Ostwald, winner of the Nobel Prize in chemistry in 1909 and an amateur painter, devised a precise number system for measuring hues and their tones (mixes with black and white). Ostwald was particularly impressed with Gustav Fechner's discovery that the minimal difference in intensity that can be distinguished by a perceiving subject bears a constant relation to the percentage of the intensity of the stimulus (Fechner's law, 1860). Ostwald designed a gray scale to have a minimal perceivable difference of value at each step (plate 6-18). He then used it to create a similar scale for every hue on a color wheel (plate 6-19). Ostwald declared that using this system it was possible to quantify (assign a number to) mixtures of color (hues mixed with black and white). He went on to propose a mathematical formula for achieving a harmonious combination of colors (plate 6-20).

Ostwald's system, published in 1916 as *Die Farbenfibel* (The color primer), was given a positive review in *De Stijl* by the Hungarian painter Vilmos Huszár, who designed

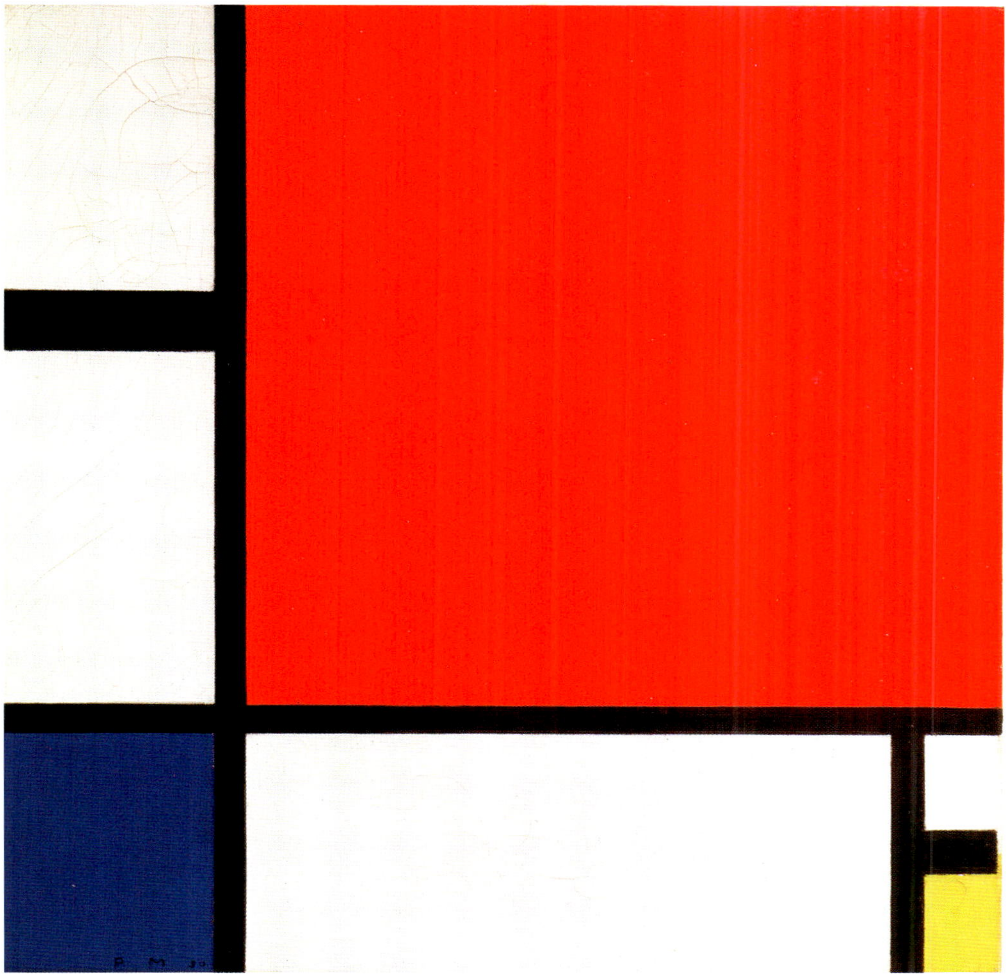

6-23. Piet Mondrian (Dutch, 1872–1944), *Composition with Red, Blue, and Yellow*, 1930. Oil on canvas, 18 × 18 in. (46 × 46 cm). Kunsthaus, Zurich. © 2015 Mondrian/Holtzman Trust.

the logo for the magazine (plate 6-22).[51] Ostwald himself published an article in *De Stijl* in 1920, "The Harmony of Color."[52] Mondrian reduced his palette to red, blue, and yellow because they were primary hues, and he aligned himself with Ostwald by stating that if black and white were added to red, blue, or yellow, the hue would still be considered a primary.[53] Following Ostwald's advice to achieve harmony by making the hues close in value (plate 6-20), Mondrian designed with many shades of gray and toned down his primaries with black and white, as in *Composition with Grid 6* (1919; see plate 6-21).[54]

During the great historical epochs, mathematics has been the basis of all the sciences and the foundation of art. When an artist uses basic geometric forms to express himself, his work is not "modern" but universal.

—Theo van Doesburg, "From Intuition to Certitude," 1930

After arriving at a similar style during World War I, Mondrian, van Doesburg, and the other De Stijl artists went their separate ways. As editor of *De Stijl*, van Doesburg would have read the manuscript of the Italian Futurist Gino Severini on four-dimensional geometry, which he serialized in 1918. Severini was trying to establish the geometric basis for Cubism and Futurism by showing how they represented a body moving through space and time.[55] Van Doesburg began to show an interest in the dynamism of Futurist art and four-dimensional geometry, and by June 1918, despite having withdrawn his invitation to Schoenmaekers to contribute to *De Stijl*, he was reading his *New Image of the World* with enthusiasm: "One of the best things seems to me his conception of time and space, and his visual representation thereof."[56] Van Doesburg also read Schoenmaekers's *Plastic Mathematics*[57] and went on to do work on geometric themes (plate 6-24). After travel

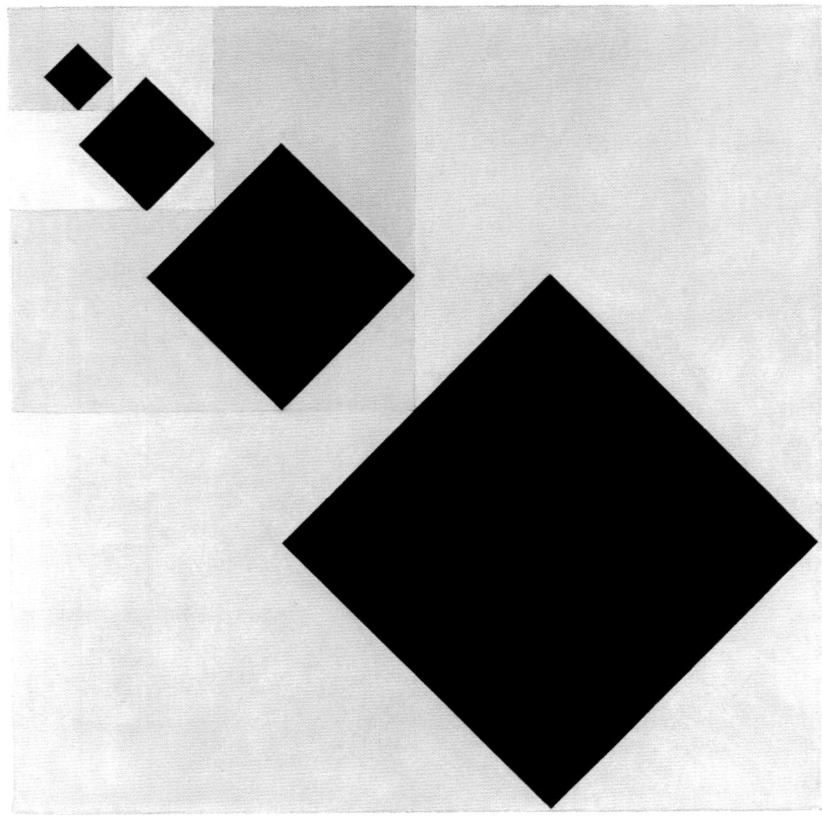

6-24. Theo van Doesburg (Dutch, 1883–1931), *Arithmetic Composition*, 1929–30. Oil on canvas, 39¾ × 39¾ in. (101 × 101 cm). Kunstmuseum, Winterthur, Permanent loan from a private collection, 2001. © Schweizerisches Institut für Kunstwissenschaft, Lutz Hartmann.

This composition is a geometric progression of doubling the area of the black squares. An arithmetic progression, such as doubling numbers in a series (as in 2, 4, 8, 16), would typically be represented on a line. Giving van Doesburg the benefit of the doubt (assuming he didn't mix up the terms), the American mathematician David Pimm has explored the ways in which the artist may have approached his geometric composition arithmetically; D. Pimm, "Some Notes on Theo van Doesburg (1883–1931) and His *Arithmetic Composition 1*," *For the Learning of Mathematics* 21, no. 2 (2001): 31–36.

Most artists are like pastry-cooks and milliners. In contrast, we use mathematical data (whether Euclidean or not) and science, that is to say, intellectual means.

—**Theo van Doesburg, "Comments on the Basis of Concrete Painting,"** ***Art Concret*, 1930**

became possible when the war ended in 1918, van Doesburg went to Germany to spread the doctrine of De Stijl.

Meanwhile, Mondrian published the first installment of his treatise, which, with a nod to Schoenmaekers, he titled "Neo-plasticism in Painting."[58] He also developed his mature Neo-plastic style, which was composed of horizontal and vertical black lines and rectangles of primary hues, as in *Composition* of 1921 (plate 6-23). Mondrian continued painting in this style, often using gray to desaturate his primaries, for the next twenty years. In each canvas he strove to achieve equilibrium of line and color and to express a universal force in nature.

Mondrian's artistic career was a spiritual journey that began when he set aside the strict Calvinist morality of his youth and adopted theosophy, which, in turn, he exchanged for a conviction that geometric abstract art could express a universal force in nature. Despite his shifting allegiances, Mondrian's art, like Brouwer's philosophy of mathematics, remained firmly rooted in German Romanticism, with its antagonism to analytic rationalism and trust in intuition.

In 1921 the Dutch Significs group made one final attempt to create an international language and establish a utopian community. By this time van Eeden had given up on psychoanalysis as providing the tools for resolving conflict because of Freud's negative attitude towards religion.[59] Van Eeden instead teamed up with the Dutch architect Jaap London, with whom he collaborated on the (unrealized) design of an ideal city (plate 6-25).[60] For his part, Mannoury was inspired by the Russian Revolution of 1917 to proclaim Lenin's Communism as the ideal social structure, and he urged his countrymen to adopt it as the

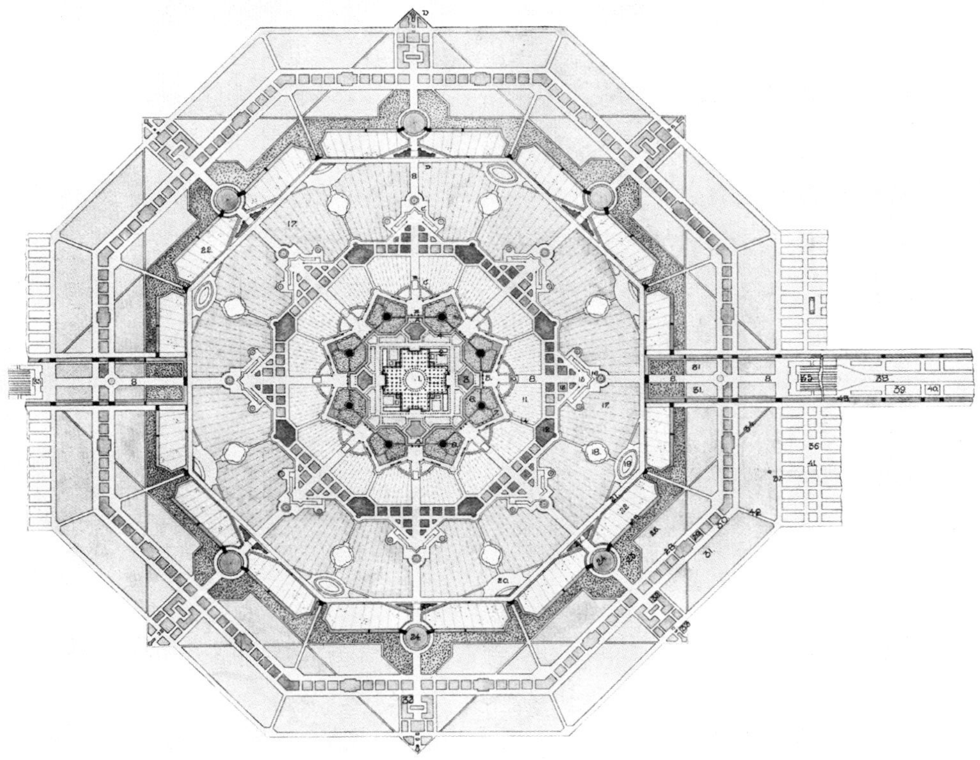

6-25. Frederick van Eeden and Jaap London, plan (left) and aerial view of an ideal city, in their *Het Godshuis in de Lichtstad* (The house of god in the city of light; Amsterdam: W. Versluys, 1921).

Reflecting Dutch interest in Buddhism, van Eeden and London laid out *Lichtstad* (city of light) in the form of a mandala, a symmetrical geometric figure used in meditation in which the central point represents pure mind/cosmic spirit. In the middle of their ideal city, van Eeden and London placed an octagon with eight domed temples, one for each of the world's great religions. In the center they built a sanctuary with a massive dome that towers over the eight historical theologies; this central sanctuary was reserved for pure contemplation.

new faith for the modern era (*Mathematics and Mysticism: A Signific Study from a Communist Point of View*, 1925).

As the war drew to a close, van Eeden and Brouwer were convinced that only intellectuals (not politicians) could achieve a lasting peace between the warring nations. So van Eeden and Brouwer travelled to The Hague in 1918 for meetings with the American consul and his staff, to whom they gave a proposal for a conference of scholars from all the belligerent and neutral nations to negotiate the terms of the German surrender.[61] When their proposal was met by blank stares, van Eeden and Brouwer gave up on their Significs project.[62]

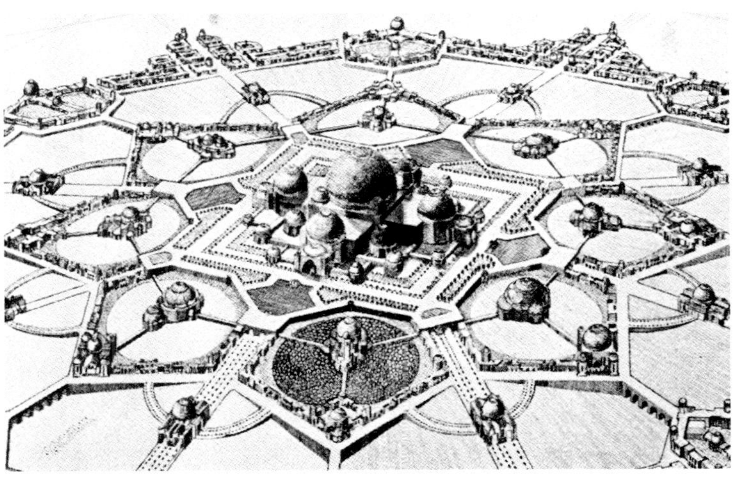

In the philosophical debates on the foundations of mathematics, Brouwer's intuitionist (constructivist) mathematics was always a minority view, yet it is still studied today.[63] Brouwer did, however, gain a large following in Göttingen and Berlin in the 1920s (see chapter 8), when the German defeat in the Great War triggered an outpouring of Romantic sentiments in the homeland of Hegel and Goethe.

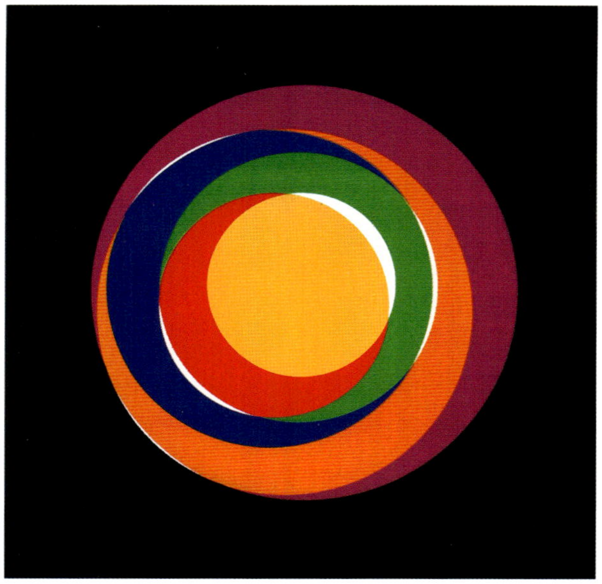
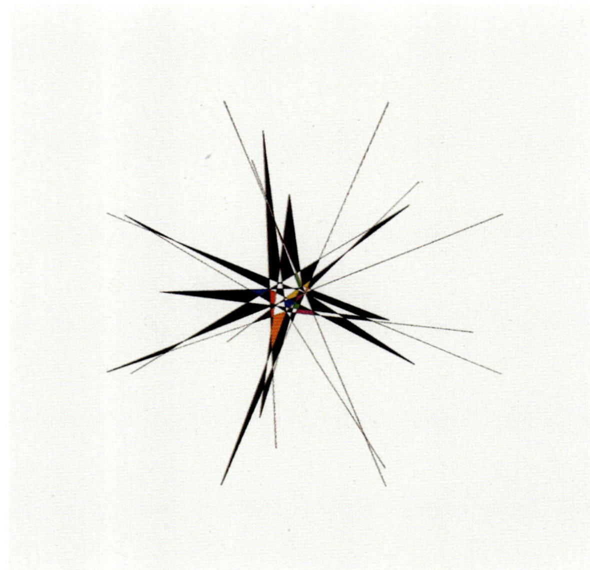
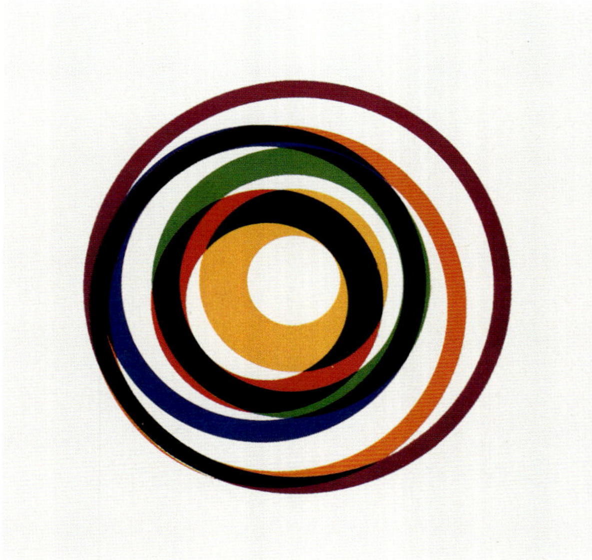
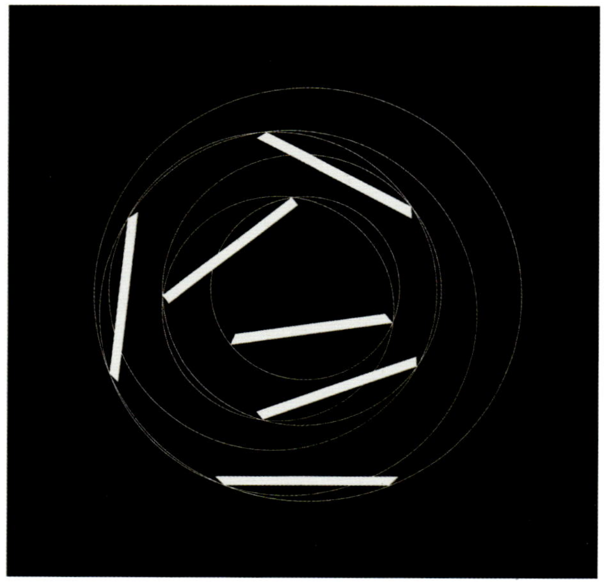
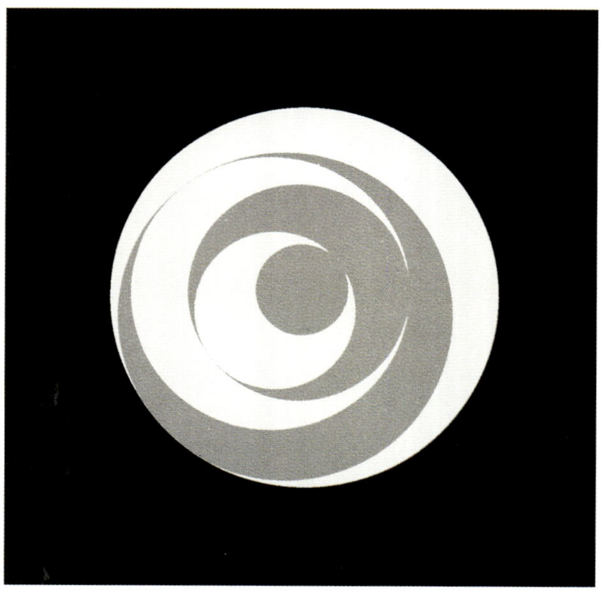
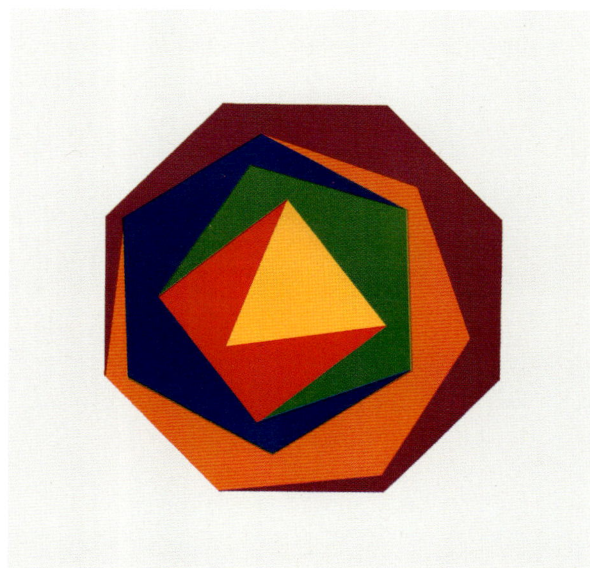

7
Symmetry

Our experience hitherto justifies us in believing that nature is the realization of the simplest conceivable mathematical ideas. I am convinced that we can discover by purely mathematical constructions the concepts and laws connecting them with each other, which furnish the key to understanding natural phenomenon. . . . Experience remains, of course, the sole criterion of the physical utility of a mathematical construction. But the creative principle resides in mathematics. In a certain sense, therefore, I hold it is true that pure thought can grasp reality, as the ancients dreamed.

—Albert Einstein, "On the Method of Theoretical Physics," 1934

SCIENTIFIC INSIGHTS into the deepest levels of the natural world are explanations based on symmetry. Since antiquity naturalists have observed bilateral symmetry in plants and animals as well as symmetrical hexagons in ice and snow. The term "symmetry" describes these shapes and forms with reference to a dividing line or plane. In the nineteenth century, scientists began to understand *why* so much of the natural world has a symmetrical structure. Using microscopes, they observed that nature's building-blocks (cells, crystals) are arranged in symmetrical patterns and come in left-right pairs.

In a wider context, the word "symmetry" refers to the property of a system that remains unchanged—is invariant—when certain operations are carried out. For example, scientists believe that the laws of nature—such as the force of gravity and the speed of light—are symmetrical in the sense that they apply equally (are invariant) throughout the universe.

The most symmetrical geometric form is a sphere (all points equidistant from a point in three-dimensional space). In the late twentieth century, scientists concluded that the universe began in perfect symmetry as a point that exploded into a sphere of plasma. As the infant universe expanded, the primordial sphere cooled, and matter condensed from the plasma to form the first particles, then atoms, gas clouds, and stars. At some point the original symmetry of the universe was broken; the resulting asymmetries appear to be the result of random shifts analogous to mutations during evolution. Today physicists are recreating samples of this primordial spherical plasma to determine the degree to which the universe retains traces of its original symmetry.

Scientists use the mathematics of group theory to describe nature's symmetry, both in the narrow sense of bilateral patterns and in the wider sense of invariance. In the early decades of the twentieth century, physicists and mathematicians in Zurich, including Albert Einstein, Andreas Speiser, and Hermann Weyl, employed group theory in their attempts to give a unified description of the forces of nature. Biologists of this era also used

7-1. Max Bill (Swiss, 1908–94), six prints from the series *Fifteen Variations on a Theme*, 1935–38 (see plate 7-18). © 2014 ProLitteris, Zurich/Artists Rights Society, New York.

group theory to describe symmetrical patterns in the chemistry of life—from molecules of glucose to DNA.

Beginning in the 1930s, Swiss artists created patterns that resonate with these mathematical descriptions of nature in terms of symmetry. Like the mathematicians and scientists, these so-called Concrete artists established basic aesthetic building-blocks—units of color and form—and arranged them using rules that preserve proportion and balance (plate 7-1). The artists described their art as "symmetrical" in the sense of being harmonious.

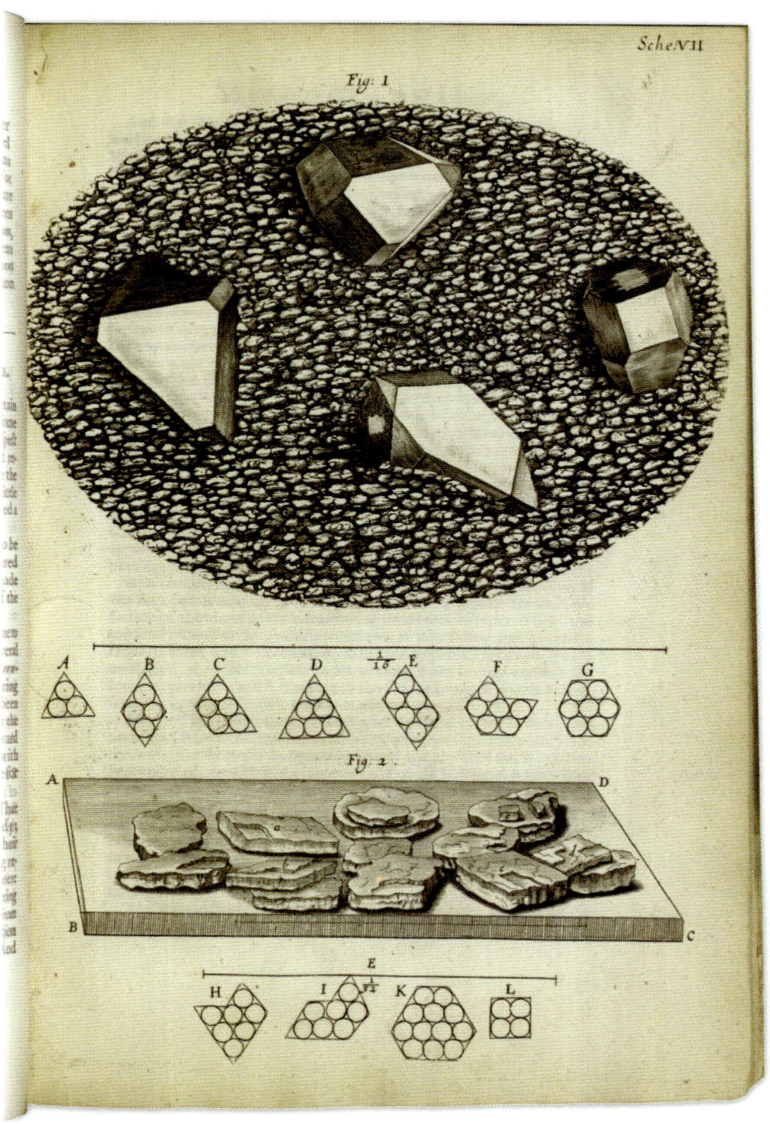

7-2. Spherical particles in crystals, in Robert Hooke, *Micrographia; or, Some Physiological Description of Minute Bodies Made by Magnifying Glasses with Observations and Inquiries Thereupon* (London: Royal Society, 1665), plate 7, fig. 1, facing 82. Rare Books and Manuscripts, The New York Public Library, Astor, Lenox, and Tilden Foundations.

This illustration accompanies Hooke's discussion "of the small Diamants [diamonds], or Sparks in flint" (82).

CRYSTALLOGRAPHY

For centuries speculation about the substructure of nature focused on the distinctive geometric form of crystals. In the 1660s the English natural philosopher Robert Hooke looked through a microscope at tiny fragments split off from rock crystals and saw that the faces of the chips had distinct geometric shapes. This led him to speculate that crystals are composed of patterns of spherical particles that are too small to be seen (plate 7-2).

At the opening of the nineteenth century the British chemist John Dalton gave a name to these hypothetical "spherical particles" when he revived the atomic theory of the pre-Socratic philosopher Democritus (see chapter 1). Dalton proposed that atoms differ in weight and combine to form molecules. A contemporary of Dalton, the French chemist René Just Haüy, suggested that the building-blocks are polyhedra rather than spheres (plate 7-3). But it was the German mineralogist Christian Weiss who had the key insight that it is not the geometric form of a crystal that distinguishes one from another but its axis of symmetry.

An axis of symmetry is an invisible line passing through a geometric form, such as a cube. The form can be described by describing its one or more axes of symmetry and stating what rotations on these axes leave its original form unchanged (invariant; see sidebar on opposite page).

Weiss symbolized the (invisible) atoms within a molecule, and molecules within a crystal, using dots (for the atoms) within an interlaced lattice structure. This model encouraged mathematicians to study crystals using Euclidean geometry and the five Platonic solids. Euclid viewed solids as stationary (not rotating), and he did not describe their symmetries. To explain a crystal as a geometric form rotating on an axis, nineteenth-century crystallographers needed a new mathematical tool: group theory.

The Rotational Symmetries of a Cube

The cube has thirteen axes of rotational symmetry. Three axes pass through the centers of the opposite faces of the cube. The form of the cube does not change (it is invariant) when it rotates on these axes to four positions (90°, 180°, 270°, or 360°). In the language of group theory, the cube has fourfold rotational symmetry about each of these three axes (for a total of twelve).

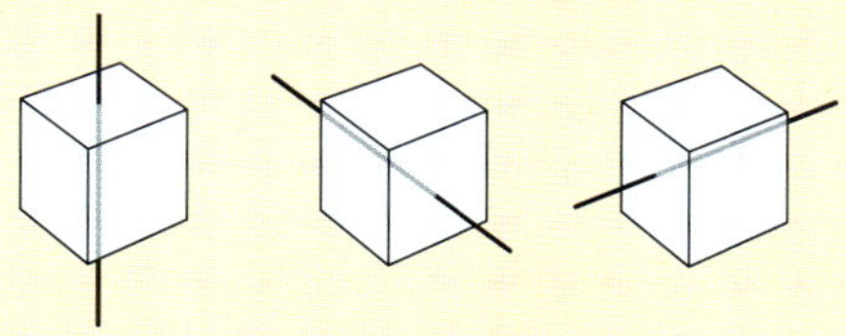

Three 4-fold axes

Four additional axes of rotational symmetry are determined by the diagonals that pierce the opposite corners of the cube. The cube has threefold rotational symmetry about each of these four axes (for a total of twelve).

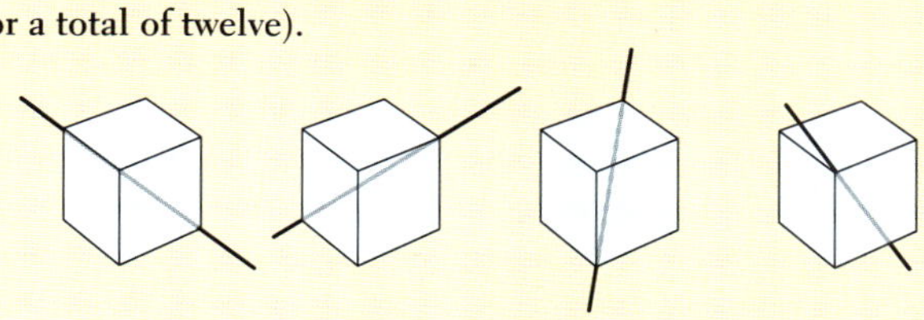

Four 3-fold axes

Six more axes of rotational symmetry are determined by axes passing through the midpoints of the edges of the cube. The cube has twofold rotational symmetry about each of these six axes (for a total of twelve).

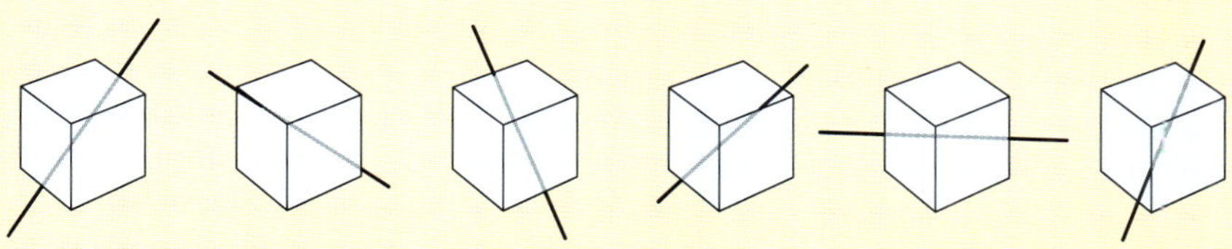

Six 2-fold axes

Taken together, there are thirty-six rotational symmetries of the cube. A sphere is the most symmetrical geometric form; it has an infinite number of axes of symmetry and it remains invariant when rotated on any one of them, to any position.

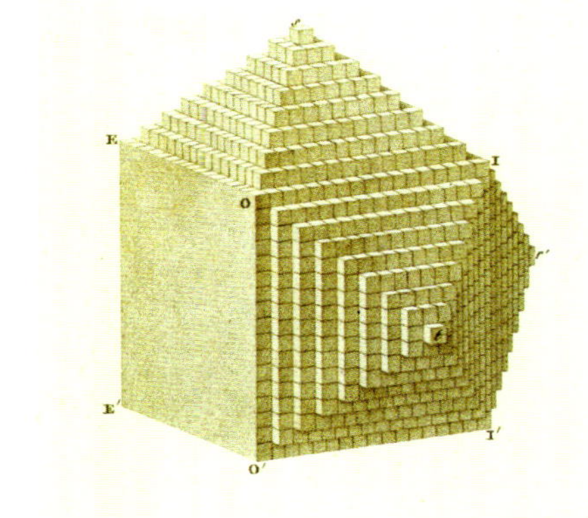

7-3. A rhombic dodecahedron built from molecular polyhedra, in René Just Haüy, *Traité de minéralogie* (Treatise on mineralogy; Paris: Louis Libraire, 1801), vol. 5, plate 2, fig. 13.

Haüy proposed that the building-blocks of crystals come in six basic forms: parallelepiped, rhombic dodecahedron, triangle-faced dodecahedron, hexagonal prism, octahedron, and tetrahedron.

GROUP THEORY

The French mathematician Évariste Galois died in a duel in 1832 at age twenty. The papers of this young scholar, including his work on group theory, which he had invented around 1830 to study solutions to certain algebraic equations, eventually ended up in the hands of Joseph Liouville, who edited and published them in 1846 as the "Galois theory."

A few decades later, while working on the classification of crystals, the French mathematician Camille Jordan recognized that Galois's algebraic group theory could describe not just crystals but *any* symmetrical system because it considers only the structure (the abstract form) of a system. Jordan published the first major work on group theory, including its application to the classification of crystals (*Treatise on Substitutions and Algebraic Equations*, 1870). After Jordan's success in classifying crystals using group theory, other geometers took up the task of describing geometric forms based on lattices in space and their symmetries.

You know, my dear Auguste, that I have explored many subjects. . . . But I am running out of time and my ideas about this immense terrain are not well enough developed.

—Évariste Galois, letter to Auguste Chevalier, written on the eve of his fatal duel, May 29, 1832

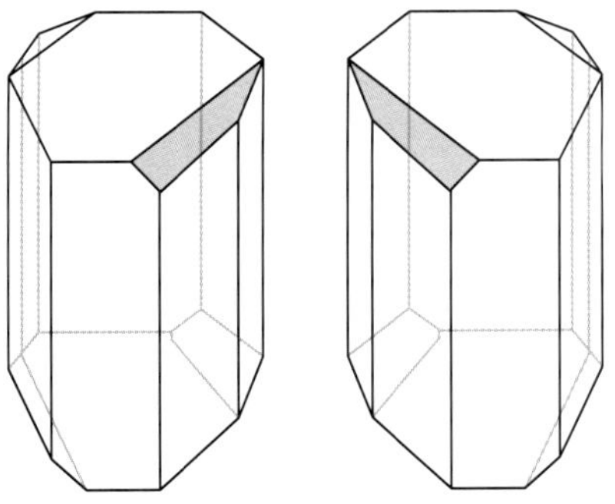

7-4. Racemic acid crystals are mirror images of each other.

Contemporary with the classification of crystals, the French chemist Louis Pasteur, who wrote his doctoral thesis on crystallography, made a major discovery about the structure of organic molecules. Pasteur investigated them in crystallized form using a beam of light that he first polarized by passing the light through a certain kind of transparent rock found in Iceland (so-called Icelandic spar) that causes the vibrations of the light-waves to become oriented in a plane. When a chemist shines a beam of polarized light through a crystal, sometimes the beam goes straight through (the chemical is "optically inactive"), but sometimes the plane of polarized light rotates to a precise degree, depending on the length of its path in the crystal and the concentration of the chemical, suggesting that the light is being turned by the invisible structure of the crystal.

In 1849 Pasteur was investigating crystals of the chemical racemic acid, a form of tartaric acid. Pasteur sent polarized light through a crystal and found that it was optically inactive. Then when examining the same crystals of racemic acid using a microscope, he noticed that they come in two asymmetrical mirror-image forms (so-called enantiomorphs), analogous to a pair of gloves for the right and left hands (plate 7-4). Pasteur meticulously separated the left- and right-hand crystals into two piles, made a solution from each, allowed them to crystallize, and then passed polarized light through them. Each rotated the light, but in opposite directions; Pasteur had discovered enantiomorphs. He correctly deduced that racemic acid contains a mixture of right- and left-hand molecules, so that when both are present, the effect of the rotated polarized light is cancelled out, making the mix appear optically inactive. By 1900 biologists were researching other pairs of mirror-image molecules and determining their roles in living systems.

Tyger, tyger, burning bright
In the forest of the night,
What immortal hand or eye
Could frame thy fearful symmetry?

—William Blake, "The Tyger," 1794

THE MOST ABSTRACT GEOMETRY: LIE'S GROUPS AND KLEIN'S ERLANGEN PROGRAM

Group theory was enlisted to restore order to mathematics after the upheaval caused by the discovery of non-Euclidean and *n*-dimensional geometries. Using the wider meaning of symmetry—the property of a system that remains unchanged by a particular operation—the late-nineteenth-century German mathematician Felix Klein showed that all forms of geometry (Euclidean, non-Euclidean, and *n*-dimensional) share features that remain invariant under certain transformations.

Klein began by extending the ideas of Poncelet, who had thought of projective geometry as the geometry of properties preserved under projective transformations ("shadows") of objects (plate 6-5 in chapter 6). Klein saw that the symmetry of different geometries could be understood as being related to different projections, which he could describe using group theory. His goal was to determine which properties are invariant under various groups of transformations. In Berlin in 1869, the Norwegian Sophus Lie joined Klein in his efforts to develop a group theory to geometry, and together they travelled to Paris in early 1870 to study with Camille Jordan. Klein focused on isolating those transformations that leave certain specific figures unchanged, defining structures such as a "Klein four-group" (see sidebar on opposite page).

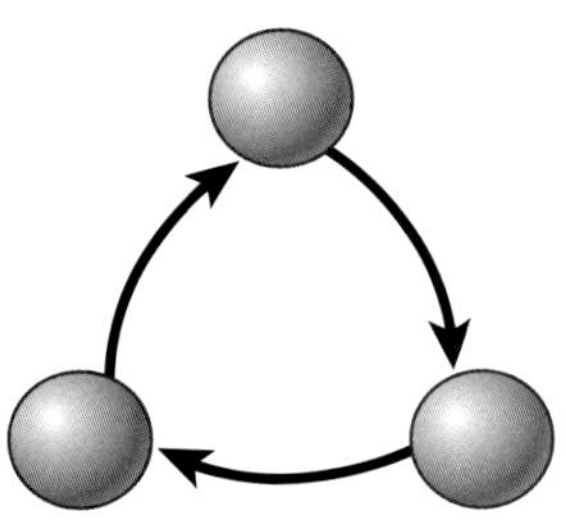

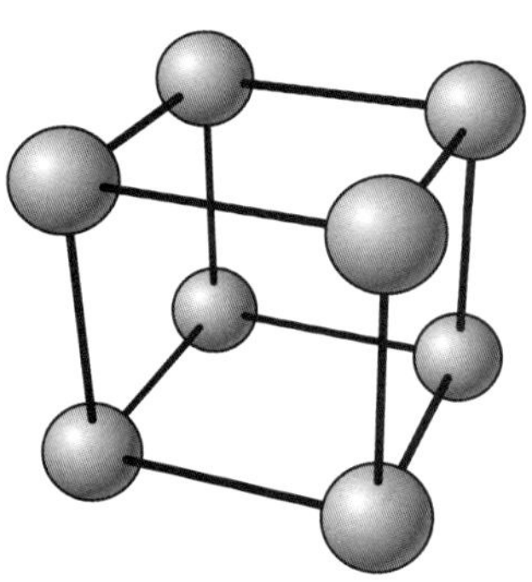

7-5. Diagrams of two systems in terms of group theory. The top system has three parts ("nodes") and one transformation that can be made in only one direction. The bottom system has eight nodes and one transformation that goes both ways.

Klein Four-Group

Group theory is used to describe the symmetry of an object or system. For example, suppose we want to measure the symmetry of a rectangle.

We do this by making a "map" of its parts and the actions ("transformations") we can perform on it that leave the rectangular shape unchanged. This map is a diagram of the rectangle's group. We make this map by doing the following:

Step 1: Identify the parts of the object that are similar. In a rectangle, the shape is determined by the position of its vertices (corners). Any geometric mapping that preserves its shape must preserve its vertices, which we number.

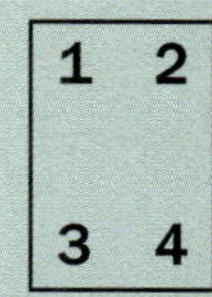

Step 2: We identify specific transformations that can be performed on the object that leave its form unchanged. In the case of this rectangle, we identify geometric transformations that send it into itself (in other words, that leave its form unchanged). Each of these corresponds to a permutation on numbered vertices.

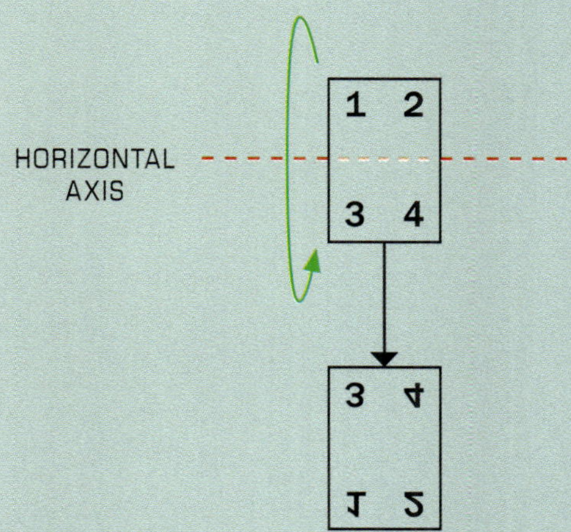

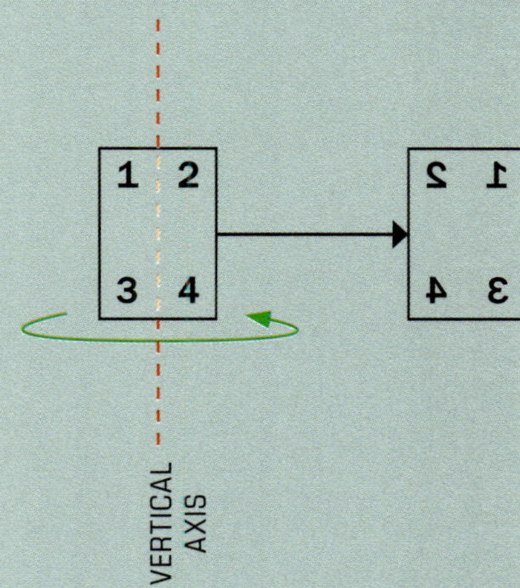

Step 3: Make a map of the possible combinations of actions:

Do nothing = Identity

Flip on its horizontal axis = HFlip

Flip on its vertical axis = VFlip

Do both (flip horizontally and then vertically, or vice versa) = HFlipVFlip

This map describes the four symmetries of the rectangle considered as a group.

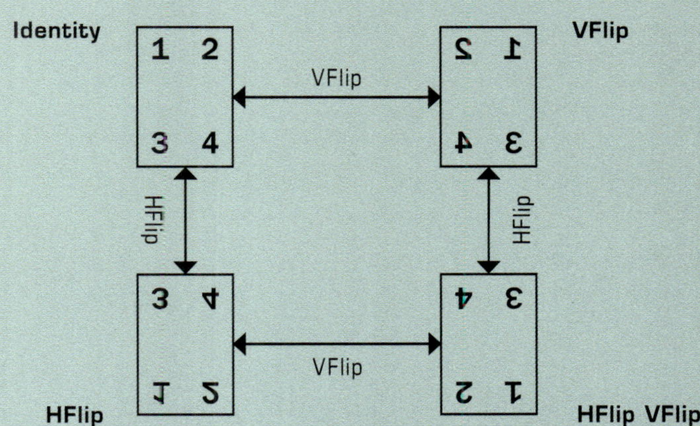

This is a diagram of the map that has been stripped of words and numbers, leaving only the bare, abstract vocabulary of group theory. The diagram has four parts ("nodes") and two transformations, one represented by dotted lines and the other by solid lines. The lines have no arrowheads, indicating that the transformation goes both ways. The transformations are deterministic (they entail no chance), they are reversible, and they are cumulative in the sense that the actions can be done repeatedly in any order, leaving the shape unchanged. In 1884 Felix Klein defined a group with four symmetries, which he called a *Vierergruppe* (German for four-group).

The outbreak of the Franco-Prussian War in July 1870 brought Klein and Lie's collaboration to an abrupt end; Klein returned to Germany, where he joined the faculty at the University of Erlangen, and Lie made his way back to Norway after a brief imprisonment in France by invading German soldiers who mistook Lie's mathematical equations for a military code. In 1871 Lie published his description of groups that allow the transformation of points from one co-ordinate system to another (a "Lie group"; *On a class of geometric transformations*, 1871). The following year Klein published his Erlangen program for the unification of geometry under its most general form (projective geometry), which he characterized by describing the invariant features of geometric figures under various groups of transformations.

SYMMETRY IN PHYSICS AND COSMOLOGY

In 1905 Einstein imagined what a light-wave would look like to an observer travelling with it at the same speed. These musings led him to propose a new axiom of science: no body can travel faster than the speed of light (about 186,000 miles per second). A mass flying through space never reaches this speed because as its velocity approaches the speed of light, the object's length is foreshortened, its mass increases, and its time scale is dilated (*Special Theory of Relativity*, 1905).[1] In 1907 the Russian-born German physicist Hermann Minkowski pointed out to Einstein that his theory implies that time must always be taken into account when giving the location of a mass travelling at great speed. Minkowski suggested joining space and time into one concept—space-time—and describing events using a four-dimensional geometry, the three spatial dimensions plus time. Adopting these suggestions, Einstein thereafter described the universe using a four-dimensional lattice of points in space-time.

Einstein's universe is symmetrical in the sense that an accurate description of the world can be given from any frame of reference. Any four-dimensional co-ordinate system can be transformed into any other set of co-ordinates, so long as the two sets of co-ordinates are moving with constant velocity relative to each other. All co-ordinate systems are interrelated in a totality—the universe. The measurements of one observer in one space-time framework can be translated into the measurements of an observer in another space-time framework. The laws of physics (and hence the observations) are invariant as one moves from one inertial (non-accelerating) four-dimensional space-time lattice to another.

Many mathematicians, including Hilbert, were intrigued by Einstein's 1905 version of the theory of relativity, which is called "special" because it describes only bodies travelling at a constant speed relative to each other. Hilbert immediately began working to extend Einstein's theory to describe frames of reference that are accelerating or decelerating relative to each other, but Einstein was the first to succeed in generalizing the theory to apply to objects that are speeding up or slowing down under the force of gravity (*General Theory of Relativity*, 1916). One conclusion of the theory of relativity is that mass and energy are interconvertible. Mass can be converted into energy, and vice versa, as Einstein declared. The conversion formula is:

$$E = mc^2$$

where E is energy, m is mass, and c^2 is the speed of light times itself.

In the late eighteenth century, the French chemist Antoine Lavoisier showed that mass is conserved in chemical reactions (*Conservation of Mass*, 1789). In the mid-nineteenth century, Hermann von Helmholtz argued for the idea of conservation of energy based on his medical studies of muscles and metabolism (*Conservation of Energy*, 1847). Einstein saw that these two previous conservation laws should be subsumed under one law because mass and energy are two aspects of a single thing—mass-energy (*Conservation of Mass-Energy*, 1905). In 1918 a colleague of Hilbert's, the German mathematician Emmy Noether, proved that these and other conservation laws can be written in group theory as transformations that leave mass, energy, and mass-energy unchanged (*Noether's Theorem*, 1918). This means that symmetry, in its wider sense of describing properties that are invariant, is a reigning principle of science.

THE UNIFICATION OF THE FORCES OF NATURE

By the time the general theory of relativity was confirmed in 1919, Einstein was trying to unify the two forces known to him: gravity, which holds the cosmos together, and electromagnetism (all forms of light, including the visible spectrum), which holds atoms together. On the assumption that gravity and electromagnetism are deterministic and that the same laws of physics apply everywhere in the universe independent of frame of reference, physicists sought an all-encompassing geometry that would allow a single field to describe gravity on a cosmic scale and electromagnetism on an atomic scale. Physicists were encouraged by the ways in which gravity and electromagnetism are similar: observation and theory indicate that the strength of the gravitational force between two bodies, such as a planet and its moon, and the strength of the electromagnetic attraction of two charged particles, such as a proton and electron, both decrease in proportion to the square of the distance between them.[2] Every mass generates a gravitational field, and every charged particle generates an electromagnetic field, although the strength of the two force fields is vastly different; the force of gravity measures $1/10^{36}$ of the force of electromagnetism. Newton gave an account of gravity as a static force that acts instantaneously throughout the universe, but Einstein described gravity as an effect of the curvature of space, with disturbances in that curvature propagating at the speed of light, as do electromagnetic effects. In other words, gravity and electromagnetism are both dynamic forces whose associated waves travel at the same speed—about 186,000 miles per second.

While Einstein was beginning his attempt at unifying these forces, two young mathematicians, the Swiss Andreas Speiser and the German Hermann Weyl, were refining group theory, the mathematical tool that Einstein and others would come to use in their attempts at unification. When Speiser and Weyl were students together at the University of Göttingen, they were kept apprised of developments in physics because their professor Hilbert was then working to generalize Einstein's theory of relativity. Both students completed doctorates in group theory, Weyl in 1908 and Speiser in 1909.[3] In 1913 Weyl joined the mathematics faculty at the Eidgenössische Technische Hochschule (ETH) in Zurich and became a colleague of Einstein, who was on the physics faculty from 1912 to 1914. Weyl, who was educated in both mathematics and physics, realized the importance of group theory for understanding the symmetries of fields, and he began working on the task of unifying gravity and electromagnetism. In 1916 Speiser returned to Switzerland to teach at the University of Zurich, and although Speiser himself did not apply group theory to physics, Weyl kept him informed about his progress on unification.

Weyl's efforts at unification were unsuccessful, but nevertheless he (correctly) saw that group theory provided physicists with the appropriate tool, which has had a lasting effect on the quest for unification. By the 1930s physicists had discovered that there are two other forces at work within the atomic nucleus: the strong nuclear force that holds the nucleus together and the weak nuclear force that causes the nucleus to decay. Using group theory as their primary mathematical tool, physicists have continued trying to unify the four forces of nature. Today unification has not been achieved, and it remains a central goal of physics.

SYMMETRY IN THE DECORATIVE ARTS

As they pursued the mathematical foundations of physics, Speiser and Weyl applied their ideas about symmetry to music and the arts. Speiser was more deeply involved with the arts than his friend and first applied group theory to the arts, after which Weyl and others followed.

Raised in a musical family in Basel, Speiser studied piano with Hans Huber, a composer and the director of the Basel conservatory, and he played piano throughout his life. As a child Speiser was also surrounded by the visual arts; for the garden of the family home, his maternal grandmother commissioned the Swiss Symbolist Arnold Böcklin to paint a series of murals.[4] From his early student days in Basel, Speiser avidly read the late-nineteenth- and early-twentieth-century scholars who were establishing art history as an academic discipline: Jakob Burckhardt, Heinrich Wölfflin, Ernst Cassirer, and Erwin Panofsky. Speiser absorbed the writings of Burckhardt, who urged his students to consider painting, sculpture, and architecture in a broad cultural context, treating epochs such as classical Greece and the Italian Renaissance as a unity and stressing that each style embodies a system of proportion (see chapter 2).

Speiser's grounding in a broad cultural perspective drew him to Burckhardt's student Heinrich Wölfflin, who became Speiser's colleague in 1924, when Wölfflin joined the art history faculty at the University of Zurich. Initially captivated by nineteenth-century theories of proportion (see chapter 2), Wölfflin also studied the psychology of his day to help him understand Renaissance and Baroque art. In his early writing Wölfflin adopted the concept of empathy current in late-nineteenth-century German psychology; the human body is symmetrical and thus a person standing before the symmetrical façade of a building experiences aesthetic pleasure by empathetic projection (*Prolegomena to a Psychology of Architecture*, 1886).[5] However, Wölfflin's most influential book, *Kunstgeschichtliche*

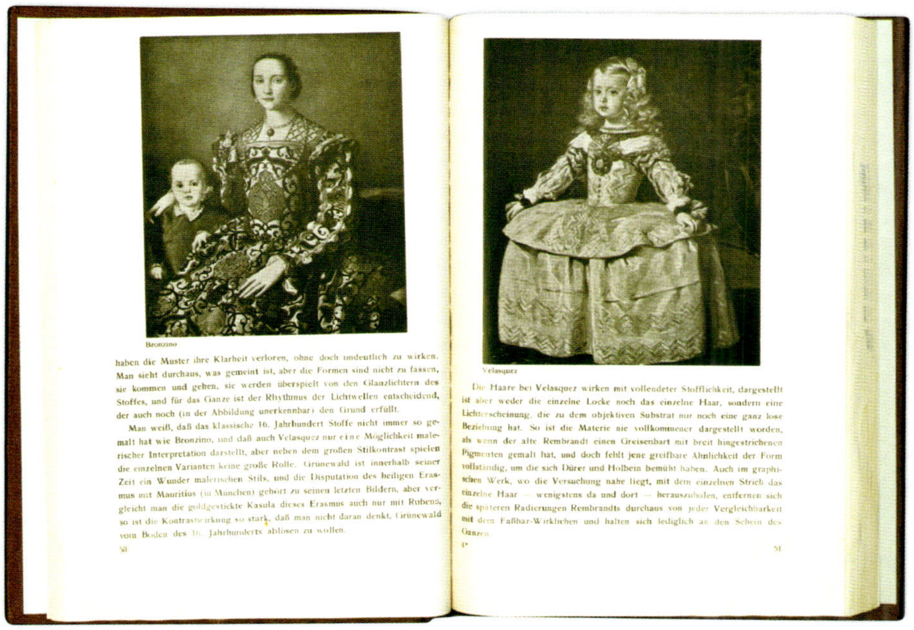

7-6. Comparison of Italian Renaissance and Baroque portraits, in Heinrich Wölfflin, *Kunstgeschichtliche Grundbegriffe* (Art history basics; Munich: Bruckmann, 1915), 50–51.

Bronzino portrayed Eleanor of Toledo (left) in a Renaissance linear style in which the figure is clearly outlined in smooth paint application, creating crisp edges between all forms. By contrast, according to Wölfflin, Velázquez used a painterly Baroque style to present Margarita of Spain (right) in a dense thicket of brushstrokes that merges seamlessly with the background.

Grundbegriffe (Art history basics, 1915; translated into English in 1932 as *Principles of Art History*), reflected the shift in focus in German psychology from the viewer's subjective feeling to an analysis of an objective pattern (a Gestalt) in the artwork. Wölfflin identified five transformations—for example, from linear to painterly—to describe changes in style from the Renaissance to the Baroque; as Wölfflin mused, "It is probably a question of the unfolding under a law, and of an effect relevant to psychology and to a rationale."[6] Remaining on the faculty at Zurich until his death in 1945, Wölfflin was a popular teacher, and Speiser regularly attended his lectures.[7] Using two projectors, Wölfflin developed what became a staple of art history lectures: showing a pair of pictures (a "slide comparison") of the same subject matter in different styles. Speiser must have enjoyed seeing a visualization of a transformation under invariance, as Wölfflin described the same subject (such as a portrait) painted in a linear style during the Renaissance (in one space-time framework) transformed by an artist using a painterly style in the Baroque (another space-time framework; see plate 7-6).[8]

At the University of Zurich Speiser taught a seminar on the philosophy of mathematics. Among the topics of discussion were the ideas of the German scholar Ernst Cassirer, who was based in Hamburg in the 1920s and early 1930s. At the opening of the twentieth century Cassirer had begun studying the formalist mathematics of Hilbert at Göttingen, and he proposed that the historian could adopt Hilbert's formalist method of analyzing abstract structure. Thus Cassirer undertook to discern Hilbert's so-called theory forms—which Cassirer called "symbolic forms"—in culture. Cassirer began with the ambitious task of analyzing the abstract scaffolding of aesthetic theories, ethics, and religious creeds, as well as the scientific worldview (*Substance and Function*, 1910). After Einstein's theory of relativity was confirmed in 1919, Cassirer adopted its core idea: an accurate description of the natural world can be given from any frame of reference (any co-ordinate system), and all co-ordinate systems are interrelated (together they comprise the universe). Cassirer

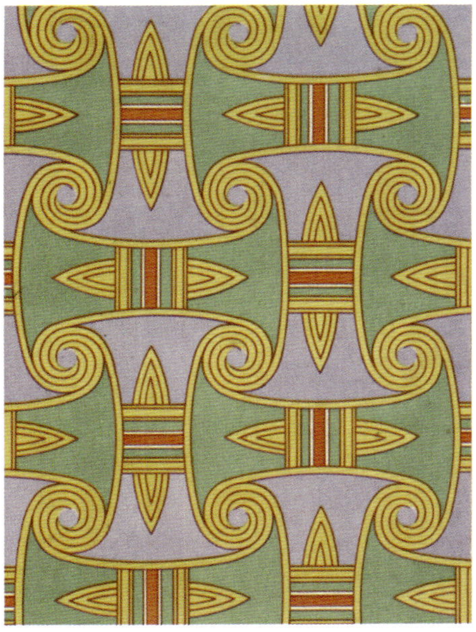

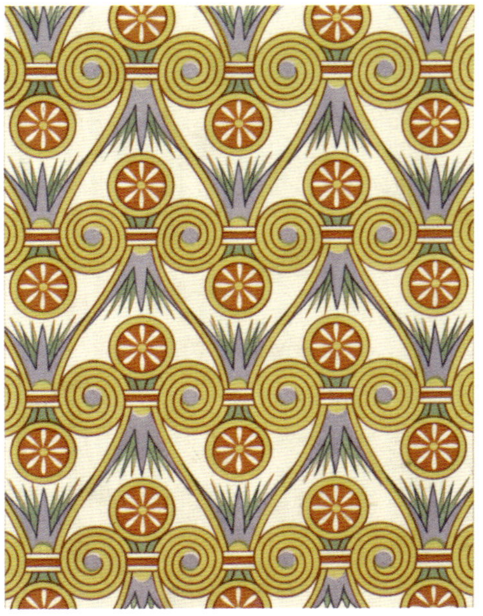

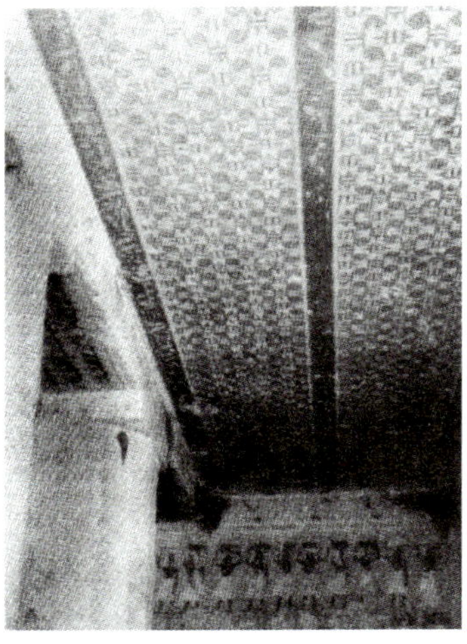

ABOVE

7-7. Ceiling patterns from the Necropolis of Thebes, Eighteenth and Nineteenth Dynasties, in Émile Prisse d'Avennes, *Histoire de l'art égyptien d'après les monuments depuis les temps les plus reculés jusqu'à la domination romaine* (Paris: Bertrand, 1878). Color lithograph. Art & Architecture Collection, Miriam and Ira D. Wallach Division of Art, Prints and Photographs, The New York Public Library, Astor, Lenox, and Tilden Foundations.

In his *Die Theorie der Gruppen von endlicher Ordnung* (Group theory of finite order), Speiser reproduced these ceiling patterns in black and white (1923, 2nd ed. 1927, 91–93). Similar patterns are visible in the fragmentary remains of the ceiling of the Tomb of Menna (see plate 1-9 in chapter 1).

LEFT

7-8. Ceilings of an Egyptian tomb, in Andreas Speiser, *Die mathematische Denkweise* (Basel, Switzerland: Birkhäuser, 1945), plate 1. Used with the permission of Springer Science and Business Media, Heidelberg.

As Speiser tells the reader of this book (118), he took this photograph on his trip to Egypt in 1928.

envisioned that he could similarly unite his aesthetic, ethical, religious, and scientific symbolic forms into an all-encompassing cultural theory of relativity.

Cassirer had an excellent resource for pursuing this goal at the University of Hamburg's Warburg Library of the Cultural Sciences. Founded by the art historian Aby Warburg, it had a rich archive of images and writings about ancient and Renaissance art, as well as the history of mythology and religion. Cassirer first presented his cultural theory of relativity in a series of lectures in the mid-1920s at the Warburg Library, which were published as his three-volume *Philosophy of Symbolic Forms* (1923, 1925, and 1929). In 1926 the young Erwin Panofsky joined Hamburg's faculty as professor of art history and became closely associated with Cassirer. The following year Panofsky, a scholar of Renaissance art, reflected Cassirer's outlook in a study of how the representation of space in painting reflects the worldview—the co-ordinates in space-time—of the artist ("Perspective as Symbolic Form," 1927).

In 1923 Speiser published a technical study of group theory for mathematicians (*Die Theorie der Gruppen von endlicher Ordnung* [The theory of groups of finite order]).

When he revised the book for a second edition in 1927, he added a chapter applying group theory to the decorative arts of Egypt and the Near East, and demonstrated that this art had real mathematical content. Speiser expanded on the theme a few years later in a book for the general public (*Die mathematische Denkweise* [The mathematical way of thinking], 1932). After describing the tiling (tessellation) of a plane with repeated shapes (of which there are seventeen possible patterns), Speiser described how Egyptian decorative patterns were the product of the repeated application of a rule—an algorithm (such as "rotate 90°")—which is a hallmark of abstract thought. He showed that the anonymous Egyptian weavers of fabrics and mats, and the creators of decorative ceilings in tombs, had discovered all seventeen possible repetitive patterns, thus demonstrating that these Egyptian craftsmen had an intuitive grasp of principles, which, since the nineteenth century, are regarded as theorems of group theory (plates 7-7 and 7-8). In other words, Egyptian geometric decorations embody mathematical patterns in the same way that Euclid's geometric diagrams in *Elements* contain the essence of his proofs. Inspired by Speiser, Weyl gave a series of lectures on symmetry and the arts at Princeton University that were published as *Symmetry* (1938).[9]

The oldest examples of surface ornament are from Egypt. We do not know whether they had a mathematical theory of groups, but their figures are certainly a geometric achievement. Today our mathematical theories are written in the form of theorems and proofs, but this is the influence of Greek mathematics. But the geometric image is the real essence of logical reasoning.

—Andreas Speiser, *Die mathematische Denkweise* (The mathematical way of thinking), 1932

Speiser travelled to Granada, Spain, to see the Islamic ornamental patterns in the architecture of the fourteenth century Moorish citadel, the Alhambra, in the company of mathematician Wolfgang Greison, who had done a mathematical analysis of Johann Sebastian Bach's *Art of the Fugue*. Greison inspired Speiser to apply group theory to Bach's fugues, and Speiser observed how Bach began each by stating a melodic pattern of notes and then he performed various transformations on the score—a mirror image, a rotation, and a reflection. The young Mozart, inspired by Bach, similarly composed music that incorporated symmetrical patterns (Speiser, *Musik und Mathematik* [Music and mathematics], 1926).

In addition to describing the patterns in Egyptian and Islamic art, Speiser applied group theory to patterns in classical architecture, including a mosaic from Pompeii, decorative foliage from the middle ages, and Islamic shrines, including a fifteenth-century mosque in Cairo (plate 7-9). He encouraged his students to apply group theory to decorative patterns, and one wrote her doctoral thesis on the Islamic tiling patterns at the Alhambra (Edith Müller, *Gruppentheoretische und Strukturanalytische Untersuchungen der Maurischen Ornamente aus der Alhambra in Granada* [Group theory and structural analysis of the Moorish ornament at the Alhambra in Granada], 1944).[10] This same Islamic tiling inspired M. C. Escher, who visited the Alhambra in 1936, to study the mathematics of tessellation and make interlocking figurative tiles of his own design (plates 7-10 and 7-11).[11]

The art of ornament contains in implicit form the oldest piece of high mathematics known to us. To be sure, the conceptual means for a complete abstract formulation of the underlying problem, namely the mathematical notion of a group of transformation, was not provided before the nineteenth century; and only on this basis is one able to prove that the seventeen symmetries already implicitly known to the Egyptian craftsman exhaust all possibilities.

—Hermann Weyl, *Symmetry*, 1938

Speiser's application of group theory to the decorative arts was adopted by anthropologists to classify decorative patterns on artifacts. Speiser was familiar with this burgeoning field from his brother, Felix Speiser, who was an anthropologist; in 1910–12 he did field work in the South Pacific on the peoples of the Vanuatu Stone Age culture of the New Hebrides Islands (today the Republic of Vanuatu; Felix Speiser, *Ethnology of Vanuatu*, 1923). Felix Speiser went on to become a professor of anthropology at the University of Basel and kept his brother apprised of developments in anthropology. The use of

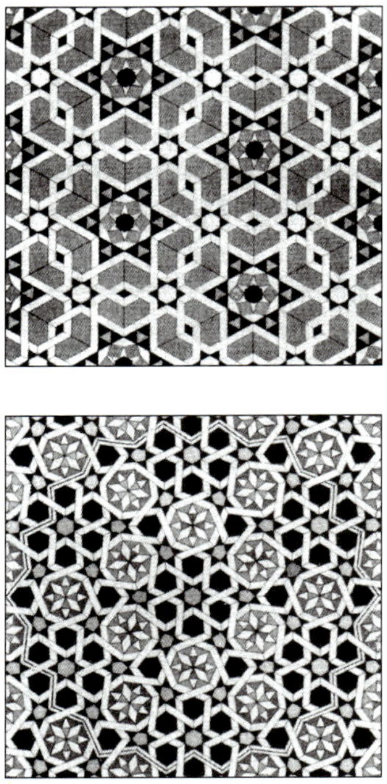

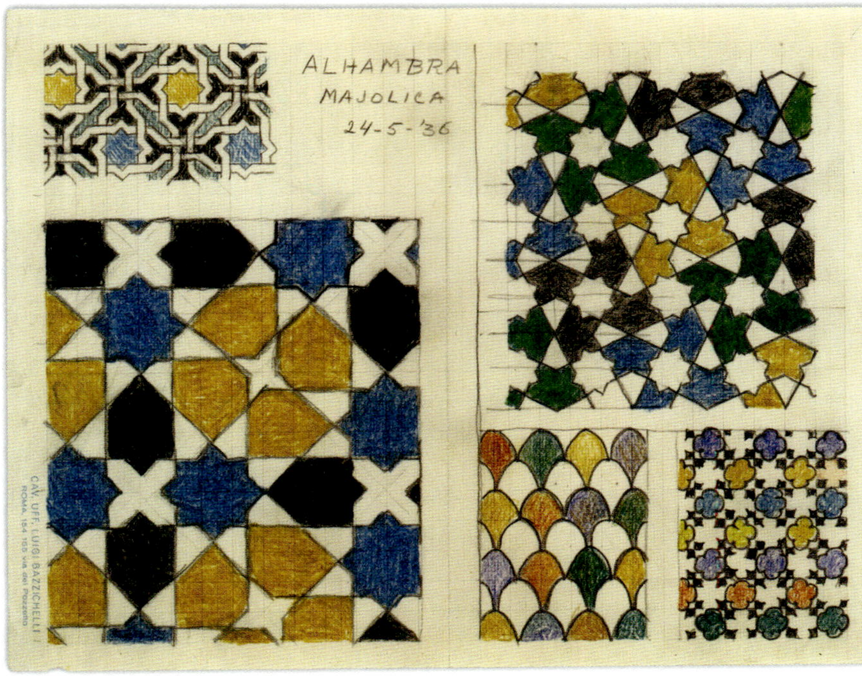

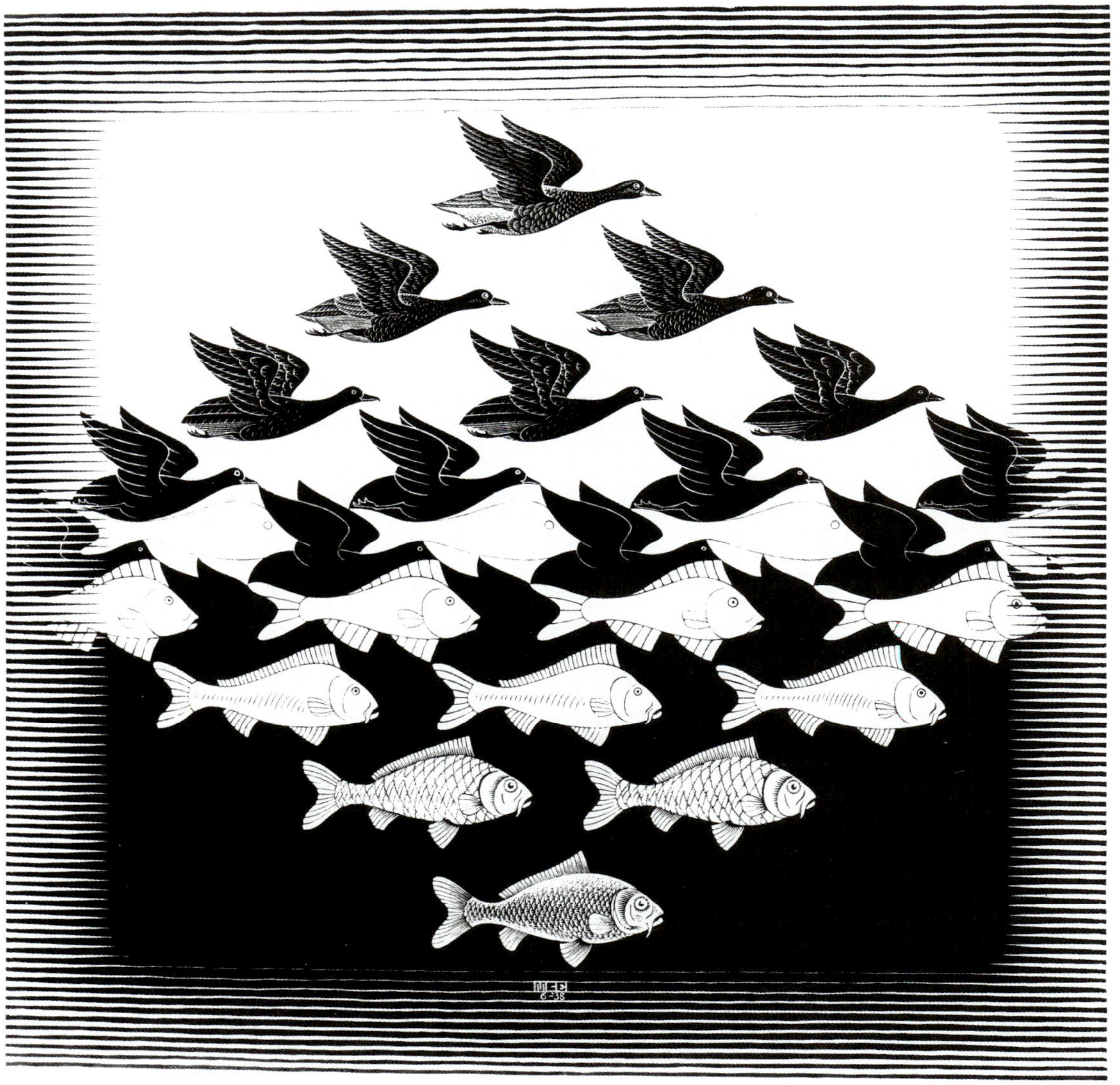

ABOVE LEFT

7-9. Islamic tile patterns, in Andreas Speiser, *Die mathematische Denkweise* (Basel, Switzerland: Birkhäuser, 1945), plate 7. Springer Science and Business Media, Heidelberg. Used with permission.

ABOVE RIGHT

7-10. M. C. Escher (Dutch, 1898–1972), drawing after the mosaics in the Alhambra, 1936. Pencil and colored crayon. Collection of Michael S. Sachs, Inc. © 2009 The M. C. Escher Company, The Netherlands. All rights reserved.

RIGHT

7-11. M. C. Escher (Dutch, 1898–1972), *Sky and Water I*, 1938. Woodcut, 17⅜ × 17⅜ in. (44.1 × 44.1 cm). © 2009 The M. C. Escher Company, The Netherlands. All rights reserved.

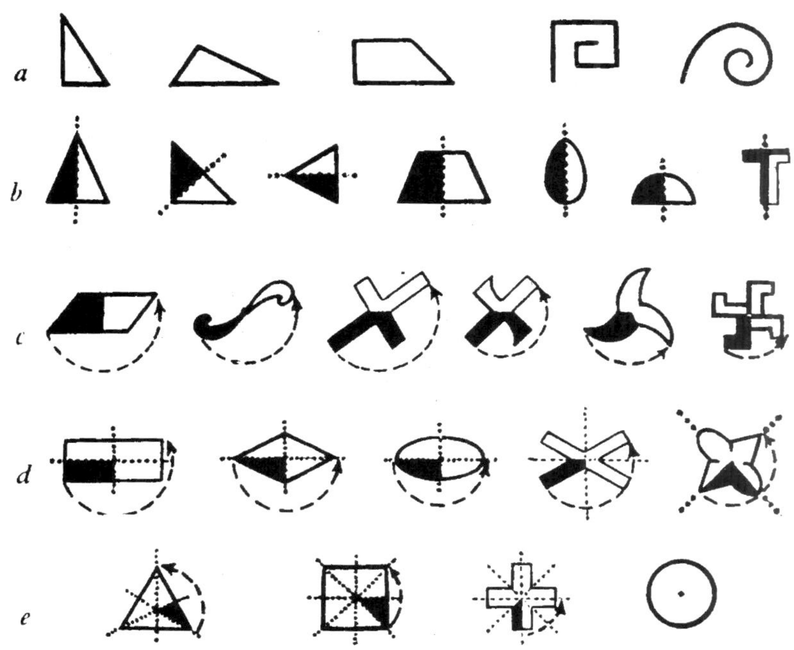

LEFT

7-12. "The Symmetry of Finite Design Illustrated by Simple Figures," in Anna O. Shepard, *The Symmetry of Abstract Design with Special Reference to Ceramic Decoration* (Washington, DC: Carnegie Institution, 1948), 218, fig. 1. Used with permission.

According to Shepard's classification system, the figures in the top row (a) are asymmetrical, those in (b) each have an axis of bilateral symmetry, those in (c) have rotational symmetry, those in (d) have both radial and biaxial symmetries, and the figures in (e) have both radial and triaxial (or multiaxial) symmetries.

BELOW

7-13. Claude Lévi-Strauss's diagram of marriage relations in *Les structures élémentaires de la parenté* (1949), translated by J. H. Bell, J. R. von Sturmer, and Rodney Needham as *The Elementary Structures of Kinship* (Boston: Beacon Press, 1969), 178. Used with permission.

With A and C representing males and B and D females, Lévi-Strauss diagramed all possible (heterosexual) marriages: (on the left) A marries B, B marries C, C marries D, and D marries A; and (on the right) since marriage is symmetrical, it is also possible that B marries A, and so on.

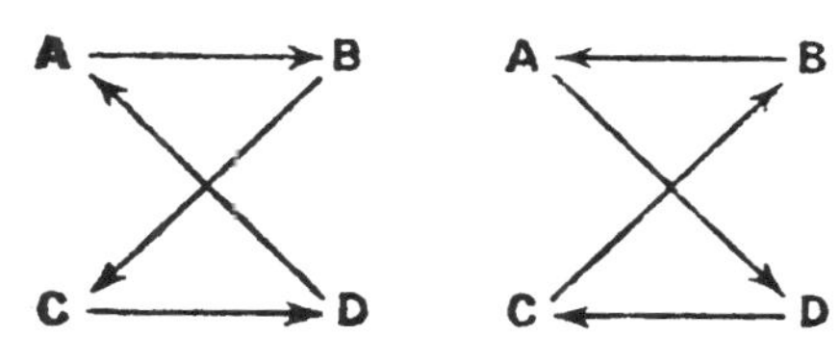

group theory to study symmetrical patterns in the decorative arts by Andreas Speiser (1927, 1932), Weyl (1938), and Müller (1944) became well known in German and English-speaking intellectual circles. In the 1940s American anthropologists used group theory to classify decorative motifs in American Indian pottery; George Brainerd analyzed the patterns in prehistoric pottery ("Symmetry in Primitive Conventional Design," *American Antiquity*, 1942), and Anna O. Shepard catalogued symmetrical patterns in her classic *The Symmetry of Abstract Design with Special Reference to Ceramic Decoration* (1948; see plate 7-12), in which she credits Speiser's pioneering work. In 1941 the French anthropologist Claude Lévi-Strauss, who was of Jewish descent, left Nazi-occupied France for the United States, where he spent the war years doing research at the New York Public Library. Analyzing field data collected by others, Lévi-Strauss adopted group theory from Brainerd and Shepard to describe patterns he found in the data, purporting to prove that societies embody the symmetrical structure of a Klein four-group (*Les structures élémentaires de la parenté*, 1949, English translation as *The Elementary Structures of Kinship*; see plate 7-13 and compare with the sidebar on page 253).

SYMMETRY OF THE MIND: GESTALT PSYCHOLOGY

Contemporary with scientific discoveries of symmetrical patterns in nature, early-twentieth-century psychologists were finding evidence that the human eye, ear, and brain have innate mechanisms for perceiving a pattern—*Gestalt* (German for "shape" or "form")—as well as an innate ability to distinguish those that are symmetrical. Some scientists even dared to speculate that the symmetries discovered by physicists, chemists, biologists, and psychologists pointed to a comprehensive wholeness and unity in nature.

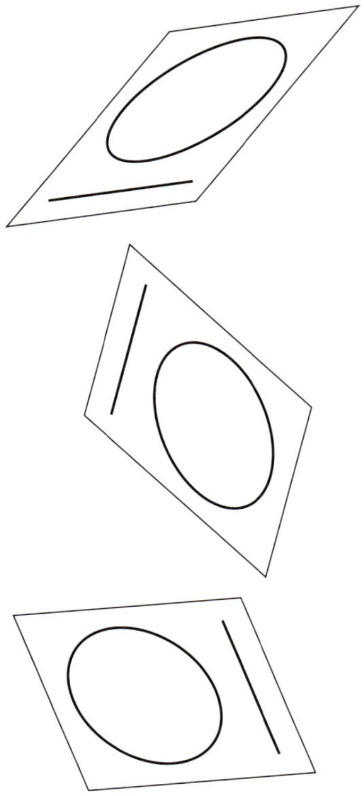

7-14. A circle and line seen from different angles.

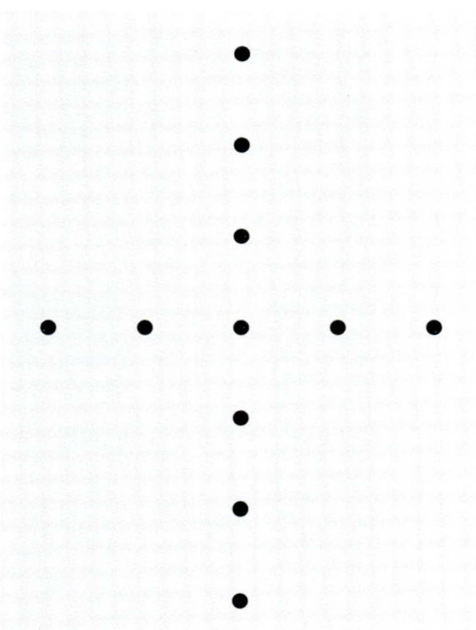

7-15. Gestalt experiment with a pattern of dots.

In the 1890s Felix Klein had asked what transformations allow an ellipse and straight line to remain invariant when they are projected from different viewpoints. Klein's contemporary, the Austrian philosopher Christian von Ehrenfels, asked a related question about the human mind. When a person looks at pictures of the same geometric form from different angles, how is it that the mind instantly recognizes that these distorted images all represent the same figure (plate 7-14)? Ehrenfels reasoned that the unification of these pictures takes place in the mind, and he set about to discover the mental faculties that perform this task. He formulated a principle of invariance whereby simple geometric objects such as circles and lines are recognized despite distortions caused by their rotation, reflection, or transformation of scale.

Ehrenfels developed his theory in opposition to the claim by his contemporary, the Austrian physicist Ernst Mach, who proposed that one perceives the world as an accumulation of discrete "atoms" of sense-data such as patches of color and instants of sound (see chapter 8). Ehrenfels argued that the eye, ear, and brain together perceive patterns of light and sound as a unity. Recognizing a melody is more than hearing the sum of the atoms of sound (the notes) because, according to Ehrenfels, the assemblage of notes is perceived as a musical pattern—a Gestalt (the melody). One recognizes the melody if it is moved up an octave or transposed into another key because the melody is invariant under these transformations ("On Gestalt Qualities," 1890).

In the 1910s and 1920s Ehrenfels's student Max Wertheimer, together with Wolfgang Köhler and Kurt Koffka, developed Ehrenfels's insight into a school of Gestalt psychology that was centered in Berlin. Using the methods of experimental psychology that had been originated in the nineteenth century by Gustav Fechner, the young Gestalt psychologists showed images to participants and asked them to describe their subjective responses. For example, when shown an array of dots (plate 7-15), most subjects reported seeing a cross (rather than eleven dots). Wertheimer took this as evidence that the mind immediately perceives the pattern, which it then analyzes into individual dots, and not the other way round, as Mach claimed.[12] Experimental psychologists such as Fechner and his disciple Wilhelm Wundt had explained optical illusions in terms of the physiology of vision first described by Helmholtz (*Handbook of Physiological Optics*, 1856–67). Wertheimer, Köhler, and Koffka extended their forbear's explanation of vision from the eye to the brain, hypothesizing (correctly) that there are innate neurological substrates for recording patterns.

When Köhler was a student at the University of Berlin, he studied not only psychology but also physics. His teacher, the physicist Max Planck, was, like Einstein, working to form a unified picture of the universe. Köhler speculated that the notion "Gestalt" provided a way to link not only psychology but also cosmology and physics into a unified vision of nature, and later, as the director of the Psychological Institute in Berlin, he worked to extend the concept of Gestalt from psychology to physics.

In the late nineteenth and early twentieth centuries, the concept of mass as Newton's point-particles moved by forces was replaced by an understanding of fields of energy: James Clerk Maxwell's electromagnetic fields (plate 7-16) and Einstein's gravitational fields. Köhler noted an analogous move in psychology from conceiving of vision in terms of points

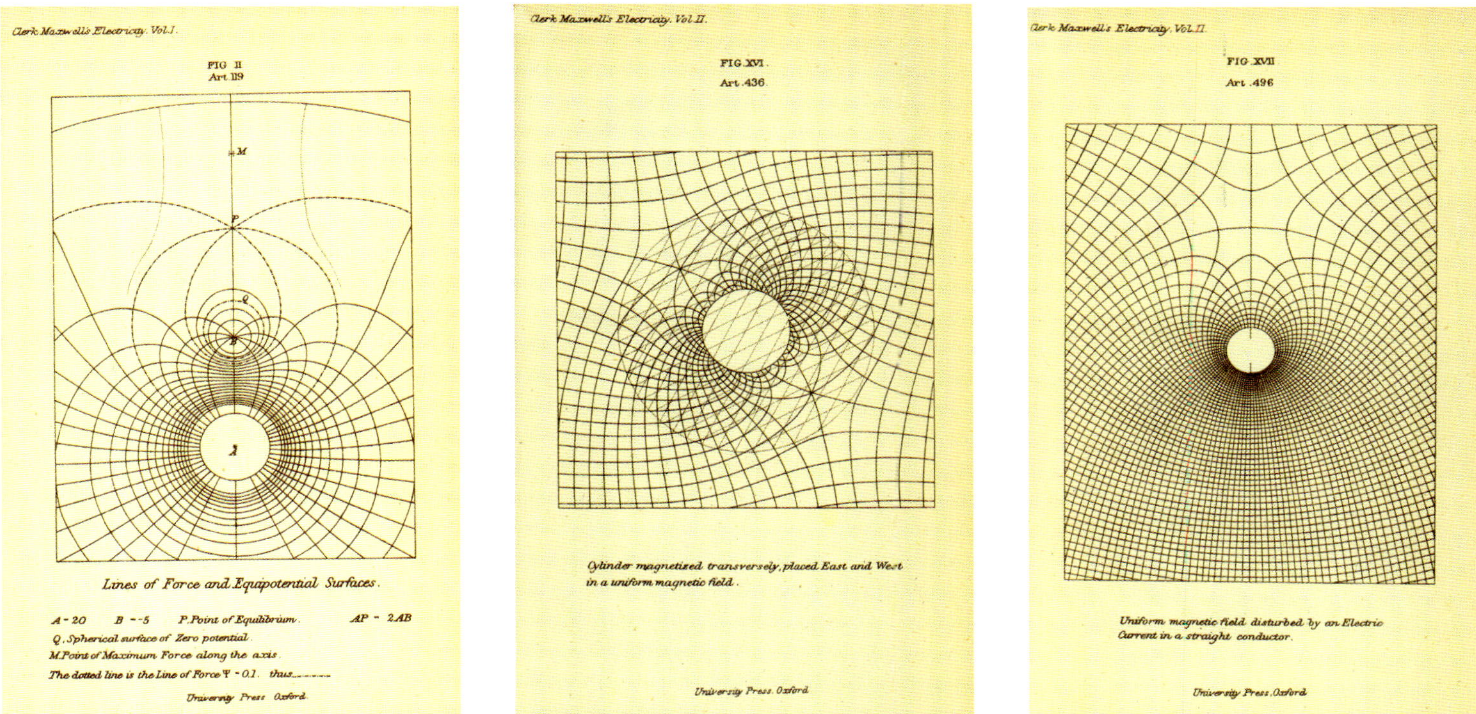

7-16. Force fields diagrams, in James Clerk Maxwell, *A Treatise on Electricity and Magnetism* (Oxford, England: Clarendon, 1873), from left to right: figs. 1:2, 2:16, and 2:17.

(Mach's sense-data) to fields (Gestalt patterns). In both physics and psychology the search for fundamental patterns had revealed a tendency for physical and psychological data to assume their simplest forms. Köhler gave Maxwell's field diagrams as an example of evenness in distribution of energy, simplicity, and symmetry (*Physical Gestalten at Rest and in a Stationary State*, 1924).

In Geneva in the early 1920s the Swiss psychologist Jean Piaget began a study of human cognition that corroborated the Gestalt principles being developed in Berlin. While doing research for his 1918 doctorate in biology, Piaget hypothesized that each organism constructs a unified worldview—a whole—from the fragmentary parts of its sensory experiences. His university studies included courses in mathematics, about which Piaget recalled that "group theory was particularly important for me with respect to the problem of unifying the whole and the parts."[13]

Piaget theorized that children have different logical and representational capacities at different ages, from infancy to adolescence, and cognitive maturation entails the progressive ability to conceive of the world more abstractly. In 1929 he published the first results of his research in the founding text of child development, *The Child's Conception of the World*. That same year Piaget joined the faculty at the University of Geneva, where he undertook an intensive ten-year study of the principal concepts of biology, mathematics, and physics with the goal of formulating a conceptual framework for the mature understanding of the world attained in adolescence.

Einstein, who was a close friend of Max Wertheimer,[14] followed the development of Gestalt psychology with interest. In 1928 Einstein suggested to Piaget that he undertake a developmental study of how children form concepts of time and speed.[15] Einstein's point was that in Newton's physics time is primary, and velocity is defined from it (speed, such as fifty miles per hour, is distance travelled divided by time). In Einstein's universe, velocity (the speed of light) is primary (it is a constant), while the passage of time is relative. In response to Einstein's suggestion, Piaget published two books in 1946 about the development in children of the ideas of time, motion, and speed.[16] Piaget's work on cognitive development has been the subject of many critiques, but his account of spatial and temporal development in children remains sound.[17] During World War II Piaget remained in neutral Switzerland, but with the rise of Nazism in the mid-1930s Wertheimer, Köhler, and Koffka left their German homeland for the United States, where the findings of Gestalt psychology were integrated into general psychology.

CONCRETE ART IN SWITZERLAND IN THE 1930S AND 1940S

Beginning in the 1930s, a group of artists in Zurich made art with symmetrical patterns in a style—Concrete Art—that was shaped by the aesthetics of De Stijl and Russian Constructivism. Representative Dutch and Russian works were regularly seen in Switzerland, such as in the large international *Die Konstruktivisten* (The Constructivists) exhibit in Basel in 1937, with works by Mondrian, Rodchenko, Tatlin, Strzemiński, and El Lissitzky. The Swiss artists also drew inspiration from Einstein's view of nature as a unified structure and Speiser's related writings on group theory.

Max Bill, Camille Graeser, Richard Paul Lohse, and Verena Loewensberg adopted the formalist aesthetic: the work of art as an autonomous system of meaning-free signs. In 1937 these artists formed the group Allianz, which exhibited together for the first time the following year in Basel (*Neue Kunst in der Schweiz*, 1938).[18] After World War II the younger artist Karl Gerstner joined the group and added algorithms for generating patterns of color (see chapter 11). Several of the Swiss Concrete artists worked as graphic designers and were educated in Gestalt psychology and the visual perception of symmetrical pattern.[19] They were familiar with Einstein, Speiser, and Weyl's popular writings on symmetry and group theory. Together these artists created a unique style that embodies symmetry. Their work is well-proportioned, often with bilateral symmetry, and some pieces display invariant features while undergoing particular transformations specified by an algorithm.

When Concrete artists introduced algorithms to generate forms, the resulting patterns had a more precise and uniform appearance than those of their German, Dutch, and Russian predecessors. For example, in Lissitzky's *Proun* (1922–23; see plate 8-8 in chapter 8), the artist arranged meaning-free rectangles, lines, and curves, creating the composition intuitively—*by eye*—based on experience. A bold black rectangle leans to the left, creating a center of interest, which is balanced by a long rust-colored rectangle and a heavy black line; curves on top are balanced by curves on the bottom. Lissitzky's placement of forms is somewhat arbitrary, in the sense that he could have repositioned the forms and the composition would still look balanced.

Trained in Germany at the Bauhaus school of design in 1927–28, Max Bill balanced his early compositions by eye (*Variation*, 1934; see plate 7-17).[20] But in the late 1930s Bill introduced algorithms to determine his compositions, as in the portfolio of prints *Fifteen*

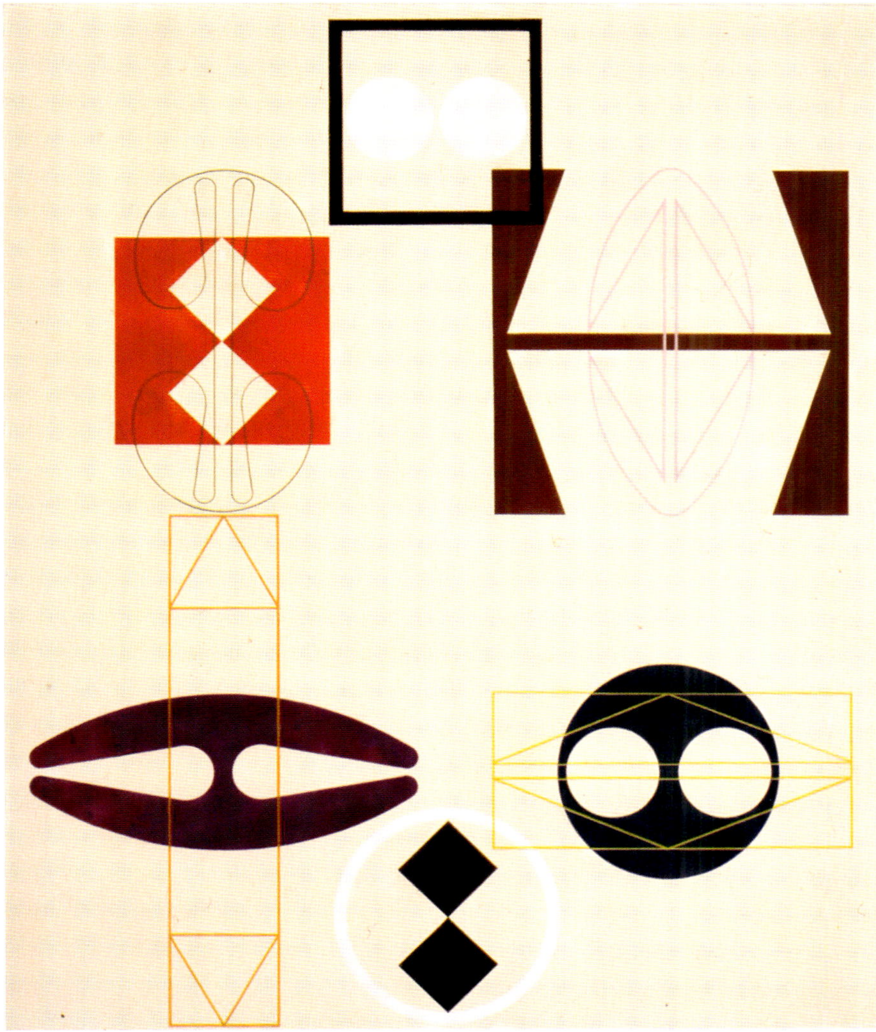

7-17. Max Bill (Swiss, 1908–94), *Variation*, 1934. Oil on canvas. © max, binia + jakob bill foundation. © 2014 ProLitteris, Zurich/Artists Rights Society, New York.

The viewer's attention is drawn to the bottom of the painting by the purple lozenge shape, the two black diamonds, and the dark blue circle with round "eyes." These shapes are balanced by their "twins" at the top of the painting: the red square with diamond-shaped holes, the black square with white "eyes," and the brown H-shape on the right.

Variations on a Theme (1938; see plate 7-18). The suite of prints resonates with themes familiar to Bill from Einstein's cosmology and Gestalt psychology: a form can be transformed from one pattern to another using rules that preserve the original (invariant) design.

Artists in Germanic culture knew the basic ideas of Einstein's space-time universe from the many lucid, popular accounts available to them, beginning with Einstein's own *Relativity: The Special and General Theory, a Popular Exposition* (1917).[21] Instruction in physics was on the Bauhaus curriculum when Bill studied there, and training in graphic design entailed lessons in Gestalt psychology. At the invitation of Walter Gropius, the director of the Bauhaus, Einstein joined the school's board of directors in 1924, and the Gestalt psychologist Karl Duncker, a colleague of Köhler in Berlin, gave a lecture on Gestalt psychology at the Bauhaus in 1929 and taught there in 1930–31.[22] This was after Bill's student days, but it indicates the interest in psychology at the school. Paul Klee, who taught the foundations course during Bill's tenure, recorded exercises using Gestalt figures in his teaching notebooks (plate 7-19).

When Bill executed *Fifteen Variations on a Theme*, he knew of Speiser through several people in the art world, including the Swiss architect Le Corbusier.[23] Speiser moved in modern art circles after his 1916 marriage to Emmy La Roche, whose brother, the Swiss

7-18. Max Bill (Swiss, 1908–94), *Fifteen Variations on a Theme*, 1935–38. Twelve lithographs, 12 × 12⅝ in. (30 × 32 cm) ea., published as a suite (Paris: Éditions des Chroniques du Jour, 1938) in *Max Bill, Maler, Bildhauer, Architekt*, ed. Thomas Buchsteiner and Otto Letze (Ostfildern-Ruit: Hatje Cantz, 2005), 228–32. Used with permission. © 2014 ProLitteris, Zurich/Artists Rights Society, New York.

Bill began this series with the theme outlined on the right, which is made from 3-, 4-, 5-, 6-, 7-, and 8-sided polygons inset slightly off center. Bill then created the 15 variations by applying rules of transformations (changing the layout into points, lines, curves, colors, etc.), while keeping the theme invariant.

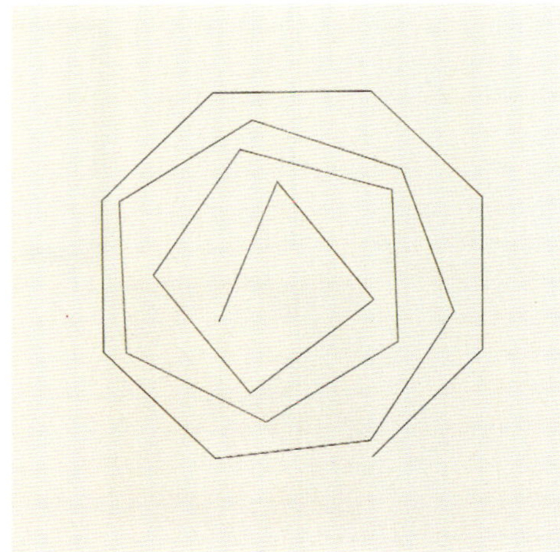

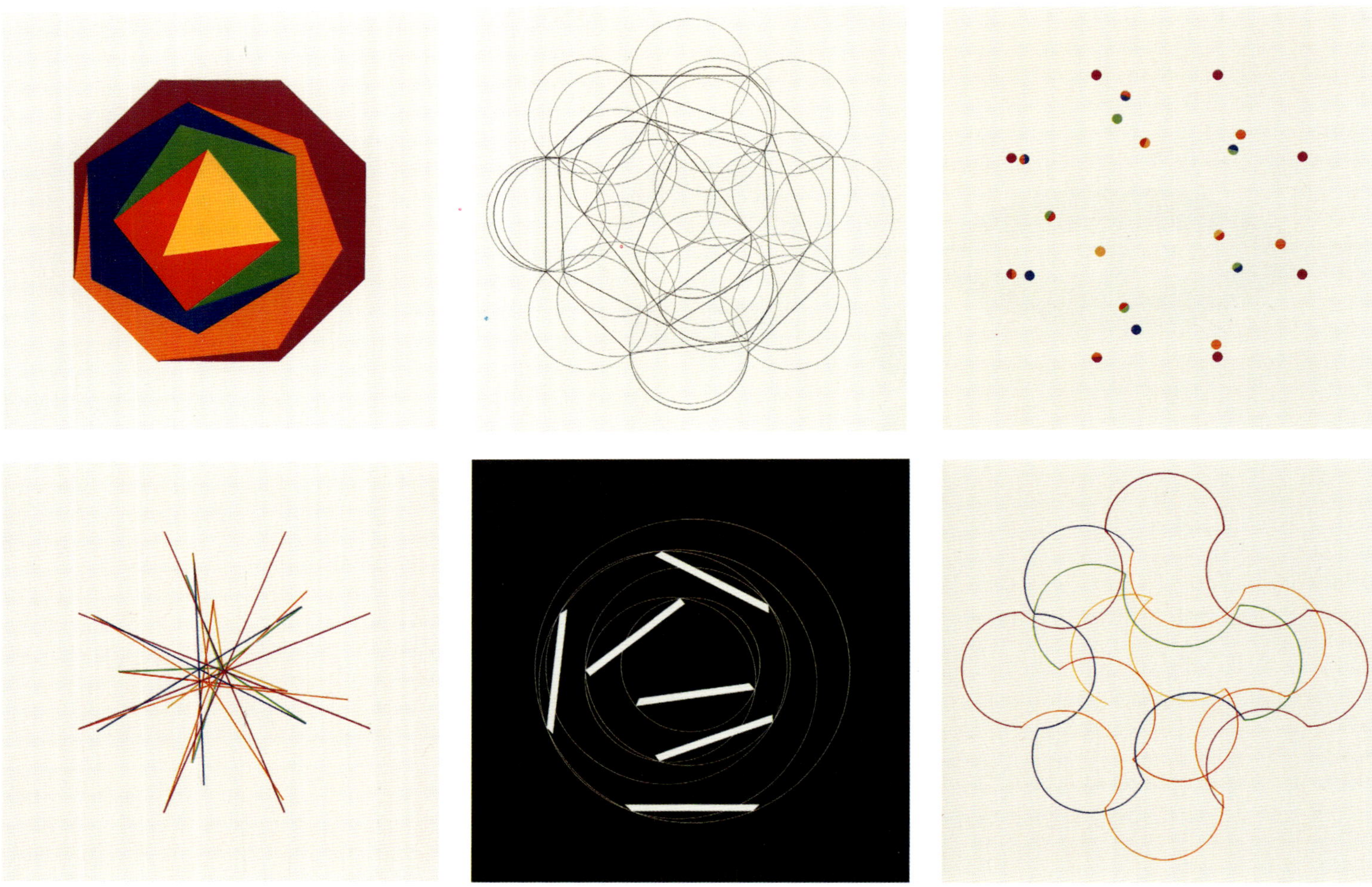

banker Raoul La Roche, was a collector of modern art in Paris. When the French government confiscated and sold German property after the outbreak of World War I, La Roche bought major Cubist paintings that had been seized from the gallery of Daniel-Henry Kahnweiler, the dealer for Picasso and Braque, and he made additional purchases from the French dealer Léonce Rosenberg, who represented Juan Gris, Fernand Léger, Amédée Ozenfant, and Le Corbusier. By the time La Roche stopped collecting in 1928, he had amassed a superb collection of about 160 works.[24] He also financed the art magazine

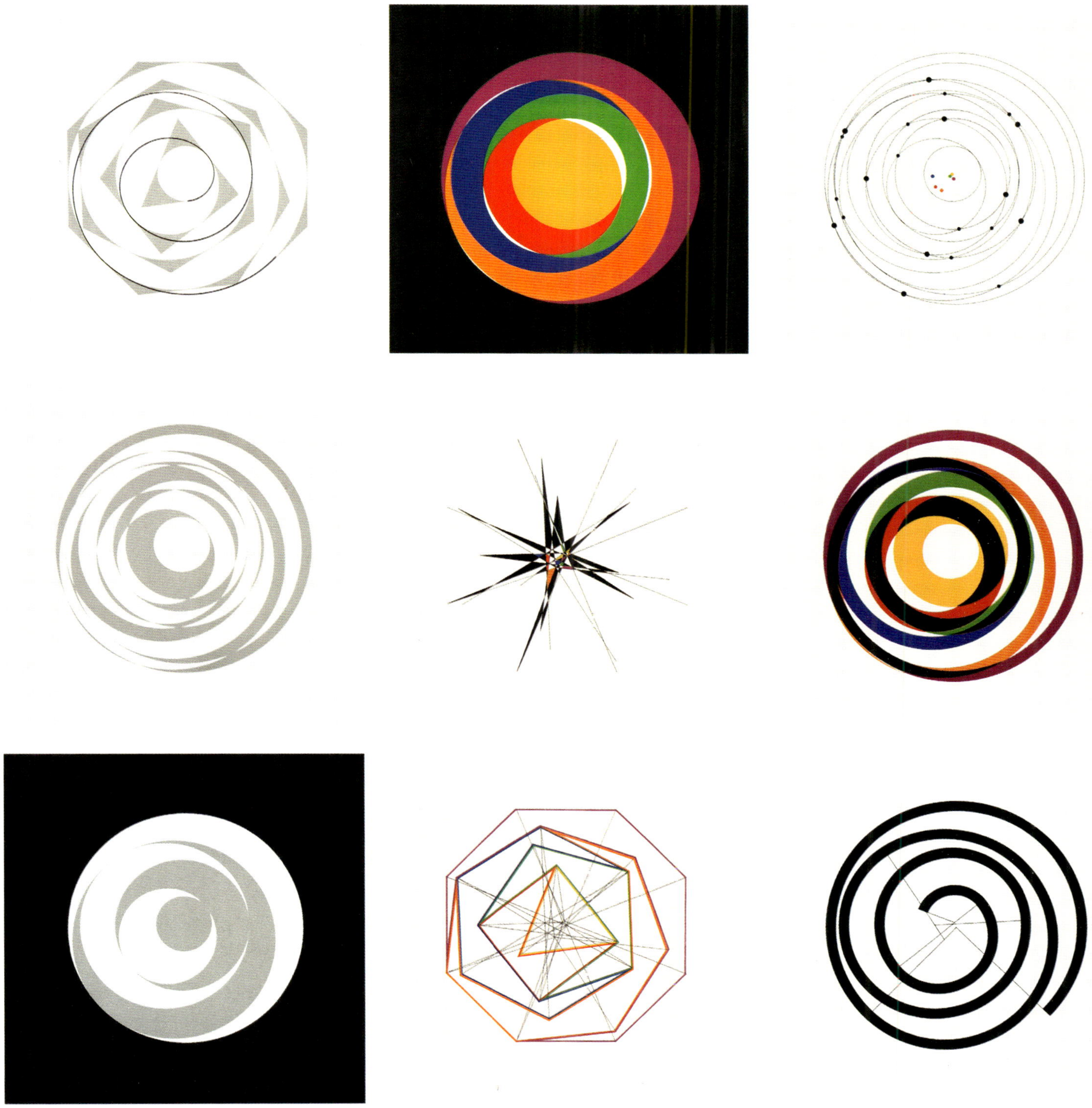

L'Esprit Nouveau, which was founded in 1920 by Le Corbusier, Ozenfant, and the Belgium poet Paul Dermée. In 1921 La Roche hired Le Corbusier to build a Paris residence to house his collection (Maison La Roche, Paris, 1923; see plate 2-34 in chapter 2).

Through La Roche, Speiser met Parisian artists, and he owned one of Braque's Cubist paintings.[25] Speiser became particularly friendly with Le Corbusier because of their shared interest in art and geometry. In recognition of Le Corbusier's work on a theory of proportion for architecture, Speiser recommended him for an honorary doctorate in

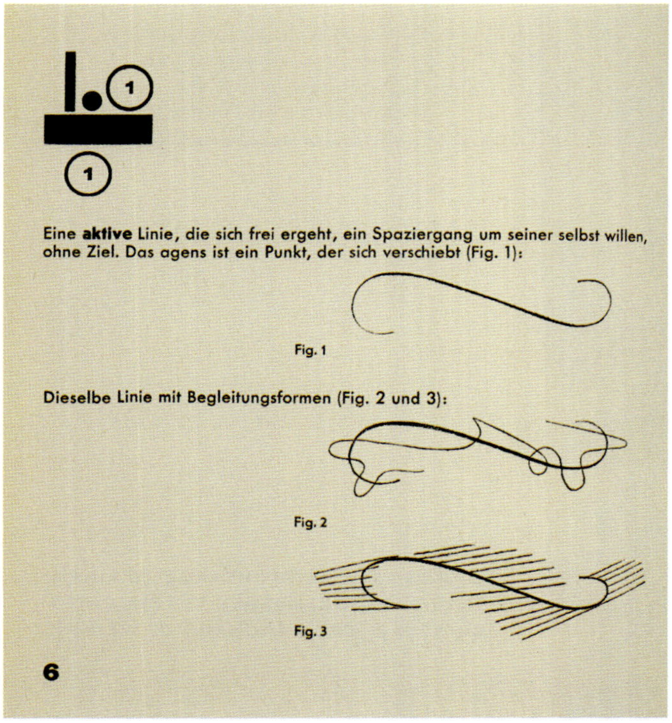

7-19. Paul Klee (Swiss, 1879–1940), *Pädagogisches Skizzenbuch* (Pedagogical sketchbook), Bauhausbücher 2 (Munich: Albert Langen, 1925), 6. Avery Architectural and Fine Arts Library, Columbia University, New York. © 2014 Artists Rights Society, New York.

Klee drew an active line (*eine aktive Linie*) (fig. 1), which is meandering on a leisurely walk, going nowhere in particular, led by a point. Then Klee demonstrated how one perceives a figure and its background together in a unified field (*Gestalt*), by showing how the form of the same active line can be obscured (fig. 2) or emphasized (fig. 3).

the philosophy of mathematics, which was awarded in 1931 by the University of Zurich.[26] At this time, Bill was editing Le Corbusier's collected works for publication, and this impressive event—an artist being awarded an honorary doctorate in mathematics—surely brought Speiser's *name* to Bill's attention. Le Corbusier was, however, a steadfast adherent to the Golden Section (see chapter 2), and he never showed any interest in Speiser's *mathematics* (group theory).[27] But Bill learned of group theory from two other colleagues who did read Speiser's publications—the Swiss psychoanalyst Adrien Turel[28] and the Swiss historian Sigfried Giedion.

In 1936 Bill organized an exhibition of Swiss artists for the Zurich Kunsthaus, *Zeitproblem in der Kunst* (The problem of time in art), which included about 160 works by forty-one artists, including Le Corbusier, Jean/Hans Arp, Sophie Täuber-Arp, Alberto Giacometti, Paul Klee, Richard Paul Lohse, Verena Loewensberg, and Bill himself. Speiser, then living in Zurich, certainly would have seen this exhibition.[29] It included Bill's *Variation* of 1934, but this painting (like another hundred in the show) was composed of simple geometric forms arranged by eye, and it lacked the distinctive appearance of symmetry that Bill achieved using algorithms to create *Fifteen Variations on a Theme*.

In *Die mathematische Denkweise* (a book dedicated to Raoul La Roche), Speiser described the ancient link between art and mathematics as culminating in Kepler's third law of planetary motion. It was in this law that the German astronomer had determined that the orbiting planets together comprise an interconnected system (the solar system), which is analogous to music (*Harmony of the World*, 1619; see the sidebar on page 63 in chapter 1). With the rise of science in the eighteenth century a split had opened between art and mathematics, which Speiser decried: "Modern art does not know about symmetry anymore."[30] When Speiser wrote *Die mathematische Denkweise* in 1932, he would have seen many styles of non-objective art, but none that were based on symmetry. He declared that to express its epoch, modern art should embody symmetry: "The group represents the principle of integral ratios which in antiquity ruled the search for the laws of nature, under the poetic name 'harmony of the spheres,' and built the basic laws of the world for Kepler. The Greeks called such a law 'logos,' which today includes the approach to nature on the macro and micro levels; it is a group concept. It allows us to chart the form of the cosmos as well as determining the possible arrangements of atoms in a crystal. Art also should be based on symmetry."[31]

Bill had been schooled in the formalist approach, and his design of *Fifteen Variations* suggests that he may have heeded Speiser's challenge by beginning to investigate the "group concept" as well as parallel ideas in Gestalt psychology. In the *Zeitproblem in der Kunst* catalogue, Bill wrote an essay in which he gave a short description of Concrete Art, which he, like Speiser, thought should express harmonic laws: "We call works of art 'Concrete' that come into being on the basis of their own inherent means and laws . . . not by abstraction from nature. Concrete painting and sculpture are composed of units of visual

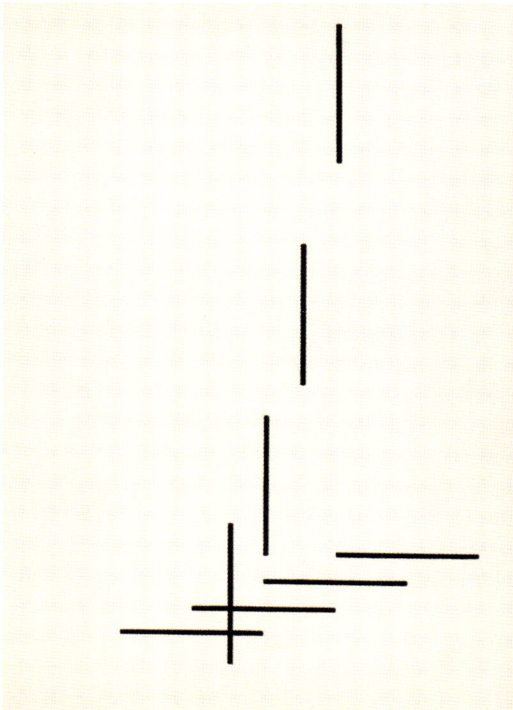

LEFT

7-20. Max Bill (Swiss, 1908–94), *Untitled*, from X = X (Zurich: Allianz, 1942), n.p. Suite of ten lithographs, 8¼ × 6 in. (21 × 15 cm) ea. Art and Architecture Collection, Miriam and Ira D. Wallach Division of Art, Prints, and Photographs. The New York Public Library, Astor, Lenox, and Tilden Foundations. © 2014 ProLitteris, Zurich/ Artists Rights Society, New York.

In the introduction to this suite of prints, Bill wrote that he based the lithographs on mathematics and geometry, yet each is "a new form with its own logic and legitimacy" (n.p.).

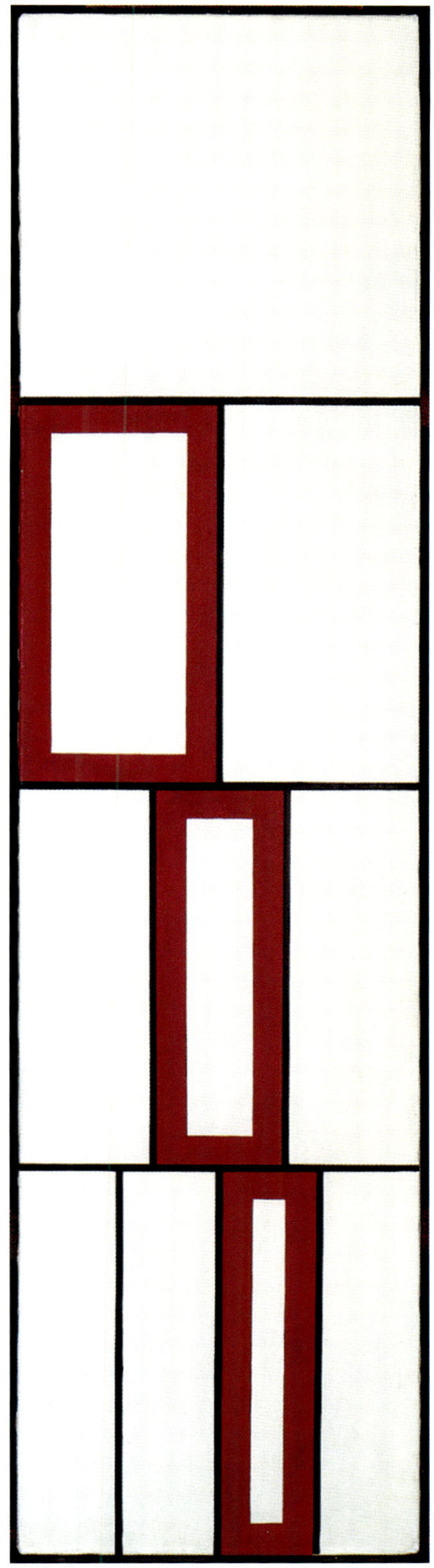

RIGHT

7-21. Max Bill (Swiss, 1908–94), *Progression with Four Squares*, 1942. Tempera on linen, 47¼ × 11¹³⁄₁₆ in. (120 × 30 cm). Kunstmuseum, Winterthur, Donated by Frawa AG, 1969. © Schweizerisches Institut für Kunstwissenschaft, Lutz Hartmann. © 2014 ProLitteris, Zurich/ Artists Rights Society, New York.

perception—colors, space, light and movement. From the shaping of these elements there springs new realities. Abstract ideas previously existing only in the mind are made visible in concrete form. Concrete Art in its ultimate outcome is the pure expression of harmonic laws and proportions."[32]

The title for the 1936 exhibit came from a catalogue essay that Bill commissioned from the young art historian Sigfried Giedion. A student of Wölfflin, Giedion had adopted his teacher's view that architecture is an expression of a culture's understanding of space and time, which Giedion discussed in his 1922 doctoral thesis *Spätbarocker und romantischer Klassizismus* (Late Baroque and romantic classicism). When Speiser and Wölffin were colleagues at the University of Zurich, Speiser too had become a student of Wölffin, and Speiser's publications in the 1930s similarly described art as a reflection of the science of its time. Giedion would certainly have known Speiser's writing because the mathematician was a close friend and colleague of his dissertation professor.

Giedion befriended Le Corbusier in Paris, and in 1928 they co-founded the influential organization dedicated to promoting modern architecture the Congrès Internationaux d'Architecture Moderne (CIAM). Giedion became a fixture of the modern art world in France and Switzerland. In the title *Zeitproblem in der Kunst*, Giedion apparently was referring to the relativity of time—the *Zeitproblem* (the "time problem") much discussed in the 1930s after the 1919 confirmation of Einstein's space-time cosmology. In a series of lectures on modern architecture at Harvard University in 1938–39 (*Space, Time, and Architecture*, 1941), Giedion explicitly stated that modern architecture is an expression of Einstein's new space-time cosmology. Thus Giedion was another figure in Bill's circle who shared Speiser's outlook.

By the early 1940s Bill was designing exclusively with algorithms such as plate 7-21,

RIGHT

7-22. Gallery 6 with Max Bill's work, as shown in the exhibition catalogue for *Konkrete Kunst* (Basel, Switzerland: Kunsthalle, 1944), 56. © 2014 ProLitteris, Zurich/Artists Rights Society, New York.

Bill's sculpture might have caught Speiser's attention because it is in the form of a Möbius strip (see plate 8-30 in chapter 8). Bill called his series *Endless Ribbon* after the symbol of infinity, ∞ , and in a 1972 interview, he recalled that he discovered the Möbius strip independently of the German mathematician August Möbius, who discovered it in 1858: "I was fascinated by a new discovery of mine, a loop with only one edge and one surface. I soon had a chance to make use of it myself. In the winter of 1935–36, I was assembling the Swiss contribution to the Milan Triennale, and there was able to set up three sculptures to characterize and accentuate the individuality of the three sections of the exhibit. One of these was the *Endless Ribbon*, which I thought I had invented myself. It was not long before someone congratulated me on my fresh and original reinterpretation of the Egyptian symbol of infinity and of the Möbius ribbon." The person who congratulated Bill on his Egyptian symbolism probably had in mind the circular symbol of a snake eating its tail, which symbolizes eternal life in the Egyptian *Book of the Dead* (ca. 1550 BC).

OPPOSITE

7-23. Camille Graeser (Swiss, 1892–1980), *Progression Red-Yellow-Blue*, 1944. Oil, 23⅝ × 13 in. (60 × 35 cm). Camille Graeser Stiftung, Zurich. © 2014 ProLitteris, Zurich/Artists Rights Society, New York.

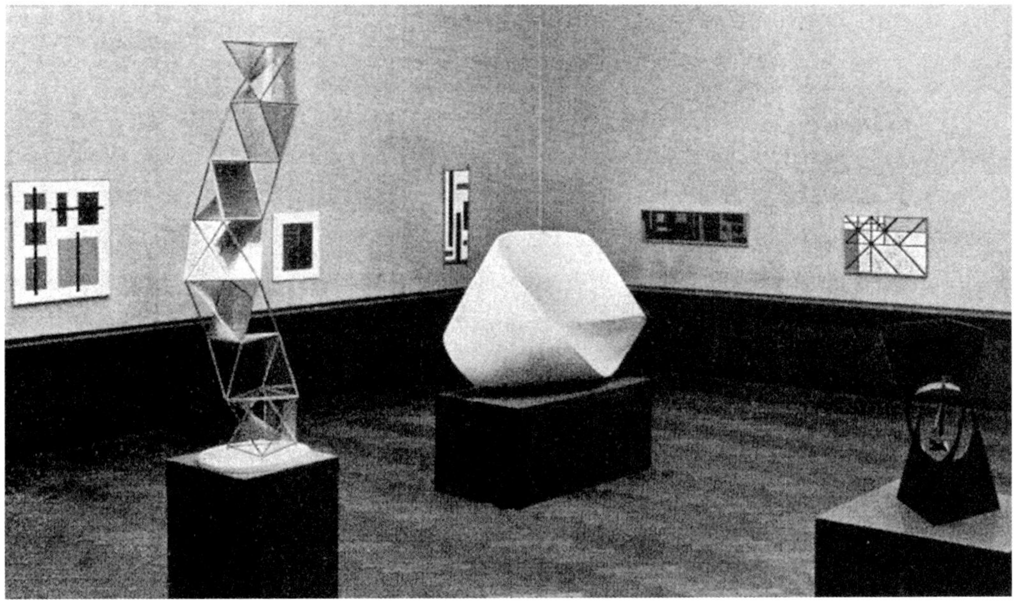

which he began with a square, and then divided it into two, three, and four parts, as well as the suite of prints with the tautological title X = X (plate 7-20). In 1944 he organized *Konkrete Kunst*, a large international survey exhibition of non-objective trends consisting of about 200 works by fifty-seven international artists: from Russia (Rodchenko, Tatlin, Strzemiński), Holland (Mondrian, van Doesberg), France (Jean Hélion, Robert Delaunay), Czechoslovakia (František Kupka), England (Henry Moore, Barbara Hepworth) and America (Alexander Calder). At the opening of the exhibition at the Kunstmuseum in Basel, the museum's director, Georg Schmidt, urged the audience to be "spiritually and intellectually receptive to the artistic expressions" of the artists.[33] Schmidt was a friend of both Speiser and La Roche (Schmidt negotiated the donation of key works from La Roche's collection to the Kunstmuseum in Basel).[34] Speiser had moved from Zurich to Basel in 1944 and certainly saw this exhibit, but most of the works were designed by eye (plate 7-22).

Following Bill's lead, Graeser, Lohse, and Loewensburg adopted design concepts from group theory and Gestalt psychology. Graeser began using simple algorithms; for example, in *Progression Red-Yellow-Blue* he doubled the area: blue is one unit, yellow is two, and red is four (1944; see plate 7-23).[35] Graeser owned a copy of Speiser's *Die mathematische Denkweise*,[36] and Weyl's *Symmetry* in its German translation of 1955.

Lohse introduced the concept of a group and operations on the elements of groups, and began describing his work with terms borrowed from group theory in the mid-1940s. To establish a group, he determined units that were commensurate with an area; in other words, the units came out even when they divided the area. He then gave precise, mechanical instructions to determine the layout of the lines, as in *Concretion I* (1945–46; see plate 7-24), which was shown in an Allianz exhibition of 1947. Lohse made other decisions informally "by eye": his choice of colors, the left-right position of the lines, and the location of the tiny squares. In other words, Lohse worked in a combination of mechanical (algorithmic) and informal (by eye) methods.[37]

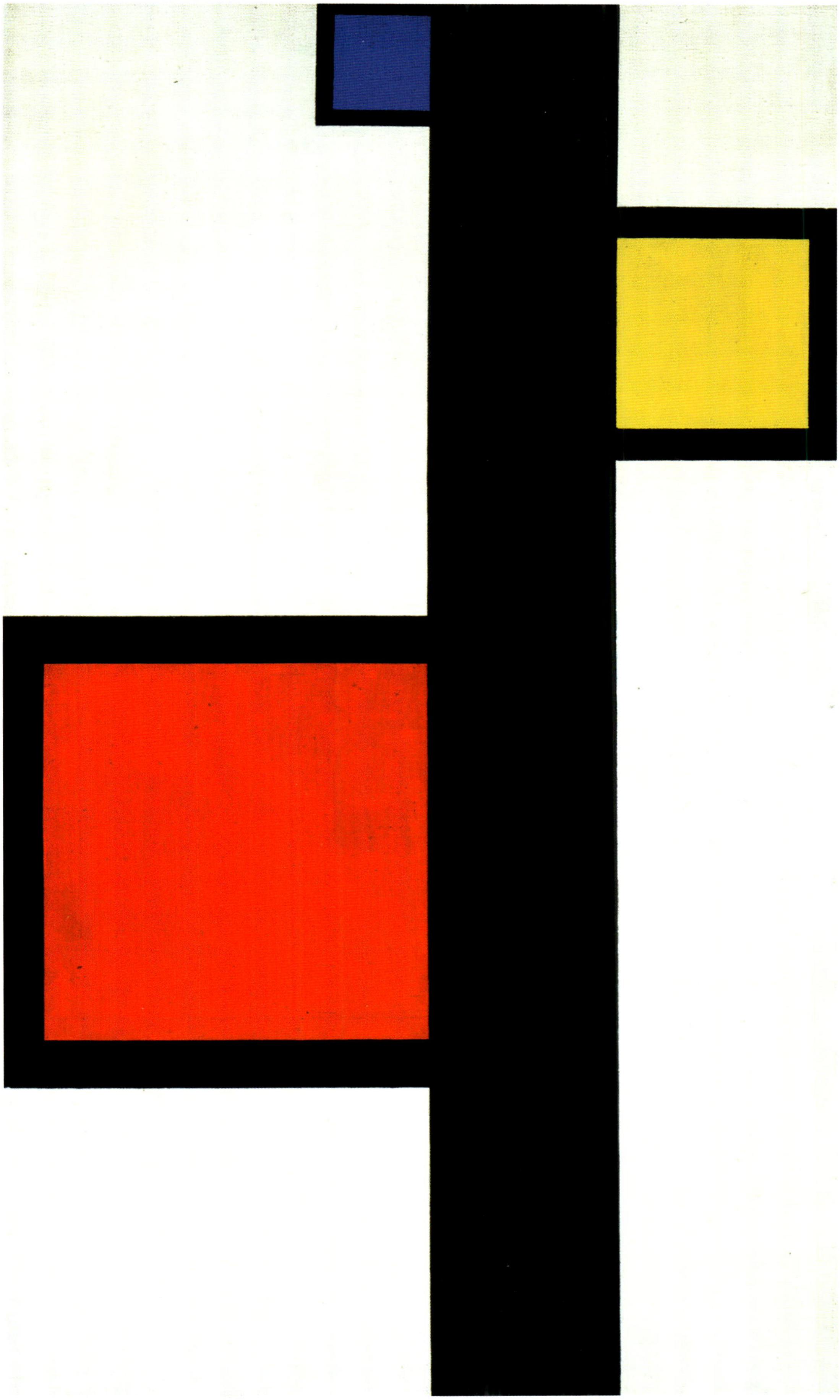

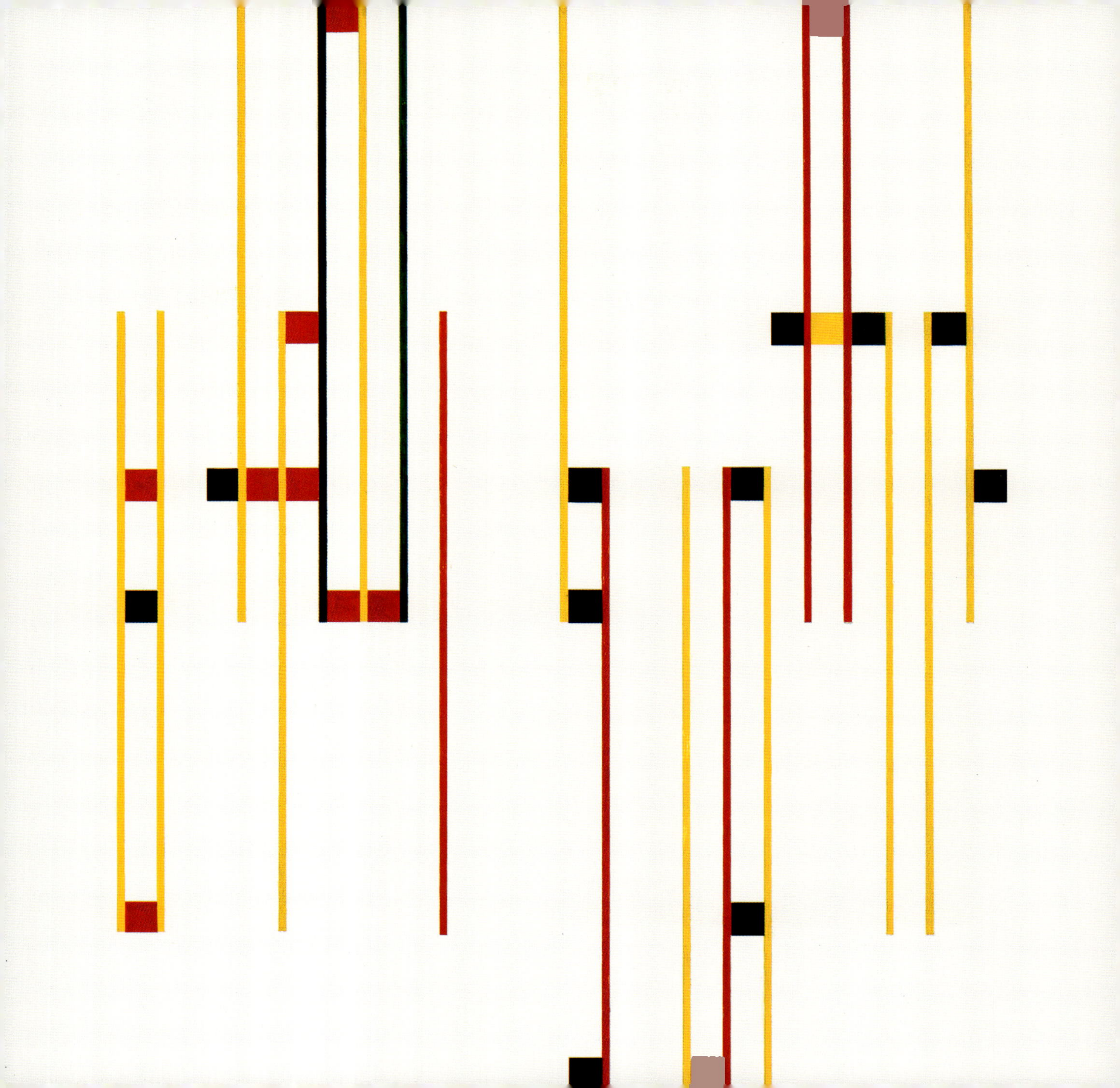

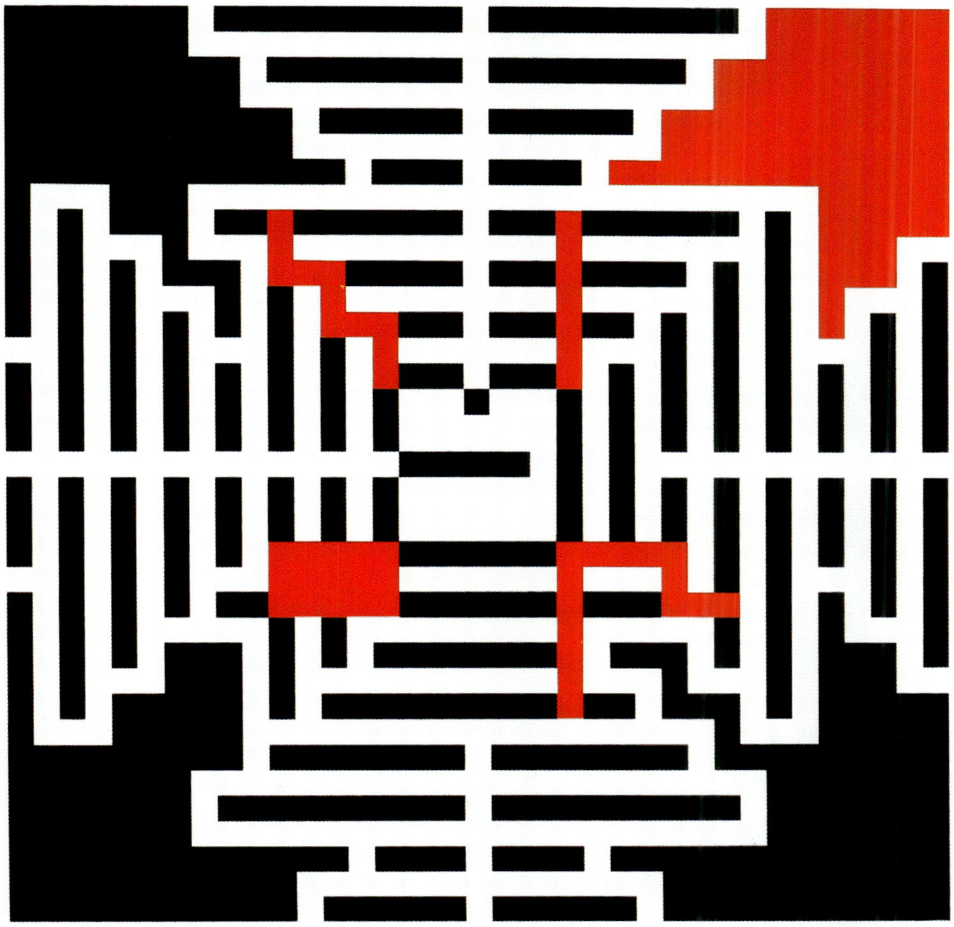

OPPOSITE

7-24. Richard Paul Lohse (Swiss, 1902–88), *Concretion I*, 1945–46. Oil on Pavatex, 27½ × 27½ in. (70 × 70 cm). © Richard Paul Lohse Foundation, Zurich. © 2014 ProLitteris, Zurich/Artists Rights Society, New York.

Concretion I measures 70 × 70 cm. After first established an underlying 10 × 10 cm grid, Lohse described how he then painted eighteen vertical lines, each extending 40 cm (four units of his grid): "After simplifying and classifying the elements of the painting, the problems of these painted groups appear naturally. They can be transformed through color, position, and dimension." "Die Entwicklung der Gestaltungsgrundlagen der konkreten Kunst" (The development of the foundation of form in Concrete Art), in the catalogue to the exhibition held October 18–November 23, 1947, *Allianz: Vereinigung moderner Schweizer Künstler* (Zurich: Kunsthaus, 1947), n.p.

LEFT

7-25. Verena Loewensberg (Swiss, 1912–86), Untitled, 1944. Oil on canvas, 23⅝ × 23⅝ in. (60 × 60 cm). Private collection. © Henrietta Coray Loewensberg, Zurich.

Lohse wrote a brief history of geometric abstract art in which he observed that Mondrian and Rodchenko had determined their compositions intuitively (by eye), as Lohse himself had in his early work: "Size and number of pictorial elements are only partially rationalized, a systematization of the latter only appears sporadically. A uniform method for the organization of the picture does not exist. The organization is based upon a subjective disposition and conception."[38] Lohse's goal was to advance the tradition of geometric abstraction beyond any reliance on intuition ("subjective disposition") by completely rationalizing (formalizing) art-making through uniform methods (algorithms).

In the mid-1940s Loewensberg also introduced regular patterns into her work (plate 7-25), but she always kept some asymmetry in the format and designed some of the elements by eye, as in this balance of the red lines and areas. Loewensberg's paintings invite the viewer to try to "figure out" the pattern like a puzzle, but the solution is elusive because her patterns are irregular.

Bill used a Klein four-group diagram to summarize the state of Concrete Art in the mid-1940s (see sidebar on page 274). He distinguished different types of Concrete painting in terms of pairs of opposite traits (two "transformations") that he arranged across horizontal and diagonal axes: geometric/non-geometric; illusionistic "pseudo-space"/a real flat

The possibility of mechanically repeating elements and facts is the signature of this epoch.

—Richard Paul Lohse, "Lines of Development," 1943–84

A Klein Four-Group of Concrete Art

Bill's caption to this diagram is: "This schema shows the four main directions in Concrete painting. It is divided into one geometric and one non-geometric half, as well as one pseudo-spatial and one two-dimensional direction. Among these trends there are a lot of gradations and personal interpretations. Nevertheless, the essential factors of Concrete painting are organized into these four main directions."

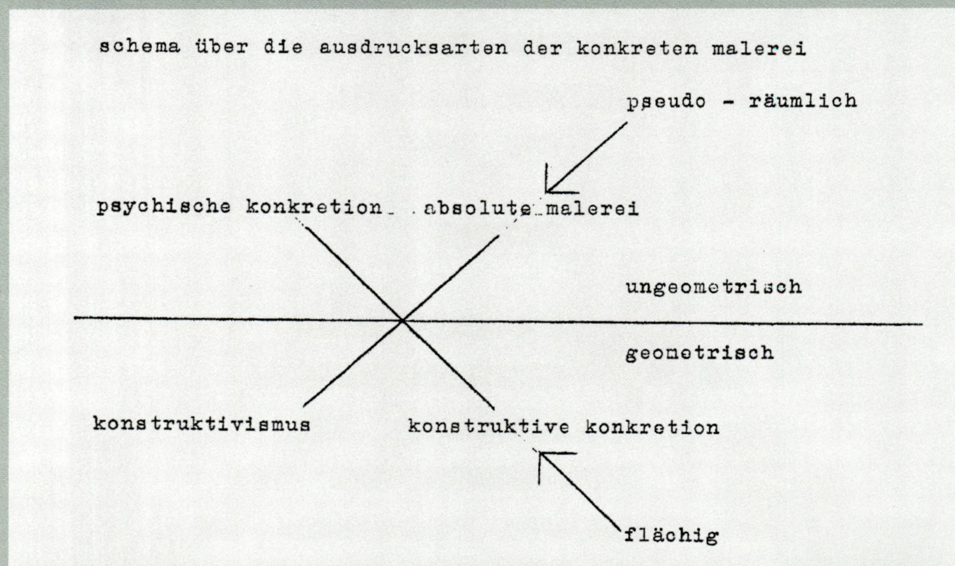

ABOVE

7-26. Max Bill (Swiss, 1908–94), "Vom Sinn der Begriffe in der neuen Kunst: Die Richtungen innerhalb der Konkrete Kunst" (The meaning of words in the new art: Directions inside Concrete Art), in *Abstrakt/Konkret* 6 [1945]: n.p. © 2014 ProLitteris, Zurich/Artists Rights Society, New York.

[Translation of the diagram]

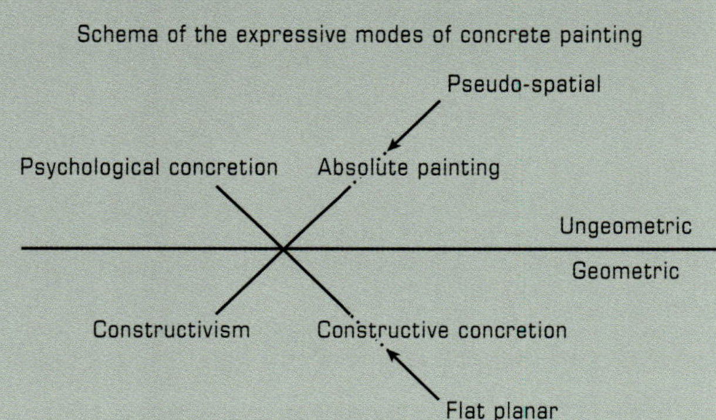

plane. Although Bill's diagram is metaphorical and lacks mathematical rigor, it embodies a group with four transformations as defined by Felix Klein.

Several artists in the circle of Concrete Art worked as graphic designers, and the bright, clear forms that they developed in their art, such as Bill's painting in plate 7-28, translated well into posters, brochures, and book covers. Lohse's cover design for a bilingual anthology of poems edited by Carola Giedion-Welcker, the wife of Sigfried Giedion (plate 7-27), is based on a square module and echoes his painting *Concretion I* (plate 7-24).[39]

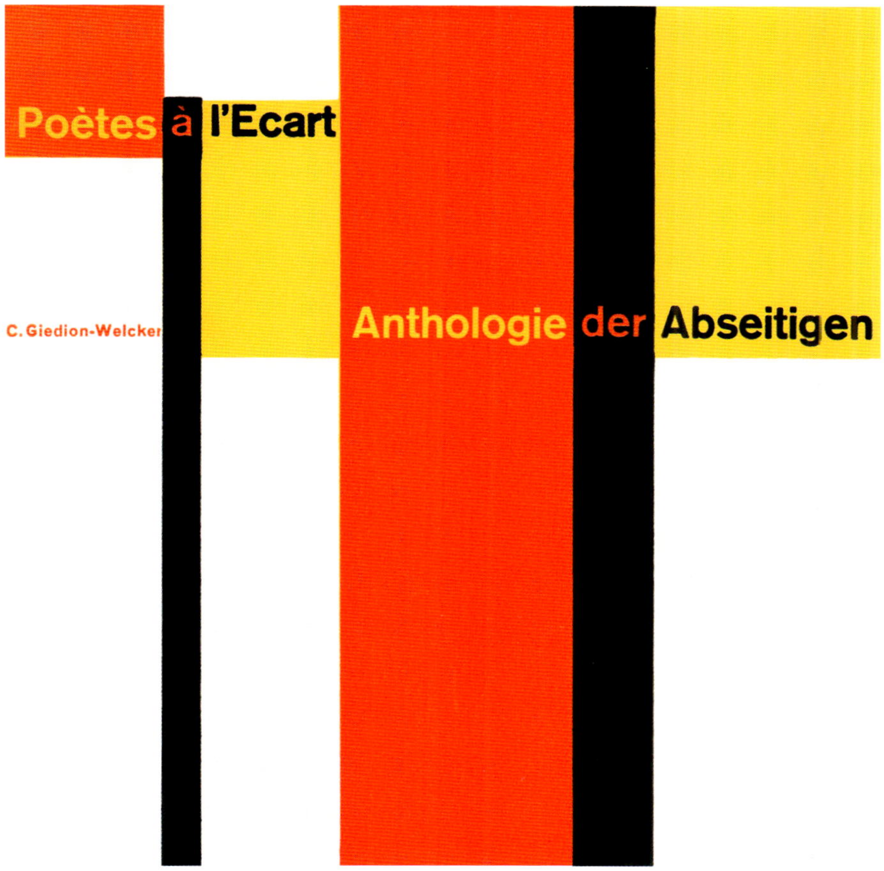

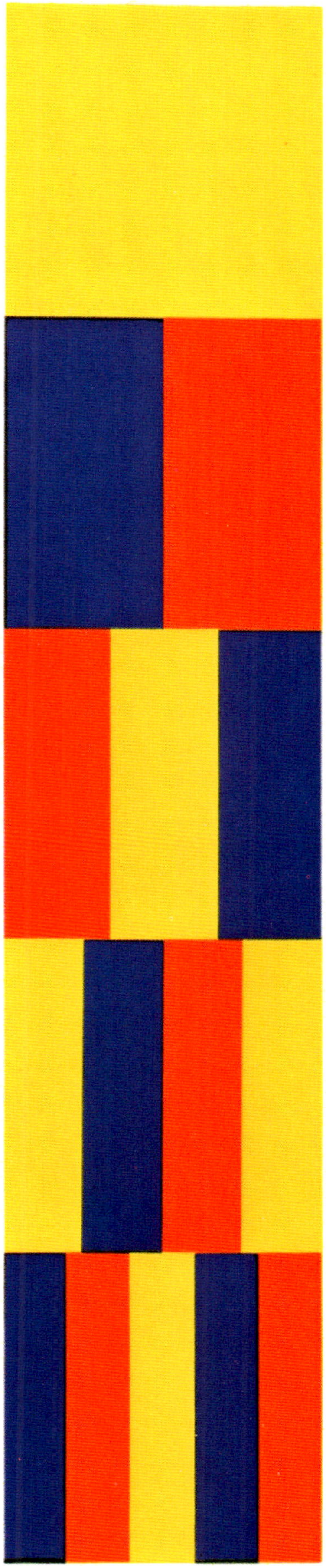

In early-twentieth-century Zurich, Einstein, Weyl, and Speiser described patterns in nature using group theory, a system that was adopted and echoed by Swiss Concrete artists because they sought the purity and order of mathematics. Their art continued to develop after World War II (see chapter 11), when the four founders were joined by an artist from the next generation, Karl Gerstner, who took the Concrete Art of Switzerland to a higher level of symmetry by adding algorithms for generating symmetrical patterns of color.

Meanwhile, the German defeat in World War I had prompted a resurgence of faith in the hallmarks of German Romanticism—feeling and individuality—and physicists in the circle of Niels Bohr declared that the subatomic realm operates by chance and uncertainty, providing a physical basis for human free will. Despite this anti-intellectual environment, Einstein held out for determinism and certainty, Hilbert dared to envision a grand tower of reason, and artists designed soaring buildings in steel and glass for imaginary cities of the future.

ABOVE

7-27. Richard Paul Lohse (Swiss, 1902–88), cover of a collection of poems edited by Carola Giedion-Welcker, *Poètes à l'écart/Anthologie der Abseitigen* (Bern, Switzerland: Benteli, 1946). © Richard Paul Lohse Foundation, Zurich. © 2014 ProLitteris, Zurich/Artists Rights Society, New York.

RIGHT

7-28. Max Bill (Swiss, 1908–94), *Progression in Five Squares*, 1942–70. © max, binia + jakob bill foundation. © 2014 ProLitteris, Zurich/Artists Rights Society, New York.

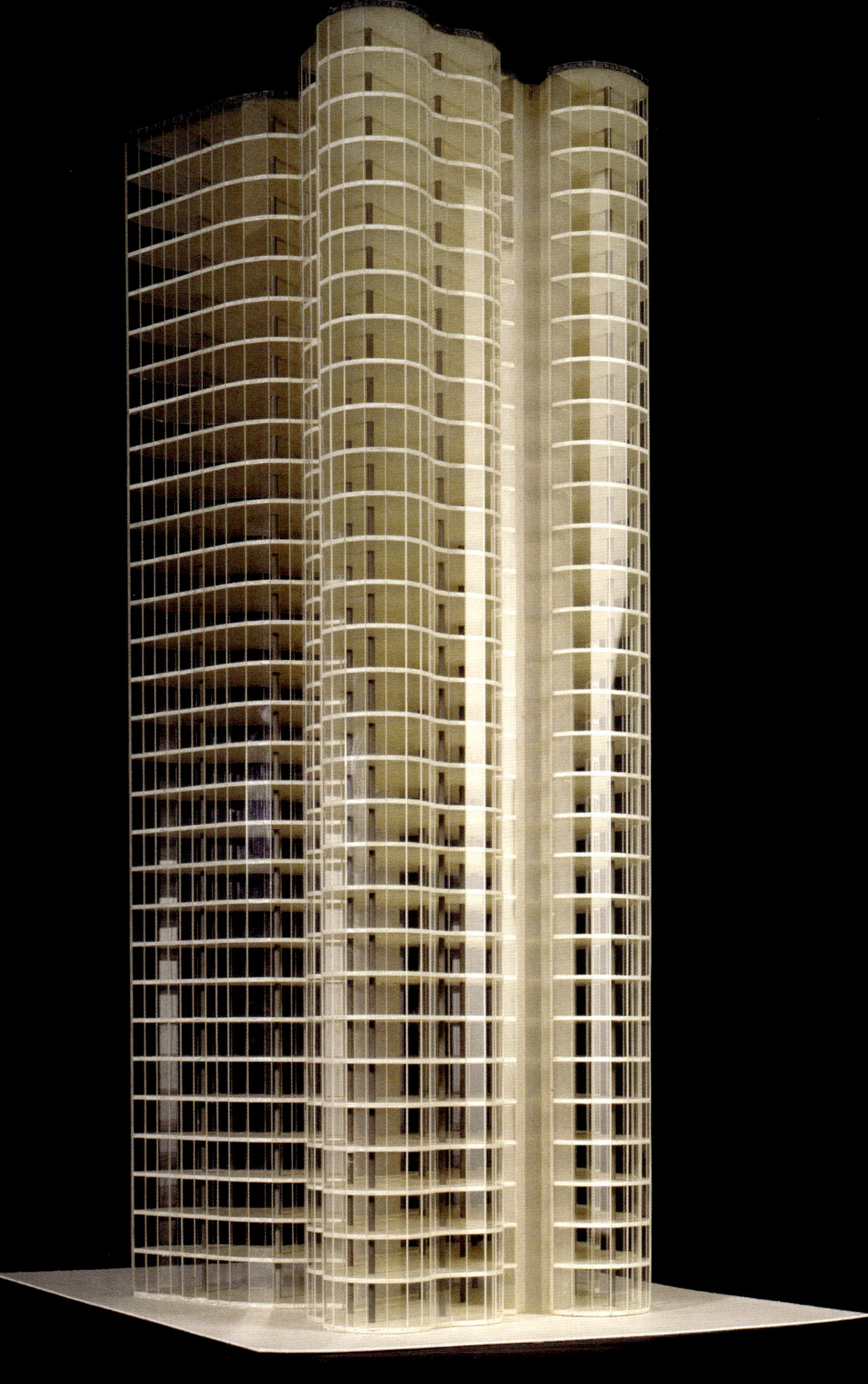

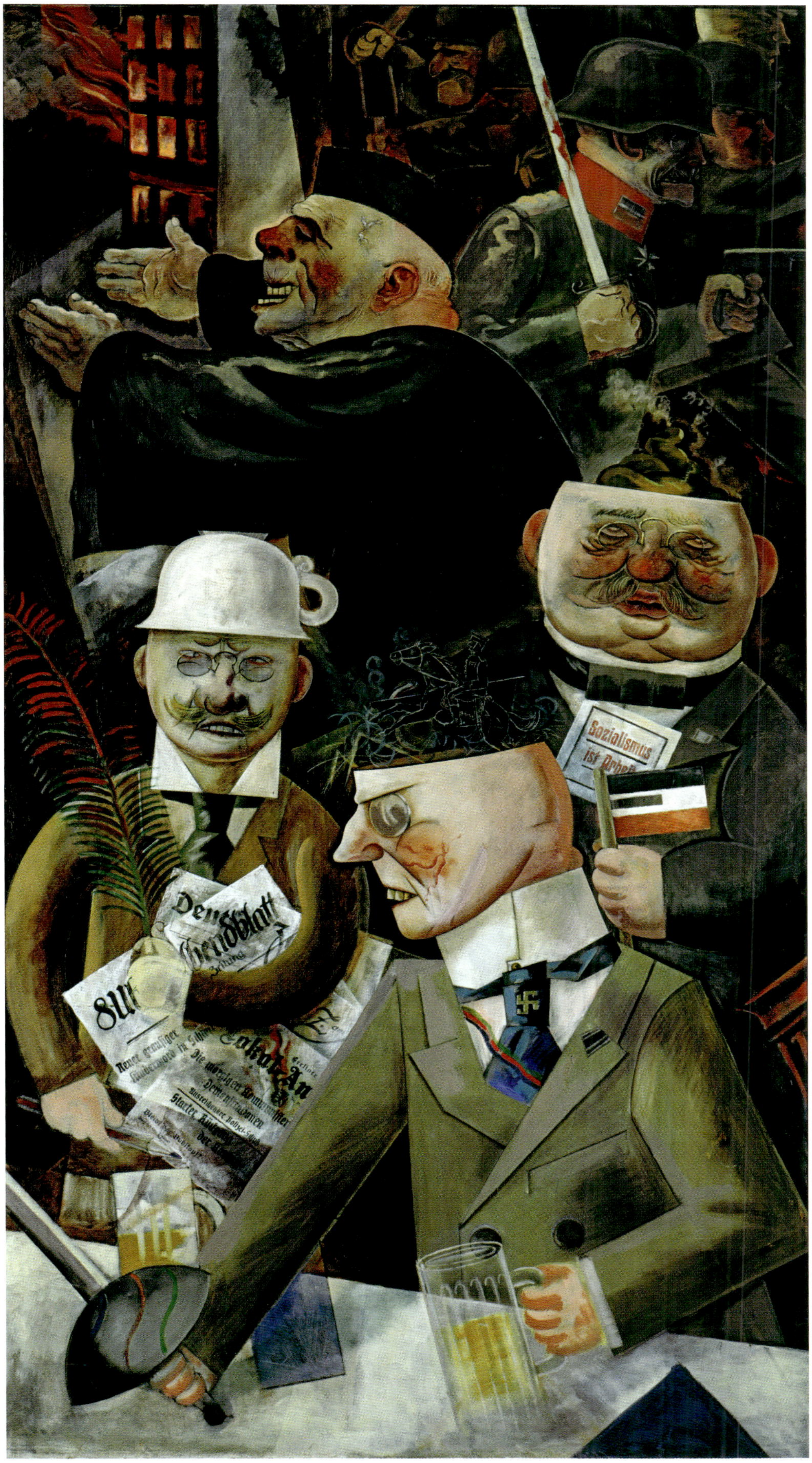

At the start of war, one hopes to win; then one expects the enemy to lose; then one is satisfied that the enemy too is suffering. In the end, one is surprised that everyone has lost.

—Karl Kraus, *Die Fackel*, October 9, 1917

8-3. George Grosz (German, 1893–1959), *Stützen der Gesellschaft* (Pillars of society), 1926. Oil on canvas, 78¾ × 42¾ in. (200 × 108 cm). Nationalgalerie, Staatliche Museen, Berlin.

Grosz symbolized the four pillars of Weimar society: the military, press, government, and church. Below, a Nazi officer—scar-faced, missing an ear, wearing a monocle and swastika—has a mind nostalgic for a bygone era before trench warfare, when the cavalry rode on horseback and men fought face-to-face. On the left a journalist, with pencil in hand and thoughts worthy of the chamber pot on his head, clutches his newspapers and a bloody palm branch, a reminder of the human cost of peace. On the right an obese, alcoholic politician, whose thoughts are like a steaming pile of dung, waves a German flag in front of the slogan *Sozialismus ist Arbeit* (Socialism is work). Above, soldiers with bloody bayonets and drawn pistols battle workers armed with spades, while a clergyman closes his eyes in prayer, blissfully unaware that the society he blesses has erupted in flames.

Contemporary with Hilbert's Program, physicists were developing quantum mechanics, and they gave two rival interpretations of the same laboratory data: the Germanic Copenhagen interpretation and the French-American de Broglie–Bohm interpretation. The two groups disagreed about the role of observation and, most fundamentally, about the nature of reality. The Copenhagen school, named for the resident city of the school's leader, the Dane Niels Bohr, declared that the natural world is ruled by chance and that reality is in the mind of the observer. Meanwhile, the Frenchman Louis de Broglie and the American David Bohm interpreted the same data as showing that the microworld of protons and electrons, like the macroworld of planets and stars, follows deterministic laws independent of human observation.

FROM PHENOMENOLOGY TO EXISTENTIALISM

As a student of mathematics and philosophy in Leipzig in the late 1870s, Edmund Husserl attended lectures by Wilhelm Wundt, the founding director of the first laboratory for experimental psychology and a disciple of Gustav Fechner. Using the laboratory methods of experimental psychology and treating logic like a branch of psychology, Husserl tried to explain how subjects do arithmetic. Gottlob Frege objected that the subject matter of psychology is the realm of ideas (which are in the head) but the subject of mathematics is the realm of numbers and forms (which, according to Platonists like Frege, exist in a mathematical-world-out-there, independent of the mind). Frege ended his review of Husserl's 1891 *Philosophy of Arithmetic*: "Reading this work had enabled me to gauge the extent of the devastation caused by the eruption of psychology into logic."[6]

Discouraged by Frege's critique and drawn to philosophy, Husserl spent the next decade developing his phenomenological approach, publishing his first study on his consciousness of numbers and logical structure, *Logical Investigations*, in 1900–1901. Husserl had based his earlier 1891 work on data collected from many research subjects, but he based his new book solely on his own consciousness of numbers. On the basis of *Logical Investigations*, Hilbert invited Husserl to join the mathematics faculty at Göttingen in 1901.

While at Göttingen, in 1905 Husserl had the key insight that to study one's own conscious experience of some phenomenon, one must step outside that experience—one must use Husserl's method of epoché, or "bracketing" the experience. Instead of focusing on a phenomenon in the world, the subject reflects on his or her own conscious experience of the world. For example, if you pause for a moment while reading this paragraph and think about yourself reading (Can you concentrate? Is your mind wandering?), you have bracketed your consciousness of—thought about yourself thinking about—the paragraph before you. When Husserl reflected on his adult experience of the counting numbers he learned as a child, he noticed that when he saw a tree in the park, he first perceived the tree and then thought about it as "one." In other words, according to Husserl, numbers are an abstraction based on perceptual experience. In general, mental experience has the form "consciousness of *x*," where *x* is some phenomenon (literally, "apparent to the senses or the mind"). Husserl was convinced that the way to understand consciousness was to examine one's mental experience of phenomena, and hence his approach came to be called "*phenomenology*." One can, theoretically, interminably bracket conscious experience in endless levels of representation (for example, the word "*tree*" can represent a tree in the park, which can, in turn, symbolize the "*tree of life*" a Darwinian metaphor for the evolution of species, and so on ad infinitum).[7] Husserl developed phenomenology to explain knowledge of mathematics, but he soon expanded the approach to explain knowledge of the natural world (in much the same way that Husserl's British contemporaries, Bertrand Russell and G. E. Moore, broadened the scope of predicate logic to form the theory of knowledge known as logical atomism). On the eve of World War I, Husserl published a general theory of consciousness in *Ideas* (1913), the founding text of phenomenology.

In the aftermath of World War I, the educated public in Germany needed philosophy to do more than describe an ideal, abstract realm in self-conscious reflection, and they turned to the *Lebensphilosophen*. Already in the nineteenth century Kierkegaard had undermined the Hegelian promise of knowledge of Absolute Spirit (the logical structure of the universe) to anyone capable of following the dialectic progression of his logic. Kierkegaard wrote an elaborate parody of Hegel's system that was an inverted dialectic designed to instill doubt and lead the reader away from absolute knowledge rather than towards it (*Either-Or*, 1843).[8] In contrast to Auguste Comte's view of society as progressing to a final positivist stage ruled by science, Nietzsche saw degeneracy all around,[9] and Fyodor Dostoyevsky described an irrational subterranean world where two and two makes five (*Notes from Underground*, 1864). In the 1920s Husserl's student and academic successor, Martin Heidegger, used his teacher's method of bracketing to describe an individual's existence in so-called existentialism (*Being and Time*, 1927). Reflecting the Romantic hostility to analytic rationality, Heidegger dismissed reason as a hopelessly inadequate guide to revealing the hidden secrets of life.[10] In the 1930s the French philosopher Jean-Paul Sartre also adopted the phenomenological method and became the leader of French existentialism (*Being and Nothingness: An Essay on Phenomenological Ontology*, 1943), which he popularized through his novels and plays such as *La nausée* (Nausea, 1938). The common method linking all versions of phenomenology and existentialism is the close study of consciousness from a first-person perspective with the goal of understanding the workings of the human mind.

I agree that two times two is four is an excellent thing, but if we're going to start praising everything, then two times two is five is sometimes also a most charming little thing.

—**The underground man, in Fyodor Dostoyevsky, *Notes from Underground*, 1864**

HILBERT'S PROGRAM

David Hilbert had spent two decades working on Einstein's theory of relativity and other topics in mathematical physics before returning to philosophical concerns after World War I. By this time axioms had been written for key areas of mathematics: Giuseppe Peano's axioms of arithmetic (1889), Hilbert's own set of axioms for geometry (1899, see chapter 4), Ernst Zermelo's axioms for set theory (1908), and Alfred North Whitehead and Bertrand Russell's axioms for logic (1910–13, see chapter 5). By the 1920s Hilbert dared to dream of establishing the foundations of mathematics at an even deeper level by finding one set of axioms for all parts of mathematics, constructing a magnificent edifice as a unified whole standing on a bedrock of certainty. Mathematicians took up Hilbert's Program in the early 1920s in an atmosphere that was very hostile to precisely the kind of grand Teutonic system they were looking for.

I mistrust all systematizers and I avoid them. The will to a system is a lack of integrity.

—**Friedrich Nietzsche, *Twilight of the Idols*, 1888**

Before the war, Brouwer had criticized Hilbert's formalist approach for allegedly making mathematics a mindless, mechanical exercise, and he invented the label "formalism" as a term of derision ("Intuitionism and Formalism," 1913; see chapter 4). In the anti-rational atmosphere after 1918, Brouwer resumed his attack, charging Hilbert with reducing mathematics to a "formula game"—a game played with formulas[11]—and he encouraged Hermann Weyl, one of Hilbert's most brilliant students, to ally himself with intuitionism. As a teenager Weyl had become convinced of the primacy of intuition from reading Immanuel Kant,[12] and during his student days at Göttingen he had come into contact with Husserl's circle of self-reflective philosophers. Weyl also shared Brouwer's interest in mysticism, and, like Brouwer (see chapter 6), Weyl read the medieval Christian mystic Meister Eckhart (a favorite of German Romantics), whose writings reinforced Weyl's introspective faith in intuition and set him on a lifelong search for a nameless spiritual force that he described as "the godhead living in impenetrable silence [*in undurchdringlichem Schweigen wohnenden Gottheit*]."[13] During the war, Weyl began to express dissatisfaction with subatomic physics and formalist logic because, in his opinion, they do not deal with "reality" and "truth": "Physics does not deal at all with the physical material and the content of reality; rather, it recognizes only its formal constitution. It has the same meaning for reality as formal logic does for the realm of truth."[14] After the war, Weyl joined Brouwer in calling for an intuitionist overhaul of mathematics (Weyl, "The New Crisis in the Foundations of Mathematics," 1921).

It is difficult to over-emphasize the significance of these events. In the third decade of the twentieth century two mathematicians [Brouwer and Weyl]—both of them of the first magnitude, and as deeply and fully conscious of what mathematics is, or is for, or is about, as anybody could be—actually proposed that the concept of mathematical rigor, or what constitutes an exact proof, should be changed!

—**John von Neumann, "The Mathematician," 1947**

Hilbert was exasperated that his star pupil had joined the ranks of Brouwer, exclaiming: "Brouwer is not, as Weyl believes, the revolution, but only another *coup* attempt [*Putschversuch*] made with a flare but using old means and doomed to failure because the ruling power is so well armed and secured by Frege . . . and Cantor."[15] Taking up another line of attack, Weyl described Hilbert's formalist mathematics as "the sort of arbitrary game in the void proposed by the more extreme branches of modern art."[16] Weyl did not state what "modern art" he had in mind, but he wrote this in the mid-1920s living in Zurich, a city where he would have encountered the geometric abstract art of Dutch De Stijl and Russian Constructivism.

There is some truth to Brouwer and Weyl's complaints; Hilbert had redefined "proof" as an entirely syntactic process. But far from being mindless, Hilbert did this because he had the philosophical sophistication to step back and reflect on the nature of the axiomatic method itself. Within his expansive view of mathematics, in which intuition played a central role, Hilbert used the axiomatic method for specific, narrow purposes. Brouwer's polarization of his "intuitionist" approach as opposed to Hilbert's "formalism" is a distortion. Hilbert himself said that metamathematics—his "proof theory"—was *based* on his intuitions about mechanical derivations: "The formula game that Brouwer so deprecates has, besides its mathematical value, an important general philosophical significance. For this formula game is carried out according to certain definite rules, in which the *technique of our thinking* is expressed. These rules form a closed system that can be discovered and definitely stated. The fundamental idea of my proof theory is none other than to describe the activity of our understanding, to make a protocol of the rules according to which our thinking actually proceeds."[17]

Contemporary with Hilbert's Program for mathematics, physicists called for a similar program for science—a systematic organization of all scientific knowledge. But physicists could not agree on what the basic building-blocks of the proposed system should be. The leading German physicist of the late nineteenth century, Max Planck, declared that the basic building-blocks are atoms, which scientists soon discovered have a subatomic structure that includes electrons, protons, and quantities ("quanta") of energy. The physicist Ernst Mach countered that the most elementary units are not physical but mental (*Contribution to an Analysis of Sensations,* 1886). Mach was educated in the tradition of German Idealism in which *ideas* are primary, and he adopted, from the experimental psychologist Gustav Fechner, the notion that simple sensations—feeling hot, seeing red—are the building-blocks of knowledge. Fechner had called them "monads" of experience that are neutral in the sense that they are neither mental nor physical but both (*Psychophysics,* 1860). In the first decade of the twentieth century, the young Bertrand Russell also adopted Fechner's neutral monads, which Russell called "sense-data" in his theory of logical atomism (see chapter 5). Mach took sense perception to be primary, and he declared that the facts of perception—sense-data (seeing red, smelling sweetness)—are the basic building-blocks of scientific knowledge (plate 8-4).

The two main attempts to find a universal language for science, both of which are mathematical at their core, developed from these two approaches. Planck, Einstein, and physicists in their circle developed quantum mechanics, which is a description of nature in terms of electrons, protons, and quanta. Followers of Mach and Russell developed logical positivism, which is a study of the language of science in terms of sense-data. The subject matter of quantum physics is the physical world (electrons, protons); it asks, for example: How many protons are in a carbon atom? The subject of logical positivism is the *language* of science (*reports* of sense-data which confirm the existence of electrons and protons); it asks questions such as: Is the word "atom" reducible to the terms "electron" and "proton"? In other words, quantum physicists practiced *science*, whereas logical positivists discussed the *philosophy* of science.

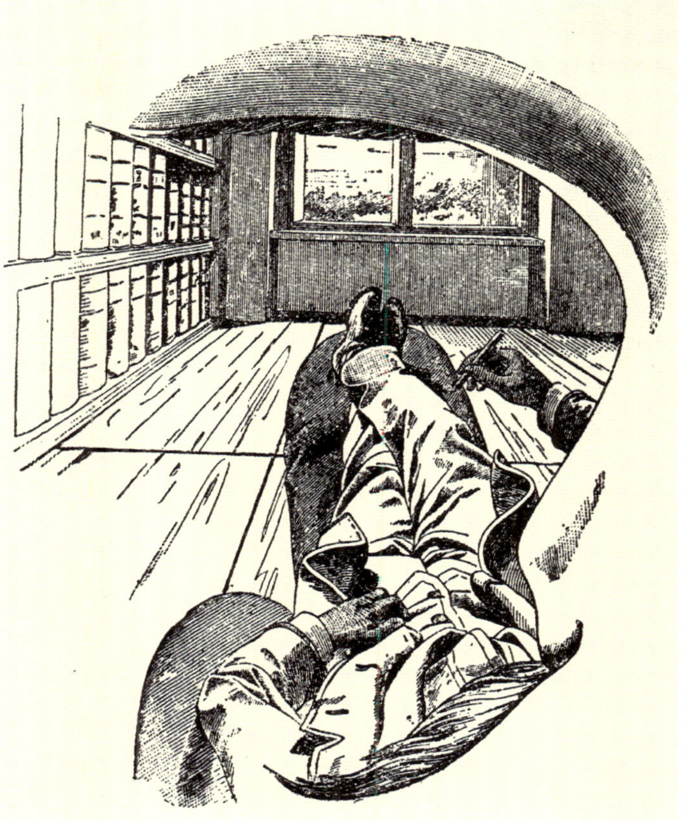

8-4. Field of vision, in Ernst Mach, *Beiträge zur Analyse der Empfindungen* (1886), translated by C. M. Williams as *Contribution to an Analysis of Sensations* (La Salle, IL: Open Court, 1897), 16, fig. 1.

This man's field of vision is filled with sense-data (sensations of light and dark, curves and straight lines, squares and rectangles), which, according to Mach, are the building-blocks of knowledge. In the preface to this book, Mach described why a physicist would be interested in the physiology of perception: "The frequent excursions that I have made into this province have all sprung from the profound conviction that the foundation of science as a whole, and of physics in particular, await their next greatest elucidations from the side of biology, and especially, from the analysis of sensations." Contrary to Mach's prediction, this is not how the *practice* of physics developed in the twentieth century, but his words foretell the future of the *philosophy* of physics as articulated by members of the Verein Ernst Mach (Association Ernst Mach), the so-called Vienna Circle, and physicists in the circle of Niels Bohr, who proclaimed the Copenhagen interpretation.

LOGICAL POSITIVISM AND THE VIENNA CIRCLE

Mach began his career working in a physics laboratory, where in the 1870s he demonstrated that a projectile moving faster than the speed of sound produces a shock wave. He described the speed of objects moving at or near supersonic velocities by defining a ratio of the speed of the projectile to the speed of sound (its "Mach number"). In addition to his practical work in physics, Mach also speculated about the philosophical topic of how to determine indisputable knowledge. He read the nineteenth-century political philosopher Auguste Comte, who, after French revolutionaries had destroyed the old monarchy, declared that the new democracy of France should be built on scientific principles that give certain ("positive") knowledge. To gain positive knowledge, Mach analyzed the logical structure of assertions about sense-data—hence Mach's twentieth-century version of Comte's positivism is called *logical* positivism.

Mach viewed science as a body of ordered knowledge. By analyzing the linguistic expressions of science, he would reveal the logical structure and interrelation of its assertions. For Mach a scientific statement is meaningful if it names an event that can be verified by observation. The assertion "the apple is red" can be verified because it names a fact—redness—that can be observed with the eyes. But Mach banned statements such as "Absolute Spirit is eternal" because it does not predict an observable event and is hence unverifiable, making the statement meaningless in a scientific context. He also threw Newton's "Absolute space" into the dustbin after Albert Michelson and Edward Morley failed in 1887 to measure the substance—æther—with which Absolute space was allegedly filled.

After World War I, mathematicians, logicians, and philosophers gathered under the banner of logical positivism to write a new universal language whose logic-based structure would unite all scientific knowledge. Centered in Vienna, home to Mach until his death in 1916, the project was officially named the Verein Ernst Mach (Association Ernst Mach) but was known as the Vienna Circle. Like their contemporaries, the Significs group in Holland, the Vienna Circle hoped that rational discourse would promote peace after the ruinous "war to end all wars."

The need for a new understanding of how everyday perception relates to physics seemed urgent because scientists had entered the subatomic realm, which can be observed only indirectly. The German physicist Moritz Schlick, a student of Planck, came to Vienna in 1922 to teach philosophy of science, and he led the weekly meetings of the Vienna Circle. In 1926 Rudolf Carnap joined the group and became its spokesman. Carnap had attended Gottlob Frege's lectures on mathematical logic at the University of Jena between 1910 and 1914, and after World War I he completed a doctoral thesis on concepts of space in mathematics, physics, and psychology.[18] Others in Schlick's inner circle included the sociologist Otto Neurath, the mathematician Hans Hahn, who was on the faculty at the University of Vienna, and his young student Kurt Gödel. The Viennese logician Ludwig Wittgenstein attended a few meetings but he kept his distance from the group. Numbering upwards of thirty people, the Vienna Circle had many differences of opinion and held no common doctrine, but they shared an enthusiasm for organizing all scientific knowledge into a system.

Since their goal was to make philosophy suitable to the scientific era, the logical positivists' greatest foes were metaphysics and psychology, which they found plagued with wooly-minded vagueness. To rid philosophy of these contaminants, Schlick, following Mach, formulated a verification principle to separate meaningful from meaningless assertions: "The meaning of a proposition can be given only by giving the rule of its verification."[19] The logical positivists were motivated by problems of interpretation in the subatomic realm—a context in which it might be useful to use this criterion. But they overreached when they applied it universally and claimed that *any* utterance that did not pass their test is meaningless. They threw away the topic of ultimate reality (the subject of metaphysics), the mind-dependent world of values (the subjects of ethics and aesthetics), and mankind's inner world (the subject of psychology). Any assertion that did not pass their verification principle, such as "The soul is immortal and able to endure all good and ill" (Plato, *Republic*, 380–367 BC), was deemed not incorrect, but meaningless.

That's not right.
It's not even wrong.

—Wolfgang Pauli, 1945

Confident that a scientific worldview would promote a civil society, in 1929 Carnap, Neurath, and Hahn published a manifesto, *The Scientific Worldview: The Vienna Circle.* They placed themselves in the philosophical tradition of the pre-Socratic philosopher Protagoras, who was the first of a group of so-called sophists who claimed that human reason can answer all questions. They were also heirs of Democritus and his follower Epicurus, who reduced everything in the universe to the mechanical interaction of inert atoms. Carnap, Neurath, and Hahn expressed confidence in the power of man's reason, and—in stark contrast to their Viennese contemporary Sigmund Freud—they optimistically banished psychology's irrational "depths": "The endeavor is to link and harmonize the achievements of individual investigators in their various fields of science. . . . Neatness and clarity are striven for, and dark distances and unfathomable depths rejected. In science there are no 'depths'; there is surface everywhere. Experience forms a complex network, which cannot always be surveyed and can often be grasped only in parts. Everything is accessible to man; and man is the measure of all things. Here is an affinity with the sophist, not with the Platonist; with the Epicureans, not with the Pythagoreans; with all those who stand for earthly being and the here and now. The scientific conception of the world knows no unsolvable riddles."[20]

Flectere si nequeo superos,
Acheronta movebo
[If I cannot sway the
higher powers, I will stir
up the underworld].

—Virgil, *Aeneid*, late first century BC, epigraph to Sigmund Freud's *The Interpretation of Dreams*, 1900

In 1928 Carnap published a detailed account of how the world of experience and science can be constructed from propositions about sense-data that are linked by logic (*Der logische Aufbau der Welt* [The logical construction of the world], 1928). Although Carnap showed how a description of sense-data (an individual's description of private conscious experience) is capable of expressing the same content as any language that describes the world-out-there, he opted to use the latter because instead of saying, "I see white. I feel polite. I sense desire," it is more convenient to say "Please pass the salt."

QUANTUM MECHANICS

At the opening of the twentieth century, physicists determined that the atom is composed of negatively charged electrons (discovered in 1897) and a positively charged nucleus (discovered in 1911). In 1899 Planck explained observations of the emission and absorption of electromagnetic radiation (light) from an opaque, non-reflecting body (a "black-body") at room temperature by supposing that it came in packets of energy with a precise quantity (*quantum* in Latin). As the energy of the light increases, so does the frequency of its electromagnetic wave, with the two always remaining in the same "constant" proportion to each other (this ratio is now called Planck's constant). Then in 1905 Einstein had the insight that light acts as a particle (a "photon") with a quantum of energy, described by the formula hv, where h is Planck's constant and v is the frequency of the light's radiation (*Photon Theory of Light*, 1905).

By the early 1920s physicists were observing that an electron does not seem to be located at a specific point (like a point particle) but rather spread out, like a cloud or a wave of matter. The Austrian Erwin Schrödinger developed an equation that describes the evolution of such waves of matter over time—a Ψ (psi) wave—that varies with the physical state of the atom (Schrödinger Equation). The following year Max Born refined this concept, saying that it is not the electron's mass (a *physical* object) that is spread out in a wave-like pattern, but rather Schrödinger's equation describes a probability distribution (a *mathematical* object) for the position of the electron. In other words, the Ψ function can be used to calculate the probability of the electron being in a given region. Thus to measure an electron, Born estimated its position at any given instant within a range of probability described by Schrödinger's Ψ wave (plate 8-5).

The mathematical formulas of quantum physics, such as Schrödinger's equation, describe the subatomic realm, just as the formulas of classical physics describe the solar system with Newton's law of gravitation and the atomic realm using Maxwell's equations for electromagnetism. First formulated in the 1920s, the formulas of quantum mechanics make predictions about events in the subatomic world that can be observed indirectly. For more than seventy-five years these mathematical descriptions of subatomic events have been repeatedly confirmed by rigorous experiment, and thus today physicists worldwide apply them to solve practical problems, such as designing solid-state electronic devices (computers, cell phones, microprocessor chips, transistors), lasers (used in barcode scanners, compact discs), and the like. The forces within the atom at the quantum level, like the force of gravity operating on a cosmic scale, are described today only by mathematics, and their underlying physical mechanisms are unknown. If tomorrow objects start falling up, not down, then physicists will question Newton's law of gravity and wonder whether classical physics is correct; if atoms stop absorbing and emitting discrete electromagnetic radiation, then physicists will doubt whether quantum mechanics is correct. But until there is observable evidence to the contrary, physicists will continue to use classical and quantum mechanics because these mathematical descriptions of the macrocosm and microcosm consistently predict facts that are confirmed by observation.

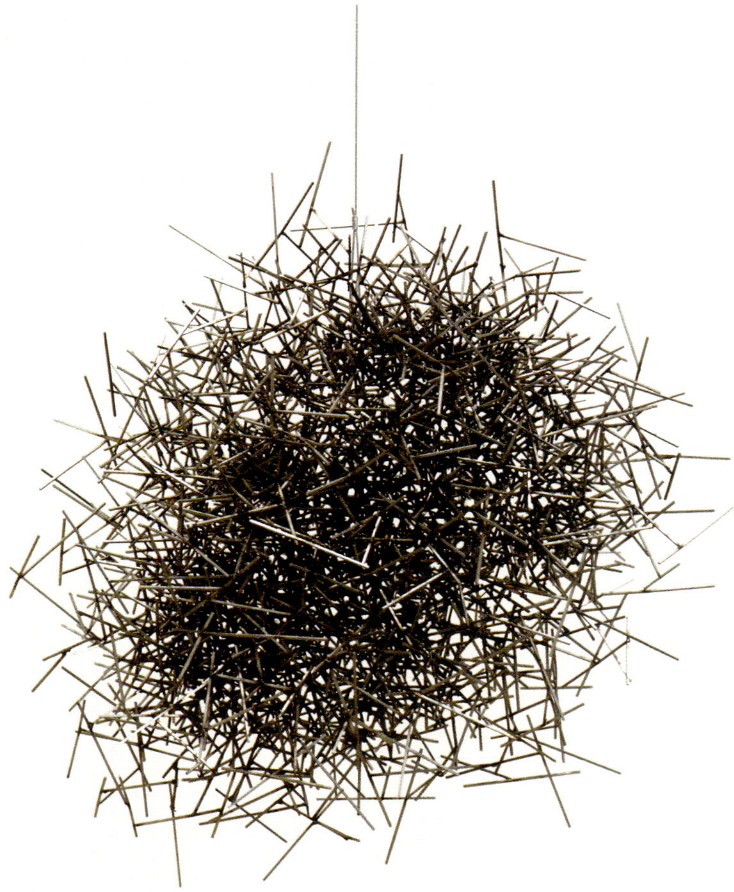

8-5. Antony Gormley (British, b. 1950), *Quantum Cloud XXXVIII (Baby)*, 2007. 2 mm square section stainless steel bar, 25½ × 25½ × 27½ in. (65 × 65 × 70 cm). © The artist. Courtesy of the artist.

Since the development of quantum mechanics in the 1920s, one of its most amazing and maddening features is the disparity between the extreme accuracy of the *practice* of quantum mechanics, which no one doubts, and the odd array of *philosophical interpretations* of quantum mechanics, on which no one can agree. One particular philosophical account of the meaning of quantum mechanics—the Copenhagen interpretation—marks a bizarre episode in the history of science.

THE COPENHAGEN INTERPRETATION

Ever since Newton announced the law of universal gravitation in 1687, scientists have worked to gain an ever more accurate description of the natural world. By the opening of the twentieth century, physicists, astronomers, and biologists felt that they were closing in on a complete picture. Then physicists, who were at the forefront of research into nature's smallest building-blocks, declared that this two-hundred-year-old scientific project was old-fashioned and naive because the natural world is ultimately unknowable. According to Niels Bohr and his circle, all scientists can ever hope to know is their subjective conscious experience—their observations—because a boundary separates them as observers from what they observe, and this crucial line between subject and object is shrouded in

8-6. Diagram of the stream of consciousness by William James in "The Stream of Thought," in James, *The Principles of Psychology* (New York: H. Holt, 1890), 1:283, fig. 33.

Making an analogy between thoughts moving in the mind and water flowing in a river, James coined the phrase "stream of consciousness," for which he proposed this model: "If we make a solid wooden frame with the sentence written on its front, and the time-scale on one of its sides, if we spread flatly a sheet of India rubber over its top, on which rectangular co-ordinates are painted, and slide a smooth ball under the rubber in the direction from 0 to 'yesterday' [9], the bulging of the membrane along this diagonal at successive moments will symbolize the changing of the thought content . . . the relative intensities, at successive moments, of the several nerve-processes to which the various parts of the thought-object correspond" (1:283).

Despite this show of confidence in his ability to measure thought, James believed that "introspective observation" (reflection on one's own thought process) was ultimately futile: "Let anyone try to cut a thought across in the middle and get a look at its section, and he will see how difficult the introspective observation of the transitive tracts is. The rush of thought is so headlong that it almost always brings us up at the conclusion before we can arrest it. Or if our purpose is nimble enough and we do arrest it, it ceases forthwith to be itself. . . . The attempt at introspective analysis in these cases is in fact like seizing a spinning top to catch the motion, or trying to turn up the gas quickly enough to see how the darkness looks" (1:244).

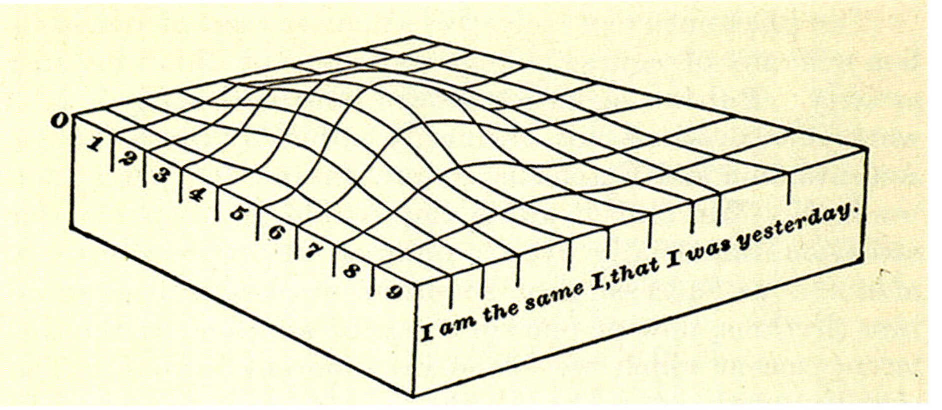

mystery and not open to scientific investigation. Furthermore, at the atomic level there is an irreparable break in the chain of cause and effect, and thus it is impossible *in principle* to give a complete description of events in the natural world. Put forth in the 1920s, the Copenhagen interpretation came to monopolize quantum mechanics and was taught as gospel in physics textbooks until around 1960, when it began its slow decline.

What caused this strange hiatus in the progress of modern science? I believe the way to make historical sense of the Copenhagen interpretation is to combine the approaches forged by historians of science Max Jammer, who described it in terms of the Idealist tradition in German philosophy and the new science of the mind (psychology),[21] and Paul Forman, who considered it within the political context of the Weimar Republic.[22] Members of the Copenhagen school were practicing physicists whose primary interest was the exciting new world of electrons and quanta being revealed to them in the laboratory. A few (not all) also had a secondary interest in the relevance of their work to traditional philosophical questions: What is certain knowledge? What is ultimate reality? But no one in Bohr's circle had advanced training in epistemology or metaphysics, so rather than formulate a consistent philosophy of science, they pieced together threads drawn from their cultural milieu: German Idealism, *Lebensphilosophie*, phenomenological and psychological theories of consciousness, and philosophical ideas emanating from the Vienna Circle.

The story begins when a young Niels Bohr enrolled in an undergraduate course in philosophy at the University of Copenhagen taught by Harald Høffding, an expert on the Danish philosopher Kierkegaard.[23] Bohr adopted Kierkegaard's anti-system attitude toward science, according to which it is impossible to conceive of nature as a completed whole because one is *part* of nature.[24] Høffding described Kierkegaard's outlook: "A [scientific] system can be conceived only if one could look back on a completed existence—but this would presuppose that one does not exist anymore!"[25] In other words, the scientist cannot be an "objective observer." Bohr was also influenced in his youth by William James, whose psychology he knew by 1905[26] (probably introduced to it by Høffding, who visited James in Cambridge, Massachusetts, in 1904[27]). From reading James, Bohr learned that an act of consciousness—the subjective experience linking him to an observed fact—is impossible to study by introspection (plate 8-6).

At the opening of the twentieth century (as we saw before) physicists had proposed two alternative building-blocks of nature: physicists, led by Planck, declared that we must build our view of nature based on *physical* entities (atoms, electrons, protons, and quanta), but philosophers, led by Mach, declared that the building-blocks are *mental* entities (ideas and sense-data). The German Idealist outlook of Mach and other logical positivists in Vienna was yet another thread in the fabric woven by Bohr and his circle, and after World War I Brouwer's *mathematical* version of German Idealism was known in the inner circle of subatomic physics in the person of Weyl (as we saw before). Following Mach, Bohr produced a new *scientific* version of German Idealism by updating the terms "monad" and "sense-data" to fit the language of the 1920s physics laboratory. According to Bohr, the basic building-blocks of the natural world are *observations*, which turn out to match observations using Schrödinger's Ψ waves (the probabilistic descriptions of the quantum state of a physical system). In other words, the building-blocks are sensations—*ideas*—in the mind of an observer.

Einstein and Schrödinger were incredulous that the Copenhagen school was shifting the focus of physics away from the *physical world* of electrons (known by their Ψ waves) to *reports of observations* of Ψ waves (which, according to Bohr, is all that electrons are; it is meaningless to talk about a physical world-out-there). As Schrödinger complained: "We are told that no distinction is to be made between the state of a natural object and what I know about it, or perhaps better, what I can know about it if I go to some trouble. Actually—so they say—there is intrinsically only awareness, observation, measurement."[28]

It may be suggested that behind the statistical universe of perception there lies a hidden "real" world ruled by causality. Such speculations seem to us—and this we stress with emphasis—useless and meaningless. For physics has to confine itself to the formal description of the relations among perceptions.

—Werner Heisenberg, "The Physical Content of Quantum Kinematics and Mechanics," 1927

The Copenhagen interpretation, in which reality (to use a quaint term) consists of human consciousness of probability waves, emanated from minds in which there was an ongoing battle between the rigorous rationality of laboratory physics and the soulful angst of *Lebensphilosophie* and the new introspective psychology. These tendencies had been present since around 1900, but they increased after 1918 when Weyl, Bohr, and Heisenberg were swept up by a tide of Romanticism—a revolt against rationalism and a celebration of intuition (per Weyl), irrationality (per Bohr[29]), and uncertainty (per Heisenberg). In other words, the gist of the Copenhagen approach *predated* the laboratory experiments that it interpreted, and beginning around 1925, Weyl, Bohr, and Heisenberg gave a *philosophical* interpretation of their laboratory data that was consistent with their Romantic worldview, culminating in 1927 in the Copenhagen interpretation of quantum mechanics.[30]

Weyl's hybrid career in mathematics and physics provides a clear example of this phenomenon: as a student at Göttingen, Weyl (as we saw before) showed an interest in the study of consciousness (phenomenology), mysticism (Eckhart), and mathematics (earning a PhD under Hilbert in 1908). During the war he began expressing dissatisfaction with physics as a tool for describing "reality" (1917); then he denounced the formalism of Hilbert (who described mathematical objects existing in a mathematical-world-out-there), and in 1918 joined Brouwer (who declared that mathematical objects exist only as ideas in the mind). Thus *before* the formulation of quantum mechanics by himself, Bohr, and others, Weyl had already dismissed the causal basis of physics on *philosophical* (not experimental) grounds, as he wrote in 1921: "It has to be said very clearly that physics in its present state can no longer support the belief that there is causality in the natural world which is based on precise laws."[31]

A man can do what he wants, but he cannot will what he wants.

—**Arthur Schopenhauer,** ***On the Freedom of the Will,*** **1839**

Historians of science have traced the origin of Bohr's notion of complementarity—the essential duality of nature—to his youthful embrace of Kierkegaard's knowledge by "qualitative dialectic" (every thesis has an antithesis, every thought is opposed by reality) summarized in the Danish philosopher's slogan "Either-Or."[32] According to Bohr, the building-blocks of nature have a dual essence, exhibiting properties of both waves and particles;[33] if he resolved to measure wave properties of an electron, then he would find waves; if he measured particle properties, then the electron would "collapse" into a particle.

Thus the philosophical and political climate in post-war Germany influenced the answers Weyl, Bohr, and Heisenberg gave to the questions: What does quantum mechanics *mean*? What does it tell us about reality?[34] Their answers reflected phenomenology, psychology, *Lebensphilosophie*, and also the rigorous logic of the Vienna Circle. Applied in the physics laboratory, the logical positivists' verification principle meant that "the electron is in location *x*" is equivalent to "a scientist shot a gamma ray at the electron, striking it in the region of *x*." Taken to its extreme (as Bohr and Heisenberg did), the verification principle seemed to imply that the electron does not *have* a location or speed until it is observed and measured, (which causes its probabilistic Ψ wave to "collapse" to a particular place and time). Thus, according to Bohr, it was pointless to talk about an objective scientist, or of the natural world existing independently of observation, as he wrote in 1927: "An independent reality in the ordinary physical sense can neither be ascribed to the phenomena nor to the agencies of observation."[35] Heisenberg adopted Schlick and Carnap's verification principle, but he also appreciated the more intuitive Viennese logician, Wittgenstein,[36] who was loosely associated with the Vienna Circle when he published his description of the limits of language (both spoken language and symbolic logic) in his *Tractatus Logico-Philosophicus*, 1921 (see chapter 9). Echoing Kierkegaard, Wittgenstein wrote that one cannot step outside the universe and describe nature as a totality,[37] ending his *Tractateus:* "Whereof one cannot speak, thereof one must be silent." Heisenberg may have applied Wittgenstein's proclamation to atomic phenomenon when he devised his so-called matrix model in which the physicist describes an electron as a statistical pattern of numbers (a mathematical object) and is not allowed to refer to (must be silent about) a "real" electron (a physical object).[38]

Heisenberg also gave a new meaning to the term "uncertainty" (*Unsicherheit*).[39] First, he argued (uncontroversially) that in order to measure an object, the observer has to interact with it in some way. For example, when counting the number of apples in a bowl, the observer looks at the apples, and light bounces off them and strikes the viewer's retinas, causing nerves to fire in the brain of the observer, who counts three apples. To observe an electron the scientist must also interact with it, but the human eye cannot see electrons because they are smaller than the wavelength of visible light. So the observer must use a beam of electromagnetic radiation with a very small wavelength, such as a gamma ray, to observe the electron. Short wavelength means high energy, so when the gamma ray interacts with the electron, the gamma ray disturbs it. The situation is analogous to a human observer determining the location of the three apples by throwing an orange at them; in marking the electron's position, the gamma ray changes the electron's momentum. When the gamma ray hits the electron, you find out where the electron *was*, but the transfer of momentum changes the path of the electron, so you have little idea of where it *is*. Measuring the position of an electron *affects* its velocity and thus one can know its momentum only approximately; similarly, measuring the momentum of an electron *affects* its position and thus one can know its position only approximately. According to Heisenberg, a degree of uncertainty is inherent in the process of measuring an electron (*Uncertainty Principle*, 1927). So far Heisenberg's principle is an undisputed truism of twentieth-century physics; the situation described above is comparable to Einstein's use of probability theory to measure particles moving in a liquid, during which the observer affects the data, and the position of each individual particle is *unknown* ("On the Movement of Small Particles Suspended in a Stationary Liquid demanded by the Molecular-Kinetic Theory of Heat" [Brownian motion paper], 1905).

But Heisenberg parted company with Einstein when he went on to give a new meaning to the word "uncertainty." Heisenberg claimed that the precise location of an electron is not only "uncertain" in the sense that it is *unknown* (because the physicist estimated its position using statistics), but also uncertain in the sense that its location is *unknowable* because the electron *lacks* a precise location until it is observed, which causes the electron (the probabilistic Ψ wave) to exist in ("collapse" to) a particular time and place. In Einstein's 1905 paper, each individual particle *has* a precise location, even though the position of all the particles moving through the liquid is unknown to the observer (whose act of measurement affects their position and who uses statistics to measure them).

From Heisenberg's new definition of "uncertain," he deduced that the data gathered by physicists is not governed by deterministic laws of cause and effect, as he concluded his 1927 paper on the Uncertainty Principle: "Quantum mechanics establishes the final failure of causality."[40] Why did Heisenberg associate statistics with the "failure of causality"—the absence of deterministic laws of cause and effect? Schrödinger's equation (the probability distribution that Heisenberg used to describe the position of the electron) *is* deterministic; the position of one electron (like the toss of one coin) involves an element of chance, but on average they obey laws of probability that *inevitably* produce a pattern, as Pascal showed in the seventeenth century. Ask any gambler.[41]

Saturday will be July 7, 2007—7/7/7. Do you believe in luck or superstition?

No. I believe in math.

—Mario DiGuiseppe, vice president, Tropicana Casino in Atlantic City, in response to Associated Press reporter Wayne Parry

By insisting that probability implies indeterminism, Heisenberg, along with Bohr, was continuing a century of associating statistics with free will, beginning with Quetelet in 1835 and continued at the fin de siècle by members of the Moscow Society of Mathematics in their work on discontinuous functions (see chapter 3). Also, Bohr had learned from Kierkegaard to reject causality. In the Danish philosopher's description of how one comes to know both oneself and ultimate reality (the Christian God, "the Highest, the Eternal"[42]), there is no deterministic cause and effect: "So long as we live we are imprisoned in becoming: hence we stand ever before the unknown, for there is no guarantee that the future will resemble the past."[43] In making a choice, a "decision of the will . . . the choice itself comes with a jerk, with a leap, in which something quite new (a new quality) is posited."[44] In announcing the Copenhagen interpretation in 1927, Bohr claimed that quantum mechanics ensures free will: "Its essence may be expressed in the so-called quantum postulate, which attributes to any atomic process an essential discontinuity, or rather individuality, completely foreign to the classical theories." Thus, according to Bohr, there is no deterministic cause and effect: "This postulate implies a renunciation as regards the causal space-time co-ordination of atomic processes . . . [and] an inherent 'irrationality.'"[45] Two years later Bohr—in a classic case of overreach—went on to write that although many have noted the "contrast between the feeling of free will, which governs the psychic life, and the apparently uninterrupted causal chain of the accompanying physiological processes," he assured libertarians that "we have learned, by the discovery of the quantum of action, that a detailed causal tracing of atomic processes is impossible . . . [because there is] a fundamental uncontrollable interference in their path. . . . We can hardly escape the conviction that in the facts that are revealed to us in the quantum theory . . . we have acquired the means of elucidating general philosophical problems."[46] Max Born echoed Bohr in a 1928 article for the German daily newspaper *Vossische Zeitung*. After summarizing classic Laplacian determinism, Born assured the public that there was a place for free will and a "higher power" in the new physics: "If the state of a closed system is known precisely at one instant, then natural laws determine the state at every later point in time. The laws of earlier physics always made this claim. Such an interpretation of nature is deterministic and mechanistic. There is no place for freedom of any kind or for the will of a higher power [*einer höheren Macht*]. . . . But recent physics has discovered new laws, supported by much empirical data, that do not conform to this deterministic schema."[47]

It is ironic that Bohr and Born, who took observation as their point of departure, should claim to find evidence in the physics laboratory for human free will (the kind of choices Kierkegaard was describing). Bohr and Born observed that electrons have "freedom" in the sense that subatomic particles in a closed system can act in a way that is not completely determined by prior conditions. But this has little to do with human "free will"—purposeful actions and decisions—which is as unobservable a phenomenon as one can imagine. When Bohr and Born linked random motion in the microworld with free will in the macroworld, they were making what philosophers call a "category mistake." Other examples of category mistakes are: "Electrons are neurotic" and "Protons are monarchists."[48] Such an error prompted Einstein to complain to Born: "I find the idea quite intolerable that an electron exposed to radiation should choose *of its own free will*, not only its moment to jump off, but also its direction."[49]

Nothing exists until it is measured.

—Niels Bohr, ca.1930

The moon is there even when I'm not looking at it.

—Albert Einstein, ca.1930

In 1927 Bohr and Einstein began their renowned debates about the philosophical meaning of quantum mechanics.[50] In the end the disagreement between Bohr/Heisenberg/Weyl/Born and Einstein/Schrödinger was over what constitutes reality. Do the moon and electrons have an independent existence, or do they only exist when observed? Einstein and Schrödinger responded that the former is the case because they wanted to keep the essence of what counted as a *physical* description in Newtonian physics. As Schrödinger stated in 1933, he introduced his wave equation because he "was faced with the difficult task of saving the *soul* of the old system."[51]

THE DE BROGLIE–BOHM INTERPRETATION

Contemporary with the Copenhagen interpretation, in 1927 the French physicist Louis de Broglie put forth an interpretation of quantum mechanics that preserved causality. De Broglie developed his theory working alone in Paris, unaffected by German Romanticism, and presented it in 1927 at the Fifth Solvay Congress in Brussels (the same gathering where Heisenberg announced his Uncertainty Principle and Bohr proclaimed that quantum mechanics ensures free will). De Broglie was followed at the podium by the Austrian physicist Wolfgang Pauli, a standard-bearer for the Copenhagen interpretation, who gave a sharply critical response to de Broglie's interpretation, effectively silencing him.[52] The causal theory languished for twenty years, during which the Copenhagen interpretation held a virtual monopoly. Professors taught generations of young physicists that it is impossible *in principle* to give a deterministic account of quantum phenomena, as the Hungarian-born mathematician John von Neumann declared in 1932: "It is therefore not, as is often assumed, a question of a reinterpretation of quantum mechanics—the present system of quantum mechanics [the laboratory data] would have to be objectively false, in order that another description of the elementary processes than the [Copenhagen] statistical one be possible."[53]

But in 1952 the American physicist David Bohm proved von Neumann wrong by giving just such "a reinterpretation of quantum mechanics" that was logically consistent, empirically adequate, and deterministic, in which physics describes an objective reality—the world-out-there—independent of an observer. His work was dismissed by most of the world's senior physicists, with the exception of de Broglie, Schrödinger, and Einstein, who encouraged Bohm. It is telling that de Broglie and Bohm were *not* German; their backgrounds were in the French and English-speaking traditions of Empiricism and Pragmatism.[54]

De Broglie and Bohm based their approach on Einstein's 1905 paper on the photoelectric effect, in which he had shown that when light interacts with matter, it does so as if it were a particle with mass (a so-called photon); indeed, light can be considered as being both a wave and a particle (a wave-particle). De Broglie (in 1924) and Bohm (in 1952) suggested that an electron *also* has dual wave-particle properties; the propagation of an electron can be treated as if it were a wave. Bohm then essentially took Schrödinger's equation for a Ψ wave (for describing the position and momentum of an electron), and gave the electron wave a particle interpretation in terms of its trajectory. The electron (considered as a particle) is guided along its path by a pilot wave, which Bohm called the particle's "quantum potential."[55] Thus, according to Bohm, electrons have a well-defined position and momentum, and they have an objective reality that does not depend on an observer.

Bohm's theory is logically consistent and empirically adequate to give a deterministic account of quantum phenomena at the level of subatomic events. Unknown to Bohm, de Broglie had proposed a similar solution at the Fifth Solvay Conference in 1927 and then gave up on it, and so in the 1950s the approach came to be called the de Broglie–Bohm interpretation.

The Double-Slit Experiment

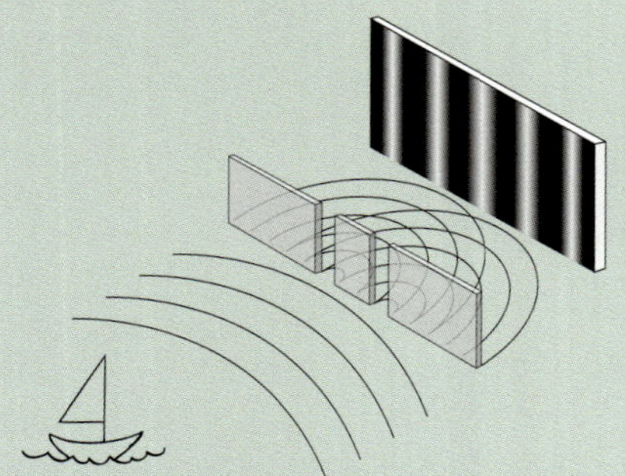

When a wave of water passes through two slits, new waves are formed. The peaks and troughs sometimes reinforce each other and sometimes cancel each other out, resulting in an interference pattern.

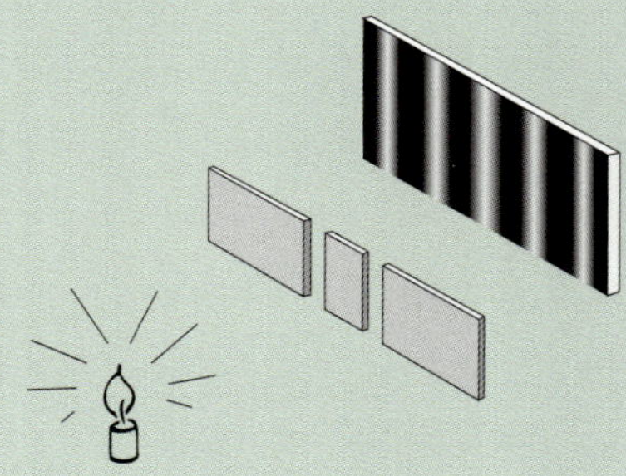

Around 1800 the British physicist Thomas Young passed light through two slits, and the interference pattern recorded on his detector screen proved that light is a wave.

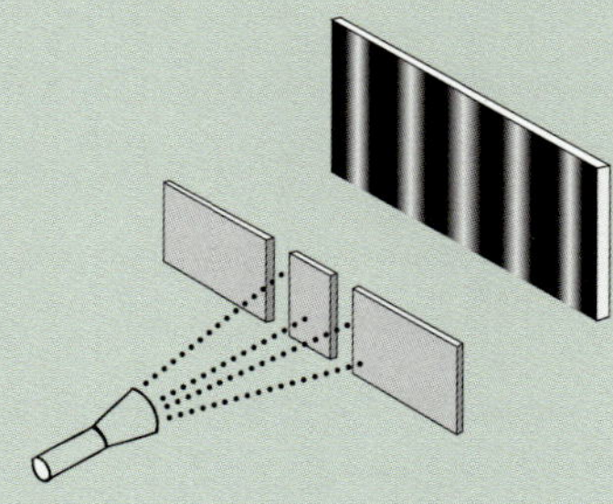

In the early twentieth century, physicists fired a stream of electrons through a double slit, and discovered that, like waves of water and light-waves, moving electrons make a wave pattern.

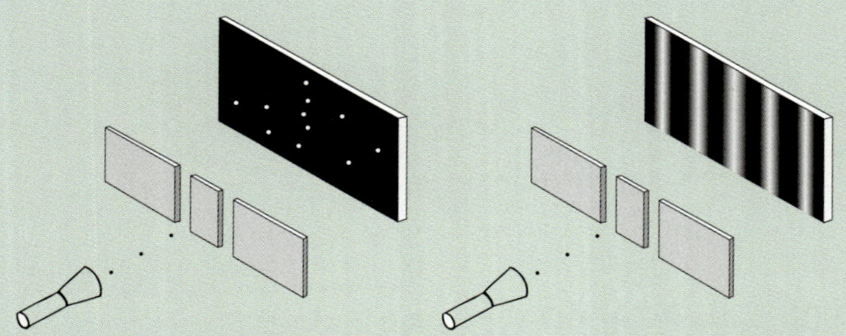

Then physicists fired electrons at the double slit one at a time (left). The cumulative recorded hits over time (right) showed an interference pattern. What was happening? Was each electron travelling through only one slit and interfering with itself? Or was each individual electron somehow travelling through both slits?

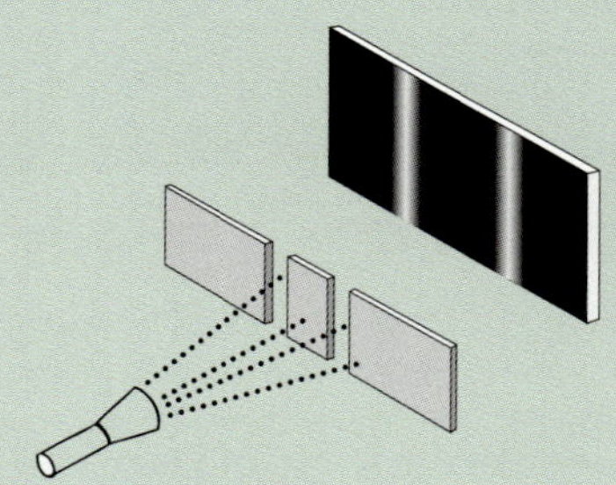

In an attempt to answer these questions, physicists put a tiny detector near the double slit and mechanically observed a stream of electrons. In this set-up, each individual electron passed through only one slit and the cumulative hits formed vertical "slit-shaped" bands, as one would expect from particles (not an interference pattern from waves, as seen above).

According to the Copenhagen interpretation, each single electron passes through both slits—it is in two places at the same time in so-called superposition. The electron has interfered with itself, yielding a new type of quantum probability that is fundamentally different from the classical probability used to measure coin tosses, where the path of each coin is an independent event. Furthermore, the act of observation has affected the path of each electron and determined whether it acts like a wave or a particle. For the American physicist Richard Feynman, this double-slit experiment was the ultimate example of the weirdness of quantum mechanics.

> We choose to examine a phenomenon [the double-slit experiment] which is impossible, absolutely impossible, to explain in any classical way, and which has in it the heart of quantum mechanics. In reality, it contains the only mystery.
>
> —Richard Feynman, *Lectures on Physics*, 1963

According to the de Broglie–Bohm interpretation, each electron (considered as a trajectory) passes through only one slit, but its associated wave passes through both; the interference pattern that develops in the wave generates a similar pattern in the trajectories guided by the wave. There is no difference between quantum and classical statistics—one probability theory governs both realms, as Einstein predicted. The apparently dramatic effect of observation (causing the interference pattern to disappear) is a result of the fact that the determination of "which slit" must involve interaction of a recording device with the system, and any such interaction affects the outcome of the experiment (in a process now termed "decoherence") without recourse to "an observer."

> While the founding fathers agonized over the question "particle" or "wave," de Broglie in 1925 proposed the obvious answer "particle" and "wave." . . . De Broglie showed in detail how [in the double-slit experiment] the motion of a particle, passing through just one of two holes in the screen, could be influenced by waves propagating through both holes. And so influenced that the particle does not go where the waves cancel out, but is attracted to where they cooperate. This idea seems to me so natural and simple, to resolve the wave-particle dilemma in such a clear and ordinary way, that it is a great mystery to me that it was so generally ignored.
>
> —John S. Bell, *Speakable and Unspeakable in Quantum Mechanics*, 1987

The de Broglie–Bohm interpretation offered an explanation for the iconic thought experiment associated with the strangeness of the quantum realm, the double-slit experiment for electrons and photons (see sidebar on opposite page). Nevertheless, during the 1950s the international physics community either ignored or actively rejected Bohm's theory. Criticism of Bohm was led by Pauli,[56] the same physicist who had spearheaded the rejection of de Broglie in 1927 and whose outlook was driven by the shifting tides of the philosophy (as opposed to the practice) of physics. In the 1930s, while teaching at the ETH in Zurich, Pauli apparently lost his positivist aversion to psychology because he gave the Copenhagen interpretation a Jungian twist. After suffering a nervous breakdown and seeking help from the Swiss psychoanalyst Carl Jung, the imperious patient did a role reversal and began analyzing the doctor's theory of archetypes, focusing on synchronicity, which Jung defined as two events that are associated in the mind not by cause and effect but by meaning.[57] Pauli agreed with Jung, who was the son of a Protestant clergyman, that the psychic ills of modern man were due to the separation of science (matter, reason) and religion (spirit, intuition; Jung, *Modern Man in Search of a Soul*, 1933). For Pauli, the Copenhagen interpretation provided the key to reuniting matter and spirit. Following in Bohr's footsteps, Pauli—in yet another case of overreach—applied quantum mechanics to ethics and psychology: "[Bohr] would like to see in 'Complementarity,' as it is manifest in physics, *a general model for the resolution of conflicts, for unifying pairs of opposites* . . . . For example, he attempted to apply this to ethics (good—evil, justice—love)" ("The Struggle for Wholeness in Physics," 1953).[58]

In the early 1950s, Pauli insisted that he rejected Bohm's challenge to the Copenhagen interpretation "for physical reasons, which have nothing to do with philosophical prejudices . . . an interpretation of quantum mechanics based on complementarity is the only one allowed [*la seule admissible*]."[59] But from a lecture Pauli gave in Mainz in 1955, it is clear that he had more than physics on his mind; Pauli was defending his (Jungian) philosophy of life (culminating in the union of science and spirit) based on Bohr's notion of complementarity. Pauli began his lecture by recounting the long history of theories of reality based on binary pairs (dark-light, male-female, odd-even), including Pythagoreans, Platonists, Taoists, Buddhists, Plotinus, Saint Augustine, Goethe, and Jung, and culminating with Bohr: "According to the formulation of Niels Bohr contemporary quantum physics has likewise come up with complementary pairs of opposites in its atomic objects such as particle-wave, position-momentum, and it must bear the freedom of the observer in mind." Pauli went on to describe how, if one can forgo rationalism, such conflicts will resolve themselves in a moment of mystical apprehension: "nothing remains but to put oneself at the mercy of these aggravated opposites and their conflicts in one way or another. It is in just that way that the researcher, more or less consciously, can tread an inner way to salvation [*Heilserkenntnis*]. Internal images, phantasies, or ideas then slowly evolve as compensation for the external situation, and they show that an approach of the poles of the pair of opposites is possible." Pauli ended by declaring that the Copenhagen interpretation amounted to a philosophy of life: "I, for my part, consider the imagined goal of an overcoming of opposites, this also including a synthesis which would comprise rational understanding as well as mystic experience of unity, to be the explicit or implicit myth of our own contemporary age."[60] In short, Pauli rejected Bohm, at least in part, because the de Broglie–Bohm interpretation challenged Bohr and Pauli's philosophy that provided a "mystic experience of unity" and was the "myth of our contemporary age."

In 1949, in the midst of Bohm's embattled research, he was subpoenaed by Senator Joseph McCarthy's House Un-American Activities Committee. During his student days at Berkeley, Bohm had been a leftist and joined a local chapter of the Communist Party. He refused to testify against others and was indicted for contempt of Congress, tried, and eventually acquitted of any crime in May 1951. During the trial Princeton University put him on paid leave and then, after his acquittal, did not renew his contract. Unable to find suitable work in the United States, Bohm moved to Brazil[61] and then to Israel, finally in 1957 settling in England, where he remained somewhat isolated and on the periphery of developments in physics for the rest of his life. Scholars have debated whether the physics community's negative reaction to Bohm was because of Cold War politics or academic politics; the weight of scholarship puts the blame on the academy.[62]

In the 1950s the Irish physicist John Stewart Bell complained bitterly that as a student he had learned from Born and von Neumann that any alternative to the orthodox Copenhagen interpretation was impossible. But then, Bell recalled, "In 1952 I saw the impossible done. It was in papers by David Bohm. Bohm showed explicitly how . . . the indeterministic description could be transformed into a deterministic one. More importantly, in my opinion, the subjectivity of the orthodox version, the necessary reference to the 'observer,' could be eliminated. Moreover, the essential idea was one that had been advanced already by de Broglie in 1927, in his 'pilot wave' picture. But why then had Born not told me of this 'pilot wave'? If only to point out what was wrong with it? Why did von Neumann not consider it? . . . Why is the pilot wave picture ignored in textbooks? Should it not be taught, not as the only way, but as an antidote to the prevailing complacency? To show us that vagueness, subjectivity, and indeterminism, are not forced on us by experimental facts, but by deliberate theoretical choice?"[63] These are good questions which have begun to be addressed in the landmark studies by Jammer and Forman and in the massive literature they prompted.[64]

QUANTUM ENTANGLEMENT

During the 1920s and 1930s Albert Einstein realized that quantum mechanics implies that under certain conditions, when a subatomic particle disintegrates it emits twin daughter particles of equal mass that have paired properties; if one particle is measured and found to have spin that is up, then the spin of the other will be measured to be down. The two particles are "entangled." According to Einstein, the twin daughter particles have their paired properties from birth, but according to Bohr, each daughter particle has only the *potentiality* of direction of spin until it is measured, and the act of measurement causes the particle to spin up or down.[65]

In an attempt to show that Bohr was wrong about the role of observation, Einstein proposed a thought experiment. Suppose that an entangled pair of particles fly off in

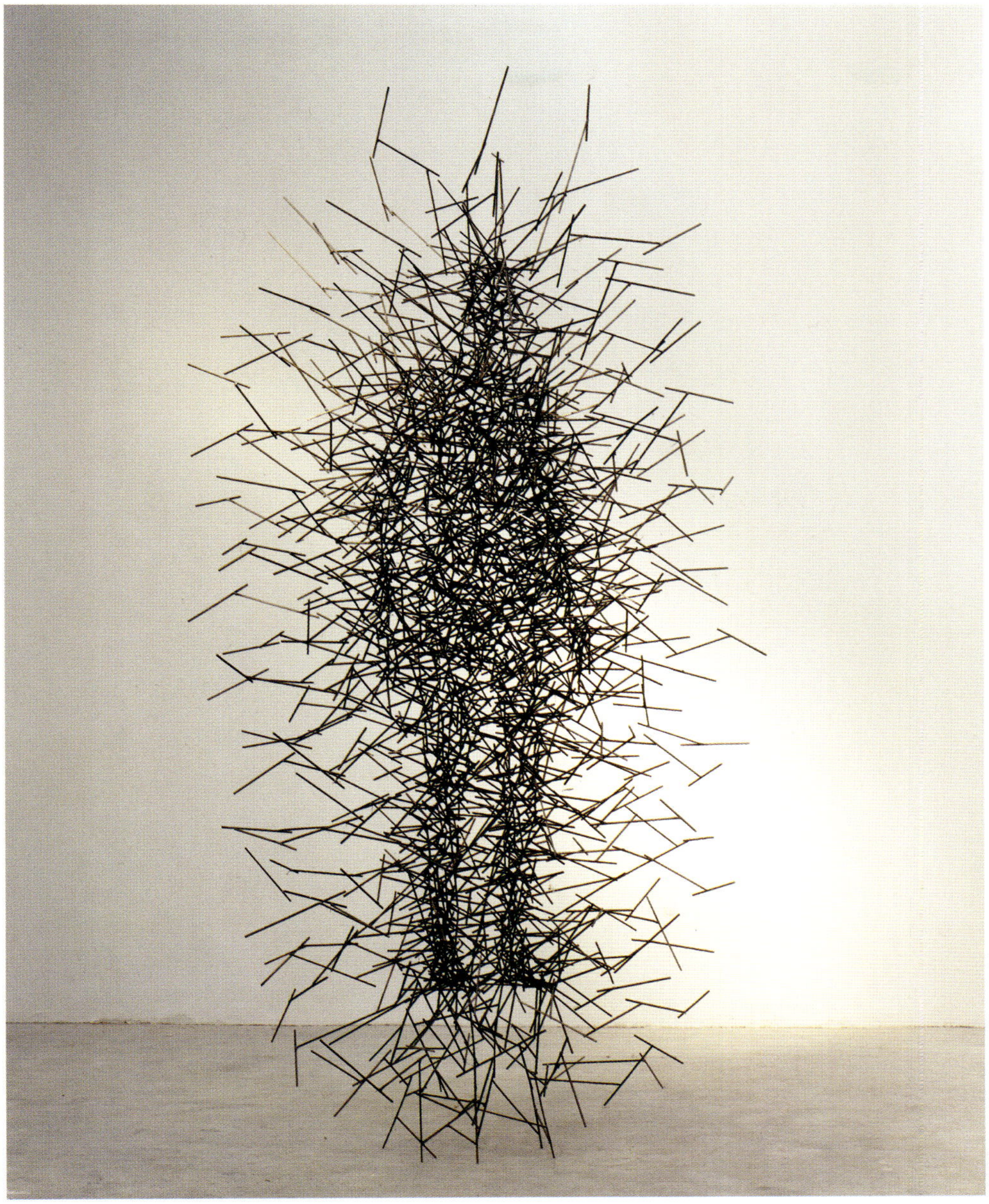

8-7. Antony Gormley (British, b. 1950), *Quantum Cloud V*, 1999. 4 mm square section mild steel bar, 107⅞ × 61 × 47 in. (274 × 155 × 119 cm). © The artist. Courtesy of the artist.

opposite directions and get very far apart to the right and left of the observer. Then imagine that an observer measures the spin of the particle on the right; suppose that its spin measures "up." According to Bohr's Copenhagen interpretation, the observer *caused* the electron to spin up and the twin on the distant left must *instantaneously* measure down. But in Einstein and Bohr's day, physicists assumed that the universe is *local* in the sense that it should be impossible to alter the property of a distant system by direct action on a local system. Assuming the locality of the universe and the theory of relativity (that nothing can travel faster than the speed of light), together with Bohr's claim that the twin photons lack spin until measured, the thought experiment ends in a physical (as distinct from a logical) paradox. Einstein formulated the paradox with his Princeton assistants Boris Podolsky and Nathan Rosen (Einstein, Podolsky, and Rosen, "Can Quantum Mechanical Description of Reality be Considered Complete?" 1935). According to Einstein, since the Copenhagen interpretation of quantum mechanics predicts a paradoxical situation, it must be incomplete. Schrödinger named Einstein's physical paradox photon "entanglement" (*Verschränkung*), and he came up with his own version of a quantum-measurement paradox (*Schrödinger's Cat*, 1935).[66]

Bohr dismissed Einstein's challenge as a senior scientist hopelessly stuck in the old-fashioned mindset of being an "objective observer" of a deterministic universe. Younger physicists considered the matter settled and moved on.

You believe in a God who plays dice, and I in complete law and order in a world which objectively exists, and which I, in a wildly speculative way, am trying to capture. . . . Even the great initial success of the quantum theory does not make me believe in the fundamental dice-game, although I am well aware that some of our younger colleagues interpret this as a consequence of senility.

—Albert Einstein, letter to Max Born, September 7, 1944

In the late 1950s Bohm and Yakir Aharonov, his collaborator in Haifa, published a new version of the Einstein–Podolsky–Rosen paradox, reformulated in terms of spin (which is the version I described before; "Discussion of Experimental Proof for the Paradox of Einstein, Rosen, and Podolsky," 1957). In the early 1960s John Stewart Bell, using the Bohm–Aharonov version of Einstein's 1935 argument, pointed out that one could infer from the mathematics that photons are entangled in some (unknown) way, and he suggested that Einstein's thought experiment could be carried out in the laboratory (*Bell Inequalities*, 1964). By the 1980s the technology had been developed to do Einstein's thought experiment, and the inexplicable nature of quantum entanglement went from being a philosophical quibble to an experimentally confirmed fact. This means that any current or future interpretation of quantum mechanics must take particle entanglement into account.[67] No physical mechanism for quantum entanglement is known; it is yet another example of a natural phenomenon that is described only by mathematics.

Einstein called particle entanglement "spooky action at a distance" because, if Bohr were right, the photon must somehow communicate its direction of spin to its distant twin moving (like a ghostly spook) faster than light. Einstein's point, however, was that Bohr was *not* right and that the apparent "spookiness" of the action disappears if you assume, as Einstein did, that the twins got their paired properties at birth (when they were created in the lab), independent of human observation. Today the term "spooky" is also used in a different sense when someone describes quantum entanglement as spooky because it (mysteriously) has no known physical cause. This is reminiscent of the bewildered response of seventeenth-century scientists to Newton's description of another "spooky action at a distance"—gravity. Today Newton's equation describing earth holding the moon in its orbit no longer seems eerily characteristic of supernatural spirits because for over three hundred years scientists have used $F = G\,(m_1 m_2/d^2)$ to predict the precise location of the moon, sun, and stars. But, like quantum entanglement, the force of gravity is described only by mathematics, and its underlying physical mechanism is unknown. (Gravitation is understood today as a phenomenon arising from the curvature of the structure of space-time, according to the general theory of relativity. Space-time is described mathematically, so we know *how* it behaves, but an explanation of *why* it behaves this way continues to elude scientists.)

Every five years the city of Kassel, Germany, hosts an international exhibition of contemporary art, Documenta, which corresponds to other huge summer art fairs, including Brazil's São Paulo Biennial, Italy's Venice Biennale, and Switzerland's Art Basel with sister exhibitions in Miami and Hong Kong. In a sign of the current interdisciplinary cultural climate (plate 8-7), in 2012 the Austrian government did not select a contemporary artist to represent its country at this major art event (as all the other participating countries did); instead they sent a physicist from the University of Vienna, Anton Zeilinger, to present five experiments demonstrating quantum entanglement on this world cultural stage. The future will tell if Zeilinger's physics lessons inspired any artists in his audience at Documenta.

THE AMBASSADOR OF RUSSIAN CONSTRUCTIVISM: EL LISSITZKY

The tension between Enlightenment rationality and Romantic expressionism was also felt in the art world during the Weimar Republic. The hub of this conflict was in the heart and mind of Walter Gropius, founding director of the Bauhaus school of design; other forceful personalities converged in Berlin, including Lissitzky, Van Doesburg, and Moholy-Nagy.

Lissitzky arrived in Berlin in 1921. His early work was inspired by Malevich's Suprematism, but his mature style was closer to Constructivism. Lissitzky grew up in traditional Jewish surroundings in Vitebsk, a town in western Russia with a large Jewish population. He had travelled to Germany before for his education in architecture and engineering, but returned home to Vitebsk when the war broke out in 1914. He supported the 1917 revolution, but as a Jew he was wary of the Bolshevik program to eliminate ethnic identity in order to create a single national culture.[68] Lissitzky made art on Jewish themes, in a loose, figurative style, such as his illustrations for Moishe Broderzon's folk tale in the booklet *The Legend of Prague* (Moscow, 1917), which is based on an entry in the sixteenth-century chronicle of the Jewish community in Prague. In August 1918 another Jewish native of Vitebsk, Marc Chagall, was appointed director of the new state-supported Popular Art Institute of Vitebsk, which had the mission of designing functional objects for the new society. The following year Chagall hired Lissitzky, and in the fall of 1919 he invited Malevich to teach at the school.[69]

Malevich arrived at the job in this small-town art school as an established Moscow artist, and he quickly organized a group of students and faculty, including Lissitzky, around the goal of applying his Suprematist style to practical objects—everything from trains to teacups. By the spring of 1920 Malevich had transformed the group into an organization, UNOVIS (an abbreviation of the Russian for "champions of new art"), and Malevich's disciples wore a Suprematist badge—a black square—on their sleeves.[70] Malevich's domineering personality and dogmatic teaching style caused a power struggle with Chagall over the curriculum; the students rallied around Malevich, and Chagall, finding himself with an empty classroom, resigned and went to Moscow.[71]

At this time Lissitzky was searching for a modern style that would harmonize with Judaism, and he chose Malevich's geometric abstraction—with its spiritual overtones—over Chagall's figurative story-telling. For Lissitzky, Suprematist art did not symbolize a traditional Jewish-Christian-Islamic Supreme (Absolute) Being, but it was part of a new vision for the future and a new way of perceiving, as he wrote in 1920: "For us Suprematism did not signify the recognition of an absolute form which was part of an already-completed universal system. On the contrary here stood revealed for the first time in all

8-8. El Lissitzky (Russian, 1890–1941), *Proun (Entwurf zu Proun S.K.)*, 1922–23. Watercolor, gouache, India ink, pencil, conté crayon, and varnish on buff paper, 8 7/16 × 11 3/4 in. (21.4 × 29.7 cm). Solomon R. Guggenheim Museum, New York, gift of the Estate of Katherine S. Dreier, 53.1343. © 2014 Artists Rights Society, New York.

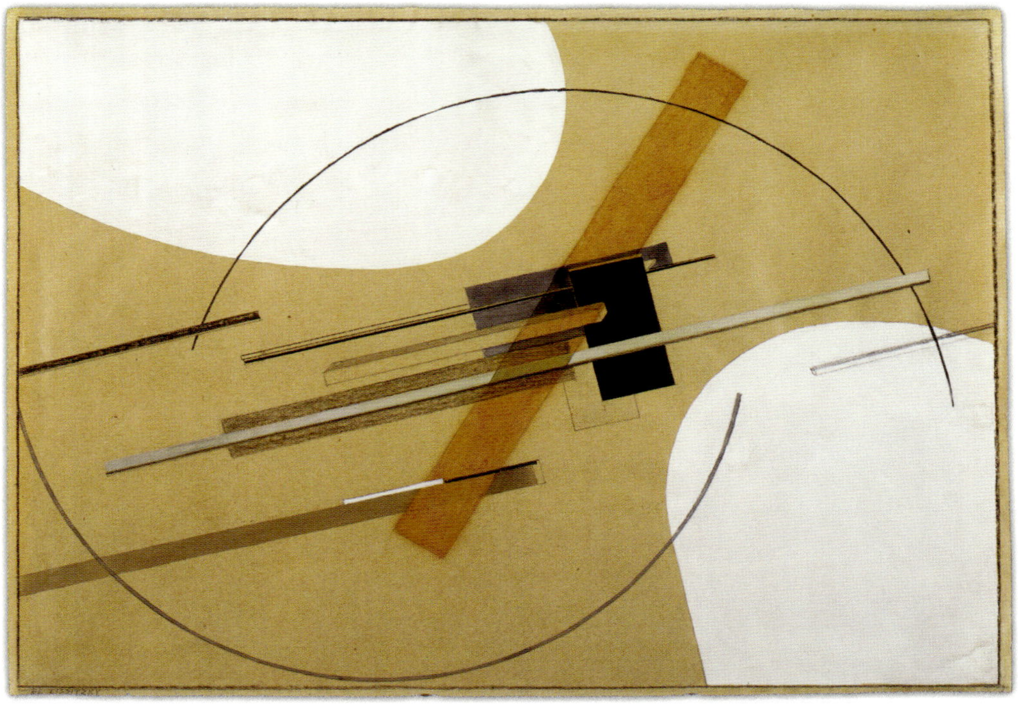

its purity the clear sign and plan for a definite new world never before experienced—a world which issues forth from our inner being and which is only now in the first stage of its formation. For this reason the square of Suprematism became known as a beacon."[72]

In response to Suprematism, Lissitzky began making non-representational designs that combined painting with architectural drawing. To name this synthesis, he coined the word "Proun" (an abbreviation of the Russian for "project for the affirmation of the new"). Like Malevich, Lissitzky composed with geometric forms, but his were much more complex and showed his training in architectural drawing (*Proun*, 1922–23; see plate 8-8). In all his projects in the 1920s—books, magazines, posters, exhibition design, and architecture—Lissitzky applied design principles he had developed in his Proun project. He designed a book for children in which he encouraged little comrades to bring two squares to life on the page (*About Two Squares, in Six Constructions: A Suprematist Tale*, 1922; see plate 8-9).[73]

Lissitzky's German education in architectural engineering included science and mathematics, and thus he was attuned to the confirmation of Einstein's general theory of relativity in 1919. Lissitzky wrote in 1920 that Newton's old Absolutes (time and space) were no more: "The absolute of all measures and standards has been destroyed. When Einstein constructed his theory of particular and general relativity, he proved that the speed with which we measure a particular distance influences the size of the unit of measure."[74] For earthlings, up-down is determined by earth's gravitational field, but for a cosmic body moving in outer space, spatial orientation changes because space is warped by the gravitational fields of massive bodies. Lissitzky symbolized space-time by designing some of his three-dimensional works to literally move through the fourth dimension of time (*Proun: Eight Positions*, 1923; National Gallery of Canada, Ottowa). In a nod to his countryman Lobachevsky, Lissitzky titled a 1925 essay he wrote on the history of mathematics "K. und

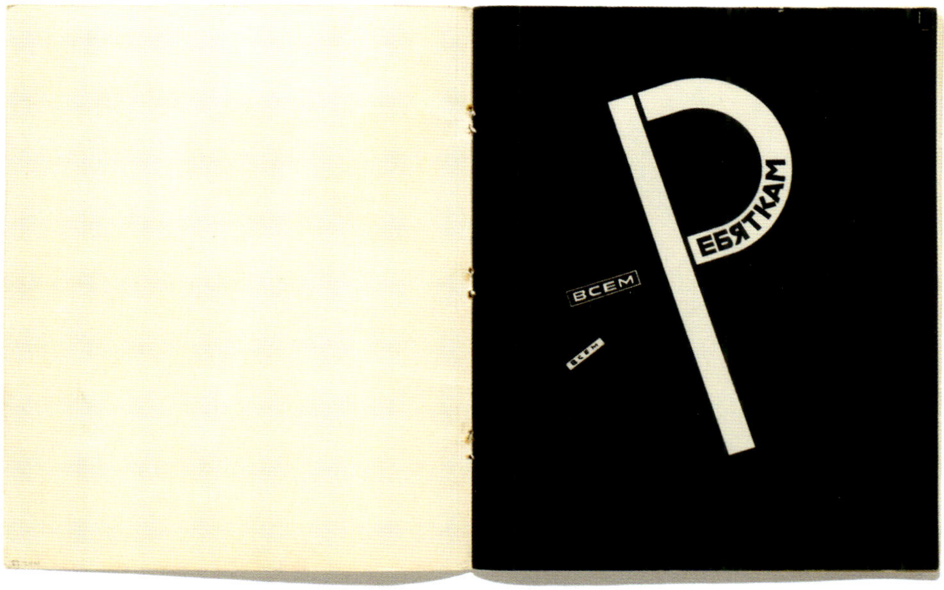

To all, all children

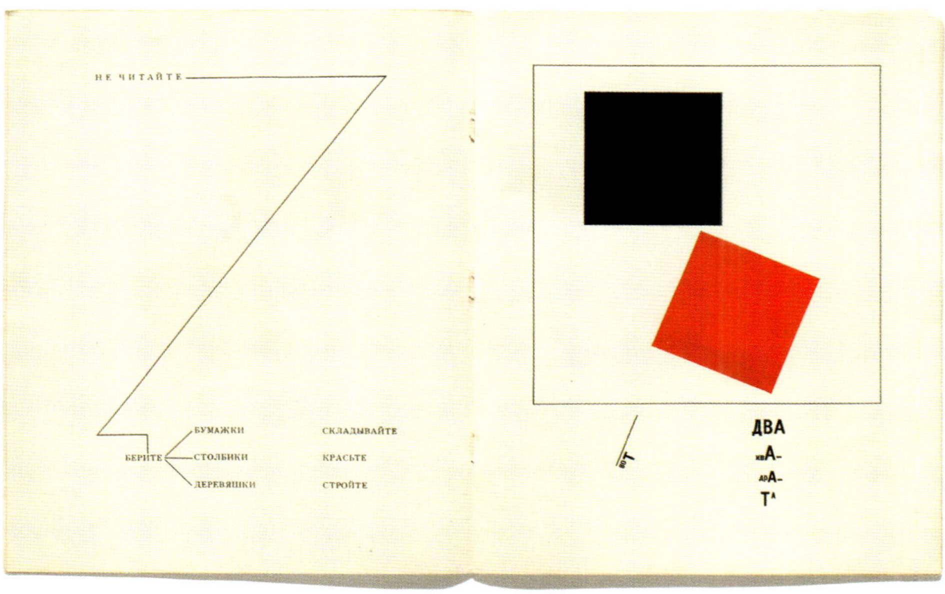

Do not read. Take paper, rods, blocks. Lay out, color, build.

Here are two squares.

8-9. El Lissitzky (Russian, 1890–1941), *Pro dva kvadrata: Suprematicheskiĭ skaz v 6-i postroĭkakh* (About two squares, in six constructions: a Suprematist tale; Berlin: Skify, 1922). Letterpress illustrations, 10 15⁄16 × 8 7⁄8 in. (27.8 × 22.5 cm) ea. Museum of Modern Art, New York, gift of the Judith Rothchild Foundation. © 2014 Artists Rights Society, New York.

Pangeometrie" (K is for the German "Kunst [Art]," and *Pangeometrie* is the title of an 1855 book by Lobachevsky).[75]

Lissitzky foresaw a new art, made collectively by anonymous artist-workers, to express the new Communism. His thinking about mathematics and art was shaped by his reading of Spengler's 1918 book, which prophesized the end of Western culture and the rise of Bolshevism.[76] The artist began the 1920–21 essay "Proun" with a quote from Spengler, and after declaring, "We are taking mathematics as the purest product of man's creativity," Lissitzky summarized the history of mathematics following Spengler. He stated a set of five axioms for his Prouns, including "Form outside material = Zero,"[77] from which he deduced that art must be made from some material (paper, wood, etc.); in other words, art cannot exist as an unembodied idea. His constructions, composed of flat, geometric planes, could be applied to architecture and design. Indeed, as a teacher in the new Communist state, Lizzitsky envisioned that Prouns would be made by his students anonymously—without attention to authorship—for collective use by the new world government. He ended his article: "We are now witnessing the birth of a new style, and it is not one of individual artists, but of a nameless creator. . . . Life is now building a new, reinforced concrete, Communist foundation for the peoples of the earth. Through the Proun we will come to build upon the universal foundation for a single world city—for all the people of the earthly globe."[78]

Within a year the UNOVIS organization had split into two groups: one wanted a meditative art with a spiritual vision, and the other favored a practical, utilitarian art at the service of society. Lissitzky, who had a foot in both camps, left Vitebsk in 1921 to become a cultural ambassador based in Berlin, with the goal of promoting Communism in the capitalist West.[79] In 1925 Malevich also left Vitebsk and returned to Moscow, where he joined Rodchenko on the faculty of VKhUTEMAS.

GEOMETRIC ABSTRACTION IN GERMANY IN THE 1920S

Theo van Doesburg arrived in Germany in 1922 and, together with Lissitzky and the German abstract filmmaker Hans Richter, announced the organization of the International Faction of Constructivists at a meeting of progressive artists in Düsseldorf. This group launched the formalist aesthetic of Constructivism as a self-conscious movement in Europe. The same year Lissitzky became co-editor with the novelist Ilya Ehrenburg of a trilingual magazine based in Berlin, Вещь/*Gegenstand*/*Objet* (Russian, German, and French for "object"). For an editorial in the first issue, Lissitzky and Ehrenburg wrote that Europe was no longer stopping the flow of ideas from Russia, and that the Constructivist aesthetic could now be brought to the West: "We stand at the outset of a great creative period. . . . The negative tactics of the 'dadaists' . . . appear anachronistic to us. Now is the time to build on land that has been cleared. . . . The fundamental feature of the present age is the triumph of the constructive method."[80] Lissitzky had developed his Proun style in Russia to express Soviet Communism, but in Berlin he described the political mission of his style more vaguely as helping people move into the future, without specifying exactly what that future should be.

After the confirmation of Einstein's space-time cosmology, van Doesburg criticized Mondrian for continuing to orient his computations up-down "relative to the static, perpendicular axis of gravity,"[81] saying that he would forge a new style, "Elementarism," that incorporated the fourth dimension of time: "While the expressive possibilities of Neo-plasticism are limited to two dimensions (the plane), Elementarism realizes the possibility of plasticism in four dimensions, in the field of space-time."[82] Van Doesburg left Amsterdam for Berlin to promote Elementarism and to publish *De Stijl* on German soil.[83] He met Moholy-Nagy, who together with the French-Germany (Alsatian) artist Jean/Hans Arp, the Austrian Raoul Hausmann, and the Russian Ivan Puni had written a manifesto, *Aufruf zur elementaren Kunst* (A call to elementary art). In consonance with van Doesburg's outlook, these artists called for an art of pure, simple forms that would express the spirit of the times. Van Doesburg published their treatise in *De Stijl* in 1921.

Moholy-Nagy arrived in Berlin in 1920. He had become familiar with the European and Russian avant-garde in Budapest through the magazine *Ma* (Hungarian for "today") while convalescing from serious injuries sustained during World War I (he served in the Austro-Hungarian army). After his discharge at age twenty-three in 1918, he took private art lessons and supported the Hungarian revolution, but he fled after the collapse of the Communist republic of Béla Kun in August 1919. He went first to Vienna but soon moved on to Berlin because "I was less intrigued with the baroque pompousness of the Austrian capital than with the highly developed technology of industrial Germany."[84] In Berlin in early 1920, Moholy-Nagy began making compositions with mechanical objects (*Bridges*, 1920–21; Gemeentemuseum, The Hague), but by 1921 he was composing with meaning-free forms in works such as *Nickel Construction* (1921; see plate 8-10), which was shown in Moholy-Nagy's first exhibition in Berlin (Galerie der Sturm, February 1922).

8-10. László Moholy-Nagy (Hungarian, 1895–1946), *Nickel Construction*, 1921. Nickel-plated iron, 14⅛ × 6⅞ × 9⅜ (35.9 × 17.5 × 23 8 cm). Museum of Modern Art, New York, gift of Sibyl Moholy-Nagy. © 2014 VG Bild-Kunst, Bonn/Artists Rights Society, New York.

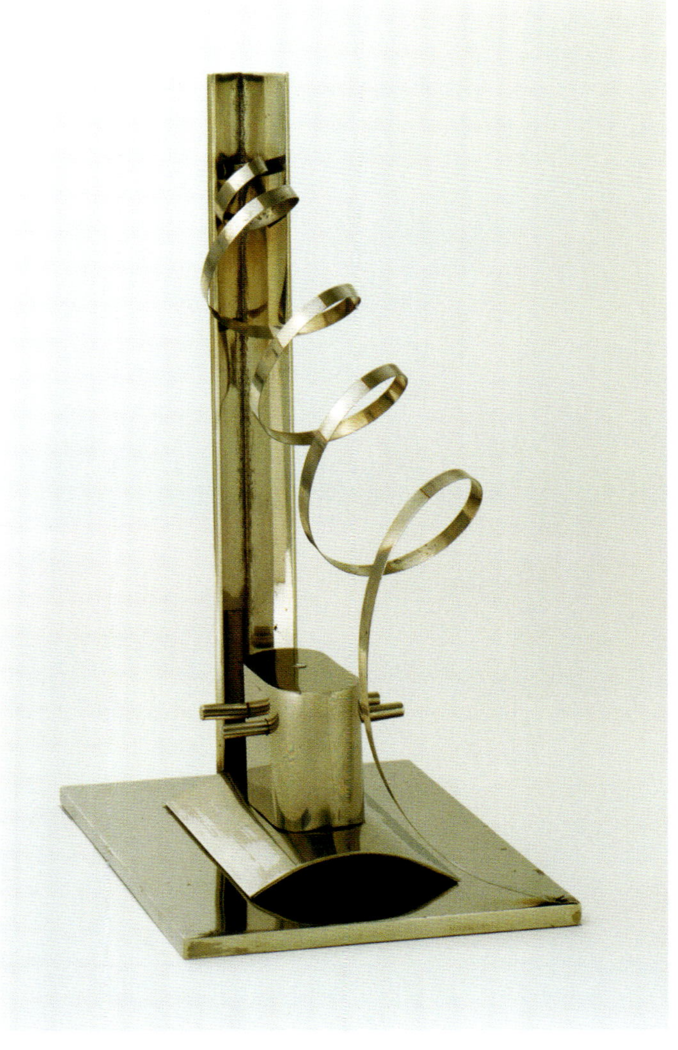

Steel and glass skyscrapers of the so-called International Style were a prime expression of the utopian mood in Berlin after World War I.[85] In the 1920s the skylines of German cities were dominated by the towers of medieval churches, from the soaring twin towers of the great cathedral of Cologne to the humble bell tower of a parish church. The German architect Bruno Taut made an analogy between the old medieval church and the new modern skyscraper—both sheathed in glass. Taut proclaimed that the spirituality of the bygone era could be revived in the modern secular age by building a glass tower—a *Stadtkrone* ("city crown")—in the center of every town, where residents could go to replenish their spirits and feel in harmony with the cosmos (plates 8-11, 8-12, and 8-13).[86]

Skyscrapers had been built since the invention of steel and the electric elevator in the mid-nineteenth century, but their steel frameworks were hidden behind non-supporting walls of brick and stone. Now German architects celebrated the new technology by exposing the building's steel skeleton beneath a skin of glass (plate 8-1, chapter frontispiece). At the Bauhaus school of design, opened in Weimar in 1919, the Romantic visionary Walter Gropius taught students a systematic, rational approach to design, but, like a medieval guild of anonymous craftsman, these students would erect buildings for the future and, like Taut's *Stadtkrone*, the new architecture would be dedicated to the spirit of the first secular, scientific age in human history.

Through Lissitzky, Ehrenburg, and others, Gropius was kept apprised of developments in Russian art education.[87] Van Doesburg met Gropius and visited the Bauhaus, liked what he saw, and moved

TOP RIGHT

8-11. Jan van Eyck (Netherlandish, ca. 1390–1441), *Die heilige Barbara* (Saint Barbara), frontispiece in Bruno Taut's *Die Stadtkrone* (Jena, Germany: Eugen Diederichs, 1919), 7. Avery Architectural and Fine Arts Library, Columbia University, New York.

Plants and animals instinctively turn towards the warmth and brilliance of sunlight. Taut used mankind's primeval response to light as the spark to rekindle spiritual life. In medieval Europe, the skyline was pierced by steeples and bell towers, as in van Eyck's image of a tower in construction, which Taut chose for the frontispiece of his 1919 book, *Die Stadtkrone*. In the Christian era Saint Barbara was imprisoned (by prejudice and dogma) in such a tower, but in the secular era mankind would be liberated (by mathematics and science) in Taut's *Stadtkrone*.

BOTTOM RIGHT

8-12. View of the *Stadtkrone* facing east, in Bruno Taut, *Die Stadtkrone* (Jena, Germany: Eugen Diederichs, 1919), 65, fig. 42. Avery Architectural and Fine Arts Library, Columbia University, New York.

As medieval worshippers saw stained glass aglow in morning light during matins in their parish church, so modern city dwellers would begin their day by facing the rising sun in their town's *Stadtkrone* (city crown).

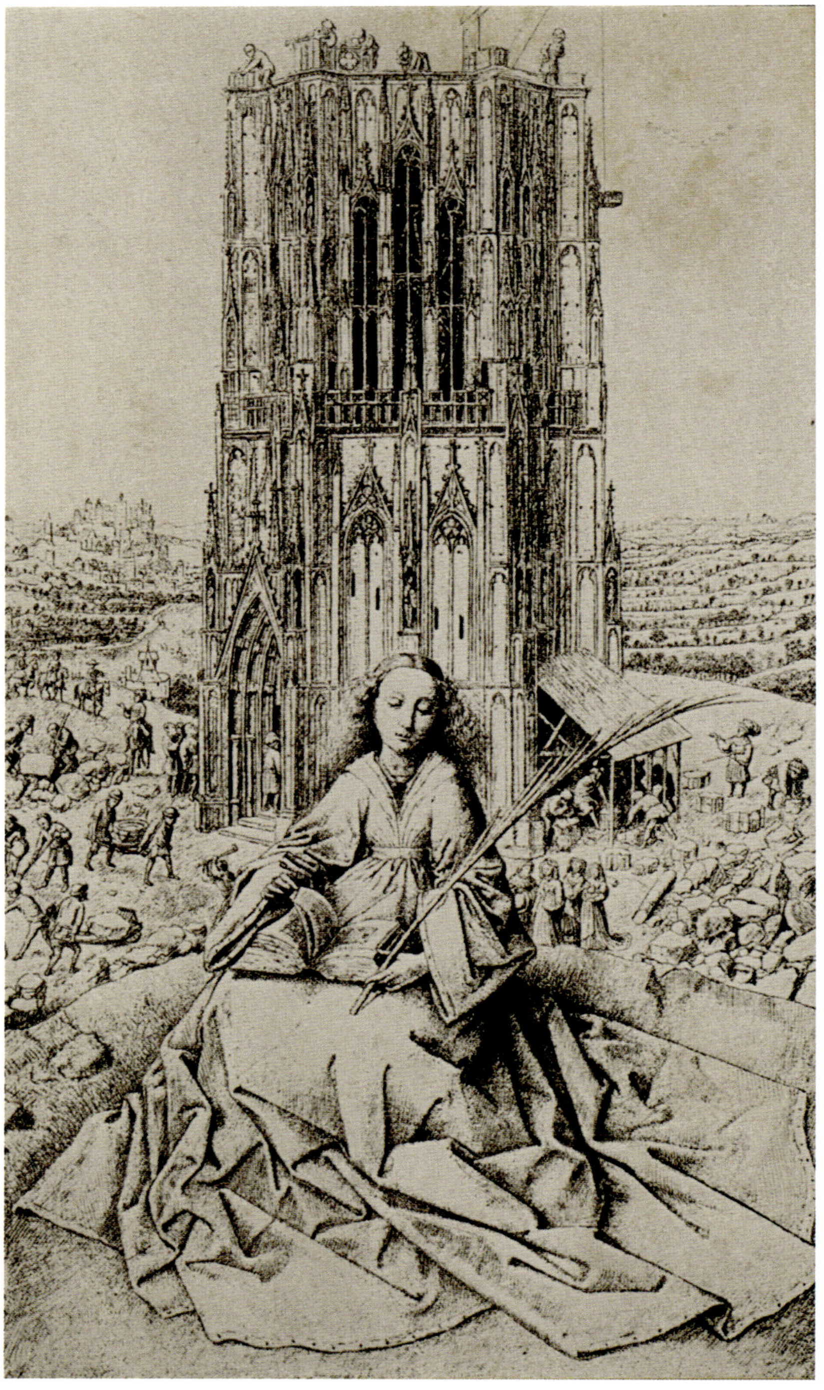

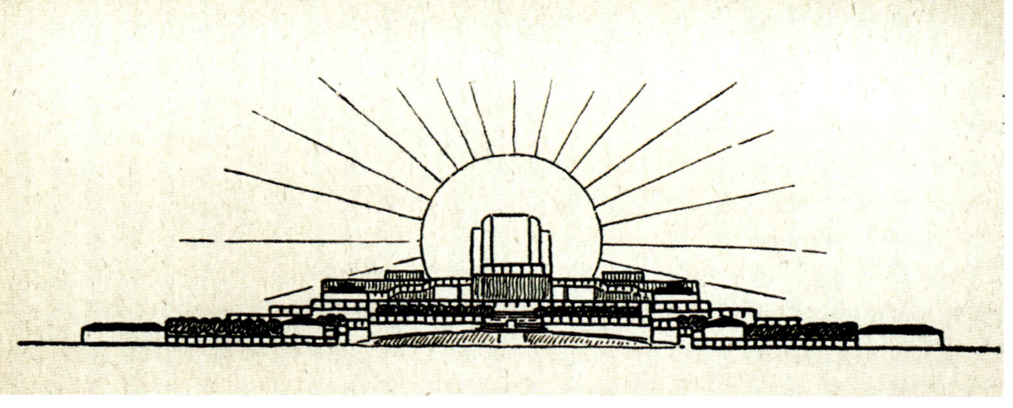

to Weimar in April 1921, hoping to land a teaching job, but Gropius was wary of van Doesburg's domineering personality and did not hire him.[88] True to his combative style, van Doesburg responded by teaching rival design classes in his Weimar apartment and giving lectures critical of what he perceived as overly Expressionist tendencies at the Bauhaus. Between 1921 and 1923 he continued to publish *De Stijl* in Weimar, calling for an art of universal abstract forms. Dismissing European and Soviet politics as "merely words," van Doesburg advocated a future shaped by a brotherhood of artists: "The International of the Mind is an inner experience that cannot be translated into words. It does not consist of a torrent of vocables but of plastic creative acts and inner intellectual force, which thus creates a newly shaped world."[89] Ironically, this self-styled prophet of an "International of the Mind" couldn't get along with other artists, as Gropius observed: "I judged him aggressive and fanatic and considered that he possessed such a narrow, theoretical view that he could not tolerate any diversity of opinion."[90]

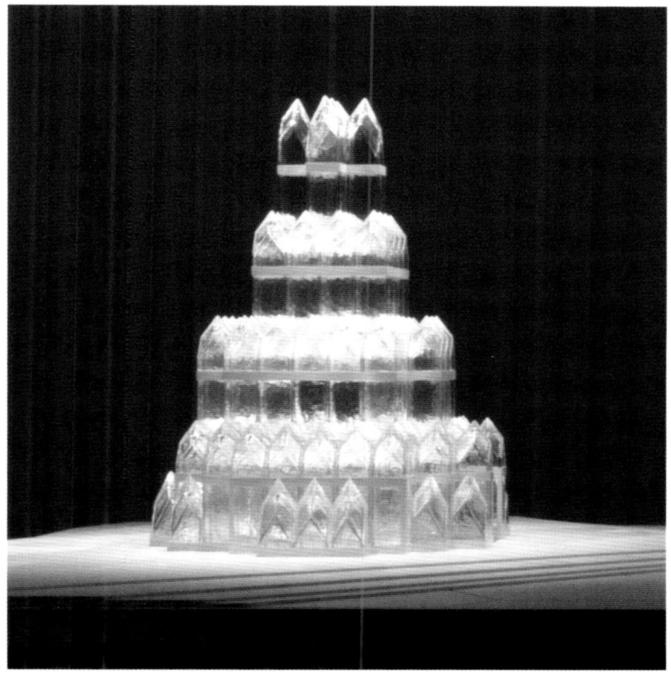

8-13. Josiah McElheny (American, b. 1966), *City-Crown*, 2007. Hand-blown glass, metal, painted wood, acrylic, electric lighting, 14 ft. (4.26 m) high, installation at the Moderna Museet, Stockholm. Courtesy of the artist and Andrea Rosen Gallery, New York, and Donald Young Gallery, Chicago, ARG# MJ2007-001.

Josiah McElheny created this tower of glass as homage to Bruno Taut's (unrealized) *Die Stadtkrone* of 1919.

As soon as Gropius opened the Bauhaus, there was an ongoing war between Enlightenment rationality and Romantic expressionism waged under the same roof (sometimes between the mind and heart of the same artist), which reflected the conflict in wider Weimar culture. The painters on staff—Paul Klee, Wassily Kandinsky, and Johannes Itten—had all been associated with the pre-war German Expressionist style Der Blaue Reiter, and their art symbolized emotions, sensations, or Absolute Spirit. Itten liven in accordance with a health movement, Mazdaznan, which was based on the teachings of Zoroaster (an ancient Persian prophet).[91] The origins of the founder of Mazdaznan are unclear; he was probably born Otto Hanisch in the mid-nineteenth century to a Russian father and German mother, and he eventually settled in Chicago, where he combined Eastern and Western doctrine in search of a universal religion. In the first decades of the twentieth century, Hanisch gave lecture tours in Germany under his Islamic pseudonym "Otoman Zar-Adhust Ha'nisch," and his disciples, including Itten and his students, followed his instructions to achieve consciousness of Absolute Spirit by eating a vegetarian diet, doing breathing exercises, and taking long walks in nature. After a brief power struggle between Itten and Gropius, Itten left and Gropius replaced him with Moholy-Nagy.[92]

Enthusiastic about technology, Moholy-Nagy endorsed the absence of the artist's hand, a point he made by telephoning a sign company and placing an order for three paintings by specifying their format and picking their hues from the company's color chart—identical images painted in three sizes: small, medium, and large (plate 8-14).[93] In an effort to promote a rational, scientific outlook, Gropius had instituted a core course in which incoming students would learn an impersonal (meaning-free) vocabulary of form and color. Moholy-Nagy's first task was to revise this foundations course. Enforcing a formalist aesthetic, he declared that students should learn a basic vocabulary of

RIGHT

8-14. László Moholy-Nagy (Hungarian, 1895–1946), *Construction in Enamel* 3 (a so-called *Telephone Picture*), 1923. Porcelain enamel on steel, 9½ × 6 in. (24 × 15 cm). Museum of Modern Art, New York, gift of Philip Johnson in memory of Sibyl Moholy-Nagy. © 2014 VG Bild-Kunst, Bonn/Artists Rights Society, New York.

BELOW

8-15. Francé's seven biotechnical elements—crystal, sphere, cone, plane, strip, rod, and spiral—were described by László Moholy-Nagy in *The New Vision: Fundamentals of Design, Painting, Sculpture, Architecture*, New Bauhaus Books, trans. Daphne M. Hoffman (New York: Norton, 1938), 46.

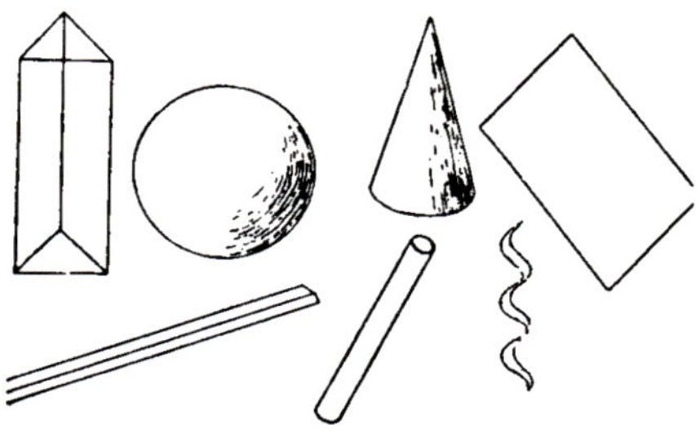

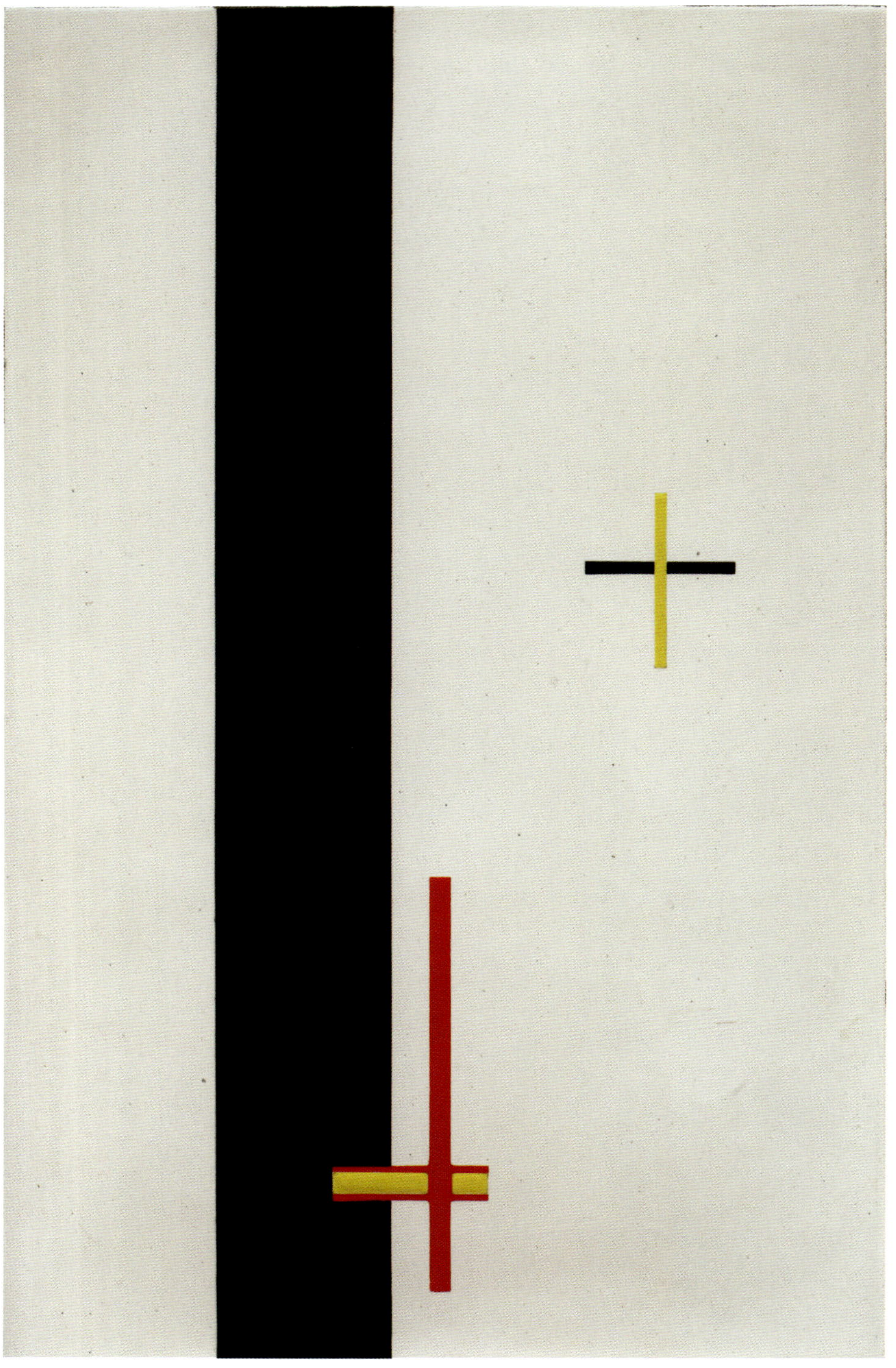

Biotechnics as a method of creative activity. The natural scientist Raoul Francé devoted himself to an intensive study of this problem. He calls his method of research and its results "biotechnics." The essential part of his teaching can be expressed in the following quotations: "Every process in nature has its necessary form. These processes always result in functional forms. . . . So man can master the powers of nature in another and quite different way from what he had done hitherto. . . . Every bush, every tree, can instruct him, advise him, and show him inventions, apparatuses, technical appliances without number."

—László Moholy-Nagy, *The New Vision*, 1929

visual elements—meaning-free forms, colors, and textures—and rules for combining them within an autonomous system. Inspired by Raoul Francé's biotechnics (plates 2-36 and 2-37 in chapter 2), Moholy-Nagy taught that seven basic forms are embodied in the natural world (plate 8-15). He used this biotechnical vocabulary in abstract sculpture such as the spiral, rods, and plane in *Nickel Construction* (plate 8-10), and he designed a functional object composed of a spiral, rod, and sphere (*Kinetic Constructive System*, 1922; see plate 8-17).[94] Francé also inspired Moholy-Nagy to introduce science photography into graphic design (plate 8-16).[95]

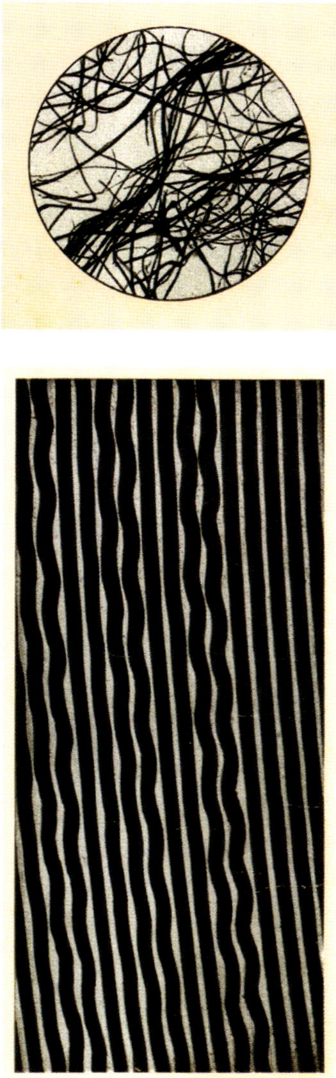

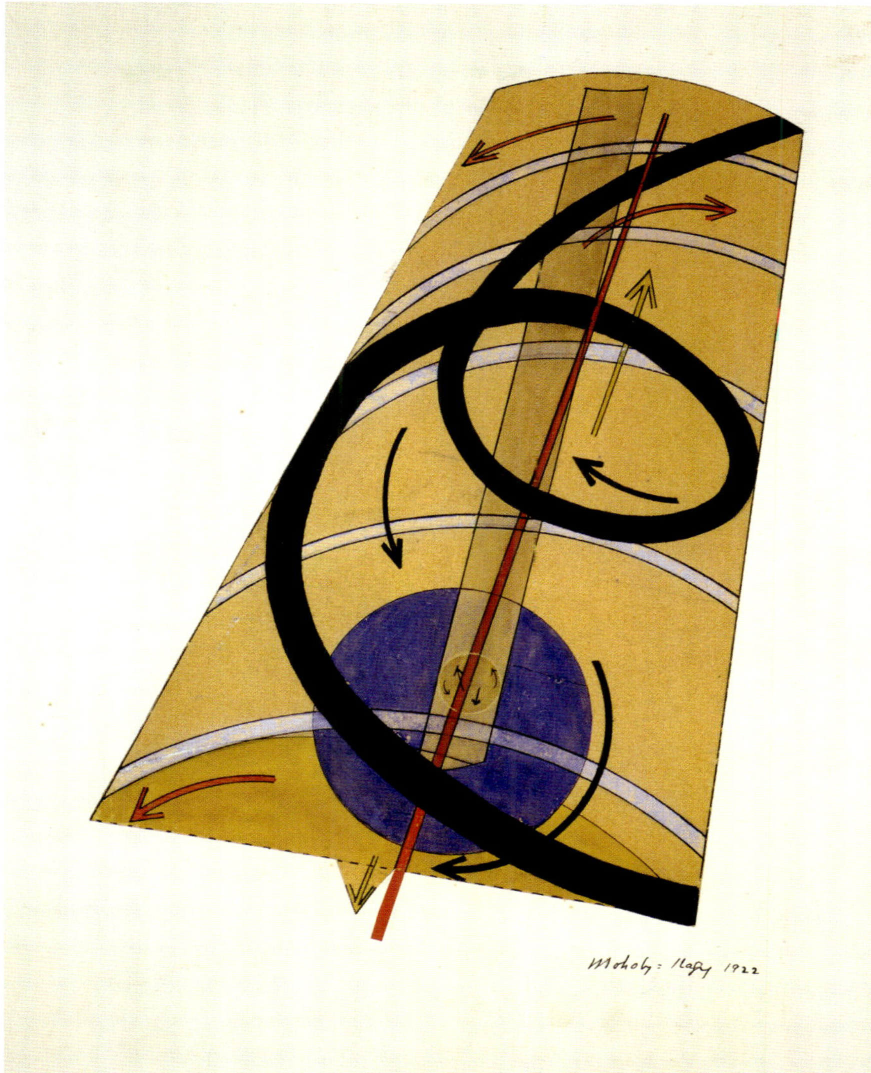

Klee and Kandinsky adapted to the guidelines of their new boss and taught students to use a universal “language of vision,” which is reflected in their Bauhaus textbooks: Klee’s *Pedagogical Sketchbook*, 1925 (plate 8-18), and Kandinsky’s *Point and Line to Plane*, 1926 (plate 8-19).[96] The formalist aesthetic extended to the Bauhaus stage and dance (plate 8-20).

Meanwhile, van Doesburg left Weimar in 1924 for Paris, where he continued publishing *De Stijl* until his sudden death from a heart attack in 1931 at age forty-seven. Meanwhile, Lissitzky contracted tuberculosis in 1923, prompting him to leave Germany for Switzerland to convalesce at a sanatorium. In March 1924 he underwent an operation to remove one lung and then, unable to get his visa renewed, he returned to the Soviet Union in 1925. Shifting his politics in response to changing times, Lissitzky became a propagandist for Stalin and, along with Rodchenko, did photomontages—combinations of geometric shapes with

ABOVE LEFT, TOP AND BOTTOM

8-16. László Moholy-Nagy (Hungarian, 1895–1946), photomicrograph of paper revealing its fibrous structure (above), and Caruso’s high C on a phonograph record (below). In *Von Material zu Architektur* (From material to architecture), Bauhausbücher 14 (Munich: A. Langen, 1929), 35, 46. © 2014 VG Bild-Kunst, Bonn/Artists Rights Society, New York.

ABOVE RIGHT

8-17. László Moholy-Nagy (Hungarian, 1895-1946), *Kinetisches konstruktives System* (Kinetic constructive system), 1922. Watercolor and black ink over pencil, collage, 24 × 19 in. (61 × 48 cm). Bauhaus-Archiv, Berlin. © 2014 VG Bild-Kunst, Bonn/Artists Rights Society, New York.

What is this? A futuristic building? An amusement-park ride? The function is unknown, but it is clear from Moholy-Nagy’s description that he wanted people to walk up and down the spiral ramp: “The structure contains an outer path mounting spirally, intended for general recreation and therefore equipped with a guard rail. Instead of steps, it is in the form of a ramp.” *The New Vision*, New Bauhaus Books, trans. Daphne M. Hoffman (New York: Norton, 1938), 186. The artist also wanted the central rod to pivot to and fro and the whole structure to be able to turn.

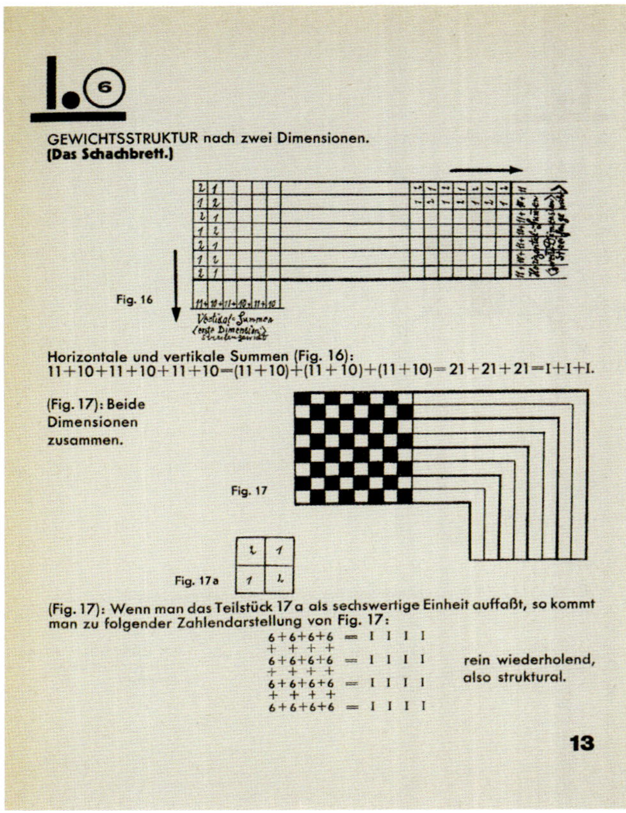

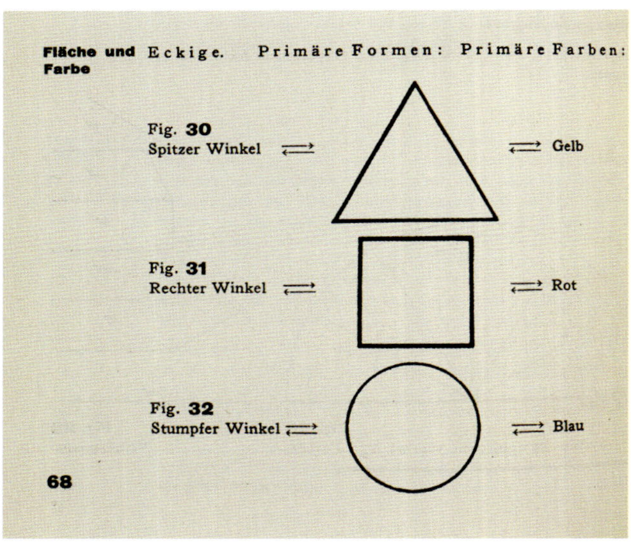

TOP LEFT

8-18. Paul Klee (Swiss, 1879–1940), *Pädagogisches Skizzenbuch* (Pedagogical sketchbook), Bauhausbücher 2 (Munich: Albert Langen, 1925), 13. Avery Architectural and Fine Arts Library, Columbia University, New York. © 2014 Artists Rights Society, New York.

In a lesson on quantitative structure (*Gewichtsstruktur*), Klee began by constructing a pattern of alternating 1s and 2s, noting that the horizontal rows and vertical columns have different visual and numerical "weights" because they sum to 11, 10, 11, 10, 11, . . . But together the rows and columns comprise a uniform pattern, which is made visible by substituting white for 1 and black for 2 to produce the chessboard (*das Schachbrett*; fig. 17). Since this grid comprises four squares with numbers that sum to 6 (fig. 17a), it is a purely repetitive (*rein wiederholend*) pattern of sixes with equal visual weight.

CENTER LEFT

8-19. Wassily Kandinsky (Russian, 1866–1944), *Punkt und Linie zu Fläche: Beitrag zur Analyse der malerischen Elemente* (Point and line to plane: Contribution to an analysis of pictorial elements), Bauhausbücher 9 (Munich: Albert Langen, 1926), 68. Avery Architectural and Fine Arts Library, Columbia University, New York. © 2014 ADAGP, Paris/Artists Rights Society, New York.

Acting like an experimental psychologist, Kandinsky circulated sheets of paper printed with a blank triangle, square, and circle. He asked Bauhaus students and staff to color in the shapes with the most "appropriate" primary colors, based on their intuitions and gut feelings. After tallying the results, Kandinsky announced a new psychological "law" that he had (allegedly) proven: Humans make an innate association of basic shape (*Primäre Formen*) to primary color (*Primäre Farben*); a pointed angle (*Spitzer Winkel*), as in a triangle, is "naturally" yellow (*Gelb*), a right angle (*Rechter Winkel*), as in a square, is red (*Rot*), and a blunt angle (*Stumpfer Winkel*), as in a circle, is blue (*Blau*). There has been no experimental confirmation of Kandinsky's law, which was based on his small sample of responses.

BELOW, LEFT TO RIGHT

8-20. Illustrations by Oskar Schlemmer (German, 1888–1943) in his "Mensch und Kunstfigur" (Man and artistic figure), *Vivos voco* 5, no. 8–9 (1926): 281–92. General Research Division, The New York Public Library, Astor, Lenox, and Tilden Foundations.

As director of the Bauhaus theater program, Oskar Schlemmer taught that the essence of dance is simple motions, performed wearing costumes made from basic geometric forms. Schlemmer described these figures (from left to right) as walking architecture (*wandelnde Architektur*; "here cubic forms are transferred to parts of the human form"; 287), the manikin (*die Gliederpuppe*; "the functional laws of the human body in relation to space . . . the club-shaped arms and legs, the spherical joints"; 288), dematerialization (*Entmaterialisierung*; "metaphysical forms of expression as symbolized by parts of the human body: the star-shape of the open hand, the ∞ of the crossed arms, the cross formed by the spine and shoulders"; 289), and a technical organism (*ein technischer Organismus*; "the laws of motion of the body . . . rotation, direction"; 288).

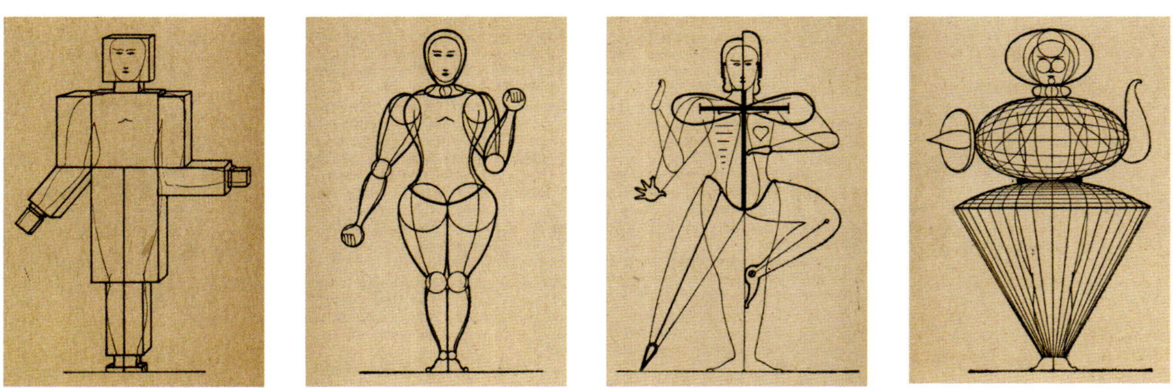

I am for beginning with multiplication tables and ABCs, because I see simplicity as the force in which any fundamental change is rooted.

—Oskar Schlemmer, "The Mathematics of Dance," 1926

photographs—as a compromise between the Russian avant-garde and social realism.[97] By the early 1930s art advisors loyal to Stalin, such as Polikarp Lebedev, were associating formalism with Western infiltration. It became politically dangerous to create an art of pure form, and abstract art ceased in Russia. Lissitzky died of tuberculosis in Moscow in 1941, and Rodchenko died in 1956 at age sixty-four, by which time Lebedev, a Communist Party functionary and outspoken supporter of social realism, had been appointed director of the State Tretyakov Gallery in Moscow in 1954.[98]

Our formalists have been imitators of Western formalism since before the Revolution.... The appearance of formalism in Soviet art is a survival of capitalism that is particularly hostile to socialism.

—Polikarp Lebedev, "Against Formalism in Soviet Art," 1936

DESIGN WITHOUT METAPHYSICS

Bauhaus architects made a more or less literal comparison between a building and a machine. At the metaphorical end of the spectrum was Gropius, who understood a building as machine-like in the sense of being functional but aspired to balance rationality with expressionism, using the practical, utilitarian aspects of a building to express the spirit of the secular age, imploring his students to create "the crystal symbol of a new faith."[99] In response to the increasing conservatism of the Weimar Republic government, in 1927 Gropius moved the Bauhaus from Weimar to Dessau and hired Hannes Meyer, an architect at the opposite, mechanistic end of the spectrum. Meyer understood a building as *literally* a machine determined by its function and geared his architectural plans for mass production, predicting that in partnership with engineers and social scientists he would make architecture an exact science. In his design for the headquarters of the League of Nations Building in Geneva, Meyer employed specialists in various scientific fields (economists, statisticians, hygienists, climatologists) to provide the data from which he "deduced" the most efficient structure of the building: "Building is not an aesthetic process. . . . The architect? He was an artist and is becoming a specialist in organization! . . . Building is only organization: social, technical, economic, mental organization."[100] Meyer's extreme functionalism caused tension with Gropius and Moholy-Nagy, who taught students a more balanced approach, mixing the function of a building with a concern for its human occupants as whole persons. In early 1928 both Gropius and Moholy-Nagy resigned, and Meyer became director of the school.

In 1929 Meyer invited Rudolf Carnap to lecture to his design students on the scientific conception of the world. Meyer shared the Vienna Circle's scientific orientation and hostility to metaphysics, which they associated with the revival of German Romanticism. In his remarks to Meyer's students Carnap declared, "I work in science and you in visible form; the two are only different aspects of one life."[101] This is true if you understand "visible form" (*sichtbare Form*) as an exact science as Meyer did, but not if you understand visual design as creating "the crystal symbol of a new faith" as Gropius did. By the late 1920s many modern architects (including Le Corbusier, J.J.P. Oud, and Mies) were already composing with a vocabulary of pure white forms, but for architects like Meyer, those squares and rectangles were contaminated with metaphysical (spiritual) associations. Meyer wanted to sterilize those pure white forms to achieve—like logical positivism—design without metaphysics.[102]

We are now in danger of becoming what we as revolutionaries opposed: a vocational training school which evaluates only the final achievement and overlooks the development of the whole man. For him there is no time, no money, no space, no concessions. . . . The spirit of construction for which I and others gave all we had—and gave it gladly—has been replaced by a tendency towards application. . . . The school today swims no longer against the current. It tries to fall in line.

—László Moholy-Nagy, letter of resignation from the Bauhaus, 1928

MATHEMATICAL MODELS

The Bauhaus was eventually closed by the Nazis in 1933, by which time most of the faculty had emigrated from Germany. Gropius and Moholy-Nagy fled to London, where an international community of émigré artists included the Russian Naum Gabo. Among the ideas exchanged and explored in the interwar era were mathematical models. The mathematics educator Felix Klein had first introduced these three-dimensional forms at the University of Göttingen in 1887 to help students visualize mathematical concepts (plate 8-21). Many universities, such as the Institute Henri Poincaré in Paris, assembled collections, and mathematical models were on public display at the Deutsches Museum in Munich (plate 8-22) and the Science Museum in London.

In London in the late 1930s, Gabo made sculpture based on mathematical models in various new media, including Lucite and nylon fiber (plates 8-23 and 8-27). Already in his *Realist Manifesto* of 1920 and in an exchange of letters with the British critic Herbert Read in London in the 1940s, Gabo declared that his sculpture symbolized the hidden mathematical structure of nature.[103]

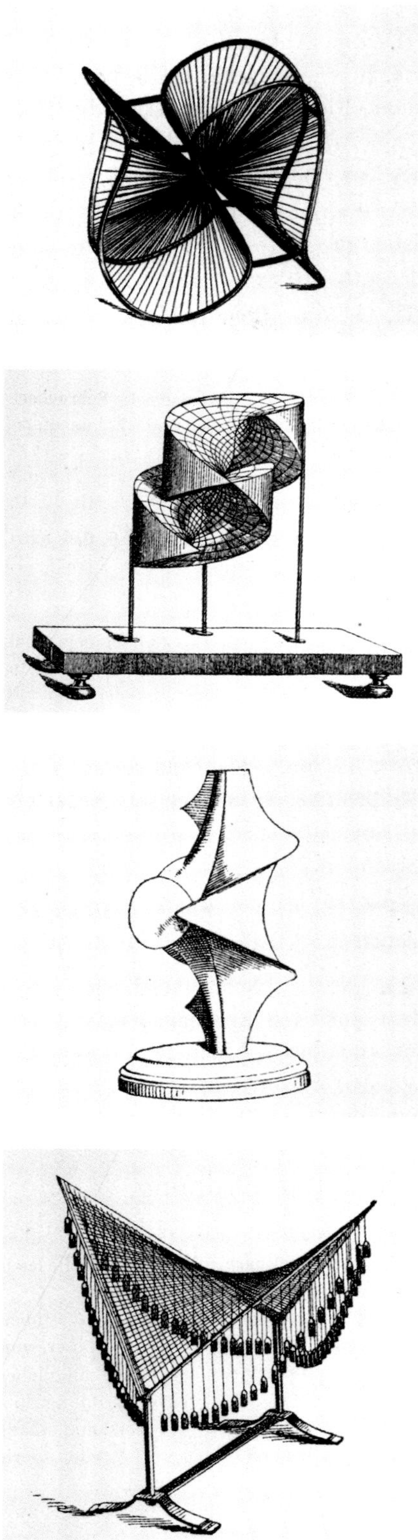

8-21. Mathematical models from Martin Schilling, *Catalog mathematischer Modelle* (Leipzig, Germany: Martin Schilling, 1911), top to bottom: 115, fig. 37; 123, fig. 83; 148, fig. 248; 144, fig. 232.

Moholy-Nagy was invited to submit a set design for the British science fiction film *Things to Come* (1936), with a screenplay written by H. G. Wells, which described the century from 1936 to 2036. Moholy-Nagy filled the set with hyperbolas, cylinders, and cones to stand for the marvels of engineering and technology that were used for good and evil in Wells's thriller (plates 8-24 and 8-28). While the émigrés in London made meaning-free sculpture to which they occasionally gave an interpretation, as in Moholy-Nagy's set design, these same mathematical models inspired British artists to make figurative sculpture. Henry Moore, who had seen the models at London's Science Museum, designed *Stringed Mother and Child* (1938; see plate 8-25) to suggest one reclining figure holding another. The strings cast shadows on the curved surface of the bronze, illustrating a principle of projective geometry: the ratio of the spaces between the strings is preserved in their projections (their shadows). Barbara Hepworth also experimented with stringed forms, as in *Pelagos* (Greek for "sea"; see plate 8-26), which she formed to suggest a crashing wave with white foam (the white paint) and ocean spray (the strings). Constructivist art in Britain culminated in the publication in 1937 of *Circle: An International Survey of Constructivist Art*, a collection of essays and artist's statements edited by Gabo, Ben Nicholson, and the architect Leslie Martin.[104]

The American Man Ray photographed mathematical models at the Institute Henri Poincaré (plate 8-29), which he then showed in an exhibition of Surrealist objects in Paris and published in the Parisian art periodical *Cahiers d'art* (1936). The founder of this influential magazine, Christian Zervos, a historian of ancient Western art, took the occasion to write an essay in which he decried artists of the 1930s who composed with geometric forms that lacked "the beauty of the world and the feelings of the heart."[105] Although Zervos did not name these cold-hearted souls, he was careful to say that they did not include Kandinsky, whose work was inspired by nature (Zervos was evidently referring to the biomorphic

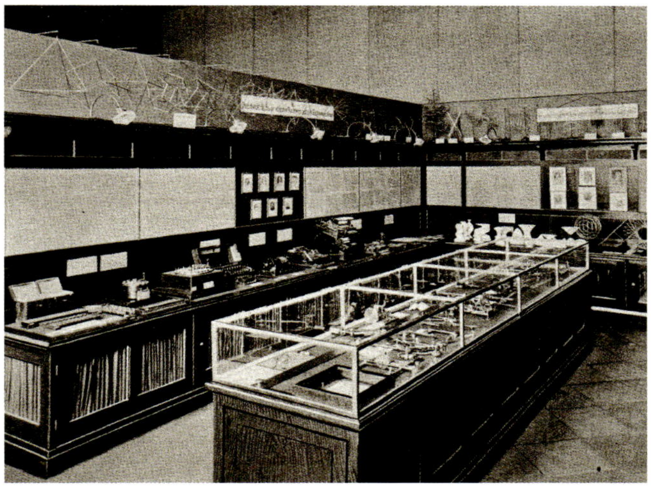

LEFT

8-22. Mathematics display from the guide book *Deutsches Museum* (Leipzig, Germany: B. G. Teubner, 1907), 62.

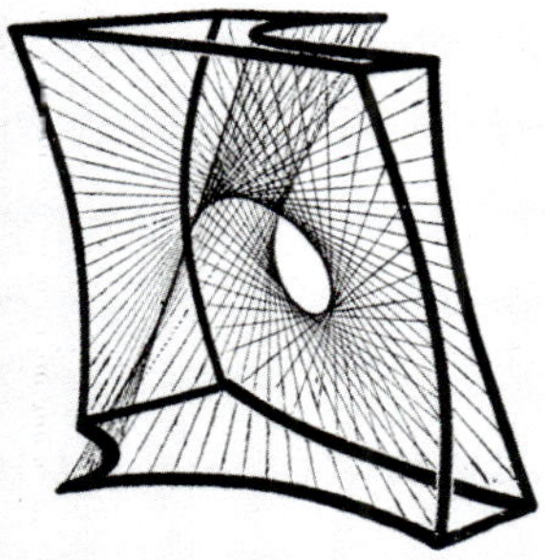

TOP RIGHT

8-23. Mathematical model, from Martin Schilling, *Catalog mathematischer Modelle* (Leipzig, Germany: Martin Schilling, 1911), 13, fig. 156.

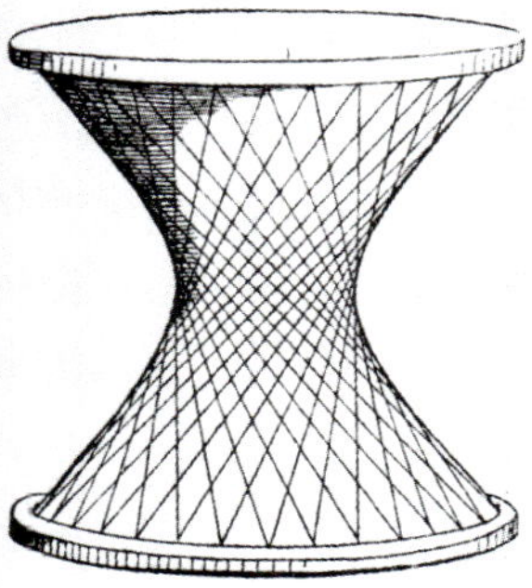

BOTTOM RIGHT

8-24. Mathematical model, from Martin Schilling, *Catalog mathematischer Modelle* (Leipzig, Germany: Martin Schilling, 1911) 112, fig. 11.

abstractions Kandinsky did in Paris in the 1930s using microscopy—for example *Entassement réglé*, 1938; Musée National d'Art Moderne, Paris). According to Zervos, geometric abstract art was an expression of a materialistic, industrial society, and he hoped that modern artists would be able to mix mathematics with human feeling—"to mathematize the soul" (*mathématiser l'âme*)—in the tradition of Pythagoras, Plato, and the Neoplatonists.[106]

BELOW LEFT

8-25. Henry Moore (English, 1898–1986), *Stringed Mother and Child*, 1938. Bronze and string, length, 4¾ in. (12.1 cm). The Henry Moore Foundation, LH 186F.

The artist was inspired by mathematical models to create this sculpture: "Undoubtedly the source of my stringed figures was the Science Museum. . . . I was fascinated by the mathematical models I saw there, which had been made to illustrate the difference of the form that is half way between a square and a circle. . . . It wasn't the scientific study of these models but the ability to look through the strings as with a bird cage and to see one form within another which excited me." "Henry Moore, Text on His Sculpture," ed. John Hedgecoe, *Henry Spencer Moore* (New York: Simon and Schuster, 1968), 105.

BELOW RIGHT

8-26. Barbara Hepworth, *Pelagos*, 1946. Wood with color and strings, 14 ½ × 15 ¼ × 13 in. (36.8 × 38.7 × 33 cm). © Tate, London.

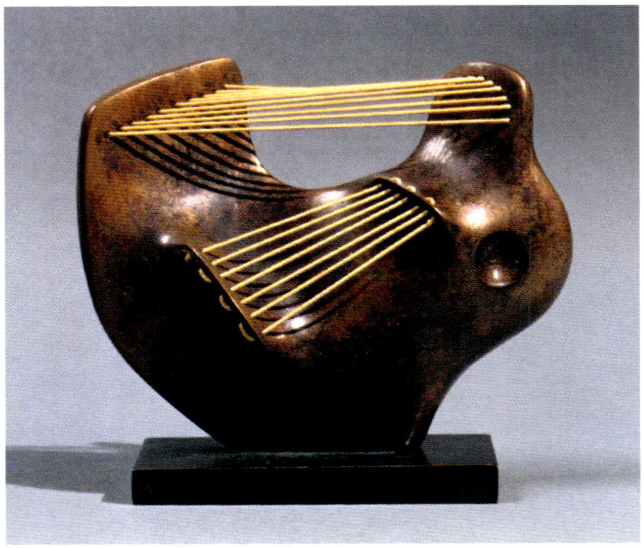

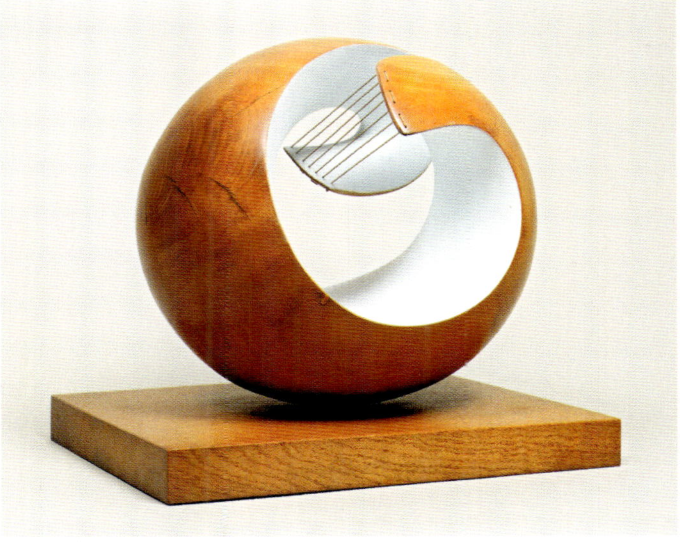

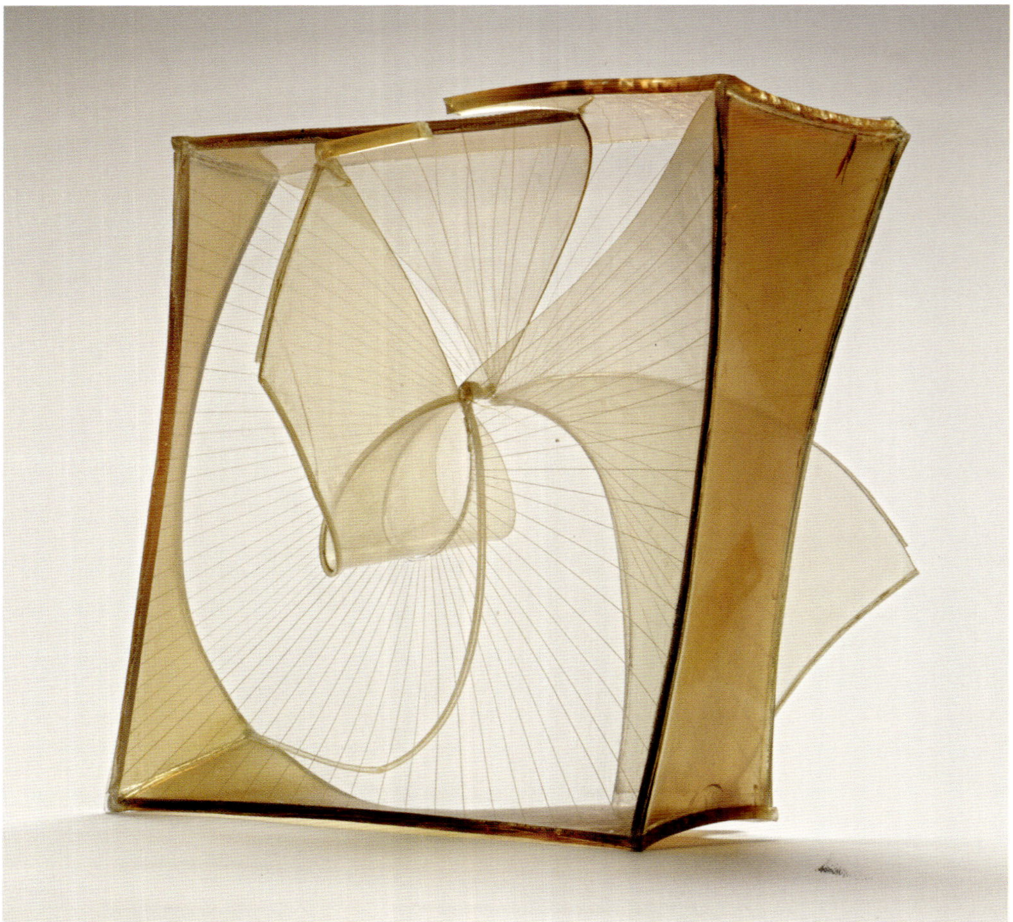

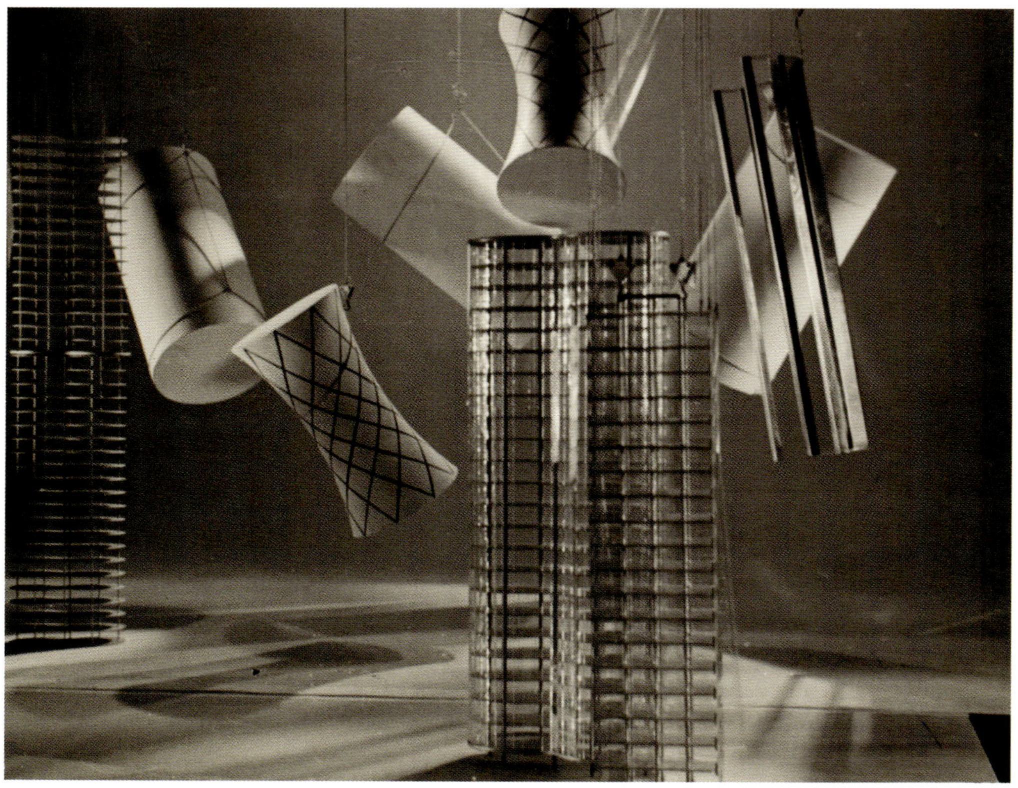

TOP LEFT

8-27. Naum Gabo (Russian, 1890–1977), *Construction in Space (Crystal)*, 1937–39. Cellulose acetate, 8¾ × 10⅝ × 7 in. (22 × 27 × 18 cm). Tate, London, inv. no. T06978. © Nina and Graham Williams.

Gabo scored the plastic surface of this work to create a pattern similar to the mathematical model on which it is based (plate 8-23).

BOTTOM LEFT

8-28. László Moholy-Nagy (Hungarian, 1895–1946), design for "special effects" in the British science fiction film *Things to Come* (London, 1936), directed by William Cameron Menzies, screenplay by H. G. Wells. Gelatin silver print, 7⅜ × 9¼ in. (18.6 × 23.4 cm). George Eastman House, Rochester, New York. © 2014 VG Bild-Kunst, Bonn/Artists Rights Society, New York.

OPPOSITE

8-29. Man Ray (American, 1890–1976), *Equation, Poincaré Institute, Paris*, 1934. Gelatin silver print, 11 13/16 × 9 3/16 in. (30 × 23.3 cm). J. Paul Getty Museum, Los Angeles. © Man Ray Trust/ADAGP/Artists Rights Society, New York.

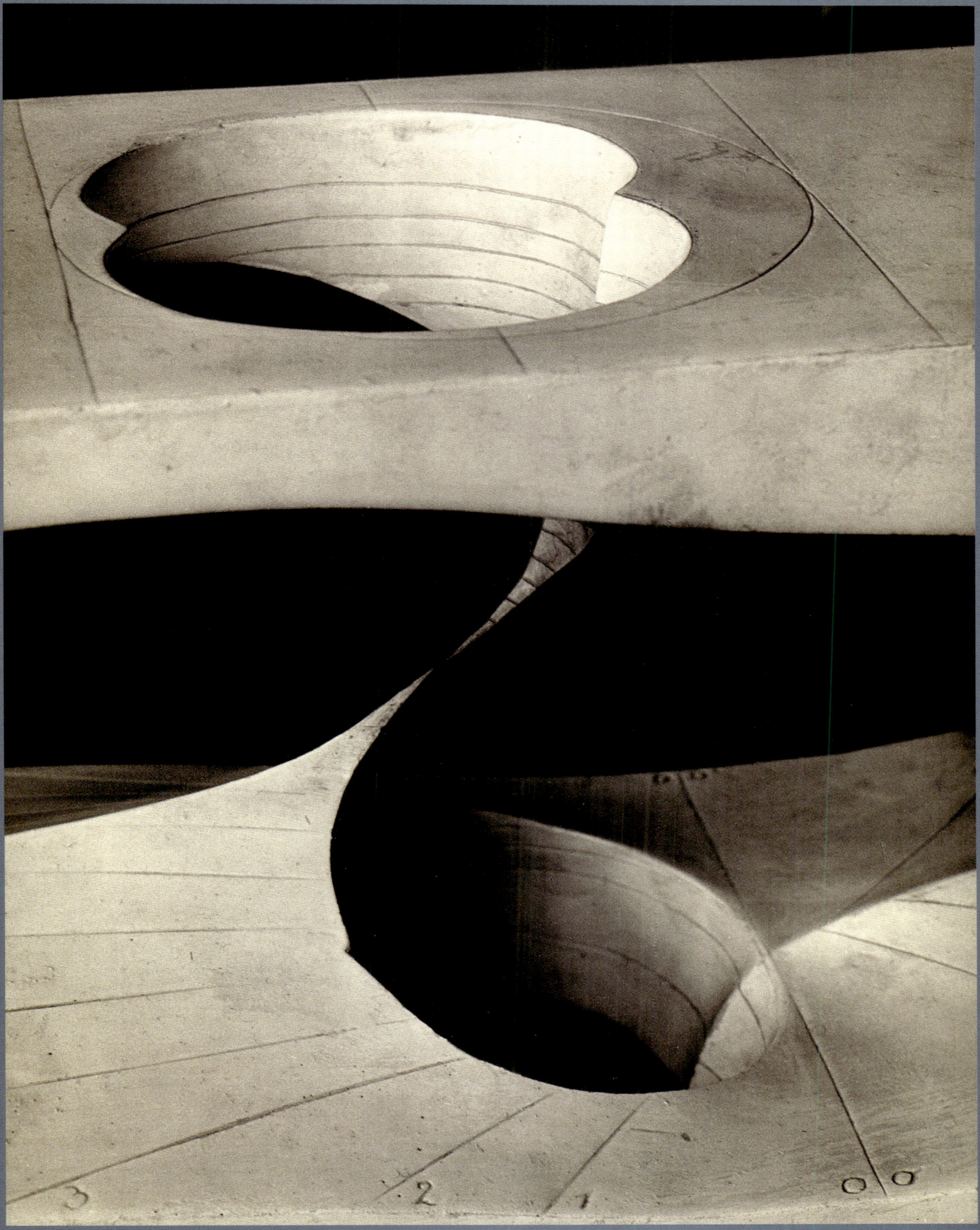
3
2
1
0 0

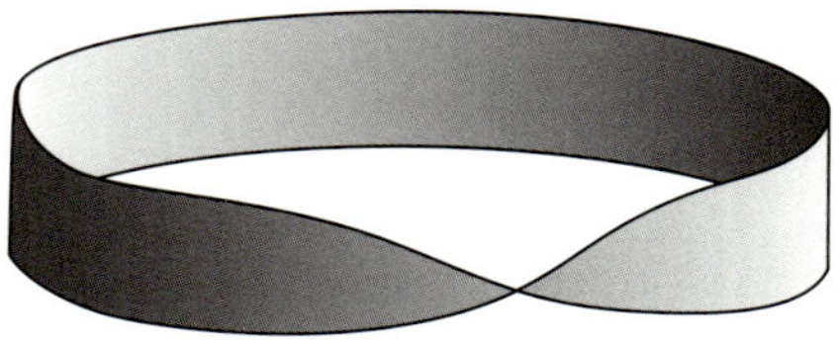

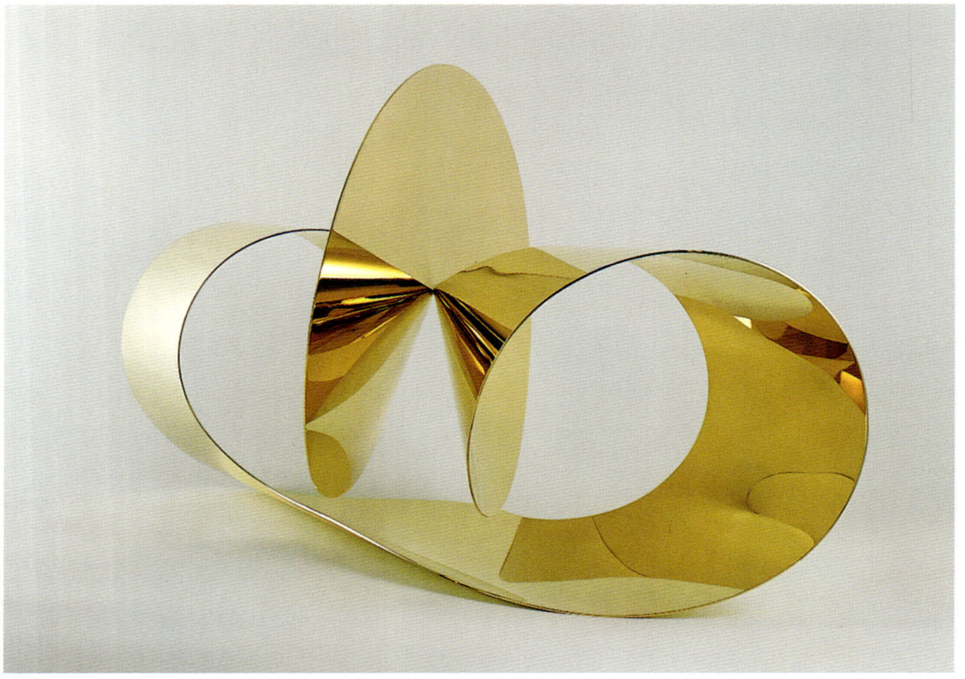

ABOVE

8-30. Möbius strip.

A Möbius strip can be made by taking a band, giving it a half twist, and affixing the ends. The result is a two-dimensional surface with only one side, a form that came to be studied as part of topology.

TOP RIGHT

8-31. Max Bill (Swiss, 1908–94), *Contour Passes Through the Center*, 1971–72. Gilt brass, 33 ½ × 16 ½ × 15 in. (85 × 42 × 38 cm). © max, binia + jakob bill foundation. © 2014 ProLitteris, Zurich/Artists Rights Society, New York.

This sculpture is a single-sided surface based on the Möbius strip, as Bill wrote: "Since 1968 there have been five new sculptures of single-sided surfaces. . . . At the moment I am working on *Contour Passes Through the Center* (1971–72); other ideas, some of them even much more complex, are still waiting to be realized according to a convincing law of form"; Max Bill, "How I Started Making Single-Sided Surfaces" (1972), trans. David Britt, in *Max Bill: Endless Ribbon, 1935–95* (Bern, Switzerland: Benteli, 2000), 89–90.

BOTTOM RIGHT

8-32. Hinke Osinga (Dutch, b. 1969) and Bernd Krauskopf (German, b. 1964), *Crocheted Lorenz Manifold*, 2004. Cotton, wire, diameter 36 in. (91.4 cm). Courtesy of the artists.

In an attempt to describe the irregular behavior of weather patterns, the American mathematician Edward Lorenz wrote the equation that describes this particular hyperbolic surface. Osinga and Krauskopf realized that Lorenz's algorithm could be translated into crochet instructions for a three-dimensional model of a Lorenz manifold. Osinga, who had learned to crochet as a child in Holland, picked up her needles, and 25,511 stitches later she had created this Lorenz manifold.

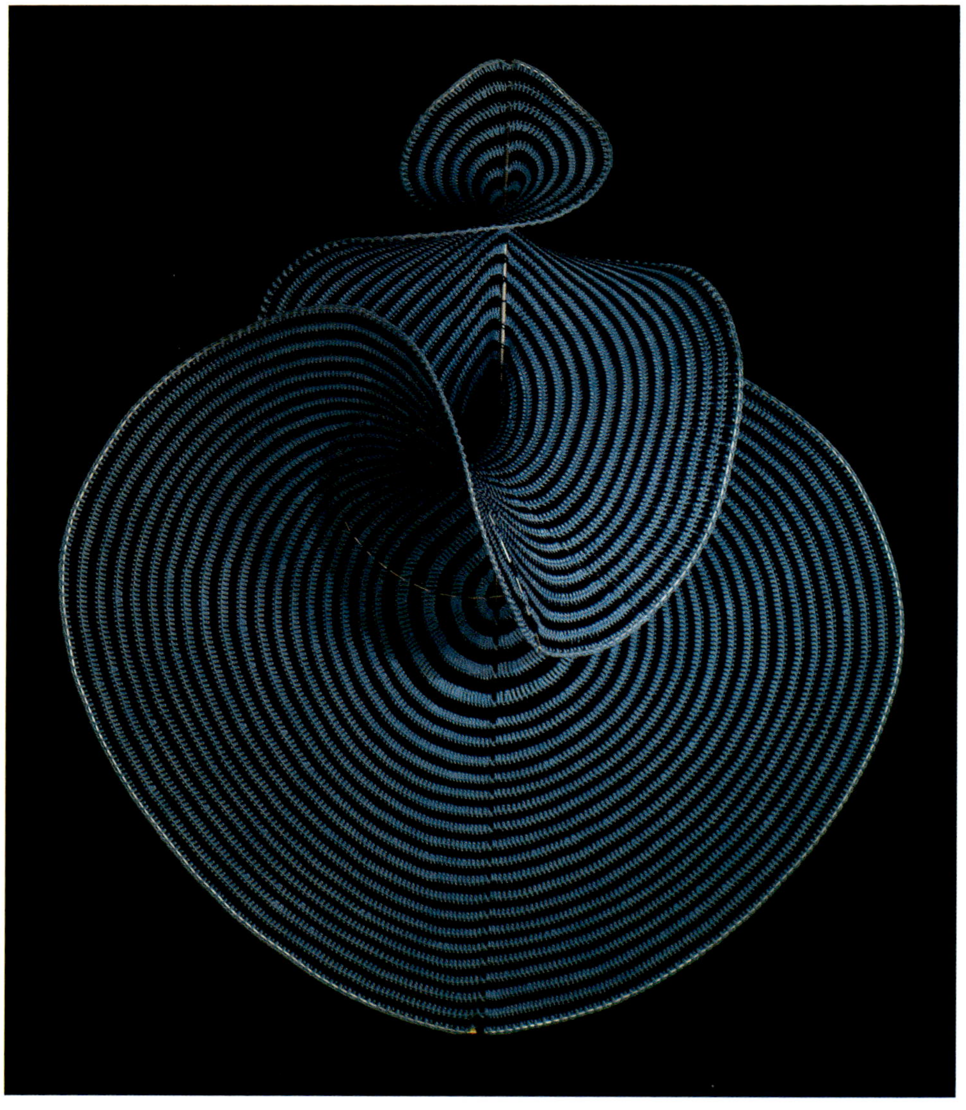

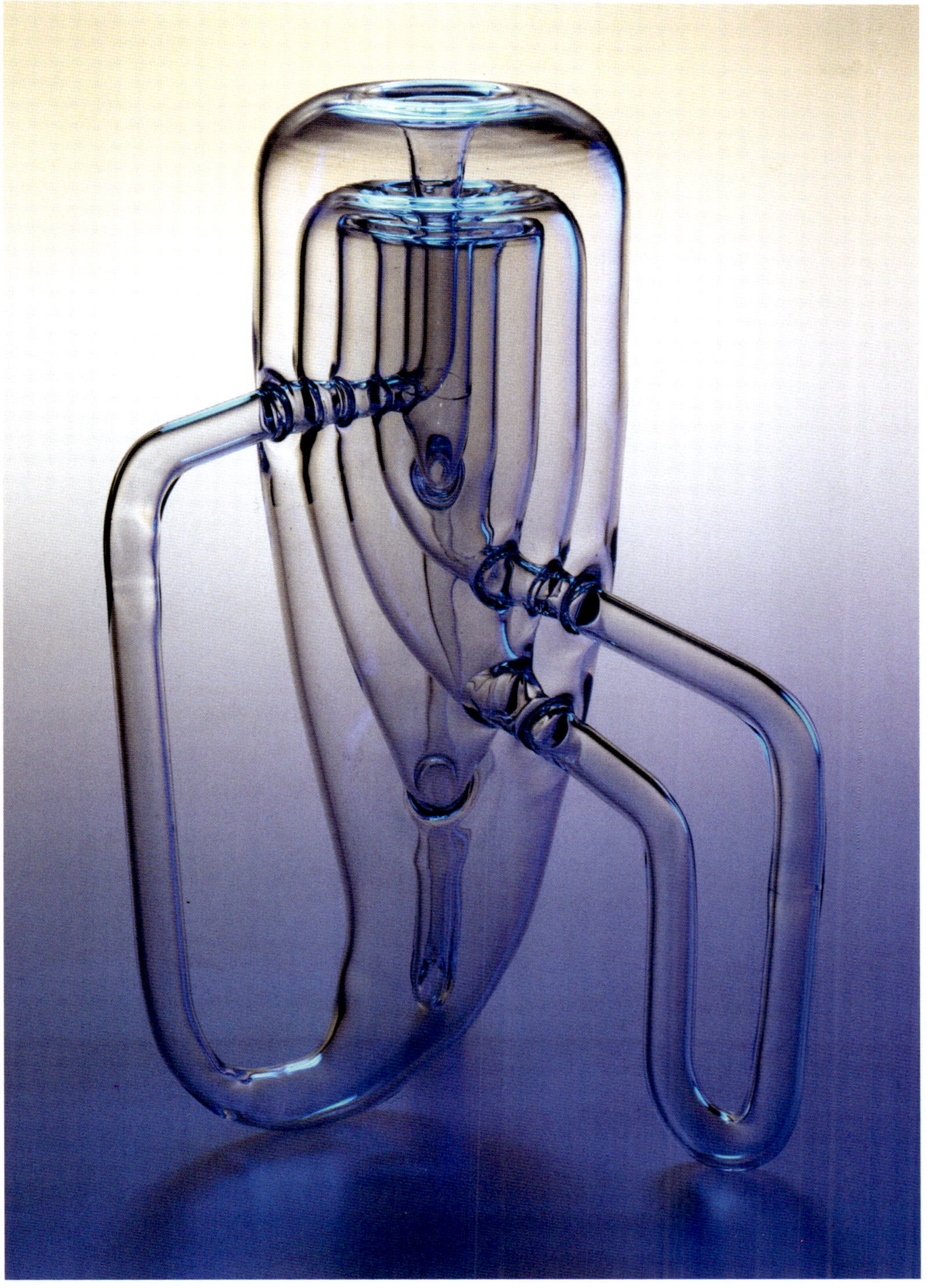

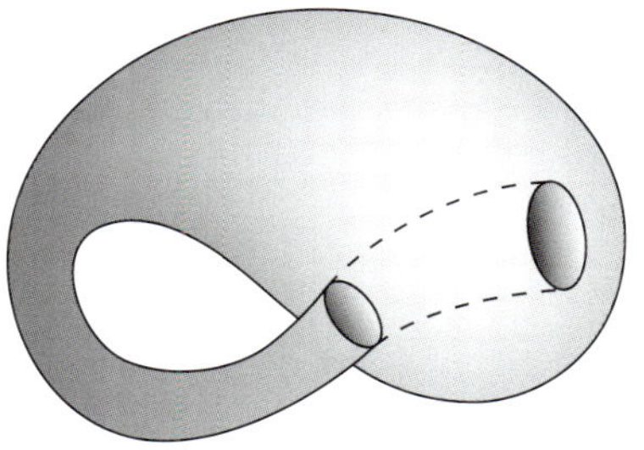

ABOVE

8-33. Klein Bottle.

If the ends of a cylinder are distorted and rejoined, you get a self-intersecting three-dimensional volume with only one surface. Although it is a closed form, such a "bottle" has no inside. This bizarre form is named for its discoverer, Felix Klein.

LEFT

8-34. Alan Bennett (British, b. 1939), *Triple Klein Bottles*, 1995. Blown glass, ca. 10 in. (25 cm) high. The Science Museum, London.

In this tour de force of glass-blowing, Alan Bennett created three Klein bottles, one inside the other. It is well known that if sliced in half on a certain plane, a Klein bottle will result in two Möbius strips. Bennett has revealed other cross-sections by slicing through his glass bottles at various angles using a diamond-edged saw; see Ian Stewart, "Glass Klein Bottles," *Scientific American* 279, no. 1 (March 1998): 100–101.

Mathematical models have continued to inspire artists such as Max Bill, who made sculptures based on a Möbius strip, a one-sided, endless shape (plates 8-30 and 8-31). The related model of a Klein bottle (plate 8-33) inspired the British glassmaker Alan Bennett to create a series of such bottles (plate 8-34). The Dutch mathematician Hinke Osinga and the German Bernd Krauskopf work on the Lorenz manifold, which is a plane in non-Euclidean, hyperbolic geometry (see plate 8-32, and the sidebar on page 155 in chapter 4). Meanwhile, the contemporary Japanese photographer Hiroshi Sugimoto turned his lens on a set of mathematical models that were acquired by the University of Tokyo around 1900 (plates 8-35 and 8-36).

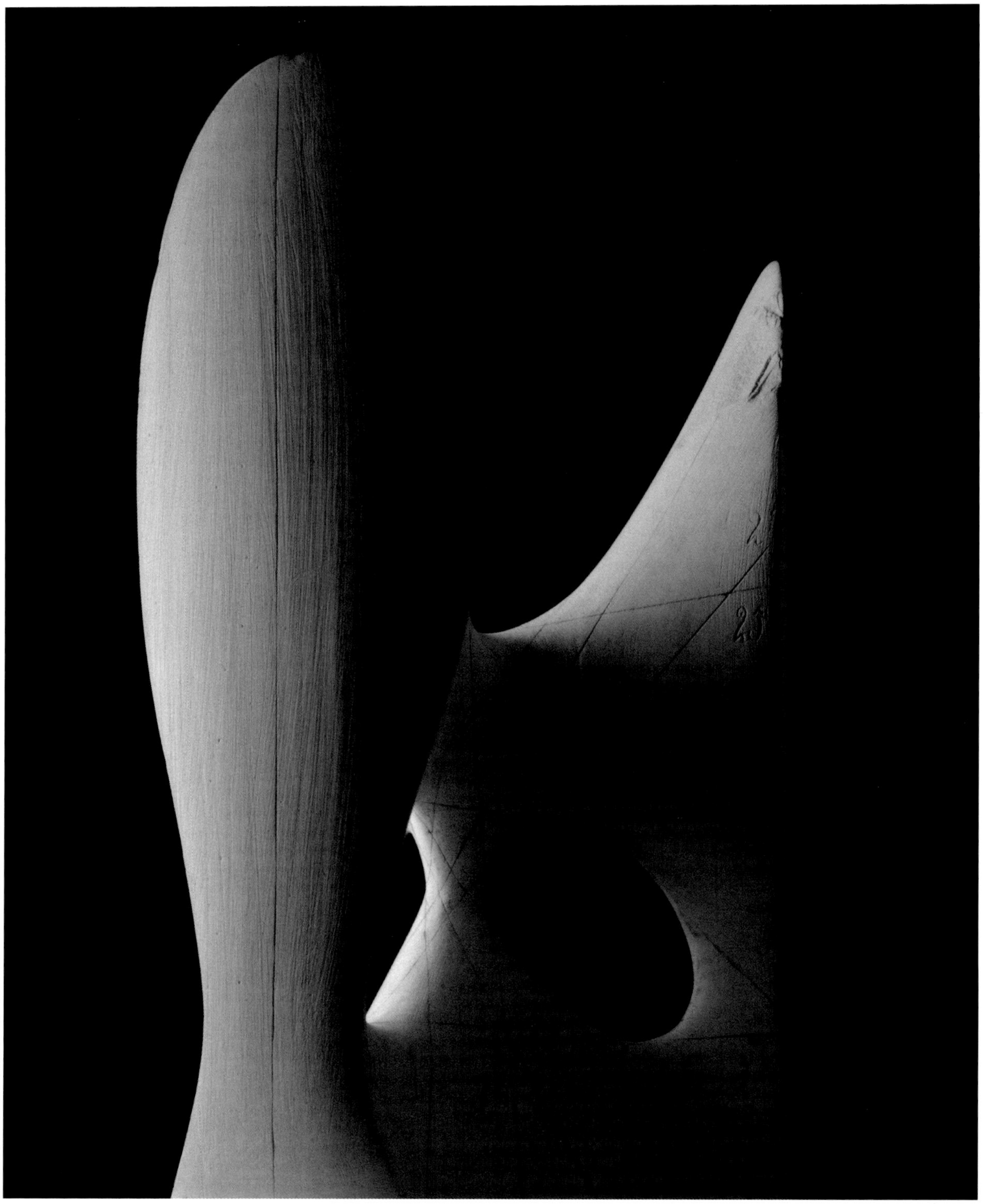

ABOVE

8-35. Hiroshi Sugimoto (Japanese, b. 1948), *Mathematical Form 0012*, 2004. Gelatin silver print, 58¾ × 47 in. (149.2 × 119.3 cm). © Hiroshi Sugimoto. Courtesy Pace Gallery, New York.

OPPOSITE

8-36. Hiroshi Sugimoto (Japanese, b. 1948), *Mathematical Form 0009*, 2004. Gelatin silver print, 58¾ × 47 in. (149.2 × 119.3 cm). © Hiroshi Sugimoto. Courtesy Pace Gallery, New York.

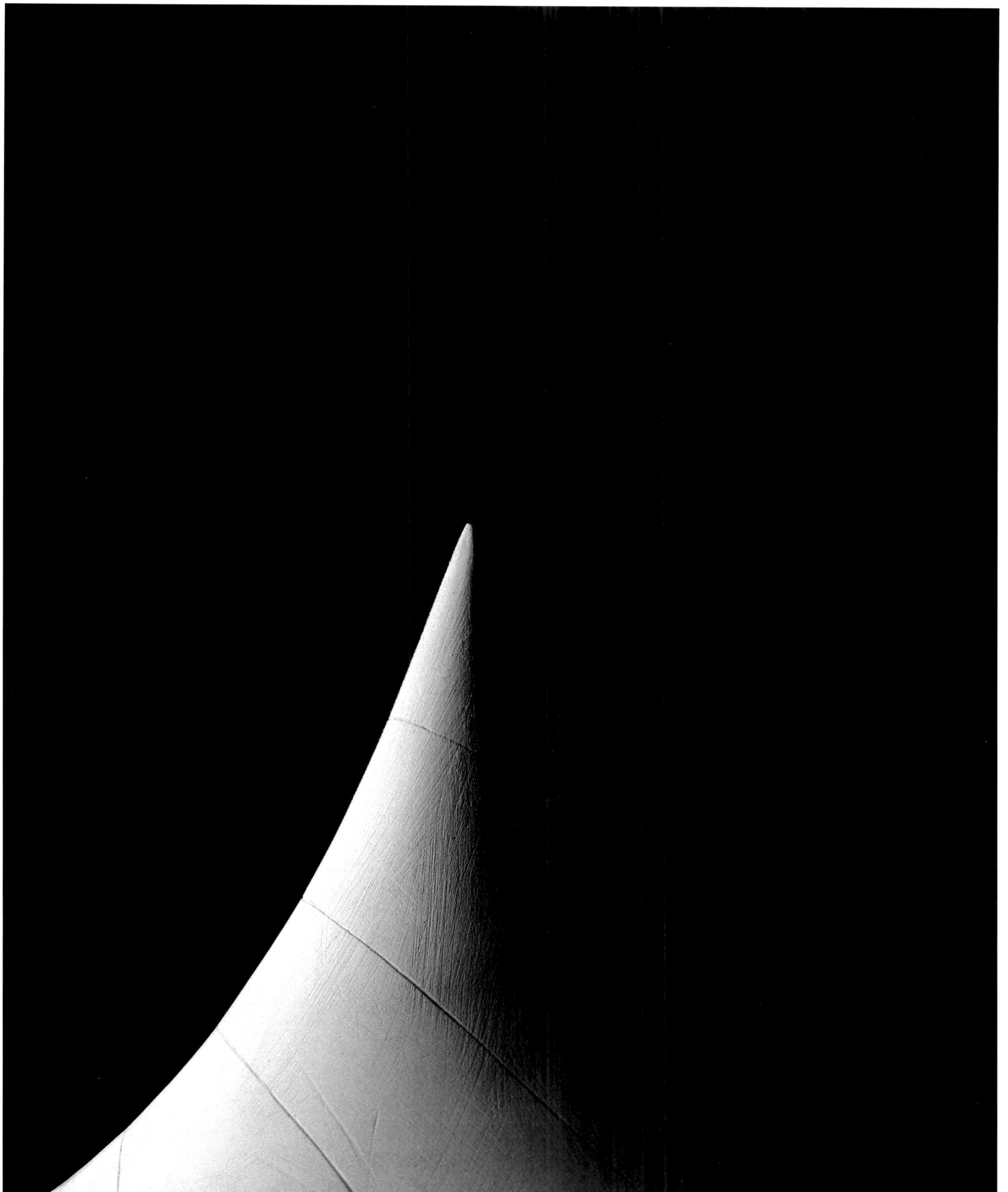

THE NEW BAUHAUS IN CHICAGO

Carnap left Vienna in 1931 to teach at the German University in Prague, but his stay was cut short by the fall of the Weimar Republic and rise of the Third Reich. With the help of the American philosopher Charles Morris from the University of Chicago and the American logician W. V. Quine from Harvard, in 1935 Carnap immigrated to the United States. From 1936 to 1952 he was a professor with Morris at the University of Chicago, whose Department of Philosophy became the international center for logical positivism. Carnap and Morris, with help from Otto Neurath, Niels Bohr, Bertrand Russell, and others, edited the writings that had been assembled in Europe by the Vienna Circle into the *International Encyclopedia of Unified Science*, the first volume of which was released by the University of Chicago Press in 1938.

Meanwhile, in 1937, Moholy-Nagy arrived in Chicago, newly hired by a group of city philanthropists to head a school of design, the "New Bauhaus" (plate 8-37). Moholy-Nagy came to know Carnap and Morris, and he invited Morris to teach philosophy at the New Bauhaus. As a follower of the American philosopher Charles Sanders Peirce, founder of semiotics, Morris felt that the logical positivists and the Bauhaus artists shared a common goal of creating a unified theory of signs: "The unity of science movement is not without significance for the arts. The approach through the theory of signs gives the possibility of a scientific aesthetics, and a language in which to talk simply and clearly about art and its relations to other components of culture. . . . The artist himself may gain release for his own activity if in an age of science and technology he can be made to see clearly the nature and importance of his own work, and to understand that in spite of irreducible and precious differences, art, science, and technology are complementary and supporting activities."[107]

The international styles of geometric abstraction that were practiced in Germany in the 1920s shared a formalist aesthetic—meaning-free colors and forms within an autonomous system—that allows for endless variations. The colorful cubes, spheres, and rectangles that today fill the world's museums of modern art and decorate the lobbies of international corporations and airports—from Chicago to Tokyo, Johannesburg to Dubai—can be traced to the formalist aesthetic, and the steel and glass skyscrapers that pierce the skylines of these cities originated in this era. But the Weimar Republic's utopian political visions, along with the grand scientific and mathematical projects of Bohr, Hilbert, and Carnap, dimmed in the 1930s. Quantum mechanics led to advances in technology, but Bohr's Copenhagen interpretation undercut a unified picture of the natural world. Furthermore, Hilbert's Program and Carnap's logical positivism were never completed, but instead led to the exposure of the limits of artificial and natural languages.

8-37. László Moholy-Nagy (Hungarian, 1895–1946), brochure for *The New Bauhaus*, Chicago, 1937–38. Bauhaus-Archiv, Berlin. © 2014 VG Bild-Kunst, Bonn/Artists Rights Society, New York.

Moholy-Nagy took the photograph and designed the typography for the cover of this brochure.

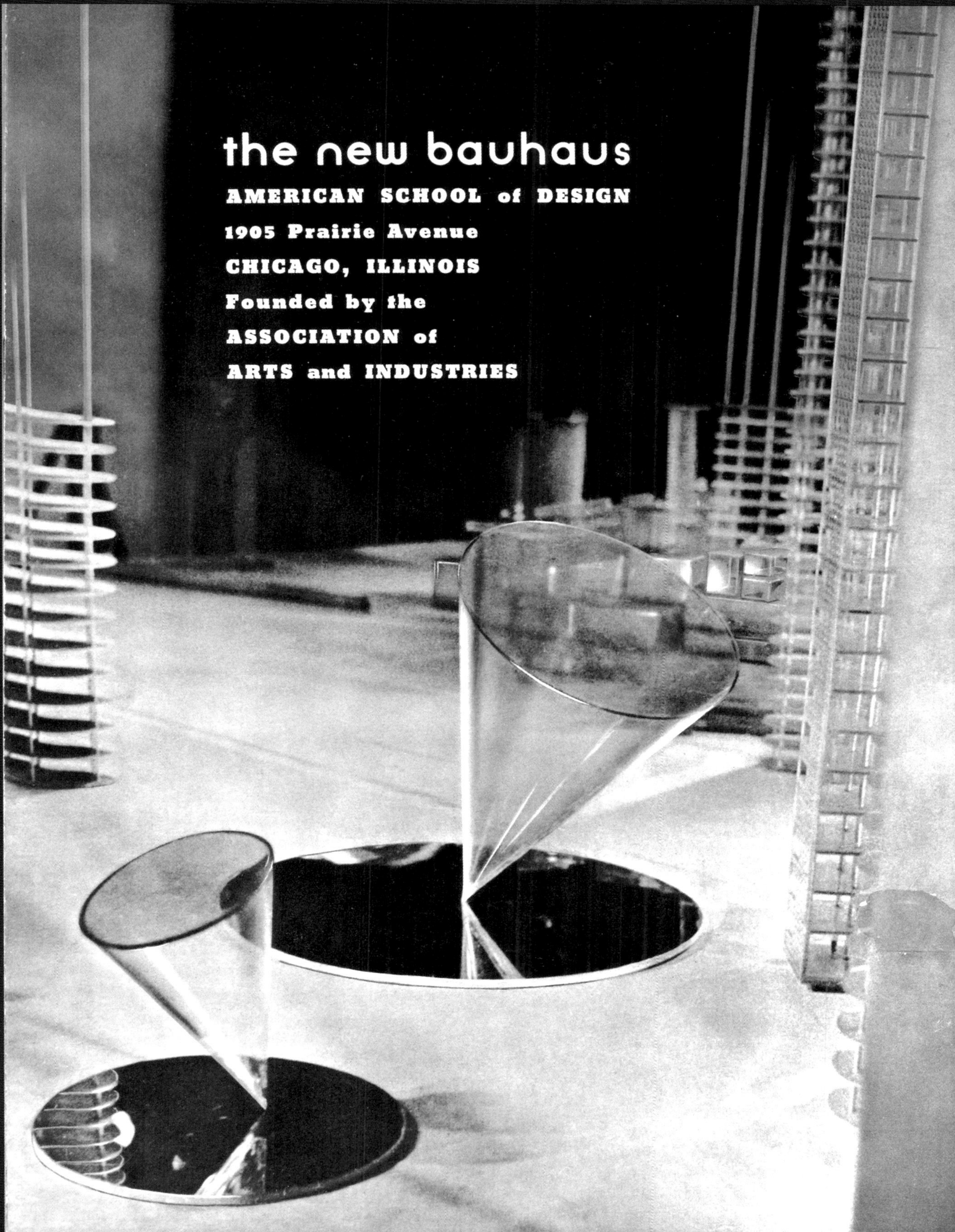
the new bauhaus
AMERICAN SCHOOL of DESIGN
1905 Prairie Avenue
CHICAGO, ILLINOIS
Founded by the
ASSOCIATION of
ARTS and INDUSTRIES

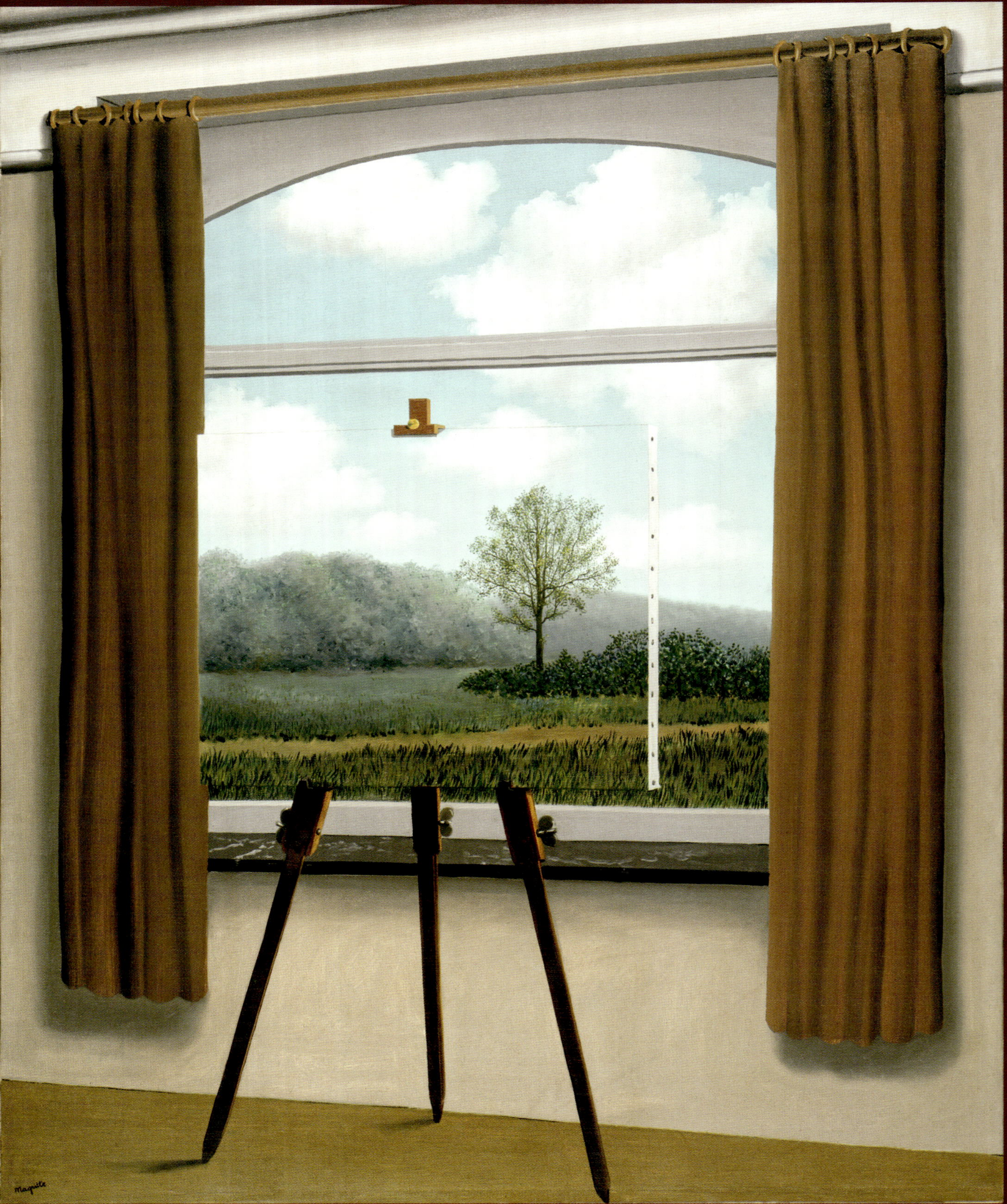

9

The Incompleteness of Mathematics

Whereof one cannot speak, thereof one must be silent.
—Ludwig Wittgenstein, ***Tractatus Logico-Philosophicus***, 1921

And what would I love except enigma?
—Giorgio de Chirico, 1910

WHEN DAVID HILBERT WROTE his new formalist axioms for geometry in the 1890s, he declared them complete in the sense that every true statement formulated in elementary geometry could be proven within his system. This is analogous to designing a chess-playing machine that could, in principle, calculate all possible positions in all possible games. Hilbert thus introduced the key distinction between a statement *in* a mathematical system—such as a theorem of geometry—and a "metamathematical" statement *about* a system—such as the assertion that a system of geometry is complete.

With the emergence of the scientific worldview, researchers commonly asked questions about their field, such as: What are the core assumptions of astronomy? Such questions are, however, peripheral to the field; if an astronomer has a discussion about the nature of astronomy, the scientist is not *doing* astronomy but engaging in the philosophy of science. Of all the sciences, only within mathematics did such questions about the nature of mathematics come to be posed *in the language of mathematics*. This is possible because mathematical methods can be applied whenever one is attempting a systematic, rational analysis of any subject matter. A mathematician who is doing metamathematics *is doing mathematics*, and self-reflection is a fundamental feature of mathematics in the modern era. As abstract (non-representational) art emerged in the late nineteenth century, artists self-consciously asked: What is the essence of art? As in mathematics, such questions came to be asked *in the vocabulary of art*, and by the early twentieth century there was "meta-art"—paintings about the essence of painting and films about filmmaking. The self-reflection inherent in the practice of both modern mathematics and modern art points to a resonance between the two fields.

In the early 1920s, during the resurgence of German Romanticism and the soaring popularity of the anti-intellectual mathematician L.E.J. Brouwer, Hilbert had challenged mathematicians to find the one set of axioms that lay at the bedrock of mathematics.

9-1. René Magritte (Belgian, 1898–1967), *La condition humaine* (The human condition), 1933. Oil on canvas, 39⅜ × 31⅞ in. (100 × 81 cm). National Gallery of Art, Washington, DC, Gift of the Collectors' Committee. © 2014 C. Herscovici/Artists Rights Society, New York.

In the Western tradition, "the human condition" is understood as curiosity that leads to suffering, as in the Greek myth of Pandora who opens a forbidden vessel from which evil escapes into the world. Reflecting modern topics in the philosophy of mind, Magritte defined the human condition as self-consciousness—the ability of the human mind to reflect on itself—which opens the door to a paradoxical mix of reality and illusion, as in this picture within a picture.

Two young Austrians, the philosopher Ludwig Wittgenstein and the mathematician Kurt Gödel, both inspired by Brouwer's critique of formal axiomatic systems, showed that there are truths that lie beyond the limits of natural and artificial languages. To the surprise of many, Hilbert's tower of mathematics founded on absolute certainty was shown to be an impossible dream. The proofs of both Wittgenstein and Gödel relied on the distinction between mathematics and metamathematics, and on the unusual ability of mathematics to describe itself in the language of mathematics. As Wittgenstein and Gödel showed, if a natural or artificial language reaches a certain (very low) level of complexity, self-reference is inherent in the system, opening the door to paradox.

Meanwhile the early-twentieth-century artists Giorgio de Chirico, René Magritte, and M. C. Escher used linear perspective—created in the Renaissance to design ordered, harmonious scenes in rational spaces—to produce irrational worlds (plate 9-1, chapter frontispiece). Although unaware of Wittgenstein and Gödel's proofs, these artists made work about enigma and paradox that resonate with these proofs because the mathematicians and artists shared common sources in the *Naturphilosophen* Hegel and Schelling, the *Lebensphilosophen* Schopenhauer, Kierkegaard, and Nietzsche, and the novelist Fyodor Dostoyevsky, who all argued that there are limits to what can be captured in abstract conceptual systems. These philosophical critiques formed part of the cultural matrix from which Wittgenstein and Gödel emerged to topple the grandest system of all—Hilbert's tower of mathematics.

METAMATHEMATICS, COMPLETENESS, AND CONSISTENCY

All such questions . . . seem to me to form an important new field of research which remains to be developed. To conquer this field we must, I am persuaded, make the concept of specifically mathematical proof itself an object of investigation, just as the astronomer considers the movement of his position, the physicist studies the theory of his apparatus, and the philosopher criticizes reason itself.

—David Hilbert, "Axiomatic Thought," 1918

The challenge of Hilbert's Program was not only to design a system with one set of axioms for all parts of mathematics, but, crucially, to prove two metamathematical statements about the system: first, that the system is *complete* (every true statement formulated in mathematics can be proven within the system), and second, that it is *consistent*, in the sense that the set of axioms can never lead to a paradox.

Questions of consistency had never arisen in mathematics before the nineteenth century because premodern mathematicians assumed that geometry and arithmetic had a paradox-free model—the world. When people used arithmetic to count bushels of rice, they applied numbers to physical objects, whose existence (or absence) is unambiguously clear: the statement "These two bushels and those two bushels make four" refers to a fact in the world-out-there. But once mathematicians began cutting their ties to the world in the mid-1800s, the question of the internal consistency of each newly invented axiomatic system loomed ever larger. If a mathematician attempted to prove the consistency of an axiomatic system by giving it an interpretation, this only moved the question to the model: Is the model consistent? Thus Hilbert challenged mathematicians to prove *absolute* consistency without resorting to a model, in other words, to prove the consistency of the system *in the language of the system itself*. Proving the absolute consistency of chess would mean developing a way to somehow make assertions about the game (statements in "metachess") using only the game's rules and vocabulary—pawns, knights, a sixty-four-square grid, and so on—and to demonstrate that the game's rules do not allow for a contradictory move, such as a pawn being both on and off the same square.

To tackle this topic, Hilbert sharpened his decades-old distinction between mathematics proper (arithmetic, geometry, logic, etc.) and metamathematics (statements *about* arithmetic, geometry, logic, etc.). He had written his 1899 geometry in a mix of words and mathematical symbols, but with the publication of Russell and Whitehead's *Principia Mathematica* in 1910–13, he had at his disposal the technical tools of predicate logic. Now it was possible to define rules of inference that operate in a completely mechanical way and to write mathematics in a formal vocabulary of symbols. For example, one could use predicate logic to assert that there is a number x which is larger than y by writing $\exists xy\ (Lxy)$, which reads "There is an x and a y such that x has the relation L to y." Furthermore and crucially, one could also use Russell and Whitehead's system to make statements in metamathematics such as "There is an x that is a proof of [a demonstration of] y"; in other words, "y is provable" by writing $\exists xy\ (Dxy)$, which reads "There is an x and a y such that x has the relation D [being a demonstration of] to y."

WITTGENSTEIN'S DECLARATION OF THE INEFFABLE

Ludwig Wittgenstein was raised in one of the wealthiest and most privileged families in Austria; he matured with a presumptuous personality and an intense work drive. In 1910, as a student of engineering at Manchester University in England, he read Bertrand Russell's 1903 *Principles of Mathematics*, which contained a summary of Frege's views about language and logic. In disagreement with those views, the twenty-one-year-old student voiced his objections in a letter to Frege, who responded by inviting Wittgenstein to come to Jena and discuss logical matters. During their 1911 visit, Frege suggested that Wittgenstein transfer to Cambridge University and study with Russell, who had just published the first volume of his landmark work, *Principia Mathematica*.

Thus in 1912—in the midst of writing volume 2 of *Principia*—Russell found himself searching for answers to penetrating questions from a new Viennese student. Their discussions ended in 1913 when Wittgenstein left Cambridge for Norway, where he spent a year in isolation, refining Russell's logical atomism.[1] As an enlistee in the Austrian army during World War I, Wittgenstein carried his manuscript in his soldier's rucksack—he fought on the Russian front in 1916 and the Italian front in 1918—completing *Tractatus Logico-Philosophicus* (Logical-philosophical treatise) while interred in a prisoner-of-war camp in Italy. Published in 1921, the slim eighty-page volume is a densely worded series of pronouncements presented like theorems, numbered in decimal notation to indicate their order within a deduction from seven axioms, the first of which is: "The world is all that is the case." Wittengenstein's *Tractatus*, its title echoing that of the Dutch philosopher Baruch Spinoza's *Tractatus Theologico-Politicus* (1670), has the deductive format of Spinoza's *Ethics* (1677), which is in turn modeled on the deductive format of Euclid's *Elements*.[2]

In *Tractatus* Wittgenstein presented a version of Russell's logical atomism, the picture theory of meaning: atomic propositions "picture" atomic facts because assertions "mirror" the logical structure of what they describe.[3] For example, when the man in plate 9-2 sees an auto, he says "auto" because his reason matches the image with a word stored in his memory. Wittgenstein was also interested in the *limits* of language, which he equated with what can be expressed in words or other symbols. Let us suppose that when the man sees the auto, he feels anxious because his authoritarian father drives a black Mercedes. His malaise is not a perception of something in the world-out-there but of his inner world, arising from his gut via neural circuitry into his brain stem. His reason can't find a word that precisely matches the sensation because it's produced by a region of the mid-brain responsible for emotion (the amygdala), which evolved long before the area of the cerebral cortex responsible for language (Broca's area). Anxiety is what psychologists call a "preverbal" mode of thought. What lies beyond the boundary of language—feelings and intuitions that cannot be put into words—Wittgenstein termed "the mystical" (*das Mystische*): "There is indeed the inexpressible. This *shows* itself; it is the mystical."[4] Wittgenstein further characterized mystical knowledge as the conviction that there is wholeness to reality—a unity: "The world and life are one—I am my world."[5] With this sentence Wittgenstein placed himself in the tradition of pantheism (nature mysticism) stretching back to the Pythagoreans and Spinoza, whose phrase *sub specie aeterni* (from the viewpoint of eternity), Wittgenstein goes on to quote:

> The contemplation of the world *sub specie aeterni* is its contemplation as a limited whole.
>
> The feeling of the world as a limited whole is the mystical feeling.[6]

According to Wittgenstein, one can talk about a part of the world but not the "limited whole." By the term "limited," he meant "bounded" or "circumscribed"; in other words, the world is a finite (not infinite) number of facts. Why is the world—this finite collection of facts—indescribable? Why must the mystic fall silent? For two reasons, one about the world and one about language.

Regarding the world, for Wittgenstein the whole world (in the sense of "the universe") is the set of every fact. In order to conceive of this set, one must "step out" of the universe and cease *being* one of the facts, as Kierkegaard had observed. In the West, this is the traditional viewpoint of a personal deity, looking down on the cosmos from beyond space and time. Indeed, Wittgenstein's vast collection of every fact recalls Cantor's "Absolute Infinity"—the entire upward hierarchy of set theory (Cantor's ascending ladder of alephs) together with everything in the universe—which, in the monotheist tradition, exists only in the omniscient mind of a Supreme Being. Thus Wittgenstein translated the ineffability associated with Western monotheism into the language of secular set theory: an ordinary, finite mortal cannot step outside the universe, and so he or she cannot describe the cosmos as a totality.

Regarding language, Wittgenstein explained why the world is indescribable by borrowing the concept of a hierarchy of languages from Frege (see the sidebar on page 207 in chapter 5). According to Wittgenstein's picture theory, the facts that compose the world are mirrored by words (assertions in a language). The structure of this language gives it the

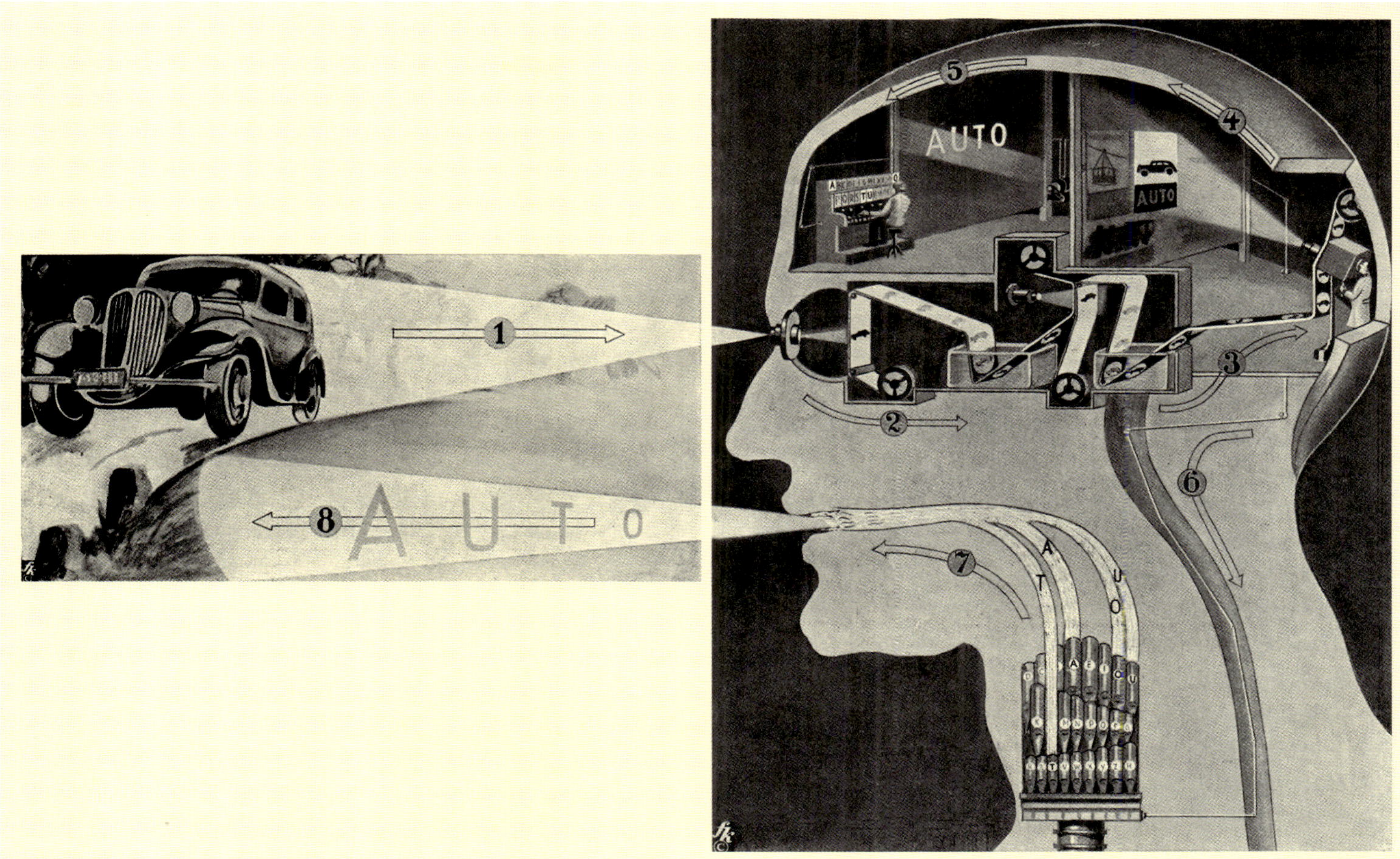

power to picture a fact, but the structure itself cannot be described *in* the language. One can, however, create a second language (a metalanguage) with an expanded domain to describe the structure of the first (plate 9-3). And there might be a third language capable of describing the structure of the second, and so on. But in order to conceive of the *entire* hierarchy of ascending languages, one must "step out" of the tower of meaning and cease speaking any of its languages. Recalling the ancient Cratylus of Athens, who stopped talking and only pointed his finger, Wittgenstein insisted that one can demonstrate the picturing relation between an assertion and a fact, but not *in* the language, disallowing the paradox posed by Magritte (plate 9-4). Wittgenstein concluded that one can conceive of the whole hierarchy but not describe it in words, ending the *Tractatus* with his seventh and final axiom: "Whereof one cannot speak, thereof one must be silent."[7]

Wittgenstein's achievement was to give a perspicacious argument that spoken language (by which he meant semantics and syntax as studied by analytic philosophy) can describe limited parts of human experience, but not intuitions about values (ethics and aesthetics) or feelings about the inner world (some of which may be describable by less rigorous, more poetic language, while others may be truly inexpressible in words).

In the *Tractatus* Wittgenstein described another traditional theme of nature mysticism: the pantheist feeling of being one with the world. Mystics describe escaping earthly bounds of mortality to unite with all of nature: the Pythagoreans transcended to a "World Soul" and Spinoza to "the Absolute—the One." Wittgenstein described his own feeling

9-2. "What goes on in our heads when we see an auto and say 'auto,'" in Fritz Kahn, *Der Mensch Gesund und Krank* (Zurich: Albert Müller, 1939).

For Russell, Wittgenstein, and Fritz Kahn (the German physician who designed this popular-science illustration), language operates in a completely mechanical (logical) way. A lens in the man's eye projects a picture of the auto onto a light-sensitive surface (his retina), which translates the image into nerve impulses that travel along his optic nerve to the visual cortex at the back of his brain. There a technician clad in a white lab coat (personifying reason) matches the picture with a word by consulting stored memories. When the technician finds a match, he projects "AUTO" to the operator of a 26-letter (A–Z) switchboard, who relays the letters to the corresponding pipes of an organ (the vocal chords in the larynx), and the pattern of sound "AUTO" emerges from the man's mouth.

9-3. Wittgenstein's tower of meaning.

In Language 1, words ("There are two dots") and images (the two dots depicted in the easel painting) represent (picture) the fact that there are two black dots below them on this page. According to Wittgenstein, this picturing relationship (between words and facts, and between images and facts) can be shown, as it is in this diagram, but it cannot be described *using Language 1* because the domain of Language 1 (the things to which its words and images refer) are facts in the world (not the relationship *between* words, images, and facts).

Language 2: *Getting outside the system in order to describe it.* The picturing relation can be described if one creates a second language (system of symbols) in which the domain is extended to include pictures of facts.

Language 3: *Building a hierarchy.* Then, a third language can be created in which the domain is extended further to include pictures of the pictures of facts, and so on.

Avoiding paradoxes of self-reference. Notice that self-reference (referring to a language in the language) is impossible in this system. Thus, for example, the (verbal) Liar paradox ("This sentence is false") cannot be formulated in Wittgenstein's tower system, nor can Magritte's (visual/verbal) conundrum *Ceci n'est pas une pipe* (This is not a pipe).

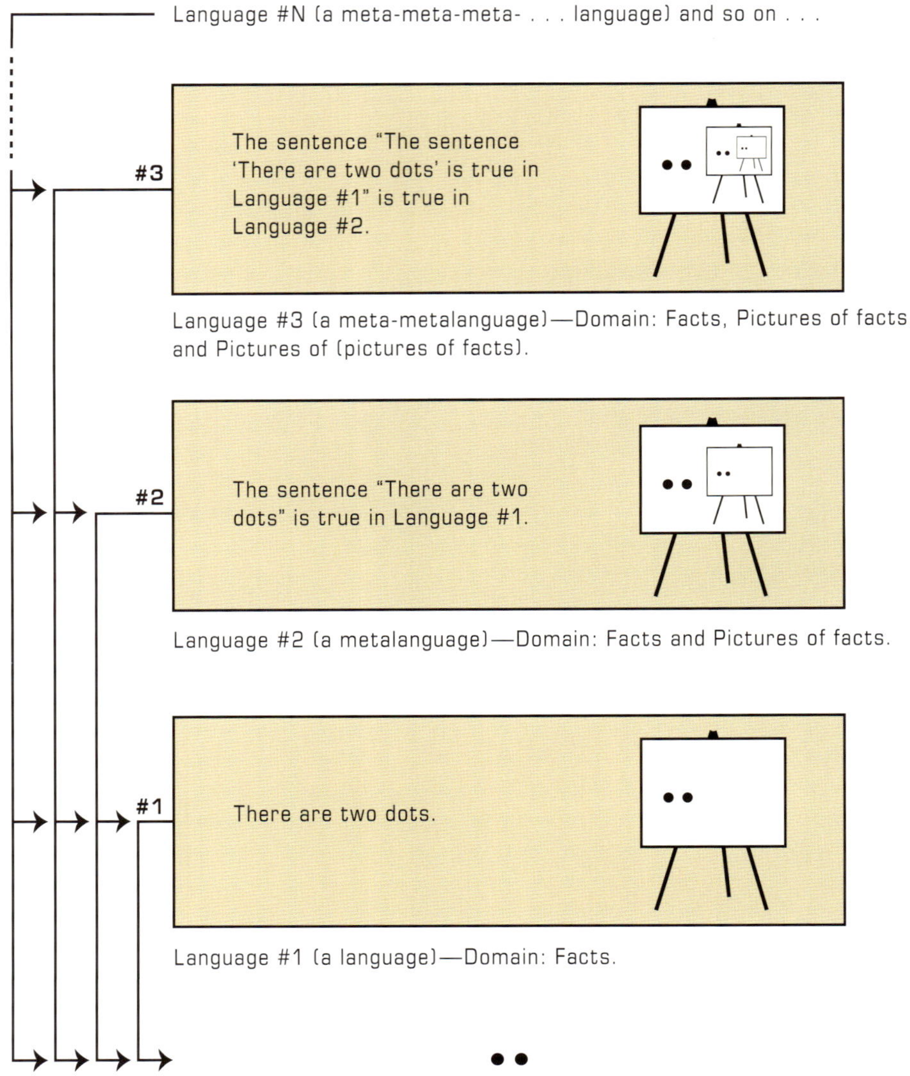

of oneness with the world as an "experience *par excellence* of absolute value" (*Notebooks*, 1914–16), and, living in a secular society, he described life as a journey of immanence (cherishing the present moment) in which he felt absolutely safe—"nothing can injure me whatever happens."[8] He described a feeling of peace even when facing the threat of death during the war: "If we take eternity to mean not infinite temporal duration but timelessness, then eternal life belongs to those who live in the present."[9] This outlook was captured by the painter Hieronymus Bosch in his portrayal of the early Christian mystic Saint Anthony, who, although surrounded by evil and dangerous tormentors, focuses on the cloudless blue sky above and its reflection in the still waters at his feet (plate 9-6).

After World War I Wittgenstein returned home to Vienna, where the Vienna Circle of mathematicians, logicians, and philosophers, led by Rudolf Carnap, were writing a new universal language for science (see chapter 8). The logical positivist group read the *Tractatus*, and Wittgenstein attended a few meetings, but he was not a regular member of the group. In the *Tractatus*, Wittgenstein had described the limits of what can be stated in a verifiable sentence, but with an attitude that was completely different from the logical positivists, whose verification principle states that the meaning of a proposition is equivalent to

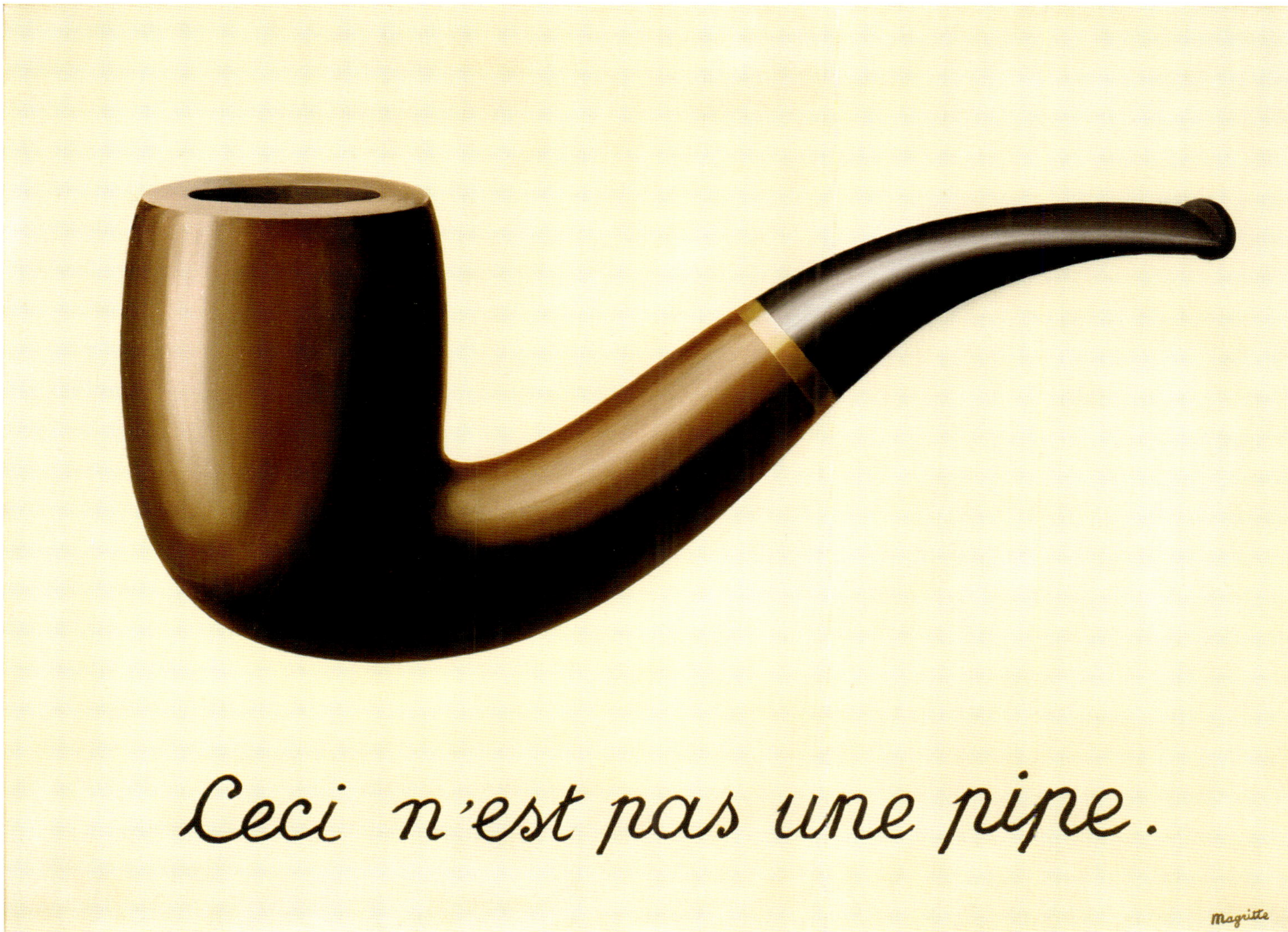

giving the rule for its verification; if there is no rule, then the proposition is meaningless. Using this criteria, Carnap dismissed metaphysics as meaningless, as if he were brushing away an annoying fly, whereas Wittgenstein concluded that "the mystical" was unverifiable (inexpressible in words or symbols) with a sense of loss. According to Wittgenstein, all metaphysical topics—as well as ethics and aesthetics—exist at the limits of the world in the sense that the world simply exists, but evaluations of what is good or beautiful come from "outside" because value judgments are mind-dependent assessments imposed on the world by humans.[10]

Feeling spent after publishing *Tractatus* in 1921, Wittgenstein left the study of logic and language and taught elementary school in rural Austria. He was drawn back into philosophy in 1928 when he attended a lecture in Vienna given by Brouwer in which the intuitionist mathematician criticized the axiomatic approach to logic and language. Brouwer's target was Hilbert's Program, but

9-4. René Magritte (Belgian, 1898–1967), *La trahison des images* (The treachery of images), 1929. Oil on canvas, 23¾ × 31 15/16 in. (60 × 81cm). Los Angeles County Museum of Art, purchased with funds from the Mr. and Mrs. William Preston Harrison Collection.

On this canvas Magritte painted a picture of a pipe and wrote a sentence that refers to that picture: *Ceci n'est pas une pipe* (This is not a pipe). This breaks the rules of Wittgenstein's tower scheme because the artist's painted pipe mirrors a fact (a pipe in the world), but his sentence refers to the relation between the painted pipe and the pipe in the world. According to Wittgenstein, the relation between picture and fact cannot be described *in* the same language (*on* the same canvas). To follow Wittgenstein's rules, Magritte would have had to create a second language (a metalanguage); he could have done this by moving *Ceci n'est pas une pipe* off the canvas and making it the title of the painting. The artist did not do this because, unlike Wittgenstein, who aspired to avoid paradox, Magritte wanted to *create* paradox to make his point that the innocent eye—the (naive) viewer's belief that a picture is identical to what it represents—opens the doors to *la trahison des images* (the treachery of images).

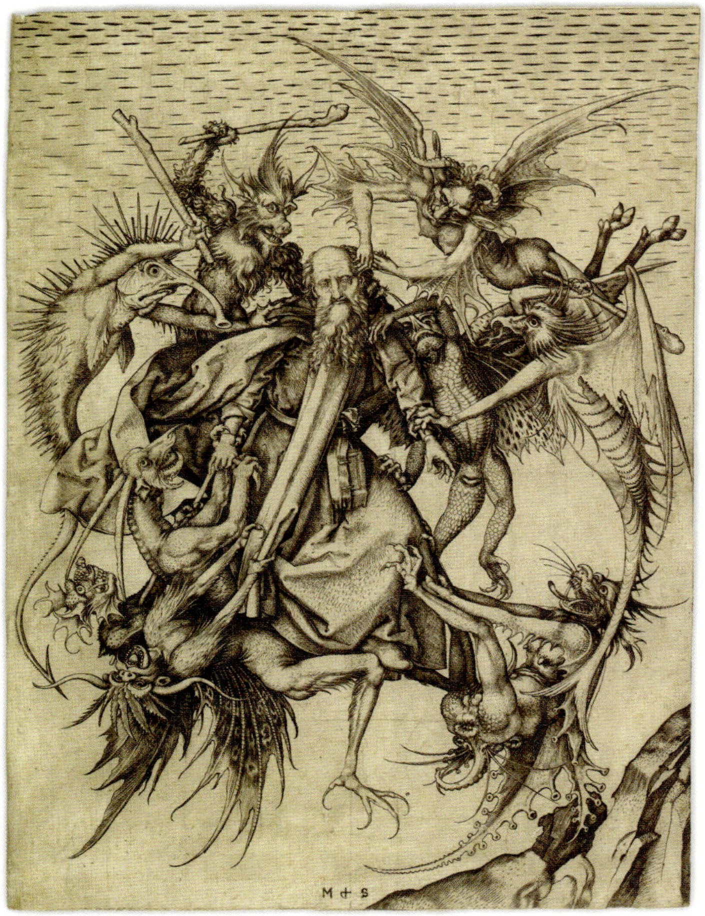

9-5. Martin Schongauer (German, 1435/50–1491), *Saint Anthony Tormented by Demons*, ca. 1470–75. Engraving, 12¼ × 9 in. (31.1 × 22.9 cm). Metropolitan Museum of Art, New York, gift of Felix M. Warburg and family, 1941.

Wittgenstein was well aware that the criticism applied to the axiomatic method that was the basis of his *Tractatus*.[11] Thus it was that Wittgenstein returned to Cambridge, completed his doctorate, and returned to philosophy.

Wittgenstein responded to Brouwer because they both did philosophy in the style of the German Romantics.[12] Russell and Carnap, on the other hand, did philosophy in the scientific style of building a structure of true propositions sustained by argument. Their theories were based on evidence, and they aspired to make philosophical progress by engaging in dialogue with their colleagues. But like the *Naturphilosophen* (and Brouwer), Wittgenstein did philosophy to achieve self-knowledge; he was on a journey that was introspective and solitary, as he wrote in the preface to *Philosophical Investigations* (published posthumously in 1953): "The philosophical remarks in this book are, as it were, a number of sketches of landscapes which were made in the course of these long and involved journeyings."[13] Wittgenstein also read the *Lebensphilosophen*,[14] and in 1947 he noted the prevalence in his own day of critics of science, seeming to leave open the possibility that they might be correct: "It isn't absurd to believe that the age of science and technology is the beginning of the end for humanity; that the idea of great progress is a delusion, along with the idea that truth will ultimately be known; there is nothing good or desirable about scientific knowledge and that mankind, in seeking it, is falling into a trap. It is by no means obvious that this is not how things are."[15]

I am not aiming at the same target as the scientists and my way of thinking is different from theirs. . . . For the place I must get to is the place where I already am.

—**Ludwig Wittgenstein, 1930**

Carnap and his colleagues, first in Vienna and later in Chicago, tried to capture all true scientific statements in a unified description of nature. Their response to Wittgenstein's demonstration that there are limits to spoken language was to dismiss topics that lay beyond the reach of science. Meanwhile, another young Austrian in their midst, the mathematician Kurt Gödel, further undermined the goals of logical positivism by proving that there are also limits to artificial language. Like Wittgenstein, Gödel was influenced by Brouwer's anti-rationalist doctrine, and according to Carnap, Gödel's proof of the incompleteness of mathematics was an expression of "the kernel of truth in the assertion made by Brouwer . . . that mathematics cannot be completely formalized."[16] The logical positivists could find ways around Wittgenstein's *Tractatus*, but Gödel's mathematical proof could not be dismissed as a philosophical quibble.

A decade after Wittgenstein pointed out the limits of spoken language, Gödel proved a parallel result for mathematics: there are truths that can be seen to be true in a system but that cannot be deduced within that system. To describe the structure of a natural language, Wittgenstein showed that one must step outside it. Gödel demonstrated that the same is true of artificial language: in a mathematical system complex enough to describe the natural numbers, self-reference is inescapable; one must step outside the first language

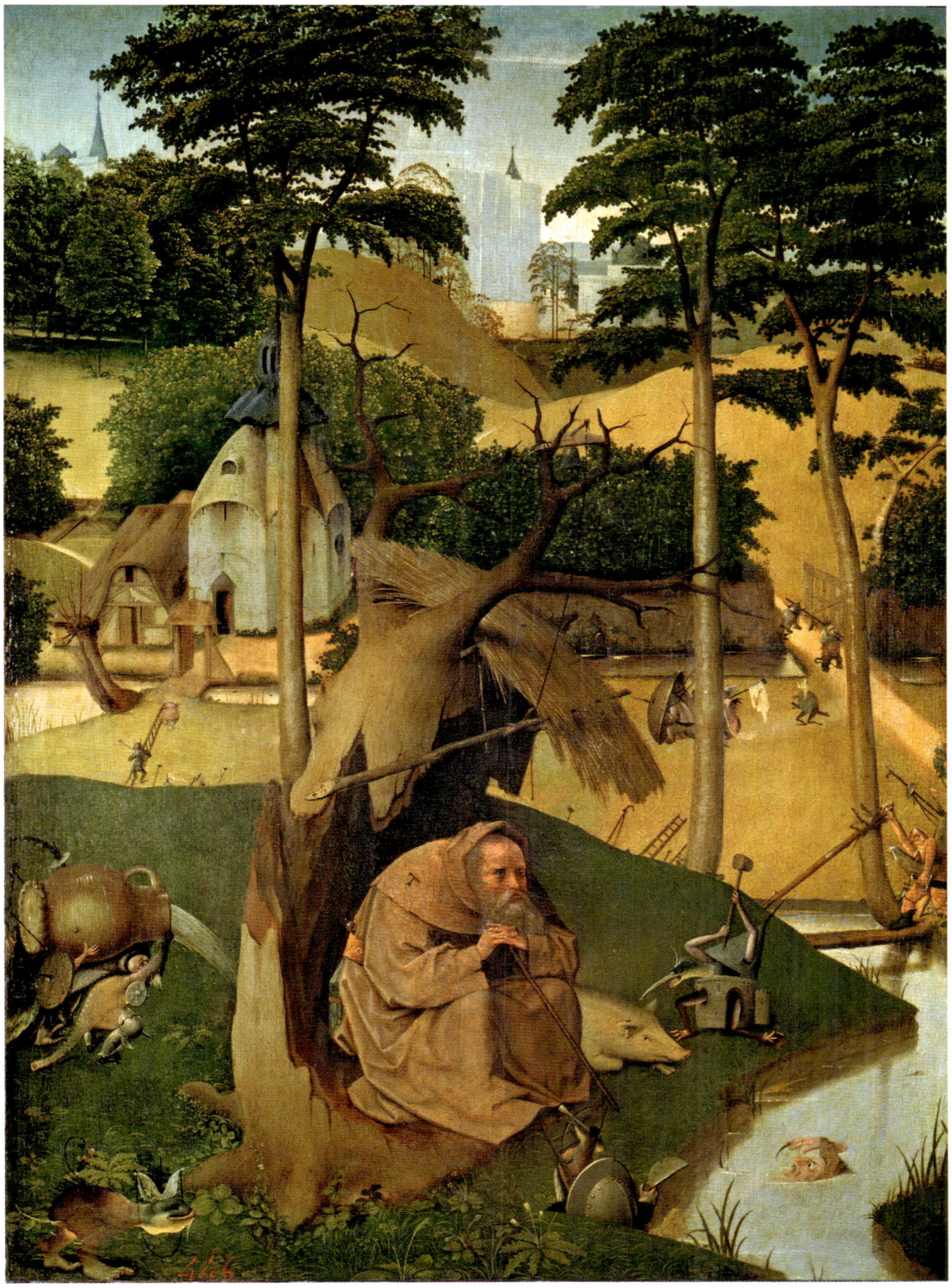

9-6. Hieronymus Bosch (Netherlandish, ca. 1450–1516), *The Temptation of Saint Anthony*, ca. 1490. Oil on wood panel, 28¾ × 20¾ in. (73 × 52.5 cm). © Museo Nacional del Prado, Madrid.

Anthony (AD 251–356) was a Christian from Alexandria who lived as a hermit in a community of Desert Fathers in the Egyptian delta, where he was tormented by devils. The lurid, erotic, terrifying details of his torment have made the temptation of Saint Anthony a favorite subject of artists such as Martin Schongauer (plate 9-5). In Bosch's unusual version, Anthony's peaceful composure is the center of interest, while the lethal threats to him lurk in less prominent places: a man in the tower on the right raises a hammer, a figure on the left pours liquid on him, a figure in front shoots a missile at him. Oblivious to such dangers, Anthony concentrates on a timeless realm beyond the reach of his attackers, symbolized by the blue sky and calm waters. In a notebook written during World War I, Wittgenstein described such intense focus: "A man who is happy must have no fear. Not even in the face of death. Only a man who lives not in time but in the present is happy" (*Notebooks*, 76, paragraph 13; July 14, 1916).

in order to describe the second. Ever since Epimenides of Crete stated the Liar paradox, mathematicians have been wary of self-reference because it can lead to paradoxes. But Gödel proved that self-reference and the potential for paradox are inherent in any axiomatic system capable of describing the integers, which is a very low level of complexity. Thus Hilbert's towering mathematical edifice and Carnap's multivolume scientific encyclopedia can never be proven complete and consistent.

Mathematicians who took up Hilbert's challenge to prove the completeness of arithmetic adopted the axioms Giuseppe Peano wrote in 1889 (see the sidebar on page 159 in chapter 4). In the first (1910) volume of *Principia Mathematica*, Russell and Whitehead wrote a set of formal axioms for propositional logic and a set for predicate logic. They were able to show that Peano's axioms for arithmetic held for a system constructed within predicate logic, thereby proving that arithmetic could be based on logic. However, to accomplish this, Russell and Whitehead had to assume an extra-logical (metaphysical) axiom, the axiom of infinity, which asserts the existence of at least one infinite set (such as the whole numbers). Russell and Whitehead went on to show how numerical patterns, and other areas of mathematics, could be brought within the scope of their axioms. In light of Russell and Whitehead's 1910 results, mathematicians engaged in Hilbert's Program turned their attention to proving the completeness and absolute consistency of the two sets of logical axioms—for propositional and predicate logic—that were in *Principia Mathematica.*

By 1920 two mathematicians, the Swiss Paul Bernays and the Russian-born American Emil Post, had independently proven the completeness of propositional logic.[17] Gödel proved in 1929 that first-order predicate logic (whose domain is individuals, such as numbers) is also complete. What about a second-order predicate logic, such as arithmetic (which is a system whose domain is expanded to include not only individuals but also sets of individuals, such as the set of whole numbers)? Gödel asked whether one could conclude from the completeness results obtained by Bernays, Post, and himself that second-order predicate logic is also complete? To everyone's surprise, Gödel proved that the answer is "No."

GÖDEL NUMBERS: FROM PROOF TO COMPUTATION

Gödel devised a mechanical procedure to translate any formula written in the formal language of arithmetic, in particular, the vocabulary of *Principia Mathematica*, into a unique number—now called its "Gödel number" (see sidebar on opposite page). He then showed that the relation between the premises and conclusion of a proof could be mapped onto an arithmetical relation between the Gödel number of the string of premises of the deduction (considered as a whole) and the Gödel number of the conclusion. Does a given conclusion follow from its premises? Gödel found a way to reduce the answer to this question to the truth or falsity of an arithmetical relationship between the Gödel number of the premises and the Gödel number of the conclusion. That is, Gödel reduced the question of whether a given conclusion follows from a given hypothesis to checking an arithmetical computation involving the Gödel numbers of the premise and conclusion.

Gödel's proof is as famous for its method as for its result because he translated the question of what is *provable* in a formal axiomatic system into what is *computable* in arithmetic. By achieving a dramatic result with this new style of proof, Gödel focused the attention of mathematicians on computation, pressing forward the development of computers.

Gödel Numbers

Gödel devised a new tool for translating formulas and statements written in a formal language, in particular that of Russell and Whitehead's *Principia Mathematica* (*PM*), into a unique number, called its Gödel number. The language of *Principia Mathematica* expresses logic and arithmetic in twelve constant signs and three kinds of variables.

To the twelve *constant signs* in *PM* assign the first twelve numbers:

Constant signs in *PM*	Numbers	The meaning that the authors of *PM* intended the signs to merit
¬	1	not
v	2	or
→	3	implies
∃	4	there exists
=	5	equals
0	6	zero
S	7	immediate successor of
(	8	left parenthesis
)	9	right parenthesis
,	10	comma
+	11	plus
·	12	times

To each *numerical variable*, associate a distinct prime number greater than 12:

x	13
y	17
z	19

To each *sentential variable*, associate the square of a prime number greater than 12:

p	132
q	172
r	192

To each *predicate variable*, associate the cube of a prime number greater than 12:

P	133
Q	173
R	193

Example: the symbols in the formula "S0 = S0" (which reads "The immediate successor of zero equals the immediate successor of zero," or, in brief, "1 = 1") would be assigned numbers as follows:

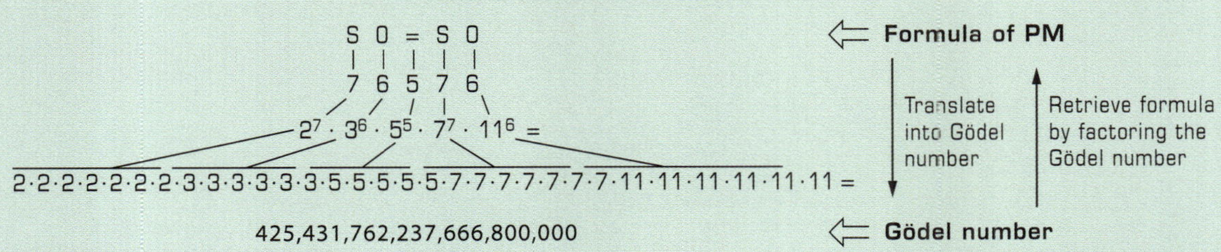

Then to determine the formula's unique Gödel number, multiply the first in the series of the prime numbers, each raised to the power of the numbers corresponding to the signs.

The great advantage of a code based on prime numbers is that, as Euclid proved (see the sidebar on page 21 in chapter 1), a composite number can be reduced to its prime factors in only one way. Thus Gödel could retrieve the original *PM* formula from its Gödel number.

GÖDEL'S PROOF OF THE INCOMPLETENESS OF MATHEMATICS

Gödel proved that the set of axioms and rules that Russell and Whitehead had written for *Principia Mathematica* is incomplete in the sense that there is a formula, which everyone would agree would have to be true (when interpreted in a model for arithmetic that satisfies a given consistent set of axioms), but that cannot be deduced from these axioms. The core of the proof is Gödel's construction of a special kind of self-referential system (plates 9-7 and 9-8). His proof is a modern variation on the Liar paradox: the absurd statement of the Cretan who said, "All Cretans are liars." One version of this ancient paradox is: "This sentence is false." This assertion refers to itself *in the language* in which the sentence is written. The difference in Gödel's proof is that he accomplished self-reference by constructing a system that links two parallel languages: the formal language of *Principia Mathematica* and Gödel numbers. In particular, Gödel gave a recipe for encoding *meta*-mathematical statements *about* formulas in *Principia Mathematica* by reference to their Gödel numbers.

Gödel then showed how to construct a sentence *Q* in the language of *Principia Mathematica* that is equivalent to an arithmetical statement whose interpretation is that *Q* is not provable in the *Principia Mathematica* system. The sentence *Q* is either true or false (it is either provable or not provable).

Consider what happens in the first alternative: If *Q* is provable (true), and given that it is logically equivalent in the *Principia Mathematica* system to an arithmetic statement that asserts that *Q* is *not* provable, then the system has a logical contradiction. Since the first alternative has led to this (unacceptable) paradox, we have to assume that *Q* is not provable.

Now consider what happens if *Q* is not provable (false). Recall that *Q* is logically equivalent (in the language of *Principia Mathematica*) to an arithmetic statement that

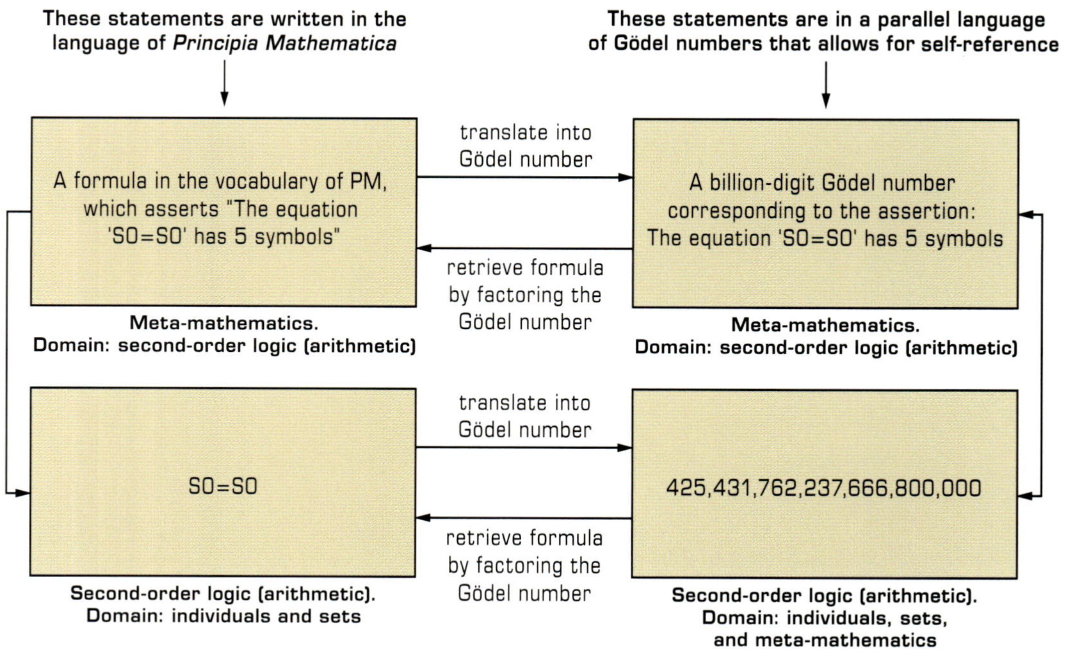

9-7. Gödel's self-referential system, shown with the statement "S0=S0" as an example.

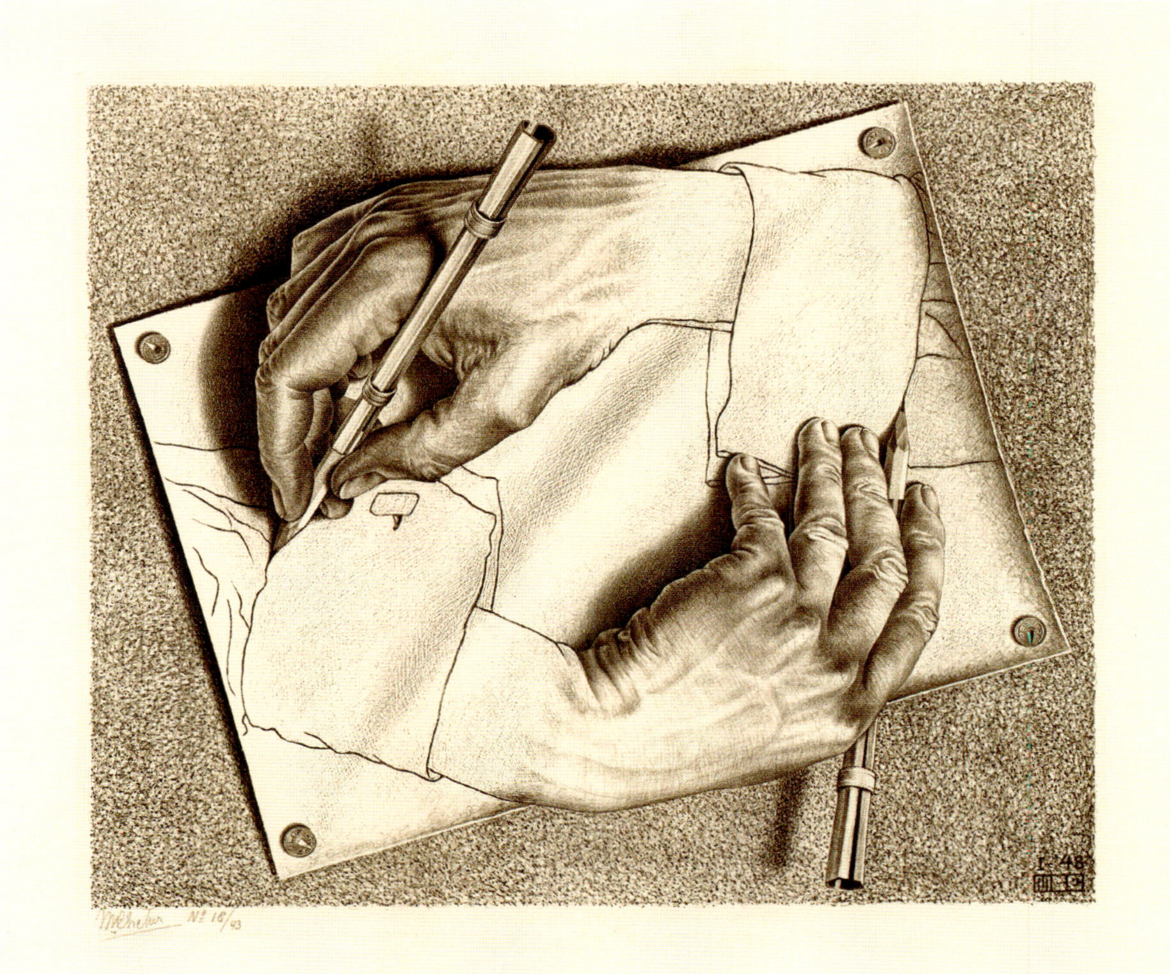

asserts that it (Q) is not provable, so the statement that Q is not provable is true. Thus Gödel has produced a true statement that cannot be proven. This means that all truths of arithmetic cannot be proven from the axioms and rules of *Principia Mathematica*, and therefore the system is incomplete. Indeed, Gödel proved that not just *Principia Mathematica* but any consistent set of axioms for arithmetic that could generate the natural numbers must be incomplete ("On Formally Undecidable Propositions of *Principia Mathematica* and Related Systems," 1931).[18]

Gödel's proof of the incompleteness of mathematics ended a half-century of attempts to rebuild mathematics on a formal axiomatic foundation and dealt a fatal blow to Hilbert's Program. For decades Hilbert, Russell, and many others had expended enormous effort to isolate the correct set of axioms for arithmetic and, ultimately, for all mathematics, never doubting that such a set existed. They all believed that the paradoxes that plagued fin-de-siècle axiomatic systems were introduced by imprecise thinking, never considering the possibility that these paradoxes were revealing limitations in the axiomatic approach itself. The Liar paradox and other earlier paradoxes did not prove incompleteness; Gödel's version of them did. Thus, contrary to expectations, Gödel proved that there is an inherent limitation to what can be demonstrated by formal proofs and within a formal system. In particular, there can be no complete and consistent set of axioms for arithmetic. This

9-8. M. C. Escher (Dutch, 1898–1972), *Drawing Hands*, 1948. Lithograph, 13 × 11 in. (33 × 28 cm).

When Gödel invented "Gödel numbers," he made arithmetic a mirror of metamathematical statements *about* arithmetic. Each contains the other. As in Gödel's new method of proof, in Escher's print there is symmetry between the "levels" that the hands represent.

suggests that the products of human reason can never be fully formalized and that new methods of mathematical proof and principles of demonstration, such as Godel's own proof, forever await discovery.

Gödel took the concept of a hierarchy of metalanguages that Wittgenstein envisioned relative to spoken language, and he exploited this idea for formal language—but with a crucial difference. Wittgenstein designed his tower of meaning to *prevent* self-reference, making it impossible to refer to a particular language *in that language* (plate 9-3). Gödel got around this restriction by showing how to mirror all legitimate processes of the metalanguage of arithmetic *in arithmetic itself* (plate 9-7).

Wittgenstein built an endlessly ascending tower. Gödel built a two-storey hall of mirrors, with the language of arithmetic spoken on the ground floor, where the axioms and theorems reside. These axioms and theorems are mirrored via Gödel numbering upstairs in the attic, where dwells metamathematics, which Gödel also reflected back downstairs into the language of arithmetic.

EARLY-TWENTIETH-CENTURY META-ART

At the opening of the twentieth century there were two key examples of what the French call *l'art pour l'art* (usually rendered in English as "art for art's sake"). The first (Cubist) example of such meta-art—art about the nature of art—was a response to the quest for the foundations of art that related to nineteenth-century physiology, especially research into how the human eye and brain form a picture of the world.[19] In his new theory of vision (*Handbook of Physiological Optics*, 1856–67) Hermann von Helmholtz proclaimed that seeing color results from light-waves striking the retina, causing nerve cells to send electrical impulses to the optical cortex at the back of the brain. After assessing the implications of Darwin's theory of evolution for the human mind, Helmholtz furthermore described vision as a lifelong learning process in which the mind continually constructs the world from visual signs. This new organic, evolutionary view of the eye and brain inspired Cézanne, whose goal was not only to record light striking his retinas (his eyes) but also to mentally structure the scene (in his brain), to "deal with nature as cylinders, spheres and cones."[20] After Cézanne's death in 1906, Picasso, Braque, and Gris continued his exploration of the question: How does a person know the world? The artists answered that the mind constructs an image of the world—and the artist creates a picture of it—by means of a multitude of signs and symbols that create illusions of reality. Making an analogy between cognition and mimesis, Picasso, Braque, and Gris made paintings that *were* imitations (they were pictures), and were *about* imitation (they were meta-art about depiction, featuring optical illusions and visual puns). For the Parisian Cubists, visual art is based on observation, and its essence is mimesis.

The second (Surrealist) example of art-about-art (meta-art) was a response to criticisms of science coming from within German philosophy that were also heeded by Wittgenstein and Gödel. This is why art by de Chirico, Magritte, and Escher resonates with the limits of language demonstrated by Wittgenstein and Gödel, even though there is no evidence that these artists had any direct knowledge of the mathematical writings.[21] De Chirico, Magritte, and Escher created paintings and prints that *are* both pictures and *about* depiction, creating paradoxical art-about-art.

De Chirico, born in Greece to Italian parents, entered Munich's Akademie der Bildenden Künste at age seventeen. There he was drawn to the art of the German Romantic Caspar David Friedrich and the Symbolists Arnold Böcklin and Max Klinger, as well as to German literature. Moving to Florence in 1910 and then to Paris in 1911, he read an Italian popularization of philosophy with chapters on Kant, Hegel, Schopenhauer, and Nietzsche.[22] De Chirico's melancholic personality drew him to the pessimistic Nietzsche, who declared that mankind had been set adrift from its ethical moorings by the rise of science and secularism in the nineteenth century. Nietzsche advocated introspection as a way to find an anchor within oneself and the careful examination of terms such as "good" and "bad," which, according to Nietzsche, are not facts in the world but human creations imposed to gain power over other people. Nietzsche declared a limit to what was knowable by reason, beyond which there were only enigmas containing elements of truth (*Beyond Good and Evil*, 1886).

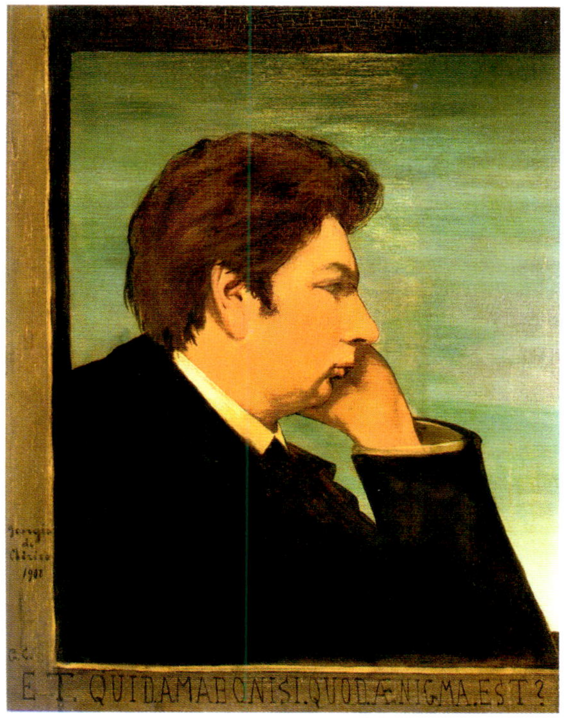

ABOVE

9-9. Giorgio de Chirico (Greek-born Italian, 1888–1978), *Self-Portrait*, 1911. Oil on canvas, 28⅜ × 21¾ in. (72.5 × 55 cm). Private collection, courtesy of the Fondazione Giorgio e Isa de Chirico, Rome. © 2014 SIAE, Rome/Artists Rights Society, New York.

De Chirico painted the brown edges (they are not part of a picture frame), on which he inscribed the Latin phrase across the bottom.

BELOW

9-10. Gustav Schultze (German, 1825–97), *Portrait of Friedrich Nietzsche*, 1882. Photograph. Klassik Stiftung Weimar, Goethe-und-Schiller Archiv, GSA 101/18.

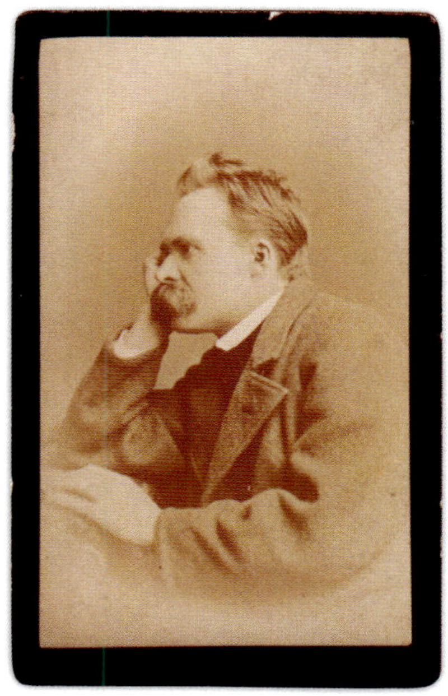

The young de Chirico identified himself with Nietzsche in a 1910 self-portrait in which he adopted the pose of the philosopher. At the bottom of the painting he wrote the Latin phrase ET QUID AMABO NISI QUOD AENIGMA EST? (And what will I love except enigma?), echoing Nietzsche's maxim, "And how could I bear to be a man if man were not also a creator and guesser of riddles and redeemer of accidents?"[23] (See plates 9-9 and 9-10.[24]) Unlike his avant-garde contemporaries in Milan, the Futurists, who were eager to burn all bridges to Italy's classical past, de Chirico wanted to breathe new life into ancient myths by rethinking them for modern culture, and he was inspired by Nietzsche's fresh interpretation of the myth of the labyrinth. Sent to the island of Crete by Apollo (human reason) to destroy the Minotaur (irrational animal impulse), the warrior Theseus entered the labyrinth carrying a thread given to him by the Cretan princess Ariadne. After slaying the beast, Theseus retraced his steps by following her thread (plate 5-1 in chapter 5); the tale is traditionally interpreted as a triumph of reason, as when Leibniz declared that his logical calculus was a thread of Ariadne. But Nietzsche inverted the usual valuation and hailed the victory of irrationality, focusing on the second half of the story. Ariadne gave Theseus the thread on the condition that he promise to take her away with him on his ship. Theseus kept his promise but then abandoned her as she lay sleeping on the island of Naxos. As Theseus sailed away, the lustful god of wine Bacchus passed by and, aroused by the sight of a sleeping young woman, possessed her (plate 9-11). According to the version of the ending favored by Nietzsche, when Ariadne woke up in the arms of the wrong man and cried out that she had been abandoned, Bacchus responded: "I am your labyrinth."[25] Faced with this turn of events, Ariadne decided to settle for Bacchus and live out her life in erotic bliss. Nietzsche preferred passionate Bacchus to cunning Theseus, but he reserved his highest praise for Ariadne for her quick turnabout in allegiance, from reason to passion.[26]

9-11. Titian (Tiziano Vecellio; Italian, ca.1488/90–1576), *Bacchus and Ariadne*, 1520–23. Oil on canvas, 69¼ × 75¼ in. (176 × 191 cm). National Gallery, London.

Leading a band of frolicking revelers, Bacchus arrives on a chariot drawn by a pair of cheetahs and discovers Ariadne, who gestures in despair towards Theseus's departed ship on the far-left horizon. At the sight of her, Bacchus leaps from his chariot and desires to make her passions soar, symbolized by the constellation—a crown of stars—high in the sky above her.

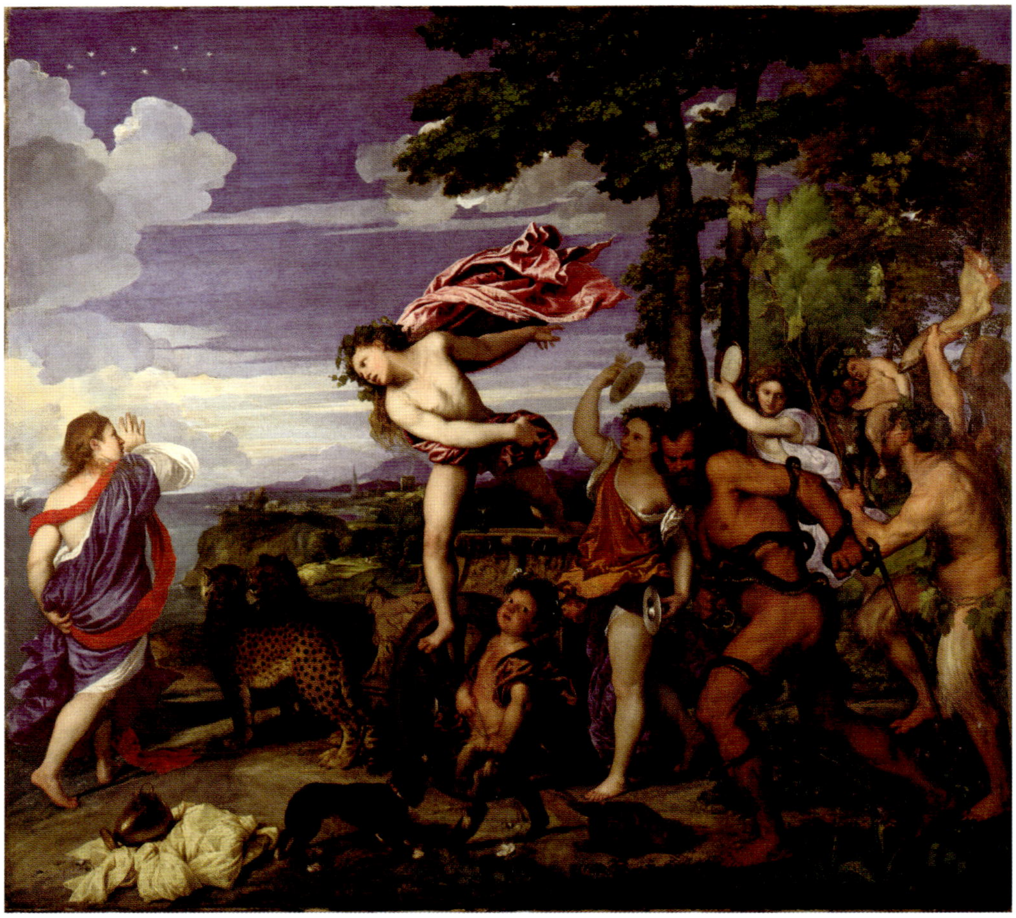

Inspired by Nietzsche's musings about Ariadne, in 1912–13 de Chirico did eight paintings featuring a white statue of sleeping Ariadne, a common subject in ancient classical art; she is shown abandoned but not yet awakened—poised between reason and passion (plate 9-13).[27] To create a disquieting metaphor for the limits of reason and imminent irrationality, de Chirico painted Ariadne in a geometric space with slight inaccuracies. In *Joys and Enigmas of a Strange Hour*, the loggia on the left and the plinth of the statue do not share a common vanishing point, which makes the statue appear to tip forward (plate 9-12). Like Gödel, de Chirico examined a symbolic system comprising a vocabulary and rules (linear perspective, the hallmark of Renaissance Italy), and then found a way—*using the rules*—to create a paradox (the loggia and the statue of Ariadne are both in, and not in, the same space).

While philosophers asked: "How does the mind represent the world?" de Chirico posed the related question: "How does a picture depict the world?" He approached his art with the self-consciousness that has characterized German philosophy ever since Kant sought certain knowledge by reflecting on the spatial and temporal scaffolding of his own mind. In *Great Metaphysical Interior* (plate 9-14), several paintings stand on easels; one is a landscape with a sanatorium; another depicts objects that appear real, as if they are stuck on the canvas' surface. On the back wall we see a cloudy sky through a window—or is it a mirror reflecting the sky in the viewer's space? As in the hierarchy of metalanguages in Wittgenstein's tower, de Chirico's painting about imitation is *in* a pictorial language (the artist's distinctive style), unable to exist outside all pictorial systems.

A picture held us captive.
And we couldn't get outside
it, for it lay in our language,
and language seemed only to
repeat it to us inexorably.

—**Ludwig Wittgenstein,**
***Philosophical Investigations*, 1953**

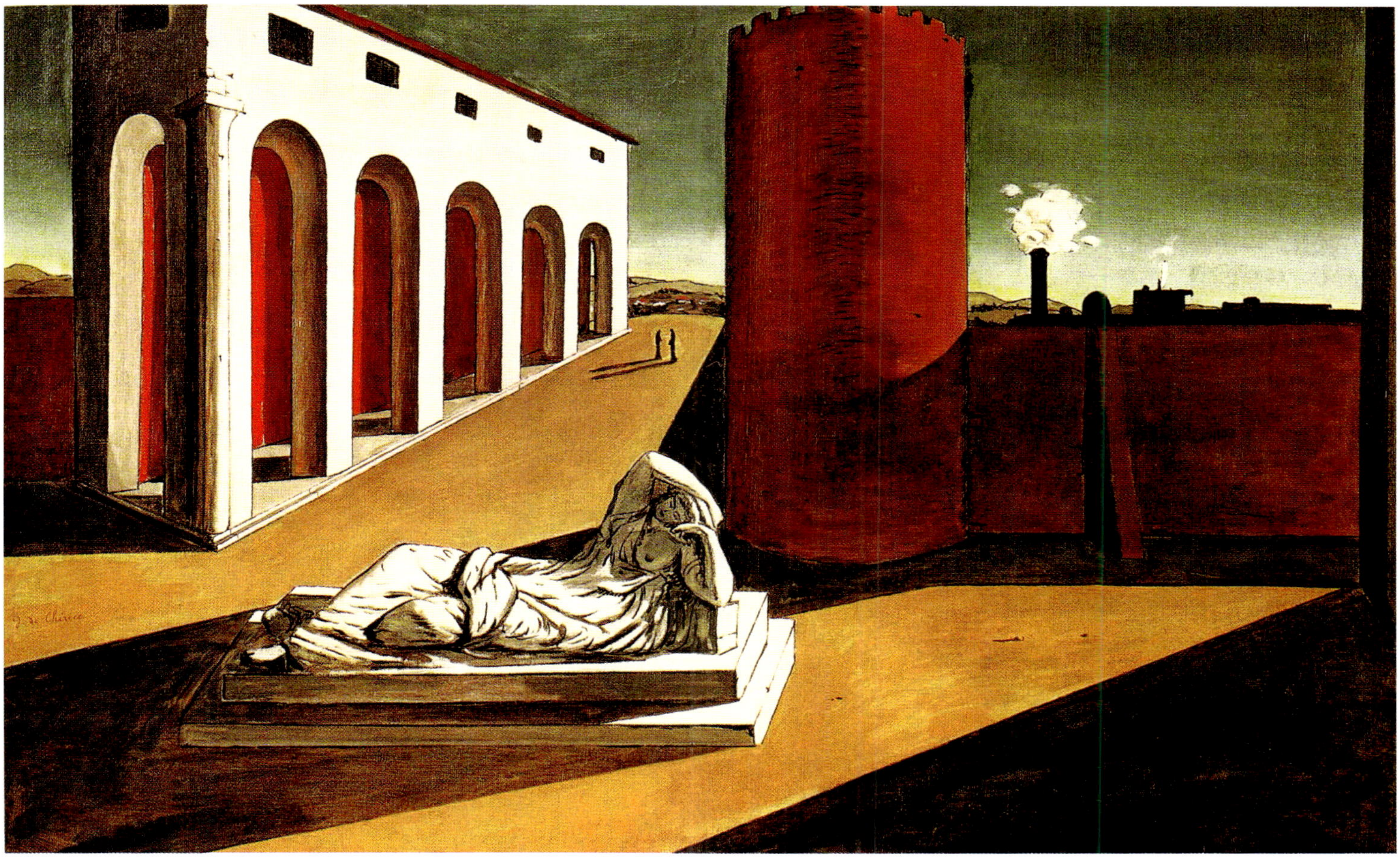

When World War I broke out in 1914, de Chirico returned from Paris to Italy, where he worked closely in Ferrara with the Futurist painter Carlo Carrà, who adopted de Chirico's style. Together with de Chirico's brother Alberto Savinio, the three formed the group Pittura Metafisica (Metaphysical painting), and after the war the publishing house Valori Plastici (Plastic values) became their mouthpiece. Through its magazine and monographs, as well as exhibitions in the early 1920s, de Chirico's style became known in northern Europe, and it was the Belgian painter René Magritte who adopted de Chirico's methods for disrupting symbolic systems in ways that were most true to their originator.

When Magritte discovered de Chirico's art in 1923, he had already survived a harrowing childhood; when he was thirteen his mother threw herself into the Sambre River, driven to suicide by her depraved husband.[28] At age eighteen Magritte entered the Académie Royale des Beaux-Arts in Brussels, and from 1916 to 1918 he experimented with Cubist and Futurist styles. Like de Chirico, he was drawn to philosophy; as a student he read a popularization of Nietzsche,[29] then Hegel and Freud.[30] Magritte found his direction as an artist when he saw a reproduction of a painting by de Chirico (*The Song of Love*, 1914; Museum of Modern Art, New York),[31] and he purchased a copy of a 1919 monograph on de Chirico that reproduced twelve other paintings, including three from the *Metaphysical Interior* series, along with commentary by French and Italian critics.[32] Magritte's strong response to the art of de Chirico is apparent in works such as *L'age des merveilles* (The age of wonders; see plate 9-15), in which oddly inanimate mechanical figures recede in space

ABOVE

9-12. Giorgio de Chirico (Greek-born Italian, 1888–1978), *The Joys and Enigmas of a Strange Hour*, 1913. Oil on canvas, 33 × 51 in. (83.7 × 129.5 cm). Private collection; courtesy of the Fondazione Giorgio e Isa de Chirico, Rome. © 2014 SIAE, Rome /Artists Rights Society, New York.

BELOW

9-13. *Sleeping Ariadne*, first–second century AD. Marble, 93 ¾ in. (238 cm) long. Roman copy of a Greek original from the third century BC. Galleria delle Statue, Musei Vaticani, Vatican City.

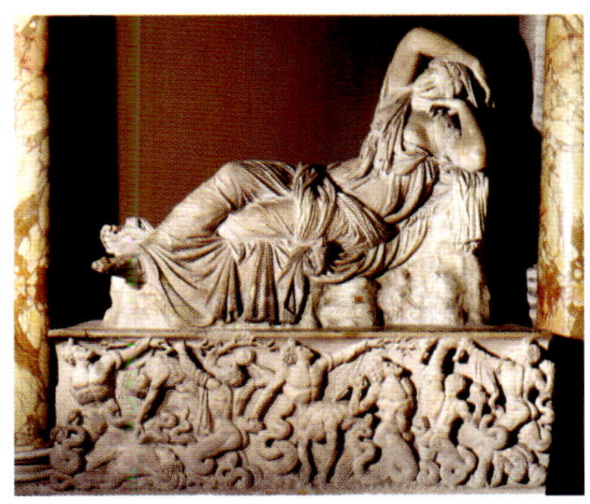

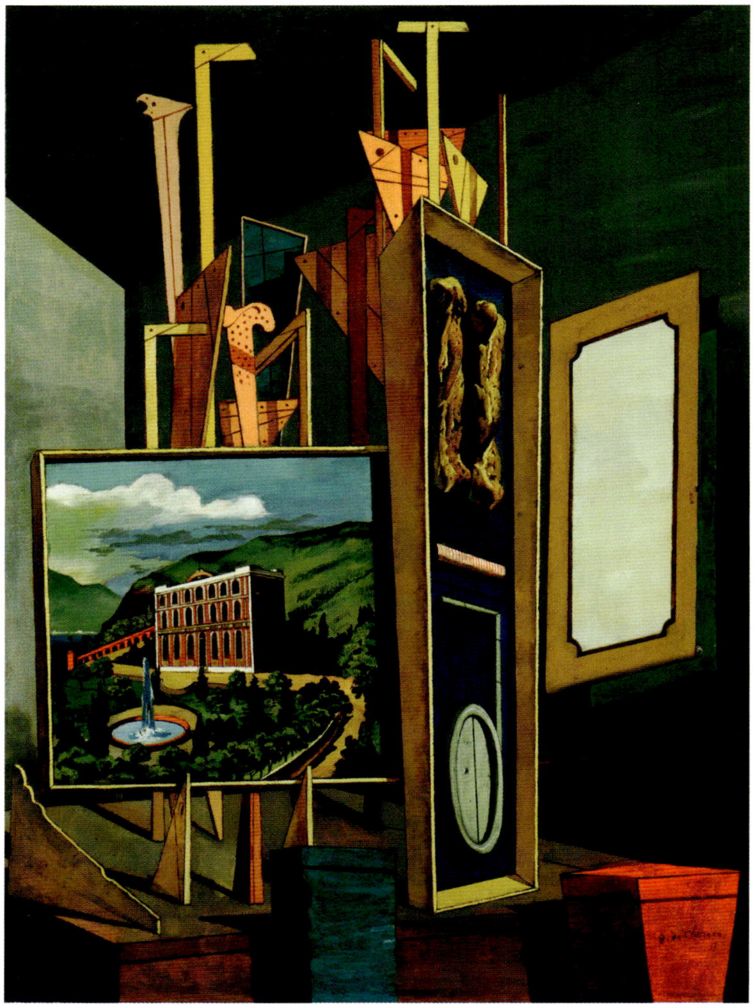

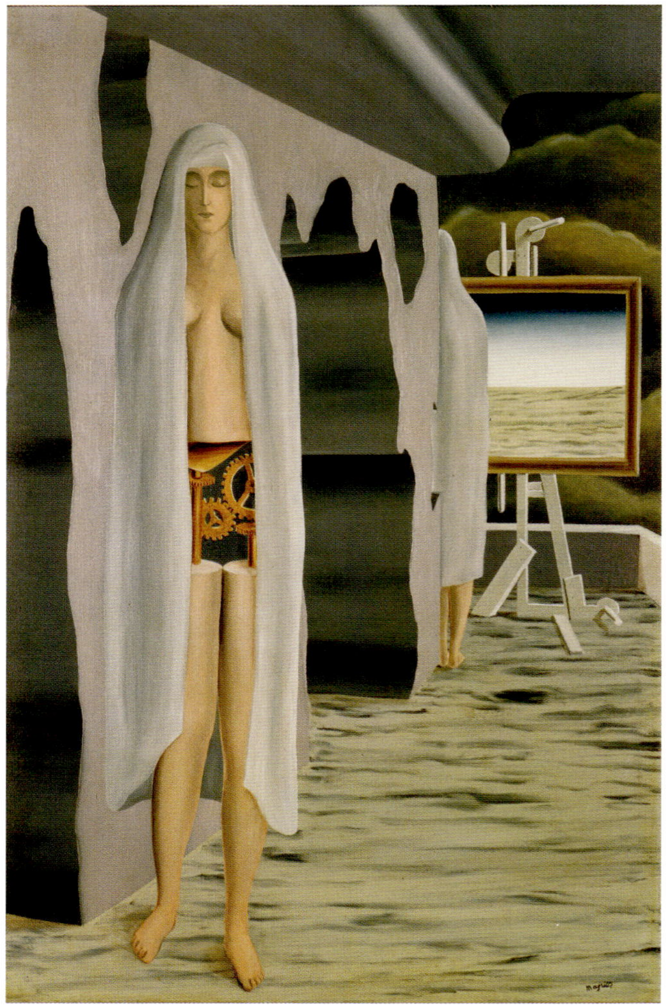

ABOVE LEFT

9-14. Giorgio de Chirico (Greek-born Italian, 1888–1978), *Great Metaphysical Interior*, 1917. Oil on canvas, 37¾ × 27¾ in. (95.9 × 70.5 cm). Museum of Modern Art, New York, gift of James Thrall Soby. © 2014 SIAE, Rome/Artists Rights Society, New York.

ABOVE RIGHT

9-15. René Magritte (Belgian, 1898–1967), *L'age des merveilles* (The age of wonders), 1926. Oil on canvas, 47¼ × 31½ in. (120 × 80 cm). Private collection. © 2014 C. Herscovici/Artists Rights Society, New York.

toward a painting-within-a painting that recalls de Chirico's *Great Metaphysical Interior*. Magritte increased the tension between the levels of representation by repeating the same foreground in both the painting and the painting-within-the-painting.

In a more discordant mix of representational levels, the landscape depicted in *La condition humaine* (The human condition) contains a picture of itself (plate 9-1, chapter frontispiece). For decades the Symbolists and the Surrealists had been consciously disrupting pictorial conventions to express the mystery of the unconscious mind. De Chirico and Magritte focused instead on the mystery of the *conscious* mind, upsetting conventions in seemingly realistic pictures to symbolize the mind's ability to reflect on itself.

In 1927 Magritte moved to Paris, where he did a series of paintings in which there are words. His interest in language may have been inspired by free association, a method used for interpreting dreams in Freudian psychoanalysis, which was the talk of Parisian café society in the 1920s. In free association the patient states whatever spontaneously comes into his or her mind when presented with a picture or word. Magritte illustrated free association in a few paintings in which he paired images with unlikely words, as they typically would be in the interpretation of a dream (for example, in *Key of Dreams*, 1930 [Private collection], the picture of a horse is paired with the word "door"). But Magritte's interest was not in the Freudian unconscious but, as for de Chirico, in expressing

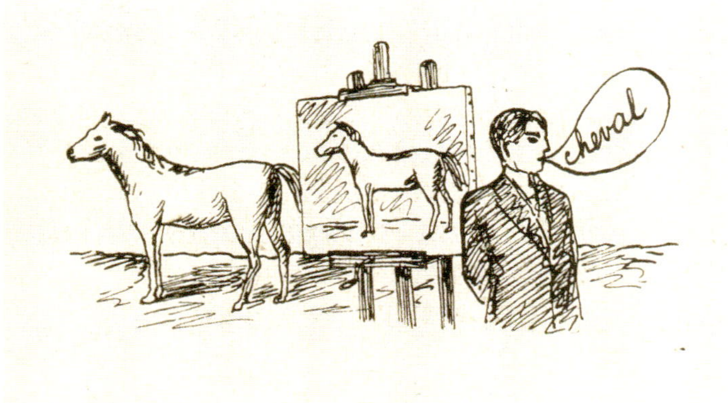

the limits of conscious thought using tools that artists had invented to depict the world realistically—silhouette, shading, overlap, and perspective.[33]

Magritte wrote an article for a Surrealist magazine comparing how words and images represent ideas in the mind and objects in the world ("Les mots et les images [Words and images]," 1929; see plate 9-16). Years later, in 1966, the French philosopher Michel Foucault published a book with a similar title, *Les mots et les choses* (Words and things; English translation as *The Order of Things*, 1970). Magritte read the book and wrote to Foucault, pointing out a difference between his own and the philosopher's understanding of resemblance. Foucault described resemblance as the relation that holds between two objects, one of which is a copy of the other; for example, a painting of a tree resembles a tree. Astutely putting his finger on a key feature of Foucault's thought, Magritte countered that resemblance is not an objective relation between two things in the world (the painting and the tree), but a subjective judgment in the mind of a viewer.[34] Magritte enclosed in his

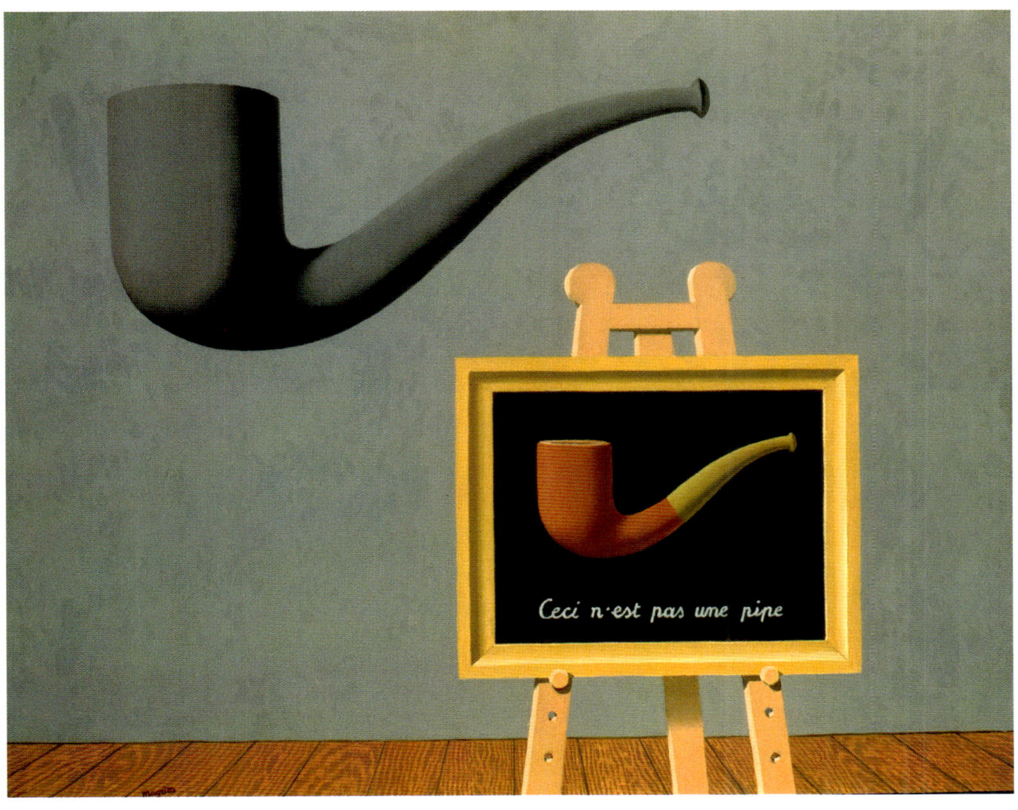

LEFT

9-16. René Magritte (Belgian, 1898–1967), *Les mots et les images* (Words and images; detail), in *La révolution surréaliste* 5, no. 12 (Dec. 15, 1929): 33. Museum of Modern Art, New York. © 2014 C. Herscovici/Artists Rights Society, New York.

A horse stands next to a picture of a horse while a man utters the word *cheval* (horse). Magritte's caption cautions: "An object never has the same function [*le même office*] as its name or its image" (33).

BELOW LEFT

9-17. René Magritte (Belgian, 1898–1967), *Les deux mystères* (The two mysteries), 1966. Oil on canvas, 25½ × 31½ in. (65 × 80 cm). Private collection. © 2014 C. Herscovici/Artists Rights Society, New York.

BELOW RIGHT

9-18. Michel Foucault's diagrams of Magritte's painting in his *Ceci n'est pas une pipe* (This is not a pipe; Montpellier: Fata Morgana, 1973), which is based on Foucault's shorter essay of the same title (*Les cahiers du chemin* 2 [1968]: 79–105). © 1973, Fata Morgana.

Foucault pointed out that there are at least three possible ways to understand the sentence *Ceci n'est pas une pipe* as it appears in Magritte's painting.

A The (painted) image (of the pipe) is not the word (*pipe*; 31).

B The word (*ceci*) is not the image (32).

C The image and the sentence *Ceci n'est pas une pipe* (taken together as an ensemble) do not form a *calligramme* (a merger of images and words, as made popular in French poetry by Guillaume Apollinaire). Foucault suggested a way to make it a *calligramme* by replacing the *u* in *une pipe* with the bowl of the pipe (33).

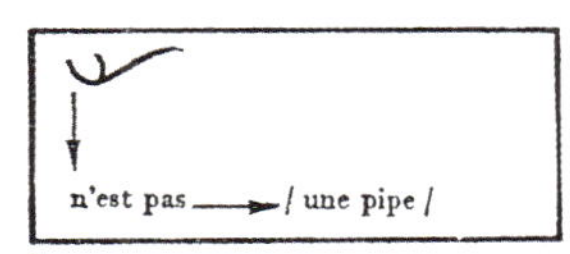

A

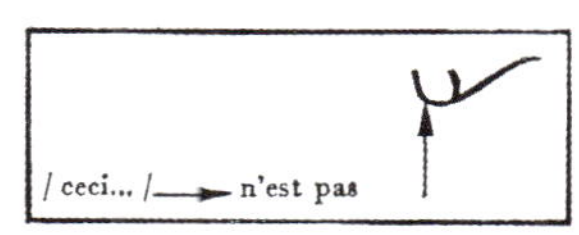

B

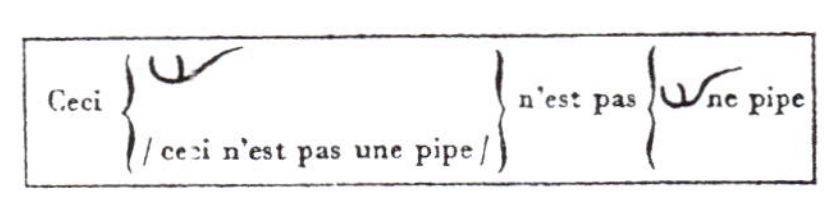

C

letter reproductions of a few of his works, including a version of the painting that includes the text "Ceci n'est pas une pipe" (This is not a pipe; plates 9-4 and 9-17). Foucault responded by publishing a short essay on Magritte in 1968 in which he analyzed "Ceci n'est pas une pipe" in the style of Ferdinand de Saussure's structural linguistics, which he expanded into a short book in 1973 (plate 9-18).[35]

A certain island G is inhabited exclusively by knights who always tell the truth and knaves who always lie. In addition, some of the knights are called "established knights" (these are knights who in a certain sense have "proved themselves").

—**Raymond Smullyan, *What is the Name of This Book?*** 1978

Meanwhile in Latin America, a contemporary of Magritte, the Argentine writer Jorge Luis Borges, incorporated Surrealist and existentialist themes in his writings. An aficionado of logical puzzles, Borges wrote about mirrors, labyrinths, and paradoxes. In his 1941 tale *La biblioteca de Babel* (The library of Babel), the narrator lives in an endless series of interlocking hexagonal rooms that are lined with bookshelves filled with volumes printed to a uniform length of 410 pages. The books are written in every possible combination of a set of twenty-five symbols, so most of the words are gibberish. The library includes a catalogue of its holdings, but there is no reader's guide to locate a book. Thus the narrator wanders through the interminable rooms searching for a book he can understand, which surely must exist since the library contains all possible combinations of the twenty-five symbols.[36] But where is that book?

IMPOSSIBLE OBJECTS

In 1958 the British mathematician Roger Penrose and his father, geneticist Lionel Sharples Penrose, discovered some impossible objects, including a staircase that one can climb forever without getting any higher (plate 9-25). Locally, each part of the figure is possible and makes sense, but as a whole the structure is not feasible, and the brain cannot make sense of it. As such, this type of optical illusion is the visual counterpart of a verbal conundrum such as the Liar paradox.

The Penroses' impossible objects inspired the Dutch printmaker M. C. Escher, who had completed his education in the early 1920s when the De Stijl aesthetic dominated the modern art world in his homeland. Unmoved by De Stijl and interested in depicting landscapes and figures, Escher settled in Italy in 1922, where he pursued his figurative style in relative isolation for over a decade. The direction of his art changed in 1936 when Escher saw the intricately patterned Islamic tiling of the Alhambra, a medieval Muslim castle and citadel in Granada, Spain (see plates 7-9 and 7-10 in chapter 7). Inspired to study the mathematics of tessellation (tiling), Escher became an illustrator of mathematical principles and went on to study many other types of mathematics, transforming abstract ideas into figurative form. He created a tower that ends in an impossible staircase, *Ascending and Descending* (1960; see plates 9-22 and 9-25), and he incorporated another figure invented by the Penroses—an impossible triangle—in his lithograph *Waterfall* (1961; see plates 9-24 and 9-27). Escher was also intrigued by the so-called crazy crate, an impossible cube of unknown origin (plate 9-26), on which he based the structure of *Belvedere* (1958; see plate 9-23). Roger Penrose came to know Escher when the Stedelijk Museum in Amsterdam organized an exhibition of the artist's work to correspond with a meeting of the International Congress of Mathematicians in 1954, which marks the beginning of Escher's great popularity with mathematicians.

After seeing this exhibition, the Canadian geometer H.S.M. "Donald" Coxeter reproduced one of Escher's tessellations in an article he wrote about non-Euclidean hyperbolic (saddle-shaped) planes (see the sidebar on page 155 in chapter 4).[37] Coxeter and Escher embarked on a lengthy and ultimately frustrating correspondence,[38] in which the artist asked for "simple explanations" of how to construct certain shapes or forms, and the mathematician responded with instructions that were incomprehensible to Escher. When the artist saw one of Coxeter's diagrams of the tiling of a (non-Euclidean) plane (which has the same pattern as Felix Klein's tiling of a hyperbolic plane in plate 11-8 in chapter 11), he adapted the pattern of curved lines to depict fish swimming in curved paths on a flat (Euclidean) plane. Escher described Coxeter's response: "I had an enthusiastic letter from Coxeter about my colored fish [*Circle Limit III*, 1959], which I sent him. Three pages of explanation of what I actually did. . . . It is a pity that I understand nothing, absolutely nothing of it."[39] Escher focused on puzzles in the mundane Euclidean world, such as his *Circle Limit* series (plate 9-19), which is about the idea of filling a figure (a circle) with motifs "getting smaller and smaller till they reach the limit of infinite smallness"[40]—just as Archimedes did in his method of exhaustion (plate 3-3 in chapter 3) and Newton in his integral calculus (see the sidebar at the bottom of page 117 in chapter 3). But Coxeter focused on his non-Euclidean research interests,[41] and in the end, each saw the other's work distorted because he viewed it through his own lens.

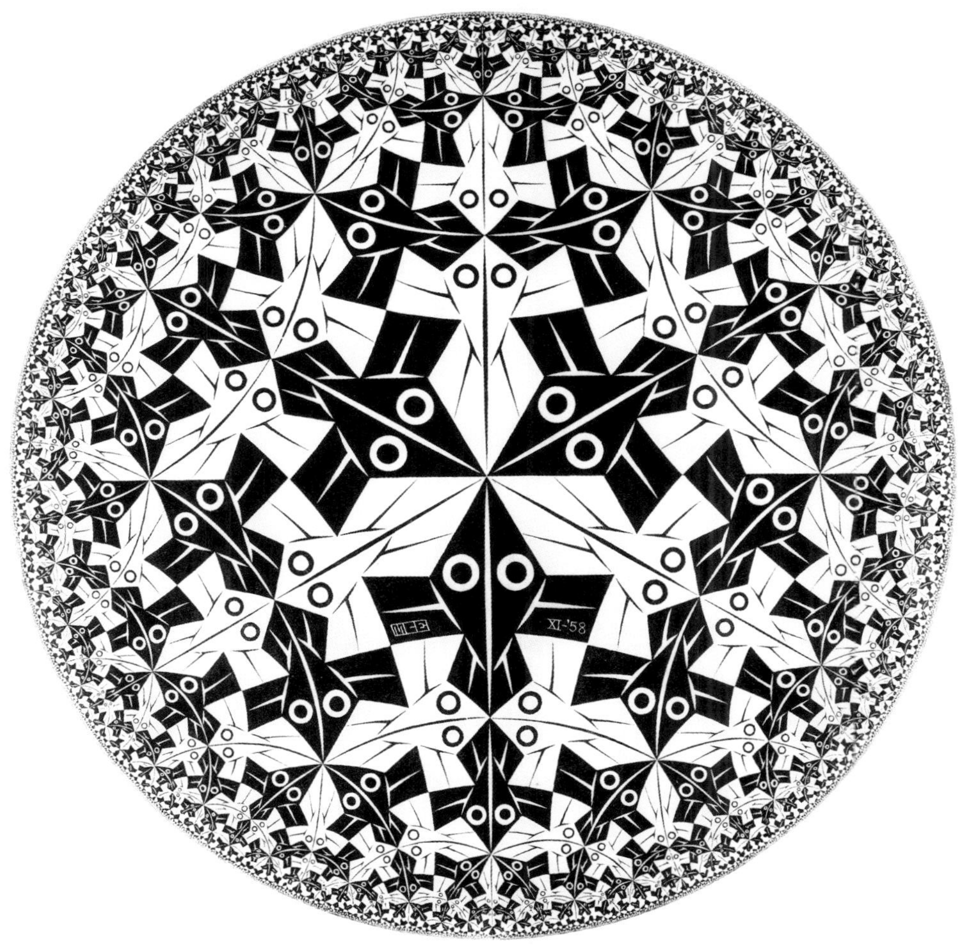

9-19. M.C. Escher (Dutch, 1898–1972), *Circle Limit I*, 1958. Woodcut, image diameter 16½ in. (41.8 cm).

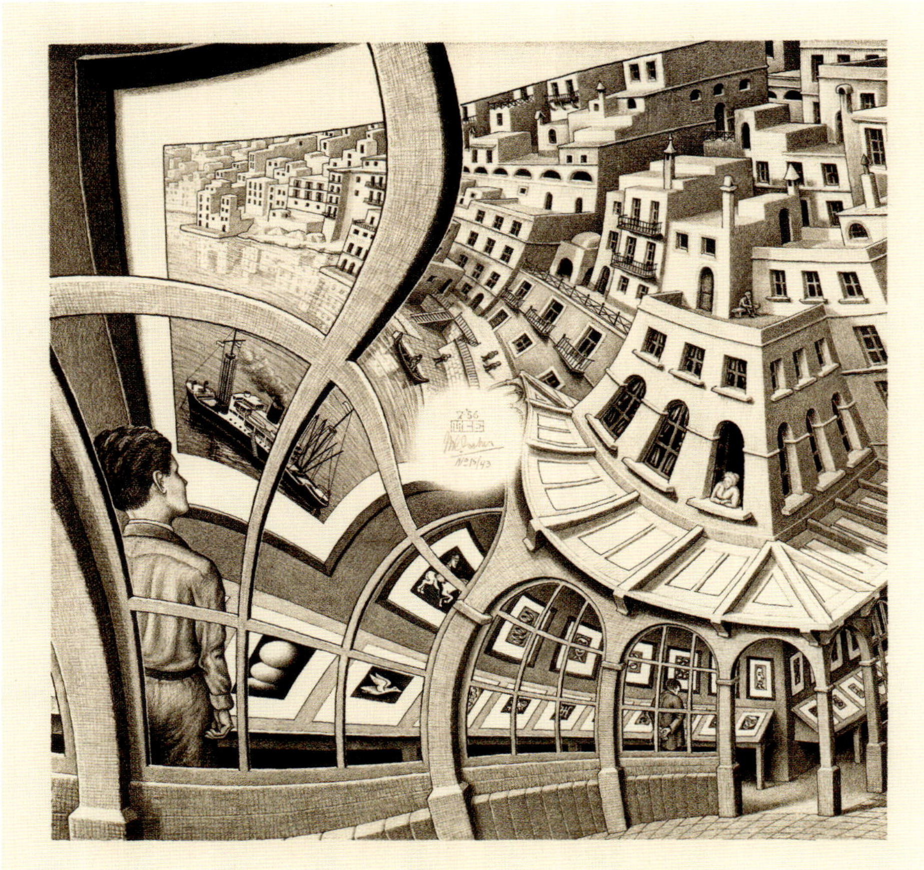

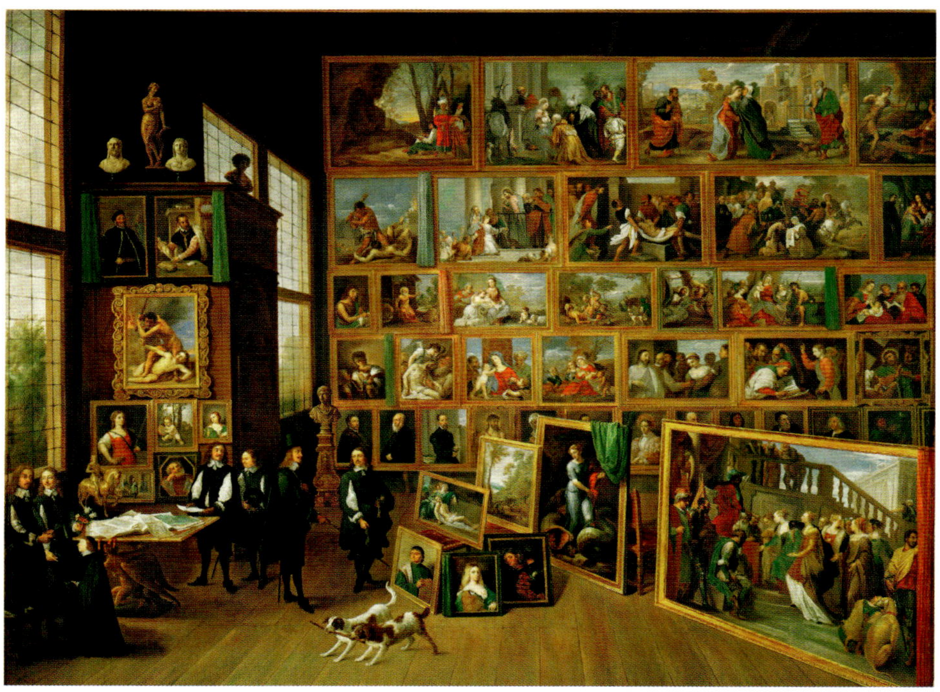

TOP LEFT

9-20. M. C. Escher (Dutch, 1898–1972), *Print Gallery*, 1956. Lithograph, 15 × 15 in. (38 × 38 cm). © The M. C. Escher Company, The Netherlands. All rights reserved.

Visitors to this print gallery enter through the arched doorway in the lower right. A boy has stopped on the far left after passing one of Escher's own prints, *Three Spheres I* (1945), located just to the right of the boy's arm in the lower row. The boy looks up at a picture of boats in a harbor. Continuing one's gaze clockwise, one sees the buildings on the waterfront leave the picture frame to become the building housing the print gallery with an arched doorway in the lower right. Two Dutch mathematicians at the University of Leiden, Bart de Smit and Hendrik W. Lenstra Jr., have suggested that Escher was unable to resolve the central section where the levels meet, which explains why he left a blank spot in the center of the print that he covered over with his monograph (his "fig leaf"). For Smit and Lenstra's reconstruction drawing of Escher's missing center, see their essay "The Mathematical Structure of Escher's *Print Gallery*," *Notices of the American Mathematical Society* 50, no. 4 (Apr. 2003): 446–51.

BOTTOM LEFT

9-21. David Teniers the Younger (Flemish, 1610–1690), *Archduke Leopold Wilhelm in His Gallery in Brussels*, ca. 1650. Oil on canvas, $48\frac{5}{8} \times 64\frac{1}{4}$ in. (123 × 163 cm). Kunsthistorisches Museum, Vienna.

OPPOSITE, TOP LEFT

9-22. M. C. Escher (Dutch, 1898–1972), *Ascending and Descending*, 1960. Lithograph, $11 \times 13\frac{3}{4}$ in. (28 × 35 cm). © 2009 The M. C. Escher Company, The Netherlands. All rights reserved.

OPPOSITE, TOP RIGHT

9-23. M. C. Escher (Dutch, 1898–1972), *Belvedere*, 1958. Lithograph, $11\frac{1}{2} \times 16\frac{1}{2}$ in. (29 × 42 cm). © 2009 The M. C. Escher Company, The Netherlands. All rights reserved.

In the lower right a boy sits holding a three-dimensional crazy crate, and on the ground is a two-dimensional sketch of one. Like Belvedere Castle in Central Park in New York, Escher's *Belvedere* features a lookout porch. Its columns, like the boards in the crazy crate, are locally consistent but globally inconsistent.

OPPOSITE, BOTTOM LEFT

9-24. M. C. Escher (Dutch, 1898–1972), *Waterfall*, 1961. Lithograph, $11\frac{13}{16} \times 15$ in. (30 × 38 cm). © The M. C. Escher Company, The Netherlands. All rights reserved.

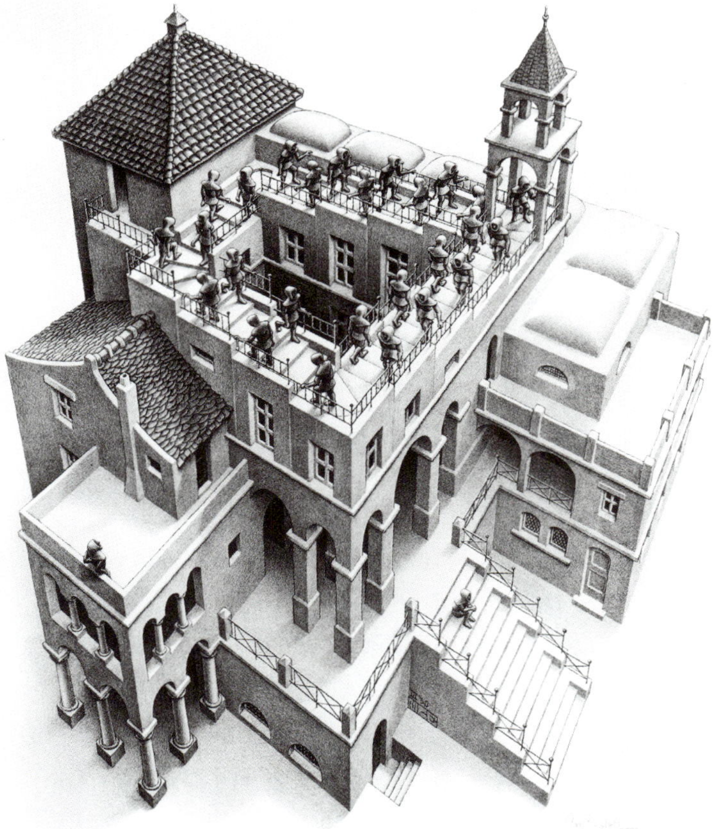

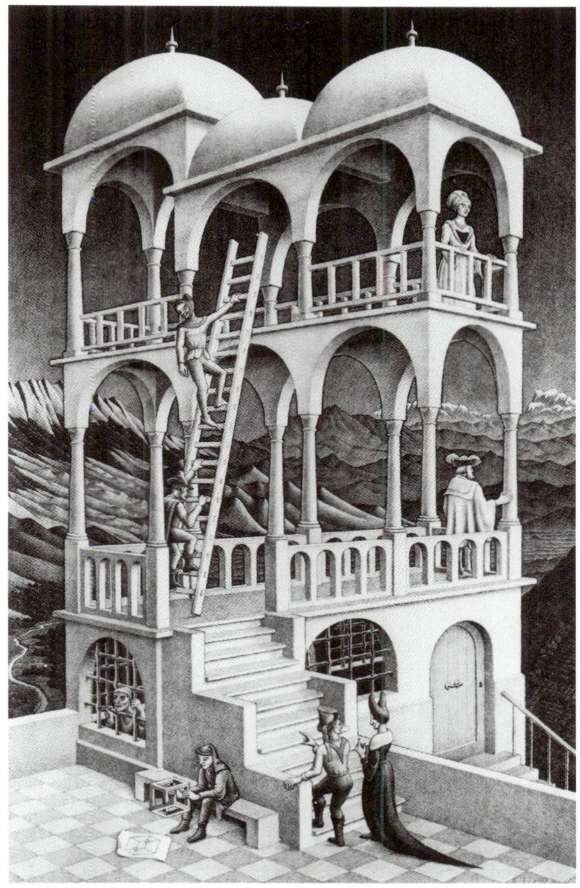

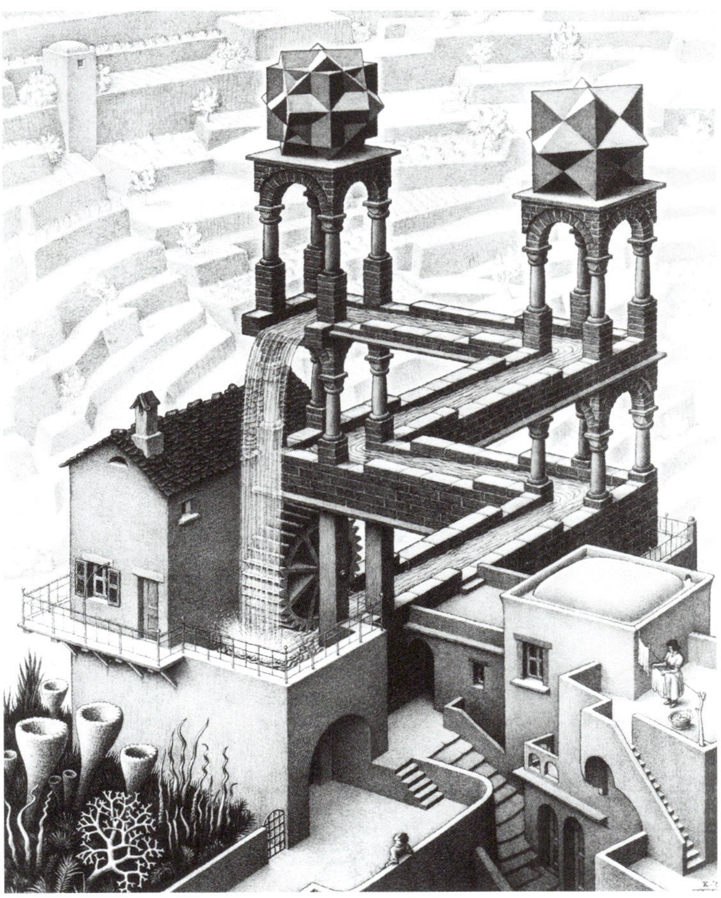

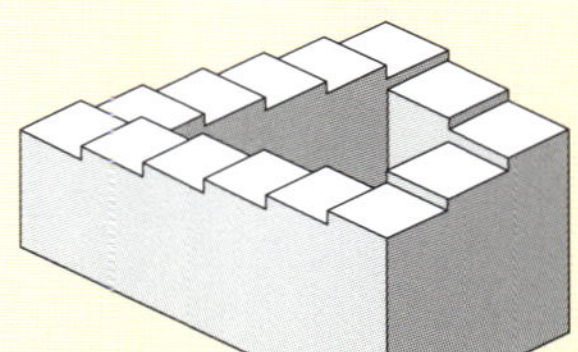

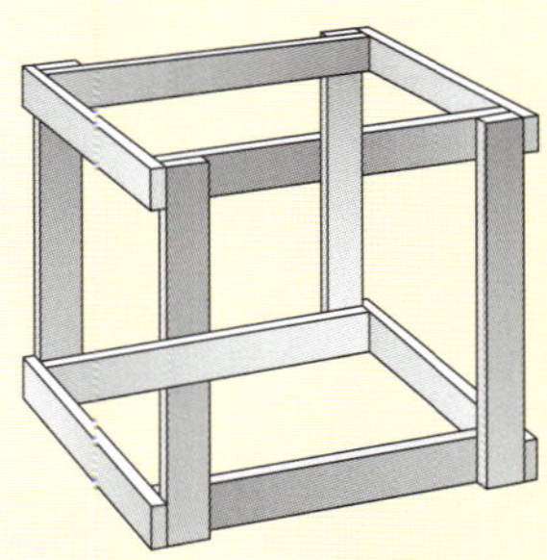

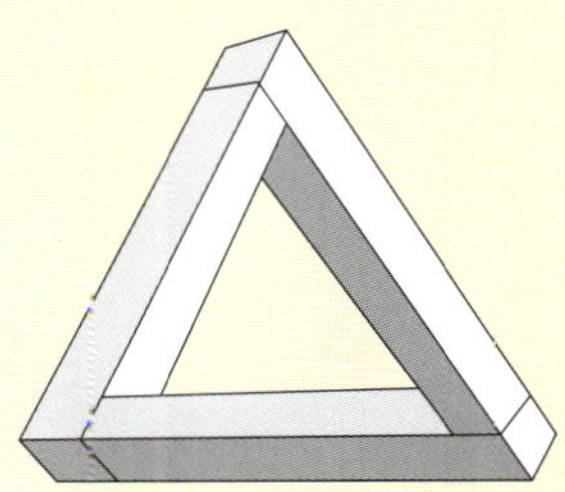

TOP LEFT

9-25. Impossible staircase.

The impossible staircase and the impossible triangle (at bottom) were first published by Roger Penrose and Lionel Sharples Penrose in their essay "Impossible Objects: A Special Type of Visual Illusion," *British Journal of Psychology* 49, no. 1 (1958): 31–33; figs. 1 and 3.

CENTER LEFT

9-26. Crazy crate.

Which board is in front and which is in back? The visual clues are locally consistent but inconsistent from one part to another, making the whole crate self-contradictory.

BOTTOM LEFT

9-27. Impossible triangle.

In addition to impossible objects, Escher and Magritte were also inspired by descriptions of mental operations from the subject's point of view given by the phenomenologist Edmund Husserl. In Husserl's discussion of the possible layers of representation in consciousness, he described a painting of a picture gallery by the seventeenth-century Dutch artist David Teniers the Younger (plate 9-21) and pointed out that when one views this painting, one is standing *in* a picture gallery of an art museum, adding yet another layer of representation to this picture of pictures.[42] The same is true when one views a self-reflective work such as de Chirico's *Great Metaphysical Interior* (plate 9-14) while standing in the Museum of Modern Art in New York. Escher's *Print Gallery* (1956; plate 9-20) offers the reflective viewer even more layers of pictures within pictures.

The evolution of culture is synonymous with the removal of ornamentation from objects of everyday use.

—Adolf Loos, "Ornament and Crime," 1908

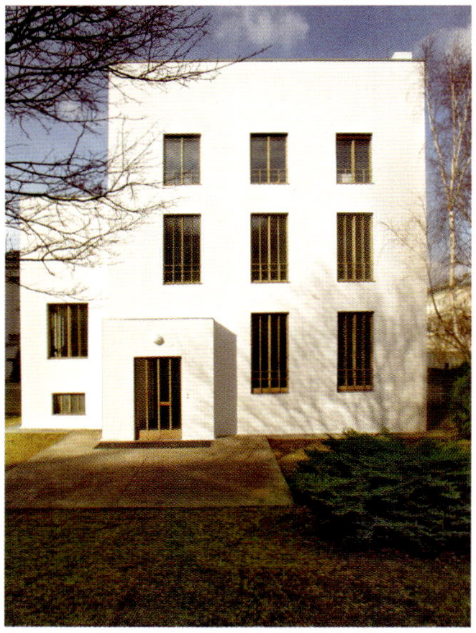

WITTGENSTEIN'S LANGUAGE GAMES

Between 1926 and 1929 the Austrian architect Paul Engelmann worked with Wittgenstein to design a town house in Vienna for his sister (plates 9-28 and 9-29). It was during this building project that Wittgenstein heard Brouwer's 1928 lecture, in response to which he returned to philosophy and developed a theory of language that can be seen as a critique of, and an informal alternative to, his earlier rigorous and mechanistic picture theory of language.[43] In his first theory Wittgenstein had presented spoken language as having a vocabulary and rules of grammar; any particular spoken language is an instance of a universal form of language. Like the edifice of mathematics, with all its diverse compartments resting on one axiomatic foundation, so the tower of language, with thousands of dialects, is, according to this approach, constructed with one universal grammar. In his second theory, Wittgenstein declared that natural language is better understood from an activity-oriented perspective in what he called a "language game." For example, a German builder and his Polish helper are surrounded by piles of blocks and beams; they communicate with a shared vocabulary of "Block" (German for "block") and "belka" (Polish for "beam"). When the builder says "Block," the helper brings him a block; when he says "belka," he gets a beam. These few words and actions constitute the "game" they play.[44] These words and actions might be their

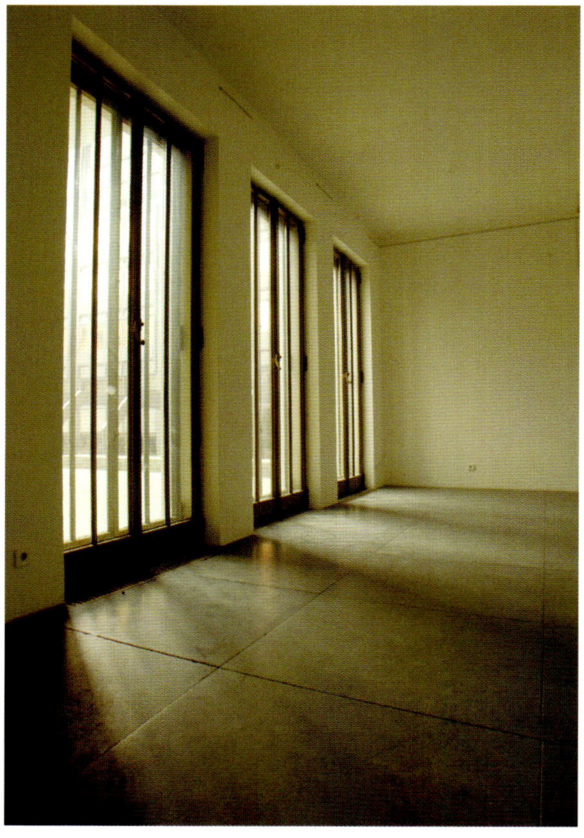

LEFT, TOP AND BOTTOM

9-28. Paul Engelmann (Austrian, 1891–1965), and Ludwig Wittgenstein (Austrian, 1889–1951), Wittgenstein House (exterior and interior), 1926–29, Vienna.

Engelmann was commissioned by Margaret Stonborough-Wittgenstein (plate 9-29) to design a house in the simple, unadorned geometric forms of the International Style practiced in Germany at the Bauhaus and in Austria by Engelmann's teacher, the architect Adolf Loos, who was known for giving pure design a moral overtone by describing decorative ornament as a crime against modern culture (plate 9-30). Wittgenstein, with his sister's encouragement, eventually took over the project from Engelmann, designing many details of the building, including the doors and windows. In the spirit of Loos's austere aesthetic, Wittgenstein forbade his sister to soften the severe interior of her home with rugs, curtains, or house plants because they would defile the purity of his design.

entire common language—a local dialect spoken in the building trade. According to Wittgenstein, all such dialects and languages have overlapping relations that he called "family resemblances," suggesting an analogy with the way members of a family can look alike because of multiple characteristics—build, temperament, hair color, etc.—although no single trait is shared by all. The unity of mathematics (the "games" one plays with numbers) is, according to Wittgenstein's later theory, also not a single rule or axiom shared by arithmetic, geometry, algebra, calculus and so on, but like a family resemblance among siblings and cousins.[45]

Changing linguistic models from an axiomatic system to a language game did not mean that Wittgenstein abandoned mathematics, because games have their own mathematical interest. During the decades that Wittgenstein reflected on language games, the mathematicians John von Neumann and Oskar Morgenstern invented game theory as a field of mathematics that studies the strategic interactions of economic players (*The Theory of Games and Economic Behavior*, 1944), and game theory has since been expanded to apply to other fields.[46] It is true, however, that Wittgenstein's shift away from logic (analysis and deduction) to ordinary language (language games) did result in a more informal approach to verbal communication. Unlike the rigorous axioms of his youth, in his later years Wittgenstein proposed a more informal approach; as he wrote, "There is not a single philosophical method, though there are indeed methods, different therapies, as it were."[47]

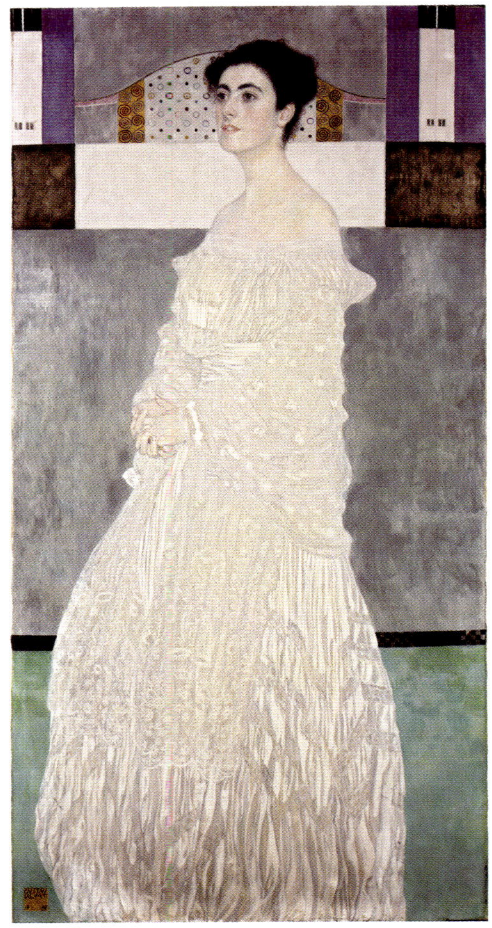

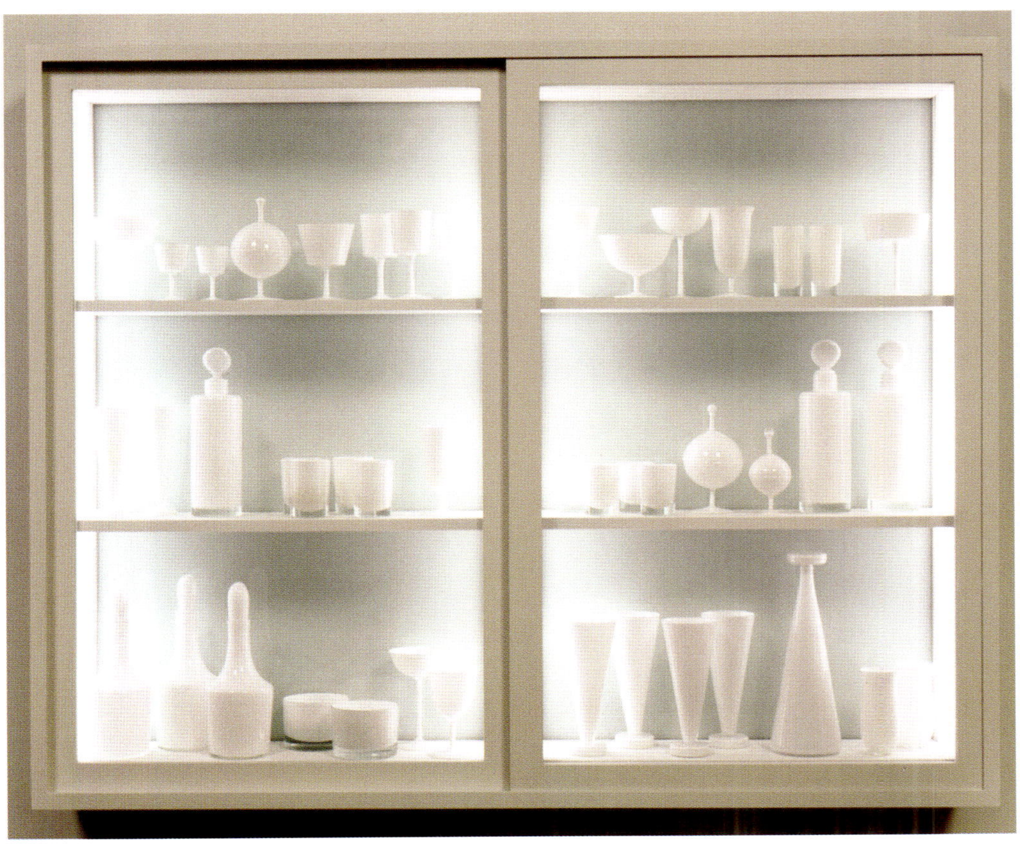

ABOVE

9-29. Gustav Klimt (Austrian, 1862–1918), *Margaret Stonborough-Wittgenstein*, 1905. Oil on canvas, 70⅞ × 35⅝ in. (180 × 90.5 cm). Neue Pinakothek, Munich.

Klimt painted this portrait of Margaret Wittgenstein on the occasion of her 1905 marriage to the American Jerome Stonborough. The couple had two sons and divorced in 1923, after which Margaret commissioned the town house.

LEFT

9-30. Josiah McElheny (American, b. 1966), *Adolf Loos's Ornament and Crime*, 2002. Blown glass, wood, glass, and electric lighting, case dimensions: 49 × 60 × 10½ in. (124.4 × 152.4 × 26.6 cm). © Josiah McElheny. Donald Young Gallery, Chicago, and Andrea Rosen Gallery, New York.

ANGLO-AMERICAN ART ABOUT LANGUAGE

After World War II, Wittgenstein's writings were widely available in translation and read in the English-speaking art world. In the 1950s the American artist Jasper Johns read Wittgenstein and began making art about symbolic systems.[48] After taking a few classes at a university and an art school, Johns essentially educated himself, displaying unusual rigor and clarity of thought. In 1954 he made his first painting of an American flag, and he went on to paint other commonplace, impersonal symbols such as numerals. In all these works the viewer senses a dislocation between his or her response to the familiar symbol and the veiled manner in which it is presented. For example, Johns executed *Numbers in Color* (plate 9-32) in colored wax over layers of newspaper that are almost invisible under the numerals.[49] Like de Chirico and Magritte, Johns focused on everyday symbols and objects, but whereas these earlier European artists had read German philosophy, Johns was inspired by the Anglo-American tradition based on Bertrand Russell's logical atomism. Johns read the early writings of Wittgenstein, which describe the picture theory of meaning: atomic propositions "picture" atomic facts because sentences "mirror" the logical structure of what they describe. In line with this idea, Johns's *Numbers* is isomorphic to numbers in the world. Johns's evocation of an emotional response through hand-craftsmanship and subtle layering of indecipherable images and texts resonates with Wittgenstein's assertion that certain truths, lying beyond language, cannot be captured by the picture theory. Finally, Johns's interest in philosophical questions about how symbolic systems work relates to Wittgenstein's later philosophy of language games, in which different languages (maps, codes, pictures, words) have family resemblances.

In *According to What?* (1963; see plate 9-31), Johns created a catalogue of different

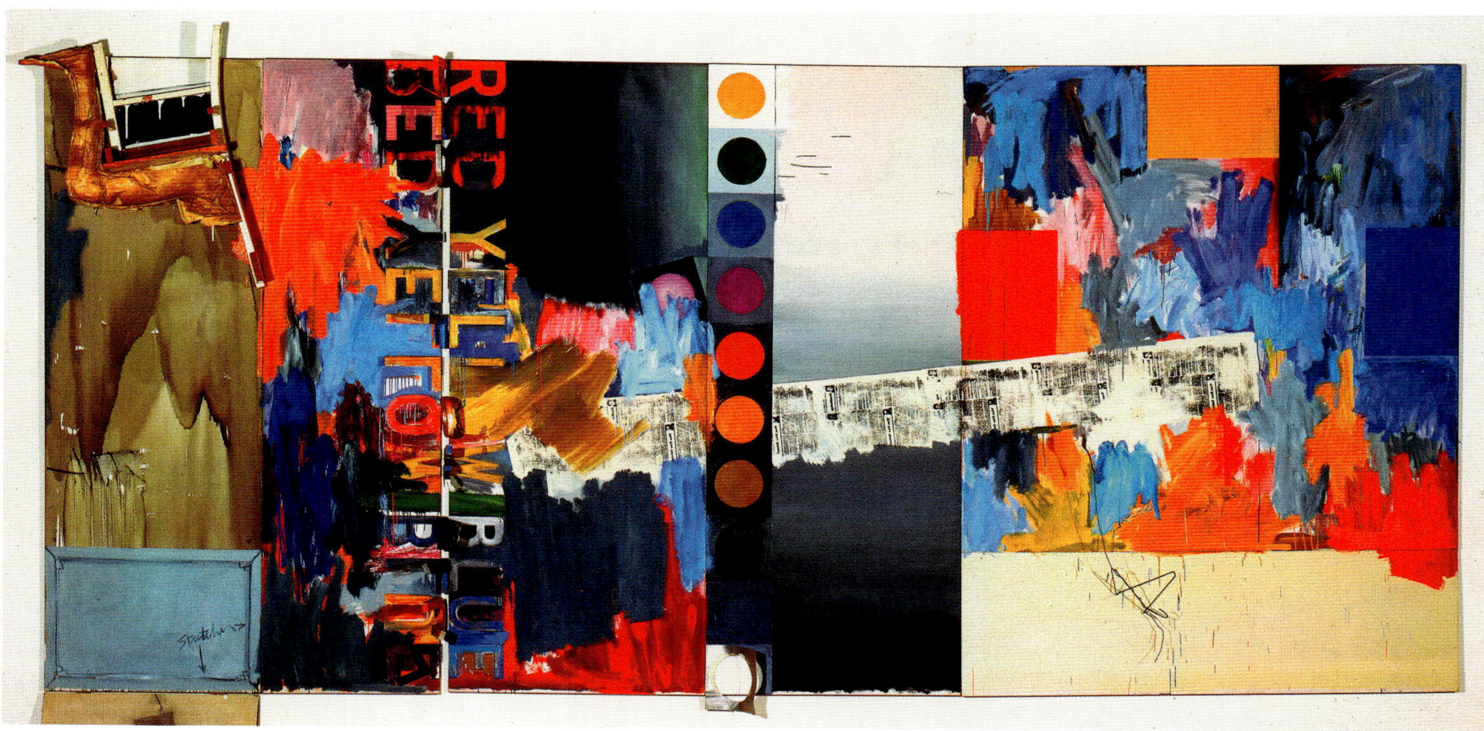

BELOW

9-31. Jasper Johns (American, b. 1930), *According to What*, 1964. Oil on canvas with objects, 88 × 192 in. (223.5 × 487.6 cm). Los Angeles County Museum of Art. © Jasper Johns/VAGA, New York.

This work shares themes with *Tu m'* (plate 11-20 in chapter 11) by Marcel Duchamp, to whom John dedicated this painting.

OPPOSITE

9-32. Jasper Johns (American, b. 1930), *Numbers in Color*, 1958–59. Encaustic and newspaper on canvas, 66½ × 49½ in. (168.91 × 125.73 cm). Albright-Knox Art Gallery, Buffalo, New York, gift of Seymour H. Knox Jr., 1959. © 2015 Jasper Johns/VAGA, New York.

Johns's pattern results from counting cyclically from zero to nine on an 11 × 11 grid, beginning with one blank in the upper left. Each number descends diagonally, and the complete 0–9 series runs horizontally across the top and bottom, and vertically down the left and right edges.

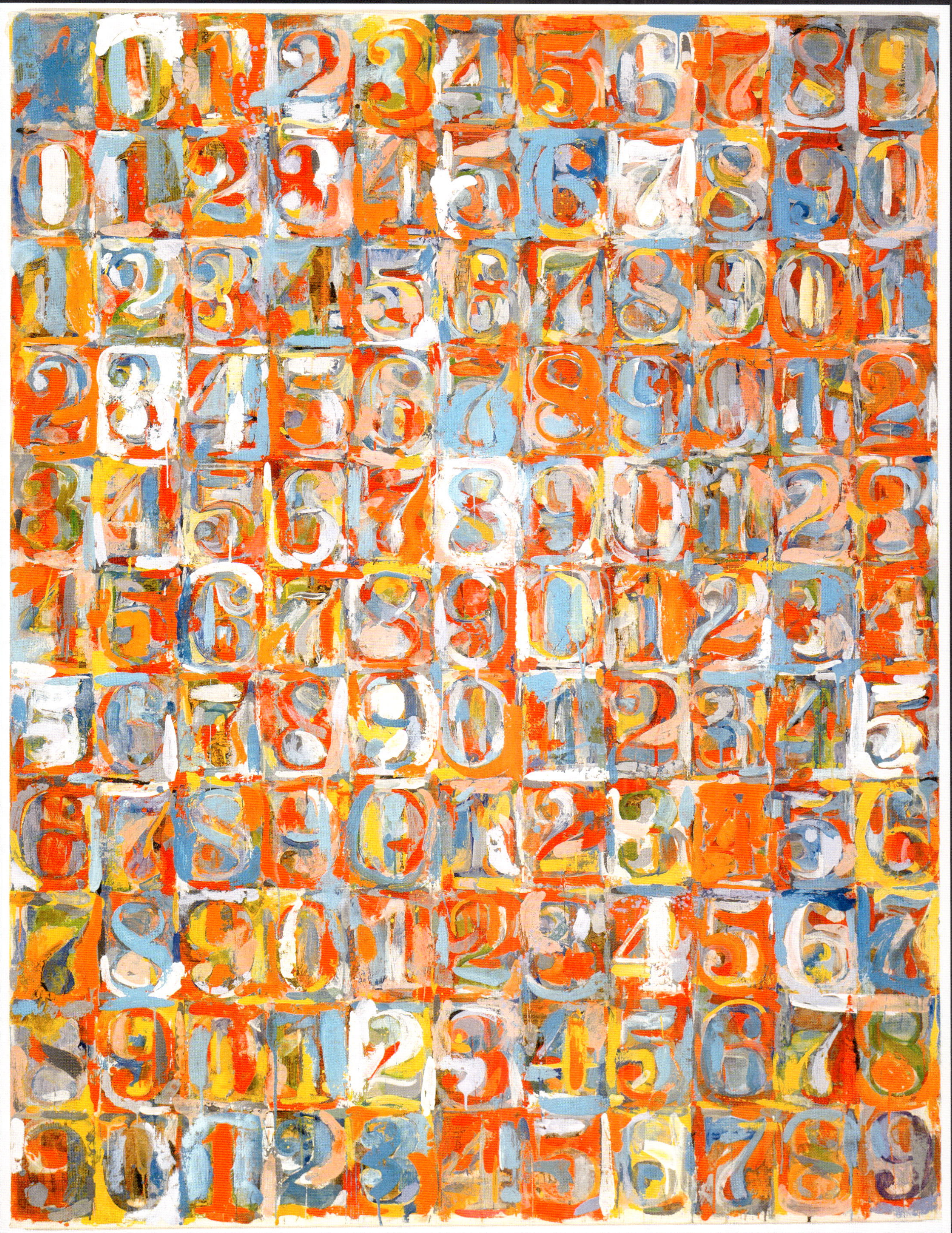

9-33. Ian Burn (Australian, 1939–93), *No Object Implies the Existence of Any Other*, 1967. Synthetic polymer paint on wood, mirror, lettering, 25⅜ × 25⅜ × 1⅛ in. (64.5 × 64.5 × 3.0 cm). Art Gallery of New South Wales, Sydney, Rudy Komon Memorial Fund 1990. © Used with permission.

In this parody of British philosophy, Ian Burn printed on glass a quotation in which the eighteenth-century Empiricist David Hume argued that one cannot connect one's idea of an object by cause and effect with one's idea of any another object (*Treatise on Human Understanding*, 1739, chap. 19). Burn then photographed the backward text in a mirror, putting Hume's empiricist dictum through multiple reflections and reversals.

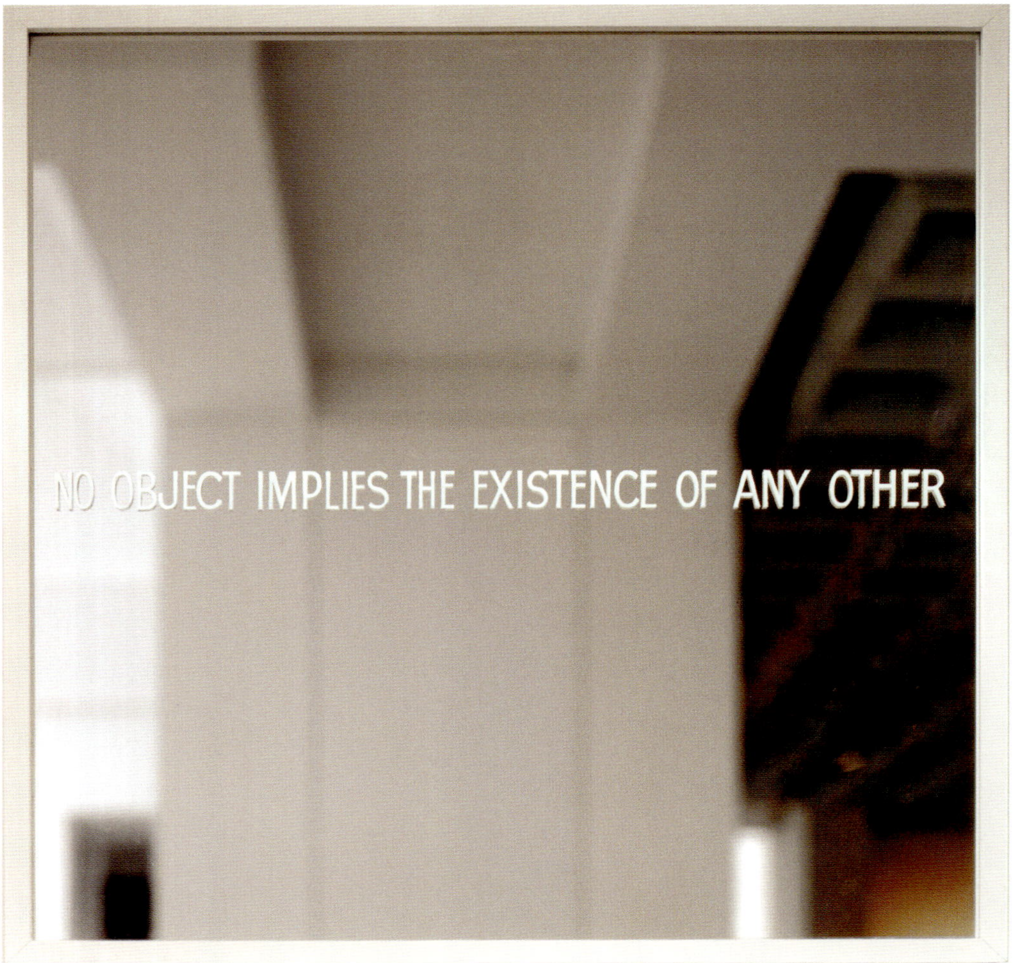

methods for visual and verbal representation: drawing, painting, printing, sculpting, and casting. Each medium creates its own symbolic system—its own language game—and to understand the different parts of the painting, one has to know what system it is recorded in. In other words, one needs to answer the question: According to what?

Johns's interest in the power and limits of signs and symbols was relayed to a small group of British artists, one of whom, Terry Atkinson, read the catalogue of a 1964 exhibition of Johns's work at Whitechapel Gallery, London. In the catalogue essay, Johns stated that he read Wittgenstein and declared, "I'm believing painting to be a language," a remark that inspired Atkinson to study Anglo-American analytic philosophy.[50] In 1968 Atkinson joined with Michael Baldwin, David Bainbridge, and Harold Hurrell to form the collective Art and Language. The group modeled itself on analytic philosophers who analyze ideas in an attempt to clarify the meaning of concepts such as "one" or "goodness." Gathering to have a dialogue in a tradition as old as Plato's Academy, the members of Art and Language met to discuss questions such as: What is art? Their discussions were not *about* art, they *were* art, according to the spokesman for the group, British art historian Charles Harrison, who described their approach: "Conversation, discussion, and conceptualization became their primary practice, as art."[51] As amateur analytic philosophers, these artists had no training in this precise and demanding field, and thus, predictably, their essays are jargon infested and lack the clarity that is the hallmark of analytic philosophy. Looking back

If Conceptual Art is pure thought, then the true Conceptual artist is someone who will remain forever unknown to us.

—Terence Parsons, remark to the author, 1980

on the development of the group, Harrison reflected: "The confusion, verbosity, prolixity, and difficulty of the writings of A & L's pre-formative and early years reflect the endeavor to articulate confusion from within confused circumstances."[52] The mouthpiece for the group was *Art-Language*, a periodical that the artists designed to look not like a glossy, hip art magazine but like *Mind*, a grey, somber journal of philosophy, which has been published by Oxford University Press since 1876.

The Art and Language group established links with other artists, including the British artist Mel Ramsden, the Australian Ian Burn (plate 9-33), and the American Joseph Kosuth. Ramsden and Burn settled in New York, where they led a discussion group, the Society for Theoretical Art and Analysis.[53] Kosuth's reading included A. J. Ayers's *Language, Truth and Logic* (1936), a lucid popularization of Anglo-American analytic philosophy.[54] Ayer wrote that assertions in logic and mathematics are valid by virtue of their form, calling them "tautologies" (for example, $A = A$ and $A = \neg\neg A$ are both tautologies).[55] Kosuth published an essay in which he explained how his art is based on analytic philosophy ("Art after Philosophy," 1969). Borrowing Ayer's terminology, Kosuth defined a work of art as a tautology: "What art has in common with logic and mathematics is that it is a tautology; i.e., the 'art idea' (or 'work') and art are the same and can be appreciated as art without going outside the context of art for varifition [*sic*; Ayer's term is 'verification']."[56] In a series of works entitled *Art as Idea as Idea*, Kosuth produced his art (his "tautologies") in the form of a photostat of the definition of a word in a dictionary, which he enlarged, mounted on board and hung on the wall. For example, one piece presents the dictionary definition of "theory" as "a mental viewing; contemplation" (*Titled [Art as Idea as Idea] The Word "Theory,"* 1967; Yale University Art Gallery). But, alas, Kosuth has concocted the kind of muddle that Frege and Russell developed modern logic to prevent. Kosuth's artwork is not an idea but the *embodiment* of an idea in a physical object (a photostat hanging on the wall). Furthermore, his artwork is not a tautology. Assuming that what Kosuth had in mind was that his artwork, entitled *The Word "Theory,"* asserts that "theory" is defined as "a mental viewing; contemplation," then this artwork/assertion is *true*, but it is not a tautology (it is not *necessarily* true by virtue of its form).

Art and Language called their work "Conceptual Art," adopting a term the American artist Sol LeWitt had coined to name art made with algorithms (see chapter 11). In 1972 the group exhibited their discussion-as-art at the international exhibition *Documenta 5* in Kassel as *Index*, a set of filing cabinets filled with transcripts of their amateur philosophy.[57]

CHINESE ART ABOUT LANGUAGE

Anglo-American art about language was widely discussed in Asia. For example, in Japan in 1974 the periodical *Geijutsu Kurabu* (Art club) published a translation of Kosuth's essay "Art after Philosophy," followed by laudatory comments by a leading Japanese critic, Minemura Toshiaki.[58] Some Asian artists, including the Japanese experimental musician Yoko Ono and the Korean video artist Nam June Paik, moved to New York, where they joined the international artists network Fluxus and pursued Anglo-American themes.

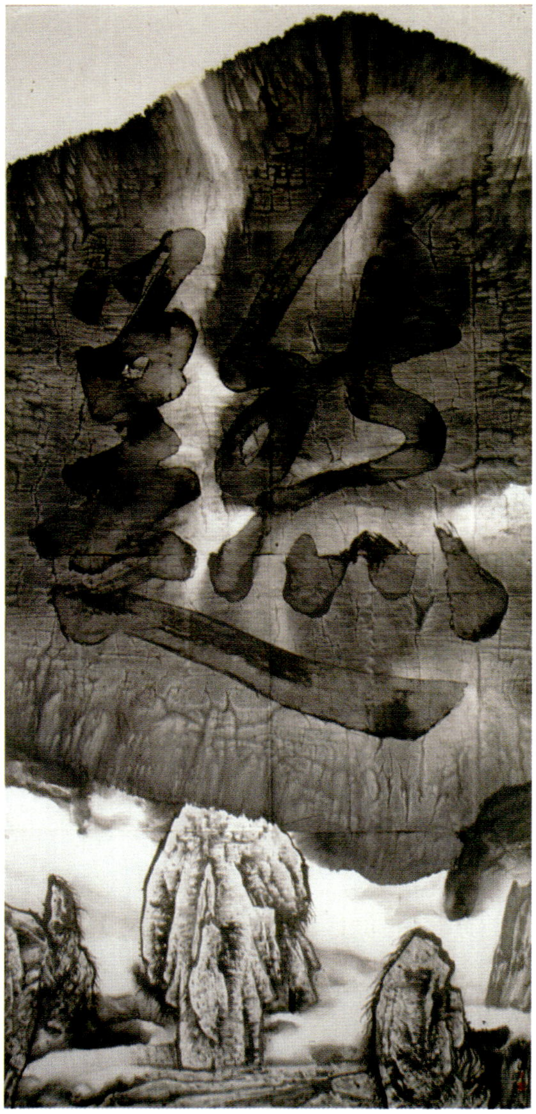

9-34. Gu Wenda (Chinese, b. 1955), *Mythos of Lost Dynasties*, C Series #5, *Transcendence*, 1996–97. Splash ink calligraphic painting, 132 × 59 in. (335.2 × 149.8 cm). © Gu Wenda.

Other Asian artists maintained close ties with their homeland and rethought topics in the philosophy of language in terms of their own languages.

In ancient China during the Warring States period (475–221 BC), scholars developed a thriving study of the semantics of their language. Chinese, the oldest living language, is written in about 3000 pictographs, each of which began as a picture that became abstracted over the centuries. But in the third century BC, the first emperor of China, the megalomaniac Qin Shi Huang, declared that only he could determine the meaning of words, thus ending the study of the philosophy of language in imperial China (221 BC–AD 1911). After the fall of the Qing dynasty and the formation of the Republic of China in 1912, intellectuals curious about Western science and mathematics invited Bertrand Russell to give a series of lectures in China, where he was well received.[59] It seemed that Western symbolic logic might take root there, but after the founding of the Chinese Communist Party in 1921 and the ascendancy of Mao Zedong, doors closed to the West, and the only Western philosophy that was allowed was Mao's version of Marxism.

The contemporary Chinese artists Gu Wenda and Xu Bing matured during Mao's Cultural Revolution (1966–76), the aim of which was abolishment of the so-called Old Four—old customs, old culture, old habits, old ideas—causing widespread destruction of ancient buildings, porcelain, silks, and manuscripts as well as the closing of universities. The decade is generally viewed today, in China and the West, as disastrous for China's cultural heritage and its modern development. As children, Gu and Xu were taught to write both in the traditional Chinese script of the literati (the scholar-bureaucrats of imperial China) and in the simplified script that was introduced in the Communist era under Mao. After Mao's death in 1976, travel restrictions eased, and Gu and Xu visited the West. Today both focus on language as the subject matter of their art.

Gu took up the study of Anglo-American philosophy of language in the 1980s, reading especially Russell and Wittgenstein,[60] and he adopted the latter's theme of the limits of language, which he symbolized with invented (and hence unreadable) characters of his own design in various historical styles of Chinese pictographs. Gu painted *Transcendence* (plate 9-34) in the style of Chinese scholarly painting as homage to the literati tradition. In the mountains above floats one of Gu's pseudo-characters, symbolizing the limits of language—Wittgenstein's "the mystical" and Gu's "myths of lost dynasties," as he called the series.

In the 1990s Gu began *The United Nations Project*, a series of installations in countries around the world in which he evokes the local languages, written in unreadable pseudo-characters, in veils woven from human hair that he has collected from barber-shops in the region. Gu named his series after the United Nations, an international organization that is devoted to unity, which is impossible in politics but, according to Gu, possible in art, using hair as the medium because it is a biological material possessed by all peoples. Differing slightly from place to place, the hair that forms the veils symbolizes the unity of mankind, "beyond knowledge, beyond nationality, beyond the borders of culture and race."[61]

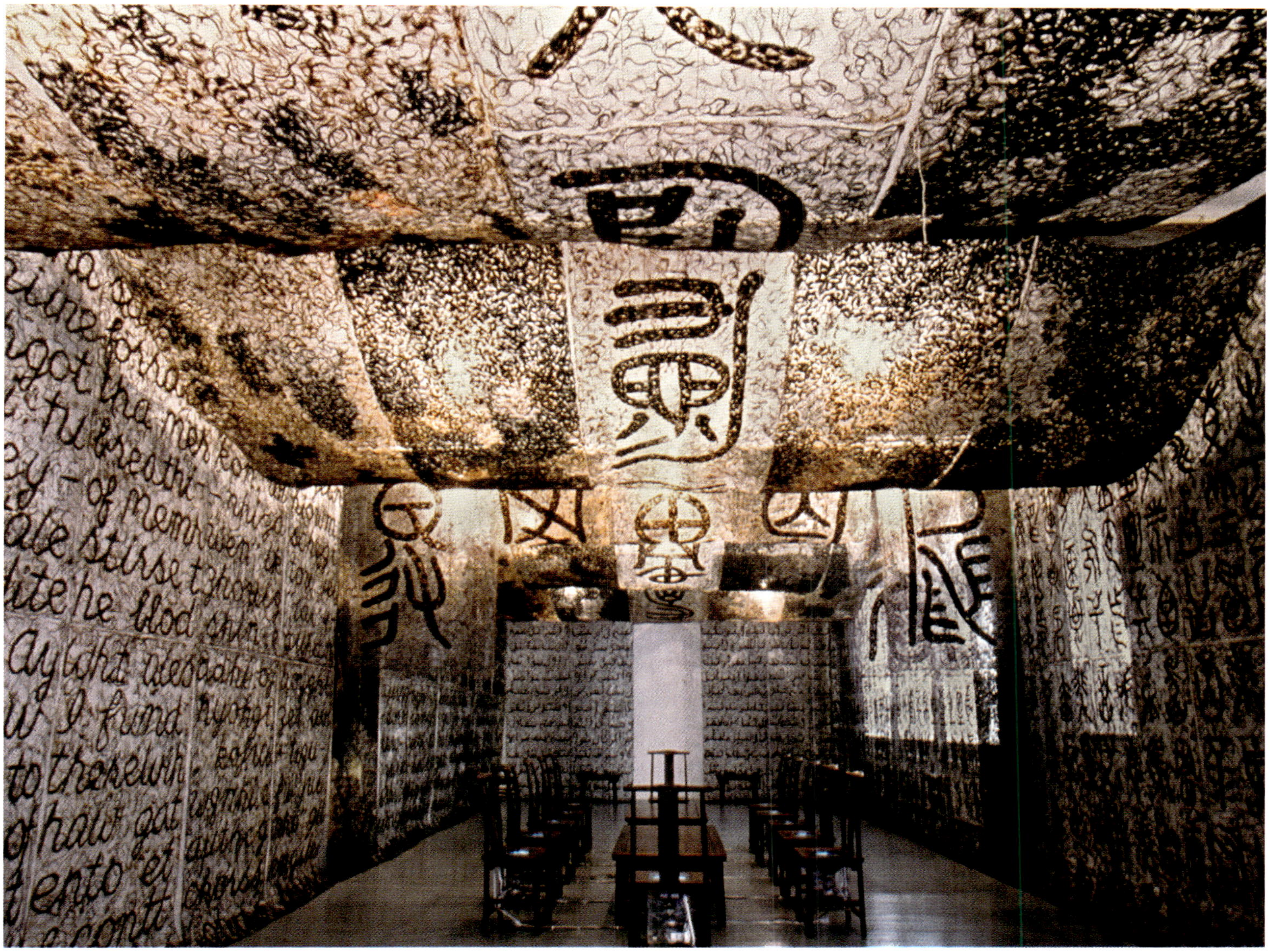

9-35. Gu Wenda (Chinese, b. 1955), *China Monument: Temple of Heaven*, from the series *The United Nations*, 1998. Installation commissioned by the Asia Society, New York, and installed in PS1 Contemporary Art Center, composed of pseudo-Chinese, Hindi, Arabic, English text made of human hair curtains collected from all over the world, 12 Ming-style chairs, 2 Ming-style tables. © Gu Wenda.

For example, in *China Monument: Temple of Heaven* (1991; see plate 9-35) the text woven into the veils on the walls and ceiling is written in four pseudo-languages, using characters from Chinese, Hindi, Arabic, and English. In the center of the room, Ming dynasty furniture is arranged for a tea ceremony, a traditional place where people come together for congenial conversation. Gu stressed the meditative atmosphere of his *China Monument*, which is named for the fifteenth-century Taoist *Temple of Heaven* in Beijing, by insetting in the seat of each chair a TV monitor with pictures of billowing clouds, so that each guest converses while metaphorically "floating on a cloud" above terrestrial concerns.[62]

Xu expressed the evolution of the Chinese language in his installation *The Living Word* (2001; see plates 9-36 and 9-37). On the gallery floor Xu arranged simplified characters from the Mao era that spell out the dictionary definition of 鸟 (Chinese for "bird"). Then these modern abstract characters return by a kind of reverse evolution, moving through the spectrum from violet characters to pictures of red birds, which are transformed (metaphorically) into living birds, who fly away, escaping the confines of language.[63]

RIGHT

9-36. Definition of 鸟 lifting off the floor in Xu Bing's *The Living Word*.

OPPOSITE

9-37. Xu Bing (Chinese, b. 1955), *The Living Word*, 2001. Mixed-media installation; carved acrylic characters, paint. Commissioned by the Arthur M. Sackler Gallery, Smithsonian, Washington, DC, 2001; shown here in its installation in the atrium of the Pierpont Morgan Gallery, New York, 2011. Xu Bing Studio and The Pierpont Morgan Library, New York.

Xu carved the Chinese characters for the definition of 鸟 (bird) in acrylic and arranged them on the floor. The artist has described this installation: "The *niao* [鸟] characters then break away from the confines of the literal definition and take flight through the installation space. As they rise into the air, the characters 'devolve' from the simplified system to standardized Chinese text and finally to the ancient Chinese pictograph based upon a bird's actual appearance. At the uppermost point in the installation, a flock of these ancient characters, in form both bird and word, soar high into the rafters toward the open windows of the space, as though attempting to break free of the words with which humans attempt to categorize and define them." Xu Bing's caption to this piece in the exhibition catalogue *Reinventing Tradition in a New World: The Arts of Gu Wenda, Wang Mansheng, Xu Bing, and Zhang Hongtu*, ed. Wang Ying and Yan Sun (Gettysburg, PA: Gettysburg College, 2004), 84.

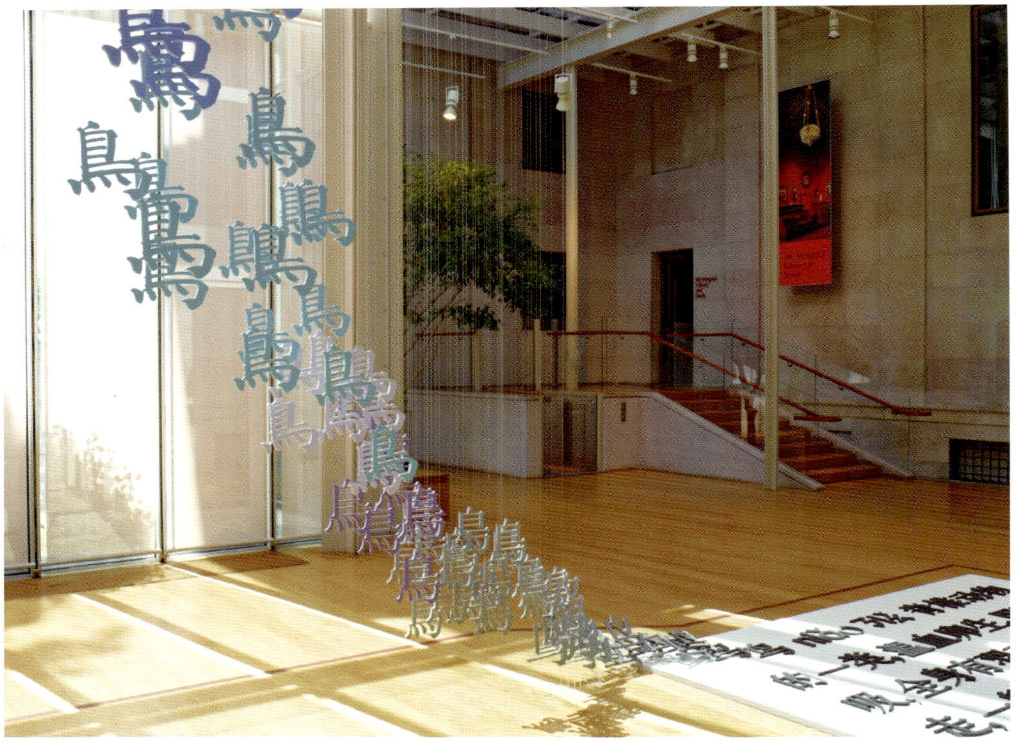

After 1931 mathematicians no longer imagined their science as a towering edifice resting on one axiomatic foundation, but more like an old, sprawling city with arithmetic and geometry residing in the Old Town at its ancient center. Over the centuries new neighborhoods have sprung up for algebra, calculus, and projective geometry, some separated by turf wars and others blended through intermarriage. Non-Euclidean geometry and set theory were initially viewed as outsiders but were eventually granted full citizenship. The latest arrival is a band of robotic refugees from engineering. Will these mindless machines be integrated into the daily life of mathematics?

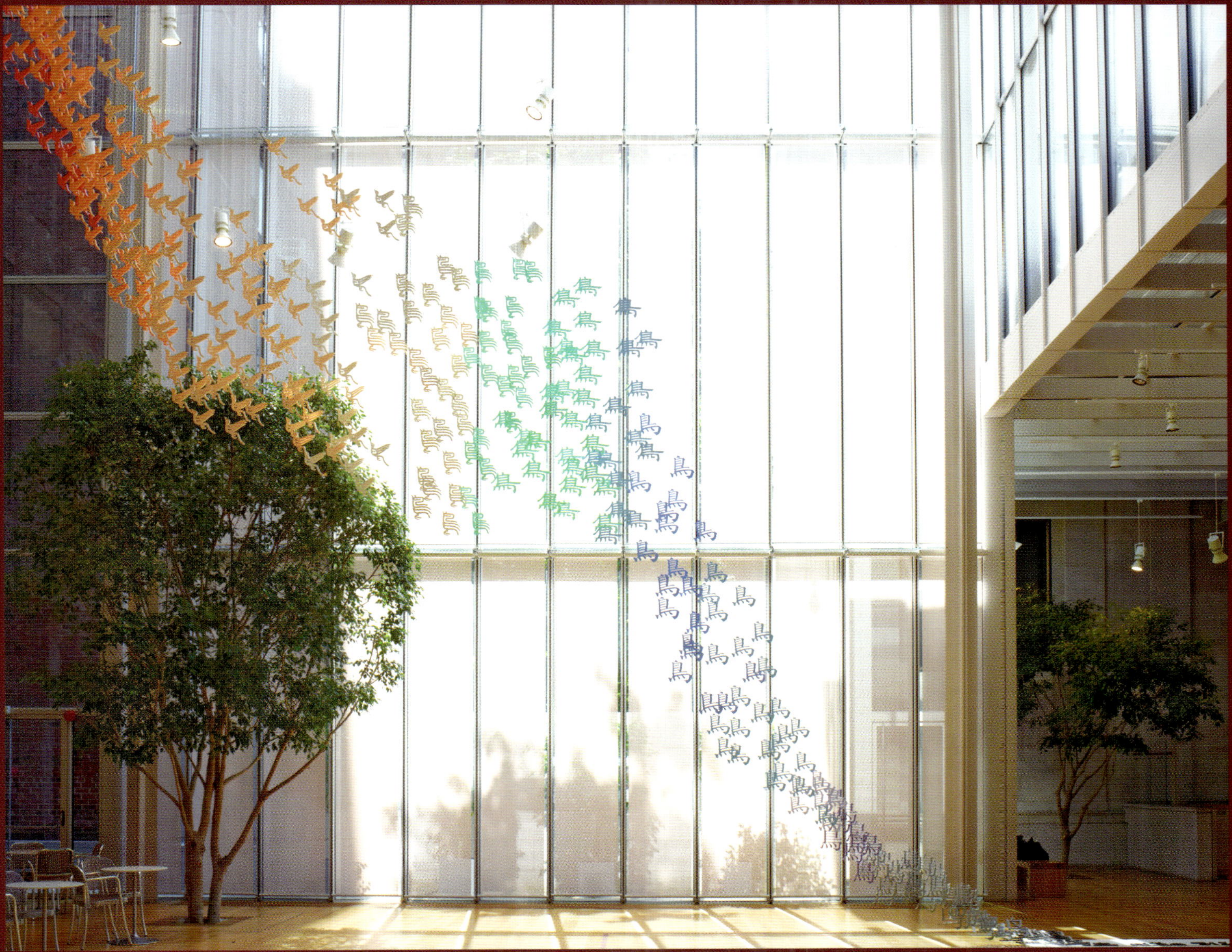

10
Computation

Only a machine can appreciate a sonnet written by another machine.
—Alan Turing, ca. 1950

In 1905 Einstein did a thought experiment: What would a light-wave look like to an observer travelling with it at the same speed? Einstein never intended to actually *make* such an observation—it's not humanly possible. In 1931 Gödel did another thought experiment: imagine a mechanical procedure to translate any formula written in mathematical symbols into a number (its "Gödel number"). Like Einstein, Gödel never intended to actually *do* arithmetic using these (immense) numbers—it's beyond the capacity of the human mind. Nevertheless, Gödel's use of such numbers to achieve his incompleteness proof led mathematicians to investigate computation further, and during World War II the British began building primitive computing machines to break Enigma, the German military code. After the war, engineers would develop those simple wartime machines into the computer industry, giving mathematicians, scientists, and artists a powerful new tool with a universal language (plate 10-1).

In 1945 the world entered the nuclear age, and many people felt a sense of urgency about finding a way to live together. The dark days of the Cold War and widespread dread of an imminent nuclear catastrophe led to renewed efforts towards improving East-West relations. Some called for a *philosophia perennis* (Latin for "eternal philosophy") to capture the truths about reality that are, they declared, found in all traditional philosophies and theologies, and they read the Japanese scholar D. T. Suzuki and the American theologian Thomas Merton. Others read psychologists Sigmund Freud and Carl Jung, who explained the brotherhood of mankind in terms of animal drives possessed by all *Homo sapiens*. Searching for eternal truths in the spirit of internationalism, intellectuals and artists reached across cultural boundaries and borrowed traditional symbols, some of which were mathematical.

10-1. Roman Verostko (American, b. 1929), detail of *The Manchester Illuminated Universal Turing Machine*, 1998 (see plate 10-2). Courtesy of the artist.

Roman Verostko created this pen-plotted drawing to celebrate the fiftieth anniversary of the first machine that embodied the logic of a Turing machine, which was built by engineers at Manchester University, where Turing worked from 1948 to 1951. After printing this sheet of computer code, the artist then "illuminated" it (burnished it with gold leaf), as monks did to their manuscripts in medieval scriptoria. The artist has stated: "These drawings, reminiscent of medieval manuscript illuminations, celebrate Alan Turing's concept of a universal Turing machine by presenting it as a valuable, precious text of our own time."

In 1928 Hilbert challenged mathematicians to solve a problem known as the *Entscheidungsproblem* (German for "decision problem"): design a mechanical (step-by-step, algorithmic) procedure that can be applied to any mathematical statement to determine whether it's true or false. The term "mechanical" had been used for decades within Hilbert's formalist camp. Then in the 1930s the young British mathematician Alan Turing focused on the concept "mechanical" and did a thought experiment: Imagine a machine with an infinitely extendable tape; the tape is divided into squares, each of which is either blank or filled with a symbol from a finite vocabulary. A tape head in the machine is capable of performing operations on the tape, such as moving to a square and changing the symbol in that square. A control mechanism in this tape head stores instructions from a finite list of rules. Once the machine has stopped its operations, the machine reads the tape and stores the state of the computation (the current operation plus the configuration of all the symbols currently on the tape; see sidebar below).

A Turing Machine

Alan Turing invented the Turing machine in 1936 to give a mechanical procedure for all possible computations. Turing modeled the structure of his (imaginary) Turing machine on a (real) teleprinter, an electromagnetic typewriter invented in the late nineteenth century in order to be able to send and receive data, such as a telegram, without the need of operators trained in Morse code. Used commonly by businesses in the 1930s, a teleprinter could also create punched tape for data storage. Turing modified the structure of a teleprinter, making the Turing machine consist of a tape that is infinitely extendable (unlike the finite tape in a teleprinter); Turing's tape is divided into squares, each of which is either blank or filled with a symbol from a finite vocabulary, which, in this example, is {0, 1}. There is a tape head that can move left or right or stay fixed at any stage, as the instructions dictate. The tape head can change the symbol, following a finite list of rules that are stored in the tape head. For example, this Turing machine has three groups of rules, A, B, and C; when the machine is carrying out one of the rules, it is in states A, B, or C. Given the symbol in the square that the tape head is above, the rule tells (1) what to change the symbol to (0 or 1); (2) where to move to (one square either left [L] or right [R]) or not to move at all (N); and (3) which state to proceed to (A, B, or C).

	State A			State B			State C		
Tape symbol	Write	Move tape head	Next state	Write	Move tape head	Next state	Write	Move tape head	Next state
0	1	R	B	0	R	A	1	N	B
1	0	L	C	1	L	C	1	L	Halt

In this example, the tape head starts at 0 in state A and takes the action dictated by the rules of state A, resulting in the moves as shown.

Once the machine halts, the current state plus the marks on the paper are its output.

The Turing machine has the basic components of a computer: an input/output device (the tape and its reader), software (the list of rules in the head), and memory (the storage of the state and the record on the paper).

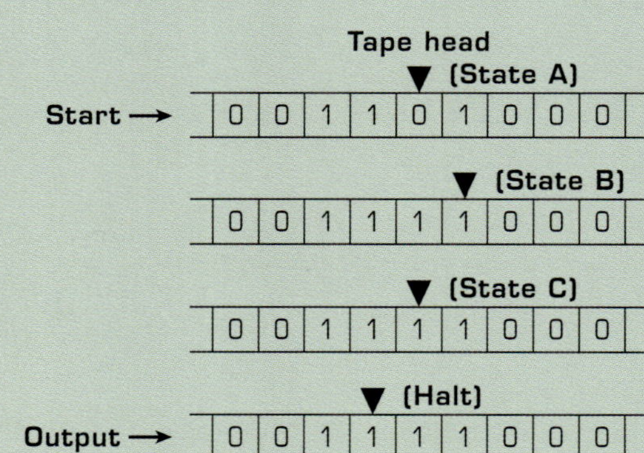

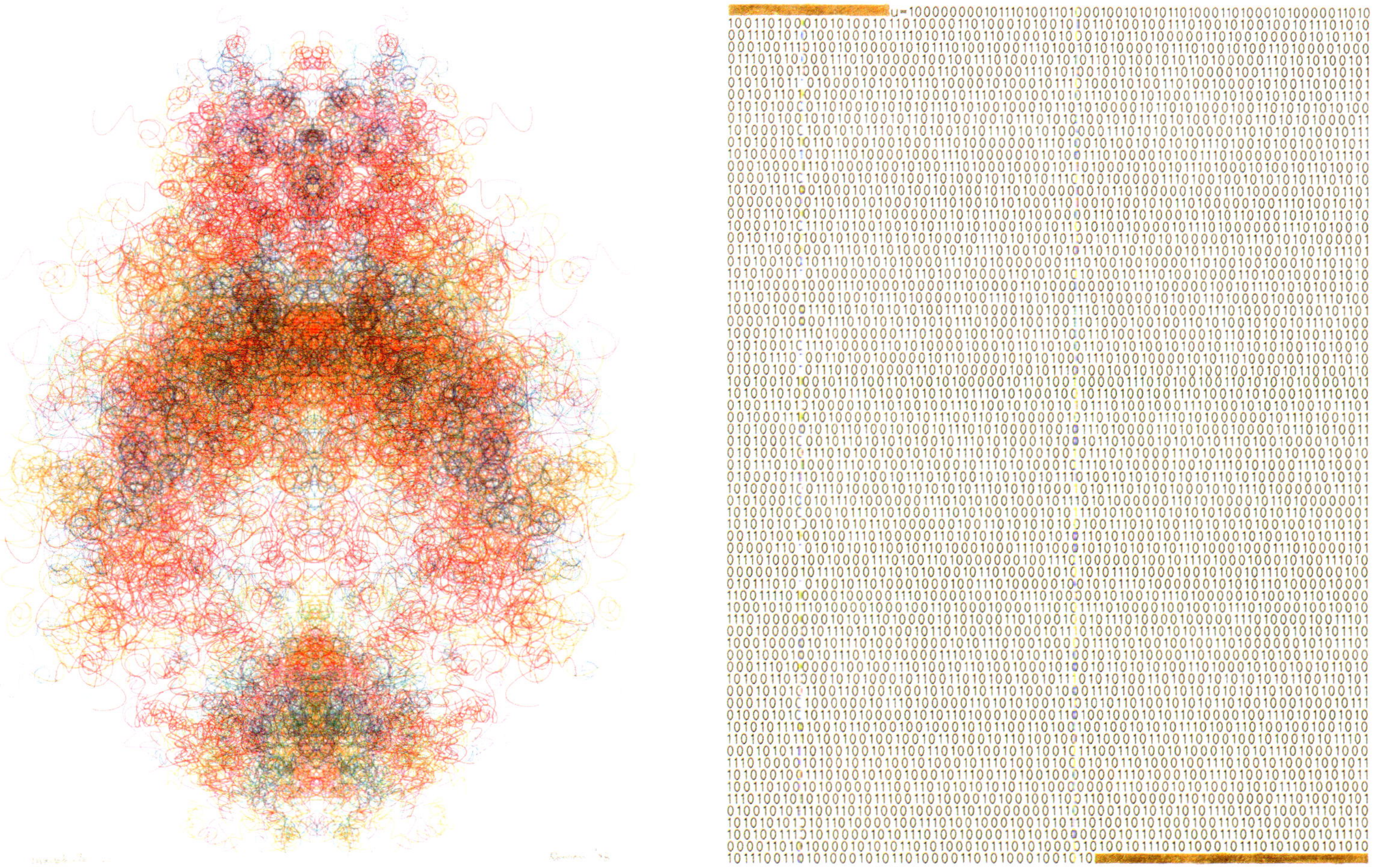

10-2. Roman Verostko (American, b. 1929), *The Manchester Illuminated Universal Turing Machine*, 1998. Pen-plotted drawing with gold leaf, no. 23 of a suite, 30 × 22 in. (76.2 × 55.8 cm). Courtesy of the artist.

To create this suite of drawings, Verostko wrote software code that would generate a family of forms. In this example (no. 23), the artist first plotted the code (on the right) and then the forms associated with the code (on the left).

With his thought experiment Turing had designed the essentials of a modern computer, with an input/output device (the tape and its reader), software (the list of rules in the tape head), and memory (the storage of the state of the computation). The Turing machine became the basis of all subsequent computers, with electrical (on-off) circuits designed using the two-valued algebra crafted by George Boole in the nineteenth century (plates 10-1 and 10-2).

Using this concept of a Turing machine, Turing was able to show that Hilbert's *Entscheidungsproblem*—a decision procedure for all mathematical statements—is impossible. Adopting methods from Gödel's 1931 incompleteness proof, Turing used a mapping of machines to numbers the way Gödel had mapped statements to numbers. Turing's key insight was that self-reference is possible when a machine (such as a Turing machine) operates on symbols because the machine is itself described by symbols (its software and memory). Indeed, a Turing machine can operate on its own description. To prove the impossibility of Hilbert's decision problem, Turing said: Suppose there is a Turing machine (think of it as the "original" machine) which, when applied to the number describing any other Turing machine, will tell whether or not that Turing machine (the "other" one) will halt (reach a decision) when fed any given number (any positive integer, some of which encode a mathematical statement). If there were such an "original" machine, it would provide a mechanical decision procedure for mathematical statements and solve

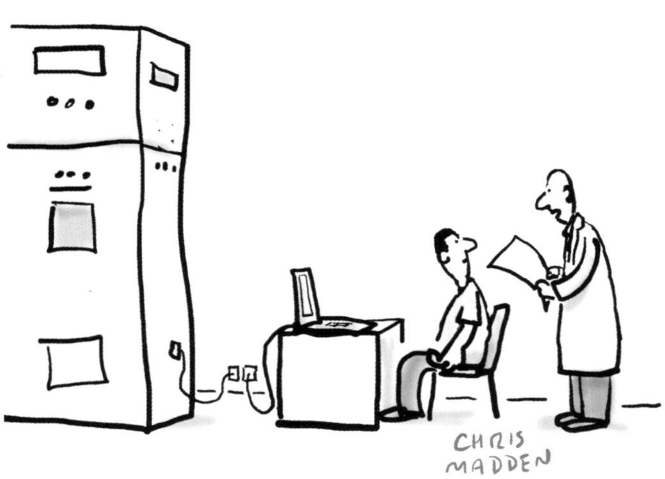

"The good news, Dave, is that the computer's passed the Turing test. The bad news is that you've failed."

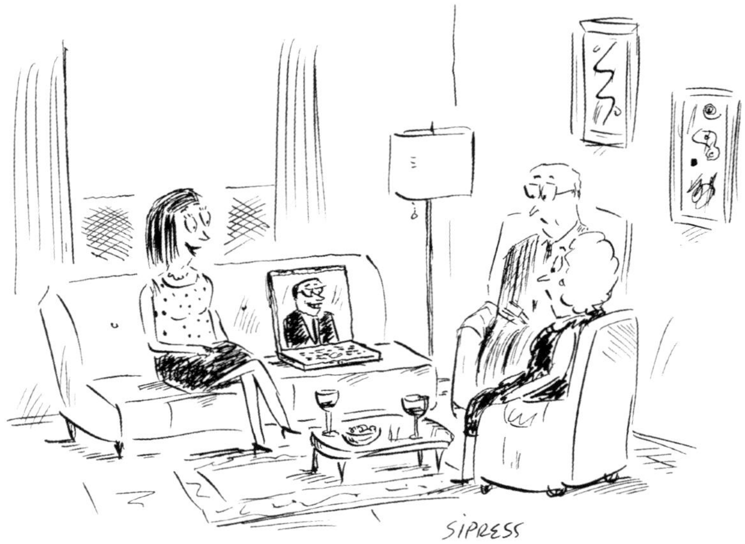

"We met online."

10-3. The Turing test.

The Turing test determines whether or not a given machine, such as a huge supercomputer (above left) or a robotic boyfriend (above right), is exhibiting intelligent behavior. According to Turing, the test is conducted by a human judge who engages in conversation in a natural language (such as English) with another human and a machine. The three are separated and cannot see each other. The exchanges are carried out in text only (the words are written but not spoken). If the judge cannot tell which participant is human and which is a machine, then the machine has passed the Turing test. Alan Turing, "Computing Machinery and Intelligence," *Mind* 59, no. 236 (1950): 433–60.

(left) Chris Madden, ca. 2000. Used with permission.

(right) David Sipress, *New Yorker*, Aug. 8, 2004. Used with permission.

Hilbert's *Entscheidungsproblem.* But Turing showed that if such a machine (the solution to the decision problem) were to exist, he could use this original to construct another Turing machine with self-contradictory properties. But a self-contradictory mechanical procedure is not logically possible, so the original machine *cannot* exist—there is no decision procedure—and the halting (decision) problem is undecidable ("On Computable Numbers, with an Application to the *Entscheidungsproblem*," 1936).

Turing did this work while a graduate student in America at Princeton University, completing a doctorate in mathematics in the spring of 1938, after which he returned home to England. When hostilities between England and Germany erupted a few months later, the British government invited Turing to join a team of mathematicians who were working to decrypt the German military code. In September 1938, Turing moved to Bletchley Park, an estate in southeast England that housed Britain's top-secret facility, and became the chief scientific figure on the team, entrusted with the responsibility of reading U-boat communication. The Germans encoded their messages using an electromechanical device—an Enigma machine—that scrambled the letters in a message. The Bletchley Park team broke the code by designing a special electromechanical device to unscramble the Enigma messages, supplying the Allies with crucial intelligence until the end of the war (see sidebar on opposite page and plate 10-4).

After 1945 Turing was recruited by British industry to develop an electronic computing machine. In his 1950 paper "Computing Machinery and Intelligence," he pioneered artificial intelligence—the mathematics of thought—and introduced a test for whether or not machines can think (the Turing test; see plate 10-3).[1]

Meanwhile John von Neumann added a significant feature to the Turing machine: the ability to store its own operating instructions (its software) so it would not have to be reprogrammed every time the machine was turned on. A stored program—software—that could be updated without resetting wiring or switches in the machine, together with a programming language, were the essential steps leading to effective general-purpose

Solving Enigma

The German Third Reich used Enigma machines to encode radio communications in an alphabet substitution code. The operator typed the letters in the message—the plaintext—onto an Enigma keyboard, and the electrical impulses from the keystrokes followed paths that were scrambled by a set of three rotors, each with twenty-six positions. Each letter from the plaintext entered the right rotor, moved left via electrical connections, and then was reflected back, and the output—the ciphertext—appeared on an Enigma lampboard. The operator then sent the message in ciphertext to a recipient who knew the settings for the three rotors and thus could decode it by a reverse process.

Code-breakers needed first to understand the structure of the machine and then to be able to determine the rotor settings for a given message. In 1932, on the eve of Hitler becoming German chancellor in January 1933, a young Polish mathematician, Marian Rejewski, got his hands on an instruction manual for an Enigma machine that had been sold to the French military by a German traitor who went by the gloomy alias Asche (German for "ashes"). Using group theory, Rejewski reconstructed the logical structure (the wiring and rotor settings) of the Enigma machine, working with fellow mathematicians Henryk Zygalski and Jerzy Różycki. As the war drew near, in July 1939 the three Polish code-breakers met their French and British counterparts in Warsaw and taught them their techniques. Two months later, during the German invasion of Poland—called the *Blitzkrieg* (German for "lightning war") because of the aerial bombardment—Rejewski, Zygalski, and Różycki were evacuated to France and Britain, where they continued to work with the Allied forces throughout the war.

The difficulty in determining the rotor settings for a given day was that the possible settings were equal to the factorial of the number of positions. The factorial of a number is the product of the integers from 1 to that number. So if the rotors each had five positions, they would have had a manageable 120 possible combinations ($1 \times 2 \times 3 \times 4 \times 5 = 120$). Factorials grow at an exponential rate; the factorial of 10 is $1 \times 2 \times 3 \times 4 \times 5 \times 6 \times 7 \times 8 \times 9 \times 10 = 3{,}628{,}800$. The Enigma machine rotors typically each had 26 positions, one for each letter of the alphabet, so it had 26 factorial possible settings, which totals 403 trillion trillion. Enigma machines were refined during the war with additional rotors and plug boards; calculations showed that a message had potentially trillions of possible encodings.

When Alan Turing reported to Bletchley Park the day after Britain declared war on Germany on September 4, 1939, he focused on a weakness in the Enigma code that the Poles had discovered: a letter could not be encrypted to itself. Turing exploited this property in designing a mechanical procedure to search for cribs—short pieces of plaintext thought to be in a ciphertext. For example, Turing found that almost every day some military commander would send the message *Keine besondere Ereignisse* ("No special occurrences," in other words, "Nothing to report"). Since all the commanders used the same rotor settings on any one day, this gave Turing a crib to search for, such as the pattern of the *e* (_*__*_*_ ___*_**_*_______*). He knew that if he found this pattern in the ciphertext, then he had probably found *e* in the plaintext. The actual search was done by machines that could scan for cribs and then rapidly test for thousands of possible rotor settings.

The Germans remained confident that Enigma was, for all practical purposes, unbreakable. But Turing, together with the team at Bletchley Park, and the Allies were able to decrypt crucial messages about battle plans and troop movements, averaging 30,000–80,000 interceptions per month. By the end of the war in 1945, the Allies could decode almost all German Enigma traffic within a day or two. After the German surrender, Admiral Dönitz (head of the U-boats) and Hermann Göring (head of the Luftwaffe) were informed that their code had been broken.

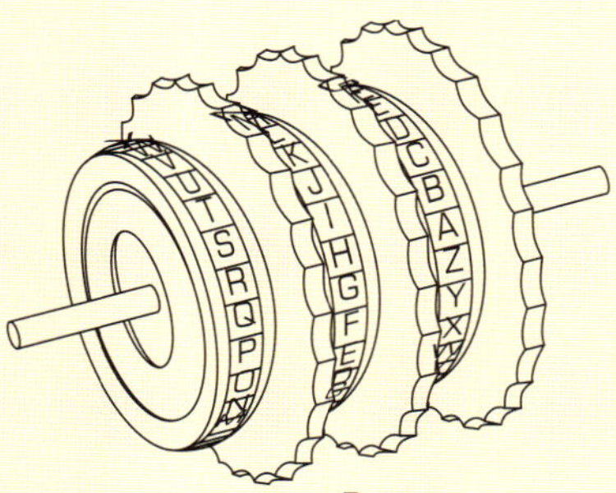

Rotors

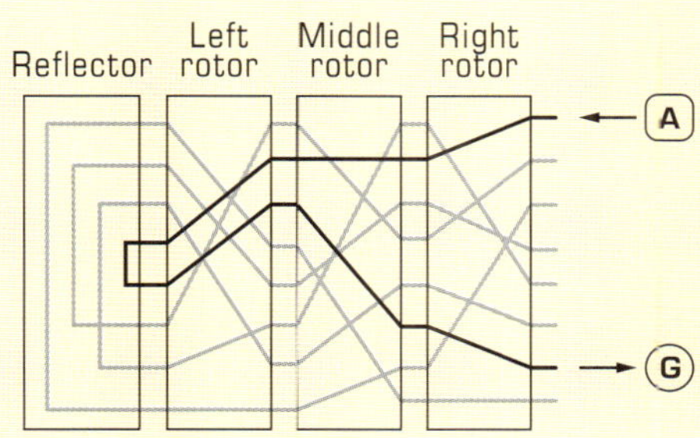

Light lines show alternate paths

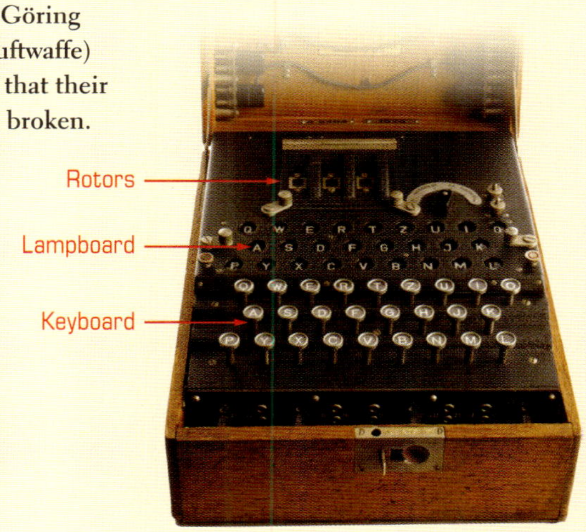

computers. Apart from Turing and von Neumann, none of the early innovators in computation—neither Leibniz, Boole, Frege, Cantor, Hilbert, nor Gödel—foresaw the application of their ideas to the construction of an all-purpose digital computer.[2]

Since its development by Turing, von Neumann, and others, computers have continued to depend on a steadily evolving set of mathematical advances, such as algorithms for the storage and retrieval of data, image processing, statistics, and so on. Computer

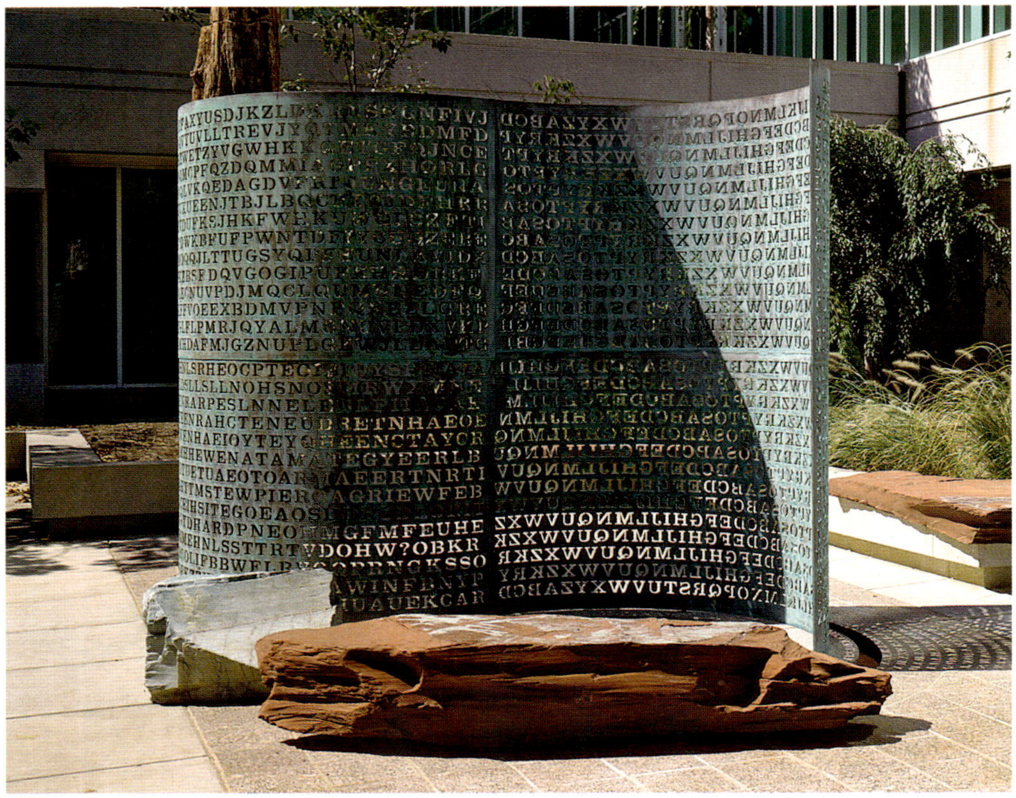

RIGHT

10-4. Jim Sanborn (American, b. 1945), *Kryptos*, 1989. Granite, quartz, lodestone, copper, encoded text, water, 12 × 20 × 10 ft. (3.6 × 6 × 3 m). Central Intelligence Agency; Langley, Virginia. Courtesy of the artist.

In 1990 Jim Sanborn won a commission for a sculpture for the new CIA headquarters in Virginia. *Kryptos* is composed of a scroll-like sheet of copper inscribed with four messages, which, in the spirit of the agency, the artist put into an alphabet code with the help of a cryptographer from the CIA. The sculpture is located in a central court where CIA code-breakers take their coffee breaks. These restless minds have succeeded in decoding the first three sections.

1. "Between subtle shading and the absence of light lies the nuance of illusion."

2. [This phrase gives the co-ordinates and location of the CIA headquarters.]

3. [A paraphrase of a quotation from the British archeologist Howard Carter, describing his opening the tomb of King Tutankhamun] "Slowly, desperately slowly, the remains of past debris that encumbered the lower part of the doorway were removed. With trembling hands I made a tiny breach in the upper left-hand corner, and then widening the hole a little, I inserted the candle and peered in."

4. The last ninety-seven letters remain undeciphered.

programmers have devised increasingly more sophisticated software, while engineers worked to make computers smaller, cheaper, and more reliable, using vacuum tubes (1940s), transistors (1960s), integrated circuits, and microprocessors (1970s), opening the door to personal computers for the office, laboratory, and studio. Today computers consist of hardware (which is fixed forever), an operating system (which is variable but is usually not changed except for updates), software (programs that can be switched on or off and that are regularly changed), and input and output devices.

AXIOMATIC APPROACHES TO MUSIC

One of the earliest aesthetic uses of the computer was in the composition and production of music. In antiquity the Pythagoreans had discovered the numerical basis of musical harmony and concluded that it was a reflection of cosmic harmony. The association of music with mathematics survived in the West into the seventeenth century, when music was still studied as part of astronomy and mathematics; polyphonic music (such as a Bach fugue) was understood as reflecting the divine order of the cosmos as discerned by Kepler and Newton. But in the eighteenth century, music separated from mathematics and composers of the First Viennese School—Mozart, Haydn, and Beethoven—began to view music as a fine art. Melodic music (such as a Mozart piano concerto or a Haydn string quartet) was seen as an expression of the composer's spirit, and in the nineteenth century the tonal music of the Romantic era (such as a Beethoven symphony) was understood as an outpouring of human emotion.

In the first decades of the twentieth century, the Viennese composer Arnold Schoenberg invented a new method of composition—atonal, twelve-tone music—that has both mathematical and expressionistic aspects. The composer first arranges twelve notes of the chromatic scale in any order, using each note only once, to form a twelve-note "basic set" of the composition (in place of the melody in tonal music). Schoenberg specified three rules to apply to the basic set to obtain its "mirror forms"[3]—the pattern of twelve notes can be played in retrograde (backward), in inversion (played with up-down reversed), or both (plate 10-5). Schoenberg then added a final rule: the original pattern and its three transformations can commence on any of the twelve notes of the basic set, resulting in a total of forty-eight possible twelve-note sequences. Using this palette of forty-eight patterns, the composer then arranges them in any order to complete the composition.

Schoenberg's twelve-tone composition method has striking similarities to a formal axiomatic system: the twelve-note basic set is like an axiom, the rules for manipulating the basic set are analogous to rules of inference, and the finished composition is like a theorem. Although music critics have often found it useful to employ mathematical tools to analyze twelve-tone compositions,[4] there is no evidence that Schoenberg, who received only a rudimentary education in mathematics, drew on contemporary sources in mathematics. In music he was largely self-taught, having studied only counterpoint with the Viennese composer Alexander von Zemlinsky. Schoenberg thought of musical "space" in terms of elementary school geometry: "The two-or-more-dimensional space in which musical ideas are presented is a unit. . . . The elements of a musical idea are partly incorporated in the horizontal plane as successive sounds, and partly in the vertical plane as simultaneous sounds."[5]

Schoenberg harkened back to the master mathematician of music, J. S. Bach, who began a fugue by stating a theme (a series of single notes). The difference between Schoenberg and Bach is that Bach's theme was tonal (it was written in a key), and Schoenberg's was atonal (his twelve-note basic set was not in a key; it made no reference to a scale). Bach then completed his fugue by applying various rules to the basic theme, some of which Schoenberg adopted.[6] Bach altered the line by retrograde, inversion, and retrograde inversion, and he staggered the line in pitch, having the same pattern start on

There is no greater perfection in music than Bach! Not Beethoven or Hayden, not even Mozart who was closest to it, ever attained such perfection.

—Arnold Schoenberg, *Preliminary Exercises in Counterpoint*, ca. 1940

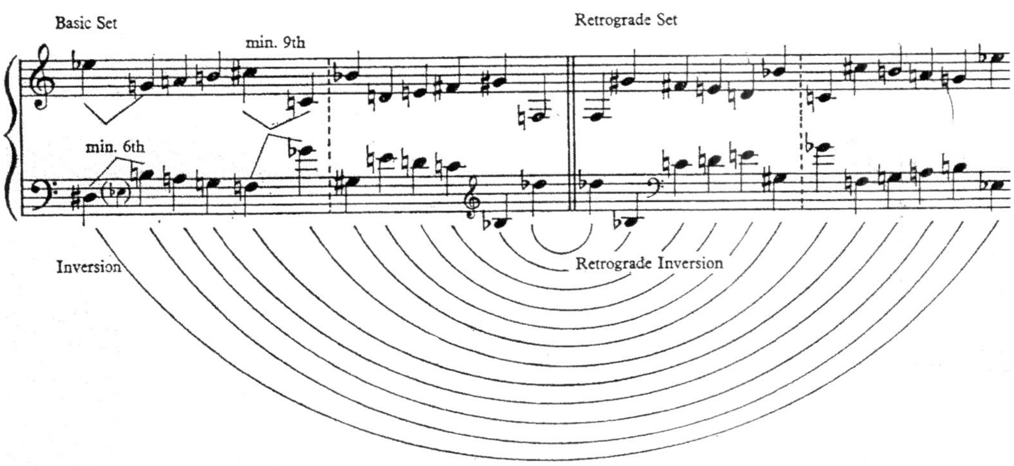

10-5. A basic set, its retrograde set, its inversion, and its retrograde inversion, in Arnold Schoenberg, "Composition with Twelve Tones (I)" (1941), in *Style and Idea: Selected Writings of Arnold Schoenberg*, ed. Leonard Stein, trans. Leo Black (London: Faber and Faber, 1941/ rpt. 1975), 214–25; the diagram is on 225.

different notes. Bach also staggered the line in time, repeating the pattern after regular intervals. Finally, Bach augmented the line in time by making it last twice as long and diminished it by making it half as long.

The expressive side of Schoenberg's music can be understood in terms of modern physiology of perception. In the nineteenth century Herman von Helmholtz discovered that hearing notes in certain ratios of whole numbers—1 : 2, 2 : 3, and 3 : 4 (an octave, a fifth, and a fourth)—as consonant, as well as hearing notes lacking these ratios as dissonant, are innate features of the human ear ("On the Physiological Causes of Harmony in Music," 1857; see chapter 4). In other words, music composed from tonal consonances (such as a Bach fugue or a Mozart minuet) embodies pure mathematics in a sensual form that resonates with the listener's inner world.[7]

In Schoenberg's day composers typically used dissonance for expressive ends, such as creating a dissonant tension that is resolved in a consonant finale. Fin-de-siècle composers such as the Russian Alexander Scriabin and the Hungarian Béla Bartók (as well as Schoenberg), had been moving away from melodic tonality and composing with atonal parts. But in 1908 Schoenberg became the first composer to completely abandon consonance and compose with *only* dissonant sounds.

Historians commonly link this pivotal moment in the history of modern music to an event in Schoenberg's personal life.[8] Schoenberg had made the acquaintance of two young Austrian painters, Oskar Kokoschka and Richard Gerstl, and in 1907 he himself seriously took up painting in an Expressionist style.[9] Schoenberg's wife, Mathilde, briefly left him in 1908 to live with Gerstl, and when she returned to her husband, Gerstl destroyed his paintings and hung himself at age twenty-five. In the following five years Schoenberg composed two monodramas (operas for one singer), *Erwartung* (Anticipation, 1909) and *Die glückliche Hand* (The delightful touch, 1910–13), in which his break with tonality was complete and the music is replete with dissonant expressions of despair (plates 10-6 and 10-7).

The work of the artist is instinctive. The conscious mind has little influence on it.

—**Arnold Schoenberg, *Theory of Harmony*, 1911**

The general public, totally cut off from the production of new music, is alienated by the outward characteristics of such music. The deepest currents present in this music proceed, however, from exactly those sociological and anthropological foundations peculiar to that public. The dissonances which horrify them testify to their own condition; for that reason alone do they find them unbearable.

—**Theodor W. Adorno, *Philosophy of Modern Music*, 1948**

In addition to expressing Schoenberg's unsettled personal life, these atonal monologues relate to topics raised by the composer's Viennese contemporary, Sigmund Freud, who in 1900 had published *Die Traumdeutung* (English translation as *The Interpretation of Dreams*). The librettist for *Erwartung*, Marie Pappenheim, was a relative of the patient Anna O (Bertha Pappenheim), whom Freud described in one of his earliest case histories (*Studies on Hysteria*, 1895). As Theodor Adorno suggested in 1949, Pappenheim wrote the libretto of *Erwartung* as the text of a dream (its uninterpreted manifest content): "The first atonal works are depositions, in the sense of psychoanalytic dream depositions [*Traumprotokollen*]."[10]

Both *Erwartung* and *Die glückliche Hand* are about seeking and not finding love, and after the outbreak of World War I Schoenberg pursued the related theological/philosophical topic of seeking and not finding truth. Although raised in a Jewish home, at age twenty-four Schoenberg had converted to Lutheranism and explored other religious outlooks, including that of the Swedish mystic and pantheist Emanuel Swedenborg. But by the 1920s his spiritual quest had led him back to Judaism, to which he officially returned in 1933 after being fired from the Prussian Academy of Art in Berlin because of his Jewish

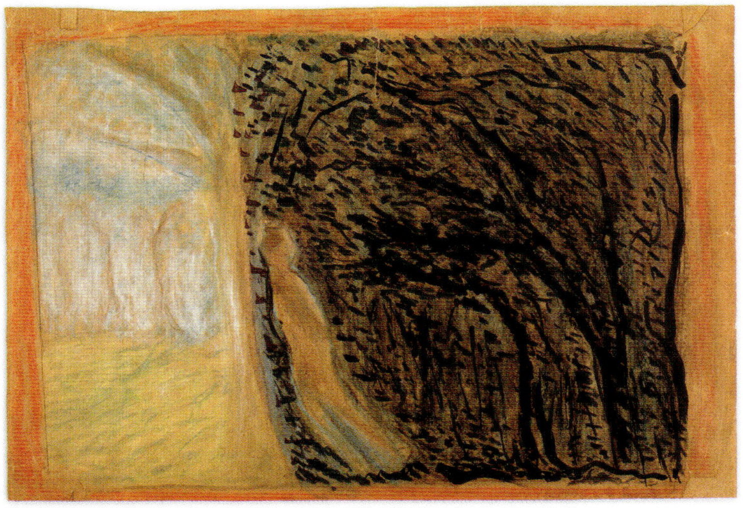

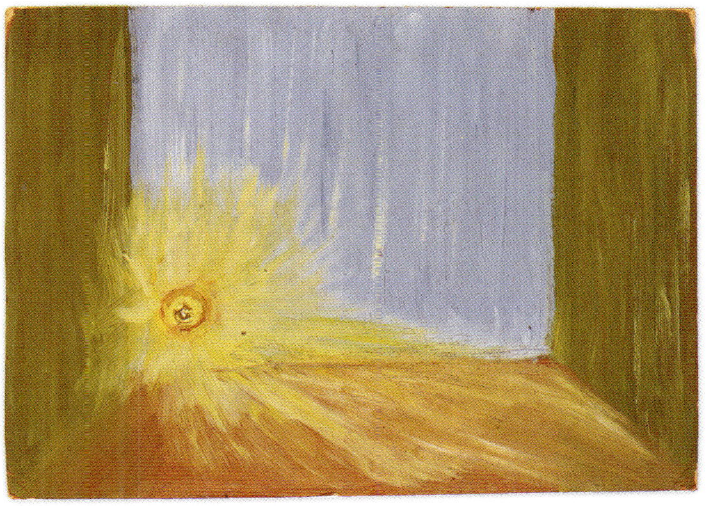

heritage. Schoenberg then immigrated to America, where he taught at the University of Southern California and UCLA until his death in 1951.[11]

In 1928 Schoenberg wrote the libretto of an opera for two voices and chorus, *Moses und Aron* (Moses and Aron), which is based on the biblical account of two brothers. Moses ascended Mount Sinai, where the God of Abraham charged him with communicating to the children of Israel a being who is "infinite, omnipresent, unperceived, and inconceivable," but Moses is inarticulate, stuttering and stumbling over his lines in the libretto. Aron has perfect diction but nothing to say because he has not been to the mountaintop. The opera's theme is the impossibility of capturing profound truth in images or words. The chorus chants the prohibition of idolatry: "You shall not make an image. For an image confines, limits, grasps what should remain limitless and unimaginable." In 1932 Schoenberg began composing the score, which he wrote in strict adherence to his twelve-tone method. In a tacit admission that there are inexpressible truths, Schoenberg did not complete the composition before his death in 1951; the unfinished opera ends with Moses singing "Oh word, thou word, that I lack."

The twelve-tone method, or "serialism" as it came to be called, was taken up by Schoenberg's disciples Alban Berg and Anton Webern, and together the three came to be known as the Second Viennese School. In the 1920s and 1930s the method was confined to their circle, and Berg and Webern, like their teacher, did not make an explicit analogy between serialism and mathematics. However, their style easily translated into machine language when computer technology became available in the 1950s, and the legacy of the twelve-tone method is in computer-assisted music.

A contemporary of the Second Viennese School, the Russian composer Joseph Schillinger, developed an approach to composition that was directly based on formalist mathematics. Schillinger studied composition at the Saint Petersburg conservatory, and after the 1917 revolution he taught at art schools in Saint Petersburg and Moscow. He analyzed a line of music into an abstract pattern by plotting it on graph paper (plate 10-8) and then manipulating the pattern according to mechanical rules (plate 10-9). After composing music using this method, Schillinger often gave it an interpretation, such as

ABOVE LEFT

10-6. Arnold Schoenberg, set design for *Erwartung* (Anticipation), no. 17, ca. 1911. Watercolor, pastel, and ink on paper, 12⅜ × 17¾ in. (31.4 × 45 cm). Arnold Schoenberg Center, Vienna, inv. no. 169.

In *Erwartung* (Anticipation, 1909), a young woman sings about fear as she wanders on a dark stage searching for her lover. After discovering his dead body, out of sheer exhaustion she reconciles herself to living without him, ending her monologue by singing: "I was seeking."

ABOVE RIGHT

10-7. Arnold Schoenberg, set design for *Die glückliche Hand* (The delightful touch), no. 18, 1910. Oil on cardboard, 8¾ × 11¾ in. (22 × 30 cm). Arnold Schoenberg Center, Vienna, inv. no. 176.

In *Die glückliche Hand*, a man sings of his love for an elusive woman who has abandoned him for a rival. In the mistaken belief that he has won her back, he is inspired to create a spectacular work of art in pure gold. But the artist soon realizes that he has created a glittering delusion—a trivial bauble in base metal—because, although the woman has returned, she does not love him.

an expression of cosmic order or a political outlook.[12] Indeed, it became Schillinger's lifelong mission to demonstrate that his formal approach to composition—the Schillinger system—could be applied not only to music but to all the arts, and he himself did paintings with patterns of symmetry, continuity, and permutation (plates 10-10 and 10-11).[13]

After the British physicists Michael Faraday and James Clerk Maxwell discovered the link between electricity and magnetism—electromagnetism—inventors applied their discovery to make the first electrical machines, such as the telegraph that sent electrical impulses in Morse code (1840s), the telephone that transmitted the pattern of sound-waves (1876), and the first phonograph that both recorded and reproduced sound (1878). Late-nineteenth-century musicians were keenly interested in the implications of these sound devices for their work, and some dreamt of building an electromagnetic machine that did not use sound-waves from the human voice or a musical instrument, but produced purely electronic sound-waves. The first such machine—the theremin—was built in Russia in 1920 by Schillinger's contemporary Léon Theremin, and subsequent music machines were designed by composers who typically had training in both music and mathematics. The underlying idea that links these many music machines to the formalist aesthetic is the concept of a piece of music as a formal axiomatic structure, with a vocabulary (notes), an axiom (a melody or a twelve-tone basic set), and algorithms (rules of transformation).

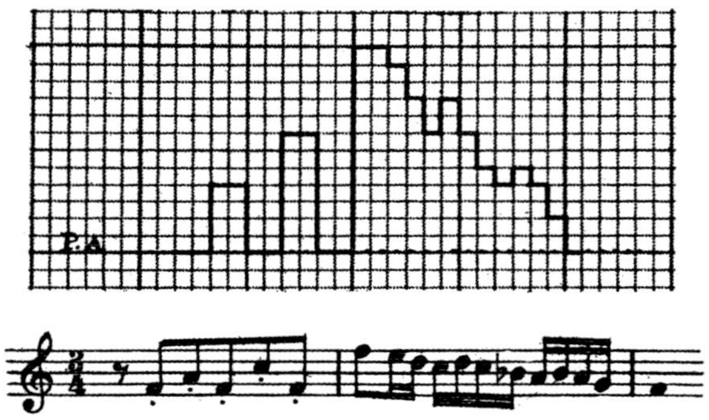

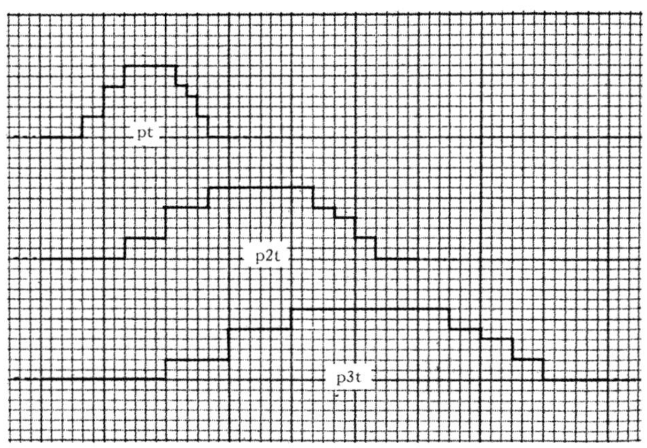

TOP

10-8. Graph of motion in Bach's two-part invention, no. 8, in Joseph Schillinger, *The Schillinger System of Musical Composition* (New York: Carl Fischer, 1941), 1:312.

In this graph Schillinger represented each sixteenth note with a horizontal square and each semitone with a vertical square.

ABOVE

10-9. Time and pitch expansion, in Joseph Schillinger, *The Schillinger System of Musical Composition* (New York: Carl Fischer, 1941), 1:212.

In 1916 Theremin graduated from the Saint Petersburg Conservatory as a cellist and then attended the Electro-Technical School for officers, completing the requirements to be a radio engineer in Tsar Nicholas II's Imperial Army. But in October 1917 Theremin deserted and joined the Bolsheviks, and for the next several years he ran the Red Army's radio station while also experimenting with translating electromagnetic wave patterns into sound-waves with the goal of designing electronic musical instruments. By 1920 he had designed the "ætherphon" (named for the mysterious atmospheric æther that allegedly filled the cosmos, but soon known as the theremin; see plate 10-12). In March 1922 Theremin demonstrated his instrument in a private meeting with Lenin, who responded enthusiastically to this example of electrification, which the leader envisioned as a key technology for the development of the Soviet Union.[14] The first orchestral work composed for an electronic solo instrument was Andrey Pashchenko's *Symphonic Mystery*, for theremin and orchestra, first performed in Saint Petersburg in 1924. Theremin gave his first U.S. recital, "Music from the Æther," at the Metropolitan Opera House in New York in 1928, and the following year he patented the instrument in the United States and RCA began commercial production.

Meanwhile, Schillinger had emigrated from Russia, and in 1929 he and Theremin set up a studio together in New York, where they collaborated on electronic devices and musical instruments for the burgeoning recorded music and film industries. In 1929

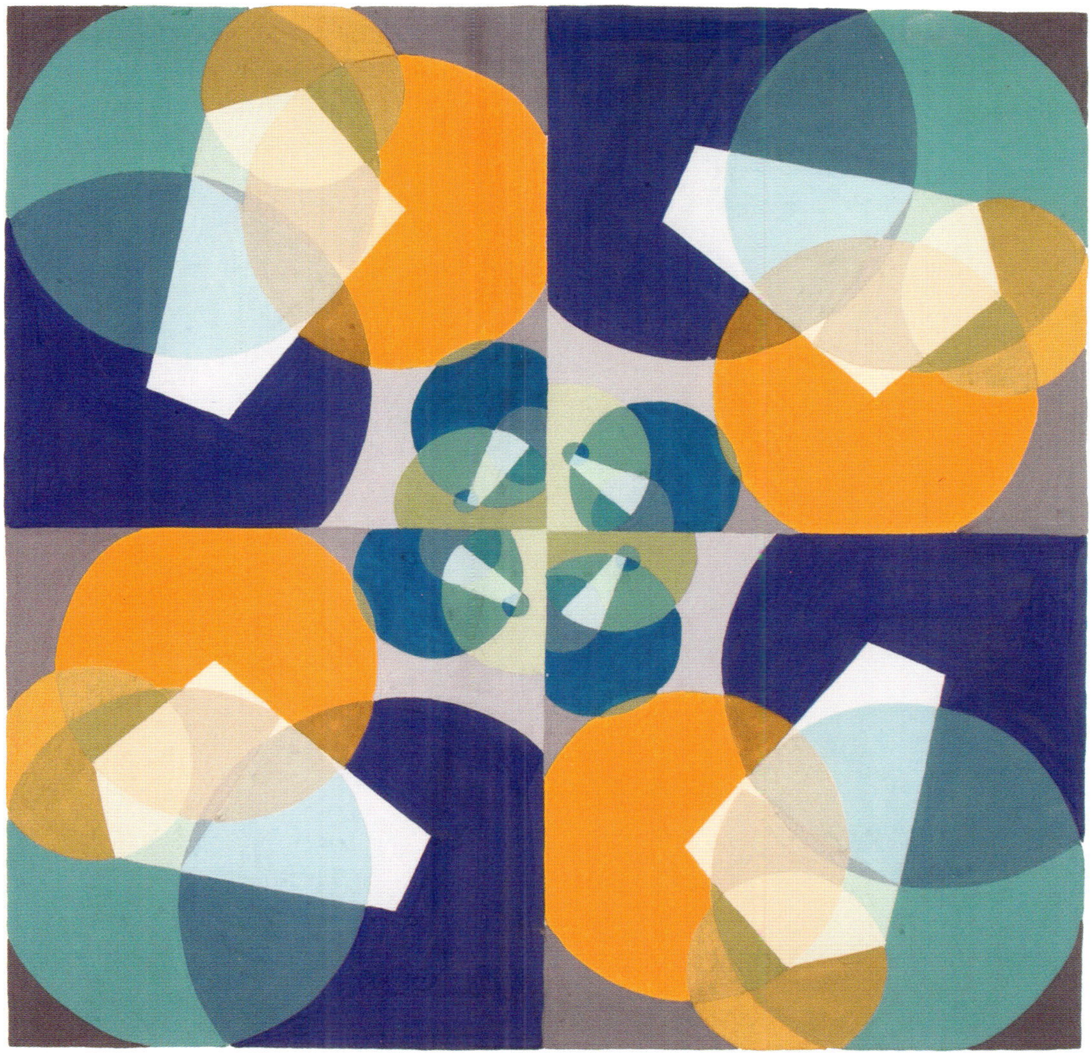

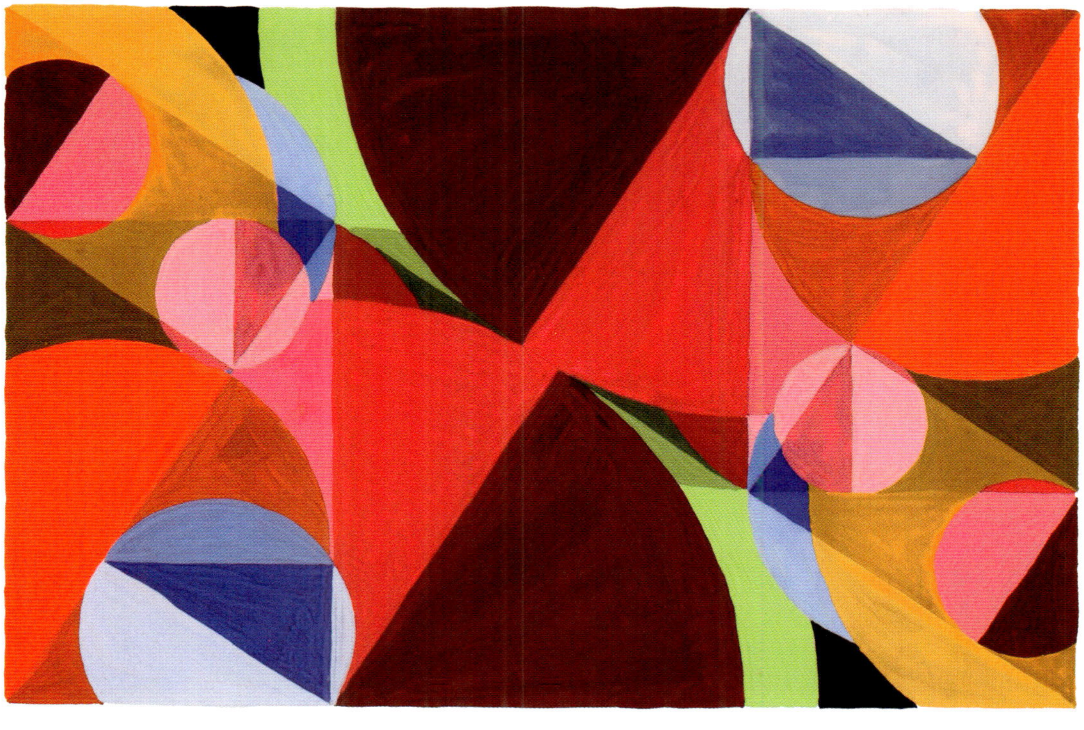

TOP RIGHT

10-10. Joseph Schillinger (Russian, 1895–1943), *Green Squares*, from the series *Mathematical Basis of the Arts*, ca. 1934. Tempera on paperboard, 11⅞ × 11¼ in. (30 × 28 cm). Smithsonian American Art Museum, Washington, DC, gift of Mrs. Joseph Schillinger.

BOTTOM RIGHT

10-11. Joseph Schillinger (Russian, 1895–1943), *Areas Broken by Perpendiculars*, ca. 1934. Opaque watercolor, 9 × 12 in. (22 × 30.5 cm). Smithsonian American Art Museum, Washington, DC, gift of Mrs. Joseph Schillinger.

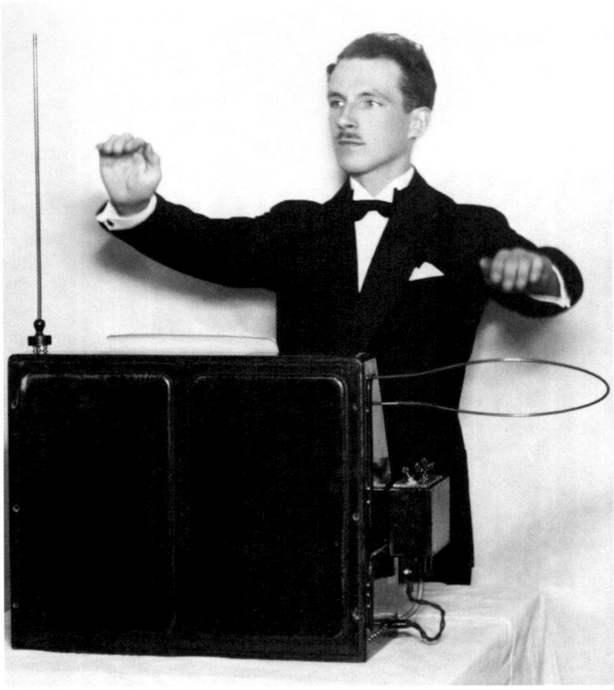

10-12. Léon Theremin performing on a theremin around 1920. Photograph. © Hulton-Deutsch Collection/Corbis.

The vertical antenna atop a theremin emits a fixed frequency of electromagnetic radiation in the radio range, resulting in a uniform wave pattern. The human body is a conductor of electricity, and so when Theremin moved his hand up and down near the antenna, it interfered with the wave pattern. These changes in wave pattern were translated by the electronic instrument into changes in the wavelength (pitch) of an audible sound, which was emitted from a loudspeaker. The result is a single tone that makes a slurring or sliding sound—a glissando—as it changes from note to note. The volume of the sound is controlled by the performer's left hand moving back and forth above a vertical loop.

After the 1917 revolution, the Russian musician Nikolai Sokoloff immigrated to America, and in 1918 he became founding music director of the Cleveland Orchestra in Ohio. There in November 1929, Sokoloff conducted the world premiere of Joseph Schillinger's *First Airphonic Suite*, with Léon Theremin as soloist on the theremin.

Schillinger composed the *First Airphonic Suite* for theremin and orchestra, and in 1932 Theremin perfected the terpsitone (named for Terpsichore, the Greek muse of dance), an electronic dance floor that was interactive; the motion of the dancers determined changes of pitch of the electronic music. Schillinger taught music, mathematics, and art history at Columbia University, where his mathematical models and painted patterns were on permanent display in the Mathematics Museum. His many private students included Tommy Dorsey, George Gershwin, and Benny Goodman. Gershwin was his student from 1932 to 1936, during which he used the Schillinger system to compose *Porgy and Bess* (1935).[15] During his eleven years in New York, Theremin was active in low-level espionage for his Soviet handlers, but despite this patriotic service, in 1938 the Soviet secret police abducted him from his New York studio and returned him to Moscow, ending his music career. After serving time in a labor camp, Theremin worked as an electrical engineer in Russia for the rest of his life.[16] In America it was primarily popular musicians who carried on the legacy of Theremin, such as the Beach Boys' composition for theremin and rock band in the 1960s.

After immigrating to America, Schoenberg taught in southern California, where a new generation picked up his serial method, developing it in the context of Cold War America. La Monte Young studied in the 1950s at Los Angeles City College with Leonard Stein, who had been an assistant to Schoenberg at UCLA; Young used the twelve-tone technique of Schoenberg and Webern in his early compositions. Swept up by the cultural upheavals of the 1960s, Young adopted Asian themes and moved to New York, where he became an originator and the guru of so-called Minimal music. To the serialist method of the Second Viennese School, Young added very long tones to create a feeling of timelessness. In Young's *Four Dreams of China* (1962), four performers produce extremely long tones that can, in theory, continue forever.[17]

As computer technology became available after World War II, the Greek architect and musician Iannis Xenakis began composing twelve-tone music with the assistance of a computer. Xenakis, who had been active in the Greek resistance during the war, fled to France in 1947 after a new conservative Greek government was installed. In Paris he got a job working in architecture for Le Corbusier, beginning as an apprentice but quickly assuming a high level of responsibility in the firm. In 1958 Le Corbusier was commissioned by a Dutch electronics company, Philips, to design their pavilion for the 1958 World's Fair in Brussels, and the architect collaborated with Xenakis, who managed the project. Le Corbusier and Xenakis gave the pavilion a dramatic geometric form composed of nine hyperbolic paraboloids (plates 10-13 and 10-14), which they built to house a multimedia spectacle highlighting electronic progress. Xenakis had already begun exploring musical sounds produced without instruments,

I'm pickin' up good vibrations
She's giving me excitations.

—Beach Boys,
"Good Vibrations," 1966

RIGHT

10-13. Le Corbusier (Charles-Édouard Jeanneret; Swiss, 1887–1965) and Iannis Xenakis (Greek, 1922–2001), Philips Pavilion, International Exposition of 1958, Brussels. Foundation Le Corbusier, Paris. © 2015 F.L.C./ADAGP, Paris/Artists Rights Society, New York.

BELOW

10-14. A hyperbolic paraboloid.

Formed from a gridded surface, a hyperbolic paraboloid is used in architecture to construct a curved surface using only straight beams. Iannis Xenakis designed the Philips Pavilion in the form of nine hyperbolic paraboloids, each of which has the curved shape of a saddle.

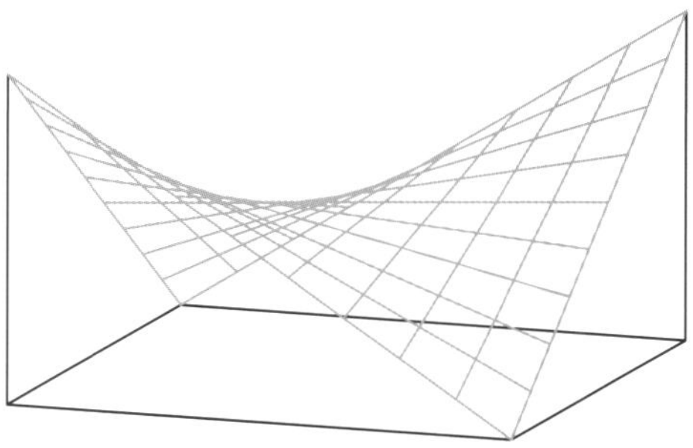

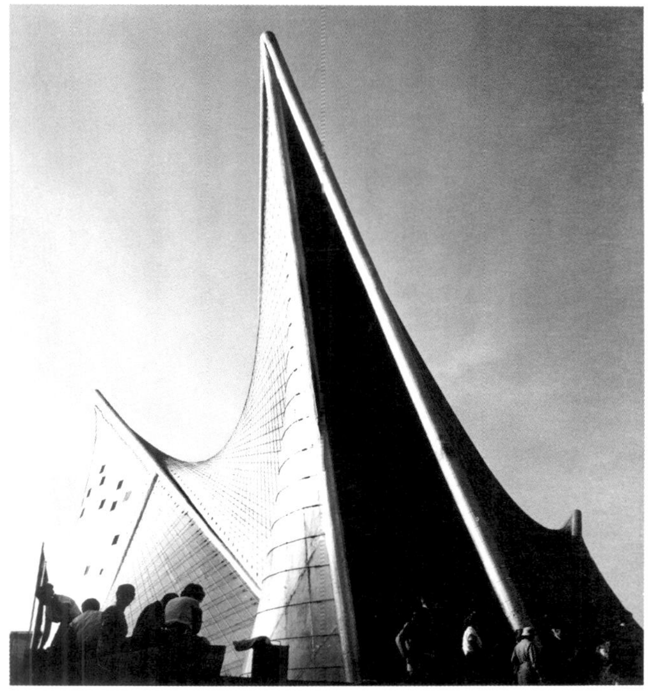

and thus it was natural for him to contribute to the Philips Pavilion project an electronic score that visitors heard as they entered the building. Xenakis left Le Corbusier's firm in 1959 to devote himself to music, becoming a leader in the new field of electronic music. In 1966 he founded the Equipe de Mathématique et Automatique Musicales (Center for mathematical and automated music) in Paris and went on to write one of the founding texts of computer-assisted composition, *Formalized Music: Thought and Mathematics in Composition* (1971).[18]

THE WORK OF ART IN THE AGE OF MECHANICAL COMPUTATION

The impact of computers on the visual arts in the second half of the twentieth century is similar to that of photography after its invention in the 1830s. The inventors of the equipment were the first to explore its aesthetic possibilities, and then artists whose styles benefited from the technology adopted the new tool. In the nineteenth century the inventor of calotype, William Fox Talbot, took photographs of landscapes and still lifes to be used as book illustrations (*The Pencil of Nature*, 1844). Soon portrait artists, such as Nadar (Gaspard-Félix Tournachon) and Julia Margaret Cameron, used the camera to achieve a likeness, as did realist painters such as Edgar Degas and Thomas Eakins.

In the 1950s and 1960s the vacuum tubes and transistors necessary to power computers made them expensive, high-maintenance, room-size machines, so the first mathematicians and computer scientists to experiment with the computer as an artistic tool worked at companies and universities that had the resources to house and maintain these massive machines.[19] In both a computer program and an algorithmic artwork, such as Max Bill's *Fifteen Variations on a Theme* (1935–38; see plate 7-18 in chapter 7), the artist applies a

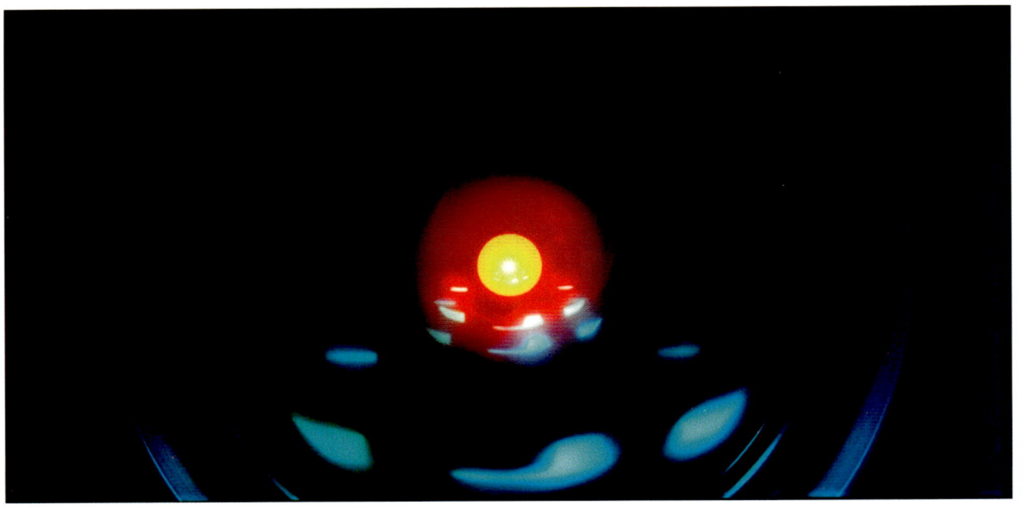

10-15. HAL 9000, a star of the film *2001: A Space Odyssey* (1968), directed by Stanley Kubrick, based on a novel by Arthur C. Clarke.

HAL was a Heuristically programmed ALgorithmic computer who had the ability to monitor all systems on the fictional spaceship *Discovery One*, as well as speak, recognize speech, read lips, recognize faces, display emotion, reason, and play chess. For audiences of *2001: A Space Odyssey* (1968), HAL passed the Turing test.

set of rules to elements (colors, shapes) and generates a set of variations. If an artist is writing a computer program and wants to allow some unpredictable surprises in the output, the artist/programmer can include a free variable, such as for color, whose value is determined by a random number generator; in other words, color is determined by chance. In computer-assisted algorithmic art, after a program with a free variable has run (ideally) all possible variations, the artist will typically pick a few art objects from the output and discard the rest.

Early criticism of computer-assisted art recalls nineteenth-century complaints about photography: it is made by a machine and thus lacks a human touch.[20] Gradually, like photography, computer art came to be appreciated for its own inherent qualities, such as the austere beauty of the visualization of an abstract idea with a level of precision that is beyond human reach (plate 10-15).

Max Bense, who was a close friend of Max Bill, led the first sustained effort to encourage engineers and artists to use the computer to produce algorithmic art. Bense completed a doctorate in physics with a thesis on Einstein's theory of relativity and quantum mechanics at the University of Bonn in 1937, but he worked in the philosophy of science and in aesthetics, two fields in which he was largely self-taught. In 1949 he joined the faculty in the philosophy department at a leading school for applied mathematics, the Technische Hochschule in Stuttgart (today the University of Stuttgart), where he remained his whole career. A prolific writer, Bense published more than eighty books on philosophy, including his multivolume *Aesthetica* (1954–65). He was a voracious reader, synthesizing midcentury intellectual trends into his grand vision of a universal aesthetics. Bense wanted to make Swiss Concrete Art "scientific" by analyzing the art of Bill and ultimately all art from all times into its elements (color, form), developing precise rules for generating art objects and then formulas for evaluating them. In 1954 Bense published the first volume of *Aesthetica*, in which, with a nod to the Swiss mathematician Andreas Speiser (see chapter 7), Bense declared his (exaggerated) thesis: symmetry is the sole cause of aesthetic pleasure. When in the 1950s the computer was introduced at the Technische Hochschule, Bense adopted it as his aesthetic model, reading works by the American mathematician Claude

Shannon, who worked as a cryptanalyst during World War II. In 1943 the British government had sent Alan Turing to Washington, DC, to update American mathematicians on the work at Bletchley Park, and for two months Turing and Shannon met regularly to discuss their common interest in the translation of speech into code. After the war Shannon described how to encrypt verbal information in order to transmit it electronically ("A Mathematical Theory of Communication," 1948). Bense adapted Shannon's methods of verbal communication for artistic communication, in what Bense termed "information aesthetics."[21]

In his later writing Bense sought a precise way to evaluate art by measuring its symmetry, and he adopted a system developed by another follower of Speiser, the American mathematician George David Birkhoff, a Harvard researcher who wrote about the mathematical foundations of Einstein's theory of relativity (*Relativity and Modern Physics*, with Rudolph E. Langer, 1923). In the early 1930s Birkhoff spent a sabbatical year studying the art and music of world cultures, and (echoing Speiser) he declared that the more symmetrical an object is, the more beautiful it is. Birkhoff gave a method for quantifying the beauty of an object by calculating the degree of symmetry of its silhouette (*Aesthetic Measure*, 1933; see plate 10-16).

In 1964 the German engineer Konrad Zuse, who in 1941 had built one of the first operational stored-program computers, designed a drawing machine—the Graphomat—to graph an equation by translating it onto a two-dimensional grid using a Cartesian co-ordinate system. Gearboxes control the movement of a pen, which is called a "plotter" because it plots points on an invisible grid as it glides across a sheet of paper. When the electric company Siemens AG in Erlangen bought a new Graphomat, the job of programming it fell to the young mathematician Georg Nees. Nees wrote programs to generate technical drawings for Siemens, and in his free time he experimented with writing programs that generate decorative patterns. As he later recalled: "When I saw figure after figure pouring from the pen, I got the shivers. I thought: 'Here is something that will never disappear.'"[22] Interested in the philosophical implications of computation, Nees began graduate work in philosophy at Erlangen and also studied with Bense at the Technische Hochschule in Stuttgart. Nees applied Bense's aesthetic theories to the generation of drawings on the Graphomat (plate 10-17), becoming the first person to receive a doctorate in computer graphics (PhD, 1969). Meanwhile, in 1964 the Technische Hochschule had also bought a Graphomat and enlisted a graduate student in mathematics, Frieder Nake, to program it. Nake also experimented with it as an aesthetic tool, writing a drawing program using a random-number generator. Nake completed his doctorate in 1967 with a dissertation on probability theory, focusing on philosophical issues regarding chance and random-number generators. Nake wrote one of the earliest computer programs designed specifically for studio artists, *Generative Aesthetics*

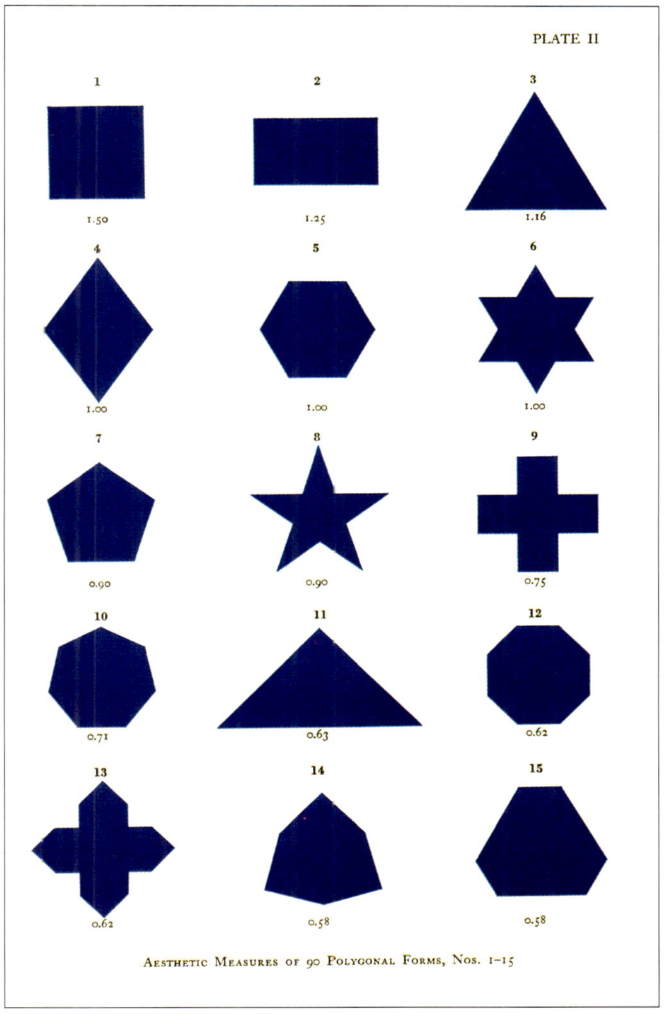

Aesthetic Measures of 90 Polygonal Forms, Nos. 1–15

10-16. George D. Birkhoff, "Aesthetic Measure of Polygonal Forms," *Aesthetic Measure* (Cambridge, MA: Harvard University Press, 1933); plate 2, facing 32.

Birkhoff reduced aesthetics to one simple formula:

$$M = O : C$$

which reads "the amount of aesthetic pleasure (M) produced by an object equals the ratio of the object's order (O) to its complexity (C)." Birkhoff described the formula: "It is the intuitive estimate of the amount of order O inherent in the aesthetic object, as compared with its complexity C, from which arises the derivative feeling of the aesthetic measure M of the different objects of the class considered" (11–12). See also George Birkhoff, "A Mathematical Theory of Aesthetics and Its Application to Poetry and Music," *Rice Institute Pamphlet* 19, no. 3 (July 1932).

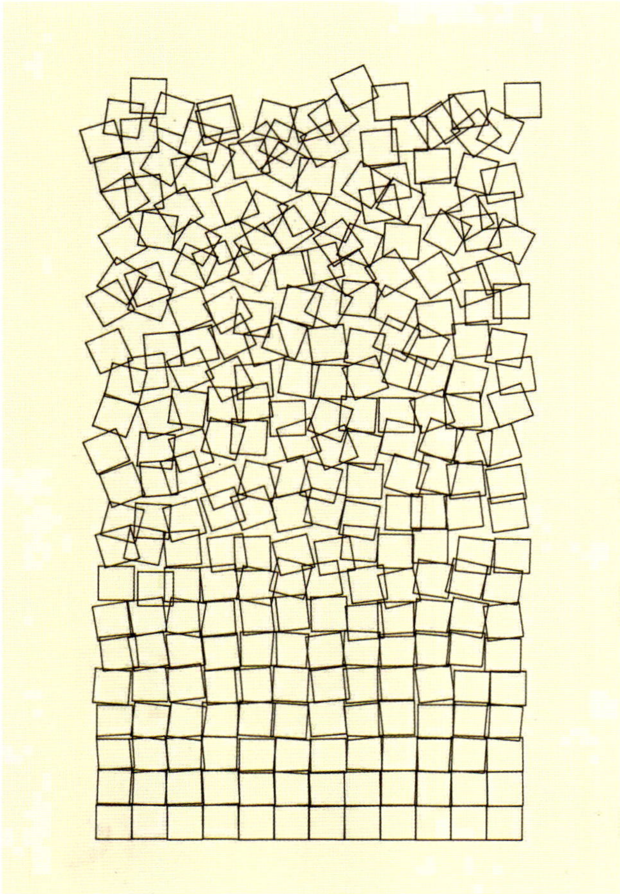

10-17. Georg Nees (German, b. 1926), *Schotter* (Gravel stones), 1965–68. Computer graphic, plotter drawing, ink on paper, 9⅟16 × 5⅛ in. (23 × 13 cm). Kunsthalle Bremen, Germany.

In 1943 the German physicist Erwin Schrödinger invented the term "negentropy" (negative entropy), which took on various meanings in biology and physics. A few years later the American pioneer of information theory Claude Shannon used the term "entropy" to name a measure of the information one is missing when one doesn't know the value of a random variable ("A Mathematical Theory of Communication," 1948). Borrowing the terms from Schrödinger and Shannon, Max Bense suggested that aesthetic objects move towards order and complete information in a process of negentropy. Nees was a doctoral student of Bense when he created *Schotter*, using a computer program that he wrote to illustrate his teacher's idea of aesthetic negentropy, moving from disorder above to order below.

I (1968). In his search for the innate aesthetic algorithms hypothesized by Bense, Nake analyzed the compositions of artists such as Paul Klee (*Hommage à Paul Klee*, ca. 1965; Kunsthalle Bremen, Germany).

Bense curated an exhibition of Nees's computer-generated drawings in 1965 for the Studio Gallery at the Technische Hochschule, and later that year Nees and Nake exhibited together in Stuttgart (*Computer-Grafik Programme*, Wendelin Niedlich Gallery, 1965). These 1965 events are considered the first exhibitions of computer art, and as Nees predicted, the computer plotter took up permanent residence in the art studios. In the mid-1960s the German artist Manfred Mohr undertook an intensive study of Bense's aesthetics, which led him to adopt an algorithmic approach and to learn computer language, with the plotter as his drawing tool (plate 10-18).[23]

In the United States in the early 1960s, engineers working in industry and at universities also began experimenting with computer-generated art objects. Like the artists in Bense's circle, A. Michael Noll, an engineer at Bell Labs in New Jersey, was interested in isolating stylistic rules. To that end, Noll analyzed paintings by artists, such as Piet Mondrian's *Composition in Line* (1916–17; see plate 6-13 in chapter 6), to create works of his own (*Computer Composition with Lines*, 1964; Kunsthalle Bremen, Germany).

The Hungarian-born György Kepes immigrated to America in 1937 to teach at the New Bauhaus in Chicago at the invitation of his countryman Moholy-Nagy. Enthusiastic about linking the arts and sciences, Kepes saw the computer as a tool that could bridge the cultural gap. To that end, in 1967 Kepes founded the Center for Advanced Visual Studies at the Massachusetts Institute of Technology, America's leading technical school, where students of contemporary art learned both artistic and computer skills. The same year in New York, a Swedish electrical engineer, Johan Wilhelm (Billy) Klüver, teamed up with the artist Robert Rauschenberg to found Experiments in Art and Technology (E.A.T.) with the goal of facilitating collaborations between ten artists (including Rauschenberg and John Cage) and thirty engineers from Bell Labs, where Klüver worked.[24] The result was a performance series, "Nine Evenings: Theater and Engineering," presented in October 1966 at the Armory in Manhattan. The 14,000 members of the audience saw the realization of ideas such as a game of tennis with rackets fitted with microphones that picked up and amplified reverberations when a ball hit the strings. Each hit also triggered a mechanism that shut off one of thirty-six ceiling lights; thus the performance ended in darkness after thirty-six hits (Robert Rauschenberg, *Open Score*, 1966).

In Zagreb in the 1960s, an art movement called New Tendencies explored the aesthetic potential of computer technology, and in 1968 its leaders launched a multilingual magazine, *Bit International*, as an international platform for computer art.[25] The same year Bense initiated an exhibition of computer art at the Institute of Contemporary Arts in London.[26] In 1970 computer-assisted art was included in the Venice Biennale, and in

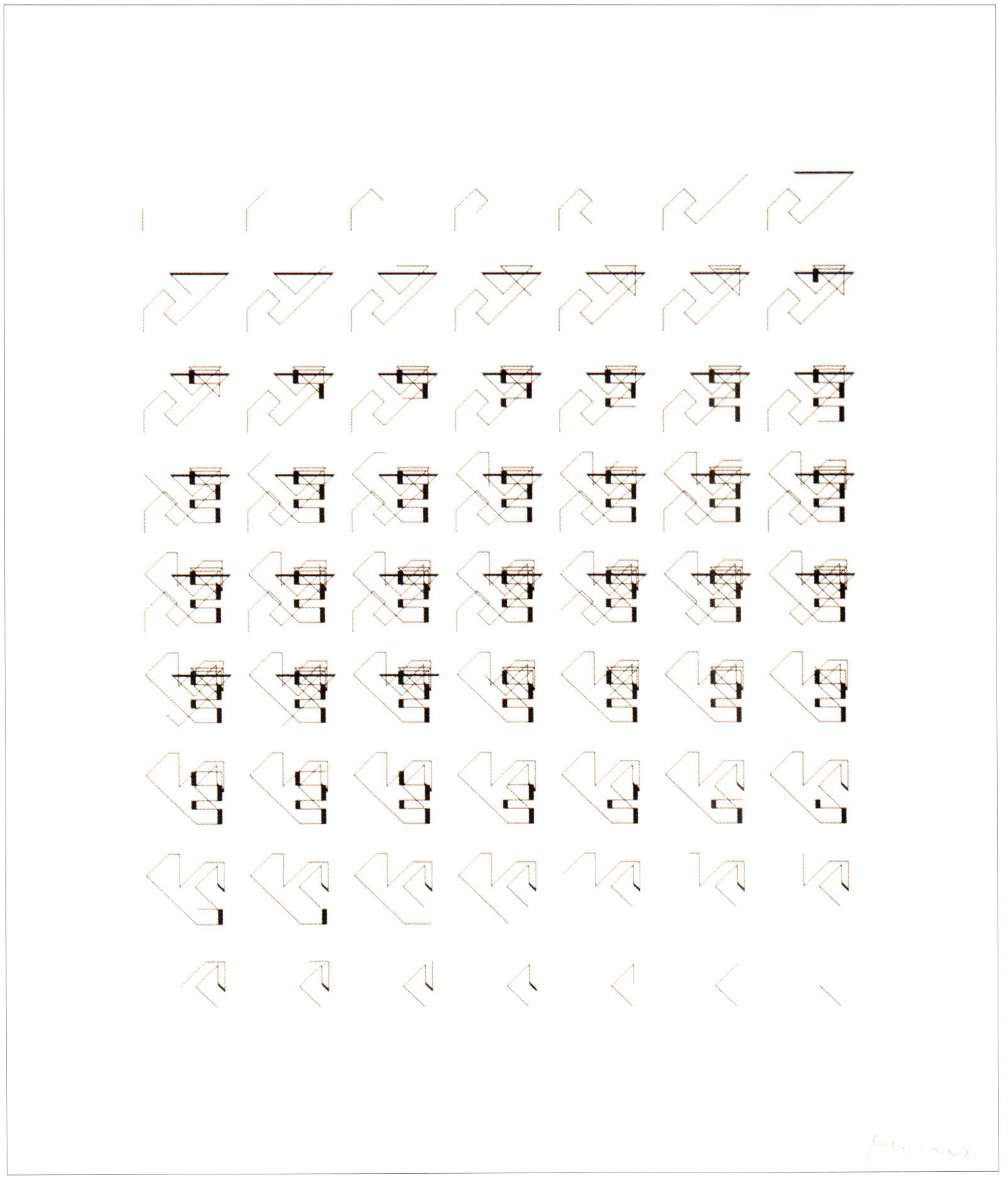

10-18. Manfred Mohr (German, b. 1938), *P-26F Inversion logique* (Logical inversion), 1970. Plotter drawing with ink on paper, 22 × 18 1/16 in. (56 × 46 cm). Courtesy of the artist.

To create this pattern, Manfred Mohr began by determining a vocabulary of symbols, including a vertical line, a diagonal, a horizontal, and so on. He then defined a grid of seven columns and nine lines, and wrote an algorithm that selected one symbol at random to put in row 1/column 1 (in the upper left). The algorithm then selected a second symbol to add (moving right) to row 1/column 2, and so on. A random process also determined the length and thickness of the lines. When the midpoint in the grid was reached (row 5/column 4), the algorithm then subtracted one symbol from the composite sign in the order it was added, so that in the last position (lower right) one element remains.

Mechanistic muses are expanding their domain to encompass every facet of creative activity.

—John R. Pierce, engineer at Bell Labs, "Portrait of the Machine as a Young Artist," 1965

1971 the Museé National d'Art Moderne in Paris hosted a one-person exhibition of plotter drawings by Manfred Mohr. By the late 1970s art made with a computer was a regular feature of art exhibitions of "new media," and with the advent of the Web in the 1990s, online institutions such as The Thing (founded 1991) and Rhizome (founded 1996) formed to present, promote, and archive digital art on the Internet, independent of traditional art museums. In today's art world the computer is as common as the camera.[27]

UNIVERSALISM

By the late nineteenth century, the universalist vision of one global outlook had developed into cross-cultural studies of philosophy and theology, marked by the first World's Parliament of Religions, which was held in conjunction with the 1893 Chicago World's Fair. William James declared that knowledge by mystical intuition was the common thread linking

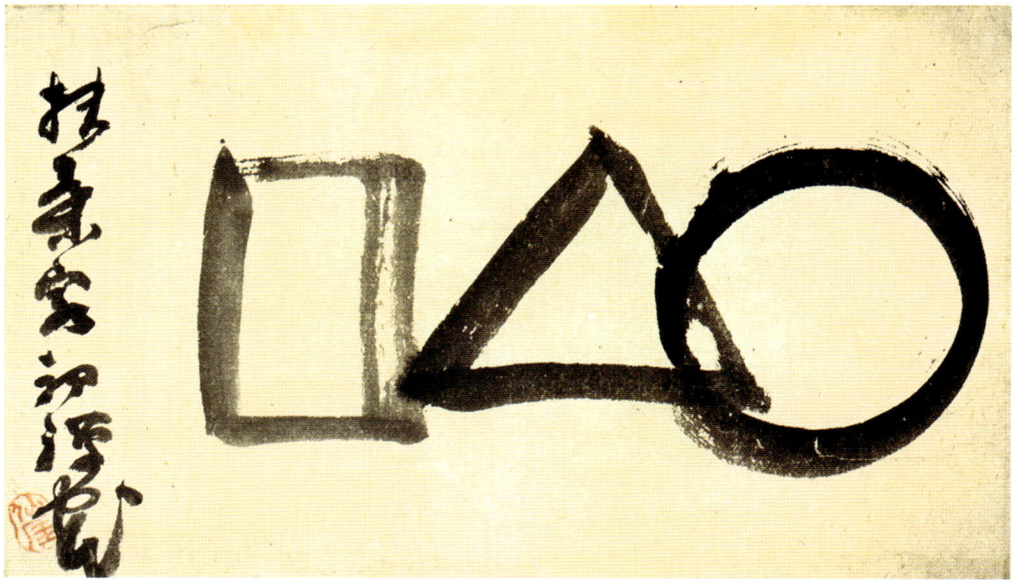

10-19. Sengai Gibon (Japanese, 1750–1837), *Untitled* (rectangle, triangle, circle), Edo period, early nineteenth century. Hanging scroll, ink on paper, 11 3/16 × 19 in. (28.4 × 48.1 cm). Idemitsu Museum of Arts, Tokyo.

The Zen Buddhist monk Sengai Gibon made this untitled ink painting; the text on the left is Sengai's signature. Suzuki cited Sengai's painting as a perfect image for contemplation, and he gave it the following interpretation: "The circle represents the infinite. . . . The triangle is the beginning of all forms. . . . Out of it comes the square. A square is a triangle doubled. This doubling process goes on infinitely and we have the multiplicity of things, which the Chinese philosopher calls 'the ten thousand things,' that is, the universe." D. K. Suzuki, *Sengai: The Zen Master* (Greenwich, CT: New York Graphic Society, 1971), 36.

world religions, and he presented a collection of accounts of mystical experience written by individuals of different faiths in his classic *Varieties of Religious Experience* (1902).

Meanwhile, the new doctors of the soul offered a secular description of the brotherhood of mankind in terms of psychology, and neurotics were offered a blend of Western psychotherapy and Eastern meditation. Charles Darwin had declared that animals have innate drives for survival and reproduction (*Origin of Species*, 1859), and, following Darwin, Sigmund Freud described how *Homo sapiens* learn to live together in civilized societies by repressing their aggressions and passions (*Civilization and Its Discontents*, 1930). Freud countered James by offering a psychological explanation of mystical experience: the pantheist longing for oneness with nature originates, according to Freud, in a memory of a time when each human *was* one with his or her surroundings—in the mother's womb and, after birth, in her arms. Freud went on to give a psychological account of other common beliefs: putting one's trust in an omnipotent personal deity (an all-powerful divine father) is, according to Freud, an infantile overvaluation of the parent that has survived into adulthood; belief in an afterlife is a delusional refusal to face mortality with stoicism (*The Future of an Illusion*, 1927).

A major figure in the search for mankind's universal outlook was the Japanese scholar D. T. Suzuki. At the 1893 World's Parliament of Religions, Suzuki met the German-born American philosopher Paul Carus, editor of the publishing house Open Court Press in LaSalle, Illinois. After the World's Fair ended, Suzuki stayed in America and worked with Carus in Illinois, beginning in 1897. Suzuki translated and Carus published a series of classic Asian texts, and Suzuki wrote popularizations of Buddhism for a Western audience. Intertwining Buddhism, Taoism, and Confucianism, Suzuki went on to become the leading spokesman for Asian philosophy in the West for half a century, dividing his time between Japan, Europe, and America, where he taught Eastern philosophy at Columbia University in New York from 1952 to 1957.

Suzuki focused on Zen Buddhism, in which the meditator's goal is to achieve a state of mind in which one is free from anger and cravings, feels in harmony with the world, and has compassion for all. This state of mind is Zen, or nirvana (D. T. Suzuki, *Zen Buddhism*, 1956). Living in New York, Suzuki merged these ancient teachings with Western psychology, emphasizing the control of anger and cravings through breathing exercises and yoga, and the attainment of feelings of harmony and compassion through psychotherapy. Combining the Eastern view that nature is a complex organic harmony with the Western view that the cosmos has a simple mathematical structure, Suzuki interpreted the geometric forms in an ink painting by the late-eighteenth-century Zen Buddhist monk Sengai Gibon as symbolizing the universe (plate 10-19). Some have criticized Suzuki for simplifying the East-West divide and for distorting Buddhist doctrine by popularizing it for a mass audience (traditional Buddhism is practiced in Asia in a monastic setting). But no one denies the immense popular response to Suzuki's spiritual/psychological alternative to atheism in the West at a time when there was widespread decline in organized religions in the Judeo/Christian/Islamic tradition. As the American interpreter of Zen Buddhism, Alan Watts, put it in 1958: "To the Westerner in search of the reintegration of man and nature there is an appeal far beyond the naturalism of Zen—in the landscapes of Ma Yuan and Sesshu, in an art which is simultaneously spiritual and secular, which conveys the mystical in terms of the natural, and which, indeed, never even imagined a break between them. Here is a view of the world imparting a profoundly refreshing sense of wholeness to a culture in which the spiritual and the material, the conscious and the unconscious, have been cataclysmically split" (plate 10-21).[28]

EAST-WEST MERGERS IN JAPANESE AND AMERICAN ART

The search for a universal outlook was expressed with a sense of urgency by artists in countries most immediately affected by the dawning of the nuclear age: Japan and the United States. Yutaka Matsuzawa was twenty-three when atomic bombs fell on his homeland in August of 1945, and he responded by composing poetry that combined Asian philosophy with Western physics, *Immortality of the Earth* (1949). After graduating from Waseda University in Tokyo with a degree in architecture, he studied philosophy with D. T. Suzuki at Columbia University in New York. Matsuzawa then returned to Japan and joined the Shingon sect of Zen Buddhism, which entails achieving nirvana by meditating on certain geometric patterns, such as a magic square that symbolizes the ancient Chinese view of reality (see plate 1-43 in chapter 1).

In 1964 Matsuzawa printed a flier, *Ψ Corpse* (plate 10-20), which, like a magic square, has symbols arranged in groups of nine and reads in part: "The text consists of 729 characters. The number is obtained by multiplying 9 to the power of 3."[29] The flier instructs the viewer to imagine that objects disappear by concentrating on the diagram. The text on the flier continues: "What makes such a horrifying thing possible? By the principle of 'omnipotence of sorts.' That is, Ψ (psi) is capable of vanishing and creating an *object*. It is an extremely dangerous truth."[30] The Ψ of Matsuzawa's title refers to Schrödinger's Ψ wave. The body of a person seeking to unite with the Brahma becomes immaterial—it is (metaphorically) a

RIGHT

10-20. Yutaka Matsuzawa (Japanese, 1922–2006), *Ψ Corpse (Pusai no shitai itai)*, 1964. Printed flier. Queens Museum of Art, New York, inv. no. 1999.6.

In the title of this flier, Matsuzawa used a Ψ (psi), the designation for Schrödinger's wave, but in Matsuzawa's text, he (inexplicably) used the Greek letter Φ (phi; for example, the upper left square, second line, third character). The numbers in the small three-by-three grid (the layout of a magic square) do not follow the traditional pattern but rather have 1 in the center and 2 below; then the counting numbers continue in a clockwise direction to 9.

OPPOSITE

10-21. Ma Yuan (Chinese, 1160–1225 AD), *Dancing and Singing: Peasants Returning from Work*, Song dynasty. Ink on silk, 75.6 × 43.7 in. (192.5 × 111 cm). National Palace Museum, Beijing, China.

In the rows of Chinese characters at the top of this painting, Ma Yuan wrote that there is much sunshine and farmers are happy. Below, peasants return from work displaying (Confucian) social harmony by singing and dancing, while experiencing (Taoist) communication with nature as a road to (Buddhist) enlightenment.

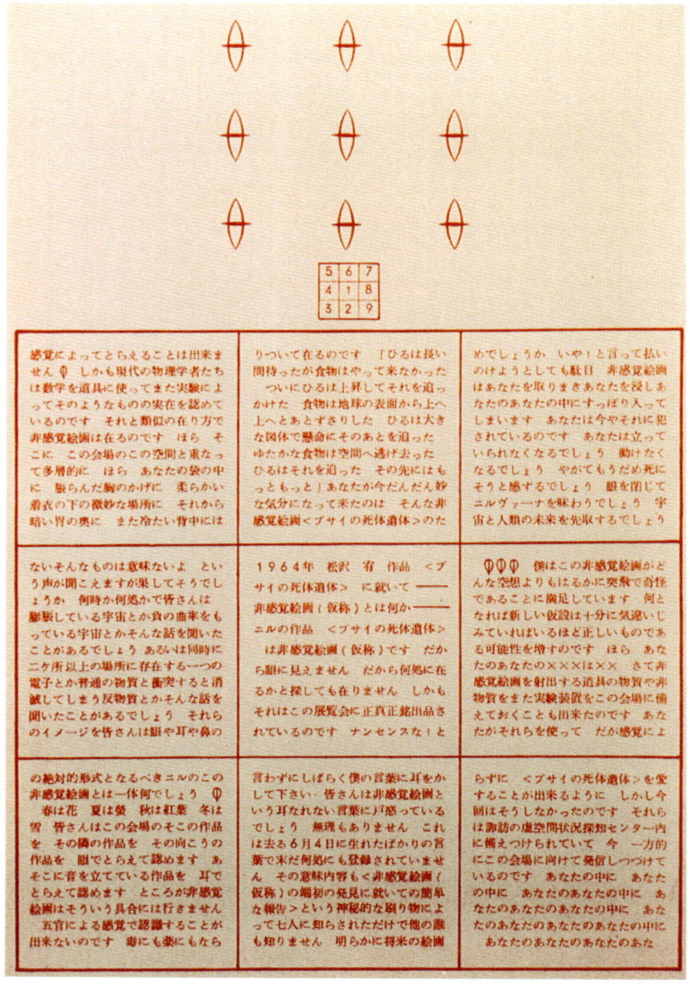

The fact that ruins receive us warmly and kindly after all, and that they attract us with their cracks and flaking surfaces, could this not really be a sign of the material taking revenge, having recaptured its original life?

—Yoshihara Jirō, *Gutai Manifesto*, 1956

"Ψ corpse"—because its atoms have become dematerialized into energy in motion (analogous to an electron as a Schrödinger Ψ wave). The doomed residents of Hiroshima who were at ground zero on the morning of August 6, 1945, were literally vaporized (the force of the uranium fission bomb broke the molecular bonds between the atoms in their bodies), turning each person into a cloud of dust—a Ψ corpse—which blew away in the wind. (There were no solid remains to bury; there are no graves for Ψ corpses). The text of Matsuzawa's *Ψ Corpse* flier goes on to describe an art exhibition of vanished objects. Viewers arrive at Matsuzawa's exhibition and enter an empty room with *chirashi* (fliers) strewn across the floor: "Therein appeared to be almost nothing. It was imperceptible whether anything or nothing was there. Thus, Nil was brimming there. In that 'void' thrown over the floor were *chirashi* fliers, printed in the color of blackened blood on paper white as bone."[31] Like Zen Buddhist monks throughout the ages, Matsuzawa eventually retreated into the mountains of Nagano Prefecture, where he lived in seclusion until his death in 2006, surrounded by his followers, who were artists and writers of the so-called Nirvana school.

By the mid-1950s the mood in Japan had shifted from despair at the destructive power of the atom bomb and Japan's defeat to a more optimistic outlook toward a future free from Japan's oppressive imperial culture and warlike past. In 1954 Yoshihara Jirō founded the Gutai Art Association based in the Osaka area, with the goal of creating

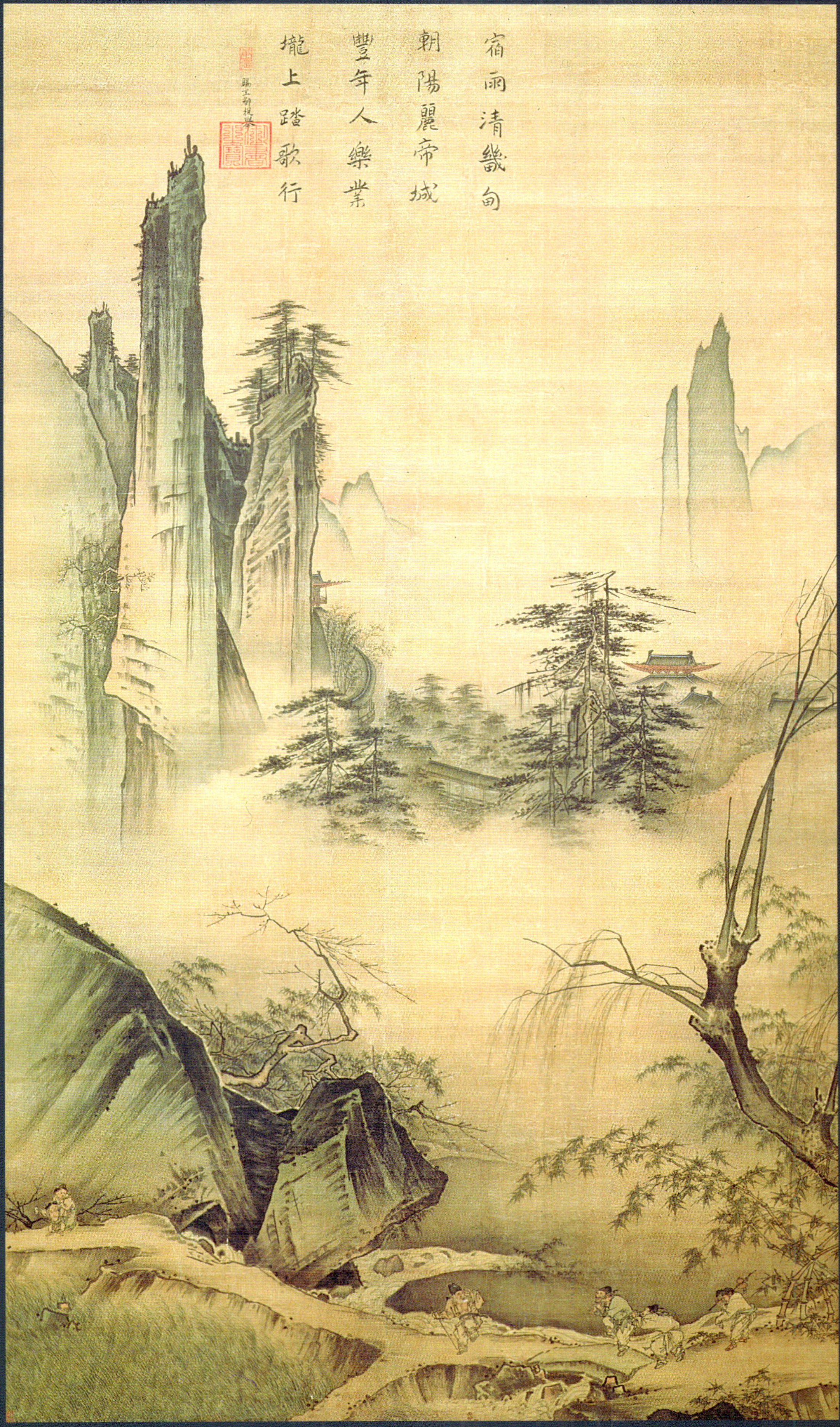
宿雨清畿甸
朝陽麗帝城
豐年人樂業
壠上踏歌行

RIGHT

10-22. Yoshihara Jirō (Japanese, 1905–72), *Red Circle on Black*, 1965. Acrylic on canvas, 71½ × 89⅜ in. (181.5 × 227.0 cm). Hyōgo Prefectural Museum of Art, Kobe. © Yoshihara Shinichiro.

OPPOSITE, TOP

10-23. Tatsuo Miyajima (Japanese, b. 1957), *U-Car (Uncertainty Car)*, 1993, LED, IC, motor, and battery. SCAI The Bathhouse, Tokyo, Japan.

Inspired by Heisenberg's Uncertainty Principle, Miyajima programmed the numbers that direct the path of two toy cars to incorporate an element of chance. The random numbers are displayed in red and yellow on top of the cars as they speed around, creating a pattern that, according to the artist, symbolizes uncertainty "by letting the counting numbers run around chaotically and freely in space."

OPPOSITE, BOTTOM

10-24. Tatsuo Miyajima (Japanese, b. 1957), *Mega Death*, 1999. LED, IC, electric wire, and sensor. Installation view at the Japan Pavilion, the 48th Venice Biennale, 1999. Courtesy of the Japan Foundation, SCAI The Bathhouse, Tokyo, Japan.

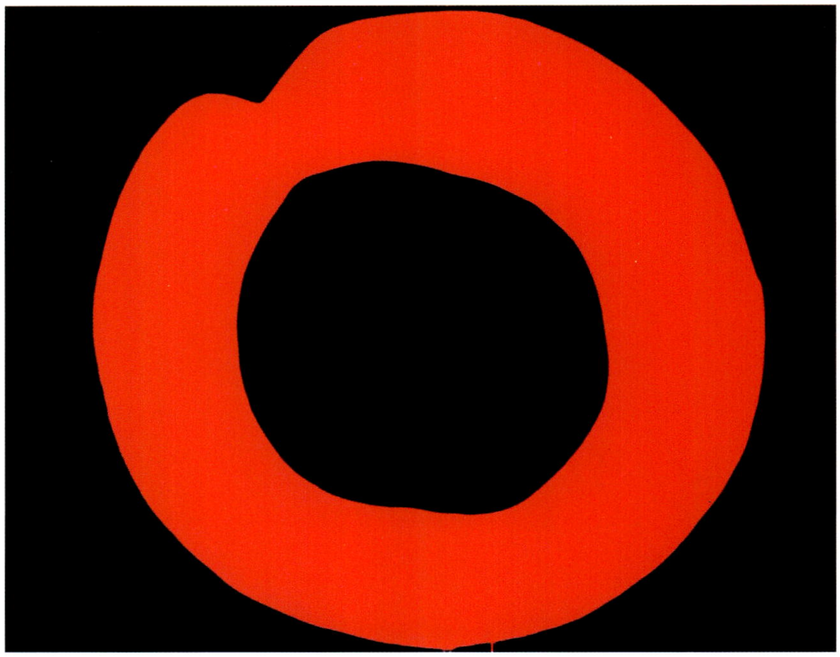

uninhibited expressions of euphoria. In the group's 1956 manifesto, Yoshihara described finding beauty in destruction, such as the rough stone of a ruined building. In the last decade of his life Yoshihara painted circles as a symbol of satori—the enlightenment of Zen Buddhism (plate 10-22).

The contemporary Japanese artist Tatsuo Miyajima intermixes themes from traditional Eastern philosophy and Western science and mathematics (plate 10-23). In *Mega Death* (plate 10-24), the twinkling numerals appear like stars in the heavens and symbolize the Buddhist/Taoist cycle of life, death, and rebirth. A human life is symbolized by numbers moving slowly from 1 to 9. When a light goes out, it symbolizes death, and when an extinguished LED turns on again, it symbolizes rebirth. An individual's death is not a tragedy but part of a natural progression, unlike the simultaneous death of many people at all stages of their lives—a "mega death"—such as the annihilation in a fraction of a second of 80,000 residents of Hiroshima, symbolized by a bank of lights going dark in unison.

Painting is special, separate, a matter of meditation and contemplation. . . . Spirituality, serenity, absoluteness, coherence. No automatism, no accident, no anxiety, no catharsis, no chance. Detachment, disinterestedness, thoughtfulness, transcendence.

—**Ad Reinhardt, 1955**

Meanwhile, in response to World War II, the New York artist Ad Reinhardt called for artists to re-examine their role in society.[32] Reinhardt was a close friend of Thomas Merton, who, after their student days at Columbia in the mid-1930s, entered a Trappist monastery in Kentucky in 1941. Merton made it his mission to combine the tradition of negative theology that was established in AD 500 by the Eastern Orthodox monk Pseudo-Dionysius (see chapter 1) with Taoism and Buddhism.[33] He initiated a dialogue with Suzuki in 1959, and the men joined forces to create a universal outlook appropriate to modern society.[34] Merton thought that much of the mystical spirituality of the early church had been lost in later Catholicism, and he urged seekers to meditate and develop self-awareness.

Reinhardt studied Asian art from 1944 to 1952 at the Institute of Fine Arts in New York, and during these years he also travelled throughout Asia, India, and the Near East. His notebooks are filled with comments about Asian thought, such as the Taoist vision of the natural world as a unified system of interrelated parts organized according to the Tao

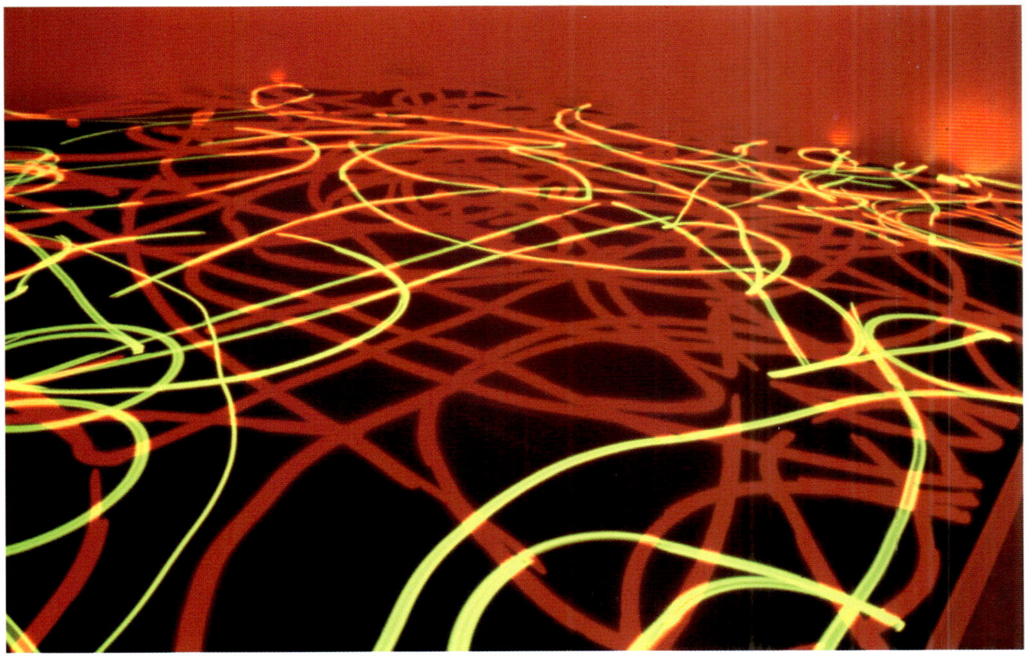

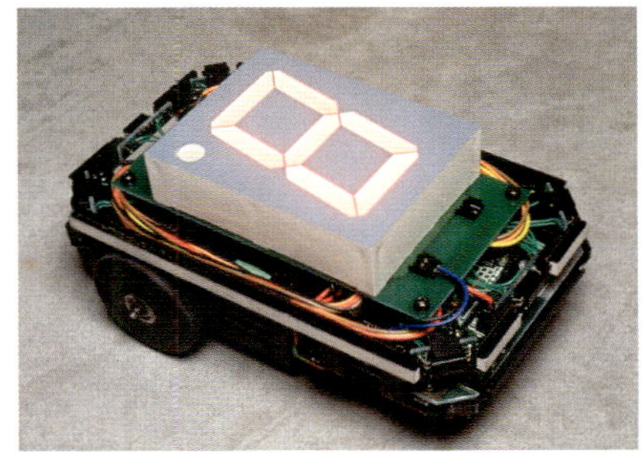

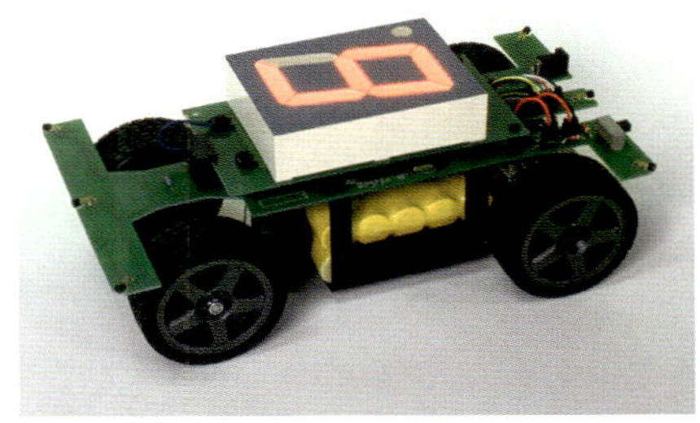

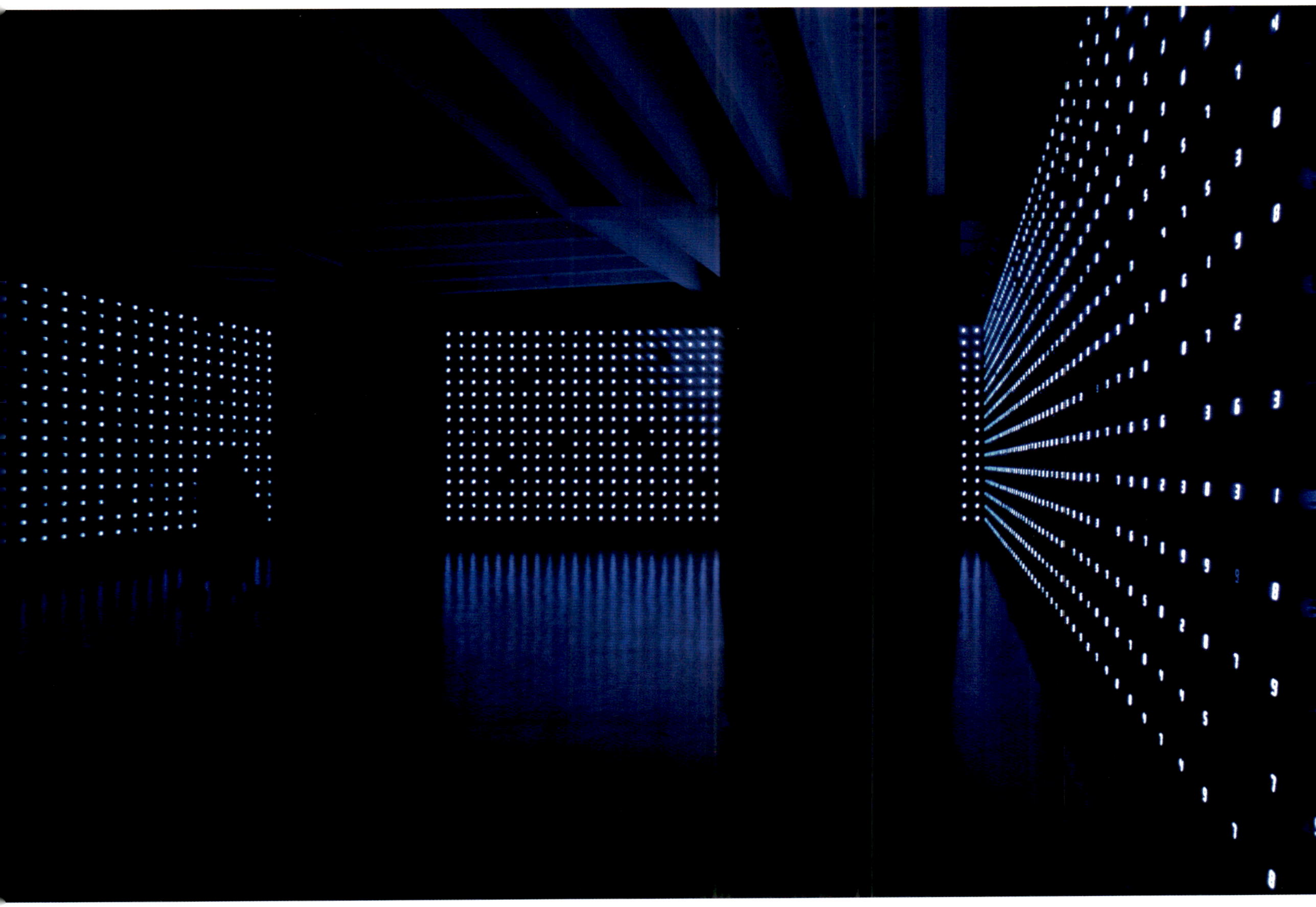

RIGHT

10-25. Ad Reinhardt (American, 1913–67), *Small Painting for T. M. (Thomas Merton)*, 1957. Oil on canvas, 10½ × 8 in. (26.7 × 20.3 cm). The Abbey of Gethsemani, Trappist, Kentucky. © 2014 Estate of Ad Reinhardt/Artists Rights Society, New York.

Based on the layout of a magic square, Reinhardt composed this painting with nine dark blue squares. Each square has a slightly different hue and surface, making the pattern barely perceptible. Merton spent much of his time in silent meditation, living as a monk in a cloistered Trappist monastery, where trivial talking is discouraged.

OPPOSITE

10-26. Mark Rothko (American, 1903–70), *Blue and Grey*, 1962. Oil on canvas, 76 × 68⅞ in. (193 × 175 cm). Foundation Beyeler, Riehen/Basel, Beyeler Collection. © 1998 Kate Rothko Prizel and Christopher Rothko/Artist Rights Society, New York.

The world is illusory only insofar as it is misinterpreted to fit our prejudices about our limited ego-selves. This simple, direct approach to reality, this unabashed apprehension of the One in the Many, of the Void in everyday life and in the ordinary world around us, is the foundation for Zen humanism in the world of today.

—Thomas Merton, "Buddhism and the Modern World," 1967

(the way)—in other words, by nature's own rules: "The Tao is through and through dark and mysterious."[35] By the late 1950s, Reinhardt was painting simple geometric shapes based on icons from Asian and Islamic cultures, which, according to Reinhardt, should be contemplated slowly and intensely over time, with the goal of purifying one's mind through meditation. In 1957 Reinhardt did a painting for Merton's monastic cell, *Small Painting for T. M. (Thomas Merton)* (see plate 10-25), which is dark blue with a barely visible pattern of nine squares arranged on a three-by-three grid, like a magic square or a cross, which struck Merton as "a most religious, devout, and latreutic small painting."[36] From 1961 until his death in 1967, Reinhardt painted a series of black squares with this same three-by-three grid format, which he described as "pure, abstract, non-objective, timeless, spaceless, changeless, relationless, disinterested painting . . . aware of nothing but art (*absolutely no anti-art*)."[37] Some of Reinhardt's American contemporaries also made simple geometric patterns as objects of contemplation, including Agnes Martin's delicate grids in oil and gold leaf (*Grey Stone II*, 1961; Fisher Landau Center for Art, Long Island City, New York), Anne Truitt's geometric sculpture based on tombstones (*Two*, 1962; Yale University Art Gallery, New Haven), and Richard Tuttle's geometric shapes in ephemeral tissue and cloth (*Cloth Octagonal* 2, 1967; Museum of Modern Art, New York).[38]

From a cave they looked
out on the world,
And struggled to understand,
And slowly the flicker of
their intelligence,
Grew and consumed the
dusk with their mind,
And the Brute-Man stood
erect and knew himself.

—Mark Rothko, "Walls of Mind: Out of the Past," 1920s

You may say I'm a dreamer
But I'm not the only one,
I hope someday you'll join us
And the world will live as one.

—John Lennon, "Imagine," 1971

Mark Rothko shared Reinhardt's interest in simple geometric shapes that exist as pure objects of contemplation (plate 10-26). Rothko read Eastern philosophy but also looked to the Western tradition after being introduced to Plato in a philosophy course as a sophomore at Yale University. He maintained an interest in Platonic thought, penning a poem about Plato's allegory of the cave in the 1920s and making frequent reference to Plato's writings in a notebook ("The Scribble Book") that he kept in the 1930s.[39] In Plato's cave the only light source is a flickering campfire, and bound prisoners see only shadows cast on the walls of the dark cave, from which they try to discern the nature of the transitory, material world. If freed from bondage, the prisoners can move around and begin an arduous ascent along a path that leads to the cave's entrance, where they get their first glimpse of the natural world in blinding sunlight (*Republic*, 514a–520a).

The American composer John Cage was a student of Suzuki at Columbia University for three years. The main idea that he came away with was that there is no separation of life and art: making a cup of tea is an aesthetic act, and every sound is musical. To make the latter point, Cage composed *4'33"* for solo piano; during the concert the pianist sits in silence with the keyboard closed for four minutes and thirty-three seconds while the audience listens to the ambient sound in the concert hall. Cage also adopted the *I-Ching* (see plates 1-46, 1-47, and 1-48 in chapter 1) as a method to compose music by tossing coins to determine the pattern and rhythm of the notes (*Music of Changes*, 1951). Cage urged his audiences to listen to his compositions with an open mind ("a mind that has nothing to do"), without preconceptions about what music is: "Where these ears are in connection with a mind that has nothing to do, then the mind is free to enter into the act of listening, hearing each sound just as it is, not as a phenomenon more or less approximating a preconception."[40] A contemporary of Cage, the Lithuanian-born American George Maciunas, who was leader of the international artist's network Fluxus, undertook an extensive study of art history at New York University in the 1950s. Maciunas became convinced that

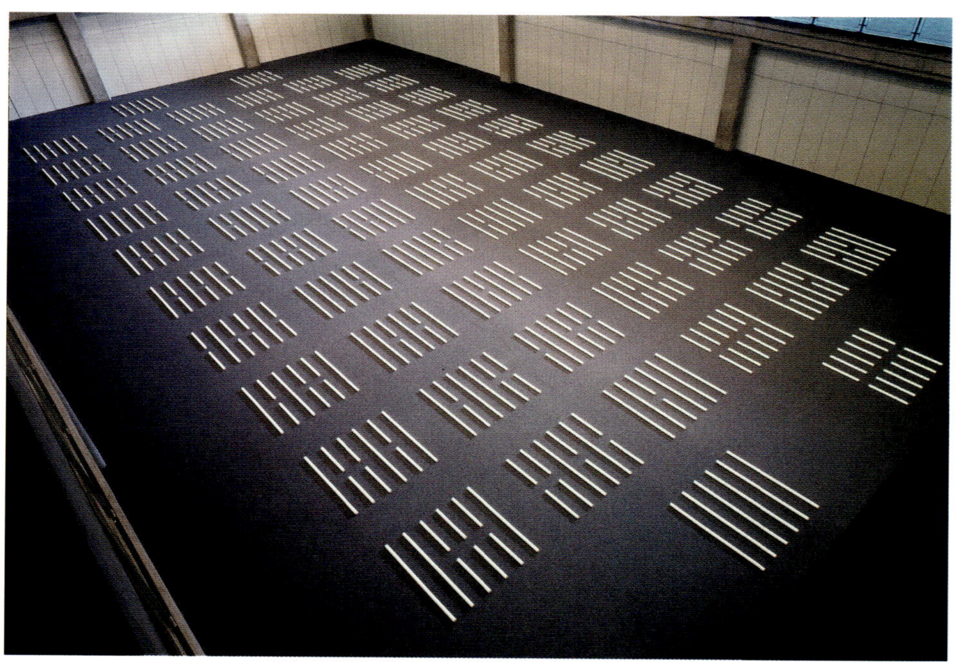

RIGHT

10-27. Walter de Maria (American, 1935–2013), *360° I-Ching/643 Sculptures*, 1981. 576 rods, wood, and lacquer. Installation at Moderna Museet, Stockholm, 1989. Collection Dia Art Foundation, New York. © Dia Art Foundation, New York.

OPPOSITE

10-28. Walter de Maria (American, 1935–2013), *The Lightning Field*, 1977. 400 stainless steel poles, diameter 2 in. (5 cm), average pole height 20 ft., 7 in. (6.27 m.), near Quemado, New Mexico. © Dia Art Foundation, New York.

The tops of the 400 poles form a flat plane, which is installed on a grid measuring 1 mile by 1 km. Presumably it is an 80 × 50 pole rectangle on a 66-ft. grid, since 80 × 66 ft. = 5280 ft. (1 mile), and 50 × 66 ft. = 3300 ft. (about 3280 ft., which is 1 km).

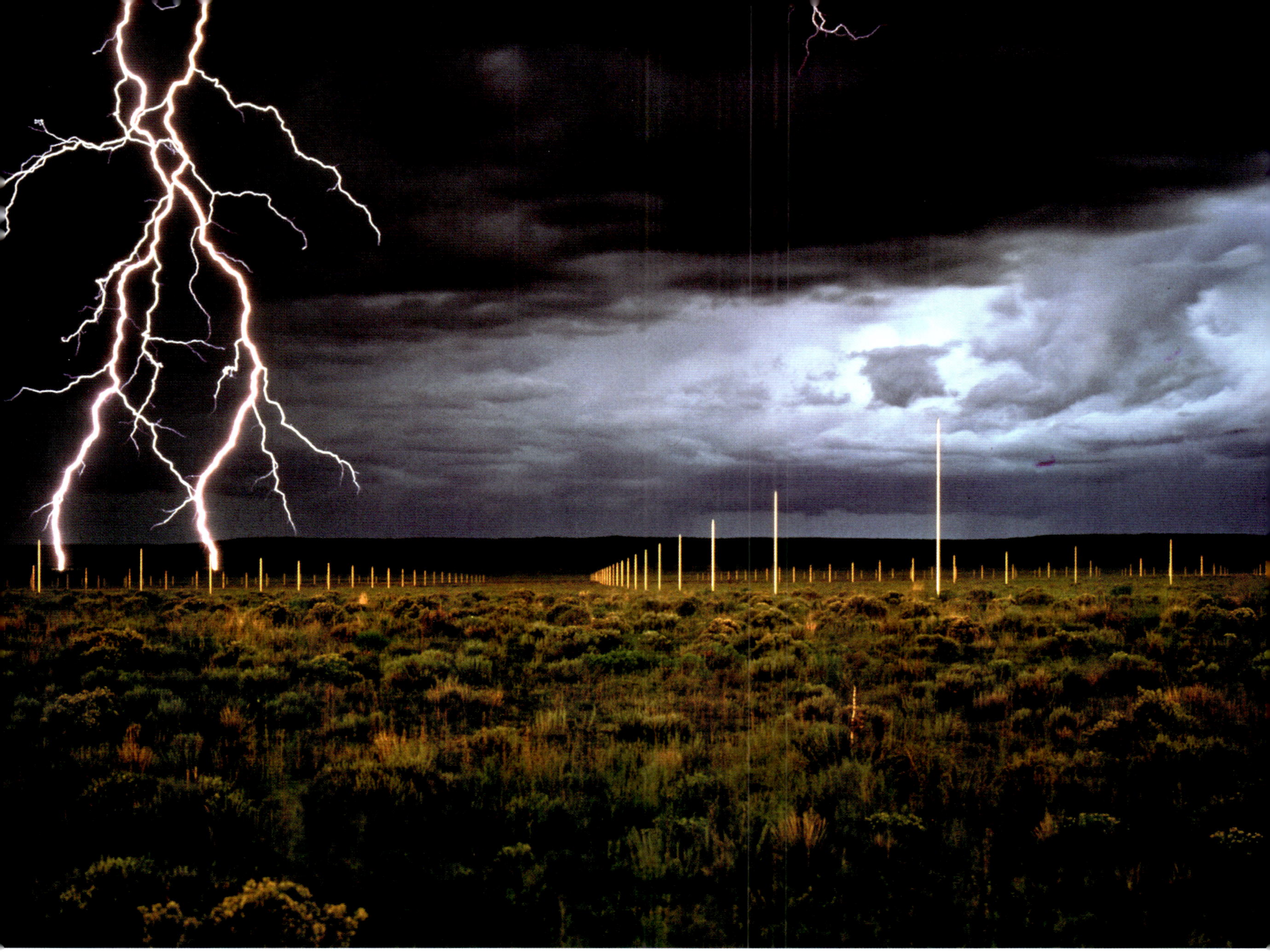

there are mathematical patterns embedded in art history, and he minutely charted these alleged patterns on a huge diagram (measuring 40 by 64 inches) of interconnected pathways among European styles from AD 600 to 1600 (*History of Art Chart*, ca. 1955–60; Gilbert and Lila Silverman Collection, Detroit).[41]

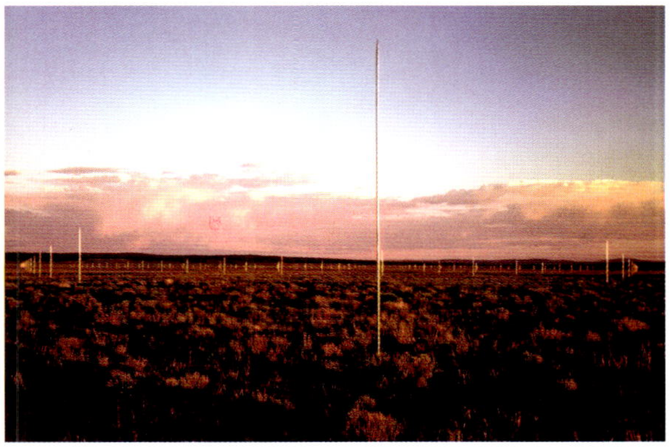

Like Reinhardt and Cage, the American artist Walter de Maria had a strong interest in Eastern thought and used numerical systems and geometric forms as objects of contemplation (plate 10-27). Visitors to his *Lightning Field*, a vast grid of metal poles stuck in a field in rural New Mexico (plate 10-28),[42] are given access to the remote site on the condition that they spend twenty-four hours in a cabin located at the edge of the grid, observing how the appearance of the metal poles changes throughout the day and night under the light of the sun and moon. Someone who meditates has a flash of insight very infrequently, if at all. By analogy, lightning (insight) strikes *Lightning Field* only during a rainstorm (brainstorm), which is very uncommon in a desert wasteland (a barren mind).

QUANTUM MYSTICISM

Members of the Cold War generation were awed and terrified—but also fascinated—by the nuclear force that the atomic bomb unleashed. Scientists and journalists responded by writing a huge post-1945 popular literature on subatomic physics. Most was written in a clear, expository style, but a few popularizers were members of the Copenhagen school, who, like Niels Bohr and Werner Heisenberg, continued to write in the promotional style they developed in the 1920s and 1930s (see chapter 8). Rather than giving equal time to alternatives as an unbiased professor would, Bohr and Heisenberg presented the Copenhagen interpretation to Cold War audiences as *the* interpretation of the subatomic realm.[43]

In 1945 the British rounded up Heisenberg, along with Hitler's other (unsuccessful) bomb builders. After detaining him in England for questioning, the British allowed Heisenberg to return to the British-occupied zone of West Germany, where he reveled in the increased prestige accorded nuclear physics, becoming director of the Kaiser-Wilhelm Institut für Physik (The Kaiser-Wilhelm institute for physics, soon renamed the Max Planck Institut für Physik). In his popular writing, Heisenberg proclaimed that the Copenhagen interpretation is the only possible meaning of quantum physics, and he presented himself as the leader of a revolution (*Physics and Philosophy: the Revolution in Modern Science*, 1959).[44]

Both Heisenberg's assertions and his style inspired members of the so-called New Age movement, who were heirs of the theosophist Helena Petrovna Blavatsky.[45] New Agers took Heisenberg's pronouncements at face value and cited them repeatedly as scientific "proof" of their magical doctrine—so-called quantum mysticism—that human consciousness causes the physical world.[46] It is only one step from Heisenberg's extreme positivism—his claim that an electron does not *have* a location or speed until it is observed—to the (absurd) conclusion that human consciousness *causes* the physical reality (the location in space and time) of the electron in the world-out-there. Heisenberg never took that step, but many occultists thought he had, based on his vaguely worded popular science writing, leading to his enormous popularity with the counter-culture of the 1960s. Occultists saw the role of observation in quantum mechanics as proof that the human mind causes the physical world, as does pure spirit (the Brahma) in Buddhism. Heisenberg's reference to "revolution" in the subtitle of his 1959 book also struck a chord in the counter-culture, as did several remarks by Bohr,[47] who selected the Taoist yin-yang symbol to represent complementarity (the dual wave-particle nature of electrons).[48] Occultists took this heedless verbiage by Bohr and Heisenberg as an oracle from the halls of science that quantum physics is the bridge—long sought by universalists—linking East and West.[49]

Capra's Tao of Physics . . . *was just physics porn—titillating but with no follow-through.*

—George Johnson, *New York Times*, June 19, 2011

New Agers also adopted Heisenberg's style of science writing as salesmanship, ratcheting up the tone in breathless claims about the association of physics to Buddhism and Taoism. Earlier theosophists never had a large following among intellectuals because they lacked academic training, but educated audiences expanded after the physicist Fritjof Capra published a book—based on his reading of Heisenberg[50]—in which he declared that human consciousness causes the physical world (*The Tao of Physics: An Exploration of the Parallels between Modern Physics and Eastern Mysticism*, 1975). Capra was born in Austria, earned a PhD in theoretical physics from the University of Vienna, and then came to America to conduct research in subatomic physics at Lawrence Berkeley National Laboratory of the University of California at Berkeley. There he joined the Fundamental Fysiks Group, which was founded in 1975 by Berkeley science graduate students who brewed an intoxicating mix of physics, meditation, hot tubs, and LSD.[51] Capra's *Tao of Physics* threw open the floodgate to a torrent of New Age books on physics; like the ancient Greeks who drank from a fountain of knowledge whose waters flowed from a sacred spring in the Pierian mountains, some intellectuals made quantum mechanics their muse, indulging in speculation that goes way beyond physics. In another example of this type of wide-eyed pseudo-science journalism, the Indian-born American physician Deepak Chopra wrote: "The physical world, including our bodies, is a response of the observer. We create our bodies as we create the experience of our world"; thus, by the power of consciousness, we can, according to Chopra, achieve immortality in an "ageless body, timeless mind."[52]

A little learning is a
dangerous thing;
Drink deep, or taste not
the Pierian spring.
There shallow draughts
intoxicate the brain,
And drinking largely
sobers us again.

—Alexander Pope,
"*An Essay on Criticism*," 1711

There is no science supporting the hypothesis that human consciousness causes the physical world. The alleged link of quantum physics to Eastern mysticism trivializes the millennia-old monastic traditions of Buddhism and Taoism.

Like New Agers, certain post-1945 poets and novelists also took the Copenhagen interpretation at face value and made an inference about the goal of literature, as American literary critic N. Katherine Hayles stated: "As Heisenberg said of science, literature is not about reality but about what we can say about reality."[53] The job of the novelist, according to Hales, is not to tell a story about a world-out-there but to reflect on the language in which the novel is written. In Paris in the 1960s, this self-conscious writing style was associated with the Nouveau Roman (new novel), and a decade later a character in an American novel by Thomas Pynchon (who studied physics as an undergraduate) solipsistically reflects on the language *of* the novel: "The new coinages seem to be made unconsciously. Is there a single root, deeper than anyone had probed, from which Slothrop's Blackwords only appear to flower separately? Or has he by way of language caught the German mania for name-giving, dividing the Creation finer and finer, analyzing, setting name hopelessly apart from named" (*Gravity's Rainbow*, 1973).[54]

The new novel doesn't
represent, it searches. And
what it searches for is itself.

—Alain Robbe-Grillet,
Pour un nouveau roman
(For a new novel), 1963

Quantum mysticism was part of the death throes of Absolute Spirit, which was mortally wounded during the destruction of Germanic culture in World War II. Capra was born and educated in Austria and Chopra in India, both countries with long traditions of mixing German Idealism and Buddhism. But in the massive shift of population after World War II, Capra and Chopra both became citizens of the United States, where their idealist seedlings withered in the native soil of pragmatism and empiricism. Today most novelists tell stories, and science journalists write in an expository style with the goal of keeping the public informed. The Copenhagen school's style of science writing as salesmanship, as well as the counter-culture's gee-whiz journalism that it inspired, have largely faded into history.[55]

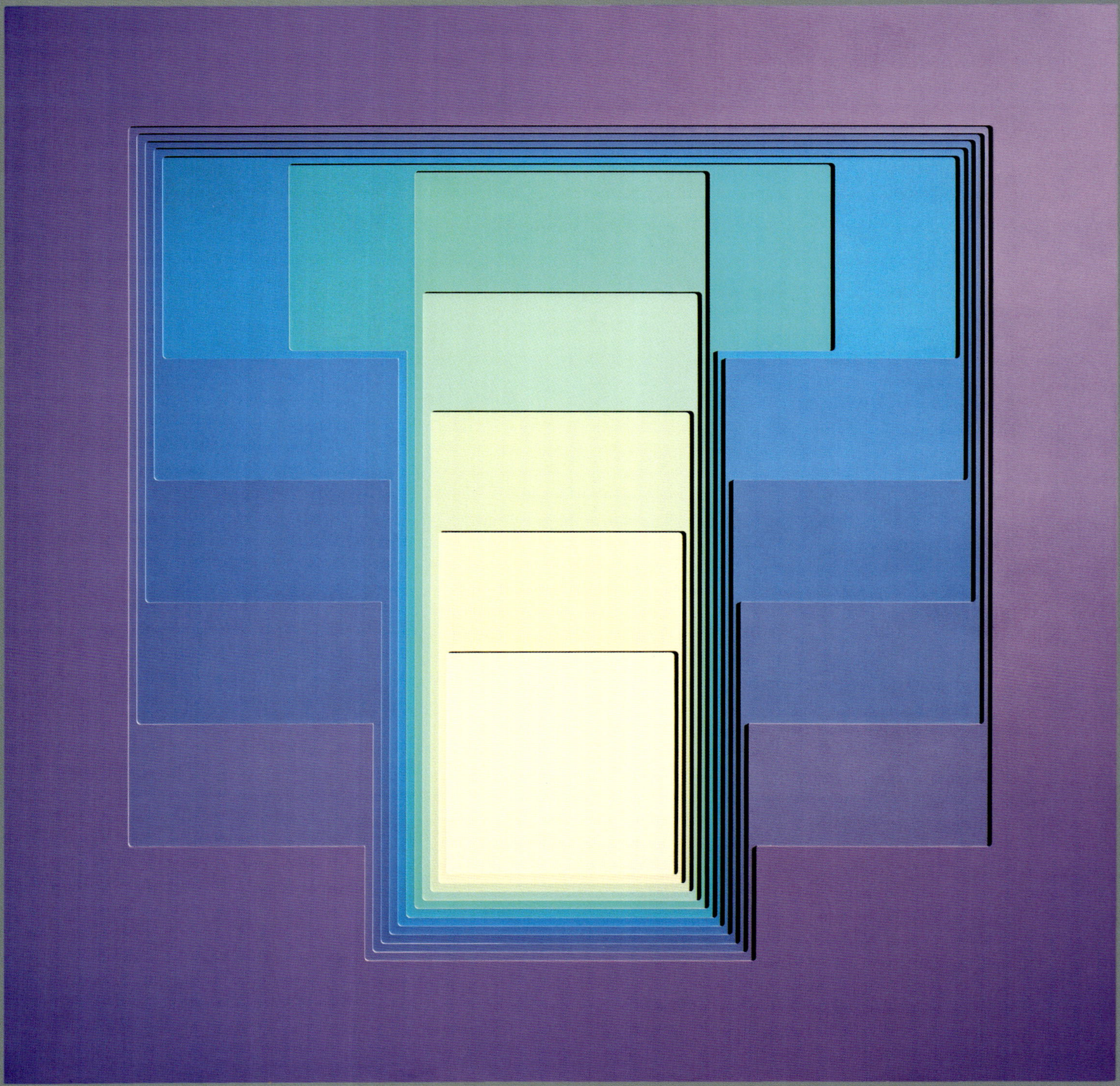

11

Geometric Abstraction after World War II

Let us now try, guided by the axiomatic concept, to look over the whole of the mathematical universe. . . . The organizing principle will be the concept of a hierarchy of structures, going from the simple to the complex, from the general to the particular.

—Nicolas Bourbaki, "The Architecture of Mathematics," 1948

I am convinced of the possibility of developing art in which the mathematical way of thinking is fundamental. . . . The art does not represent the surface of things, but rather the underlying cosmic structure. The art presents the overall view of the world that we have today. The art does not depict nature, but rather creates new systems that make visible the underlying patterns.

—Max Bill, "The Mathematical Way of Thinking in the Art of our Time," 1949

THE VIBRANT CENTERS of modern mathematics and quantum physics—Göttingen, Berlin, Vienna, and Copenhagen—were reduced to intellectual rubble during World War II. When David Hilbert retired from the University of Göttingen in 1930, his successor was Hermann Weyl, but after Hitler was named chancellor of Germany in 1933, Weyl, who had a Jewish wife, fled to America and joined Einstein, who had already found refuge at the Institute for Advanced Studies in Princeton, New Jersey. Two of Weyl's colleagues at Göttingen, Emmy Noether and Max Born, were among the first Jewish professors to be expelled from German universities; Noether went to America and taught at Bryn Mawr College, and Born went to Cambridge University. In 1935 Rudolf Carnap, whose pacifist convictions put him at risk in Nazi Germany, found a haven at the University of Chicago. After Hitler ordered the arrest of Danish Jews on October 1, 1943, Niels Bohr fled Nazi-occupied Denmark for the United States, where he joined the Manhattan Project to build the atom bomb.

Mathematics and science had been propelled by faith in rationality, objectivity, and universalizable knowledge, but these Enlightenment ideals were war casualties in defeated, divided Germany. In the city of Ulm, the Hochschule für Gestaltung was established in the 1950s to revive the Bauhaus curriculum in West Germany; the school attracted many foreign students but few Germans. Meanwhile artists in the carpet-bombed cities of East Germany—Dresden, Leipzig, and Berlin—made Neo-Expressionist art about a mad world in which mankind balances on a burning bridge over a primordial abyss (plate 11-2).

Although the war undermined confidence for all who suffered losses, in countries that were not utterly destroyed, artists continued to express confidence in the Enlightenment ideals—albeit in a somewhat tattered state—by creating abstract art that was geometric

11-1. Karl Gerstner (Swiss, b. 1930), *Color Sound 66: Introversion*, 1998. Nitrocellulose lacquers on phenolic resin panels, 46¾ × 46¾ in. (119 × 119 cm). Courtesy of the artist.

The artist overlaid twelve colored planes with T-shaped holes to produce this uniform "step" from one hue to the next. To achieve these intense colors, Gerstner developed a palette of ten primary pigments of great purity with the help of the psychologist Max Lüscher and the chemist Hans Gärtner of the Swiss pharmaceutical corporation Ciba-Geigy (now Novartis), which is based in Basel.

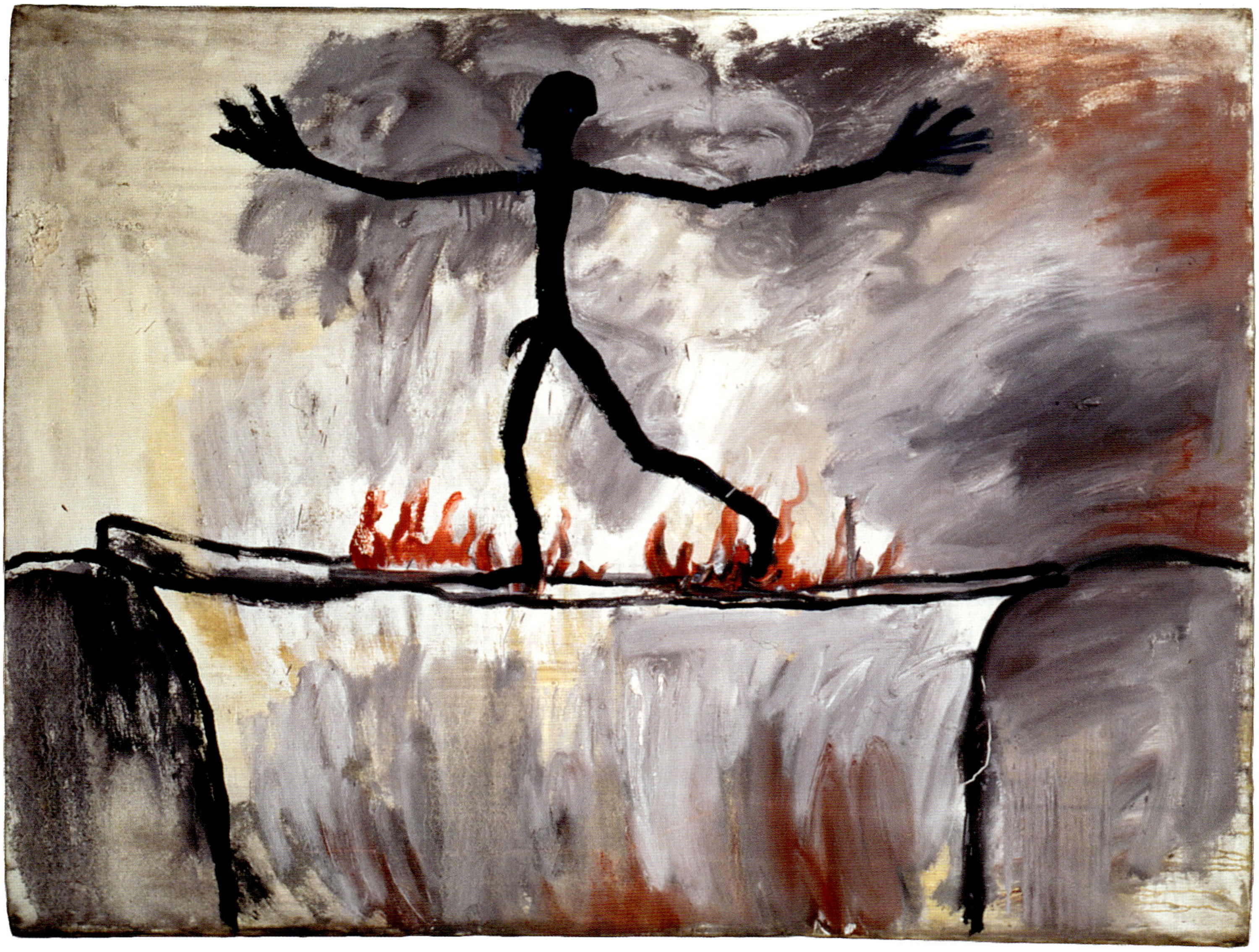

11-2. A. R. Penck (German, b. 1939), *Der Übergang* (The crossing), 1963. Oil on linen, 37 × 47¼ in. (94 × 120 cm). Courtesy of the artist, Ludwig Forum, Aachen, and Ludwig Stiftung, Aachen. © 2014 VG Bild-Kunst, Bonn/Artists Rights Society, New York.

and orderly, executed with detachment, and expressed the power of reason. Geometric art was made in Switzerland, which remained neutral during the war; in France, which was occupied but not heavily bombed; in England, which fought off an invasion; and in the Americas (United States, Canada, and Latin America), which sent troops but did not have war on their soil. In these lands geometric abstract art flourished after 1945, resonating with an international culture that had gained enormous respect for the exact sciences—especially physics—after the world entered the nuclear age. In Zurich, pre-war styles of Concrete Art moved to a higher level of symmetry when the young Karl Gerstner added precise procedures for determining not only form but also color. In Paris a group of mathematicians adopted the collective name Nicolas Bourbaki and imported David Hilbert's formalism from Germany, thereby throwing off the intuitionist mantle that had settled on French mathematics. Following Bourbaki's lead, French authors pursued a style of literary formalism under the banner of Oulipo, and Parisian painters developed Optical Art. In London, members of Systems Art broke with the British realist tradition and created an art of pure form and color. Latin American Constructivists in São Paulo, Rio de Janeiro,

Buenos Aires, and Caracas created styles that associated geometric abstraction with orderly social progress. In New York, Conceptual artists in the circle of Sol LeWitt devised rigorous rules to create mathematical objects, while Minimal artists pursued a reductionist agenda. Although symmetry reigns as the queen of science, in 1957 Mother Nature revealed a startling exception—an asymmetry in the heart of nature—inspiring Robert Smithson to create his asymmetrical modernist icon, *Spiral Jetty* (1970).

CONCRETE ART IN SWITZERLAND AFTER WORLD WAR II

Switzerland was surrounded on her three borders by Hitler's Germany, Mussolini's Italy, and Nazi-occupied France, and Concrete artists interpreted their pristine canvases as symbolizing order in the midst of chaos. After the war they designed functional objects to promote a peaceful life in a more harmonious European society.[1]

Max Bill dominated Swiss architecture and graphic design after 1945, always stressing the continuity between fine and applied art, as he wrote in 1950: "Designers who realize new forms are consciously or unconsciously reacting to trends in contemporary art because it is in art that the intellectual and spiritual currents of every epoch find their visible expression" (plate 11-4).[2] In the early 1950s Bill moved to Ulm to help the designer Otl Aicher and his wife Inge Scholl establish an art school, the Hochschule für Gestaltung, as a memorial to Scholl's younger siblings, Hans and Sophie Scholl, who were beheaded by the Nazis in 1943 for a non-violent act of resistance—distributing pacifist leaflets—at the university in Munich where they were students. As founding director of the Hochschule, which opened in Ulm in 1953, Bill modeled its curriculum after the Bauhaus and featured "good form" (plate 11-3). Bill knew the term from Gestalt psychology, where "good form" denotes a simple form that is often symmetrical,[3] and in the post-war years, Bill gave it the political connotation of promoting an ethical, orderly environment.[4] Despite defeated Germany's loss of esteem, Bill's strong reputation in the international art world attracted foreign students from forty countries around the globe. But relatively few local students came, a sign that young German artists had lost confidence in the rational basis of the orderly styles taught at Ulm.[5] Bill put his efforts into reviving the craft-based approach of his alma mater, but other Ulm faculty members, especially Argentine painter Tomás Maldonado and Dutch architect Hans Gugelot, taught technologies suitable for mass production. In protest against this shift in the school's mission, Bill resigned and returned to Zurich in 1956.[6]

In designing a pattern, for example, one has a feeling how successive parts should follow one another; one knows what a "good" continuation is, how "inner coherence" is to be achieved, etc.; one recognizes a resultant "good Gestalt" simply by its own "inner necessity."

—Max Wertheimer, "Laws of Organization in Perceptual Forms," 1923

After the war Andreas Speiser released a second 1945 edition of his 1932 book *Die mathematische Denkweise* (The mathematical way of thinking), in which he stressed that, although group theory was not invented until the nineteenth century, the search for the fundamental laws of nature had always been a search for symmetrical patterns—the building-blocks from which nature is made. Bill's good friend Adrien Turel wrote to Speiser in 1949, praising the book, at the very time that Bill was finishing his own treatise, "Die mathematische Denkweise in der Kunst unserer Zeit" (The mathematical way of thinking in the art of our time), in which Bill reflected not only Speiser's title but also the mathematician's thesis that the search for the underlying structure of the natural world has revealed the symmetry of nature.[7] After noting, as Speiser had, that Bach composed using symmetrical patterns, Bill declared that Swiss Concrete Art is based on mathematics, which he defined as both a part of human thought (innate logic) and a way of understanding the world (axioms of science) as well as an exploration of basic mathematical reality: "I am convinced of the possibility of developing art in which the mathematical way of thinking is fundamental. . . . The mainspring of visual art is geometry, which is the arrangement of forms on a surface or in space. Mathematics is a basic method of thought and a principal means by which we conceive the world around us. Mathematics is also a science of the relationships of object to object, group to group and motion to motion."[8]

Bill then drew the important distinction between art *as* mathematics and art *inspired by* mathematics. The latter is created by a process analogous to the practice of mathematics: "Contemporary art that is based on mathematical principles is not an illustration of a mathematical formula, and the artists hardly use pure mathematics. Rather the artists build up patterns of rhythms and relationships, based on laws that have their own aesthetic origin. As such the art is analogous to the development of mathematics, where fresh advances were conceived based on the thoughts of one of its pioneers."[9]

Besides exploring Speiser's ideas about the philosophy of mathematics, Bill also noted the application of mathematics to physics and cosmology. In the 1947 Allianz exhibit that included Lohse's early use of group theory (see plate 7-24 in chapter 7), Bill had exhibited a painting that he named *Infinite and Finite* (1947), borrowing terms from popularizations of Einstein's cosmology, in which space has finite volume yet is boundless because it curves like the surface of a sphere. In his essay, Bill gave examples of mathematical ideas, including infinity and a two-dimensional object with only one side (a Möbius strip). He listed mathematical structures such as ancient geometry and modern curved space-time, concluding that mathematical patterns underlie the natural world and hence mankind's conception of order: "The following are things that do not seem to have any bearing on our daily needs but nevertheless they are of the greatest significance: the mystery of mathematical problems, the inexplicableness of space, the nearness or distance of the infinite, a space beginning on one side and ending in a different form on another side that is congruent with the first; limitations without fixed boundary lines; the manifold which is nonetheless a unity; a uniformity which is changed by the presence of a single stress; a force field; parallel lines that meet and the infinite that turns back upon itself; and then again the square, in all its stability, the straight line untouched by relativity, and an ellipse on which a straight line can be formed at any point. These are the elementary patterns underlying all human order and they are present in all the structures that we recognize."[10]

Bill went on to state that the mathematical patterns embedded in the cosmos are the content of Concrete Art: "These patterns supply contemporary art with a new content. This art is not formalism for which it is often mistaken, and this art does not symbolize beauty. Rather the forms embody thoughts, ideas and cognitions. The art does not represent the surface of things, but rather the underlying cosmic structure. The art presents the overall view of the world that we have today. The art does not depict nature, but rather creates

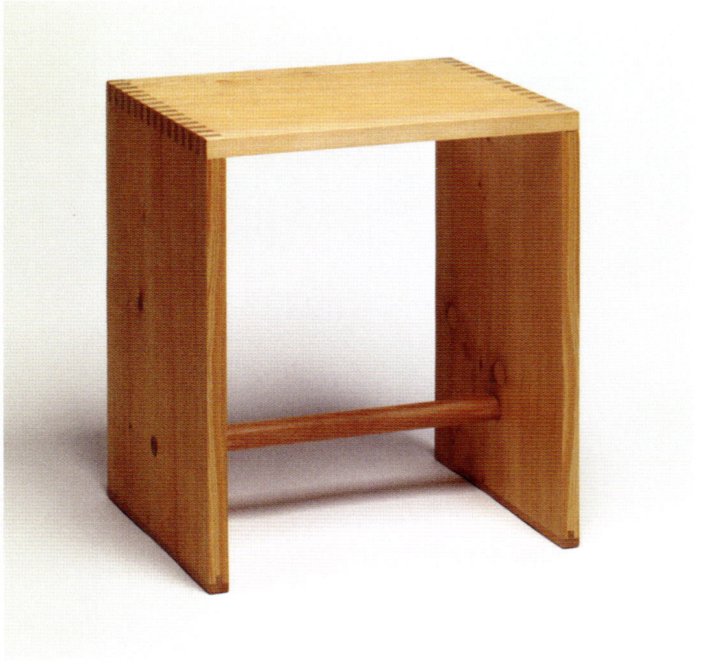

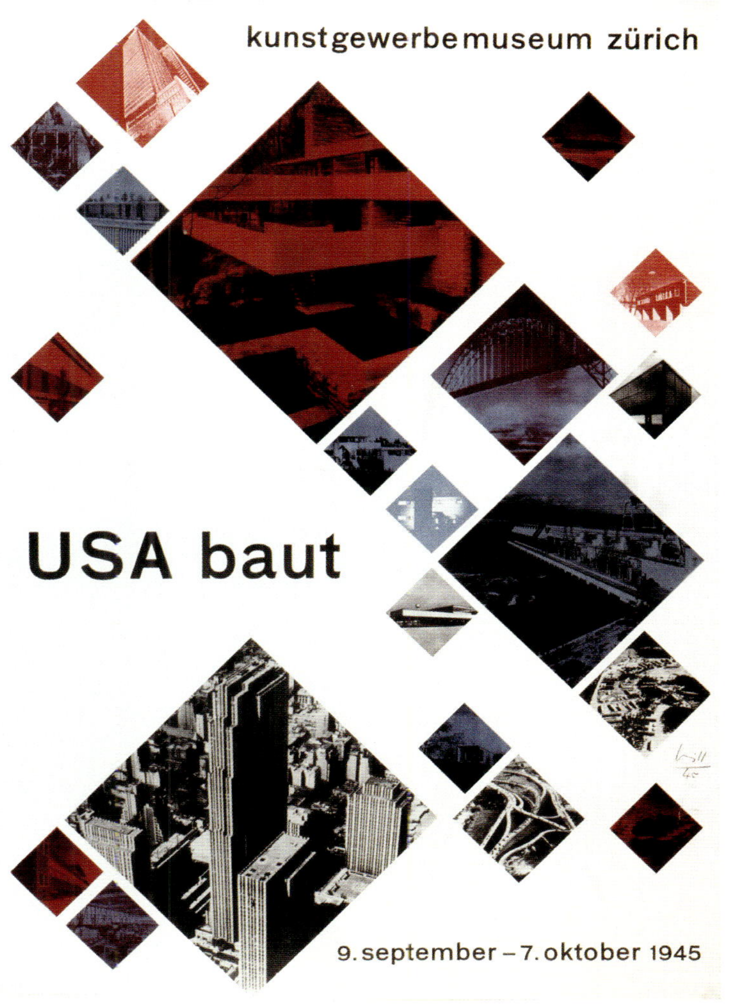

ABOVE

11-3. Max Bill (Swiss, 1908–94), *Ulmerhocker* (Ulm stool), 1954. Plywood. Museum für Gestaltung Zürich, Designsammlung. © ZHdk. © 2014 ProLitteris, Zurich/Artists Rights Society, New York.

RIGHT

11-4. Max Bill (Swiss, 1908–94), exhibition poster for *USA baut* (USA builds), Kunstgewerbemuseum, Zurich, 1945. Museum für Gestaltung Zürich, Plakatsammlung. © ZHdk. © 2014 ProLitteris, Zurich/Artists Rights Society, New York.

new systems that make visible the underlying patterns."[11] Thus, following Speiser, Bill proclaimed that Concrete Art expresses the understanding of nature as embodying mathematical patterns, which forms the basis of astronomy and physics: "The difference between the traditional concept of art and that which we have described here is of the same nature as the difference between the laws of Archimedes and those of modern astrophysics."[12]

In 1951 Bill and Andreas Speiser were speakers, along with Rudolf Wittkower, the renowned scholar of Italian Renaissance and Baroque art, at a conference on the topic of proportion in art and architecture held in Milan.[13] This is the earliest date that one can say with confidence that Bill met Speiser, although they had known each other's work since the mid-1930s. Bill responded to Speiser's thinking because, like the mathematician, he viewed "the mathematical way of thinking" in a broad cultural context, including pure mathematics (the perfect sphere), mathematics applied to nature (the cosmos as a space-time matrix), and mathematics applied to design (a chair with good form). The city of Ulm was the birthplace of Einstein, who had created the most exotic theories, and it was where Bill taught the most practical applied mathematics—designing sensible, inexpensive architecture and design to help rebuild post-war Germany. When Bill learned that Einstein's family home was being torn down to make room for a new building, he rushed

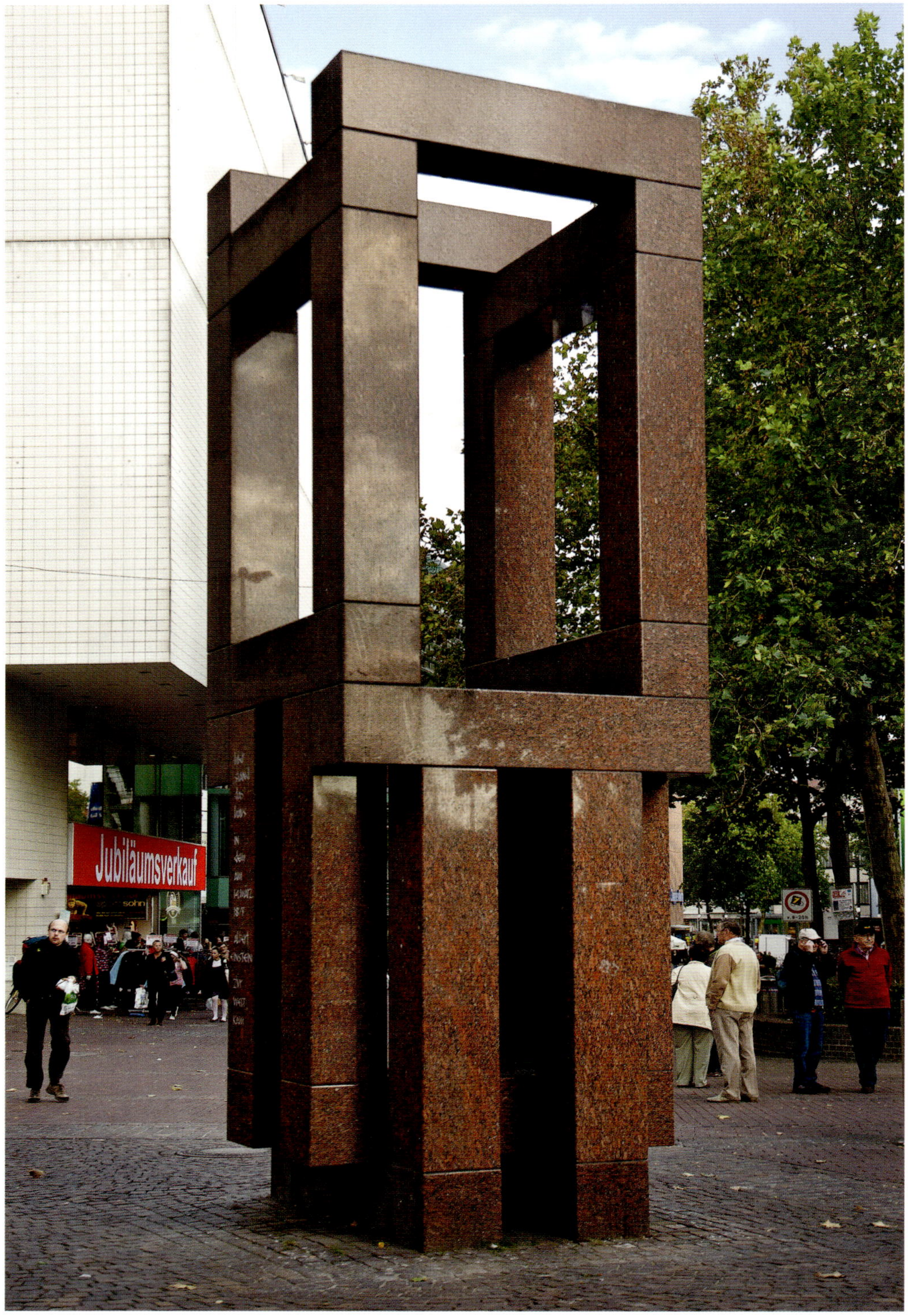

11-5. Max Bill (Swiss, 1908–94), *Monument to Einstein*, 1979–82. Granite, 19 ft. 8 in. (6 m) high. Ulm, Germany. © 2014 ProLitteris, Zurich/Artists Rights Society, New York.

to the site and salvaged the stone threshold, dragged it home, and gave it a place of honor in his garden.[14] Years later Bill designed a sculpture in memory of the father of space-time cosmology and convinced the city fathers of Ulm to put it on permanent public display to honor their most famous son (*Monument to Albert Einstein*, 1984; see plate 11-5).

In the early 1950s, when the Swiss Concrete artists ranged in age from Verena Loewensberg in her forties to Camille Graeser in his seventies (see chapter 7), the group was joined by Karl Gerstner, an artist in his twenties who shared their interest in algorithms,

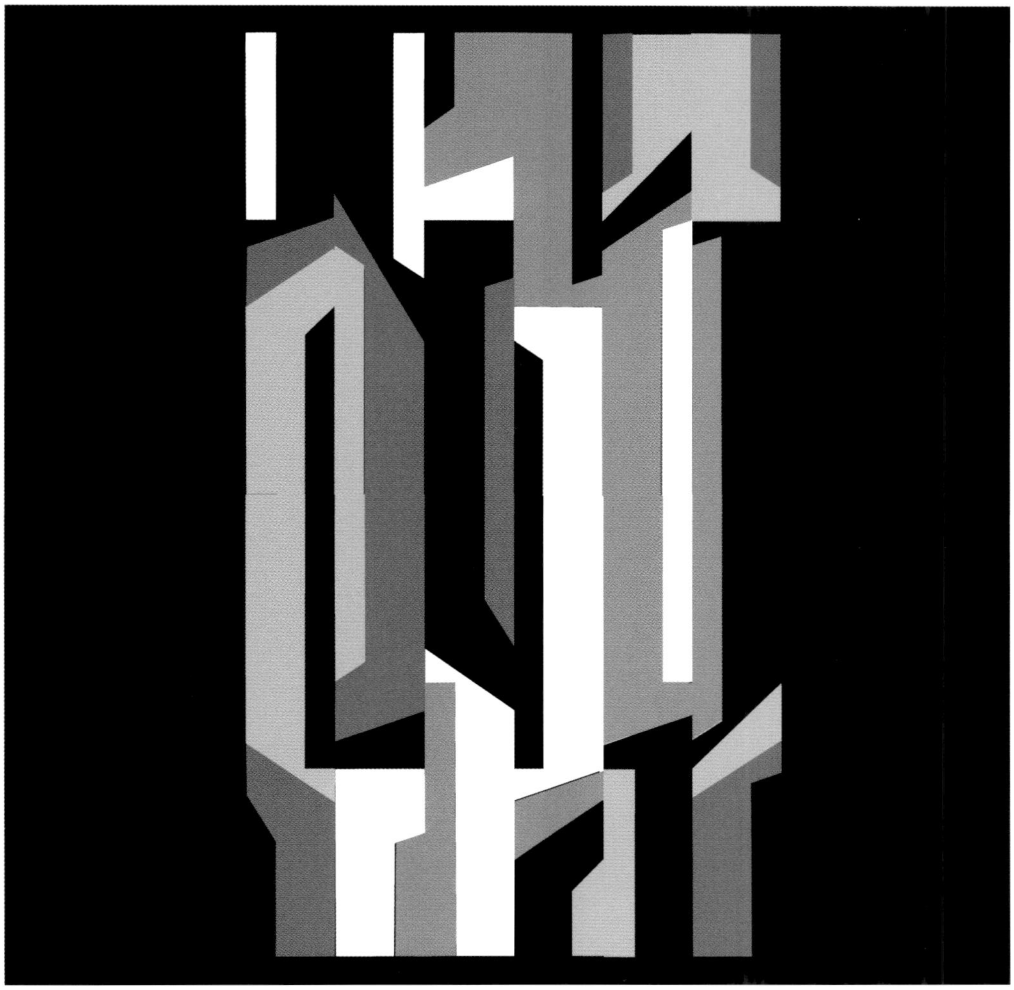

11-6. Karl Gerstner (Swiss, b. 1930), *Aperspective 1: The Endless Spiral of a Right Angle*, 1952–56. Synthetic resin paints on 12 Plexiglas panels, 3½ × 17¾ in. (9 × 45 cm) ea., secured with magnets onto iron over a base of black Plexiglas, 39⅜ × 39⅜ in. (100 × 100 cm). Courtesy of the artist.

visual perception, and the new cosmology.[15] In 1952 Gerstner created *Aperspective 1*, a work with twelve black and white units fixed to magnets so that the viewer can rearrange them (plate 11-6). The title reminds the viewer that in Einstein's cosmology an accurate description of the natural world can be given from any frame of reference (any coordinate system), and thus mankind lacks a unique perspective—humans are "aperspective." Resonating with Einstein's cosmos as a bounded infinity, Gerstner's twelve units can be repositioned within a fixed framework. Throughout his career Gerstner viewed his patterns of form and color as internally consistent aesthetic systems that are symbolic of the underlying unity and wholeness of the natural world.

Gerstner began experimenting with rules for generating forms, as the older generation of Concrete artists had done. Looking at their oeuvre with a fresh eye, he noticed that Bill, Lohse, Graeser, and Loewensberg all used algorithms to determine the form of their work but that they chose their color by eye. Gerstner determined to expand the scope of Swiss Concrete Art by incorporating precise rules for determining color, with the aim of creating a unified style (plate 11-1, chapter frontispiece).

In one of his first experiments, *Blue Excentrum* (1956; see plate 11-7), Gerstner laid out five concentric circles that create a pattern of vertical bands, which he colored from white to black and from low- to high-intensity blue. Gerstner could calculate the

RIGHT

11-7. Karl Gerstner (Swiss, b. 1930), *Blue Excentrum*, 1956. Baked enamel on aluminum turnable discs, diameter 23⅝ in. (60 cm). Courtesy of the artist.

OPPOSITE

11-8. Frontispiece to Andreas Speiser, *Die Theorie der Gruppen von endlicher Ordnung* (The theory of groups of finite order; 1927), 4th ed. (Basel, Switzerland: Birkhäuser, 1956). Springer Science and Business Media, Heidelberg. Used with permission.

This diagram was drawn by Felix Klein to illustrate the tiling of a hyperbolic (saddle-shaped) plane (as shown in the sidebar on page 155 in chapter 4). Speiser added the color.

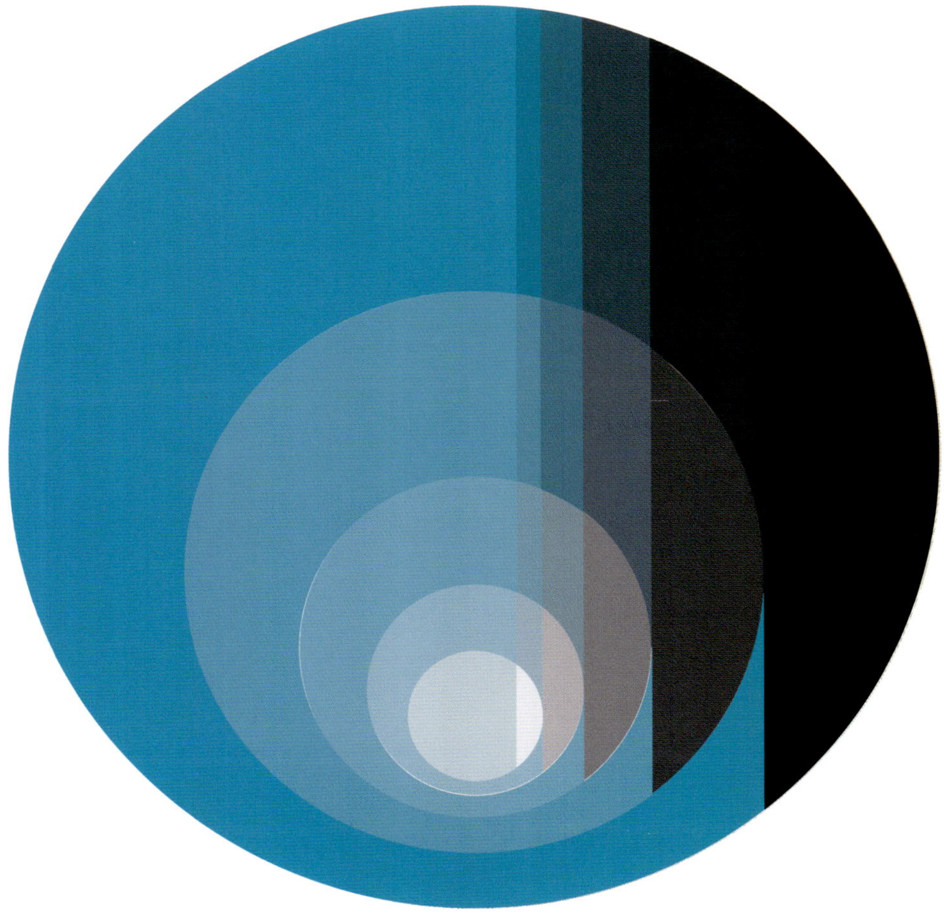

If we put any object inside a kaleidoscope, a beautiful pattern emerges. Turning the kaleidoscope causes amazing effects. The colors shine brighter and the dead matter appears full of life. In this mysterious transformation of a scrap of material into a beautiful ornament, we have an example of the force of art. The aesthetic effect does not originate in the object, but in the symmetries.

—**Andreas Speiser,**
Die mathematische Denkweise
(The mathematical way of thinking), 1932

progression of circles precisely because the diameter of a circle is an objective measure, but he could only approximate the progression of the shades of gray and blue. The measurement of a circle is a timeless truth (it is part of geometry), but the measurement of a painted surface is imprecise because pigments vary (they are part of chemistry) and color perception is subjective (it is part of physiology and psychology). Nevertheless, Gerstner determined to obtain color progressions with maximum precision.

In 1956 Gerstner devised a modular system—a movable palette with 196 hues in 28 groups—for experimenting with progressions that link form with color (plates 11-9 and 11-10). At this point Gerstner met Speiser, then seventy-one years old, who enthusiastically responded to the artist's paintings because they resonate with group theory.[16] The older Concrete artists had already read Speiser's essays, and now Gerstner also began referring to his artwork using the mathematician's terms: groups, permutations, algorithms, and invariance.

The following year Speiser released the fourth (1956) edition of his original 1927 book on group theory. Perhaps inspired by meeting Gerstner, Speiser added a new frontispiece—a colorful diagram of a non-Euclidean space (plate 11-8). Speiser also added a new ending to the book, in which he expressed the hope that such diagrams might "open up new possibilities" for modern artists.

In 1957 Gerstner organized the exhibition *Kalte Kunst? Zum Standort der heutigen Malerei* (Cold art? On the state of today's painting), with works by Bill, Graeser, Lohse, and Loewensberg, together with himself and other young artists. The exhibition catalogue

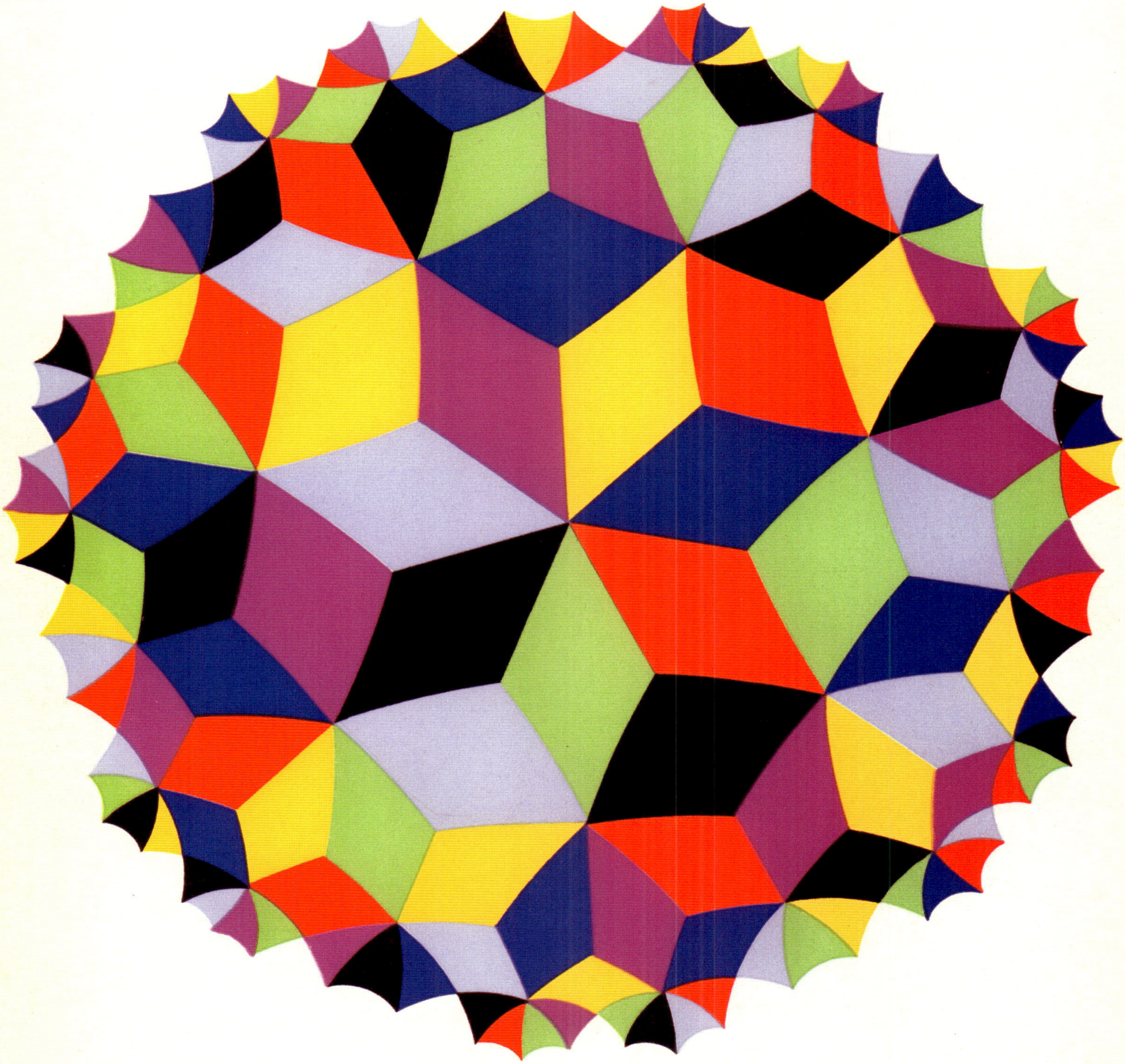

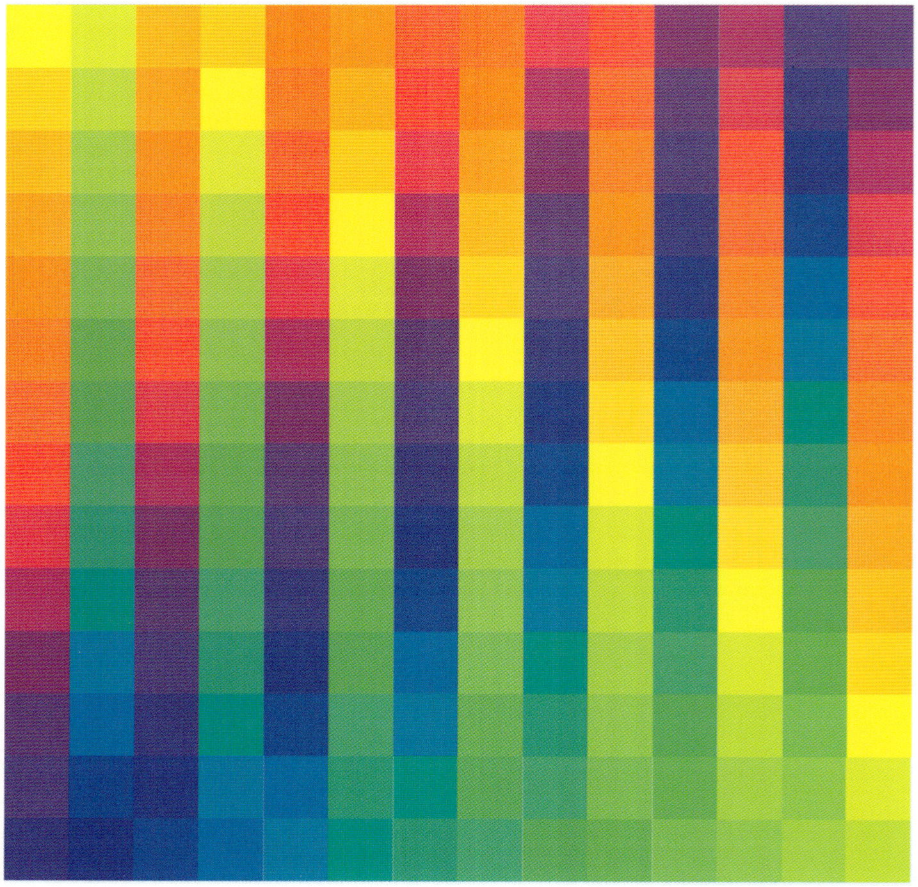

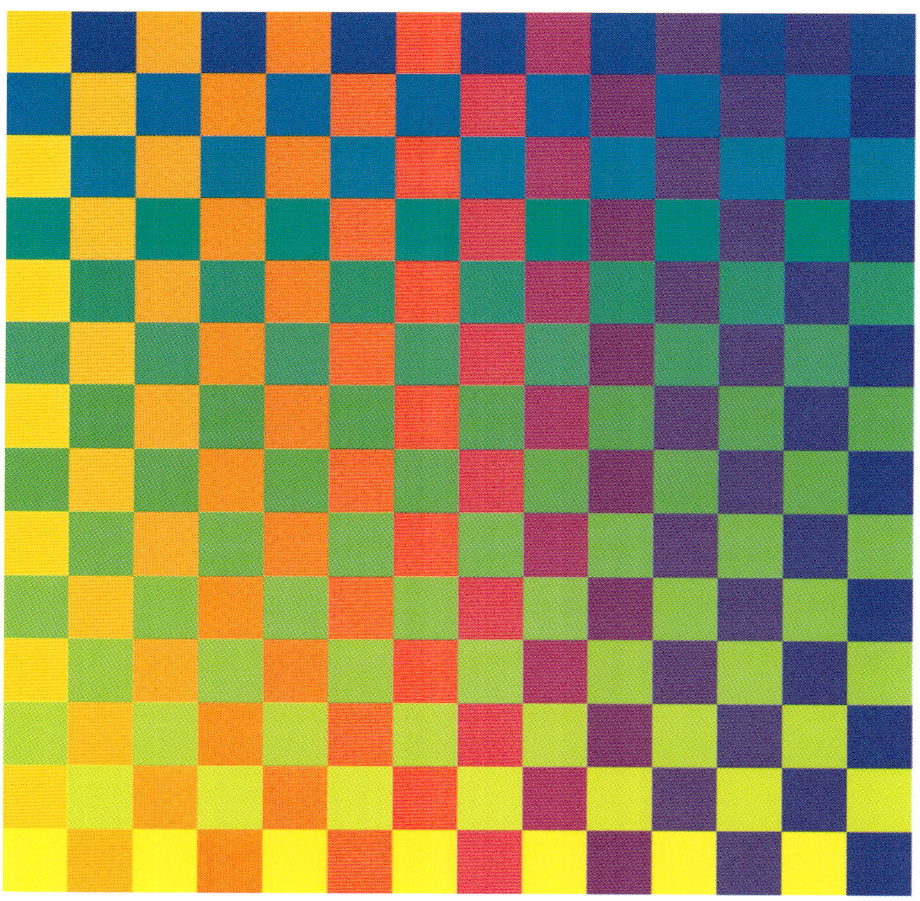

TOP LEFT

11-9. Karl Gerstner (Swiss, b. 1930), *Polychrome of Pure Colors*, 1956–58. Printer's ink on cubes of Plexiglas, 1¼ × 1¼ in. (3 × 3 cm) ea., fixed in a chrome-plated metal frame, 18⅞ × 18⅞ in. (48 × 48 cm) ea. Courtesy of the artist.

Gerstner's palette of 196 squares has 28 groups with 7 squares each. Shown here are two of myriad possible arrangements.

BOTTOM LEFT

11-10. Karl Gerstner (Swiss, b. 1930), *Polychrome of Pure Colors*, 1956–58. Printer's ink on cubes of Plexiglas, 1¼ × 1¼ in. (3 × 3 cm). ea., fixed in a chrome-plated metal frame, 18⅞ × 18⅞ in. (48 × 48 cm) ea. Courtesy of the artist.

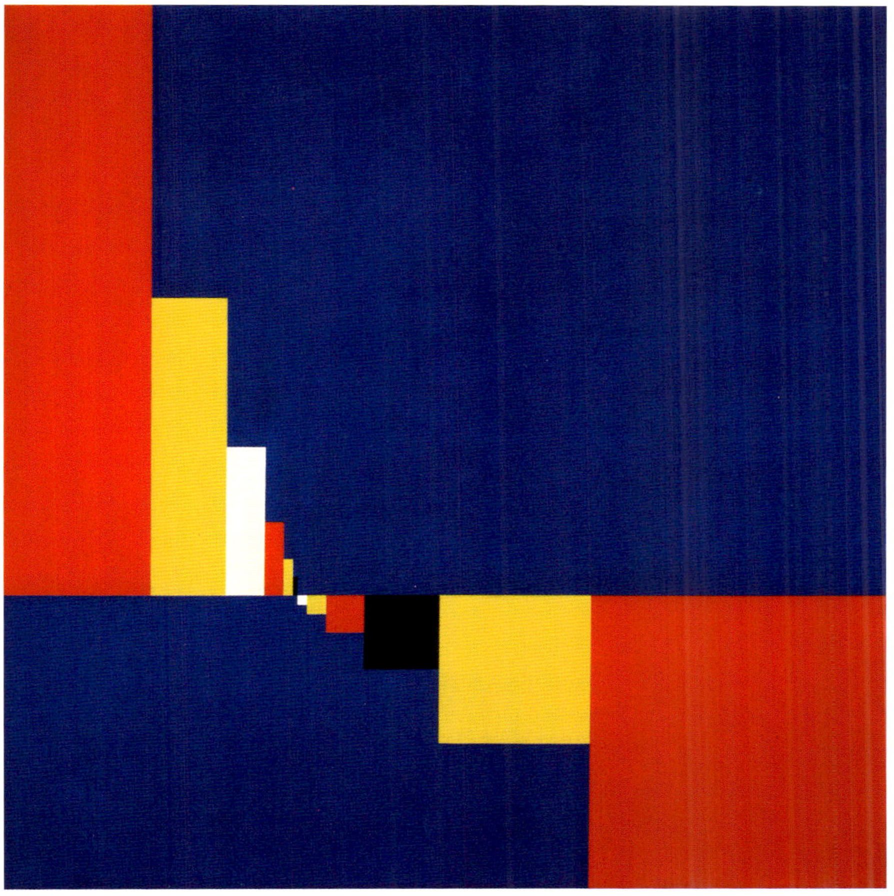

11-11. Camille Graeser (Swiss, 1892–1980), *Equivalence on the Horizontal*, 1957–58. Oil, 37⅞ × 37¼ in. (96 × 94.5 cm). Kunstmuseum, Winterthur. Camille Graeser Stiftung, Zurich. © 2014 ProLitteris, Zurich/Artists Rights Society, New York.

provides a good overview of Concrete Art in the late 1950s. Gerstner included construction diagrams of many of the paintings, which he drew with the approval of each artist. The catalogue bibliography includes Speiser's *Die mathematische Denkweise* (1932) and Weyl's *Symmetry* (1938). Each of the paintings by the older generation used algorithms to establish a pattern of forms. For example, in *Equivalence on the Horizontal* of 1957–58 (plate 11-11), Graeser put basic units at the center of the pattern (the small vertical black rectangle and the white square) and then generated the composition by doubling the areas of these units. In *Red Square* of 1946 (plate 11-12), Bill established the center of interest—a red square within a white square (turned 90°)—and then filled in the white background following a recipe-like formula (plate 11-13). Gerstner showed *Yellow Exentrum* (a variation of *Blue Exentrum*; plate 11-7), which introduced his idea of integrating the progression of form with a progression of color.

Gerstner and Speiser remained good friends until the mathematician's death in 1970, after which Gerstner made a series, *Color Lines*, inspired by Speiser's application of group theory to Islamic tiling (plate 11-14). Gerstner described the work: "And so it is to Andreas Speiser that I owe my thanks for the inspiration that led to *Color Lines*. . . . A pattern of Islamic ornament consists of a set of polygons which is turned, reflected, and displaced. Through its multifold intersections, simple symmetries take shape."[17] Gerstner developed the layout for *Color Lines* by tracing one line through a pattern of Islamic tiles, creating what he called a "visual labyrinth."

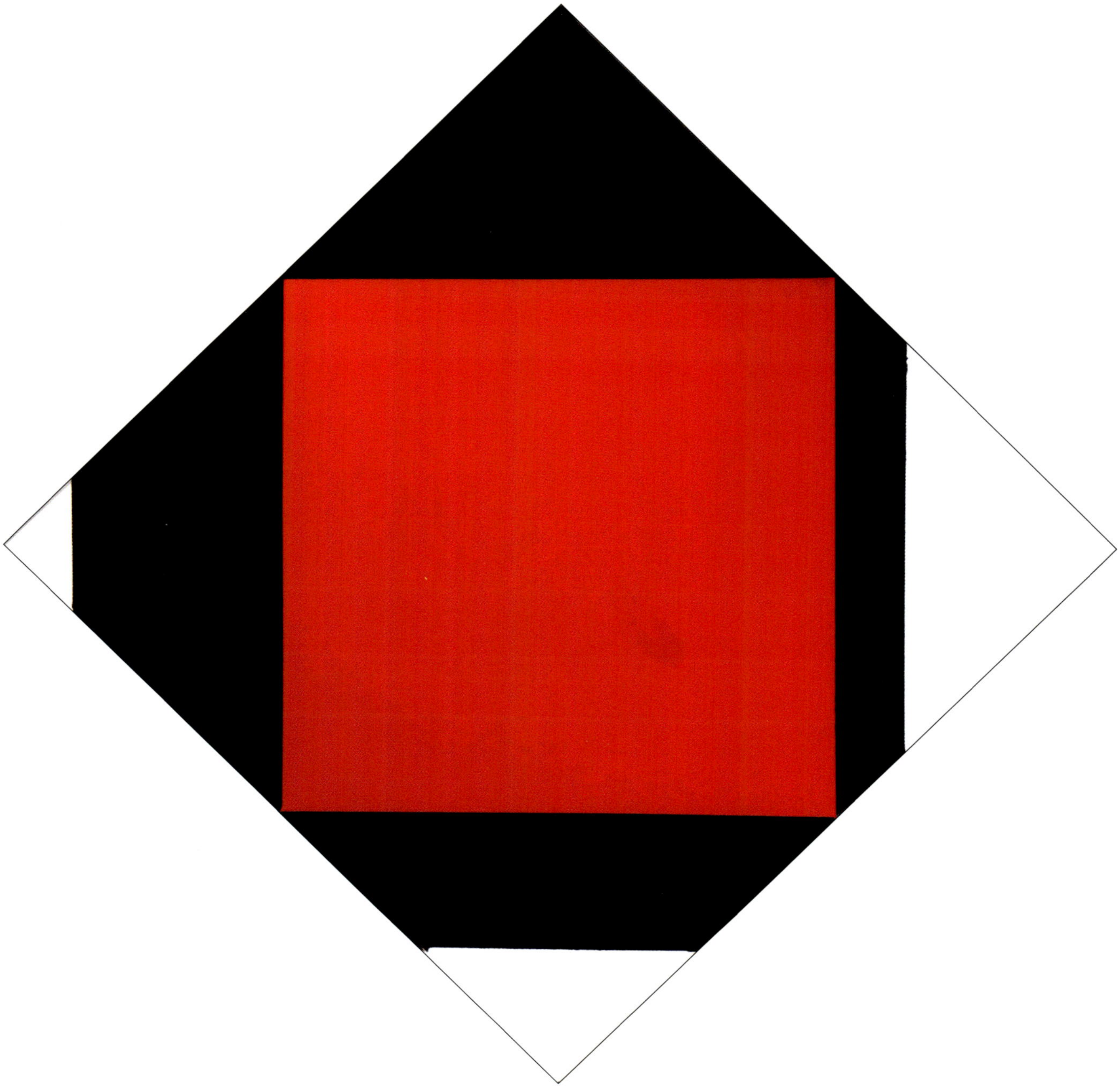

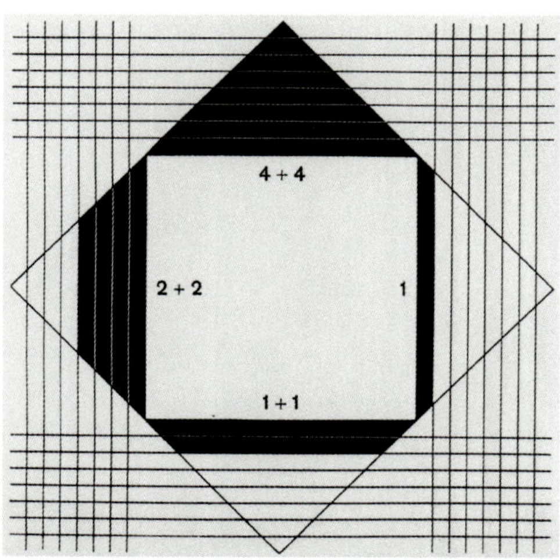

Graeser, Bill, and Loewensberg never adopted rules for determining color. Graeser continued in his style of the 1950s for several decades, with works such as *Red-Green Volume* of 1974 (plate 11-15), in which a square, though dislodged, retains a strict order. Loewensberg continued her use of slightly irregular repeating forms in almost monochromatic compositions (plate 11-16), and Bill explored geometric forms in graphic art (plate 11-18). Only Lohse eventually adopted from Gerstner the idea of using algorithms for color, completing his first painting with color progressions in the mid-1960s.[18] Lohse continued working in this style until his death in 1988, creating complex illusions (plate 11-17). For his part, Gerstner has continued to explore the integration of rigorous procedures for form and color that he pioneered (plate 11-19).

OPPOSITE TOP

11-12. Max Bill (Swiss, 1908–94), *Red Square*, 1946. Oil on canvas, 31½ in. (80 cm) diagonal. (This is a copy, made by Bill, of a 70 cm–diagonal painting of 1946 that was lost.). © max, binia + jakob bill foundation. © 2014 ProLitteris, Zurich/ Artists Rights Society, New York.

OPPOSITE BOTTOM

11-13. Layout of Max Bill's *Red Square* (1946) in Karl Gerstner, ed., *Kalte Kunst? Zum Standort der heutigen Malerei* (Teufen, Switzerland: Arthur Niggli, 1957), fig. 21.

Karl Gerstner drew this diagram of the layout of Bill's *The Red Square* (plate 11-12) for the catalogue of the exhibition *Kalte Kunst?*

TOP RIGHT

11-14. Karl Gerstner (Swiss, b. 1930), *Color Lines c-15/1-02*, 2000. Acrylic paint on Cibatool profile bars (in low relief) in frames varnished black, 41 × 41 in. (104 × 104 cm). Courtesy of the artist.

BOTTOM RIGHT

11-15. Camille Graeser (Swiss, 1892–1980), *Red-Green Volume 1:1*, 1974. Acrylic on canvas, 47¼ × 47¼ in. (120 × 120 cm). Camille Graeser Stiftung, Zurich. © 2014 ProLitteris, Zurich/ Artists Rights Society, New York.

FOLLOWING SPREAD, LEFT

11-16. Verena Loewensberg (Swiss, 1912–86), Untitled, ca. 1978. Oil on canvas, 29½ × 29½ in. (75 × 75 cm). Private Collection. © Henrietta Coray Loewensberg, Zurich.

FOLLOWING SPREAD, RIGHT

11-17. Richard Paul Lohse (Swiss, 1902–88), *Thirty Systematic Color Series with Red Diagonals*, 1943–1970. Oil on canvas, 65 × 65 in. (165 × 165 cm). © Richard Paul Lohse Foundation, Zurich. © 2014 ProLitteris, Zurich/Artists Rights Society, New York.

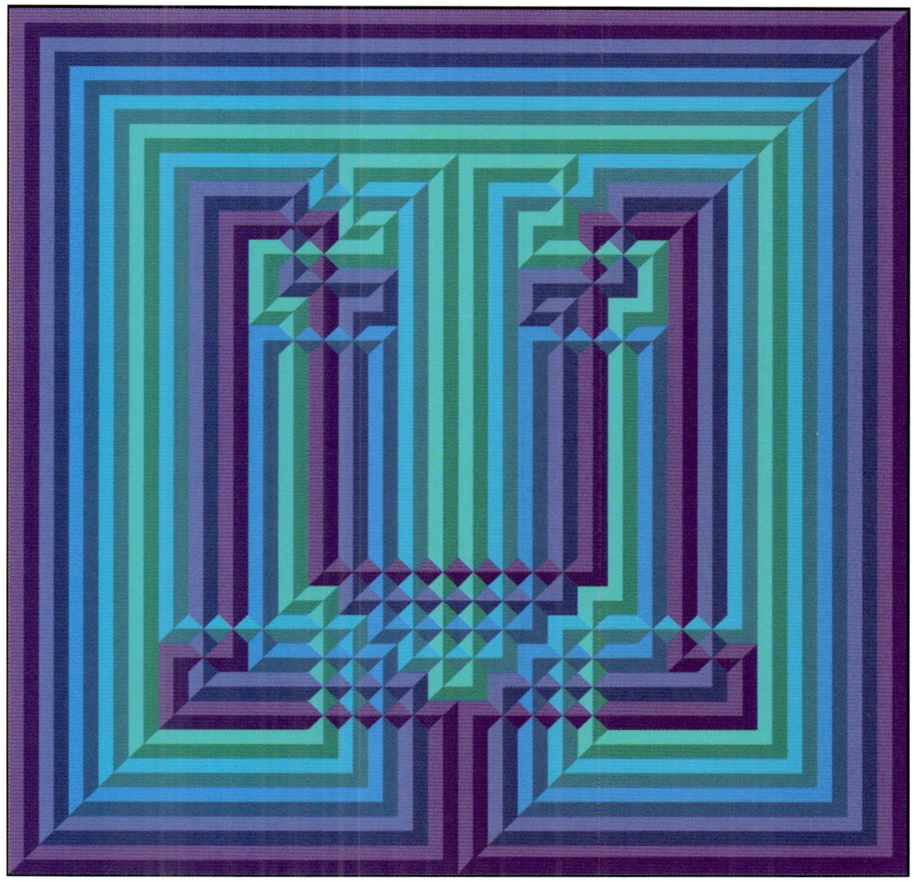

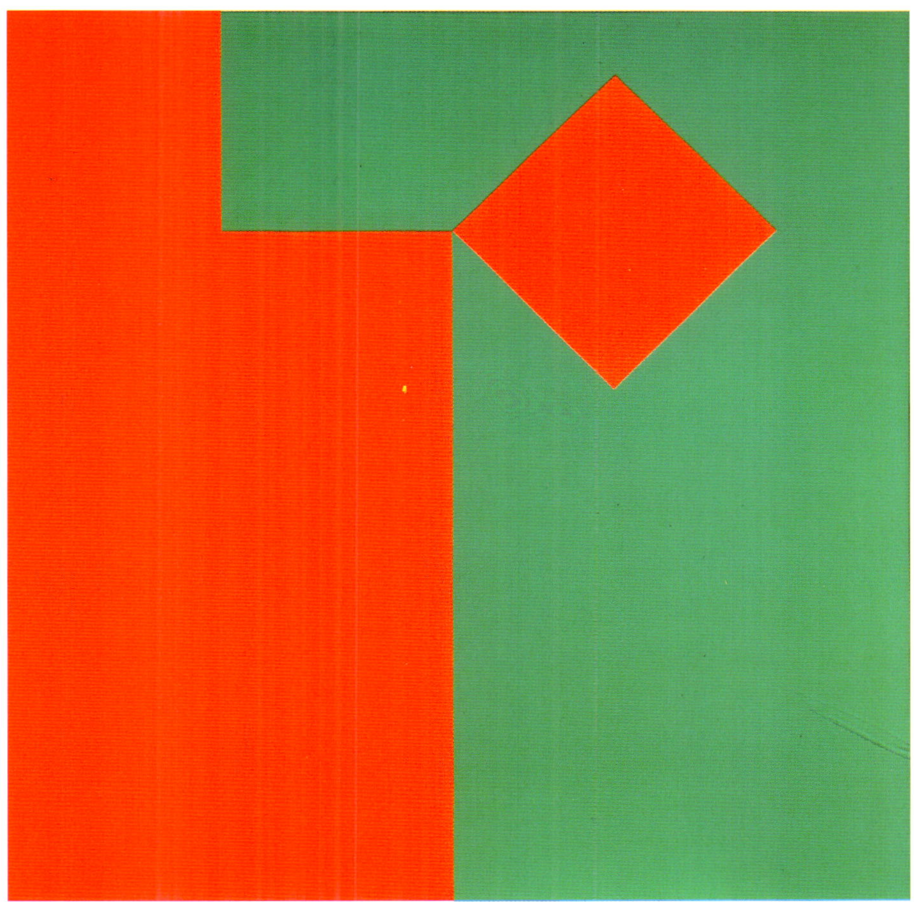

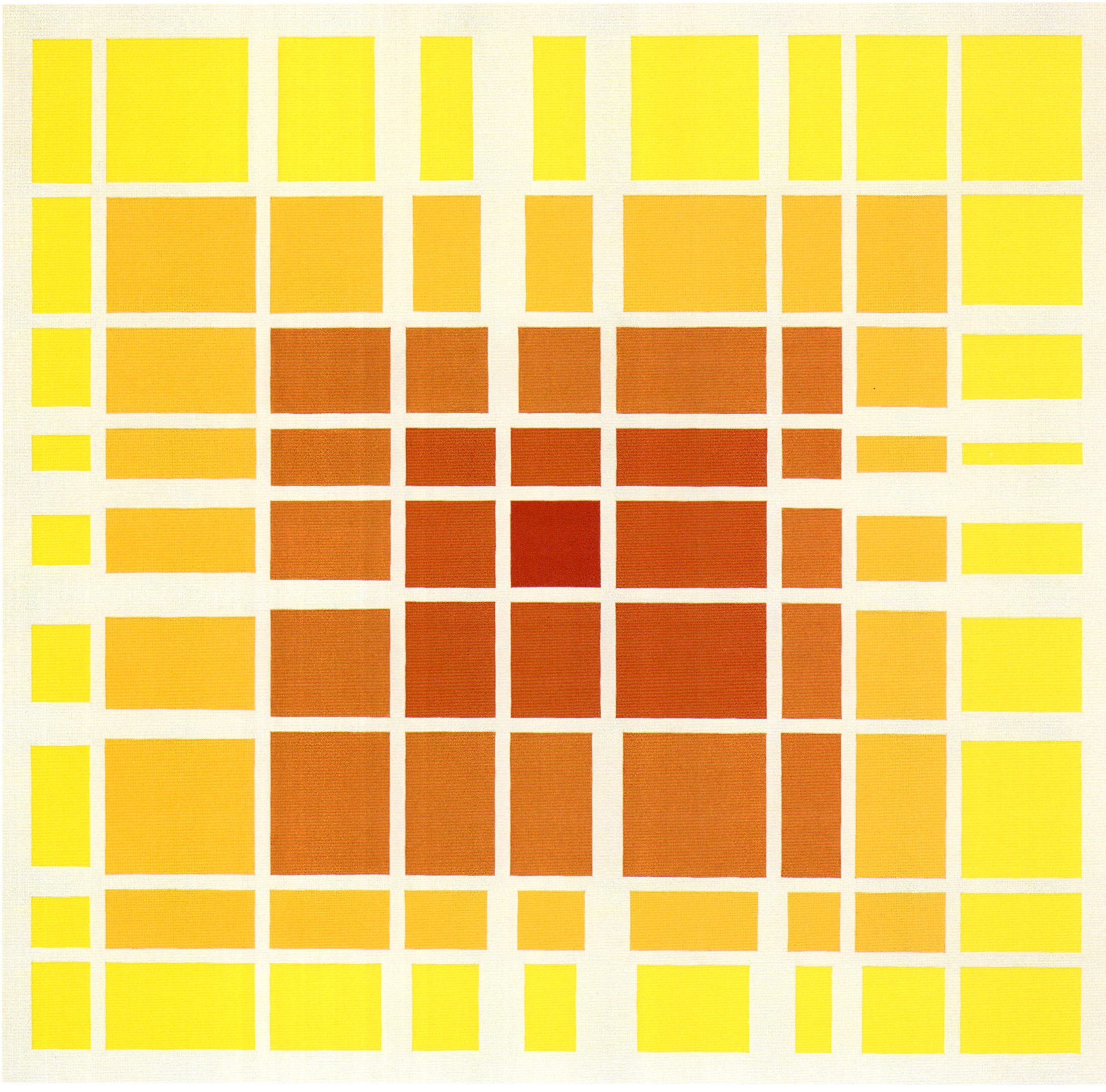

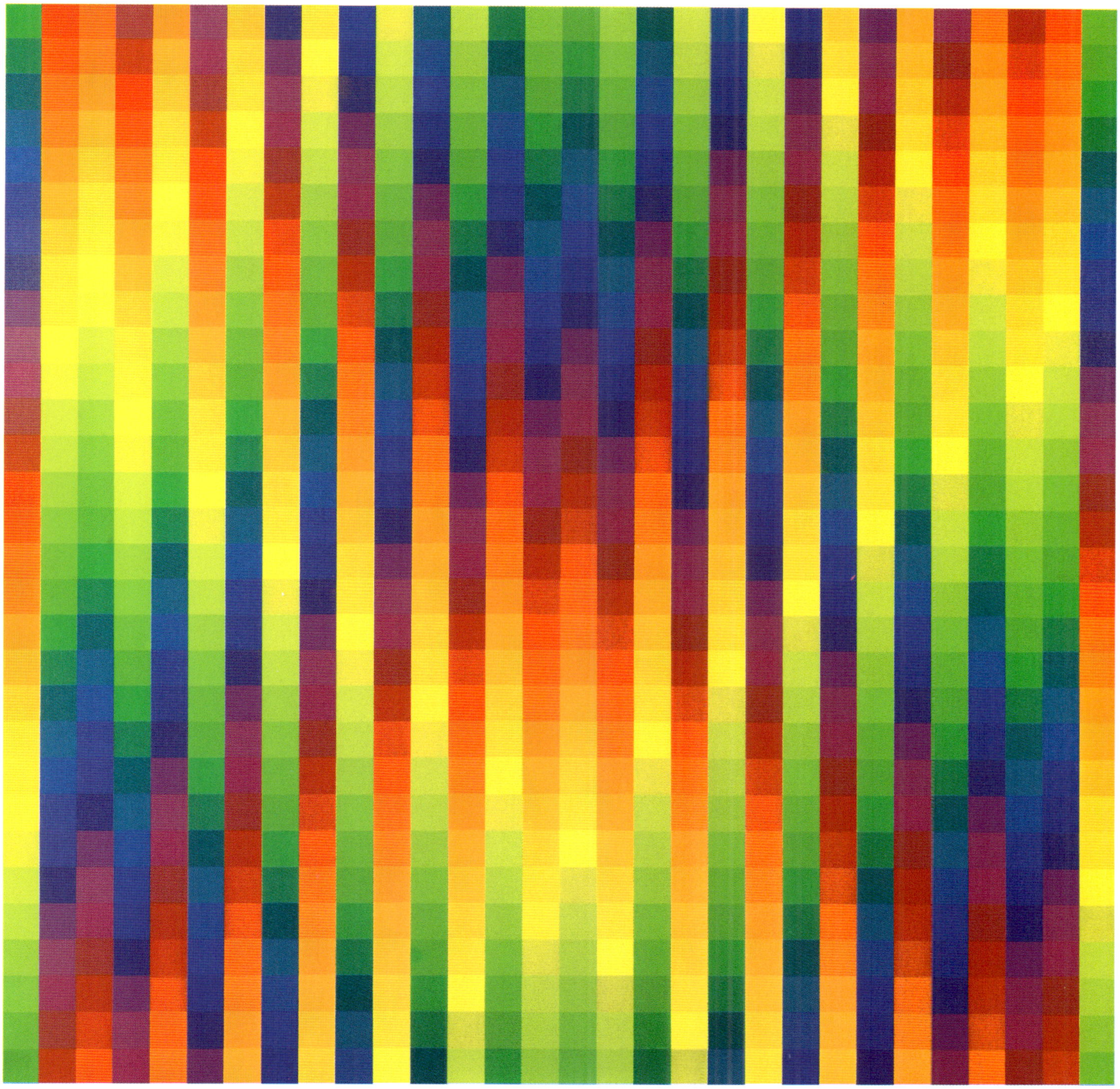

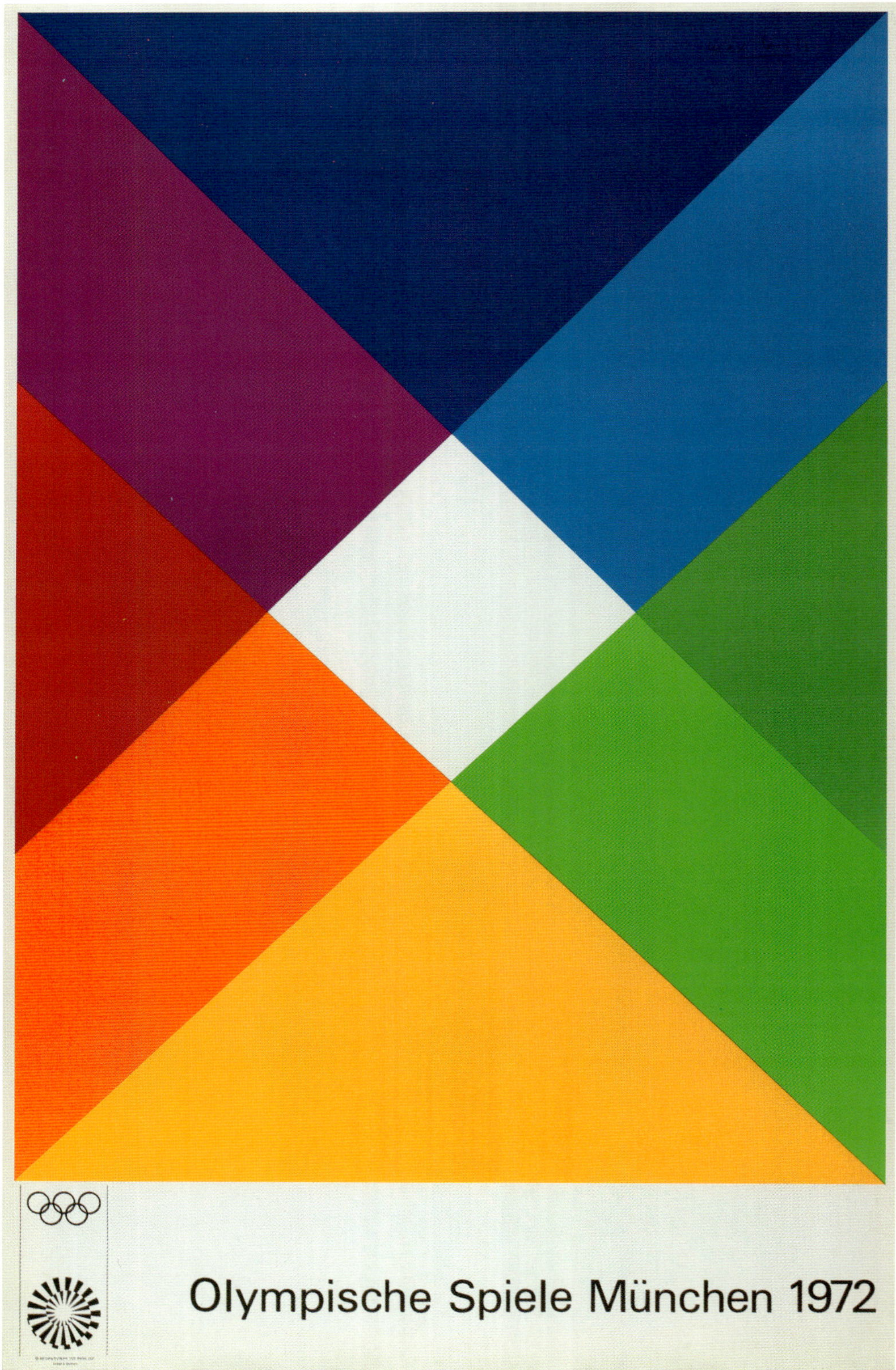

LEFT

11-18. Max Bill (Swiss, 1908–94), poster for the Olympische Spiele München (Munich Olympic games), 1972. Museum für Gestaltung Zürich, Plakatsammlung. © ZHdk. © 2014 ProLitteris, Zurich/ Artists Rights Society, New York.

OPPOSITE

11-19. Karl Gerstner (Swiss, b. 1930), *Color Spiral Icon x65b*, 2008. Acrylic on aluminum, diameter 41 in. (104 cm). Collection of Esther Grether, Basel, Switzerland.

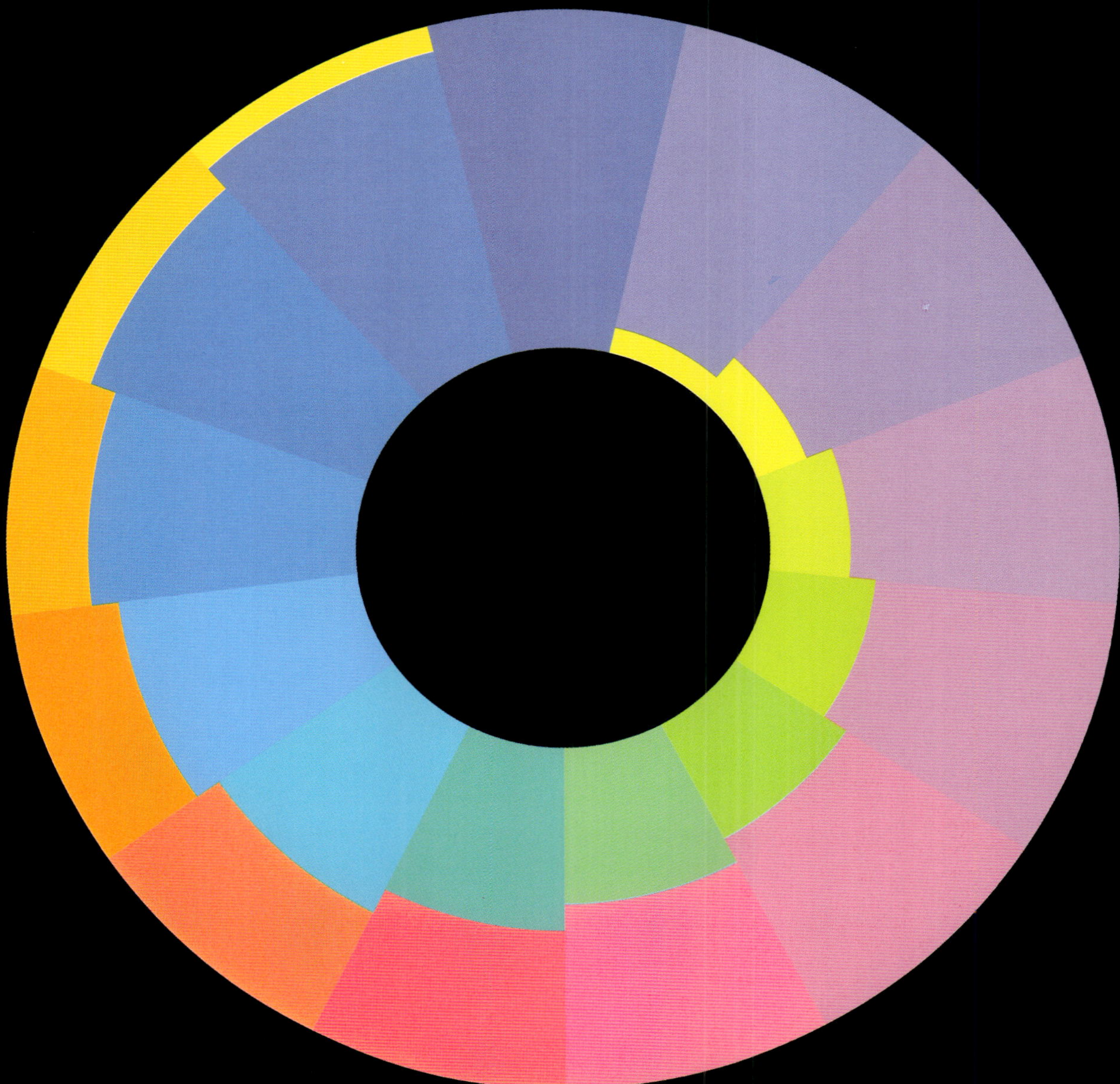

FRENCH FORMALIST MATHEMATICS: NICHOLAS BOURBAKI

At the center of our universe are found the great types of structures, . . . they might be called the mother structures. . . . Beyond this first nucleus are structures that might be called multiple structures. They involve two or more of the great mother structures not in simple juxtaposition (which would not produce anything new) but combined organically by one or more axioms which set up a connection between them. . . . Farther along we come finally to the theories properly called particular. In these the elements of the sets under consideration, which in the general structures have remained entirely indeterminate, obtain a more definitely characterized individuality.

—**Nicolas Bourbaki, "The Architecture of Mathematics," 1948**

At the opening of the twentieth century, mathematics in France was dominated by Henri Poincaré, who took an intuitive approach similar to L.E.J. Brouwer (see chapter 6). Wary of German axiomatic methods, Poincaré urged his French colleagues to follow their intuitions. Blurring the distinction between mathematical axioms and scientific theories, Poincaré declared that the axioms of geometry do not describe a timeless mathematical realm but the space of the world around us. Thus, according to Poincaré, the axioms of geometry are mere conventions—arbitrary, changeable rules.

Reading these declarations in Poincaré's several popularizations, the Surrealist Marcel Duchamp made an analogy between the conventions of geometry and art; just as Euclidean and non-Euclidean geometries are equally valid in mathematics, so, according to Duchamp, Impressionism and Dadaism are equally valid in art.[19] Aesthetics is not a question of true or false, right or wrong, but of capricious opinion. Duchamp summarized this outlook in *Tu m'* (plate 11-20), in which he presented a selection of pictorial conventions: a color chart, a realistic hand (executed by a sign painter), shadows cast by three of his ready-mades (*Bicycle Wheel, Corkscrew, Hat Rack*), and outlines of his irregular meter-sticks (*Three-Standard Stoppages*) traced in the lower left.[20] What does it all mean? Each viewer can arrive at a convenient interpretation by supplying the title's missing verb—*Tu m'aimes* (You love me), *Tu m'ennuies* (You bore me), and so on.

After Poincaré's death in 1912 and the release in 1916 of Ferdinand de Saussure's *Cours de linguistique générale* (Course in general linguistics), a formalist/structuralist approach began to gain ground in France. In the 1930s a group of young mathematicians associated with the École Normale Supérieure in Paris drafted a plan to overhaul French mathematics by introducing formal axiomatic methods as practiced at Göttingen under David Hilbert.[21] Inspired by the avant-garde spirit of Surrealist poets and artists, the young

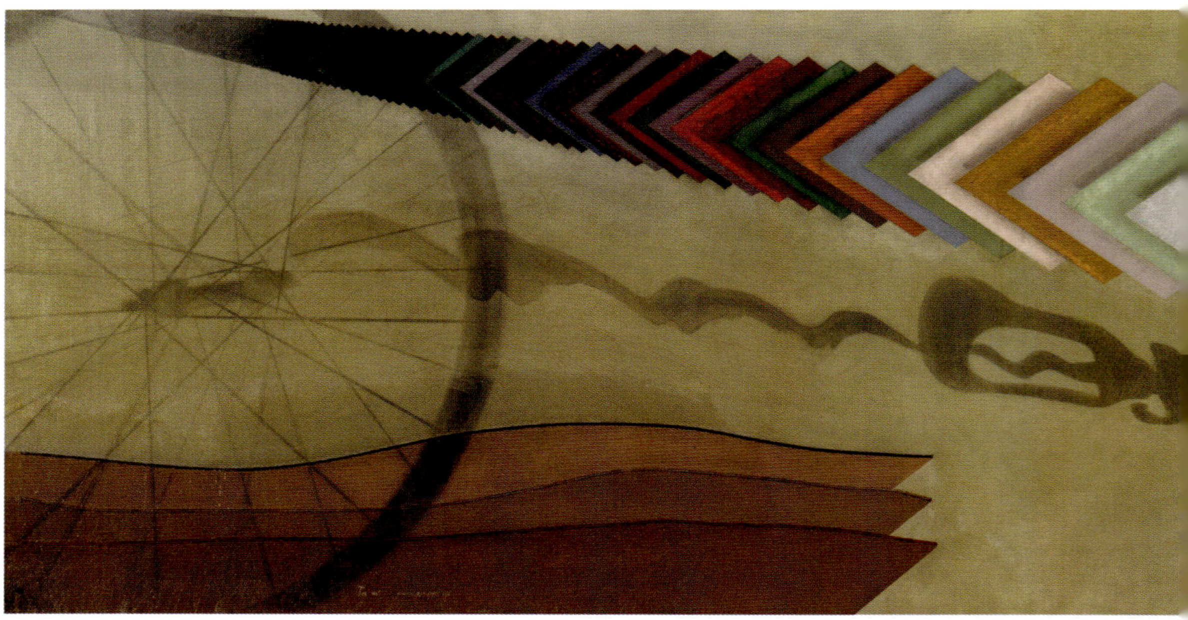

11-20. Marcel Duchamp (French, 1887–1968), *Tu m'*, 1918. Oil on canvas, with bottle-brush, safety pins, and bolt, 27½ × 119⁵⁄₁₆ in. (69.8 × 303 cm). Yale University Art Gallery, gift of the Estate of Katherine S. Dreier. © 2014 Succession Marcel Duchamp/ADAGP, Paris/Artists Rights Society, New York.

Duchamp emphasized the illusory nature of depiction by affixing several real objects to the illusionistic surface of this painting; three (real) safety pins span a painted picture of the (unreal) tear in the center-right of the fabric, from which protrudes a (real) bottle-brush.

Parisian mathematicians met in secret, as if they were conspirators. To convey their hope that French intuitionism would be "conquered" by German formalism, the group adopted the pseudonym "Nicolas Bourbaki," the surname adopted from an anti-hero, a French general who lost battles against Germanic troops during the Franco-Prussian War (1870–71).

Borrowing Hilbert's building metaphor, Bourbaki described mathematics as a vast hierarchy of structures composed of sets ("The Architecture of Mathematics," 1948). But Boubaki went beyond Hilbert in the group's commitment to formalism, or structuralism (*structuralisme*) as it is called in France. Hilbert's goal was to apply his "theory-forms" in the real-world-out-there (such as in the physics laboratory), but Bourbaki argued that pure mathematics should be done in ivory-tower isolation from science and society. Also, Hilbert applied formalist methods to a body of mathematics that *already existed*, such as Euclid's geometry, in order to reveal its abstract structure. Bourbaki *designed from scratch* uninterpreted mathematical structures, each of which was not the means to an end (not applied to any subject matter), but an end in itself.[22]

Beginning in 1939, Bourbaki published a multivolume overview of mathematical concepts, written in a rigorous, axiomatic format, and named *Éleménts de mathématique* (Elements of mathematics) in homage to Euclid's *Elements*. Boubaki's encyclopedia is known for its extremely abstract, arid style, with no appeal to intuition, no interpretation or models, and few figures or diagrams to illustrate the geometric proofs. In their zeal to promote formalism, in 1948 Bourbaki instituted a series of popular monthly public lectures that coincided with the publication of a pamphlet describing the mathematical concepts covered in each lecture. Bourbaki also led a movement to supplant rote education (such as memorizing multiplication tables) in elementary school mathematics with creative thinking, hands-on projects, and thought-provoking questions. In the 1960s France was the first country to introduce set theory in kindergarten by teaching children that numbers are sets, a curriculum reform ("new math") that soon spread throughout the developed world.[23]

You can't take three from two,
Two is less than three,
So you look at the four
in the tens place.
Now that's really four tens,
So you make it three tens,
Regroup, and you change
a ten to ten ones,
And you add them to the
two and get twelve,
And you take away
three, that's nine.
Is that clear?

—Tom Lehrer, "New Math," 1965

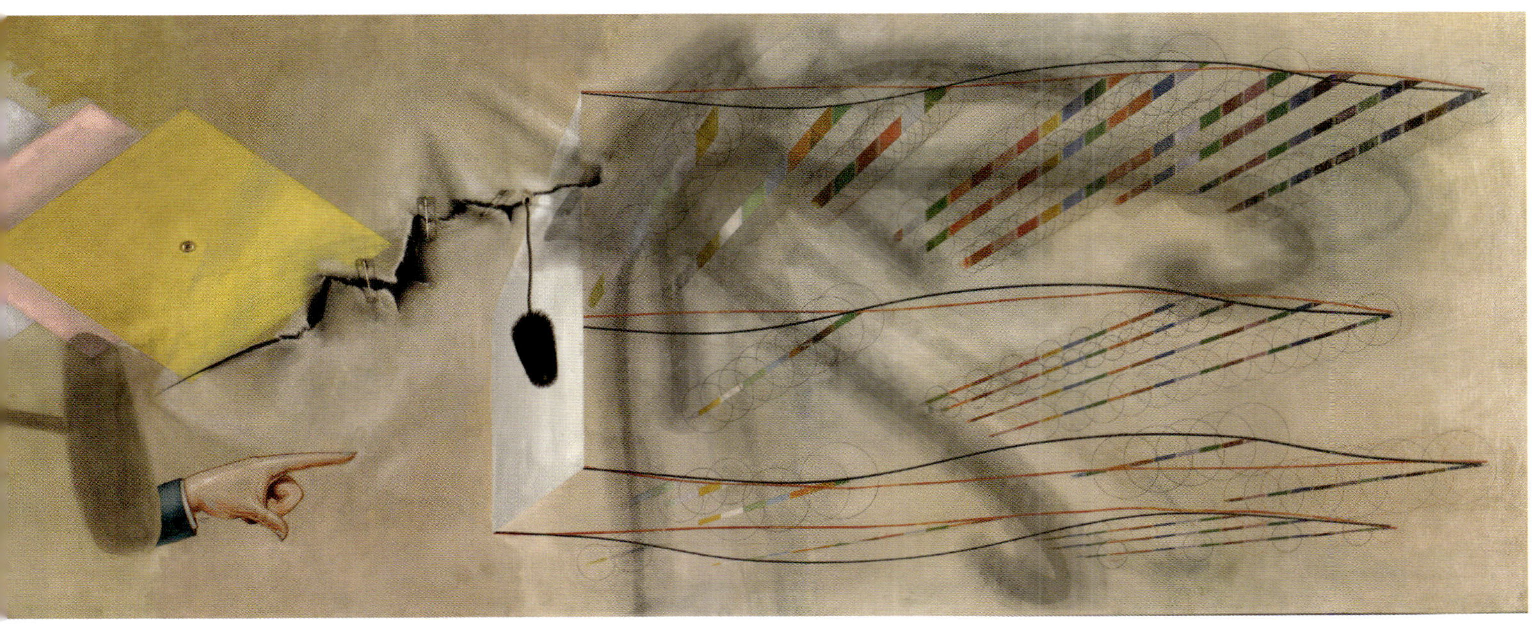

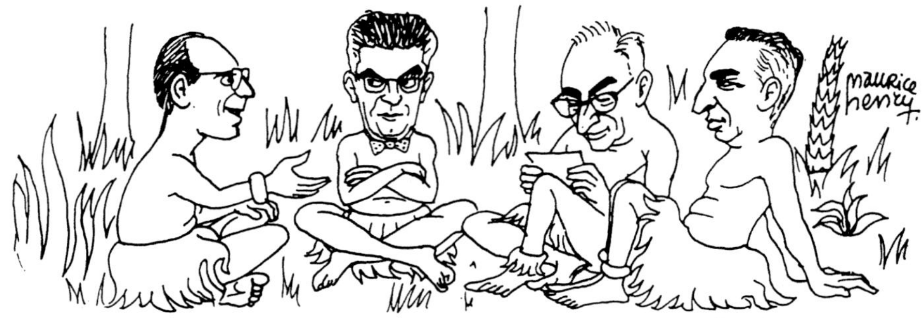

11-21. Maurice Henry, *Le déjeuner sur l'herbe structuraliste* (The structuralist luncheon on the grass), *La quinzaine littéraire*, July 1, 1967. Used with permission.

With a nod to the scantily clad bathers in Édouard Manet's *Le déjeuner sur l'herbe* (The luncheon on the grass; 1862–63; Musée d'Orsay, Paris), the French cartoonist Maurice Henry exposed the structuralist camp in native attire after they adopted methods (such as the Klein four-group) from cultural anthropology (see the sidebar on page 446): (left to right) Michel Foucault, Jacques Lacan, Claude Lévi-Strauss, and Roland Barthes.

This mathematics of man . . . is resolutely determined to break away from the hopelessness of the "great numbers"—the raft to which the social sciences, lost in an ocean of figures, have been helplessly clinging; its ultimate object is no longer to plot progressive and continuous movements in monotonous graphs. . . . We are less concerned with the theoretical consequences of a 10 percent increase in the population in a country having 50 million inhabitants than with the change in structure occurring when a "two-person household" becomes a "three-person household."

—**Claude Lévi-Strauss, "The Mathematics of Man," 1954**

Although Bourbaki argued that pure mathematics should not be applied to mundane affairs, the anthropologist Claude Lévi-Strauss, who had already adopted group theory while in America during World War II (see chapter 7), returned to Paris in the late 1940s and advocated using mathematical methods in the social sciences. Lévi-Strauss adopted methods from Saussure's structural linguistics—not from Bourbaki—but he nevertheless insisted that linguistic structuralism was in the spirit of Bourbaki.[24] Other French intellectuals followed Lévi-Strauss's lead and introduced mathematical methods from structural linguistics into their respective fields, becoming *post*structuralists: the psychoanalyst Jacques Lacan, the philosopher Michel Foucault, and the literary critic Roland Barthes, each focused on the abstract structure of their respective disciplines, beginning in the 1950s with Lacan's "The Function and Field of Speech and Language in Psychoanalysis" (1953; see plate 11-21). Unlike these poststructuralists, a literary group of French writers—Oulipo—adopted mathematical methods directly from Bourbaki.

FRENCH LITERARY FORMALISM: OULIPO

In the mid-twentieth century the French literary scene underwent a transition from intuition-based writing techniques to formal methods in the spirit of Bourbaki. The intuitionist protocols belonged to the Surrealists, who were intrigued by the neurologist Pierre Janet's research on the unconscious mind as revealed by automatic speech (*L'automatisme psychologique*, 1889). In 1919 the Surrealist poets André Breton and Philipe Soupault began experimenting with automatic writing by suspending conscious control and letting their words flow in unedited stream-of-consciousness (*Les champs magnétiques*, 1919). In the first manifesto of Surrealism, Breton defined the new style as "psychic automatism in a pure state" (*Manifeste du surréalisme*, 1924). Artists devised analogous automatic techniques, such as letting a pen wander aimlessly across a sheet of drawing paper (Jean/Hans Arp, *Automatic Drawing*, 1917–18; Museum of Modern Art, New York).

François Le Lionnais and Raymond Queneau experimented with chance verse in their youth in the 1920s and 1930s. Le Lionnais went on to become a chess master after studying mathematics and engineering, and Queneau became a novelist who signed an anti-Breton pamphlet (*Une cadavre* [A corpse], 1930). Feeling camaraderie with Bourbaki's anti-establishment spirit, Le Lionnais and Queneau joined forces in the 1960s and hit upon the idea of adopting the formalist approach—especially set theory—to reinvigorate literature. They formed the group Ouvroir de Littérature Potentielle (Workshop of potential literature), which they nicknamed "Oulipo." Oulipo created the "potential" for literature in the sense that their methods produced only structures; it was up to the creative writer to fill in the framework with words.[25]

The sonnet was a favorite of Le Lionnais and Queneau because it relies on a formal structure; for instance, a fourteen-line sonnet by William Shakespeare has the rhyme pattern ABAB CDCD EFEF GG (plate 11-22). In 1961 Queneau published the manifesto of Oulipo in the form of ten sonnets, each printed on a right-hand page of a book and sliced into fourteen strips, one for each sonnet line (plate 11-23). In his preface Queneau described his book: "Above all it is a sort of machine to make poems." By flipping the strips

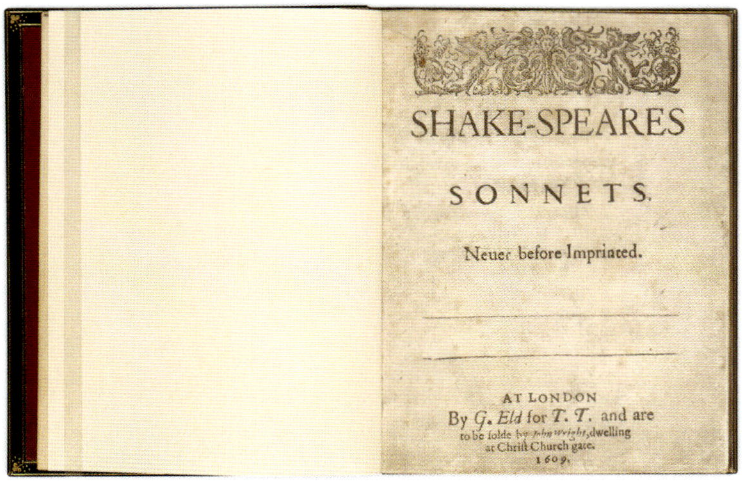

SHAKE-SPEARES

SONNETS.

Neuer before Imprinted.

AT LONDON
By G. Eld for T. T. and are
to be solde by iohn wright, dwelling
at Christ Church gate.
1609.

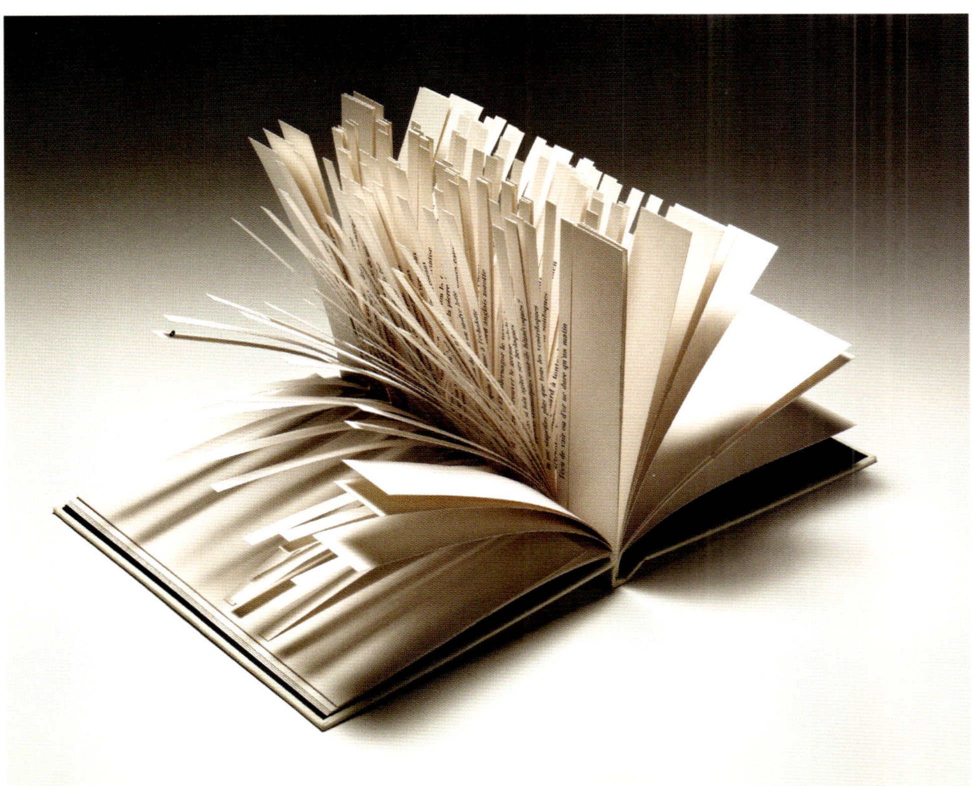

TOP LEFT

11-22. William Shakespeare, *Sonnets* (London, 1609). Beinecke Rare Book and Manuscript Library, Yale University.

Sonnet 147

My love is as a fever, longing still
For that which longer nurseth the disease,
Feeding on that which doth preserve the ill,
The uncertain sickly appetite to please.
My reason, the physician to my love,
Angry that his prescriptions are not kept,
Hath left me, and I desperate now approve
Desire is death, which physic did except.
Past cure I am, now reason is past care,
And frantic-mad with evermore unrest;
My thoughts and my discourse as
madmen's are,
At random from the truth vainly express'd;
For I have sworn thee fair and thought
thee bright,
Who art as black as hell, as dark as night.

BOTTOM LEFT

11-23. Raymond Queneau (French, 1903–76), *Cent mille milliards de poèmes* (One hundred thousand billion poems; Paris: Gallimard, 1961).

Queneau's book of sonnets recalls Wolfgang Amadeus Mozart's composition *Musical Dice Game* (published posthumously in 1792), which is a minuet in sixteen measures, of which Mozart specified only the final measure. The performer determines the other fifteen by rolling a pair of dice to come up with a sequence of fifteen numbers, each assigned to a measure in a set written by Mozart. There are more than 100 trillion possible combinations of sixteen measures, each minuet in perfect form.

to the right or left, the reader can view 10^{14} (100,000 billion) sonnets, each of which is structurally correct and makes some sense (*Cent mille milliards de poèmes* [One hundred thousand billion poems], 1961).

In another play on the sonnet form, the Oulipo poet Jacques Bens wrote what he called "irrational sonnets." The fourteen lines of his sonnets have a rhyme pattern AAB, C, BAAB, C, CDCCD, resulting in a line grouping of three, one, four, one, and five (31415), which are the first digits of π (pi). Oulipo members invented numerous other clever literary rules, such as $M + n$, where M is any kind of word (*mot* in French) such as a noun, adjective, or verb, and n is any positive number (*nombre*). The procedure is as follows: in a given text, such as an essay by Voltaire, the author replaces each occurrence of the chosen kind of word (such as each noun) with the *n*th word in a specified dictionary (such as the *Le petit Larousse*, 1955 edition). In yet another witticism, Queneau responded to a remark by David Hilbert, "It must be possible to replace in all geometric statements the words *point, line, plane* by *table, chair, mug*" (see chapter 4), by replacing *point*, *line*, and *plane* in Hilbert's *The Foundations of Geometry* (1899) with *word*, *phrase*, and *paragraph* to get a set of axioms for literature (*The Foundations of Literature after David Hilbert*, 1976).[26]

FRENCH OPTICAL ART

This experimental, programmed painting seems to respond to two needs: first, the need of the public that wants to be part of the "creation" of art, in order to demystify it and better understand it; second the need of aestheticians for new material, those men of science who are both mathematicians and psychologists.

—François Morellet, manifesto of Groupe de Recherche d'Art Visuel, 1962

OPPOSITE

11-24. François Morellet (French, b. 1926), *Random Distribution of 40,000 Squares, Following the Even and Odd Numbers in a Phone Book*, 1960. Oil on canvas, 39⅜ × 39⅜ in. (100 × 100 cm). Courtesy of the artist. © 2014 ADAGP, Paris/Artists Rights Society, New York.

When World War II ended, many European cities were in disarray and New York had not yet emerged as an art capital. The center of the international art world remained in Paris, where artists worked in late versions of Surrealism and early styles of gestural abstraction, called Tachisme (from *tache*, "stain"). The geometric abstraction of Art Concret had never died out, and some artists in Paris in the 1950s were also familiar with the basic concepts of formalist mathematics from the Bourbaki's monthly series of public lectures.

In 1944 Denise René opened her Paris gallery with an international group of artists associated with French Art Concret and Swiss Concrete Art (*Art Concret*, 1945). Two years later, the painter Auguste Herbin founded the international Salon des Réalités Nouvelles, and for a decade the salon's group exhibitions and related book series, *Réalités nouvelles*, provided the largest forum for abstract art in Europe. Another French voice for geometric abstraction emerged in 1949 with the founding of the magazine *Art d'aujourd'hui*.

The young French artist François Morellet happened upon the approach to art he was seeking while travelling to São Paulo in 1950. A retrospective of Max Bill's work had just closed when Morellet arrived in the city, but artists were still talking about it. This second-hand encounter with Swiss Concrete Art set him on the path of logic and precision. By the mid-1950s Morellet was using algorithms to generate all his work, and he was back in Paris, where Bourbaki's monthly lectures on structure were the talk of the art world. Morellet had been raised Roman Catholic, but as an adult he explored Buddhist meditation practices and read the Armenian theosophist G. I. Gurdjieff.[27] For Morellet, making art according to a regular, repetitive process—like meditation—produced a regular pattern when he did not assert his ego and just passively performed the task. After initially being uncomfortable with chance, Morellet used random patterns once he began

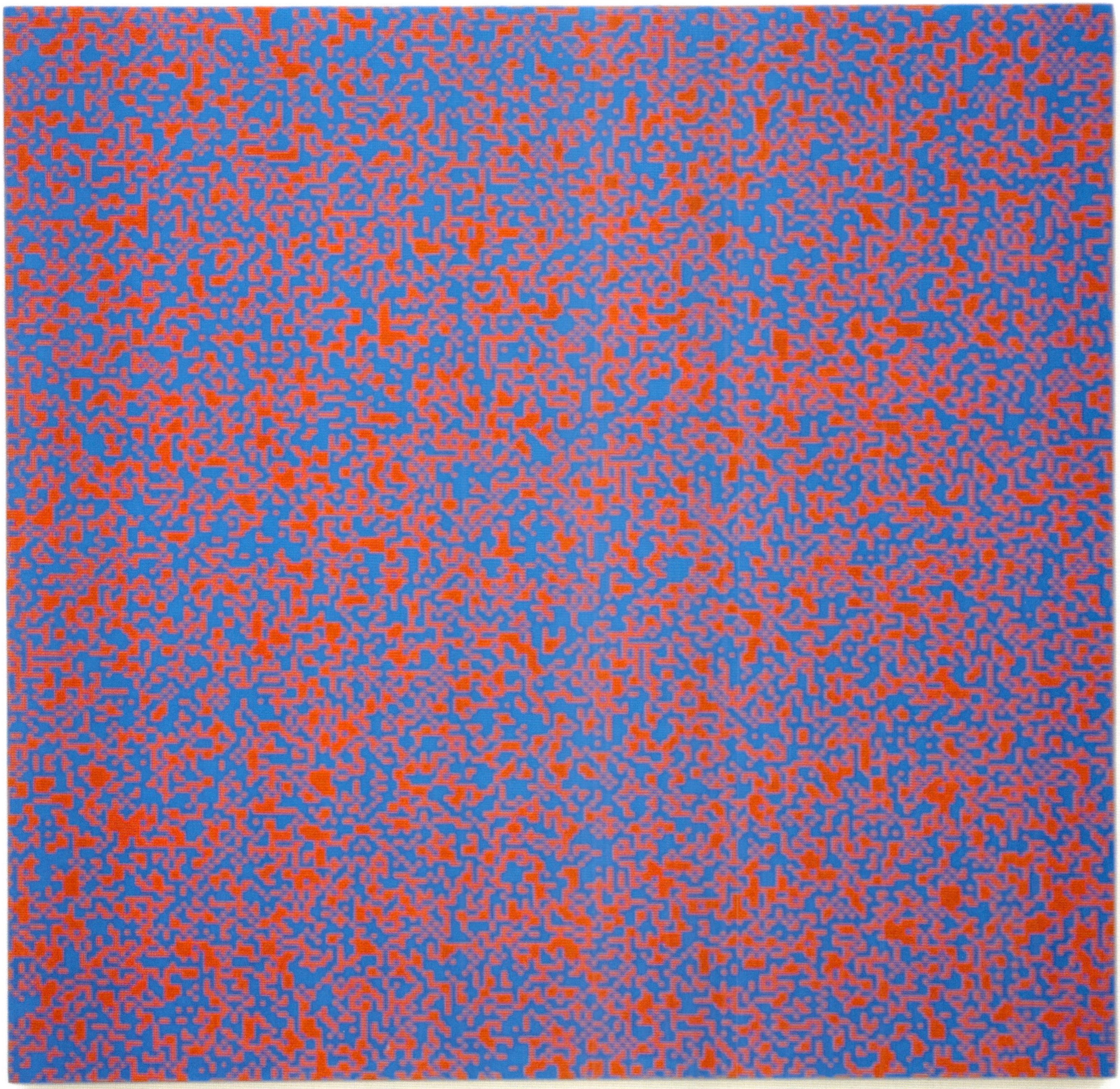

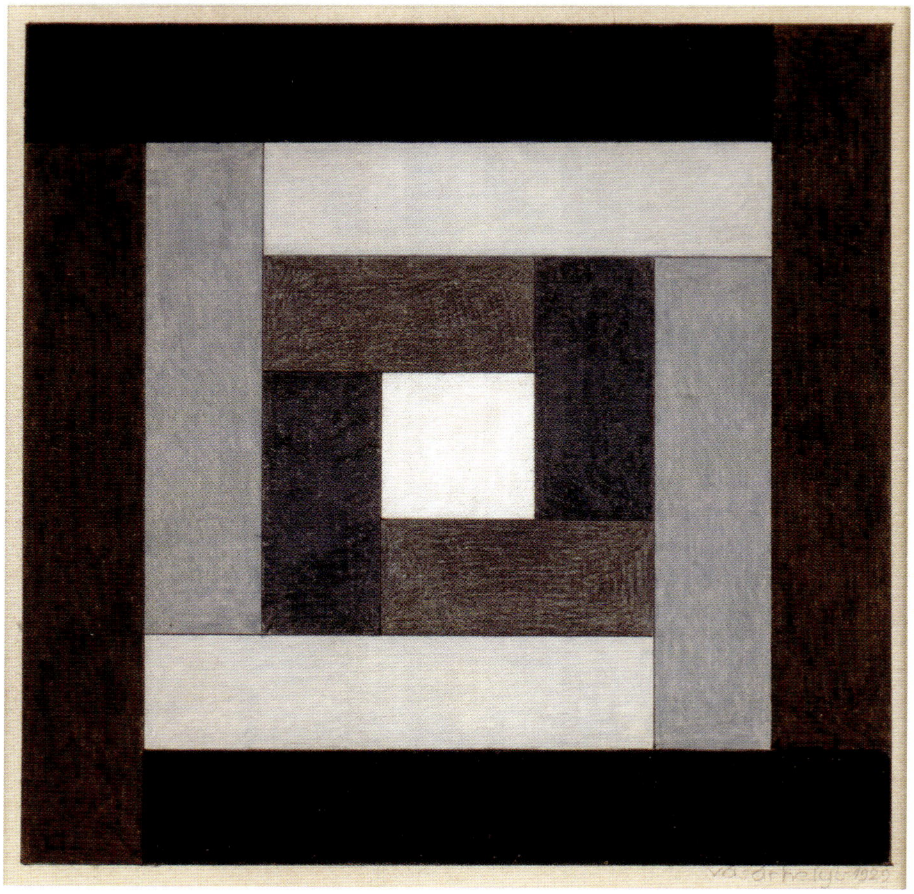

We are finished with the "nature" of Romanticism; our nature is biochemistry, astrophysics, and wave mechanics. We affirm that all creation of man is formal and geometric like the secret structure of the universe.

—Victor Vasarely, 1952

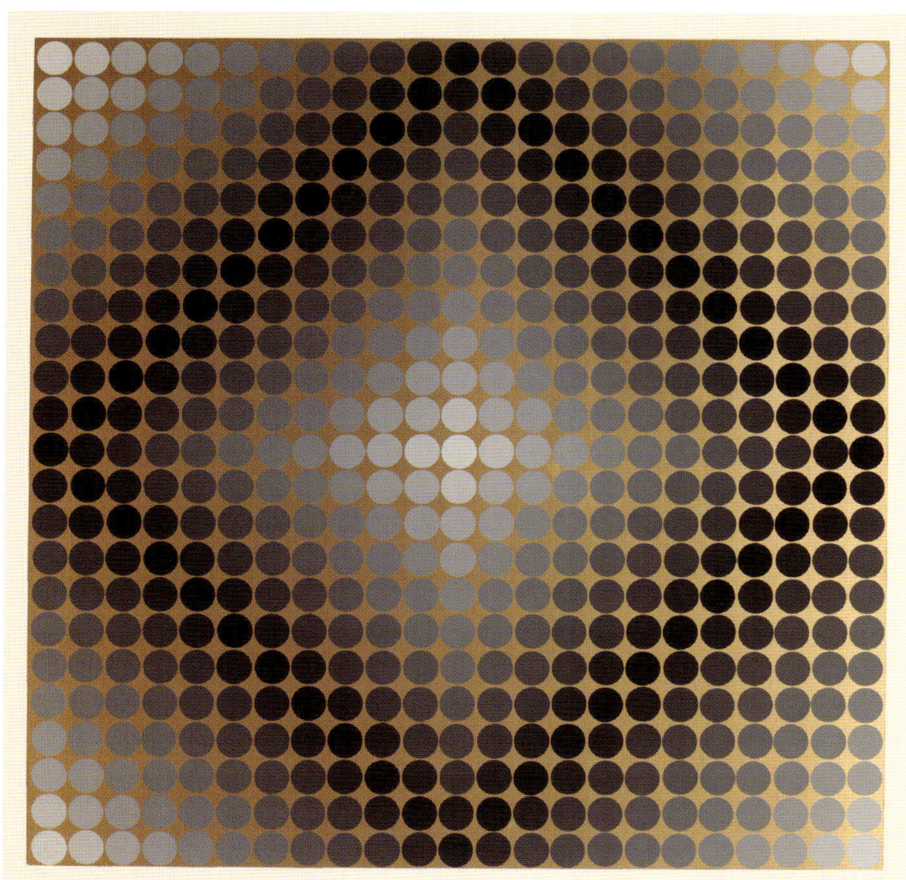

TOP LEFT

11-25. Victor Vasarely (Hungarian-born French, 1908–97), *Bauhaus Study A*, 1929. Oil on board, 9 × 9 in. (23 × 23 cm). Collection Michèle Vasarely, formerly Musée de Gordes, France. © 2014 ADAGP, Paris/Artists Rights Society, New York.

BOTTOM LEFT

11-26. Victor Vasarely (Hungarian-born French, 1908–97), *Untitled*, ca. 1960. Screenprint, $23\frac{5}{8}$ × $23\frac{5}{8}$ in. (60 x 60 cm). National Gallery of Victoria, Melbourne, Australia.

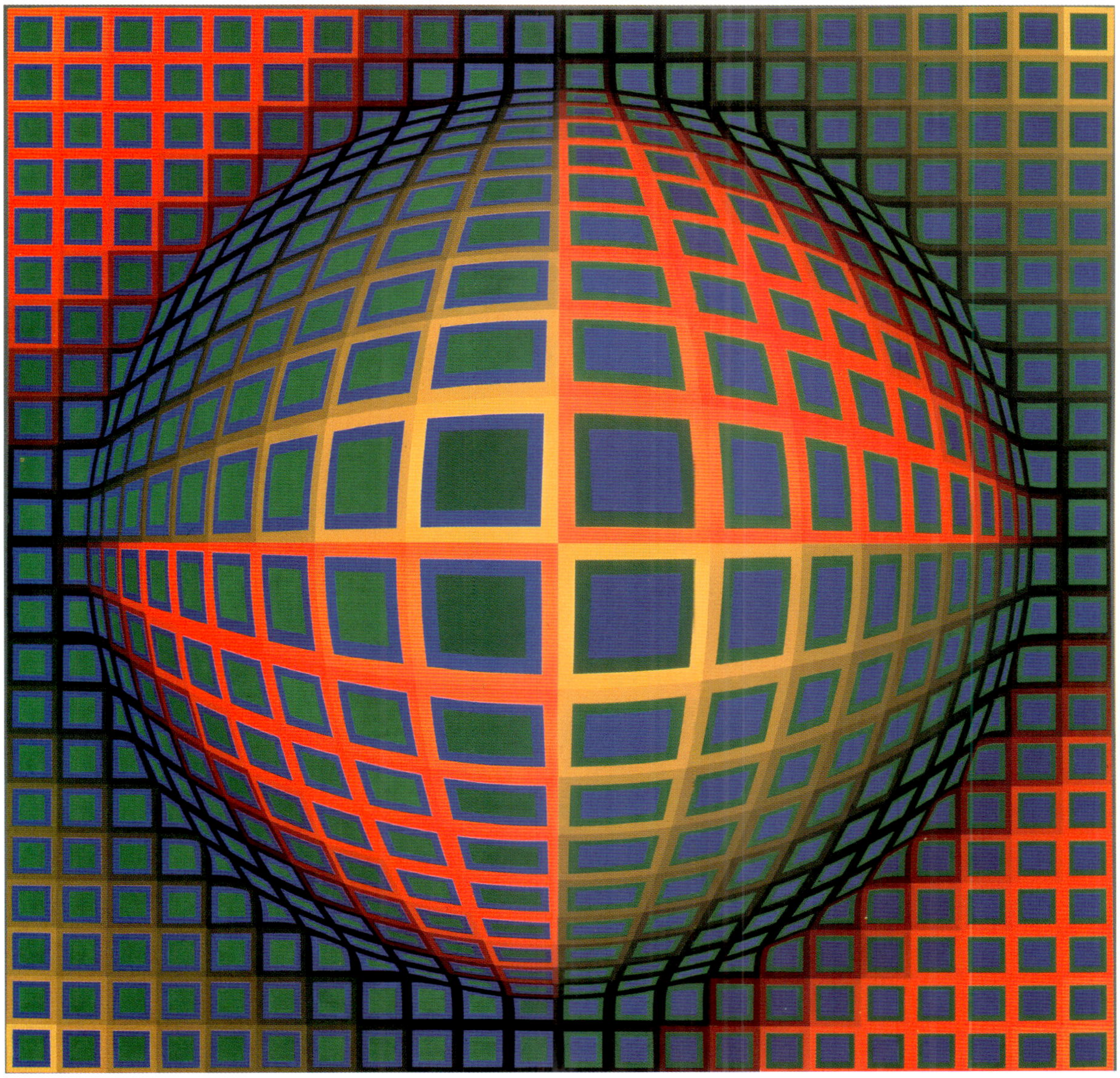

11-27. Victor Vasarely (Hungarian-born French, 1908–97), *Vega-Nor*, 1969. Oil on canvas, 78¾ × 78¾ in. (200.02 × 200.02 cm). Albright-Knox Art Gallery, Buffalo, New York, gift of Seymour H. Knox Jr. 1969. © 2014 ADAGP, Paris/ Artists Rights Society, New York.

generating them in a systematic way, such as rolling dice. For example, in *Random Distribution of 40,000 Squares* (plate 11-24), he arranged red and blue squares in a grid according to the pattern of odd and even numbers in a telephone book.

Among the artists Morellet knew in Paris was the Hungarian Victor Vasarely, who had studied in Budapest in the late 1920s at the Mühely (Hungarian for "workshop"), an art school patterned after the Bauhaus curriculum, where he learned the basic language of vision (plate 11-25). Vasarely worked in Paris as a graphic designer for a decade while searching for his own style in experiments that culminated in the 1950s in black and white designs dedicated to Malevich (*Tribute to Malevich*, 1954; Universidad Central de Venezuela, Caracas). But whereas the Suprematist Malevich had interpreted his forms as steps on the path to Absolute Spirit, Vasarely aimed to express "the internal geometry of nature" (*Yellow Manifesto*, 1955). Vasarely also considered the viewer's perception to be the essence of the aesthetic experience: "Movement does not rely on composition nor a specific subject, but on the apprehension of the act of looking, which by itself is considered as the only creator." Vasarely curated the exhibition *Le mouvement* (Movement,

1955; Galerie Denise René, Paris), launching Optical Art and Kinetic Art with works by himself, Jesús Rafael Soto, Yaacov Agam, Jean Tinguely, and others. In the 1960s Vasarely introduced color into his work and went on to combine color patterns with the apparent motion of his warped surfaces to create dramatic optical illusions (plates 11-26 and 11-27).

Although Optical Art became international, its base remained in Paris. The Argentines Julio Le Parc, Horacio García Rossi, and Francisco Sobrino joined with Morellet, Vasarely's son Jean-Pierre Vasarely (known as Yvaral), and the Frenchman Joël Stein to form the Groupe de Recherche d'Art Visuel (Group for research on visual art) in 1961. As the name implies, the artists took a scientific approach to studying optical effects, and, like Vasarely, they employed principles from Gestalt psychology, focusing on optical and kinetic effects of form and color. They denounced aesthetic elitism and declared that their art was democratic in the sense that the viewer's response was caused by innate features of the eye (such as the structure of the retina) that are shared by all humans. The viewer completed the art by looking at the dazzling patterns, and according to Morellet, the viewer's participation "finally gave a social meaning to our geometry."[28] Since 1976 the French artist Bernar Venet has taken a more formalist approach, painting mathematical symbols as typographic marks in his compositions, focusing on patterns made by the symbols as opposed to their meaning (which is unknown to him). In his series *Saturation* (plate 11-28) Venet created a multilayered palimpsest of superimposed mathematical symbols.[29]

THE DEATH OF BOURBAKI

Since Bourbaki was an imaginary person, he was supposed to be immortal, but, alas, he grew old. After a strike by students and workers swept France in May 1968, intellectuals lost interest in abstract structures—whether from Bourbaki or Lévi-Strauss—and demanded a philosophy that related to society. By the early 1970s there was a rapid decline of poststructuralism in the general culture in France and a related decline in Bourbaki, which the public had come to associate with structural linguistics (although they have different historical origins).

The ordered structure blew up.

—**Michel Serres, *Hermès IV*, 1977**

Another cause of Bourbaki's demise was the poststructuralists' overextension of the boundaries of their intellectual terrain; since Bourbaki was yoked to poststructuralism, it was swept away by the rage provoked by certain poststructuralists.[30] Structuralism had begun in the 1910s when Saussure declared that words in a natural (spoken) language have no fixed definitions—their meaning arises from their use; terms like "truth" and "law" are defined within a cultural context. The poststructuralists Lévi-Strauss, Lacan, Foucault, and Barthes were motivated by problems of interpretation of cultural phenomena (as studied by anthropologists, psychoanalysts, historians, and literary critics)—contexts in which it might be useful to apply Saussure's linguistic tools. But French poststructuralists overreached when they generalized Saussure's *verbal* linguistics to apply to *all* signs and declared that the theorems of mathematics and the laws of science are cultural conventions that do not describe a world-out-there. By making such claims, the poststructuralists produced a local commotion in academia that set off an avalanche of criticism from the halls of science.[31]

OPPOSITE

11-28. Bernar Venet (French, b. 1941), *Saturation* 3, 2002. Acrylic on canvas, 6 × 6 ft. (183 × 183 cm). Private collection, Seoul, South Korea. © 2014 ADAGP, Paris/Artists Rights Society, New York.

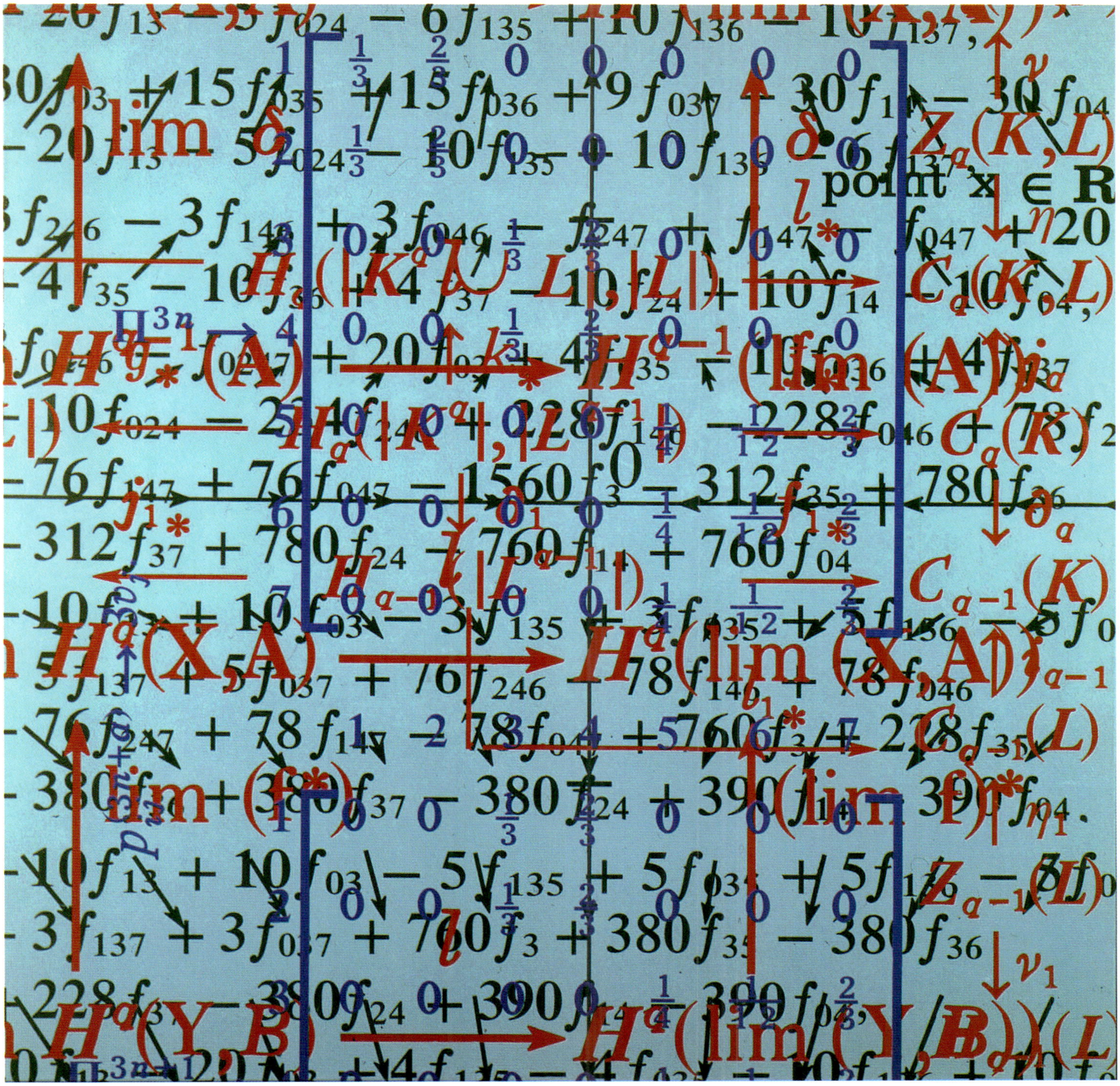

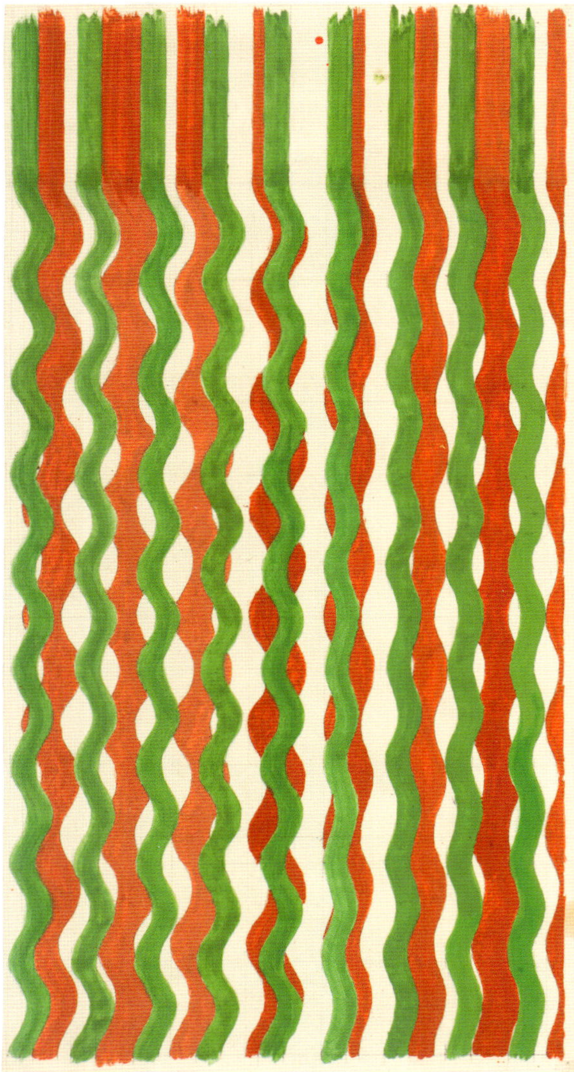

BRITAIN

In London, the stage was set for the emergence of British Process Art when critic Herbert Read held the exhibition *Growth and Form* in 1951. Based on the British botanist D'Arcy Wentworth Thompson's classic analysis of biological processes (*On Growth and Form*, 1917; see plate 2-42 in chapter 2), Read made an analogy (not found in Thompson) between a growing organism and a "growing" work of art—one in the process of being made. In the mid-1950s the painter Michael Kidner, along with the younger artists Jean Spencer and Malcolm Hughes, echoed this thinking in their so-called Systems Art, in which the process by which an artwork is created became its subject.[32] They wrote algorithms and then systematically produced works, such as *Phased Brown and Green Wave* (plate 11-29), which are pure geometric abstraction devoid of narrative content.

In the mid-1960s the British artist Bridget Riley developed her geometric painting style in relative isolation, unaware of the nascent group of Optical artists in Paris. She was catapulted onto an international stage in 1965 by the inclusion of her work alongside paintings by Vasarely and Morellet in the *Responsive Eye* exhibition at the Museum of Modern Art in New York (plate 11-31).[33] Riley works more intuitively than her French counterparts, not drawn to precise formulas, executing her paintings by hand. Simon Thomas, who studied visual arts at the Royal College of Art in London in the 1980s, went on to create sculpture with striking geometric patterns, serving as artist-in-residence at the University of Bristol, both in the department of physics (1993–95) and mathematics (2002; plates 11-30 and 11-32).

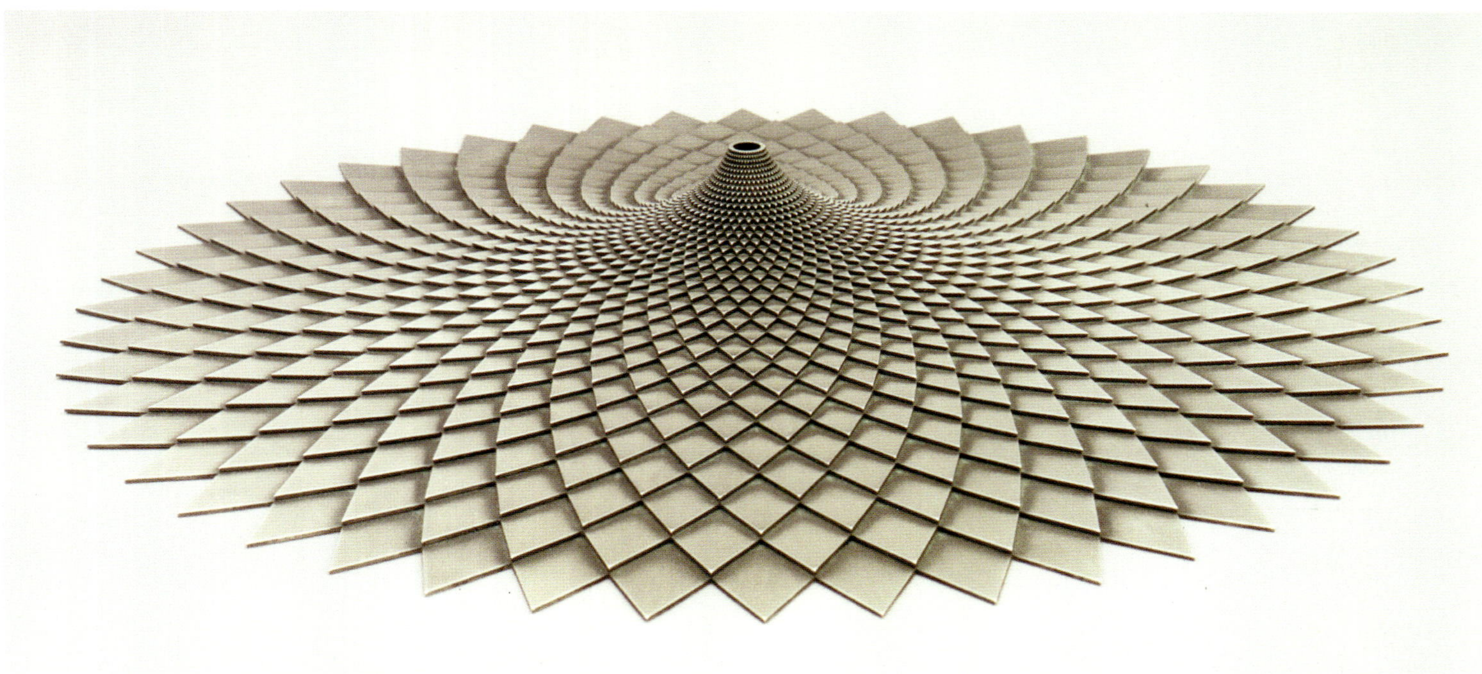

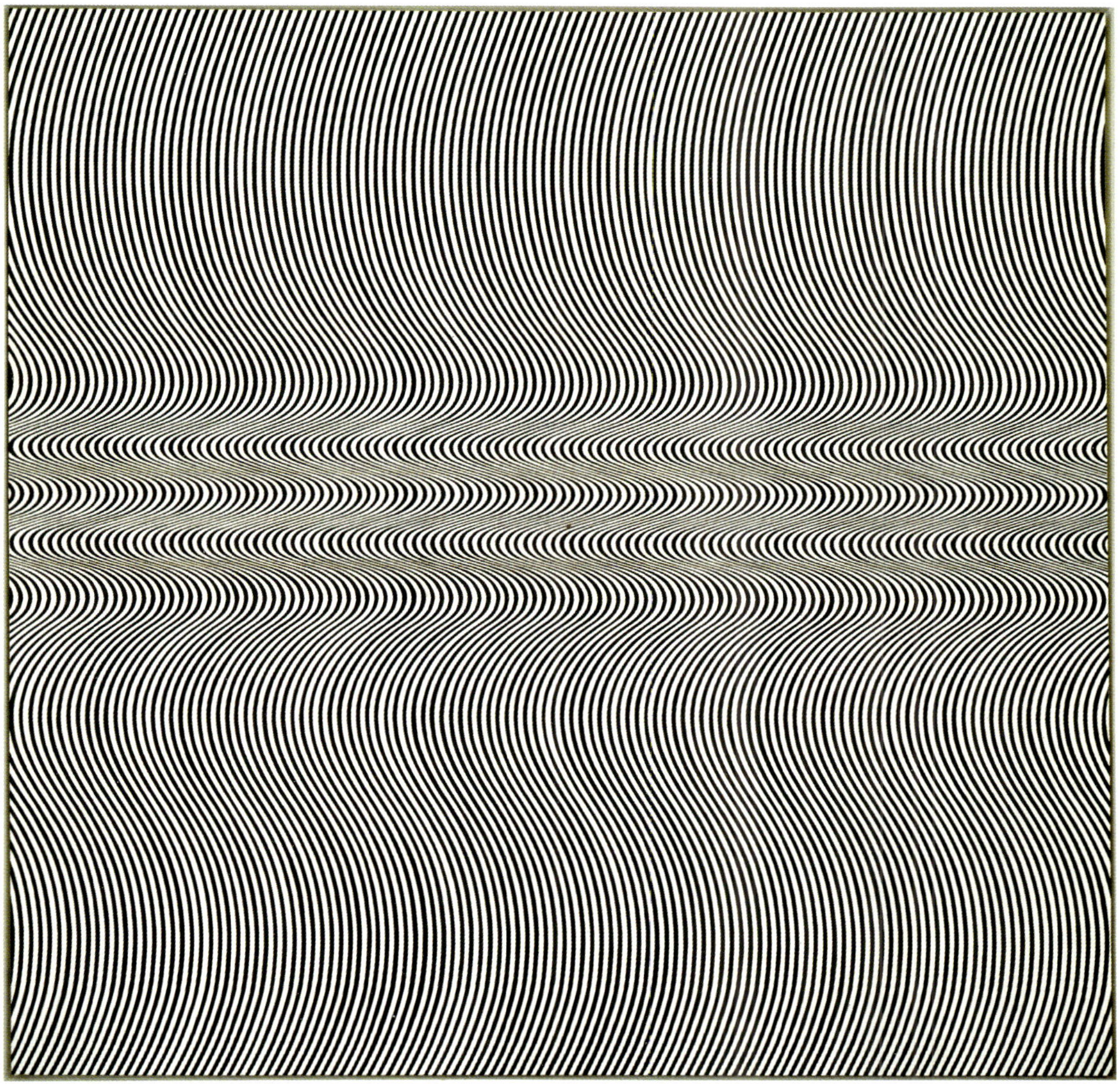

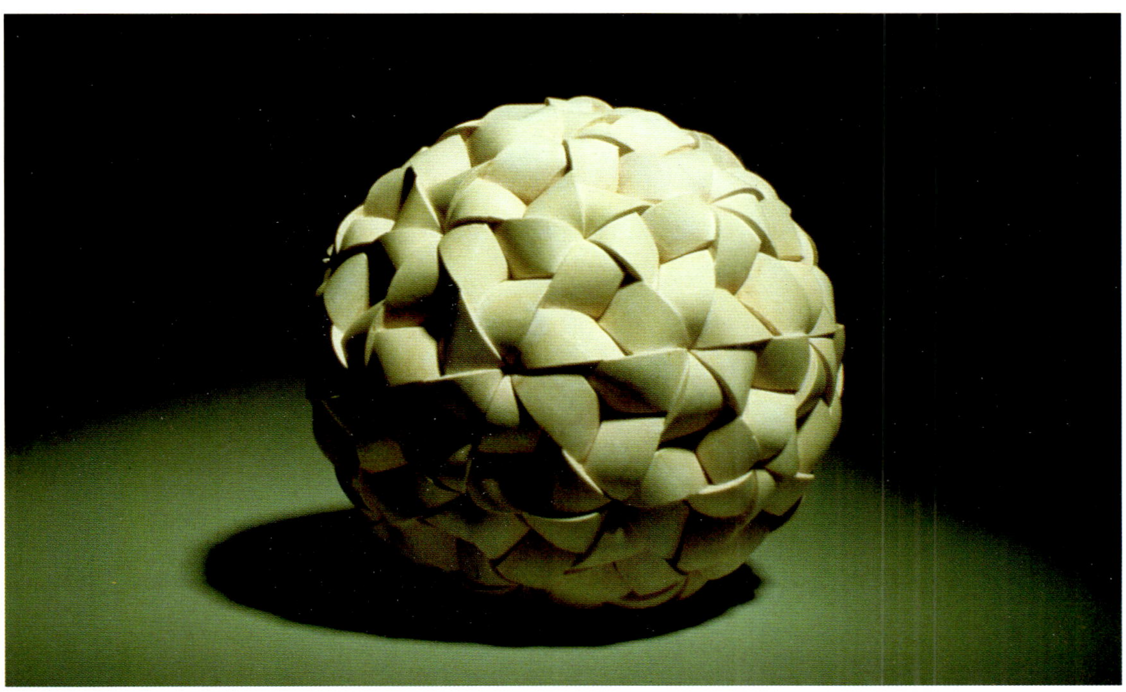

OPPOSITE TOP

11-29. Michael Kidner (English, 1917–2009), *Phased Brown and Green Wave*, 1964. Gouache and pencil on paper, 22 × 11½ in. (56 × 29 cm). © Tate, London.

OPPOSITE BOTTOM

11-30. Simon Thomas (British, b. 1960), *Planeliner*, 2005. Bead blasted stainless steel, 23⅝ in. (60 cm) diam. × 2¼ in. (5.55 cm) high. Courtesy of the artist.

The artist has described this sculpture: "The geometric foundation of this sculpture is a simple equation. The combination of a few proportioned elements set together in a consistent system can generate an event of resounding clarity."

TOP LEFT

11-31. Bridget Riley (English, b. 1931), *Current*, 1964. Synthetic polymer paint on composition board, 58⅜ × 58⅞ in. (148.1 × 149.3 cm). Museum of Modern Art, New York, Philip Johnson Fund. © Bridget Riley 2015. All rights reserved, courtesy Karsten Schubert, London.

BOTTOM LEFT

11-32. Simon Thomas (British, b. 1960), *Crystal Orb*, ca 2010. Crystal, 9¾ in. (25 cm) diam. Courtesy of the artist.

This pattern is based on the division of a spherical surface into equilateral triangular faces, which are themselves divided into thirds.

LATIN AMERICA

Many Latin American artists were attracted to the precise styles and geometric aesthetics of European Constructivism and Concrete Art, and politicians in São Paulo, Rio de Janeiro, Buenos Aires, and Caracas associated geometric abstract art with orderly social progress. In Uruguay, artists were inspired by abstract art brought back from France by Joaquín Torres García. As a young man, Torres García acquired a deep knowledge of the Constructivist aesthetic while studying in Europe. During the 1920s he used the Golden Section to give geometric structure to his compositions, but after meeting Theo van Doesburg he adopted the grid format of De Stijl, although he rejected their palette of primary colors and used muted earth tones (plate 11-33).[34] Returning to Uruguay in 1934, Torres García proposed a Latin American version of Constructivism—Universalismo Constructivo—in which artists filled the grid with international signs and symbols, including some from pre-Columbian cultures (plate 11-34). To that end in 1935 Torres García and his followers formed the Asociación Arte Constructivo in his hometown of Montevideo,[35] where in 1943 Torres García opened an art workshop, El Taller Torres García.

11-33. Joaquín Torres García (Uruguayan, 1874–1949), *Wood Planes of Color*, 1929. Oil on wood, 17 × 8 in. (43.2 × 20.3 cm). Carmen Thyssen-Bornemisza Collection on deposit at Museo Thyssen-Bornemisza, Madrid. © 2014 Alejandra, Aurelio, and Claudio Torres.

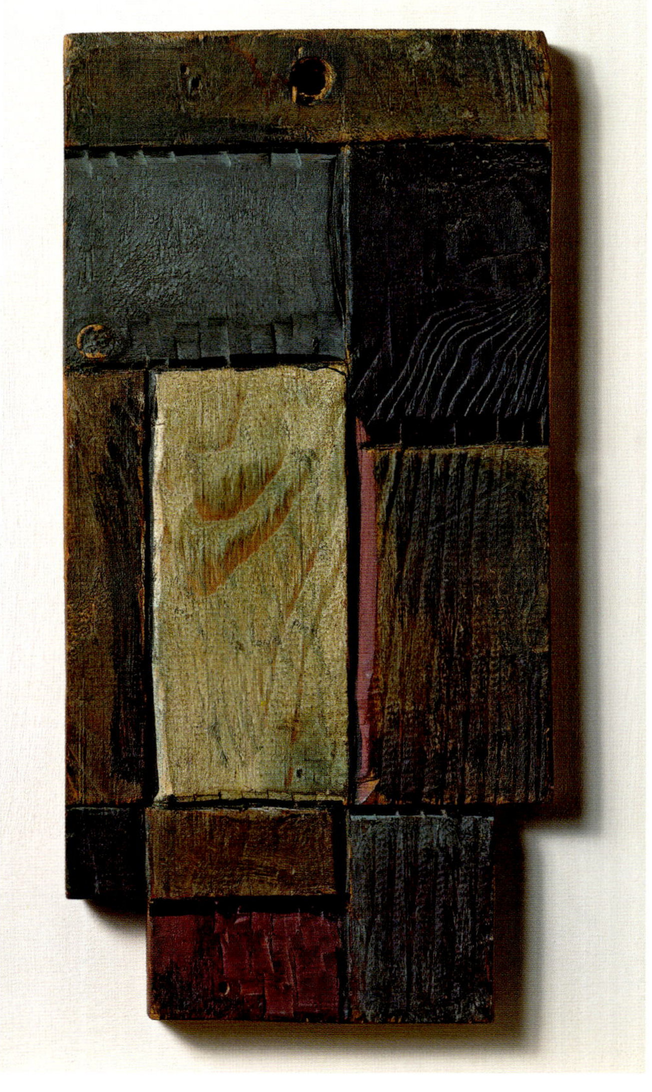

Gonzalo Fonseca was a native of Montevideo who studied at the Taller Torres García in the 1940s and in Europe in the 1950s, pursuing the ideals of Universal Constructivism by creating stone sculptures that resonate with both the pre-Columbian and European classical traditions. Fonseca carved sculpture with forms that fit together like pieces of a puzzle (plate 11-35).

In Brazil in the early 1940s, logic was added to the curriculum at the University of São Paulo, where the faculty included the young American logician W. V. Quine, who had completed a doctorate as a student of Alfred North Whitehead at Harvard University in 1932. There was strong government support for geometric abstract art, which suited the new industrial landscape of Brazil. In 1948 government-sponsored museums of modern art opened in São Paulo and Rio de Janeiro, and in 1951 the government launched an international art fair, the São Paulo Biennial. The founding director of the Museu de Arte Moderna in São Paulo was the Belgian art critic Léon Degand, whose first curatorial effort was an international survey of ninety-five art works, (*Do figurativismo ao abstracionismo* [From figurative to abstract art], 1949), followed by a retrospective of Max Bill (1950). At the first São Paulo Biennial of 1951, which included several Swiss Concrete artists, Max Bill won the grand prize, thereby insuring his importance for a generation of Brazilian artists, some travelling to Ulm for their education.[36]

In 1952, São Paulo artists led by Waldemar Cordeiro formed the Grupo Ruptura, which adopted the Constructivist style.[37] Cordeiro was raised in Italy by his Italian mother and Brazilian father,

The idea of structure is the cornerstone of Constructivism, and also of painting: to paint is to construct.

—**Joaquín Torres García,**
***Universalismo Constructivo*, 1944**

LEFT

11-34. Joaquín Torres García (Uruguayan, 1874–1949), *Universal Composition*, 1937. Oil on board, 42½ × 33½ in. (108 × 85 cm). Musée National d'Art Moderne, Centre Georges Pompidou, Paris.

Universal Composition includes symbols from pre-Columbian culture (a sun, a man, a fish, and a star), from classical Greece (the letters alpha and omega, a temple), and from science and mathematics (a triangle, pentagon, compass, balance, and numbers).

BELOW

11-35. Gonzalo Fonseca (Uruguayan, 1922–97), *For the Geometrician's Grave*, 1970. Tennessee pink marble, 12 × 20 × 5 in. (30.4 × 50.8 × 12.7 cm). Courtesy of Caio Fonseca.

Two slabs of stone are laced together with leather straps to form a book that opens to reveal "pages" inset with geometric forms that can be dislodged from their niches. The open book suggests a tombstone "for the geometrician's grave."

and after studying art in Rome he moved to São Paulo in 1945 at age twenty. He advocated creating art in complete objectivity, giving visual form to an algorithm, as in *Visible Idea*, in which black and white linear patterns, each the inverted mirror image of the other, trace out corresponding curves (1956; see plate 11-36).

The most dramatic expression of the Constructivist aesthetic in Brazil, however, was the design and construction of the country's new capital, Brasília, by the architect Oscar Niemeyer and city-planner Lúcio Costa. Building a new capital city in the barren center of Brazil, then home to a few cattle ranchers, was the utopian vision of the politician Juscelino Kubitschek. Born in poverty in Brazil to a Czech mother and Polish father, Kubitschek matured with an ambitious sense of social mission. He won the Brazilian presidency in 1956 with the slogan "Fifty years of progress in five," and soon after he took office he commissioned Niermeyer to create a city of the future with uniform housing and mass transit that would symbolize the equality of citizens. Niermeyer designed the buildings—completed in just four years—in steel, glass, and reinforced concrete in the International Style, giving the city a strikingly unified appearance (plate 11-37).[38]

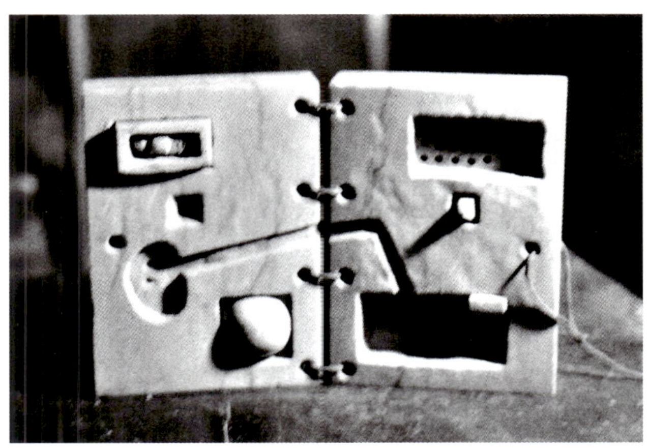

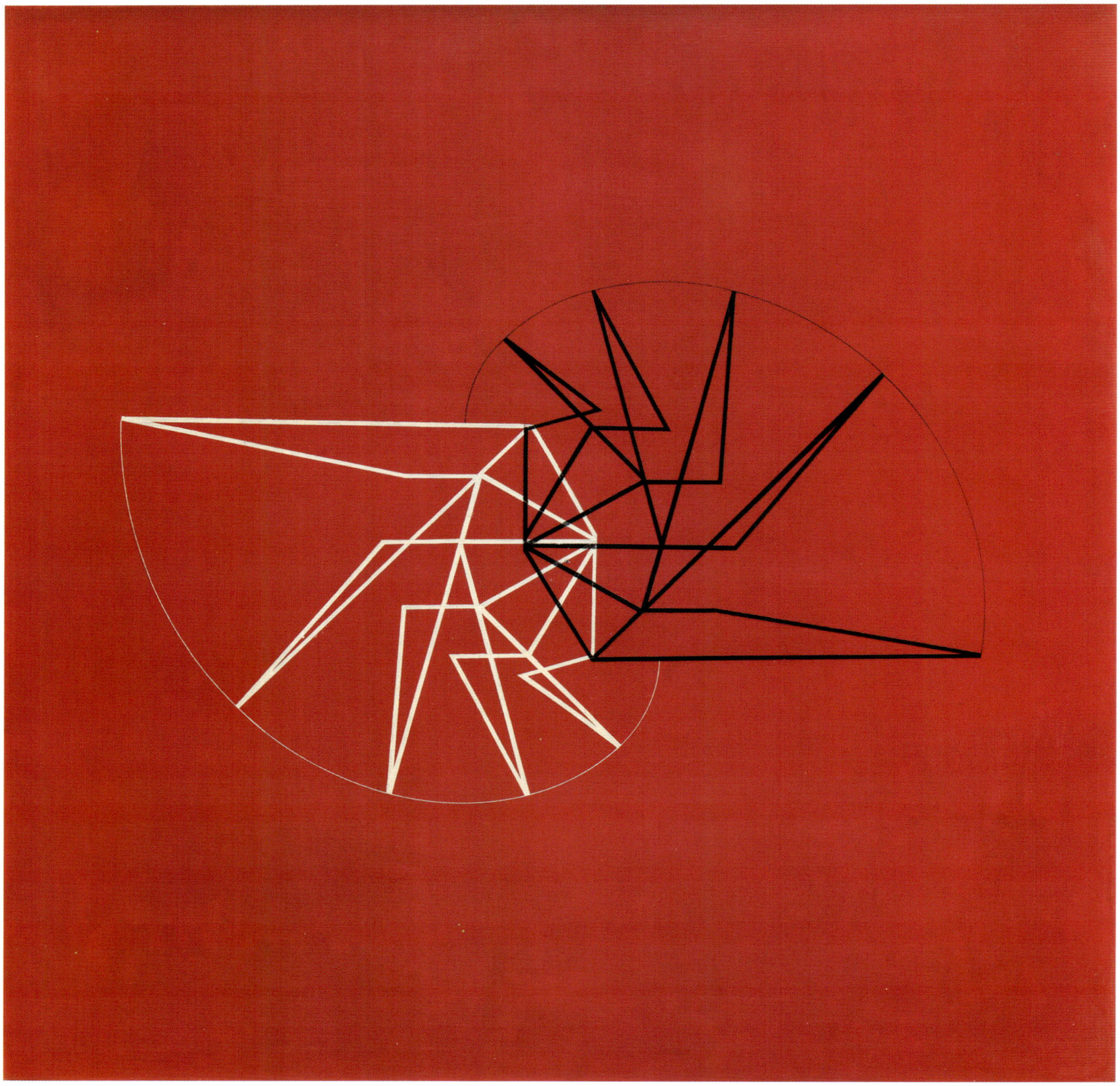

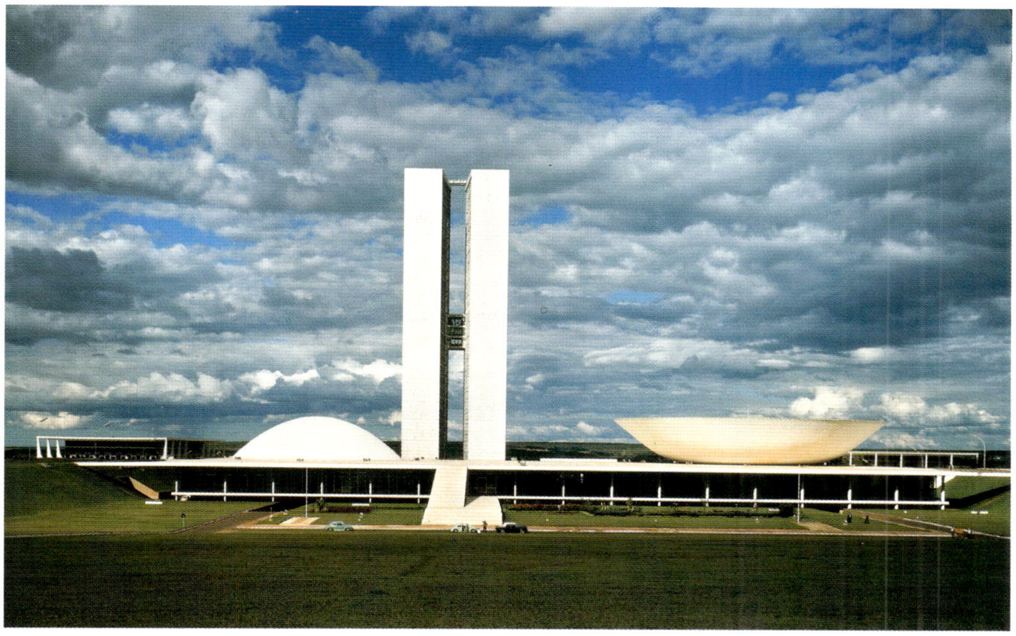

In the late 1950s a group of Brazilian artists in Rio dared to propose a formalist style in which the forms were not meaning-free but imbued with emotional expression—Neo-Concrete Art. The poet Ferreira Gullar wrote their manifesto: “Art cannot be merely illustrations of a priori concepts. . . . Such statements might lead one to believe that Neo-Concrete artists want to shun objectivity and lose themselves in subjective chaos. But in fact, we seek a kind of deeper objectivity resulting from the intimate integration of material with mankind’s feeling and mind.”[39] In a typical example, Hélio Oiticica arranged precise rectangles on a grid but shifted their orientation so they would not quite line up, giving the impression that the squares are moving and shifting (*Metaesquema* [Metascheme], 1959; see plate 11-38).

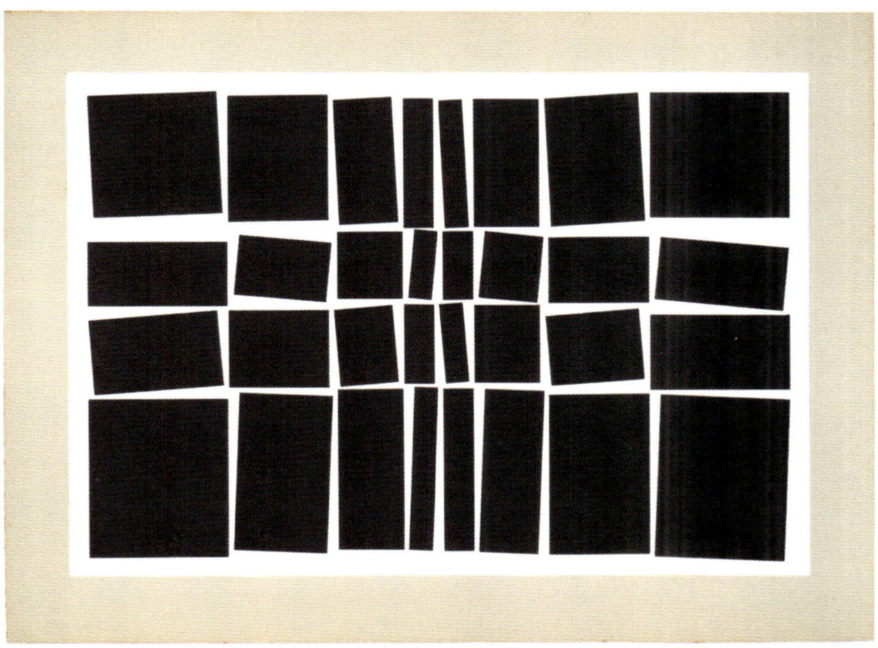

OPPOSITE

11-36. Waldemar Cordeiro (Italian-born Brazilian, 1925–73), *Visible idea*, 1956. Acrylic on plywood, 23⁹⁄₁₆ × 23⅝ in. (59.9 × 60 cm). Private collection. © Waldemar Cordeiro Estate.

TOP LEFT

11-37. Oscar Niemeyer (Brazilian, 1907–2012), National Congress of Brazil, located on the Three Powers Plaza, Brasília, designed 1956, opened 1960.

Niemeyer designed the plaza to house the executive, judiciary, and legislative branches of Brazil’s government. The buildings in this photograph are home to the legislature (the Congress): the semi-hemisphere on the left houses the senators and the one on the right holds the representatives, connected by two rectangular office towers.

BOTTOM LEFT

11-38. Hélio Oiticica (Brazilian, 1937–80), *Metaesquema* (Metascheme), 1958. Gouache on board, 19⅞ × 26¾ in. (50.5 × 68 cm). Museum of Modern Art, New York, purchased with funds given by Patricia Phelps de Cisneros in honor of Paulo Herkenhoff. © 2009 Projeto Hélio Oiticica.

RIGHT

11-39. Gyula Kosice (Hungarian-born Argentine, b. 1924), *Over Relief,* 1948. Luminous glass tubes mounted on wooden box, acrylic, Plexiglas, 25½ × 14¾ × 7 in. (65 × 37.5 × 17 cm). Courtesy of the artist.

OPPOSITE

11-40. Tomás Maldonado (Argentine, b. 1922), *Composition 208,* 1951. Oil on canvas, 19¾ × 19¾ in. (50.2 × 50.2 cm). Private collection. © Tomás Maldonado.

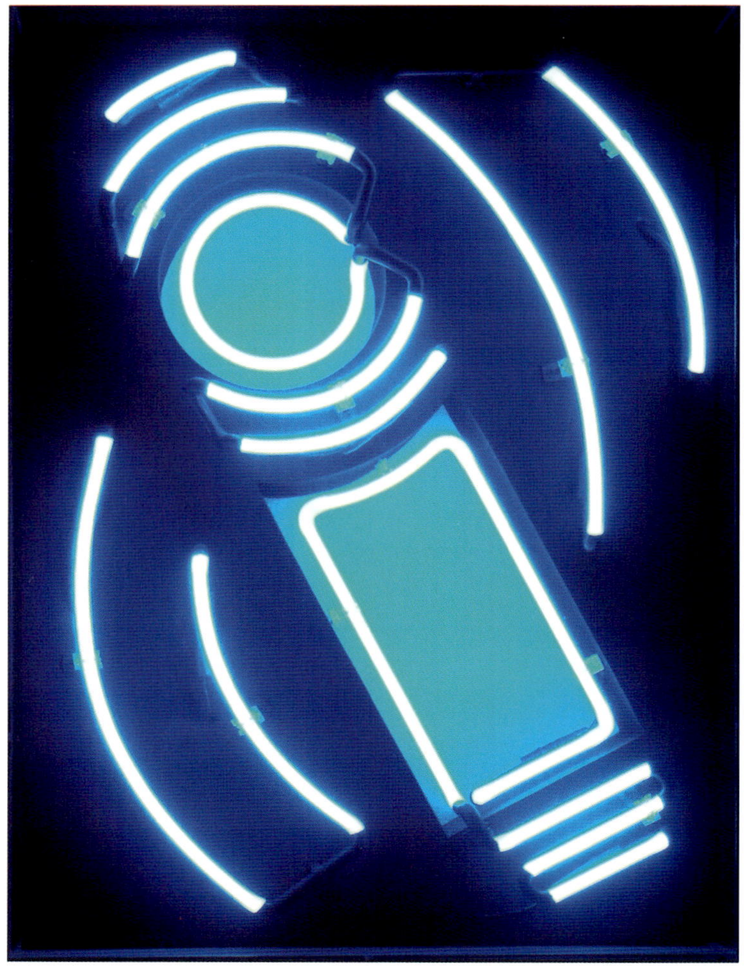

Scientific aesthetics will replace the old speculative, idealist aesthetics. There is no reason now to carry on a dialogue about the nature of the Beautiful. The metaphysics of the Beautiful has died of exhaustion. The physics of beauty is what is needed now.

—Tomás Maldonado et al., *Inventionist Manifesto*, 1946

On the eve of World War I, Germany, which had colonialist intentions, sent several German physicists to Buenos Aires to create a major center for theoretical physics at the Universidad Nacional de La Plata, which remained active until the early 1930s.[40] After graduating from La Plata, Mario Bunge completed a PhD in physics at Simon Fraser University in Canada, and then returned to Argentina to teach physics and philosophy at his alma mater. In the 1950s, Bunge organized several groups that read Gottlob Frege and Bertrand Russell, becoming one of the first advocates for analytic philosophy and logic in Argentina.[41] Argentine intellectuals hoped that education in mathematics and science would bring economic and social stability, and Argentine artists expressed confidence in reason in precise Constructivist styles. There were two groups of abstract artists, Madí and the Asociación Arte Concreto-Invención, and together in 1944 they published a single issue of the journal *Arturo*, which catalyzed a decade of creative output in Buenos Aires. The organizer of Madí, the painter Gyula Kosice, declared in the group's manifesto that the artists of Madí would "construct objects with timeless, eternal value."[42] Interested in physics and technology, Kosice began making objects from gas-filled glass tubes in 1946, and he interpreted the glowing light as symbolizing immaterial energy (plate 11-39).

Artists in the Asociación Arte Concreto-Invención created meaning-free, physical ("concrete") objects, such as Tomás Maldonado's *Composition 208* (1951; see plate 11-40). Internal divisions running parallel to the edges of the painting emphasize its literal

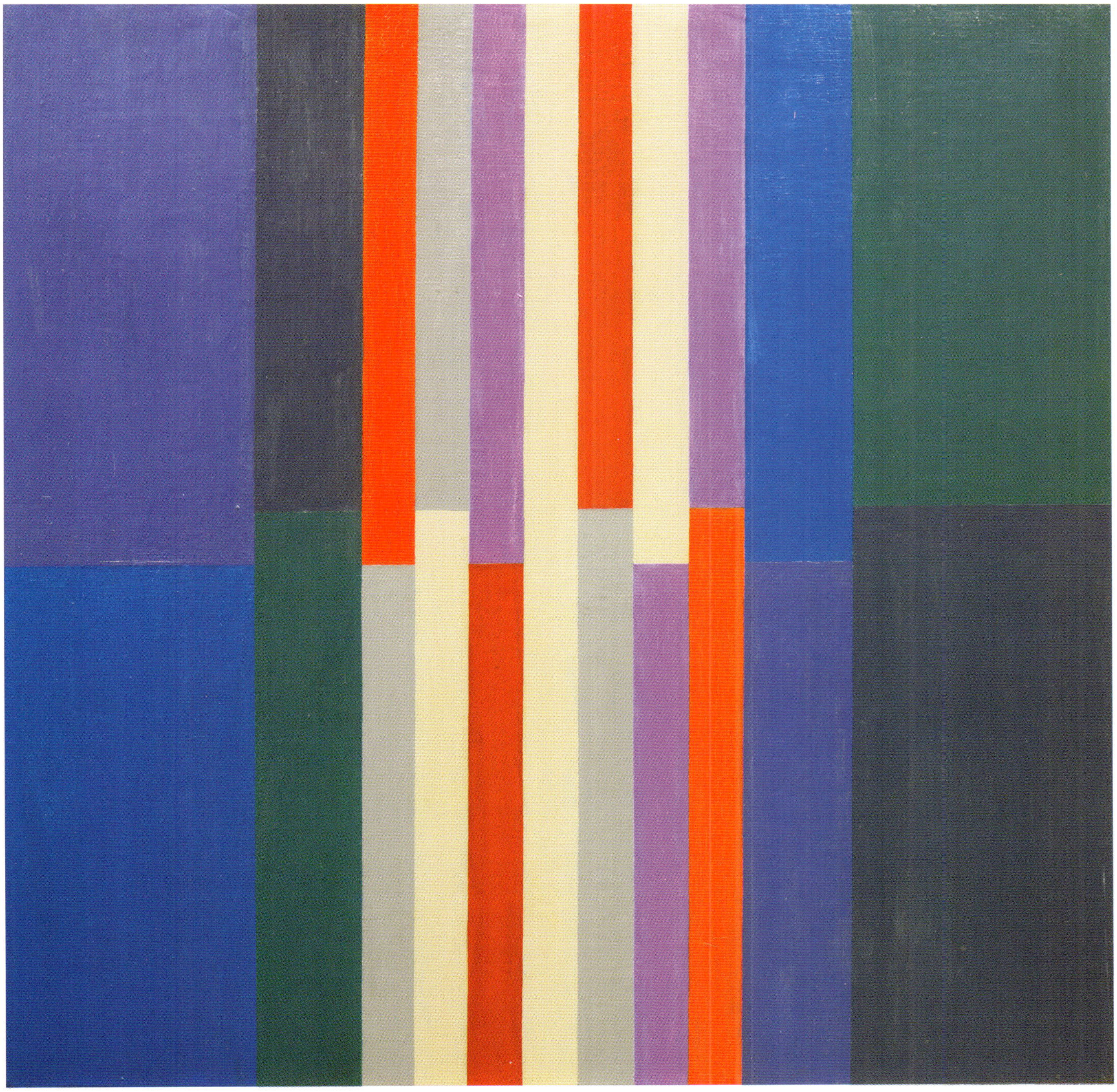

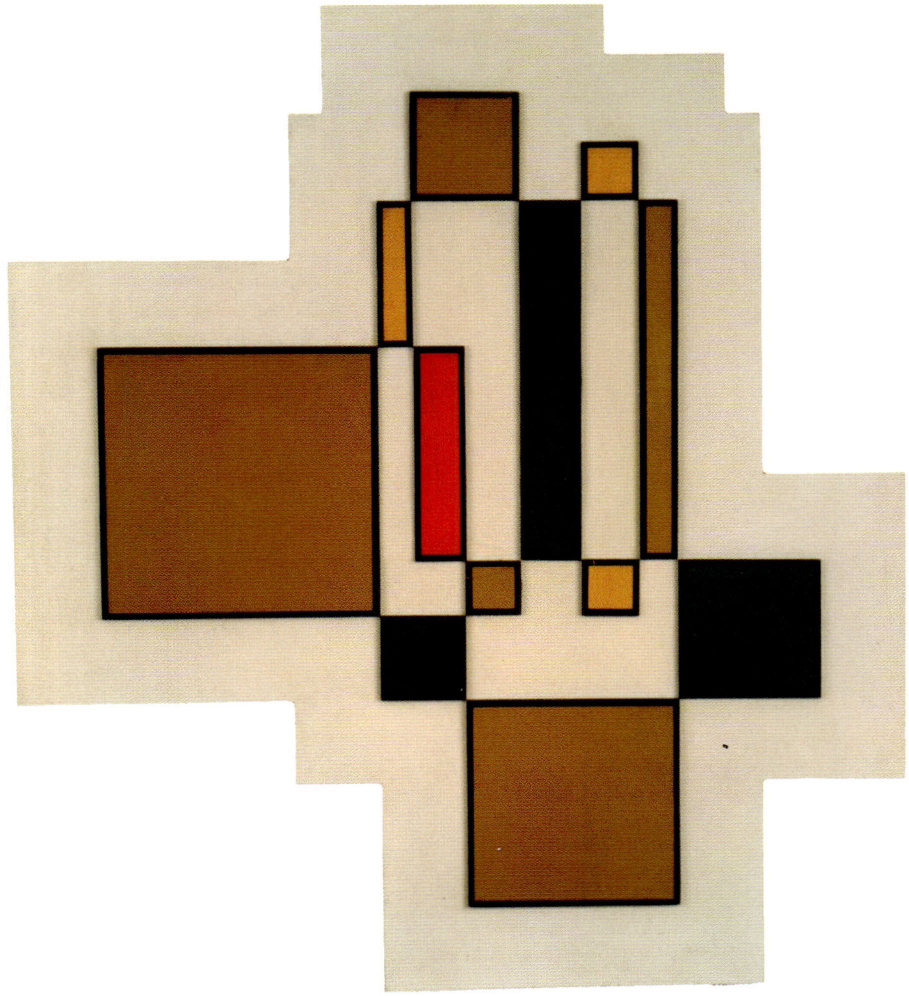

11-41. Rhod Rothfuss (Uruguayan, 1920–69), *Acento rojo* (Red accent), 1957. Lacquer on panel mounted to panel, 27½ × 24⅜ in. (69.8 × 61.9 cm). Collection of Diane and Bruce Halle, Phoenix, Arizona.

(actual physical) size and shape. Maldonado wanted to create a completely flat image in *Composition 208*, but viewers tend to see the red, purple, and beige lines floating in front of the darker green and blue "background." Echoing the formalist master narrative (see chapter 4), Maldonado's goal was to rid painting of illusionism: "The biggest lacks in non-representational art were caused by its failure to achieve either a new composition or the definitive removal of the illusory. Thus we began by breaking with the traditional format of the painting."[43] In an attempt to achieve "the definitive removal of the illusory," Maldonado invented the shaped canvas in the early 1950s, working with Rhod Rothfuss, a Uruguayan who lived in Argentina. For example, viewers are less inclined to see Rothfuss's *Acento rojo* (plate 11-41) as a "window" into an illusory world because they notice the irregular edges of the canvas, which remind them that they are looking at a physical object.[44] In 1948 Maldonado travelled to Europe, where he met Max Bill, and the two artists formed a lasting bond over their shared interests. Maldonado joined the faculty of the Hochschule für Gestaltung in Ulm in 1954, after which he divided his time between Buenos Aires and Ulm, serving as director of the Hochschule from 1964 to 1966.[45]

In Buenos Aires in 1964, the philosopher Thomas Moro Simpson, who was a follower of Bunge, published one of the first books in Spanish about logic and analytic philosophy—*Formas lógicas, realidad y significado* (Logical forms, reality and significance; 1964). But

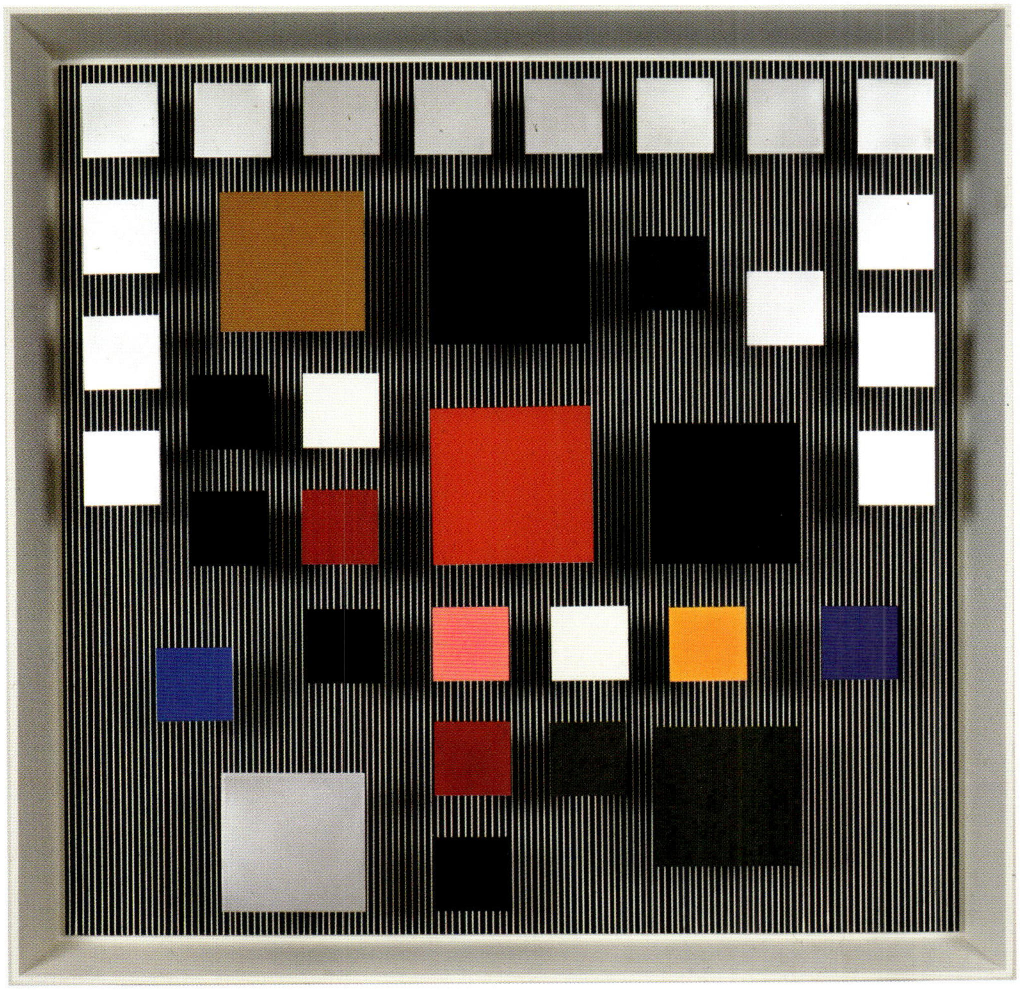

11-42. Jesús Rafael Soto (Venezuelan, 1923–2005), *Ambivalence in the Color Space, no. 21*, 1981. Acrylic, metal panels, wood elements, mounted on board, 41½ × 41½ × 6¼ in. (105 × 105 × 16 cm). Atelier Soto and Sicardi Gallery. © Sicardi Gallery, Houston, Texas. © 2014 ADAGP, Paris/Artists Rights Society, New York.

artistic and academic freedom ended in Argentina in 1966 when a coup installed a military dictatorship that decimated cultural life in the decades-long Guerra Sucia (Spanish for "Dirty war") of state-sponsored killing of Argentine civilians. Bunge was exiled to Montreal; Maldonado settled first in Europe and then the United States. Many younger artists, such as the painter César Paternosto, left Argentina for the United States. After moving to New York in 1967, Paternosto found his voice in the 1970s when he incorporated patterns from pre-Columbian textiles into his abstract formats (plates 11-43 and 11-44).[46] The political turmoil in Argentina subsided with the return of civilian rule in 1983, but many, including Bunge, Maldonado, and Paternosto, remained expatriates.

Venezuela became wealthy in the early twentieth century because of its huge oil resources, but the art establishment in Caracas remained old-fashioned. Progressive Venezuelan artists travelled to Paris, where they formed the émigré group Los Disidentes (The dissidents), including Jesús Soto and Carlos Cruz-Diez. During the 1952–58 rule of the military dictator Marcos Pérez Jiménez, the country underwent rapid commercial and cultural growth, and the visionary architect Carlos Raúl Villanueva designed the new campus of the Universidad Central de Venezuela in Caracas. At Villanueva's request, Soto and Cruz-Diez returned home from Paris in 1952 to create public art for the campus. Villanueva also commissioned sculpture from Alexander Calder and murals from Victor

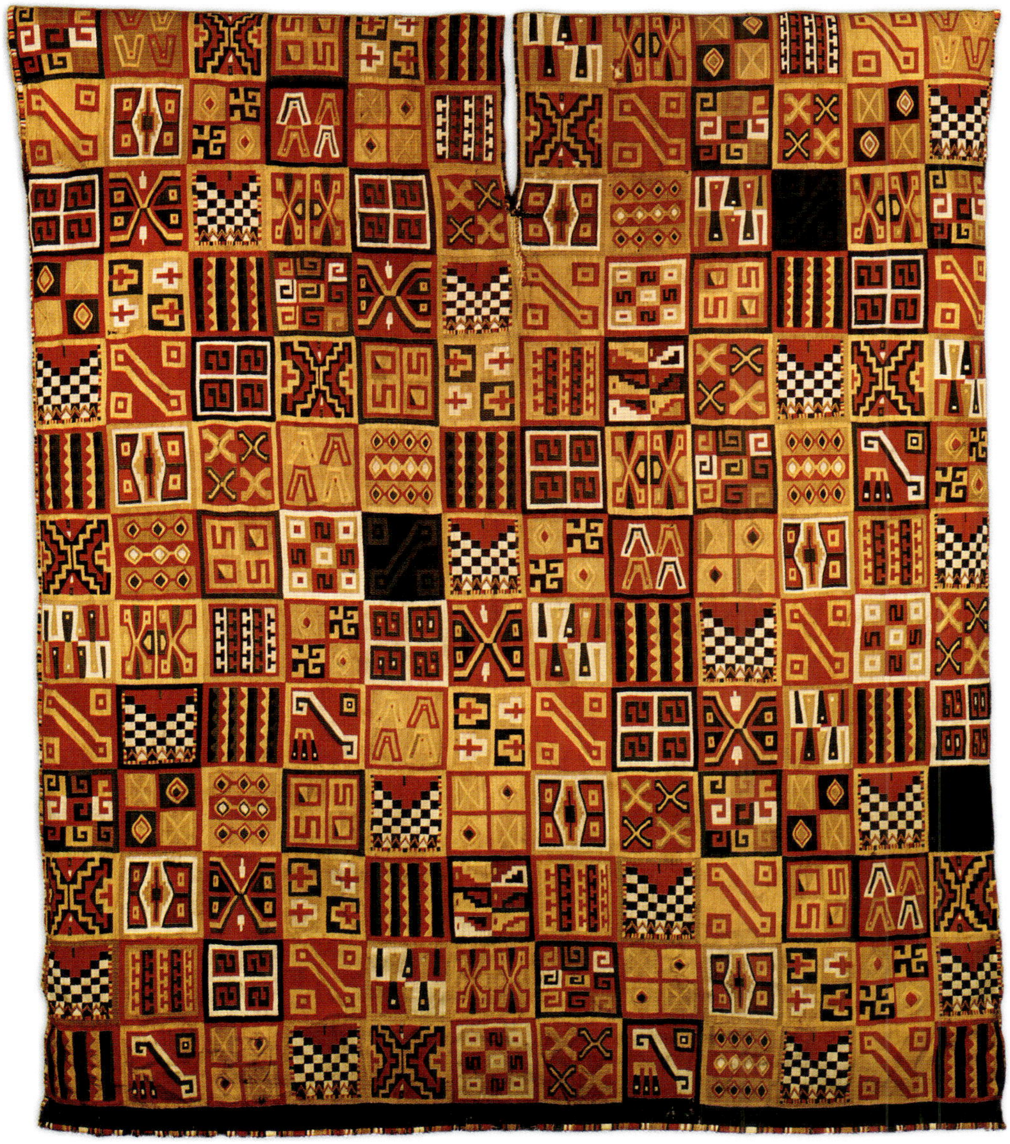

OPPOSITE

11-43. César Paternosto (Argentine, b. 1931), *T'oqapu*, 1982. Acrylic on canvas, 66 × 66 in. (167.6 × 167.6 cm). Courtesy of the artist and the Museo de Arte Contemporáneo Esteban Vicente, Segovia, Spain.

In pursuit of a universal Constructivist style, Paternosto combined geometric abstraction with Andean symbolism in paintings such as this. *T'oqapu* means "garment of the kings" in Quechua, the Inca language.

LEFT

11-44. T'oqapu tunic, from Peru. Inca, AD 1450–1540. Wool and cotton, 35½ × 30⅜ in. (90.2 × 77.15 cm). © Dumbarton Oaks, Pre-Columbian Collection, Washington, DC.

From the thirteenth to the early sixteenth centuries, the Incas built an empire high in the Andes Mountains, living in villages such as Machu Picchu at an elevation of 9000 feet. They had no written language but recorded information in knotted ropes (*quipu*; see plate 12-10 in chapter 12) and in patterns woven into their textiles. The small rectangular geometric units (*t'oqapu*) in this tunic may have been a code for social rank; only high-ranking Incas wore garments adorned with these patterns.

Vasarely because their geometric style fit the industrial, technological vision that Pérez Jiménez had for Venezuela.[47] Other government commissions followed, and Venezuela became a vibrant center for contemporary art in the Constructivist style.[48]

Soto designed artworks that are kinetic in the sense that they move or their appearance changes when the viewer moves. For example, the floating planes in a relief sculpture by Soto cast shadows that shift with the viewer's position (plate 11-42). Soto also suspended colored metal tubes to create volumes that the viewer can walk around and through (*Penetrables*; see plate 11-45). In Cruz-Diez's series *Physichromies*, the artist made reliefs with vertical strips which he painted different colors on their two sides, so that a moving viewer sees the colors change (plate 11-46).

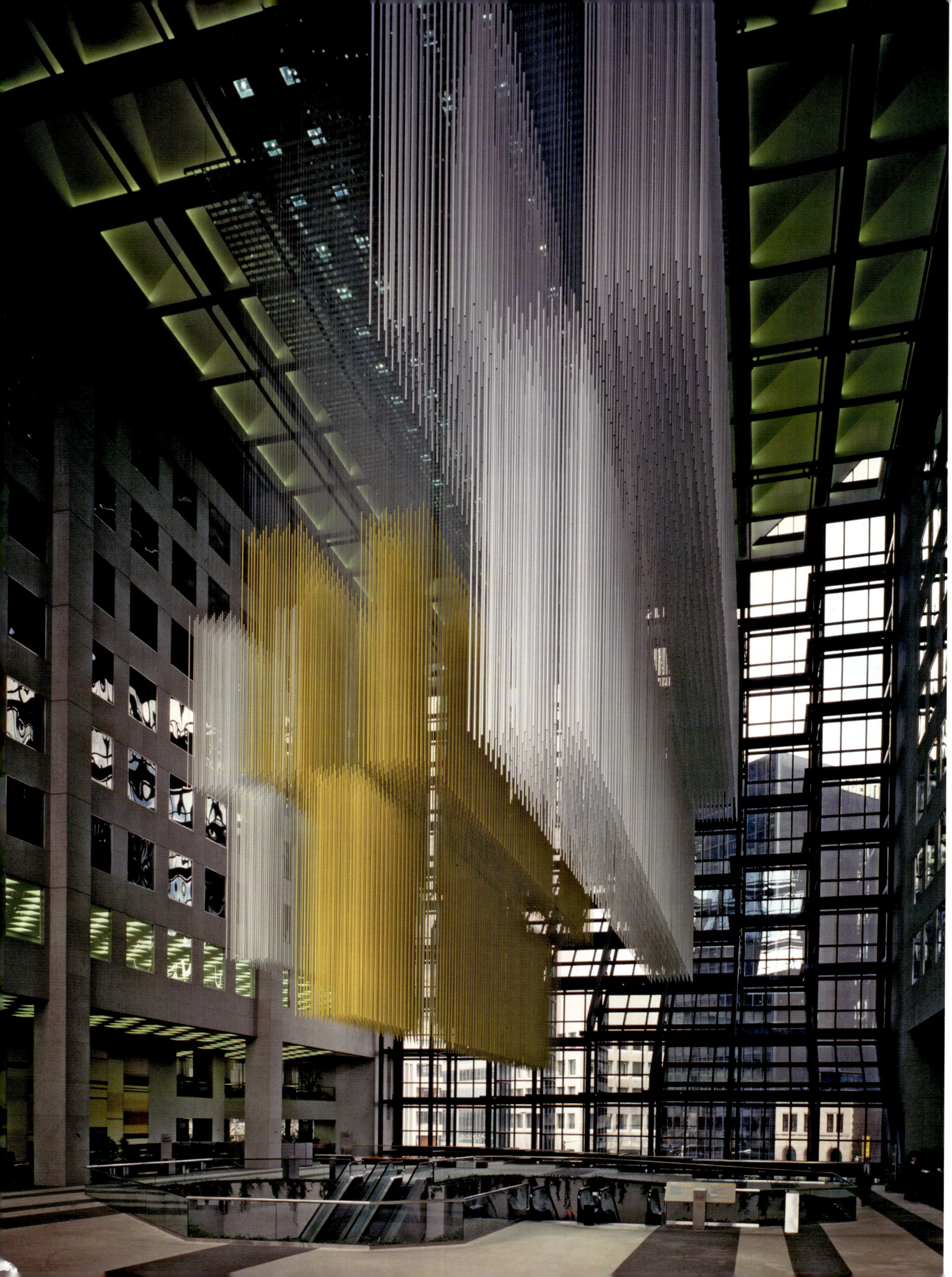

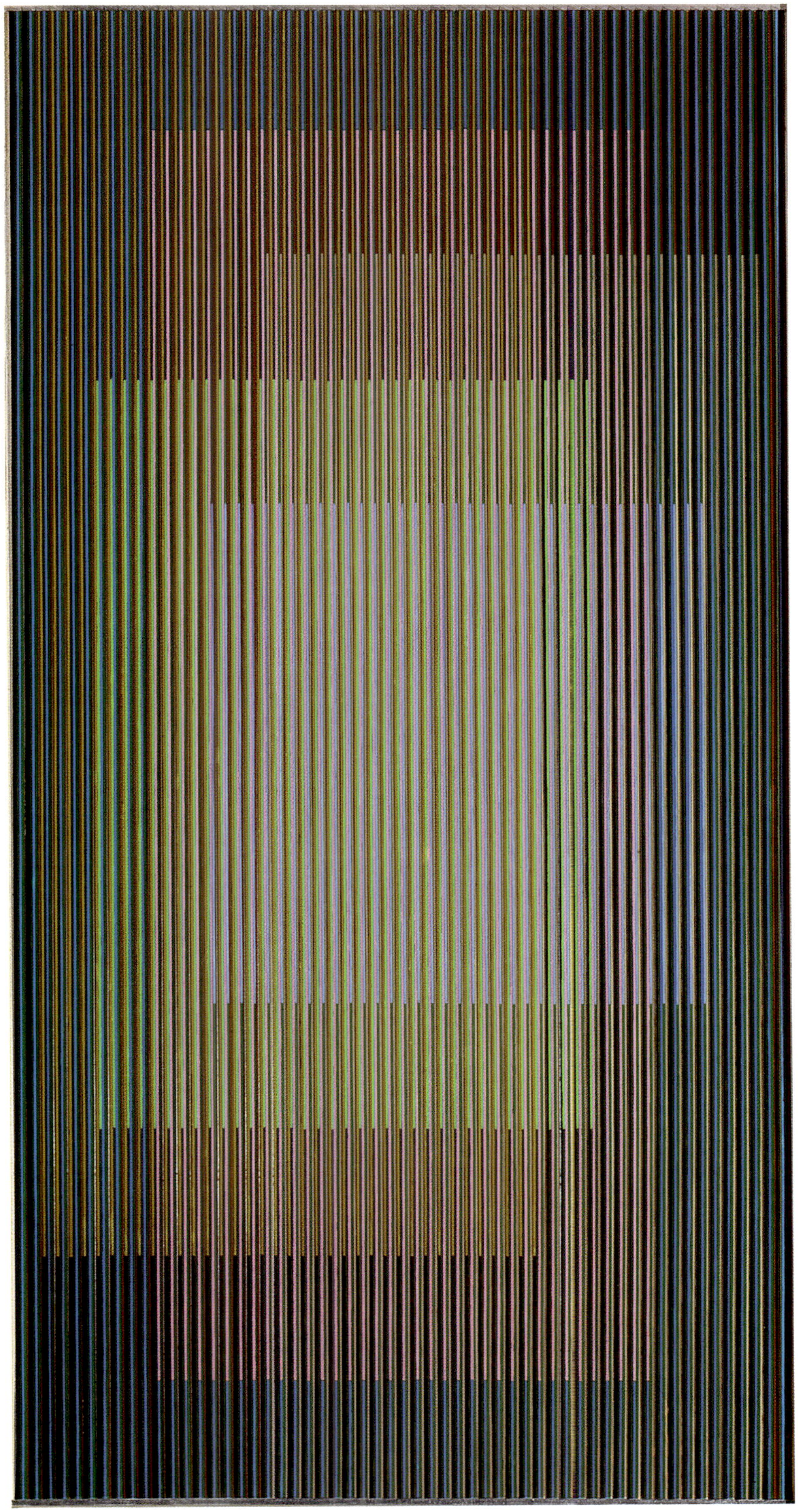

OPPOSITE

11-45. Jesús Rafael Soto (Venezuelan, 1923–2005), *Virtual Volume*, from the series *Penetrables*, 1977, in the lobby of the Royal Bank of Canada, Toronto. Atelier Soto and Sicardi Gallery. © Sicardi Gallery, Houston, Texas. © 2014 ADAGP, Paris/Artists Rights Society, New York.

RIGHT

11-46. Carlos Cruz-Diez (Venezuelan, b. 1923), *Physichromie no. 394*, 1968. Vinyl paint, plywood, cardboard, plastic, and metal, 47¾ × 24½ × 2½ in. (121.3 × 62.2 × 6.4 cm). Blanton Museum of Art, University of Texas at Austin, gift of Irene Shapiro, 1986.

NORTH AMERICA

Two approaches to geometric abstraction arose in America in the late 1950s: an algorithmic approach led by Sol LeWitt and an attempt to reduce art to its literal (physical) essence by so-called Minimalists.[49] One who followed the algorithmic path was George Rickey, who studied painting in Paris in the 1930s, entering the Cubist circle of Fernand Léger and Amédée Ozenfant. Working as a gunnery instructor during World War II, Rickey honed his skills with ball bearings and steel. After the war he attended Naum Gabo's lectures at the New Bauhaus in Chicago, and learned the Constructivist aesthetic. By the 1950s Rickey had developed his mature style: non-motorized, kinetic sculpture in simple lines and planes. In *Three Red Lines* (1966; see plate 11-47), triangular steel blades pivot in response to air currents—from a breeze to a gale—balanced by internal counterweights as they sweep out three large arcs. In the late 1960s Rickey published a detailed history of Russian Constructivism, the German Bauhaus, and international post-1945 geometric sculpture (*Constructivism: Origins and Evolution*, 1967).

BELOW

11-47. George Rickey (American, 1907–2002), *Three Red Lines*, 1966. Painted steel and hydraulic devices, 37 ft. (11.2 m) high. Hirshhorn Museum and Sculpture Garden, Washington, DC. © Estate of George Rickey/VAGA, New York.

OPPOSITE

11-48. Charles Biederman (American, 1906–2004), *Work no. 27, Red Wing*, 1968–69. Painted aluminum relief, 41 × 29¼ × 8 in. (104 × 74.2 × 20.3 cm). Minneapolis Institute of Arts, gift of funds from Mr. and Mrs. John E. Andrus III in memory of Faye Cole Andrus, inv. no. 74.53. © Charles Biederman.

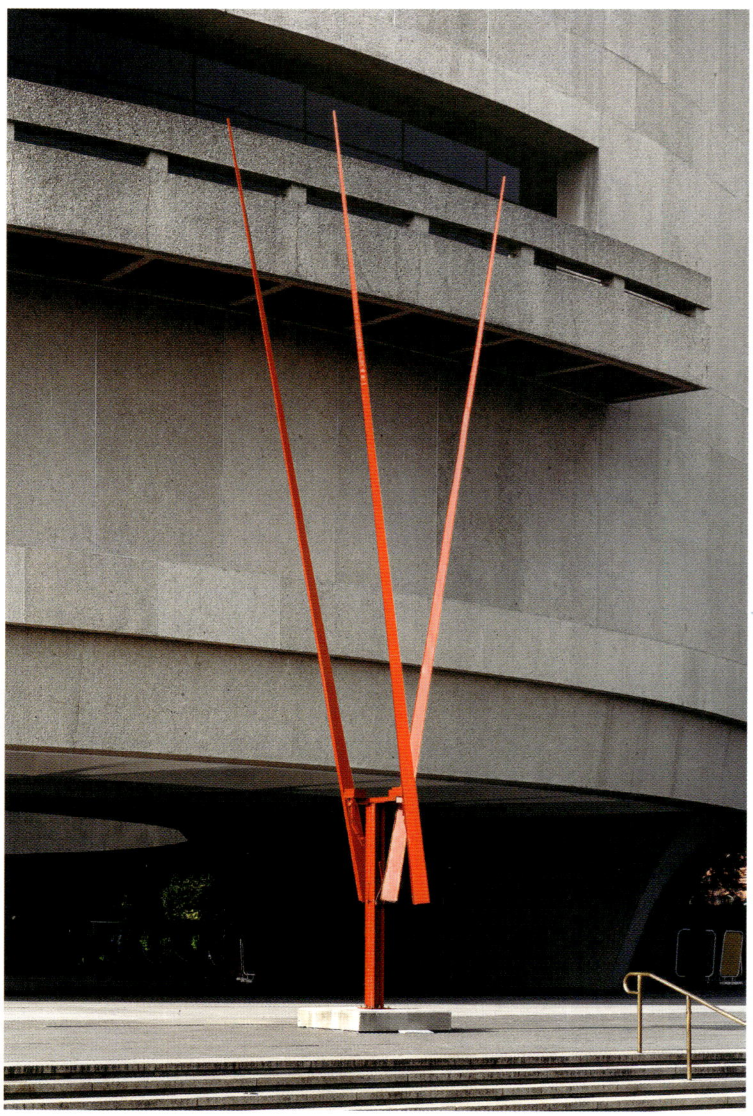

As an American in Paris in the 1930s, Charles Biederman made reliefs inspired by Piet Mondrian's grid paintings (such as plate 6-23 in chapter 6). Feeling alienated from the art world in France, Biederman returned to his homeland, where he also felt estranged from the art communities in New York and Chicago. In 1942, at age thirty-six, he gave up city life and moved to a farm near Red Wing, Minnesota, where he lived as a hermit for the rest of his life. By the 1950s he had arrived at his mature style: brightly colored horizontal and vertical geometric forms, centrally clustered and projecting from a monochrome plane (plate 11-48). After reading a popularization of physics by David Bohm (*Causality and Chance in Modern Physics*, 1957), Biederman wrote to Bohm in March 1960, commencing a decade-long correspondence about their mutual interest in creativity in art and science.[50] Like Biederman, Bohm was an exile of sorts, somewhat isolated in England after being forced out of American academia during the McCarthy era (see chapter 8). Bohm entered into dialogue with Biederman at a time when he was expanding his thinking beyond physics to the broader topic of *order* (including order in culture), about which he developed a holistic philosophy in the second half of his life (*Wholeness and the Implicate Order*, 1980).[51]

The young American Ellsworth Kelly came to Paris in 1948 and experimented with the regular format favored by Concrete artists (*Window*, 1949; Musée National d'Art Moderne, Paris). After visiting the studio of Jean/Hans Arp, who

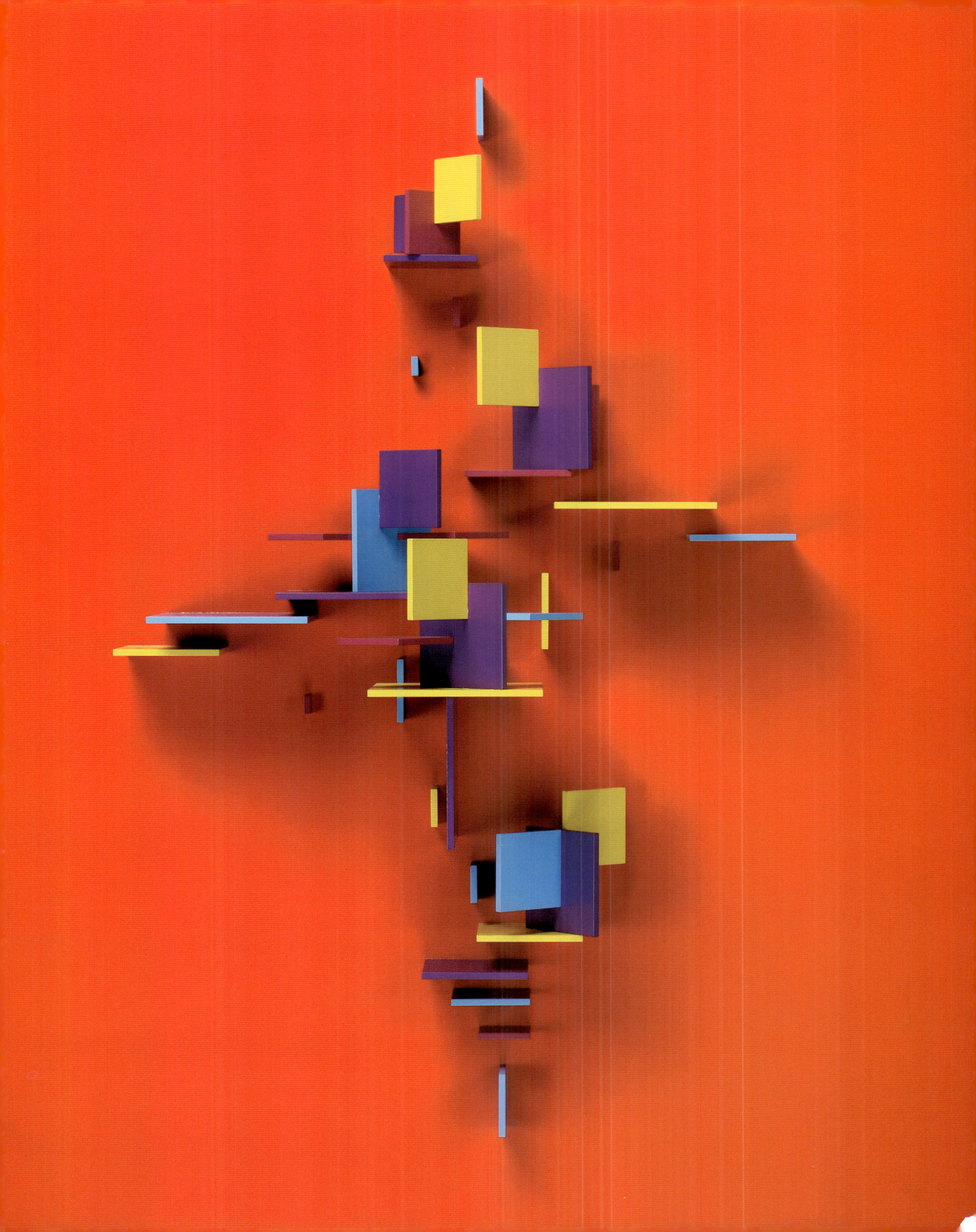

RIGHT

11-49. Ellsworth Kelly (American, b. 1923), *Spectrum Colors Arranged by Chance VI*, 1951. Cut-and-pasted color-coated paper and pencil on black paper, 37¼ × 37¼ in. (94.6 × 94.6 cm). Museum of Modern Art, purchased with funds given by the Edward John Noble Foundation, the Herbert and Nanette Rothschild Fund, and Mrs. Pierre Matisse. © Ellsworth Kelly.

OPPOSITE

11-50. Frank Stella (American, b. 1936), *Die Fahne hoch*! 1959. Enamel on canvas, 121½ × 73 in. (308.6 × 185.4 cm). Whitney Museum of American Art, New York; gift of Mr. and Mrs. Eugene M. Schwartz and purchased with funds from the John I. H. Baur Purchase Fund, the Charles and Anita Blatt Fund, Peter M. Brant, B. H. Friedman, the Gilman Foundation, Inc., Susan Morse Hilles, The Lauder Foundation, Frances and Sydney Lewis, the Albert A. List Fund, Philip Morris Inc., Sandra Payson, Mr. and Mrs. Albrecht Saalfield, Mrs. Percy Uris, Warner Communications Inc., and the National Endowment for the Arts, 75.22. © 2014 Frank Stella/Artists Rights Society, New York.

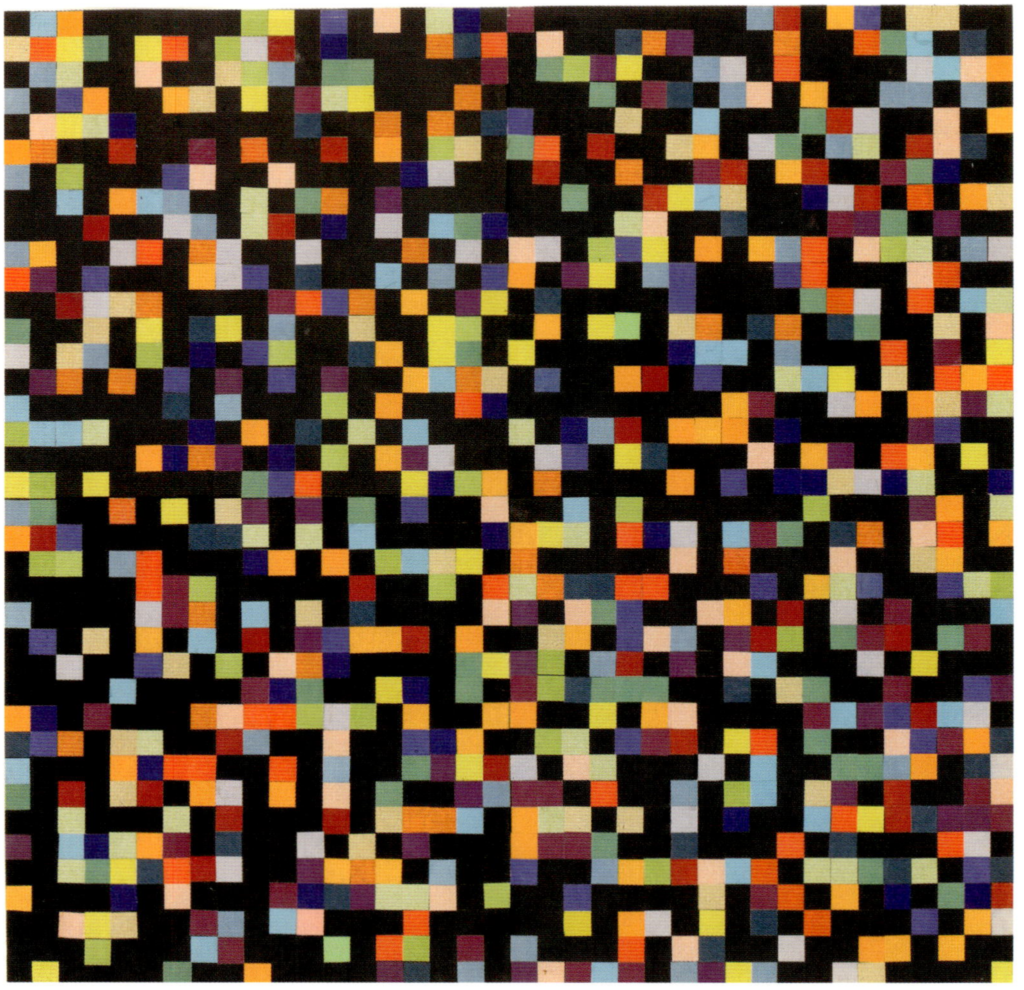

showed the young artist his collage *Squares Arranged According to the Laws of Chance* (1916–17; Museum of Modern Art, New York), Kelly experimented with chance techniques within a grid format (plate 11-49).[52] When Kelly returned to New York in 1954, he was another voice for the Constructivist aesthetic of making art following algorithms. The use of a methodical, step-by-step approach is evident in work of the 1950s by Jasper Johns. In *Numbers in Color* (1958–59; see plate 9-32 in chapter 9), Johns placed the numbers zero to nine in a regular pattern, following the internal logic of a rule. It was in this spirit that the young Frank Stella moved from the gestural abstraction of the New York school to methodically painting black stripes in a symmetrical, uniform pattern, thus creating the monochrome series *Black Paintings*, which includes *Die Fahne hoch!* (1959; see plate 11-50). Once Stella established this system, his execution was predetermined: the stripes run parallel to the edges of the canvas, and the width of the stripes equals the width of the stretcher bars. Stella followed *Black Paintings* with the series *Aluminum Paintings*, which had the same striped format in aluminum paint. To prevent an illusionistic reading of the rectangular canvas, he notched its edges, much as Maldonado and Rothfuss had shaped their canvases in the early 1950s (plate 11-41). Stella gave the paintings in his *Black Series* names that evoke death and disaster; *Die Fahne hoch!* (The flag on high!) was the anthem of the Nazi party from 1930 to 1945, and *Bethlehem's Hospital* (1959; Glenstone) refers to

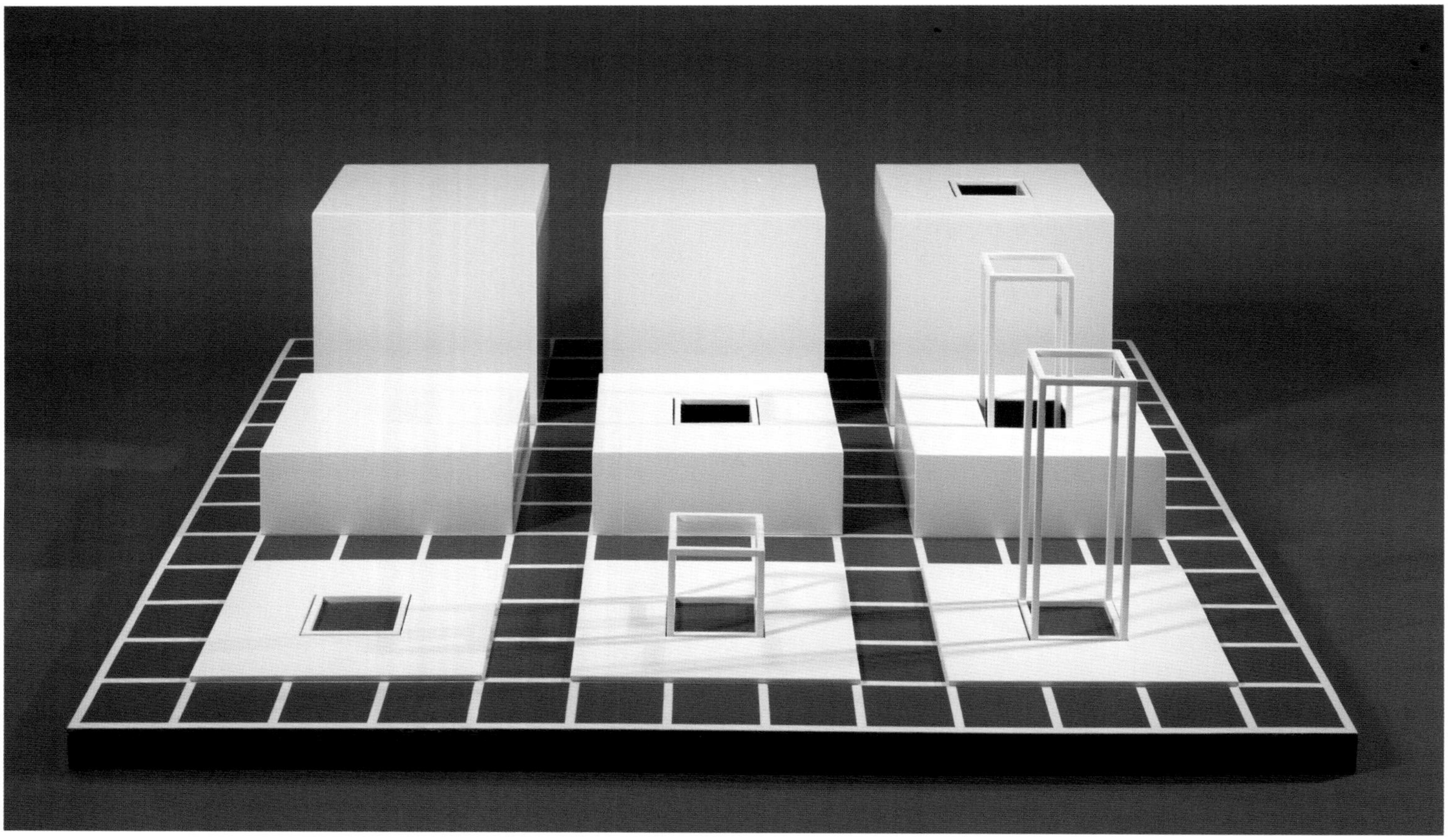

ABOVE

11-51. Sol LeWitt (American, 1928–2007), *Serial Project no. 1 set C*, 1966–69. Baked enamel on aluminum, 9 × 33 × 33 in. (22.9 × 83.8 × 83.8 cm). The Pace Gallery, New York. © 2014 The LeWitt Estate/Artists Rights Society, New York.

OPPOSITE

11-52. Sol LeWitt (American, 1928–2007), *Wall Drawing 56*, 1970 (as shown as a detail of *Four Square Composite*, 1971, silkscreen). Black pencil. LeWitt Collection, Chester, Conneticut. © 2014 The LeWitt Estate/Artists Rights Society, New York.

LeWitt's algorithm was to divide a square into four equal parts and then, beginning in the upper left, draw superimposed lines: vertical, horizontal, right-left diagonal, and left-right diagonal.

When an artist uses a conceptual form of art, it means that all the planning and decisions are made beforehand and the execution is a perfunctory affair.

—**Sol LeWitt, "Paragraphs on Conceptual Art,"** ***Artforum*****, 1967**

Bethlehem Hospital in London, one of the world's oldest lunatic asylums, whose nickname "bedlam" is synonymous with the confusion and uproar of madness.[53] Stella gave more dispassionate titles to works in the *Aluminum Paintings* series, using names of historical places and persons, such as Avicenna (a medieval Persian polymath; 1960; Menil Collection, Houston).

Sol Lewitt was the most methodical of the American algorithmic artists (plates 11-51 and 11-52). Beginning in the early 1960s, he produced artworks by establishing a vocabulary, operations, and then carrying out the permutations. Indeed, for LeWitt the algorithm *is* the artwork; execution was secondary and could be done by a team of assistants. In 1967 LeWitt coined the term "Conceptual Art" to name art made with algorithms.[54] LeWitt stressed the simplicity of the structures used by artists: "Conceptual art doesn't have anything to do with mathematics or philosophy. It's simple arithmetic."[55] He also emphasized the intuitive nature of working with numbers and forms: "Conceptual artists are mystics rather than rationalists. They leap to conclusions that logic cannot reach."[56]

Many artists in LeWitt's circle explored algorithmic processes. Carl Andre built his *Elements Series* (1960) from nine units of wood, each 12 by 12 by 36 inches, in a methodical process that recalls Aleksandr Rodchenko's assemblage of wood modules in his *Spatial Construction* series (1920–21; see plates 4-14 and 4-15 in chapter 4). Using group theory transformations, Mel Bochner drew a diagram of the triaxial rotation of a cube (plate 11-53; compare the sidebar on page

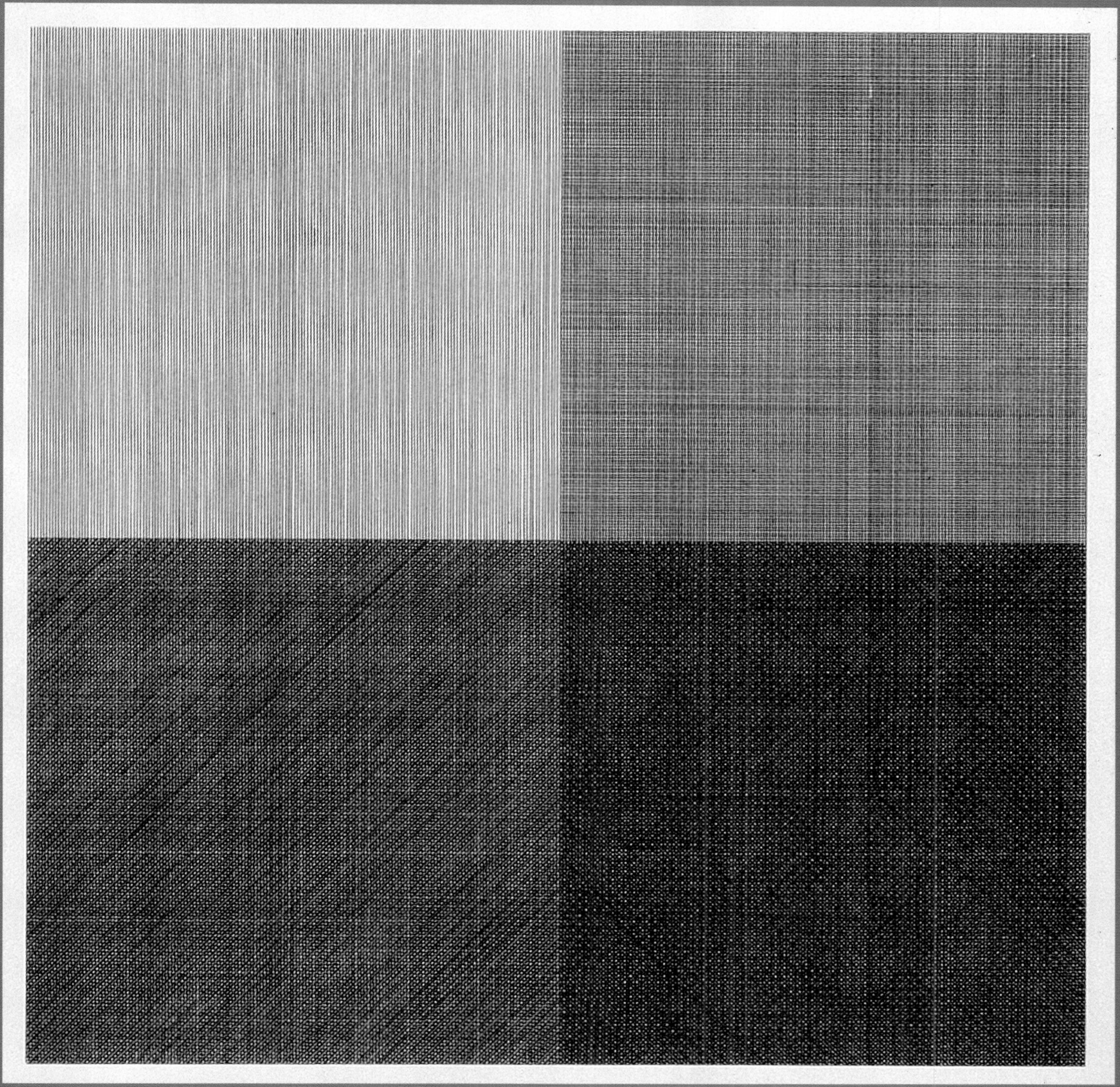

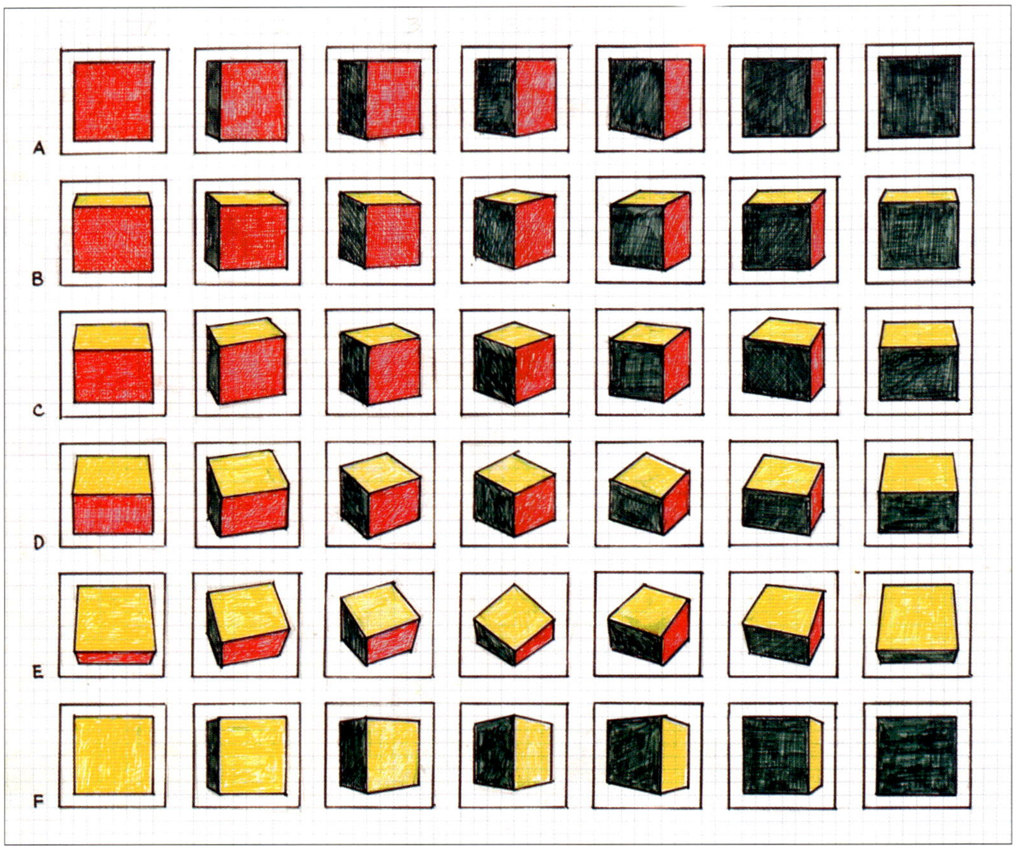

RIGHT

11-53. Mel Bochner (American, b. 1940), *Projected Plan for 42 Color Photographs: Tri-axial Rotation of a Cube*, 1966. Pencil on paper. Courtesy of the artist.

OPPOSITE

11-54. Margaret Kepner (American, b. 1945), *Magic Square 25*, 2010. Archival inkjet print, 18 × 18 in. (45.7 × 45.7 cm). Courtesy of the artist.

The artist Margaret Kepner based this print on a magic square of order 25. To represent the numbers in the magic square, which range from 0 to 624, Kepner devised a visual base 5 system: four concentric squares serve as the 1, 5, 25, and 125 places, while shades of grey stand for the numerals 0 to 4. When Kepner put the numbers 0–624 into the visual code, she got the 625 unique symbols (nested squares in shades of grey) which she needed to make the pattern. As in any magic square, each row, column, and main diagonal sums to the same number; in this case the "magic constant" is 7800. This particular magic square of order 25 contains 25 mini-magic squares of order 5; the rows, columns, and diagonals of these mini-magic squares sum to the magic constant 1560. In addition, certain other groups of 5 squares, such as the plus sign shapes and X shapes highlighted in red, add up to 1560.

251 in chapter 7) and arranged sets of stones on the floor of an art gallery (plate 3-16 in chapter 3). Bochner coined the term "serial attitude" to describe the working method of artists who repeatedly apply a set of rules ("The Serial Attitude," 1967).[57] Occasionally an American artist with significant training in mathematics, such as the printmaker Margaret Kepner, has devised a pattern that moves beyond the simple arithmetic and geometry of the artists in LeWitt's circle (plate 11-54).

Other American artists adopted an algorithmic approach to the geometry of the cosmos. In 1957 the Soviet Union launched *Sputnik*, the first man-made object to orbit earth, beginning a decade of intense space exploration. In response to the popular fascination with outer space, Americans made Earth Art—vast observatories in the deserts of the American Southwest.[58] NASA built high-tech reflector telescopes and sleek, shiny spacecraft to conquer outer space. Earth Art was low-tech outdoor sculpture, which alluded to ancient monuments that were used as naked-eye (no-telescope) observatories. Located in the wilderness, such Earth Art encouraged viewers to assume a reflective state of mind and contemplate their place in the cosmos. Nancy Holt's *Sun Tunnels* (1973–76, Great Basin Desert, Utah; see plate 11-55), which is oriented to the summer and winter solstices, is reminiscent of prehistoric monuments such as Stonehenge in England.[59] The many earth-star alignments in the Egyptian pyramids inspired Charles Ross to create *Star Axis* (begun 1971, New Mexico; see plate 11-56), which is aligned with the earth's axis and oriented to the North Star.[60]

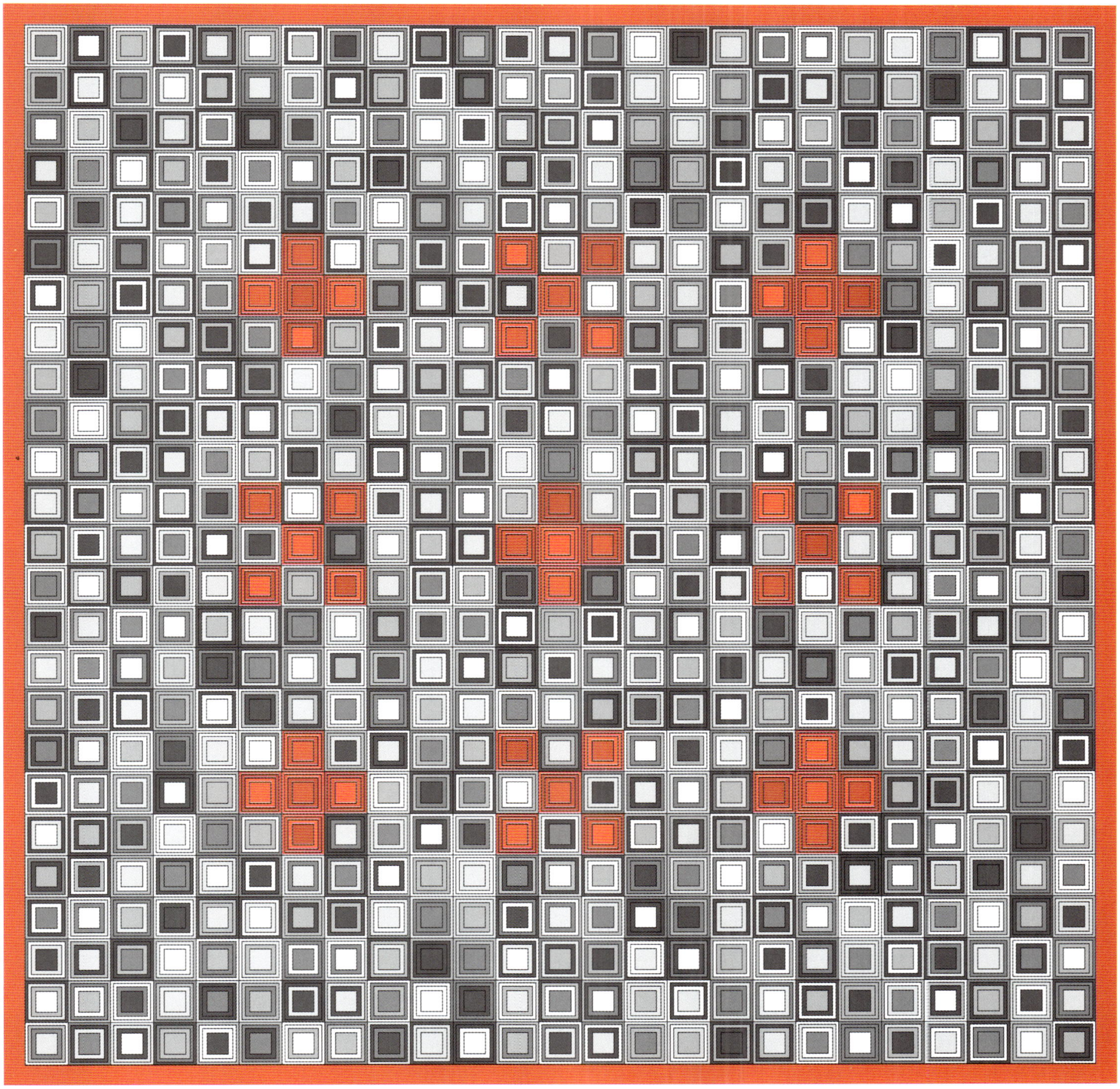

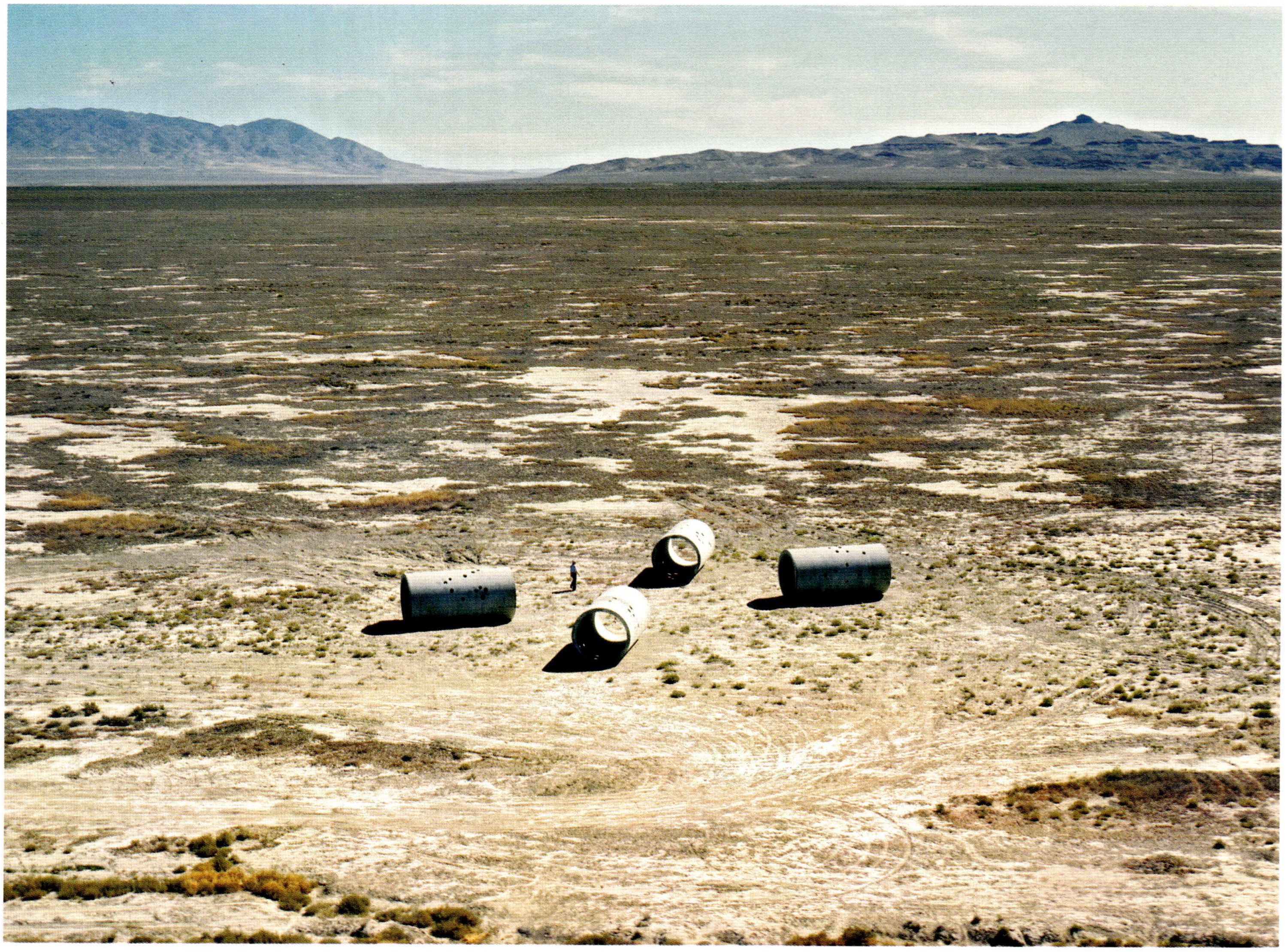

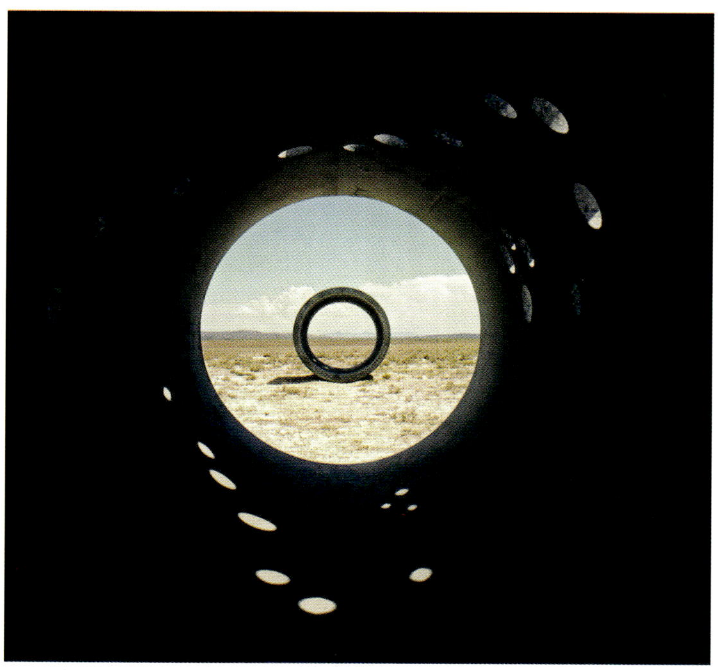

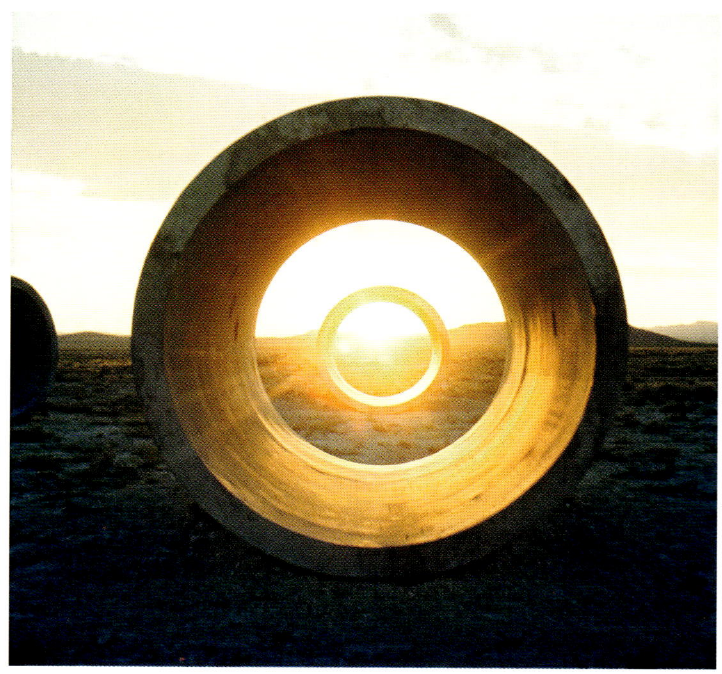

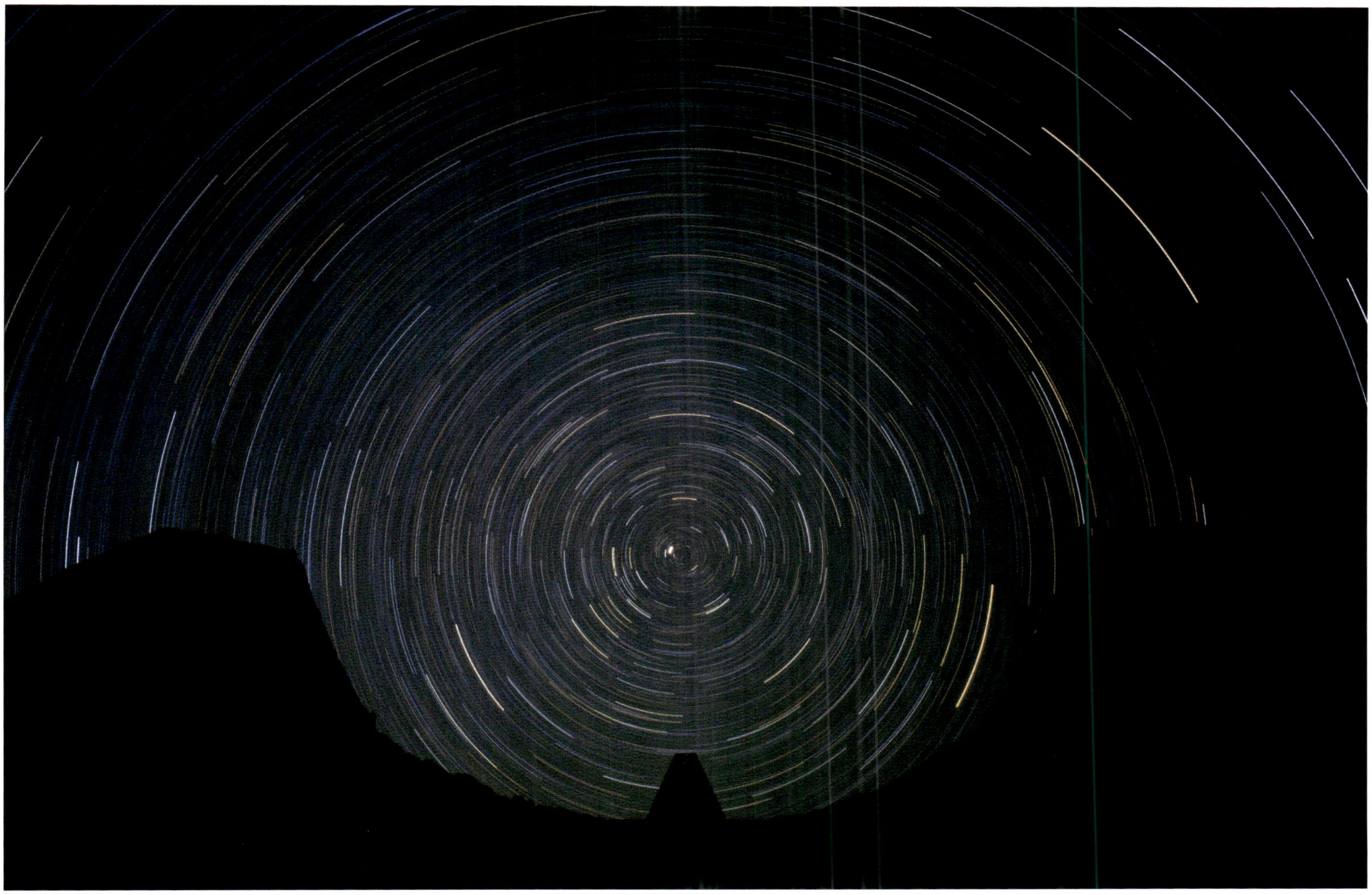

OPPOSITE PAGE

11-55. Nancy Holt (American, 1938–2014), *Sun Tunnels*, 1973–76. Concrete pipes, each with an outer diameter of 9 ft. 3½ in. (2.83 m). Great Basin Desert, Utah. Courtesy of the artist. © Nancy Holt/VAGA, New York.

Holt's sculpture *Sun Tunnels* consists of four huge cement pipes aligned to points on the compass. The photograph in the lower right shows sunrise on the summer solstice. Into the tops of the pipes Holt cut holes, through which sunlight falls in the pattern of the constellations Draco, Perseus, Columba, and Capricorn. The artist has described the experience of standing inside a pipe during the day (lower left) and seeing a constellation at her feet: "It's an inversion of the sky/ground relationship—bringing the sky down to earth."

ABOVE AND RIGHT

11-56. Charles Ross (American, b. 1937), Star Tunnel and Hour Chamber in *Star Axis*, begun 1971. New Mexico. © Charles Ross 2011. © Charles Ross/VAGA, New York.

The best time to see *Star Axis* is on a clear night. The photograph above was taken from the bottom of a staircase at the entrance to the Star Tunnel using a time-lapse exposure, resulting in the "blurred" stars sweeping out circles around the still point—Polaris, the North Star. As the earth rotates, it wobbles slightly, slowly changing the orientation of its axis, which describes an approximate circle in the sky every 26,000 years. Earth's north polar axis now points to Polaris, but over time its axis will point to a different "North Star" (or no star at all). The viewer ascends the Star Tunnel on a stairway that is parallel to the axis of the earth; the years of the 26,000-year cycle are engraved on the stairs, and a circle of sky framed in an aperture at the top of the stairs represents the position of the North Star for the year engraved on each stair. Ross has said: "The intent of *Star Axis* is to give us awareness of the motion of the universe in relation to ourselves. By walking through this work, one will be able to directly experience the entire 26,000-year cycle of polar precession."

At the top of the stairs the viewer enter a triangular Hour Chamber (right) and views the night sky through a triangular window. Throughout the night, the stars appear to rotate in a counter-clockwise, circular pattern; Ross calculated the triangular opening such that stars move from the triangle's left to right edge in exactly one hour. Polaris is at the apex of the triangle—it is the "star axis."

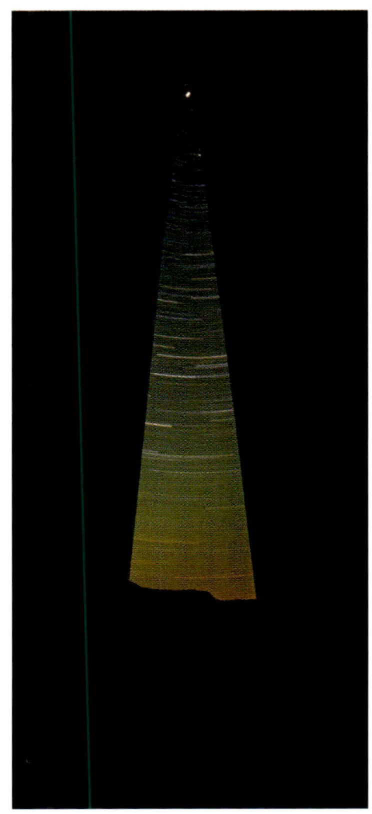

AMERICAN MINIMAL ART

In modern thought, once the foundations of a field have been axiomatized, work ceases because the job is done. For example, after Giuseppe Peano and David Hilbert wrote formal axioms for arithmetic and geometry in the 1890s, later mathematicians rarely revisited the topic. When they did, it was to suggest some amendment to Peano and Hilbert's original axioms, or a change in notation.[61] Similarly, after Aleksandr Rodchenko and Władysław Strzemiński enacted the formalist master narrative and reduced art to its essence in monochrome canvases and simple geometric solids, abstract artists from the 1920s onward assumed these artistic "axioms" as *premises;* in other words, meaning-free color and form became the vocabulary of art. There are many styles of geometric abstraction—De Stijl, Suprematism, Concrete Art, Optical Art—because artists can arrange color and form into myriad patterns. But there is only one style—Russian Constructivism—that reduced art to its (physical, formalist) essence, because this reductionist process has a natural end-point. Or more accurately, there *was* only one such style until the 1960s, when American artists re-enacted the formalist master narrative and once again reduced art to its formalist essence in monochrome canvases and simple geometric solids—Minimal Art. Why did these artists replay the narrative when all the participants—artists and critics—knew the ending?[62]

American artists returned to the topic of the foundations of art because they were swept up in a wave of enthusiasm for new beginnings in the New World after World War II. Life in Europe was still in turmoil, and New York was emerging as a vibrant intellectual center filled with refugees. In this context Jackson Pollock, Mark Rothko, Adolph Gottlieb, and Barnett Newman created a style of gestural abstraction. Critic Harold Rosenberg described the expressive content of these paintings in poetic prose, while his rival, Clement Greenberg, stressed their formal properties. By the late 1950s and early 1960s, the Cold War was ending, and the next generation of American artists moved away from the pathos of their elders and painted in more cool-headed styles of geometric abstraction. By 1960 many refugees had returned to Europe, but by then it was clear that the center of the art world had shifted from Paris to the shores of the new political and economic superpower, the United States. New York artists and critics—energized by the feeling that they were part of a fresh start—asked anew questions about the foundations of art.[63] Critical debates raged in the art press throughout the 1960s, driven by the passionate conviction that it was of utmost importance to discover the essence of art; the answer could be revealed at the opening of an upcoming gallery exhibition or in the next issue of *Artforum*.

Post-war America occupied a different historical context than revolutionary Russia, and thus—despite the fact that the Constructivists and Minimalists reduced art to the same elements (color and form)—American artists and critics made a meta-art claim that one never heard in Moscow, Saint Petersburg, and Kazan. The Russian Constructivists lived during the political upheaval of the Russian revolution and its aftermath, and after reducing art to meaning-free signs, they then applied those colors and forms to practical design tasks for the new Communist state. The American Minimal artists lived after the atomic bomb and the destruction of Russian and Germanic culture by Stalin and Hitler's fascism,

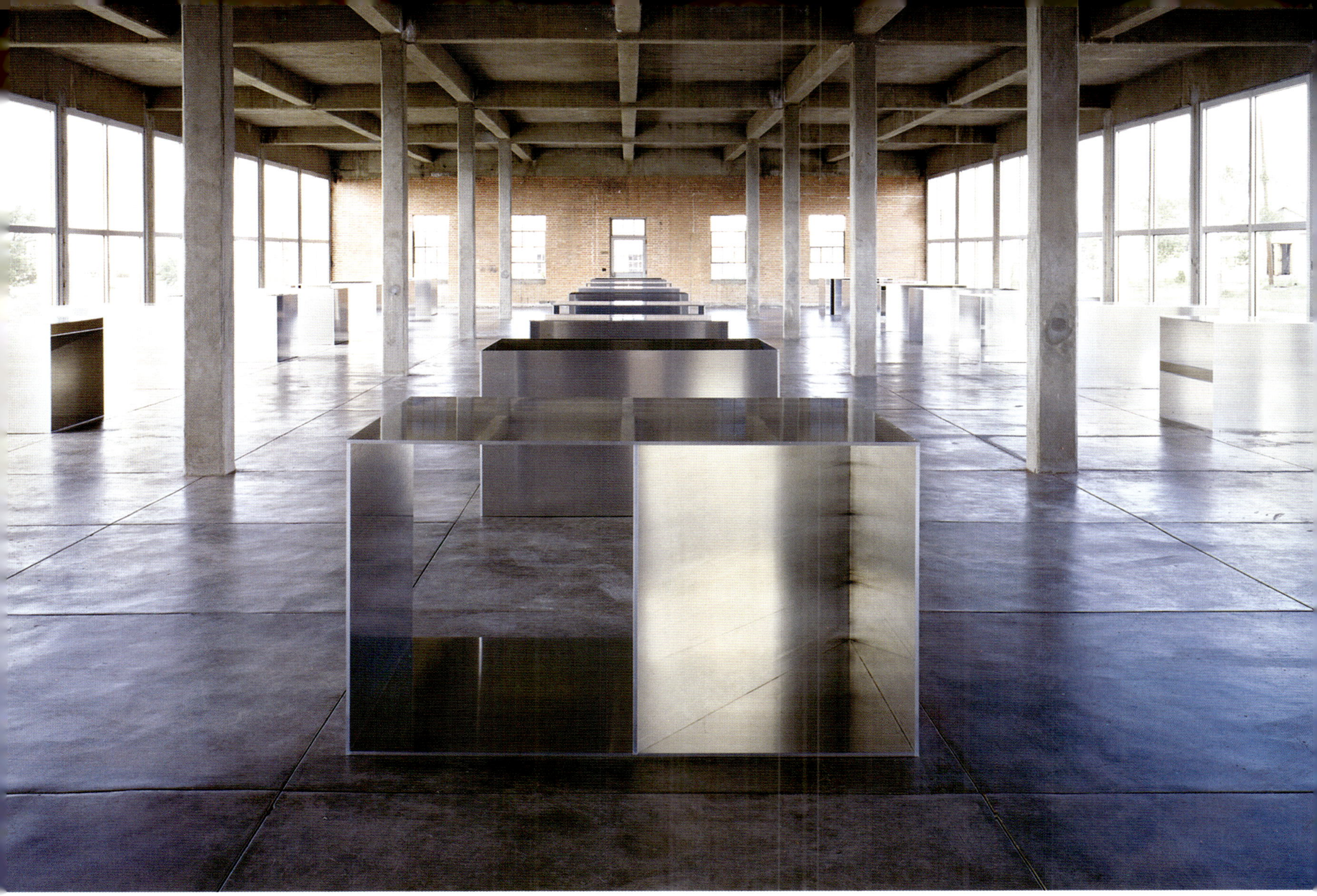

after which Western culture entered an era of cynicism and dread during the Cold War and the volatile 1960s. Rodchenko and Tatlin had reduced visual art to its physical essence (color on canvas, form in steel) as the *means* to an end, but American Minimal artists went one step further and declared that the physical, formalist essence of art is *all it ever can be* and that color and form should not be applied to some meaning-filled practical task or interpreted in terms of some symbolism or given a transcendental reference. In doing this, the American focus on "just the facts" was part of the unmasking of metaphysical certitudes that occurred throughout Western culture after 1945, and it also reflected a more general anti-metaphysical atmosphere that pervades modern secular thought. The goal of certain (not all) artists associated with Minimalism was to pursue what I will call a "Minimalist Program" to make painting and sculpture devoid of meaning (in the many senses of that word). Rodchenko, Tatlin, and other early formalist artists *never* did this (see chapter 4).[64] Like the fin-de-siècle formalist mathematicians led by Hilbert, early-twentieth-century abstract artists in Russia, Poland, Germany, Austria, and Switzerland—even when they made art in factory-style anonymity—developed a meaning-free vocabulary as a *tool*, a visual *language*, with which to communicate important and profound content, as art always has.

American artists who pursued the Minimalist Program were not a fixed group; key representatives were the sculptors Donald Judd (plate 11-57), Robert Morris, Dan Flavin,

11-57. Donald Judd (American, 1928–94), *Untitled*, 1982–86. 100 cubes in mill aluminum. Collection of the Chinati Foundation, Marfa, Texas. © Judd Foundation/VAGA, New York.

The form of a work of art derives from two fundamental premises: the material *(colors, sounds, words) and the* construction, *through which the material is organized in a coherent whole, acquiring its artistic logic and profound meaning.*

—**Nikolai Tarabukin, *Towards a Theory of Painting*, 1916** and Carl Andre, but the literature on Minimalism typically cites dozens of others, some of whom I discuss elsewhere in this book (because, in my opinion, their work falls outside the Minimalist Program), including Frank Stella (plate 11-50), Sol LeWitt (plates 11-51 and 11-52), and Mel Bochner (plates 3-16, 11-53, and 11-61). The Minimalists Judd, Morris, Flavin, and Andre had divergent, continually shifting outlooks, but those who followed the Minimalist Program were united in their aspiration to make art without meaning: they "sought to purge themselves of allusion,"[65] there was a "literalist evacuation of content,"[66] and they made abstract art that "refuses to signify."[67] The 1960s critics included the senior defender of Abstract Expressionism, the formalist Clement Greenberg, and his young followers Barbara Rose and Rosalind Krauss, along with many other writers, including the artists themselves, many of whom were academically trained. Like any fluid ideas in a live intellectual climate, the artistic and philosophical concepts put forth often intermingled, and for the purpose of this book it is not important to keep them all straight. I will focus on one topic relevant to the history of mathematics and art—meaning-free signs—and argue that the Minimalist Program was unattainable because, as critics noted from the outset, signs *signify*.[68]

Most accounts of Minimal Art begin with Stella's *Black Paintings*, but Stella didn't pursue the Minimalist Program; he gave allusive titles to his monochrome *Black Paintings* and soon moved on to polychrome figure-ground illusions in series such as *Concentric Squares* (1962–63) and *Irregular Polygons* (1966–67). Although his brush with the Minimalist Program was short-lived, during that brief encounter Stella could talk the talk: "I always get into arguments with people who want to retain the old values in painting—the humanistic values that they always find on the canvas. If you pin them down, they always end up asserting that there is something there besides the paint on the canvas. My painting is based on the fact that only what can be seen there *is* there. It really is an object. . . . What you see is what you see."[69]

Malevich, Kandinsky, and Mondrian, in different ways, universalized their art by theory, but in New York there is little reliance on Platonic or Pythagorean mysteries.

—Lawrence Alloway, "Systemic Painting," 1966

This amalgamated philosopher-artist-draughtsman-aristocrat seems to lead us to an Academy once again, a sort of Platonic Academy—minus Plato.

—Brian O'Doherty, "Minus Plato," 1966

Unlike Stella's short-lived declaration of Minimalism, his contemporaries the sculptors Donald Judd and Robert Morris put forth the Minimalist Program with lasting enthusiasm, each reducing their sculpture to simple rectangular solids that were shown in Gallery 5 of the Jewish Museum in New York in the 1966 exhibition *Primary Structures* (plate 11-58). Judd described his cubic sculptures shown in plate 11-58 as meaning-free physical objects—"specific objects."[70] Judd was an academically trained spokesman for the Minimalist Program; he earned a bachelor's degree in philosophy from Columbia University, where he also completed the coursework for a master's degree in art history before becoming a widely published art critic. Lacking studio training, Judd had his sculpture factory-fabricated. Judd remained consistent to the Minimalist Program throughout his long career, continuing to have cubes fabricated for decades, culminating in his installation of one hundred aluminum cubes in the 1980s (plate 11-57).

In contrast to Judd's meaning-free art, Morris declared that his two L-shaped sculptures in plate 11-58 were meaning-free *props* for art, which was an *event*—the viewer's perception of these sculptures while walking around in Gallery 5 of the Jewish Museum.[71] In fact, Morris's first sculpture (*Column*, 1961) *was* a prop that he made for his performance

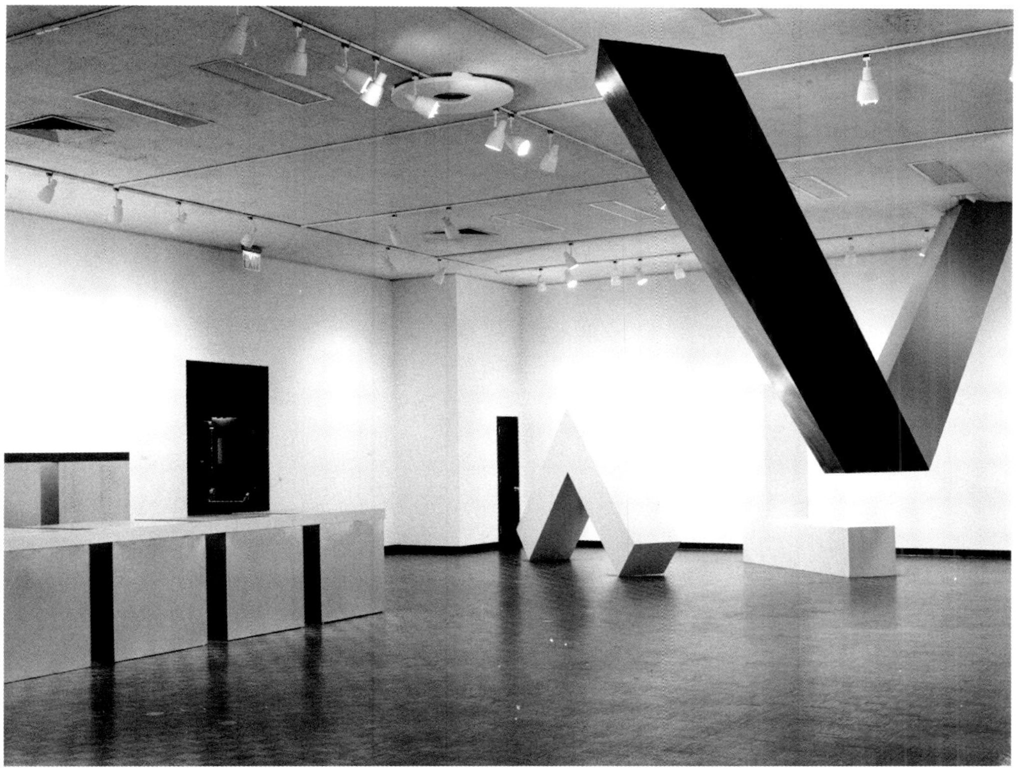

11-58. Installation view of Gallery 5 of the exhibit on *Primary Structures: Younger American and British Sculptors* at the Jewish Museum in New York, 1966. The Jewish Museum, New York. © Judd Foundation/VAGA, New York.

On view in Gallery 5 were (from left to right) Donald Judd's *Untitled*, 1966 (the four free-standing cubes), Robert Morris's *Untitled (L Beams)*, 1965, and (suspended from the ceiling) Robert Grosvenor's *Transoxina*, 1965. In a belated acknowledgement of the internationalism of post-1945 geometric abstraction, in 2014 the director of the Jewish Museum, Jens Hoffmann, organized *Other Primary Structures* to include "art with an apparently Minimalist aesthetic that was created contemporaneously with *Primary Structures*, between about 1960 and 1970, in parts of the world long considered peripheral or underdeveloped. It includes works by well-known [Latin American] artists such as Hélio Oiticica, Lygia Clark, and Gego, as well as lesser-known artists from Africa, East Asia, Eastern Europe, and the Middle East. . . . Japan and South Korea." *Other Primary Structures* (New York: Jewish Museum, 2014), n.p.

with the Judson Dance Theater in Greenwich Village. As a dancer, Morris was especially sensitive to the motion of viewers, and as an articulate writer (having earned a master's degree from the City University of New York), he used research done by Gestalt psychologists and phenomenologists, especially the writings on vision and motion by the French philosopher Maurice Merleau-Ponty, *Phénoménologie de la perception* (Phenomenology of perception, 1945), which had appeared in English translation in 1962.[72] Echoing themes that run throughout this technical literature, Morris stated that when we see something such as a cube, "the nature of this belief and the way it is formed involve perceptual theories of 'constancy of shape' and 'tendencies toward simplicity.' Kinesthetic cues, memory traces, and physiological factors regarding the nature of binocular parallax vision and the structure of the retina and brain. . . . Simplicity of shape does not necessarily equate with simplicity of experience."[73] Morris soon moved beyond the stark simplicity of the solids in his L beams, introducing illusory surfaces (*Untitled* [*Mirrored Cubes*], 1965) and, in the late 1960s and early 1970s, works made from felt cloth that droops unpredictably, which he described in his essay "Anti-Form" (*Artforum*, April 1968).

Other artists in *Primary Structures* did not adhere to the Minimalist Program but gave their art meaningful associations. Judd and Morris did not title their sculptures in Gallery 5, but Robert Grosvenor called his dark sculptural V-form suspended from the ceiling *Transoxina* (plate 11-58), the ancient name for the homeland in central Asia of Iranian nomadic tribes, according to the epic poem *Shahnameh* (Persian for "The Books of Kings") written by Hakim Abu'l-Qasim Ferdowsi Tusi between circa AD 977 and 1010.

Some artists in the *Primary Structures* exhibition saw themselves as being in dialogue with their Russian forebears, especially Rodchenko, Tatlin, and Malevich, whose *White on White* (1918; see plate 4-10 in chapter 4) had been on permanent display in New York since 1935, when the founding director of the Museum of Modern Art, Alfred H. Barr Jr., acquired it for his fledgling museum.[74] After reading Camilla Grey's *The Great Experiment: Russian Art, 1863–1922* (published in 1962), Dan Flavin made homage to Tatlin's Tower (1919; see plate 4-30 in chapter 4) in a series of wall pieces that he commenced in 1963 (plate 11-59), and it was in this series that Flavin, following Tatlin, also began activating the corner (see Tatlin's *Counter-Corner Relief*, 1914–15; plate 4-8 in chapter 4).[75] When Gyula Kosice first made art with luminous glass tubes in 1943 (plate 11-39), he hand-molded the tubes, but Flavin used standard (off-the-shelf) fluorescent light fixtures. Robert Morris also acknowledged his Russian sources in 1966 when he observed: "Tatlin was perhaps the first artist to free sculpture from representation and establish it as an autonomous form both by the kind of image, or rather non-image, and by his literal use of materials."[76] Other artists in *Primary Structures* set up a "rather ridiculous American vs. European polarity"[77] by distancing themselves from the European roots of abstract art. In 1964 Judd stated: "I'm more interested in Neo-plasticism and Constructivism than I was before, perhaps, but I was never influenced by it, and I'm certainly influenced by what happens in the United States rather than by anything like that."[78] Stella and Judd were especially hostile to their archrival, Max Bill, because he was held in such high esteem in Europe and Latin America. As Stella stated in 1964: "I find all that European geometric painting—sort of post-Max Bill school—a kind of curiosity—very dreary. . . . European geometric painters really strive for what I call relational painting. The basis of their whole idea is balance. You do something in one corner and you balance it with something in the other corner. Now the 'new painting' is being characterized as symmetrical. . . . It's non-relational. In the newer American painting we strive to get things in the middle."[79] Judd added: "Those [compositional] effects tend to carry with them all the structures, values, feelings, of the whole European tradition. It suits me fine if that's all down the drain."[80]

I believe that art is shedding its vaulted mystery for a common sense of keenly realized decoration. Symbolizing is dwindling—becoming slight. We are pressing downward toward no art—a mutual sense of psychologically indifferent decoration.

—Dan Flavin, "Some Remarks," *Artforum*, 1966

Minimal Art coalesced into a style in many gallery exhibitions of the mid-1960s. The movement peaked in 1968 when it was christened "the 'real' art of today" by the then-most-powerful museum of modern art in the world, the Museum of Modern Art in New York, which organized a major exhibition of work by thirty-one American artists, *The Art of the Real: USA, 1948–1968*, and sent it on a triumphal tour of Europe, where it was shown at the Grand Palais in Paris, the Kunsthaus in Zurich, and the Tate in London. In the exhibition catalogue the curator, Eugene Goossen, was the mouthpiece for the Minimalist Program: "The 'real' of today as posited in this new art has nothing to do with metaphor, or symbolism, or any kind of metaphysics. . . . the spectator is not given symbols, but facts."[81] Goossen provoked derisive howls on both sides of the Atlantic[82] when he went on to claim that Abstract Expressionism and Minimal Art are unique products of American history, and that they have no historical basis in the art of Europe or Russia.[83]

11-59. Dan Flavin (American, 1933–96), *Monument for V. Tatlin*, 1969. Fluorescent light fixtures, 96 × 35 × 5 in. (243.8 × 77.5 × 12.7cm). Museum of Contemporary Art, Los Angeles, gift of Lenore S. and Bernard A. Greenberg. © 2014 Stephen Flavin/Artists Rights Society, New York.

Ai Weiwei paid homage to Tatlin's Tower as a symbol of hope by making his *Fountain of Lights* from thousands of sparkling crystals, their myriad tiny rainbows creating a magical atmosphere (see plate 4-31 in chapter 4). In contrast, Dan Flavin made his *Monument for V. Tatlin* from fluorescent light fixtures that emit a cool light, creating a warehouse atmosphere of "psychologically indifferent decoration."

The leading critic of the Minimalist Program, Clement Greenberg, took statements by Stella, Judd, and Morris at face value and rejected Minimal Art because it was meaning-free. Greenberg's formalism was rooted not in the European formalist/structuralist tradition of David Hilbert, Ferdinand de Saussure, and the Russian linguist Roman Jakobson (see

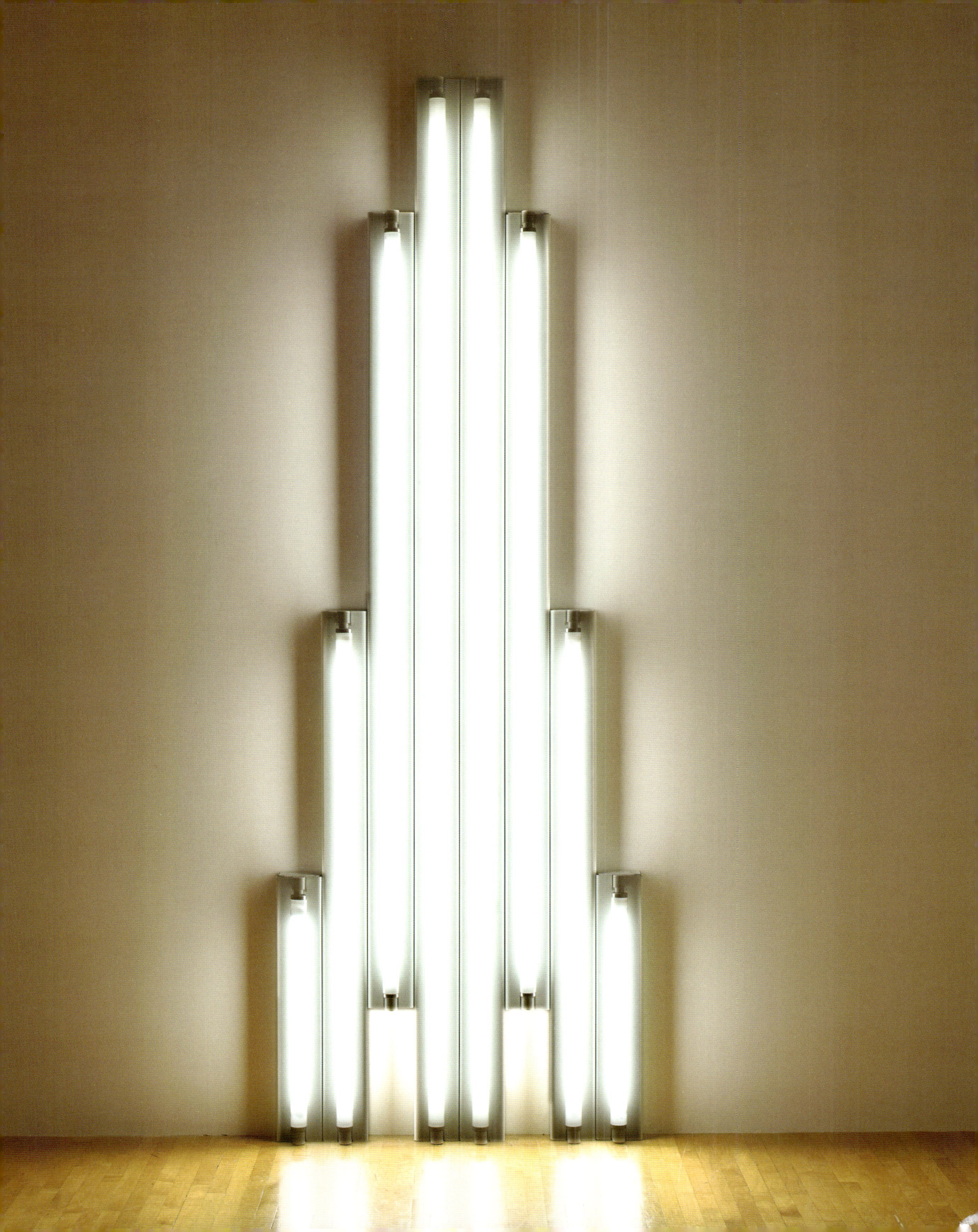

chapter 4), but in the Anglo-American logical tradition that had been adopted by Roger Fry, who combined Bertrand Russell's logic with John Ruskin's more intuitive approach (see chapter 5).[84] Like Fry (and Ruskin), Greenberg relied on his immediate, subjective intuition to make an aesthetic judgment and to discover in art the "emotional elements of design" for which Fry's colleague, Clive Bell, had coined the term "significant form."[85] Greenberg intuited emotional elements in the brushed forms of Rothko and dripped paint of Pollock, but the rectangles and cubes of Stella, Judd, Morris, and Andre left him cold because they were devoid of "significant form"—form that *signifies*, elements that convey *emotion*. Thus regarding the objects in Gallery 5 of the Primary Structures exhibition (in plate 11-58), according to Greenberg's definition neither Judd nor Morris's geometric solids (nor the two black doors, for that matter) are truly art: "Minimal works are readable as art, as almost anything is today, including a door, a table, or a blank sheet of paper . . . it would seem that a kind of art nearer the condition of non-art could not be envisioned or ideated at this moment."[86] Many audiences, echoing Greenberg, complained that the line between art and non-art was erased when meaning-free art was fabricated (by an anonymous workman) from non-art material, such as a stack of bricks (plate 11-60).

From the outset, younger formalists who were followers of Greenberg said "Not so fast!" Barbara Rose, who was then a graduate student in art history at Columbia University, found that Minimalism encouraged in the viewer a kind of meditative state that was long associated with monastic traditions in which the mystic goes on a spiritual journey—a *via negativa*—concentrating on nothing to discover everything (see chapter 1). In her 1965 essay "ABC Art," which was one of the earliest attempts to put Minimal Art in historical context, Rose positioned it in the mystical tradition: "The art I have been talking about is obviously a negative art of denial and renunciation. Such protracted asceticism is normally the activity of contemplatives or mystics. . . . Like the mystic, in their work these artists deny the ego and the individual personality, seeking to evoke, it would seem, that semi-hypnotic state of blank consciousness of meaningless tranquility and anonymity which both Eastern monks and yogis and Western mystics . . . sought."[87]

As the activity of the mystic must end in a via negativa, *a theology of God's silence, a craving for the cloud of unknowningness beyond knowledge and the silence beyond speech, so must art tend toward anti-art, the elimination of the "subject" (the "object," "the image"), the substitution of chance for intention, and the pursuit of silence.*

—**Susan Sontag, "The Aesthetics of Silence," 1967**

The following year Rosalind Krauss, then a graduate student in art history at Harvard, wrote that she found in Minimal Art other kinds of meaning—sensual allusions and spatial illusions. When Krauss looked at a Judd sculpture, she saw a meaning-filled sign that was "startlingly sensuous, almost voluptuous" ("Allusion and Illusion in Donald Judd," 1966).[88] Describing an untitled work fabricated from brushed aluminum and violet enamel (1965; Whitney Museum of American Art, New York), Krauss wrote: "Judd's own criticism would seem to accept only art which eschews both allusion and illusion. Yet his sculpture derives its power from a heightening of illusion—although not of pictorial illusion but of lived illusion."[89] To define "lived illusion" Krauss, like Morris, invoked the phenomenologist Merleau-Ponty's description of how a moving viewer perceives an object.[90] Describing a mural by Kenneth Noland (*Across Center*, ca. 1965), which measures four feet high by twenty feet wide (so one has to walk back and forth to see the whole painting), Krauss wrote: "The sensation thus produced, that one cannot know absolutely the nature of the shape of the painting, that one's view is always adumbrated, that the work in its

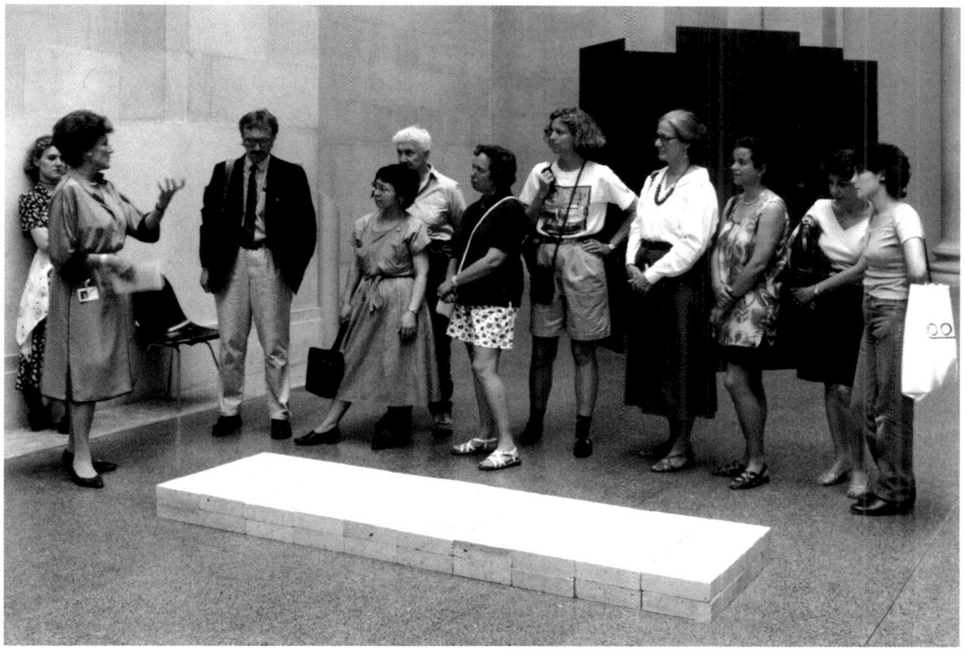

11-60. Carl Andre (American, b. 1935), *Equivalent VIII*, 1966. 120 bricks, in two 60-brick layers. © 2011 Tate, London.

For his *Equivalent* series, Carl Andre stacked 120 standard building bricks in two 60-brick layers in different ("equivalent") configurations (such as a line of 60 bricks, rectangles of 2 × 30 bricks, 3 × 20 bricks, 4 × 15 bricks, 5 × 12 bricks, 6 × 10 bricks). In 1976 the Tate in London purchased *Equivalent VIII* (the 6 × 10 configuration) and a critic for *The Times* (February 15, 1976) initiated a raucous public debate by asking "Is it art?" This photograph shows *Equivalent VIII* being presented by a brave tour guide during the 1976 controversy. For a description of the incident see James Meyer, *Minimalism* (London: Phaidon, 2000), 190–91, and for the artist's viewpoint see Carl Andre, "The Bricks Abstract," *Arts Monthly* 1 (Oct. 1976): 24.

entirety is highly illusive, throws one more surely and more persuasively onto an immediate experience of color alone."[91] Krauss concluded that Judd and Noland would not—*could* not—achieve the Minimalist Program: "From within the context of an increased sensuosity neither artist will desert meaning."[92]

In addition to the emotional and perceptual meanings of Minimal Art as described by Rose and Krauss, all art has political meaning within its larger social context, as described by art historian Anna C. Chave. The monastic aura described by Rose resulted, according to Chave, from the way the art was displayed—simple objects in sparse settings viewed in respectful silence—but this presentation in a cathedraled space was not arranged by the artists but rather by the wealthy patrons who commissioned such installations.[93] According to Chave, few Minimalists fit Rose's description of artists who "deny the ego and the individual personality," and those who did were women—Agnes Martin, Anne Truitt, and Eva Hess. Most Minimalists were males like Judd and Morris who, according to Chave, used materials to convey a machismo aesthetic of power and domination.[94] Furthermore, on the topic of scale, historians concur that when it comes to Minimalist sculpture, size matters.[95]

Another kind of meaning of Minimal Art is economic. Who paid for the hundred aluminum cubes in plate 11-57? Judd had them fabricated with money from the Dia Art Foundation, the brainchild of the German art dealer Heiner Friedrich and his wife, the American Philippa de Menil. Born in Berlin in 1938, Friedrich grew up during the Nazi era and later recalled: "My early experience of total destruction made me want to create the permanence of indestructible properties, particularly the creative work of artists."[96] After directing art galleries in Munich, Cologne, and New York, he met de Menil, an heiress to the Texas-based Schlumberger oil-services fortune, and together they founded Dia in 1974. The inspiration for it came to Friedrich while visiting Padua, Italy, where he was overwhelmed by the early Renaissance Scrovegni Chapel (known as the Arena Chapel), the interior of which contains fresco cycles on the lives of Jesus and Mary painted by Giotto

11-61. Mel Bochner (American, b. 1940), *Language Is Not Transparent*, 1970. Chalk on paint on wall, 72 × 48 in. (182.8 × 121.9 cm). Courtesy of the artist.

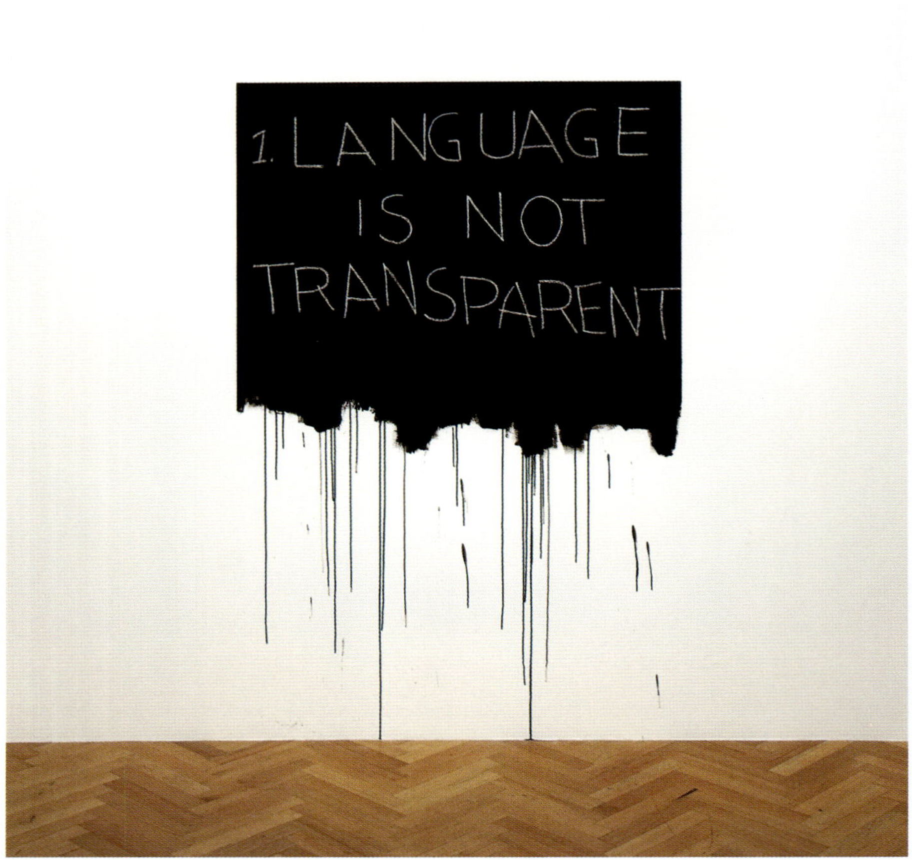

(ca. AD 1305). According to Friedrich, the Arena Chapel "became for me the true insight for the unfolding and development of Dia"[97] because the frescoed interior was conceived and executed by one artist, it is preserved in perpetuity for pilgrims, and it was paid for by a businessman, Enrico Scrovegni, whose family fortune was made in banking. (Scrovegni commissioned the chapel to atone for the sin of usury—lending money at interest.)

With access to de Menil's immense fortune, Friedrich, in all sincerity, aspired to be a patron in the tradition of the Scrovegni and Medici banking families; he chose artists, including Judd, and offered them unlimited support to realize their artistic vision; each "chosen one" was given a stipend, a studio, and the promise of a future one-man museum dedicated to his work. For a decade Judd was given a stipend that reached $17,500 per month in 1981, along with studio space and the construction of a museum in Marfa, Texas, where the Dia Foundation spent $5 million purchasing and refurbishing property for Judd, including two farms and Fort Russell (a former Army post on 340 acres of land), which is today home to Judd's art museum (the Chinati Foundation), where his hundred aluminum cubes (plate 11-57) are on permanent display.[98]

I'm totally uninterested in European art and I think it's over with.

—**Donald Judd, 1964**

In addition to being convinced of the power of patronage, Friedrich came away from Padua with a spiritual vision of the ability of art to transport the viewer to a transcendental realm.[99] This revelation was lost on Judd, but it was pursued by other of Friedrich's chosen ones: Dan Flavin,[100] La Monte Young, and Walter de Maria (see chapter 10). By the early 1980s, Dia was hemorrhaging money on multiple cultural projects when the oil boom hit and Schlumberger stock plummeted. In 1985 Friedrich and de Menil were forced to sell art at auction and cut back on their support of artists. Judd, feeling betrayed by Friedrich ("I distrusted him from the beginning. The only problem is that I didn't distrust him enough"[101]), threatened a lawsuit and, in an out-of-court settlement, won custody of all his art, an additional $2 million, and ownership of the Texas properties.[102] The moral of this tale: the financial meaning of Judd's one hundred aluminum cubes is that their extravagant display of wealth and power is an (unintended) symbol of Western capitalism in the 1980s.

On a more philosophical level, the Minimalist Program was doomed because definitions of art, like standards of ethics, are not given a priori but are formulated within a culture. Although Judd earned an undergraduate degree in philosophy at Columbia,[103] it is evident from the views expressed in his published writings that he did not master semantics (theory of meaning). However, he did take a course on logical positivism, whose spokesman, Rudolf Carnap, had a just-the-facts, anti-metaphysical attitude that is a good philosophical fit for Judd's outlook. But in Anglo-American philosophical circles in the 1960s discussions of meaning were framed in terms of Carnap's polar opposite in the Vienna Circle, Ludwig Wittgenstein, whose picture theory of language (*Tractatus Logico-Philosophicus*, 1921) and writing on language games (*Philosophical Investigations*, 1953) were read by Judd's contemporaries Jasper Johns and Mel Bochner (plate 11-61).[104] Wittgenstein's basic point about language games is that the meaning of words and images is a consequence of a "game" people play with verbal and visual symbols within a social context. Other artists in Judd's generation, such as Robert Morris, had also learned from Freud's psychoanalysis and Merleau-Ponty's phenomenology that human consciousness makes endless symbolic associations with all it encounters, making it impossible for an artist to void meaning from an object by decree.[105] With any symbol, such as Judd's cubes, everyone is free to play a language game (per Wittgenstein) or to free-associate (per Freud) and *give* it meaning, or—if the symbol does not hold the viewer's interest—to turn away and let the mind wander elsewhere.

Judd never wavered from his positivist outlook, presiding over his cubes in Marfa until his death in 1994, but in the late 1960s Morris, Flavin, and others abandoned the Minimalist Program and went their separate ways. The critical world was also changing; Greenberg's lambasting of Minimalism had cost him his following, and younger critics turned their backs on the British formalism of Roger Fry along with Greenberg. The other tradition—Russian formalism/French structuralism—had also been present in America in the person of Roman Jakobson, who, in the aftermath of the Russian Revolution, had left Moscow for Prague, from which he fled to Scandinavia in the late 1930s. Being of Jewish descent, he narrowly escaped the advancing Nazi army and came to America in 1941. Jakobson eventually joined the faculty of the Department of Slavic Languages at Harvard from 1949 to 1965, after which he held an honorary chair at MIT until his death in 1982. The formalist/structuralist tradition of the linguists Jakobson, Jan Baudouin de Courtenay, and Ferdinand de Saussure developed in the first decades of the twentieth century from the same cultural context as Russian Constructivism, and thus (unlike Fry's formalism) it was a promising idea to add some linguistic concepts to the art historian's tool kit with which to analyze Minimalism because it was in the Constructivist tradition. In the introduction to her 1977 book *Passages in Sculpture*, Krauss did just that and adopted the methods of structural linguistics, together with a phenomenological approach to motion (the "passages" of the title).[106] Other American art critics and art historians also adopted linguistic and phenomenological tools, prompting a shift in the 1970s to theory as opposed to aesthetics (traditional academic philosophy of art), which since the Enlightenment has been based on Kant's *Critique of Judgment* (1790). This anti-aesthetic climate in 1970s Anglo-American art criticism was, in part, a consequence of the rejection of Greenberg, who gave Kant the kiss of death by praising him as an authority.[107]

In 1976 Krauss co-founded the journal *October* (named for Sergei Eisenstein's 1928

The Klein Four-Group: From Anthropology to Psychoanalysis and the Arts

In 1872 the German mathematician Felix Klein invented the four-group diagram as part of his effort to unify Euclidean and non-Euclidean geometries by isolating those transformations that leave geometric figures unchanged (see the sidebar on page 253 in chapter 7). One hundred years later, Klein's diagram began appearing on pages written by members of the *October* group, having travelled the following circuitous route. In the 1940s, George Brainerd and Anna O. Shepard introduced group theory into anthropology, and were the first to use the Klein four-group diagram to analyze decorative patterns on artifacts (see plate 7-12 in chapter 7). As a Jewish refugee in New York during World War II, the anthropologist Claude Lévi-Strauss adopted group theory from these Americans and used it to analyze field data based on his assumption that societies have a symmetrical structure corresponding to Klein's four-group (see plate 7-13 in chapter 7). After 1945 Lévi-Strauss returned to Paris, where the Lithuanian linguist Algirdas Greimas developed the diagram as a tool with which to analyze linguistic meaning, on his assumption (adopted from Saussure) that language has a symmetrical structure. Moreover (going beyond Saussure), Greimas made the additional (big) assumption that linguistic structure corresponds to Klein's four-group (*Sémantique structurale*, 1966; translated by Daniele McDowell, Ronald Schleifer, and Alan Velie as *Structural Semantics*, 1983). Greimas renamed the Klein four-group the "Semiotic Square." Under this alias the diagram entered French intellectual circles in the 1960s and was imported to American academia in the 1970s, on the (very big) assumption that the meanings residing in the unconscious mind (per Lacan), a painting (per Jakobson), and a sculpture (per Krauss) all have the same four-part symmetrical structure that Felix Klein discerned in Euclidean and non-Euclidean geometries.

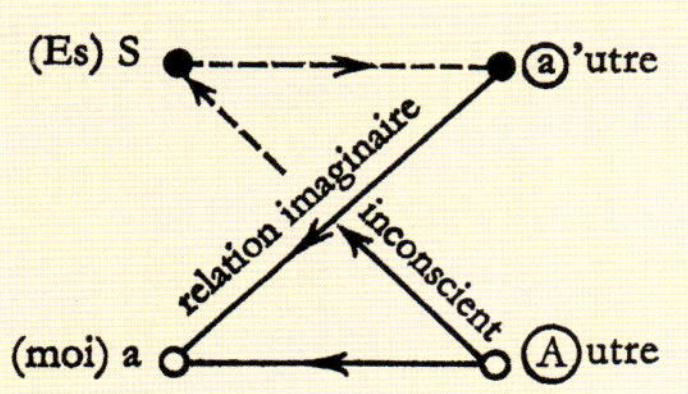

Psychoanalysis

11-62. Jacques Lacan's diagram of the relations between the id (*Es*), the ego (*moi*) and the other (*autre*) in *Écrits* (Paris: Éditions du Seuil, 1966), 53.

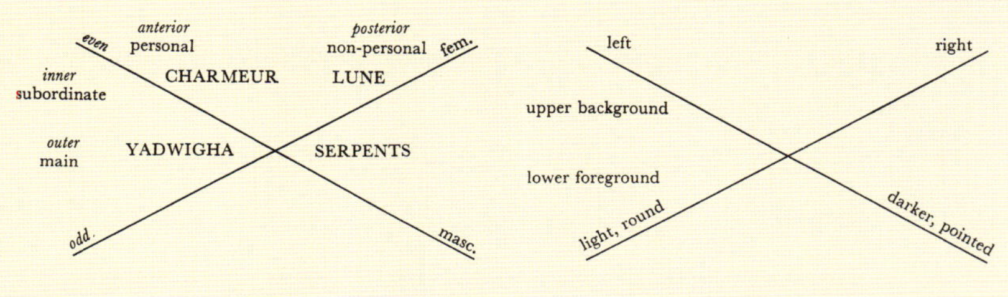

Literature

11-63. Roman Jakobson's diagram of the grammatical subjects (left) and the pictorial composition (right) of Henri Rousseau's painting *The Dream* (1910) in "On the Verbal Art of William Blake and Other Poet-Painters," *Linguistic Inquiry* (1970): 334. Used with permission. Echoing Max Bill's 1945 application of the Klein four-group to the visual arts, Jakobson, who was then on the linguistics faculty at Harvard, used it to analyze a painting in this essay, which he dedicated to his friend, the American art historian Meyer Shapiro, who was then on the art history faculty at Columbia University.

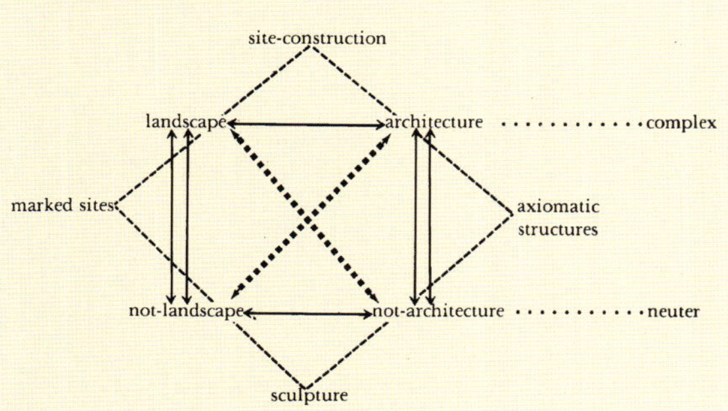

Visual Art

11-64. Rosalind Krauss's diagram of types of sculpture in "Sculpture in the Expanded Field," *October*, 8 (1979): 38. Used with permission.

film about the October Revolution of 1917),[108] which became the mouthpiece for Russian formalism and the structuralist ideas of Saussure and his French-speaking heirs, led by Claude Lévi-Strauss and including the *post*structuralists Michel Foucault, Jacques Lacan, and Roland Barthes, who adopted as their logo Felix Klein's four-group diagram (see sidebar on opposite page), which the Lithuanian linguist Algirdas Greimas renamed the "Semiotic Square."[109] When the *October* group, which included Krauss's students Hal Foster and Benjamin H. D. Buchloh, imported structuralism into American academia in the mid-1970s, it had already gone into free fall in France, where the search for abstract structure seemed irrelevant after May 1968. As Americans read French structural linguistics in the 1970s, an anti-essentialist mood settled on the New York art world, and the Minimalist Program, which assumes that art has a fixed physical essence, came to seem passé.

ASYMMETRY: ROBERT SMITHSON

A contemporary of the Minimalists, Robert Smithson, viewed geometry not as minimal but maximal—brimming with meaning. As a young artist, he asked the old questions: Where does everything come from? How did we get here? This led him to study philosophy and theology: "I was always interested in origins and primordial beginnings . . . I guess this was always haunting me until about 1959 and 1960 when I got interested in Catholicism through T. S. Eliot."[110] In these years Smithson painted mythological, sexual, and religious subjects in an organic, expressionist style (*Green Chimera with Stigmata* of 1961; plate 11-65). Smithson shifted his attention from religion to science after reading the science

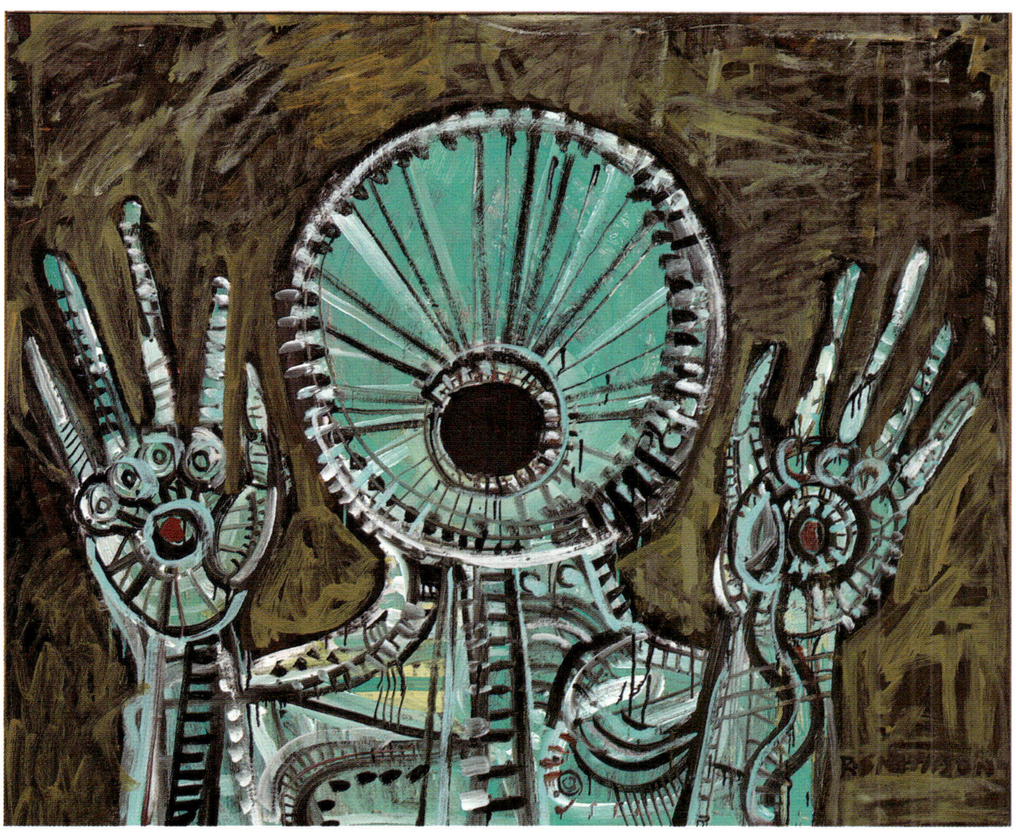

11-65. Robert Smithson (American, 1938–73). *Green Chimera with Stigmata*, 1961. Oil on canvas, 47¾ × 57 in. (121.2 × 144.7 cm). © The Holt-Smithson Fundation/VAGA, New York. Courtesy James Cohen Gallery, New York and Shanghai

Smithson invented this composite classical and Christian image: a fire-breathing, lion-headed monster of Greek mythology, a Chimera, has hands marked with the stigmata, the wounds of the crucified body of Jesus.

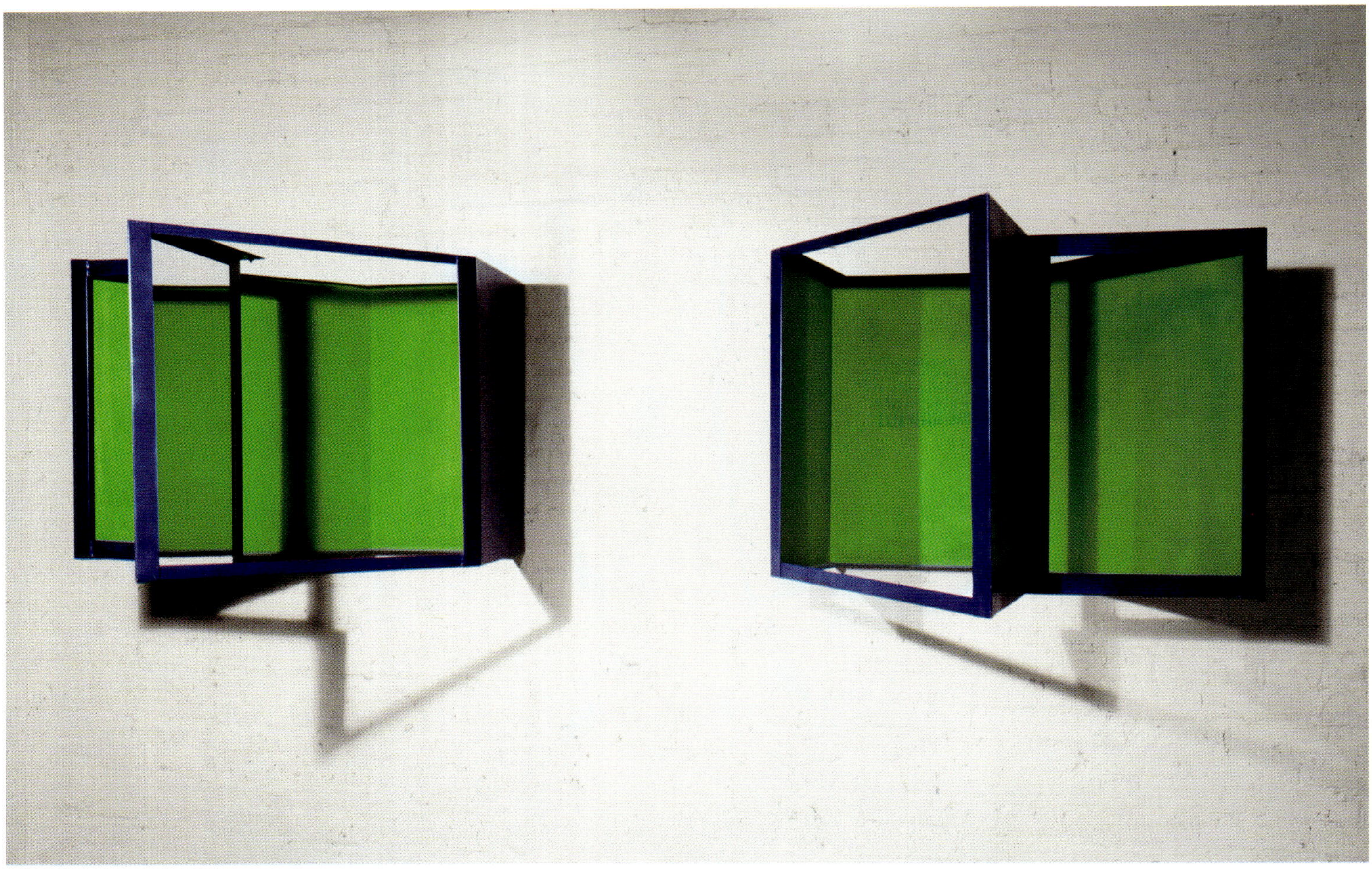

11-66. Robert Smithson (American, 1938–73), *Enantiomorphic Chambers*, 1965/reconstruction of 2003. Steel and mirrors, two components, each 24 × 30 × 31 in. (61 × 76.2 × 78.7 cm). Location of the original is unknown; the reconstruction is in the collection of the Museet for Samtidskunst, Oslo. © The Holt-Smithson Foundation/VAGA, New York. Courtesy James Cohen Gallery, New York and Shanghai.

journalist Martin Gardner's *Ambidextrous Universe* (1964), a book that describes the origin of the cosmos in terms of left-right ("mirror") symmetry, inspiring Smithson to explore these themes in a vocabulary of simple geometric forms.[111] For example, Smithson made a pair of cubes with mirrored surfaces that suggest molecules that exist in two asymmetrical, mirror images (*Enantiomorphic Chambers*, 1964; see plate 11-66 and compare plate 7-4 in chapter 7). Smithson studied geology and showed a special interest in crystals.[112]

When Smithson changed his focus in 1964, he nonetheless continued asking the same questions. He read the Austrian writer Anton Ehrenzweig, who studied art, psychoanalysis, and Gestalt psychology in Vienna before fleeing Nazi-occupied Austria in 1938 and settling in London, where he joined fellow refugees, the Austrian Ernst Gombrich and the German Rudolf Arnheim. A holistic thinker, Ehrenzweig sought an aesthetics that merged psychoanalysis and Gestalt psychology by revealing the "hidden order" of art as an expression of the (Freudian) unconscious and the (Gestalt) unity of opposite forms (male-female, up-down, left-right; *The Hidden Order of Art*, 1967). Ehrenzweig described the ability of geometric abstract art to express the unconscious mind: "Though abstract art is already degenerating today into mannerisms, its origin in deeply unconscious layers of the mind is beyond question. . . . In our abstract art there is a dramatic short circuit between its high sophistication and love of geometry on the one hand and an almost oceanic lack of differentiation obtaining in its matrix in the unconscious mind."[113] In 1967

Ehrenzweig gave an account of one of the degenerate "mannerisms," which (although he didn't name names) describes Minimal Art: "Abstraction becomes truly empty whenever it is dissociated from an unconscious matrix. It will then turn into a vacuous 'generalization.' Empty generalizations can be hurled with such facility because they have cut themselves loose from their anchorage in the deep."[114]

The stark geometry and large scale of Minimal Art gives it an aura of profundity, but if you scratch the surface you hit bottom. Smithson wanted to take a deep dive below the "vacuous" literal surface to reach Ehrenzweig's "unconscious levels" and symbolize primordial forms in nature. Smithson described his 1964 move into abstraction: "I was interested in this kind of archetypal gut situation that was based on kind of primordial needs and the unconscious depths."[115] His search for these qualities brought him face-to-face with two key events in the science and mathematics of his day: NASA's manned mission to the moon and the discovery by physicists at Columbia University in New York of an asymmetry in the bedrock of nature, the so-called fall of parity.

In October 1968 the United States launched its first manned expedition to the moon, followed by manned missions in December 1968 and March 1969. The goal of these widely publicized launches was to conquer "the last frontier," echoing earlier colonialist American expeditions to remote lands. In a 1969 interview Smithson expressed his opinion that the political spectacle of the American-Soviet space race trivialized the awesomeness of the cosmos. He expressed admiration for the seventeenth-century mathematician Blaise Pascal, who wrote, "The eternal silence of these infinite spaces terrifies me," and whose thinking reflected the fifteenth-century mystic, the German Catholic cardinal Nicholas of Cusa. As Smithson said, "My understanding of the order of the universe is based on the Pascalian idea that the universe is a sphere whose circumference is everywhere and whose center is nowhere. . . . The need to put the men on the moon is essentially a personal, humanistic approach . . . anthropomorphic. If they thought about the infiniteness of space like Pascal where they actually are terrified, you know you can just sort of be terrified and have a tragic view."[116] Smithson felt that the American space program was too narrowly focused on mankind, whereas he focused on the vastness of space—the sublime of the Romantic landscape painters—like Cusa, Pascal, and the leading modern mathematician of infinity, Georg Cantor, whose publication on transfinite numbers Smithson owned.[117] Many modern astronomers, including Edwin Hubble, were also deeply romantic about their scientific exploration of the cosmos.[118]

In 1964 (the year Smithson began working with mirrors) American and Soviet scientists used the surface of the moon as a mirror, off which to bounce light for measuring the earth-moon distance ("Moon Is Employed as a 'Space Mirror,'" *New York Times*, March 6, 1964). Although some light is reflected off the moon's rocky surface, much of it is absorbed. More light would be reflected if there were a better mirror on the surface of moon, and the plan was for American astronauts to place a special kind of mirror there.

With the newspapers filled with giddy speculation about whether the moon landing would be attempted in May or June of 1969, on April 15, 1969, Smithson set off with a group of twelve-inch-square mirrors on his own expedition to a remote land, the Yucatán

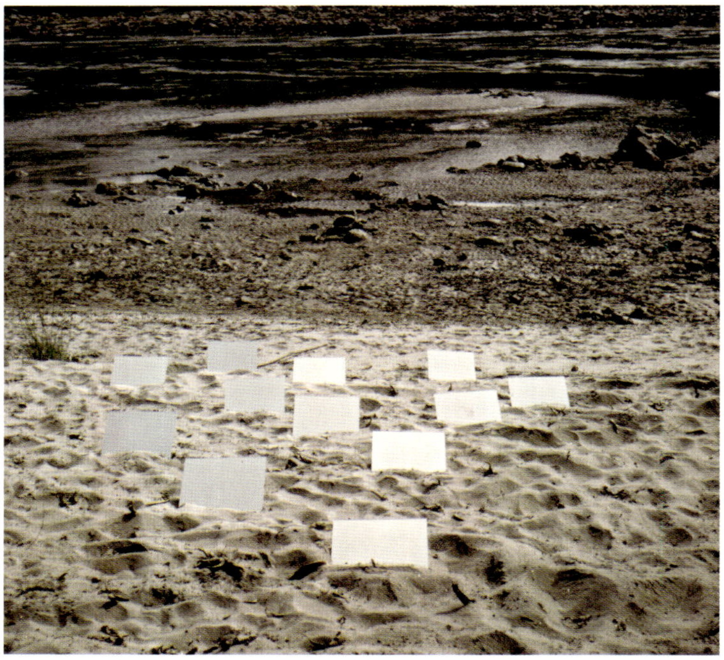

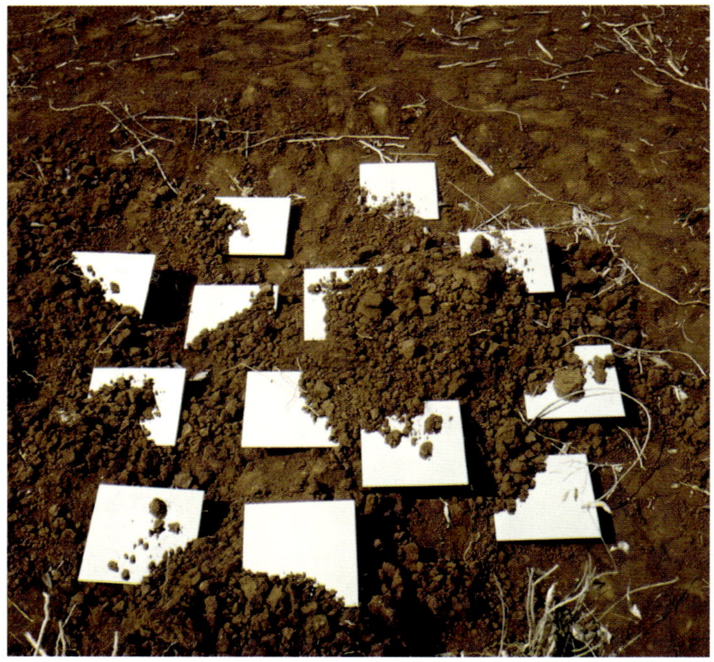

ABOVE LEFT

11-67. Robert Smithson (American, 1938–73), *Yucatan Mirror Displacement (6),Yucatan, Mexico,* 1969. Number six in the series of nine 126-format chromogenic-development transparencies. Solomon R. Guggenheim Museum, New York. © The Holt-Smithson Fundation/ VAGA, New York. Courtesy James Cohen Gallery, New York and Shanghai.

ABOVE RIGHT

11-68. Robert Smithson (American, 1938–73), *Yucatan Mirror Displacement (2), Yucatan, Mexico,* 1969. Number two in the series of nine 126-format chromogenic-development transparencies. Solomon R. Guggenheim Museum, New York. © The Holt-Smithson Fundation/ VAGA, New York. Courtesy James Cohen Gallery, New York and Shanghai.

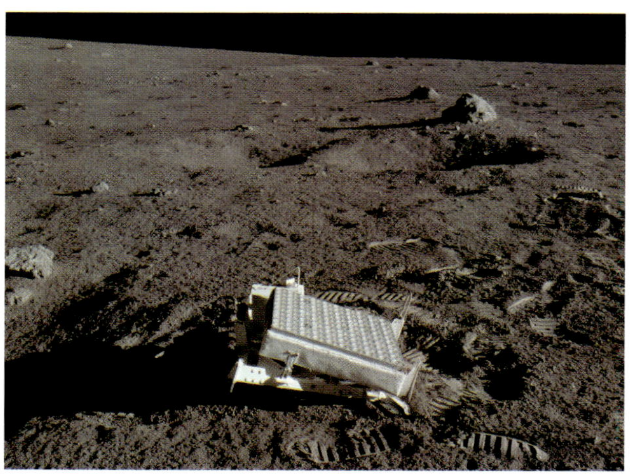

11-69. Lunar laser retroreflector, 2 × 2 ft. (61 x 61 cm) panel with 100 mirrors. Lunar and Planetary Institute of NASA, image AS14-67-9386.

The first Lunar laser retroreflector was placed on the moon by *Apollo 11* astronauts on July 21, 1969. Later explorers aboard *Apollo 14*, *Apollo 15*, and the Soviet *Lunokhod* (Moonwalker) *1* and *2*, also placed retroreflectors on the moon.

peninsula in Mexico. There, in nine locations, he arranged his mirrors in a grid pointed at the sky (plates 11-67 and 11-68). A few months later, on July 21, 1969, American astronauts landed on the moon, where they placed a twenty-four-inch-square grid of one hundred small mirrors pointed at the earth (plate 11-69). NASA's hundred mirrors are called "retroreflectors" because they return a beam of light along its path of approach. Each retroreflector is a transparent cube of glass backed by three mutually perpendicular mirrors that form a corner (they are "corner-reflector" mirrors). From a telescope on earth, an astronomer shoots a laser beam at the lunar retroreflector, and the light returns into the same telescope. By measuring the time it takes the light to make the round trip, multiplying by the speed of light and dividing by two, the astronomer can measure the distance between the terrestrial telescope and the surface of the moon to an accuracy of a few centimeters (about an inch).

The same side of the moon always faces the earth, with which it is in gravitational lock, and thus NASA's lunar corner-reflector mirrors always point towards the earth, bouncing light back and forth between earth and moon. In contrast, Smithson's mirrors point skyward, reflecting sunlight as the earth rotates and beaming it out into "the infiniteness of space" like a beacon.[119] While American and Soviet scientists competed to see who could be the first to plant their nation's flag on the moon, Smithson commented that his cosmos—his Yucatán—was not a place to aggressively conquer but to passively experience as an exotic "other," an "elsewhere" filled with mystery: "Elsewhere. That's how the Yucatán runs. The Yucatán is elsewhere."[120]

Smithson's shift from making expressionistic paintings on religious themes to site-specific mirror sculptures on cosmic themes was prompted by his reading of Gardner, who discusses mirrors and symmetrical structures in nature. But the motivation and centerpiece of Gardner's *Ambidextrous Universe* was not symmetry but asymmetry—the fall of parity—the surprising counter-example to nature's grand symmetrical scheme.

As we saw in chapter 7, symmetry became a fundamental principle of science after researchers confirmed the conservation of *mass* (Antoine Lavoisier, 1789), of *energy* (Hermann von Helmholtz, 1847), and of *mass-energy* (Albert Einstein, 1905). Furthermore, Emmy Noether showed how, using group theory, many conservation laws can be understood as transformations that leave the mass, energy, and mass-energy of the universe unchanged, which points to a deep symmetry in nature (*Noether's Theorem*, 1918). A similar relation holds between symmetries and conserved quantities in the subatomic realm. In particular, the symmetry of mirror reflection in the quantum realm yields the principle of the conservation of *parity*, a quantity that takes the values of even or odd, which are assigned to all subatomic particles.

After entering the quantum realm and confirming the left-right symmetry of subatomic interactions in the 1930s, scientists expected that symmetry would be respected throughout nature. But to everyone's surprise, in the late 1950s scientists discovered an exception to mirror symmetry.

There was a troubling puzzle in particle physics concerning two particles called theta and tau. They seemed identical except that one disintegrated to a collection of particles whose parity was even, while the other disintegrated to a collection of particles whose parity was odd. Two Chinese-born American theoretical physicists, Tsung-Dao Lee and Chen Ning Yang, took up the theta-tau puzzle and had the bold insight that things would be simpler if these two particles were actually the same particle disintegrating in two different ways.[121] But for that to happen, conservation of parity would have to be thrown out, which would imply that in the subatomic world mirror symmetry does not always hold.[122] Lee and Young first made this proposal in 1956, when such a breaking of symmetry was almost unthinkable. By the end of 1957 it was an experimentally confirmed fact.

The first experimental confirmation was by Chien-Shiung Wu, who was a colleague of Lee at Columbia and an expert on beta decay, a process in which atomic nuclei decay by emitting electrons (so-called beta particles). These processes involve the weak nuclear force, which also mediates the decay of the theta and tau particles—the puzzling pair that led Lee and Yang to question the conservation of parity. Lee and Yang observed that a spinning nucleus defines a direction in space; call that direction "north." Then they did a thought experiment and imagined reflecting that nucleus in a vertical mirror; this reverses the direction of the spin, so the reflection of the spinning nucleus is a nucleus whose "north" points south (plate 11-70). Consider the distribution of beta particles (electrons) emitted by this nucleus when it decays. If the distribution is not fifty-fifty up and down, then it can't look like the distribution of beta particles from the mirrored nucleus, because

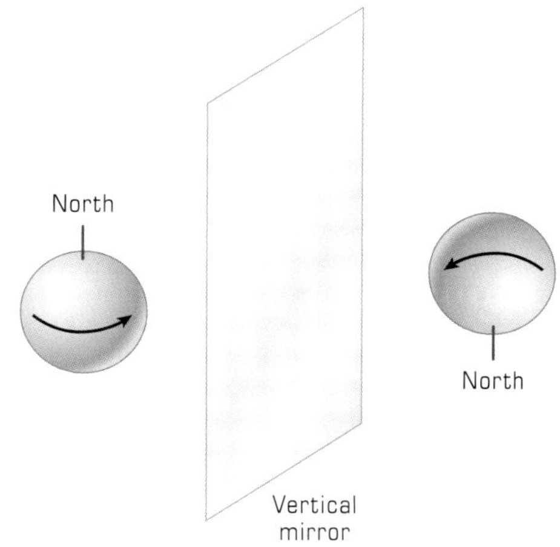

11-70. A spinning atomic nucleus and its mirror image.

Reflection in the vertical mirror changes the counter-clockwise rotation of the atomic nucleus shown on the left, to the clockwise rotation in the reflected image on the right, but does not reverse up and down. Since the spinning mass is positively charged, it generates a magnetic field, whose north pole points upward on the left but downward in the mirrored version on the right. Conservation of parity is equivalent to invariance under mirror symmetry. So if parity is conserved, the behavior of these mirrored nuclei will be the same.

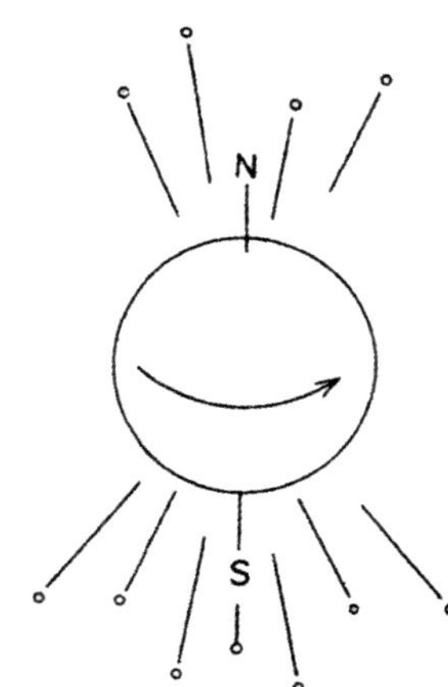

11-71. A spinning cobalt-60 nucleus, in Martin Gardner, *The Ambidextrous Universe* (New York: Basic Books, 1964), 243, fig. 62.

In this diagram of a spinning cobalt-60 nucleus, the nuclear magnetic field has its north pole pointing upward. Experiment shows that electrons are more likely to be flung out from its south pole, which means that parity is not conserved.

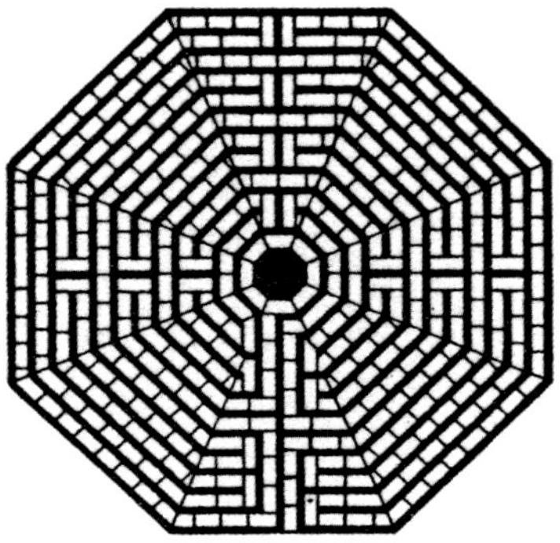

11-72. Plan of the labyrinth in Amiens Cathedral, France, in Robert Smithson, "The Quasi-Infinities and the Waning of Space," *Arts Magazine* 41, no. 1 (1966): 28–31; the plan is on 29.

the "north" of the mirrored nucleus points south, whereas the mirror image leaves the geometrical direction unchanged.

Wu and her colleagues experimented using radioactive cobalt atoms cooled close to absolute zero; to achieve this low temperature they used the equipment at the National Bureau of Standards in Washington, DC. Once cooled, the atoms were put in a strong magnetic field to orient their north-south axes. By January 1957 their data showed that the beta particles were emitted preferentially downward in this setup (plate 11-71). This meant that parity is not conserved in weak interactions, and mirror symmetry does not hold for weak interactions. In other words, the radioactive decay of cobalt-60 is a counter-example to nature's symmetry, which is amazing.

The discovery that parity is not conserved in some radioactive decay underscores the central role of mathematics in scientific exploration. Lee and Yang's hypothesis was that some weak interactions might violate parity conservation sometimes, and they had to think up a situation in which the decay of a nucleus was governed by weak forces; the decay of cobalt-60 fit the bill. To design and carry out her experiment on radioactive cobalt, Wu was guided by Lee and Yang's mathematical description of the decay of cobalt to know where to look.

Gardner was inspired to write *Ambidextrous Universe* by the *absence* of symmetry—the startling and mysterious fall of parity; Smithson's *Spiral Jetty* (1970; see plate 11-73) is asymmetrical—its spiral shape cannot be superimposed on its mirror image. When Smithson created this work, he was on a quest for primordial, archetypal symbols, and he may have found one when he read Gardner's dramatic description of the fall of parity. Radioactive beta decay is a phenomenon that is described only by mathematics in the sense that there is no physical description of a mechanism that accounts for the observed experimental data: beta particles flying out in an asymmetrical pattern. The phenomenon is a tremendous mystery.

Everyone expected Albert Michelson and Edward Morley to detect a motion of the earth relative to a fixed "æther." It was the negative result of this test that was so upsetting. Everyone expected Madam Wu to find a left-right symmetry in the process of beta decay. Nature sprung another surprise.

—**Martin Gardner,**
***Ambidextrous Universe*, 1964**

After completing *Spiral Jetty*, Smithson commented that he felt that there had been a "tug of war going on between the organic and the crystalline" inside himself. When asked which had won, he responded, "I think they kind of met—a kind of dialectic occurred later on, so both areas were resolved."[123] The (expressionistic, primordial) organic and the (cerebral, geometric) crystalline are combined in *Spiral Jetty*, which is a well-marked path to a special place, reminiscent of another asymmetrical path followed by a religious pilgrim—the labyrinth inside the entrance to a medieval cathedral—which Smithson described as one "through which the mind will pass in an instant."[124] The labyrinth to which he refers (plate 11-72) has a single path that, like *Spiral Jetty*, leads inexorably to its center (it is not a maze designed to confuse). Cathedral labyrinths are a miniature version of a pilgrimage to Jerusalem, a voyage that medieval parishioners hoped to make once in life. After travelling to the remote site on the Salt Lake, the art pilgrim walks to the center and gazes at earth and sky. It can be a moment of transcendence when, as Ehrenzweig wrote, "there is a dramatic short circuit between its high sophistication and love of geometry on the one hand and an almost oceanic lack of differentiation."[125]

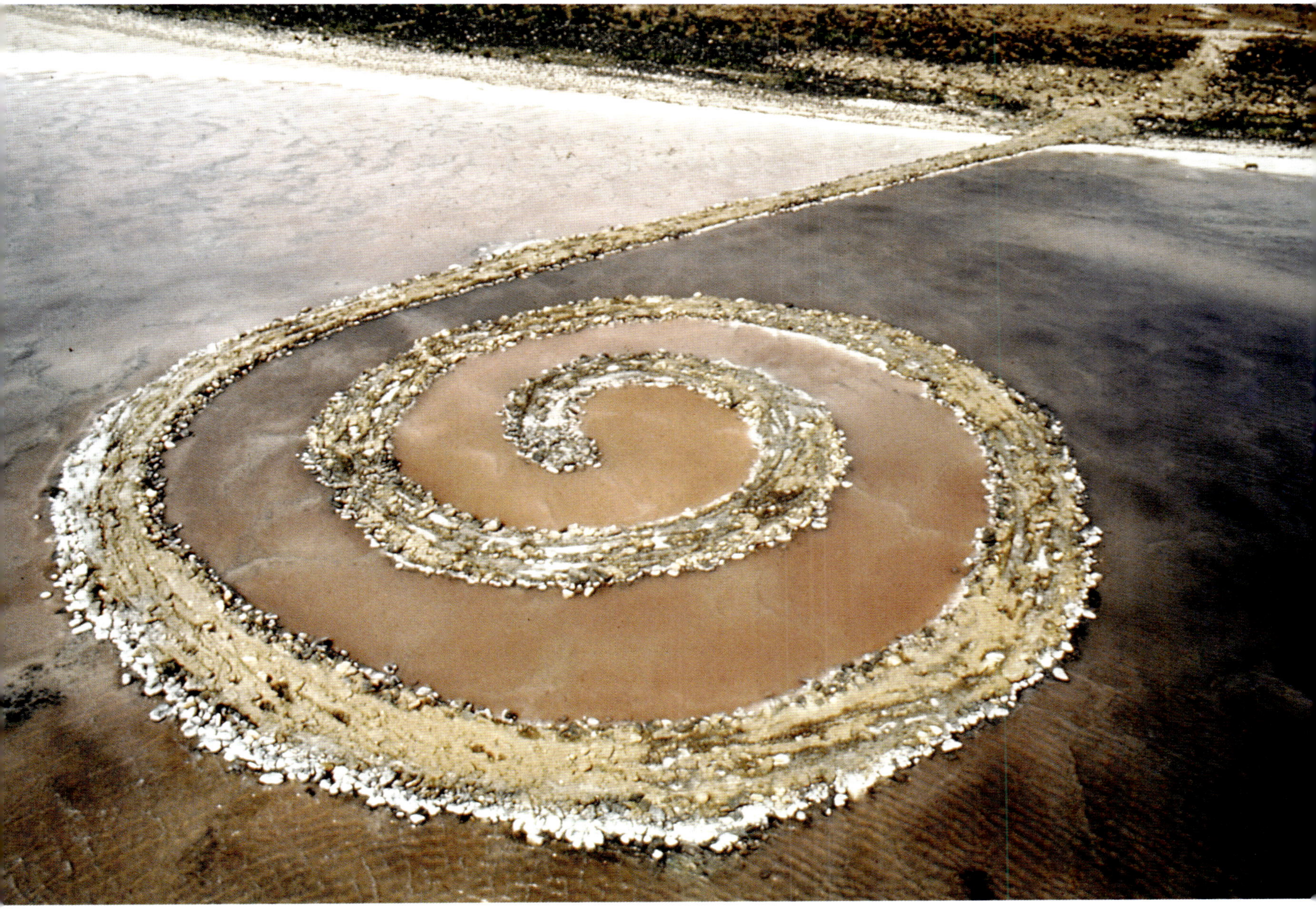

American art historian Jennifer L. Roberts has interpreted Smithson's *Spiral Jetty* as an expression of four-dimensional space—the occult realm of the Absolute—which the Russian theosophist Peter Ouspensky described in his 1911 book *Tertium Organum* (see chapter 3).[126] In fact, there is no historical evidence for this.[127] Rather, Smithson went from a youthful fascination with religion to a mature engagement with a wide range of intellectual topics, including psychology, literature, philosophy, and not antiquated pseudo-science but the *real* science and mathematics of his generation, which is filled with speculation about the traditional theological/philosophical questions that captivated Smithson. For insightful commentary on these profound issues, the artist did not need to look further than Gardner, who in a discussion of the "numinous," the "awareness of a *mysterium tremendum*" (Latin for "tremendous mystery"), described his feeling about the mathematical-world-out-there: "I, too, believe in this 'wholly other' realm, a realm in which our universe is an infinitesimal island, and I call myself a mystic in the Platonic sense."[128]

11-73. Robert Smithson (American, 1938–73), *Spiral Jetty*, 1970. Mud, salt crystals, rocks, water, 1500 ft. (457.2 m) long and 15 ft. (4.57 m) wide, extending into the Great Salt Lake, Utah. Dia Center for the Arts, New York. © The Holt-Smithson Foundation/VAGA, New York. Courtesy James Cohen Gallery, New York and Shanghai.

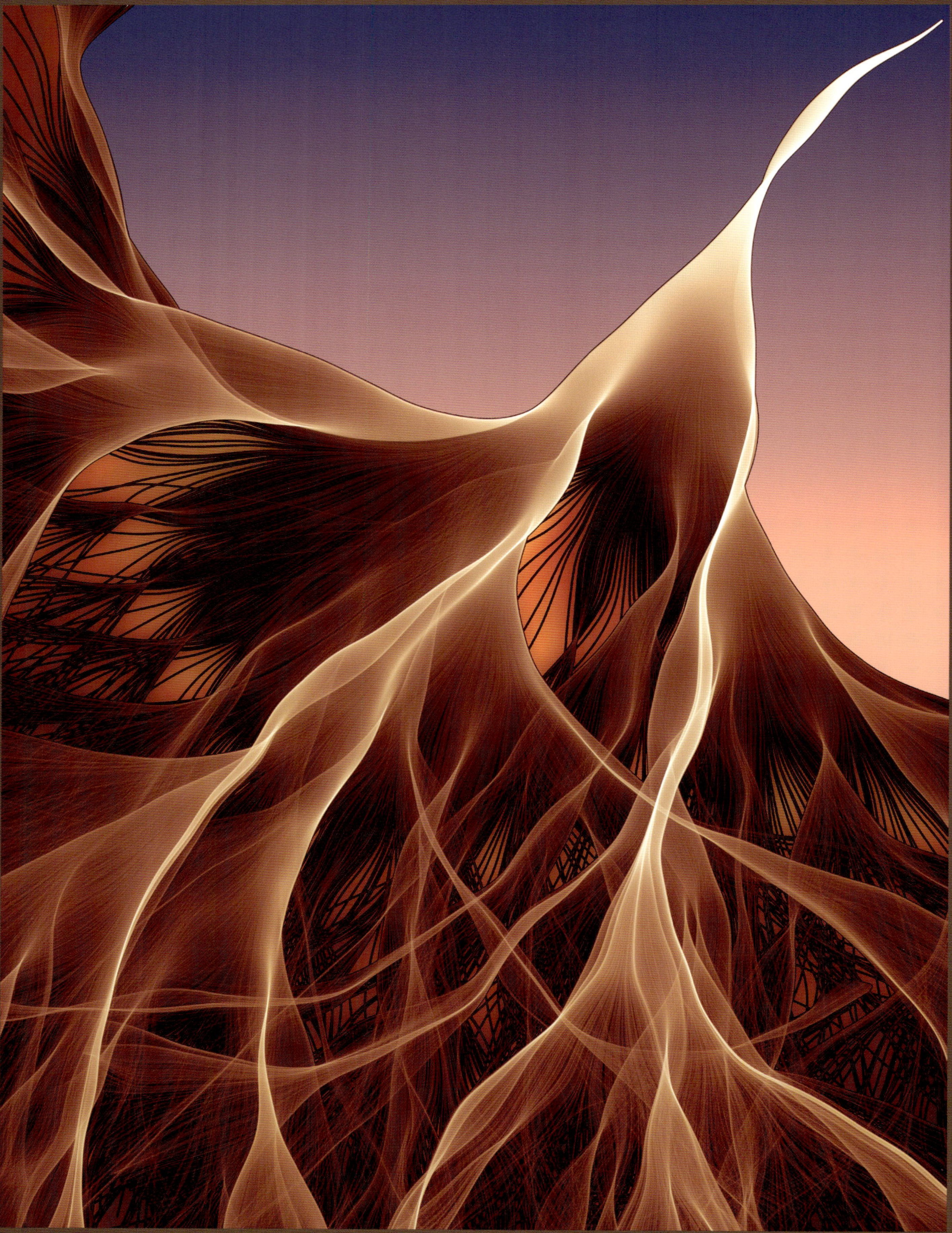

12
Computers in Mathematics and Art

Ever since Euclid, mathematical proofs have served a dual purpose: certifying that a statement is true and explaining why it is true. In the future those two epistemological functions may be divorced. In the future, the computer assistant may take care of the certification and leave the mathematician to look for an explanation that humans can understand.
—Dana Mackenzie, "What in the Name of Euclid Is Going On Here?" *Science*, 2005

WHEN COMPUTERS BECAME available in the 1950s, mathematicians wondered whether these mindless machines might someday not only calculate sums at lightning speeds but also take over some task in actually proving a theorem. This happened in 1976 when the American mathematician Kenneth Appel and the German Wolfgang Haken announced that, with the assistance of a computer, they had proven a conjecture—the four-color theorem—that had been unsolved for more than a century. Other computer-assisted proofs followed, launching a debate about the role of computers in mathematics and what constitutes a proper proof. By the 1980s, scientists were employing computers to collect and analyze massive data, the mathematician Benoît Mandelbrot had invented fractal geometry, and the computer had entered the studio, where artists created computer graphics, digital photography, and computer animation.

PROOFS BY EXHAUSTION

Kenneth Appel and Wolfgang Haken proved the four-color theorem using a so-called proof by exhaustion: the statement to be proved is divided into a finite number of cases and then each is proved separately, thereby "exhausting" the cases. For example, Euclid used a proof by exhaustion to show that there is an infinite number of prime numbers (see the sidebar on page 22 in chapter 1); a whole number is either prime or composite, so he had two cases to prove. If there are just a few cases, they can be proved by hand as Euclid did, but if there are hundreds or thousands of cases, it becomes impossible to do all the calculations by hand in a lifetime. This changed with the invention of the computer.

The four-color theorem began as a speculation in 1852 when a British youth, Francis Guthrie, was coloring in a map of the counties of England, and he noticed that he only needed four colors to make all adjacent counties a different hue. He asked his

12-1. Eric J. Heller (American, born 1946), *Exponential Electron Flow*, ca. 2000. Digital print. Courtesy of the artist.
The physicist and artist Eric J. Heller created this image by first launching (from the upper right) a flow of electrons, which fan out and form branches as indirect effects of travelling over bumps. He then took his (black and white) digital recording of electron flow, put it into Photoshop, and added the color.

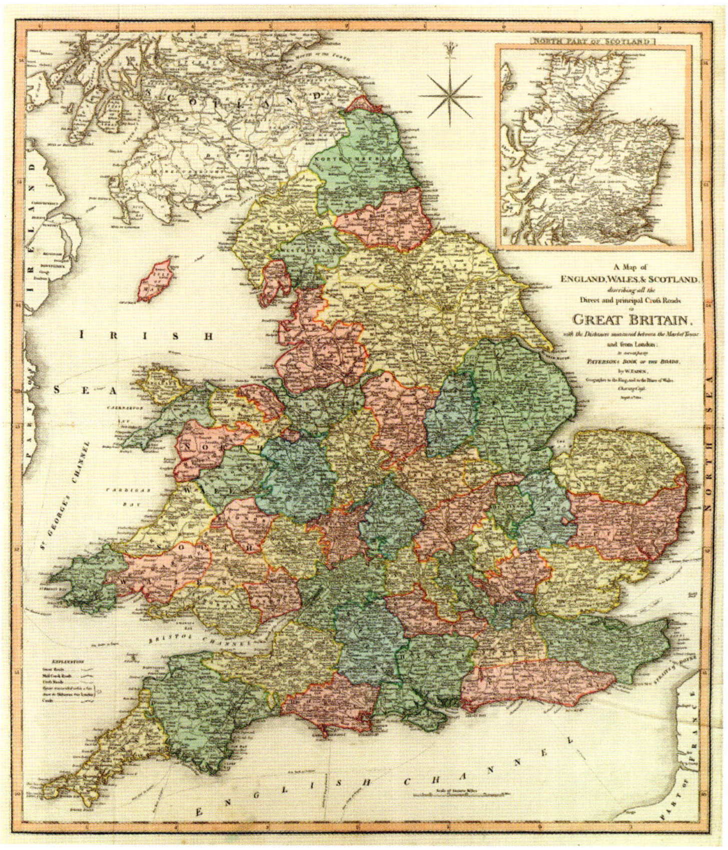

12-2. A map of England, Scotland, and Wales in William Faden, *Paterson's Book of the Roads* (London: William Faden, 1801), 7. David Rumsey Map Collection, inv. no. 2014.007. © 2000 Cartography Associates.

The counties of England are colored yellow, pink, blue, and green.

mathematician brother, Frederick, whether four colors suffice to color any map, and the conjecture entered the mathematics literature, becoming one of the best-known unsolved problems of the last hundred years. Intuitively the theorem states that given any division of a plane into contiguous regions (such as a map), the regions can be colored using at most four colors such that no two adjacent regions have the same color (plate 12-2).

The difficulty in proving the four-color theorem is that it applies to *any* division of the plane (any possible map), so it could have millions of regions with any imaginable shapes. Appel and Haken's method for proving this result was to show that there can be no simplest counter-example—no irreducibly five-colorable map. Appel and Haken began in 1972 by first dividing the theorem into cases, determining that all possible maps belong to one of 1936 sets of cases. Then they wrote a program to test that four colors suffice for each case (each archetypal division of the plane) and that there was not a map requiring five. Over the next four years, the computer applied their algorithm, carrying out billions of calculations in 1200 hours of computer time, while the mathematicians themselves analyzed many cases by hand.

Appel and Haken created considerable controversy when they announced that they had proven the four-color theorem with computer assistance.[1] Some objected that such a proof is so long that it is impossible for the human mind to follow it. Indeed, the proof can be checked only by another computer using special proof-checking software, which was done by the British mathematician Georges Gonthier in 2004, after which Appel and Haken's proof was generally accepted by the mathematics community. Some argued that the algorithm (the computer program) written by the mathematicians should be considered the proof, and the computer's part as a mere mechanical calculation. Also, a computer is a physical object made from electrical and mechanical parts, which can be defective. Others raised aesthetic objections, arguing that any proof by exhaustion—with or without computer assistance—lacks the simplicity and brevity of an elegant proof (the nickname for a proof by exhaustion is "brute force proof").[2]

In 1998 a second famous, even older mathematical problem was solved with a computer-assisted proof by exhaustion. In 1611 Johannes Kepler wrote a book on snow, which he imagined as composed of tiny spheres of ice, and he asked: "What is the densest arrangement of spheres in space?" He answered by describing the most efficient way to pack the same-sized spheres so that they take up the least space (achieve the highest density). Assuming the first layer in a large crate is a grid-like sheet of spheres, should the spheres in the next layer be stacked directly on top of those below (A in plate 12-3), or should each sphere in the second layer be set into the hollows made by the four spheres beneath it, as grocers stack oranges (B in plate 12-3)? Kepler believed that the latter method achieved the highest density, but he could not prove it. The difficulty in proving the conjecture is that it applies to *all* possible arrangements (regular and irregular) of spheres, not only a pile of neatly stacked oranges but also the countless ways of tossing the spheres into a box haphazardly and then shaking the box to make them settle. A proof by exhaustion would have to show that the orange-pile method was best by proving that *all* the other cases have lower densities. In 1953, just as computer technology was becoming available, the Hungarian mathematician László Fejes Tóth showed that the density of all the cases could be determined in a finite (although very large) number of calculations, and so in principle Kepler's conjecture could be solved using a proof by exhaustion. After engineers developed computers with enough speed, programming power, and memory to do the massive number of calculations, the American mathematician Thomas Hales developed a multistage strategy to prove Kepler's conjecture that included writing and running the programs. In 1998 Hales announced that he had proven Kepler's conjecture; he posted his entire proof on the Internet, including 250 pages of his explanatory text and three gigabytes of the computer's number crunching.

The tale of the reception of Hale's proof gives a glimpse into the impact that the computer and the Internet are having on the culture of mathematics.[3] Mathematicians all over the world could immediately download the proof and start checking it, speeding up the response to the proof by the mathematics community, which did not have the Internet in 1976 when the proof of the four-color theorem was announced. The editor of the *Annals of Mathematics*, which is a joint publication of Princeton University and the Institute for Advanced Studies in Princeton, and one of the most prestigious mathematics journals in the world, took the unusual step of contacting Hales and offering to publish a summary of his proof, but only after it was reviewed by an independent panel of referees. After four years of intensive work, the referees, led by the world's expert on the topic, the Hungarian mathematician Gábor Fejes Tóth (the son of László Fejes Tóth), announced in 2003 that the validity of the proof was 99 percent certain but that they could not verify every last one of the billions of computer calculations. It is a mark of the difference between mathematics and science that 99 percent certain was not good enough for Robert MacPherson, editor of the *Annals*, who wrote to Hales: "The news from the referees is bad, from my perspective."[4] Hales responded by announcing a project to write a proof of Kepler's conjecture that is completely formalized in the sense that it does not rely at all on intuition (in the mind of someone reading it) but has all the logical steps explicitly stated so that the proof can be both written and read by a machine. To accomplish this massive task, Hales assembled an international team of collaborators who are connected by the Internet, and he estimates that the team will complete the formal proof in about twenty years. At that time Hales's proof of Kepler's conjecture will be able to be checked by automated proof-checking software, and this will remove any lingering doubts.[5] Meanwhile on a practical level, the mathematics of the efficient packing of spheres, developed by Hales and his collaborators, is used today by engineers working in nanotechnology (where the "spheres" are atoms) to design the storage of information on compact disks and the compression of data for easy transmission around the globe.

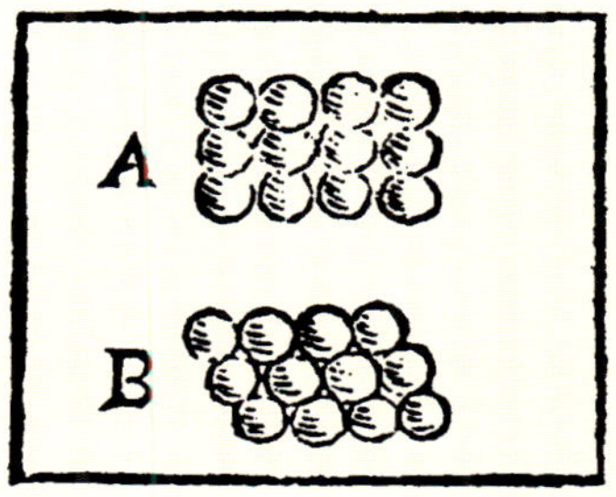

12-3. Johannes Kepler, *Strena, seu, de Nive Sexangula* (On the six-cornered snowflake; Frankfurt-am-Main, Germany: Godfrey Tampach), 1611.

In his short treatise on snow, Kepler suggested that the way to pack spheres ("spherical pellets") into a container and achieve maximum density is shown in diagram (B): "Not only is every pellet touched by its four neighbors in the same plane, but also by four above and four below, and so throughout one will be touched by twelve, and under pressure spherical pellets will become rhomboid. This arrangement will be more comparable to the octahedron and pyramid. The packing will be the tightest possible, so that in no other arrangement could more pellets be stuffed into the same container." Johannes Kepler, *Strena, seu, de Nive Sexangula*, trans. C. Hardie (Oxford, England: Clarendon, 1966), 15. No mathematician was able to prove Kepler's conjecture until the computer arrived and gave the needed assistance.

COMPUTER VISUALIZATION

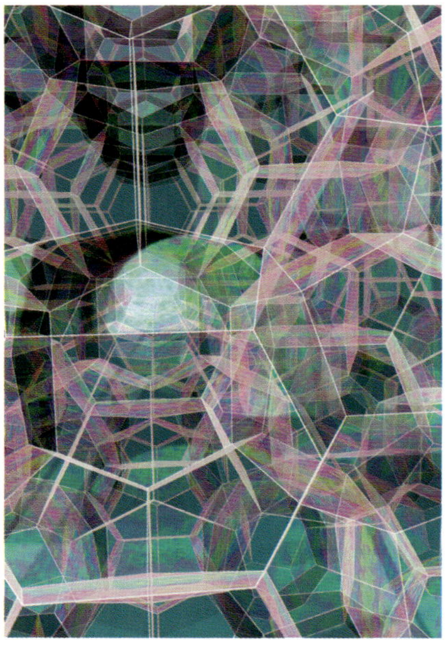

In addition to assisting mathematicians in proofs by exhaustion, computers have also been enlisted to investigate the geometry of space. In 1887 the Irish mathematician Lord Kelvin asked what would be the partition of a space into cells (such as soap bubbles) that is *optimal* in the sense that the walls of the cells have the least surface area. In other words, each wall of every bubble would be a minimal surface—one for which any distortion, no matter how small, increases its area. Kelvin conjectured that the optimal partition would be into equal-volume cells of a single uniform shape, a honeycomb formed with fourteen-sided polyhedra (with six square faces and eight hexagonal faces)—a truncated octahedron, which is known as a Kelvin structure. But in 1993 an Irish physicist, Denis Weaire of Trinity College Dublin, together with his graduate student Robert Phelan, found a better solution by examining computer simulations of foam. They discovered a structure composed of two cells with different forms (although they have equal volume)—a Weaire–Phelan structure (plates 12-4 and 12-5)—which has 0.3 percent less surface area than the one proposed by Kelvin. Architects used the Weaire–Phelan structure to design the translucent, foamy walls of the National Aquatics Center (the "Water Cube"; see plate 12-6) that was built adjoining the National Stadium (the "Bird's Nest") for the 2008 Olympics in Beijing.

In the late 1960s the American mathematician Thomas Banchoff began using computer graphics to model (visualize) spaces of four and higher dimensions. The computer produces an image on a two-dimensional screen. To model a three-dimensional object (such as a building), each point in three-dimensional space is specified as a triple of numbers. Then the computer can rotate this three-dimensional matrix and display any two-dimensional projection of the building as seen from any viewpoint in three-dimensional space. Similarly, a four-dimensional object can be specified by a quadruple of numbers, and the computer can rotate the four-dimensional matrix and display a two-dimensional image of it on the computer screen for a given viewpoint in

ABOVE

12-4. John M. Sullivan (American, b. 1963), *Foamy Partition: Weaire–Phelan*, 1995–99. Computer graphic, 22 × 33 in. (55.8 × 83.8 cm). Courtesy of the artist.

John Sullivan, a professor of mathematics at the Technische Universität in Berlin, was part of the team that proved that the Weaire–Phelan structure beats Kelvin's conjecture by being a least-area way to fill space with equal-volume cells, although they could not prove that this foam is truly optimal. John M. Sullivan, "Comparing the Weaire–Phelan Equal-Volume Foam to Kelvin's Foam," *Forma* 2, no. 3 (1996): 233–42. After using his computer to visualize the Weaire–Phelan space, Sullivan then transformed the mathematical pattern (shown here from *inside* the foam) into an artistic computer graphic by adding color and manipulating the design, creating this print of a foamy partition inspired by the Weaire–Phelan structure.

BELOW LEFT

12-5. A Weaire–Phelan structure.

BELOW RIGHT

12-6. *National Aquatic Center* (the "Water Cube"), 2008, Beijing, China. Architects: the Australian firm PTW, China State Construction Engineering, and China Construction Design International.

The Water Cube adjoins the Bird's Nest, which is visible on the right.

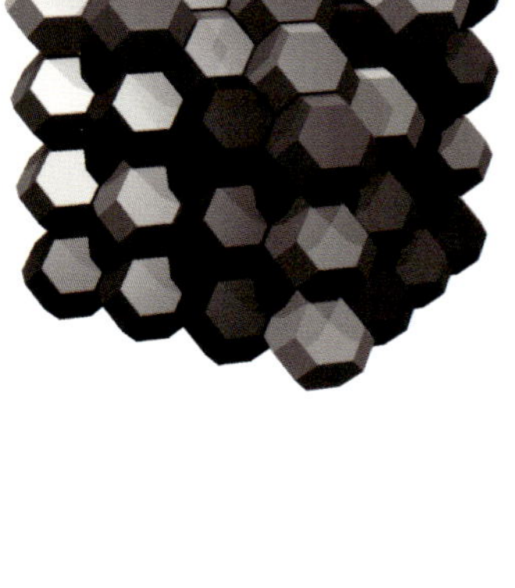

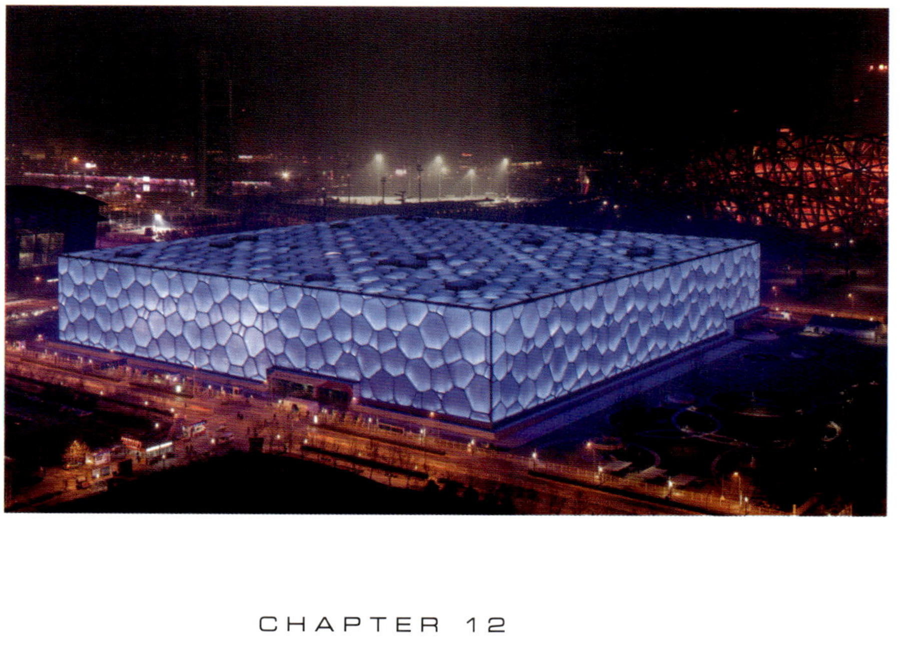

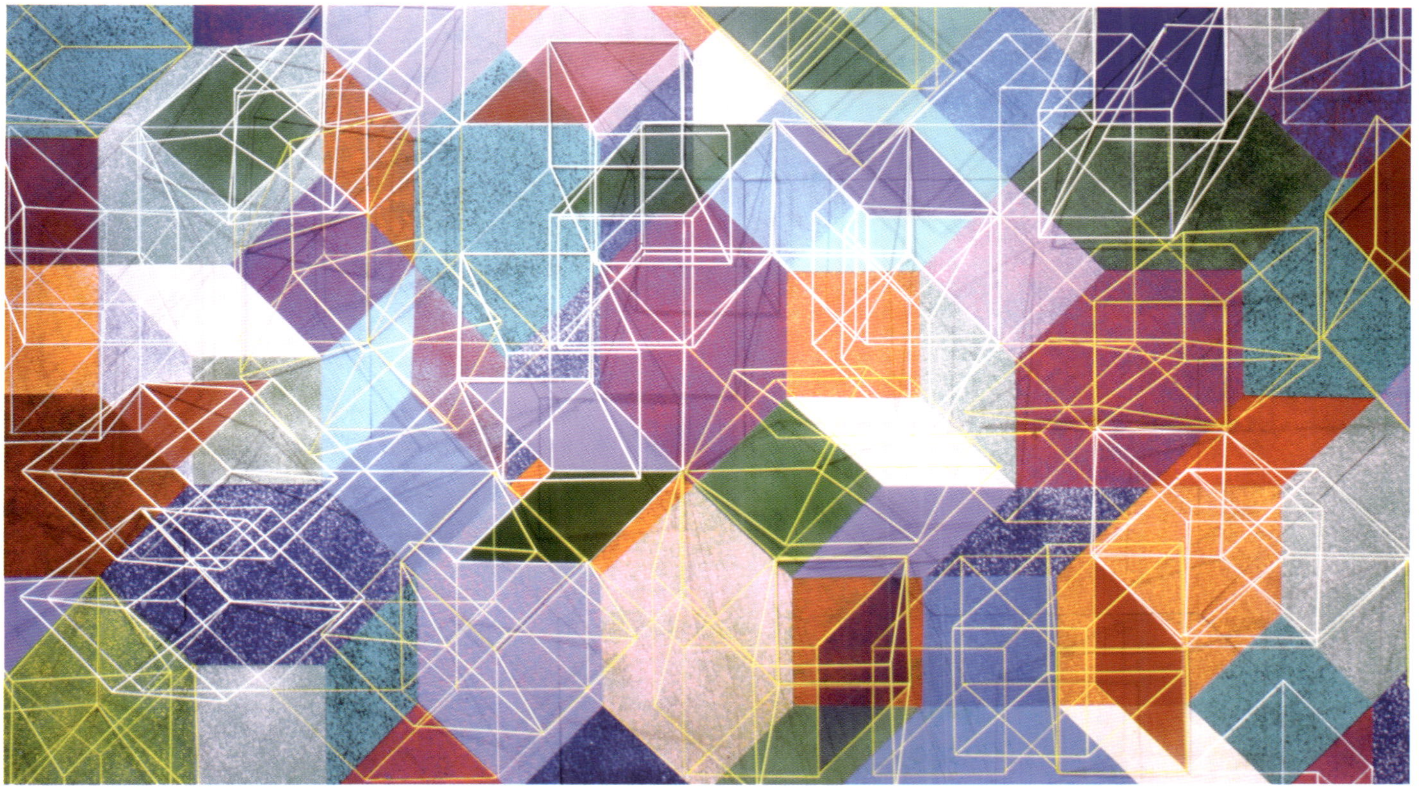

three-dimensional space (plate 12-7), and so on for higher dimensions. In *Flatland* (1884), Edwin A. Abbott envisioned fanciful encounters with beings that have higher dimensions (plate 4-2 in chapter 4), and today Abbott's fairy tale has become a reality using computer visualization, as Banchoff has observed (*Beyond the Third Dimension: Geometry, Computer Graphics, and Higher Dimensions*, 1990). Tony Robbin was a founding member of the Pattern Painting group in New York in the 1970s when he saw Banchoff's models and realized that, using computer graphics, he could visualize four-dimensional objects. Since then Robbin has painted cubes and other solids in four-dimensional space, but without the occult associations of his early-twentieth-century predecessors (plate 12-8).

But by far the most dramatic and influential use of the computer as a tool to visualize the geometry of space began in the 1970s, when a mathematician working at the computer corporation IBM in upstate New York developed fractal geometry (see "Fractal Geometry" on page 482 in this chapter).

ABOVE

12-8. Tony Robbin (American, b. 1943), *Fourfield* (detail), 1980–81. Oil on canvas with wire, 8 ft. 5 in. × 28 ft. × 15 in. (2.56 × 8.9 m × 38 cm). Courtesy of the artist.

To visualize an object that exists in a higher spatial dimension, it is common to imagine it as casting a "shadow" on a surface in a lower dimension. For example, in linear perspective (see the sidebar on pages 78–79 in chapter 2) the projection of a three-dimensional cube onto a two-dimensional picture plane can be thought of as the cube's shadow. Robbin's aim in *Fourfield* (of which this image is a detail) is to evoke the appearance of four-dimensional cubes (so-called hypercubes) for a viewer who exists in three-dimensional space. The artist first painted two-dimensional planes in colorful patterns, and then over them mounted three-dimensional structures made from colored rods; the form of both the planes and the transparent volumes are derived from hypercubes. The artist's goal is to give the viewer a glimpse into hyperspace in the following way; as one walks back and forth in front of this 28-ft.-wide mural, one looks through a maze of colorful rods at a constantly changing pattern of colored planes, volumes, and shadows.

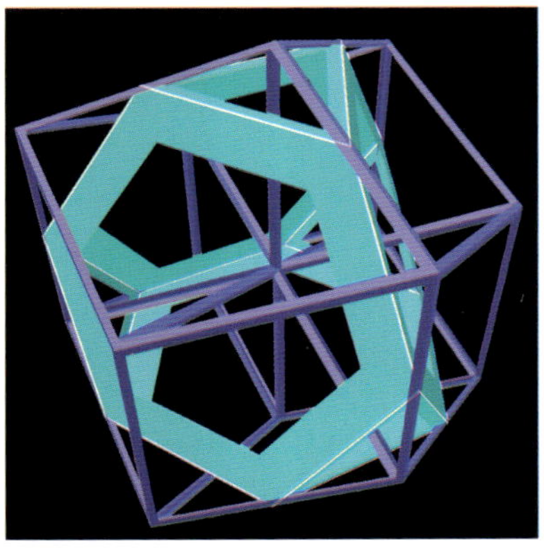

LEFT

12-7. Davide Cervone (American, b. 1962), *Hypercube*, 2002. Still from an interactive version (2002) of the computer animation *Hypercube: Projections and Slicing* (1978), by Thomas Banchoff and Charles M. Strauss.

Davide Cervone, a collaborator of Thomas Banchoff, made this image as part of a 2002 interactive version of the 1978 film *Hypercube: Projections and Slicing*, which Banchoff made in cooperation with the computer scientist Charles M. Strauss to illustrate a four-dimensional cube moving through three-dimensional space.

KNOTS

Knots have been used throughout history for practical purposes and to form decorative patterns (plates 12-9, 12-10, 12-11, 12-12, and 12-13). Nineteenth-century mathematicians began to classify knots by counting the number of times the strands cross over, then in the first decade of the twentieth century knots came to be studied within topology (see chapter 6). As L.E.J. Brouwer showed, planes can have any number of dimensions, and so topologists studied multidimensional knots. Early-twentieth-century mathematicians also began to use group theory to study tessellation, such as Islamic tiling patterns, many of which include knots (see plate 7-9 in chapter 7). Mathematicians studied knots as closed loops within three-dimensional space (plate 12-15), with the goal of isolating features of a knot that remain invariant as it is transformed from one knot to another.

The topology of knots has recently found applications in physics, where the strands are strings of energy, each hypothesized to be vibrating on a plane in multidimensional

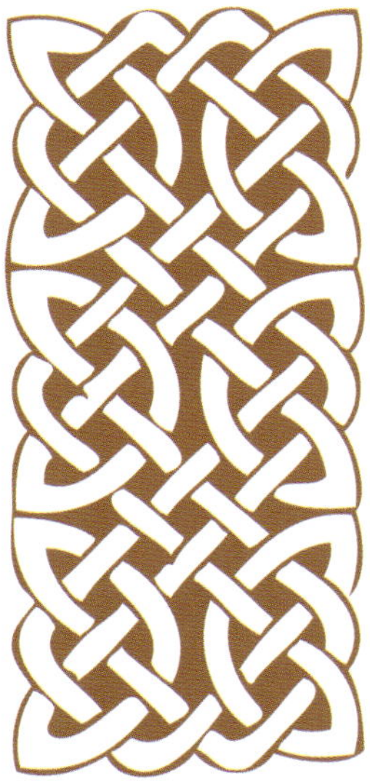

ABOVE

12-9. A Celtic knot is a single strand in a closed loop—a so-called endless knot—which is the type most studied by mathematicians (see plate 12-15). The strand of an endless knot follows an interlaced path, passing alternatively above and below itself, until returning to its starting point and beginning again. This figure is composed of two endless knots interwoven with each other.

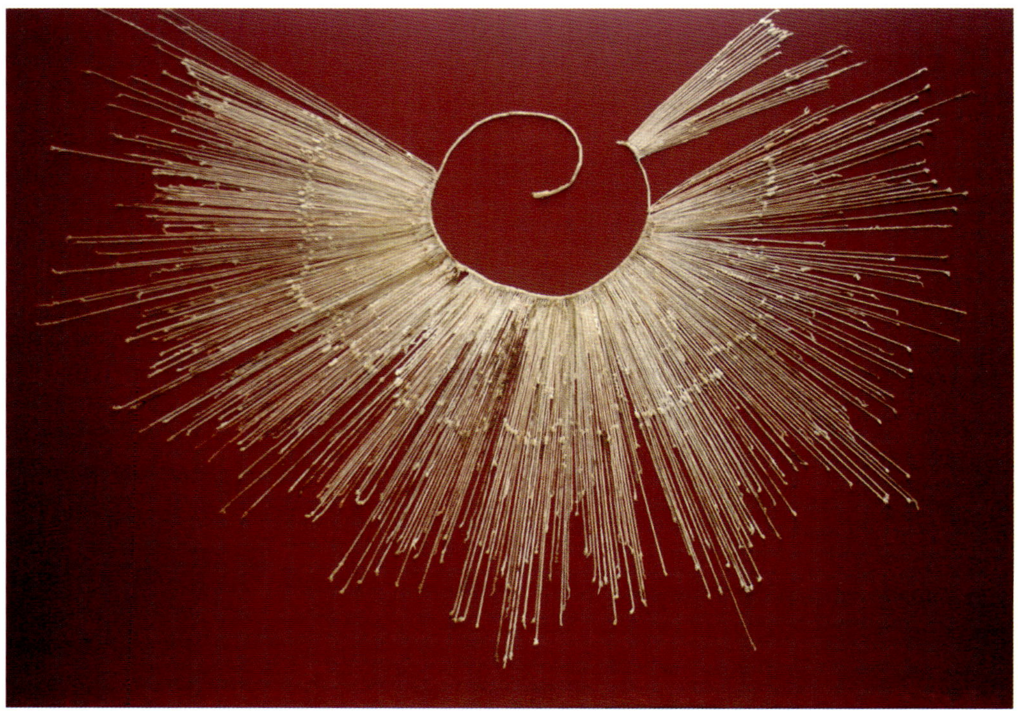

RIGHT

12-10. *Quipu*, Inca era, ca. AD 1400–1532, found at the Laguna de los Cóndores, Peru. Cotton cord, pendant strings average about 15 in. (38 cm) long. Centro Mallqui, Museo Leymebamba, Peru.

The Inca gave knots numerical-logical meaning by recording information in knotted ropes such as this *quipu* (Quechua for "knot"). Runners carried numerical messages encoded in *quipu* throughout the Inca Empire, which stretched from Ecuador to Chile, with its heart at Cusco in the Andes of southern Peru. This *quipu* is shown with its primary cord in a curve and its knotted pendants radiating outward; it was carried by runner with the primary cord rolled into a spiral. The Inca's decimal system of counting and their method of symbolizing large numbers has been codified by historians of mathematics Marcia Ascher and Robert Ascher in *Mathematics of the Incas: Code of the Quipu* (Ann Arbor: University of Michigan Press, 1981), and their accounting methods have been explored by anthropologists Gary Urton and Carrie J. Brezine in "Khipu Accounting in Ancient Peru," *Science* 309 (Aug. 12, 2005): 1065–7.

OPPOSITE

12-11. Chi-rho page, *Book of Kells*, ca. AD 800. Illuminated manuscript on vellum. The Board of Trinity College, Dublin, MS 58 fol. 34r.

Created by Celtic monks around AD 800, this manuscript was in the Abbey of Kells north of Dublin during the late medieval and early modern era. The book contains the four Gospels of the New Testament, written by the apostles Matthew, Mark, Luke, and John. Matthew began his account by stating the genealogy of Christ, after which this page occurs to mark the beginning of his narrative of the life of Christ (Matthew 1:18). Christ's name in Greek—ΧΡΙΣΤΟΣ—is abbreviated here with the letters X (chi) and *P* (rho)—XP. The large illuminated *P* dominates the page, with the small X in the lower right, both lavishly decorated with knot work.

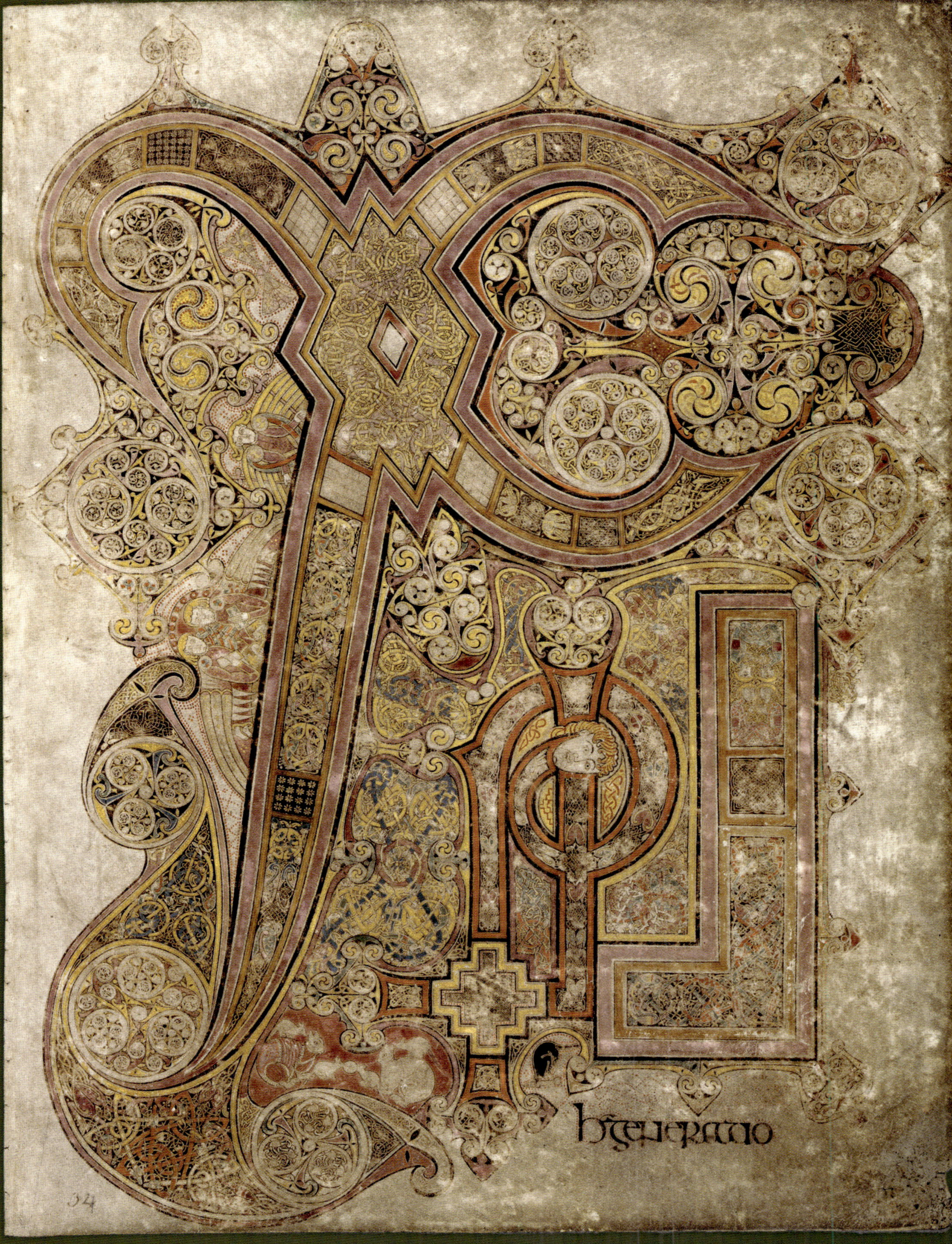
h generatio

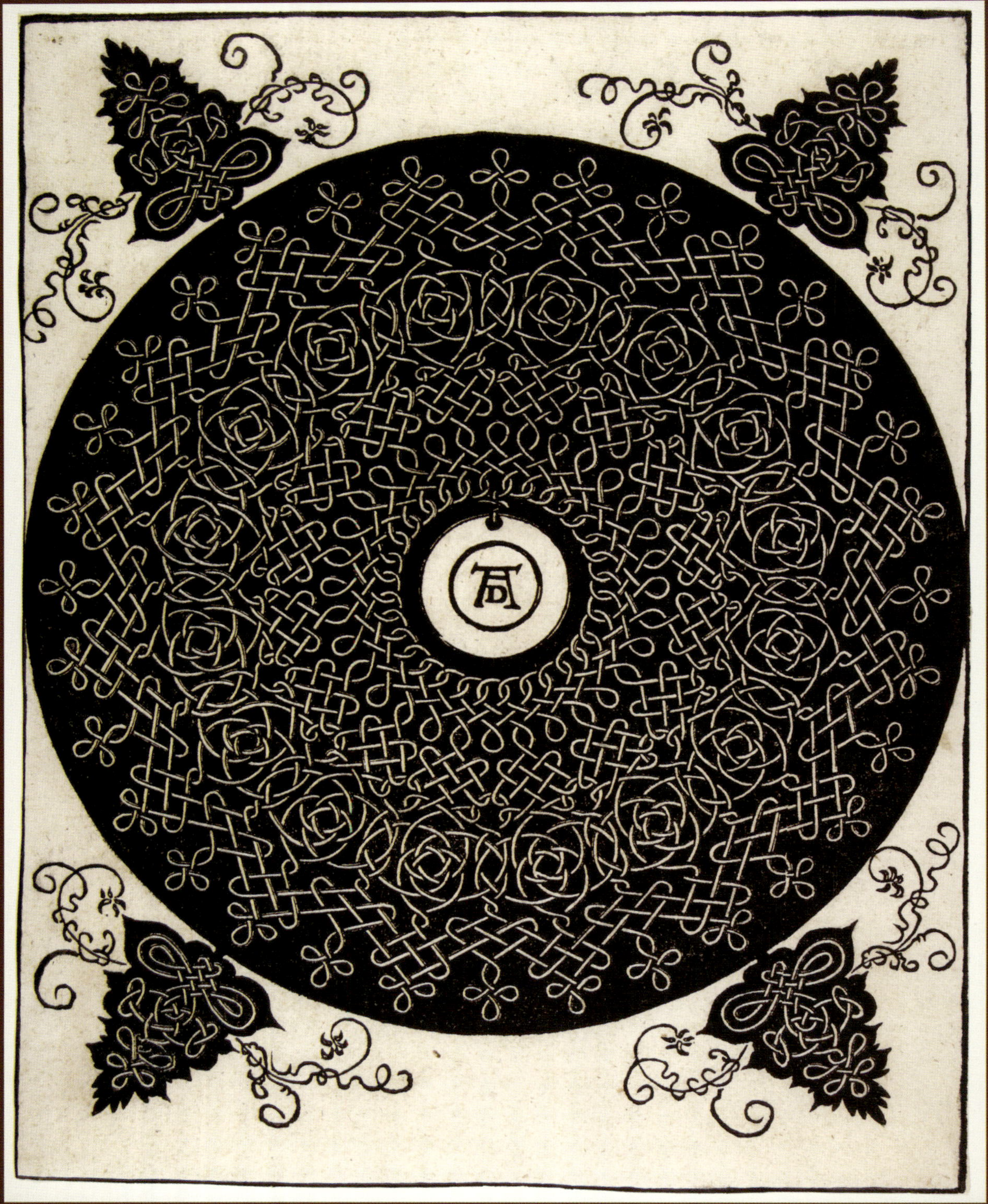

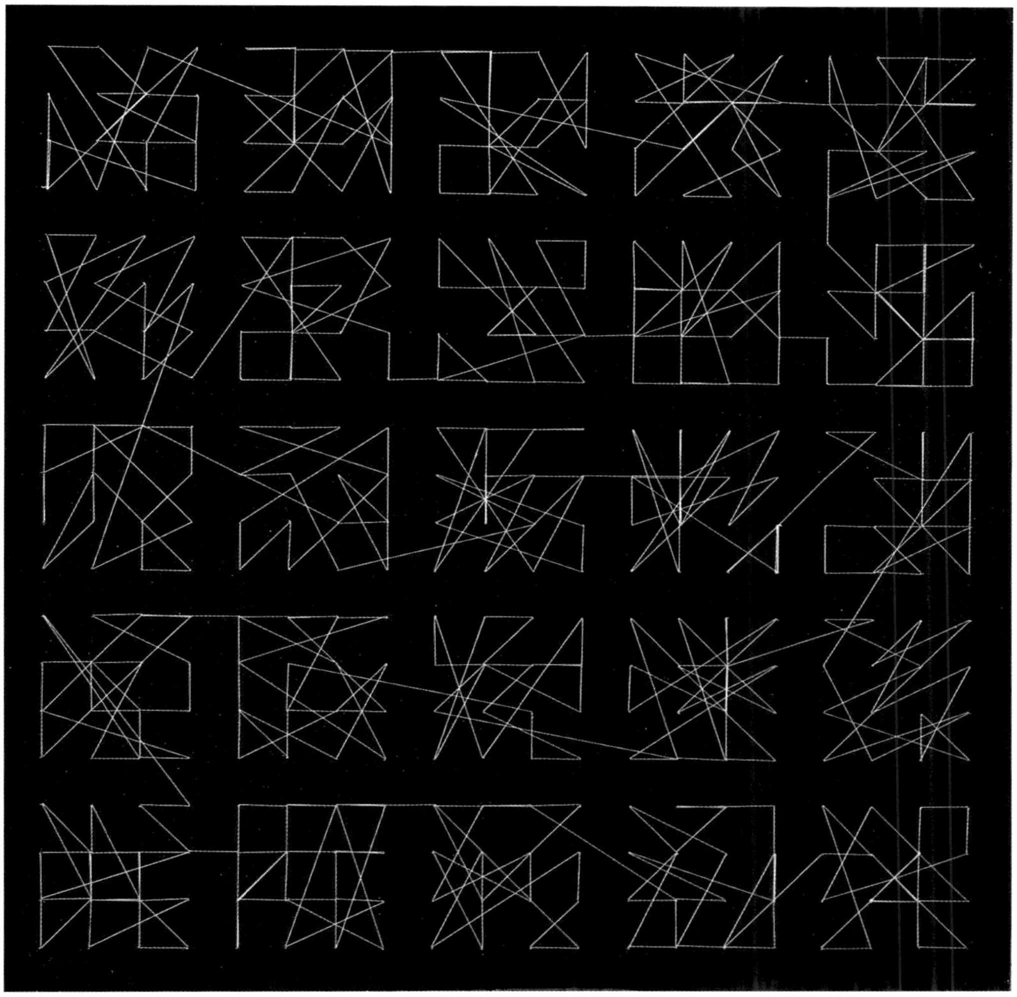

OPPOSITE

12-12. Albrecht Dürer (German, 1471–1528), *Embroidery Pattern with Round Medallion in Its Center; Third Knot* from the series *Six Knots*, n.d. Woodcut, 10¾ × 8 7/16 in. (27.3 × 21.4 cm). Metropolitan Museum of Art, New York, George Khuner Collection, gift of Mrs. George Khuner, 1975.

During the Renaissance, German printers achieved a detailed illustration such as this by engraving it in metal. Only Dürer dared cut such an intricate pattern in relief in wood. His image is a copy of an anonymous engraving of a drawing by Leonardo da Vinci, which is lost.

LEFT

12-13. Vera Molnar (Hungarian-born French, b. 1924), *Homage to Dürer: 400 Needles Crossed by a Thread*, 1989/2004. Acrylic on untreated cotton cloth, cut needles, approximately 131 ft. (40 m) thread (knotted once), 33⅛ × 33⅛ in. (84 × 84 cm). Digital Art Museum, Berlin.

Vera Molnar emigrated from Hungary to Paris, where in 1968 she became one of the first artists to experiment with a computer. Always drawn to mathematics, Molnar composed this pattern of 400 needles on a large 5 × 5 grid made from twenty-five smaller 4 × 4 grids, each with sixteen needles. After piercing the fabric with the needles, she threaded a single strand through them and "crossed" the thread, making it into a knot in homage to Albrecht Dürer's *Six Knots*.

space. Knot theory also has applications in the study of proteins, each of which is composed of a unique chain of amino acids (of which there are twenty). Once the linear chain of amino acids is complete, the protein assumes its final three-dimensional form in milliseconds as the chain of acids ties itself into a knot (plate 12-14) or folds like origami (see the following). When biologists began studying proteins in the early twentieth century, they found it impractical to analyze their structure calculating with paper and pencil because as each of the twenty different amino acids is added to the chain, the possible form of the protein increases dramatically. For example, a protein that is a chain of five amino acids can twist and fold into an astronomical 20^5 (3,200,000) possible structures.[6] (A typical protein is a chain of not five but hundreds; the blood protein hemoglobin is a chain of 574 amino acids). The invention of the computer put the modeling of proteins within reach, and today interdisciplinary teams of mathematicians, computer scientists, and molecular biologists work together to unlock the door to this central mystery of modern biology—protein folding.

TOP RIGHT

12-14. Julian Voss-Andreae (German, b. 1970), *Cycloviolacin*, 2007. Powder-coated steel maquette, 30 × 34 × 24 in. (76 × 86 × 61 cm). Courtesy of the artist.

To make this violet-colored sculpture of a protein, the German artist Julian Voss-Andreae symbolized amino acids as bars with lozenge-shaped holes. The chain of amino acids has crossed over itself and formed a knot, which is the final form of the protein cycloviolacin. This protein was discovered in a violet native to Australia in 2006 by David Craik, a biologist at the University of Queensland Brisbane who searches the jungles and forests of his homeland for proteins that might have a practical use for humans, such as fighting a disease; David Craik, B. Chen, M. L. Colgrave, and C. N. Wang, "Cycloviolacin H4, a Hydrophobic Cyclotide from *Viola hederaceae*," *Journal of Natural Products* 69, no. 1 (2006): 23–28.

BOTTOM RIGHT

12-15. Knots as closed loops within three-dimensional space.

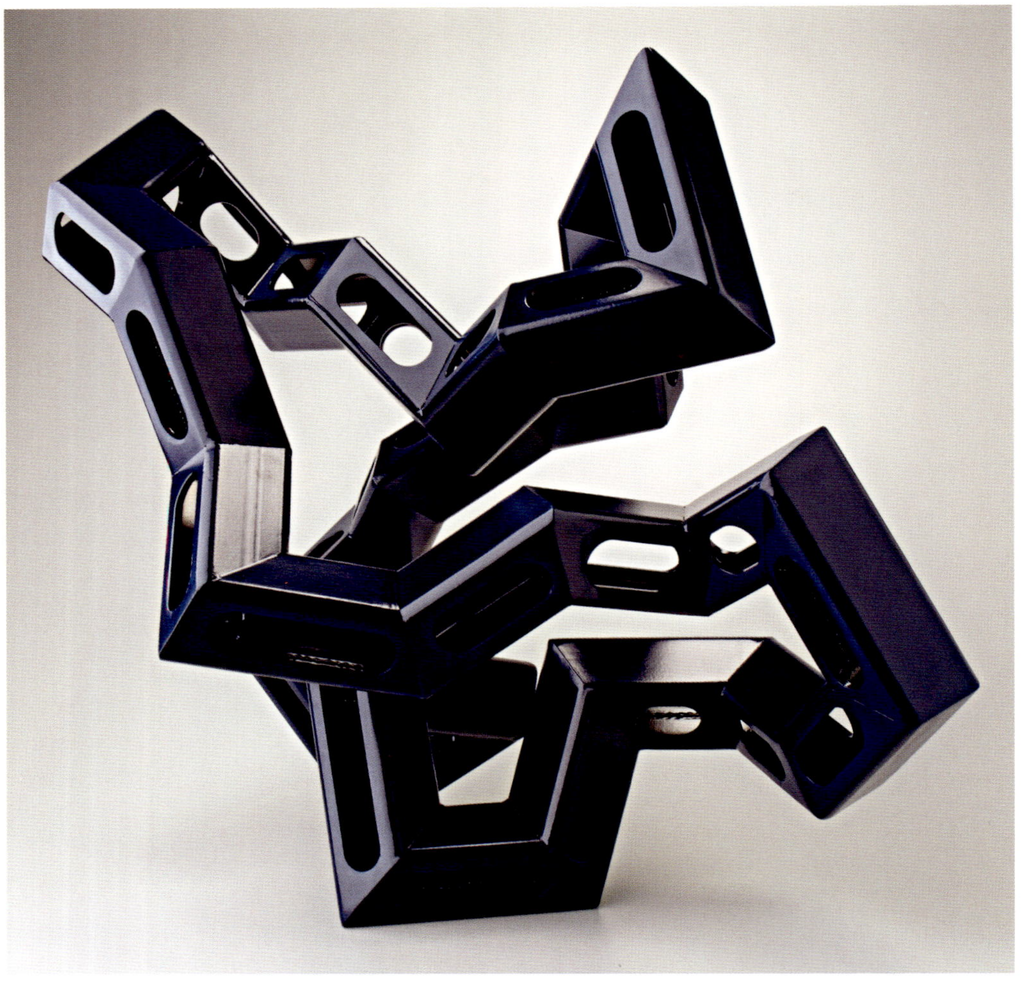

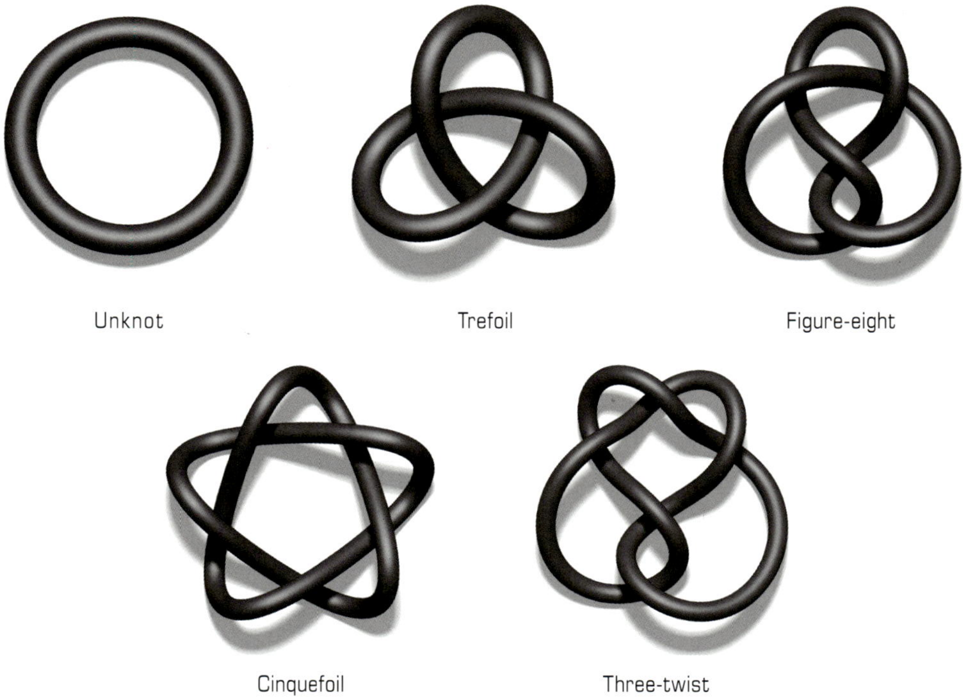

NETWORKS

The study of networks began with a problem posed in the eighteenth century by the Swiss mathematician Leonhard Euler. The city of Königsberg in Prussia (today Kaliningrad in Russia) straddles the Pregel River, in the middle of which there are two islands. Seven bridges connect the islands to each other and to the mainland (plate 12-16). Euler asked: Is there a route through the city that crosses each bridge only once? Euler developed a technique to analyze the problem by making a graph of the network of bridges, and he proved that the answer is "No" (*The Seven Bridges of Königsberg*, 1735).

With the development of railroads in the nineteenth century, the topic of finding an optimal route for a journey was of practical interest, and it also became the theme of a parlor game. The topic entered the mathematics literature in 1930, when the Viennese mathematician Karl Menger described it as the "messenger problem" (*das Botenproblem*) of finding an optimal delivery route. It was soon dubbed the "travelling salesman's problem": given a list of cities and the distances between each pair, find the shortest route that visits each city once and returns to the city of origin (plate 12-18). The mathematics of networks developed dramatically in the late twentieth and early twenty-first centuries in response to multiple applications, including the routing of airliners and designing circuitry for microchips. In 1989 the Internet was up, and in 1996 the search engine Google was launched, creating vast virtual landscapes that need to be mapped to optimize routing (plates 12-17, 12-19, 12-20, 12-21, 12-22, and 12-23).

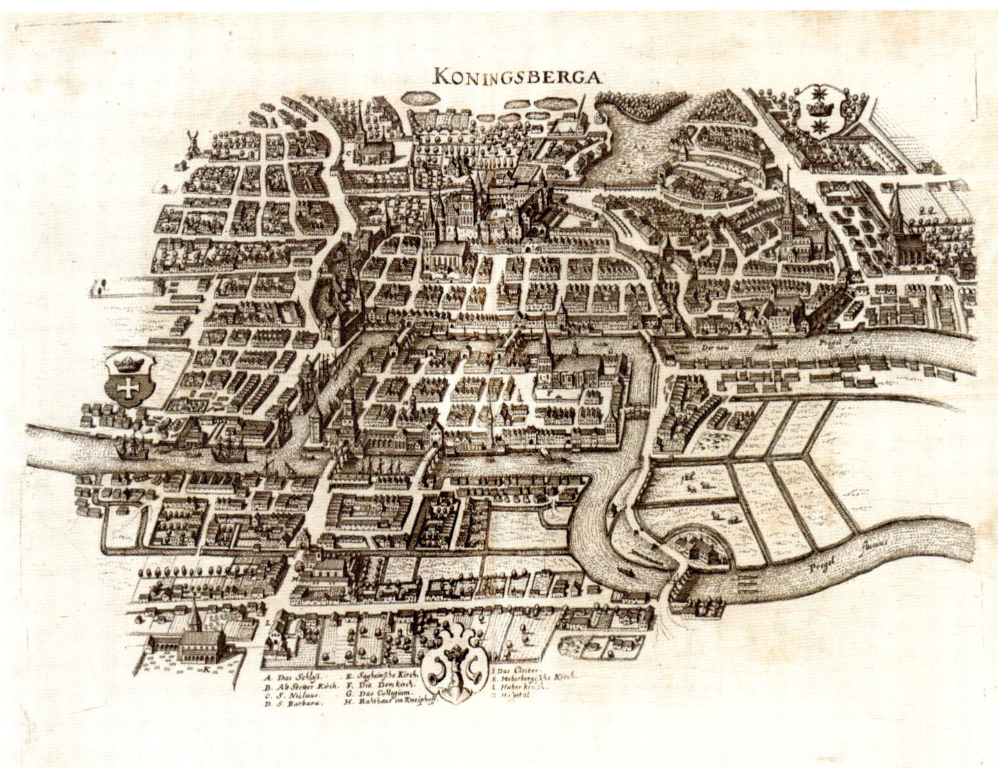

12-16. *The Seven Bridges of Königsberg*, map by Martin Zeiller (German, 1589–1661) in *Topographiae Prussiæ et Pomerelliæ* (Frankfurt am Main, Germany: M. Merian, 1649), between 30 and 31. Rare Books Division, The New York Public Library, Astor, Lenox, and Tilden Foundations.

TOP RIGHT

12-17. Gerard Caris, (Dutch, b. 1925), *Polyhedral Net Structure #2*, 1972. Steel wire, cardboard, linen, acrylic, 59¾ × 59¼ × 43⁵⁄₁₆ in. (152 × 150 × 110 cm). Zentrum für Kunst und Medientechnologie, Karlsruhe. © 2014 Pictoright Amsterdam/Artists Rights Society, New York.

BOTTOM RIGHT

12-18. Robert Bosch (American, b. 1963), *Knot?* 2006. Digital print, 34 × 34 in. (86.3 × 86.3 cm). Courtesy of the artist.

The American mathematician Robert Bosch drew this continuous line based on the solution to a 5000-city instance of the travelling salesman problem. From a distance, the print appears to depict a black cord against a grey background in the form of a Celtic knot (see plate 12-9). But on close inspection the apparent "grey" is actually a continuous white line moving on top of a black background. The white line never crosses over itself—it is a network rather than a knot—and so the answer to the title is "Not."

OPPOSITE

12-19. Naheed Raza (British, b. 1980), *Mile of String*, 2009. String, ca. 20 in. (50 cm) diam. Courtesy of the artist.

Naheed Raza created this sculpture while artist-in-residence in the mathematics department of University College London, where she collaborated with Steven Bishop, whose research is on complex systems and chaos theory. According to Raza, "*Mile of String* is a three-dimensional drawing made from a single line of string one mile long, manipulated to such an extent that it holds itself in a complex fractal-like form entirely under its own tension. Although able to trace the outer contour, the majority of the form lies hidden within cavernous spaces inaccessible to the fingertips. It evokes our desire of touch through its highly tactile nature while simultaneously denying our ability to grasp it in its entirety because of the seemingly infinite detail. It is reminiscent of spatial chaos, Einstein's warped space-time and the enigmatic structures of DNA and proteins which coil and supercoil in ways that elude our contemporary understanding."

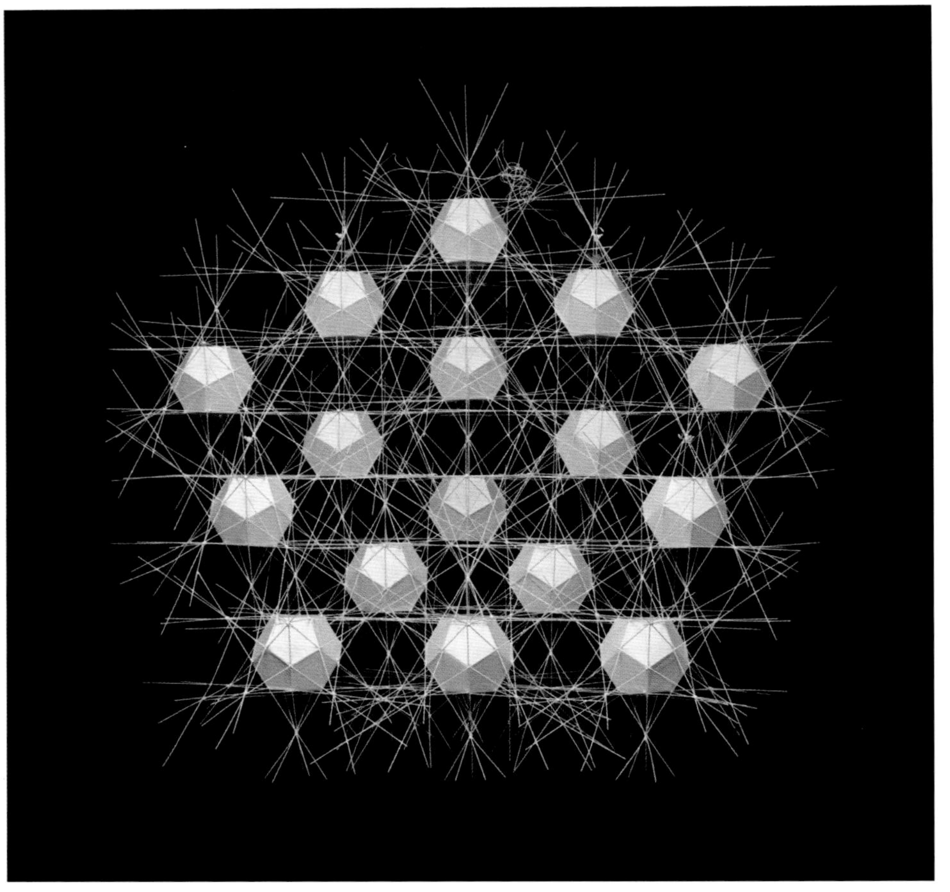

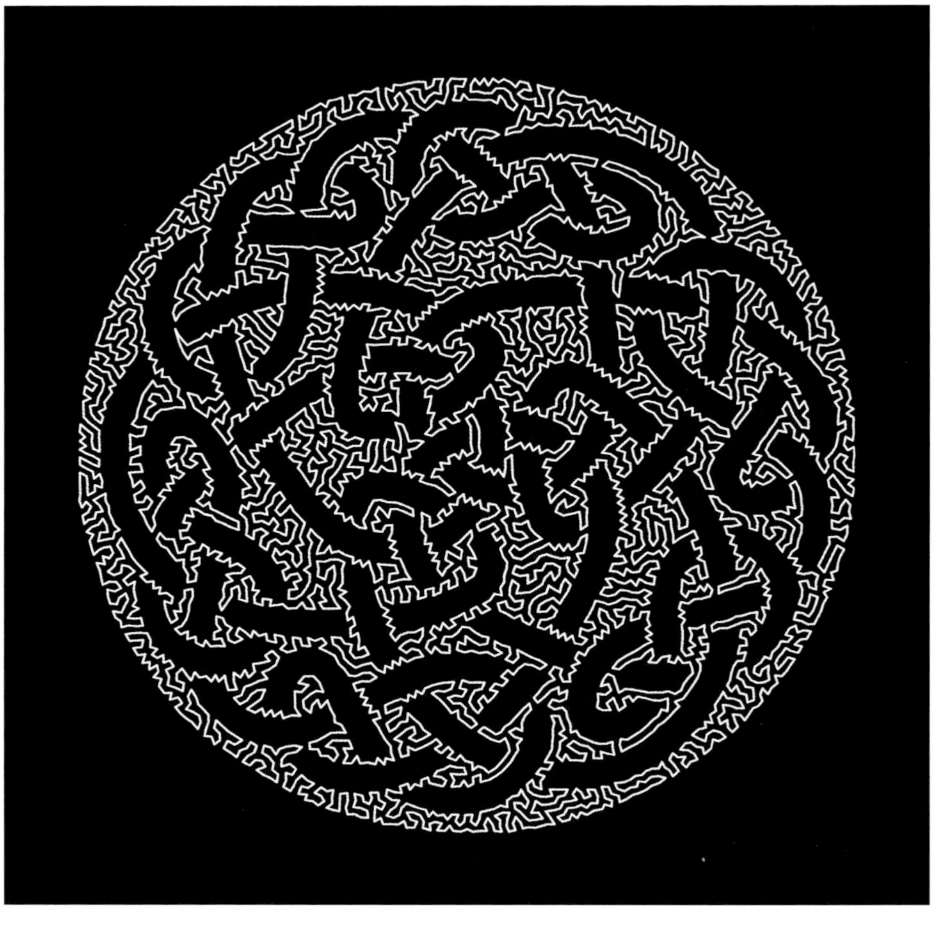

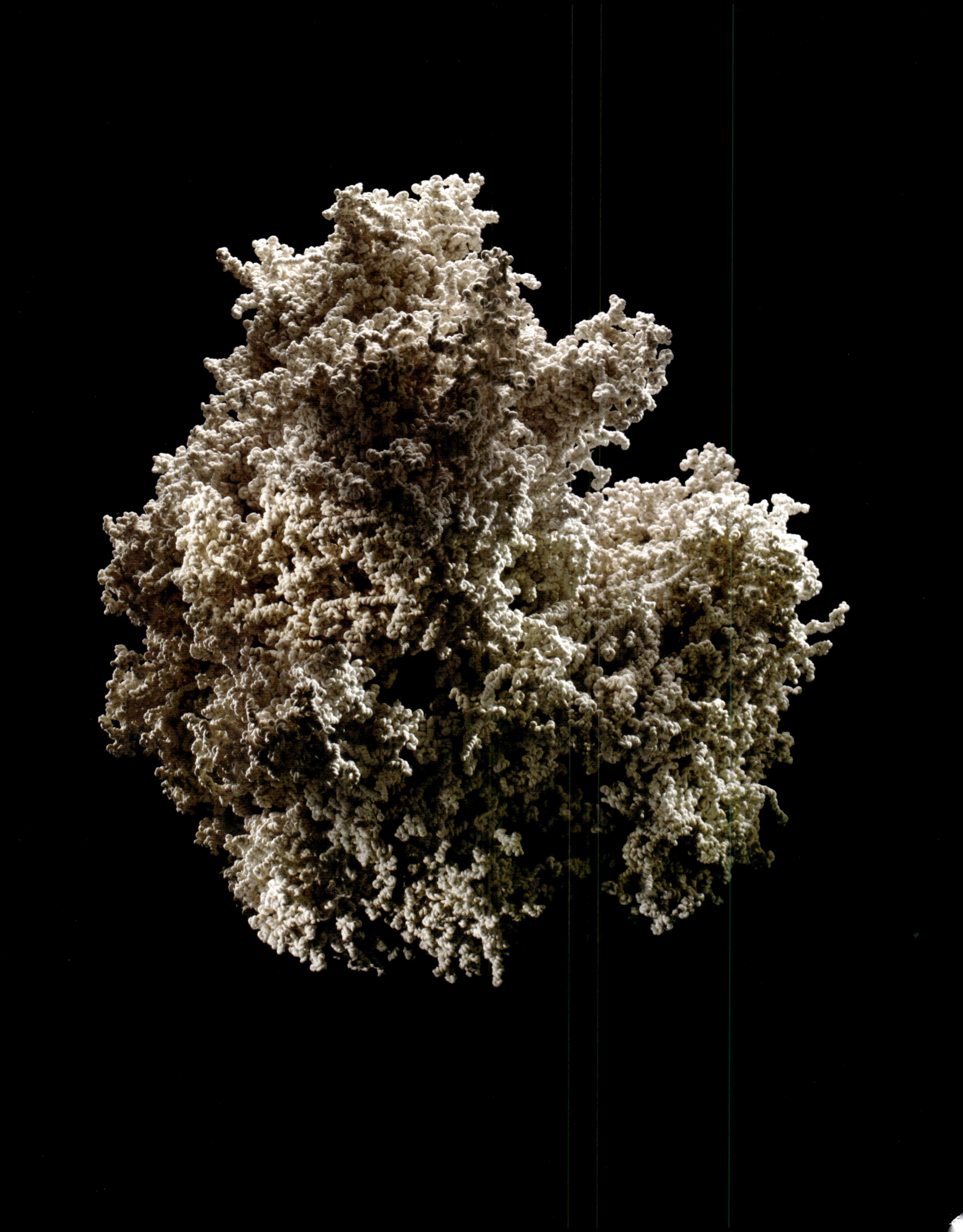

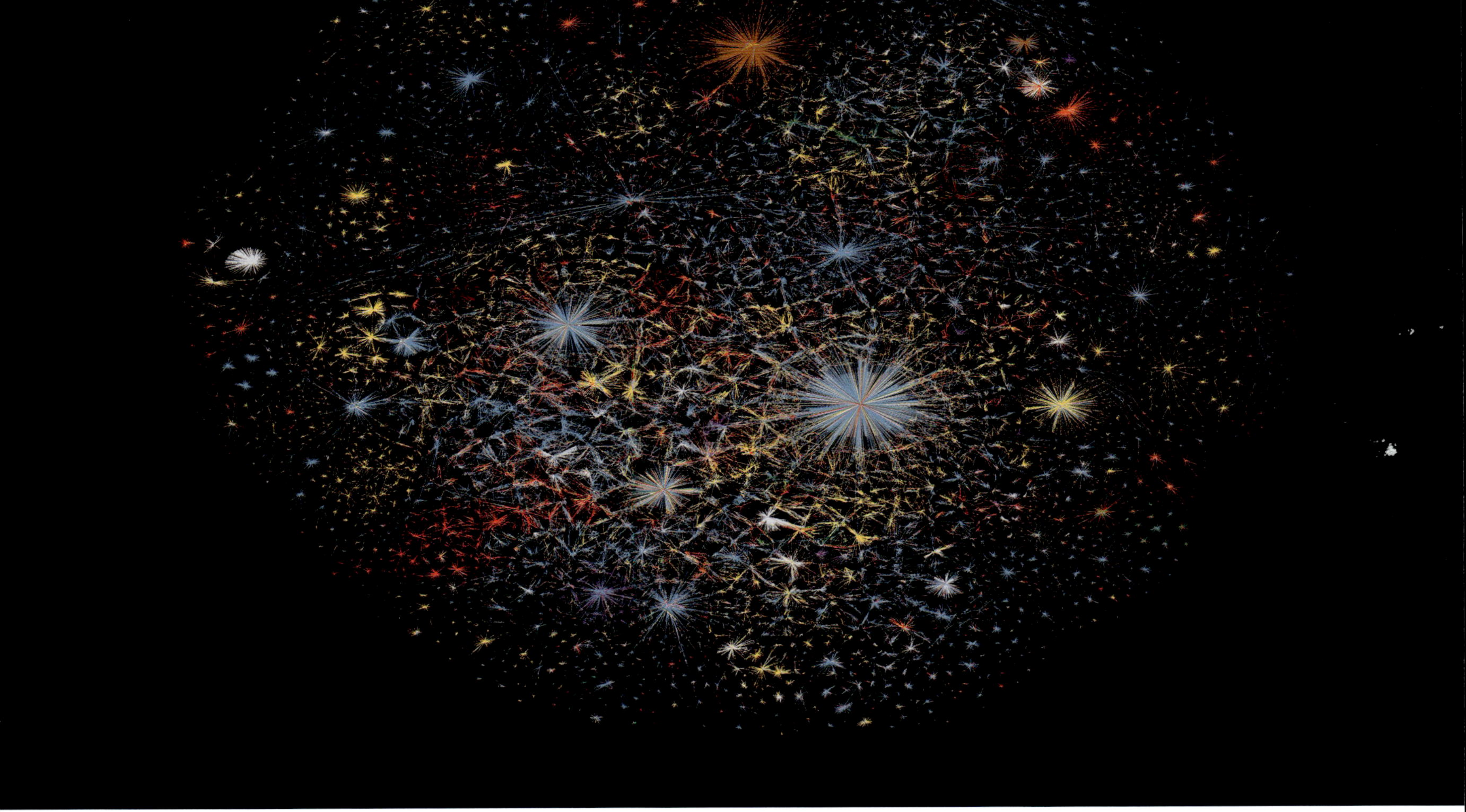

ABOVE

12-20. Map of the Internet, 2012. Lumeta, Somerset, New Jersey.

This map is the result of a long-term project devoted to collecting routing information and using it to generate tree-like paths on the Internet. The image depicts the shortest outgoing routes from a test computer at Lumeta, a company that maps the Internet, in Somerset, New Jersey, to each of more than 450,000 registered or announced nets on the Internet. Each node can represent a handful of computers on a small network or a large company with thousands of hosts. Each intermediate node is a router. This map shows only the shortest path to each destination, not every path.

BELOW

12-21. Charles M. Schultz, *Peanuts*, © 1963. Peanuts Worldwide LLC. Dist. by Universal Uclick. Used with permission. All rights reserved.

A googol is 10^{100} (1 followed by 100 zeros). It has no particular mathematical interest other than being a very large, but finite, number. The name of the Internet search engine Google originated as a misspelling of "googol." The mission of the company is "to organize the world's information and make it universally accessible and useful." In other words, Google organizes a very large, but finite, amount of information.

Charles Schultz surely meant for Schroeder to say that the chances of him and Lucy getting married were one to a googol, but he got his words backwards. If the odds of something happening are a googol to one, then for all practical purposes it's a certainty. If only Lucy had been on her toes about ratios, she could have seized the moment.

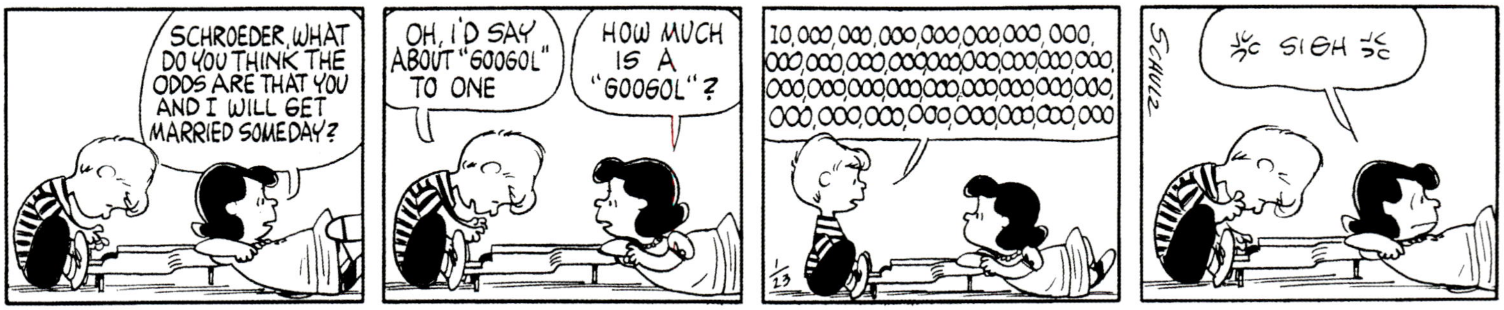

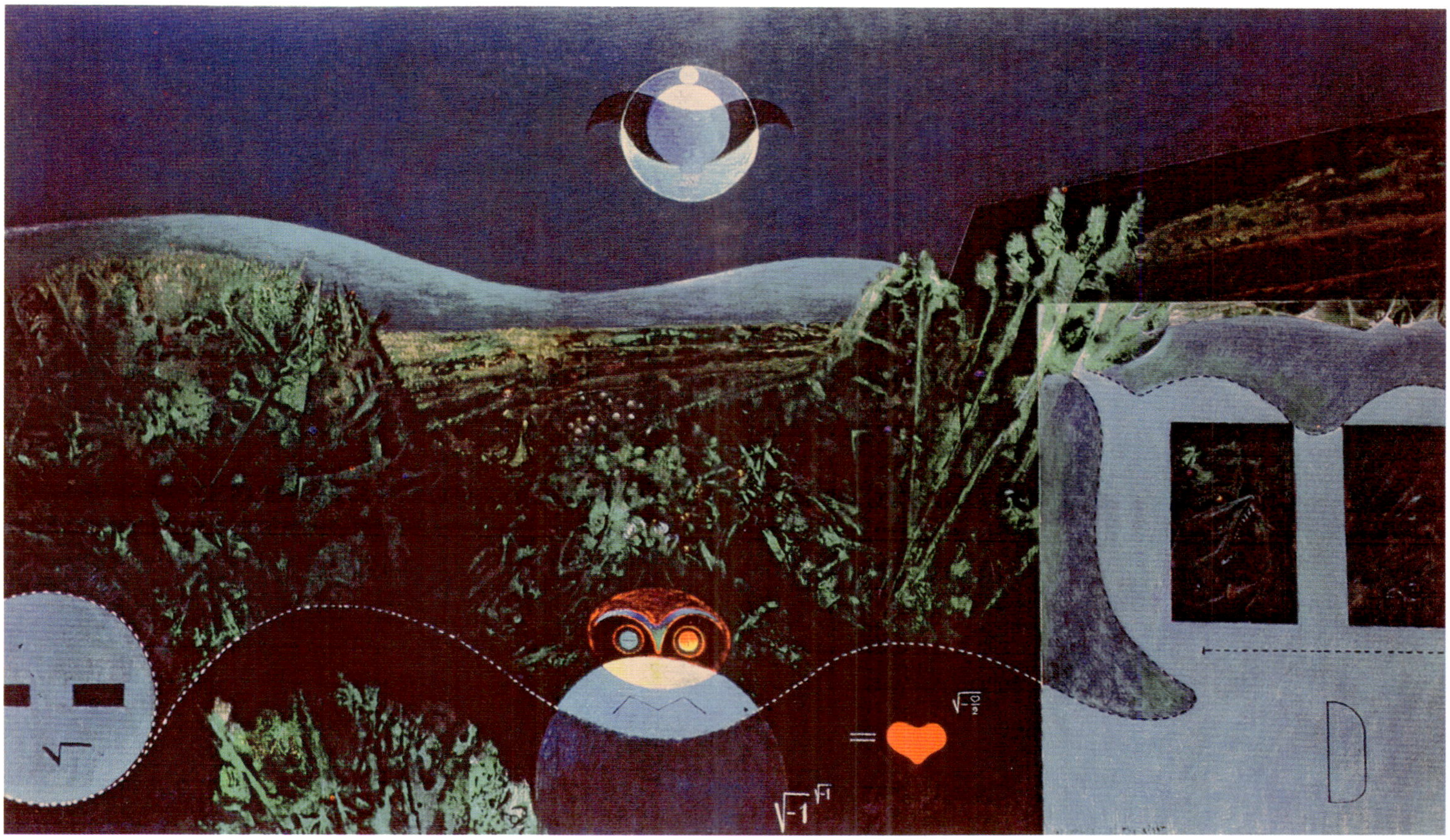

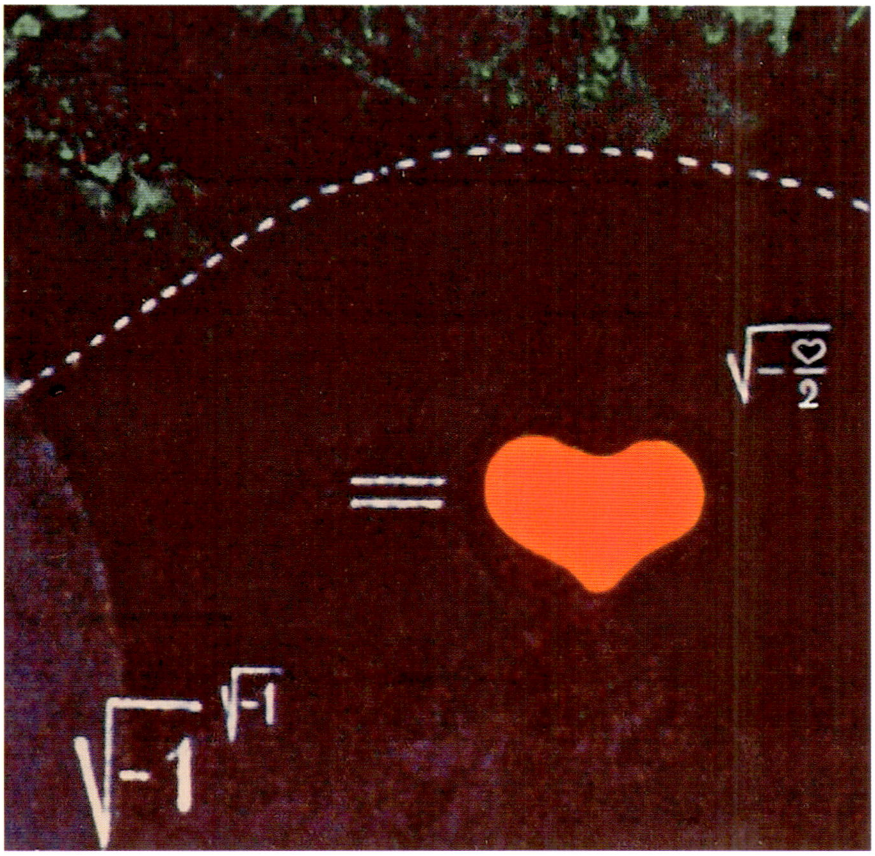

ABOVE AND LEFT (DETAIL)

12-36. Max Ernst (German, 1891–1976), *Phases of the Night*, 1946. Oil on canvas, 36 × 64 in. (91.5 × 162.5 cm). © 2014 ADAGP, Paris/Artists Rights Society, New York.

In 1946 Max Ernst married the American artist Dorothea Tanning, and the couple settled near Flagstaff, Arizona. It was there that Ernst did this painting, in which he equates romance (the red heart) with an imaginary number, the square root of –1, which Ernst wrote as $\sqrt{-1}$. Seen by moonlight in the desert under the gaze of owl-like creatures, the imaginary realm is multiplied—the imaginary number is raised to the power of an imaginary number—and, to balance the equation, love also soars—the heart is raised to a power whose terms are love (the outlined heart) and a "couple" (the 2). The title may allude to the phase of a wave or other rhythmic oscillation, since, as a groom, Ernst would want the "phases of the night" to be in sync.

FOLLOWING SPREAD, LEFT

12-37. Jean Claude Meynard (French, b. 1951), *Excès* (Excess), 2001. Digital pattern on Plexiglas, 47¼ × 47¼ in. (120 × 120 cm). Courtesy of the artist.

FOLLOWING SPREAD, RIGHT

12-38. Sylvie Donmoyer (French, b. 1959), *Reflections on a Menger Sponge*, ca. 2010. Digital print, 12 × 12 in. (30.4 × 30.4 cm). Courtesy of the artist.

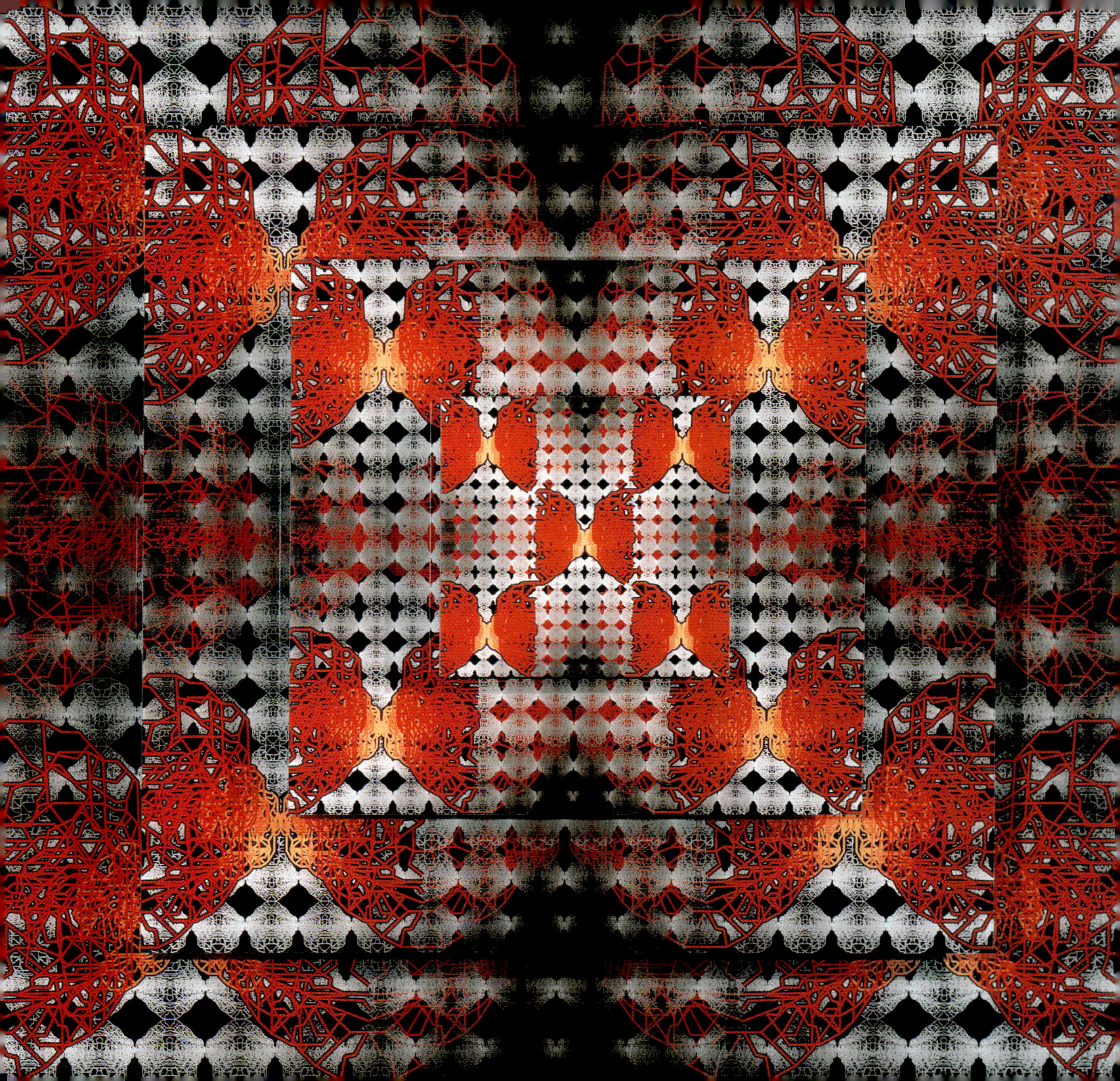

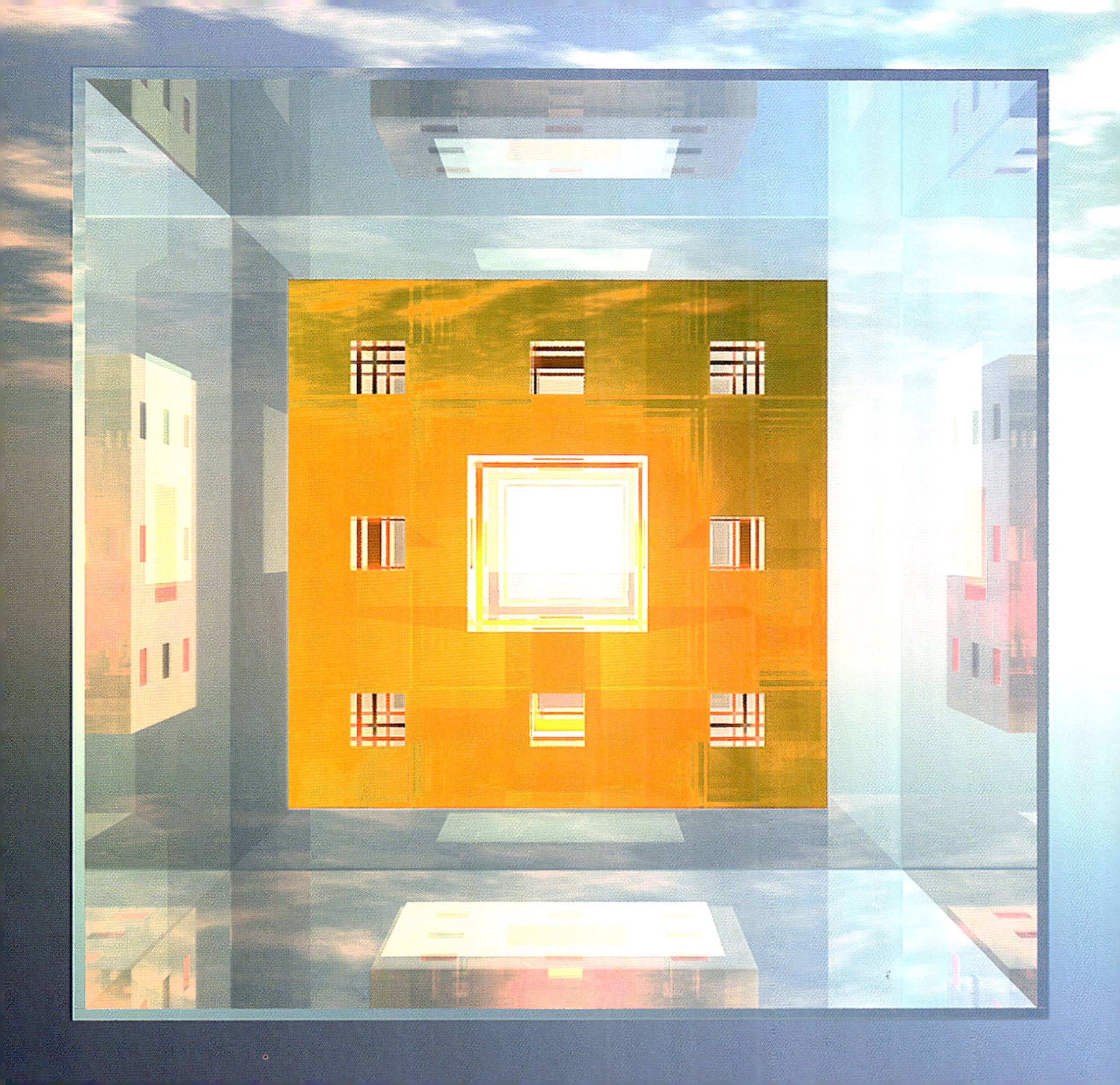

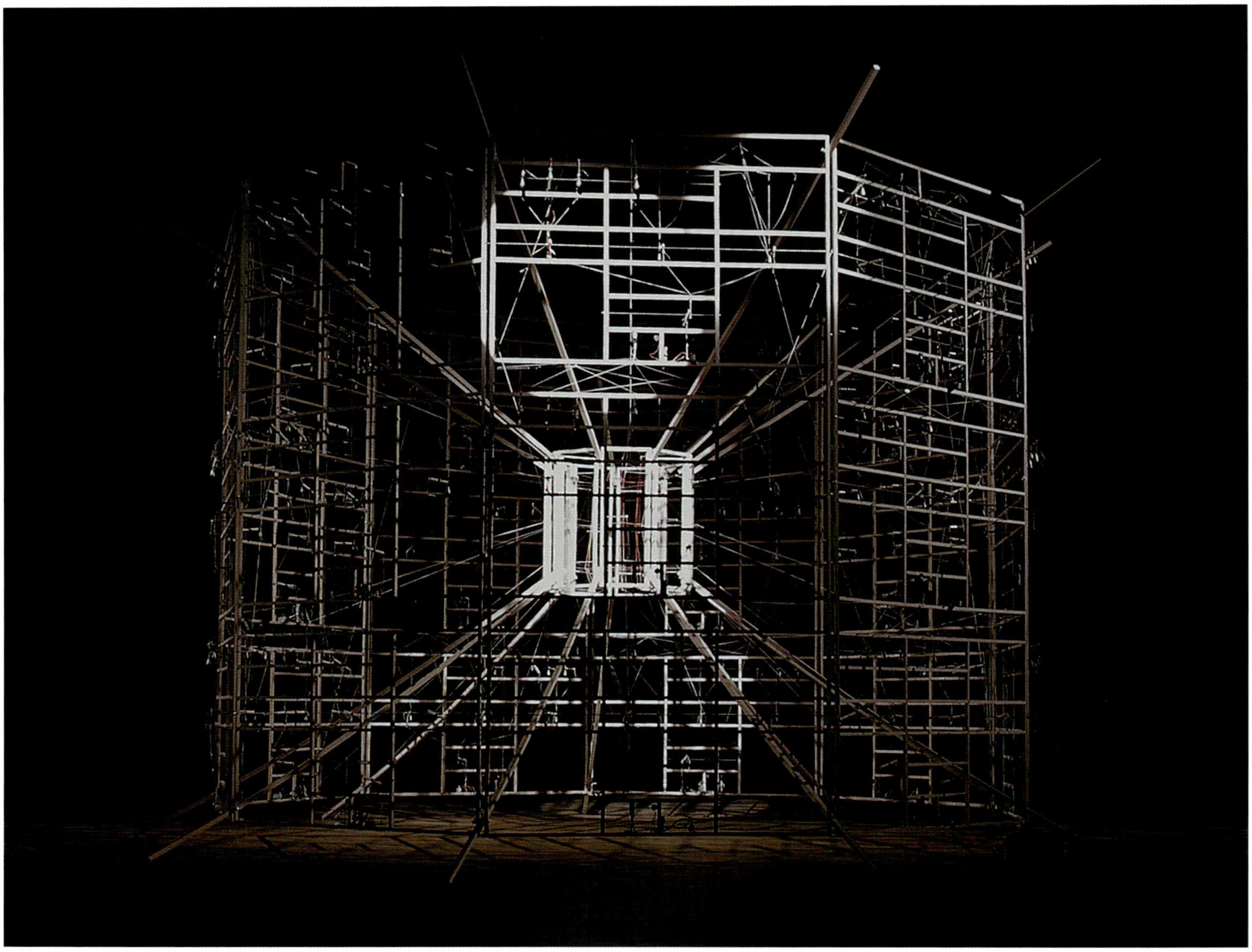

12-39. Ralf Baecker (German, b. 1977), *Rechnender Raum* (Calculating space), 2007. Strings, beechwood, elastic band, aluminum parts, custom electronics, servos, switches. Courtesy of the artist. © Ralf Baecker.

After studying computer science, Ralf Baecker completed his education at the Kunsthochschule für Medien in Cologne. The artist built this large kinetic sculpture as a fully operational computer made of sticks, wire, and metal plumbs. Baecker's *Rechnender Raum*, named in of honor of Konrad Zuse's classic text, carries out its computations on its perimeter where the moving sticks and pulleys can be seen, and then sends its calculations to the inner cube, which is inaccessible to the viewer. Although the structure appears to reveal its workings, the *Rechnender Raum* guards its calculations as a secret within.

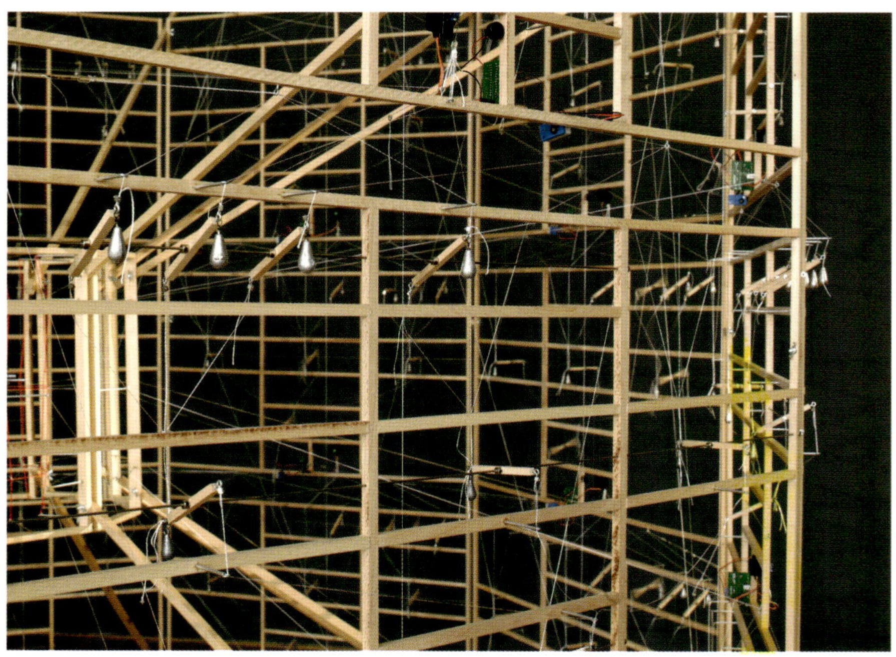

By the 1960s computer scientists had confirmed that certain simple recursive rules (the ones that do not preserve self-similarity) can lead to very complex (chaotic) behavior in a system and that such dynamic systems are very sensitive to their initial conditions, as Julia had suspected. Although the terms "complex" and "chaotic" are used interchangeably to designate recursive systems that lack self-similarity, the outcome of repeatedly applying such an algorithm is completely deterministic. The word "chaotic" refers to the *apparent* randomness of the (deterministic) pattern.

The German Konrad Zuse (inventor of the Graphomat; see chapter 10) and the American Edward Fredkin suggested in the 1960s that if space is composed of discrete points and time ticks off in seconds, then, in theory, the whole universe could be modeled as one big recursive system.[11] Zuse described the cosmos as a three-dimensional lattice of points (*Rechnender Raum* [Calculating space], 1969), which inspired the contemporary German artist Ralf Baecker (*Rechnender Raum* [Calculating space], 2007; see plate 12-39).

To the British mathematician John Horton Conway, recursive algorithms suggested organic processes such as cell division. In English the word "cell" refers to both a *living cell* that is the basic unit of organisms and a *computer cell* that is the basic unit of a cellular automaton. Exploiting this coincidence, Conway invented a computer game with the punny name *The Game of Life* (1970). (A cellular automaton is a computer system consisting of a grid of cells that change according to a set of rules that determine the new state of each individual cell as a function of its current state and the current state of its neighboring cells.) The player sets the initial conditions of the cellular automaton and writes simple "transition rules" that govern the passage of information about the computer cell color (either black or white) from a cell to its neighbors. With each tick of the clock, the cells follow the rules and turn black or white, looking uncannily like living microorganisms seen through a microscope. Even the most simple transition rules create patterns that are impossible to predict (plate 12-40), inspiring the American artist Leo Villareal to create upscale versions of *The Game of Life* (plates 12-43 and 12-44). After noticing that weather patterns are similarly sensitive to very slight changes in the initial values of variables, the American mathematician Edward Lorenz (creator of the Lorenz manifold; see plate 8-32 in chapter 8) performed one of the first applications of a complex model to weather prediction ("Does the Flap of a Butterfly's Wings in Brazil Set Off a Tornado in Texas?" 1972).

Following in the footsteps of Zuse, Fredkin, Conway, and others, the physicist and mathematician Stephen Wolfram, the designer of *Mathematica* (a computer program widely used by mathematicians and scientists), argued that the natural world can be modeled on a computer because natural objects (snowflakes, plants, animals) arise when simple rules (crystallization, cell division) are repeatedly applied to basic units. Natural objects are, according to Wolfram, best described by giving the rule (the line of computer programming) that generates them (*A New Kind of Science*, 2002). Wolfram illustrated his publication with graphic examples of simple rules that produce complex patterns, making them widely known and available to artists (plates 12-41 and 12-42).

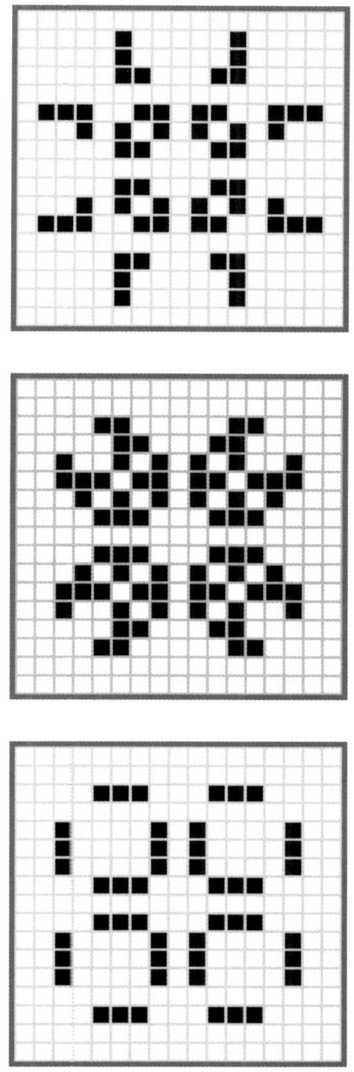

12-40. John Horton Conway (British, b. 1937), *The Game of Life*, 1970. Cellular automaton.

The Game of Life is a cellular automaton made up of an infinite two-dimensional grid of cells, each of which is dead (white) or alive (black). The player sets an initial pattern of living (black) cells and then watches as the universe evolves. In this example, the first generation is created by each cell simultaneously responding to its eight neighbors as follows:

Any live cell with fewer than two live neighbors dies.

Any live cell with two or three live neighbors lives on.

Any live cell with more than three live neighbors dies.

Any dead cell with exactly three live neighbors becomes a live cell.

Future generations are created as the rules are applied again and again in a regular rhythm, like a ticking clock. This example shows an initial symmetrical pattern and two generations that appear to oscillate as they evolve, creating a pulsing pattern (like a jellyfish in the ocean or a pulsar in outer space).

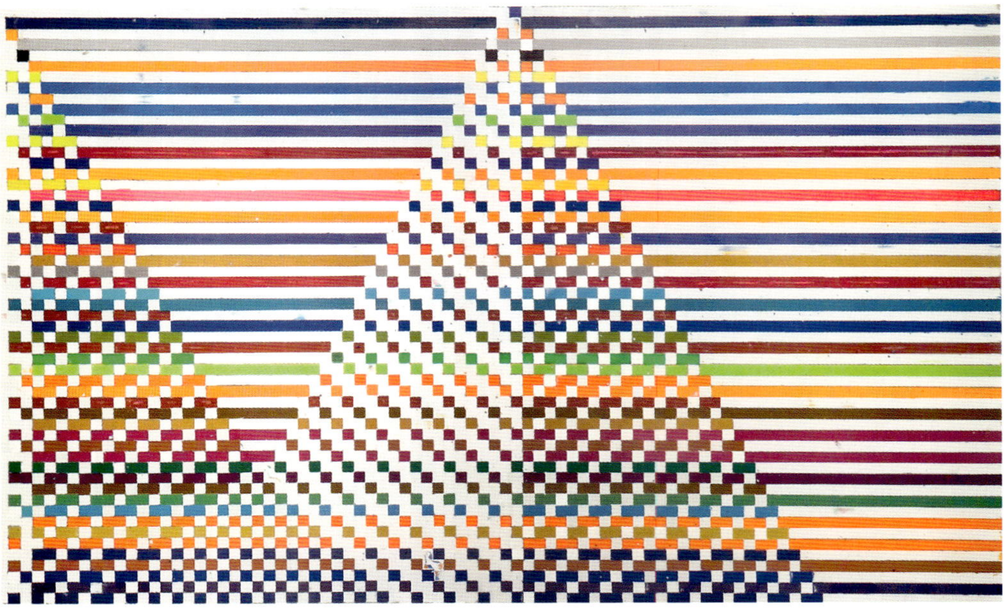

12-41. Richard Purdy (American, b. 1956), *198*, 2005. Encaustic on wood, 20 × 32 in. (50.8 × 81.2 cm). Nancy Hoffman Gallery, New York.

Inspired by Stephen Wolfram's *A New Kind of Science* (2002), Richard Purdy used a simple rule for cellular automata associated with the binary form of the number 198 to generate the pattern in this artwork, which he hand-painted in thick impasto.

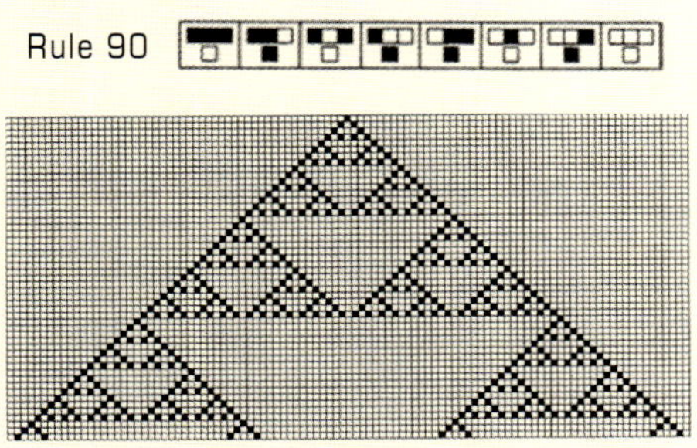

Application of Rule 90 for 50 steps, beginning with one black square in the top row center.

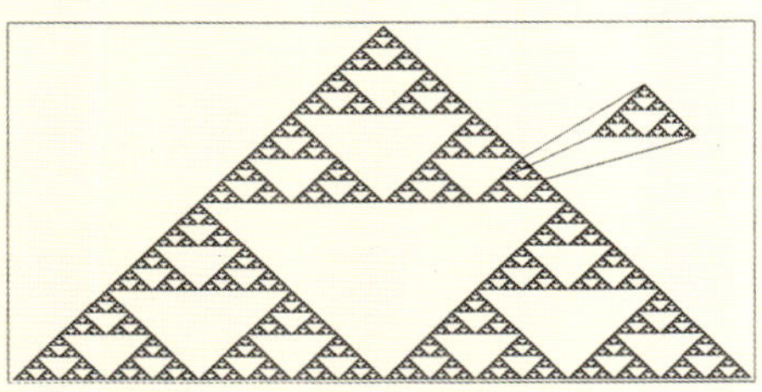

Application of Rule 90 for 500 steps. The rule generates a fractal pattern, which is self-similar.

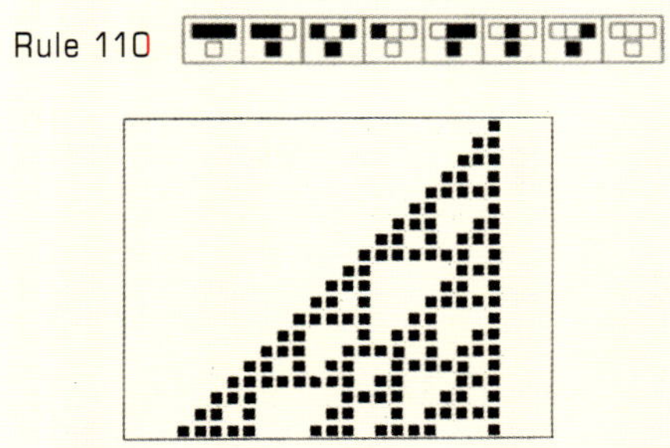

Application of Rule 110 for 20 steps.

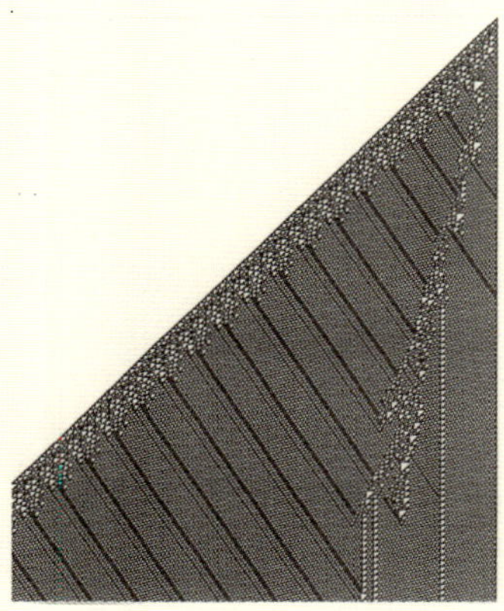

Application of Rule 110 for 700 steps.

12-42. Stephen Wolfram, *A New Kind of Science* (Champaign, IL: Wolfram Media, 2002), 25, 26, 32, and 33. © Stephen Wolfram.

Rules: The top row shows a pattern of colors (black or white) of three cells (center, left, and right), which determines the color of the center cell in the next row.

OPPOSITE, TOP

12-43. Leo Villareal (American, b. 1967), *Multiverse*, 2008. White LEDs, custom software, electric hardware, 200 ft. (60.96 m) long. Site-specific installation: National Gallery of Art, Washington, DC. Courtesy of the artist and Gering & López Gallery, New York.

Like John Conway's *The Game of Life*, Villareal's *Multiverse* is a grid of cells. Conway's end product is a handheld computer game. Dramatically increasing the scale, Villareal's grid is spread across existing architecture—a pedestrian-moving walkway—and his cellular points of light are LEDs. Pedestrians glide through a metaphorical multiverse, a concept in contemporary cosmology inspired by the possibility that, like the parallel paths on this walkway, there are parallel universes.

OPPOSITE, BOTTOM

12-44. Leo Villareal (American, b. 1967), *The Bay Lights*, 2013. White LEDs, custom software, electric hardware, 1.8 miles (2.9 km) long. Site-specific installation on the San Francisco–Oakland Bay Bridge, California. Courtesy of the artist and Gering & López Gallery, New York.

To form the grid of a cellular automaton, Villareal positioned white LEDs along the vertical steel support cables that connect the deck of the bridge to the suspension cables at the top. The pattern of lights is determined by a computer system consisting of the initial condition of the cells (the array of LEDs, each either on or off), and rules that determine the new state of each cell over time. The result is a series of abstract patterns that ascend and descend the cables as they move across the bridge. Villareal wrote the rules to reflect patterns in the weather, tides, and bridge traffic. Visible from the San Francisco waterfront, the blinking lights are positioned to be invisible to bridge motorists to avoid distraction.

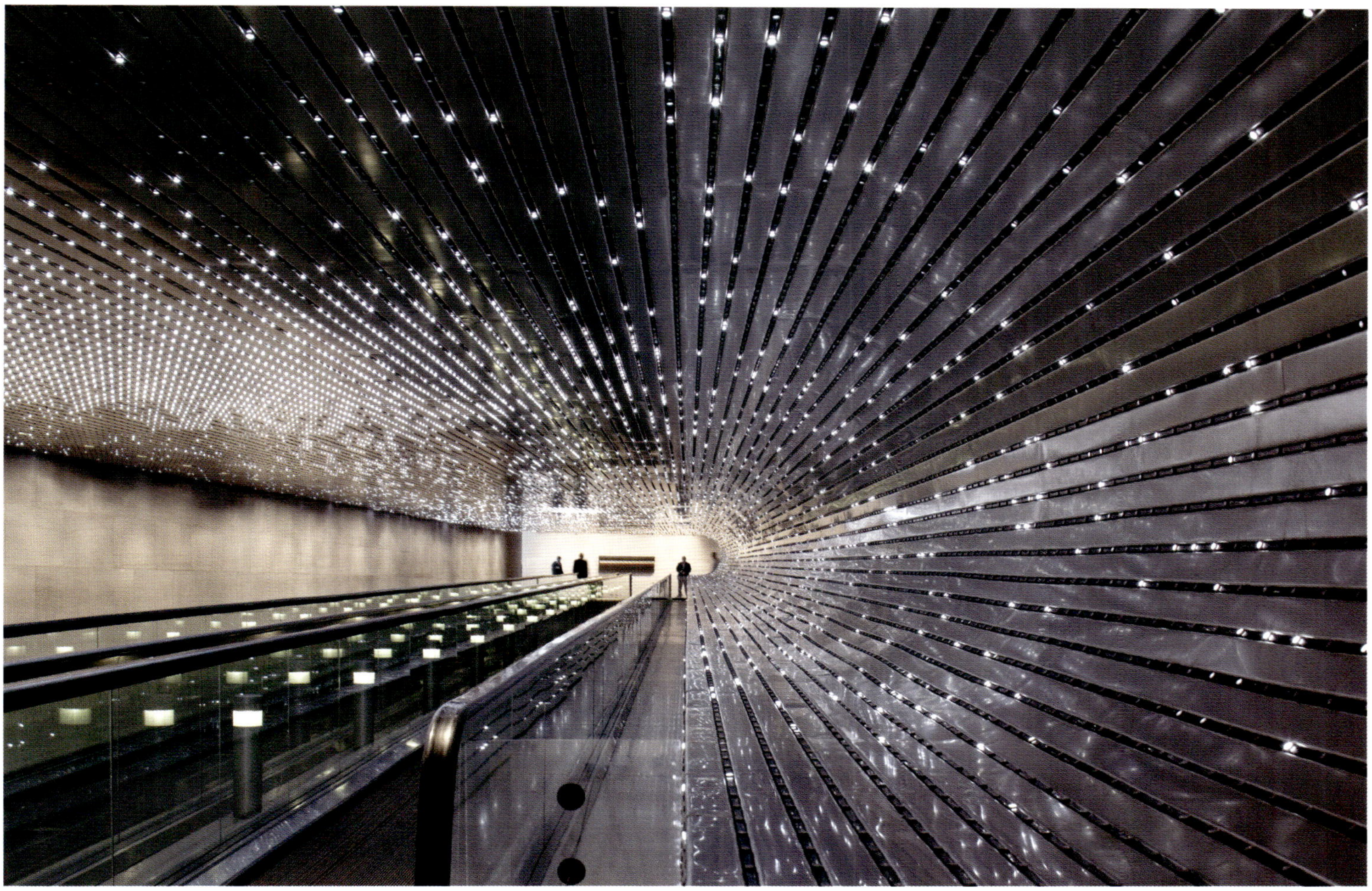

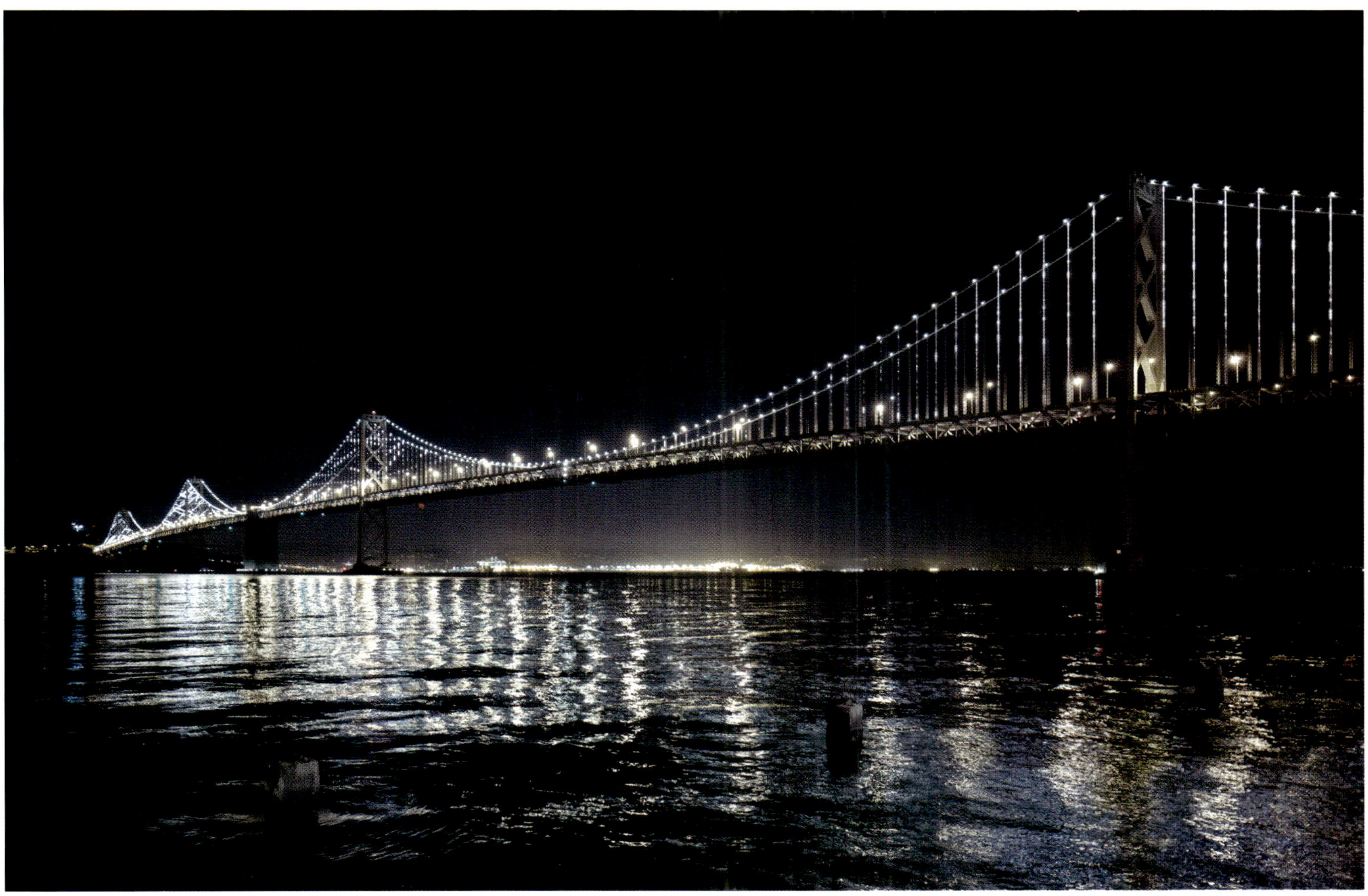

FRACTAL GEOMETRY

In the 1970s Benoît Mandelbrot developed a new kind of geometry, which he named "fractal" (from the Latin *fractus* for "broken," as in a fragment) to describe complex patterns in dynamic, organic systems that result from the repeated application of recursive algorithms. Mandelbrot was born into a Jewish family in Warsaw. In 1936 his family fled the Nazis to Paris, where Benoît Mandelbrot's uncle, Szolem Mandelbrot, a professor of mathematics at the Collège de France and a member of the Bourbaki group, took charge of his young nephew's education. After the outbreak of World War II, the family relocated to the south of France to the small farming town of Tulle, where Mandelbrot spent his adolescence escaping detection as a Jew in Nazi-occupied France during 1940–44. Mandelbrot remembers this as a formative experience of learning to live by his wits and being careful whom he trusted. After the war he returned to Paris and studied at the École Polytechnique during 1945–47 with Gaston Julia, completing a PhD in mathematics at the Université de Paris (the Sorbonne) in 1952.

Mandelbrot found himself increasingly unsympathetic to the strict formalism of the Bourbaki school that dominated French mathematics. In contrast to Boubaki's abstract formalism, Mandelbrot saw pictures—visualizations of geometric structures. So Mandelbrot left France and relocated to America in 1958. Now thirty-four, married and with a son, he moved to upstate New York to work for IBM at its Watson Research Laboratory, where he studied self-similar random processes and algorithms, linking these two in a deep way. He considered an odd fact about measurement that had been pointed out by

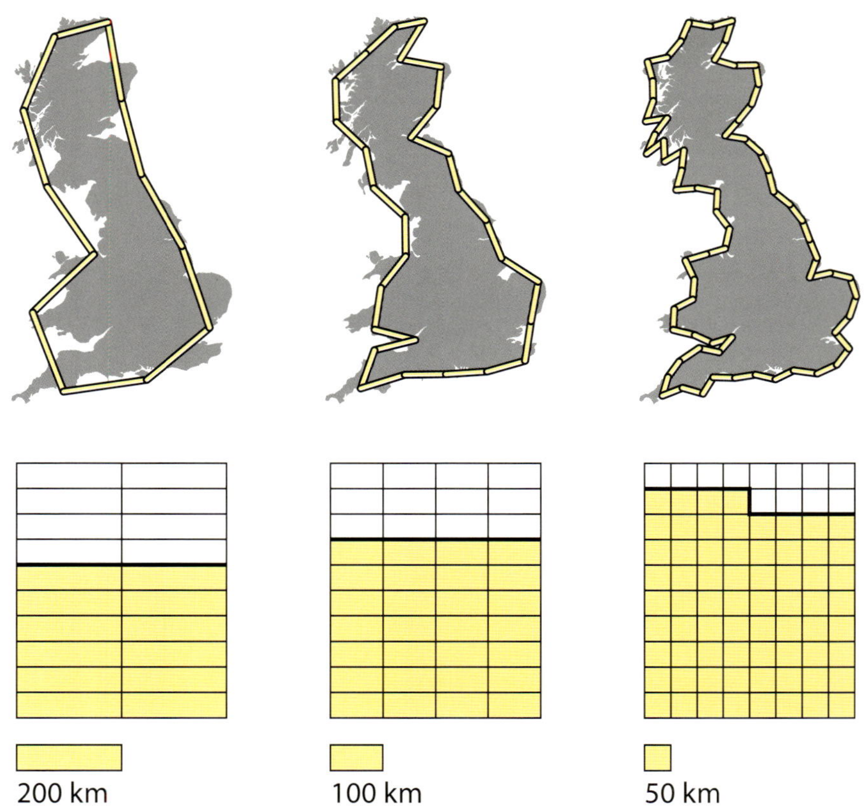

12-45. Measuring the coastline of Britain with different units.

Lewis Fry Richardson observed that the length of an irregular shape increases without limit as the unit decreases. For example, if you measure the coastline of Britain with a 200-km ruler and then cut the ruler in half, the measurement increases significantly.

If the unit is 200 km, then the coastline is 12 units long and its length is 2400 km.

If the unit is 100 km, then the coastline is 28 units long and its length is 2800 km.

If the unit is 50 km, then the coastline is 68 units long and its length is 3400 km.

the British mathematician Lewis Fry Richardson: the length of a country's border changes with the unit used (plate 12-45). In a landmark paper, "How Long Is the Coast of Britain?" (1967), Mandelbrot suggested that mathematicians needed to rethink measurement when the object in question is irregular, and he proposed that the Koch curve pointed the way to measure the coastline of an island nation such as Britain. The Koch snowflake encloses a bounded finite area, but it has an infinite perimeter (plate 12-46). According to Mandelbrot, one should not try to measure the *length* of a coastline or of Koch's snowflake, but its *roughness* (its degree of complexity).

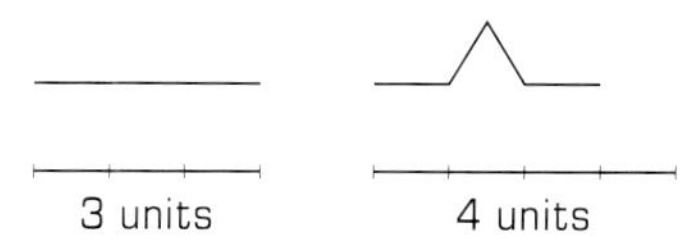

12-46. The infinite perimeter of the Koch snowflake.

Each side of the Koch snowflake is divided into three units of equal length. At each iteration, the three units are replaced by four, increasing the length ad infinitum.

As a student in the 1940s, when Mandelbrot heard Julia describe a recursive process, he tried to imagine what it would look like if he could plot all the successive stages of the function on a grid and show its progression toward the limiting shape. This was tantamount to making graphs of all the (countless) iterations of a recursive function on transparent sheets of paper, and then overlaying the layers to see what shape they make together. Mandelbrot adopted computers as the tool with which to carry out this task; by the 1970s computer technology had advanced enough so that he could run Julia's recursive algorithms thousands of times and plot the resulting numbers into points on a graph. The resulting image is a visualization of patterns of change in a dynamic system over time (plate 12-47).

Julia had written hundreds of recursive algorithms to describe many different patterns of change, from slow motion to rapid change, from a slight perturbation to an immense shift. Mandelbrot unified this work, describing all Julia's patterns of change in one master equation that describes the so-called Mandelbrot set. Working with the assistance of computer programmers, he then plotted on a graph the result of repeatedly applying his master equation, creating a detailed pattern of self-similar curves when viewed at smaller and smaller scales (plate 12-48). As such, the Mandelbrot set is the archetypal image of fractal geometry, visualizing dynamic patterns of change—much as the circle is the archetypal image of Euclidean geometry, visualizing motionless, timeless symmetry. Mandelbrot launched his fractal effort in France in 1975 with *Les objets fractals: Forme, hasard, et dimension*, which he expanded in an English edition of 1977 as *Fractals: Form, Chance, and Dimension*.

The Mandelbrot set is an excellent example of an abstract object that exists in the Platonic realm of set theory, as the British mathematician Roger Penrose described it: "Have you ever seen those pictures produced by computers, the object known as the Mandelbrot set? It's as if you are travelling to some distant world. You turn on your sensing device and see this incredibly complicated configuration, with all sorts of structure to it, and you try to figure out what it is. . . . Very elaborate and impressive! Yet from just seeing the equations, nobody had the remotest conception that they would produce patterns of this nature. Now these landscapes aren't conjured up out of someone's imagination; everyone sees the same pattern. You're exploring something with a computer, but it's not dissimilar from exploring something with experimental apparatus."[12]

Like Euclidean geometry, fractal geometry is embodied in the natural world, in the simple, repeated processes that have been applied over and over again to form our

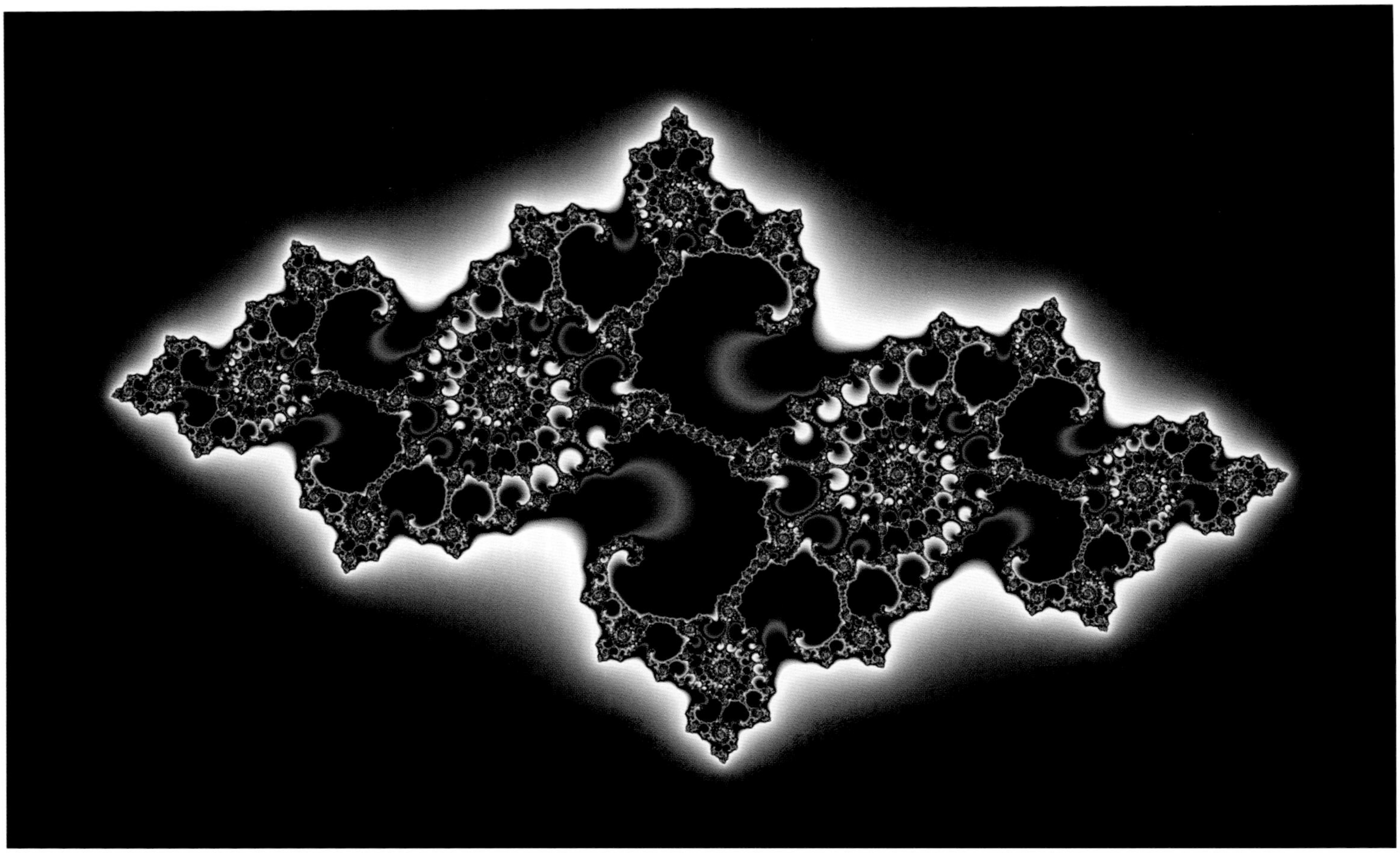

ABOVE

12-47. A Julia set.

OPPOSITE

12-48. This computer graphic shows part of a mathematical object discovered by Mandelbrot, the so-called Mandelbrot set (upper left). The computer zooms in like a microscope, seeing the set in progressively finer detail.

universe and life forms. Nature is composed of so-called emergent systems in the sense that complex patterns—billowing clouds, flowing liquids—arise ("emerge") from a multiplicity of simple interactions (plates 12-49 and 12-50). A mountain evolves over millennia by repeated uplifting of the underlying rock and erosion, resulting in a form that is dense and intricate. According to Mandelbrot, a mountain is best described by stating the simple, recursive algorithms that made it (plates 12-51 and 12-52).

Some prominent mathematicians took a dim view of fractal geometry, and Mandelbrot responded to his critics by taking the unusual step of publishing a revised edition of his book aimed at a general educated audience (*The Fractal Geometry of Nature*, 3rd ed., 1982). Writing in the self-promotional style of science journalism initiated by Heisenberg in the 1920s (see chapter 8), Mandelbrot presented fractal geometry to the public as describing "forms that Euclid leaves aside,"[13] filling his book with illustrations demonstrating the mathematical structure of "amorphous" trees, mountains, and clouds. Although the book didn't silence his critics,[14] it generated great public interest in fractal geometry.[15] *Les objets fractals* was released in Paris at a time when Bourbaki's formalist approach was in decline, and the French philosopher Jean-François Lyotard heralded Mandelbrot as a mathematician who had contributed to the creation of a new "postmodern science [*la science postmoderne*]."[16] Historians of science lambasted Lyotard for uncritically accepting Mandelbrot's exaggerated claims about the revolutionary nature of fractal geometry.[17]

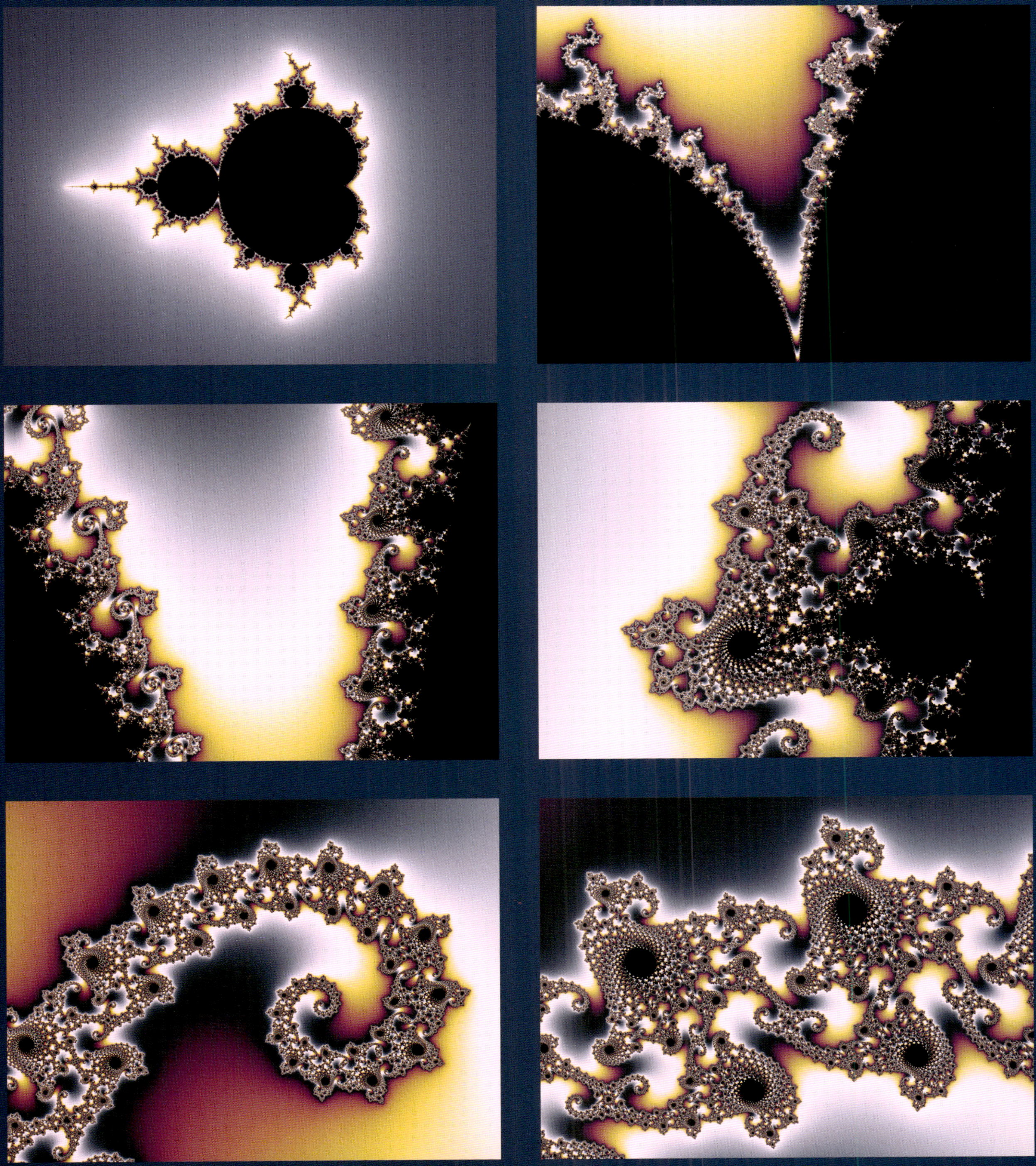

RIGHT

12-49. Keith Tyson (English, b. 1969), *Cloud Choreography: Clouds in Your Coffee*, 2009. Oil on aluminum, diameter 42 in. (122 cm). © Keith Tyson. Courtesy Pace Gallery, New York.

OPPOSITE

12-50. Keith Tyson (English, b. 1969), *Cloud Choreography: Halifax*, 2009. Acrylic on aluminum, 78 × 78 in. (198 × 198 cm). © Keith Tyson. Courtesy Pace Gallery, New York.

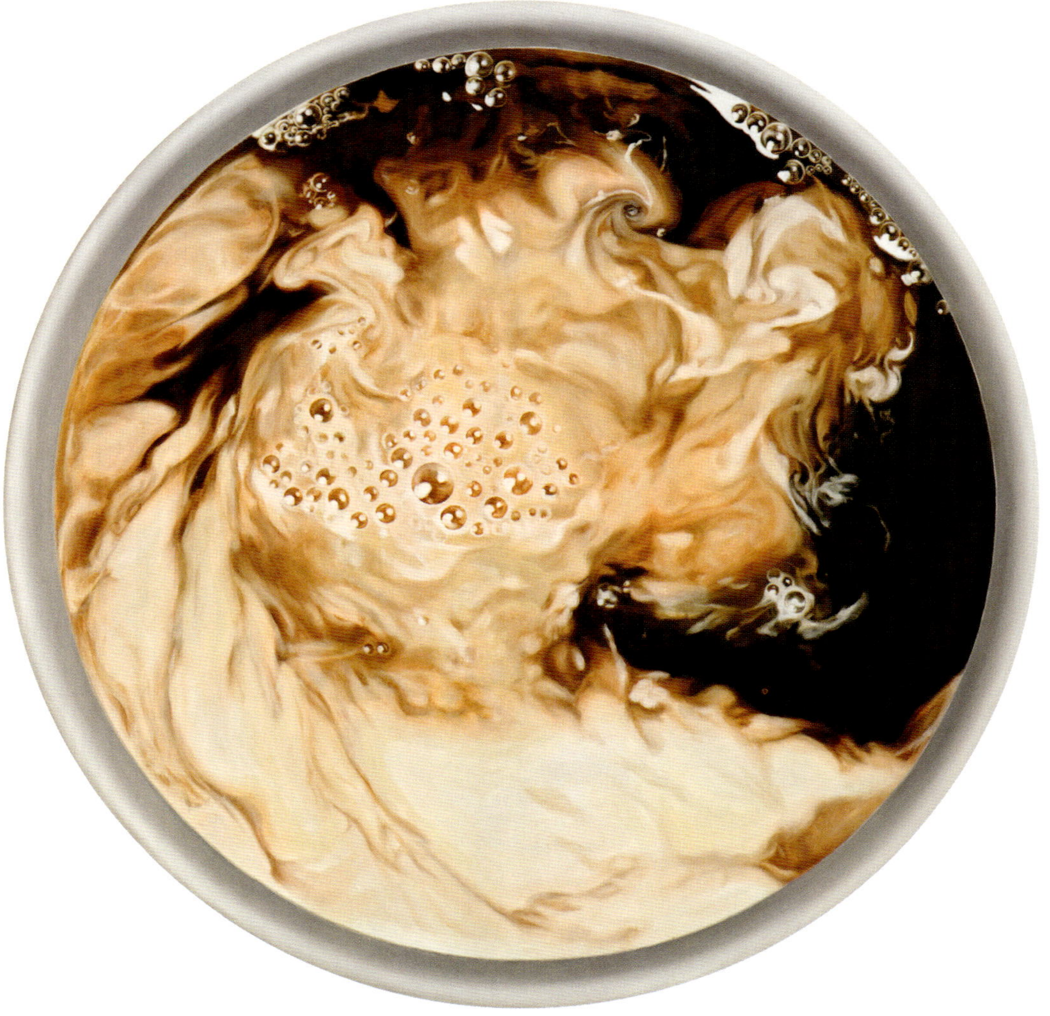

Mandelbrot's promotional style was adopted by the German mathematician Heinz-Otto Peitgen, who curated an exhibition of computer graphics of fractals and photographs of fractal patterns in nature, dramatically titled *Schönheit im Chaos* (Beauty in chaos), for the Goethe-Institut (a German cultural center) in Munich in 1985. After touring Germany, the exhibition was displayed at local chapters of the Goethe-Institut in forty countries around the world, generating enormous popular interest. For example, after the exhibition was shown at the Goethe-Institut in New York (under its English title, *Frontiers of Chaos*), the New Museum of Contemporary Art responded by organizing an exhibition of contemporary art about fractals, curated by Alice Yang (*Strange Attractors: Signs of Chaos*, 1989). (An "attractor" is a set towards which a dynamic system evolves over time; if the dynamic system is chaotic, its attractor will have a fractal structure, which mathematicians call "strange." It bears repeating that, in the context of fractal geometry, the term Peitgen and Yang have latched onto—"chaos"—refers to the apparent randomness of patterns that result from applying completely ordinary [not strange], orderly [not unpredictable] recursive algorithms.)

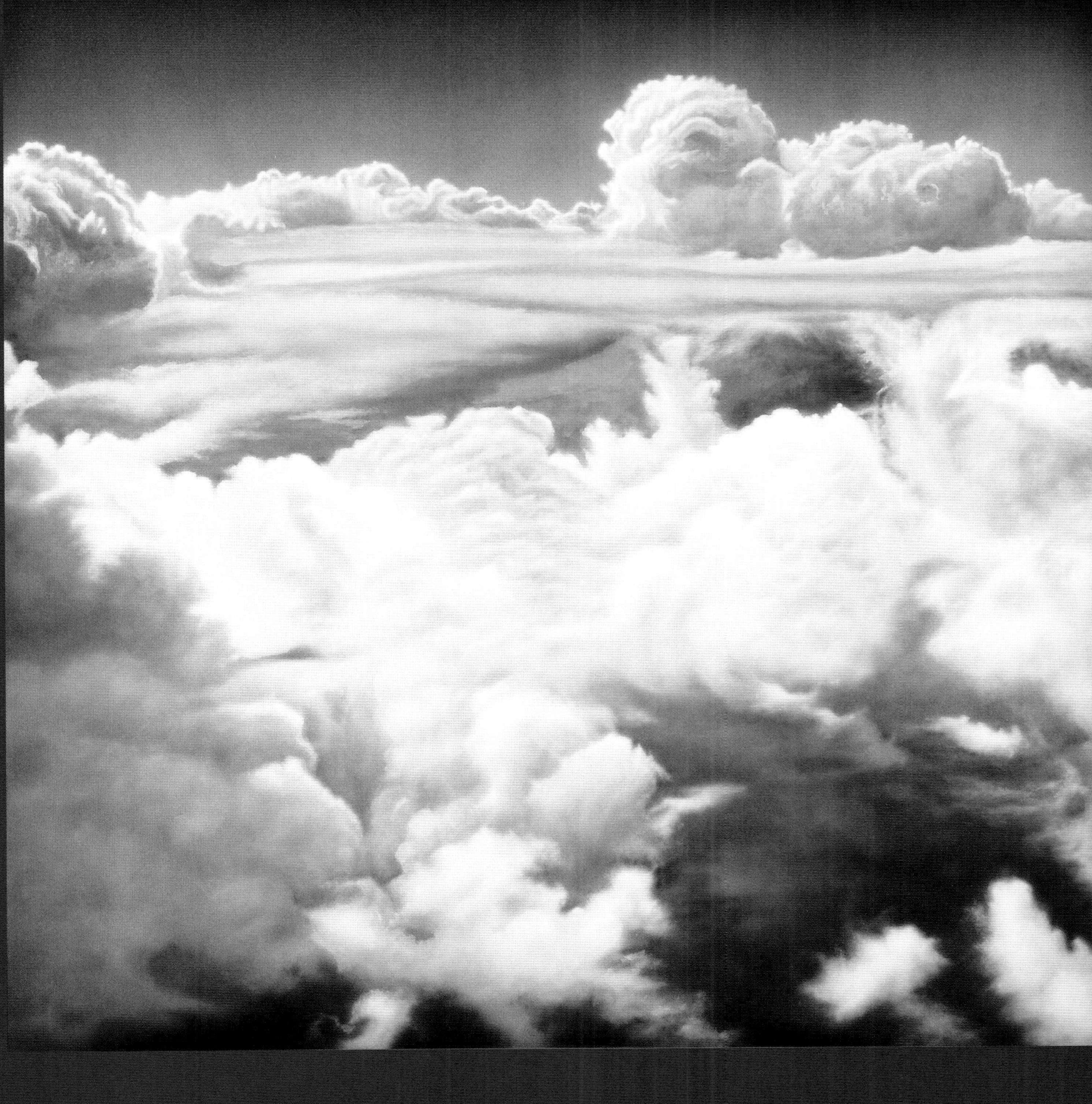

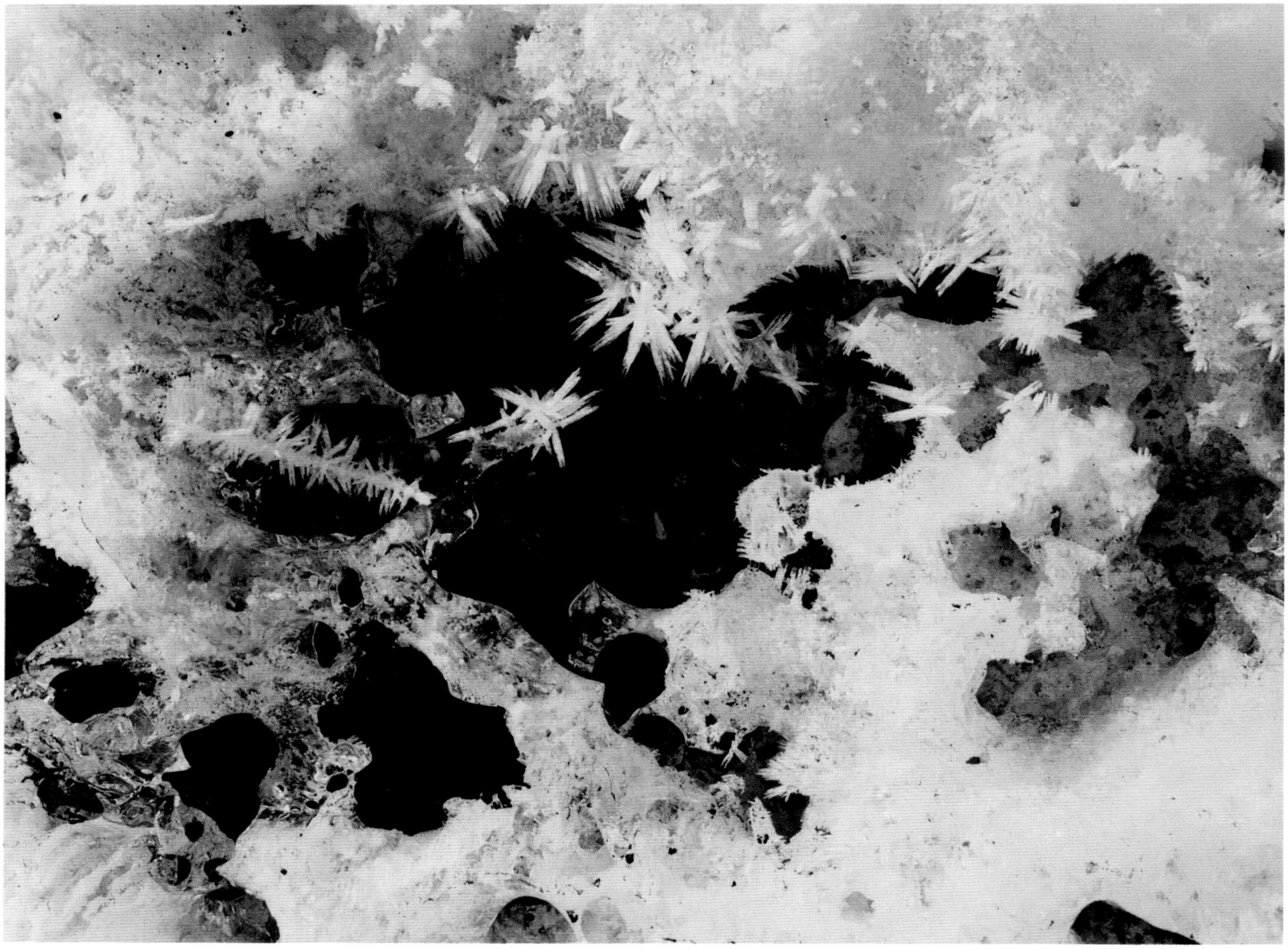

ABOVE

12-51. Eliot Porter (American, 1901–90), *Ice Crystals, Black Island, McMurdo Sound, Antarctica,* December 17, 1975. Dye inhibition print (Kodak dye transfer), 8½ × 10¼ in. (21.59 × 26 cm). © 1990 Amon Carter Museum, Fort Worth, Texas, Bequest of the artist, 1990. 51.521.2.

OPPOSITE

12-52. Charles Brown (American, b. 1954), *Monte Rosa, Switzerland,* 2011. Inkjet print, 20 × 20 in. (50.8 × 50.8 cm). Courtesy of the artist.

Clouds are not spheres, mountains are not cones,
coastlines are not circles, and bark is not smooth,
nor does lightning travel in a straight line.

—**Benoît Mandelbrot,**
***The Fractal Geometry of Nature*, 1982**

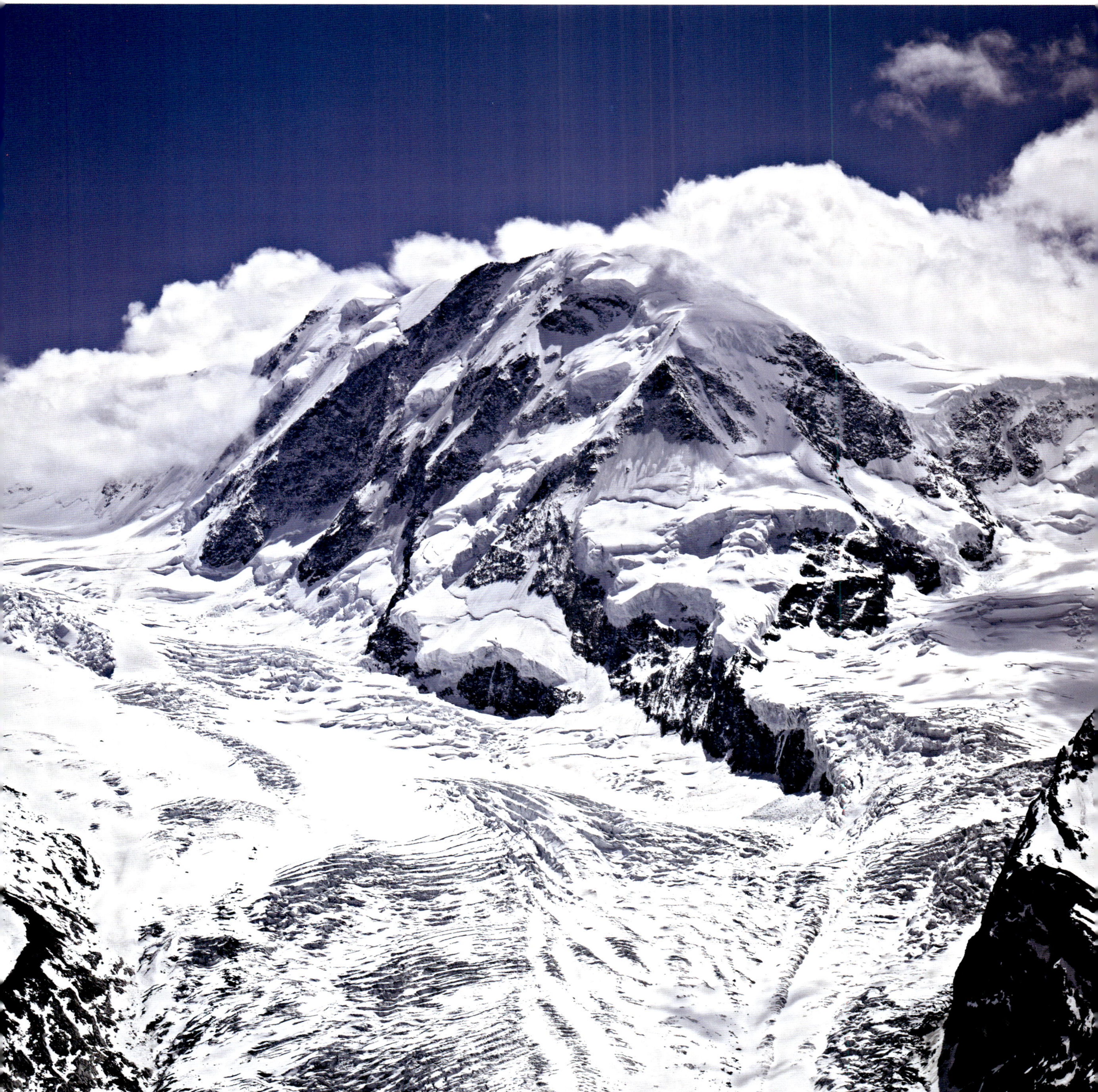

RECURSIVE ALGORITHMS IN SCIENCE, TECHNOLOGY, AND THE ARTS

Fractal geometry offers deep insights into the natural world, and its new tools have been employed in science, technology, and the arts. Like a branching tree, the human cardio-vascular system has a fractal structure (plate 12-53). Medical imaging using ultrasound gives a good picture of the overall structure of blood vessels in an organ such as the kidney or liver, but the vascular structures of a tiny tumor within the organ of a living patient are too small to see. The medical researcher Peter N. Burns has found that the network of blood vessels in a healthy organ has a uniform, ordered fractal pattern, but that the blood vessels in cancerous tissue have a rough, disordered fractal pattern (plate 12-54). Today Burns and his colleagues are working to develop a mathematical model of healthy and pathological vascular networks, on the hypothesis that the model will allow them to use ultrasound images of the large-scale structure of blood vessels in an organ to predict whether or not the organ contains a tiny, invisible malignant tumor.[18]

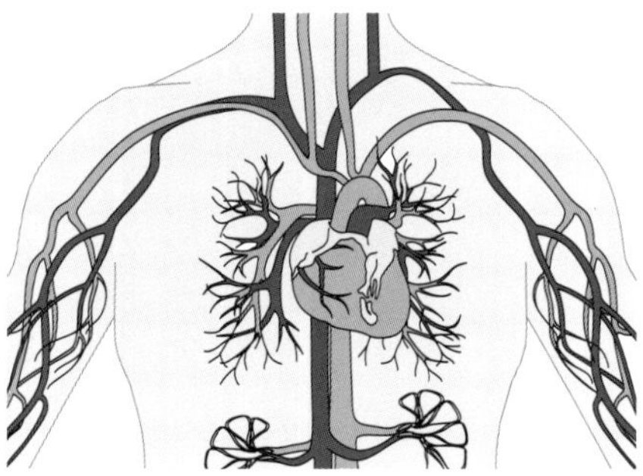

12-53. Human cardio-vascular system.

In the 1990s the American radio astronomer Nathan Cohen discovered that an antenna with the fractal form (plate 12-55) could be smaller and receive a wider range of frequencies than an antenna made in the dipole ("rabbit ears") form with straight lines and segments of a circle. Mobile phones and laptop computers provide multiple features—telephone, e-mail, global positioning systems—each of which is sent on a separate radio frequency and thus needs to be received by its own antenna. Cohen proved that the most efficient shape to receive multiple frequencies is an antenna in a fractal pattern such as a Sierpiński carpet, which is now the standard form of antennae embedded in smartphones and other wireless devices.[19]

A

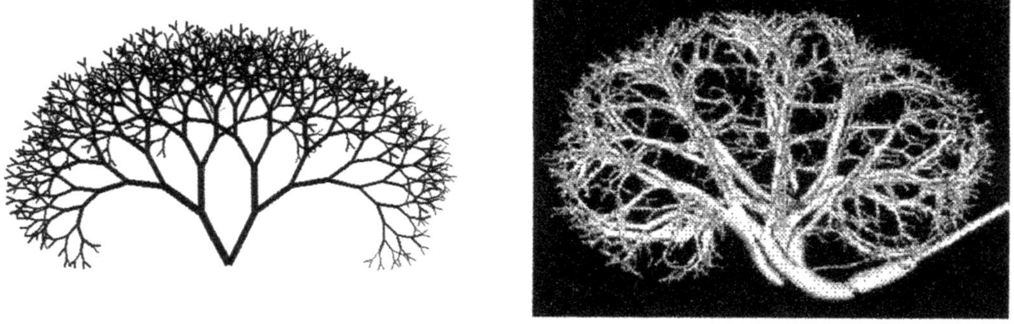

B

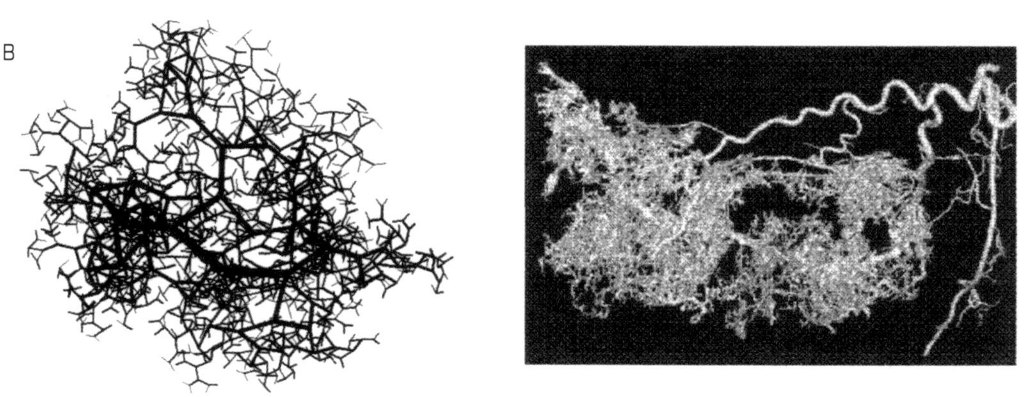

12-54. Fractal blood-flow patterns in healthy and pathological tissue, in Raffi Karshafian, Peter N. Burns, and Mark R. Henkelman, "Transit Time Kinetics in Ordered and Disordered Vascular Trees," *Physics in Medicine and Biology* 48 (2003): 3225–37; images are on 3231–2. Used with permission.

A Computer-generated kidney vascular tree model (left) and the kidney of a healthy rabbit.

B Computer-generated tumor vascular tree model (left) and a cancerous tumor (a neuroblastoma) in a mouse.

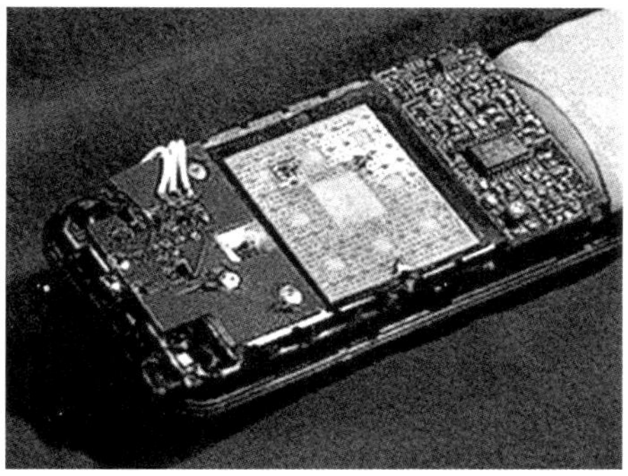

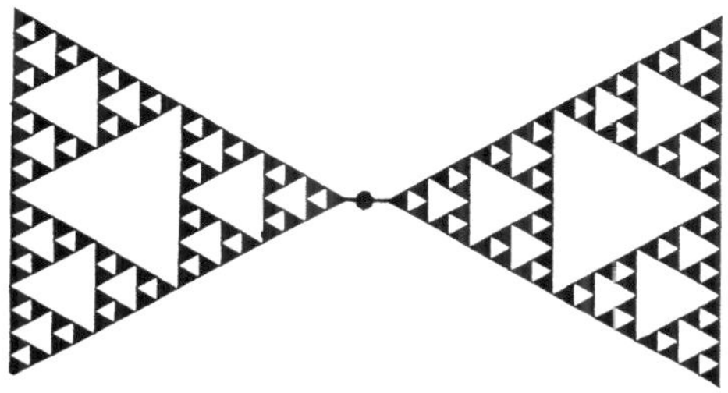

12-55. (Left) Fractal antenna in a cell phone in the form of a square Sierpiński carpet (see plate 12-33). Courtesy Fractal Antenna Systems, Bedford, Massachusetts.

(Right) Fractal antenna in the form of a triangular Sierpiński carpet, in Nathan Cohen and Robert G. Hohlfeld, "Self-Similarity and the Geometric Requirements for Frequency Independence in Antennae," *Fractals* 7, no.1 (1999): 79–84. © 2011 World Scientific Publishing Company. Used with permission.

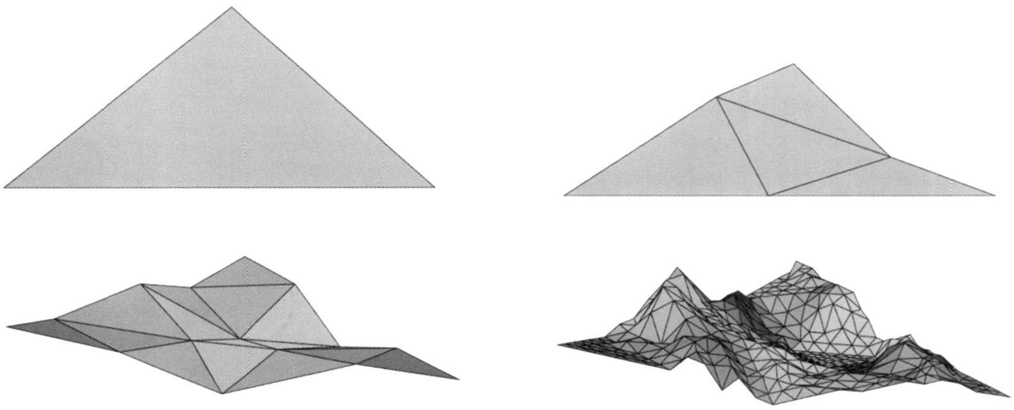

12-56. Equipped with a computer, an animator can model a mountain such as this in seconds by repeatedly dividing the sides of a large triangle (conceived as a pyramid on a three-dimensional grid) into smaller triangles, using a program that introduces random changes at each iteration. This process produces a realistic mountain by essentially mimicking the simple, repetitive processes (sedimentation, crystallization) that nature uses to make a mountain. Earlier hand-drawn cartoons simulated depth by overlapping flat planes, and motion by moving left and right on a two-dimensional grid. This computer simulation of a mountain was created in three dimensions, so the animator can show the mountain from the viewpoint of someone moving *through* the space.

BELOW

12-57. "Planets at Different Resolutions," in Alain Fournier, Don Fussell, and Loren Carpenter, "Computer Rendering of Stochastic Models," *Communications of the ACM* (Association for Computing Machinery) 25, no. 26 (1982): 381, fig. 17. Used with permission.

Fractal geometry can be used to model any natural object that results from some random (stochastic) process. This illustration shows landmasses on a planet going from rougher (at lower resolution) to more detailed (at higher resolution) as a recursive algorithm is applied more and more times. The authors suggest that the rough version is good for a long shot and the more detailed for a close-up. Carpenter and his colleagues published this article in 1982, the year that Carpenter used this technique to create an artificial planet in *Star Trek II*. The three young animators concluded their article with the understatement: "We suggest that recognition of the importance of the stochastic properties of the real world will lead to greatly increased flexibility in the modeling techniques used in computer graphics" (383).

When moving pictures began being made in the 1890s, animators created cartoons by drawing each frame by hand, requiring hundreds of pictures for a short scene. To keep the task manageable, cartoonists outlined animals and landscapes in simple shapes (plate 12-58). When computers became available to graphic artists in the 1970s, a new generation of animators adopted the old vocabulary, writing programs that would move simple, flat shapes across a two-dimensional plane (the screen of the monitor). In 1978 one young animator, Loren Carpenter, realized the potential of fractal geometry for computer graphics. To draw a complex object in three-dimensional space, Carpenter began by drawing a mountain as a simple geometric form—a pyramid—and then divided each of its triangular sides into smaller triangles, repeating this process over and over again (plate 12-56). In 1980 Carpenter joined Lucasfilm

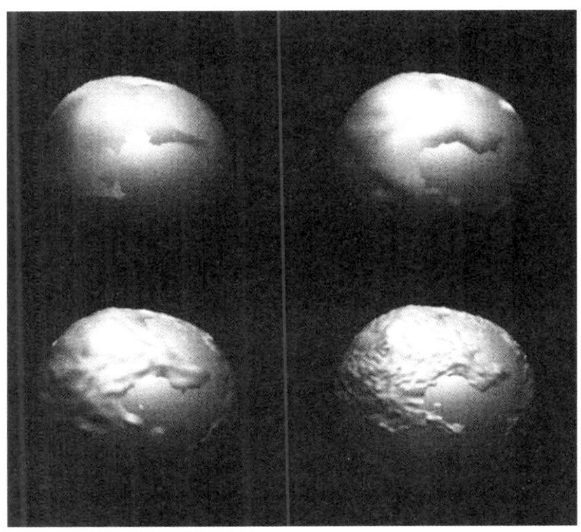

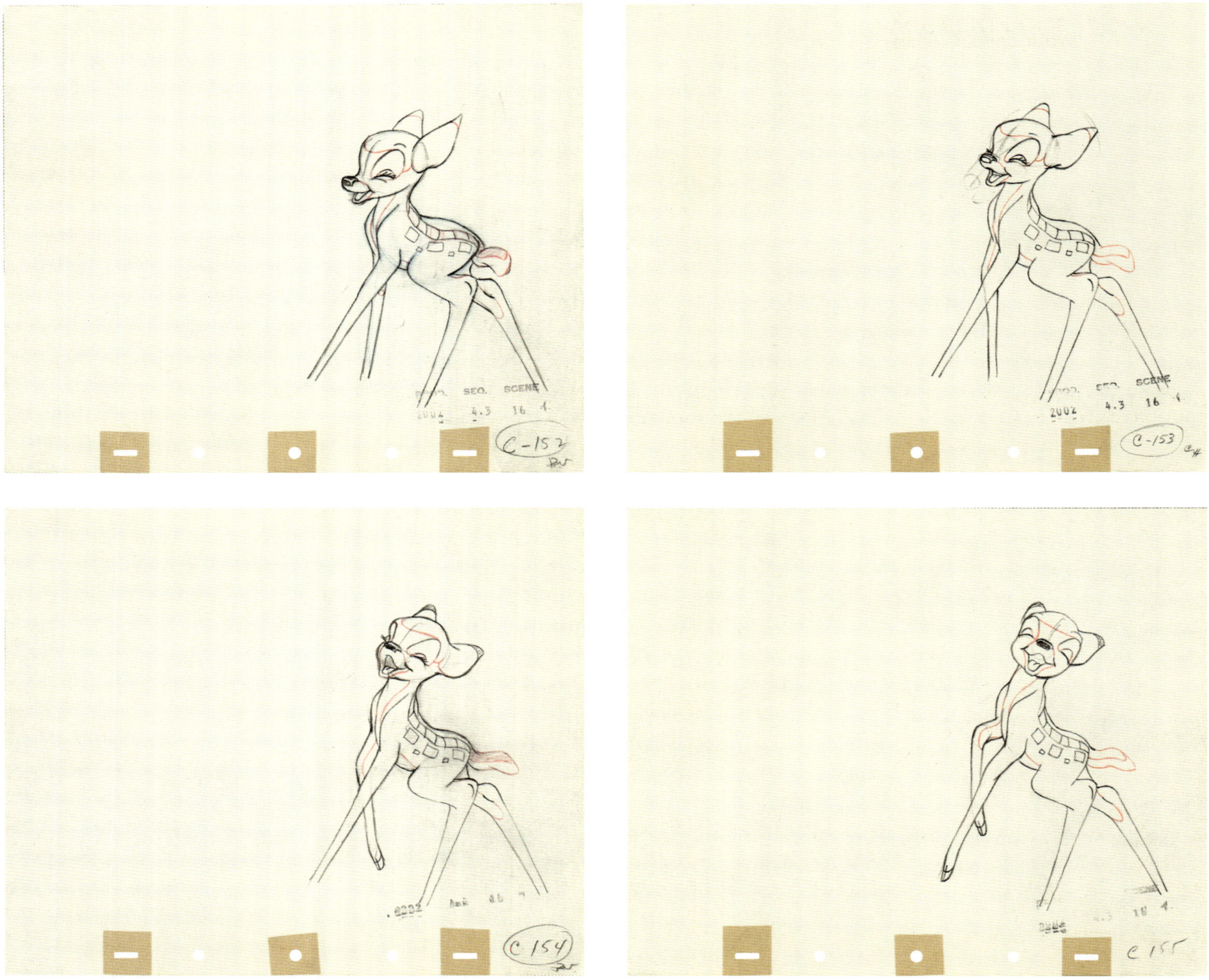

12-58. Frank Thomas (American, 1912–2004), animation drawings for the film *Bambi*, 1942. Pencil on animation paper. © Disney.

To animate this sequence showing Bambi walking, Frank Thomas stacked up translucent animation paper on a special desk that had a light source behind a glass disc, allowing him to see through the paper. The paper was held in place by the peg holes at the bottom of each sheet. Thomas would flip the drawings to make sure he was getting the correct movement and expression, while maintaining the look of the character as determined by the director. Thomas executed only the key drawings, and assistants filled in the in-between drawings, each slightly different from its neighbors.

The plot of Walt Disney's 1942 film *Bambi* was based on a 1923 novel by the Austrian Felix Salten, *Bambi*: *Eine Lebensgeschichte aus dem Walde* (Bambi: A life in the woods). Written during a resurgence of German Romanticism after World War I, Salten described an idyllic life in nature, where a great prince of the forest protects his fellow animals from being shot by hunters. With the rise of Nazism in the 1930s, Salten, who was from a prominent Viennese Jewish family, found himself—like Bambi—being hunted in his homeland; in 1936 Hitler banned the novel *Bambi* in Germany, and after the Nazi annexation of Austria in 1938, Salten fled Vienna for Zurich, where he died in exile in 1945. Meanwhile, Disney released *Bambi* during World War II, where the coming-of-age story of a brave young deer (portrayed as more adorable than his melancholy Austrian counterpart) won the hearts of American audiences, whose sons were overseas fighting to protect their homeland from a deadly enemy.

in San Francisco, where he created an artificial planet for *Star Trek II: The Wrath of Khan* (1982), which contained the first completely computer-generated sequence in a feature film (plate 12-57). The computer graphics community rapidly adopted fractal geometry, which revolutionized special effects and gave animators the tools to construct creatures that appear to be alive, such as the dinosaurs in *Jurassic Park* (1992; see plate 12-59).

Mathematicians, engineers, and scientists typically use computer graphics programs to visualize their data. If they are so inclined, some go on to manipulate the imagery in a metaphorical direction, moving from doing mathematics or science to making art.[20] Since the 1980s it has become common for professional technical organizations and corporations—especially groups of mathematicians, physicists, and computer scientists—to exhibit their members' computer-generated imagery and include discussion of art within their professional contexts. The American mathematician George W. Hart took up woodcarving in the 1990s to explore polyhedral forms. When new computer

12-59. *Jurassic Park* (1993), directed by Steven Spielberg, based on a 1990 novel by Michael Crichton. © Universal Studios Licensing LLC.

After spending decades digging up fossil bones, two American paleontologists (played by Sam Neill and Laura Dern) find themselves at the feet of a living, breathing *Brachiosaurus*. On the left is the billionaire entrepreneur (in white, played by Richard Attenborough) who built Jurassic Park, a prehistoric theme park populated with cloned dinosaurs. In earlier dinosaur films such as *Godzilla* (Japanese, 1954), the extinct creature was played by a man in a dinosaur costume; this creature was created with the assistance of a computer. *Jurassic Park* won the 1993 Academy Award for best special effects.

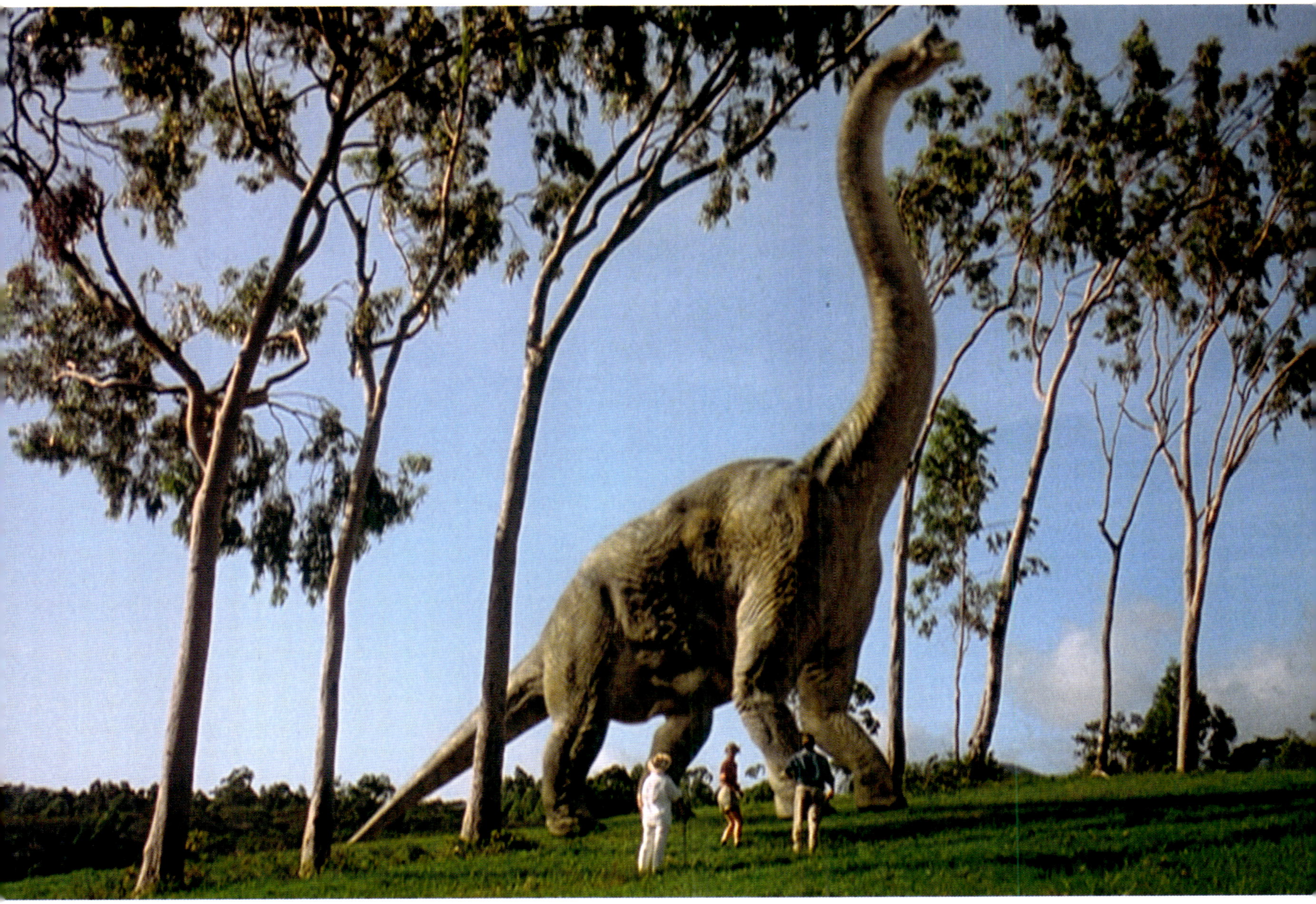

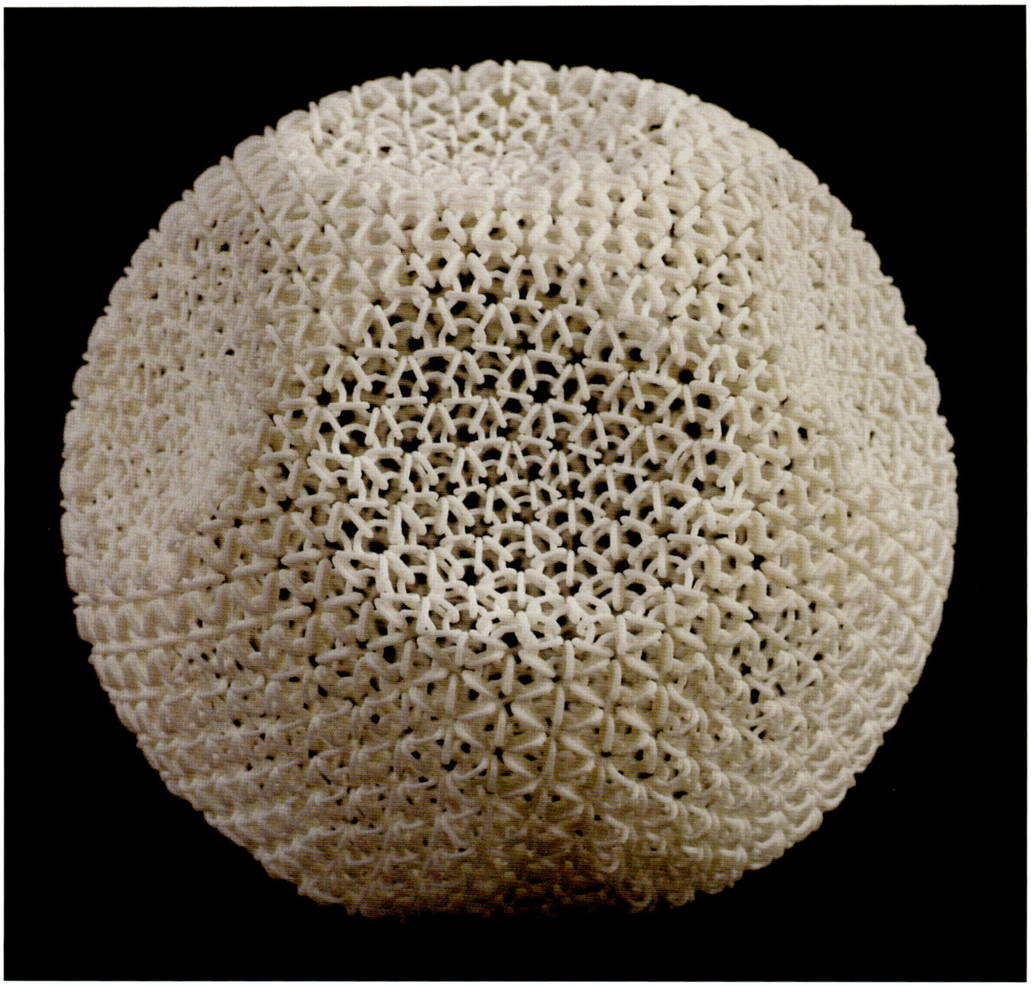

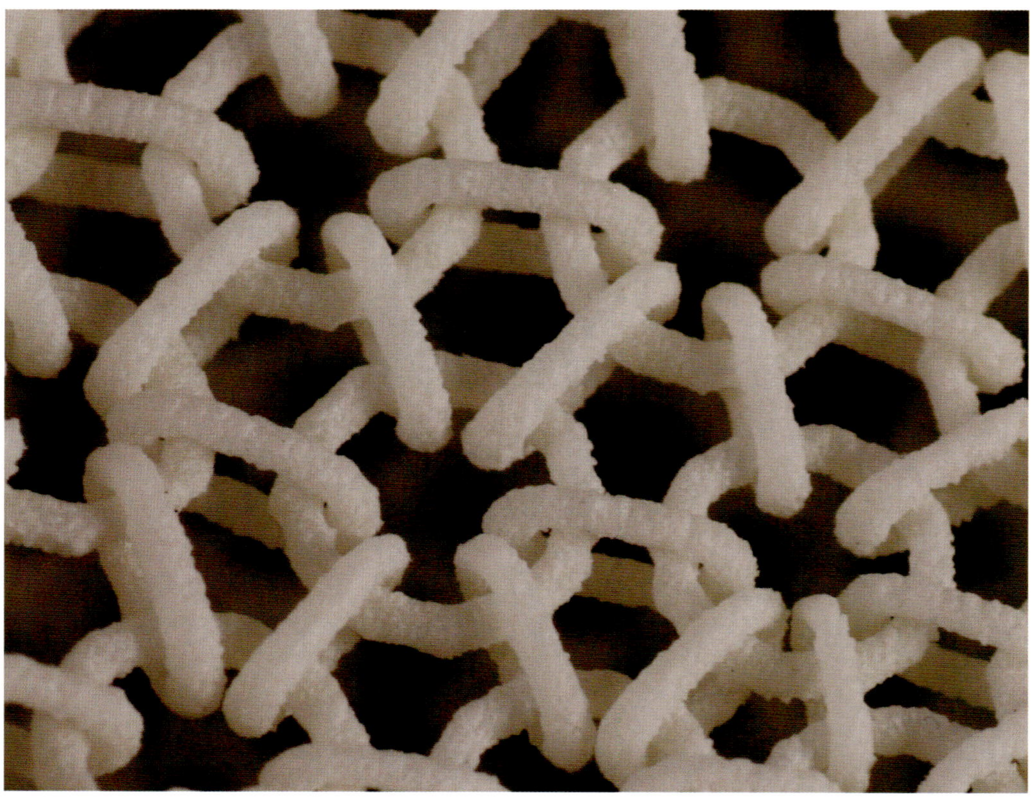

LEFT, TOP AND BOTTOM (DETAIL)

12-60. George W. Hart (American, b. 1955), *Ball and Chain*, 2009. Nylon, three-dimensional printing, diameter 6 in. (15.2 cm). Courtesy of the artist.

This chain-link sphere was made by a manufacturing process (three-dimensional printing), in which successive layers of very thin material (nylon) were laid down and slowly built up into the final form. The job of this artist (who is also a computer scientist) was to compose a digital file describing a chain-link sphere, which the manufacturer then fabricated. As the object is turned, sections collapse under its own weight.

OPPOSITE

12-61. Yoichiro Kawaguchi (Japanese, b. 1952), *Festival*, 1991. Computer graphic. Courtesy of the artist.

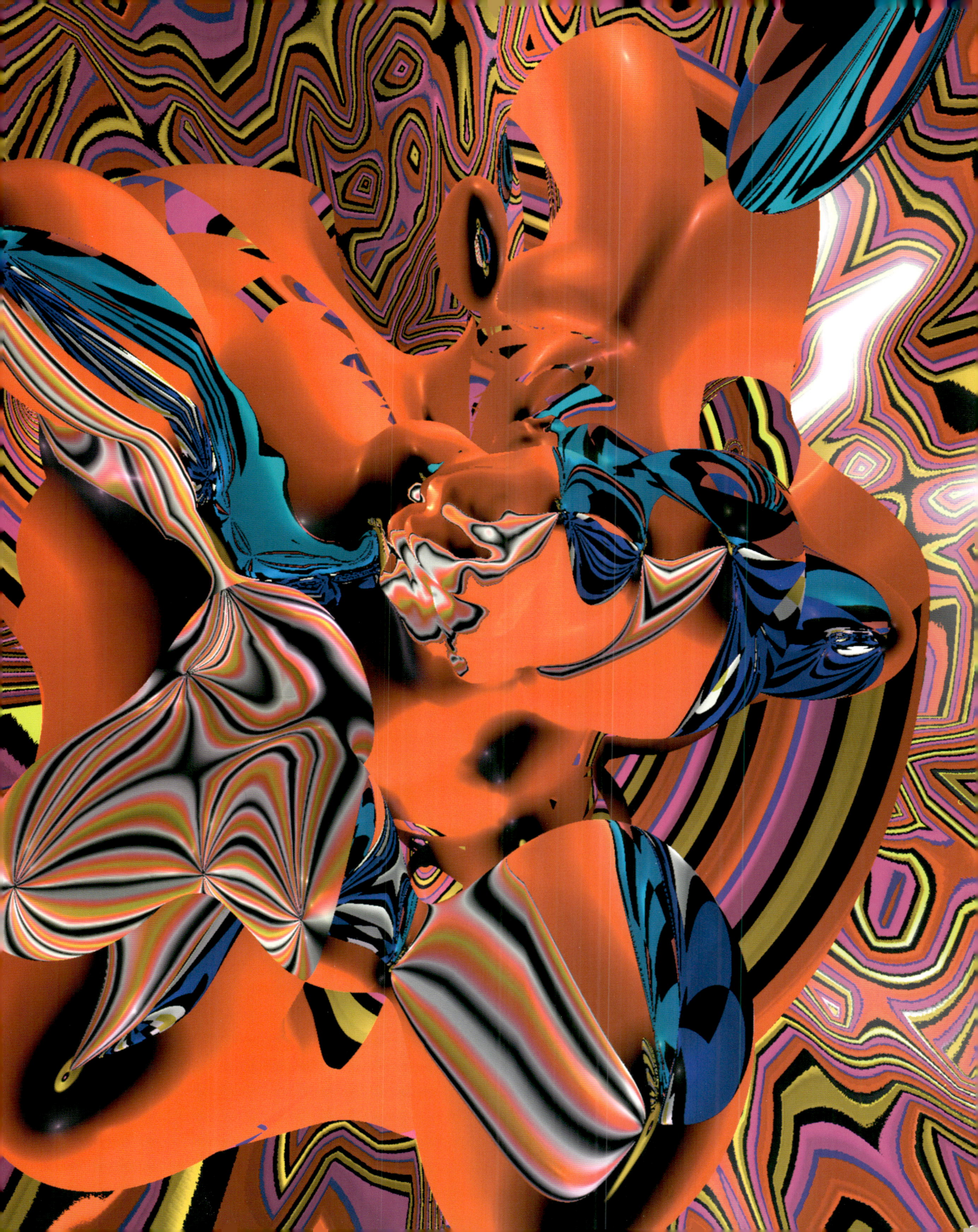

OPPOSITE, TOP

12-62. Martin Dörbaum (German, b. 1971), *Insert I (Col de la Forclaz)*, from the series *Objekt klein* A, 2005. Chromira print Aludibond/Diasec, 29½ × 52¾ in. (75 × 134 cm). Courtesy of the artist.

OPPOSITE, BOTTOM

12-63. Martin Dörbaum (German, b. 1971), *Tea Leaf Prophecy*, from the series *30° N–120° E*, 2008. Iris print on Somerset Velvet paper, 13 × 19¾ in. (33 × 50 cm). Courtesy of the artist.

graphic technologies became available, Hart explored more complex polyhedrons by "printing" three-dimensional objects that are impossible for him to carve by hand, such as this sphere made from chain-link (plate 12-60).[21]

When personal computers became available in the 1980s, a mathematician or artist with a specific professional task in mind had to begin by learning to program the machine. The Japanese computer animator Yoichiro Kawaguchi wrote programs in order to create abstract art with evolving organic forms (plate 12-61), as the artist wrote in 1982: "The unimaginable form, color, and way of moving of the flora and fauna of the natural world give me ideas for my work in computer graphics."[22] Then computer scientists began designing off-the-shelf products such as the computational software programs Mathematica (first version released in 1988 by Wolfram Research) and Photoshop (a graphics editing program, the first version of which was released by Adobe in 1990). Such programs, which are continually updated and improved, together with computer graphics printers, have made the computer an aesthetic tool that's easy to use, with wide applications. The German artist Martin Dörbaum creates "camera-less" photography by designing a virtual three-dimensional world in his computer, which he then "photographs" by transposing the three-dimensional space to a two-dimensional plane. This process allows Dörbaum to photograph moving objects at night (plates 12-62 and 12-63), which is impossible using a traditional camera because little available light means long exposure time.

Eric J. Heller is on the physics and chemistry faculty at Harvard University, where he studies rogue waves on large and small scales. A rogue wave (freak wave, killer wave) can occur far out at sea on the surface of the deep ocean. Unpredictable, these large waves threaten even ocean liners. Similarly, when a wave of electrons flows through a semiconductor, a freak wave can suddenly arise where circuits cross and threaten the smooth functioning of the device. In addition to doing physics, Heller makes graphic art by taking his experimental data on wave patterns, putting it into Photoshop, and using the editing program to refine the image and add color (plate 12-1, chapter frontispiece).

In the late twentieth century the computer was gradually adopted by mathematicians, scientists, and artists, giving them powerful new tools.

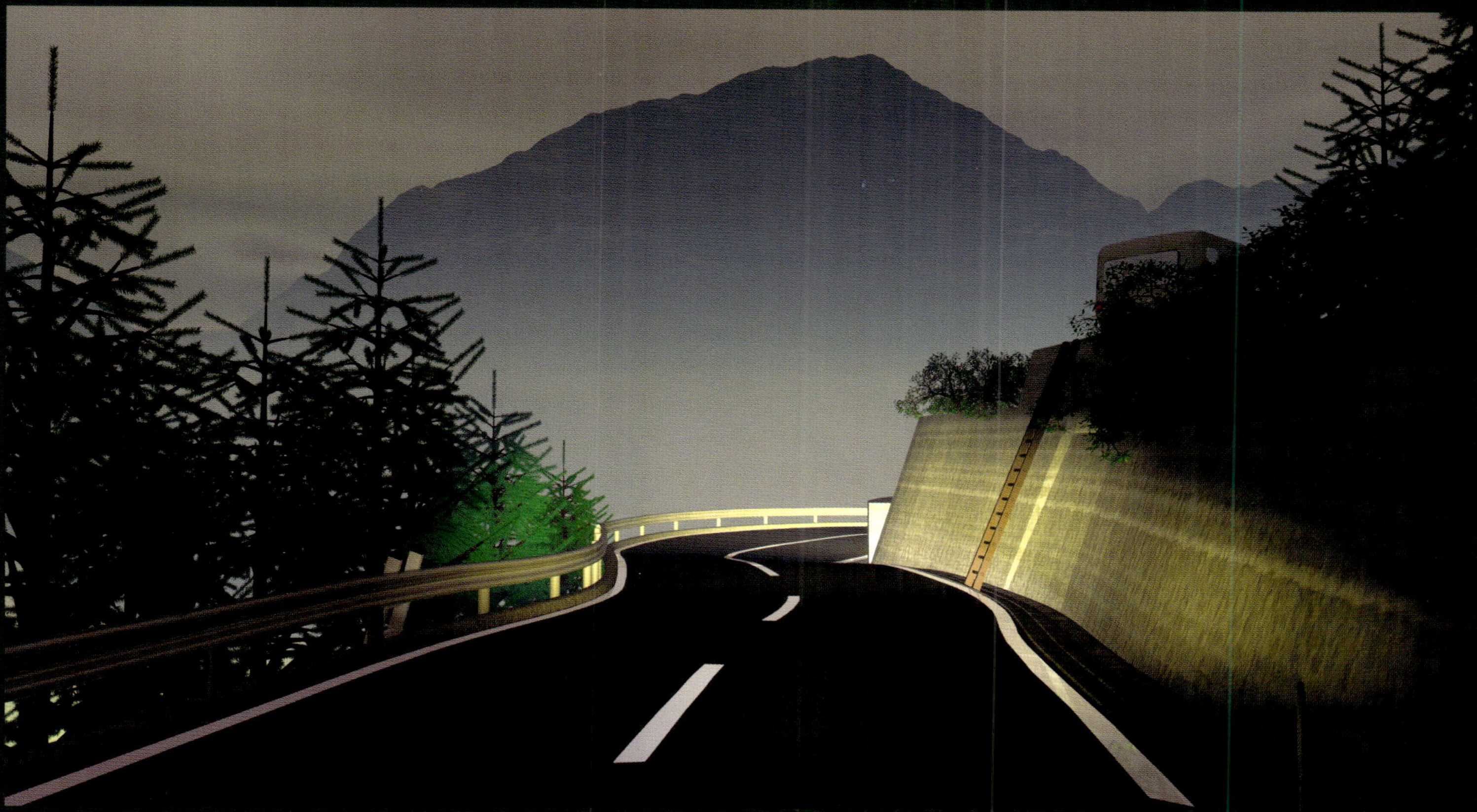

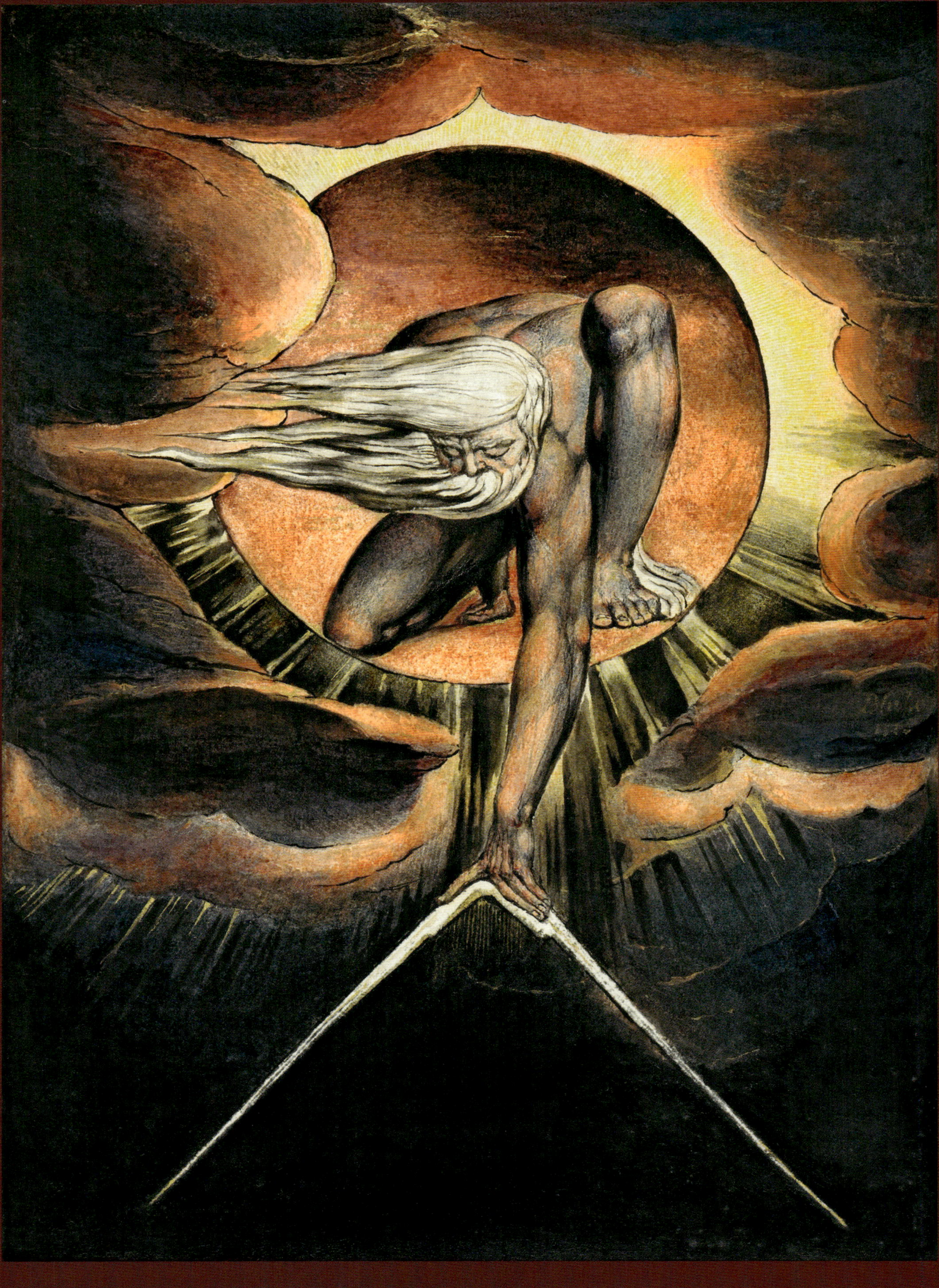

13
Platonism in the Postmodern Era

The equations that really work in describing nature with the most generality and the greatest simplicity are very elegant and subtle. . . . Now there must be skeptics out there who will tell you that these beautiful equations might have nothing to do with nature. That's possible, but it's uncanny that they are so graceful and that they capture so much of what we already know about physics while shedding so much light on theories that we already have.

—Edward Witten, "Viewpoints on String Theory," 2003

In the years after World War II, the Germanic vision of the unity of mathematics and the wholeness of nature was lost, its philosophical underpinnings dislodged by the destruction of scholarly communities throughout Germanic lands. Today the writings of Kant and Hegel continue to be studied in academia but as historical documents; German Idealism is no longer a living philosophy debated in the coffeehouses of Berlin, Vienna, and Copenhagen.

Two German Jews, Theodor Adorno and Max Horkheimer, were the first to give a penetrating analysis of the social and cultural consequences of this loss of a unified worldview, a condition that came to be called "postmodernism."[1] After spending the war years in exile, Adorno and Horkheimer returned to their homeland and described a world in the radioactive aftermath of the atomic bomb: "Enlightenment, understood in the widest sense as the advance of thought, has always aimed at liberating human beings from fear and installing them as masters. Yet the wholly enlightened earth is radiant with triumphant calamity."[2] Adorno and Horkheimer asked whether Enlightenment reason, which had propelled such phenomenal advances in mathematics and science, had contributed to the overwhelming assertions of power at Auschwitz and Hiroshima. Are mankind's scientific pursuits driven by a blind desire to dominate the earth and its inhabitants? Echoing the words of Dostoyevsky's underground man, Adorno and Horkheimer argued that notions of abstract truth and logic—two plus two equals four—are not neutral and eternal but express a worldview and are therefore instruments of class struggle (plate 13-1).

13-1. William Blake (English, 1757–1827), *Ancient of Days*, frontispiece (1824) to *Europe, a Prophecy*, 1794. Etching, watercolor, and gouache on paper, 9¼ × 6⅝ in. (23.4 × 16.8 cm). Whitworth Art Gallery, The University of Manchester.

Romantic poet William Blake warned that science and mathematics give men power that can lead to disaster, urging a restricted view of the scope and limits of human reason. Echoing these sentiments as well as Western creation stories, Blake's mythological narrative *Europe, A Prophecy* (1794) begins with this illustration of the Ancient of Days, a character Blake invented to embody reason and law. His opposite is the prophet Los, the embodiment of irrationality and imagination, who ominously ends the poem by sending his sons to war:

And with a cry that shook all nature to the utmost pole
Call'd all his sons to the strife of blood.

Beginning in the early 1950s, Adorno and Horkheimer worked at the Institute for Social Research in Frankfurt am Main, where they adopted critical self-reflection as a method to point a way out of the rubble of what remained of Western culture. The school was known as the Frankfurt School of Critical Theory; during the volatile 1960s Adorno and Horkheimer mounted an attack against logical positivism, claiming that Rudolf

If the Führer wants it,
two and two makes five.

—Hermann Göring, ca. 1938

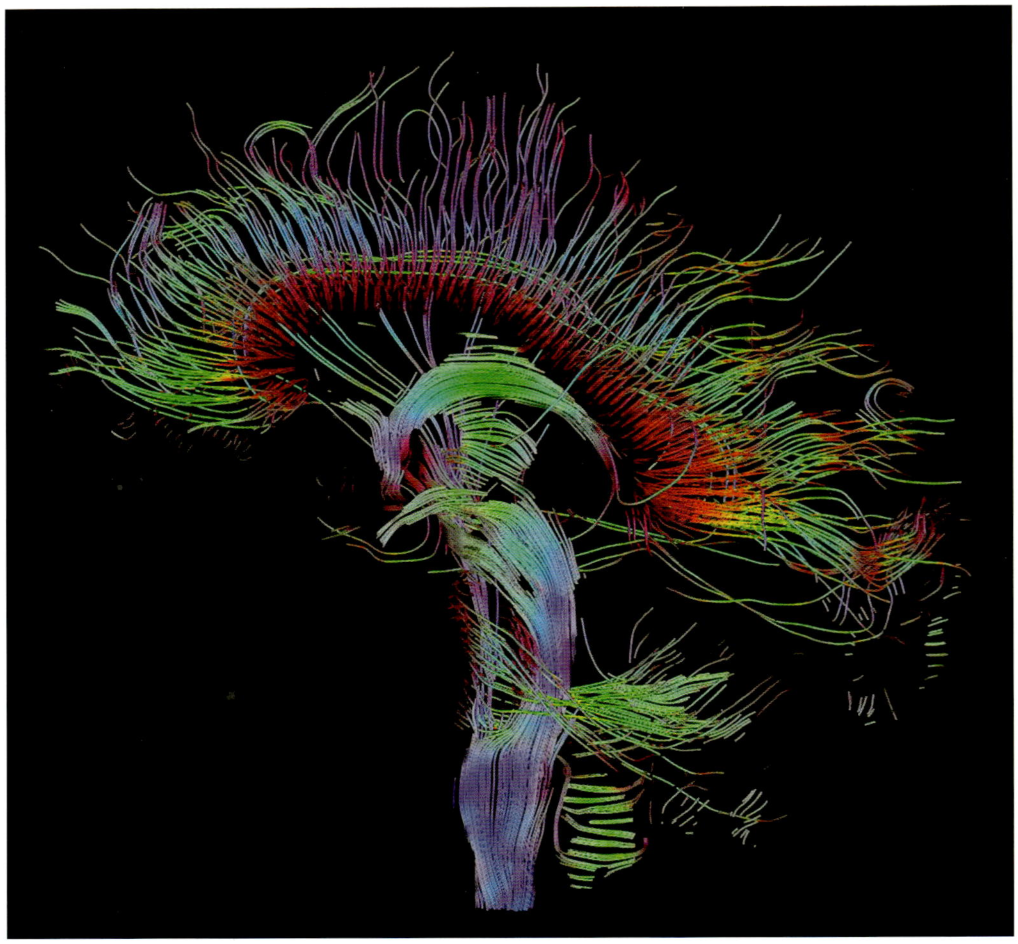

13-2. Nerve fibers intersecting the midsagittal plane in a living human brain, reconstructed using diffusion magnetic resonance imaging. Thomas Schultz, University of Bonn, Germany.

The German computer scientist Thomas Schultz made this image of nerve fibers that intersect a vertical ("midsagittal") plane that divides the brain into right and left hemispheres. The image reveals neural pathways passing through the thalamus, a symmetrical structure in the center of the brain, which acts as a relay station between the lower brain and the cortex, with nerve fibers projecting out in all directions. Schultz made the image with a technique related to magnetic resonance imagery (MRI), based on the principle that any spinning, electrically charged sphere (a proton in a hydrogen atom or a planet in the solar system) creates a magnetic field, acting like a bar magnet spinning on a north-south, positive-negative axis. If part of a person's body (in this case the head) is put inside a large circular magnet that generates a strong magnetic field, the polar axis of the proton in the nucleus of each hydrogen atom (harmlessly) snaps into orientation with the positive and negative regions of the magnetic field. If a radio wave is then passed through the person's head, it momentarily knocks all the tiny atomic magnets off their alignment. After the wave has passed, each proton resonates—reorients itself—with the magnetic field. As each proton moves, it emits a radio signal indicating its position, acting as a tiny radio transmitter. Recording these signals requires a supercomputer since the human brain has 100 billion nerve cells, every cell containing billions of hydrogen atoms, each sending a radio signal. The radiologist uses the data to produce an image—a magnetic resonance image—that is essentially a map of the density of hydrogen atoms in the soft tissue. Schultz used the related technique of diffusion MRI that records the movement of hydrogen atoms (in water molecules) as they move (diffuse) through tissue, clinging to nerve fibers.

Carnap's narrow focus on only what is the case (just the facts) disconnected the language of the exact sciences from society, where what was needed was to imagine what is *not* the case and envision a better world.

In the 1950s the center of mathematics and science shifted from Europe to the United States, where scientific researchers shunned the more theoretical Germanic approach. In Princeton, Boston, New York, and Chicago, international communities of mathematicians and scientists with diverse backgrounds gathered and asked the old questions: How does mankind know the world? What is everything made of? In the 1950s biologists transformed debates about human consciousness from philosophy to neuroscience when they discovered a pattern of brain waves linked to dreaming, so-called REM (rapid eye movement) sleep, and developed the first drug (chlorpromazine, trade name "Thorazine") that is effective against a major mental illness (schizophrenia). By the 1980s they had invented non-invasive imaging tools—positron emission tomography (PET) and magnetic resonance imaging (MRI; plate 13-2)—with which to observe the workings of the living, conscious human brain; these imaging tools are as important to neurology as the telescope is to astronomy. Young researchers also transformed cosmology from philosophy to astrophysics by gathering data using rocket-launched telescopes that look back in time from above earth's atmosphere (plate 13-3) and particle accelerators that probe the subatomic realm.

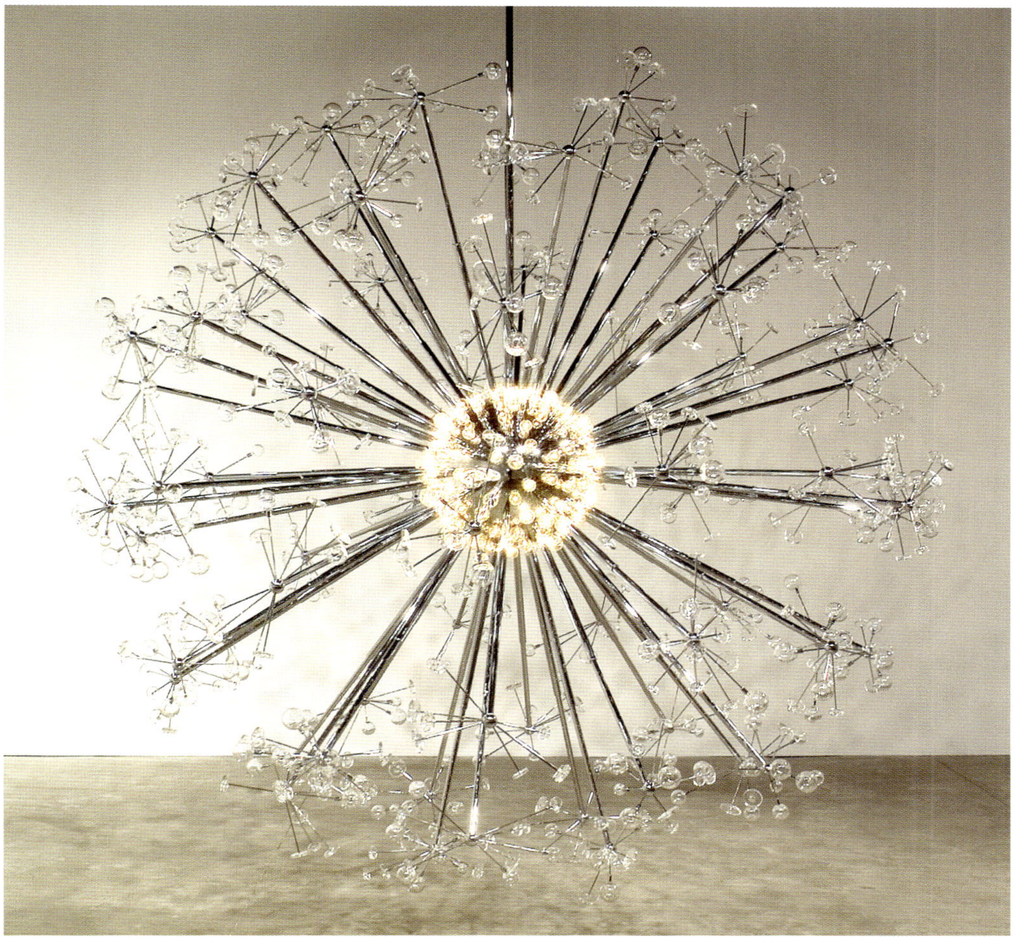

13-3. Josiah McElheny (American, b. 1966), *Last Scattering Surface*, 2006. Hand-blown glass, chrome-plated aluminum, electric lighting, steel cable and rigging, diameter 10 ft. (3 m), hung 6 in. (15.2 cm) above the floor. © Josiah McElheny. Courtesy Donald Young Gallery, Chicago, and Andrea Rosen Gallery, New York, ARG# MJ2006-007.

In collaboration with cosmologist David Weinberg, Josiah McElheny made this spherical sculpture to represent the creation of the universe in the Big Bang when, from a point of infinite density, primordial plasma formed and began expanding. In the early years of the new universe, all forms of electromagnetic radiation (including visible light) could not move freely in the dense plasma but were scattered. But after about 380,000 years, the primordial plasma had expanded and thinned enough to allow light to shine forth and travel in a straight path (symbolized in the sculpture by the chrome rods that emit a flash of light as they emerge from the darkness of the inner sphere). This process is analogous to how an observer on earth can't see the sun on a cloudy day; sunlight can't pass through the cloud but is dispersed in various directions, bouncing off water droplets that compose the vapor, until the light reaches the cloud's edge—the "last scattering surface"—and then light shines forth through clear air in a straight path to earth. Just as an observer on earth can see only light that has emerged from the last scattering surface of an (opaque) cloud, so astronomers can see only the light streaming out from the last scattering surface of the (opaque) primordial plasma. Cosmologists have made a map of that light by recording the photons that emerged from the last scattering surface of the cosmos when it was 380,000 years old. (The data was gathered by a satellite launched by NASA in 2001, the Wilkinson Microwave Anisotropy Probe—WMAP). The map revealed slight variations in the intensity of this ancient light (symbolized in the sculpture by the dappled light on the surface of the inner sphere), caused by variations in the density and temperature of the expanding matter. Over the eons, gravity pulled together these concentrations of matter to form the stars and galaxies of the universe today (symbolized by the sparkling clusters of glass on the boundary of the sculpture), 13.8 billion years after the Big Bang.

In today's global culture there are many conflicting opinions about the nature of postmodernism, but most would agree that the outlook has two features. First, terms like "truth" and "certainty" do not name objective entities but are meaningful only within a system of human thought. Mathematics is the science in which the concept of certainty is most deeply rooted in its history; for millennia mathematicians in the Platonic tradition have described a realm of eternal, perfect abstract objects that exist outside time and space, which mankind knows with certainty. Given its unique history, mathematics has been largely immune to postmodern critiques, but some postmodern thinkers, like those in the Frankfurt school, have raised doubts about the truth and certainty of mathematics.

Second, the postmodern outlook holds that there is no unity of mathematics or wholeness of nature—no single, autonomous structure that incorporates all mathematical and scientific knowledge. Already in the early twentieth century Gödel's incompleteness proof had dashed hopes for a complete axiomatic structure for pure mathematics, and the Copenhagen interpretation of quantum physics had erected roadblocks to a unified mathematical description of nature, launching debates about the role of observation in measurement that continue to this day.

> *Everything flows and nothing abides.*
>
> **—Heraclitus, late sixth century BC**

THE STANDARD MODEL AND THE SEARCH FOR SUPERSYMMETRY

After the general theory of relativity was confirmed in 1919, Einstein began trying to unify gravity and electromagnetism, and he continued working towards this elusive goal until his death in 1955. Meanwhile, in the 1920s physicists discovered two other fundamental forces that operate within the atomic nucleus: the strong and weak nuclear forces (plate 13-4). Physicists today use group theory—the mathematics of symmetry—in their ongoing attempts to unify these four forces because group theory can describe transformations of scale while keeping the laws of physics invariant. In 1954 the Chinese-born American physicist Chen Ning Yang and the American Robert L. Mills described the strong force in terms of Lie groups, and in the 1960s the Americans Steven Weinberg and Sheldon Glashow, together with the Pakistani Abdus Salam, were able to unify electromagnetism and the weak force using group theory.

In the late nineteenth and early twentieth centuries, scientists had discovered the subatomic particles that comprise ordinary matter: the electron (1897), proton (1911), and neutron (1932). Then in the 1950s, equipped with powerful particle accelerators, they began discovering dozens and dozens of other, less common subatomic particles. In 1964 the Japanese physicist Kazuhiko Nishijima and the American physicist Murray Gell-Mann independently proposed that this growing multitude of particles could be organized into a simple schema by realizing that there is a lower level of matter—composed of quarks and leptons—which exhibits a symmetrical pattern that can be displayed on a chart, the Standard Model, which shows the generations of matter that are known today (plate 13-5).[3]

As the monolithic Copenhagen school began to crumble after World War II, several rival interpretations of quantum mechanics emerged. In the 1950s, the American physicist Hugh Everett proposed a so-called many-worlds interpretation, in which he hypothesized that, in addition to the familiar world-out-there, there are other similar worlds (in the sense of other universes) that exist in parallel in the same space and time. When a quantum experiment is performed, such as shooting an electron with, say, a fifty-fifty chance of being absorbed by an atom, both outcomes occur—the electron is absorbed in one universe and passes by in an alternative universe. So one observer detects the absorption of the electron in his or her world, while in a parallel universe his or her double detects nothing. Everett's motivation for proposing the existence of other worlds was that this so-called multiverse restores determinism to the microworld ("Relative State Formulation of Quantum Mechanics," 1957; see plates 13-6 and 12-43 in chapter 12).

By the 1970s there was extensive work being done on the philosophical foundations of quantum mechanics,[4] and in the 1980s a new generation of physicists was renewing efforts to give a complete description of the natural world in a seamless picture including both quantum and classical mechanics. Physicists developed the de Broglie–Bohm interpretation into today's Bohmian mechanics, along with another half-dozen rival interpretations, including string theory.[5]

String theorists hypothesize that subatomic particles such as quarks and electrons

OPPOSITE

13-4. Matthew Ritchie (English, b. 1964), *The Four Forces*, 2008. Oil and marker on linen, 88 × 77 in. (223.5 × 195.6 cm) ea. © Matthew Ritchie. Courtesy Andrea Rosen Gallery, New York.

The artist has symbolized the four forces of nature, clockwise from the upper left: electromagnetism, the strong and weak nuclear forces, and gravity.

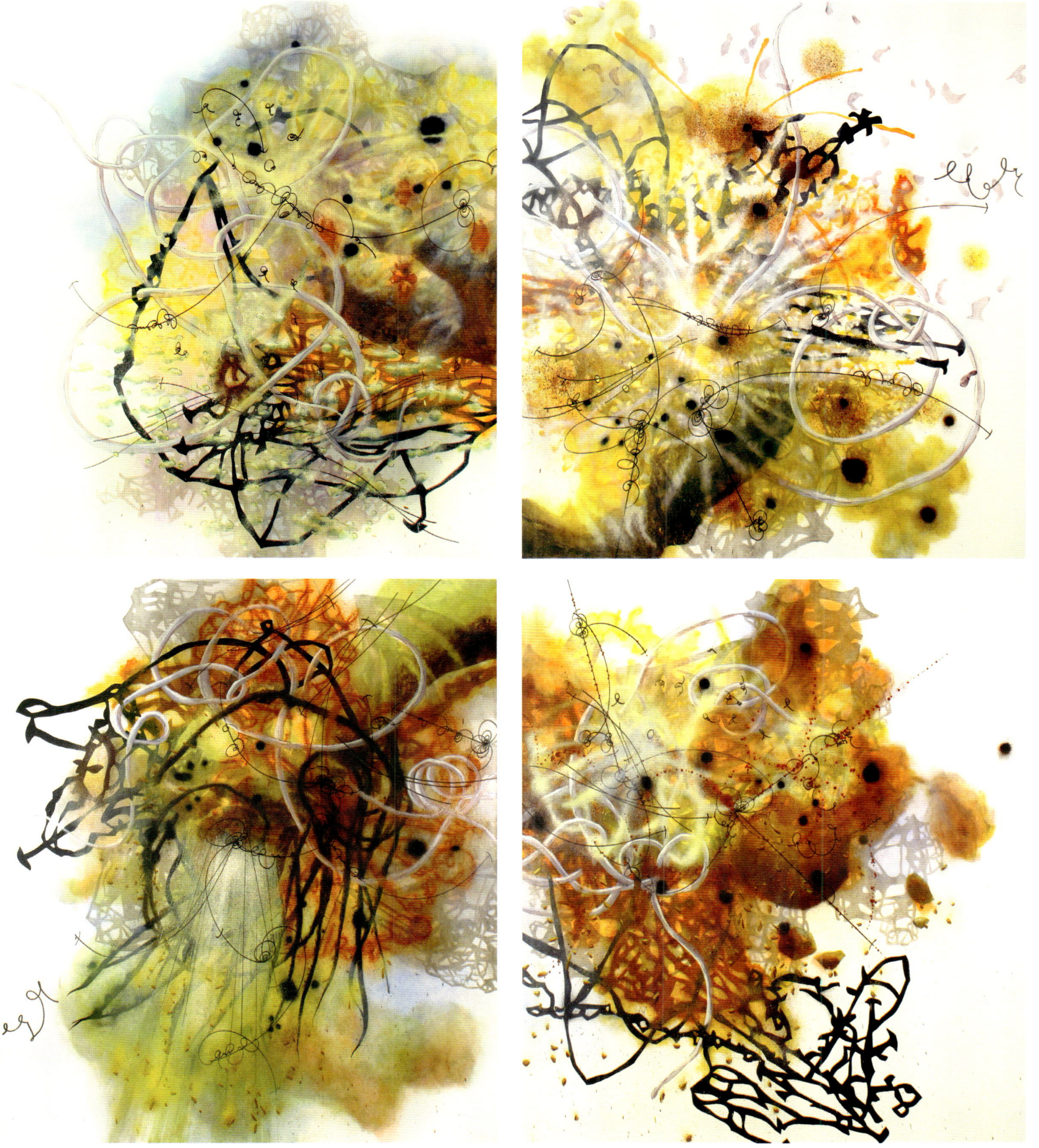

comprise an even lower level of mass-energy in the form of one-dimensional lines ("strings"). Their proposal that these hypothetical strings vibrate in as many as ten spatial dimensions over time has sparked renewed interest in higher dimensions of space. There are many examples in the history of science of researchers beginning with some unexplained data (such as the movement of the planets through the night sky), and first describing the data mathematically and then formulating a theory to explain the patterns they find in it (such as Kepler's laws of planetary motion). Then they deduce (predict) an observable outcome with which they can confirm (or refute) their theory. Because the hypothetical strings are alleged to exist at such a small scale, there *is* no unexplained data—strings have never been observed—but only a mathematical description of them. (Physicists estimate that the size of one string is to a hydrogen atom as one hydrogen atom is to the solar system.) Thus string theorists, such as the American physicist Edward Witten, write a mathematical description of strings that experimental physicists can use to design an experiment, so that they can observe a result that will confirm (or rule out) the existence of strings. Historically scientists have typically adopted an existing piece of mathematics to describe their data (such as Kepler's adoption of Apollonius's conic section to describe the elliptical orbits of the planets). But Witten and his colleagues have reversed the normal order by writing the mathematics, which has then been adopted by mathematicians for other purposes. In 1990 Witten became the first physicist to win the Fields Medal (the mathematician's equivalent of the Nobel Prize).[6]

In 1930 the British physicist Paul Dirac argued, based on the principle of symmetry, that familiar negatively charged electrons should have complementary "twins" with a positive charge—"positrons." The American physicist Carl Anderson's observations of positrons in 1932 confirmed the symmetry of nature at a deep level. String theory predicts, furthermore, that the universe has so-called supersymmetry: for every known species

This web of time—the strands of which approach one another, bifurcate, intersect or ignore each other through the centuries—embraces every possibility. We do not exist in most of them. In some you exist and not I, while in others I do, and you do not, and in yet others both of us exist. In this one, in which chance has favored me, you have come to my gate. In another, you, crossing the garden, have found me dead. In yet another, I say these very same words, but am an error, a phantom.

—Jorge Luis Borges, *The Garden of Forking Paths*, 1941

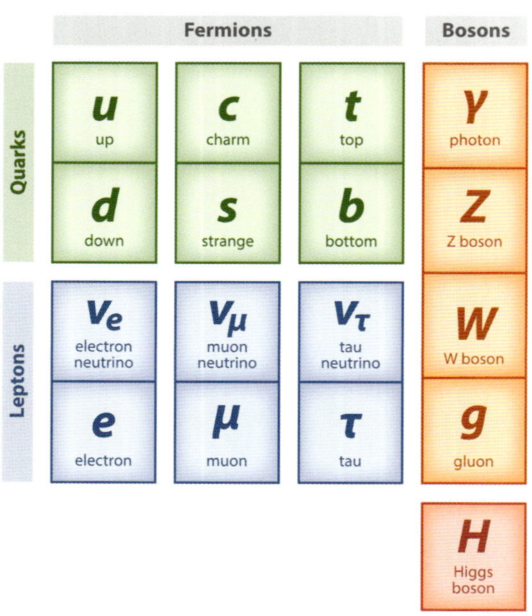

13-5. The Standard Model of elementary particles.

The Standard Model is a theoretical chart describing how the known elementary particles are organized and how they interact with one another via the forces. This table shows the three generations of matter, including six quarks and six leptons. There are also four bosons that are force-carrying particles, each of which transmits an indivisible unit—a "quantum"—of energy. Additionally, there is one boson (named for the British physicist Peter Higgs) that does not transmit a force but is part of the process that gives mass to particles.

The lightest quarks in the Standard Model, designated "up" and "down," combine in groups of three to make up the protons and neutrons in the nucleus of every atom. The groups of three reminded a physicist of a line from James Joyce's *Finnegan's Wake*, "Three quarks for Muster Mark," which led to the name "quarks." Together with electrons, the threesome quarks make up all elements on the periodic table. Quarks are bound together by the strong nuclear force, carried by gluons, to form protons (which have a positive charge) and neutrons (which are electrically neutral). Negatively charged electrons surround the nucleus, forming an atom held together by the attraction of the opposite charges of its protons and electrons. Photons (particles of light) carry this electromagnetic force. (The force of gravity is not represented on this chart.)

Thus familiar atoms are made from just a few of the elementary particles in the Standard Model. The other particles were abundant in the early universe but are rare today. To observe them physicists must either create the rare particles in particle accelerators, or they can study cosmic rays in earth's atmosphere, hoping to observe a rare particle that happens to fly in from outer space.

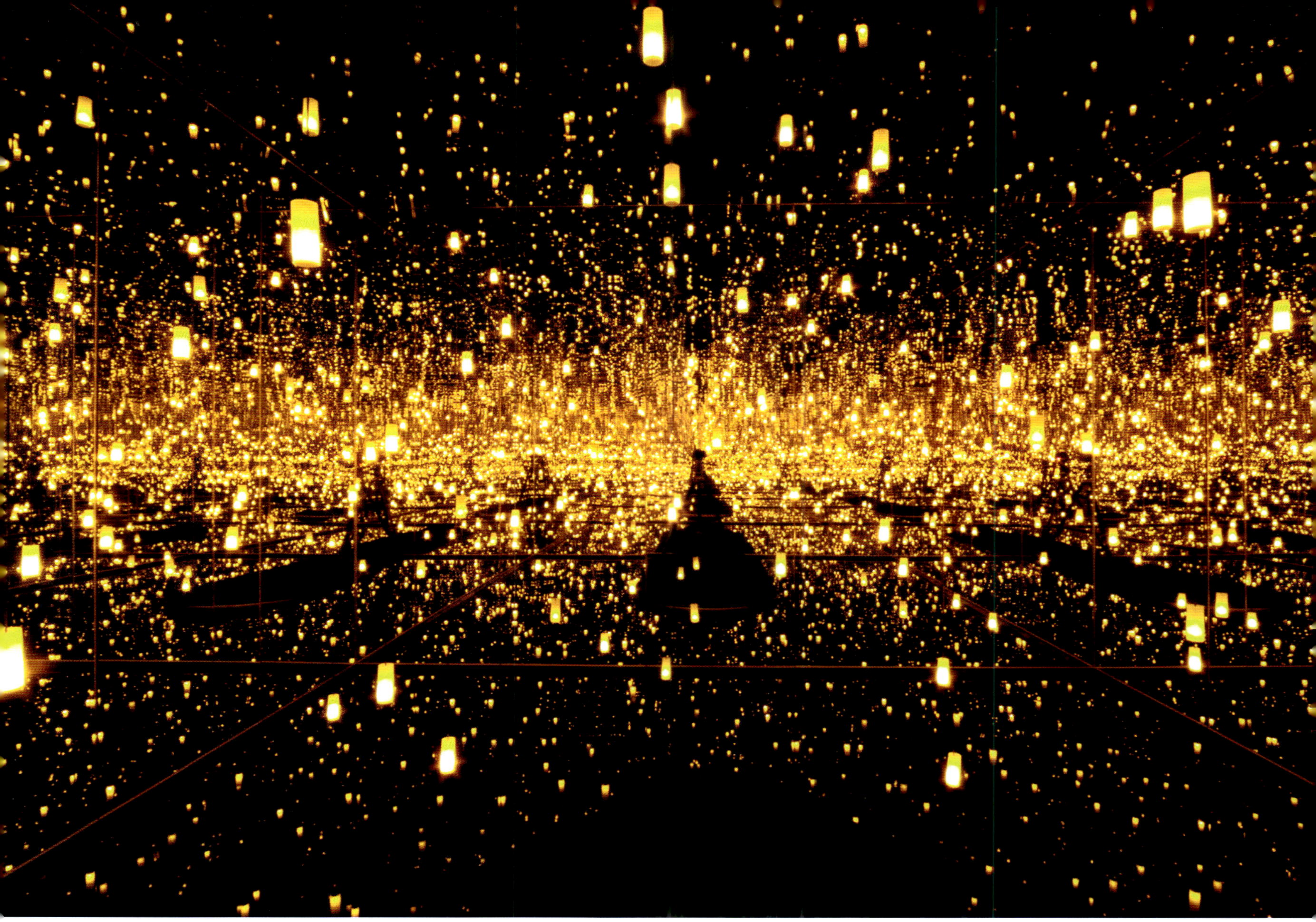

of particle (everything on the Standard Model), there exists a partner species. Physicists hypothesize that these partners, which are rare today, were common in the early universe. Researchers hope to be able to observe the partner particles by recreating the kind of plasma of which the infant universe was composed. This plasma is made in a particle accelerator by accelerating two steams of protons, going in opposite directions, and crashing them into each other, causing some head-on collisions that yield the primordial plasma (plate 13-7). Physicists have made such plasma before at accelerators such as the Fermi National Accelerator Laboratory near Chicago, and they believe that the partner series has so far eluded detection because they are more massive than their known counterparts. Today the most powerful accelerator in the world is the Large Hadron Collider (LHC) operated by CERN (Conseil Européen pour la Recherche Nucléaire/European Council for Nuclear Research), an organization located on the French-Swiss border near Geneva, founded in 1954 by twelve European countries to explore the peaceful use of nuclear energy. In March 2010 the LHC achieved a collision that surpassed the energy of all previous colliders. A hadron (such as a proton) is a composite particle made up of three quarks; the LHC has the power to accelerate streams of protons to 99.999999 percent of the speed of light before colliding them. String theorists predict that the LHC will create

13-6. Yayoi Kusama (Japanese, b. 1929), *Aftermath of Obliteration of Eternity*, 2009. Installation, mixed media. Courtesy of the artist. © Yayoi Kusama.

One hundred and fifty lamps are suspended above water in a darkened hall of mirrors. Reflected endlessly, the light from each lamp creates another world with slight variations in angle and scale.

The task of the physicist is to see down to the underlying, very simple, symmetric reality.

—**Steven Weinberg, interview of 1985**

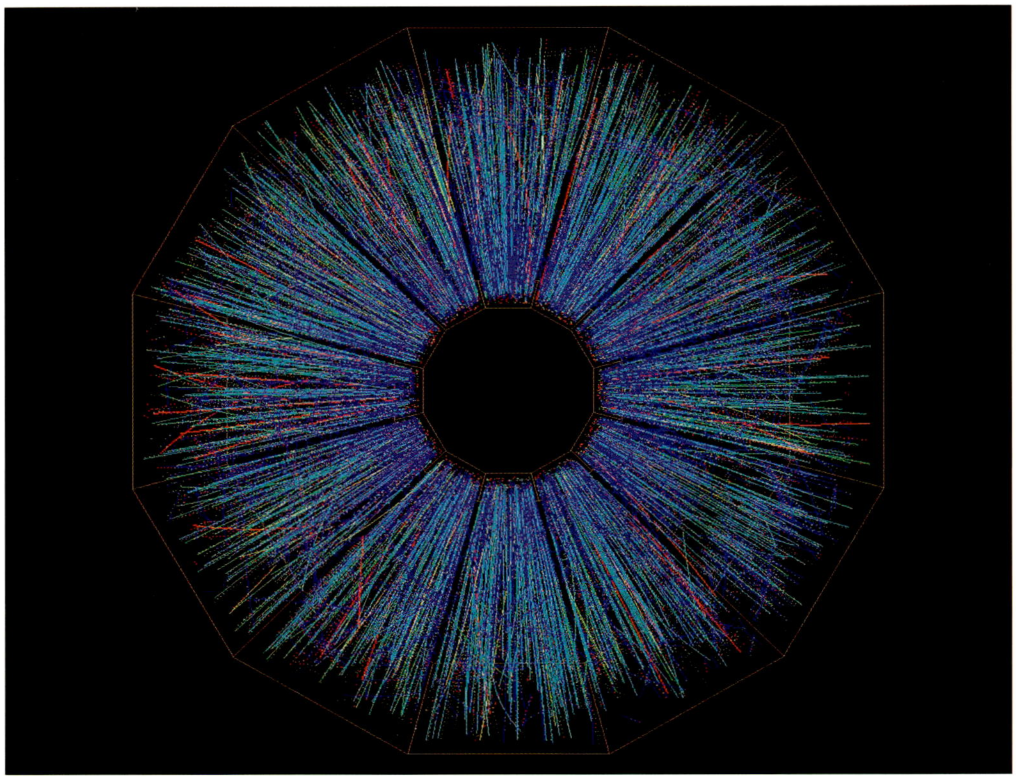

13-7. Nucleus-to-nucleus head-on collision of gold. Solenoidal Tracker at RHIC (STAR) Collaboration, Relativistic Heavy Ion Collider (RHIC), Brookhaven National Laboratory on Long Island, New York.

The Relativistic Heavy Ion Collider (RHIC) at Brookhaven National Laboratory can create matter as hot and dense as in the first microseconds after the Big Bang. RHIC does this by colliding so-called heavy ions, the nuclei of large atoms such as gold (with its 197 protons and neutrons). In a head-on collision close to the speed of light, nuclei break apart into more elementary particles, which scientists can then study using a detector such as STAR. Based on the principle of a solenoid, STAR consists of a set of concentric cylinders through which electricity flows, creating a magnetic field along its axis that lies along the path of the accelerator's beam. Specializing in comprehensive detail, STAR locates the path of up to 6000 particles and clears its memory fast enough to record the details of 1000 collisions per second.

the partner particles and thereby confirm the supersymmetry of nature. In addition to unifying the forces of nature and confirming supersymmetry, another goal of string theory is to account for the values of about twenty numbers—including the speed of light, the mass of an electron, and the force of gravity—that make the universe the way it is.

"WE ARE STARDUST"

Einstein's quest for the unification of the forces of nature, which led to the Standard Model, is an example of the deeply felt conviction that the natural world has a wholeness with which humans are one. This outlook has found expression throughout history in various forms, including Pythagorean pantheism, Spinoza's faith in "God/Nature," the German Romanticism of the *Naturphilosophen*, and Einstein's belief that nature embodies mathematical patterns, which he called his "cosmic religiosity."[7] Like any passion, such an emotion can provide the energy to pursue a goal, or it can get out of control and, as Freud said, the horse can throw the rider.[8] This has happened repeatedly in modern times when surges of Romantic sentiments have thrown reason out of the saddle.

Today there is a new manifestation of this quest for unity, despite postmodern proscription against universalizable truths. Many biologists and astrophysicists have observed that certain scientific breakthroughs made after World War II can be interpreted as implying a unity of human beings with all life forms and with the cosmos. The discovery of the double-helix structure of DNA in the cell nucleus (1953) and the subsequent mapping of mitochondrial DNA gave conclusive evidence that—on a molecular level—members of the species *Homo sapiens* are linked to each other and to all earthly organisms on a tree of life. The biologist E. O. Wilson has observed that this gives mankind a secular creation myth—"the epic of evolution" (*On Human Nature*, 1978; see plate 13-8).[9]

We are stardust.
Billion-year-old carbon.
We are golden.

—**Joni Mitchell, "Woodstock," 1969**

The post-1945 era also witnessed the first successful experiments with large-scale nuclear fusion,[10] which confirmed the origin of the elements in a primordial explosion—the Big Bang. All ninety-two elements on the periodic table were made in this explosion that created the cosmos and in subsequent stellar fusion (fusion occurs when the nuclei of two atoms fuse together to form a larger nucleus). Hydrogen and helium (with one and two protons, respectively) were made in the aftermath of the Big Bang. The nuclei of light elements such as carbon (six protons), nitrogen (seven protons), and calcium (twenty protons) fused in the thermonuclear core of stars. Once the fusion process forms atoms as large as iron (with twenty-six protons), the star becomes unstable, and if it has enough mass, it collapses in an explosion, a supernova (plate 13-9), which produces stupendous energy capable of fusing—in a fraction of a second—the nuclei of all the heavy elements, from cobalt to uranium (with twenty-seven and ninety-two protons, respectively). The presence of all ninety-two elements on earth is conclusive evidence that our solar system formed from the debris of a supernova, which means that—on an atomic level—humans are linked to the cosmos in the sense that all the atoms in their bodies are *literally* stardust.

The nitrogen in our DNA, the calcium in our teeth, the iron in our blood, the carbon in our apple pies were made in the interiors of collapsing stars. We are made of starstuff.

—**Carl Sagan, host of *Cosmos: A Personal Voyage*, 1980**

A related outlook exists in the mathematics world among those, such as Martin Gardner, who are philosophical Platonists. For them, the embodiment of numbers and geometric forms in the natural world implies that—on the level of arithmetic and geometry—there is a unity (an isomorphism) between the physical world-out-there and an immaterial mathematical-world-out-there.[11]

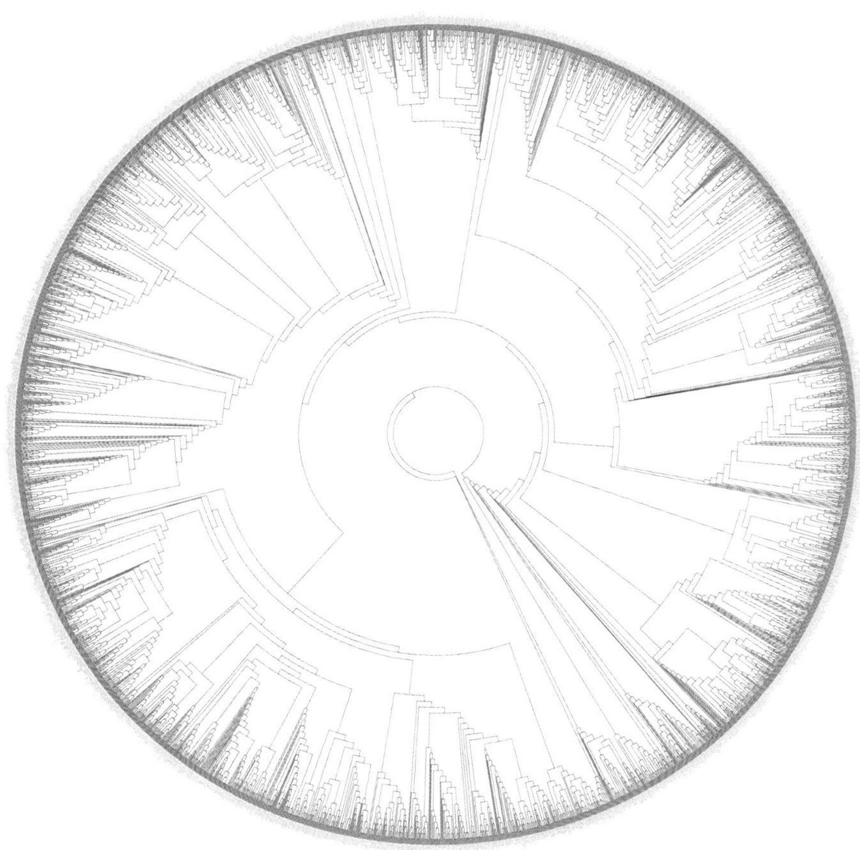

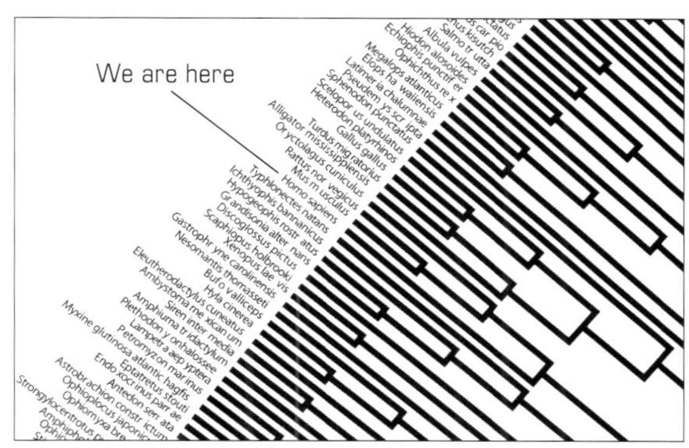

13-8. Radial tree of life, 2003. David Hillis, Derrick Zwickl, and Robin Gutell, University of Texas at Austin.

Biologists made this tree by analyzing genetic material from a sampling of about 3000 species out of an estimated nine million thought to exist today on earth, of which 1.7 million have been formally described and named. The 3000 samples came from the major biological groups: archaea (the simplest in form, with no internal membrane), bacteria (the largest group), and eukaryota (cells with a nucleus, including plants, animals, and fungi). The center represents the root of the tree (the last universal ancestor of all life on earth), and the branches show the interconnection of the lineages that end in 3000 outer twigs.

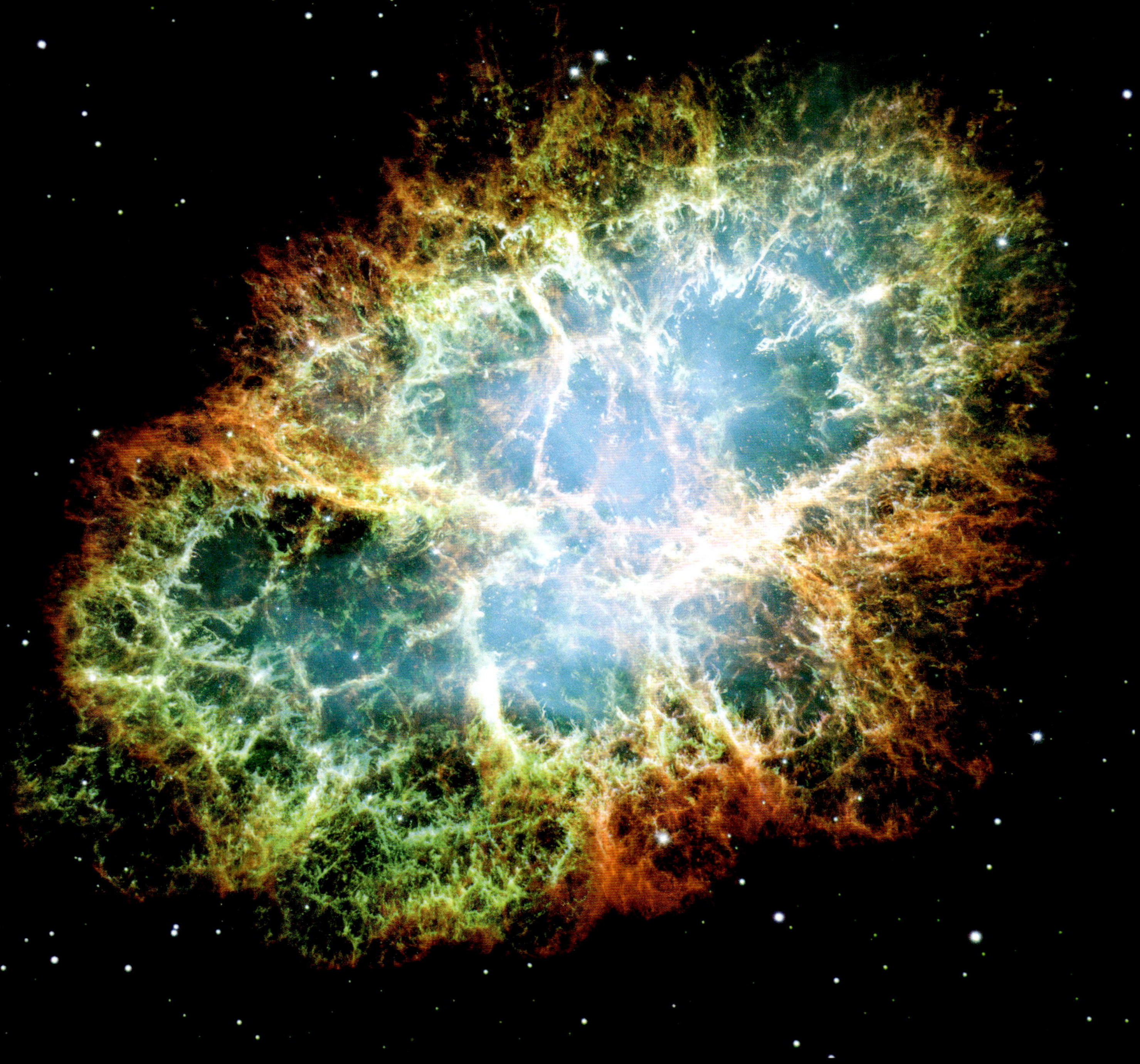

We are in the universe and the universe is in us.

—Neil deGrasse Tyson,
host of *Cosmos: A Spacetime Odyssey*, 2014

"TRUTH" TRUTH AND CERTAINTY

Mathematics is the science that, throughout its long history, has been most closely associated with truth and certainty, and thus it remained largely outside the postmodern debates about cultural relativism that swept other fields after World War II. In the early twentieth century Brouwer and Poincaré argued that arithmetic and geometry are mind-dependent human inventions, with Poincaré stressing that geometry is a social convention (see chapter 6). In Brouwer and Poincaré's day this was a minority outlook, but in the post-1945 climate of cultural relativism, many more mathematicians have described mathematics as mind-dependent. For example, the Hungarian-born American mathematical physicist Eugene Wigner found it inexplicable that a human invention (mathematics) is embedded in nature ("The Unreasonable Effectiveness of Mathematics in the Natural Sciences," 1960). From this perspective, the fact that the moon embodies a sphere should astonish us as much as if a botanist were to discover that the underground network of tree-roots in Central Park is isomorphic to the New York City subway system. Another proponent of mathematical relativism, Morris Kline, professor of mathematics at New York University from 1938 to 1978, taught generations of students that mathematics is a human invention. In the introduction to his textbook, which has the anxiety-producing title *Mathematics: The Loss of Certainty*, Kline declared: "It is now apparent that the concept of a universally accepted, infallible body of reasoning—the majestic mathematics of 1800 and the pride of man—is a grand illusion."[12] What caused this appalling loss to the culture of mankind? Kline lays the blame at the feet of David Hilbert and Bertrand Russell because they held a debate about the foundations of mathematics: "The disagreements about the foundations of the 'most certain' science are both surprising and, to put it mildly, disconcerting. The present state of mathematics is a mockery of the hitherto deep-rooted and widely reputed truth and logical perfection of mathematics."[13] One is relieved to read on and discover that Kline gives no historical evidence to support his devastating pronouncement. Indeed, the history of the foundational debates points to the opposite conclusion; Hilbert and Russell were motivated to isolate the correct foundations of mathematics because they believed in its eternal truth and certainty (see chapters 4 and 5).

In a 1981 book on the history of mathematics, *The Mathematical Experience*, two American mathematicians, Philip J. Davis and Reuben Hersh, similarly describe mathematics not as a science but as one of the humanities: "Mathematics is not the study of an ideal, preexisting nontemporal reality. Neither is it a chess-like game with made-up symbols and formulas. . . . Mathematics is similar to an ideology, a religion, or an art form; it deals with human meanings, and is intelligible only within the context of culture."[14] This thesis makes it difficult to explain why two and two makes four in all cultures throughout history, a question these authors do not address.

Since the 1980s mathematicians concerned with the certainty of their science have focused on what constitutes a proper proof, addressing issues such as the role of intuition, the use of computers, and the required degree of rigor. The use of computers to achieve certainty in proofs that require multiple applications of an algorithm (and thus costly computer time) has raised concern that certainty has a high price tag in today's digital world. In 1994 the Israeli mathematician Doron Zeilberger predicted that the rigorous logic of Euclid's axiomatic method

OPPOSITE

13-9. Crab Nebula, NGC 1952. Six light-years wide, located 6500 light-years from earth in the constellation Taurus. NASA, ESA, Jeff Hester and Allison Loll, Arizona State University.

The Crab Nebula is the remnant of a massive star that exploded, causing what appeared in the night sky as a new star—a supernova—on AD July 4, 1054; it was recorded by Chinese and Japanese astronomers (Europe was in the Dark Ages). When the star collapsed, it crushed its atomic nuclei together, forming an ultradense core, which is essentially a huge atomic nucleus measuring about 18 miles (30 km) in diameter. When the protons and neutrons fused, the protons lost their charge so that the massive core is solid neutrons, a "neutron star." Great mass means strong gravity, so light can escape the core only at its magnetic poles, from which it ejects twin beams, like a lighthouse beacon, as it spins. (The Crab Nebula's neutron star has been spinning thirty times a second since AD 1054.) During the explosion, atoms also spewed outward, sending forth into the cosmos the heavy elements that fused during the cataclysm. The colors of the outer part of this remnant indicate the makeup of the clouds of gas that remain today, including green for sulfur and red for oxygen. The blue light in the center comes from electrons whirling around the spinning core.

will become a thing of the past. In the future, according to Zeilberger, mathematicians who do computer-assisted proofs will settle for a high degree of probability at little or no cost and not require absolute certainty that can be achieved only with a large amount of computer time: "As absolute truth becomes more and more expensive, we would sooner or later come to grips with the fact that few non-trivial results could be known with old-fashioned certainty. Most likely we will wind up abandoning the task of keeping track of price altogether and complete the metamorphosis to non-rigorous mathematics."[15] The American mathematician George E. Andrews responded: "Through the summer of 1993, I was desperately clinging to the belief that mathematics was immune from the giddy relativism that has pretty well destroyed a number of disciplines in the university. . . . Then came an article by my friend and collaborator Doron Zeilberger . . . a first-rate mathematician. Thus one expects that his futurology is based on firm ground. So what is his evidence for this paradigm shift? It was at this point that my irritation turned to horror."[16] Given the reaction to Zeilberger by Andrews and others, one can rest assured that a brave new world devoid of "old-fashioned certainty" is not imminent.

In the current atmosphere of rapidly changing technologies and shifting attitudes towards truth, what stands out in the recent literature on proof theory is the unstated assumption (it goes without saying) that mathematics has a deeply rooted certainty. The triumph of modern mathematics is that it is capable of revealing and proving—with certainty—its own limitations. It is above all this quality of exactness and precision that continues to inspire artists to adopt mathematical methods and concepts.

Mathematicians are vigilant in protecting and preserving the certainty of their science, which gives mathematics its unique status in modern culture, where the interplay between mathematics and the natural world underlies all science and technology. This quality of being determined, fixed, and unfailing is grounded in two Platonic assumptions that have been the *philosophical* basis of mathematics for two and a half millennia: mathematics describes immaterial, abstract objects that exist outside time and space (in the sense that they are embodied in, but do not interact with, the natural world), and mankind knows these objects by contemplation. Reflecting the anti-metaphysical mood that pervades modern thought, many mathematicians adopt these Platonic assumptions to guide their practical work, without pursuing the philosophical implications. The more reflective person may consider the question "Where does mathematics come from?" unanswerable, but nevertheless an awe-inspiring mystery.

The certainty of mathematics is also grounded in the *practice* of mathematics in the sense that it is known with certitude—subjective certainty. Mathematical concepts and proofs have been subjected to repeated, rigorous examination by millions of people. Indeed, throughout its long history mathematics has offered glimpses into a whole ocean of truth, as Newton mused late in his life: "I do not know what I may appear to the world, but to myself I seem to have been only like a boy playing on the sea-shore, and diverting myself in now and then finding a smoother pebble or a prettier shell than ordinary, whilst the great ocean of truth lay all undiscovered before me."[17]

13-10. Eric J. Heller (American, b. 1946), *Transport* 2, ca. 2000. Digital print. Courtesy of the artist.

This image records the movement of electrons launched from the center in all directions, after which they fanned out and formed branches.

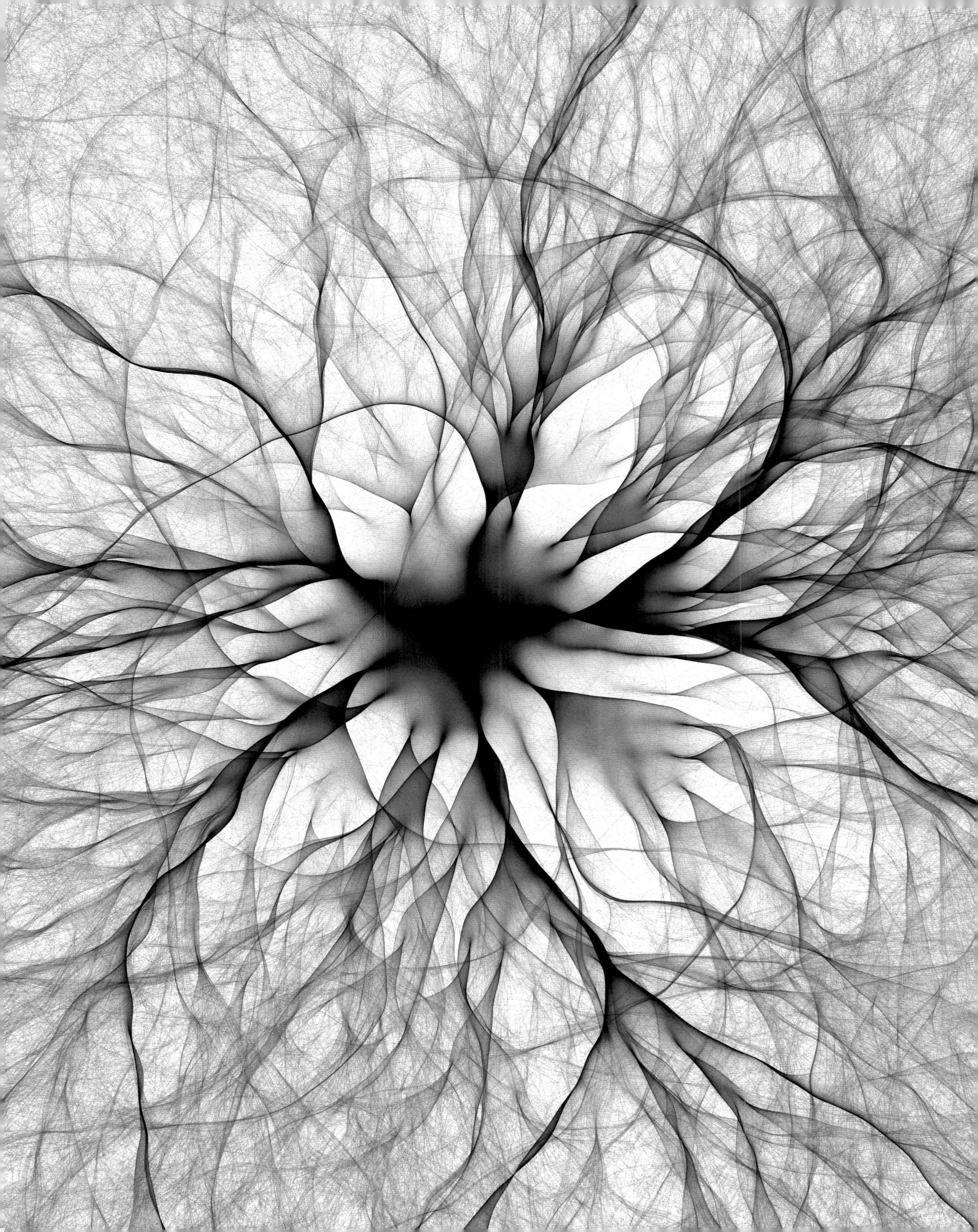

Notes

1. ARITHMETIC AND GEOMETRY

1. On the cognition of numbers in animals and humans, see Stanislas Dehaene's book *The Number Sense: How the Mind Creates Mathematics* (New York: Oxford University Press, 1997), and his essay "Single-Neuron Arithmetic," *Science* 297, no. 5587 (Sept. 6, 2002): 1562–1653.
2. See Thomas Wynn, "The Intelligence of Later Acheulean Hominids," *Man* 14, no. 3 (1979): 371–91; Wynn discusses symmetrical tools using the long-held estimates of their dates ranging from 1.4 to 1.6 millions years ago. Those dates were recently pushed back to 1.76 million years ago after the discovery of symmetrical hand axes made by *Homo erectus* in Kenya; see Christopher J. Lepre et al., "An Earlier Origin for the Acheulian," *Nature* 477 (Sept. 1, 2011): 82–85. "Acheulian" is an alternative spelling of "Acheulean."
3. The perception of two-dimensional shapes is processed in the lower temporal lobe, which is located in the center of the brain, whereas the perception of three-dimensional form is based in the parietal lobe, which is in the cerebral cortex; for a description of experimental data on the evolution of the perception of shape and form, see Stephen M. Kosslyn, *Image and Brain: The Resolution of the Imagery Debate* (Cambridge, MA: MIT Press, 1994), and Thomas Wynn, "Evolutionary Developments in the Cognition of Symmetry," in *Embedded Symmetries: Natural and Cultural*, ed. Dorothy K. Washburn (Albuquerque: University of New Mexico Press, 2004), 27–46.
4. See M. Kohn and S. Mithen, "Handaxes: Product of Sexual Selection?" *Antiquity* 73 (1999): 518–26.
5. This theory is put forth by anthropologist Ellen Dissanayake in her book *Homo Aestheticus: Where Art Comes from and Why* (New York: Free Press, 1992).
6. On the discovery of the flute shown in plate 1-6, see Nicholas J. Conrad, Maria Malina, and Susanne C. Münzel, "New Flutes Document the Earliest Musical Tradition in Southwestern Germany," *Nature* 460, no. 7256 (2009): 737–40.
7. Historians generally conclude from problem 51 of the Rhind papyrus that the Egyptians had a method for determining the area of any particular triangle, but it is not known if they had a proof of their method; see Edward J. Gillings, *Mathematics in the Time of the Pharaohs* (Cambridge, MA: MIT Press, 1972), 138–39.
8. *Proclus, A Commentary on the First Book of Euclid's Elements* (fifth century AD), 65. 3–7, 352.13, trans. Glenn R. Morrow (Princeton, NJ: Princeton University Press, 1992), 52, 113. No writing by Thales survives; his thought is reconstructed from testimony of the historian Herodotus (ca. 484–424 BC), Plato, and Aristotle, whose student Eudemus wrote a history of mathematics that was lost, but not before Proclus, with a copy of Eudemus's text at hand, wrote this commentary on Euclid's *Elements* in the fifth century AD.
9. For this and other theorems attributed to Thales, see Thomas Heath, *A History of Greek Mathematics* (Oxford, England: Clarendon, 1921), 1:130–37.
10. See Charles H. Kahn, *Anaximander and the Origins of Greek Cosmology* (New York: Columbia University Press, 1994).
11. On the symbolism of this biblical image of creation, see Katherine H. Tachau, "God's Compass and *Vana Curiositas*: Scientific Study in the Old *Bible Moralisée*," *Art Bulletin* 80, no. 1 (1998): 7–33.
12. Protagoras is quoted by Diogenes Laërtius in *Lives of Eminent Philosopher* (early third century AD), 9:2, trans. R. D. Hicks (Cambridge, MA: Harvard University Press, 1950), 465.
13. Critias, *Fragment I*, in August Nauck, *Tragicorum Graecorum fragmenta* (1856), trans. Frederich Solsen, in *Plato's Theology* (Ithaca, NY: Cornell University Press, 1942), 35.
14. According to Socrates, Anaxagoras's theory of nature was well known in Athens, where copies sold for a mere drachma; Plato, *Apology of Socrates* (ca. 395–87 BC), 26d–e, trans. Michael C. Stokes (Warminster, England: Aris and Phillips, 1997), 61.
15. Anaxagoras fled to the city of Lampsacus in Asia Minor, where he organized a school of philosophy. For a description of the trial of Anaxagoras see A. E. Taylor, "On the Date of the Trial of Anaxagoras," *Classical Quarterly* 11 (Apr. 1917): 81-87; and J. Mansfield, "The Chronology of Anaxagoras's Athenian Period and the Date of his Trial," *Mnemosyne* 32 (1979): 39–60, 33 (1980): 17–95.
16. Plato, *Apology of Socrates*, 26d, trans., 61.
17. The Pythagorean theorem was attributed to Pythagoras from the fourth century BC to the mid-twentieth century, when the classical scholar Walter Burkert showed that it and other mathematical achievements long associated with his name were the work of his followers; see Burkert, *Lore and Science in Ancient Pythagoreanism*, trans. E. Minar (from the first German edition of 1962; Cambridge, MA: Harvard University Press, 1972). Recent scholarship has shown that it is impossible to determine precisely who wrote the numerology and who did the mathematical proofs ascribed to Pythagoras, so in this book I refer to the group collectively as the "Pythagoreans." The ancient Greeks certainly recognized that Pythagoras and his followers all shared beliefs about numbers, mathematics, and cult practices. In a section of Burkert's book entitled "The Later Non-Aristotelian Tradition and Its Sources, Speusippus, Xenocrates, and Heraclides Ponticus" (53–83), he demonstrates that after the death of Plato, the two leading figures in his Academy, Speusippus, who was Plato's nephew and immediate successor as head of the academy, and Xenocrates, deliberately created the deception that Pythagoras in the sixth century, rather than the Pythagoreans in the fourth century (Plato's contemporaries), originated the Pythagorean theorem and other mathematical ideas (the mathematical structure of music, the harmony of the spheres) in order to establish an early lineage for and give ancient authority to their late master's mathematical cosmology. In the end, Pythagoras and his followers constitute one body of ancient thought—the Pythagoreans. For current views of this complex interweaving of ancient texts, see Charles Kahn, *Pythagoras and the Pythagoreans: A Brief History* (Indianapolis, IN: Hackett: 2001); and Christoph Riedweg, *Pythagoras: His Life, Teaching and Influence*, trans. Steven Redall (Ithaca, NY: Cornell University Press, 2007).
18. See Christiane Sourvinou-Inwood, "Festivals and Mysteries: Aspects of the Eleusinian Cult," in *Greek Mysteries: the Archeology and Ritual of Ancient Greek Secret Cults*, ed. Michael B. Cosmopoulos (London: Routledge, 2003), 25–49.
19. For a compendium of all fragments by Archytas, a discussion of their authenticity, and a translation, see Carl A. Huffman, *Archytas of Tarentum: Pythagorean, Philosopher, and Mathematician King* (Cambridge: Cambridge University Press, 2005).
20. In the *Republic* (530d), Plato quoted a sentence from Archytas's writing on music (from Fragment 1), which he described as part of Pythagorean harmonics. Plato then went on to tell the myth of Er, in which the soldier described his vision of the structure of the universe as eight concentric spheres that hold the fixed stars, the five (known) planets, the sun, and moon, and accompanying the movement of the spheres, the Sirens sing eight notes at constant pitch forming a scale. *Republic* (380–67 BC), 616–17, *The Republic of Plato*, trans. F. M. Cornford (Oxford, England: Clarendon, 1941/rpt. 1955), 345–46.
21. In *Timaeus* (366–360 BC), 35a–36b, Plato described the division of the World Soul into harmonic intervals that were known as the Pythagorean diatonic; *Plato's Cosmology: The* Timaeus *of Plato*, trans. F. M. Cornford (London: Kegan Paul, Trench, Trübner, 1937), 54–57.
22. Plato, *Republic* (380–67 BC), 620, trans. 350. As Plato described in *Laws* 10: "When a change [in the soul] is more significant and more unjust, the movement is into the depths and the places said to be below, which people call 'Hades.' . . . O child or young man, who believes that he is neglected by the gods: the one who becomes more vicious is conveyed to the vicious souls, while the one who becomes better is conveyed to the better souls, in life and in very death, to experience and to do what is appropriate for like to do to like. From this judgment of the gods neither you nor anyone else who's become luck-

less will ever boast of having escaped"; 904c–d, trans. Robert Mayhem in *Plato, Laws* 10 (Oxford, England: Clarendon, 2008), 36.

23. For attempts to reconstruct Polykleitos's lost canon, see J. J. Pollitt, "The *Canon* of Polykleitos and Other Canons," in *Polykleitos, the Doryphoros, and Tradition*, ed. Warren G. Moon (Madison: University of Wisconsin Press, 1995), 19–24.
24. For discussion of the ontological status of mathematical objects in Plato's dialogues, see M. F. Burnyeat, "Plato on Why Mathematics Is Good for the Soul," in *Mathematics and Necessity: Essays in the History of Philosophy*, ed. T. Smiley (Oxford: Oxford University Press, 1999), 1–81, esp. sec. 7, "The Metaphysics of Mathematical Objects," 33–35. Since the time of Aristotle, scholars have commonly referred to mathematical objects as "intermediaries," meaning that they are an indirect way of talking about Plato's Forms. In this essay Burnyeat argues that, in the end, Plato leaves their ontological status unclear (34).
25. Plato, *Timaeus* (366–60 BC), 29e, trans. 33.
26. The order of the cosmos proves that it is "governed by a marvelous intelligence and wisdom"; *Philebus* (360–47 BC), 28d–30d, *The Dialogues of Plato*, trans. B. Jowett (Oxford, England: Clarendon, 1871), 3:175–78.
27. On atomism, see Andrew Pyle, *Atomism and Its Critics: Problem Areas Associated with the Development of the Atomic Theory of Matter from Democritus to Newton* (Bristol, England: Thoemmes Press, 1995).
28. Divine Reason, according to Plato, must have a soul, and therefore the living cosmos had a World Soul. Assuming that there was an analogy between musical harmony and the harmony of the soul, Plato constructed the World Soul according to ratios in Pythagorean harmonic theory, using Philolaus's ratios (the Pythagorean diatonic scale; *Timaeus*, 36a–b).
29. Aristotle, *Physica* (fourth century BC), *The Works of Aristotle*, trans. R. P. Hardie and R. K. Gaye, ed. W. D. Ross (Oxford, England: Clarendon, 1930/rpt. 1970), 2: bk. 2, 193b–94b.
30. Aristotle, *Metaphysica*, (fourth century BC), *The Works of Aristotle*, trans. and ed. W. D. Ross (Oxford, England: Clarendon, 1908/rpt. 1972), 8: bk. delta, 1072b.
31. On Euclid's place within Greek philosophy, see Ian Mueller, *Philosophy of Mathematics and Deductive Structure in Euclid's Elements* (Cambridge, MA: MIT Press, 1981).
32. *The Thirteen Books of Euclid's Elements*, trans. Thomas Heath, 2nd ed. (Cambridge: Cambridge University Press, 1926), 1:153–55. All quotations from Euclid in this book are from Heath's translation.
33. Pythagorean triples are sets of three numbers that make the Pythagorean theorem come out true, such as: 3-4-5; 5-12-13; and 9-12-15. In 1900 a German historian of mathematics, Moritz Cantor, speculated that Egyptians surveyors (such as those shown in plate 1-9) may have knotted their ropes in a pattern of Pythagorean triples in order to form a right triangle when laying out a plot of farmland; Moritz Cantor, *Vorlesungen über Geschichte der Mathematik* (1900; rpt. New York: Johnson, 1965), 1:105–6. Cantor acknowledged that there was no ancient evidence to support his speculation, but historians have heedlessly presented it as a fact. For a list of historians who have rejected Cantor's speculation, see Richard J. Gillings, "The Pythagorean Theorem in Ancient Egypt," *Mathematics in the Time of the Pharaohs* (Cambridge, MA: MIT Press, 1972), appendix 5. Additionally, historians have long thought that around 1800 BC neighbors of the Egyptians in Babylon (present-day Iraq) devised a method for generating Pythagorean triples. A key piece of evidence for this conjecture is a Babylonian clay tablet with columns of numbers on it (Plimpton 322 in the G. A. Plimpton Collection, Columbia University Library, New York), which was excavated in the nineteenth century. In 2002 Eleanor Robson showed that this interpretation of Plimpton 322 is almost certainly incorrect; see Robson, "Words and Pictures: New Light on Plimpton 322," *American Mathematics Monthly* 109 (Feb. 2002): 105–19.
34. See Otto Neugebauer, *The Exact Sciences in Antiquity* (Copenhagen: E. Munksgaard, 1951), 96.
35. While Babylonian mathematics was long seen as a precursor to that of Greece, in today's climate of postcolonial studies, there is an impulse to see the uniqueness of each culture's mathematics to order society and understand the world. For example, in 2010 two historians of science, Alexander Jones and Christine Proust, titled their exhibition of cuneiform tablets: *Before Pythagoras: The Culture of Old Babylonian Mathematics*, implying that the Babylonians discovered key elements of Greek mathematics (Institute for the Study of the Ancient World, New York University, Nov. 12 –Dec. 16, 2010). Eleanor Robson takes this approach in *Mathematics in Ancient Iraq: A Social History* (Princeton, NJ: Princeton University Press, 2008), tracing mathematics in the Middle East from its earliest beginnings in the fourth millennium BC to the end of indigenous Mesopotamian culture in the second century BC, when cuneiform writing was slowly abandoned.
36. The principal ancient sources for Eudoxus's model are Aristotle's *Metaphysics* (fourth century BC), bk. lambda, ch. 8, and Simplicius (AD 490–560) in his commentary *On Aristotle on the Heavens*, 491–97. On the basis of these sources, the late-nineteenth-century Italian historian Giovanni V. Schiaparelli did a detailed reconstruction of Eudoxus's model that was accepted as correct for over a century (*Le Sfere omocentriche de Eudosso, de Callip ed di Aritotele*, 1875). Eudoxus was convinced that the cosmos consisted of twenty-seven concentric, crystalline spheres. The stars were embedded ("fixed") in the immobile outer sphere within which transparent spheres carried the planets Saturn, Jupiter, Mars, Mercury, Venus (in that incorrect order), the sun, moon, and Earth at the center. Aristotle endorsed the principle of Eudoxus's model, which he expanded to include a total of 55 spheres. What was unique about Eudoxus's model, according to Schiaparelli's reconstruction, was his way of accounting for each planet's daily, annual, and retrograde motions by using four spheres to drive each of the planets; the sun and moon each required three, for the grand total of twenty-seven. In a series of articles in the last two decades, historians have thrown the details of Schiaparelli's nineteenth-century model in doubt by questioning, ultimately, the reliability of Simplicius. See Henry Mendell, "Reflections on Eudoxus, Callipus and their Curves: Hippopedes and Callippopeds," *Centaurus* 40 (1998): 177–275; Henry Mendell, "The Trouble with Eudoxus," *Ancient and Medieval Traditions in the Exact Sciences*, ed. Patrick Suppes, J. Moravcsik, and Henry Mendell (Stanford, CA: CSLI, 2000), 59–138; Ido Yavetz, "On the Homocentric Spheres of Eudoxus," *Archive for History of Exact Sciences* 15 (1998): 67–114; and Ido Yavetz, "On Simplicius' Testimony Regarding Eudoxan Lunar Theory," *Science in Context* 16 (2003): 319–30.
37. The Library of Alexandria sustained damage during invasions of Alexandria by Roman emperors Julius Caesar (48 BC) and Aurelian (third century AD); it was ruined further after Emperor Theodosius, who supported Christianity in the collapsing Roman Empire, decreed that pagan buildings be destroyed (391 AD), and met its final end after the Arab conquest of Egypt (642 AD). See Mostafa El-Abbadi, *The Life and Fate of the Ancient Library of Alexandria* (Paris: Unesco/UNDP, 1990), esp. 145–78.
38. For the history of translations of Euclid's *Elements*, see Heath, *Greek Mathematics*, 1:206–12. For the history of the translations of Ptolemy's *Almagest*, see *Ptolemy's Almagest*, trans. G. J. Toomer (New York: Springer, 1984), 2–3.
39. Diophantus of Alexandria (third century AD) is sometimes called "the father of algebra," but he did not use general methods that took him from some to all, as Thomas Heath concluded in his study of algebraic concepts in Greek mathematics: "We do not find in Diophantus' work statements of method put generally as book-work to be applied to examples. Thus we do not find the separate rules and limitations for the solution of different kinds of equations systematically arranged, but we have to seek them out laboriously from the whole of his work, gathering scattered indications here and there, and to formulate them in the best way that we can." Thomas Heath, *Diophantus of Alexandria: A Study in the History of Greek Algebra* (Cambridge: Cambridge University Press, 1885/rpt. 1910), 58.
40. On the Platonic mix of morality and mathematics, see M. F. Burnyeat, "Platonism and Mathematics: A Prelude to a Discussion" (1987), in *Mathematics and Metaphysics in Aristotle/Mathematik und Metaphysik bei Aristoteles*, ed. Andreas Graeser (Bern, Switzerland: Paul Haupt, 1987), 213–40; Andrew Barker, "Ptolemy's Pythagoreans, Archytas, and Plato's Conception of Mathematics," *Phronesis* 39, no. 2 (1994): 113–35; and Burnyeat, "Plato on Why Mathematics Is Good for the Soul," 1–81.
41. *The Seventh Letter*, 341 CD, in Plato, *Phaedrus and the Seventh and Eighth Letters*, trans. Walter Hamilton (Harmondsworth, England: Penguin, 1973), 136.

42. Speusippus, Fragments 44 (Aristotle, *Metaphysica*, 1091 A29 –B3) and 46 (Aristotle, *Metaphysica*, 1075 A31–B1), in Leonardo Tarán, *Speusippus of Athens* (Leiden, the Netherlands: Brill, 1981), 150–51. Fragments 44 and 46 are translated in Aristotle, *Metaphysica* (fourth century BC) 8: bk. 3 (see n. 30).
43. Plato, *Republic* (380–67 BC), 500c, trans. 204.
44. Aristotle, *Nicomachean Ethics* (fourth century BC), bk. 10, 1177, line 30 to 1178, line 2, in *Works*, 9 (see n. 30).
45. *Parmenides* (380–67 BC) 142a, in *Plato and Parmenides*, trans. F. M. Cornford (London: Kegan Paul, Trench, Trübner, 1939), 129.
46. After quoting Plato's description of the One as "beyond being" (*Republic*, 509b), Plotinus goes on to concur with Plato's inference in *Parmenides* that the One is ineffable: "This phrase 'beyond being' does not mean that it is a particular thing—for it makes no positive statement about it—and it does not say its name, but all that it implies is that it is 'not this.' But if this is what the phrase does, it in no way comprehends the One: it would be absurd to try to comprehend that boundless nature"; Plotinus, *Ennead* (ca. AD 253–70), 5:6, trans. A. H. Armstrong (Cambridge, MA: Harvard University Press, 1984), 173.
47. Proclus wrote: "So the soul, exercising her capacity to know, projects on the imagination, as on a mirror, the ideas of the figures; and the imagination, receiving in pictorial form these impressions of the ideas within the soul, by their means affords the souls an opportunity to turn inward from the pictures and attend to herself. . . . She wants to penetrate within herself to see the circle and the triangle there, all things without parts and all in one another, to become one with what she sees and enfold their plurality, to behold the secret and ineffable figures in the inaccessible places and shrines of the gods, to uncover the unadorned divine beauty and see the circle more partless than any center, the triangle without extension, and every other object of knowledge that has regained unity." *Proclus, Commentary*, 113.
48. For the parallels between the medieval and modern debates, see W. V. Quine, "On What There Is," in *From a Logical Point of View* (Cambridge, MA: Harvard University Press, 1953), 1–19, esp. 14–15.
49. Augustine, *De doctrina christiana* (On Christian doctrine; begun ca. 396 and finished 426), in *On Christian Doctrine*, trans. D. W. Robertson (Indianapolis, IN: Bobbs-Merrill, 1958), 75.
50. Augustine, *De ordine* (On Order) (AD 386), 16:44, in *On Order*, trans. Silvano Borruso (South Bend, IN: St. Augustine's Press, 2007), 107–9.
51. "Augustine: Now, then, have you ever seen with the eyes of the body such a point, or such a line, or such width? Evodius: No, never. These things are not bodily. Augustine: But if, by some sort of remarkable affinity of realities, bodily things are seen with bodily eyes, it must be that the soul by means of which we see these incorporeal things is not a body, nor like a body." Saint Augustine, *De quantitate animae* (The greatness of the soul; AD 387–88), 13:22, in *St. Augustine: The Greatness of the Soul*, trans. Joseph M. Colleran, (Westminster, MD: Newman, 1950), 39.
52. Augustine, *De ordine*, 15:42, trans. 105-7.
53. Augustine, *Quantitate*, 16:27, trans. 46.
54. Needham summarized his view of so-called ecumenical natural philosophy as follows: "The standpoint here adopted assumes that in the investigation of natural phenomenon all men are potentially equal, and that the ecumenism of modern science embodies a universal language that they can all comprehensibly speak, that the ancient and medieval sciences (though bearing an obvious ethnic stamp) were concerned with the same natural world and could therefore be subsumed into the same ecumenical natural philosophy, and that this has grown, and will continue to grow among men, *pari passu* with the vast growth of organization and integration in human society, until the coming of the world co-operative commonwealth which will include all peoples as the waters cover the sea." Joseph Needham, "The Roles of Europe and China in the Evolution of Ecumenical Science" (1964), in *Science and Civilization in China* (Cambridge: Cambridge University Press, 2004), 7: part 2, 42. Needham began his career as a biologist who sought ways to merge his scientific, mechanistic worldview with religion, which for him meant panpsychism (a nineteenth-century version of monadology), or as Needham put it, "a modification of Spinoza"; see Needham, "Mechanistic Biology and the Religious Consciousness," in *Science, Religion, and Reality*, ed. Joseph Needham (New York: Macmillan, 1925), 219–58; the quote is on 259.
55. *The Nine Chapters on the Mathematical Art: Companion and Commentary*, trans. Shen Kangshen, John N. Crossley, and Anthony W.-C. Lun (Oxford: Oxford University Press; Beijing: Science Press, 1999), and *Les neuf chapitres: Le classique mathématique de la Chine ancienne et ses commentaries*, trans. Karine Chemla and Guo Shuchun (Paris: Dunod, 2004).
56. Chemla has presented her revision of Chinese mathematics in a series of publications, including collections of essays by herself and others, *La valeur de l'exemple: perspectives chinoises*, ed. Karine Chemla (Saint-Denis, France: Presses Universitaires de Vincennes, 1997), and *Divination et rationalité en Chine ancienne*, ed. Karine Chemla, D. Herper, and M. Kalinowsi (Saint-Denis, France: Presses Universitaires de Vincennes, 1999). She has summarized the core of her position in the essay "Generality above Abstraction: The General Expressed in Terms of the Paradigmatic in Mathematics in Ancient China," *Science in Context* 16 (2003): 413–58.
57. Trans. J. Needham and W. Ling in Needham, *China*, 3:23 (see n. 54). The translation of the Gougu theorem has become an issue in recent debates about the distinction between the particular and the general in Chinese mathematics. In the 1999 English translation of *The Nine Chapters*, Shen Kangshen, John N. Crossley, and Anthony W.-C. Lun translated *gougu* as "right-angled-triangle." In a review of the translation, Mary Tiles complained: "The translation obscures the fact that there is some debate about whether the Chinese operated using the concepts of angle and triangle. 'Go' and 'gu' refer to the two arms of a carpenter's square, not directly to the geometrical right-angled triangle. Angular measures are not discussed." *Philosophy East and West* 52, no. 3 (2002): 386–89; the quote is on 388.
58. Especially after Sabetai Unguru's 1975 essay in which he called for a cultural history of mathematics, "On the Need to Rewrite the History of Greek Mathematics," *Archive for History of Exact Sciences* 15 (1975): 67–114. For a sample of the broad range of post-Needham scholarship, see several commemorative collections of essays, which are not necessarily laudatory of his work: *Chinese Science: Explorations of an Ancient Tradition*, ed. Shigeru Nakayama and Nathan Sivin (Cambridge, MA: MIT Press, 1973), commemorating Needham's seventieth birthday, and two memorial volumes after his death in 1995, *Beyond Joseph Needham: Science, Technology, and Medicine in East and Southeast Asia*, ed. Morris Low (Chicago: University of Chicago Press, 1999), which focused on Asian science other than Chinese; and *Situating the History of Science: Dialogues with Joseph Needham*, ed. S. Irfan Habib and Dhruv Raina (New York: Oxford University Press, 1999). For an overview of non-Western mathematics, see Marcia Ascher, *Mathematics Elsewhere: An Exploration of Ideas across Cultures* (Princeton, NJ: Princeton University Press, 2002).
59. For statements of this viewpoint, see Joseph Needham, "Human Law and the Laws of Nature," in Needham, *China*, 2:518–83; A. C. Graham, "China, Europe, and the Origins of Modern Science," in *Chinese Science; Explorations of an Ancient Tradition*, ed. Shigeru Nakayama and Nathan Sivin (Cambridge, MA: MIT Press, 1973), 45–69, esp. 55–58; and in the same volume, Kiyosi Yabuuti, "Chinese Astronomy: Development and Limiting Factors," 91–103, esp. 92–94.
60. The cultural historians D. L. Hall and R. T. Ames have described a distinct concept of order in Confucianism, contrasting Eastern "aesthetic order" with Western "logical order"; see *Thinking Through Confucius* (Albany, NY: State University of New York Press, 1987), esp. "The Primacy of Aesthetic Order," 132–38.
61. Chuang Tzu, *The Writings of Chuang Tzu*, bk. 14, The Revolution of Heaven (ca. 300 BC), in *The Tao tê ching; The Writings of Chuang-tzŭ; The Thâi-shan Tractate of Actions*, trans. James Legge (Taipei: Book World, 1891/1963 rpt.), 393.
62. On the possible relationships between Eastern religions and Greek mystery cults, see F. M. Cornford, "Tao, Rta, and Asha," in *From Religion to Philosophy: A Study in the Origins of Western Speculation* (1912; Mineola, NY: Dover, rpt. 2004), 172–77; and Wilhelm Halbfass, "The Philosophical View of India in Classical Antiquity," in *India and Europe: An Essay in Understanding*, n.t. (translation of *Indien und Europa, Perspektiven ihrer geistigen Begegnung*, 1981; Albany, NY: State University of New York Press, 1988), 2–23. On Neoplatonism and Indian philosophy, see Paul Hacker, "Cit and Nous," *Neoplatonism and Indian Thought*, ed. R. Baine Harris (Norfolk, VA: International Society for Neoplatonic

Studies, 1982), 161–80; and for an overview of the scholarly literature on Plotinus, see in the same volume, Albert M. Wolters, "A Survey of Modern Scholarly Opinion on Plotinus and Indian Thought," 293–308. For an overview of the scholarly literature on possible connections between ancient Christianity and Buddhism, see Zacharias P. Thundy, *Buddha and Christ: Nativity Stories and Indian Traditions* (Leiden, the Netherlands: Brill, 1993), 1–17. Halbfass, Hacker, Wolters, and Thundy all doubt that there could have been an exchange of ideas between representatives of Eastern thought and mystery cults. Only the great classical scholar F. M. Cornford is cautiously optimistic.

63. "There are Greek writers, some of an earlier, some of a later date, who have borrowed this doctrine [transmigration of souls] from the Egyptians," Herodotus, *The Persian Wars* (fourth century BC), bk. 2:123, in *The Greek Historians: The Complete and Unabridged Historical Works of Herodotus*, trans. and ed. George Rawlinson and Francis R. B. Godolphin (New York: Random House, 1942), 1:141.
64. "Accordingly Clitarchus [a historian of Alexander the Great], in his twelfth book, says that the gymnosophists despise death"; Diogenes Laërtius, *Lives and Opinions of Eminent Philosophers* (third century AD), I:6, trans. C. D. Yonge (London: Henry G. Bohn, 1853), 8.
65. "In his twenty-eighth year he [Plotinus] felt the impulse to study philosophy and was recommended to the teachers in Alexandria. . . . He became eager to make acquaintance with the Persian philosophical discipline and that prevailing among the Indians. As Emperor Gordian was preparing to march against the Persians, he joined the army and went on the expedition." Porphyry, *Vita Poltini* (Life of Plotinus; early third century AD), 3:5ff, trans. A. H. Armstrong, in *Plotinus* (Cambridge, MA: Harvard University Press, 1966), 9.
66. Pseudo-Dionysius the Areopagite, *On Divine Names* (ca. AD 500), V: 5–6, in *The Works of Dionysius the Areopagite*, trans. John Parker (Merrick, NY: Richwood, 1897–99/rpt. 1976), 1:77–78.
67. Pseudo-Dionysius the Areopagite, *The Heavenly Hierarchy* (ca. AD 500), I:3, in *Works of Dionysius*, 2:3.
68. Ibid., 2:1.
69. Thierry of Chartres, *Tractatus de sex dierum operibus* (twelfth century), in *Commentaries on Boethius by Thierry of Chartres and Others of His School*, ed. Nikolaus M. Haring (Toronto: Pontifical Institute of Medieval Studies, 1971), 568, sec. 30; trans. Peter Ellard in *The Sacred Cosmos: Theological, Philosophical, and Scientific Conversations in the Early Twelfth Century School of Chartres* (Scranton, NY: University of Scranton Press, 2007), 15.
70. Thierry, *Tractatus*, 568, sec. 31, trans. 108–9.
71. The fifth-century mystic Pseudo-Dionysius was the titular patron saint of the Abbey of Saint-Denis because there was a mix-up in the identity of three persons: (1) Dionysius the Areopagite, who was converted by Saint Paul in the first century AD (Acts 17:34); (2) Saint Denis, a bishop of Paris who was martyred in circa AD 250; (3) Pseudo-Dionysius, an (anonymous) Orthodox monk who lived in the fifth century AD, when he adopted the pseudonym "Dionysius the Areopagite" (in order to give himself a first-century pedigree). The Abbey of Saint-Denis was founded in the seventh century, and by the early ninth century the legends of Dionysius (1) and Denis (2) began to be intermixed, which bestowed on the Parisian Denis the prestige of having been converted by Saint Paul. The two historical persons eventually became identified as one, aided by the coincidence that the name "Denis" derives from "Dionysius." The disambiguation of the three persons was done in the 1940s by Erwin Panofsky, who gave the following account: in 814 the Byzantine Emperor Michael II gave an original Greek manuscript of the fifth-century mystic Pseudo-Dionysius (3) to the son of Charlemagne, Louis the Pious, who believed it to have been written by the convert of Saint Paul, Dionysius/Denis (1/2), and he so deposited it in the library of the Abbey of Saint-Denis, where it was translated into Latin. This is the book that Suger read. Suger's mistaken identity of the author is, however, irrelevant to the unmistakable impact it had on him. Panofsky saw Suger as heir to the Cluniac view that the sanctuary should be splendidly decorated to reflect heaven, and he used the writing of Pseudo-Dionysius to combat criticism from the Cistercians, who advocated a plainer interior; see the writings of Abbot Suger in *On the Abbey of St.-Denis and Its Art Treasures*, trans. and ed. Erwin Panofsky (Princeton, NJ: Princeton University Press, 1946). Most historians, myself included, concur with Panofsky's view of Abbott Suger. For arguments against the influence of Pseudo-Dionysius's writings on Suger, see Peter Kidson, who wrote, "Suger was not in any serious sense a follower of the Pseudo-Dionysius," in "Panofsky, Suger, and St-Denis," *Journal of the Warburg and Courtauld Institutes* 50 (1987), 1–17; and Lindy Grant, who also downplays the writings of Pseudo-Dionysius in *Abbot Suger of St-Denis: Church and State in Early Twelfth-Century France* (London: Longman, 1998), 270–71.
72. Suger, *On the Abbey*, 73-75.
73. "Thus, when—out of my delight in the beauty of the house of God—the loveliness of the many-colored stones has called me away from the external cares, and worthy meditation has induced me to reflect, transferring that which is material to that which is immaterial, on the diversity of the sacred virtues: then it seems to me that I see myself dwelling, as it were, in some strange region of the universe which neither exists entirely in the slime of the earth nor entirely in the purity of Heaven; and that, by the grace of God, I can be transported from this inferior to that higher world in an anagogical manner." Suger, *On the Abbey*, 63–65.
74. This is the intriguing suggestion of the medieval historian Sumner McNight Crosby in "Crypt and Choir Plans of St. Denis," *Gesta: International Center of Medieval Art* 5 (1966): 4–8.
75. Aristotle, *Nicomachean Ethics* (fourth century BC), bk. 5, sec. 7, trans. Martin Oswald (Indianapolis: Bobbs-Merrill, 1962), 131.
76. Thomas Aquinas, *Summa Theologiae* (A treatise on theology; 1265–73), trans. Fathers of the English Dominican Province (Westminster, MD: Christian Classics, 1911/rpt. 1981), 2:105.
77. See David Topper and Cynthia Gillis, "Trajectories of Blood: Artemisia Gentileschi and Galileo's Parabolic Path," *Woman's Art Journal* 17 (Spring–Summer 1996), 10–13.
78. Johannes Kepler, *Harmonices Mundi* (Harmony of the world; 1619), 5:7, trans. E. J. Aiton, A. M. Duncan, and J. V. Field (Philadelphia: American Philosophical Society, 1997), 446.
79. The gravitational constant is a physical constant. Scientists believe that physical constants (such as gravity and the speed of light) are the same throughout the universe and they do not change over time. Gravity has a fixed numerical value that is based on the measurement of the attraction between two uniform spheres of known mass. It is especially difficult to measure because the force of gravity is so weak. Newton did not know the strength of the force when he described universal gravitation, but he was confident that it would be determined one day. The gravitational constant was first measured by the British scientist Henry Cavendish (*Philosophical Transactions*, 1798).
80. For example, in 1662 Robert Boyle wrote the *mathematical* formula stating that there is an inverse relation between the volume and pressure of a gas ($vp = c$, the volume [v] times the pressure [p] is always the same [it equals a constant, c]). For example, if the volume is cut in half, then the pressure doubles, assuming the temperature remains the same. Two hundred years later, when atomic theory was developed, scientists gave a *physical* description of gas that explained *why* volume and pressure have this relation. Gas molecules in a container move randomly in all directions, bouncing off the walls; when they are squeezed into a volume that is half the size, they hit the walls twice as often, doubling the pressure.
81. Isaac Newton, *Philosophiae Naturalis Principia Mathematica* (Mathematical principles of natural philosophy; 1687), the quote is in the "General Scholium" that Newton added to the second (1713) translation from Latin by Andrew Motte in 1729, translation revised by Florian Cajori, ed. R. T. Crawford, in *Sir Isaac Newton's Mathematical Principles of Natural Philosophy and His System of the World* (Berkeley: University of California Press, 1947), 547.
82. Ibid.
83. Charles Darwin, *The Descent of Man and Selection in Relation to Sex* (1871), 2nd ed. (New York: Appleton, 1883), 623.
84. For scholarly opinions on the Galileo affair, see the essays collected in *The Church and Galileo*, ed. Ernan McMullin (Notre Dame, IN: University of Notre Dame Press, 2005).
85. Johannes Kepler, *Harmonices Mundi*, 146.
86. Galileo Galilei, *The Assayer* (1623), in *Discoveries and Opinions of Galileo*, trans. Stillman Drake (Garden City, NY: Doubleday, 1957), 237–38.

87. Galileo Galilei, *Dialogue Concerning the Two Chief World Systems* (1632), trans. Stillman Drake, (New York: Modern Library, 2001), 118. Galileo's claim formed part of the Inquisition's indictment of him, which read in part: "That he wrongly asserts and declares a certain equality between the human and the divine intellect in the understanding of geometric matters" (Special Commission Report on the *Dialogue*, Sept. 1632), trans. Maurice A. Finocchiaro in *The Galileo Affair: A Documentary History* (Berkeley: University of California Press, 1989), 222.
88. On Galileo's influence see, for example, Anthony Blunt, who wrote "Galileo's writings were much studied in intellectual circle's during Borromini's youth, and it is, I believe, from him that Borromini derived his conception of nature," in *Borromini* (London: Penguin, 1979), 47; and, more recently, John Hendrix, *The Relation between Architectural Forms and Philosophical Structures in the Work of Francesco Borromini in Seventeenth-Century Rome* (Lewiston, NY: Mellen, 2002), 45, 93, and 121. Kepler's influence on Borromini has also been suggested; see John G. Hatch, "The Science behind Francesco Borromini's Divine Geometry," *Visual Arts Publications* 4 (Jan. 1, 2002): 127–39. On the underlying geometry of the plan and dome, see Julia M. Smyth-Pinney, "Borromini's Plans for Sant'Ivo alla Sapienza," *Journal of the Society of Architectural Historians* 59, no. 3 (Sept. 2000): 312–37.
89. Newton, *Principia*, 544. See also James E. Force, "Newton's God of Dominion: The Unity of Newton's Theological, Scientific, and Political Thought," in *Essays on the Context, Nature, and Influence of Isaac Newton's Theology*, ed. James E. Force and Richard H. Popkin (Dordrecht, the Netherlands: Kluwer, 1990), 75–102.

2. PROPORTION

1. Vitruvius, *On Architecture* (first century BC), 3.1.3 in *Vitruvius, Ten Books on Architecture*, trans. Ingrid D. Rowland (Cambridge: Cambridge University Press, 1999), 47.
2. See Judith V. Field, *Piero della Francesca: A Mathematician's Art* (Oxford: Oxford University Press, 2005), 122–23.
3. Ratios were used as a tool in Greek mathematics before they appeared in the *Elements*, but Euclid was the first to offer a *theory* of proportion, which is often ascribed to Eudoxus. The Euclid/Eudoxus theory of proportion rests on the definition "being in the same ratio" (*Elements*, book 5). See Ken Saito, "Phantom Theories of pre-Eudoxean Proportion," *Science in Context* 16, no. 3 (2003): 331–47.
4. Campanus made the comment about a theorem (*Elements*, book 14, proposition 10) that was considered to be by Euclid in Campanus's day, although it is not today. But this doesn't matter because he could have made the remark about any of Euclid's (genuine) theorems that entail the ratio. The Latin original of the quotation is from Pacioli's edition of Campanus's translation (Venice, 1509), 137v.; the English translation is by Albert van der Schoot in his essay "The Divined Proportion," in *Mathematics and the Divine: A Historical Study*, ed. T. Koetsier and L. Bergmans (Amsterdam: Elsevier, 2005), 655–72; the quote is on 662.
5. The Renaissance biographer Vasari complained that Pacioli's (unattributed) printing of Piero's *De quinque corporibus regularibus* was plagiarism: "Maestro Luca, claiming the authorship of these books, had them printed as his own, for they had fallen into his hands after the death of Piero"; Giorgio Vasari, *Lives of the Most Eminent Painters, Sculptors, and Architects* (1550/enlarged 1568), trans. Gaston du C. de Vere (London: Philip Lee Warner, 1912–15), 3:22. In 1509, books had been printed in multiple copies for only about fifty years; giving Pacioli the benefit of the doubt, the confusion in authorship may be because standards of use were still unclear. However, another sixteen-century author, Daniele Barbaro, did credit Piero for the parts of *De prospectiva pingendi* that he borrowed for his book *La pratica della perspettiva* (1569). Pacioli also printed (unattributed) many problems from Piero's *Trattato d'abaco* in his *Summa de Arithmetica* (1494).
6. For an exhaustive chronological listing of references to Euclid's ratio in mathematical literature, see appendix I of Roger Herz-Fischler, *A Mathematical History of Division in Extreme and Mean Ratio* (Waterloo, Ontario: Wilfrid Laurier University Press, 1987), 164–70.
7. Adolf Zeising, *Neue Lehre von den Proportionen des menschlichen Körpers, aus einem bisher unerkannt gebliebenen, die ganze Natur und Kunst durchdringenden morphologischen Grundgesetze entwickelt und mit einer vollständigen historischen Uebersicht der bisherigen Systeme begleitet* (A new system of proportions of the human body, which was until now unrecognized, and of the fundamental morphological law which permeates all of nature and art; Leipzig, Germany: R. Weigel, 1854), v.
8. See Roger Fischler's discussion of the incorrect use of statistical data in studies that claim to find the Golden Ratio in various objects: "How to Find the 'Golden Number' without Really Trying," *Fibonacci Quarterly* 19 (1981): 406–11.
9. For a summary of false historical claims about the Golden Section, see George Markowsky, "Misconceptions about the Golden Ratio," *College Mathematics Journal* 23 (1992): 2–19. On pseudo-science related to the Golden Ratio, see Martin Gardner, "The Cult of the Golden Ratio," *Skeptical Inquirer* 18, no. 3 (1994): 243–47.
10. A century later scholars are still spending their research time trying to prove (or disprove) Fechner's (nonexistent) assertion that the 1.618 ratio gives aesthetic pleasure. For example, Tony Collyer and Alex Pathan, professors of physics and engineering (respectively) at Sheffield Hallam University in England, argue for the positive in "The Pyramids, the Golden Section, and 2π," *Mathematics in School* 29, no. 5 (2000): 2–5. Susan T. David and John C. Jahnke, two psychologists at Miami University in Oxford, Ohio, argue for the negative in "Unity and the Golden Section: Rules for Aesthetic Choice?" *American Journal of Psychology* 104, no. 2 (1991): 257–77.
11. See, for example, Friedrich Röber, who purported to demonstrate that ancient architects used the Golden Ratio to determine the angle of inclination of the pyramids at Giza and the proportions of the cella of the Parthenon in Athens; *Die aegyptischen Pyramiden in ihren ursprünglichen Bildungen: nebst einer Darstellung der proportionalen Verhältnisse im Parthenon zu Athen* (Egyptian pyramids in their original buildings, next to a presentation of the proportional relations of the Parthenon in Athens; Dresden: Woldemar Türk, 1855). Little is known about Röber; for a detailed history of attempts to determine geometric patterns in the Pyramid of Khufu at Giza, see Roger Herz-Fischler, *The Shape of the Great Pyramid* (Waterloo, Ontario: Wilfrid Laurier University Press, 2000).
12. According to Thiersch, "We are searching for a law that is compatible with a multiplicity of forms and has been proven under various conditions. A step towards finding such a law has been taken by the German thinker Zeising who singled out the Golden Section. . . . By examining the successful works of all times, we find that in every building a basic form is repeated, and the individual parts are always in this basic form. There are infinitely many different forms, which may in and of themselves be neither beautiful nor ugly. The harmony is achieved by the repetition of the main figure in its subdivisions." "Die Proportionen in der Architektur," in *Handbuch der Architektur*, Josef Durm, ed. (Darmstadt, Germany: Diehl, 1883), 38–77; the quote is on 39.
13. Jacob Burckhardt to August Thiersch, Sept. 27, 1883, in Burckhardt, *Briefe*, ed. Max Burckhardt (Basel, Switzerland: Schwabe, 1974), 8:156.
14. Heinrich Wölfflin, "Zur Lehre von den Proportionen," *Deutsche Bauzeitung* 23, no. 46 (1889): 278. Thiersch's response to Wölfflin's addition of reverse regulating lines was unenthusiastic, warning that they made the analysis of architecture more complicated mathematically (*mathematisch verwickelter*) and were somewhat arbitrary (*einer gewissen Willkür*); *Deutsche Bauzeitung* 23, no. 55 (July 6, 1889): 328. Undeterred, Wölfflin continued using reverse regulating lines.
15. Heinrich Wölfflin, *Renaissance und Barock* (Munich: T. Ackermann, 1888), 55. In the original German edition of this book, Wölfflin acknowledged August Thiersch as his source in a footnote (55, note 2), which was omitted from the English translation by Kathrin Simon, *Renaissance and Baroque* (Ithaca, NY: Cornell University Press, 1966).
16. Plate 2-25 is one of sixteen of Thiersch's Renaissance diagrams that Burckhardt reproduced, each with the credit line "*Nach A. Thiersch*" (After A. Thiersch). These credit lines were omitted from the diagrams in the English translation by James Palmes, *The Architecture of the Italian Renaissance*, ed. Peter Murray (Chicago: University of Chicago Press, 1985), 70–76.
17. Burckhardt also cited both Zeising's *Neue Lehre von den Proportionen* (1854) and Thiersch's "Die Proportionen in der Architektur" (1883); *Geschichte der Renaissance in Italien*, 3rd ed., ed. Heinrich Holtzinger (Stuttgart, Germany: Ebner and Seubert, 1891), 98–99; English transla-

tion as *The Architecture of the Italian Renaissance*, trans. James Palmes, ed. Peter Murray (Chicago: University of Chicago Press, 1985), 70. Burckhardt had published the first edition of this book in 1868. The second 1878 edition is the last one that was only by his hand (he died in 1897), but the editor of the third 1891 edition, Heinrich Holtzinger, specifically stated in his preface that the new section on proportion in the third edition was written by Burckhardt and that it was at Burckhardt's request that Thiersch's diagrams were added; *Renaissance in Italien* (1891), vi.

18. Burckhardt, *Renaissance in Italien*, 98–99, trans. 70.
19. For example, Wölfflin credited Thiersch in a lecture about German Renaissance architecture that he delivered in 1914 to the Munich Academy, which he included in a collection of essays published in 1941; Heinrich Wölfflin, *Gedanken zur Kunstgeschichte* (Basel, Switzerland: Schwabe, 1941), 115.
20. Heinrich Wölfflin, *Kunstgeschichtliche Grundbegriffe* (1915), 6th ed. (Munich: Bruckmann, 1923), 199–202.
21. Dietrich Neumann, "Teaching the History of Architecture in Germany, Austria, and Switzerland: '*Architekturgeschichte*' vs. '*Bauforschung*,'" *Journal of the Society of Architectural Historians* 3 (2002): 370.
22. Elam is head of graphic design at the Ringling School of Art and Design, Sarasota, Florida.
23. See Harald Siebenmorgen, *Die Anfänge der "Beuroner Kunstschule," Peter Lenz und Jakob Wüger, 1850-1875: Ein Beitrag zur Genese der Formabstraktion in der Moderne* (Sigmaringen, Germany: Jan Thorbecke, 1983), 162.
24. Lenz's letters to Blessing are quoted by Siebenmorgen in *Beuroner Kunstschule*, 180.
25. Charles Henry, "Introduction à une esthétique scientifique," *Revue contemporaine* 2 (Aug. 25, 1885): 441–69. In this essay Henry discusses Zeising and Fechner's writing on proportion (444) together with Pacioli's on the divine Golden Section (453).
26. Charles Henry, "Correspondance" (letter to the editor), *Revue philosophique* 29 (1890): 332–36; the quote is on 332. On Seurat's disinterest in the Golden Section, see Roger Herz-Fischler, "An Examination of Claims regarding Seurat and 'The Golden Number,'" *Gazette des Beaux-Arts* 125 (1983): 109–12.
27. Édouard Schuré, *Les grands initiés: Esquisse de l'histoire secrète des religions* (Paris: Perrin, 1889).
28. Maurice Denis recorded his and Sérusier's contact with Beuron in his *Journal* (Paris: La Colombe, 1957), 1:191ff. He also discussed the topic in *Paul Sérusier, sa vie, son oeuvre* (Paris: Floury, 1942), 74–93. This monograph on Sérusier by Denis was bound together with (published as pages 37–112 of) Paul Sérusier's *ABC de la peinture* (1921); *Suivi d'une étude sur la vie et l'œuvre de Paul Sérusier par Maurice Denis* (Paris: Floury, 1942). On the relationship of the Nabis and artists of Beuron, see Annegret Kehrbaum, *Die Nabis und die Beuroner Kunst* (Hildesheim, Germany: Olms, 2006).
29. Denis, *Sérusier*, 76.
30. This despite the fact that Lenz's adoption of the Golden Section dates to the 1860s–1870s, and he was not using it much by the 1890s, when the Nabis artists visited him in Beuron. See, for example, Lenz's book *Ästhetik der Beuroner Schule* (Vienna: Braumüller, 1898; French edition as *L'Esthétique de Beuron*, trans. Paul Sérusier [Paris: Bibliothèque de l'Occident, 1905]), in which Lenz does not mention the Golden Section.
31. Maurice Denis, "Définition du néo-traditionnisme" (1890), in *Théories, 1890–1910: Du symbolisme et de Gauguin vers un nouvel ordre classique* (Paris: Bibliothèque de l'Occident, 1912), 1.
32. Paul Sérusier, *ABC de la peinture* (Paris, Floury, 1921/rpt. 1942), 15–20.
33. Ghyka recalled their meeting in his memoir, *The World Mine Oyster: Memoirs* (London: Heinemann, 1961), 302–3.
34. Salvador Dalí, *50 secretos "mágicos" para pintar* (Barcelona: Luis de Caralt, 1951), English translation as *50 Secrets of Magic Craftsmanship*, trans. Haakon Chevalier (New York: Dial Press, 1948), 5.
35. Dalí's library, which is preserved at the Centre d'Estudis Dalinians, Fundació Gala-Salvador Dalí, Figueres, Spain, includes the following books by Ghyka: *Esthétique des proportions dans la nature et dans les arts* (The aesthetic of proportions in nature and the arts; 1927); *Essai sur le rythme* (Essay on rhythm; 1938), *The Geometry of Art and Life* (1946), which is the English translation of *Le Nombre d'or* (1931); and Ghyka's *A Practical Handbook of Geometrical Composition and Design* (1952).
36. Dalí owned a copy of Pacioli in Spanish translation, *La divina proporción* (Buenos Aires: Losada, 1946). Evidence that the artist read Pacioli closely is that he reproduced Leonardo's illustrations for *Divina Proportione* (1509), which do not appear in Ghyka's books. Also, for the title page of *50 Secretos*, Dalí used Pacioli's geometric design of an alphabet, which was published in *Divina Proportione* but is absent from Ghyka's books.
37. Ghyka assured Dalí that the dodecahedron did indeed symbolize the universe: "Regarding the questions that you put to me on the subject of solids corresponding to the macrocosm and microcosm: for the macrocosm it is clearly the *Dodecahedron* that Plato mentioned in *Timaeus* as the model used by the Grand Designer ('God arranging with Art' [*'le Dieu arrangeant avec art'*]) of the Cosmos." Ghyka, undated letter (in French) to Dalí, 1940s, Centre d'Estudis Dalinians, Fundació Gala-Salvador Dalí, Figueres, Spain.
38. Le Corbusier's copy of Ghyka's *Esthétique des proportions dans la nature et dans les arts* (Aesthetics of proportion in nature and the arts, 1927) is preserved at the Foundation Le Corbusier, Paris. Roger Herz-Fischler has established the chronology of Le Corbusier's use of the Golden Section at the Villa at Garches on the basis of Le Corbusier's drawings and writings; Herz-Fischler, "Le Corbusier's 'Regulating Lines' for the Villa at Garches (1927) and Other Early Works," *Journal of the Society of Architectural Historians* 43 (1984): 53–59. See also Roger Herz-Fischler, "The Early Relationship of Le Corbusier to the 'Golden Number,'" *Environment and Planning B* 6 (1979): 95–103. A few years later in his second book on the Golden Section, Ghyka featured Le Corbusier's Villa at Garches and hailed August Thiersch's "law of analogy" (*loi de l'analogie*); Matila Ghyka, "Le Corbusier and P. Jenneret, Regulating Lines, Villa at Garches" (1927), in Matila Ghyka, *L'nombre d'or: Rites et rythmes pythagoriciens dans le développement de la civilisation occidentale* (Paris: Gallimard, 1931), illustration between 156 and 157; Ghyka's praise for Thiersch is on 11.
39. For a detailed account of the several mathematicians who were involved in producing the Modulor, see Judi Loach, "Le Corbusier and the Creative Use of Mathematics," *British Journal for the History of Science* 31 (1998): 185–215.
40. See Carla Marzoli, curator, *Studi sulle proporzioni: Mostra bibliografica*, exh. cat. (Milan: La Bibliofila, 1951); and Rudolf Wittkower, "International Congress on Proportion in the Arts," *Burlington Magazine* 94, no. 587 (1952): 52, 55. A few years later Wittkower wrote an essay on the history of proportion in which he distanced himself from proponents of the Golden Section and described "the bankruptcy of the Milan meeting" (210) at which Le Corbusier displayed his Modulor; "The Changing Concept of Proportion," *Daedalus* 89, no. 1 (1960): 199–215.
41. Le Corbusier, *L'Unité d'habitation de Marseille* (Souillac, France: Mulhouse, 1950), 26, 44.
42. For an exposition of phyllotaxis, see John H. Conway and Richard K. Guy, "Phyllotaxis," in *The Book of Numbers* (New York: Springer, 1996), 113–24.
43. But tenacious myths die slowly; in 2001 Duke University awarded a doctorate in the history of music for a dissertation on J. S. Bach's (alleged) use of the Golden Ratio. Tushaar Power, *J. S. Bach and the Divine Proportion* (PhD diss., Duke University, 2001).

3. INFINITY

1. For a description of the origins of abstract art in the scientific worldview, see Lynn Gamwell, *Exploring the Invisible: Art, Science, and the Spiritual* (Princeton, NJ: Princeton University Press, 2002).
2. The earliest extant text that names Pythagoras as the person who discovered that the square root of 2 is irrational is *On the Pythagorean Life*, which was written in the late third century AD by the Neoplatonist Iamblichus, who also relayed that one Pythagorean brother who dared to speak of irrational numbers outside the cult community was taken out to sea and tossed overboard. Lacking an earlier source, it is impossible to separate fact from fiction in the discovery of this example of irrationality; see Walter Burkert, *Lore and Science* (see chap. 1, n. 17), 454–65. Taking another approach, the historian D. H. Fowler

has argued that the very lack of extant mention of the square root of 2 before the third century AD is evidence that discovery of incommensurate ratios was an incidental event in Plato and Euclid's time and did not cause great consternation; see *The Mathematics of Plato's Academy: A New Reconstruction* (Oxford, England: Clarendon, 1999), 356–69.

3. Four-dimensional geometries had been developed in the eighteenth century by the French mathematician Joseph-Louis Lagrange, who spent his career trying to improve Newton and Leibniz's calculus. Lagrange pointed out that if one is describing a point moving through three-dimensional space, such as a cannon ball shot at an enemy's tower, one needs to know not only the ball's location in space, but also its point in time. So he suggested adding a fourth "temporal" variable to the three spatial variables, giving the location of an event in three-dimensional space at a moment in time (*Theory of Analytic Functions*, 1797).
4. For example, Arthur Cayley, "On Some Theorems of Geometry of Positions" (1846) in David Eugene Smith, *A Source Book in Mathematics* (New York: McGraw Hill, 1929), 527–29; and Hermann Grassman, *Die lineale Ausdehnungslehre* (1844), trans. Mark Kormes, in Smith, *Source Book*, 684–96.
5. See Edith Dudley Sylla, "The Emergence of Mathematical Probability from the Perspective of the Leibniz-Jacob Bernoulli Correspondence," *Perspectives on Science* 6, nos.1 and 2 (1998): 41–76.
6. See Desmond MacHale, "Early Mathematical Work," in *George Boole: His Life and Work* (Dublin: Boole Press, 1985), 44–72.
7. For the history of attempts to prove the continuum hypothesis, see Paul J. Cohen, *Set Theory and the Continuum Hypothesis* (New York: W. A. Benjamin, 1966).
8. Nicholas of Cusa, *De Docta Ignorantia* (On learned ignorance; 1440), sec. 5, in *Nicholas of Cusa on Learned Ignorance: A Translation and Appraisal of De Docta Ignorantia*, trans. Jasper Hopkins (Minneapolis: Banning Press, 1985), 51. In 1902 Bertrand Russell thought that he could map the totality of all sets (the universe of sets) into the system of ordinal numbers (number referring to order; first, second, third . . .), which would have been Cusa's "Absolute Maximality." But Russell soon realized that he was wrong; see Bertrand Russell, "Recent Work in the Philosophy of Mathematics," *International Monthly* 4 (1901): 83–101; reprinted as "Mathematics and the Metaphysicians" in Bertrand Russell, *Mysticism and Logic: And Other Essays* (London: G. Allen and Unwin, 1917), 74–96, esp. 88–89.
9. Ibid., sec. 33, trans. 62.
10. Ibid., sec. 63, trans. 75–76.
11. On Nicholas and Bruno's belief in other worlds, see Steven J. Dick, *Plurality of Worlds: The Extraterrestrial Life Debate from Democritus to Kant* (Cambridge: Cambridge University Press, 1982), esp. 23–43 and 61–105.
12. In the second (1713) edition of *Principia*, Newton added to his "Definitions" a section of notes, entitled "Scholium," in which he distinguished between the everyday, common conceptions of time and space as they relate to "sensible objects," as opposed to mathematical concepts of time and space: "Absolute, true, mathematical time . . . [and] Absolute space, in its own nature, without relation to anything external"; Newton, *Principia*, 6 (see chap. 1, n. 81).
13. Titus Lucretius Carus, *De rerum natura* (first century BC), trans. Cyril Bailey (Oxford, England: Clarendon, 1947), bk. 2, lines 216–93: "But that the very mind feels not some necessity within in doing all things, and is not constrained like a conquered thing to bear and suffer, this is brought about by the tiny swerve (*clinamen*) of the first-beginnings in no determined direction of place and at no determine time" (lines 289–93).
14. Bouvet to Leibniz, Nov. 4, 1701, in *Leibniz Korrespondiert mit China: Der Briefwechsel mit den Jesuitenmissionaren (1689–1714)*, ed. Rita Widmaier (Frankfurt am Main, Germany: Vittorio Klostermann, 1990), 147–70.
15. Gottfried Leibniz, *Remarks on Chinese Rites and Religion* (1708), trans. Henry Rosemont and Daniel J. Cook (LaSalle, IL: Open Court, 1994), sec. 9, 73–74.
16. In addition to Joachim Bouvet, Leibniz also met the Jesuit missionary Claudio Filippo Grimaldi, who was on leave from Beijing, after which Leibniz corresponded regularly with Grimaldi and other clerics about Chinese philosophy; see Franklin Perkins, *Leibniz and China: A Commerce of Light* (Cambridge: Cambridge University Press, 2004), 114ff.
17. G. W. Leibniz, *Discourse on the Natural Theology of the Chinese* (1716), sec. 48, in *Writings on China*, trans. Daniel J. Cook and Henry Rosemont (LaSalle, IL: Open Court, 1994), 116.
18. G.W.F. Hegel, "Oriental Philosophy" (1816), in *Hegel's Lectures on the History of Philosophy*, trans. E. S. Haldane and Frances H. Simson (London: Routledge and Kegan Paul, 1974), 1:125. On Platonic themes in German Romanticism, see Douglas Hedley, "Platonism, Aesthetics, and the Sublime at the Origins of Modernity," in *Platonism at the Origins of Modernity: Studies on Platonism and Early Modern Philosophy*, ed. Douglas Hedley and Sarah Hutton (Dordrecht, the Netherlands: Springer, 2008), 269–82.
19. On German Romantic views of India, see Halbfass, "Hegel" and "Schelling and Schopenhauer," in *India and Europe*, 84–99 and 100–20 (see chap. 1, n. 62).
20. There is a vast literature on this and related occult topics in early modern art, beginning with the ground-breaking 1983 book by Linda Dalrymple Henderson, *The Fourth Dimension and Non-Euclidean Geometry in Modern Art* (Princeton, NJ: Princeton University Press, 1983); 2nd rev. ed. (Cambridge, MA: MIT, 2013). See also Henderson's summary of the topic in "The Image and Imagination of the Fourth Dimension in Twentieth-Century Art and Culture," *Configurations* 17, nos. 1–2 (Winter 2009): 131–60.
21. On Cantor's psychological and philosophical (as distinct from his mathematical) motivations for developing set theory, see I. Grattan-Guinness, "Psychology in the Foundations of Logic and Mathematics: The Cases of Boole, Cantor, and Brouwer," in his *History and Philosophy of Logic* (1982), 3:33–53; and José Ferreirós, "The Motives behind Cantor's Set Theory: Physical, Biological, and Philosophical Questions," *Science in Context* 17, no. 2 (2004): 49–83.
22. Georg Cantor, *Grundlagen einer allgemeinen Mannigfaltigkeitslehre: Ein mathematisch-philosophischer Versuch in der Lehre des Unendlichen* (Foundations of a general theory of manifolds: A mathematico-philosophical investigation into the theory of the infinite; 1883), trans. and ed. William Ewald, in *From Kant to Hilbert: A Source Book in the Foundations of Mathematics* (Oxford, England: Clarendon, 1996), 2:878–920; the quote is on 896. In endnote 6 to this essay (918), Cantor stated his affinity with Plato, Spinoza, and Leibniz.
23. Ibid., 893.
24. The details of Cantor's mental illness are described by his biographer Joseph Warren Dauben in *Georg Cantor: His Mathematics and Philosophy of the Infinite* (Cambridge, MA: Harvard University Press, 1979), esp. 136 and 284.
25. Mittag-Leffler made this request to Cantor in an unpublished letter, Nov. 14, 1884, which is in the archive of the Institut Mittag-Leffler, Djursholm, Sweden; Dauben, *Cantor*, 126, 331, note 17.
26. Georg Cantor to Gösta Mittag-Leffler, Sept. 22, 1884, in *Georg Cantor: Briefe*, ed. H. Meschkowski and W. Nilson (Berlin: Springer, 1991), 202.
27. Ibid.
28. "Ich nenne im Anschluß an Leibniz die einfachen Elemente der Natur, aus deren Zusammensetzung in gewissem Sinne die Materie hervorgeht, *Monaden* oder *Einheiten*" (Referring to Leibniz, I call the simple elements of nature which built the compounds of matter *monads* or *units*); Georg Cantor, "Über verschiedene Theoreme aus der Theorie der Punktmengen in einem n-fach ausgedehnten stetigen Raume Gn," *Acta Mathematica* 7 (1885): 105–24, in *Gesammelte Abhandlungen mathematischen und philosophischen Inhalts*, ed. E. Zermelo (Berlin: Julius Springer, 1932), 261–76; the quote is on 275.
29. Georg Cantor to Wilhelm Wundt, Oct. 16, 1883, *Briefe*, 142.
30. Cantor to Mittag-Leffler, Nov. 16, 1884, *Briefe*, 224.
31. Ibid. Cantor made a similar statement in a paper published the following year: "Auf diesem Standpunkt ergibt sich als die erste Frage, woran aber weder Leibniz noch die Späteren gedacht haben, welche *Mächtigkeiten* jenen beiden Materien in Ansehung ihrer Elemente, sofern sie als Mengen von *Körper-* resp. *Äthermonaden* zu betrachten sind, zukommen; in dieser Beziehung habe ich mir schon vor Jahren die *Hypothese* gebildet, daß die *Mächtigkeit* der *Körpermaterie* diejenige ist, welche ich in meinen Untersuchungen die erste Mächtigkeit nenne, daß dagegen die *Mächtigkeit* der *Äthermaterie* die *zweite* ist." (From this perspective arises the first question, which neither Leibniz nor his followers have thought about, of which *cardinalities* accord to both materials, considering their

elements, as long as they are seen as quantities *of corporeal monads* and *æther monads*. In this regard I have formed the *hypothesis* that the *cardinality of the corporeal material* is the one I call the first cardinality in my papers, and the *cardinality of the æther material* is therefore *second*.) Cantor, "Über verschiedene Theoreme," 276.

32. Benoît B. Mandelbrot, *The Fractal Geometry of Nature* (San Francisco: W. H. Freeman, 1977/rev. ed. 1982), 1.
33. Benoît B. Mandelbrot, "Fractal Events and Cantor Dusts," in *Fractal Geometry*, 74–82.
34. For the Cantor-Gutberlet correspondence, see Georg Cantor, "Mitteilungen zur Lehre vom Transfiniten" (1887), in *Gesammelte Abhandlungen*, 396–98.
35. "Über im absoluten Geiste ist immerdar die ganze Reihe im actualen Bewußtsein"; C. Gutberlet, "Das Problem des Unendlichen," *Zeitschrift für Philosophie und philosophische Kritik* 88 (1886): 179–223; the quote is on 206.
36. In the introduction to his key 1887 publication on infinity, Cantor included a quote from a letter from Cardinal Franzelin in which the Jesuit endorsed Cantor's concept of Absolute Infinity; Georg Cantor, "Mitteilungen zur Lehre vom Transfiniten" (1887), in *Gesammelte Abhandlungen*, 378–439; the quote is on 399–400.
37. I have been unable to locate this oft-quoted remark in Kronecker's published papers or memorabilia. In his obituary (Apr. 20, 1892) Kronecker's friend, the American mathematician Henry B. Fine, reported that Kronecker made a similar remark: "'God made numbers and geometry,' I once heard him say, 'but man [made] the functions;'" "Kronecker and his Arithmetical Theory of Algebraic Equations," *Bulletin of the New York Mathematical Society* (today *Bulletin of the American Mathematical Society*) 1 (1892): 183. On the anxiety expressed in the mathematics community over the increasing abstraction of mathematics, which peaked around 1900, see Jeremy J. Gray, "Anxiety and Abstraction in Nineteenth-Century Mathematics," *Science in Context* 17, no. 1–2 (2004): 23–47.
38. On the Moscow Society of Mathematicians, see Alexander Vucinich, *Science in Russian Culture* (Palo Alto, CA: Stanford University Press, 1963), 2:352–56.
39. Nikolai Bugayev, "Les mathématiques et la conception du monde au point de vue philosophie scientifique" (1897), trans. from Russian (n.t.), *Verhandlungen des ersten internationalen Mathematiker-Kongresses in Zürich von 9 bis 11 August 1897*, ed. Fernand Rudio (Leipzig, Germany: Teubner, 1898), 221.
40. Ibid., 217. See also S. S. Demidov, "N. V. Bougaiev et la création de l'école de Moscou de la théorie des fonctions d'une variable réelle," *Mathemata, Boethius series: Texte und Abhandlungen zur Geschichte der exakten Wissenschaft* 12 (Stuttgart, Germany: Steiner, 1985), 651–73.
41. See Vucinich, *Science in Russian Culture*, 2:512, note 45.
42. Charles E. Ford, "Dmitrii Egorov: Mathematics and Religion in Moscow," *Mathematical Intelligencer* 13, no. 2 (1991): 28.
43. For a biographical sketch of Florensky, see Nicoletta Misler's introduction to Pavel Florensky, *Beyond Vision: Essays on the Perception of Art*, trans. Wendy Salmond, ed. Nicoletta Misler (London: Reaktion, 2002), 13–28.
44. See Graham Priest and Richard Routley, "The History of Paraconsistent Logic," in *Paraconsistent Logic: Essays on the Inconsistent*, ed. Graham Priest, Richard Routley, and Jean Norman (Munich: Philosophia, 1989), 3–75.
45. Pavel Florensky, "Letter Two: Doubt," in *The Pillar and Ground of the Truth* (1914), trans. Boris Jakim (Princeton, NJ: Princeton University Press, 1997), 24.
46. Pavel Florensky, "Letter Four: The Light of Truth," in *The Pillar*, 67.
47. Luzin described Plotinus as a "mystic . . . who is no stranger to deep logical work required for a real worldview" (Luzin to Florensky, Apr. 12, 1909), trans. Loren Graham and Jean-Michel Kantor, in their book *Naming Infinity: A True Story of Religious Mysticism and Mathematical Creativity* (Cambridge, MA: Harvard University Press, 2009), 93.
48. In a discussion of numbers, including aleph-one (which is Cantor's second level of infinity [II Class]), Luzin wrote: "Let us concern ourselves with psychology. We, in our mind, consider natural numbers *objectively existing*. We, in our mind, consider the *totality of all* natural numbers *objectively existing*. We, finally, consider the *totality of all transfinites* of the II Class *objectively existing*"; trans. Graham and Kantor, in "Appendix: Luzin's Personal Archives," in *Naming Infinity*, 207. Graham and Kantor claim that Florensky, Luzin, and their Name Worshipping sect of Orthodoxy, believed that the existence of God is mind-dependent; for them, as Graham and Kantor write, "The name of God *is* God" (11). I disagree with Graham and Kantor's description of the Name Worshipper's ontology. Following the historian of mysticism Bernard McGinn, I would describe these Orthodox mystics as chanting the name "Jesus Christ" to evoke the *presence* of God. When they ceased chanting, God was no longer *present to them*, but God *continued to exist* independent of them, and, indeed, outside time and space for eternity. On the concept of "presence" in the Christian mystical tradition, see Bernard McGinn, "The Nature of Mysticism: A Heuristic Sketch," in his *The Presence of God: A History of Western Christian Mysticism* (New York: Crossroad, 1991), xiii–xx. Mind-dependent deities, like mind-dependent mathematical objects, can lead to contradictions (for example, consider the ontological status of God when one monk is chanting and—concurrently—another is not). The astute Florensky and Luzin would certainly have foreseen such paradoxes. The absence of any discusssion of such paradoxes in their writings is evidence that they did not hold the view that Graham and Kantor attribute to them.
49. As Luzin wrote: "We want the following: having assumed that we face the objectively existing totality of all natural and transfinite numbers of the II Class, we connect with each of the transfinites of the II Class a definition, a 'name'—and moreover uniformly for all those transfinites we are considering"; trans. Graham and Kantor, "Luzin's Archives," *Naming Infinity*, 207.
50. For a description of the emergence of abstract art in Germanic culture, see Lynn Gamwell, "German and Russian Art of the Absolute: A Warm Embrace of Darwin," in *Exploring the Invisible*, 93–109 (see n. 1).
51. On the role of mathematics in Russian modernism, see Anke Niederbudde, *Mathematische Konzeptionen in der russischen Moderne: Florenskij, Chlebnikov, Charms* (Munich: Otto Sagner, 2006). The author focuses on the topics of infinity, perspective, measurement, and the ontological status of numbers and signs in the work of Pavel Florensky, Velimir Khlebnikov, and the absurdist poet Daniil Kharms.
52. As a neurologist, Kulbin was familiar with the experimental psychology that supported this premise. Kulbin's clinical publications include: *Chuvstvitelnost: Ocherki po psikhometrii i klinicheskomu primeneniiu eia dannykh* (Sensation: Studies in psychometry and the clinical application of its data; Saint Petersburg, 1903).
53. Kruchenykh made this statement about the origins of *zaum* in a 1959 interview with Nikolay Khardzhiev, which is translated by Gerald Janecek in his book *Zaum: The Transrational Poetry of Russian Futurism* (San Diego, CA: San Diego State University, 1996), 49.
54. On the translation of this coined word, see Gerald Janecek, "*Zaum*: A Definition," in *Zaum*, 1–3.
55. P. D. Ouspensky, *Tertium Organum: The Third Canon of Thought, a Key to the Enigmas of the World* (1911), trans. Claude Bragdon and Nicholas Bessaraboff (New York: Alfred A. Knopf, 1968), 236.
56. Kruchenykh, "New Ways of the Word" (1913), trans. Anna Lawton and Herbert Eagel, in *Russian Futurism through Its Manifestos, 1912–1928* (Ithaca, NY: Cornell University Press, 1988), 70. Instead of the name that I'm writing as "Peter Ouspensky," the translators of Kruchenykh's essay use the alternative spellings "P. Uspensky" (in the quotation, 70) and "Petr Uspenskii" (in the endnote to the quotation, 309).
57. Kruchenykh, "Declaration of the Word as Such" (1913), in Lawton and Eagel, *Russian Futurism*, 68.
58. Gerald Janecek has assembled examples of *zaum* poetry in his encyclopedic study *Zaum*.
59. In his book *Kazimir Malevich and the Art of Geometry* (New Haven, CT: Yale University Press, 1996), the British art historian John Milner discusses Malevich's possible relationship to many different kinds of geometry, including Pythagorean diagrams, the Fibonacci series, four-dimensional geometries, systems of proportion based on the Golden Section, and diagrams used by medieval alchemists and modern theosophists. Despite the illustrations of many historical geometric figures, as well as Milner's own diagrams of the geometric structure of some of Malevich's paintings, the author writes in vague generalities and nowhere explains exactly how any of the illustrated geometric figures

relate to Malevich's art. Furthermore, the author repeatedly alludes to an important hidden meaning that Malevich used geometry to symbolize, but Milner never divulges the secret to the readers of his book.

60. In Andréi Nakov's *Kazimir Malewicz: Catalogue raisonné* (Paris: Adam Biro, 2002), see works F390–F481 (ca. 1913–15), with titles that include the terms "transrational" and "alogical." In 1915 there are many references to the theosophists' "fourth dimension" (*Painterly Realism of a Boy with a Knapsack—Color Masses in the Fourth Dimension*, 1915), after which, as seen in S421–S455 (ca. 1915–18), the titles now include the term "cosmic." On Malevich and theosophy, see Jean Clair, "Malévitch, Ouspensky, et l'espace néo-platonicien," in *Malévitch, 1878–1978*, ed. Jean-Claude Marcadé (Lausanne, Switzerland: L'âge d'homme, 1979), 15–30. Clair argues that Plato's myth of the cave (*Republic*, bk. 7), which Ouspensky discusses at length in *Tertium Organum*, underlies the outlook of other theosophists such as Charles Hinton, as well as Malevich.
61. On the relation of Russian icons to the Russian avant-garde, see Margaret Betz, "The Icon and Russian Modernism," *Artforum* 15, no. 10 (1977): 38–45; R. Milner-Gulland, "Icons and the Russian 'Modern Movement,'" *Icons 88: To Celebrate the Millennium of the Christianization of Russia, an Exhibition of Russian Icons in Ireland*, ed. Sarah Smyth and Stanford Kingston (Dublin: Veritas, 1988), 85–96; and Andrew Spira, *The Avant-Garde Icon: Russian Avant-Garde Art and the Russian Icon Painting Tradition* (Hampshire, England: Lund Humphries, 2008).
62. Kazimir Malevich, *Chapitre de l'autobiographie du peintre* (Chapter from the painter's autobiography; 1933), trans. from Russian to French by Dominique Moyen and Stanislas Zadora, in Marcadé, *Malévitch 1878–1978*, 164 (see n. 60). In one of Florensky's many essays mixing aesthetics with theology, he argued that if the reverse perspective in a Russian Orthodox icon looks naive to us, it is we who are naive because we assume that the artist was trying to achieve a naturalistic representation of the everyday world. According to Florensky, icon painters were trying to achieve a realistic depiction of a supernatural realm where space is irrational—"reversed"; see Pavel Florensky, "Reverse Perspective" (first published posthumously in Russian in 1967) in the collection of his essays, *Beyond Vision*, 201–72 (see n. 43).
63. For a description of this brief episode in the style of Malevich, see Christina Lodder, "The Transrational in Painting: Kazimir Malevich, Algoism, and *Zaum*," *Forum for Modern Language Studies: The International Avant-Garde* 32, no. 2 (1996): 119–36. For a comparison of the algoism of the Russian avant-garde and the nonsense of the British logician Lewis Carroll, author of *Alice's Adventure in Wonderland* (1865; first Russian translation as *Sonia v Tsarsivw Dive* [Sonia in the kingdom of wonder], 1879), see Nikolai Firtich, "Worldbackwards: Lewis Carroll, Aleksei Kruchenykh and Russian Algoism," *Slavic and East European Journal* 48, no. 4 (2004): 593–606.
64. Malevich, *Chapitre de l'autobiographie*, 168.
65. From the report of a meeting to plan *Victory over the Sun* that was held in 1913 by Kruchenykh, Malevich, and Matyushin; they published the report as "Pervy Vserossiysky Siezd Bayachey Budushchego (poetov-futuristov) [First Russian Congress of the poets-futurists]," *Zhurnal za 7 dney* (Pb), no. 28 (1913): 606; trans. Gerald Janacek in *Zaum*, 111.
66. Aleksei Kruchenykh, *Victory over the Sun* (1913), act 1, scene 1, in *Victory over the Sun: The World's First Futurist Opera*, trans. Rosamund Bartlett, ed. Rosamund Bartlett and Sarah Dadswell (Exeter, England: University of Exeter Press, 2012), 26. This book includes a facsimile edition of the original Russian libretto together with an English translation.
67. Ibid., scene 4, trans. 36.
68. Ibid. John Bowlt has suggested that Malevich's invented marks (black squares and circles) are not merely metaphors but depictions of a total solar eclipse in keeping with theme of the libretto; see Bowlt, "Darkness and Light: Solar Eclipse as a Cubo-Futurist Metaphor," in *Victory over the Sun*, trans. Bartlett, 65–77.
69. See Charlotte Douglas, *Swans of Other Worlds: Kazimir Malevich and the Origins of Abstraction in Russia* (Ann Arbor, MI: UMI Research Press, 1980), 3; and Charlotte Douglas, *Kazimir Malevich* (New York: Abrams, 1994), 21–22. Andréi Nakov, who compiled the artist's *catalogue raisonné*, has also complained that Malevich's nasty habit of dating works from when he *conceived* of a painting rather than when he actually *did* the painting, has thwarted the efforts of scholars to establish an accurate chronology; *Malewicz: catalogue raisonné*, 37.
70. Malevich's Russian manuscript was translated by A. von Riesen into German and published as Bauhausbücher 11 under the title *Die gegenstandslose Welt* (Munich: Albert Langen, 1927). In 1959 Howard Dearstyne did the English translation from German (the Russian manuscript having been lost) as *The Non-Objective World* (Chicago: Paul Theobald, 1959), 68, 76. Charlotte Douglas has pointed out that when the original Russian text survived, Malevich's term *oshchushchenie* should have been translated "sensation" rather than "feeling," as it often is, giving the incorrect impression that Malevich was concerned with emotion; *Swans of Other Worlds*, 57–58. I suggest, however, that the artist's vast writings are so poetic that they defy pinning down meanings of terms, and at times Malevich certainly seems to be writing about how he feels (emotionally) as opposed to what he senses (perceptually). When Malevich did write about sensation, he focused on physical stimuli that were subliminal (below conscious awareness), with the goal of developing his mental facility (his intuition) to perceive such sensations and then represent (symbolize) them on his canvas. On the physiological sensations symbolized by Malevich, see Christina Lodder, "Man, Space, and the Zero of Form: Kazimir Malevich's Suprematism and the Natural World," in *Meanings of Abstract Art: Between Nature and Theory*, ed. Paul Crowther and Isabel Wünsche (New York: Routledge, 2012), 47–63.
71. Pavel Florensky, *Iconostasis*, trans. Donald Sheehan and Olga Andrejev (Crestwood, NY: St. Vladimir's Seminary Press, 1996), 63.
72. On Malevich's sources in Schopenhauer, see the introduction to Kazimir Malevich, *The World as Non-Objectivity: Unpublished Writings 1922–25*, trans. Xenia Glowacki-Prus and Edmund T. Little, ed. Troels Andersen (Copenhagen: Borgen, 1976), 7–10.
73. Malevich, *Non-Objective World*, 68.
74. Malevich, *World as Non-Objectivity*, 354.
75. Alexander Benois, "Posledniaia futuristicheskaia vystavka," *Rech'* (Jan. 9, 1916) 3, translated by Jane A. Sharp in her essay "The Critical Reception of the 0,10 Exhibition: Malevich and Benau," in *The Great Utopia: The Russian and Soviet Avant-Garde, 1915–1932* (New York: Guggenheim Museum, 1992), 39–52; the quote is on 42. "Alekandr Benau" is another spelling of the name "Alexander Benois."
76. Malevich painted versions of *Black Square* in 1915, 1920, 1924, 1929, and 1930; Nakov, *Malewicz: catalogue raisonné*, 37. He also did more than fifty variations of *Black Square* in many different contexts: see Nakov's catalogue raisonné numbers S-113 to S-174 on 205–18.
77. See the exhibition catalogue, *Das schwarze Quadrat: Hommage an Malewitsch*, ed. Hubertus Gassner (Ostfildern, Germany: Hatje Cantz, 2007). The exhibition included a grisly homage to Malevich by the contemporary Slovenian art group IRWIN, who restaged Malevich's lying-in-state (with one artist playing dead) in their installation *Corpse of Art* (2003; 184–85, fig. 121). For a discussion of Malevich's *Black Square* as the origin of monochrome painting, see Yve-Alain Bois, "Malevitch, le carré, le degré zero," *Macula* 1 (1976): 28–49. For the record, however, Malevich's painting is not monochrome but a black square on a white background; the trophy for creating the first monochrome painting goes to Aleksandr Rodchecko (see chap. 4).
78. Adolphe Quetelet, *Sur l'homme et le développement de ses facultés, ou Essai de physique sociale* (Paris: Bachelier, 1835), 12. On the widespread use of probability theory to defend free will in the nineteenth century, see Theodore M. Porter, "Statistical Law and Human Freedom," in *The Rise of Statistical Thinking, 1820–1900* (Princeton, NJ: Princeton University Press, 1986), 151–92.
79. The quote is from Quetelet's preface to the 1842 English translation of his *Essai de physique sociale* (1835) as *A Treatise on Man and the Development of His Faculties*, trans. R. Knox (Edinburgh: William and Robert Chambers, 1842), x.
80. Pavel A. Nekrasov, *Filosofiia i logika nauki o massovikh proiavleniiakh chelovecheskoi deiatelnosti (Peresmotr osnovanii sotsialnoi fiziki Ketle)* (The philosophy and logic of the study of mass phenomena of human activity [A review of the social physics of Quetelet]; Moscow: Universitetskaia tipografiia, 1902). See also Eugene Seneta, "Statistical Regularity and Free Will: L.A.J. Quetelet and P. A. Nekrasov," *International Statistical Review/Revue Internationale de Statistique* 71, no. 2 (2003): 319–34.

81. See the review by the Russian mathematician Dmitrii M. Sincov (German transcription as Sintzow) of Nekrasov's *Filosofiia* (1902) in the Berlin journal *Jahrbuch über die Fortschritte der Mathematik* 33 (1902): 236. Sincov also reviewed responses to Nekrasov's thesis by a few other Russians, summarizing their common opinion: "The mathematical proof of metaphysical doctrines such as the freedom of the will is quite impossible [*ganz unmöglich*]"; *Jahrbuch über die Fortschritte der Mathematik* 34 (1903): 66.
82. See Ayda Ignez Arruda, "On the Imaginary Logic of N. A. Vasil'év" (1977), in *Non-Classical Logic, Model Theory and Computability*, ed. Ayda Ignez Arruda, N. da Costa, and R. Chuanqui (Amsterdam: North Holland, 1977), 3–24; and Valentin A. Bazhanzov, "The Fate of One Forgotten Idea: N. A. Vasiliev and His Imaginary Logic," *Studies in Soviet Thought* 39, nos. 3–4 (1990): 333–42.
83. Friedrich Engels, *Dialectics of Nature* (1883), trans. and ed. Clemens Dutt (New York: International Publishers, 1940), 309.
84. For the details of Florensky's exile and execution, see Graham and Kantor, *Naming Infinity*, 144–45 (see n. 47). See also Eugene Seneta, "Mathematics, Religion, and Marxism in the Soviet Union in the 1930s," *Historia Mathematica* 31, no. 3 (2004), 337–67.
85. However, one practitioner of the new science of the mind, the Chilean psychoanalyst Ignacio Matte Blanco, has argued the (far-fetched) thesis that the Freudian unconscious mind is an infinite set in the sense that if one assumes that the infant looks to the mother as a source of all possible knowledge, then "we attribute to the mother's breast all the potentialities or same cardinal number of power as that attributed to the class of all possible knowledge"; *The Unconscious as Infinite Sets* (London: Duckworth, 1975), 180.
86. Friedrich Schleiermacher, *On Religion: Speeches to Cultured Despisers* (1799), trans. Richard Crouter (Cambridge: Cambridge University Press, 1988), 139–40.

4. FORMALISM

1. For the differences between German and French science and their expression in the arts, see Lynn Gamwell, "The French Art of Observation," and "German and Russian Art of the Absolute," in Gamwell, *Exploring the Invisible*, 57–109 (see chap. 3, n. 1).
2. Plato, *Timaeus* (366–60 BC), 33b, in *Dialogues*, trans. 3: 452 (see chap. 1, n. 26).
3. John Ruskin, *Modern Painters* (London: Smith, Elder, 1846), 2:193.
4. For discussions of formalism that focus on British criticism, see Arnold Isenberg, "Formalism," in *Aesthetics and the Theory of Criticism* (Chicago: University of Chicago Press, 1973), 22–35; and Richard Wollheim, "On Formalism and Pictorial Organization," *Journal of Aesthetics and Art Criticism* 59, no. 2 (2001): 127–37.
5. L.E.J. Brouwer, "Intuitionism and Formalism" (1912), trans. Arnold Dresden, *Bulletin of the American Mathematical Society* 20, no. 2 (1913): 81–96; the quote is on 83.
6. For a history of attempts to prove the parallel postulate, see Boris A. Rosenfeld, "The Theory of Parallels," in *A History of Non-Euclidean Geometry: Evolution of the Concept of Geometric Space*, trans. Abe Shenitzer (New York: Springer, 1988), 35–109.
7. For a history of the discovery of non-Euclidean geometries, see Marvin Jay Greenberg, *Euclidean and Non-Euclidean Geometries: Development and History* (New York: W. H. Freeman, 1993), esp. 869–74. For a facsimile reproduction of Bolyai's original publication (as the appendix to Farkas Bolyai's textbook), together with commentary, see Jeremy J. Gray, *János Bolyai, Non-Euclidean Geometry, and the Nature of Space* (Cambridge, MA: Burndy Library, 2004).
8. Bolyai chose to study Euclid's fifth postulate in the form of Playfair's axiom: "Through any point lying outside a line, many (an infinite number of) lines can be drawn through the point that do not intersect that given line." See Harolde E. Wolfe, *Introduction to Non-Euclidean Geometry* (Bloomington, IN: Indiana University Press, 1941), 24.
9. Immanuel Kant, *Critique of Pure Reason* (1781), A84/B116 to A92/B124, trans. Norman Kemp Smith (London: Macmillan, 1929), 120–25.
10. Ibid.
11. Hermann von Helmholtz, "On the Origin and Meaning of Geometrical Axioms" (1876), trans. Edmund Atkinson, in *Ewald, Kant to Hilbert*, 2:668–70 (see chap. 3, n. 22).
12. The episode was recorded by Helmholtz's friend and biographer, Leo Königsberger, in *Hermann von Helmholtz* (1905), trans. Frances A. Welby (New York: Dover, 1905/rpt. 1965), 254–67. Riemann had submitted the essay as his *Habilitationsschrift* (similar to a PhD dissertation) to his professors at Göttingen in 1854.
13. Hermann von Helmholtz, "Über die Tatsachen, die der Geometrie zugrunde liegen" (On the facts underlying geometry; 1868), in Helmholtz, *Wissenschaftliche Abhandlungen* (Leipzig, Germany: Johann Ambrosia Barth, 1883), 2:618–39; the quote is on 619.
14. As Thomas Heath has stated: "Euclid preferred to assert as a postulate, directly, the fact that all right angles are equal; and hence his postulate must be taken as equivalent to the *invariability of figures*, or, what is the same thing, the *homogeneity of space*." Heath, *Greek Mathematics*, 1:375 (see chap. 1, n. 9).
15. This is the judgment of Heath, *Greek Mathematics*, 1:375.
16. David Hilbert, *Grundlagen der Geometrie* (Foundations of geometry; 1899), trans. E. J. Townsend as *The Foundation of Geometry* (LaSalle, IL: Open Court; London: Kegan Paul, Trench, Trübner, 1902/rpt. 1962), 4.
17. Hermann Weyl gives this quote in his obituary of Hilbert, "David Hilbert and His Mathematical Work," *Bulletin of the American Mathematical Society* 50 (1944), 612–54; the quote is on 635. Weyl names Otto Blumenthal as his source for Hilbert's remark, and Blumenthal also recounted the anecdote and stated that Hilbert made such remarks as early as 1891; in David Hilbert, *Gesammelte Abhandlungen* (Berlin: Springer, 1970), 3:403.
18. Hilbert, *Grundlagen der Geometrie*, 4–5.
19. Gottlob Frege, "The Concept of Number," in his *Die Grundlagen der Arithmetik* (1884), trans. J. L. Austin as *The Foundations of Arithmetic* (Oxford, England: Blackwell, 1953), 67–99.
20. Cantor, *Grundlagen einer allgemeinen Mannigfaltigkeitslehre*, 896 (see chap. 3, n. 22).
21. David Hilbert to Gottlob Frege, Nov. 7, 1903, in Frege, *Philosophical and Mathematical Correspondence*, trans. Hans Kaal, ed. Brian McGuinness (Oxford, England: Blackwell, 1980), 52. See also José Ferreirós, "Hilbert, Logicism, and Mathematical Existence," *Synthese* 170, no. 1 (2009): 33–70, esp. 55–59.
22. The series of publications on the topic of Platonism in mathematics includes W. V. Quine, "Success and Limits of Mathematics" (1978), in *Theories and Things* (Cambridge, MA: Harvard University Press, 1981), 148–55; and Hilary Putnam, "What Is Mathematical Truth?" in *Mathematics, Matter, and Method: Philosophical Papers*, 2nd ed. (Cambridge: Cambridge University Press, 1979), 1:60–78. For an in-depth discussion of Quine-Putnam's Platonism versus Benacerraf's anti-Platonism, see Mark Balaguer, *Platonism and Anti-Platonism in Mathematics* (Oxford: Oxford University Press, 1998).
23. Plato, *Seventh Letter*, 136 (see chap. 1, n. 41).
24. Paul Bernays, "Über den Platonismus in der Mathematik," in *Abhandlungen zur Philosophie der Mathematik* (Darmstadt, Germany: Wissenschaftliche Buchgesellschaft, 1976), 62–78; the quote is on 65.
25. See Peter van Inwagen, "The Nature of Metaphysics," in *Contemporary Readings in the Foundations of Metaphysics*, ed. Stephen Laurence and Cynthia Macdonald (Oxford, England: Blackwell, 1998), 11–21.
26. Bertrand Russell, "Reflections on my Eightieth Birthday" (1952), in his *Portraits from Memory and Other Essays* (London: George Allen and Unwin, 1956), 53.
27. Peter Renz, "Mathematical Proof: What It Is and What It Ought to Be," *The Two-Year College Mathematical Journal* 12, no. 2 (1981): 83–103; the quote is on 101.
28. Davies goes on: "This is not surprising since Platonism grew out of the Pythagorean mystery religion, in which mathematics played a key role." See E. Brian Davies, "Let Platonism Die," *European Mathematical Society Newsletter* 64 (June 2007): 24–25; the quote is on 24.
29. For an overview of the literature on the ontology of abstract objects, see Harty Field, "Mathematical Objectivity and Mathematical Objects," in Laurence and Macdonald, *Foundations of Metaphysics*, 387–403 (see n. 25).
30. David Corfield, *Towards a Philosophy of Real Mathematics* (Cambridge: Cambridge University Press, 2003).
31. Johann Friedrich Herbart, *Über philosophisches Studium* (1807), in *Sämtliche Werke* (Leipzig, Germany: Leopold Voss, 1850), 2:373–463.

32. Riemann's extensive notes on Herbart's philosophy, including a summary of Herbart's *Über philosophisches Studium* (1807), are preserved in the Riemann Archive at Göttingen; see Erhard Scholtz, "Herbart's Influence on Bernhard Riemann," *Historia Mathematica* 9 (1982), 413–40. For further details about Riemann's interest in Herbart, see Detlef Laugwitz, "The Role of Herbart's Philosophy," in Laugwitz, *Bernhard Riemann, 1826–1866: Turning Points in the Conception of Mathematics*, trans. Abe Shenitzer (Basel, Switzerland: Birkhäuser, 1999), 287–92; and Jeremy Grey, *Plato's Ghosts: The Modernist Transformation of Mathematics* (Princeton, NJ: Princeton University Press, 2008), 83–86 and 91–93.
33. For the epigraph to his seminal *Grundlagen der Geometrie* (Foundations of geometry; 1899), Hilbert chose the following quote from Immanuel Kant: "All human knowledge begins with intuitions, thence passes to concepts and ends with ideas" (*Critique of Pure Reason*, 1781).
34. Kant described all value judgments (both aesthetics and ethics) as subjective: "Two things fill the mind with ever new and increasing admiration and reverence, the more often and steadily one reflects on them: *the starry heaven above me and the moral law within me.* I do not need to search for them as though they were veiled in obscurity or in the transcendent region beyond my horizon; I see them before me and connect them immediately with the consciousness of my existence." Immanuel Kant, *Critique of Practical Reason* (1788), trans. Mary Gregor (Cambridge: Cambridge University Press, 1997), 133.
35. Hilbert made this statement in his lecture to the International Congress of Mathematics in Paris in 1900, in which he challenged the mathematics community to solve twenty-three key problems in the coming century; David Hilbert, "Mathematical Problems" (1901), trans. Mary Winston Newsom, *Bulletin of the American Mathematical Society* 8 (July 1902): 437–79; the quote is on 478.
36. Ibid., 479.
37. Johann Friedrich Herbart, *Kurze Encyklopädie der Philosophie aus praktischen Gesichtspunkten entworfen* (Halle, Germany: Schwetschke, 1831), sec. 72, 124–25.
38. Hermann von Helmholtz, "On the Physiological Causes of Harmony in Music" (1857), in his *Popular Lectures on Scientific Subjects*, trans. E. Atkinson (London: Longmans, Green, 1893/rpt. 1904), 53–93; the quote is on 92.
39. For example, Hilbert stated: "It is an error to believe that rigor in the proof is the enemy of simplicity. On the contrary we find it confirmed by numerous examples that the rigorous method is at the same time the simpler and the more easily comprehended." Hilbert, "Mathematical Problems," 441. Indeed, in his 1900 formulation of twenty-three key problems for the twentieth century, Hilbert originally included a twenty-fourth problem *about* simplicity—finding a method to determine that a proof is in its simplest form—but he omitted it from his speech and his published list. The problem lay hidden until the historian of mathematics Rüdiger Thiele recently rediscovered it among Hilbert's notes; see Rüdiger Thiele, "Hilbert's Twenty-Fourth Problem," *American Mathematical Monthly* 110 (Jan. 2003): 1–24.
40. See Alexander Vucinich, "Probability Theory," in his *Science in Russian Culture*, 2:336–43 (see chap. 3, n. 38).
41. For nineteenth-century parallels between organic evolution and the "evolution" of language, see Stephen G. Alter, *Darwinism and the Linguistic Image: Language, Race, and Natural Theology in the Nineteenth Century* (Baltimore: Johns Hopkins University Press, 1999).
42. Today in dictionaries that give etymology, reconstructed words from the Indo-European mother tongue are marked with an asterisk.
43. Baudouin wrote: "Phonetic laws can be compared with such 'laws' as apply to meteorological generalizations." "Phonetic Laws" (1910), trans. Edward Stankiewicz, in *A Baudouin de Courtenay Anthology: The Beginnings of Structural Linguistics* (Bloomington: Indiana University Press, 1972), 276.
44. It was Aleksei Kruchenykh's coined words that especially interested Baudouin; see Gerald Janecek, "Baudouin de Courtenay versus Kruchenykh," *Russian Literature* 10 (1981): 17–30.
45. Andrey Bely, "Lyrical Poetry and Experiment" (1909), in *Selected Essays of Andrey Bely*, trans. Steven Cassedy (Berkeley: University of California Press, 1985), 222–73. (All the essays in this collection were in Bely's 1910 book *Symbolism*.) Bely refers to these diagrams as both "geometric figures" (260) and "statistical figures" (265).
46. For a history of the application of statistics to poetry in the tradition of Bely, see the Czech critic Jiří Levý, "Mathematical Aspects of the Theory of Verse" (1969), in *Statistics and Style*, ed. Lubomír Doležel and Richard Bailey (New York: Elsevier, 1969), 95–112.
47. Velimir Khlebnikov, epigraph to "Artists of the World!" (1919), in *Collected Works of Velimir Khlebnikov*, trans. Paul Schmidt (Cambridge, MA: Harvard University Press, 1987), 1:364.
48. Khlebnikov, *Collected Works*, 1:365.
49. Ibid., 1:365, 367.
50. Velimir Khlebnikov, *The Burial Mound of Sviatagor* (1908), in *Collected Works*, 1:234. Sviatagor is the name of a Russian mythical hero. I was unable to locate this essay in the original Russian; Khlebnkov surely intended "geometry" for what the translator rendered as "geomeasure." For a discussion of Khlebnikov's use of Lobachevsky's geometry as a mythic symbol, see Henryk Baran, "Xlebnikov's Poetic Logic and Poetic Illogic," in *Velimir Chlebnikov*, ed. Nils Åke Nilsson (Stockholm: Almqvist and Wiksell, 1985), 7–25. Baran contrasts Khlebnikov's early use of antithetical pairs of opposite terms (his "poetic illogic") with his more resolved constructions (his "poetic logic") after October 1917 and World War I.
51. For a comprehensive study of the art of Tatlin and Rodchenko, see Christina Lodder's *Russian Constructivism* (New Haven, CT: Yale University Press, 1983).
52. David Burliuk, "Cubism (Surface—Plane)" (1912), in *Russian Art of the Avant Garde: Theory and Criticism, 1902–1934*, trans. John Bowlt (New York: Viking, 1976), 70.
53. Ibid., 70, 73.
54. Ibid., 77.
55. For a detailed catalogue of Tatlin's counter-reliefs, see *Vladimir Tatlin: Retrospektive*, ed. Anatolij Strigalev and Jürgen Harten (Cologne, Germany: DuMont, 1993), inv. nos. 340–62 on 245–553, and inv. nos. 391–92 on 257–58.
56. Viktor Shklovskii, "On *Faktura* and Counter-Reliefs" (1920), trans. Eugenia Lockwood, in *Tatlin*, ed. Larissa Alekseevna Zhadova (New York: Rizzoli, 1988), 341–42; the quote is on 341. On *faktura*, see also Benjamin H. D. Buchloh, "From *Faktura* to Factography," *October* 30 (Autumn 1984): 82–119, esp. 85–95; and Maria Gough, "*Faktura*: The Making of the Russian Avant-Garde," *RES: Anthropology and Aesthetics* 36 (Autumn 1999): 32–59.
57. Sergei K. Isakov, "On Tatlin's Counter-Reliefs" (1915), trans. Eugenia Lockwood, in Zhadova, *Tatlin*, 333–35; the quote is on 334.
58. For an overview of Rodchenko's career, see Magdalena Dabrowski, "Aleksandr Rodchenko: Innovation and Experiment," in *Aleksandr Rodchenko*, ed. Magdalena Dabrowski, Leah Dickerman, and Peter Galassi (New York: Museum of Modern Art, 1998), 18–49.
59. For a description of the debates, see Christina Lodder, "Towards a Theoretical Basis: Fusing the Formal and Utilitarian," in *Russian Constructivism*, 73–108. In 1919 Rodchenko was already active in the revolutionary group *Zhivskul'ptarkh*, a name formed by combining the words *"zhivopis"* (painting), *"skulptura"* (sculpture), and *"arkhitektura"* (architecture); he designed for this group an information kiosk, "The Future is Our Only Goal" (1919). For a description of *Zhivskul'ptarkh*, see Kestutis Paul Zygas, *Form Follows Form: Source Imagery of Constructivist Architecture, 1917–25* (Ann Arbor, MI: UMI Research Press, 1981), 14–23. On Rodchenko's kiosk see Victor Margolin, *The Struggle for Utopia: Rodchenko, Lissitzky, Moholy-Nagy, 1917–1946* (Chicago: University of Chicago Press, 1997), 16–20.
60. Aleksandr Rodchenko, "The Famous Theorem of Cantor" (1920) in *Aleksandr Rodchenko: Experiments for the Future: Diaries, Essays, Letters, and Other Writings*, trans. Jamey Gambrell, ed. Alexander N. Lavrentiev (New York: Museum of Modern Art, 2005), 102. Rodchenko cites his source for Cantor's theorem as A. Solonovich, "Equation of the World Revolution," *Klich*, no. 3 (Moscow, 1917). Planets, suns, and galaxies are distinct physical objects, and thus, even though there are many of them, they are countable. Rodchenko seems unaware that, on the other hand, the points on a line segment are *uncountably* many.
61. Malevich, *Non-Objective World*, 68 (see chap. 3, n. 70).
62. Today many of Rodchenko's *Spatial Constructions* are lost, but in the 1920s he made photographs and drawings of all of them. These documents are reproduced in *Alexander Rodchenko: Spatial Constructions/ Raumkonstruktionen*, trans. Michael Eldred and Gerlinde Weber-

Niesta, ed. Krystyna Gmurzynska and Mahias Rastorfer, (Ostfildern, Germany: Hatje Cantz, 2002), which served as the catalogue of an exhibition of Rodchenko's *Spatial Constructions* at the Wilhelm Lehmbruck Museum in Duisberg. With the permission of the Rodchenko estate, in 2002 the lost originals were reproduced in a limited edition, which are illustrated in this publication.

63. Rodchenko's three monochrome paintings (*Red, Blue,* and *Yellow*) are described (incorrectly) as a triptych by Benjamin H. D. Buchloh in his essay "The Primary Colors for the Second Time: A Paradigm Repetition of the Neo-Avant-Garde," *October* 27 (Summer 1986): 41–52. James Meyer makes the same mistake when he describes the "tripartite structure" of Rodchenko's "monochrome triptych *Pure Colors: Red, Blue, and Yellow*" in *Minimalism* (London: Phaidon, 2000), 19. If one assumes that Rodchenko's goal was to reduce painting to its simplest form, then his end-point could not have been a triptych (which has three parts). He first showed his three monochromes in the exhibition, 5 × 5 = 25 (Moscow, 1921), which was an exhibition of five artists, each represented by five works, for a total of twenty-five works of art on display. In the catalogue Rodchenko listed his work as *Red* (1921), *Yellow* (1921), *Blue* (1921), *Line* (1920), and *Square* (1921), for a total of five. See the facsimile of the original Russian exhibition catalogue (n. p.), which is bound separately but boxed together with John Milner, *The Exhibition 5 × 5 = 25: Its Background and Significance* (Budapest: Helikon, 1992). For a history of monochrome painting spawned by Rodchenko's 1921 paintings, see Thierry de Duve, "The Monochrome and the Blank Canvas," in *Kant after Duchamp* (Cambridge, MA: MIT Press, 1996), 199–279.
64. Rodchenko made this statement about his 1921 monochrome paintings in a memoir that he wrote in 1940 on the occasion of the tenth anniversary of the death of Mayakovsky, "Working with Mayakovsky" (1940), in *Rodchenko, Experiments for the Future,* 214–30; the quote is on 228.
65. Nikolai Tarabukin, *Ot mol'beria k mashine* (From the easel to the machine; 1923), trans. Christina Lodder, in *Modern Art and Modernism: A Critical Anthology*, ed. Francis Frascina and Charles Harrison (New York: Harper & Row, 1982), 139. This essay and four others by Tarabukin (including "Pour une théorie de la peinture," which was written in 1916 and published in Moscow in 1923), are available in French translation by Gérard Conio in his *Dépassements constructivistes: Taraboukine, Axionov, Eisenstein* (Lausanne, Switzerland: L'âge d'homme, 2011).
66. For example, in 1919 Rodchenko wrote a rather gloomy artist's statement: "The collapse of all 'isms' in painting was the beginning of my ascent. To the toll of the funeral bells of colorist painting, the last 'ism' is laid to eternal rest here, the last hope and love collapse, and I leave the house of dead truths." "Rodchenko's System," in the catalogue of the 10th *State Exhibition: Non-Objective Creation and Suprematism* (Moscow, 1919), in *Rodchenko, Experiments for the Future,* 84. Also, the excavation of foundations can entail demolition and thus create a negative aura, as Tarabukin observed: "When Manet's canvases first appeared about sixty years ago at Parisian exhibitions and inspired a complete revolution in the artistic world of Paris of the time, the first stone was removed from the foundation of painting. Until recently we were still inclined to see the whole subsequent development of painterly forms as a progressive process towards the perfection of those forms. In the light of most recent developments, we now perceive this, on the one hand, as a steady destruction of the painterly organism into its constituent elements, and on the other, as a gradual degeneration of painting as a typical art form." Tarabukin, *Ot mol'beria k mashine,* 135.
67. On Strzemiński and Kobro's search for unity see the French historian Yve-Alain Bois, "Strzemiński and Kobro: In Search of Motivation," in *Painting as Model* (Cambridge, MA: MIT Press, 1990), 123–55. Bois describes their "motivation" as their search for unity, which doomed them to commit the original sin of modernism—essentialism—against which the Polish historian Andrzej Turowski defends the artists, especially Kobro, in his essay "Theoretical Rhythmology, or the Fantastic World of Katarzyna Kobro," trans. Alina Kwiatkowska, in *Katarzyna Kobro, 1898–1951* (Łódź, Poland: Museum Sztuki; Leeds, England: Henry Moore Institute, 1998), 83–88.
68. Władysław Strzemiński, "L'art moderne en Pologne" (1934), trans. Antoine Budin, in Władysław Strzemiński and Katarzyna Kobro, *L'Espace uniste: Écrits du constructivisme polonaise,* (Lausanne, Switzerland: L'âge d'homme, 1977), 148.
69. All quotes in this paragraph are from Władysław Strzemiński, "B = 2; to read . . ." (1924), trans. Joanna Holzman, Piotr Graff, and Michael Trevelyans, in *Constructivism in Poland, 1923 to 1936* (Cambridge, England: Kettle's Yard Gallery, 1973), 33–36.
70. Henri Poincaré, review of David Hilbert, *Grundlagen der Geometrie* (1899), in *Bulletin des sciences mathématiques* 26 (1902): 249–72; the quote is on 252.
71. Freeman Dyson, *The Scientist as Rebel* (New York: New York Review Books, 2006), 9.
72. In a 1990 study of the role of mathematics in the emergence of modern culture, the historian of mathematics Herbert Mehrtens described Hilbert as a pure formalist; *Modern-Sprache-Mathematik: Eine Geschichte des Streits um die Grundlagen der Disziplin und des Subjekts formaler Systeme* (Frankfurt am Main, Germany: Suhrkamp, 1990). Mehrtens's thesis is that modern mathematics emerged from a conflict between those who are modern (*die Moderne*), led by Hilbert, for whom the subject of mathematics is a meaning-free language (the "*Sprache*" of the title), and those who oppose modernism (*die Gegenmoderne*), counter-modernists led by Felix Klein, for whom mathematics was about ideal, abstract objects, a battle which, according to Mehrtens, the modernists won. I suggest that Hilbert was *both* one of Mehrtens's modernists (when working on foundational topics) and one of Mehrtens's counter-modernists in the sense that Hilbert was a set-theoretic Platonist, for whom, as we saw before (on pages 161-62), consistency *entails* the existence of abstract objects.
73. Hilbert's contributions to mathematical physics are detailed by Lewis Pyenson in "Physics in the Shadow of Mathematics: The Göttingen Electron-Theory Seminar of 1905," *Archive for History of Exact Sciences* 21, no. 1 (1979): 55–89; and Leo Corry in *David Hilbert and the Axiomatization of Physics (1898–1918): From* Grundlagen Der Geometrie *to* Grundlagen Der Physik (Dordrecht: Kluwer, 2004). Corry focuses on Hilbert's interest in isolating the mathematical content of physics and of avoiding possible contradictions in theoretical physics.
74. For examples of negative criticism of Rodchenko's work in the 5 × 5 = 25 exhibition, written by Russian critics in the 1920s, see Aleksandr Lavrent'ev, "On Priorities and Patents," in Dabrowski, *Rodchenko,* 58 (see n. 58).
75. On the the mix of theory and practice in Moscow between 1920 and 1926, see Maria Gough, *The Artist as Producer: Russian Constructivism in Revolution* (Berkeley: University of California Press, 2005). In great detail, Gough describes the mix of theory and practice as a tale of "formalism, functionalism, and failure" (191).
76. Rodchenko to Stepanova, May 4, 1925, in *Rodchenko, Experiments for the Future,* 168–69 (see n. 60). See also Christina Kiaer, "Rodchenko in Paris," in *Imagine No Possessions: The Socialist Objects of Russian Constructivism* (Cambridge, MA: MIT Press, 2005), 198–240.
77. On the concept and design of Tatlin's *Monument to the Third International*, see Norbert Lynton, *Tatlin's Tower: Monument to Revolution* (New Haven, CT: Yale University Press, 2009), 81–106. On the symbolism of the spiral in Russian culture, as well as Tatlin's possible source in Khlebnikov's numerology—his "rhythms of time"—see Christina Lodder, "Tatlin's Monument to the Third International as a Symbol of Revolution," in *The Documented Image: Visions in Art History*, ed. Gabriel Weisberg and Laurinda Dixon (Syracuse, NY: Syracuse University Press, 1987), 275–88.
78. Nikolai Punin, *Pamyatnik tret'ego internatsionala* (Saint Petersburg: Otdela Izobrazitel' nylch Iskusstv, N.K.P., 1920), trans. Christina Lodder, in her essay, "Tatlin's Monument," 279.
79. Charlotte Douglas, "Kazimir Malevich," in *Kazimir Malevich*, ed. Phyllis Freeman (New York: Abrams, 1994), 34.
80. Paul Bernays, "On Platonism in Mathematics" (1935), in *Philosophy of Mathematics: Selected Reading,* ed. Paul Benacerraf and Hilary Putnam (Englewood Cliffs, NJ: Prentice Hall, 1964), 274–86.
81. For example, discussing working mathematicians and physicists who refer to abstract entities, Rudolf Carnap described each as being "like a man who in his everyday life does with qualms many things which are not in accord with the high moral principles he professes on Sundays." "Empiricism, Semantics, and Ontology" (1950), in *Philosophy of Mathematics,* 214. In this same article Carnap, although acknowledg-

ing the existence of abstract objects, bristled at being called a "Platonic realist" by W. V. Quine (250, n. 6) because this would imply that he accepted "Plato's metaphysical doctrine of universals [Forms]."

82. David Hilbert, "Über das Unendliche" (On the infinite), *Mathematische Annalen* 95, no. 1 (1926): 161–90; the quote is on 170.

5. LOGIC

1. Plato, *Cratylus* (ca. 380–67 BC), in *Dialogues*, trans. 2:260 (see chap. 1, n. 26).
2. Ibid., 2:265–67.
3. Aristotle, *Metaphysica*, 8:9 (see chap. 1, n. 30).
4. Leibniz was left with a massive classification of data, which he planned to assemble into a multivolume encyclopedia. He worked on this task intermittently for decades, trying in vain to get learned societies to give their expert advice and monarchs to pay for the production of the encyclopedia, which he left unfinished at his death in 1716. On Leibniz's encyclopedia project, see Maria Antognazza, *Leibniz: An Intellectual Biography* (Cambridge: Cambridge University Press, 2009), 92–100, 233–62, and 529–31.
5. Leibniz used this Latin phrase in a 1677 letter to the French scholar Jean Gallois, which is reprinted in part in Louis Courturat, *La Logique de Leibniz* (Hildesheim, Germany: Olms, 1901/rpt. 1961), 90, n. 3.
6. Lewis Carroll, *The Game of Logic* (London: Macmillan, 1887), xiii. The board game consists of a small card (ca. 4 in. by 6 in.), with red and grey counters (paper disks ½ in. diameter), in an envelope printed with the same title page as the book, but with the date 1896. The board game was sold with the book (an envelope containing the game slips into the back cover of the book). For an illustrated description of the game see Robin Wilson's biography, *Lewis Carroll in Numberland: His Fantastical, Mathematical, Logical Life* (New York: Norton, 2008), 175–83.
7. Frege's own symbolism, which he called his "conceptual notation," was not adopted by other logicians. In this book I use symbols devised by Giuseppe Peano, which became the basis for all later notations after Bertrand Russell and Alfred North Whitehead used it in their *Principia Mathematica*, 1910–13.
8. On the thesis that philosophical questions are questions about language, see Richard M. Rorty, "Metaphysical Difficulties of Linguistic Philosophy," in *The Linguistic Turn: Recent Essays in Philosophical Method*, ed. Richard M. Rorty (Chicago: University of Chicago Press, 1967), 1–39.
9. Russell letter to Frege (1902), in *From Frege to Gödel: A Source Book in Mathematical Logic, 1879–1931*, ed. Jean van Heijenoort (Cambridge, MA: Harvard University Press, 1967), 124–25.
10. Frege published the modified axiom as an appendix to volume two of *Grundgesetze der Arithmetik* (Basic laws of arithmetic; Jena, Germany: H. Pohle, 1903). Although Russell had shown that Frege's system was inconsistent, Frege's theorem (proving the reduction of arithmetic to logic) did not use the flawed axiom and was thus unaffected.
11. On Russell's attempt to base mathematics in logic, see C. W. Kilmister, "A Certain Knowledge? Russell's Mathematics and Logical Analysis," in *Bertrand Russell and the Origins of Analytical Philosophy*, ed. Ray Monk and Anthony Palmer (Bristol, England: Thoemmes, 1996), 269–86.
12. G. E. Moore, *Principia Ethica* (Cambridge: Cambridge University Press, 1903/rpt. 1929), 188.
13. In the reductionist spirit of the era, the American logician Henry M. Sheffer showed that the primitive concepts "or" and "not" could be reduced to "nor" (written | and called the "Sheffer stroke"), and the French logician Jean Nicod showed that the logical calculus could be written with only one rule and one axiom, both expressed using the Sheffer stroke. H. M. Sheffer, "A Set of Five Independent Postulates for Boolean Algebras, with Application to Logical Constants," *Transactions of the American Mathematical Society* 14 (1913): 481–88; and Jean Nicod, "A Reduction in the Number of Primitive Propositions of Logic," *Proceedings of the Cambridge Philosophical Society* 19 (1916–19): 32–41.
14. For arguments that these three axioms are not purely logical axioms, see Rudolf Carnap, "The Logistic Foundations of Mathematics" (1931), in *The Philosophy of Mathematics; Selected Reading*, ed. Paul Benacceraf and Hilary Putnam (Cambridge: Cambridge University Press, 1983), 41–52.
15. For the details of this approach, see David Bostock, *Russell's Logical Atomism* (Oxford: Oxford University Press, 2012).
16. On Fry's relation to France see Mary Ann Caws and Sarah Bird Wright, "Roger Fry's France," in *Bloomsbury and France: Art and Friends* (Oxford: Oxford University Press, 2000), 303–25.
17. See Russell's account of Fry and the Apostles in *The Autobiography of Bertrand Russell* (Boston: Little, Brown, 1967–69), 1:84. Fry gave a paper to the Apostles entitled "Do We Exist?" in which he criticized idealists who, like McTaggart, believe they can be sure only of their own ideas. Thus knowledge is unattainable "except with regard to themselves," to which Fry added the aside, "I think Bertie [Bertrand Russell] was once one." Fry then went on to state his own view: "In fact it seems to me more reasonable to regard the ego as a configuration of sensations and memories which has not much more solidity and coherence than the configuration of atoms in a molecule." Thus Fry allied himself with the Russell/Moore view of knowledge as a construction from sense-data ("a configuration of sensations"), borrowing Russell's atomic metaphor. Fry's unpublished manuscript is referred to as "one of his Apostle papers" by Christopher Green in his introduction to *Art Made Modern: Roger Fry's Vision of Art*, ed. Christopher Green (London: Courtauld Institute of Art, 1999), 16. The undated manuscript is catalogued in the Modern Archive of King's College, Cambridge, under the heading "Papers for the Apostles, 1887–89," inv. no. "Fry AI" (per a letter to the author from Christopher Green, May 1, 2004). But the paper must be later than 1887–89, because Fry knew Russell only after 1890; so the most plausible date for the paper is the late 1890s, when logical atomism was the subject of lively debate between Russell and Moore, or soon after their 1903 publications, which brought the analytic outlook to public attention (Russell's *Principles of Mathematics* was reviewed in *Times Literary Supplement* on Sept. 18, 1903).
18. For example: in the 1890s Fry designed a dress for Alys (Russell, *Autobiography*, 1:115); although Alys initially condemned her sister's marriage, by 1903 Berenson was reading Russell's philosophy (Berenson letter to Russell, Mar. 22, 1903, in Russell, *Autobiography*, 1:291); in 1904 the Berensons were guests in the Russell home (Russell letter to Lowes Dickenson, July 20, 1904, in Russell, *Autobiography*, 1:289); in 1905 the Russells were house guests of the Frys (Russell, *Autobiography*, 1:272). Fry and Russell remained in the same social circle after their marriages dissolved; during 1910–11 they had concurrent affairs with the wife of the British politician Philip Morrell, the indefatigable Ottoline Morrell (see Frances Spaulding, *Roger Fry: Art and Life* [Berkeley: University of California Press, 1980], 141–43), and in the early 1920s Fry painted Russell's portrait (*Portrait of Bertrand Russell, Earl Russell*, ca. 1923, National Portrait Gallery, London).
19. On Fry's long association with the magazine, see Caroline Elam, "'A More and More Important Work': Roger Fry and *The Burlington Magazine*," *Burlington Magazine* 145, no. 1200 (2003): 142–52.
20. In a letter of Aug. 3, 1908 to Clive Bell, Woolf described her determination to understand *Principia Ethica*: "I am climbing Moore like some industrious insect, who is determined to build a nest on the top of a Cathedral spire"; Virginia Woolf, *The Letters of Virginia Woolf*, ed. Nigel Nicolson (New York: Harcourt Brace Jovaovich, 1975), 1:340; see additional references to Moore on 347, 352, and 375. Woolf was acquainted with Russell; in a letter of August 12, 1908 to Vanessa Bell, Woolf described turning down an invitation to visit with "Mr. and Mrs. Bertie Russell" because they possess "too much elderly brilliance for my taste" (Woolf, *Letters*, 1:351; also see references to Russell on 1:358 and 1:365).
21. Anon., "The New Symbolic Logic," *Times Literary Supplement*, 321.
22. Bertrand Russell, "Mysticism and Logic" (1901), in *Mysticism and Logic*, 1–32 (see chap. 3, n. 8).
23. For a comparison of Ruskin and Fry, see Jacqueline V. Falkenheim, *Roger Fry and the Beginnings of Formalist Art Criticism* (Ann Arbor, MI: UMI Research Press, 1980), 52–54. Falkenheim confines her discussion of the origins of British "formalist art criticism" to the art world (Ruskin, Walter Pater, and James McNeill Whistler), and she does not mention developments in logic (Russell and Whitehead).
24. Roger Fry, "An Essay in Aesthetics" (1909), in his *Vision and Design* (London: William Clowes, 1920), 11-25; the quote is on 14–15.

25. Roger Fry, "Mantegna as a Mystic," *Burlington Magazine* 8, no. 32 (1905): 87–98; the quote is on 90–91.
26. Ibid., 98.
27. Roger Fry, "The Post-Impressionists," *Manet and the Post-Impressionists* (1910), in *A Roger Fry Reader*, ed. Christopher Reed (Chicago: University of Chicago Press, 1996), 81–85; the quote is on 83. On Fry's mix of formalism and mysticism in his approach to Cézanne, see Maud Lavin, "Roger Fry, Cézanne, and Mysticism," *Arts Magazine* 58 (Sept. 1983): 98–101. For Fry's association with Edward Carpenter, the English socialist and student of Hindu mysticism, see Linda Dalrymple Henderson, "Mysticism as the 'Tie that Binds': The Case of Edward Carpenter and Modernism," *Art Journal* 46, no. 1 (1987): 29–37.
28. Roger Fry, "An Essay in Aesthetics" (1909), in *Vision and Design*, 11–25; the quote is on 25. The British analytic philosopher Richard Wollheim has argued that Fry described "form" as if it were an objective property of an artwork. But, according to Wollheim, this entails an inconsistency because Fry's discovery of "form" was a subjective experience for him. Thus Fry's "form" cannot be the basis of his objective art criticism. See Wollheim, "On Formalism and Pictorial Organization," *Journal of Aesthetics and Art Criticism* 59, no. 2 (2001): 127–37.
29. Roger Fry, *Second Post-Impressionist Exhibition* (London: Ballantyne, 1912), 14.
30. Roger Fry, "An Essay in Aesthetics" (1909).
31. Clive Bell, *Art* (London: Chatto and Windus, 1914), 8.
32. Ibid., 25.
33. Roger Fry, book review of Clive Bell, *Art*, 1914, in Reed, *Roger Fry Reader*, 128.
34. William James, "The Stream of Thought," *The Principles of Psychology* (New York: H. Holt, 1890), 1:224–90.
35. In 1905 Freud wrote: "And so it happens that anyone who tries to make him well is to his astonishment brought up against a powerful resistance, which teaches him that the patient's intention of getting rid of his complaint is not so entirely and completely serious as it seemed. A man of letters, who incidentally is also a physician—Arthur Schnitzler—has expressed this piece of knowledge very correctly in his *Paracelsus*." "Fragment of an Analysis of a Case of Hysteria" (1905), *Standard Edition of the Complete Psychological Works of Sigmund Freud*, trans. James Strachey (London: Hogarth Press, 1953–74), 7:44.
36. The extensive literature on Virginia Woolf and British analytic philosophy includes S. P. Rosenbaum, "The Philosophical Realism of Virginia Woolf" (1971), in *English Literature and British Philosophy*, ed. S. P. Rosenbaum (Chicago: University of Chicago Press, 1971), 316–56; the Finnish analytic philosopher Jaakko Hintikka has contributed "Virginia Woolf and Our Knowledge of the External World," *Journal of Aesthetics and Art Criticism* 38 (Fall 1978–80): 5–14; see also Deborah Esch, "'Think of a kitchen table': Hume, Woolf, and the Translation of an Example," in *Literature as Philosophy: Philosophy as Literature*, ed. Donald G. Marshall (Iowa City: University of Iowa Press, 1987), 272–76; and Ann Banfield, *The Phantom Table: Woolf, Fry, Russell and the Epistemology of Modernism* (New York: Cambridge University Press, 2000).
37. Virginia Woolf, *To The Lighthouse* (1927; New York: Harcourt, Brace and World, 1955), 125.
38. Ibid., 127.
39. Ibid., 128.
40. Henry Moore interview by James Johnson Sweeney, "Henry Moore," *Partisan Review* 14, no. 2 (1947): 180–85; the quote is on 182.
41. An undated note, probably from the late 1950s, in *Henry Moore: Writings and Conversations*, ed. Alan Wilkinson (London: Lund, Humphries, Aldershot, 2002), 114.
42. Henry Moore, "Contemporary English Sculptors: Henry Moore," *Architectural Association Journal* (1930), in *Henry Moore on Sculpture: A Collection of the Sculptor's Writings and Spoken Words*, ed. Philip James (London: Macdonald, 1966), 57.
43. See A. M. Hammacher, *The Sculpture of Barbara Hepworth*, translated from Dutch by James Brockway (New York: Abrams, 1968), 15.
44. On Henry Moore's sculptural legacy, see Christa Lichtenstern, *Henry Moore: Work, Theory, Impact*, translated from German by Fiona Elliot and Michael Foster (London: Royal Academy of Arts, 2008), 286–402.
45. See Alan G. Wilkinson, "The 1930s: Constructive Forms and Poetic Structures," in *Barbara Hepworth: A Retrospective*, ed. Penelope Curtis and Alan G. Wilkinson (London: Tate, 1994), 31–77; and Norbert Lynton, "The 1930s: London," in his *Ben Nicholson* (London: Phaidon, 1993), 76–171.
46. Theo van Doesburg, "Vers la peinture blanche" (1929), in *Art Concret* (Apr. 1930): 11. Nicholson began a 1934 artist's statement by quoting the British physicist Arthur Eddington's statement that the natural world is mind-dependent: "The universe we live in is the creation of our minds. . . . If we are to know anything about that nature it must be through something like a religious experience." Nicholson added: "As I see it, painting and religious experience are the same thing, and what we are all searching for is the understanding and realization of infinity—an idea which is incomplete, with no beginning, no end and therefore giving to all things for all time." *Unit 1: The Modern Movement in English Architecture, Painting, and Sculpture*, ed. Herbert Read (London: Cassell, 1934), 89.
47. Read did the first English translation of Worringer's *Formproblem der Gotik* (1912) as *Form in Gothic* (1927), the subject of which is abstraction during the medieval era of war and famine.
48. See especially Herbert Read, *The Meaning of Art* (London: Faber and Faber, 1931), 148–53.
49. Roger Fry, review of Herbert Read's *Art Now* (1933), *Burlington Magazine* 64 (1934): 242, 245; the quote is on 242.
50. Eliot recorded that he spent Christmas vacation of 1914 engrossed in Russell's *Principia Mathematica*. Robert Sencourt, *T. S. Eliot: A Memoir* (London: Garnstone, 1971), 49.
51. Russell, *Autobiography*, 2:9–10 (see n. 17).
52. Arguing that abstract ideas (such as the concept "love") are just as real as physical objects (apples and oranges), Eliot cited Russell: "If we are allowed to accept certain remarks of Pascal and Mr. Bertrand Russell about mathematics, we believe that the mathematician deals with objects—if he will permit us to call them objects—which directly affect his sensibility." "The Perfect Critic," in Eliot, *The Sacred Wood: Essays on Poetry and Criticism* (London: Methuen, 1920/rpt. 1950), 9. In Eliot's review of Russell's *Mysticism and Logic: and Other Essays* (1917), the poet praised Russell's "wonderfully lucid introduction to the logical point of view." "Style and Thought," *Nation* (Mar. 23, 1918): 768–70; the quote is on 768.
53. T. S. Eliot, "Hamlet and His Problems," in *Sacred Wood*, 95–103; the quote is on 100.
54. T. S. Eliot, "Commentary," *Criterion* 6 (1927): 291.
55. The termination of the friendship has been documented by the literary historian Robert H. Bell: in the summer of 1917 Russell made amorous advances towards Vivien Eliot, which he described in his 1917 letters to Colette O'Neil, wife of the Irish actor Miles Malleson, who, unlike Vivien, returned Russell's affections. By early 1919 T. S. Eliot and Vivien Eliot had asked Russell in writing not to contact them. "Bertrand Russell and the Eliots," *American Scholar* 52 (1983): 309–25.
56. On Russell's impact on Eliot's "The Waste Land," see Keith Green, "'These fragments I have shored against my ruins': Russell and Modernism," in *Bertrand Russell, Language and Linguistic Theory* (London: Continuum, 2007), 144–61.
57. In 1962 Hugh Kenner, a Canadian literary scholar, observed that the history of late-nineteenth- and early-twentieth-century literature "is closely parallel to the history of mathematics during the same period." "Art in a Closed Field" (1962), in *Learners and Discerners: A Newer Criticism*, ed. Robert Sholes (Charlottesville: University Press of Virginia, 1964), 110–33; the quote is on 112. Kenner described Joyce's *Ulysses* as one example of "works of art as patterns gotten by selecting elements from a closed set and then arranging them inside a closed field" (114). He also stated that "the dominant intellectual analogy of the recent age is drawn . . . from general number theory" (122). Beyond these general remarks, Kenner didn't say *what* development in "number theory" (arithmetic) he had in mind; perhaps he was referring to the axiomatization of arithmetic by Gottlob Frege in *Grundgesetze der Arithmetik* (1893).
58. James Joyce, *A Portrait of the Artist as a Young Man* (New York: B. W. Huebsch, 1916), 241.
59. Joyce used "mosaic" to describe his corrected galleys in a letter to his editor Joyce to Weaver, Oct. 7, 1921, in *Letters of James Joyce*, ed. Stuart Gilbert (New York: Viking, 1957), 172.
60. Joyce displayed an interest in axiomatic (non-Euclidean) mathematics in his notes for *Ulysses*, in which he refers to Nikolai Lobachevsky and

Bernhard Riemann. *Joyce's* Ulysses *Notesheets in the British Museum*, ed. Phillip E. Herring (Charlottesville: University Press of Virginia, 1972), 474, notesheet "Ithac – 13," lines 87–88.

61. Joyce made references to Bertrand Russell's *Introduction to Mathematics* (1919) in notesheet "Page 30 'Ithaca,'" lines 26–38, in *Joyce's Notes and Early Draft for* Ulysses: *Selections from the Buffalo Collection*, ed. Phillip E. Herring (Charlottesville: University Press of Virginia, 1977), 109–11. The scholarly literature on mathematics in Joyce's *Ulysses* includes Richard E. Madtes, *The "Ithaca" Chapter of Joyce's* Ulysses (Ann Arbor, MI: UMI Research Press, 1983), esp. chap. 2, "The Rough Notes"; Patrick A. McCarthy, "Joyce's Unreliable Catechist: Mathematics and the Narration of 'Ithaca,'" *English Literary History* 51, no. 3 (1984): 605–18; Joan Parisi Wilcox, "Joyce, Euclid, and 'Ithaca,'" *James Joyce Quarterly* 28, no. 3 (1991): 643–49; Mario Salvadori and Myron Schwartzman, "Musemathematics: The Literary use of Science and Mathematics in Joyce's *Ulysses*," *James Joyce Quarterly* 29, no. 2 (1992): 339–55; and for a summary of the literature on the topic, see T. J. Rice, "Appendix A: Joyce, Mathematics, and Science," in *Joyce, Chaos and Complexity*, ed. T. J. Rice (Urbana: University of Illinois Press, 1997), 141–44.
62. James Joyce, *Ulysses* (1922; Paris: Shakespeare and Co., 1926), 660.

6. INTUITIONISM

1. Ralph Waldo Emerson, "The Transcendentalist," 1841 lecture in Boston; in *The Collected Works of Ralph Waldo Emerson*, ed. Robert E. Spiller and Alfred R. Ferguson (Cambridge, MA: Belknap, 1971) 1:201 and 207.
2. See Frederik van Eeden, "The Theory of Psycho-Therapeutics," *Medical Magazine* 1, no. 3 (1892): 232–57. According to van Eeden, he and van Renterghem used the same techniques as Charcot, Liébeault, and Bernheim, but the Dutchman called their method "suggestive psycho-therapy" and avoided the term "hypnotism" because of its negative association with stage performers, as in a circus (233). Van Eeden described their method as following "the simple and lucid principle of *leading the mind of the patient to effect his own cure under the influence of suggestion, sustained by sleep*" (234).
3. See Albert Willem van Renterghem, "L'evolution de la psychothérapie en Hollande," in *Deuxième Congrès Internationale de L'Hypnotisme, Paris, 1900*, ed. Edgar Bérillon and Paul Farez (Paris: Vigot, 1902), 54–62. After noting that there were numerous doctors in Holland who used methods developed in Nancy, Van Renterghem predicted that the Nancy school would become "la science officielle dans mon pays" (the official science of my country; 62).
4. For a detailed description of the lectures given at the congress, see Henri F. Ellenberger, *The Discovery of the Unconscious; the History and Evolution of Dynamic Psychiatry* (New York: Basic Books, 1970), 758–61.
5. For example, an American correspondent in Paris reported that Édouard Manet "is less mad than the other maniacs of Impressionism." Anon, *Art Journal* 6 (1880): 189.
6. Frederik van Eeden, "Vincent van Gogh (November 1890)," in his *Studies* (Amsterdam: W. Versluys, 1894–97), 2:100–108; the quote is on 108. Two decades later, van Eeden was still referring to van Gogh's art as a pathological symptom, such as "de decadenten als van Gogh en de Franschen" (the decadents like Van Gogh and the French); van Eeden, entry of Jan. 8, 1909, *Dagboek: 1878–1923*, ed. H. W. van Tri (Culemborg, the Netherlands: Tjeenk Willink-Noorduijn, 1971), 2:952.
7. See Frederik van Eeden, "A Study of Dreams," *Proceeding of the Society for Psychical Research* 67, no. 26 (1913): 413–61.
8. Frederik van Eeden, *Happy Humanity* (Garden City, NY: Doubleday, Page, 1912), 89. Van Eeden wrote this book in English for an American audience after giving a lecture tour in America in the mid-1890s.
9. Mannoury's method of doing mathematics is known secondhand from his student, Brouwer, who adopted it. Their method is described by Brouwer's biographers; see Walter P. van Stigt, *Brouwer's Intuitionism* (Amsterdam: Elsevier, 1990), esp. "Brower's Philosophy," 147–92; and Dirk van Dalen, *Mystic, Geometer, Intuitionist: The Life of L.E.J. Brouwer* (Oxford, England: Clarendon, 1999), esp. "Mathematics and Mysticism," 41–79.
10. L.E.J. Brouwer, *Life, Art, and Mysticism* (1905), in *L.E.J. Brouwer, Collected Works*, ed. A. Heyting (Amsterdam: North Holland, 1975), 1:6.
11. Frederik van Eeden, *Redekunstige grondslag van verstandhouding* (Logical foundation of understanding; 1897), in his *Studies*, 3: 5–84 (see n. 6).
12. Brouwer wrote a commentary on van Eeden's book *The Joyous World* for his student newspaper (*Studenten-weekblad Delft*, Oct. 6, 1904), which is quoted by his biographer van Dalen in *Mystic, Geometer, Intuitionist*, 64.
13. Brouwer's biographers concur that Brouwer consulted the writings of Böhme and Eckhart when he was writing *Life, Art, and Mysticism*. See Walter P. Stigt, *Brouwer's Intuitionism* (Amsterdam: North Holland, 1990), 30; and van Dalen, *Mystic, Geometer, Intuitionist*, 68. Although Brouwer does not cite Böhme or Eckehart by name, his language reflects their mystical tradition; see especially the section "Transcendent Truth" in *Life, Art, and Mysticism* in *Brouwer, Collected Works*, 1:89. Brouwer also uses the term "karma" throughout this 1905 text to refer to his spiritual self, and decades later, in a discussion of how to attain wisdom, he gave a lengthy quote from a Hindu sacred text, *Bhagavad-Gita* (ca. 800 BC); see L.E.J. Brouwer, "Consciousness, Philosophy and Mathematics" (1948), in *Brouwer, Collected Works*, 1:486.
14. L.E.J. Brouwer, "Consciousness, Philosophy, and Mathematics" (1948), in *Brouwer, Collected Works*, 1:480–84; the quote is on 480.
15. Ibid., 480. On Brouwer's belief that the perception of time is the cornerstone of intuitionist mathematics, see Grattan-Guinness, "Psychology in the Foundations of Logic and Mathematics," 43–46 (see chap. 3, n. 21).
16. Van Eeden published Freud's letter in the weekly periodical *De Amsterdammer* (Jan. 17, 1915); there is an English translation in *Freud, Standard Edition*,14:301–2 (see chap. 5, n. 35). Freud's 1915 essay "Thoughts for the Times on War and Death" is in the *Standard Edition*, 14:274–300.
17. Van Eeden met Welby in 1892 during his visit to England for a psychiatry conference (van Eeden, *Dagboek*, 1:500; see n. 6). They corresponded about their common interest in the triadic structure of signs until Welby's death in 1912. For van Eeden's view of Welby, see *Happy Humanity*, 84–87 (see n. 8).
18. On the relation of Peirce and Welby, see *Semiotic and Significs: The Correspondence between Charles S. Peirce and Lady Victoria Welby*, ed. Charles S. Hardwick and James Cook (Bloomington: Indiana University Press, 1977).
19. Van Eeden met Schoenmaekers in 1904; entry of July 3, 1904, *Dagboek*, 2:592. In 1905 van Eeden wrote an essay on Schoenmaekers's writings (entry of Sept. 27, 1905, *Dagboek*, 2:625); they met for conversation, and van Eeden appreciated Schoenmaekers's inquisitive mind (entry of Mar. 14, 1906, *Dagboek*, 2:647). After this initial contact with van Eeden, Schoenmaekers spent 1911–12 in America enrolled at the Meadville Unitarian Theological Seminary in Pennsylvania, and then he returned to Holland and settled in Laren, near van Eeden's Walden and Brouwer's hut, and entered van Eeden's circle (entry of July 5, 1915, *Dagboek*, 3:1145).
20. Van Eeden, entries of Oct. 22 and Nov. 27, 1915, *Dagboek*, 3:1466 and 1471.
21. L.E.J. Brouwer, "Intuitive Significs," the introduction to a lecture given by van Eeden on Mar. 13, 1918, trans. Walter P. van Stigt; appendix 4 in van Stigt, *Brouwer's Intuitionism*, 416–17; the quote is on 417 (see n. 9).
22. Van Eeden, entry of July 27, 1915, *Dagboek*, 3:1451.
23. Ibid., entry of July 5, 1915, *Dagboek*, 3:1450.
24. Ibid., entry of Dec. 12, 1893, *Dagboek*, 1:264. Van Eeden bought Toorop's *Les Rôdeurs* (The prowlers), 1891; Musée Kröller-Müller, Otterlo. The drawing is reproduced in Victorine Hefting, *Jan Toorop, 1858–1928: Impressionniste, symboliste, pointilliste* (Paris: Institut Néerlandais, 1977), n.p., cat. no. 44.
25. Robert P. Welsh has determined that *Passion Flower* (ca. 1901) may have been shown in the Spoor-Mondrian-Sluyters 1909 exhibition; see Welsh, *Catalogue Raisonné of the Naturalistic Works (until Early 1911)*, vol. 1 of *Piet Mondrian: Catalogue Raisonné* (Munich: Prestel, 1998), 1:213–14.
26. *Devotion* (1908) was definitely included in the Spoor-Mondrian-Sluyters 1909 exhibition, per Welsh, *Catalogue raisonné*, 1:418–19.
27. C. L. Dake, "Schilderkunst: drie avonturiers in het Stedelijk Museum," *De Telegraaf* (Jan. 8, 1909), trans. Hans Janssen and Joop M. Joosten,

in "1908–1910," in *Mondrian, 1892–1914: The Path to Abstraction*, ed. Hans Janssen (Fort Worth, TX: Kimbell Art Museum, 2002), 128–38.

28. Van Eeden, entry of Jan. 8, 1909, *Dagboek*, 2:952 (see n. 6).
29. "Mondrian is a straighforward case of acute decline" (*Mondriaan een typisch geval van eenvoudig acuut verval*), in Frederik van Eeden, "Gezondheid en Verval in Kunst," *Op de Hoogte: Maandschrift voor de Huiskammer* 6, no. 2 (Feb. 1909): 79–85; the quote is on 84. On van Eeden's views of modern art, see Michael White, "'Dreaming in the Abstract': Mondrian, Psychoanalysis and Abstract Art in the Netherlands," *Burlington Magazine* 148, no. 1235 (2006): 98–106. I disagree with Michael White's suggestion that the source of van Eeden's view of modern art as pathology was the German political writer Max Nordau's *Entartung* (Degeneration) of 1892. Van Eeden was much too sophisticated to swallow Nordau's narrow-minded anti-intellectualism, and all the historical evidence points to van Eeden being a follower (as was Nordau) of Cesare Lombroso, best-selling author of the *L'uomo di genio* (Man of genius), whom van Eeden met when they gave papers together at the Exposition Universelle in Paris in 1889. Lombroso was a respected pillar in van Eeden's community, holding professorships in psychiatry and medical jurisprudence at the University of Turin for more than twenty years, but Nordau was held in low regard by van Eeden's peers, including Sigmund Freud and William James. During Freud's stay in Paris in 1886, the young neurologist got a letter of introduction to Nordau, but during their meeting Freud "found him vain and stupid and did not cultivate his acquaintance." Ernest Jones, *The Life and Work of Sigmund Freud* (New York: Basic Books, 1953), 1:188. James described Nordau as a follower of Lombroso who pushed his master's thesis to an extreme: "One disciple of the school, indeed, has striven to impugn the value of works of genius in a wholesale way (such as works of contemporary art, namely, as he is unable to enjoy, and they are many) by using medical arguments," to which James added a footnote to "Max Nordau, in his bulky book entitled *Degeneration*"; William James, *The Varieties of Religious Experience: A Study in Human Nature* (1902), in his *Writings, 1902–1910* (New York: Library of America, 1987), 24. In addition to Lombroso, van Eeden was also inspired by the American John ("truth-to-nature") Ruskin, and he endorsed Ruskin's observation that when an artist does not accurately copy natural color, this is an indication of decadence (*indicator der decadence*) and noticeable decline (*verval merkbaar*); van Eeden, entry of Jan. 8, 1909, *Dagboek*, 2:952.
30. On the rapid adoption of French and Austrian psychoanalysis in Holland, see Ilse N. Bulhof, "Psychoanalysis in the Netherlands," *Comparative Studies in Society and History* 24, no. 4 (1982): 572–88. Van Eeden never wrote about Mondrian's art after his conversion to Freudian aesthetics; for discussion of Mondrian through a psychoanalytic lens, see Peter Gay, *Art and Act: On Causes in History—Manet, Gropius, Mondrian* (New York: Harper & Row, 1976), in which this cultural historian, who holds a certificate in psychoanalysis, suggested that one of the "causes in history" is an artist's personal life, especially his or her sex life or, in Mondrian's case, the lack thereof. Thus Mondrian's reclining horizontal/female lines and erect male/verticals can, according to Gay, be understood as an expression of his repressed sexuality (210–26). For another psychoanalytic perspective, see Pieter van der Berg, "Mondrian: Splitting of Reality and Emotion," in *Dutch Art and Character: Psychoanalytic Perspectives on Bosch, Breughel, Rembrandt, Van Gogh, Mondrian*, ed. Joost Baneke et al. (Amsterdam: Swets and Zeitlinger, 1993), 117–20. The art historian Donald Kuspit, who, like Gay, holds a certificate in psychoanalysis, discussed two Dutchmen with opposite types of personalities, Van Gogh and Mondrian, as examples when comparing the views that art is a sublimated sexual instinct (per Freud) or a transitional object (per the British pediatrician Donald Winnicott). Kuspit, "Art: Sublimated Expression or Transitional Expression? The Examples of Van Gogh and Mondrian," *Art Criticism* 9, no. 2 (1994): 64–80.
31. See Michel Seuphor [pseud. of F. L. Berckelaers], *Piet Mondrian: Life and Work* (New York: Abrams, 1956), 53. Reflecting back on his career after achieving his signature style, Mondrian wrote: "It is in my work that I am something, but compared to *The Great Initiates*, for example, I am nothing." Mondrian to Seuphor, 1934, quoted in Seuphor, *Mondrian* (1956), 58.
32. Rudolf Steiner, *Theosophy* (1909), trans. Elizabeth Douglas Shields (Chicago: Rand McNally, 1910), 211–12.
33. Mondrian owned a paraphrase of the lectures, which he kept with him his whole life; his copy, which is missing its title page, is known in the literature under various titles, including *Dutch Lectures*; see R. P. Welsh, "Mondrian and Theosophy," in *Piet Mondrian, 1872–1944* (New York: Guggenheim Museum, 1971), 39. The edition Mondrian owned was probably Rudolf Steiner, *Theosofie* (Amsterdam: Theosofische Uitgevers Maatschappij, 1909).
34. Steiner, *Theosophy*, 178–94.
35. Seuphor, *Mondrian*, 54–58.
36. Helena Petrovna Blavatsky, *Isis Unveiled: A Master-Key to the Mysteries of Ancient and Modern Science and Theology* (New York: Bouton, 1877), 2:270.
37. Mondrian wrote, "While it [Cubism] achieves greater unity than the old art, because its composition has strong plastic expression, Cubism loses unity by following the fragmented character of natural appearance. . . . [But in my case] by learning to perceive nature more and more purely, painting came to abstraction." "Neo-plasticism in Painting," *De Stijl* 1 (1917): 3, reprinted in *De Stijl: Extracts from the Magazine*, trans. R. R. Symonds, ed. Hans L. C. Jaffé (London: Thames and Hudson, 1970), 54, 88.
38. Mondrian made this statement in a sketchbook from which his friend Michel Seuphor (pseud. of Fernand Berckelaers) quoted in "Piet Mondrian, 1914–18," *Magazine of Art* 45, no. 5 (1952): 217. In 1952 Harry Holtzman, the executor of Mondrian's estate, owned this and one other sketchbook, which were later published as *Two Mondrian Sketchbooks, 1912–1944*, ed. Robert P. Welsh and J. M. Joosten (Amsterdam: Meulenhoff, 1969).
39. James Clerk Maxwell, "On Faraday's Lines of Force" (1856), *Transactions of the Cambridge Philosophical Society* 10 (1864): 30.
40. Van Eeden, entry of July 15, 1915, *Dagboek*, 3:1445.
41. M.H.J. Schoenmaekers, *Beginselen der Beeldenden Wiskunde* (Bussum, the Netherlands: van Dishoek, 1916), 56.
42. In 1914 the theosophy journal *Theosophia* rejected an early version of the essay "Neo-plasticism in Painting"; see Carel Blotkamp, *Mondrian: Art of Destruction* (London: Reaktion, 1995), 107.
43. Piet Mondrian, "A Dialogue on Neo-plasticism," *De Stijl* 2, no. 5 (1919), in Jaffé, *De Stijl: Extracts*, 124.
44. Theo van Doesburg, "Kunst-kritiek," *Eenheid* (Nov. 6, 1915), quoted in *De Stijl: The Formative Years, 1917–1922*, trans. Charlotte I. Loab and Arthur L. Loab, ed. Carel Blotkamp et al. (Cambridge, MA: MIT Press, 1986), 8.
45. Van Doesburg to Anthony Kok, Feb. 7, 1916, in *Theo van Doesburg. 1883–1931*, ed. Evert van Straaten (The Hague: Staatsuitgeverij, 1983), 56.
46. For a discussion of the contents of van Doesburg's essay, see White, "Mondrian, Psychoanalysis," 103–4 (see n. 29).
47. Van Doesburg to Anthony Kok, Feb. 7, 1916, trans. Michael White, quoted in White, "Mondrian, Psychoanalysis," 104.
48. Mondrian to Steiner, Feb. 23, 1921. Mondrian wrote the letter in French, and it was translated into German and included in Rudolf Steiner, *Wenn die Erde Mond wird: Wandtafelzeichnungen zu Vorträgen 1919–1924*, ed. Walter Kugler (Cologne, Germany: DuMont, 1992), 151.
49. Mondrian to van Doesburg, Spring 1917, trans. Carla van Spluntren, quoted in Rudolf W. D. Oxenaar, "Van der Leck and De Stijl: 1916–1920," in *De Stijl, 1917–1933: Visions of Utopia*, ed. Mildred Friedman (Oxford, England: Phaidon, 1982), 73.
50. An article by Oud appeared in the first issue of *De Stijl*; J.J.P. Oud, "The Monumental Townscape," *De Stijl* 1, no. 1 (1917), in Jaffé, *De Stijl: Extracts*, 95–96 (see n. 37).
51. Vilmos Huszár, "Iets over *Die Farbenfibel* van W. Ostwald," *De Stijl* 1, no. 10 (1918): 113–18. Books recommended by the editor of *De Stijl* included three by W. Ostwald: *Die Farbenfibel (1917)* and *Die Harmonie der Farben (1918)*, in *De Stijl* 2, no. 6 (1919): 72; and *Mathematische Farbenlehre (1918)*, in *De Stijl* 3, no. 1 (1919): 12.
52. Wilhelm Ostwald, "Die Harmonie der Farben," *De Stijl* 3, no. 7 (1920): 60–62.
53. Piet Mondrian, "Neo-plasticism in painting," *De Stijl* 1, no. 3 (1917): 29–30, in Jaffé, *De Stijl: Extracts*, 55.
54. Ostwald's system was used by designers for only about a decade, ending with the chemist's death in 1932. His system proved too limited

because it classified the mix of a hue only with black and white and not with other hues.

55. Gino Severini, "La peinture de l'avant garde," *De Stijl* 1 (1917), published in segments 18ff, 27ff, 45ff, 59ff, 94ff, and 118ff.
56. Theo van Doesburg to Anthony Kok, June 22, 1918, quoted in Blotkamp, *DeStijl: The Formative Years*, 30 (see n. 44).
57. Ibid.
58. Piet Mondrian, "Neo-plasticism in Painting," *De Stijl* 1 (1917): 29. The title of Schoenmaekers's book, *Beginselen der Beeldenden Wiskunde*, is usually translated from Dutch to English as *Plastic Mathematics*. The Dutch word *beeldenden* has the meaning "creating form," so "plastic mathematics" means mathematical entities (lines, points, planes) that create form; *beeldenden* is usually translated into German as *Gestaltung* ("form") and French as *plastique* ("plastic"). Mondrian's phrase *De nieuwe beelding*, which in English is rendered "Neo-plasticism," similarly means the creation of new form in its usual translations as *die neue Gestaltung* (German) and *le néo-plasticisme* (French).
59. In Freud's earliest essay on the psychology of religion, "Obsessive Actions and Religious Practices" (1907; Dutch translation, 1914), he described repetitive religious rituals as a symptom of a "universal obsessional neurosis." What Freud saw as mature stoicism in the scientific spirit (facing death without delusions of immortality), van Eeden perceived as cynicism. Van Eeden recorded many opinions about Freud in his *Dagboek* (see n. 6). As early as July 31, 1910, van Eeden was complaining that Freud had a brutish lack of feeling for his patients' "higher or finer feelings [*fijnere of hoogere gevoelens*]," in other words, for their religious beliefs (2:1111).
60. On the balance between science and religion in this project, see Luc Bergmans, "Science and the House of God in the City of Light," in *Utopianism and the Sciences, 1880–1930*, ed. M. G. Kemperink and Leonicke Vermeer (Leuven, Belgium: Peeters, 2010), 144–57.
61. Van Eeden, entries of May 22, 1918, and July 24, 1918, *Dagboek*, 4:1671–72 and 4:1689–90.
62. After van Eeden and Brouwer abandoned the project in 1926, it was continued by the die-hard Mannoury; see Luc Bergmans, "Gerrit Mannoury and His Fellow Significians on Mathematics and Mysticism," in *Mathematics and the Divine: A Historical Study*, ed. Teun Koetsier and Luc Bergmans (Amersterdam: Elsevier, 2005), 550–68.
63. On the state of intuitionism today, see the essays collected in *One Hundred Years of Intuitionism, 1907–2007*, ed. Markus van Atten (Basel, Switzerland: Birkhäuser, 2008).

7. SYMMETRY

1. Time dilation means that as the mass approaches the speed of light, its "proper time" (the ticking of its personal clock) is slowed.
2. In Einstein's day scientists believed that gravity was a force that only attracts, whereas electromagnetism both attracts and repels. But in 1998 astronomers detected signs that some unknown antigravity effect—the so-called dark force—seems to be accelerating the expansion of the cosmos, which suggests that gravity also repels.
3. For a discussion of the interrelated work of Speiser and Weyl, see Patricia Radelet–de Grave, "Andreas Speiser (1885–1970) et Hermann Weyl (1885–1955), scientifiques, historiens et philosophes des sciences," *Revue philosophique de Louvain* 94, no. 3 (Aug. 1996): 502–35.
4. The information about Speiser's life in this and the following paragraph is based on an interview (Mar. 9, 2008, in Arlesheim, Switzerland) by the author with Andreas Speiser's nephew, the mathematician and physicist David Speiser, whose work is in group theory and elementary particle theory, and his wife, Ruth Speiser, who is the daughter of Hermann Weyl.
5. The ideas of artistic sympathy and empathy were introduced by Robert Vischer in *On the Optical Sense of Form: A Contribution to Aesthetics* (1873) and developed by Theodor Lipps, who founded an institute of experimental psychology in Munich in 1894. On the interrelations of Vischer, Lipps, and Wölfflin, see Harry Francis Mallgrave and Eleftheries Ikonomou, "The Psychology of Form and Style Transformations: Heinrich Wölfflin and Adolf Göller," in *Empathy, Form, and Space: Problems in German Aesthetics, 1873–1893*, trans. Harry Francis Mallgrave (Santa Monica, CA: Getty Center for the History of Art, 1994), 39–56.
6. Wölfflin wrote this sentence in a 1942 autobiographical statement that is in the archives of the Österreichische Akademie der Wissenschaften, Vienna. The sentence is rendered in French by an anonymous translator in *Relire Wölfflin*, ed. Joan Hart (Paris: Musée du Louvre, 1995), 148. On Wölfflin's relation to Gestalt psychology, see Friedrich Sander, "Gestaltpsychologie und Kunsttheorie," in *Ganzheitspsychologie: Grundlagen, Ergebnisse, Anwendungen*, ed. Friedrich Sander and Hans Volkelt (Munich: C. H. Beck'sche, 1962), 383–403.
7. Speiser recalled "the hours I have spent among a crowd of students . . . listening to Wölfflin's lecture, checking the clock from time to time, not with unpleasant feelings but hoping only that the time is not yet up, as I feared." Andreas Speiser, *Die mathematische Denkweise* (Zurich: Rascher, 1932), 96.
8. Rosalind Kraus has noted the similarity of Wölfflin's approach to an analysis of symmetrical systems, although she does not relate Wölfflin's art history methodology to the mathematics of group theory: "If we take Wölfflin as the father of art historical structuralism, we realize that his 'principles of art history' amount to a code. A set of bipolar opposites that constitute the formal language or system out of which art writes itself as history. These oppositions—line/color, closed/open, planarity/recession, unity/multiplicity, clear/indistinct—comprise the code in relation to which style can be either manifested or perceived." Krauss, "Representing Picasso," *Art in America* 68, no. 10 (Dec. 1980): 90–96; the quote is on 93.
9. Hermann Weyl, *Symmetry* (Baltimore: Washington Academy of Sciences, 1938); the book was reprinted as *Symmetry* (Princeton, NJ: Princeton University Press, 1952).
10. Continuing Müller's analysis, in 1987 the Spanish mathematician José María Montesinos identified examples of all seventeen possible tiling patterns on a two-dimensional plane at the Alhambra; see Montesinos, *Classical Tessellation and Three-fold Manifolds* (Berlin: Springer, 1987).
11. On Escher's exploration of tiling patterns, see Doris Schattschneider, *M. C. Escher: Visions of Symmetry* (New York: Abrams, 2004), esp. 44–52 and 342–46.
12. Max Wertheimer, "Untersuchungen zur Lehre von der Gestalt," *Psychologie Forschung* 4 (1923): 301–50; an abstracted version of this essay was translated by Willis D. Ellis as "Laws of Organization in Perceptual Forms," in *A Source Book of Gestalt Psychology* (London: Kegan Paul, Trench, Trübner, 1938), 71–88.
13. Jean Piaget, an autobiographic statement, "Jean Piaget," in *History of Psychology as Autobiography*, ed. Edwin G. Boring et al. (Worcester, MA: Clark University Press, 1952), 4: 242–43.
14. On their relationship, see Arthur I. Miller, "Albert Einstein and Max Wertheimer: A Gestalt Psychologist's View of the Genesis of Special Relativity Theory," *History of Science* 13, no. 2 (1975): 75–103.
15. The exchange is described by John H. Flavell in *The Developmental Psychology of Jean Piaget* (Princeton, NJ: Van Nostrand, 1963), 259.
16. Jean Piaget, *Le développement de la notion de temps chez l'enfant* (The development of the notion of time in the child; Paris: Presses Universitaires de France, 1946) and *Les notions de mouvement et de vitesse chez l'enfant* (The ideas of motion and speed in the child; Paris: Presses Universitaires de France, 1946).
17. Piaget's legacy is summarized in Harry Beilin, "Piaget's Enduring Contribution to Developmental Psychology," *Developmental Psychology* 28 (1992): 191–204; and John H. Flavell, "Piaget's Legacy," *Psychological Science* 7, no. 4 (July 1996): 200–203. On the importance of Piaget's work for understanding mathematical cognition, see Gisele Lemoyne and Mireille Favreau, "Piaget's Concept of Number Development: Its Relevance to Mathematics Learning," *Journal for Research in Mathematical Learning* 12, no. 3 (1981): 179–96.
18. On Allianz, see Rudolf Koella, "El grupo de artistas Allianz y los Concretos de Zurich," in *Suiza Constructiva*, ed. Patricia Molins (Madrid: Museo Nacional Reina Sofía, 2003), 54–57.
19. On the relation of Concrete Art to graphic design in Switzerland, see Richard Hollis, *Swiss Graphic Design: The Origins and Growth of an International Style, 1920–1965* (New Haven, CT: Yale University Press, 2006). The graphic art career of Lohse is documented in *Richard Paul Lohse Konstruktive Gebrauchsgrafik*, ed. Richard Paul Lohse Foundation with Christof Bignens and Jörg Stürtzebecher (Ostfildern-Ruit, Germany: Hatje Cantz, 1999).

20. On Bill's Bauhaus training, see Jakob Bill, *Max Bill am Bauhaus* (Bern, Switzerland: Benteli, 2008).
21. On popularizations of Einstein's theory of relativity, see Gerald Holton, "Einstein's Influence on Our Culture," in *Einstein, History, and Other Passions* (Woodbury, NY: American Institute of Physics, 1995), 3–21.
22. The course on Gestalt psychology was taught by Karlfried von Dürckheim of Leipzig; Hannes Meyer to Mayor Hesse of Dessau, Aug. 16, 1930, in Hans M. Wingler, *The Bauhaus: Weimar, Dessau, Berlin, Chicago* (Cambridge, MA: MIT Press, 1969), 163–65.
23. The French historian Mars Ducourant has stated that Bill's use of Speiser's group theory dates to 1946, but he gives no historical evidence to support this date. "Art, sciences et mathématiques: De la Section d'Or à l'Art Concret," in *Art Concret* (Paris: Espace de l'Art Concret, 2000), 45–54, see esp. 52. I date Bill's first use of group theory a decade earlier based on evidence I give in the text.
24. La Roche's collection today forms the core of the modernist holdings of the Kunstmuseum, Basel.
25. Speiser recalled a conversation with Braque: "The painter Braque once told me of having observed in his youth that pictures in perspective always draw attention to the background, away from the observer. He then had the idea of using perspective in the reverse direction, toward the observer. This effect can be seen in his pictures as early as 1905, and thus, through his collaboration with Picasso, arose the so-called Cubism, in which objects project out of the picture toward the foreground in planes. Now this is evidently a mathematical principle, and one that has been extremely fruitful"; Andreas Speiser, "Symmetry in Science and Art," *Daedalus* (Winter 1960): 191–98; the quote is on 198.
26. LeCorbusier was awarded the degree under the description "the brilliant creator [*dem genialen Schöpfer*] and the designer of spatial forms and mathematical laws for modern architecture." *Universität Zürich: Bericht über das akademische Jahr, 1932–33* (Zurich: Orell Füssli, 1933), 63. Although Speiser encouraged Le Corbusier's search for ideal proportions, he discouraged the architect's interest in the Golden Section. When Le Corbusier conjectured that the Golden Ratio (about 1.618) is embodied in the (alleged) spiraling orbits of the planets, Speiser informed the architect that Johannes Kepler had determined that planets travel in elliptical orbits (Kepler's *First Law of Planetary Motion*, 1609); Speiser to Le Corbusier, June 13, 1954, Foundation Le Corbusier, Paris.
27. In Le Corbusier's correspondence with Speiser spanning 1928 to the 1950s, and in his voluminous writings, the architect never expressed interest in group theory; Foundation Le Corbusier, Paris.
28. Turel was Bill's close friend and intellectual companion from the mid-1930s. Bill included in the 1936 Zurich exhibit a piece by the young sculptor Hanns Welthi, the brother of Turel's wife; *Zeitprobleme in der Schweizer Malerei und Plastik*, catalogue of an exhibition held June 13–July 22, 1936 (Zurich: Kunsthaus, 1936), 40. Turel corresponded with Speiser and had read Speiser's chapter on decorative patterns ("On Symmetry in Ornament") in his *Die mathematische Denkweise* (a book which Speiser dedicated to Raoul La Roche); Turel to Speiser, Nov. 28, 1949, Adrien Turel Stiftung, Zurich, MS 25. Turel's letter was written on the occasion of the release of the second (1945) edition of Speiser's book, but in this and his subsequent correspondence with Speiser, Turel exhibits a long-time interest in Speiser's ideas; see esp. Turel to Speiser, June 27, 1952, and Mar. 20, 1955. Speiser's perspective on the history and philosophy of mathematics and the arts was central to Turel's intellectual pursuits.
29. Author interview with David Speiser, Mar. 9, 2008.
30. Speiser, *Die mathematische Denkweise*, 16 (see n. 7).
31. Ibid., 21.
32. Max Bill, "Konkrete Gestaltung," in *Zeitprobleme in der Schweizer Malerei und Plastik* (Zurich: Kunsthaus, 1936), 9.
33. Schmidt's remarks are recorded in the periodical *Abstrakt/Konkret* 1 (1944): n.p.
34. Bill, in his capacity as director of an advertising design business, was also a colleague of Schmidt, whom Bill often hired as a free-lance copywriter; author interview with Jakob Bill (Max Bill's son), Mar. 10, 2008, Zurich. Schmidt's brother, the architect Hans Schmidt, had graduated from the Bauhaus and was a friend of Bill.
35. Graeser's algorithmic method is manifest in his many drawings and diagrams for his finished paintings; see *Camille Graeser: Vom Entwurf zum Bild: Entwurfszeichnungen und Ideenskizzen, 1938–1978*, ed. Richard W. Gassen and Vera Hausdorff (Cologne, Germany: Wienand, 2009).
36. Author interview with Vera Hausdorff, conservator of the Camille Graeser-Foundation, Mar. 11, 2008, Zurich.
37. On Lohse's techniques to achieve his color and form (other than his use of group theory), see Hans Joachim Albrecht, "Farbensinn und konstruktive Logik: Color Sense and Constructive Logic," trans. Maureen Oberli-Turner, in *Richard Paul Lohse: Drucke: Dokumentation und Werkverzeichnis/Prints: Documentation and catalogue raisonné*, ed. Johanna Lohse James and Felix Wiedler (Ostfildern, Germany: Hatje Cantz, 2009), 34–45.
38. Richard Paul Lohse, "Standard, Series, Module: New Problems and Tasks of Painting," in *Module, Proportion, Symmetry, Rhythm*, ed. Gyorgy Kepes (New York: George Braziller, 1966), 142.
39. This point is made in the layout of the catalogue raisonné of Lohse's graphic design, *Lohse: Konstruktive Gebrauchsgraphik (see* n. 19). The catalogue begins with six pairs of Lohse's fine and graphic works of art, printed on facing pages (including *Concretion I* [12] and his cover for Giedion-Welcker's *Poètes à l'écart* [13]), showing their similarity (8–19).

8. UTOPIAN VISIONS AFTER WORLD WAR I

1. Felix Klein, "Festrede zum 20 Stiftungstage der Göttinger Vereinigung zur Förderung der angewandten Physik und Mathematik," *Jahresbericht der Deutschen Mathematiker-Vereinigung* 27 (1918): 217–28; the quote is on 217.
2. Oswald Spengler, *The Decline of the West* (1918), trans. Charles Francis Atkinson (New York: Knopf, 1957), 1:21.
3. Ibid., 1:85.
4. Ibid., 1:88.
5. On changing visions of "utopia" in the sense of interpretations of a present life with an eye to a better future, see Fredric Jameson, "Utopia as Method, or the Uses of the Future," in *Utopia/Dystopia: Conditions of Historical Possibility*, ed. Michael D. Gordin, Helen Tilley, and Gyan Prakash (Princeton, NJ: Princeton University Press, 2010), 21–44.
6. Gottlob Frege, review of Husserl, *Philosophie der Arithmetik* (1891), in *Zeitschrift für Philosophie und philosophische Kritik* 103 (1894): 313–32; the quote is on 332; translated by Hans Kaal as "Review of E. G. Husserl, *Philosophie der Arithmetik I*," in *Collected Papers on Mathematics, Logic, and Philosophy*, ed. Brian McGuinness (New York: Blackwell, 1984), 195–209; the quote is on 209.
7. This feature of Husserl's thought made him a favorite of the Surrealist painter René Magritte; see Caroline Joan S. Picart, "Memory, Pictoriality, and Mystery: (Re)presenting Husserl via Magritte and Escher," *Philosophy Today* 41 (1997): 118–26.
8. On Kierkegaard's critique of Hegel, see David L. Rozema, "Hegel and Kierkegaard on Conceiving the Absolute," *History of Philosophy Quarterly* 9, no. 2 (1992): 207–24.
9. For Nietzsche's critique of Comte, see Patrik Aspers, "Nietzsche's Sociology," *Sociological Forum* 22, no. 4 (2007): 474–99.
10. For a description of this atmosphere in German academia, see Fritz K. Ringer, *The Decline of the German Mandarins: The German Academic Community, 1890–1933* (Cambridge, MA: Harvard University Press, 1969), esp. 200–52 and 367–449.
11. The essays by Brouwer on this topic are reprinted, along with Hilbert's responses, in the collection *From Brouwer to Hilbert: The Debate on the Foundations of Mathematics in the 1920s*, ed. Paolo Mancosu (New York: Oxford University Press, 1998).
12. Weyl wrote: "In all cases, this process of confirmation (and not the proof) remains the ultimate source from which knowledge derives its authority; it is the 'experience of truth.'" Weyl, *Das Kontinuum* (1918), translated by Stephen Pollard and Thomas Bole as *The Continuum* (Kirksville, MO: Thomas Jefferson University Press, 1987), 119.
13. Hermann Weyl described his spiritual journey in retrospect in "Erkenntnis und Besinnung (ein Lebensrückblick)" (1954), in *Gesammelte Abhandlungen*, ed. Komaravolu Chandrasekharan (Berlin: Springer, 1968), 4:631–49; the quote is on 647; translated by T. L. Saaty and F. J. Weyl as "Insight and Reflection (a Review of My Life)" (1954), in *The Spirit and Uses of the Mathematical Sciences* (New York:

McGraw-Hill, 1955), 281–301; the quote is on 298. Elsewhere in this same essay Weyl wrote: "Of all spiritual experiences, the ones which brought me greatest happiness were the study of Hilbert's magnificent *Report on the Theory of Algebraic Numbers* as a young student in 1905, and in 1922 the reading of Eckhart. . . . Here I finally found for myself the entrance into the religious world" (299).

14. Hermann Weyl, *Raum, Zeit, Materie: Vorlesungen über allgemeine Relativitätstheorie* (1918), 2nd ed. (Berlin: Springer, 1919), 227.
15. David Hilbert, "Neubegründung der Mathematik" (New foundation of mathematics; 1922), in Hilbert, *Gesammelte Abhandlungen*, 3:157–77; the quote is on 160 (see chap. 4, n. 17). For a description of the shifting attitudes towards Brouwer in Weimar Germany, see Dennis E. Hesseling, *Gnomes in the Fog: The Reception of Brouwer's Intuitionism in the 1920s* (Basel, Switzerland: Birkhäuser, 2003); see esp. 222–24 on Brouwer's relation to Weyl.
16. Hermann Weyl, "The Current Epistemological Situation in Mathematics" (1925–27), trans. Benito Müller, in Mancosu, *From Brower to Hilbert*, 140.
17. David Hilbert, "Die Grundlagen der Mathematik" (The foundations of mathematics; 1927), in van Heijenoort, *From Frege to Gödel*, 475 (see chap. 5, n. 9).
18. Rudolf Carnap, *Der Raum: Ein Beitrag zur Wissenschaftslehre* (Space: A contribution to scientific theory; Berlin: Reuther and Reichard, 1922).
19. Moritz Schlick, "Meaning and Verification," *Philosophical Review* 44 (1936), reprinted in Schlick, *Gesammelte Aufsätze, 1926–1936* (Vienna: Gerold, 1938), 341.
20. Rudolf Carnap, Hans Hahn, and Otto Neurath, *Wissenschaftliche Weltauffassung: Der Wiener Kreis* (A scientific worldview: The Vienna Circle; 1929), n.t., in Otto Neurath, *Empiricism and Sociology* (Dordrecht, the Netherlands: Reidel, 1973), 299–318; the quote is on 306.
21. Max Jammer, *The Conceptual Development of Quantum Mechanics* (New York: McGraw-Hill, 1966), esp. 166–80; and *The Philosophy of Quantum Mechanics: The Interpretations of Quantum Mechanics in Historical Perspective* (New York: Wiley, 1974).
22. Paul Forman, "Weimar Culture, Causality, and Quantum Theory, 1918–1927: Adaptation by German Physicists and Mathematicians to a Hostile Intellectual Environment," *Historical Studies in the Physical Sciences* 3 (1971): 1–115; and Forman, *"Kausalität, Anschaulichkeit,* and *Individualität,* or How Cultural Values Prescribed the Character and Lessons Ascribed to Quantum Mechanics," in *Society and Knowledge*, ed. Nico Stehr and Volker Meja (New Brunswick, NJ: Transaction, 1984), 333–47.
23. Harald Høffding was also a close friend of Bohr's father Christian Bohr, a professor of physiology at the university. For the details of Bohr's knowledge of Kierkegaard, see Jammer, *Conceptual Development of Quantum Mechanics*, 172–76.
24. Although, according to Kierkegaard, one can formulate an abstract, logical system (of mathematics) as a completed whole. On Kierkegaard's distinction between knowledge of mathematics and of science, see Harald Høffding, *Søren Kierkegaard som filosof* (1892), translated into German by Albert Dorner und Christof Schrempf as *Søren Kierkegaard als Philosoph* (Stuttgart, Germany: Frommann, 1896), 67; and Harald Høffding, "Søren Kierkegaard," in *A History of Modern Philosophy*, trans. B. E. Meyer (London: Macmillan, 1908), 2:285–89, esp. 2:287.
25. "An das System könnte man erst denken, wenn man auf die abgeschlossene Existenz zurückblicken könnte—das würde aber voraussetzen, daß man nicht mehr existierte!" Høffding in *Kierkegaard som filosof*, 69.
26. Bohr recollected this in a 1962 interview, which is quoted at length by Gerald Holton in "The Roots of Complementarity," *Daedalus* 99 (Fall 1970): 1015–55; the quote is on 1034–35. For the details of Bohr's knowledge of James, see Jammer, *Conceptual Development of Quantum Mechanics*, 176–79.
27. William James to F.C.S. Schiller, Oct. 26, 1904, in *The Letters of William James*, ed. Henry James (Boston: Atlantic Monthly Press, 1920), 2:216. James wrote the preface to Høffding's *The Problems of Philosophy* (1905), which was Galen M. Fisher's English translation of Høffding's *Filosofiske problemer* (1902).
28. Erwin Schrödinger, "The Present Situation in Quantum Mechanics" (1935), trans. John D. Trimmer, in *Quantum Theory and Measurement*, ed. John Wheeler and Woyciech Hubert Zurek (Princeton, NJ: Princeton University Press, 1983), 152–67; the quote is on 157. Schrödinger wrote his Schrödinger equation to describe a wave of matter over time, which is a physical object in the natural world-out-there; but for Born the equation described a probability distribution, which is an abstract object in the mathematical-world-out-there. Today physicists and philosophers of science are again describing Schrödinger's Ψ waves as entities in the natural world; see *The Wave Function: Essays on the Metaphysics of Quantum Mechanics*, ed. Alyssa Ney and David Z. Albert (Oxford: Oxford University Press, 2013).
29. Niels Bohr, "The Quantum Postulate and the Recent Developments in Atomic Theory" (1927), n.t., *Nature* 121, no. 3050, (1928): 580–90, reprinted in Bohr, *Atomic Theory and the Description of Nature* (Cambridge: Cambridge University Press, 1934), 52–91; the word is on 54.
30. See Forman, "Weimar Culture," 1–115; and Forman, *"Kausalität, Anschaulichkeit, and Individualität,"* 333–47. Forman has written: "With extraordinary suddenness the German mathematics community began to feel how insecure were the foundations upon which the entire structure of mathematical analysis rested, how dubious the methods by which the structure had been erected. Now, with quasi-religious enthusiasm, considerable numbers of German mathematicians rallied to L.E.J. Brouwer's standard, calling for a complete reconstruction of mathematics, a redefinition of the enterprise, which, appropriately enough, went under the name 'intuitionism.'" Forman, "Weimar Culture," 60.
31. Hermann Weyl, *Raum, Zeit, Materie: Vorlesungen über allgemeine Relativitätstheorie* (1918), 4th ed. (Berlin: Springer, 1921), 283. Weyl ended this 1921 edition on a Romantic note: "Whoever looks back over the ground that has been traversed . . . must be overwhelmed by a feeling of freedom won—the mind has cast off the fetters that held it captive. He must feel transfused with the conviction that reason is not only a human, a too human, makeshift in the struggle for existence, but that, in spite of all disappointments and errors, it is yet able to follow the intelligence which has planned the world, and that the consciousness of each one of us is the center at which the One Light and Life of Truth comprehends itself in Phenomena. Our ears have caught a few of the fundamental chords from that harmony of the spheres of which Pythagoras and Kepler once dreamed" (284). Forman has described Weyl as follows: "The quantum theory was for Weyl a *post factum* rationalization for a position whose adoption represented an actualization of his own intellectual/emotional proclivities by contact with a *Zeitgeist* of the corresponding character." Paul Forman, "The Reception of an Acausal Quantum Mechanics in Germany and Britain," in *Weimar Culture and Quantum Mechanics*, ed. Cathyrn Carson, Alexei Kojevnikov, and Helmuth Trischler (London: Imperial College Press, 2011), 221–60; the quote is on 226.
32. On Kierkegaard's concept of knowledge by dialectic, see Høffding, "Søren Kierkegaard," 2:28 (see n. 24). On Bohr's sources in Kierkegaard, see Jammer, *Conceptual Development of Quantum Mechanics*, 172–76. In his detailed study of the origin of the concept of "complementarity," the historian Gerald Holton has concurred with Jammer: "Now it would be absurd as it is unnecessary to try to demonstrate that Kierkegaard's conceptions were directly and in detail translated by Bohr from their theological and philosophical context to a physical context. Of course, they were not. All one should do is permit oneself the open-minded experience of reading Høffding and Kierkegaard through the eyes of a person who is primarily a physicist—struggling, as Bohr was, first with his 1912–13 work on atomic models, and again in 1927." Holton, "Roots of Complementarity," 1042 (see n. 26).
33. Bohr's term *complementarity* is the source of many of the philosophical puzzles of quantum mechanics that physicists educated in the Copenhagen interpretation continue to argue about to this day. One way to view the current impasse is that it is a philosophical disagreement about words. There is no *logical* reason why the subatomic realm must be described using a distinction that originated to describe the macroworld, as Steven Weinberg (Nobel Laureate in Physics, 1979) has written: "So irrelevant is the philosophy of quantum mechanics to its use, that one begins to suspect that all the deep questions about the meaning of measurement are really empty, forced upon us by language, a language that evolved in a world governed very nearly by classical physics." *Dreams of a Final Theory: The Search for the Fundamental Laws of Physics* (New York: Pantheon, 1992), 84.

34. Forman has described this moment: "My conclusion is that there was little connection between quantum mechanics and the philosophical constructions placed on it, or the world-view implications drawn from it. The physicists allowed themselves, and were allowed by others, to make the theory out to be whatever they wanted it to be—better, whatever their cultural milieu obliged them to want it to be." Forman, "*Kausalität, Anschaulichkeit*, and *Individualität*," 344.
35. Bohr, "Quantum Postulate"; the quotations are on 580 and 54, respectively (see n. 29). On Heisenberg's changing attitudes towards objectivity, see Cathyrn Carson, "Objectivity and the Scientist: Heisenberg Rethinks," *Science in Context* 16, nos. 1–2 (2003): 243–69, esp. 247–49.
36. According to Max Jammer, *Conceptual Development of Quantum Mechanics*, 198.
37. "The contemplation of the world *sub specie aeterni* [from the viewpoint of eternity] is its contemplation as a limited whole." Ludwig Wittgenstein, *Tractatus Logico-Philosophicus* (6.45), trans. C. K. Ogden (London: Kegan Paul, Trench, Trübner, 1922), 187.
38. This is the hypothesis of Max Jammer in *Conceptual Development of Quantum Mechanics*, 180 and 197–200.
39. In his original 1927 paper in which he introduced the term, Heisenberg used the word *Ungenauigkeit* (indeterminacy) throughout the body of his text and introduced *Unsicherheit* (uncertainty) only in an addendum. "Über den anschaulichen Inhalt der quantentheoretischen Kinematik und Mechanik," *Zeitschrift für Physik* 43, no. 3–4 (1927): 172–98; *Unsicherheit* (uncertainty) occurs on 197 and 198. When Carl Eckart and Frank C. Hoyt translated a textbook by Heisenberg (*The Physical Principles of the Quantum Theory, 1930*), they rendered both *Ungenauigkeit* and *Unsicherheit* as "uncertainty," which became the standard term in English.
40. Heisenberg, "Über den Inhalt der quantentheoretischen Mechanik," 197, translated by John A. Wheeler and W. H. Zurek as "The Physical Content of Quantum Kinematics and Mechanics" in Wheeler and Zurek, *Quantum Theory and Measurement*, 62–86; the quote is on 83 (see n. 28).
41. For an overview of determinism in physics, see John Earman's *Primer on Determinism* (Dortrecht: Reidel, 1986), and his recent update "Aspects of Determinism in Modern Physics," in *Philosophy of Physics*, ed. Jeremy Butterfield and John Earman (Amsterdam: North-Holland, 2007), 2:1369–1434. Some members of the Copenhagen school have argued that coin tosses (in classical mechanics) are independent events, whereas electron paths (in quantum mechanics) are intertwined events. But in the 1950s the de Broglie–Bohm interpretation showed that one probability theory governs both realms, as Einstein predicted would be the case; according to Einstein, "Assuming the success of efforts to accomplish a complete physical description [of quantum mechanics], the statistical quantum theory would, within the framework of future physics, take an approximately analogous position to the statistical mechanics within the framework of classical mechanics." Einstein, "Reply to Criticisms" (1949), in *Albert Einstein: Philosopher-Scientist*, ed. Paul Arthur Schilpp (LaSalle, IL: Open Court, 1949/rpt. 1970), 672.
42. Høffding, "Kierkegaard," 2:287 (see n. 24).
43. Ibid., 2:286–87.
44. Ibid., 2:287. Jammer has noted that "Høffding's discussion of the problem of knowledge seems to foreshadow certain conceptual traits in later quantum mechanics"; Jammer, *Conceptual Development of Quantum Mechanics*, 173.
45. Bohr, "Quantum Postulate"; the quotations are on 580 and 53–54, respectively (see n. 29).
46. Niels Bohr, "The Quantum of Action and the Description of Nature" (1929), in Bohr, *Atomic Theory*, 92–101; the quotes are on 100–1 (see n. 29). Bohr also assured the pious that physics has applications in biology and that quantum mechanics makes it impossible to give a physiochemical explanation of the origin of life: "With regard to the more profound biological problems, however, in which we are concerned with the freedom and power of adaption of the organism in its reaction to external stimuli [we found that] the same conditions . . . which determine the limitation of the causal mode of description in the case of atomic phenomena [means that] the very problem of the distinction between the living and the dead escapes comprehension." Niels Bohr, "The Atomic Theory and the Fundamental Principles Underlying the Description of Nature" (1929), in Bohr, *Atomic Theory*, 102–19; the quotes are on 118–19. Bohr is here echoing Ernst Mach's "profound conviction that the foundation of science as a whole, and of physics in particular, await their next greatest elucidations from the side of biology" (see plate 8-4 in chapter 8).
47. Max Born, "Gibt es physikalische Kausalität?" *Vossische Zeitung*, Apr. 12, 1928, 9.
48. For the logic of category mistakes, see Bernard Harrison, "Category Mistakes and Rules of Language," *Mind* 74, no. 295 (1965): 309–25.
49. Albert Einstein to Max Born, Apr. 29, 1924, in *The Born-Einstein Letters: Correspondence between Albert Einstein and Max and Hedwig Born from 1916 to 1955*, trans. Irene Born (London: Macmillan, 1971), 82. Today, despite the fracturing of the Copenhagen monolith, the link between indeterminacy and free will remains a serious theme of research in mathematics, as manifest in the proof offered in 2006 by the British mathematician John Conway and his Belgian colleague Simon B. Kochen that if humans have free will, then the position of certain elementary particles must be indeterminate: "If indeed there exist any experimenters with a modicum of free will, then elementary particles must have their own share of this valuable commodity." John Conway and Simon B. Kochen, "The Free Will Theorem," *Foundations of Physics* 36, no. 10 (2006): 1441–73; the quote is on 1441. This free will theorem prompted a refutation in 2010 by a team of four mathematicians and physicists, who argued that Conway's and Kochen's theorem applies only to deterministic models of the world. Sheldon Goldstein, Daniel V. Tausk, Roderich Tumulka, and Nino Zanghi, "What Does the Free Will Theorem Actually Prove?" *Notices of the American Mathematical Association*, Dec. 2010, 1451–53.
50. For a description of the Bohr-Einstein debates, see David Lindley, *Uncertainty: Einstein, Heisenberg, Bohr, and the Struggle for the Soul of Science* (New York: Doubleday, 2007); and Manjit Kumar, *Quantum: Einstein, Bohr, and the Great Debate about the Nature of Reality* (New York: Norton, 2008). For discussion of Bohr's philosophy of science by professional philosophers, see *Niels Bohr and Contemporary Philosophy*, ed. Jan Faye and Henry J. Folse (Dordrecht, the Netherlands: Kluwer, 1994), esp. Don Howard's essay "What Makes a Classical Concept Classical? Toward a Reconstruction of Niels Bohr's Philosophy of Physics" (201–29), in which he writes: "There was a time, not very long ago, when Niels Bohr's influence and stature as a philosopher of physics rivaled his standing as a physicist. But now there are signs of growing despair—much in evidence during the 1985 Bohr centennial—about our ever being able to make good sense out of his philosophical views" (201).
51. Erwin Schrödinger, "The Fundamental Idea of Wave Mechanics," Nobel Lecture, Dec. 12, 1933, in *Nobel Lectures, Physics 1922–1941* (Amsterdam: Elsevier, 1965), 305–16; the quote is on 309.
52. For a description of de Broglio's presentation at the Solvay conference and Pauli's response, see Guido Bacciagaluppi and Antony Valentini, *Quantum Theory at the Crossroads: Reconsidering the 1927 Solvay Conference* (Cambridge: Cambridge University Press, 2009). The authors argue that Pauli's objection to de Broglie's theory was incorrect (212–20).
53. John von Neumann, *Mathematical Foundations of Quantum Mechanics* (1932), trans. Robert T. Beyer (Princeton, NJ: Princeton University Press, 1955), 325.
54. On this point see Forman, "Acausal Quantum Mechanics," 221–60 (see n. 31).
55. For the details of Bohm's model, see James T. Cushing, *Quantum Mechanics: Historical Contingency and the Copenhagen Hegemony* (Chicago: University of Chicago Press, 1994), 42–75; and David Z. Albert, *Quantum Mechanics and Experience* (Cambridge, MA: Harvard University Press, 1994), 134–79.
56. Pauli's critical paper is "Remarques sur le problème des paramètres caché dans la méchanique quantique et sur la théorie de l'onde pilote" (1952), in *Louis de Broglie: Physicien et Penseur* (Paris: Michel, 1953), 33–42.
57. See Jung and Pauli's joint publication *Naturerklärung und Psyche* (The explanation of nature and psyche; 1952), translated with modesty as *The Interpretation of Nature and the Psyche*, with essays by C. G. Jung, "Synchronicity: An Acausal Connecting Principle," trans. R.F.C. Hall; and Wolfgang Pauli, "The Influence of Archetypal Ideas on the

Scientific Theories of Kepler," trans. Priscilla Silz (London: Routledge and Kegan Paul, 1955). The relation of Pauli and Jung is documented in their correspondence; see *Wolfgang Pauli und C.G. Jung: Ein Briefwechsel, 1932–1958*, ed. C. A. Meier (Berlin: Springer, 1992). Pauli maintained his interest in Jungian thought until his death in 1958; he was a neighbor of Max Bill in Zurich and spoke of Jung on his frequent visits to the artist's home in the late 1940s and 1950s. Author interview with Jakob Bill, Mar. 10, 2008, Zurich.

58. Wolfgang Pauli, "Das Ganzheitsstreben in der Physik" (The struggle for wholeness in physics; 1953), an essay sent by Pauli in a letter to the Swiss physicist Markus Fierz (1912–2006), professor of physics at Basel University, in early 1953 (Pauli Letter Collection at CERN, Geneva, inv. no. PLC 0092107), in Kalervo Vihtori Laurikainen, *Beyond the Atom: the Philosophical Thought of* Wolfgang Pauli, translated from Finnish by Carol Westerlund (Berlin: Springer, 1988). German original of Pauli's letter is on 90–93; the quote is on 91; English translation of the quote by Eugene Holman is on 207.
59. Pauli, "Remarques," 42 (see n. 56).
60. Wolfgang Pauli, "*Wissenschaft und das abendländische Denken*" (Science and Western thought; 1955), in Laurikainen, *Wolfgang Pauli*, 96–103. The original German quote is on 209–15; the English translation by Eugene Holman is on 213–15.
61. Bohm accepted a teaching job at the University of São Paulo that was arranged by the Brazilian physicist Jayme Tiomno, who had completed a doctorate in physics at Princeton in 1950 under Bohm and John Wheeler. Thus it was that Bohm went to Brazil in October 1951, and—unnerved that U.S. officials confiscated his passport on his arrival at the São Paulo airport—he became a citizen of Brazil, where he found support for his research and a welcoming atmosphere. Although Marxism was officially banned in Brazil, many intellectuals, such as Oscar Niemeyer, were members of the Brazilian Communist Party, as was the Brazilian physicist Mario Schönberg, who was a colleague of Bohm at the University of São Paulo. Bohm travelled to Argentina for talks with Mario Bunge. On Bohm's life, see F. David Peat, *Infinite Potential: The Life and Times of David Bohm* (Reading, MA: Helix, 1997).
62. As Jammer described it: "In the early 1950s [there was] an almost unchallenged monocracy of the Copenhagen school in the philosophy of quantum mechanics"; Jammer, *Philosophy of Quantum Mechanics*, 250 (see n. 21). On the rejection of Bohm because of academic politics, see Olival Freire Jr., "Science and Exile: David Bohm, the Cold War, and a New Interpretation of Quantum Mechanics," *Historical Studies in the Physical and Biological Sciences* 36, no. 1 (2005): 1–34. For the minority view that Bohm was rejected because of the McCarthyist climate, see Russell Olwell, "Physical Isolation and Marginalization in Physics: David Bohm's Cold War Exile," *Isis* 90, no. 4 (1999): 738–56.
63. John Stewart Bell, *Speakable and Unspeakable in Quantum Mechanics* (Cambridge: Cambridge University Press, 1987), 160.
64. For recent perspectives on the Forman thesis, both pro and con, see Carson, *Weimar Culture and Quantum Mechanics* (see n. 31).
65. For a description of the philosophical debate between Bohr and Einstein, see Jammer, *Philosophy of Quantum Mechanics*, esp. 109–58.
66. See Erwin Schrödinger, "Die gegenwärtige Situation in der Quantenmechanik," *Naturwissenschaften* 23 (Nov. 1935): 807–12; translated by John D. Trimmer as "The Present Situation in Quantum Mechanics," in *Wheeler and Zurek, Quantum Theory and Measurement* (see n. 28).
67. See George Johnson, *A Shortcut through Time: The Path to the Quantum Computer* (New York: Knopf, 2003), esp. 42–50 and 141–54.
68. On Lissitzky's quest for a Jewish style, see John Bowlt, "From the Pale of Settlement to the Reconstruction of the World," and Ruth Apter-Gabriel, "El Lissitzky's Jewish Works," both essays are in *Tradition and Revolution: The Jewish Renaissance in Russian Avant-Garde Art*, 1912–1928, ed. Ruth Apter-Gabriel (Jerusalem: Israel Museum, 1987), 43–60 and 101–24; and Margolin, *Struggle for Utopia*, 22–37 (see chap. 4, n. 59).
69. See John Bowlt, "Malevich and His Students," *Soviet Union* 5, part 2 (1978), 258–59.
70. For a photograph of a student wearing a black square, see Aleksandra Shatskikh, *Vitebsk: the Life of Art*, trans. Katherine Foshko Tsan (New Haven, CT: Yale University Press, 2007), 138, fig. 111. On Malevich's group in Vitebsk, see Christina Lodder, "International Constructivism and the Legacy of UNOVIS in the 1920s: El Lissitzky, Katarzyna Kobro and Władysław Strzemiński" (2003), in *Constructivist Strands in Russian Art 1914–1937* (London: Pindar, 2005), 537–58.
71. The details of Chagall's departure are recounted by Shatskikh in *Vitebsk*, 108–47.
72. El Lissitzky, "Suprematism in World Construction," *UNOVIS* 1 (1920), in Sophie Lissitzky-Küppers, *El Lissitzky: Life, Letters, Text*, (London: Thames and Hudson, 1968/rev.ed.1980), 331.
73. There is a facsimile edition of this children's book: El Lissitzky, *Pro dva kvadrata* (About two squares; 1922; Cambridge, MA: MIT Press, 1991), which has (printed on vellum overlaying the storybook pages), an English translation by Christiana van Manen. For other Russian avant-garde children's books, see Eveny Steiner, *Stories for Little Comrades: Revolutionary Artists and the Making of Early Soviet Children's Books*, trans. Jane Ann Miller (Seattle: University of Washington Press, 1999), which illustrates El Lissitzky's 1928 sketches for an unpublished primer on numbers, *Four Arithmetic Operations* (34–39).
74. El Lissitzky, "Proun" (1920–21), *De Stijl* 5–6 (1922), trans. John E. Bowlt, reprinted in *El Lissitzky: Ausstellung* (Cologne, Germany: Galerie Gmurzynska, 1976), 63; this text was written in 1920–21 and given as a lecture at the Moscow Institute of Artistic Culture on Oct. 23, 1924. On mathematics in Lissitzky's work, see Yve-Alain Bois, "Lissitzky, Malevich, et la question de l'éspace," in *Suprematisme* (Paris: Galerie Jean Chauvelin, 1977), 29–46; and Esther Levinger, "El Lissitzky's Art Games," *Neohelicon* 14, no. 1 (Dec. 1987): 177–191.
75. For discussion of Lissitzky's "K. und *Pangeometrie*," see Alan C. Birnholz, "Time and Space in the Art and Thought of El Lissizsky," *Structurist*, no. 15–16 (1975–76): 89–96; and Yve-Alain Bois, "From ¬ ∞ to 0 to + ∞: Axonometry, or Lissitzsky's Mathematical Paradigm," in *El Lissitzky, 1890–1941: Architect, Painter, Photographer, Typographer*, ed. Caroline de Bie et al. (Eindhoven, the Netherlands: Municipal van Abbemuseum, 1990), 27–33.
76. Many Russian intellectuals distanced themselves from the German Spengler because of his right-wing politics, but he was nevertheless very popular in Russia. On the Russian art critic Nikolai Tarabukin's love/hate attitude towards Spengler, see Maria Gough, "Tarabukin, Spengler, and the Art of Production," *October* 93 (Summer 2000): 78–108.
77. El Lissitzky, "Proun" (1920–21), 67 (see n. 74).
78. Ibid., 67 and 70.
79. El Lissitzky's political mission is described by Christina Lodder in "El Lissitzky and the Export of Constructivism," in *Situating El Lissitzky: Vitebsk, Berlin, Moscow*, ed. Nancy Perloff and Brian Reed (Los Angeles: Getty Research Institute, 2003), 27–46.
80. El Lissitzky and Ilya Ehrenburg, "The Blockade of Russia Moves Towards Its End" (1922), in *The Tradition of Constructivism*, ed. Stephen Bann (New York: Viking, 1974), 53–57; the quote is on 55. On the history of uses of the word "*constructivism*" in modern art, see Bann's introduction to this book, xxv–xlix. On the internationalization of Constructivism, see *Von Kandinsky bis Tatlin/From Kandinsky to Tatlin: Constructivism in Europe*, ed. Kornelia von Berswordt-Wallrabe (Schwerin, Germany: Staatliches Museum, 2006).
81. Theo van Doesburg, "Elementarism: Fragment of a Manifesto," *De Stijl* 7, no. 78 (1926–27), in Jaffé, *De Stijl: Extracts*, 213–16; the quote is on 214 (see chap. 6, n. 37).
82. Ibid.
83. See Doris Wintgens Hötte, "Van Doesburg Tackles the Continent: Passion, Drive, and Calculation," in *Van Doesburg and the International Avant-Garde: Constructing a New World*, ed. Gladys Fabre and Doris Wintgens Hötte (London: Tate, 2009), 10–19. On the role of the *De Stijl* magazine in Van Doesburg's international ambitions, see Krisztina Passuth, "*De Stijl* and the East-West Avant-Garde: Magazines and the Formation of International Networks," also in *Van Doesburg* (2009), 20–27.
84. László Moholy-Nagy, *The New Vision: Abstract of an Artist* (New York: Wittenborn, 1946), 70.
85. On the several concepts of "utopia" in the mid-war era, see *Central European Avant-Gardes: Exchange and Transformation, 1910–1930*, ed. Timothy Benson (Los Angeles: Los Angeles County Museum of Art, 2002), and *Modernism 1914–1939, Designing a New World*, ed. Christopher Wilk (London: Victoria and Albert Museum, 2006).

86. Bruno Taut, *Die Stadtkrone* (Jena, Germany: Diederichs, 1919), 68.
87. On art education in Soviet Russia and the Weimar Republic, see Christina Lodder, "The VKhUTEMAS and the Bauhaus," in *The Avant-Garde Frontier: Russia Meets the West 1910–1930*, ed. Gail Harrison Roman and Virginia Hagelstein Marquardt (Gainesville: University Press of Florida, 1992), 196–240.
88. On van Doesberg's sojourn in Germany see Sjarel Ex, "'De Stijl' und Deutschland, 1918–1922: Die Ersten Kontakte," in *Konstruktivistische Internationale Schöpferische Arbeitsgemeinschaft, 1922–1927: Utopien für eine*, ed. Bernd Finkeldey et al. (Stuttgart, Germany: G. Hatje, 1992), 73–80. See especially van Doesburg's postcard to Antony Kok with a picture of the Henry van der Velde building, Staatliche Hochschule für bildende Kunst, Weimar, the place where Gropius had founded the Bauhaus in 1919. Van Doesburg metaphorically renamed the Hochschule by writing "DE STIJL" across the building's façade pictured on the postcard (76).
89. Theo van Doesburg, "Towards a Newly Shaped World," in Joost Balijeu, *Theo van Doesburg* (New York: Macmillan, 1974): 113–14; the quote is in 114.
90. Walter Gropius to the Italian architect Bruno Zevi, Nov. 3, 1952, in Bruno Zevi, *Poetica dell'architettura neo-platica* (Turin: Einaudi, 1953/ rpt. 1974): 229–30; English trans. in Balijeu, *Theo van Doesburg*, 41.
91. On Itten's adoption of Mazdaznan, see Magdalena Droste, *The Bauhaus 1919–1933, Reform and Avant-Garde* (Cologne, Germany: Taschen, 2006), 25; and Éva Forgács, *The Bauhaus Idea and Bauhaus Politics*, trans. John Bátki (Budapest: Central European University Press, 1995), 51.
92. On the Expressionist and Constructivist tendencies in the early years of the Bauhaus, see Magdalena Droste, "Aneignung und Abstoßung: Expressive und Konstruktive Tendenzen am Weimar Bauhaus," in *Bauhaus-Ideen um Itten, Feininger, Klee, Kandinsky: Vom Expressiven zum Konstruktiven*, ed. Brigitte Salmen (Murnau, Germany: Schlossmuseum Murnau, 2007), 11–31. On Gropius's attempt to balance his utopian vision with practice, see Peter Müller, "Mental Space in a Material World: Idealized Reality in the Weimar Director's Office," in *Bauhaus: A Conceptual Model*, ed. Bauhaus Archiv Berlin, Stiftung Bauhaus Dessau, and Klassik Stiftung Weimar, n.t. (Ostfildern-Ruit, Germany: Hatje Cantz, 2009), 153–56.
93. On these *Telephone Pictures*, see Brigid Doherty, "László Moholy-Nagy's Constructions in Enamel, 1923," in *Bauhaus, 1919–1933: Workshops for Modernity*, ed. Leah Dickerman and Barry Bergdoll (New York: Museum of Modern Art, 2009), 130–33.
94. In the 1920s the Russian artists El Lissitzky and Naum Gabo were also fluent in Francé's biotechnical vocabulary. After reading Francé's book *Bios: Die Gesetze der Welt* (1923), El Lissitzky wanted to write to the author and arrange to meet him; El Lissitzky to Sophie Lissitzky-Küppers, Mar. 10, 1924, in Lissitzky-Küppers, *El Lissitzky*, 46 (see n. 72). Gabo also expressed interest in the biomorphic writings of Francé and Ernst Kállai; see Martin Hammer and Christina Lodder, *Constructing Modernity: The Art and Career of Naum Gabo* (New Haven, CT: Yale University Press, 2000), 282–83.
95. See Oliver A. I. Botar, "László Moholy-Nagy's New Vision and the Aestheticization of Scientific Photography in Weimar Germany," *Science in Context* 17, no. 4 (2004): 525–56.
96. On Kandinsky's use of mathematics in *Point and Line to Plane* (1926), see Christopher Short, "The Role of Mathematical Structure, Natural Form, and Pattern in the Art Theory of Wassily Kandinsky: The Quest for Order and Unity," in *Meanings of Abstract Art: Between Nature and Theory*, ed. Paul Crowther and Isabel Wünsche (New York: Routledge, 2012), 64–80.
97. See Boris Groys, *The Total Art of Stalinism: Avant-Garde, Aesthetic Dictatorship, and Beyond*, trans. Charles Rougle (Princeton, NJ: Princeton University Press, 1992). See also Margarita Tupitsyn, who has argued that there is a tension in El Lissitzky's post-1925 propaganda between the Stalinist restraint of free action and the avant-garde loose experimentation with typeface and layout; *El Lissitzky: Beyond the Abstract Cabinet; Photography, Design, Collaboration* (New Haven, CT: Yale University Press, 1999). Yve-Alain Bois has suggested that the shifting political winds in Russia caused El Lissitzky to present space differently in his early and later work; (early) Suprematist Lissitzky critiqued the illusions of linear perspective in his *Proun*, employing an axonometric perspective that lacks a unique viewpoint, but (late) Stalinist Lissitzky accepted illusion and made propagandist photomontages. "El Lissitzky: Radical Reversibility," *Art in America* 76, no. 4 (1988): 161–81.
98. For the Soviet view of Western art during the Stalinist era, see the collection of documents *Russian and Soviet Views of Modern Western Art: 1890s to Mid-1930s*, trans. Charles Rougle, ed. Ilia Dorontchenkov (Berkeley: University of California Press, 2009), esp. 305–7.
99. Walter Gropius, *Program of the Staatliche Bauhaus in Weimer* (1919), trans Wolfgang Jabs and Basil Gilbert, in Wingler, *Bauhaus*, 31 (see chap. 7, n. 22).
100. Hannes Meyer, *Bauhaus: Zeitschrift für Bau und Gestaltung Schriftleitung* (Dessau, Germany: Bauhaus, 1928), 4:12–13.
101. Rudolf Carnap, "*Wissenschaft und Leben*" a lecture delivered at the Bauhaus in Dessau, Germany, October 15, 1929. Carnap's handwritten notes for the lecture are preserved in the Rudolf Carnap Papers (RC 110-07-49), Archives for Scientific Philosophy, University of Pittsburgh. Quoted by permission of the University of Pittsburgh. All rights reserved.
102. The details of the relation between Meyer's Bauhaus and the Carnap's Vienna Circle have been chronicled by Peter Galison in "Aufbau/ Bauhaus: Logical Positivism and Architectural Modernism," *Critical Inquiry* 16 (Summer 1990): 709–52. Galison argues that the two groups had *parallel* goals (logicians removing metaphysics/architects removing ornament). But by 1929 ornament was long gone in avant-garde architecture. In fact, Carnap and Meyer had the *same* goal: obliterate metaphysics.
103. See Naum Gabo and Antoine Pevsner, *The Realist Manifesto* (1920), in Herbert Read and Leslie Martin, *Gabo: Constructions, Sculpture, Paintings, Drawings, Engravings* (Cambridge, MA: Harvard University Press, 1957), which includes Gabo's original Russian text (150) with English translation (151–52) and "Constructive Art: An Exchange of Letters between Naum Gabo and Herbert Read," *Horizon* 10, no. 55 (1944): 60–61.
104. See Jane Beckett, "*Circle*: The Theory and Patronage of Constructivist Art of the Thirties," in *Circle: Constructive Art in Britain 1934–40*, ed. Jeremy Lewison (Cambridge, England: Kettle's Yard Gallery, 1982), 11–19.
105. Christian Zervos, "Mathématiques et l'art abstrait," *Cahiers d'art* (1936): 4–20); the quote is on 4.
106. Ibid., 6.
107. Charles Morris, "Science, Art, and Technology," *Kenyon Review* 1, no. 4 (1939): 409–23; the quote is on 422–23.

9. THE INCOMPLETENESS OF MATHEMATICS

1. This same year, the steel tycoon Karl Wittgenstein died and left his son Ludwig an immense fortune. Preferring an ascetic lifestyle, the young Wittgenstein gave it all away.
2. The title of the first German edition of Wittgenstein's treatise was *Logisch-Philosophische Abhandlung* (Logical-philosophical treatise), ed. Wilhelm Ostwald, *Annalen der Naturphilosophie* 14 (1921). Reflecting British taste for Latin titles, the book's first edition, with the German text and English translation by C. K. Ogden on facing pages, was named *Tractatus Logico-Philosophicus* (London: Kegan Paul, Trench, Trübner, 1922).
3. In 1918–19 Russell himself published a series of essays, "Philosophy of Logical Atomism," in which he acknowledged his debt to Wittgenstein. But Wittgenstein's own version of the picture theory of language has proven more influential.
4. Ludwig Wittgenstein, *Tractatus Logico-Philosophicus* (6.522), trans. 187. In 1901 Russell had published an essay about mysticism in which he described how the mathematician knows an ultimate order of reality by mystical intuition, in other words, by pure cognition ("Mysticism and Logic" [1901], in *Mysticism and Logic*, 1–32 [see chap. 3, n. 8]). It is unclear whether or not Wittgenstein read Russell's essay before composing his *Tractatus*. For a comparison of Russell and Wittgenstein's approaches to the topic, see Brian McGuiness, "The Mysticism of the *Tractatus*," *Philosophical Review* 75 (1966): 305–28.
5. Wittgenstein, *Tractatus Logico-Philosophicus* (6.21 and 6.3), trans. 151.
6. Ibid. (6.545), trans. 187.
7. Ibid. (7), trans. 89.

8. Ludwig Wittgenstein, "A Lecture on Ethics" (1929–30), *Philosophical Review* 74, no. 1 (1965): 3–12; the quote is on 8.
9. Wittgenstein, *Tractatus* (6.4311), trans.185.
10. On Wittgenstein's defense of metaphysics against attacks by positivists, see Christopher Hoyt, "Wittgenstein and Religious Dogma," *International Journal for Philosophy of Religion* 61, no. 1 (2007): 39–49.
11. On the relation of Wittgenstein's philosophy to that of Brouwer in 1928, see Mathieu Marion, "Wittgenstein and Brouwer," *Synthese* 137, no. 1/2 (2003): 103–27. See also Hesseling, *Gnomes in the Fog*, esp. 190–98 (see chap. 4, n. 15).
12. On Wittgenstein's roots in German Romanticism, see M. W. Rowe, "Wittgenstein's Romantic Inheritance," *Philosophy* 69, no. 269 (1994): 327–51. Although there is no evidence that Wittgenstein read Hegel and Schelling, he definitely read the writings of Goethe, which expresses their Romantic philosophy. See M. W. Rowe, "Goethe and Wittgenstein," *Philosophy* 66, no. 257 (1991): 283–30.
13. Ludwig Wittgenstein, *Philosophische Untersuchungen/Philosophical Investigations* (1953), trans. G.E.M. Anscombe, 2nd ed. (Oxford, England: Blackwell, 1958/rpt. 1998), n.p. Wittgenstein began this book with a quotation from Saint Augustine (*Confessions*, 1:8) about the nature of language.
14. In notes collected and published posthumously, Wittgenstein made frequent reference to the *Lebensphilosophen*; see Wittgenstein, *Vermischte Bemerkungen: Culture and Value*, trans. Peter Winch, ed. G. H. von Wright (Oxford, England: Blackwell, 1980/2nd ed. 1997), references to Nietzsche, 9, 59; Kierkegaard, 31, 32, 38, 53; and to Schopenhauer, 19, 26, 34, 36, 71.
15. Wittgenstein, *Culture and Value*, 56e.
16. Rudolf Carnap, *The Logical Syntax of Language* (1934), trans. Amethe Smeaton (New York: Harcourt, Brace, 1937), 222.
17. Paul Bernays, "Axiomatische Untersuchung des Aussagen-Kalküls der *Principia Mathematica*," *Mathematische Zeitschrift* 25 (1926): 305–20; and Emil L. Post, "Introduction to a General Theory of Elementary Propositions," *American Journal of Mathematics* 43 (1921), 163–85.
18. For an exposition of Gödel's proof, see Ernst Nagel and James R. Newman's classic text *Gödel's Proof* (New York: New York University Press, 1958). For a biography of Gödel, see John W. Dawson, *Logical Dilemmas: The Life and Work of Kurt Gödel* (Wellesley, MA: A. K. Peters, 1997).
19. For a discussion of the origins of Cubism summarized in this paragraph, see Lynn Gamwell, "Looking Inward: Art and the Human Mind," in *Exploring the Invisible*,129–47 (see chap. 3, n. 1).
20. Cézanne to Emile Bernard, Apr. 15, 1904, in *Paul Cézanne, Correspondance*, ed. John Rewald (Paris: B. Grasset, 1978), 296.
21. The theme of self-reference in mathematics and art is the premise of Douglas Hofstader's book *Gödel, Escher, Bach: An Eternal Golden Braid* (New York: Basic Books, 1979). If one were to approach this topic historically and replace the (unhistorical) "eternal golden braid" for a particular time and place (early-twentieth-century Germanic culture), then the trilogy would be Gödel, Magritte, Schoenberg.
22. De Chirico read Giovanni Papini, *Il crepuscolo dei filosofi* (The crucible of philosophy): *Kant, Hegel, Schopenhauer, Comte, Spencer, Nietzsche* (Milan: Società Editrice Lombarda, 1906). Papini was a young writer on art and literature who, a few years later, wrote about Futurism for the avant-garde magazine *Lacerba*.
23. Friedrich Nietzsche, *Ecce Homo* (1888), in *Basic Writings of Nietzsche*, trans. Walter Kaufmann (New York: Modern Library, 1968), 764.
24. On the theme of enigma in the art and literature of de Chirico and his brother Andrea de Chirico (who changed his name to Alberto Salvinio in 1909), see Keala Jewell, *The Art of Enigma: The De Chirico Brothers and the Politics of Modernism* (University Park: Pennsylvania State University Press, 2004).
25. This line occurs in Nietzsche's poem "Ariadne's Lament" (1889), in *The Portable Nietzsche*, trans. Walter Kaufmann (New York: Penguin, 1968), 345.
26. Nietzsche wrote: "Man [in the sense of "mankind"; he is referring here to Ariadne] is to my mind an agreeable, courageous, inventive animal that has no equal on earth; it finds its way in any labyrinth." Friedrich Nietzsche, *Beyond Good and Evil* (1886), in *Writings of Nietzsche*, trans. Kaufmann, 426.
27. De Chirico's series on Ariadne is documented by Michael R. Taylor in *Giorgio de Chirico and the Myth of Ariadne* (London: Merrell, 2002).
28. For a psychoanalytic consideration of the effects of Magritte's early life on his art, see Ellen Handler Spitz, "Testimony through Painting," in her *Museums of the Mind: Magritte's Labyrinth and Other Essays* (New Haven, CT: Yale University Press, 1994), 26–36; and the French psychoanalyst Jacques Roisin, *Ceci n'est pas une biographie de Magritte* (Brussels: Alice, 1998), esp. "Les eaux profondes" (Deep waters), 56–76.
29. Magritte read Elie Faure's *Les constructeurs* (Paris: G. Crès, 1914), with essays on Dostoyevsky, Nietzsche, and Cézanne.
30. René Magritte and the critic Louis Jean Scutenaire discussed Hegel, Nietzsche, and Freud in their co-authored essay "L'Art Bourgeois," *London Bulletin*, no. 12 (Mar. 15, 1939): 13–14.
31. On this key 1923 date, see *René Magritte: Catalogue Raisonné*, ed. David Sylvester (London: Philip Wilson, 1992), 1:39.
32. Magritte acquired the booklet *12 opere di Giorgio de Chirico* (Rome: Valori Plastici, 1919), with brief remarks by writers including Guillaume Apollinaire, Carlo Carrà, Maurice Raynal, André Salmon, Ardengo Soffici, and Louis Vauxcelles. In his remarks, Soffici compared de Chirico's use of geometry to the quattrocento master of linear perspective, Paolo Uccello, n.p.
33. On Magritte's subversion of linear perspective, see Jean Clair, "Seven Prolegomenae to a Brief Treatise on Magrittian Tropes," *October* 8 (Spring 1979): 89–110.
34. Magritte to Foucault, May 23, 1966, in René Magritte, *Écrits Completes*, ed. André Blavier (Paris: Flammarion, 1979), 639–40.
35. Michel Foucault, *Ceci n'est pas une pipe* (This is not a pipe; 1968; Montpellier: Fata Morgana, 1973). For other discussion of Magritte through the lens of phenomenology/existentialism, see Martin Jay, "In the Empire of the Gaze: Foucault and the Denigration of Vision in Contemporary French Thought," in *Foucault: A Critical Reader*, ed. David Couzens Hoy (Oxford, England: Blackwell, 1986), 175–204, in which Jay questions Foucault's interpretation of Magritte regarding the distinction between resemblance and similitude; and Gary Shapiro, "Pipe Dreams: Eternal Recurrence and Simulacrum in Foucault's Ekphrasis of Magritte," *Word and Image* 13, no. 1 (1997): 69–76. On Magritte from the perspective of the founder of semiotics, Charles Sanders Peirce (1839–1914), see André de Tienne, "Ceci n'est-il pas un signe? Magritte sous le regard de Peirce," in *Magritte au Risque de la Sémiotique*, ed. Nicole Everaert-Desmedt (Brussels: Facultés Universitaires Saint-Louis, 1999). For discussion of Magritte in terms of British analytic philosophy, especially late Wittgenstein, see Suzi Gablik, *Magritte* (Greenwich, CT: New York Graphic Society, 1970).
36. The American mathematician William Goldbloom Bloch has described the mathematics behind Borges's tale in *The Unimaginable Mathematics of Borges' Library of Babel* (Oxford: Oxford University Press, 2008).
37. H.S.M. Coxeter, "Crystal Symmetry and Its Generalization," *Transactions of the Royal Society of Canada* 51, ser. 3, sec. 3 (1957): 1–13.
38. The encounter of Escher and Coxeter is documented by Doris Schattschneider in "Coxeter and the Artists: A Two-Way Inspiration," in *The Coxeter Legacy*, ed. C. Davis and E. Eller (Providence, RI: American Mathematical Society/Fields Institute, 2006), 255–80.
39. Escher to his son George, May 28, 1960, reproduced in F. H. Bool, J. B. Kist, J. L. Locher, and F. Wierda, *M. C. Escher: His Life and Complete Graphic Work* (New York: Abrams, 1982), 100–101.
40. Escher to Coxeter, Dec. 5, 1958; quoted by Coxeter in his essay "The Non-Euclidean Symmetry of Escher's Picture *Circle Limit III*," *Leonardo* 12 (1979): 19.
41. See, for example, Coxeter, "Escher's *Circle Limit III*," 19–25 and 32.
42. Edmund Husserl, *Ideen zu einer reinen Phänomenologie und phänomenologischen Philosophie* (Halle, Germany: Max Niemeyer, 1913/rpt. 1922), 211. David Teniers did ten paintings of picture galleries to document the art collection of his patron, the Archduke Leopold Wilhelm, who was governor of the southern Netherlands, which is today Belgium. In addition to these paintings, Teniers also produced an engraved illustrated catalogue, the *Theatrum Pictorium* (Theater of painting) of the 243 most prized works, selected from Leopold Wilhelm's total collection of about 1300 Renaissance and Baroque works, which today is the core collection of the Kunsthistorisches Museum in Vienna. To my knowledge, there was not a Teniers painting of a picture gallery in the collection of the Gemäldegalerie in Dresden when Husserl wrote *Ideen*, so he must have seen a work that was on loan from

another collection in Europe, where all the Teniers gallery paintings reside. For the history of Teniers documentation of the Leopold Wilhelm collection, see Ernst Vegelin van Claerbergen, ed., *David Teniers and the Theater of Painting* (London: Courtauld Institute Art Gallery, 2006); an inventory of the ten gallery paintings is on page 65.

43. After hearing Brouwer's 1928 lecture, Wittgenstein wrote extensively about the philosophy of mathematics from 1929 to 1944. Philosophers have generally assumed that Wittgenstein adopted Brower's anti-Platonist stance in 1928, but recent scholarship is shepherding him back to the Platonist flock; see Hilary Putnam, "Was Wittgenstein *Really* an Anti-Realist about Mathematics?" in *Wittgenstein in America*, ed. Timothy McCarthy and Sean C. Stidd (Oxford, England: Clarendon, 2001), 140–94.
44. Wittgenstein, *Philosophical Investigations* (sec. 2–3), 6 (see n. 13).
45. Ibid. (sec. 67), 36.
46. On the historical context of the development of game theory, see Giorgio Israel and Ana Millán Gasca, "The Theory of Games: A New Mathematics for the Social Sciences," in *The World as a Mathematical Game: John von Neumann and Twentieth Century Science*, translated from Italian by Ian McGilvay (Basel, Switzerland: Birkhäuser, 2009), 128–33. For an overview of applications of game theory to politics, the stock market, and much more, see Avinash K. Dixit, Susan Skeath, and David H. Reiley Jr., *Games of Strategy*, 3rd ed. (New York: Norton, 2009).
47. Wittgenstein, *Philosophical Investigations* (sec. 133), 57.
48. On Johns and Wittgenstein, see Peter Higginson, "Jasper's Non-Dilemma: A Wittgensteinian Approach," *New Lugano Review* 10 (1976): 53–60. On Johns's use of words, see Esther Levinger, "Jasper Johns's Painted Words," *Visible Language* 23, nos. 2–3 (1989): 280–95; and Harry Cooper, "Speak, Painting: Word and Device in Early Johns," *October* 127 (Winter 2009): 49–76.
49. On Johns's use of numbers, see Roberta Bernstein, "Numbers," in *Jasper Johns: Seeing with the Mind's Eye* (San Francisco: San Francisco Museum of Modern Art, 2012), 44–55.
50. According to the historians of Art and Language, Charles Harrison and Fred Orton, *A Provisional History of Art and Language* (Paris: E. Fabre, 1981), 20.
51. Ibid., 22.
52. Ibid., 21.
53. Ramsden and Burn recorded the minutes of a meeting of the Society for Theoretical Art and Analysis in *Art-Language* 1, no. 3 (June 1970): 1.
54. Kosuth quoted extensively from a 1950 edition of Ayer in his essay "Art after Philosophy," *Studio International* 178, no. 915 (Oct. 1969): 134–37, 178; and no. 916 (Nov. 1969): 160–61.
55. In addition to Ayer, another source for Kosuth was Marcel Duchamp, who declared "Everything is tautology" in an interview published in 1967: "Viennese logicians have worked out a system in which everything is, as far as I understand, a tautology, that is, a repetition of premises. In mathematics, it goes from a very simple theorem to a very complicated one, but everything is in the first theorem. So, metaphysics is tautology; religion is tautology; everything is tautology, except black coffee because the senses are in control!" Pierre Cabanne, *Entretiens avec Marcel Duchamp* (Paris: Belfond, 1967), 204.
56. Kosuth, "Art after Philosophy," 136. Harrison was assistant editor of *Studio International* and arranged for Kosuth to have this platform, from which he became a spokesman for Conceptual Art.
57. Art and Language continues to exist as a group, consisting of Charles Harrison, Michael Baldwin, and Mel Ramsden; for one of their recent essays, which details their family quarrels with the *October* group, see Art and Language, "Voices Off: Reflections on Conceptual Art," *Critical Inquiry* 33, no. 1 (2006): 113–35.
58. *Geijutsu Kurabu*, no. 8 (Apr. 1974): 42–67; see Reiko Tomii, "Concerning the Institutionalism of Art: Conceptualism in Japan," *Global Conceptualism: Points of Origin, 1950s–1980s*, ed. Luis Camnitzer, Jane Farver, and Rachel Weiss (New York: Queens Museum of Art, 1999), 16, 27, n. 6 and n. 10.
59. On Russell and China, see Eric Hayot, "Bertrand Russell's Chinese Eyes," *Modern Chinese Literature and Culture* 18, no. 1 (2006): 132–39.
60. Simon Leung and Janet Kaplan, "Pseudo-Languages: A Conversation with Wenda Gu, Xu Bing, and Jonathan Hay," *Art Journal* 58, no. 3 (Fall 1999): 90.
61. Ibid.
62. See Gao Minglu, "Seeking a Model of Universalism: The *United Nations* Series and Other Works," in *Wenda Gu: Art from Middle Kingdom to Biological Millennium*, ed. Mark H. C. Bessire (Cambridge, MA: MIT Press, 2003), 20–29.
63. On Xu Bing's use of the Chinese language, see Liu Yuedi, "Calligraphic Expression and Contemporary Chinese Art: Xu Bing's Pioneer Experiment," in *Subversive Strategies in Contemporary Chinese Art*, ed. Mary Bittner Wiseman and Liu Yuedi (Leiden, the Netherlands: Brill, 2011), 87–108.

10. COMPUTATION

1. Turing's career was cut short when he committed suicide in 1954 at age forty-one after being convicted by a British court of homosexual acts. The British government offered Turing the choice between jail or libido-reducing injections of estrogen. Turing, who was a world-class marathon runner (his best time was 2 hours, 36 minutes), chose to undergo the injections from which he suffered dire physical side effects. Two years later Turing died of cyanide poisoning. Homosexuality was decriminalized in Britain in 1967, and in 2013 the British government granted Turing a posthumous pardon. For the life and death of Turing, see the biography by Andrew Hodges, a professor of mathematics at Oxford University, who emerged in the 1970s as a civil rights activist during the gay liberation movement; Hodges, *Alan Turing: the Enigma* (New York: Simon and Schuster, 1983), which was the basis for the film *The Imitation Game* (2014), directed by Morten Tyldum. For a summary of the many artistic responses to the central irony of Turing's life—a man destroyed by the country he saved—see Michael Olinick, "Artists Respond to Alan Turing," *Math Horizons* 19, no. 4 (Apr. 2012): 5–9.
2. This point is stressed by Martin Davis in the epilogue to his history of the mathematics (as opposed to the technology) of computers, *The Universal Computer: The Road from Leibniz to Turing* (New York: Norton, 2000), 209.
3. Arnold Schoenberg, "Composition with Twelve Tones (I)" (1941), in *Style and Idea: Selected Writings of Arnold Schoenberg*, trans. Leo Black, ed. Leonard Stein (London: Faber and Faber, 1941/rpt. 1975): 214–25; the phrase is on 225.
4. For example, the American composer Milton Babbitt (1916–20), who was on the Princeton University faculty of both music (beginning in 1938) and mathematics (beginning in 1943), applied set theory to analyze twelve-tone music composed by others and to create his own compositions. See Milton Babbitt, "Some Aspects of Twelve-Tone Composition," *The Score* 12 (June 1955): 53–61.
5. Schoenberg, "Composition with Twelve Tones," 220.
6. See Rudolf Stephan, "Schoenberg and Bach," trans. Walter Frisch, in *Schoenberg and His World*, ed. Walter Frisch (Princeton, NJ: Princeton University Press, 1999), 126–40.
7. See Edward Rothstein, *Emblems of Mind: The Inner Life of Music and Mathematics* (New York: Random House, 1995).
8. See Joan Allan Smith, *Schoenberg and His Circle* (London: Collier Macmillan, 1986), 174; and Allen Shawn, *Arnold Schoenberg's Journey* (New York: Farrar, Straus, and Giroux, 2002), 44–47, 93. Shawn writes: "It is not hard to take this chapter in Schoenberg's personal life into account when encountering both the high anxiety quotient and the specific literary themes embodied in his musical works from the period 1909–1913" (47).
9. On Schoenberg's visual art, see *Arnold Schoenberg, das bildnerische Werk/Arnold Schoenberg, Paintings and Drawings*, ed. Thomas Zaunschirm (Klagenfurt, Austria: Ritter, 1991).
10. Theodor Adorno, *Philosophy of New Music* (1949), trans. Robert Hullot Kentor (Minneapolis: University of Minnesota Press, 2006), 35. Adorno also wrote: "The monologue *Erwartung* has as its heroine a woman who, at night and at the mercy of all night's terrors, searches for her lover, only to find him murdered. She is consigned to the music as an analytical patient to the couch"; 37. See Lewis Wickes, "Schoenberg, *Erwartung*, and the Reception of Psychoanalysis in Musical Circles in Vienna until 1910–1911," *Studies in Music* 23 (1989): 88–106; and Alexander Carpenter, "Schoenberg's Vienna, Freud's Vienna: Re-examining the Connections between the Monodrama *Erwartung* and the early history of Psychoanalysis," *Musical Quarterly* 93, no. 1 (2010), 144–81.

11. On Schoenberg's return to Judaism and his late work, see Kenneth H. Marcus, "Judaism Revisited: Arnold Schoenberg in Los Angeles," *Southern California Quarterly* 89, no. 3 (2007): 307–25.
12. In 1927 the Soviet state commissioned Schillinger to compose a work entitled *October* to mark the tenth anniversary of the Russian revolution.
13. For Schillinger's intermixing of science, mathematics, and music, see Warren Brodsky, "Joseph Schillinger (1895–1943): Music Science Promethean," *American Music* 21, no. 1 (Spring 2003): 45–73. Schillinger's teaching notes were published posthumously as *The Mathematical Basis of the Arts* (New York: Philosophical Library, 1948).
14. On Theremin's meeting with Lenin, see Albert Glinsky, *Theremin: Ether Music and Espionage* (Urbana: University of Illinois Press, 2000), 28–31.
15. See Paul Nauert, "Theory and Practice in *Porgy and Bess*: The Gershwin-Schillinger Connection," *Musical Quarterly* 78, no. 1 (1994): 9–33.
16. On Theremin's bizarre and dramatic life, see Glinsky, *Theremin*.
17. Heiner Friedrich, the founding director of the Dia Foundation in New York, offered to fund Young's perpetual performances at a building (6 Harrison Street) in the SoHo district of New York. Dia renovated the building, named the *Dream House*, and lavishly funded it as the residence and performance space of Young and his partner Marian Zazeela from 1979 to 1985. The music stopped after Dia withdrew funding in 1985 due to a drop in the value of its stock. See Phoebe Hoban, "Medicis for a Moment," *New York Magazine* 18, no. 46 (1985): 56–57.
18. See James Harley, *Xenakis: His Life in Music* (New York: Routledge, 2004).
19. For a history of computer art focusing on these early years, see Barbara Nierhoff-Wielk, "*Ex machina*—the Encounter of Computer and Art: A Look Back," in *Ex Machina—Frühe Computergrafik bis 1979* (Bremen, Germany: Kunsthalle Bremen, 2007), 20–57.
20. See Grant Taylor, "Soulless Usurper: Reception and Criticism of Early Computer Art," in *Mainframe Experimentalism: Early Computing and the Foundations of Digital Arts*, ed. Hannah B. Higgins and Douglas Kahn (Berkeley: University of California Press, 2012), 17–37.
21. See Christof Klütsch, "Information Aesthetics and the Stuttgart School," in Higgins and Kahn, *Mainframe Experimentalism*, 65–89. Bense also borrowed from the field of linguistics, which by the 1950s was adopting a computer model for language. In 1957 the American Noam Chomsky hypothesized that language has the structure (syntax) it does because human beings are born with an innate set of algorithms—a universal grammar—for producing language (*Syntactic Structures*, 1957). Bense turned Chomsky's theory of generative grammar into what he called "generative aesthetics," and while Chomsky and his students at MIT researched (with considerable success) innate linguistic algorithms, Bense and his students in Stuttgart tried (with little success) to discover innate aesthetic algorithms. On the fate of Max Bense's "information aesthetics" in Germany, see Claus Pias, "Hollerith 'Feathered Crystal': Art, Science, and Computing in the Era of Cybernetics," trans. Peter Krapp, *Grey Room* 29 (Fall 2007): 110–33. This essay includes the transcript of an infamous face-off that was staged in 1970 between Bense, whose cheerful enthusiasm for technology led him to adopt the upbeat slogan *Programmierung des Schönen* (Programming the beautiful), and the diametrically opposed Joseph Beuys, whose doleful mission was to make art that would help heal the wounds suffered by German society during World War II (113–14).
22. Georg Nees, "Künstliche Kunst: Wie man sie verstehen kann" (Artificial art: How one can understand it), in *Georg Nees, Künstliche Kunst: Die Anfänge* (Bremen, Germany: Kunsthalle Bremen, 2005), n.p.; quoted in *Ex Machina*, 428.
23. A recent exhibition of work by contemporary artists inspired by Bense suggests that he still holds sway in German intellectual circles; see *Bense und die Künste* (Karlsruhe, Germany: Zentrum für Kunst und Medientechnologie, 2010).
24. On the roles of Kepes and Klüver in post-1945 debates about art and technology, see Anne Collins Goodyear, "Gyorgy Kepes, Billy Klüver, and American Art of the 1960s: Defining Attitudes towards Science and Technology," *Science in Context* 17, no. 4 (2004): 611–35.
25. On computer art in Zagreb, see Margit Rosen, ed., *A Little-Known Story about a Movement, a Magazine, and the Computer's Arrival in Art: New Tendencies and Bit International, 1961–1973* (Cambridge, MA: MIT Press, 2011).
26. *Cybernetic Serendipity: The Computer and the Arts*, curated by Jasia Reichardt (London: Institute of Contemporary Art, 1968); see also *White Heat Cold Logic: British Computer Art, 1960–1980*, ed. Paul Brown, Charlie Gere, Nicholas Lambert, and Catherine Mason (Cambridge, MA: MIT Press, 2009). Bense himself curated *Computerkunst—On the Eve of Tomorrow* for Kubus in Hanover, 1969.
27. For overviews of computer art, see Christiane Paul, *Digital Art* (London: Thames and Hudson, 2003), Rachel Greene, *Internet Art* (London: Thames and Hudson, 2004), and Wolf Lieser, *Digital Art: Neue Wege in der Kunst* (Potsdam, Germany: H. F. Ullmann, 2010).
28. Alan Watts, "Square Zen, Beat Zen, and Zen," *Chicago Review* 12, no. 2 (1958): 3–11; the quote is on 5–6.
29. Matsuzawa distributed his 1964 *Ψ Corpse* flier to visitors gathered for his *Han Bunmei Ten* (Anti-civilization exhibition; 1965); translated from Japanese by Reiko Tomii, in Camnitzer, Farver, and Weiss, *Global Conceptualism*,19 (see chap. 9, n. 58).
30. Ibid.
31. Ibid.
32. On Reinhardt's political views, see the artist's undated notes in the section "Art and Politics" in Ad Reinhardt, *Art as Art: The Selected Writings of Ad Reinhardt*, ed. Barbara Rose (New York: Viking, 1975), 171–81.
33. On the relevance of negative theology to a secular society, see the Belgian Catholic theologian Louis Dupré, "Spiritual Life in a Secular Age," *Daedalus* 111 (1982): 21–31.
34. Merton described their relationship in "Wisdom in Emptiness: A Dialogue by Daisetz T. Suzuki and Thomas Merton," in *Zen and the Birds of Appetite* (New York: New Directions, 1968), 99–138.
35. Undated notes by Ad Reinhardt in *Art as Art*, 108. In another page of undated notes, Reinhardt assembled quotations on "Black" that reads "'The Tao is dim and dark' (Lao Tzu, fourth century BC) . . . 'The divine dark' (Meister Eckhardt, fourteenth century AD) 'Dark night of the soul' (of Saint John of the Cross, sixteenth century AD)" (98).
36. Merton to Reinhardt, Nov. 23, 1957, Thomas Merton Study Center, Bellarmine College, Louisville, Kentucky. This and other letters are reproduced by Joseph Mashek in his "Five Unpublished Letters from Ad Reinhardt to Thomas Merton and Two in Return," *Artforum* 17 (Dec. 1978): 23–27; the quote is on 24.
37. Ad Reinhardt, "The Black Square Paintings" (1961), in *Art as Art*, 82–83; the quote is on 83. On the meaning of Reinhardt's art, see Yve-Alain Bois, "The Limit of Almost," in *Ad Reinhardt* (New York: Rizzoli, 1991), 11–33. On Reinhardt's attitude towards religion, see Michael Corris, "Neither Sacred nor Secular," in *Ad Reinhardt* (London: Reaktion, 2008), 86–91.
38. On the impact of Asian art on American artists of Reinhardt's generation, see Bert Winter-Tamaki, "The Asian Dimensions of Post-war Abstract Art: Calligraphy and Metaphysics," 145–97, and Alexandra Monroe, "Buddhism and the Neo-Avant-Garde: Cage Zen, Beat Zen, and Zen," 199–273, in *The Third Mind: American Artists Contemplate Asia*, ed. Alexandra Monroe (New York: Guggenheim Museum, 2009).
39. James Breslin reproduced Rothko's poem from "The Scribble Book," which is in the Mark Rothko Archives, in his biography *Mark Rothko: A Biography* (Chicago: University of Chicago Press, 1993), 44. Breslin also recounts that Rothko's friend Sally Avery recalled that he made frequent references to Plato in the 1930s (244). David Anfam has relayed a similar recollection by Buffie Anderson about Rothko's interest in Plato in the 1940s; David Anfam, *Mark Rothko: The Works on Canvas: Catalogue raisonné* (New Haven, CT: Yale University Press, 1998), 98.
40. John Cage, "Composition as Process" (1958), *Silence: Lectures and Writings* (Middletown, CT: Wesleyan University Press, 1961), 18–57; the quote is on 23.
41. See Astrit Schmidt-Burkhardt, "Mapping Art History," in *Maciunas's Learning Machines: From Art History to a Chronology of Fluxus* (Berlin: Vice Versa, 2003), 13–15 and 85–113. An earlier example of such a chart is Alfred J. Barr's "Diagram of Stylistic Evolution from 1890 to 1935" on the book cover of *Cubism and Abstract Art* (New York: Museum of Modern Art, 1935).
42. De Maria was one of the art patron Heiner Friedrich's "chosen ones" (see chap. 11), and the installation of *Lightning Field* was funded by

the Dia Foundation; see Hoban, "Medicis for a Moment," 56 (see n. 17).

43. On the origin in the 1920s–1930s of attitudes expressed by German physicists in post-1945 popular science journalism, see Cathryn Carson "Who wants a Postmodern Physics?" in *Science in Context* 8, no. 4 (1995): 635–55, esp. 644ff.
44. For example, Heisenberg wrote: "All opponents of the Copenhagen Interpretation do agree on one point. It would, in their view, be desirable to return to the reality concept of classical physics or, to use a more general philosophical term, to the ontology of materialism. They would prefer to come back to the idea of an objective real world whose smallest parts exist objectively in the same sense as stones and trees exist, independently of whether or not we observe them. This however is impossible or at least not entirely possible because of the nature of the atomic phenomena." *Physics and Philosophy: The Revolution in Modern Science* (London: George Allen and Unwin, 1959), 115. This book contains the Gifford Lectures on the intellectual history of physics that Heisenberg gave (in English) in late 1955 to early 1956 at St. Andrews University in Scotland.
45. On the links between theosophy and the New Age movement, see Olav Hammer, *Claiming Knowledge: Strategies of Epistemology from Theosophy to the New Age* (Leiden, the Netherlands: Brill, 2001). Back in the 1920s, Max Planck had complained that occultism was on the rise: "In our age, when so much is happening to promote progress, the belief in miracles in various forms of occultism, spiritualism, theosophy, and all the numerous shadings of these names, penetrates the educated and uneducated public more than ever, despite the stubborn defensive efforts mounted by the scientific side"; Max Planck, "Kausalgesetz und Willensfreiheit" (1923), in *Vorträge und Erinnerungen* (Darmstadt, Germany: Wissenschaftliche Buchgesellschaft, 1949/rpt. 1975), 139–68; the quote is on 162–63. Following in the footsteps of the founder of theosophy, Helena Petrovna Blavatsky, the post–World War I theosophists sought the authority of science to support their occultism. Austrian theosophist Rudolf Steiner called for the release of a dynamic life force from the soul of every German (*Appeal to the German People and the Civilized World, 1919*), prompting the German physicist Max van Laue to complain that Steiner was preaching "the 'spiritualization' of science" (*'Vergeistigung' der Naturwissenschaft*); Max van Laue, "Steiner und die Naturwissenschaft," *Deutsche Revue* 47 (1922): 41–49; the quote is on 41. In a 1922 book with the scientific-sounding title *Consciousness of the Atom* (New York: Lucifer Press), the American theosophist Alice Bailey vowed to present "the testimony of science as to the relation of matter and consciousness" (5), and during World War II she launched the New Age movement (*Discipleship in the New Age*, 1944).
46. For a detailed description of this phenomenon, see Victor J. Stenger, *Physics and Psychics: The Search for a World Beyond the Senses* (Amherst, NY: Prometheus, 1990); and Stenger, *The Unconscious Quantum: Metaphysics in Modern Physics and Cosmology* (Amherst, NY: Prometheus, 1995).
47. For example, in 1937 Bohr wrote: "For a parallel to the lesson of atomic theory regarding the limited applicability of such customary idealizations [the distinction between an observer who records a measurement and the measured event], we must in fact turn to quite different branches of science, such as psychology, or even to that kind of epistemological problems with which already thinkers like Buddha and Lao Tzu have been confronted, when trying to harmonize our positions as spectators and actors in the great drama of existence." Niels Bohr, "Biology and Atomic Physics" (1937), in *Niels Bohr: Collected Works*, ed. Finn Aaerud (Amsterdam: Elsevier, 1999), 10:49–62; the quote is on 60. Occultists typically did not notice that Bohr hastened to add: "The recognition of an analogy in the purely logical character of the problems which present themselves in so widely separate fields of human interest does in no way imply acceptance in atomic physics of any mysticism foreign to the true spirit of science" (60).
48. In 1947 the Danish government honored Bohr by giving him an award (a knighthood), and for the occasion Bohr designed a coat of arms, for which he chose the yin-yang symbol, together with the Latin dictum *contraria sunt complementa* (opposites are complementary) to symbolize the dual wave-particle nature of subatomic particles. Bohr's biographer Abraham Pais has written that Bohr pursued his interest in philosophy later in life, after his 1920s–1930s work on complementarity, and that his work on atomic theory was not influenced by philosophy (Eastern or Western); see Pais, *Niels Bohr's Times: in Physics, Philosophy, and Polity* (Oxford, England: Clarendon, 1991), 424.
49. Fritjof Carpra reproduced Bohr's coat of arms in *The Tao of Physics: An Exploration of the Parallels between Modern Physics and Eastern Mysticism* (Berkeley, CA: Shambhala, 1975), 144.
50. Capra cites the authority of Heisenberg throughout *Tao of Physics*, quoting him directly on 10, 18, 28, 45, 50, 53, 67, 140, and 264.
51. For a history of the Fundamental Fysiks Group, see David Kaiser, *How the Hippies Saved Physics: Science, Counterculture, and the Quantum Revival* (New York: Norton, 2011).
52. Deepak Chopra, *Ageless Body, Timeless Mind: The Quantum Alternative to Growing Old* (New York: Random House, 1993), 5.
53. N. Katherine Hayles, *The Cosmic Web: Scientific Field Models and Literary Strategies in the Twentieth Century* (Ithaca, NY: Cornell University Press, 1984), 84. On the impact of the Copenhagen interpretation on literature, see Susan Strehle, *Fiction in the Quantum Universe* (Chapel Hill, NC: University of North Carolina Press, 1992); Maureen DiLonardo Troiano, *New Physics and the Modern French Novel* (New York: P. Lang, 1995); and Elisabeth Emter, *Literatur und Quantentheorie: Die Rezeption der modernen Physik in Schriften zur Literatur und Philosophie deutschsprachiger Autoren, 1925–1970* (Berlin: Walter de Gruyter, 1995). During the Cold War, many literary figures were inspired by the Copenhagen interpretation as promoted in the 1950s by the founder of existentialism, Martin Heidegger. The philosopher had followed developments in physics since 1927, the year his book *Being and Time* was released in Germany and Heisenberg published his Uncertainty Principle. In 1935 Heidegger and Heisenberg met for lengthy conversations at the philosopher's retreat in Todtnauberg; on the relationship of Heidegger and Heisenberg, see Cathyrn Carson, "Modern or Anti-modern Science? Weimar Culture, Natural Science, and the Heidegger-Heisenberg Exchange," in Carson, *Weimar Culture and Quantum Mechanics*, 523–42 (see chap. 8, n. 31). Later that year Heidegger gave a lecture praising the philosophical erudition of his new scientific soul mate: "The leading minds of atomic physics today, Niels Bohr and Heisenberg, think philosophically through and through and for that reason alone create new ways of posing questions and above all hold out in the realm of what bears questioning." Martin Heidegger, *Die Frage nach dem Ding: Zu Kants Lehre von den transzendentalen Grundsätzen* (Tübingen, Germany: Max Niemeyer, 1962), 51; English trans. W. B. Barton and Vera Deutsch as *What Is a Thing?* (Chicago: Henry Regnery, 1967), 67. Heidegger remained a leading voice for existentialism in the 1950s, even though his professional activities were restricted because of his Nazi leanings; see *The Heidegger Case: On Philosophy and Politics*, ed. Tom Rockmore and Joseph Margolis (Philadelphia: Temple University Press, 1992). Heidegger presented Heisenberg to the Cold War generation as the spokesman for science and mathematics in public lectures such as "The Question of Technology" (1953). The event, at which Heisenberg also gave a lecture, "The Picture of Nature in Modern Physics" (1953), is described by Cathryn Carson in "Science as Instrumental Reason: Heidegger, Habermas, Heisenberg," *Continental Philosophy Review* 42, no. 4 (2010): 483–509.
54. Thomas Pynchon, *Gravity's Rainbow* (New York: Viking, 1973), 391.
55. But, alas, not completely; a recent presentation of quantum mysticism as fact is the 2004 documentary-style American film *What the Bleep Do We Know?* directed by William Arntz, Betsy Chasse, and Mark Vicente, which grossed $15 million at the box office. Physicist Lisa Randall has described quantum mysticism as "the bane of scientists" in her book *Knocking on Heaven's Door: How Physics and Scientific Thinking Illuminate the Universe and the Modern World* (New York: Ecco, 2011), 10.

11. GEOMETRIC ABSTRACTION AFTER WORLD WAR II

1. Richard Paul Lohse described his work as "democratic" in the sense of being composed of equal units arranged in a non-hierarchical order: "The serial principle is a radical democratic principle." Hans Joachim Albrecht et al., *Richard Paul Lohse: Modulare und serielle Ordnungen, 1943–84/Ordes modulaires et sériels, 1943–84/Modular and Serial Orders, 1943–84* (Zurich: Waser, 1984), 142. Lohse's anti-fascist senti-

ments were stoked by his marriage from 1936 to 1939 to the German painter Irmgard Burchard, who fled to Zurich in 1934. In response to the Nazi exhibition of modern art, *Entartete Kunst* (Degenerate art), which opened in Munich in 1937, Burchard organized a 1938 exhibition of twentieth-century German art at the Burlington Gallery, London. Supportive of her political activism, Lohse designed an announcement for the exhibition, although it was not printed and survives only as a draft (per author's conversation with the artist's daughter, Johanna Lohse James, director of the Richard Paul Lohse Stiftung, Mar. 12, 2008, in Zurich). On the social meaning of Lohse's prints, see Felix Wiedler, "Die soziale Substanz innerhalb des Multiplikativen/The Social Substance within the Multiplicative Aspect," trans. Jane Thorley Wiedler, in *Lohse: Drucke*, 46–62 (see chap. 7, n. 37).

2. Max Bill, *Form: Eine Bilanz über die Formentwicklung um die Mitte des XX. Jahrhunderts/A Balance Sheet of Mid-Twentieth-Century Trends in Design/Un bilan de l'évolution de la forme au milieu du XXe siècle* (Basel, Switzerland: Karl Werner, 1952), 11.
3. Max Wertheimer introduced the term *gute Gestalt* ("good Gestalt") in "Untersuchungen zur Lehre von der Gestalt," *Psychologische Forschung*, ed. K. Koffka, W. Köhler, M. Wertheimer et al. (Berlin: Springer, 1923), 4:326.
4. In 1949 the editors of the design magazine *Werk* made *gute Form* ("good form") the theme of an issue, to which Max Bill contributed "Schönheit aus Funktion und als Funktion" (The beauty of function and [beauty] as a function) *Werk* 36, no. 8 (1949): 272–74. This same year Bill organized the exhibition *Die gute Form* (Zurich: Kunstgewerbemuseum, 1949). See Claude Lichtenstein, "Theorie und Praxis der guten Form: Max Bill und das Design," in *Max Bill: Aspekte seines Werkes* (Sulgen, Switzerland: Niggli, 2008), 144–57.
5. See Herbert Lindinger, "Ulm: Legend and Living Idea," in *Ulm Design, 1953–1968: The Morality of Objects*, trans. David Britt, ed. Herbert Lindinger (Cambridge, MA: MIT Press, 1990), 9–13. The Ulm student body was around half foreign, as contrasted with 10% foreign in other German colleges of the era. Students came to Ulm from Algeria, Argentina, Austria, Belgium, Brazil, Canada, Chile, Columbia, Finland, France, Great Britain, Greece, Hungary, India, Indonesia, Israel, Japan, Mexico, Netherlands, New Zealand, Norway, Peru, Poland, South Africa, South Korea, Sweden, Switzerland, Thailand, Trinidad, Turkey, United States, Venezuela, Vietnam, and Yugoslavia.
6. After only moderate success in linking the Hochschule with industry and confronted with student demands in the wake of the May 1968 strikes in France that swept Europe, the Ulm school collapsed financially and closed in 1968.
7. Turel to Speiser, Nov. 28, 1949, Adrien Turel Stiftung, Zentralbibliothek Zürich, MS 25.
8. Max Bill, "Die mathematische Denkweise in der Kunst unserer Zeit" (The mathematical way of thinking in the art of our time), in *Antoine Pevsner, Georges Vantongerloo, Max Bill* (Zurich: Kunsthaus, 1949), n.p.; the essay also appeared in the magazine *Werk* 36, no. 3 (1949).
9. Ibid.
10. Ibid.
11. Ibid.
12. Ibid.
13. The conference was on the occasion of a major exhibition on the history of proportion, featuring mathematical diagrams from Leon Battista Alberti's *Ten Books on Architecture* (completed 1452, printed 1485) to Le Corbusier's *Modulor* (1946). See Marzoli, *Studi sulle proporzioni* (see chap. 2, n. 40).
14. The incident was recalled by Bill's second wife, Angela Thomas, in the catalogue for an exhibition she curated; *Max Bill* (Studen, Switzerland: Fondation Saner, 1993), 36. The Swiss art historian Margaret Staber has also recalled that when she was Bill's student at Ulm in the later 1950s, he gave her a small book about Einstein and Freud, *Ein Briefwechsel A. Einstein—Sigmund Freud, Warum Krieg?* (Paris: Internationales Institut für geistige Zusammenarbeit, 1933); author's interview with Staber, Mar. 12, 2008, Zurich.
15. See *Der Geist der Farbe: Karl Gerstner und seine Kunst*, ed. Henri Stierlin (Stuttgart, Germany: DVA, 1981); English edition as *The Spirit of Colors: The Art of Karl Gerstner*, trans. Dennis Q. Stephenson, ed. Henri Stierlin (Cambridge, MA: MIT Press, 1981). Like the older artists, Gerstner worked as a graphic designer; see *Karl Gerstner: Rückblick auf 5 × 10 Jahre Graphik Design*, ed. Manfred Kröplien (Ostfilden-Ruit, Germany: Hatje Cantz, 2001); English edition as *Karl Gerstner: Review of 5 × 10 Years of Graphic Design*, trans. Tas Skorupa and John St. Southward, ed. Manfred Kröplien (Ostfilden-Ruit, Germany: Hatje Cantz, 2001).
16. Author's interview with Karl Gerstner, Mar. 8, 1908, in Schönenbuch, Switzerland.
17. Karl Gerstner, "Sketches for the Color Lines," *24 Facsimile Pages from a Sketchbook* (Zurich: Editions Pablo Stähli, 1978), n.p.
18. Lohse claimed to have begun using color progressions decades earlier (in other words, before Karl Gerstner) by dating his painting in plate 11-17 to 1943–70. Another example is Lohse's giving the date 1950–69 to *Six Continuous Color Bands with Equal Quantities* (Collection of the Foundation H. and R. Rupf, Kunsthalle Bern). The unusually long span of these dates (twenty-six and nineteen years, respectively) is because Lohse dated his work from when he had the first idea to make the painting, to when (decades later) he actually executed the artwork. Since I can only examine documents (not read minds), I date Lohse's paintings to when he made them. Specifically, I base the date of the commencement of Lohse's use of color progressions to the *completion* date of his paintings with color progressions, which is no earlier than the mid-1960s. Catalogues of paintings that were exhibited during his lifetime, such as a 1967 survey of about thirty works spanning his career from 1942 to 1967, include no work with color progressions (*Richard Paul Lohse*, exhibition at Galerie Denise René, Paris, Nov. 4–Dec. 4, 1967). Additional evidence for a mid-1960s date comes from Lohse's graphic art (in which he consistently employed the same patterns of color and form as in his fine art). Color progressions do not appear in Lohse's graphic art before the mid-1960s. See *Richard Paul Lohse: Catalogue raisonné* (Ostfildern-Ruit, Germany: Hatje Cantz, 1999), vol. 1 (graphic art). Lohse did drawings for paintings with color progressions that he dated much earlier than the completed paintings, which were not exhibited until 1985, three years before the artist's death. Thus there is no way to independently confirm the pre-1960 date that he assigned to these drawings; see Richard Paul Lohse, *Zeichnungen: Dessins, 1935–1985: Hans-Peter Riese, Friedrich W. Heckmanns, Richard Paul Lohse* (Baden, Germany: LIT, 1986).
19. Several popularizations by Poincaré were well known in early-twentieth-century Paris: Henri Poincaré, *La science et l'hypothèse*, 1902), *Science et mèthode* (1904), and *La Valeur de la science* (1908). On Duchamp and Poincaré, see Craig Adcock, "Conventionalism in Henri Poincaré and Marcel Duchamp," *Art Journal* 44, no. 3 (1984): 249–58. Duchamp was a cynic who playfully distorted scientific and mathematical symbols to create nonsensical conundrums; see Craig Adcock, *Marcel Duchamp's Notes from the Large Glass: An N-Dimensional Analysis* (Ann Arbor, MI: UMI Research Press, 1983); and Linda Dalrymple Henderson, *Duchamp in Context: Science and Technology in the Large Glass and Related Works* (Princeton, NJ: Princeton University Press, 1998).
20. For an analysis of the pictorial conventions in Duchamp's *Tu m'*, see Karl Gerstner's diagrams of the painting's key components; Gerstner, *Marcel Duchamp: Tu m'*, trans. John S. Southard (Ostfildern-Ruit, Germany: Hatje Cantz, 2001).
21. For Bourbaki's sources in Hilbert, see Leo Corry, *Modern Algebra and the Rise of Mathematical Structure* (Basel, Switzerland: Birkhäuser, 1996). After describing Hilbert's formalist program and its impact on Bourbaki's metamathematical vision of the nature of mathematics as a structuralist hierarchy, Corry argues that Bourbaki did not carry out its vision in the axiomatic systems that the group actually designed. The historian of mathematics J. S. Bell has suggested that the source of Bourbaki's structuralism was not Hilbert's formalism but rather French structuralist linguistics, but he gives no historical evidence to support this hypothesis; Bell, "Category Theory and the Foundations of Mathematics," *British Journal of the Philosophy of Science* 32 (1981): 349–58.
22. For the difference between Hilbert's and Bourbaki's conceptions of the modern axiomatic method, see Leo Corry, "The Origins of Eternal Truth in Modern Mathematics: Hilbert to Bourbaki and Beyond," *Science in Context* 10, no. 2 (1997): 253–96.
23. For an illustrated history of Boubaki, see Maurice Mashaal, *Bourbaki: A Secret Society of Mathematicians*, translated from French by Anna Pierrehumbert (Providence, RI: American Mathematical Society, 2006).

24. The historical basis of this common observation is detailed by David Aubin in "The Withering Immortality of Nicolas Bourbaki: A Cultural Connector at the Confluence of Mathematics, Structuralism, and the Oulipo," *Science in Context* 10, no. 2 (1997): 297–342. Aubin does not claim that Bourbaki directly influenced linguistic structuralism, or vice versa, but rather he introduces the notion of a "cultural connector" to describe Bourbaki's role in French culture. Nevertheless, the French mathematician and historian Jean-Michel Kantor has vigorously disputed any historical link between Lévi-Strauss's linguistic structuralism and Bourbaki, claiming that Lévi-Strauss and intellectuals in his circle cited Bourbaki to give their social science a mathematical pedigree; see Kantor, "Bourbaki's Structures and Structuralism," *Mathematical Intelligencer* 33, no. 1 (2011): 1.
25. On the formation of Oulipo, see Warren Motte's introduction to *Oulipo: A Primer of Potential Literature*, trans. and ed. Warren Motte (Lincoln: University of Nebraska Press, 1986), 1–22. Le Lionnais and Queneau formed Oulipo as a subcommittee of the College of Pataphysics, a large group of artists and writers, including Marcel Duchamp, Max Ernst, Lucio Fontana, and Joan Miró, who were followers of the inventor of pataphysics, the satirist Alfred Jarry. In 1898 Jarry had defined "pataphysics" as "the science of imaginary solutions," stressing the inability of scientists to give a complete physical description of reality by reason alone. Jarry considered the ancient atomist Lucretius's *clinamen* (the random swerving of atoms during collisions; see chapter 3) as a fundamental aspect of poetical creativity. For Jarry's sources in Lucretius, see Andrew Hugill, *'Pataphysics: A Useless Guide* (Cambridge, MA: MIT Press, 2012), 15–16; and Steve McCaffery, *Prior to Meaning: The Protosemantic and Poetics* (Chicago: Northwestern University Press, 2001), in which the author writes: "Jarry's pataphysical strategies similarly involve the *method* of the *clinamen* without an attendant atomistic ontology" (24; see also 20–22).
26. Queneau's *Les fondements de la littérature d'après David Hilbert* (1976) was published as fascicle no. 3 in the *La bibliothèque oulipienne* (Paris: Éditions Seghers, 1990), 1:35–48.
27. Gurdjieff was a teacher of Peter Ouspensky in Russia from 1915 to 1918, but after the revolution of October 1917, conditions in Russia became inhospitable to spiritual teachings, so Gurdjieff went to Fontainebleau-Avon and established the Institute for the Harmonious Development of Man in 1923, eventually settling in Paris, where he taught until his death in 1949. Morellet was drawn to Gurdjieff's view, which was recorded by his student Ouspensky, that artists should take a mathematical approach to art with the goal of affecting the viewer's intellect and soul. See Peter Ouspensky, *In Search of the Miraculous*, n.t. (New York: Harcourt, Brace, 1949), 27.
28. François Morellet, "Discours de la méthode," in *François Morellet: Discours de la méthode* (Mainz, Germany: Galerie Dorothea van der Koelen, 1996), 6.
29. See Thierry Lenain and Thomas McEvilley, *Bernar Venet* (Paris: Flammarion, 2007).
30. On the linking of Bourbaki and structuralism, and the demise of both, see Aubin, "Bourbaki: A Cultural Connector," 297–342 (see n. 24).
31. For a critique of French poststructuralist assertions that science has no claim to objective reality, see Alan Sokal and Jean Bricmont, *Impostures intellectuelles* (Paris: Odile Jacob, 1997); English edition as *Fashionable Nonsense: Postmodern Philosophers' Abuse of Science* (London: Profile Books, 1998); and Steven Weinberg, *Facing Up: Science and Its Cultural Adversaries* (Cambridge, MA: Harvard University Press, 2001). For a study of the relationship of French poststructuralism to the history of modern mathematics, see Vladimir Tasić, *Mathematics and the Roots of Postmodern Thought* (Oxford: Oxford University Press, 2001). For a lucid exposition of the murky arguments put forth by Tasić, who is a Serbian novelist, see the review by American mathematician Michael Harris, who follows poststructuralist debates from his perch on the mathematics faculty at the University of Paris; *Notices of the American Mathematical Society* 50, no. 4 (Aug. 2003): 790–99.
32. In 1966 a British art critic, Lawrence Alloway, independently coined the term "Systemic Art" to refer to art done in New York, where he had lived since 1961; see Lawrence Alloway, "Introduction," in *Systemic Painting* (New York: Solomon R. Guggenheim Museum, 1966), 11–21.
33. Riley's *Current*, 1964 (plate 11-31), was featured on the cover of the catalogue of *The Responsive Eye* (New York: Museum of Modern Art, 1965), which was curated by William C. Seitz. On the reception of Riley's work in this 1965 exhibition, as well as the viewer's sensation of movement when looking at *Current*, see Pamela M. Lee, "Bridget Riley's Eye/Body Problem," in *Chronophobia: On Time in the Art of the 1960s* (Cambridge, MA: MIT Press, 2004), 154–214.
34. When van Doesburg met Torres García in 1928, he was directing the French group of non-objective artists *Art Concret*, having left Germany in a huff because he was passed over for a teaching job by Gropius. The petulant Dutchman clashed with his Uruguayan disciple, and in 1930 Torres García broke with him and helped found the rival Parisian non-objective group, *Cercle et Carré*. See *Antagonistic Link: Joaquín Torres García, Theo van Doesburg*, ed. Jorge Castillo (Amsterdam: Institute of Contemporary Art, 1991).
35. See the artist's manifesto, which was reproduced as a 77-page hand-lettered artist's book: Joaquín Torres-García, *La tradición del hombre abstracto: Doctrina Constructivista* (Montevideo, Uruguay: 1938).
36. For example, Mary Vieira (Brazilian, 1927–2001; at Ulm 1952–54), and Geraldo de Barros (Brazilian, 1924–1998; at Ulm in the 1950s). On the Constructivism in Brazil, see *Arte Constructiva no Brazil/Constructive Art in Brazil*, ed. Aracy Amaral (São Paulo, Brazil: DBA Melhoramentos, 1998). On Max Bill in Latin America, see María Amalia García, "Max Bill and the Map of Argentina: Brazilian Concrete Art," in *Building on a Construct: The Adolpho Leirner Collection of Brazilian Constructive Art*, ed. Héctor Elea and Mari Carmen Ramirez (Houston: Museum of Fine Arts, 2009), 53–68.
37. See Héctor Elea, "Waldemar Cordeiro: From Visible Ideas to the Invisible Work," in *Building on a Construct*, ed. Elea and Ramirez, 128–55.
38. See Guilherme Wisnik, "Brasília: Die Stadt als Skulptur/Brasília: the City as Sculpture," in *Das Verlangen nach Form: Neoconcretismo und zeitgenössische Kunst aus Brasilien* (Berlin: Akademie der Künste, 2010), 77–83 (German); 276–80 (English). Although Max Bill's Concrete Art was a source for Niemeyer's vocabulary of geometric forms, Bill found Niemeyer's architecture cold and inhumane—"antisocial barbarity" (280).
39. *Manifesto neoconcreto* (1959), excerpts in Ferreira Gullar, *Etapas da arte contemporânea: Do cubismo ao neoconcretismo* (São Paulo, Brazil: Nobel, 1985), 242–43.
40. See Lewis Pyenson, *Cultural Imperialism: German Expansion Overseas, 1900–1930* (New York: P. Lang, 1985), 139–246.
41. See Jorge J. E. Gracia, "Philosophical Analysis in Latin America," *History of Philosophy Quarterly* 1, no. 1 (1984): 111–22.
42. Gyula Kosice, "Del manifesto de la escuela," *Arte madí universal* 0 (1947): n.p.
43. Tomás Maldonado, "Lo abstracto y lo concreto en el arte moderno," *Arte Concreto* 1 (1946): 5–7; the quote is on 7. On Maldonado's vision for geometric art in Argentina, see Omar Calabrese, "Tomás Maldonado, le arti e la cultura come totalità/Tomás Maldonado, the Arts and Culture as a Totality," English trans. Dominique Ronayne, in *Tomás Maldonado* (Milan: Skira, 2009), 12–31.
44. For a description of the first shaped canvases, see Rhod Rothfuss, "El marco: Un problema de la plástica actual" (The frame: A problem for literal painting), *Arturo* 1, no. 1 (1944): n.p.
45. On Maldonado's contribution to the curriculum at Ulm, see William S. Huff, "Albers, Bill, e Maldonado, il corso fondamentale della scuola di design di Ulm (HfG)/Albers, Bill, and Maldonado, the Basic Course of the Ulm School of Design (HfG)," English to Italian translation by the Language Consulting Congressi, Milan, in *Tomás Maldonado* (Milan: Skira, 2009), 104–21.
46. César Paternosto has argued that ancient Andean cultures had an abstract vision, discernable in their stonework and textiles, of which Western abstract art is a recent expression; see Paternosto, *Piedra abstracta: La escultura inca, una visíon contemporánea* (Abstract stone: Inca sculpture, contemporary vision; Buenos Aires: Fondo de Cultura Económica, 1989).
47. For the contemporary Venezuelan artist, Alessandro Balteo Yazbeck (born 1972), who makes art about this institutionalization of geometric abstraction in Caracas in the 1950s, see Kaira M. Cabañas, "If the Grid Is the New Palm Tree of Latin American Art," *Oxford Art Journal* 33, no. 3 (2010): 365–83.
48. See Francine Birbragher-Rozencwaig, "La pintura abstracta en Venezuela, 1945–1965," in *Embracing Modernity: Venezuelan Geometric*

Abstraction, ed. Francine Birbragher-Rozencwaig and Maria Carlota Perez (Miami: Frost Art Museum, 2010), 9–14. On Soto's role in this era, see *Soto: Paris and Beyond, 1950–1970*, ed. Estrellita B. Brodsky (New York: Grey Art Gallery, New York University, 2012).

49. Nana Last makes a similar distinction and argues, as I do, that whereas the algorithmic route branches in endless directions, the reductionist path has a single end-point; see Last, "Systematic Inexhaustion," *Art Journal* 64, no. 4 (2005): 110–21.
50. See David Bohm and Charles Biederman, *Bohm-Biederman Correspondence: Creativity in Art and Science*, ed. Paavo Pylkkanen (New York: Routledge, 1999).
51. Olival Freire Jr. has argued that the evolution of Bohm's thought about physics in the 1960s is best recorded in his letters to Biederman; see Friere, "Causality in Physics and in the History of Physics: A Comparison of Bohm's and Forman's Paper," in *Weimar Culture and Quantum Mechanics*, ed. Carson et al., 397–411, esp. 404–9 (see chap. 8, n. 31). With no encouragement from Bohm, New Agers have used his writings about wholeness for their own ends; for example, "Nonetheless, all eastern religions (psychologies) are compatible in a very fundamental way with Bohm's physics and philosophy. All of them are based on the experience of a pure, undifferentiated reality which is that-which-is." Gary Zukav, *Dancing with Wu Li Master: An Overview of the New Physics* (London: Rider/Hutchison, 1979), 326.
52. On Kelly's use of chance, see Yve-Alain Bois, "Kelly in France: Anti-composition in the Many Guises," in *Ellsworth Kelly: The Years in France, 1948–1954*, ed. Mary Yakush (Washington, DC: National Gallery of Art, 1992), 9–36, esp. 23–26.
53. On the titles of Stella's series *Black Paintings*, see Brenda Richardson, "Titles," in *Frank Stella: The Black Paintings* (Baltimore: Baltimore Museum of Art, 1976), 3–11. See also Anna C. Chave, "Minimalism and the Rhetoric of Power," *Arts Magazine* 64, no. 5 (1990): 44–63.
54. Sol LeWitt, "Paragraphs on Conceptual Art," *Artforum* 5, no. 10 (1967): 79–83; the quote is on 80.
55. Ibid.
56. Sol Lewitt, "Sentences on Conceptual Art," *Art-Language* 1, no. 1 (1969): 11.
57. "Serial order is a method, not a style"; so Mel Bochner began his essay "The Serial Attitude," *Artforum* (Dec. 1967): 73–77; the quote is on 73. In his erudite survey of the history of the serial attitude, Bochner illustrated the magic square of order four in Albrecht Dürer's *Melencolia I* (1514; shown in the background of plate 2-16 in chapter 2) and included quotations from the American philosopher Josiah Royce (1855–1916), the American psychologist J. J. Gibson (1904–79), as well as Schoenberg, Wittgenstein, and Sol LeWitt. The historian Peter Lowe has suggested that Theo van Doesburg also took a serial attitude when he made *Arithmetic Composition* (plate 6-24 in chapter 6); "La composition arithmétique—un pas vers la composition sérielle dans la peinture de Théo van Doesburg," in *Théo van Doesburg*, ed. Serge Lemoine (Paris: Philippe Sers, 1990), 228–33. For discussion of mathematical themes in Bochner's work, see the essays collected in *Mel Bochner: Thought Made Visible*, ed. Richard Field (New Haven, CT: Yale University Art Gallery, 1995), 75–106.
58. On non-American Earth Art inspired by developments unrelated to the American-Soviet space race, see Mel Gooding, *Song of the Earth* (London: Thames and Hudson, 2002); and *Ends of the Earth: Land Art to 1974*, ed. Philipp Kaiser and Miwon Kwon (Los Angeles: Los Angeles Museum of Contemporary Art, 2012), which features Earth Art from Great Britain, Germany, the Netherlands, Iceland, Israel, Japan, as well as the United States.
59. See *Nancy Holt: Sightlines*, ed. Alena J. Williams (Berkeley, CA: University of California Press, 2011).
60. See Thomas McEvilley and Klaus Ottmann, *Charles Ross: The Substance of Light* (Santa Fe, NM: Radius, 2012).
61. An example is Henry M. Sheffer and Jean Nicod's simplification of the notation used in logic (see chap. 5, n. 13).
62. The key text on post-1945 recycling of the art of Rodchenko and others is *Theory of the Avant-Garde* (1974), in which the German literary critic Peter Bürger argued that the early-twentieth-century "historical avant-garde" art made original points—for example, the Russian Constructivists asserted that color and form are the physical essence of visual art—but that the post-1950 "neo-avant-garde" movements (such as Minimal Art) are redundant: "The concept of the historical avant-garde movements used here applies primarily to Dadaism and early Surrealism but also and equally to the Russian avant-garde after the October revolution. Partly significant differences between them notwithstanding, a common feature of all these movements is that they do not reject individual artistic techniques and procedures of earlier art but reject that art in its entirety, thus bringing about a radical break with tradition." Peter Bürger, *Theory of the Avant-Garde* (1974), trans. Michael Shaw (Minneapolis: University of Minnesota Press, 1984), 109, n. 4. Bürger argues that what makes a style "avant-garde" is that it makes a "radical break" with the past, and that this can be done only once. The art historian Benjamin H. D. Buchloh has countered that neo-avant-garde artists live in in another time and place, and therefore they are not being repetitive but doing creative recycling that expresses another era. Thus, according to Buchloh, Yves Klein's *Red, Blue, Yellow* (ca. 1951) is not a rehash of Rodchenko's *Red, Blue, Yellow* (1921); Buchloh, "Primary Colors for the Second Time" (see chap. 4, n. 63). While Buchloh's retort may hold up for certain examples (although I wouldn't put money on Yves Klein), Bürger has a point when one considers that any reductionist agenda, be it Hilbert's Program or Russian Constructivism, has a (single) end-point. The consuming goal of reductionists is to *find* that end-point.
63. In James Meyer's history of this era, *Minimalism: Art and Polemics in the Sixties* (New Haven, CT: Yale University Press, 2001), the author gives a detailed year-by-year summary of the art, the exhibitions, and the critical debates.
64. James Lawrence has contrasted the void of meaning in Minimal Art with the complex content of the Russian avant-garde in "Back to Square One," in *Rethinking Malevich*, ed. Charlotte Douglas and Christina Lodder (London: Pindar, 2007), 294–313, esp. 307–13.
65. Meyer, *Minimalism*, 184.
66. Ibid., 185.
67. Ibid., 187.
68. For an overview of theories of meaning, see David Lewis, "General Semantics," *Synthese* 22, nos. 1–2 (1970): 18–67, in which this philosopher of language introduces the topic by stating that theories of meaning encompass two topics: "First, the description of possible languages or grammars as abstract semantic systems whereby symbols are associated with aspects of the world; and second, the description of the psychological and sociological facts whereby a particular one of these abstract semantic systems is the one used by a person or population" (19).
69. Bruce Glaser, "Questions to Stella and Judd" (1964), an interview with Frank Stella and Donald Judd that was broadcast on WBAI in New York in Feb. 1964, edited by Lucy Lippard, and published in *Art News* 65 (Sept. 1966): 55–61; the quote is on 58.
70. "It isn't necessary for a work to have a lot of things to look at, to compare, to analyze one by one, to contemplate. The thing as a whole, its quality as a whole, is what is interesting." Donald Judd, "Specific Objects," *Arts Yearbook* 8 (1965), in Donald Judd, *Complete Writings, 1959–1975* (Halifax: Nova Scotia College of Art and Design, 1975), 181–89; the quote is from 187.
71. "In the simpler regular polyhedrons such as cubes and pyramids one need not move around the object for the sense of the whole, the Gestalt, to occur. One sees and immediately 'believes' that the pattern within one's mind corresponds to the existential fact of the object"; Robert Morris, "Notes on Sculpture: Part I," *Artforum* 4, no. 6 (Feb. 1966): 42–44; the quote is on 44. At the time of this writing Morris was completing an MA thesis on Brancusi at Hunter College, which may explain his academic tone.
72. On the impact of Merleau-Ponty on Minimal Art, see Alex Potts, "The Phenomenological Turn," in *The Sculptural Imagination* (New Haven, CT: Yale University Press, 2000), 207–34. On Merleau-Ponty's place in recent art history, see Brendan Prendeville, "Merleau-Ponty, Realism, and Painting: Psychophysical Space and the Space of Exchange," *Art History* 22 (Sept. 1999): 364–88; and Amelia Jones, "Meaning, Identity, Embodiment: The Uses of Merleau-Ponty's Phenomenology in Art History," in *Art and Thought* (Oxford, England: Blackwell, 2003), 71–90.
73. Morris, "Notes on Sculpture: Part I," 44; the terms "constancy of shape" and "tendencies toward simplicity" are from the Gestalt psychologist Wolfgang Köhler.

74. Barr had a strong interest in the Russian avant-garde after travelling to Moscow in 1927–28. After acquiring Malevich's *White on White* in 1935, he made it a centerpiece of his landmark exhibition *Cubism and Abstract Art* (Mar. 2–Apr. 19, 1936). For the details of the acquisition and display of this painting, see Sybil Gordon Kantor, *Alfred H. Barr, Jr. and the Intellectual Origins of the Museum of Modern Art* (Cambridge, MA: MIT Press, 2002), 181–83. In 1980 Stella conceded that Malevich's *White on White* was "an unequivocal landmark" that "kept us going, as a focus of ideas"; Stella interview, with Maurice Tuchman, in "The Russian Avant-Garde and the Contemporary Artist," in *The Avant-Garde in Russia, 1910-30* (Los Angeles: Los Angeles County Museum of Art, 1980), 120.
75. Until this time Constructivism not well known because it was suppressed in both the USSR and the United States; see Benjamin H. D. Buchloh, who uses the example of the Russian-born Naum Gabo's reception in his homeland and the West, before and after World War II; Buchloh, "Cold War Constructivism," in *Reconstructing Modernism: Art in New York, Paris, and Montreal, 1945–1964*, ed. Serge Guilbaut (Cambridge, MA: MIT Press, 1990), 85–112.
76. Morris, "Notes on Sculpture: Part I," 43. Other artists in the Minimalist circle who expressed interest in the Russian tradition include Carl Andre, who mused, "Frank Stella is a Constructivist. He makes paintings by combining identical, discrete units." Carl Andre and Hollis Frampton, *12 Dialogues, 1962–63*, ed. Benjamin H. D. Buchloh (Halifax: Nova Scotia College of Art and Design, 1980), 37. The art historian Maria Gough has expanded on this point in her essay on the early (1958–62) "literalist" work of Stella and Andre in "Frank Stella is a Constructivist," *October* 119 (Winter 2007): 94–120. For Andre, the constructed sculpture of Rodchenko and Tatlin was "a great alternative to the semi-Surrealist work of the 1950s such as Giacometti's and the late Cubism of David Smith"; Andre, interview with Tuchman, "Russian Avant-Garde and the Contemporary Artist," 120. On the relation of Constructivism to Minimalism, see also Hal Foster, "Some Uses and Abuses of Russian Constructivism," in *Art into Life: Russian Constructivism, 1914–1932*, ed. Richard Andrews (New York: Rizzoli, 1990), 241–53. LeWitt recalled: "If you had to find a historical precedent, you had to go back to the Russians . . . the area of main convergence between the Russians and the Americans in the 1960s was the search for the most basic forms." LeWitt, interview with Tuchman, "Russian Avant-Garde and the Contemporary Artist," 119.
77. According to Elisabeth C. Baker in "Judd the Obscure," *Artnews* 67, no. 2 (1968): 45.
78. Glaser, "Questions to Stella and Judd," 58 (see n. 69).
79. Ibid., 55.
80. Ibid., 56. A decade later Judd was still dismissing the Russians; see Donald Judd, "On Russian Art and Its Relation to my Work," in Judd, *Complete Writings*, 114–18 (see n. 70).
81. Eugene Goossen, *The Art of the Real: USA, 1948–1968* (New York: Museum of Modern Art, 1968), 7 and 11.
82. Although the exhibition catalogue for *The Art of the Real* had a lengthy bibliography, Philip Leider, editor-in-chief of *Artforum*, complained that Goossen's essay was "altogether remarkable for being written as if not a single item of criticism in the entire bibliography had ever been *read* by Mr. Goossen"; Leider, "Review of *The Art of the Real*, Museum of Modern Art," *Artforum* 7, no. 1 (Sept. 1968): 65. In Paris, Marcelin Pleynet, editor of *Tel Quel*, described the complex philosophical issues at stake in the Minimalist Program, the subtleties of which escaped Goossen, especially Minimal artists' "naive attempt" (*cette tentative naïve*) to void meaning from an object; "Peinture et 'réalité,'" *L'enseignement de la peinture* (Paris: Seuil, 1971), 163–85. Furthermore, many historians noted that "minimalist" (reductionist) tendencies were not exclusive to American art of the 1960s; see, for example, *Minimalism in Germany: The Sixties/Minimalismus in Deutschland: Die 1960er Jahre*, ed. Renate Wiehager (Ostfildern: Hatje Cantz, 2012). The chauvinism of Goossen prompted complaints that the wealthy trustees of MOMA used exhibitions for their own political agendas; see Michael Kimmelman, "Revisiting the Revisionists: The Modern, Its Critics, and the Cold War," in *The Museum of Modern Art at Mid-Century: At Home and Abroad*, ed. John Szarkowski (New York: Museum of Modern Art, 1994), 38–55.
83. Ibid., 7–8.
84. To my knowledge, Greenberg never cited Roman Jakobson or Ferdinand de Saussure in his writings; Greenberg's many discussions of Roger Fry in his *Collected Essays and Criticism* (Chicago: University of Chicago Press, 1986–1995) include "Review of an Exhibition of Hans Hofmann and a Reconsideration of Mondrian's Theories" (1945), in 2:18; "Review of *Eugène Delacroix: His Life and Work* by Charles Baudelaire" (1947), 2:156; "T. S. Eliot: The Criticism, the Poetry" (1950), 3:66; "Cézanne and the Unity of Modern Art" (1951), 3:84; "The Early Flemish Masters" (1960), 4:102. Greenberg often disagreed with Fry, but "even at his most dogmatic, Fry usually has hold of the truth somewhere" (4:102). Greenberg did not always distinguish the anxious, self-doubting Fry from the bombastic Clive Bell, a common oversight made by readers of British criticism. For example, Victor Burgin and Charles Harrison treat Fry and Bell as a unified voice and toss them together into the dustbin of art history; see Frances Spalding, "Roger Fry and His Critics in a Postmodernist Age," *Burlington Magazine* 128, no. 1000 (1986): 490.
85. Bell, *Art*, 8 (see chap. 5, n. 31).
86. Clement Greenberg, "Recentness of Sculpture," in *American Sculpture of the Sixties*, ed. Maurice Tuchman (Los Angeles: Los Angeles County Museum of Art, 1967), 24–26; the quote is on 25. For Greenberg's response to the shifting political winds in post-1945 America (as manifest in his many writings and revisions), see Francis Frascina, "Institutions, Culture, and America's 'Cold War Years': The Making of Greenberg's *Modernist Painting*," *Oxford Art Journal* 26, no. 1 (2003): 71–97.
87. Barbara Rose, "ABC Art," *Art in America* 53, no. 5 (1965): 57–69; the quote is on 69. Her spiritual interpretation of Minimalism was adopted by many critics and historians; even James Meyer, who scrupulously takes the literalists at their word, echoes Rose when he states: "Minimal art communicates precisely in its 'lack of communication.'" Meyer, *Minimalism*, 187 (see n. 63). Meyer is here quoting Theodor Adorno, for whom silence can speak volumes: see Adorno, *Aesthetic Theory*, trans. C. Lenhardt (London: Routledge and Kegan Paul, 1984), 7. In an essay for the catalogue of an exhibition of Minimal Art in Japan, Lucy Lippard distinguished Asian art (ego-less and contemplative) from New York Minimalism (moralistic and puritanical); "The Cult of the Direct and the Difficult," *Two Decades of American Painting* (Tokyo: National Museum of Modern Art, 1966), 10–12 (in Japanese translation); reprinted in Lippard, *Changing: Essays in Art Criticism* (New York: Dutton, 1971), 112–19.
88. Rosalind Krauss, "Allusion and Illusion in Donald Judd," *Artforum* 4, no. 9 (May 1966): 25–26; the quote is on 25.
89. Ibid., 26.
90. "The lived perspective, that which we actively perceive, is not a geometric or photographic, one." Merleau-Ponty, "Cézanne's Doubt," which in 1964 became available in English translation in *Sense and Nonsense*, trans. Hubert L. Dreyfus and Patricia Allen Dreyfus (Evanston, IL: Northwestern University Press, 1964), 9–25; the quote is on 14. In 1964 the literary critic Susan Sontag had declared that when an artist or author makes a work intended to be factual and literal, it should not be interpreted because doing so would not be in the spirit of the author's intention: "The function of criticism should be to show *how it is what it is*, even *that it is what it is*, rather than to show *what it means*." Sontag, "Against Interpretation" (1964), in *Against Interpretation and Other Essays* (New York: Farrar, Straus, and Giroux, 1966), 3–14.
91. Krauss, "Allusion and Illusion," 26.
92. Ibid., 26. The artist Robert Smithson similarly described Judd's work as filled with illusion in his 1965 essay "Donald Judd," in *Writings of Robert Smithson*, ed. Nancy Holt (New York: New York University Press, 1979), 21–23. Smithson observed, "An uncanny materiality inherent in the surface engulfs the basic structure. . . . The important phenomenon is always the basic lack of structure at the core of 'facts.' The more one tries to grasp the surface tension, the more baffling it becomes" (22–23). Another critic who found Judd's work illusory was Elisabeth C. Baker, who described the "slithery reflections coming off the surfaces" and noted, "Even the mathematical schemes are so simple as to leave nothing special when you know what they are. On the other hand, the works project enigma." "Judd the Obscure," 45 (see n. 77).
93. On the spiritual vision of major patrons of Minimal Art, including Heiner Friedrich, Philippa de Menil, and Giuseppe Panza, see Anna

C. Chave, "Revaluing Minimalism: Patronage, Aura, and Place," *Art Bulletin* 90, no. 3 (2008): 466–86.

94. Chave, "Minimalism and the Rhetoric of Power," 44–63 (see n. 53).
95. As in the exhibition *Scale as Content: Ronald Bladen, Barnett Newman, Tony Smith* at the Corcoran Gallery of Art (Oct. 7, 1967, through Jan. 7, 1968), which featured Ronald Bladen's black painted aluminum sculpture in the form of an X that measured 22 ft. high by 24 ft. wide (*The X*, 1967–68), Barnett Newman's 26-ft.-high *Broken Obelisk* (1967), and Tony Smith's *Smoke* (1967), which was 24 × 48 × 34 ft. See Lucy Lippard, "Escalation in Washington," *Art International* 12, no. 1 (1968), reprinted in Lippard, *Changing*, 237–54 (see n. 87).
96. Interview of Heiner by Phoebe Hoban, quoted in "Medicis for a Moment," 54 (see chap. 10, n. 17).
97. From an interview with Michael Kimmelman, quoted in "The Dia Generation," *New York Times Magazine*, Apr. 6, 2003.
98. See Hoban, "Medicis for a Moment," 57–58; and Marianne Stockebrand, *Chianti: The Vision of Donald Judd* (Marfa, TX: Chianti Foundation, in association with Yale University Press, 2010).
99. Friedrich and Philippa de Menil are two examples of wealthy patrons who commissioned installations with a cathedral aura. They were married in 1979 in a Sufi (Muslim) ceremony, and they supported a Sufi community in upstate New York. Another major patron of Minimal Art, the Italian Giuseppe Panza, whom Friedrich advised, had Old World religious aspirations; see Anna C. Chave, "Revaluing Minimalism" (see n. 93). On the Dia and Panza collections, see Rosalind Krauss, "The Cultural Logic of the Late Capitalist Museum," *October* 54 (Fall 1990): 3–17.
100. Flavin, who was raised Catholic and attended parochial schools, began making sculptural "icons" in 1961 by wiring a light bulb onto an ordinary object, which he described: "My icons differ from a Byzantine Christ held up in majesty; they are dumb—anonymous and inglorious." Notebook entry dated Aug. 9, 1962; this notebook is accessible only through Flavin's later quotations from it, as in *Dan Flavin: Three Installations in Fluorescent Light* (Cologne, Germany: Kunsthalle Köln, 1973), 83. After being chosen by Friedrich and collected by Panza, Flavin did colorful light installations that suited the spiritual aspirations of these collectors in the cavernous spaces of the Dia Foundation and Panza's Italian Villa Varese, which today houses a large collection of Flavin's work. See *Dan Flavin: The Complete Lights, 1961–1996,* ed. Michael Govan and Tiffany Bell (New York: Dia Art Foundation, 2004). As noted before, Anna C. Chave has argued that spiritual associations with light were made by the patrons, not the artists; "Revaluing Minimalism."
101. Interview of Judd by Phoebe Hoban, quoted in "Medicis for a Moment," 58.
102. The settlement took the form of the non-profit Chianti Foundation, under the directorship of Donald Judd; see Judd's obituary, *New York Times*, Feb. 13, 1994; and Kimmelman, "Dia Generation" (see n. 97).
103. Judd earned a BA in philosophy in 1953 from Columbia University, where he took courses on metaphysics, epistemology, Plato, Descartes, Spinoza, and logical positivism; see the chronology in *Donald Judd*, ed. Nicholas Serota (New York: D.A.P., 2004), 247; and David Raskin, *Donald Judd* (New Haven, CT: Yale University Press, 2012), 130, note 29.
104. On Johns's reading of Wittgenstein, see chap. 9, n. 49. Mel Bochner's *On Certainty: The Wittgenstein Illustrations* (1991) is one of several of the artist's works inspired by Wittgenstein.
105. Judd also didn't learn phenomenology, even though Arthur Danto, who studied with Merleau-Ponty in Paris from 1949 to 1950, joined the philosophy faculty at Columbia University in 1951 (during Judd's undergraduate days).
106. Krauss stated: "The history of modern sculpture coincides with the development of two bodies of thought, phenomenology and structural linguistics, in which meaning is understood to depend on the way any form of being contains the latent experience of its opposite; simultaneity always containing an implicit experience of sequence." Rosalind Krauss, *Passages in Modern Sculpture* (Cambridge, MA: MIT Press, 1977), 4–5. Krauss then went on to combine these approaches in her writing; see David Carrier's intellectual biography, *Rosalind Krauss and American Philosophical Art Criticism: From Formalism to Beyond Postmodernism* (Westport, CT: Praeger, 2002).
107. See Diarmuid Costello, "Greenberg's Kant and the Fate of Aesthetics in Contemporary Art Theory," *Journal of Aesthetics and Art Criticism* 65, no. 2 (2007): 217–28. In another essay Costello has called for a reconsideration of Kant's philosophy of art; Costello, "Kant after LeWitt; Towards an Aesthetics of Conceptual Art," in *Philosophy and Conceptual Art*, ed. Peter Goldie and Elisabeth Schellekens (Oxford, England: Clarendon, 2007), 92–115. For a sympathetic assessment of Greenberg as well as a positive view of Kant's "pure aesthetic judgment" (updated to proclaim "This is art" rather than Kant's outmoded Enlightenment judgment "This is beautiful"), see Thierry de Duve, *Clement Greenberg entre les Lignes* (Paris: Dis Voir, 1996).
108. According to Annette Michelson, Douglas Crimp, and Joan Copec, in their introduction to *October: The First Decade, 1976–86* (Cambridge, MA: MIT Press, 1987), ix.
109. Following in the footsteps of Lévi-Strauss, who used group theory to analyze *anthropological* meaning, the Lithuanian linguist Algirdas Julien Greimas adopted the Klein four-group diagram as a tool with which to analyze *linguistic* meaning, in *Sémantique structurale: recherche de méthode* (Paris: Presses Universitaires de France, 1966), translated into English as *Structural Semantics: An Attempt at a Method* (Lincoln: University of Nebraska Press, 1984). Greimas developed the square further in 1968 with Francis Rastier; the two betrayed their ignorance of the history of the diagram by stating that the semiotic square "makes it possible to compare the [linguistic] model . . . to structures called the Klein group in mathematics and the Piaget group in psychology." "The Interaction of Semiotic Constraints," *Yale French Studies* 41 (1968), reprinted in *On Meaning: Selected Writings in Semiotic Theory* (Minneapolis: University of Minnesota Press, 1987), 49–50. Indeed the origin of the semiotic square in the Klein four-group is largely lost in poststructuralist literature, such as the standard source *Semiotics: The Basics* (London: Routledge, 2007), in which author Daniel Chandler states (incorrectly) that "the semiotic square is adapted from the 'logical square' of opposition of scholastic philosophy" (106). Nowhere in this 307-page college textbook on linguistics do students learn about Felix Klein or group theory.
110. Interview of Robert Smithson by Paul Cummings, published as "Interview with Smithson for the Archives of American Art/Smithsonian Institution" (1972), in Holt, *Writings of Smithson*, 148 (see n. 92).
111. Smithson owned Martin Gardner's *Ambidextrous Universe* (New York: Basic Books, 1964); see "Catalogue of Robert Smithson's Library: Books, Magazines, Records," in *Robert Smithson* (Los Angeles: Museum of Contemporary Art, 2004), 249–63. Several art historians have cited *Ambidextrous Universe* as a catalyst for Smithson; see Ann Reynolds, *Robert Smithson: Learning from New Jersey and Elsewhere* (Cambridge, MA: MIT Press, 2003), 252, fn. 111; Jennifer L. Roberts, *Mirror-Travels: Robert Smithson and History* (New Haven, CT: Yale University Press, 2004), 52–53; Thomas Crow, "Cosmic Exile: Prophetic Turns in the Life and Art of Robert Smithson," in *Robert Smithson* (Los Angeles: Museum of Contemporary Art, 2004), 52; Linda Dalrymple Henderson, "Space, Time, and Space-time: Changing Identities of the Fourth Dimension in Twentieth-Century Art," in *Measure of Time*, ed. Lucinda Barnes (Berkeley: University of California, 2007), 87–101.
112. On Smithson's interest in crystals, see Larisa Dryansky, "La carte cristalline: Cartes et cristaux dans l'oeuvre de Robert Smithson," *Les cahiers du musée national d'art moderne* 110 (2009–10): 62–85.
113. Anton Ehrenzweig, *The Hidden Order of Art: A Study in the Psychology of Artistic Expression* (Berkeley: University of California Press, 1967), 128–29.
114. Ibid.
115. Oral history interview with Robert Smithson, July 14–19, 1972, Archives of American Art, Smithsonian Institution.
116. Interview with Smithson by Dennis Wheeler, in Smithson Papers, Archives of American Art, New York (1969), interview 2, reel 3833, frame 1132; in *Robert Smithson: The Collected Writings*, ed. Jack Flam (Berkeley: University of California Press, 1996), 210–11.
117. Smithson owned a Dover reprint of Cantor's 1887 paper "Contributions to the Theory of the Transfinite"; see "Catalogue of Smithson's Library" (see n. 111).
118. See Edwin Hubble, *The Realm of the Nebula* (New Haven, CT: Yale University Press, 1936). Smithson owned a 1958 edition of the Hubble book; see "Catalogue of Smithson's library."

119. In the context of a discussion of double vision, the art historian Jennifer L. Roberts made a similar observation about Smithson's *Yucatán Mirror-Displacement* series: "What is striking about the mirrors is the care with which Smithson has installed each so that its face parallels the others, as if the array were a single delicate instrument, tuned to receive a specific frequency or to observe a specific quadrant of the heavens." Jennifer L. Roberts, "Landscapes of Indifference: Robert Smithson and John Lloyd Stephens in Yucatán," *Burlington Magazine* 82, no. 3 (2000): 556. I suggest that Smithson did not arrange the mirrors "*as if* the array were a single delicate instrument," but rather his arrangement *was* such an instrument.
120. Interview of Smithson by Dennis Wheeler, Smithson Papers, Archives of American Art, New York (1970), interview 4, reel 3833, frame 1177; in Flam, *Smithson: Writings*, 230.
121. Tsung-Dao Lee and Chen Ning Yang, "Question of Parity Conservation in Weak Interactions," *Physical Review* 104 (1956): 254–58.
122. See Gardner, "The Fall of Parity," *Ambidextrous Universe*, 237–53 (see n. 111).
123. Interview of Smithson by Paul Cummings, published as "Interview with Smithson for the Archives of American Art/Smithsonian Institution" (1972), in Holt, *Writings of Smithson*, 290 (see n. 92).
124. Robert Smithson, "The Quasi-Infinities and the Waning of Space," *Arts Magazine* 41, no. 1 (Nov. 1966): 29.
125. Ehrenzweig, *Hidden Order of Art*, 128–29 (see n. 113).
126. According to Roberts, "What the mystical-spatial fourth dimension was also able to do for Smithson was to suggest a cool, hard space beyond limited anthropomorphic perception, a space wherein the seemingly fatal contradictions and paradoxes that he explored in his religious paintings might be resolved." *Mirror-Travels: Robert Smithson and History*, 56 (see n. 111). In 2007, Linda Dalrymple Henderson, author of *The Fourth Dimension and Non-Euclidean Geometry in Modern Art* (1983), lent her support to Robert's interpretation of *Spiral Jetty*; Henderson, "Space, Time, and Space-Time," 98–99 (see n. 111).
127. Smithson and Ouspensky are separated by a half-century of advances in neuroscience, which hastened the decline of theosophy (see "The End of the Absolute" in chap. 3). What evidence is there that Smithson based his art on this old-fashioned mode of thought? According to Roberts: "Although there is no evidence that Smithson read Ouspensky's original text, he would have been indirectly familiar with his ideas through his other readings on the subject. He was certainly reading Kazimir Malevich (who had been heavily influenced by Ouspensky)" (*Mirror-Travels*, 54). Smithson's "other readings" included a chapter entitled "The Fourth Dimension" in Gardner's *Ambidextrous Universe*, which, according to Roberts, sparked the artist's interest in "a 'transcendental world' where asymmetries might be resolved" (*Mirror-Travels*, 52). But Smithson's reading of this chapter is not evidence that the artist believed in occult hyperspace; indeed, the evidence points to the opposite conclusion. As a science journalist, Gardner was a crusader against pseudo-science who *debunked* the occult. His chapter "The Fourth Dimension" is about the *science* of *n*-dimensional space. Gardner also mentions (without endorsing) occult interpretations of four-dimensional space, the "notion that departed souls inhabit a higher space" (*Ambidextrous Universe*, 173; see n. 111), and a colorful character who believed that he could make contact with an occult four-dimensional realm, the "German astronomer and physicist Johann Carl Zöllner, a remarkably stupid fellow" (173). If one assumes that Smithson trusted Gardner's plain-speaking voice, then the artist certainly knew the difference between science and pseudo-science, but Roberts blurs this crucial distinction when she refers to "Smithson's embrace of what Martin Gardner called '4-space'" (*Mirror-Travels*, 53). Gardner described four-dimensional space as a mathematical object (a topic in *science*), but in this context Roberts is using the term "4-space" in reference to "the discourse of hyperspace" in occult literature (a topic in *pseudo-science*). What did Gardner think of Ouspensky? In his 1952 book *Fads and Fallacies in the Name of Science*, Gardner described Ouspensky (who did not graduate from high school) not as a "Russian mathematician" (per Roberts, *Mirror-Travels*, 54) but as a disciple of the Russian theosophist George Burdjieff, whose texts were "almost as unreadable as Madame Blavatsky's writings" (Gardner, *Fads and Fallacies* [New York: Dover, 1952/rpt. 1957], 215). Was writing by Ouspensky any more intelligible? Roberts describes Ouspensky as an "eloquent proponent" of hyperspace, who wrote a "highly influential tract on the fourth dimension" (*Mirror-Travels*, 54). The bewildered reader of Roberts's lengthy quotations from Ouspensky will be relieved to learn that the level-headed Gardner thought the theosophist's speculations were, to quote Wolfgang Pauli, "not even wrong." As Gardner put it: "Ouspensky's speculations are so mixed with esoteric revelation, and so far removed from science, that I have not discussed them in this book" (*Fads and Fallacies*, 215).
128. Martin Gardner, *The Whys of a Philosophical Scrivener* (New York: W. Morrow, 1983), 330; see also 326–42.

12. COMPUTERS IN MATHEMATICS AND ART

1. For example, the American philosopher Thomas Tymoczko complained that the computer-assisted proof of the four-color theorem "introduces empirical experiments into mathematics [and] . . . raises for philosophy the problem of distinguishing mathematics from the natural sciences." Tymoczko, "The Four-Color Theorem and its Philosophical Significance," *Journal of Philosophy* 76, no. 2 (1979): 57–83; the quote is on 58.
2. See Robin Wilson, *Four Colors Suffice: How the Map Problem was Solved* (Princeton, NJ: Princeton University Press, 2002).
3. On the impact of electronic media on proof theory, see Arthur Jaffe, "Proof and the Evolution of Mathematics," *Synthese* 111 (1997): 133–46.
4. Report by Robert MacPherson, 2003, posted on Hales's Web site, Flyspeck Fact Sheet: http://code.google.com/p/flyspeck/wiki/FlyspeckFactSheet (May 10, 2015).
5. See Thomas C. Hales, "Historical Overview of the Kepler Conjecture," *Discrete and Computational Geometry* 36, no. 1 (2006): 5–20; and Tomaso Aste and Denis Weaire, *The Pursuit of Perfect Packing* (Bristol, PA: Institute of Physics, 2000).
6. A protein is made of a chain of amino acids; there are twenty different amino acids. Thus, in a chain composed of five amino acids, there are twenty possibilities for each of the five positions, so the number of possible proteins with five amino acids is $20 \times 20 \times 20 \times 20 \times 20 = 20^5 = 3{,}200{,}000$. Each of these 3,200,000 five-amino-acid proteins is a self-contained system with parameters that may vary independently. For example, each of the five amino acids has about 300 possible rotations relative to its neighbor, and each has a side chain, which can bond at about 150 different angles. These parameters expand a protein's possible folds exponentially. See Lila M. Gierasch and Jonathan King, ed., *Protein Folding: Deciphering the Second Half of the Genetic Code* (Washington, DC: American Association for the Advancement of Science, 1990).
7. See Jane S. Richardson and David C. Richardson, "The Origami of Proteins," in Gierasch and King, *Protein Folding*, 5–16.
8. Quoted by Margaret Wertheim in "Scientist at Work: Erik Demaine, Origami as the Shape of Things to Come," *Science Times* section of the *New York Times*, Feb. 15, 2005.
9. Georg Cantor, "Über unendliche, lineare Punktmannigfaltigkeiten," (On infinite linear point-manifolds [sets]), *Mathematische Annalen* 21 (1883): 545–91.
10. On the sadly under-researched topic of mathematics and eroticism, see H. von Hug-Hellmuth, "Einige Beziehungen zwischen Erotik und Mathematik" (Some relationships between eroticism and mathematics) *Imago* 4 (1915): 52–68. Von Hug-Hellmuth was a Viennese psychoanalyst and his editor at *Imago* was Sigmund Freud. Von Hug-Hellmuth began his essay by commenting that the sexual drive is capable of sublimation in many forms and went on to discuss the symbolism of numbers and forms in Pythagorean and Platonic cosmology. On the impact of physics on Max Ernst, see Gavin Parkinson, "Quantum Mechanics and Particle Physics: Matta, Wolfgang Paalen, Max Ernst," in *Surrealism, Art, and Modern Science: Relativity, Quantum Mechanics, Epistemology* (New Haven, CT: Yale University Press, 2008), 145–76, esp. 168–72.
11. Konrad Zuse, *Rechnender Raum: Schriften zur Datenverarbeitung* (Braunschweig: Vieweg, 1969), translated as *Calculating Space* (MIT Technical Translation AZT-70-164-GEMIT; Cambridge, MA: MIT, 1970); and Edward Fredkin, "Digital Mechanics," *Physica D. Nonlinear Phenomena* 45 (1990): 254–70.

12. Interview with Roger Penrose in *Omni* 8 (June 1986): 67–73; the quote is on 70.
13. "Compared with Euclid . . . nature exhibits not simply a higher degree but an altogether different level of complexity. The number of distinct scales of length of natural patterns is for all practical purposes infinite. The existence of these patterns challenges us to study those forms that Euclid leaves aside as being 'formless,' to investigate the morphology of the 'amorphous.'" Mandelbrot, *Fractal Geometry*, 1 (see chap. 3, n. 32).
14. For example, the American physicist Leo P. Kadanoff complained that Mandelbrot's fractal geometry lacked a secure theoretical base, without which "much of the work on fractals seems somewhat superficial and even slightly pointless." Kadanoff, "Fractals: Where's the Physics?" *Physics Today* 39, no. 2 (1986): 6–7. The American mathematician Steven G. Krantz wrote: "As to the assertion that they [fractal graphic images] provide a glimpse of a new science or the language for developing a new analysis of nature, I would say that any contribution that fractal theory has made in this direction has been accidental. In short, the emperor has no clothes." Krantz, "Fractal Geometry," *Mathematical Intelligencer* 11, no. 4 (1989): 12–16; the quote is on 16; in the same issue see Mandelbrot's reply (17–19).
15. Mandelbrot, *Fractal Geometry*. The public response to the book included the appearance of a fractal on the cover of *Scientific American* 253, no. 2 (Aug. 1985), along with an article by A. K. Dewdney, who wrote: "A computer microscope zooms in for a close look at the most complicated object in mathematics" (16–24). In 1986 a lavishly illustrated book was published by two professors at the University of Bremen, the mathematician Heinz-Otto Peitgen and the physicist Peter Richter, *The Beauty of Fractals* (Berlin: Springer, 1986), and in 1987 the science journalist James Gleick published the best-selling *Chaos: Making a New Science* (New York: Viking, 1987).
16. "En s'intéressant aux indécidables, aux limites de la précision du contrôle, aux quanta, aux conflits à information non complète, aux *'fracta,'* aux catastrophes, aux paradoxes pragmatiques, la science postmoderne fait la théorie de sa propre évolution comme discontinue, catastrophique, non rectifiable, paradoxale." (By concerning itself with undecidables, the limits of precise control, quanta, conflicts in incomplete information, *'fracta,'* catastrophes, and pragmatic paradoxes, postmodern science is theorizing its own evolution as discontinuous, catastrophic, nonrectifiable, and paradoxical). Jean-François Lyotard, *La condition postmoderne: Rapport sur le savoir* (Paris: Minuit, 1979), 97.
17. For withering criticisms of Lyotard's *science postmoderne*, see Jacques Bouveresse, *Rationalité et cynisme* (Paris: Minuit, 1984), 125–30; and Alan Sokal and Jean Bricmont, *Impostures intellectuelles* (Paris: Odile Jacob, 1997), 123–26. For an attempt to make sense of *la science postmoderne* see Amy Dahan Dalmedico, "Chaos, Disorder, and Mixing: A New Fin-de-Siècle Image of Science?" in *Growing Explanations: Historical Perspectives on Recent Science*, ed. M. Norton Wise (Durham: Duke University Press, 2004), 67–94.
18. See Raffi Karshafian, Peter N. Burns, and Mark R. Henkelman, "Transit Time Kinetics in Ordered and Disordered Vascular Trees," *Physics and Medicine and Biology* 48 (2003): 3225–37.
19. See Nathan Cohen and Robert G. Hohlfeld, "Self-Similarity and the Geometric Requirements for Frequency Independence in Antennae," *Fractals* 7, no. 1 (1999): 79–84.
20. For a historical overview of the computer graphic print as an artwork, from the 1950s to 2008, see Debora Wood, *Imaging by Numbers: A Historical View of the Computer Print* (Evanston, IL: Mary and Leigh Block Museum of Art, 2008).
21. In another art-related development of computation, in 1999 the American physicist Richard P. Taylor was the lead author of two articles that were published in peer-reviewed science journals, in which Taylor reported that he had discovered fractal patterns in Jackson Pollock's drip paintings, which were made by pouring streams of paint from a can unto a horizontal canvas lying on the floor. According to Taylor, these fractal patterns are unique to Pollock's drip paintings and could be used to authenticate his work; Richard P. Taylor, Adam P. Micolich, and David Jonas, "Fractal Analysis of Pollock's Drip Paintings," *Nature* 399, no. 6735 (June 3, 1999): 422; and Richard P. Taylor, Adam P. Micolich, and David Jonas, "Fractal Expressionism," *Physics World* 12 (Oct. 1999): 25–28. The art world ignored Taylor until 2006, when the Pollock/Krasner Foundation hired him to use fractal analysis to determine whether certain paintings were fakes or the real thing, prompting a lively debates about fine art methods of authentification and connoisseurship, which are summarized by Claude Cernuschi, Andrzej Herczynski, and David Martin in "Abstract Expressionism and Fractal Geometry," in *Pollock Matters*, ed. Ellen G. Landau and Claude Cernuschi (Chestnut Hill, MA: McMullen Museum of Art, 2007), 91–104. As these authors relate, the weight of opinion among art world professionals has been negative towards Taylor's (alleged) discovery, but the authors end on an optimistic note about the future research on fractal patterns: "Art historians might even feel encouraged to draw upon such scholarship" (101).
22. Yoichiro Kawaguchi, "A Morphological Study of the Form of Nature," *Computer Graphics* 16 (1982): 223–42.

13. PLATONISM IN THE POSTMODERN ERA

1. Following the practice of cultural historians, I use the term *"modern"* to refer to the revival during the Renaissance of the classical outlook, which is based on the conviction that nature embodies patterns that mankind can discern, and Galileo and Kepler's application of mathematics to nature that spurred the Enlightenment of Newton and the German Idealism of Kant and Hegel. By "postmodern," I mean the era after World War II when this classical vision was lost.
2. Theodor W. Adorno and Max Horkheimer, *Dialectic of Enlightenment: Philosophical Fragments* (1947), trans. Edmund Jephcott, ed. Gunzelin Schmid Noerr (Palo Alto, CA: Stanford University Press, 2002), 1. See Jay Bernstein, "Adorno on Disenchantment: The Skepticism of Enlightenment Reason," in *German Philosophy since Kant*, ed. Anthony O'Hear (Cambridge: Cambridge University Press, 1999), 305–28.
3. For most physicists, the classification of subatomic particles into octets is known as the Standard Model, but Gell-Mann, who won the Nobel Prize in Physics in 1969 for his role in devising the chart, had the poor judgment to encourage "quantum mysticism" (see chap. 10) by calling the chart the "Eightfold Way," in an allusion to the Eightfold Way of Buddhism, which is a path to enlightenment that Gell-Mann, who is Jewish, never followed; see Murray Gell-Mann and Yuval Ne'eman, *The Eightfold Way* (New York: W. A. Benjamin, 1964), a collection of papers including Gell-Mann's "The Eightfold Way: A Theory of Strong Interaction Symmetry," *California Institute of Technology Synchrotron Laboratory Report CTSL-20* (1961), 11–57.
4. For a summary of the alternative interpretations of quantum mechanics through the 1970s, see Olival Freire Jr., "The Historical Roots of 'Foundations of Physics' as Fields of Research, 1950–1970," *Foundations of Physics* 34, no. 11 (2004): 1741–60.
5. For a summary of alternative interpretations of quantum mechanics as of around 2000, see Brian Greene, *The Fabric of the Cosmos: Space, Time, and the Structure of Reality* (New York: Vintage, 2004), 202–16. For discussion of Bohmian mechanics, see *Bohmian Mechanics and Quantum Theory*, ed. James Cushing, Arthur Fine, and Sheldon Goldstein (Dordrecht, the Netherlands: Kluwer, 1996); and Detlef Dürr, Sheldon Goldstein, and Nino Zanghi, *Quantum Physics without Quantum Philosophy* (New York: Springer, 2013).
6. This unusual situation prompted two American mathematicians, Arthur Jaffe and Frank Quinn, to propose a merger of mathematics with theoretical physics, by, for example, changing the accepted mathematical norms of rigor to allow for the publication in mathematics journals of work by physicists, who typically write in a less formal style. Jaffe and Quinn also suggested that mathematicians adopt a division of labor that is used in physics, dividing their efforts into theoretical mathematics (the initial, speculative stage of mathematical discovery) and experimental mathematics (confirmation of the speculation with a rigorous proof). See Arthur Jaffe and Frank Quinn, "Theoretical Mathematics: Towards a Cultural Synthesis of Mathematics and Theoretical Physics," *Bulletin of the American Mathematical Society* 29, no. 1 (1993): 1–13. Jaffe and Quinn's proposal prompted a reflection on the distinction between mathematics and physics in a broad historical perspective by Israel Kleiner and Nitsa Movshowitz-Hadar in their "Proof: A Many-Splendored Thing," *Mathematical Intelligencer* 19, no. 3 (1997): 16–26.

7. Albert Einstein, "Religion and Science," *New York Times Magazine*, Nov. 9, 1930, section 6.
8. Freud used the metaphor to describe the ego riding the forces of the id: "The functional importance of the ego is manifested in the fact that normally control over the approaches to motility devolves upon it. Thus in its relation to the id it is like a man on horseback, who has to hold in check the superior strength of the horse; with this difference, that the rider tries to do so with his own strength while the ego uses borrowed forces. The analogy may be carried a little further. Often a rider, if he is not to be parted from his horse, is obliged to guide it where it wants to go; so in the same way the ego is in the habit of transforming the id's will into action as if it were its own." Freud, "The Ego and the Id" (1923), in *Standard Edition*, 19:25 (see chap. 5, n. 35).
9. "The evolutionary epic is mythology in the sense that the laws it addresses here and now are believed but can never be definitely proved to form a cause-and-effect continuum from physics to the social sciences, from the world to all other worlds in the visible universe and backward through time to the beginning of the universe. . . . The evolutionary epic is probably the best myth we will ever have." E. O. Wilson, *On Human Nature* (Cambridge, MA: Harvard University Press, 1978), 192 and 201.
10. The first successful test of nuclear fusion on a large scale was a multi-megaton detonation (hydrogen bomb) on November 1, 1952.
11. Martin Gardner, *Philosophical Scrivener*, 300; see also 326–42 (see chap. 11, n. 128).
12. Morris Kline, *Mathematics: The Loss of Certainty* (New York: Oxford University Press, 1980), 6.
13. Ibid. For other examples of relativist approaches to mathematics, see *New Directions in the Philosophy of Mathematics*, ed. Thomas Tymoczko (Boston: Birkhäuser, 1986); the editor (who is a philosopher) states in the introduction to this volume that the essays he has collected all focus on the production and communication of mathematics as a fluid, ever-changing body of knowledge (as opposed to a body of timeless, certain truths), in an approach he calls "quasi-empiricist" (xvi). For another relativist viewpoint, see Philip Kitcher, "Mathematical Naturalism" (1988), in *History and Philosophy of Modern Mathematics*, ed. William Aspray and Philip Kitcher, Minnesota Studies in the Philosophy of Science 11 (Minneapolis: University of Minnesota Press, 1988), 293–325.
14. Philip J. Davis and Reuben Hersh, *The Mathematical Experience* (Boston: Birkhäuser, 1998), 410.
15. Doron Zeilberger, "Theorems for a Price: Tomorrow's Semi-Rigorous Mathematical Culture," *Mathematical Intelligencer* 4, no. 4 (1994): 11–14; the quote is on 14.
16. George E. Andrews, "The Death of Proof? Semi-Rigorous Mathematics? You've Got to Be Kidding!" *Mathematical Intelligencer* 4, no. 4 (1994): 16–18; the quote is on 16. Nonetheless, the Scottish mathematician Jonathan Borwein has pointed out that results with a high degree of probability are already accepted in an approach called "experimental mathematics"; see Borwein et al., "Making Sense of Experimental Mathematics," *Mathematical Intelligencer* 18, no. 4 (1996): 12–18.
17. Newton is quoted by David Brewster (1781–1868) in his *Memoirs of the Life, Writings, and Discoveries of Sir Isaac Newton* (Edinburgh: T. Constable, 1855), 2:407.

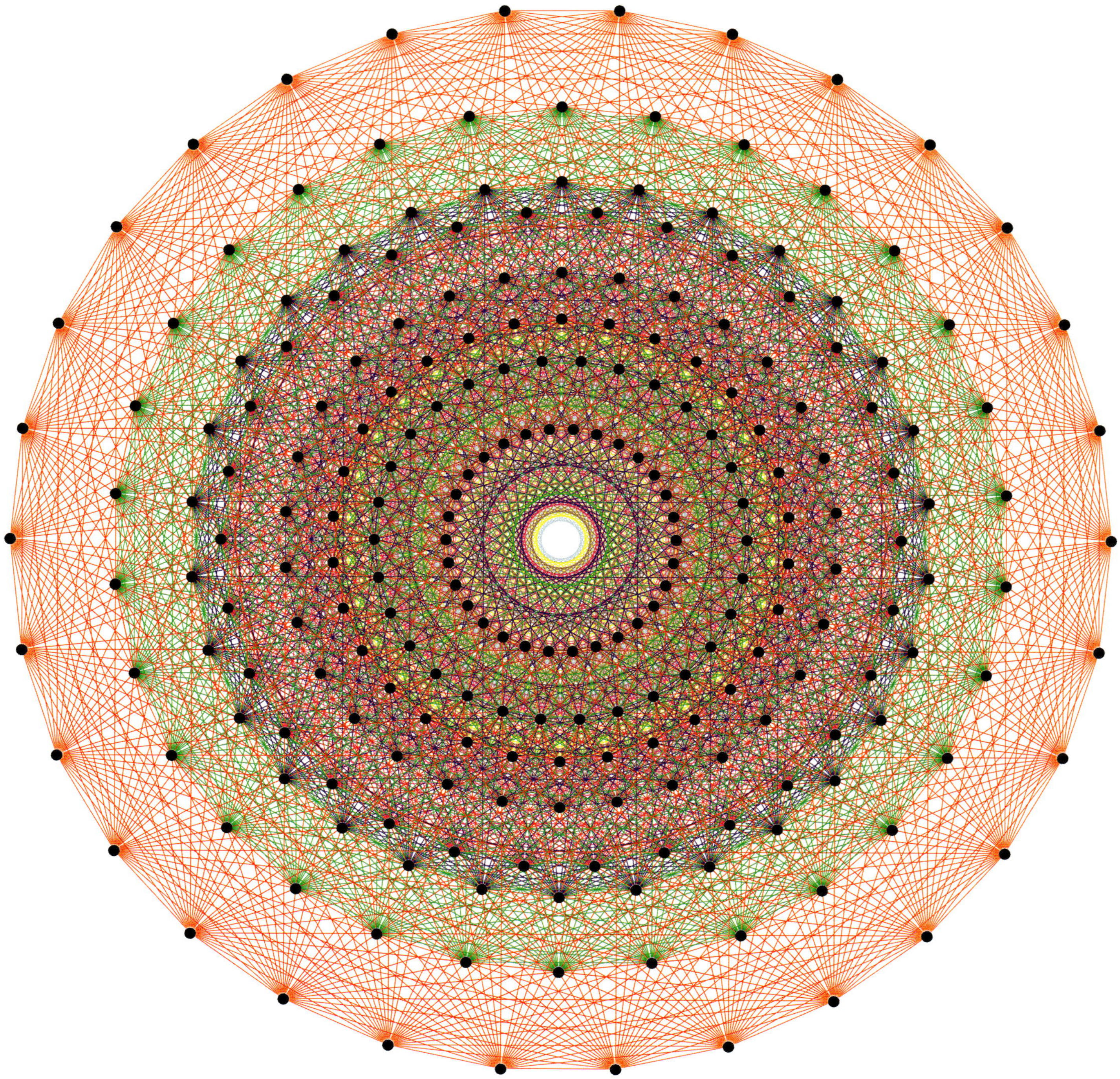

0-6. A computer-generated image of the root system of the symmetry group E-8, 2007.

E-8 is a 248-dimensional group of symmetries of a 57-dimensional object. In March 2007 an international team of mathematicians announced that they had completed a computation to describe all the ways that E-8 can act as a symmetry group. The symmetry group E-8 has attracted attention recently because it has applications in string theory. There is hope that E-8 will provide a way to incorporate the Standard Model into string theory. Mathematicians usually work alone or with a few colleagues, but because of the complexity of the calculation, this project required four years of work by a team of eighteen mathematicians and seventy-seven hours of time on a supercomputer. The team's core group consisted of Jeffrey Adams (University of Maryland), Dan Barbasch (Cornell University), Fokko du Cloux (University of Lyon), Marc van Leeuwen (University of Poitiers), John Stembridge (University of Michigan), Peter Trapa (University of Utah), and David Vogan (MIT).

Acknowledgments

In 2002 Princeton University Press published my book on art and science (*Exploring the Invisible: Art, Science, and the Spiritual*), and Vickie Kearn, executive editor for mathematics, suggested I write a book on art and mathematics. I thank her for giving me this good idea and for shepherding the book through its development. Since my education was in the fine arts, I had to learn the mathematics, and I was fortunate to have the guidance of Peter Renz, mathematics editor of the book. Suzanne Kotz, the trenchant editor of my previous books, served as art editor.

While writing this book I taught courses on the history of mathematics to art students attending the School of Visual Arts in New York. Nothing clarifies one's thinking more than struggling to find words to communicate a complex concept to bright students. I am grateful to my students for their enthusiasm about mathematics, and to the provost Jeff Nesin for his leadership in the interdisciplinary education of artists. In my day-to-day teaching, I received steadfast support from Tom Huhn, chairman of art history, and Robert Milgrom, co-chairman of humanities and sciences.

I benefitted from criticism of my manuscript by the five anonymous readers recruited by the Press and from the four readers I enlisted: comments on art history came from Jean Becker; on theology from Rolf-Walter Becker; on physics from Robert Pompi; and on logic from John M. Vickers. I received encouragement from Peter Gelker, Mary Jane Harris, Ann Lehman Katz, Howard Katz, Ed Marquand, Elizabeth Meryman, and Robert Simon. The foreword was written with insight and panache by astrophysicist Neil deGrasse Tyson.

My research on the history of group theory and Concrete Art took me to Zurich, where I profited from conversations with David Speiser, nephew of Andreas Speiser, and David's wife, Ruth Speiser, the daughter of Hermann Weyl; Jakob and Chantal Bill, son and daughter-in-law of Max Bill; Johanna Lohse James, daughter of Richard Paul Lohse; Vera Hausdorff, director of the Camille Graeser Archive; Karl Gerstner; and Margaret Staber, art historian. I am also grateful to Victor Belyakov for research assistance in Moscow.

Many librarians aided me, especially the research staffs of New York Public Library and the Zentralbibliothek Zürich; Jennifer Tobias, Museum of Modern Art Library, New York; Arnaud Dercelles, Foundation Le Corbusier, Paris; and Carme Ruiz, Centre d'Estudis Dalinians, Fundació Gala-Salvador Dalí, Figueres, Spain.

The translations from Latin were done by Jonathan Canning, from Russian and Polish by Silvia Vassileva-Ivanova, from Dutch by Bert van Manen, from Chinese by Xingwei Huang, Hong Mei Jin, Hui Chen Ou Yang, and Lisi Zhang, and from Japanese by Haruka Song. I am responsible for the translations from French, Spanish, and German, the latter with assistance from Silvia Vassileva-Ivanova.

The key task of visualizing abstract concepts in the one-hundred-and-two mathematics diagrams was done superbly by Dimitri Karetnikov of Princeton University Press and the staff of Umbra Studio in New York. Space does not permit me to describe all the help I got assembling the 444 art images that appear on these pages. I do, however, want to acknowledge the extraordinary generosity of the grandson of Aleksandr Rodchenko, Aleksandr Lavrent'ev of Moscow, who gave me vintage photographs of his grandfather's work, and the unstoppable determination of the architectural photographer Achim Bednorz of Cologne, who travelled to Chartres, France, to photograph for this book the rare statues of Pythagoras, Euclid, and Aristotle that adorn the west façade of the Cathedral. Other individuals who understood the importance and power of images include Hattula Moholy-Nagy, daughter of László Moholy-Nagy; Lawrence Schoenberg, son of Arnold Schoenberg; Kumiko Matsuzawa, the grand-daughter of Yutaka Matsuzawa; Maria Kozlovskaya and Maria Mednikova of the Institute of Archaeology, Moscow; Ana Quinones of the Russian Archives; Wencke Clausnitzer-Paschold of the Bauhaus-Archiv, Berlin; Sando Zwiesele of the Benedictine Abbey of St. Martin of Beuron, Germany; Thilo von Debschitz of Q, Wiesbaden; Angelika Harthan, of Galerie Angelika Harthan, Stuttgart; Sidse Buck of the Louisiana Museum of Modern Art, Humlebæk, Denmark; Johanna Descher of the École des Ponts, Champs-sur-Marne, France; Stefan Binnewies and Josef Pöpsel of the Capella Observatory, Crete; Hans Huang of the National Palace Museum, Taiwan; Shinichiro Yoshihara of the Hyogo Prefectural Museum of Art, Kobe; John Moffett of the Needham Research Institute, Cambridge, England; Abbot Elias Dietz of the Trappist Abbey of Gethsemani, Kentucky; Eric Heller of the Physics Department, Harvard University; Julie S. Heath of Warner Brothers; Margaret Adamic of Disney; Chris Focht of the State University of New York at Binghamton; and in New York City, astronomer Brian Abbot of the Hayden Planetarium at the American Museum of Natural History; computer animator Chris Anne Lindo; and Robbi Siegel of Art Resource.

I am grateful to the staff of the Corinne Goldsmith Dickinson Center for Multiple Sclerosis at Mount Sinai Hospital in New York, where I am a patient, for helping me keep enough of my circuits firing so I could carry out this project.

Jason Snyder designed the dramatic cover of this book and its elegant interior pages, on which he skillfully merged art images, math diagrams, and text into a seamless whole. Production was done by the expert staff of Princeton University Press, under the watchful eyes of production editors Karen Carter, Ellen Foos, and Ali Parrington. The endpapers were secured with help from Marge Salik of Talas, and the printing of the book in China was co-ordinated by the amazingly efficient Alex Cheng of Imago USA.

Credits

PHOTO CREDITS

Key to abbreviations

ArtRes	Art Resource, New York
Bridgeman	Bridgeman Images
CF	Chris Focht
Escher	www.mcescher.com
Met	Metropolitan Museum of Art, New York
MOMA	Museum of Modern Art/Scala/Art Resource, New York
Tate	Tate, London
Vatican	Photographic Archives of the Vatican Museums

Permission to reproduce illustrations is provided by the owners or sources as listed in the captions. Additional photography credits are as follows (numerals in boldface refer to plates).

1-7 Bridgeman; **1-5** Asia Society, New York; **1-9** Met/ArtRes; **1-12** © Sandro Vannini/Corbis; **1-17** Scala/ArtRes; **1-20** Michael Christoph; **1-23**, **1-30** CF; **1-31** Zev Radovan, Jerusalem; **1-41** Scala/ArtRes; **1-50** Norihiro Ueno; **1-54**, **1-55** Achim Bednorz, Cologne; **1-57** Sonia Halliday; **1-52**, **1-59** Achim Bednorz, Cologne; **1-60** P. Zigrossi, Vatican; **1-67** Alinari/ArtRes; **1-79** Achim Bednorz, Cologne; **2-1** Scala/ArtRes; **2-3** René-Gabriel Ojéda © RMN-Grand Palais/ArtRes; **2-9** Scala/ArtRes; **2-10** Scala/Ministero per i Beni e le Attività culturali/ArtRes; **2-12** Stapleton Collection/Bridgeman; **2-14** © National Gallery, London/ArtRes; **2-17** Ari Magg, Reykjavik; **2-15** Bridgeman; **2-31** Adélaïde Beaudoin © RMN-Grand Palais/ArtRes; **2-33** National Gallery of Art, Washington, DC; **2-34** Olivier Martin-Gambier; **2-39** Eija Hartemaa; **2-40** © Tate; **2-41** © Tim Fitzharris/Minden Pictures/ Corbis; **2-42** © 145/Burazin/Ocean/Corbis; **2-43** Spencer Hansen, Bryan Tarnowski; **3-1** Tom Loonan; **3-35** © Malcolm Lightner/Corbis; **3-36** © Mike ver Sprill; **4-5** Dagli Orti/ArtRes; **4-9**, **4-10** MOMA; **4-20** Tom Van Eynde; **4-23** CF; **4-27** © Tate; **4-28**, **4-32** MOMA; **5-2** Alfredo Dagli Orti/ ArtRes; **5-11** Courtesy of the Getty's Open Content Program; **5-14** Biff Henrich; **5-15**, **5-18** © Tate; **6-4** CF; **6-12** MOMA; **6-18**, **6-19**, **6-20** CF; **6-23** Erich Lessing/ArtRes; **6-25**, **7-6** CF; **7-11** Escher; **7-12**, **7-13**, **7-16** CF; **8-3** Joerg P. Anders/bpk Berlin/ArtRes; **8-2** Jean-Claude Planchet © CNAC/ MNAM/Dist. RMN-Grand Palais/ArtRes; **8-5**, **8-7** Stephen White, London; **8-10** MOMA; **8-13** Åsa Lundén; **8-14** MOMA; **8-27** © Tate; **9-4** Museum Associates/LACMA/ArtRes; **9-5** Met/ArtRes; **9-6** Scala/ArtRes; **9-8** Escher; **9-10** Klassik Stiftung Weimar; **9-11** © National Gallery, London/ArtRes; **9-13** © Vatican; **9-15** Herscovici/ArtRes; **9-14** MOMA; **9-17** ADAGP/ArtRes; **9-19**, **9-20** Escher; **9-21** Erich Lessing/ArtRes; **9-22**, **9-23**, **9-24** Escher; **9-30** Tom Van Eynde; 9-29 bpk, Berlin/ArtRes; 9-31 Erich Lessing/ArtRes; **9-32** Biff Henrich; **9-33** Jennie Carter; **9-36**, **9-37** Graham S. Haber; **10-13** Ralph Alberto; **10-16** CF; **10-21** Bridgeman; **10-24** Shigeo Anzai; **10-26** Robert Bayer, Basel; **10-27** Tord Lund; **10-28** John Cliett; **11-2** ©Anne Gold, Aachen; **11-3** Museum für Gestaltung Zürich, Designsammlung, Jaggy and U. Romito, ©ZHdk; **11-4** Museum für Gestaltung Zürich, Plakatsammlung, ©ZHdk; **11-5** Nadja Wollinsky/Stadtarchiv Ulm; **11-23** CF; **11-26** Bridgeman; **11-28** Archives Bernar Venet, New York; **11-31** MOMA; **11-34** Philippe Migeat © CNAC/MNAM/Dist. RMN-Grand Palais/ArtRes; **11-36** Mark Morosse; **11-37** De Agostini Picture Library/ Bridgeman; **11-46** Rick Hall; **11-48** Minneapolis Institute of Arts; **11-49** MOMA; **11-50** © Whitney Museum of American Art; **11-51** Ellen Page Wilson; **11-52** R. J. Phil; **11-58** Rudolf Burckhardt © The Jewish Museum, New York/ArtRes; **11-57** Florian Holzherr, 2002; **11-62**, **11-63**, **11-64** CF; **11-65** © Christie's Images/Bridgeman; **11-73** Gianfranco Gorgoni; **12-2** www.davidrumsey.com; **12-6** Eric Gregory Powell; **12-10** Gary Urton; **12-12** MOMA; **12-17** John Baggen; **12-23** Andreas von Lintel; **12-41**, **12-42** James Ewing; **13-3** Tom Van Eynde; **13-4** Jeremy Lawson.

DIAGRAM CREDITS

Credit for the mathematics diagrams is as follows (numerals in boldface refer to plates; SB refers to sidebars).

6–7, **12–56**: Chris Lindo; SB on 29, **1–44**, **1–45**, **1–63**, **1–65**, **1–71**, **2–6**, **3–2**, **3–3**, **3–6**, **3–7**, SB on 114, SB on 116, SB on 117, **3–8**, **3–18**, **3–23**, SB on 155, **5–4**, **5–5**, **5–8**, **6–5**, SB on 251, **7–4**, **7–5**, **7–14**, SB on 274, **8–30**, **8–33**, SB on 331, **9–25**, **9–26**, **9–27**, SB on 356, SB on 359, **10–14**, **11–70**, **12–40**, **12–53**, **13–5**: Princeton University Press; **1–11**, **1–13**, **1–14**, **1–19**, **1–22**, **1–24**, **1–25**, **1–26**, **1–27**, **1–28**, **1–29**, **1–32**, **1–36**, **1–37**, SB on 39, **1–48**, **1–58**, **1–72**, SB on 63, **1–73**, **1–74**, **1–78**, SB on 78–79, SB on 88, **2–24**, **2–25**, SB on 102, **2–41**, **2–42**, **3–5**, **3–12**, **3–13**, **3–14**, **3–15**, **3–17**, **4–3**, SB on 207, SB on 253, SB on 294, **9–3**, **9–7**, **12–31**, **12–32**, **12–33**, **12–35**, **12–46**, **12–47**, **12–48**: Umbra Studio, New York.

MUSIC LYRICS CREDITS

Credit for the music lyrics that appear in the margins of the pages listed below.

380: "Imagine," words and music by John Lennon © 1971 (renewed) Lenono Music, All Rights Administered by Downtown DMP Songs/Downtown Music Publishing LLC, All Rights Reserved, Used by Permission, Reprinted by permission of Hal Leonard Corporation; 366: "Good Vibrations," words and music by Brian Wilson and Mike Love, © 1966 Irving Music, Inc., Copyright renewed, All Rights Reserved, Used by Permission, Reprinted by permission of Hal Leonard Corporation; 403: "New Math," words and music by Tom Lehrer, who gave permission for eternity in the known and unknown universes, courtesy of Maelstrom Music; 507: "Woodstock," words and music by Joni Mitchell, © 1969 (renewed) Crazy Crow Music, All Rights Administered by SONY/ATV Music Publishing, 8 Music Square West, Nashville, TN 37203, All Rights Reserved.

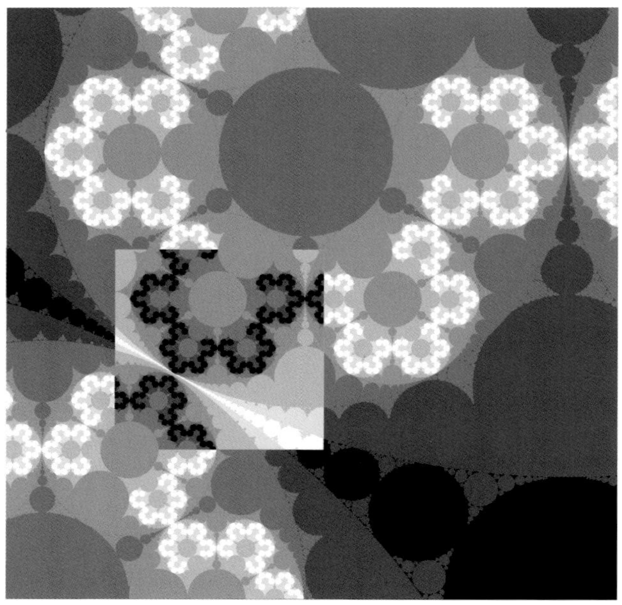

Karl Gerstner (Swiss, b. 1930), *Color Fractal 6.10: Self-Referential Black and White*, 1984-90. Acrylic on aluminum, 59 x 59 in. (150 x 150 cm). Courtesy of the artist.

Index

Page numbers in *italic* refer to a plate or sidebar

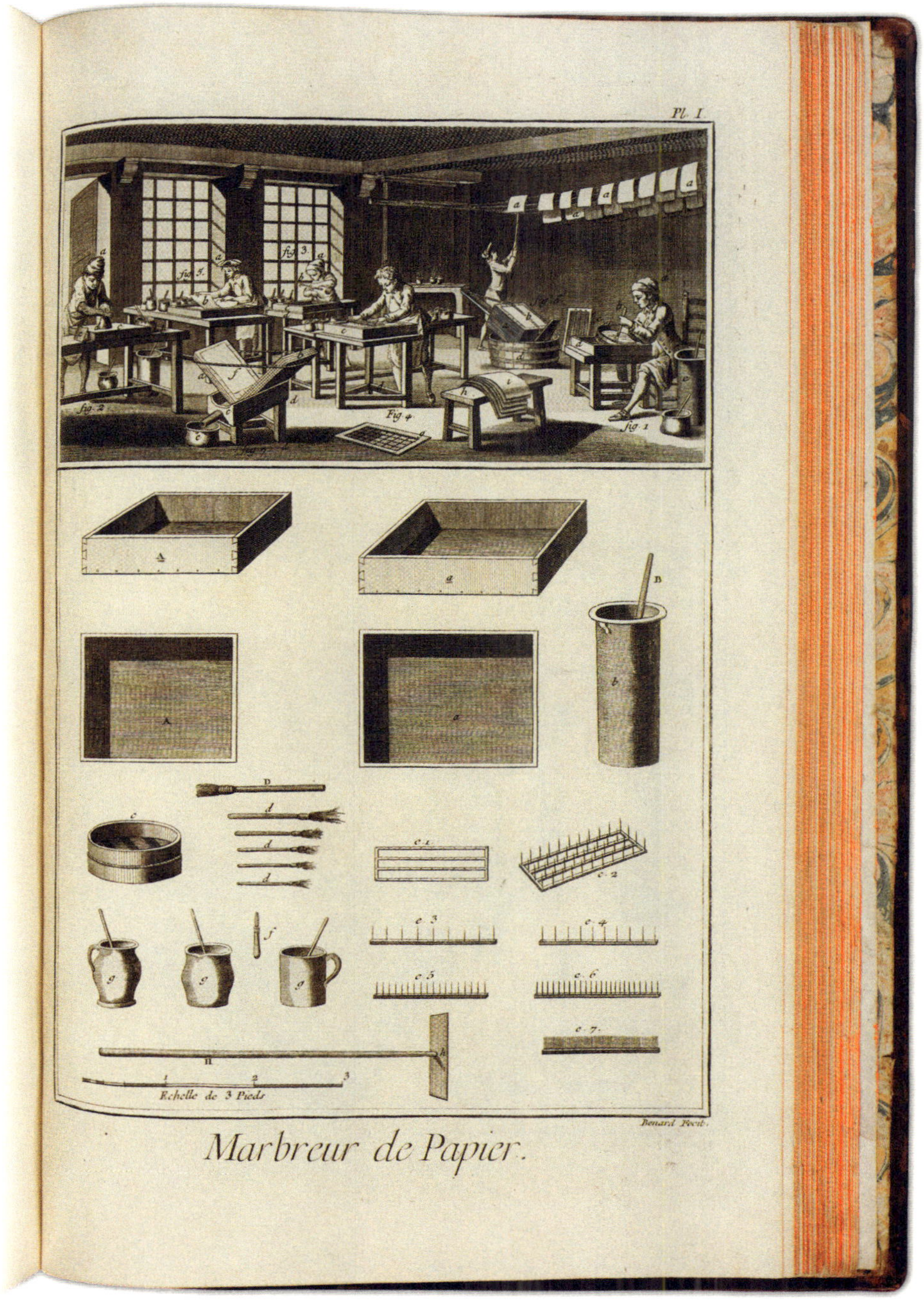

ENDPAPERS

Renato Crepaldi (Brazilian, born 1973), *Grey French Shell*, 2015. Marbled paper. Used with permission.

The endpapers of this book are a photographic reproduction of a piece of marbled paper by the contemporary artist Renato Crepaldi, who uses the traditional methods illustrated in the eighteenth-century engraving from *l'Encyclopedie* (shown above), co-edited by Denis Diderot and the mathematician Jean le Rond d'Alembert, known for his work on regular, repeating wave patterns (*d'Alembert's Equation*, 1747). The kind of irregular, non-repeating (but self-similar) shapes in Crepaldi's endpapers were long considered too complex to be analyzed mathematically. But after the invention of the computer and fractal geometry, mathematicians today study the patterns formed by flowing liquids in paper marbling. See Shufang Lu, Aubrey Jaffer, Xiaogang Jin, Hanlin Zhao, and Xiaoyong Mao, "Mathematical marbling," *IEEE Computer Graphics and Applications* 32, no. 6 (Nov.–Dec. 2012): 26–35.

ABOVE

Marbreur de Papier (Paper marbling), in Denis Diderot and Jean le Rond d'Alembert, *l'Encyclopedie* (Paris: Briasson, David, Le Breton, Durand, 1768), 4:275. Beinecke Rare Book and Manuscript Library, Yale University.

Paper marbling is a method of flowing pigment across a wet surface to produce patterns that often look like cut and polished marble. First, water is poured into a shallow wooden tray, and colored pigments are mixed to a thin, viscous consistency of gouache or opaque watercolor. Then using one of the whisks (the six little bundles of broom straw, shown horizontally), the artist applies paint to the surface of the water, on which it floats. In the endpapers for this book, Crepaldi achieved the look of scattered stones by letting drops of paint fall rapidly from a small whisk. Artists create other patterns by combing the floating pigment (various rakes and combs are shown in the lower right of the engraving) or sprinkling powder using a sieve (shown to the left of the whisks). When the desired look has been achieved, the artist lays a sheet of absorbent paper on the surface and captures the floating design. As such, each marbled paper is unique.